Main groups

8A
18

| | | | | | | | | 2
He
Helium
4.002602 |

Nonmetals

| 3A
13 | 4A
14 | 5A
15 | 6A
16 | 7A
17 |

| 5
B
Boron
10.811 | 6
C
Carbon
12.0107 | 7
N
Nitrogen
14.00674 | 8
O
Oxygen
15.9994 | 9
F
Fluorine
18.9984032 | 10
Ne
Neon
20.1797 |

| 1B
11 | 2B
12 |

10

| 13
Al
Aluminum
26.981538 | 14
Si
Silicon
28.0855 | 15
P
Phosphorus
30.973761 | 16
S
Sulfur
32.066 | 17
Cl
Chlorine
35.4527 | 18
Ar
Argon
39.948 |

| 28
Ni
Nickel
58.6934 | 29
Cu
Copper
63.546 | 30
Zn
Zinc
65.39 | 31
Ga
Gallium
69.723 | 32
Ge
Germanium
72.61 | 33
As
Arsenic
74.92160 | 34
Se
Selenium
78.96 | 35
Br
Bromine
79.904 | 36
Kr
Krypton
83.80 |

| 46
Pd
Palladium
106.42 | 47
Ag
Silver
107.8682 | 48
Cd
Cadmium
112.411 | 49
In
Indium
114.818 | 50
Sn
Tin
118.710 | 51
Sb
Antimony
121.760 | 52
Te
Tellurium
127.60 | 53
I
Iodine
126.90447 | 54
Xe
Xenon
131.29 |

| 78
Pt
Platinum
195.078 | 79
Au
Gold
196.96655 | 80
Hg
Mercury
200.59 | 81
Tl
Thallium
204.3833 | 82
Pb
Lead
207.2 | 83
Bi
Bismuth
208.98038 | 84
Po
Polonium
[208.98] | 85
At
Astatine
[209.99] | 86
Rn
Radon
[222.02] |

| 110
Ds
Darmstadtium
[271] | 111
Rg
Roentgenium
[272] | 112
*
[277] | 113
[284] | 114
[289] | 115
[288] | 116
[292] |

| 63
Eu
Europium
151.964 | 64
Gd
Gadolinium
157.25 | 65
Tb
Terbium
158.92534 | 66
Dy
Dysprosium
162.50 | 67
Ho
Holmium
164.93023 | 68
Er
Erbium
167.26 | 69
Tm
Thulium
168.93421 | 70
Yb
Ytterbium
173.04 | 71
Lu
Lutetium
174.967 |

| 95
Am
Americium
[243.06] | 96
Cm
Curium
[247.07] | 97
Bk
Berkelium
[247.07] | 98
Cf
Californium
[251.08] | 99
Es
Einsteinium
[252.08] | 100
Fm
Fermium
[257.10] | 101
Md
Mendelevium
[258.10] | 102
No
Nobelium
[259.10] | 103
Lr
Lawrencium
[262.11] |

Element 112 has a proposed name of Copernicium which is, at the time of this publication, under review by IUPAC.

ANALYTICAL CHEMISTRY
AND
QUANTITATIVE ANALYSIS

David S. Hage
University of Nebraska, Lincoln

James D. Carr
University of Nebraska, Lincoln

Prentice Hall

Boston Columbus Indianapolis New York San Francisco Upper Saddle River
Amsterdam Cape Town Dubai London Madrid Milan Munich Paris Montréal
Toronto Delhi Mexico City São Paulo Sydney Hong Kong Seoul Singapore Taipei Tokyo

Publisher: Daniel Kaveney
Editor in Chief, Chemistry and Geosciences: Nicole Folchetti
Marketing Manager: Erin Gardner
Assistant Editor: Laurie Hoffman
Editorial Assistant: Kristen Wallerius
Marketing Assistant: Nicola Houston
Managing Editor, Chemistry and Geosciences: Gina M. Cheselka
Project Manager, Science: Maureen Pancza
Art Project Managers: Connie Long and Ronda Whitson
Design Director: Jayne Conte
Cover Designer: Suzanne Duda
Senior Manufacturing and Operations Manager: Nick Sklitsis
Operations Specialist: Maura Zaldivar
Cover Image Specialist: Karen Sanatar
Photo Research Manager: Elaine Soares
Composition/Full Service: GEX Publishing Services
Cover Photograph: Image courtesy of Rich Irish, Landsat 7 Team, NASA GSFC; Data provided by EROS Data Center.

Credits and acknowledgments borrowed from other sources and reproduced, with permission, in this textbook appear on the appropriate page within the text.

Library of Congress Cataloging-in-Publication Data
Hage, David S.
  Analytical chemistry and quantitative analysis / David S. Hage and James D. Carr.—1st ed.
      p. cm.
  Includes bibliographical references and index.
  ISBN-13: 978-0-321-59694-9 (alk. paper)
  ISBN-10: 0-321-59694-3 (alk. paper)
  1. Chemistry, Analytic—Quantitative. I. Carr, James D. II. Title.
  QD271.7.H34 2011
  543—dc22
                            2009052011

Printed in the United States
10 9 8 7 6 5 4 3 2 1

Prentice Hall
is an imprint of

ISBN 10:   0-321-59694-3
ISBN 13:   978-0-321-59694-9

www.pearsonhighered.com

CONTENTS

PREFACE

The purpose of this text is to acquaint the student with basic laboratory techniques for chemical analysis and the appropriate selection and use of these methods. This includes such items as the proper use and maintenance of balances, laboratory glassware, and notebooks, as well as mathematical tools for the evaluation and comparison of experimental results. Basic topics in chemical equilibria are reviewed and used to help demonstrate the principles and proper use of classical methods of analysis like gravimetry and titrations. Students are also introduced to common instrumental techniques, such as spectroscopy, chromatography and electrochemical methods. One important change made from other texts is that we have tried to organize and weight this material in a manner that better reflects the relative importance of these methods in today's analytical laboratories.

The chapters in this textbook are arranged into several groups with common themes. This design makes it possible to easily flow from one topic to the next in a variety of ways. For instance, students who require training in chemical equilibria and related calculations can learn about this topic in Chapters 6 through 10, while students who already have a solid background in this area can move on to later chapters that deal with techniques such as gravimetry or titrations. An instructor who wishes to discuss some instrumental methods before classical methods can use the first set of chapters to provide a general background on chemical analysis, followed by a discussion on electrochemistry, spectroscopy, or chromatography. We believe this format gives the instructor the greatest flexibility in using this text as either a one-semester introduction to analytical chemistry or as part of a more traditional two-semester sequence that begins with quantitative analysis and later moves on to instrumental methods. This format makes this textbook a flexible but practical tool that can be utilized to provide foundation coursework in analytical chemistry, in line with recent ACS guidelines outlined in a 2008 report entitled "Undergraduate Professional Education in Chemistry: ACS Guidelines and Evaluation Procedures for Bachelor's Degree Programs."

An underlying goal that we had in writing this textbook was to give students an appreciation of the role analytical chemistry has played in the development of science and continues to play in everyday life. To do this, we use real examples in each chapter to help illustrate the principles that are being discussed. This format also meets recent guidelines that encourage the use of problem- or inquiry-based learning. Special sections are also included as sideboxes to indicate important developments in the history of chemical analysis and/or common applications of analytical chemistry to real-world problems. In illustrating the methods in this textbook, we have gone beyond the standard inorganic and organic analyses that are common in many texts and have included examples from fields that range from environmental science, pollution monitoring and industrial processes to pharmaceutical science, food testing, and clinical analyses. In doing this, it is our hope that students who read this textbook will come away with a view of analytical chemistry as an important, living, and ever-changing science. These students should also have a greater appreciation of how the creation and use of methods for chemical analysis are important in the scientific process of discovery.

One key difference between this textbook and others is the way in which the students learn about each topic. For example, many of the chapters begin with an opening scenario in which the student is presented with a problem or group of problems that require the use of a particular analytical method. The student is then introduced to the method and guided through a series of topics that are needed for him or her to understand and use this technique. This format allows us to cover the same topics as other quantitative analysis texts but employs a more student-friendly style than the traditional topic-oriented approach. Another advantage of this format is that it will help students to more easily see the value and utility of each topic as it is presented. This is reinforced by exercises that are scattered throughout the text and by related homework problems that appear at the end of each chapter. Most of these problems can be solved using elementary algebra; however, sections are also included at the end of each chapter with "Challenge Problems," some of which involve the use of spreadsheets and all of which allow the student to address the chapter's material on a more in depth level. There is also a section at the end of each chapter entitled "Topics for Discussion and Reports" which provides the instructor and students with opportunities to explore material and methods that are related to those presented in the chapter but that are normally not covered in a traditional course on quantitative analysis. The "Challenge Problems" and "Topics for Discussion and Reports" are designed to develop the abilities of a student in inquiry-based and open-ended investigations into the area of analytical chemistry. Within these sections there are also many opportunities for writing, critical thinking and reasoning in topics related to chemical analysis.

SUPPLEMENTS

The supplemental materials listed below are available to support instructors and students as they use this textbook.

- Solutions Manual. This resource provides detailed answers for problems that appear at the end of each chapter.
- Test Bank. This resource can be used by instructors as a source of exam questions, complete with solutions.
- PowerPoint lecture slides. Each chapter of the book will be accompanied by a set of PowerPoint slides that can be used directly or after customization by an instructor to their individual preferences.
- Extended homework assignments. One extended homework assignment per chapter is available. These assignments are based on a key component in each chapter (e.g., describing the fraction of species for an amino acid during the discussion of acid-base reactions).
- In-class work sheets. Each chapter in the book is accompanied by a chapter summary hand-out with key focus points for one major topic. These worksheets will have blanks and/or problems to discuss and work on in class.
- Podcasts. Each chapter is accompanied by a podcast that summarizes the most important points in the chapter and highlights the points students should take away from the chapter.

The Solutions Manual has been developed by the authors, and the remaining supplemental materials have been prepared by Dr. Charles W. (Bill) McLaughlin of Montana State University. Please contact your local Pearson representative for more details about the supplements program that accompanies this textbook.

ACKNOWLEDGEMENTS

Any effort such as writing a textbook is not just the result of work by the authors, but of many who contribute through their support, suggestions or insight. We would first like to thank our families for their support and help during this project. Jill, Ben, Brian and Bethany all helped in many ways as D.S. Hage worked on his portions of this project, and Rosalind gave her support to J.D. Carr as he worked on his sections. Many long days and hours were spent on putting this textbook together. The help by all of these family members with providing the time and support needed for this writing was crucial and is greatly appreciated. Their help with proofreading, preparing graphs, and acquiring photos is also appreciated. Jill is particularly acknowledged for all her help in editing and proofreading during the creation of this textbook.

Many students also have provided feedback, comments and assistance through the development of this textbook. These students include (in alphabetical order): Jeanethe Anguizola, John Austin, Omar Barnaby, Sara Basiaga, Raychelle Burks, Jianzhong Chen, Sike Chen, Mandi Conrad, Abby Jackson, Jiang Jang, Krina Joseph, Liz Karle, Ankit Mathur, Annette Moser, Mary Anne Nelson, Corey Ohnmacht, Efthimia Papastavros, Erika Pfaunmiller, Shen Qin, John Schiel, Matt Sobansky, Sony Soman, Stacy Stoll, David Stoos, Zenghan Tong, Michelle Yoo, and Hai Xuan. The input from these current and future teachers and leaders in analytical chemistry was greatly appreciated as we tried to create a textbook that could be effectively used by such individuals in the classroom.

There are also many current and former colleagues who have contributed in various ways to this textbook. The input and efforts by Carlos Castro-Acuna, Paul Kelter, and Jody Redepenning in the early phases of this project are acknowledged. We also thank Richard Stratton for his encouragement and support during the early phases of this project. Valuable comments on information on specific topics were also received from Daniel Armstrong, Chad Briscoe, Ronald Cerny, Carrie Chapman, Barry Cheung, William Clarke, Patrick Dussault, Don Johnson, Rebecca Lai, Robert Powers, Peggy Ruhn, Ed Schmidt, and John Stezowkski.

We thank the publishing professionals at Pearson who helped us turn our ideas and manuscript into this final textbook. In particular, we thank Publisher Dan Kaveney, Assistant Editor Laurie Hoffman, and Production Project Manager Maureen Pancza. Micah Petillo and the staff at GEX Publishing Services, were also a valuable part of this process. We thank Dr. William C. Wetzel (Thomas More College) for his valuable input and efforts in checking the accuracy of this textbook. We also thank Dr. Bill McLaughlin for his input during this project and the excellent supplementary materials that he has developed that we believe will greatly increase the impact this book will have in helping students learn about analytical chemistry.

Finally, we would like to thank all of the reviewers who were part of assembling the text. We are grateful to the following chemists for reviewing all or part of the manuscript: Lawrence A. Bottomley (Georgia Institute of Technology), Heather A. Bullen (Northern Kentucky University), James Cizdziel (University of Mississippi), Darlene Gandolfi (Manhattanville College), James G. Goll (Edgewood College), Harvey Hou (University of

Massachusetts, Dartmouth), Elizabeth Jensen (Aquinas College), Mark Jensen (Concordia College, Moorhead), Irene Kimaru (St. John Fisher College), Abdul Malik (University of South Florida), Stephanie Myers (Augusta State University), Niina J. Ronkainen-Matsuno (Benedictine University), Brian E. Rood (Mercer University), Clayton Spencer (Illinois College), Cynthia Strong (Cornell College), Matthew A. Tarr (University of New Orleans), Jason R. Taylor (Roberts Wesleyan College), Lindell Ward (University of Indianapolis), William C. Wetzel (Thomas More College), and Xiaohong Nancy Xu (Old Dominion University).

David S. Hage

James D. Carr

ABOUT THE AUTHORS

David S. Hage is a professor of analytical and bioanalytical chemistry in the Department of Chemistry at the University of Nebraska, Lincoln. He received his B.S. in chemistry and biology from the University of Wisconsin, La Crosse, his Ph.D. in analytical chemistry from Iowa State University, and he was a postdoctoral fellow in clinical chemistry at the Mayo Clinic. He is a full professor at the University of Nebraska, Lincoln.

Dr. Hage is the author of over 145 research publications, reviews and book chapters. He recently edited a book entitled the *Handbook of Affinity Chromatography* (Taylor Francis) and is a coauthor on the textbook *Chemistry: An Industry-Based Introduction* (CRC Press). He received the 1995 Young Investigator Award from the American Association for Clinical Chemistry and the 2005 Excellence in Graduate Education Award from the University of Nebraska, Lincoln. He was made a Bessey Professor of Chemistry in 2006 at the University of Nebraska.

James D. Carr is a professor of analytical chemistry in the Department of Chemistry at the University of Nebraska, Lincoln. He received his B.S. in chemistry from Iowa State University and his Ph.D. in chemistry from Purdue University. He was then a postdoctoral fellow at the University of North Carolina, Chapel Hill. He is a full professor at the University of Nebraska, Lincoln.

Dr. Carr is the author of approximately 50 research publications and articles. He is the coauthor of *Chemistry: A World of Choices* (McGraw-Hill), a liberal arts general chemistry textbook. He is also the author or coauthor of several versions of general chemistry and quantitative analysis lab manuals and study guides (gen chem only). He has won several teaching awards, including the University of Nebraska Distinguished Teaching Award in 1981; University of Nebraska Recognition Awards for Contributions to Students in 1992, 1993, 1994, 1995, and 2000; and the University of Nebraska Outstanding Teaching and Instructional Creativity Award in 1996. He is a member of the University of Nebraska, Lincoln Academy of Distinguished Teachers and received the Distinguished Teacher Award from the Nebraska Teaching Improvement Council in 2001.

ANALYTICAL CHEMISTRY AND QUANTITATIVE ANALYSIS

Chapter 1

An Overview of Analytical Chemistry

Chapter Outline

1.1 INTRODUCTION: THE CASE OF THE MYSTERIOUS CHEMIST

We turned down a narrow lane and passed through a small side door, which opened into a wing of the great hospital. It was familiar ground to me, and I needed no guiding as we ascended the bleak stone staircase and made our way down the long corridor, with its vista of white-washed wall and dun-colored doors. Near the further end a low, arched passage branched away from it and led to the chemical laboratory. This was a lofty chamber, lined and littered with countless bottles. Broad, low tables were scattered about, which bristled with retorts, test-tubes, and little Bunsen lamps, with their blue, flickering flames.

*   There was only one student in the room, who was bending over a distant table absorbed in his work. At the sound of our steps he glanced around and sprang to his feet with a cry of pleasure. "I've found it! I've found it!" he shouted to my companion, running toward us with a test-tube in his hand. "I have found a reagent which is precipitated by hemoglobin and nothing else." Had he discovered a gold mine, greater delight could not have shown upon his features.*

*   "Dr. Watson—Mr. Sherlock Holmes," said Stamford, introducing us.*[1]

In this passage from the 1887 story "A Study in Scarlet," Sir Arthur Conan Doyle describes the first meeting between Dr. John H. Watson and the great, fictional detective Sherlock Holmes. Holmes is best known for his use of careful observation and deduction as tools for solving crimes. But he also relied heavily on chemical analysis for providing important clues in some of his cases. In the preceding excerpt, Holmes is working on a new method for confirming blood stains (see Figure 1.1). This was a problem often encountered by law officials at the time of this story because there was no reliable means for proving whether a spot on a suspect's clothing was blood or a stain from another source, like mud, rust, or food. The method developed by Holmes overcame this problem by specifically looking for hemoglobin, the protein in red blood cells that produces their color.[2] In modern laboratories, chemical tests not only can confirm whether a stain is blood but can determine if the blood is from a human or an animal and if it came from a particular victim or suspect.[3–5]

    Blood stain analysis is just one of many examples of how chemical tests are used to solve everyday problems. Other examples include techniques for monitoring pollutants in air or water and methods for detecting bacteria or contaminants in our food. Chemical measurements are also important in various industries for determining the quality or purity of their products. This includes companies that produce food, textiles, drugs, plastics, and metals. In addition, chemical analysis plays an important role in forensic science and clinical

FIGURE 1.1 Sherlock Holmes at work in his laboratory. Although Holmes was a fictional character, he was one of the earliest advocates in the use of analytical chemistry for crime scene investigations. (Original artwork by Sidney Paget, who drew illustrations for many of the early stories about Holmes.)

testing, and is a vital component of research in biology, biochemistry, medicine, and materials science. In fact, almost every day your life is probably affected in some way by chemical analysis.

The field of chemistry that deals with the use and development of tools and processes for examining and studying chemical substances is known as **analytical chemistry**. A fairly simple, but comprehensive, definition of analytical chemistry is "the science of chemical measurements."[6–8] The purpose of this book is to introduce you to common techniques for identifying, measuring, and characterizing chemicals or mixtures of chemicals. As you go through this book, you will learn how each of these methods works and how each is used to address various real-world problems. The underlying principles behind each technique will also be presented and you will see how knowledge of these principles can guide you in the correct choice and use of such methods.

1.2 THE HISTORY OF CHEMICAL ANALYSIS

1.2A Origins of Chemical Analysis

The earliest use of chemical testing dates back to ancient times. This can be illustrated in the analysis of precious metals like gold and silver. We have known since the

beginning of recorded history how to refine these metals, creating a need for methods that can determine the purity of the final product. This analysis was accomplished by using a small-scale version of the process for obtaining silver from lead ore—the use of fire to remove silver from lead and other metals. To perform this *fire assay*, a portion of the gold or silver was weighed, combined with lead, and melted in a furnace. A blast of air was then used to convert the lead and metal impurities to solid metal oxides, which could easily be removed from the surface of the melted silver or gold. The difference in mass before and after this treatment was then used to determine the original silver or gold's purity (see Figure 1.2). There are several references to this method in the Bible.[9–13] This assay is also mentioned in tablets sent between 1350 and 1375 B.C.E. from King Burraburiash of Babylon to Pharaoh Amenophis IV of Egypt, in which the Babylonian king complains about the quality of some gold that was sent to him by the Pharaoh.[9,12]

Another early example of chemical analysis is a method supposedly developed by the Greek mathematician Archimedes. Archimedes (287–212 B.C.E.) had been asked by King Hiero II of Syracuse to determine whether goldsmiths had cheated the king by mixing silver with the gold he had given them to make a crown for use as a ceremonial wreath (see Figure 1.3). After thinking about how to answer this question without damaging the crown, Archimedes developed an approach in which he compared the amount of water displaced by the crown and an equal mass of pure gold.[9,10] According to legend, Archimedes came up with this scheme as he lowered himself into a bath and watched water spill over the sides. Once he realized this effect could be used to examine the gold content of the crown, he is said to have jumped from the bath and exclaimed "Eureka!" (which means "I have found it!"), giving us an expression that has now become associated with scientific discovery.

In the years spanning from the time of the Roman empire through the Middle Ages, other chemical measurements were developed for looking at the quality of water, metals, medicines, and dyes.[9] It wasn't until the Renaissance, however, that these techniques became important in the systematic study of nature. During this era the phrase "chemical analysis" was first coined for describing such measurements. This term was suggested by Robert Boyle in his 1661 book *The Skeptical Chymist*.[9–11] Boyle was a nobleman who helped popularize the careful use of experiments for studying the physical properties and composition of matter, thus setting the stage for modern chemistry. In fact, it was by using this approach that he was able to develop what is now known as "Boyle's law," which describes the relationship between the pressure and volume of a fixed number of moles of a gas at a constant temperature.

For many years thereafter, chemical analysis was regarded as simply a tool and not as a field of study in its own right. This situation changed in the late 1700s, when a Swedish scientist named Torbern Bergman began to

FIGURE 1.2   The *fire assay*, a technique that has been used since ancient times for determining the purity of gold and silver. This technique is also called *cupellation*, a name derived from this method's use of a special container known as a *cupel*. (This image is from a book written in 1540 by Vannoccio Biringuccio entitled *De la Pirotechnia*, the earliest book published in Europe on the subject of metallurgy.)

systematically organize existing methods of chemical analysis according to the substances they were used to examine. His work was published between 1779 and 1790 as a collection of five volumes entitled *Opuscula physica et chemica*. It is this event that some think represents the beginning of chemical analysis as a distinct branch of chemistry. As Bergman's work became better known, other books on this topic also began to appear. One of

these was a textbook written by C. H. Pfaff in 1821 (*Handbuch der analytischen Chemie*) in which the name "analytical chemistry" was used for this new field of science.[9]

1.2B Chemical Analysis in the Modern World

During our industrial age, the use of chemical analysis has continued to grow and is now an important part of almost every aspect of our lives. A few of these applications are shown in Figure 1.4. They range from forensic science to biotechnology, agriculture, and materials science. Chemical analysis is also widely used in commercial applications, including the testing of foods, metals and other manufactured products. You will see many examples of such applications in this book, as we discuss the various approaches used for chemical measurements.

Much of the analytical chemistry that is performed daily has a very practical and applied aspect to it. Many of the chemical measurements made in industrial laboratories are concerned with determining the composition or properties of a product or raw material to ensure this item is satisfactory for sale or further use (an application known as *quality control*). Other examples involve the use of chemical analysis in hospitals for patient testing or in environmental laboratories that monitor the quality of our air, food, and water.

Another important application of analytical chemistry is its use in studying the world around us. This application includes cancer research, the discovery of new drugs, and the development of new synthetic materials, to name a few. The use of analytical methods for such work is often a two-way process, because the need

FIGURE 1.3   A golden wreath (or "crown") from the late second to early first century B.C.E. This wreath is probably similar to the crown that was examined by Archimedes. (Reproduced with permission, courtesy of the Benaki Museum Athens, © 2010.)

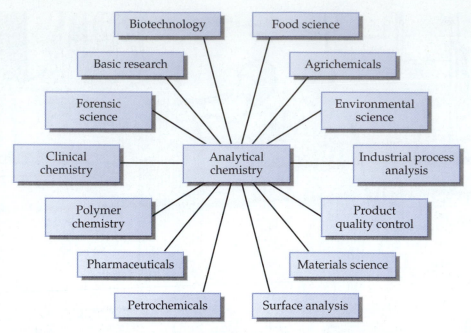

FIGURE 1.4   Common applications of analytical chemistry in today's world.[7]

for more detailed chemical information promotes the development of new techniques, which leads to further research made possible by the ability to obtain more data about a sample. As an example, the techniques of infrared spectroscopy (IR) and mass spectrometry (MS) both appeared in the early 1900s as research tools for characterizing specific properties of atoms and molecules. During World War II, however, there was a large growth in the synthesis and use of polymers. This growth led to a need for techniques that could analyze polymers and other compounds, such as the use of IR for examining functional groups and MS for determining the masses and structures of molecules. These developments, in turn, made it easier to synthesize and study other types of chemicals.

Because analytical chemistry impacts so many areas, it is not surprising that contributions to this field have been made by individuals from a variety of backgrounds. Table 1.1 lists the Nobel Prizes that have been awarded for research resulting in new or improved analytical methods. This list includes people from fields such as chemistry, physics, and medicine, and reflects the importance of chemical analysis in all these areas of science.

1.3 GENERAL TERMS USED IN CHEMICAL ANALYSIS

1.3A Sample-Related Terms

Now that we have seen the role analytical chemistry plays in our world, we need to define a few terms we will use throughout this book to discuss this topic. The first terms we will consider are those used to describe the material we wish to characterize. In most situations, it is not desirable or practical to look at all of the material of interest, so we instead take a smaller, representative portion for study. An example would be when a nurse or doctor takes a sample of blood to determine the amount of a particular drug that is present in your body. The portion of material taken for analysis is referred to as the **sample**.[14] Ideally, we would like this sample to be as representative as possible of the rest of the material to be examined. Ways for meeting this goal will be described in Chapter 5 when we discuss approaches for acquiring a chemical sample.

Within most samples there is a large variety of substances present. The entire group of substances that makes up a sample is called the sample **matrix**. The particular substance we are interested in measuring or studying in the sample is known as the **analyte**.[14] In some cases, the analyte might be an atom, molecule, or ion, while in others it might be a larger substance such as a polymer, virus particle, or cell. The technique used to examine the analyte should produce a signal that is related to the presence of this analyte in the sample. Although we are not always interested in looking at other components in the sample, we still have to consider these components as we choose and use an analysis method. This is the case because not all analysis methods are compatible with all types of samples. In addition, some sample components in the matrix may cause an error in the final result if they have not been properly dealt with before or during the analysis.

Another way we can classify an analyte is in terms of its relative contribution to the overall sample. At one extreme we have analytes that make up a significant portion of the sample. The term **major component** (or *major constituent*) is used to refer to such substances, especially if they make up more than 1% of the sample.

TABLE 1.1   Nobel Prizes Awarded for Developments in Chemical Analysis

Year and Area of Award	Awardees	Area of Study
1915–Physics	Sir William L. Bragg and Sir William H. Bragg	X-ray crystallography
1922–Chemistry	F. W. Aston	Mass spectrometry
1923–Chemistry	Fritz Pregl	Microanalysis of organic compounds
1930–Medicine	Karl Landsteiner	Blood typing
1930–Physics	Sir Chandrasekhara Venkata Raman	Raman spectroscopy
1943–Chemistry	George De Hevesy	Radioactive tracers
1948–Chemistry	Arne Wilhelm Kaurin Tiselius	Electrophoresis
1952–Chemistry	A. J. P. Martin and Richard L. M. Synge	Partition liquid chromatography
1952–Physics	Felix Bloch and Edward M. Purcell	Nuclear magnetic resonance spectroscopy
1953–Chemistry	Frits Zernike	Phase-contrast microscopy
1959–Chemistry	Jaroslav Heyrovsky	Polarography
1960–Chemistry	Willard Frank Libby	Carbon-14 dating
1977–Medicine	Rosalyn Yalow	Radioimmunoassays
1978–Medicine	Daniel Nathans, Werner Arber, and Hamilton O. Smith	Genetic studies with restriction enzymes
1980–Chemistry	Walter Gilbert and Frederick Sanger	DNA sequencing
1981–Physics	Nicolaas Bloembergen and Arthur L. Schawlow	Laser spectroscopy
1982–Chemistry	Sir Aaron Klug	Crystallographic electron microscopy
1985–Chemistry	Herbert A. Hauptman and Jerome Karle	Direct methods for determining crystal structures
1986–Physics	Gerd Binning and Heinrich Rohrer	Scanning tunneling microscopy
1986–Physics	Ernst Ruska	Electron microscopy
1991–Chemistry	Richard R. Ernst	High-resolution nuclear magnetic resonance spectroscopy
1993–Chemistry	Kary B. Mullis	Polymerase chain reaction
1999–Chemistry	Ahmed Zewail	Femtosecond spectroscopy
2002–Chemistry	John B. Fenn and Koichi Tanaka	Soft desorption ionization for mass spectrometry
2002–Chemistry	Kurt Wüthrich	Nuclear magnetic resonance spectroscopy for 3-dimensional studies of biological macromolecules
2003–Medicine	Paul C. Lauterbur and Sir Peter Mansfield	Magnetic resonance imaging
2005–Physics	John L. Hall and Theodor W. Hänsch	Laser-based precision spectroscopy
2008–Chemistry	Osama Shimomura, Martin Chalfie, and Roger Y. Tsien	Discovery of green fluorescent protein

Source: This information was obtained from *The Nobel Prize Internet Archive* (www.almaz.com/nobel).

For instance, a gold bar that is 99% pure would have gold as its major component. A substance present at lower levels, such as 0.01–1% of the total sample, is called a **minor component**. Likewise, a substance present at a level below 0.01% (100 parts-per-million) is known as a **trace component**. Table 1.2 illustrates these concepts using the composition of dry air as an example.[15] Such a classification scheme is important because the relative amount of an analyte in a sample is often a key factor in determining what techniques can be used for examining this analyte. This type of classification has lead to a division of methods according to whether they are used for **major component analysis, minor component analysis**, or **trace analysis**.[16]

1.3B Method-Related Terms

The Analytical Process.   A second group of terms we need to define concern the method being used to characterize our sample. Some words we have already used to describe this include *assay, analysis*, and *determination*. To illustrate this, we could say in our opening example that Sherlock Holmes was developing an *assay* for the *determination* of hemoglobin in

TABLE 1.2   Types of Sample Components Based on Relative Amount in the Sample*

Type of Sample Component	Relative Amount in Sample	Example: Composition of Dry Air (Without Water Vapor)
Major Component	1–100%	Nitrogen (78.1%), Oxygen (20.9%)
Minor Component	0.01–1%	Argon (0.9%), Carbon Dioxide (0.03%)
Trace Component	< 0.01% (100 ppm)	Neon (18.2 ppm), Helium (5.2 ppm), Methane (2 ppm), Krypton (1.1 ppm), Hydrogen (0.5 ppm), Nitrogen Dioxide (0.5 ppm), Xenon (0.09 ppm)

* All values are expressed in terms of the volume of gas per unit volume of air (v/v). The abbreviation *ppm* in this table stands for "parts-per-million," where 1 ppm = 0.0001% (see Chapter 3). The ranges given in the table for major, minor, and trace component analysis are only approximate and vary slightly depending on the technique and type of sample that is being examined.

Source: This information was obtained from the *CRC Handbook of Chemistry and Physics, 81st Ed.*, CRC Press, Boca Raton, FL, 2000.

blood or that he was conducting a hemoglobin *analysis*. Each of these terms concerns the general act of examining the sample and its analyte. The approach used to perform this assay is the **analytical method** or "analytical technique." Again going back to the opening example, we can say Holmes was using the *analytical method* of selective precipitation to determine whether hemoglobin was present in his sample. The entire group of operations used for the analysis is known as the *procedure* or *protocol*.

As shown in Figure 1.5, there are many steps in the overall procedure for a chemical analysis. You first need to determine what question is being asked about the sample and to identify the information that will be needed to answer this question. In the work by Holmes, the general question was "Is the sample a

blood stain?", which he sought to answer by looking for hemoglobin. The second step is to select an appropriate sample. To select a sample you must consider the nature of the material being examined, the types of analytes to be measured, and the distribution and suspected levels of these analytes within the material. For a blood stain, this process would involve locating a sample at the scene of a crime and obtaining a representative portion for analysis.

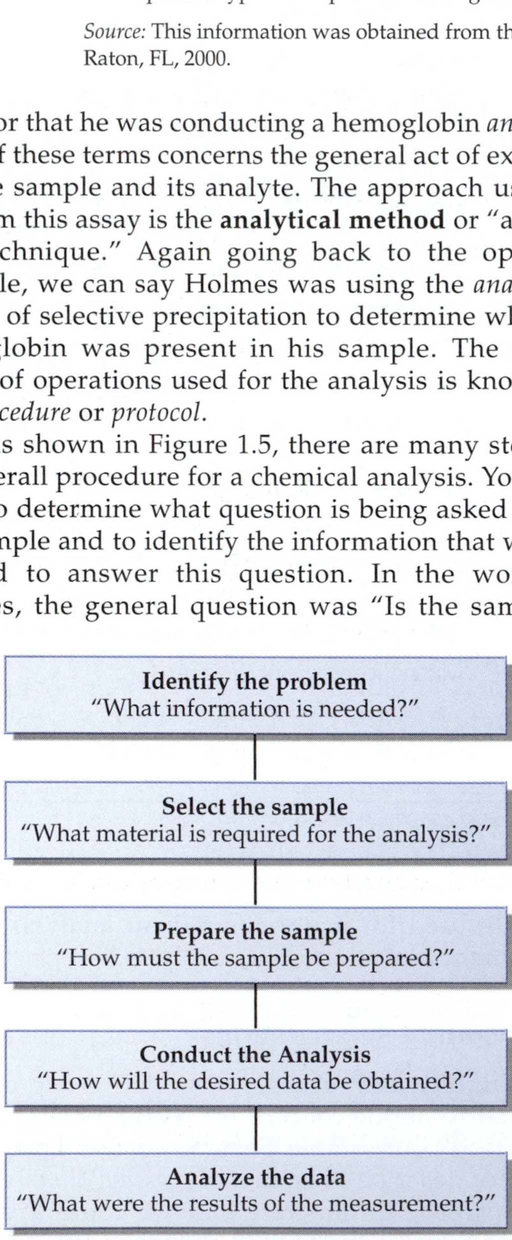

The third step in an analysis is sample preparation. The degree of preparation needed will depend on the sample's complexity, the types of analytes being examined, and the measurement method. For the technique used by Holmes, sample preparation probably involved placing a small part of the stained material into a container to which reagents could be added. The fourth and fifth steps in an analysis are the actual examination of the sample and use of these results for chemical measurement or characterization. In the technique developed by Holmes, these steps were represented by the addition of a reagent to the stained material and the observation of whether a precipitate was formed by the presence of hemoglobin.

Types of Analytical Methods.   The large number of chemicals and samples that occur in our world means we also need many different methods for their measurement or characterization. Some common types of analytical techniques are listed in Figure 1.6. These techniques can be placed into three categories: classical methods, instrumental methods, and separation methods. **Classical methods** were the first analytical techniques developed and produce a result by using experimentally determined quantities such as a mass or volume, along with the use of atomic or molecular masses and well-defined chemical reactions.[17,18] An example of a classical method is gravimetric analysis (discussed in Chapter 11), which is

Identify the problem
"What information is needed?"

Select the sample
"What material is required for the analysis?"

Prepare the sample
"How must the sample be prepared?"

Conduct the Analysis
"How will the desired data be obtained?"

Analyze the data
"What were the results of the measurement?"

FIGURE 1.5   The general steps in a procedure for chemical analysis.

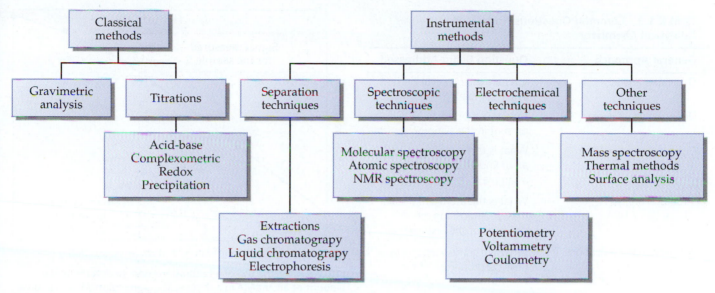

FIGURE 1.6   General categories of analytical techniques.

based on measuring the mass of a chemical product that either contains or is related to the analyte. The fire assay for gold is an example of a gravimetric method. Another such method is a titration (see Chapters 12, 13, and 15), in which a chemical substance is measured by determining the volume or amount of a well-defined reagent that is needed to react with this analyte. Most classical methods are performed as manual techniques; however, some are conducted in modern laboratories with the aid of automated systems.

An **instrumental method** uses an instrument-generated signal for detecting the presence of an analyte or determining the amount of an analyte in a sample. There are many instrumental methods, ranging from electrochemical methods (which make use of the production or consumption of electrons by chemicals) to spectroscopic methods (which use electromagnetic radiation to characterize or measure analytes). Other techniques in this category are mass spectrometry, thermal methods, and approaches for surface analysis. Instrumental techniques were developed long after classical methods of analysis but are used in most of today's chemical measurements.[19,20]

A **separation method** is an approach used to remove one type of chemical from another. A separation method is often needed when the goal is to examine a chemical or group of chemicals in a complex sample. Chemical separations can be used as part of either a classical method or an instrumental method to isolate an analyte from a sample, remove interfering chemicals, or place the analyte in an appropriate matrix for further study. Some separation methods are carried out manually (for instance, an extraction), while others require special equipment and are considered "instrumental methods" (as occurs for gas chromatography and high-performance liquid chromatography). Separation techniques are quite common and make up an important part of modern methods for chemical analysis.[19,20]

1.4 INFORMATION PROVIDED BY CHEMICAL ANALYSIS

Each chemical assay has its own unique set of requirements, but we can sort these methods into general categories based on the type of information they provide about a sample (see Table 1.3). Most chemical assays involve a comparison between the sample and a material known to contain the analyte of interest (known as the **standard**). This comparison provides a means for the positive identification or measurement of the analyte in a sample.

The first type of chemical measurement that might be carried out is a **qualitative analysis**. The goal here is to simply determine whether a particular analyte is present in a sample. An example would be the fictitious assay for hemoglobin that was developed by Sherlock Holmes for the detection of blood. In methods for qualitative analysis we are not necessarily interested in how much of the analyte is present, although there must be a certain minimum amount present for its detection. Instead, we only care if the compound of interest is present above this minimum level. This approach is also sometimes called a **screening assay** and is often used to help decide whether further tests should be conducted on a sample.

Another question that can be asked is "How much of the analyte is present in the sample?" This question is answered by using **quantitative analysis**. The goal here is to measure, **quantitate**, or **quantify** (to provide a numerical value for) the actual amount of analyte in a sample. Such an approach is utilized when it is necessary to determine the concentration of an analyte or its contribution to the overall composition of a sample. For instance, quantitative analysis would be used by a food company to measure the protein, carbohydrate, and fat content in a product. This type of analysis would also be used by a hospital laboratory to determine whether a drug given to a patient is within the proper range for treatment of a disease.

TABLE 1.3   Common Questions Addressed by Analytical Chemistry

General Approach	Question Being Addressed
Qualitative analysis	Is a particular analyte present in the sample?
Quantitative analysis	How much of the analyte is present in the sample?
Chemical identification	What is the identity of an unknown chemical in a sample?
Structural analysis	What is the atomic/molecular mass, composition, or structure of the analyte?
Property characterization	What are some chemical or physical properties of the analyte?
Spatial analysis	How is the analyte distributed throughout a sample?
Time-dependent analysis	How does the amount of an analyte or a property of the analyte change over time?

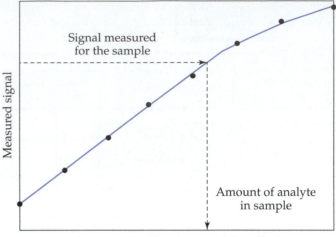

FIGURE 1.7   The use of a calibration curve to determine the amount of an analyte in a sample. The signal plotted in this curve is determined by using standards containing known amounts of the analyte. The experimental results obtained for the standards are represented by the solid dots, and the solid line is the best-fit curve that passes through these results. Further information on calibration curves and their properties can be found in Chapter 5.

Quantitative analysis is probably the most common type of analytical chemistry that is conducted on a routine basis. A quantitative analysis can be used directly on a sample or may follow an earlier screening assay. Although the response of some methods (such as a gravimetric analysis or titration) can be used directly to determine the amount of an analyte in a sample, most quantitative methods require the use of standards for this purpose. This task is accomplished by making a plot of the signals given by a method for standards that contain known amounts of the analyte. This process is referred to as **calibration** and gives a graph known as a **calibration curve** (see Figure 1.7). When a sample is later examined by the same method, the signal it produces is compared to the calibration curve and used to determine the amount of analyte that must have been present in the sample to give such a response.

Besides using analytical chemistry to measure a substance, it is also often necessary to identify a substance in a sample. This application, known as **chemical identification**, might be used by a chemist to identify a potential drug candidate that has been isolated from a plant, or it might be used by an environmental scientist to determine the nature of a new pollutant found in a water or soil sample. One way chemical identification can be carried out is by comparing the unknown compound's behavior in an analytical method to that observed for standard samples of known chemicals. Another way chemical identification might be accomplished is by using techniques that provide direct clues regarding the compound's composition and structure.

Two types of testing that are closely related to chemical identification are **structural analysis** and **property characterization**. In structural analysis, the goal is to determine features such as the mass, composition, functional groups, or structure of the analyte. Structural analysis might provide a detailed description of a chemical or help identify an unknown substance. In property characterization, measurement of some specific chemical or physical property of the analyte is desired. Property characterization of a material might involve examining how the material interacts with light or electrons, its ability to react with other chemicals, or its color, crystal shape, and mechanical strength. Like structural analysis, property characterization can be conducted with either standard samples of known chemicals or with unknown compounds that are to be identified through their measured properties.

Many materials have compositions that are different from one section of their matrix to the next. In these cases, the method of *spatial analysis* can be used to provide more detailed information about the material's composition. Spatial analysis deals with determining how a particular analyte is distributed throughout a matrix by examining small sections of the material, thus allowing chemical information to be obtained from different regions. This type of analysis is valuable when you are examining a *heterogeneous material* (that is, a material with a composition that varies from one point to the next within its structure). One example is *surface analysis,* which is used in areas such as the semiconductor industry during the production of storage media and computer chips.

Numerous samples that are studied by chemical analysis are taken from systems that change over time. For instance, if a doctor measures the amount of glucose in your blood, a sample taken shortly after you have eaten will give a much higher result than one that is taken just after you awake in the morning. This change in concentration can be studied by using *time-dependent (temporal) analysis*, which examines how the amount of one or more analytes varies as a function of time. Changes in an analyte over long periods of time can often be examined by using the same methods that are employed for quantitative analysis. However, for shorter periods of time more specialized techniques may be required.

| EXERCISE 1.1 | What Information Is Required from a Chemical Analysis? |

What general type of chemical test (for instance, qualitative analysis, quantitative analysis, etc.) is needed in each of the following situations?

a. An assay of drinking water to determine if the concentration of a particular pollutant is within legal limits
b. Studies to determine the nature of an unknown toxin in a food sample
c. The location of a specific type of protein within a cell

SOLUTION

(a) This is an example of a quantitative analysis, because the amount of a specific chemical is to be measured. (b) These studies will involve some form of chemical identification. This may be done by comparing an isolated sample of the unknown compound to known samples of standard chemicals, or by performing structural analysis or property characterization on the toxin to provide clues as to its identity. (c) Spatial analysis is required for this application because it is necessary to look at the protein's distribution and location in the cell. This analysis might be carried out in a microbiology or biochemistry laboratory to identify a cell or to provide clues as to a protein's function within the cell.

1.5 OVERVIEW OF TEXT

In this chapter, we have had our first glimpse at the field of analytical chemistry. We have discussed the origins of this field and considered some of its applications in today's world. We have also considered some general terms used within analytical chemistry and discussed the types of information that can be obtained by chemical analysis.

In the following chapters we will revisit many of these concepts and look more deeply at the techniques used in chemical analysis. We will begin in Chapters 2–5 by going over some of the basic tools employed in any analytical method, such as techniques for good laboratory practices, the preparation of solutions and the statistical treatment of experimental results. We will then study classical methods of chemical analysis, such as gravimetric analysis and titrations. Chapters 6–10 include a review of the principles behind chemical equilibrium, followed by a discussion in Chapters 11–13 of how these principles are employed in classical methods.

The focus in the remaining chapters will be on common approaches for instrumental analysis. This section will begin with a discussion of electrochemical methods (Chapters 14–16), followed by spectroscopic methods (Chapters 17–19) and techniques for chemical separations (Chapters 20–23). Throughout this book, you will also be introduced to a variety of other techniques (such as mass spectrometry and nuclear magnetic resonance (NMR) spectroscopy) which are important tools for the modern analytical chemist. In each case, you will learn about various applications for these methods and see how these methods impact our everyday lives.

Key Words

Analyte *4*	Instrumental method *7*	Property characterization *8*	Separation method *7*
Analytical chemistry *2*	Major component *4*	Qualitative analysis *7*	Standard *7*
Analytical method *6*	Major component analysis *5*	Quantitate (quantify) *7*	Structural analysis *8*
Calibration *8*	Matrix *4*	Quantitative analysis *7*	Trace analysis *5*
Calibration curve *8*	Minor component *5*	Sample *4*	Trace component *5*
Chemical identification *8*	Minor component	Screening assay *7*	
Classical method *6*	analysis *5*		

Other Terms

Analysis (assay,	Heterogeneous material *8*	Spatial analysis *8*	Time-dependent (temporal)
determination) *5*	Procedure (protocol) *6*	Surface analysis *8*	analysis *9*
Fire assay (cupellation) *2*	Quality control *3*		

Questions

INTRODUCTION AND HISTORY OF CHEMICAL ANALYSIS

1. Define the terms "analytical chemistry" and "chemical analysis."

2. What was the first use of chemical analysis? How have the uses of analytical chemistry changed from ancient to modern times?

3. Describe some general applications of analytical chemistry in the modern world.

4. What is the relationship between research in analytical chemistry and research in other fields, such as medicine, environmental science, or biology?

GENERAL TERMS USED IN CHEMICAL ANALYSIS

5. What is meant in analytical chemistry by the term "sample" (when used as a noun)? How is this related to or different from the terms "analyte" and "matrix"?

6. Identify the sample, analyte, and matrix in each of the following situations.
   (a) Estimation of the amount of sulfur in coal
   (b) Analysis of the drug content in a tablet by a pharmaceutical company
   (c) Measurement of carbon monoxide in fumes emitted by an industrial plant

7. Explain the difference between major, minor, and trace components within a sample.

8. Determine whether each of the following substances in common household items is an example of a major, minor, or trace sample component.
   (a) The amount of protein and fat in a portion of 95% lean beef (5% fat)
   (b) The amount of aspirin (acetylsalicylic acid) in a 250 mg nonprescription tablet that contains 80 mg of this drug
   (c) The vitamin C in an orange, which typically contains 50–60 mg vitamin C per 100 g total mass

9. What are the five general steps in any type of chemical analysis?

10. Explain what is meant in analytical chemistry by a "classical method." What is meant by an "instrumental method"? How do these two types of methods differ?

11. Discuss why a separation method might be used as part of a chemical analysis. What are some examples of separation methods?

INFORMATION PROVIDED BY CHEMICAL ANALYSIS

12. What is a standard? What is a calibration curve? How is each of these used in chemical analysis?

13. Compare and contrast the information that is provided by each of following types of general analytical methods.
    (a) Qualitative analysis vs. quantitative analysis
    (b) Structural analysis vs. property characterization
    (c) Spatial analysis vs. temporal analysis

14. What general approach (for instance, qualitative analysis, quantitative analysis, etc.) is needed in the following situations?
    (a) An analysis of samples from athletes to determine whether they are using performance-enhancing drugs
    (b) Identification of an unknown compound from a plant that is believed to have antitumor properties
    (c) Measurement by a pharmaceutical company of the actual amount of a drug that is present in one of their products
    (d) Location of the point where a pollutant is entering a river

CHALLENGE PROBLEMS

15. Trace analysis can be divided into many subcategories, depending on how small of an amount of analyte must be detected and on the size of the sample that is being used.[14,16]
    (a) Look up the definitions for each of the following terms and explain how they differ from each other: microtrace analysis, nanotrace analysis, and picotrace analysis.
    (b) Differentiate between what is meant by "microtrace analysis" and "ultratrace analysis" in chemical testing.
    (c) Identify several examples of analytes and samples that would fit into the various categories listed in (a) and (b).

16. Look up a research paper that discusses the development or use of an analytical method. Identify each of the following factors within the paper.
    (a) The sample, analyte, and matrix being examined
    (b) The type of assay, analytical method, and procedure being used
    (c) The general type of analytical method (classical or instrumental method)
    (d) The type of question that is being asked
    (e) The general analysis format (qualitative analysis, quantitative analysis, etc.)

17. A general indication of the role analytical chemistry plays in our world can be seen by references that are made to such testing in literature and popular media. An example is the opening excerpt in this chapter from "A Study in Scarlet." Locate another example from a book, movie, or television show (for example, see Reference 21). Identify the types of analytes that are being measured and the approach that is used in your example. Determine whether the analytical method is a real technique or a fictional one.

18. The development of improved electronics led to a huge growth in the development of new instrumental methods in the 1940s and 1950s. Similar growth occurred with the introduction of personal computers during the 1970s and 1980s.[19,20] What current trends and recent advances do you think will be important in the future development of chemical analysis?

TOPICS FOR DISCUSSION AND REPORTS

19. The modern analysis of precious metals such as gold and silver often combines classical methods of chemical analysis with more modern instrumental methods.[9] Report on how such assays are currently performed.

20. Obtain more information on one of the individuals listed in Table 1.1 and write a report on the contribution that person made to chemical analysis. Discuss how this development impacted the person's field of research or other areas of science.

21. The availability of reliable, commercial equipment for performing instrumental or classical methods has often been a key step in determining how quickly a new analysis technique sees widespread use. Several people and companies who have played important roles in the past development of such equipment are discussed in References 19 and 20. Obtain information on one of these individuals or companies and discuss how that person's work contributed to the field of analytical chemistry.

22. The journal *Analytical Chemistry* is an important source of reviews and research articles on methods for chemical measurements. Reference 22 describes how this journal has changed over the past century. Examine this article and discuss how this journal has reflected the changes in chemical analysis over the past 100 years.

23. Select an article from a current newspaper or magazine that discusses a topic in which chemical analysis was used to provide key information. Describe the type of chemical analysis that was conducted and the type of information it provided. Also, discuss how this information was used in the article.

References

1. A. C. Doyle, "A Study in Scarlet," *Beeton's Christmas Annual*, 1887.
2. S. M. Gerber, "A Study in Scarlet: Blood Identification in 1875," *Chemistry and Crime: From Sherlock Holmes to Today's Courtroom*, S. M. Gerber, Ed., American Chemical Society, Washington, DC, 1983, Chapter 3.
3. F. M. Gdowski, "Bloodstain Analysis-Case Histories," *Chemistry and Crime: From Sherlock Holmes to Today's Courtroom*, S. M. Gerber, Ed., American Chemical Society, Washington, DC, 1983, Chapter 7.
4. L. Kobilinsky, "Bloodstain Analysis-Serological and Electrophoretic Techniques," *Chemistry and Crime: From Sherlock Holmes to Today's Courtroom*, S. M. Gerber, Ed., American Chemical Society, Washington, DC, 1983, Chapter 8.
5. C. S. Tumosa, "The Detection and Species Identification of Blood—A Bibliography of Relevant Papers from 1980 to 1995," *Forensic Science Review*, 8 (1996) 74–90.
6. R. W. Murray, "Analytical Chemistry: The Science of Chemical Measurements," *Analytical Chemistry*, 68 (1991) 271A.
7. J. Tyson, *Analysis: What Analytical Chemists Do*, Royal Society of Chemistry, London, 1988.
8. M. Valcarcel, "A Modern Definition of Analytical Chemistry," *Trends in Analytical Chemistry*, 16 (1997) 124–131.
9. S. Kallmann, "Analytical Chemistry of the Precious Metals: Interdependence of Classical and Instrumental Methods," *Analytical Chemistry*, 56 (1984) 1020A–1027A.
10. F. Szabadvary, *History of Analytical Chemistry*, Pergamon Press, New York, 1966.
11. G. D. Christian, "Evolution and Revolution in Quantitative Analysis," *Analytical Chemistry*, 66 (1995) 532A–538A.
12. J. O. Nriagu, "Cupellation: The Oldest Quantitative Chemical Process," *Journal of Chemical Education*, 62 (1985) 668–674.
13. Examples of citations for the fire assay in the Bible include Numbers 31:22, 1st Peter 1:7, and Revelation 3:18.
14. H. M. N. H. Irving, H. Freiser, and T. S. West, *Compendium of Analytical Nomenclature: Definitive Rules—1977*, Pergamon Press, New York, 1977.
15. *CRC Handbook of Chemistry and Physics, 81st ed.*, CRC Press, Boca Raton, FL, 2000.
16. H. A. Laitinen, "History of Trace Analysis," *Journal of Research of the National Bureau of Standards*, 93 (1988) 175–185.
17. C. M. Beck II, "Classical Analysis: A Look at the Past, Present and Future," *Analytical Chemistry*, 63 (1991) 993A–1003A.
18. C. M. Beck II, "Classical Analysis: A Look at the Past, Present and Future," *Analytical Chemistry*, 66 (1994) 224A–239A.
19. J. Poudrier and J. Moynihan, "Instrumentation Hall of Fame," *Made to Measure: A History of Analytical Instrumentation*, J. F. Ryan, Ed., American Chemical Society, Washington, DC, 1999, pp. 10–38.
20. J. T. Stock, "A Backward Look at Scientific Instrumentation," *Analytical Chemistry*, 65 (1993) 344A–351A.
21. C. A. Lucy, "Analytical Chemistry: A Literary Approach," *Journal of Chemical Education*, 4 (2000) 459–470.
22. D. Noble, "From Wet Chemistry to Instrumental Analysis: A Perspective on Analytical Science," *Analytical Chemistry*, 4 (1994) 251A–263A.

Chapter 2

Good Laboratory Practices

Chapter Outline

2.1 INTRODUCTION: A QUESTION OF QUALITY

It is still referred to as the "Trial of the Century." Even after more than a decade, the O.J. Simpson case is often cited as an example of what can go wrong in a forensic investigation. This case began on June 17, 1994, with a low-speed chase of a white Ford Bronco by Los Angeles Police. What followed was the arrest of O. J. Simpson as a suspect in the murder of his ex-wife, Nicole Brown, and her friend Ronald Goldman.[1,2] Part of the evidence presented at the resulting trial was a set of DNA tests conducted on blood and hair samples taken from the victims, the murder scene, and Simpson's vehicle and clothing (see Figure 2.1). As part of this analysis, the samples were first processed by a method known as the *polymerase chain reaction* (PCR) (see Box 2.1). Although some DNA matches were found, it was argued at the trial that the original samples had been obtained by individuals who were not properly trained in collecting such materials. Other evidence suggested the samples had been contaminated during their collection and handling, and that the DNA samples had undergone significant degradation prior to analysis. It was thus decided by the jury that the test results did not provide any conclusive evidence in this case.[1,2]

Although it is unfortunate when a laboratory produces questionable results, this example does indicate the importance of making sure that proper methods are being used when you handle or analyze a sample. To accomplish this goal, every laboratory needs to have a set of procedures in place to ensure its work is conducted in a safe and meaningful manner. This is true regardless of whether you are working in a commercial laboratory, university, government agency, or industrial facility. The rules and procedures that are used to guide work in these settings are referred to as **good laboratory practices (GLPs)**.[4,5] In this chapter, we will discuss some basic practices that should be followed in all laboratories and see why these practices are especially important in analytical chemistry.

2.1A What Are Good Laboratory Practices?

Good laboratory practices can be thought of as a set of guidelines that promotes proper work and conduct within the laboratory. There are many things covered by this definition. For instance, laboratories in the United States that test or develop food additives and drugs are required by the Food and Drug Administration to have guidelines in place for the training of personnel, the selection and care of facilities and equipment, sample handling and analysis, and the recording and reporting of data. Similar guidelines are used by the U.S. Environmental Protection Agency for laboratories and companies that work with insecticides and herbicides, as well as by other countries in the regulation of drug companies or industries that routinely deal with chemical hazards.[6,7]

One of the primary purposes of good laboratory practices in analytical chemistry is to give confidence that the final results are a valid representation of a sample. In

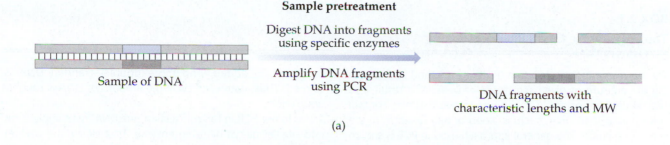

Sample pretreatment

Digest DNA into fragments using specific enzymes

Amplify DNA fragments using PCR

Sample of DNA

DNA fragments with characteristic lengths and MW

(a)

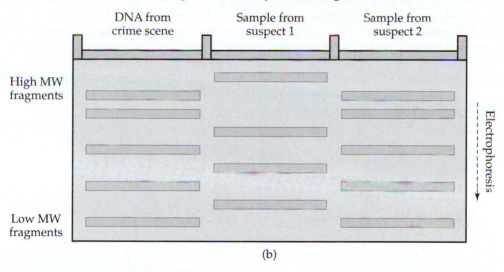

Separate and analyze DNA fragments

DNA from crime scene

Sample from suspect 1

Sample from suspect 2

High MW fragments

Low MW fragments

Electrophoresis

(b)

FIGURE 2.1   The analysis and comparison of DNA in forensic samples. The method of electrophoresis, which is used here to separate and compare DNA fragments from various samples, is discussed in more detail in Chapter 23. Abbreviation: MW, molecular weight (molar mass).

reaching this goal, it is important to consider all of the steps in an analysis and to have proper procedures in place for each step. Many of the problems in the DNA results for the O. J. Simpson trial could have been eliminated if the individuals obtaining the evidence had been adequately trained in DNA sample collection. Similarly, the problems with sample contamination and degradation could have been minimized if better procedures had been in place for the storage and handling of DNA samples.

2.1B Establishing Good Laboratory Practices

Good laboratory practices require that there be well-defined methods describing how routine work should be carried out in a laboratory. This goal is accomplished by using a **standard operating procedure** (or **SOP**). A standard operating procedure is a specific set of instructions that describes how a particular task should be performed. An SOP might be a document that describes a procedure for synthesizing a chemical, a safety protocol, or a method for calibrating an instrument. In a teaching setting, an SOP is often a written description of an experiment that is provided to students by their instructor or given in a laboratory manual.

An example of an SOP is given in Figure 2.3. This SOP describes the proper method for using an electronic balance, a common laboratory device that we will learn about in Chapter 3. As you can see from this example,

the described procedure is specific in its directions and can easily be understood by anyone who must conduct the procedure. The same features should be present in any well-written SOP.

EXERCISE 2.1	**Developing a Standard Operating Procedure**

Write a short SOP for the evacuation of your laboratory during a fire. Share your procedure with others and use their feedback to rewrite and improve your SOP.

SOLUTION

The fire evacuation procedure will be specific for your laboratory, but should include an indication of where fire alarms are located, what they sound like, how individuals should exit the laboratory, where they should go, and actions to avoid as they exit (for example, taking the time to retrieve notebooks or personal items).

You can see from this last exercise that many of the standard operating procedures for a laboratory can be quite specific for that setting, such as the evacuation route you would take during a fire. However, there are also procedures that will be found in most laboratories. Examples include rules involving safety, the handling of chemicals, the use of a laboratory notebook, and the reporting of data.

BOX 2.1
The Polymerase Chain Reaction

The polymerase chain reaction (PCR) is a method for increasing the amount of given sequences of DNA. As a result, this technique is important in analyzing samples that may contain trace amounts of DNA, such as blood or hair taken from the scene of a crime. The general approach used in PCR is shown in Figure 2.2. The original DNA or sections of this DNA is first divided into its two separate strands (step 1). Small pieces of DNA known as "primers" that bind to specific regions of these strands are then added and act as a starting point for their replication (step 2). Enzymes and nucleotides are also added, which allow the two pieces of single-stranded DNA to be converted into two pieces of double-stranded DNA (step 3). This ends one PCR cycle. In the next cycle, the DNA is again separated into single strands, more primers and

reagents are added, and more double-stranded DNA is formed. Thus, with each cycle the amount of DNA is roughly doubled.[3]

To use PCR in forensic testing, the DNA in a sample is cut into smaller pieces by using enzymes that cleave the DNA at well-defined sequences that vary in their location from person to person, as shown in Figure 2.1. These pieces are then amplified by PCR and later separated by their difference in size by utilizing the method of electrophoresis (see Chapter 23). The pattern that is obtained for the DNA sample is then compared to those found with other individuals or samples. For instance, the unknown in Figure 2.1 has the best match with sample from suspect number 2 because the patterns obtained for their DNA fragments are identical.[2]

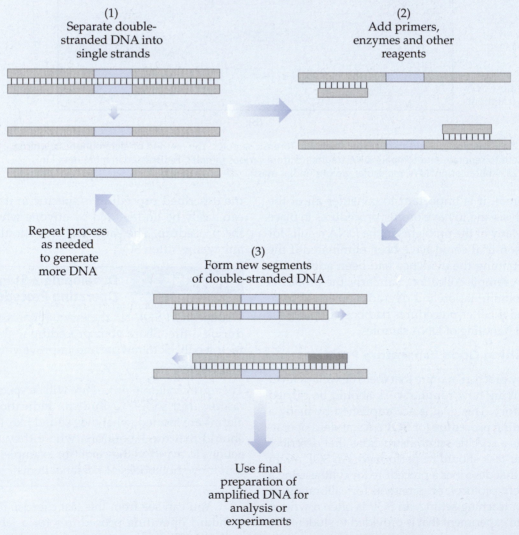

(1)
Separate double-stranded DNA into single strands

(2)
Add primers, enzymes and other reagents

Repeat process as needed to generate more DNA

(3)
Form new segments of double-stranded DNA

Use final preparation of amplified DNA for analysis or experiments

FIGURE 2.2   Basic steps in the amplification of DNA by the polymerase chain reaction.

STANDARD OPERATING PROCEDURES
DOC. NO. 03.02.009
ISSUE 10

TITLE: Electronic balance

PURPOSE/SUBJECT: To accurately weigh and display the correct
                                  amount of a given substance

A. BALANCE CONDITIONS:

The balance will be located in a place so as not to be affected by temperature,
wind currents, sunlight, or other external factors that would change the
accuracy of the balance.

B. BALANCE OPERATION:

1. If possible, the balance will be left on all the time. If the balance must be
    turned off , allow 30 minutes to warm up.

2. Place a weigh container in the middle of the pan and close the doors (where
    applicable). Allow balance to equilibrate.

3. Tare the balance according to manufacturer's directions.

4. Weighing:

    a. Keep the balance doors (where applicable) open only long enough to add
        the material you are weighing to the weigh container.

    b. Accurate readings of weight are taken only when the doors are closed
        (where applicable).

    c. Carefully transfer the material to be weighed into the weigh container.
        Do not return excess material to the original container.

    d. Remove any spills on the balance before taking final weight.

    e. Clean the balance with a lint-free cloth or brush and close doors after use.

    f. Documentation of balance identification number should be recorded in
        the study notebook.

C. SERVICE AND CALIBRATION:

1. The accuracy of the balance is checked and recorded by authorized in-house
    personnel twice a month or as needed if the balance is moved.

2. The balance is serviced every six months by a contract service.

D. REFERENCE STANDARD BALANCE PROCEDURES:

1. One analytical balance will be designated for the purpose of weighing
    reference standards.

2. This balance will have its performance monitored by verification with
    external weights. The verification procedure will be performed every day
    that reference standards are being weighed and documented in a log book.

E. SAMPLE WEIGHING:

1. Analytical or top loading balances are used for weighing samples. This
    balance will have its performance monitored by verification with external
    weights. The verification procedure will be performed every day that
    samples are being weighed and documented in a log book.

FIGURE 2.3   Example of a standard operating procedure (SOP) for using an electronic balance.

2.2 LABORATORY SAFETY

Probably the most important part of good laboratory practices is to make sure your experiments are performed in a safe manner. A chemical analysis laboratory is usually a safe place to work because you are often dealing with small amounts of chemicals and the reactions being employed are often not hazardous. And yet, even in this setting it is possible for accidents to occur. In the chemical laboratories of large companies, accidents lead to an average of one hour of work being missed for every 400,000 employee hours. In teaching laboratories, it has been estimated that the number of accidents may be 100–1000 times greater than in industry.[6]

One reason for the lower rate of accidents in industrial laboratories is that these laboratories are staffed with trained professionals and not with students who are still learning how to handle chemicals. Another factor is that industrial laboratories follow rigid guidelines that help create a safe working environment. The SOPs used to promote safety in a laboratory are referred to as the *chemical hygiene plan* (CHP).

2.2A Common Components of Laboratory Safety

There are many things that can be included under the category of "laboratory safety." These range from the use of a well-designed work area, to the availability of safety equipment and adequate training of people in the laboratory.

Table 2.1 gives some common safety features found in chemical laboratories. These features include safety showers, eye washes, room exits, emergency phones, fire extinguishers, first aid kits, equipment for handling chemical spills, and facilities for the handling, storage, and disposal of chemicals. Each of these items either helps prevent your exposure to chemicals or minimizes any damage if there is such exposure.

Some laboratories have additional features that may be included in this list. One example would be the use of "biohazard" containers in a facility that works with biological materials, like the one that analyzed DNA samples for the O. J. Simpson trial. Another example would be the special precautions taken in a laboratory that works with radioactive substances, where employees need to routinely monitor their work areas for radiation levels and wear badges that determine the amount of radioactivity to which they have been exposed.

2.2B Identifying Chemical Hazards

Definition of a Chemical Hazard. To work safely in a laboratory, you must be able to determine in advance whether each chemical you use has any hazards associated with it. This allows you to adopt appropriate methods for handling the substance and for minimizing the risk a chemical might pose to you or others in the laboratory. A **chemical hazard** (or **hazardous chemical**) can be defined

---

TABLE 2.1   Common Safety Features Found in the Chemical Laboratory

Protective Eye Wear	Laboratory eye wear can range from safety glasses (which protect the eyes from flying materials, including chemicals) to safety goggles (which are more rugged and protect a greater area around the eye) and face shields. Specialized glasses or goggles may be required for work with lasers or ultraviolet light sources.
Fire Extinguishers	Most chemical laboratories will have OSHA Class A, B, or C fire extinguishers that use a chemical such as monoammonium phosphate to smother fires. Other chemical-based fire extinguishers may contain carbon dioxide, which is especially good against flammable liquids, or halons (such as CF_3Br or CF_2BrCl), which work well with electrical equipment.
Eye Washes, Showers, and Fire Blankets	Many laboratories contain special eye washes and showers that are designed to flush the eye or body with large amounts of water in case of a chemical spill. Another feature that is usually present is a fire blanket, which is intended to be wrapped around a person's body if his or her clothing is on fire. All of these items should be in locations that are easy to access in an emergency.
Laboratory Clothing	Laboratory coats are used to avoid any spilled chemicals coming in contact with your body. Special gloves may be required when handling hot substances or those that are hazardous or corrosive in nature. Laboratory workers should not wear sandals or open-toed shoes, which would expose the feet to chemicals or sharp items (like broken glass) that might be on the floor. Other clothing should be an appropriate length to protect the arms and legs from a fire or spill. Workers with long hair should have this hair restrained to keep it from accidentally coming in contact with chemicals or fire.
Proper Training	All persons in the laboratory should be trained in the proper handling, storage, and disposal of the chemicals they will be using. They should also know how to safely and properly handle the laboratory glassware and equipment they will employ. There should be advanced planning and training of laboratory workers in how to handle emergencies such as fires or chemical spills. Everyone working in the laboratory should be familiar with the location of emergency telephones and with the proper routes to take during an emergency evacuation.
Other Features	All safety features and laboratory exits should be clearly marked and easy to access. Individuals who are trained in first aid and the cleanup of spilled chemicals should be available at or near the laboratory setting. A first aid kit and equipment for handling chemical spills should be available in the laboratory, as well as a clearly marked disposal area for broken glassware.

as "any chemical that is a physical or a health hazard." Chemicals that are physical hazards include those that are explosive, highly reactive, or flammable. Health hazards are chemicals that may be toxic or corrosive, produce cancer or birth defects, or cause damage to specific parts of the body, such as the lungs, skin, or eyes.[7–10] It is quite common for a single chemical to present several different types of hazards (for example, being both flammable and toxic). A variety of terms are used to describe these hazards and health effects, including "carcinogen," "corrosive," and "irritant." The definitions of these and related terms can be found in Table 2.2. Further information on how a chemical is determined to be a hazard is provided in Box 2.2.

Chemical Hazard Symbols and Labels. Besides being familiar with the terms used to describe hazardous chemicals, you should also pay close attention to labels on chemical containers. Appropriate warning labels for chemical hazards are required in the United States for use by all manufacturers, distributors, and importers of chemicals. These warnings may be in the form of words, pictures, or symbols (see Figure 2.5) and indicate the dangers that are associated with a chemical, such as its flammability or ability to produce injuries.

One system used to identify chemical hazards is the **National Fire Prevention Association (NFPA) label.** This label is usually drawn as a diamond with four colored

---

TABLE 2.2   Terms for Describing Chemicals That Are Physical Hazards or Health Hazards

Term	Definition	Examples
Physical hazards		
Flammable	A material that is easily ignited	Gasoline, diethyl ether
Explosive	A substance that can cause a sudden, violent chemical reaction with the release of gas and heat	Trinitrotoluene (TNT), nitroglycerine
Oxidizer	A substance that readily yields oxygen to support the combustion or oxidation of other chemicals	Potassium permanganate, organic peroxides
Water-reactive	A chemical that will react with water to become flammable or give off large quantities of flammable or toxic substances	Potassium or sodium metal
Compressed gas	A gas that is held in an enclosed container at an elevated pressure	Cylinders of hydrogen or oxygen gas
Health hazards		
Radioactive	A material that emits ionizing radiation	Radon gas
Acute toxin	A chemical that causes a harmful effect after a single exposure	Sodium cyanide
Chronic toxin	A chemical that causes a harmful effect after long-term exposure	Benzo[a]pyrene (a carcinogen)
Poison	A substance that can kill, injure, or impair a living organism	Arsenic compounds
Biohazard	A biological substance that presents a health hazard	AIDS virus
Etiological agent	A microorganism or related toxin that can cause human disease	Salmonella (common in food poisoning)
Irritant	A noncorrosive chemical that causes reversible inflammation (swelling and redness) on contact with living tissues	1-Propenylsulfenic acid (found in onions)
Corrosive	A chemical that causes the destruction of living tissue at the site of contact	Strong acids and bases, such as hydrochloric acid and sodium hydroxide
Allergen	A substance that may produce an allergic reaction	Aflatoxins (found in peanuts)
Asphyxiant	A chemical that interferes with transport of oxygen in the body	Carbon monoxide, natural gas
Carcinogen	A substance that causes cancer	Benzene, carbon tetrachloride
Reproductive toxin	An agent that creates damage to the reproductive system, such as a substance that causes a change in DNA (a *mutagen*) or the production of non-hereditary birth defects (a *teratogen*)	Ethanol, mercury, and lead compounds

BOX 2.2
Determining the Safety of Chemicals

All chemicals have some risk associated with them. This is true for even "safe" chemicals like sodium chloride and glucose, which can lead to an increase in blood pressure or promote the development of diabetes when present at high levels in the circulation for long periods of time. But how do we determine the dangers that a specific chemical will have? This question is the subject of the field of *toxicology*, which is the study of how chemicals affect living organisms. To determine the safety of a chemical, there are several questions to consider. For example, how much of the substance must be present to create an effect and what will be the nature of this response? Once we know the answers to these questions, we can determine how to minimize our risk when handling such agents.

There are several approaches used to identify hazardous chemicals. One strategy is to compare the structure of a chemical with those of other substances that have known effects on the body. Another approach that can sometimes be used is to look at the effects in people who have been exposed to a specific chemical through their lifestyle (as would occur with smoking or illegal drug use), an accident (such as a chemical spill), or environmental factors (for instance, exposure to lead paint in a home). But for new chemicals, and especially those used in foods and drugs, chemical hazards are identified using animals or other living systems as models for human exposure. These animal studies are often conducted using rats or mice as the subjects, with the agent of interest being given to the animal by the most likely route of exposure (mouth, skin, lungs, etc.) and the animal later being examined for any response.[11-13]

Although animal studies are an extremely important component of chemical safety testing, they are expensive and time-consuming. There is also a growing concern regarding the ethics of this approach, leading to an increase in the development of non-animal-based methods.[13] A good example is the *Ames test* (see Figure 2.4), which uses microorganisms to test for the ability of a chemical to produce mutations in DNA.[14,15] Once a hazardous chemical has been identified, it is next necessary to characterize the risk that would be associated with handling this substance. This process involves determining the amount of chemical that is necessary to provoke a reaction, as well as the risk of a single concentrated dose (an *acute exposure*) versus a repeated, long-term dose (*chronic exposure*). It is also valuable to determine what organs or parts of the body will be affected by the chemical.[11-13] This information allows appropriate safety measures to be developed for handling this agent, as would then be included in its material safety data sheet.

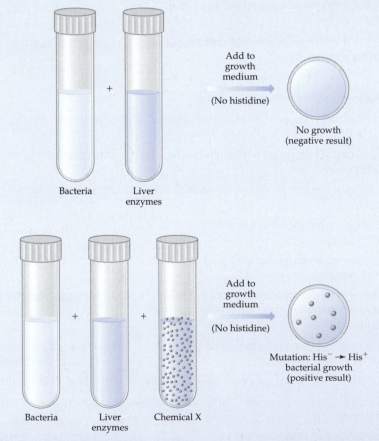

FIGURE 2.4 The *Ames test* for chemical hazards. This technique uses a strain of the microorganism *Salmonella enterica* (formerly *S. typhimurium*), which has a defective gene for an enzyme that synthesizes the amino acid histidine, giving "His⁻" bacteria. However, the action of some chemicals can mutate this bacteria's DNA and change it to a "His⁺" form that can grow in a culture with no histidine present. Such growth indicates the added chemical is a possible cancer-causing agent. A suspension of liver enzymes (usually from a rat) is usually added to the bacteria to also produce and test metabolites generated by the chemical.

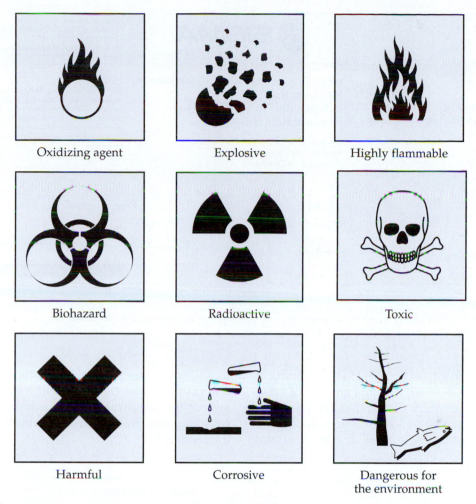

FIGURE 2.5 Common symbols for chemical hazards. These symbols are shown in color in the center of this book.

areas, as demonstrated in Figure 2.6 (see color version in the center of this book). The blue region at the left of the diamond represents the overall health risk of the chemical. The red region at the top indicates the compound's general level of flammability. The yellow region to the right represents the chemical's ability to react with other substances. The white region at the bottom provides other information, such as the reactivity of the chemical with water or its relative ability to oxidize other compounds. The blue, red, and yellow regions each contain a number between zero and four, where zero represents the safest compounds and four represents those with the greatest danger. Sometimes this information is provided without the diamond shape, but the meaning of the 0–4 rankings is the same regardless of how these values are presented.

EXERCISE 2.2 | **Identifying Chemical Hazards**

The NFPA label in Figure 2.6 is for ethidium bromide, a chemical that might be found in a laboratory that conducts DNA testing. The label from a bottle of ethidium bromide is also provided on the next page. Based on the NFPA label and the label from this chemical's container, what types of physical or health hazards must be considered when you are working with ethidium bromide?

SOLUTION
The NFPA label for ethidium bromide indicates that this chemical should pose no significant threat in terms of its flammability or reactivity. The same label, however, also shows that this chemical is a level 3 health hazard. The label from the container gives more specific information by showing that ethidium bromide is a toxic substance that can cause heritable genetic damage and act as an irritant to the eyes, respiratory system, or skin. Precautions to follow during the handling of this chemical are also listed on the container label.

2.2C Sources of Information on Chemicals

Material Safety Data Sheets. It is important to consider the possible hazards of each chemical you may deal with in a laboratory. There are several sources of information to help in this task. Probably the most complete set of information on a chemical can be found in its **material safety data sheet** (or **MSDS**). An MSDS is a set of one or more sheets that must be sent with each chemical that is produced or provided by a manufacturer. An example of an MSDS is given in Appendix A. Other examples can be found on the Internet or obtained from chemical suppliers.

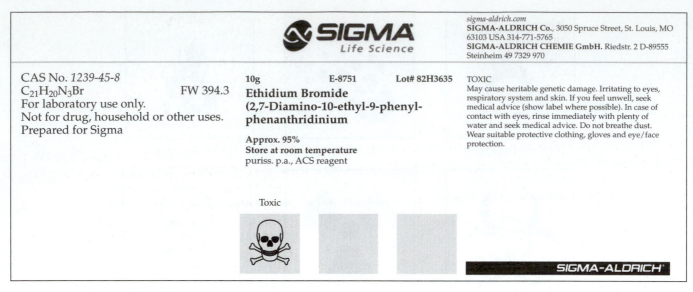

sigma-aldrich.com
SIGMA-ALDRICH Co., 3050 Spruce Street, St. Louis, MO 63103 USA 314-771-5765
SIGMA-ALDRICH CHEMIE GmbH. Riedstr. 2 D-89555 Steinheim 49 7329 970

CAS No. *1239-45-8*
$C_{21}H_{20}N_3Br$ FW 394.3
For laboratory use only.
Not for drug, household or other uses.
Prepared for Sigma

10g   E-8751   Lot# 82H3635

**Ethidium Bromide
(2,7-Diamino-10-ethyl-9-phenyl-phenanthridinium**

**Approx. 95%
Store at room temperature**
puriss. p.a., ACS reagent

Toxic

TOXIC
May cause heritable genetic damage. Irritating to eyes, respiratory system and skin. If you feel unwell, seek medical advice (show label where possible). In case of contact with eyes, rinse immediately with plenty of water and seek medical advice. Do not breathe dust. Wear suitable protective clothing, gloves and eye/face protection.

SIGMA-ALDRICH

Figure for Exercise 2.2
A label from a container of ethidium bromide, an agent used in labeling and detecting DNA. (Reproduced with permission, courtesy of Sigma-Aldrich.)

Even a quick glance at the MSDS in Appendix A shows that this resource contains a large amount of information. For instance, an MSDS should describe the chemical and physical properties of a chemical, including the ability of this chemical to create a fire or explosion, and its ability to react with other chemicals. Other items you will find listed in an MSDS will include the health hazards of the chemical and procedures to dispose of this substance or to follow after someone has been exposed to the chemical.

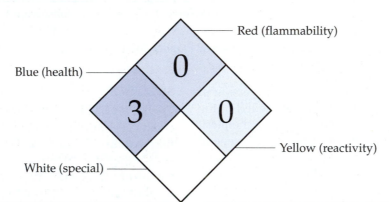

Red (flammability)		
0		Not combustible
1		Combustible if heated
2	Warning	Combustible liquid flash point of 100° to 200° F
3	Caution	Flammable liquid flash point below 100° F
4	Danger	Flammable gas or extremely flammable liquid

Blue (health)		
0		No unusual hazard
1	Caution	May be irritating
2	Warning	May be harmful if inhaled or absorbed
3	Warning	Corrosive or toxic. Avoid skin contact or inhalation
4	Danger	May be fatal on short exposure. Specialized protective equipment required

Yellow (reactivity)		
0	Stable	Not reactive when mixed with water
1	Caution	May react if heated or mixed with water but not violently
2	Warning	Unstable or may react violently if mixed with water
3	Danger	May be explosive if shocked, heated under confinement or mixed with water
4	Danger	Explosive material at room temperature

White (special)	
W	Water-reactive
Oxy	Oxidizing agent

FIGURE 2.6   The National Fire Prevention Association (NFPA) labeling system. This particular NFPA label is for the chemical ethidium bromide. A horizontal line is sometimes used with symbols in the white (special) category; for instance, a "W" with a line through it means a chemical is water-reactive.

EXERCISE 2.3   Using a Material Safety Data Sheet

Determine the physical hazards and potential health risks of benzene by using the MSDS that is provided for this chemical in Appendix A.

SOLUTION

The MSDS shows that benzene is a flammable liquid, with an NFPA flammability rating of level 3. The health risks of benzene include the fact that it is an irritant to the respiratory tract, skin, and eyes. This chemical is also a carcinogen and mutagen.

Other Resources.   There are many other sources of information that can be used to learn about the risks of chemicals. A summary of the physical and chemical properties of common compounds can be found in the *CRC Handbook of Chemistry and Physics*[16] or the Merck Index.[17] Information on the health effects of specific chemicals can also be found in the *Sigma-Aldrich Library of Chemical Safety*,[18] *Sax's Dangerous Properties of Industrial Materials*,[19] and *A Comprehensive Guide to the Hazardous Properties of Chemical Substances*.[20] Other useful texts that deal with laboratory safety and chemical hazards include References 7–10.

2.2D Proper Handling of Chemicals

Minimizing Exposure to Chemicals.   Whenever you are working with chemicals, it is wise to minimize your exposure to any substances that are possible hazards. To do this, it is important to be familiar with the various ways chemicals can enter the body.[7, 10] The first way is through inhalation, which can be a problem when you are working with a gas or volatile liquid. An inhaled chemical can damage the mouth, throat, and lungs or pass on to the blood and other parts of the body. For instance, inhaled ether can quickly enter the blood and lead to dizziness, disorientation, or unconsciousness. The inhalation of small solid particles is also a problem, because these particles can lodge in the lungs and form a long-term irritant, as occurs with asbestos fibers. The inhalation of chemicals can be greatly reduced by working with volatile substances in a ventilation hood. If needed, you can also wear a mask to prevent breathing in dust or small particles.

Contact with the skin or eyes is the second way you can be exposed to chemicals. This is the most common type of chemical exposure, so it is essential that you *always* use eye protection and have adequate clothing when you are in a laboratory. In addition, wear suitable gloves when working with hazardous chemicals and make sure to properly cover all cuts and abrasions, which can provide easy routes for chemical entry. Furthermore, clean your work area after you are finished with an experiment. This habit will avoid exposing others to substances you have been using and is especially important

when you are using equipment or computers that may be operated by others.

Ingestion is the third way you might be exposed to a chemical. This route is not as common as exposure though the skin and eyes, but it can occur. For example, a small amount of a chemical on your hands or within the air may enter your body as you eat. It is for this reason that you should *never* eat or drink within a laboratory. Applying cosmetics, smoking, and using gum or chewing tobacco in a laboratory is also discouraged. For the same reason, you should always wash your hands after handling chemicals and before leaving a laboratory.

The final possible route of chemical exposure is through injection, which can occur when you are dealing with chemicals and sharp objects such as needles or blades. An accidental injection is not common but can be serious in that it may place chemicals directly into the bloodstream. The best guard against injection is to carefully use sharp devices and to dispose of these devices (along with broken glassware) in a sturdy container, typically labeled "sharps." This practice not only protects everyone in your lab but ensures that those who later remove these items will be able to do so safely.

EXERCISE 2.4   Identifying Precautions for Handling a Chemical

Based on the MSDS in Appendix A, what precautions for protection and minimizing exposure should you follow when handling benzene?

SOLUTION

The recommended precautions are stated under Section 8.0 of this MSDS (Exposure Controls/Personal Protection). Benzene is a volatile compound, so adequate ventilation is required when you are working with this substance and appropriate respiratory protection is needed. Gloves, eye protection, and adequate clothing are also recommended to avoid contact of this chemical with the eyes and skin. Washing your hands after handling this substance is a further precaution that should be taken to avoid accidental ingestion.

Chemical Storage.   Another factor to consider as part of laboratory safety is the way in which chemicals are stored. The goal is to keep the chemicals in a stable form that will not present a hazard or cause any interactions with nearby substances. There are several general guidelines to aid in this storage process.[7–10] Flammable or explosive chemicals should be kept in special metal cabinets that have been designed for such substances. Volatile chemicals should be kept in well-ventilated areas like a vented cabinet or hood. Gas cylinders should be securely fastened to a wall or laboratory bench, and water-reactive substances must be kept in a dry, moisture-free environment. It is also essential to keep chemicals that might react with each other in separate areas, such as acids and bases, or oxidizing agents and reducing agents. Some substances, like many biological materials, may require refrigeration. Others, such as possible carcinogens, radioactive substances, or toxins, should be kept in their own well-labeled areas.

One thing that is essential whenever you are using or storing chemicals is to make sure the chemical containers are properly labeled. The label you place on a chemical container should include (1) the name of the chemical, (2) the date it was prepared, received, and/or opened, and (3) the name of the person who prepared or used the chemical. Abbreviations for chemical names should not be used. Proper labels will help others know exactly what substances are present in a container. These labels will also help you avoid errors in experiments due to chemical mix-ups and make it easier for you to store and later dispose of the chemicals in an appropriate manner.

Chemical Disposal. Another question to consider as part of laboratory safety is "What should be done with any used chemicals after an experiment has been completed?" The procedures for disposing or handling used chemicals are referred to as *laboratory waste management*. Proper laboratory waste management goes beyond dealing with just the excess products or reagents you might produce during an experiment. This management also includes chemicals that have reached their expiration date or that may have changed from their original composition.

It is important to always know in advance how you should dispose of a chemical. Inappropriate disposal can lead to contamination of the environment or the creation of a possible fire or explosion hazard. The Occupational Safety and Health Administration (OSHA) and the U.S. Environmental Protection Agency both require that all chemicals be disposed of in a safe and responsible manner.[7–10, 21] Information on how this should be done can be found in a chemical's material safety data sheet. As an example, the MSDS for benzene in Appendix A says that this chemical should be disposed of (where allowed by law) by burning it in a chemical incinerator. When carried out properly, this process converts all of the benzene (C_6H_6) into water and carbon dioxide, two nontoxic chemicals.

The disposal of chemicals is both an environmental and legal issue. It is for this reason that many companies, colleges, and universities have special offices to deal with governmental agencies and to collect and dispose of excess chemicals. Even though most laboratories do not dispose of their own chemicals, there are a few general principles that all laboratory workers should follow. For instance, before a material is prepared for disposal, it is necessary to identify which chemicals in the material are hazardous or have special disposal requirements. This identification can be accomplished by looking at resources like the material safety data sheets for the chemicals. If you are dealing with more than one substance, it is essential to separate the chemicals into groups based on their reactivity

and composition. And finally, all excess chemicals should be placed into appropriate storage containers and enough information should be provided on the content of these containers so that correct disposal procedures can be selected.

2.3 THE LABORATORY NOTEBOOK

Another part of good laboratory practices is the need for scientists to maintain a complete and accurate record of the work they conduct in the laboratory. They do this by using a **laboratory notebook**. The laboratory notebook is a record of the procedures that were used by a scientist in an experiment, the results that were obtained, and the conclusions that were reached from the experiments. The laboratory notebook also plays a vital role when these results are eventually communicated to others in articles or reports, and can be used to establish when a particular experiment was conducted, as might be needed when a scientist applies for a patent.

The practice of using notebooks to describe experiments and observations goes back many centuries. One famous example is the work of Leonardo da Vinci. During his lifetime (1452–1519), da Vinci conducted research and recorded observations in his notebooks on topics spanning from architecture, engineering, and mechanics to geology, chemistry, and biology. To protect his writings, he would write his entries backwards so they could only be read with a mirror. This practice meant many of his discoveries were not widely known until long after his death. In contrast to this, modern scientists use the laboratory notebook as a tool to clearly communicate results and procedures for use by others.

2.3A Recommended Notebook Practices

Although there are no specific requirements on what should be in every laboratory notebook, there are some general guidelines that should always be followed (see Table 2.3).[22–25] The purpose behind each of these guidelines becomes clear when you consider what the ultimate goals are in keeping a laboratory notebook. One goal is to provide a detailed and accurate account of your research that others can use in their own work. This is the reason why it is important to record data directly into the notebook and to keep a thorough record of your experiments and results. A second purpose of the notebook is to present a record that is easy to understand by others. The use of clear headings, a table of contents, and a consistent format for entries helps this to be achieved. A third goal for keeping a notebook is to provide a record of how and when your experiments were conducted. A record like this is important in academic and industrial laboratories as proof of when research was conducted for establishing intellectual property rights (as needed for obtaining patents) or in documenting work for later use in scientific reports and publications. These are just some reasons why it is essential for a laboratory worker to always date and sign all entries in a notebook.

TABLE 2.3   Recommended Practices for Keeping a Laboratory Notebook

General Notebook Properties	The notebook should contain only permanently bound pages. All pages should be numbered in advance and used in sequential order, with no blank pages being left in between.
Entry Format	All entries should be easy to read and made in permanent ink (*not* with a pencil or water-soluble marker). The entry for each experiment should include a short statement of the experiment's purpose, a description of the methods and conditions that were used, and a statement of the results that were obtained. Each of these sections should have a clear, concise heading. In addition, there should be a table of contents at the beginning of the notebook in which the title, dates, and pages of each experiment are listed.
Items to Include in an Entry	Chemical structures, tables, and diagrams that result from or describe the experiments should be included. You should also include a description of how you prepared your reagents and samples (including masses, volumes, and types of equipment that you used). Any abbreviations that you utilize should be defined within the notebook. You should provide examples of any calculations that were carried out during the experiment, including those employed during the preparation of samples and reagents. If you are using a procedure that was developed by someone else, a reference to this source should be given. If you have changed the procedure, the changes you made should be described in detail.
When to Make an Entry	*Always* record data directly into your notebook. When you are measuring the mass of a substance, *do not* write the measured values on a loose piece of paper and later transfer them into the notebook. This also applies to other items, such as sketches of equipment and tables of data that you generate during an experiment.
Graphs and Printouts	Graphs and printouts from computers or instruments can be taped, pasted, or stapled within the notebook. A brief description of what the data represent and how the data were obtained should be noted. Each attached item should be signed and dated so that part of the signature and date is on the attached material and part overlaps onto the notebook page. For large numbers of graphs and printouts, a separate notebook or file may be required. However, these items should still be signed and dated, and their location should be clearly described within the main laboratory notebook.
Entry Validation	Sign and date all notebook entries as they are being made. In addition, you should sign and date the bottom of each notebook page once it has been completed. The notebook should also be examined and signed on a regular basis by at least one other person in your laboratory, such as your supervisor or instructor.
Handling Errors	Any errors or changes that are made in the notebook (for example, an incorrect description of a procedure or an incorrectly written data point) may be noted by crossing these out with a single line, with the date and your initials written next to the correction. *Never* erase or totally remove an earlier entry from a notebook.
Dealing with Empty Spaces	*Never* tear or remove any pages from a laboratory notebook. A well-kept notebook will not have any empty pages between entries. If there is any unused space at the end of a page, you can mark this off by placing a large "X" through it.

An example of a good notebook entry is shown in Figure 2.7. This example illustrates many of the principles we have already discussed. This entry is written in a style that is easy to read and follow, and the description of the experiment is sufficiently complete to allow someone else to reproduce the experiment. Maintaining a laboratory notebook in this fashion takes diligence and practice, but such efforts will make this notebook a valuable aid in your laboratory work.

2.3B Electronic Notebooks and Spreadsheets

An issue of growing interest in chemical laboratories is the use of computers to maintain an *electronic laboratory notebook* (or *ELN*).[26–28] An ELN is a digital record of a laboratory experiment in which text can be combined with graphs, structures, images, and other computer-based sources of information. Such a system has the potential to offer much greater flexibility than a standard notebook in terms of data analysis, report generation, and the communication of results. Electronic notebooks, however, also have potential disadvantages, such as being subject to data loss due to computer viruses or computer failure. In addition, there are security issues that can result if hackers enter into the notebook. These problems can be minimized by routinely making backup copies of the notebooks and by taking appropriate measures to prevent their unauthorized use. Another potential issue occurs if the ELN files can easily be updated or are subject to tampering. However, efforts have been made in recent years to overcome this problem and make ELN entries more suitable as a permanent record of laboratory work.

Many industrial laboratories use electronic notebooks as part of a *laboratory information management system* (*LIMS*). An LIMS is a computer software package for collecting data from instruments in a laboratory and processing this information into a suitable form for a report.[29] When dealing with electronic notebooks, an LIMS can be modified to produce a printed and verifiable copy of an entry as it is generated. In a smaller laboratory the same result can be obtained by printing a copy of each page from the electronic notebook and permanently attaching these copies into a binder. Thus, at least for the present, the paper version of the laboratory notebook still holds an important place in the chemical laboratory.[27, 28]

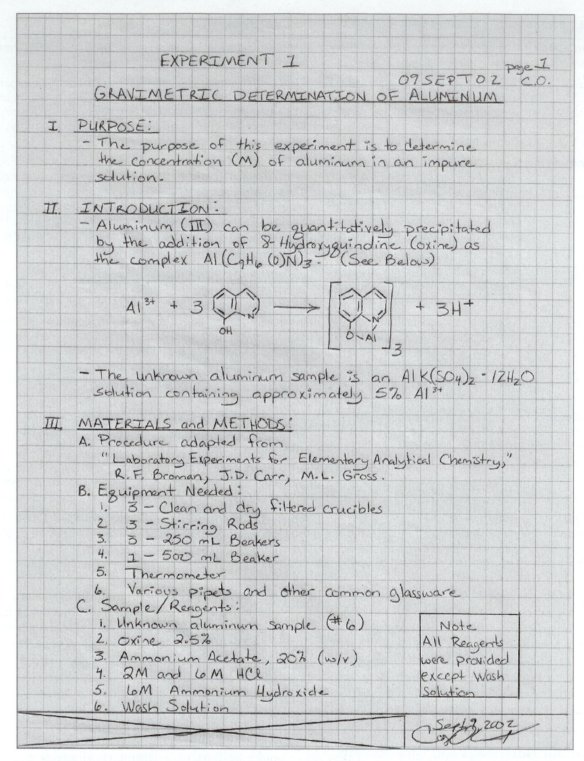

FIGURE 2.7A   Page from a well-kept student laboratory notebook. The correct structure for the product of the described method is Al(oxine)₃. (Reproduced with permission from C. Ohnmacht.) *(continued on next page)*

Another computer-based tool that can be used with either a traditional or electronic notebook is a **spreadsheet**. This is a program used to record, analyze, and manipulate data. A spreadsheet is a valuable tool in automatically carrying out repetitive calculations. An example of a spreadsheet is shown in Figure 2.8. In this particular spreadsheet, a set of results has been placed in a table, where the numbers or individual entries can be used in calculations or to make graphs. These graphs, tables, and calculations can later be placed within a notebook and used to prepare reports. You will be given the opportunity throughout this book to work with spreadsheets in solving problems related to analytical chemistry. If you are not already familiar with how to use spreadsheets, a step-by-step guide can be found in Appendix C. Further help can be found in sources like References 30–33.

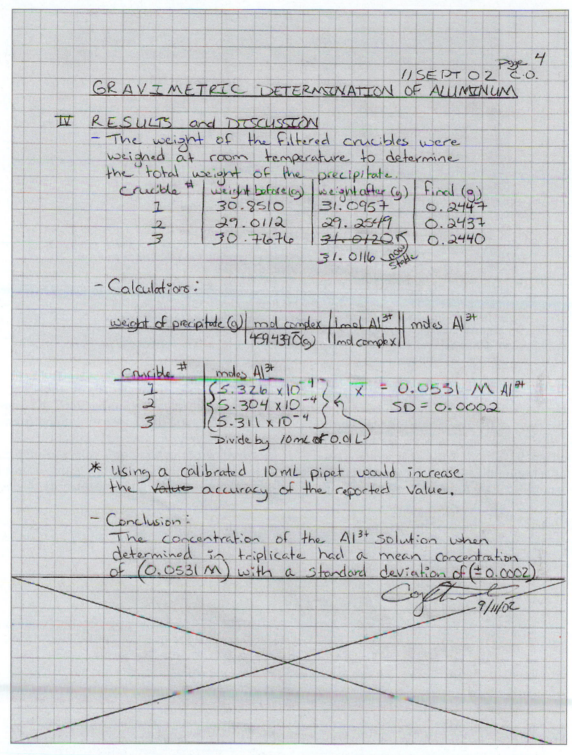

FIGURE 2.7B   Page from a well-kept student laboratory notebook. (Reproduced with permission from C. Ohnmacht.) *(continued from previous page)*

2.4 REPORTING EXPERIMENTAL DATA

2.4A The SI System of Measurements

Because they must share results with others throughout the world, scientists have long recognized the need for a standard set of units in which they can report data. This goal is accomplished through the **SI system** of measurements, also known as the *International System of Units* or the *Système Internationale d'Unités*. The SI system provides a set of uniform standards for describing mass, length, time, and other measurable quantities. This system traces its beginnings to the late 1700s in France and was adopted by other nations, including the United States, in the *Treaty of the Meter*, which was signed in Paris on May 20, 1875.

Fundamental and Accepted SI Units.   All measurements in the SI system can be described by a small group of base units that are related to physical constants or

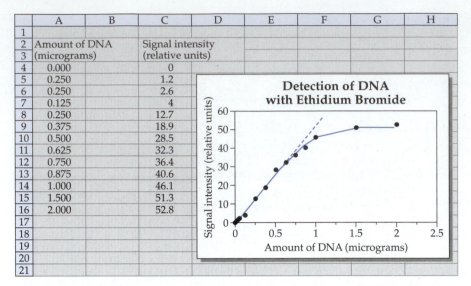

	A	B	C	D	E	F	G	H
1								
2	Amount of DNA		Signal intensity					
3	(micrograms)		(relative units)					
4	0.000		0					
5	0.250		1.2					
6	0.250		2.6					
7	0.125		4					
8	0.250		12.7					
9	0.375		18.9					
10	0.500		28.5					
11	0.625		32.3					
12	0.750		36.4					
13	0.875		40.6					
14	1.000		46.1					
15	1.500		51.3					
16	2.000		52.8					
17								
18								
19								
20								
21								

FIGURE 2.8   An example of a spreadsheet used to analyze data for the quantitative measurement of DNA in samples based on the fluorescence of ethidium bromide in the presence of the DNA. The dashed line shows the best-fit response to the results for samples containing 0 to 0.625 μg DNA, while the solid line shows a smoothed curve for the entire data set. Methods for finding the best-fit line for a linear relationship are discussed in Chapter 4.

well-defined values.[34,35] These *fundamental SI units* are given in Table 2.4. You are probably familiar with many of these units, such as the *meter* as a unit of length, the *kilogram* as a unit of mass, and the *second* as a unit of time. Another fundamental SI unit is the *mole*, which is used in chemistry for counting the number of members of a substance (for example, there are 6.02×10^{23} molecules of

H_2O in one mole of water). The other base units in the SI system are the *kelvin* for temperature, the *ampere* for electrical current, and the *candela* for luminous intensity.

The seven fundamental units of the SI system can be combined to obtain other values known as *derived SI units* (see Table 2.5). For instance, the SI unit for electric charge (the coulomb, C) is the quantity of electricity that

TABLE 2.4   Fundamental Units of the SI System

Measured Quantity	Base Unit	Symbol	Definition of Unit
Length	meter	m	A meter is defined as the distance traveled by light in a vacuum in 1/299,792,458 of a second.
Mass	kilogram	kg	A kilogram is defined as the mass of a platinum–iridium cylinder that is kept as the international standard for the kilogram at the International Bureau of Weights and Measures in Sèvres, France.
Time	second	s	A second is defined as the amount of time equal to 9,192,631,770 periods of the radiation corresponding to the transition between the two hyperfine levels of the ground state of cesium-133.
Amount of substance	mole	mol	A mole is defined as the number of individual entities of a substance that is equal to the number of carbon atoms in 0.012 kg of carbon-12.
Temperature	kelvin	K	The kelvin temperature scale assigns the lowest possible temperature (absolute zero) a value of 0 K, and the triple point of water (where the gas, solid, and liquid forms of water all exist in equilibrium) a value of 273.16 K.
Electric current	ampere	A	One ampere is defined as the constant current that produces a force of 2×10^{-7} newton per meter of length when maintained in two straight parallel conductors of infinite length and negligible circular cross section that are placed one meter apart in a vacuum.
Luminous intensity	candela	cd	A candela is the luminous intensity measured in a given direction from a source that emits monochromatic radiation with a frequency of 540×10^{12} hertz, and that has a radiant intensity of 1/638 watt per steradian in the observed direction.

TABLE 2.5   Derived and Accepted SI Units with Special Names

Measured Quantity	Base Unit (Symbol)	Relationship to Other SI Units
Frequency	hertz (Hz)	1 Hz = 1/s
Force	newton (N)	$1\ N = 1\ m \cdot kg/s^2$
Pressure	pascal (Pa)	1 Pa = 1 N/m
Energy	joule (J)	$1\ J = 1\ N \cdot m$
	electron-volt (eV)	$1\ eV = 1.60218 \times 10^{-19}\ J$
Electric charge	coulomb (C)	$1\ C = 1\ A \cdot s$
Power	watt (W)	1 W = 1 J/s
Electric potential	volt (V)	$1\ V = 1\ W/A = 1\ J/(A \cdot s)$
Electric resistance	ohm (Ω)	$1\ \Omega = 1\ V/A$
Temperature	degree Celsius (°C)	°C = K − 273.15
Time	minute (min)	1 min = 60 s
	hour (h)	1 h = 3,600 s
	day (d)	1 d = 86,400 s
Volume	liter (L)	$1\ L = 10^{-3}\ m^3$
Mass[a]	unified atomic mass unit (u)	$1\ u = 1.66054 \times 10^{-27}\ kg$

[a]In many biologically related fields (and particularly in work with proteins), the "dalton" (symbol, Da) is commonly used in place of the unified atomic mass unit, where 1 Da = 1 u.

flows for one second at a current of one ampere ($1\ C = 1\ A \cdot s$). Other derived units we will use in this book are the volt, the joule, and the degree Celsius. The SI system also allows the use of some common units that are related to but not directly derived from the fundamental SI units. These related units are called *accepted SI units*. Examples in analytical chemistry include the liter as a measure of volume, minutes and hours as units of time, the electron volt as a unit of energy, and the unified atomic mass unit as a measure of mass.

Converting Units.   Although all scientific measurements should ideally be reported using the SI system, you will often encounter cases where numbers are reported in other units. Common examples in the United States are the use of pounds, gallons, inches, feet, and miles. There are many older scientific papers that have reported data in units now discouraged by the SI system. Thus, it is important that you be able to convert between these units and SI units. A list of conversion factors that are available for this purpose can be found in many resources, including the *CRC Handbook of Chemistry and Physics*.[16]

To illustrate how you can make such a conversion, let's consider a physical constant that we will be using throughout this text—the speed of light in a vacuum. In English units, this speed is commonly given as 186,000 miles per second. In the SI system, this same constant would be given in meters per second. To make this change, you need to obtain the conversion factor

between miles and meters (1 mile = 1609 m) and then use this factor as a ratio in the following relationship.

$$\text{Speed of light (m/s)} = (186{,}000\ \text{mile/s}) \cdot (1609\ \text{m/mile})$$
$$= 299{,}000{,}000\ \text{m/s} \qquad (2.1)$$

The result is a value that is now expressed in the appropriate SI units.

Notice in this conversion that we recorded and compared the units on our numbers to make sure the final result was expressed in the desired fashion. This process is known as **dimensional analysis**, and it is a tool we will use throughout this book to check calculated results. To carry out dimensional analysis, you first need to write down the units on all the numbers in your calculation, as shown in Equation 2.1. Next, if there are any numbers being multiplied or divided by each other, you can cross out any common units that appear in both the numerators and denominators of these terms. This was done in the previous example by crossing out the unit of "mile" in the upper and lower portions of the right-hand terms in Equation 2.1. You also need to make sure that all numbers being added or subtracted have the same units (which was not required in our "speed of light" example that only used multiplication). If you have set up your equations and numbers in the right manner and used appropriate conversion factors, your final answer should also have the correct units.

EXERCISE 2.5   Converting Between Units of Temperature

One area in which conversions are often needed in science is in the reporting and recording of temperature. In the United States, temperatures are commonly described in degrees Fahrenheit (°F) while other countries generally use degrees Celsius (°C), where °C = (°F − 32) · (5/9). When a chemist analyzes samples at 77°F, what is the temperature in °C? What units must be associated with "32," "5," and "9" in the temperature conversion equation to make this calculation valid?

SOLUTION

Let's first use dimensional analysis to examine the given conversion equation. The term (°F – 32) must have both numbers in degrees Fahrenheit to give a valid result. The purpose of this term is to adjust for the different reference points used in the Celsius and Fahrenheit scales for the freezing point of water (32°F versus 0°C). Next, we have the ratio 5/9, which corrects for the different sizes of these scales, where a change of five degrees Celsius corresponds exactly to nine degrees Fahrenheit, or 5°C/9°F. Placing these units into the temperature conversion equation gives the more complete expression °C = (°F − 32°F) · (5°C/9°F), which we can then use as follows.

$$(77°F − 32°F) \cdot (5°C/9°F) = \mathbf{25°C}$$

Note that all of the units now work out to give us the final desired answer in degrees Celsius.

SI Prefixes. As our understanding of nature has increased, so has our need to describe bigger and smaller quantities in measurements. This need can be met by using *scientific notation*. For instance, the speed of light could be written in scientific notation as 2.998×10^8 m/s, or approximately 3×10^8 m/s. Another option in the SI system is to use a prefix along with the main measurement unit to represent various factors of 10 (10^3, 10^6, 10^9, and so on). A list of these prefixes and their abbreviations is provided in Table 2.6. These prefixes are used by adding them to the SI unit of interest. As an example, a speed of 3×10^8 m/s could also be written as 3×10^{10} centimeters per second (cm/s), 3×10^5 kilometers per second (km/s), or 0.3 gigameters per second (Gm/s).

The same approach can be used with all of the other units given in Tables 2.4 and 2.5, with the exception of the kilogram. For mass measurements, the gram is used in place of the kilogram as the base unit to which a prefix is added. The prefix you choose is usually in the same general range as the number you are describing. This is why the prefix giga- was chosen to describe the speed of light, because 3×10^8 fell near 10^9. However, more common prefixes can also be selected, as was done when we gave the speed of light in units of cm/s or

TABLE 2.6   Prefixes Used in the SI System

Prefix Name (symbol)	Meaning
yotta- (Y)	10^{24} (1 septillion)
zetta- (Z)	10^{21} (1 sextillion)
exa- (E)	10^{18} (1 quintillion)
peta- (P)	10^{15} (1 quadrillion)
tera- (T)	10^{12} (1 trillion)
giga- (G)	10^9 (1 billion)
mega- (M)	10^6 (1 million)
kilo- (k)	10^3 (1 thousand)
hecto- (h)	10^2 (1 hundred)
deka- (da)	10^1 (ten)
deci- (d)	10^{-1} (1 tenth)
centi- (c)	10^{-2} (1 hundredth)
milli- (m)	10^{-3} (1 thousandth)
micro- (μ)	10^{-6} (1 millionth)
nano- (n)	10^{-9} (1 billionth)
pico- (p)	10^{-12} (1 trillionth)
femto- (f)	10^{-15} (1 quadrillionth)
atto- (a)	10^{-18} (1 quintillionth)
zepto- (z)	10^{-21} (1 sextillionth)
yocto- (y)	10^{-24} (1 septillionth)

km/s. Not all the prefixes in Table 2.6 are common in analytical chemistry, although many that describe moderate to small quantities are often employed in this field, especially during trace analysis.

EXERCISE 2.6   SI Prefixes and DNA Analysis

One objection made in the O.J. Simpson trial was that the DNA samples had been contaminated. This issue was of concern because the samples had been processed by PCR, which allows even a single copy of contaminant DNA to be converted into a large amount of material. Ideally, the amount of DNA will double each time PCR is carried out on a sample. Thus, if you start with only one molecule of DNA (where 1 molecule = 1.66×10^{-24} mol) you get up to 2^{15} molecules of DNA after 15 cycles of PCR or 2^{30} molecules after 30 cycles. Using the prefixes in Table 2.6, determine how many moles of DNA would be present after 15 and 30 PCR cycles if you begin with a single copy of DNA.

SOLUTION

We are told that after 15 PCR cycles the amount of DNA can increase by as much as 2^{15} (or 32,768-fold). By combining this information with the fact that one molecule = 1.66×10^{-24} mol (or **1.66 ymol**), we get the following maximum moles of DNA that could be present after 15 cycles.

mol duplicated DNA

$$= (32{,}768) \cdot (1.66 \times 10^{-24} \text{ mol DNA})$$

$$= 5.44 \times 10^{-20} \text{mol} = \textbf{54.4 zmol or 0.0544 amol}$$

Using the same approach, after 30 cycles there could be 1.78×10^{-15} mol, or **1.78 fmol**, of DNA.

2.4B Significant Figures

Another use of good laboratory practices in data reporting is to make sure you have the correct number of digits when you record results. There are two principles to remember when doing this: (1) all measurements have some degree of uncertainty in them, and (2) the uncertainty will determine the number of digits you can use when reporting a result. The total number of digits that we can use to reliably report a result is known as the number of **significant figures** for that value.

Recording Results. To illustrate what is meant by "significant figures," let's examine two types of devices you might use to measure temperature (see Figure 2.9). The first of these is an electronic thermometer, which has a **digital display**. In this case, the measured temperature has been converted into a number with a well-defined number of digits. To read this type of result you would generally record all of the digits shown (in this case, 25.1°C) because the manufacturer of this instrument has previously determined that the last displayed digit (the number "1" to the right of the decimal) is the first value in which there is some degree of uncertainty. Even though it would be possible to build a display that gives more digits, doing so would only add numbers that have random variations in their values.

The second type of device you might use is one with an **analog display**. This display has a signal that has not yet been converted to a distinct number, but is instead displayed on a continuous scale of values, such as the display on a standard mercury thermometer. In recording a number from such a device, you must find the two values on the display that are just above and below the measured response. In Figure 2.9 the measured temperature is somewhere between 25.0 and 25.2°C. Although we do not know the exact temperature beyond this point, we can estimate how far it lies between these two values to provide one more significant figure in our recorded value. When we make this estimate, we get a number that lies about 0.1 units above 25 degrees, again giving a temperature of 25.1°C.

The same approach employed for reading an analog display should be used to read a result from a graph. Significant figures are also essential to consider when you prepare a graph, in that you should select both a scale and grid size that allow your data to be displayed with the most possible significant figures. An example of a graph that has been prepared in this manner is given in Figure 2.10. This figure also illustrates some good laboratory practices for plotting data points and labeling a graph so that the final plot will be easy for others to understand.

There are a few simple rules that allow you to tell how many significant figures are or should be in a number.[36] A review of these rules, as well as those for rounding a value to the correct number of digits, can be found in Appendix A. For instance, a temperature reported as 25°C implies that we have two significant figures, while a temperature given as 25.1°C implies that we have three significant figures. The more significant figures you can give for a number, the easier it will be for you to analyze and compare your data to reference values or results obtained by others. It is important, however, to avoid including nonsignificant (random) figures in a result because this will make the number appear more reliable than it really is.

Combining Results.   When you are using a series of numbers to calculate other values, you need to remember that the uncertainty (or number of significant figures) in your final answer will be affected by the uncertainty in all the numbers that are used in the calculation. The rules to follow when determining the number of significant figures in a calculated result are also provided in Appendix A. These rules cover many basic operations, such as addition, subtraction, multiplication, division, and the process of taking the logarithm or antilogarithm of a number. An example of this process is given in the following exercise. We will revisit this issue in Chapter 4 under the topic of "error propagation," when we later discuss the origin of these rules and see a more exact approach for determining the uncertainty in a calculated value.

EXERCISE 2.7 **Determining the Significant Figures in a Molar Mass**

Ethidium bromide ($C_{21}H_{20}BrN_3$) is used in analytical laboratories to label and detect DNA. Calculate the molar mass (or "molecular weight" in units of g/mol) for this compound, using the correct number of significant figures in your answer.

SOLUTION

The average atomic masses for C, H, Br, and N (as listed in the periodic table) are 12.01115, 1.00797, 79.909, and 14.0067 g/mol, respectively. By using these values and

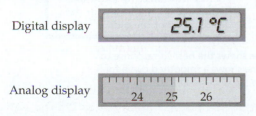

| Digital display | 25.1 °C |

| Analog display | 24 25 26 |

FIGURE 2.9 Examples of digital and analog displays.

the known number of atoms in one molecule of ethidium bromide, you can obtain the following molar mass for this chemical.

Carbon:	21 · (12.01115 g/mol)	= 252.23415
Hydrogen:	20 · (1.00797 g/mol)	= 20.1594
Bromine:	1 · (79.909 g/mol)	= 79.909
Nitrogen:	3 · (14.0067 g/mol)	= + 42.0201
Molar mass:		394.32265
		= **394.323 g/mol**

Several rules for significant figures were used in this example. Because 21, 20, 1, and 3 are integers and have unlimited significant figures, the products of the atomic masses and these integers should have the same number of significant figures as the atomic masses. When these products were added, their sum was rounded off to have the same number of digits past the decimal as the product with the fewest number of digits in this location (79.909 for bromine, which had three digits to the right of

the decimal). This process gave a final answer of 394.323 g/mol, with a total of six significant figures (three to the left and three to the right of the decimal).

Notice in the last exercise that we did not round to the correct number of significant figures until we obtained the final answer. This was done to avoid **rounding errors**, which are errors produced when a number is rounded off too early in a calculation. If rounding errors are not dealt with properly, they can result in the loss of valid significant figures in a calculated result. One way you can avoid rounding errors is to allow each value in the calculation to carry at least one nonsignificant figure until the final result is obtained. These additional nonsignificant figures are known as **guard digits**. An underline was used to identify the guard digits in the last exercise (see the "5" in the calculated mass of 252.23415 for the carbon in one mole of ethidium bromide as an example). This is the same practice we will follow throughout this book whenever guard digits are required.

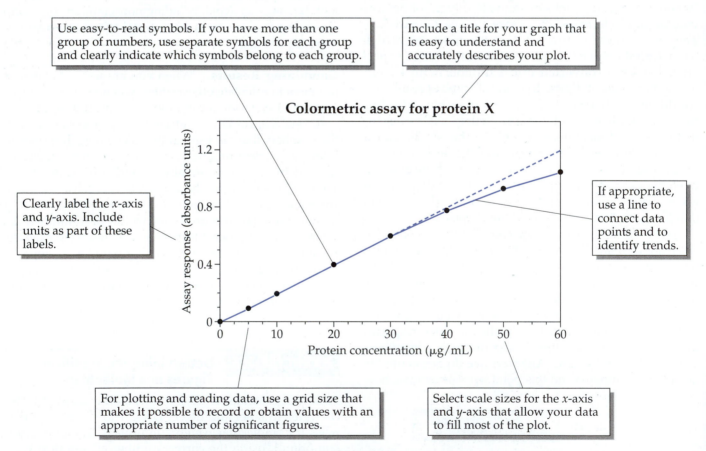

Use easy-to-read symbols. If you have more than one group of numbers, use separate symbols for each group and clearly indicate which symbols belong to each group.

Include a title for your graph that is easy to understand and accurately describes your plot.

Clearly label the x-axis and y-axis. Include units as part of these labels.

If appropriate, use a line to connect data points and to identify trends.

For plotting and reading data, use a grid size that makes it possible to record or obtain values with an appropriate number of significant figures.

Select scale sizes for the x-axis and y-axis that allow your data to fill most of the plot.

FIGURE 2.10   Good laboratory practices for preparing a graph. The dashed line shows the best-fit response to the results for samples containing 0 to 30 μg/mL of protein X, while the solid line shows a smoothed curve for the entire data set. Methods for finding the best-fit line for a linear relationship are discussed in Chapter 4.

Key Words

Other Terms

Questions

INTRODUCTION

1. What is meant by "good laboratory practices"? Why are good laboratory practices important in chemical laboratories?
2. What is a "standard operating procedure"? How is this related to good laboratory practices?
3. What are some standard operating procedures that are used in your laboratory?
4. Write a standard operating procedure for each of the tasks listed below.
   (a) Operation of an eyewash
   (b) Disposal of used needles or sharp objects in your laboratory
   (c) Storage of concentrated hydrochloric acid
5. Use the Internet to obtain standard operating procedures for each of the following tasks.
   (a) Calibration of a volumetric flask
   (b) Preparation of a chromic acid solution
   (c) Use of a volumetric pipet

LABORATORY SAFETY

6. What is a "chemical hygiene plan"? What things are usually included in such a plan?
7. What are some common safety features found in a modern laboratory? What is the purpose of each safety feature?
8. Familiarize yourself with the safety equipment in your laboratory, and draw a map that shows the location of this equipment. How does this equipment compare to the list in Table 2.1?
9. What is a "chemical hazard"? What types of substances are considered chemical hazards?
10. Define each of the following terms regarding chemicals that are physical hazards.
    (a) Flammable
    (b) Explosive
    (c) Oxidizer
    (d) Radioactive
    (e) Water-reactive
    (f) Compressed gas
11. Define each of the following terms related to chemicals that are health hazards.
    (a) Acute toxin
    (b) Poison
    (c) Biohazard
    (d) Carcinogen
    (e) Irritant
    (f) Corrosive
    (g) Asphyxiant
    (h) Etiological agent
    (i) Reproductive toxin
    (j) Allergen
    (k) Mutagen
    (l) Teratogen
12. Using a chemical catalog or other resource, determine what types of hazards (if any) are present for each of the following compounds.
    (a) Hydrogen peroxide
    (b) Potassium hydroxide
    (c) Sodium chloride
    (d) Carbon tetrachloride
    (e) Acetonitrile
    (f) Hydrogen gas
13. What is an NFPA label? What chemical or physical properties are described in this type of label?
14. The NFPA symbols for acetonitrile and sodium borohydride are shown in the following figure. What do these symbols tell you about the chemical or physical properties of these chemicals?

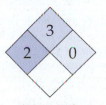

Acetonitrile
(CH_3CN)

Sodium borohydride
($NaBH_4$)

15. What is a "material safety data sheet"? What information does this provide about a chemical?

16. Use the Internet or other resources to locate the material safety data sheet for methylene chloride (CH_2Cl_2), a common solvent used in products like paint thinner. From this information, what can you say about the health and safety risks of this chemical?

17. Sodium bicarbonate is commonly used in analytical laboratories to control the acid/base properties of a solution. This chemical is also used in homes as "baking soda" for cooking and as an ingredient in toothpaste and air fresheners. Obtain a copy of the material safety data sheet for this compound from the Internet or another source. What does this information tell you about the safety of using sodium bicarbonate in these applications?

18. What are the four major ways in which chemicals can enter the body? What precautions can you take to minimize each of these routes of chemical exposure?

19. What problems could result in each of the following situations?
    (a) A technician works with ethyl ether, a highly volatile chemical, outside of a hood.
    (b) A student fails to clean an area after spilling a small amount of hydrochloric acid.
    (c) A scientist leaves some needles on the bench after using them to prepare samples.
    (d) A worker fails to label a new chemical reagent before leaving the laboratory.
    (e) Some graduate students order a pizza and have it delivered to their laboratory.

20. Use the Internet or other resources to determine how the following chemicals should be stored.
    (a) Hexane
    (b) Sodium iodate
    (c) Zinc metal

21. Below is list of several common laboratory chemicals. What storage conditions are needed for each of these chemicals and which can be safely stored together?
    (a) n-Hexane
    (b) Benzaldehyde
    (c) Nitric acid
    (d) Phosphoric acid
    (e) Sodium hydroxide

22. What is "laboratory waste management"? How is laboratory waste management conducted in your own laboratory? What is your role in this process?

23. Using material safety data sheets or other resources, determine how each of the chemicals in Problem 21 should be handled as part of laboratory waste management.

THE LABORATORY NOTEBOOK

24. What is the role of the laboratory notebook in today's world? Give three specific examples of why it is important to keep thorough and up-to-date entries in such a notebook.

25. What are some recommended practices for keeping a laboratory notebook? Which of these practices are used in your laboratory?

26. What is an "electronic laboratory notebook"? What are the advantages and disadvantages of using an electronic laboratory notebook?

27. What is a "spreadsheet"? What are some ways that a spreadsheet can be used along with a laboratory notebook?

REPORTING EXPERIMENTAL DATA

28. Explain what is meant by the "SI system of measurements"? Why is such a system used in science?

29. What is meant by the term "fundamental SI unit"? List each of the fundamental SI units and state what types of quantities they are used to measure.

30. What is meant by the terms "derived SI unit" and "accepted SI unit"? Give two specific examples for each of these two types of units.

31. Convert each of the following values into fundamental, derived, or accepted SI units. In some cases it may be necessary to use another resource to find an appropriate conversion unit.
    (a) 22,489 ft
    (b) 5.68 atm
    (c) 130 lb
    (d) 120 mile/hr
    (e) 2200 cal
    (f) 25.0 gal

32. Convert each of the following temperatures into the requested units.
    (a) A typical body temperature of 98.6°F in °C and K
    (b) The value of absolute zero (0 K) in °C and °F
    (c) A standard temperature of 20°C in K and °F
    (d) The temperature of a freezer (–20°C) in °F and K

33. State each of the following values using SI prefixes.
    (a) 2.58×10^{-11} g
    (b) 125×10^{-6} L
    (c) 150,000 g/mol
    (d) 589×10^{-9} m
    (e) 600×10^{6} Hz
    (f) 25,000 V

34. What is the difference between an analog display and a digital display? Explain the proper way to read and record a result from each of these two types of displays.

35. Figure 2.11 shows a scale from an older instrument used in absorbance spectroscopy. Estimate the values represented by the needle on both the upper and lower sides of this scale (percent transmittance and absorbance).

FIGURE 2.11   A reading on the analog output of a Spectronic 20 absorbance spectrometer.

36. What good laboratory practices should be followed when you are preparing a graph?

37. What is meant by "significant figures" in science? Why is this an important part of good laboratory practices?

38. How many significant figures are in each of the numbers shown below?
    (a) $F = 9.64853415 \times 10^4$ C/mol
    (b) $m/z = 183.2280$ u
    (c) $-\log(\% T) = 1.238$
    (d) 1 km = 0.62137 mile
    (e) 1 in = 2.54 cm
    (f) $[OH^-] = 6.00 \times 10^{-7}$ M

39. Round each of the following numbers to three significant figures.
    (a) $[Na^+] = 1.525$ M
    (b) $-\log a_{H+} = 7.463$
    (c) $h = 6.6260688 \times 10^{-34}$ J·s
    (d) $t = 5.515$ ns
    (e) $10^{-pCa} = 8.370$
    (f) $K_a = 0.1650$

40. A student needs to add the following numbers: 52.7866, 34.0988, and 14.1146. What is the sum of these values after each of these three numbers has been rounded to five, four, three, or two significant figures?

41. Round off each of the following values to the stated number of significant figures.
    (a) 8.854×10^{-12} to two significant figures
    (b) 1.283×10^{-9} to three significant figures
    (c) 6.735 to three significant figures
    (d) 3.049×10^{15} to two significant figures

42. The value of π (3.14159265358...) is currently known to over a trillion decimal places. What approximate numbers would you use for this constant if you rounded it to three, four, or five significant figures?

43. Using the atomic masses listed in the periodic table at the front of this text, determine the molar mass of glucose ($C_6H_{12}O_6$). Express your final answer using the correct number of significant figures. Show how this value changes when you express it using three, four, or five significant figures.

44. Write the answers to the following calculations using the correct number of significant figures.
    (a) 107.868 + 35.4527
    (b) 2.5898 − 0.133 − 0.003517
    (c) 98.4/99.976
    (d) $\log(2.01 \times 10^{-6})$
    (e) antilog(−2.891)
    (f) $10^{-6.82}$

45. Write each of the following answers with the correct number of significant figures.
    (a) 189.032 + 153.02 − 32.0861
    (b) $(1.053 \times 10^{-5}) \cdot (3.56 \times 10^{-8})/(0.48)$
    (c) (0.9323/0.184) + 4.8520
    (d) $0.998 \cdot (18.99840 + 12.0107)$
    (e) $6.82 + \log(0.1235)$
    (f) $0.238 \cdot 10^{-4.221}$

46. What is "dimensional analysis"? How is this used when you are performing a calculation?

47. Report the results of the following calculations using the correct number of significant figures. Use dimensional analysis to determine the final units for your answers.

    (a) Mass of Cu in 5 g of $CuSO_4 \cdot H_2O$ =
    $$(5.000 \text{ g } CuSO_4 \cdot H_2O) \cdot \frac{(1 \text{ mol } CuSO_4 \cdot H_2O)}{(177.63 \text{ g } CuSO_4 \cdot H_2O)}$$
    $$\cdot \frac{(1 \text{ mol Cu})}{(1 \text{ mol } CuSO_4 \cdot H_2O)} \cdot \frac{(63.55 \text{ g Cu})}{(1 \text{ mol Cu})}$$

    (b) Titrated concentration of NaOH in a sample
    $= (95.8 \text{ mL} - 25.3 \text{ mL}) \cdot (1 \text{ L}/1000 \text{ mL}) \cdot (0.105 \text{ mol/L HCl})$
    $\cdot (1 \text{ mol NaOH}/1 \text{ mol HCl})/(0.500 \text{ L NaOH})$

    (c) Molar mass of $C_{12}H_{22}O_{11}$
    $= (12 \text{ mol C/mol } C_{12}H_{22}O_{11}) \cdot (12.01115 \text{ g C/mol C}) +$
    $(11 \text{ mol O/mol } C_{12}H_{22}O_{11}) \cdot (15.9994 \text{ g O/mol O})$
    $+ (22 \text{ mol H/mol } C_{12}H_{22}O_{11}) \cdot (1.00794 \text{ g H/mol})$

    (d) Density of a lead cylinder
    $$= (23.2850 \text{ g} - 0.0165 \text{ g})/[(2.52 \text{ cm}) \cdot \pi \cdot (0.51 \text{ cm})^2]$$

48. Report the results of the following calculations using the correct number of significant figures. Use dimensional analysis to determine the final units for your answers.
    (a) $n = PV/RT = [(2.50 \text{ atm}) \cdot (3.15 \text{ L})]/[(0.0821 \text{ L} \cdot \text{atm}/(\text{mol·K})) \cdot (273.15 \text{ K} + 25.0 \text{ K})]$
    (b) $\Delta G°_{AgCl}$
    $= -2.303 \cdot (8.314 \text{ J}/(\text{mol} \cdot \text{K})) \cdot (298 \text{ K}) \cdot \log(1.0 \times 10^{10})$
    (c) $-\log(\gamma_{Ca^{2+}}) = \dfrac{[0.51 \cdot (+2)^2 \cdot (0.10)^{\frac{1}{2}}]}{[1 + (0.10)^{\frac{1}{2}}]}$
    (d) Percent (w/w) of sulfur in H_2SO_4
    $$= 100 \cdot \frac{(1 \text{ mol S}/1 \text{ mol } H_2SO_4) \cdot (32.066 \text{ g S}/1 \text{ mol S})}{(98.078 \text{ g } H_2SO_4/1 \text{ mol } H_2SO_4)}$$

49. The relationship for temperatures expressed in degrees Celsius and Kelvin was given in Table 2.5 as °C = K − 273.15. What units must be associated with the number "273.15" in this equation to make it valid? Based on dimensional analysis, what other values and units must be present for this relationship to be correct?

50. Two values used for the ideal gas law constant (R) are 8.314 J/(mol · K) and 1.987 cal/(mol · K). Prove that these two numbers are equivalent by using dimensional analysis and appropriate conversion factors.

51. What is a "rounding error"? How can these errors be minimized or avoided?

52. Based on the definition of the meter in Table 2.4, the true speed of light in a vacuum is 299,792,458 m/s. This value is close to but not exactly the same as what was obtained in Equation 2.1. What do you think the reason is for this difference?

53. Insulin is a two-chain polypeptide hormone that helps the body to regulate its blood sugar levels. The empirical formula for insulin from humans is $C_{257}H_{383}N_{65}O_{77}S_6$. Calculate the molar mass of this hormone using two significant figures for the atomic masses of carbon, hydrogen, nitrogen, oxygen, and sulfur. Repeat this calculation several times using three, four, five, and six significant figures for each of these atomic masses. What can you conclude from your results?

CHALLENGE PROBLEMS

54. Most chemical laboratories are equipped with OSHA Class A, B, or C fire extinguishers. Identify the types of fire extinguishers in your laboratory. Discuss how they work and describe the types of fires and situations in which each should be used.

55. The health effects of some chemicals can be directed primarily at certain organs or tissues in the body. Using Reference

7–10 or other sources, determine what parts of the body are affected by each of the following types of chemical hazards. Provide a specific example of each and discuss the health problems that they cause.

(a) Neurotoxin
(b) Nephrotoxin
(c) Hepatotoxin
(d) Hematopoietic toxin

56. "Flammable" is a term that covers a wide range of physical hazards. Related but more specific terms that are sometimes used are given below.[7–10] Define each of these terms and give a specific example of a substance in each category.

(a) Combustible (inflammable)
(b) Spontaneously combustible
(c) Pyrophoric

57. The following data were obtained for standards analyzed by atomic absorption spectroscopy for the measurement of copper in water.

Copper Concentration in Standard (mg/L)	Assay Signal (Absorbance Units)
0.0	0.005
5.0	0.109
10.0	0.206
20.0	0.415
30.0	0.616
40.0	0.809
50.0	1.035

Prepare a spreadsheet for these results similar to the one shown in Figure 2.8. Use this spreadsheet to prepare a calibration curve for these results, following the good laboratory practices given in this chapter for graph preparation.

TOPICS FOR DISCUSSION AND REPORTS

58. The use of scientific data in the courtroom can become complicated when a relatively new method, such as the DNA tests employed in the O. J. Simpson trial, is used to provide evidence. The current guidelines U.S. judges use to determine whether such evidence can be utilized is based on the case of *Daubert* v. *Merrell Dow Pharmaceuticals*.[37,38] Locate information on this trial and report on how it has affected the use of scientific results in the courtroom.

59. Forensic testing, as used for the O. J. Simpson trial, is just one of many applications for PCR. Other examples include its use in food testing, clinical analysis, and paleontology. Obtain more information on these applications of PCR and discuss your findings.

60. Maintaining a documented "chain of custody" for samples is an important aspect of modern forensic testing. Locate more information on this topic. Discuss what is meant by a "chain of custody" and describe the procedures that are used by law enforcement agencies and laboratories with regards to this topic. In what other areas do you think a chain of custody might be essential during sample handling and analysis?

61. Scientific fraud that involves the use of imaginary or falsified data is an extremely rare event, but it does occur from time to time. Obtain information on an example of scientific fraud from past newspapers or magazines. Discuss what role, if any, laboratory notebooks played in detecting or confirming this fraud.

62. Although the chemical laboratory is normally a safe place to work, accidents do occasionally occur. Get information on a recent accident that has occurred in an industrial or academic laboratory. Describe why the accident happened and discuss what could have been done to prevent it.

63. The use of animals for chemical testing is a topic of ongoing debate.[11–13] Obtain more information on both animal-based methods and alternative systems for chemical testing. Discuss the advantages or disadvantages of each approach.

64. Changes to the SI system are made every few years at an international meeting known as the *General Conference on Weights and Measures* (or *Conférence Générale des Poids et Mesures*). One recent change that has been made is in the definition of the kilogram, which until present has been the only SI unit that was still defined based on a physical object.[39] Report on how the definition of the kilogram is being changed so that it is instead based on an unvarying physical property of nature. Discuss how the definitions of some other SI units, like the meter and liter, have also changed over the years.

References

1. F. M. Schmalleger, *Trial of the Century: People of the State of California vs. Orenthal James Simpson*, Prentice Hall, Englewood Cliffs, NJ, 1996.

2. B. S. Weir, "DNA Statistics in the Simpson Matter," *Nature Genetics*, 11 (1995) 365–368.

3. H. A. Erlich, Ed., *PCR Technology: Principles and Applications for DNA Amplification*, Oxford University Press, New York, 1997.

4. A. T. Sullivan, "Good Laboratory Practices and Other Regulatory Issues: A European View," *Drug Development Research*, 35 (1995) 145–149.

5. R. A. Nadkarni, "ISO 9000: Quality Management Standards for Chemical and Process Industries," *Analytical Chemistry*, 65 (1993) 387A–395A.

6. The Laboratory Safety Workshop, Natick, MA.

7. A. K. Furr, Ed., *CRC Handbook of Laboratory Safety*, 5th ed., CRC Press, Boca Raton, FL, 2000.

8. Forum for Scientific Excellence, *Concise Manual of Chemical and Environmental Safety in Schools and Colleges*, Vols. 1–5, J. P. Lippincott, New York, 1990.

9. S. K. Hall, *Chemical Safety in the Laboratory*, CRC Press, Boca Raton, FL, 1994.

10. National Research Council, *Prudent Practices in the Laboratory*, National Academy Press, Washington, DC, 1995.

11. A. A. Blumberg, "Risks and Chemical Substances," *Journal of Chemical Education*, 71 (1994) 912–918.

12. A. C. Huggett, B. Schilter, M. Roberfroid, E. Antignac, and J. H. Koeman, "Comparative Methods of Toxicity Testing," *Food and Chemical Toxicology*, 34 (1996) 183–192.

13. A. M. Goldberg and J. M. Frazier, "Alternatives to Animals in Toxicity Testing," *Scientific American*, 261 (1989) 24–30.

14. B. N. Ames, F. D. Lee, and W. E. Durston, "An Improved Bacterial Test System for the Detection and Classification of

Mutagens and Carcinogens," *Proceedings of the National Academy of Sciences USA*, 70 (1973) 782–786.

15. J. McCana and B. N. Ames, "Detection of Carcinogens as Mutagens in the *Salmonella*/microsome test: Assay of 300 Chemicals: Discussion," *Proceedings of the National Academy of Sciences USA*, 73 (1976) 950–954.

16. D. R. Lide, Ed., *CRC Handbook of Chemistry and Physics*, 83rd ed., CRC Press, Boca Raton, FL, 2002.

17. P. Heckelman, A. Smith, and M. J. Oneil, Eds., *The Merck Index*, 13th ed., Merck & Co., Rahway, NJ, 2001.

18. R. E. Lenga, Ed., *Sigma-Aldrich Library of Chemical Safety*, 2nd ed., Sigma-Aldrich Chemical Co., Milwaukee, WI, 1988.

19. R. J. Lewis, Sr., *Sax's Dangerous Properties of Industrial Materials*, 8th ed., Van Nostrand Reinhold, New York, 1992.

20. P. A. Patnaik, *A Comprehensive Guide to the Hazardous Properties of Chemical Substances*, Van Nostrand Reinhold, New York, 1992.

21. ACS Task Force on Waste Management, *Laboratory Waste Management: A Guidebook*, American Chemical Society, Washington, DC, 1994.

22. H. M. Kanare, *Writing the Laboratory Notebook*, American Chemical Society, Washington, DC, 1985.

23. J. S. Dood and M. C. Brogan, Eds., *The ACS Style Guide: A Manual for Authors and Editors*, American Chemical Society, Washington, DC, 1986.

24. I. Krull and M. Swartz, "Laboratory Notebook Documentation," *LC · GC*, 15 (1997) 1122–1129.

25. J. R. Wagner, "Purpose and Proper Maintenance of a Laboratory Notebook," *Tappi Journal*, 77 (1994) 130–132.

26. R. E. Dessy, "Electronic Lab Notebooks: A Shareable Resource," *Analytical Chemistry*, 67 (1995) 428A–433A.

27. M. C. Fitzgerald, "The Evolving, Fully Loaded, Electronic Laboratory Notebook," *Chemical Innovation*, 30 (2000) 2–3.

28. R. A. Dabek and J. Orndorff, "Laboratory Notebooks: A Medieval Artifact in an Electronic World," *Chemtech*, March (1999) 6–12.

29. G. A. Gibbon, "A Brief History of LIMS," *Laboratory Automation and Information Management*, 32 (1996) 1–5.

30. S. Copestake, *Excel 2002 in Easy Steps*, Barnes & Noble, Warwickshire, UK, 2003.

31. E. J. Billo, *Microsoft Excel for Chemists*, 2nd ed., Wiley, New York, 2001.

32. R. De Levie, *How to Use Excel in Analytical Chemistry and in General Scientific Data Analysis*, Cambridge University Press, Cambridge, UK, 2001.

33. D. Diamond and V. Hanratty, *Spreadsheet Applications in Chemistry Using Microsoft Excel*, Wiley, New York, 1997.

34. B. N. Taylor, Ed., *The International System of Units (SI)*, NIST Special Publication 330, National Institute of Standards and Technology, Gaithersburg, MD, 1991.

35. *Correct SI Metric Usage*, United States Metric Association.

36. J. F. Kenney and E. S. Keeping, "Significant Figures," in: *Mathematics of Statistics, Part 1*, 3rd ed., Van Nostrand, Princeton, NJ, 1962, pp. 8–9.

37. C. A. Kuffner, Jr., E. Marchi, J. M. Morgado, and C. R. Rubio, "Capillary Electrophoresis and Daubert: Time for Admission," *Analytical Chemistry*, 68 (1996) 241A–246A.

38. D. T. Case and J. B. Ritter, "Disconnects between Science and the Law," *Chemical & Engineering News*, 78(7) (2000) 49–60.

39. S. K. Ritter, "Redefining the Kilogram," *Chemical & Engineering News*, 86(21) (2008) 43.

Chapter 3

Mass and Volume Measurements

Chapter Outline

3.1 INTRODUCTION: J. J. BERZELIUS

J. J. Berzelius has been called the greatest experimental chemist of all time. This Swedish scientist was the leading figure in chemistry during much of the first half of the 1800s (see Figure 3.1A). Working with no more than a few students at a time, he was responsible for discovering several elements, determining the atomic weights of 50 elements, and devising the system of elemental symbols (H for hydrogen, O for oxygen, and so on) that we use to this day. He also determined the chemical composition of approximately 2000 minerals, put together the first comprehensive list of atomic weights, coined the terms "protein" and "catalyst," and invented many pieces of glassware that are still found in modern laboratories.[1–3]

A key feature of Berzelius's work was his insistence on the use of good laboratory techniques and careful measurements (see Figure 3.1B). The result is that all of the atomic masses he reported (when normalized to the value he used for oxygen) are within 1% of the values listed in the modern periodic table.[1,2] This level of accuracy is particularly impressive because instrumental methods of analysis, as we now know them, did not exist at his time. Instead, Berzelius used classical analysis methods that employed known chemical reactions and mass or volume measurements of the reactants or products.

Mass and volume determinations are still important in today's laboratories, where they form the basis of many

FIGURE 3.1A Jöns Jakob Berzelius (1779–1848), an early pioneer in analytical chemistry and one of the founders of modern chemistry. A balance that was used by Berzelius is shown in the back of his portrait. The portrait above shows Berzelius in 1843 and is from a painting by O. J. Soedermark. (Reproduced with permission and courtesy of the Royal Swedish Academy of Sciences.)

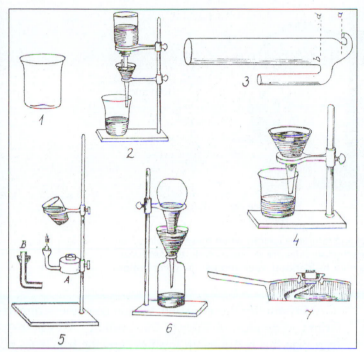

FIGURE 3.1B   The drawings of the equipment shown above are from an 1836 book entitled *Lehrbuch der Chemie* by Berzelius, and include several items of glassware that Berzelius used in his research.

reference methods and are used to prepare samples and reagents for other analytical techniques. In this chapter we discuss mass and volume measurements and the procedures to follow in these measurements. We also consider how mass and volume can be combined to describe the content of samples and reagents.

3.2 MASS MEASUREMENTS

Mass is one of the most fundamental properties of matter and is defined as the quantity of matter in an object. This quantity is typically determined in a laboratory by using a **balance**, which is a precision weighing instrument used to measure small masses.[4] The balance is the earliest known measurement device. References to balances and weights can be found in cultures that range from ancient Egypt to China.[5] The device in the painting of Berzelius in Figure 3.1A is typical of what most people imagine when they think of a balance, in which two weighing pans are held by a beam on either side of a central fulcrum. The name "balance" comes from the Latin word *bilanx*, which means "having two pans."[4] The operation of such a device is relatively simple. The object to be measured is placed on one of the two pans while objects of a known mass are added to the other side. When the two sides are level, the mass of the object is determined by adding the masses of the weights on the opposite side.

In analytical chemistry, balances and mass measurements are used for many purposes. Balances are used to measure samples for analysis, to weigh chemicals for the preparation of reagents, and to determine the amount of a product that results from a reaction. The advantages of

mass measurements are that they can be made quickly, accurately, and reproducibly with relatively simple and inexpensive equipment. The only requirements are that you have enough material to examine and that the mass of this material is sufficiently stable to be measured.

3.2A The Determination of Mass

Weight versus Mass.   The process of determining either the mass or weight of a substance is often called **weighing**, but the terms "mass" and "weight" actually refer to quite different things. *Mass* refers to the quantity of matter in an object. We learned in Chapter 2 that the base SI unit for mass is the *kilogram*, but mass can also be expressed in related units like the gram and milligram. **Weight** differs from mass in that it is a measure of the pull of a force on an object.[4] In most cases, gravity is the main force acting on an object as it is weighed, but other forces can also play a role.

The difference between mass and weight can be illustrated if you consider what would happen if you were weighed on both the Earth and the Moon. Your mass would be the same in each location because your body would contain the same amount of matter. However, your weight on the Moon would be about 1/6 of that on Earth because the Moon's gravity is only 16.7% as strong as the Earth's. Even if you move from the equator to the North or South poles on Earth your weight will differ by as much as 0.5% at sea level. This change occurs because the Earth is not a perfect sphere, a feature that causes your weight and distance from the Earth's center (which affects gravitational attraction) to vary at different locations. Similar changes occur as you move up or down in altitude, but none of these factors will alter your mass.[4,6,7]

Converting Weight to Mass.   It is important to know the difference between mass and weight because when you place an object on a laboratory balance you are measuring the net force of gravity plus other forces on this object (or the object's *weight*). What we would really like to determine, however, is the object's *mass*. The way a balance makes this conversion is by comparing the object to a reference weight with a known mass. If the object and reference weight are placed on opposite sides of the balance, they will have the same forces acting on them from the surrounding medium and will be experiencing the same local gravitational field. Thus, the difference in force between the two sides should be directly related to their difference in mass.

The forces that act on an object when it is on a balance are illustrated in Figure 3.2. First, there is the force of gravity that is pulling downward on the object. This force is described by Equation 3.1.

$$\text{Force due to gravity} = m_{obj} \cdot g \qquad (3.1)$$

where m_{obj} is the mass of the object and g is the *gravitational acceleration constant*, which is a measure of

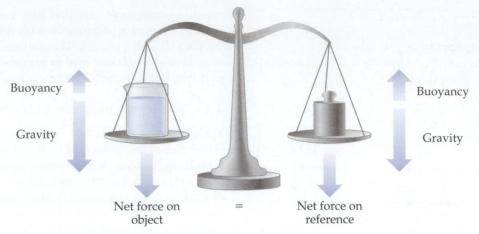

FIGURE 3.2   The forces that act on an object and reference weight when these are placed on opposite sides of an equal-arm balance. When the two sides are at equal levels, the net force acting on the measured object and reference must be identical, even though the contributions from gravity and buoyancy may not be the same on each side.

the pull of gravity at a particular location. A similar relationship can be written for the reference weight, where the force due to gravity will be equal to $m_{ref} \cdot g$.

Gravity is not the only force that determines an object's weight. Another important factor is **buoyancy**. Buoyancy is a force that works against gravity when you are weighing an object.[4] This force occurs whenever an object is surrounded by air or any medium other than a vacuum. If an object is less dense than its surrounding medium, this medium will displace the object to a region of lower density, causing the object to rise. This effect is why a helium balloon floats into the sky. The size of this force when we are weighing an object is related to the gravitational pull on the medium (usually air) that is displaced when the object and reference weight are placed on the balance. The size of this force is given by Equation 3.2, where the negative sign indicates that the force due to buoyancy is working against the force of gravity.

$$\text{Force due to buoyancy} = -m_{air} \cdot g \qquad (3.2)$$

In this relationship, the mass of displaced air can be determined from the air's density (d_{air}) and the volume of the object that displaced this air ($m_{air} = d_{air} \cdot V_{obj}$). Another way of determining the displaced air's mass is to use the density of air and the mass and density of the weighed object, where $m_{air} = m_{obj} \cdot (d_{air}/d_{obj})$.

When the object and reference weight are at the same level on an equal-arm balance, the overall forces acting on the object and reference weight must also be the same. We can describe this situation by combining Equations 3.1 and 3.2 to produce the following new relationship.

Weight of Object   **Weight of Reference**

$$m_{obj} \cdot g - m_{obj} \cdot (d_{air}/d_{obj}) \cdot g = m_{ref} \cdot g - m_{ref} \cdot (d_{air}/d_{ref}) \cdot g \qquad (3.3)$$

To simplify this equation, we can combine common terms and divide both sides by the constant g. We can then rearrange this relationship to show how the true mass of the object (m_{obj}) is related to its apparent mass, as represented by the mass of reference (m_{ref}).

$$m_{obj} = m_{ref} \cdot \frac{[1 - (d_{air}/d_{ref})]}{[1 - (d_{air}/d_{obj})]} \qquad (3.4)$$

Equation 3.4 indicates that the mass of the measured object will indeed be directly related to the value of m_{ref}. It is this principle that allows the mass of this object to be measured on a balance. However, this equation also indicates that m_{ref} and the actual mass of our object are not necessarily the same value. This difference occurs because the ratio $[1 - (d_{air}/d_{ref})]/[1 - (d_{air}/d_{obj})]$ also appears in Equation 3.4. This ratio represents the different buoyancy effects that are acting on the object and reference weight. Although this ratio is often ignored in routine mass measurements, it must be considered when highly accurate masses are required. We will come back to this topic in Section 3.2C when we learn how to adjust mass values for buoyancy effects. (See Box 3.1 and Figure 3.3 for further discussion of how force measurements can be used in chemical analysis.)

3.2B Types of Laboratory Balances

Mechanical Balances and Electronic Balances.   One type of laboratory balance is a mechanical balance. This device uses a mechanical approach for determining mass. An example is the equal-arm (or two-pan) balance, in which a sample and reference weights are placed on opposite sides of a beam that is held across a central fulcrum. This type of balance has been used for thousands of years and was found in analytical laboratories through the mid twentieth century, but it is not common in modern laboratories.[7] Another type of mechanical balance is the *substitution balance*, or *single-pan mechanical balance*. This device has a

BOX 3.1
Atomic Force Microscopy

There are other ways besides the use of laboratory balances in which scientists make use of force measurements to obtain information on a chemical sample. One important example is the method of *atomic force microscopy* (or *AFM*). Atomic force microscopy is a high-resolution method that is capable of imaging individual atoms or molecules on the surface of a sample. This technique was first developed in 1986 and has since become one of the most valuable tools for examining and manipulating chemicals at the nanoscale.[8–10]

The instrument that is used to perform AFM is called an *atomic force microscope* (see Figure 3.3). The basic design of this instrument includes a small cantilever with a sharp tip (called the "probe") that is passed over the surface of the sample. The end of this tip has a width on the order of nanometers, which is comparable to the size of atoms and small molecules. As this tip is passed gently over the sample, forces between the

tip and sample will cause a slight deflection in the tip and cantilever. These deflections are typically measured by using a laser beam that is aimed at the cantilever and deflected onto a detector array that monitors any small changes in the position of this laser beam. This information is then used to create an image of the surface of the sample.

Although modern AFM instruments are capable of atomic-scale resolution, they are commonly utilized in examining materials on a slightly larger scale, as shown in Figure 3.3. A big advantage of AFM is that, unlike some other high-resolution imaging methods, this technique can be used with both conductive and nonconductive samples. AFM can also be employed with samples that are in air or in liquids and that are hard or soft in nature. These properties have made AFM an important tool in fields that range from materials science to biomedical research.[9,10]

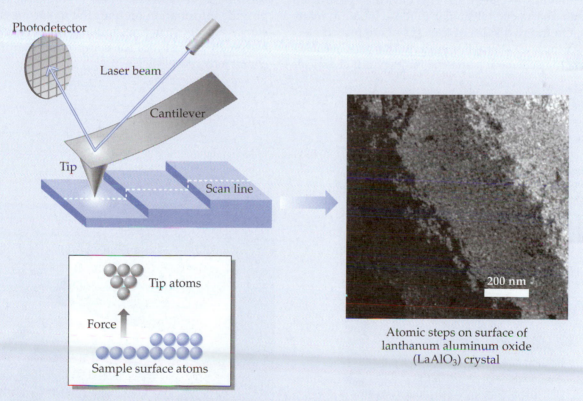

Atomic steps on surface of
lanthanum aluminum oxide
($LaAlO_3$) crystal

FIGURE 3.3   The basic operation of atomic force microscopy (AFM) and an image of the surface of a lanthanum aluminum oxide crystal that was obtained by this technique. (Reproduced with permission and courtesy of C. L. Cheung, University of Nebraska.)

single pan placed on one side of a beam along with a set of removable weights. The other side of the beam is connected to a fixed counterweight. When no sample is present, both sides of the beam are in balance. When a sample is placed onto the pan, the positions of the two sides are disturbed and some weights on the sample side are removed to restore these positions. The mass of these weights is then used to determine the mass of the sample.

The most popular type of balance in modern chemical laboratories is the **electronic balance** (see Figure 3.4). This balance uses an electronic mechanism to determine the mass of an object, which is accomplished by attaching the sample pan to one or more bars that are held between the two ends of a permanent magnet. When a sample is placed onto the pan, the bars are pushed downward. A position sensor signals the balance to apply a current through the bars, which produces an electromagnetic force that causes the bars to again move upward. The size of the applied current that is needed to move the bar and sample pan to their original positions is then measured, providing a value that is proportional to the mass of the sample.[4,7] The low cost and ease of use of electronic balances have made them popular in analytical laboratories.[4,7,11] With these balances, it is possible to correct electronically for the mass of a sample container and for variations in the instrument's response with changes in temperature. Most electronic balances are also able to perform automatic calibration with built-in reference weights and can interface directly with computers for data collection.

Other Mass-Measuring Devices.   There are a variety of other instruments that use mass measurements for chemical analysis. One example is a *quartz crystal microbalance (QCM)*. Rather than being used to measure relatively large amounts of material, a QCM is employed as a sensor for trace amounts of chemicals. This device is constructed from a thin quartz crystal similar to those found in many watches. On the two sides of the crystal are placed electrodes that apply an alternating current. This current causes the quartz to oscillate at a specific frequency. If chemicals adsorb to the surface of the crystal, as might occur when it contains a coating that favors such binding, the mass of the crystal will change. This adsorption process can produce a measurable change in the frequency at which the crystal is vibrating, allowing the mass of deposited material to be measured. The maximum amount of a substance that can be measured by this device is around a few hundred micrograms and it can be used to examine changes in mass as small as a fraction of a nanogram.[12]

Another mass-measuring device that we will encounter throughout this book is a *mass spectrometer*. Unlike laboratory balances, which estimate the total mass of a sample, a mass spectrometer measures the masses of *individual* atoms or molecules. This type of measurement is accomplished by first converting the atoms or molecules into gas-phase ions, which are then separated and analyzed based on their mass and charge. The result is a graph known as a *mass spectrum*, in which the amount of each detected ion is plotted versus its *mass-to-charge ratio* (*m/z*) (see Figure 3.5). This plot can provide information on both the molar mass of the substance and its structure, making mass spectrometry an extremely useful tool for compound identification and characterization.[13–15]

Analytical balance

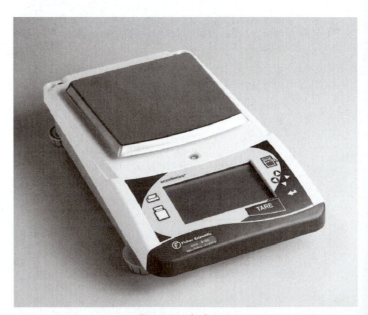

Precision balance

FIGURE 3.4   Some examples of electronic balances. The balance with the enclosed weighing compartment is known as an **analytical balance**, and the balance that has the open weighing area is called a **precision balance**, or *top-loading balance*. (These images are for the AB-S/FACT Class Analytical Balance and accuSeries II Electronic Toploading balance and are reproduced with permission and courtesy of Mettler Toledo and Fisher Scientific, respectively.)

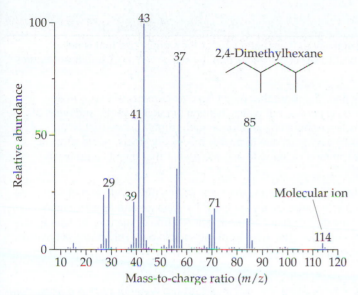

FIGURE 3.5   An example of a mass spectrum.

3.2C Recommended Procedures for Mass Measurements

Selection and Use of Balances.   Most laboratories have several types of balances, so a question you often need to ask is "What is the best balance for my particular mass measurement?" An important thing to consider is the *maximum load* (or *capacity*) of each balance, which refers to the largest mass a balance can reliably measure. Another feature to consider is the *readability*, which is the smallest division in mass that can be read on the balance's display.[4,7] A third factor used to compare balances is their *resolution*, which is determined by dividing the capacity of a balance by its readability.[7]

$$\text{Resolution} = \frac{\text{Capacity}}{\text{Readability}} \qquad (3.5)$$

As you can see from this relationship, the resolution is a direct measure of how many distinct masses can be determined by a particular balance (1.0000 g vs. 1.0001 g, and so

on). Most laboratory balances have a resolution of at least 10,000 and some have resolutions as high as 20 million.

Table 3.1 lists the capacities, readabilities, and resolutions of various laboratory balances. These devices can be placed into one of two general groups based on their readability and design. A balance that has an enclosed weighing compartment (for greater stability and accuracy) and that can provide mass measurements to within at least 0.1 milligram is known as an **analytical balance**. If the balance has an open weighing area, it is called a **precision balance** (or *top-loading balance*).[4,7] Of the various balances that are listed in Table 3.1, the standard precision balance and macroanalytical balance are the ones most often seen in analytical laboratories, due to their good readabilities and their ability to handle the range of masses needed for most routine assays.

After you have selected a balance, you next must make sure you are familiar with its proper use and care. Using good laboratory practices with the balance will help ensure the balance is operating correctly and is providing the best possible mass measurements. Table 3.2 summarizes a few guidelines that pertain to the care of any balance, including the handling of samples and chemicals on a balance and the proper recording of mass measurements.

Weighing Methods.   Another choice you must make before making a mass measurement is the technique you will employ during the weighing process. There are two common approaches for weighing.[6] The first approach is *direct weighing*, which is carried out by simply placing an object on the balance pan and recording its mass from the display. This approach is commonly used for inert and solid objects, such as reference weights. It is important to remember that chemicals should *never* be placed directly onto a balance. Instead, the chemicals should be weighed while they are in or on a container (for instance, a piece of weighing paper, a plastic weighing boat, or a beaker). This precaution protects the balance from exposure to the chemical, but it also makes it difficult to use direct weighing for chemicals on most mechanical balances.

TABLE 3.1   Common Types of Balances

Analytical Balances	Capacity (g)	Readability (g)	Resolution
Macroanalytical balance	50–400	0.0001	$(0.5 - 4) \times 10^6$
Semimicrobalance	30–200	0.00001	$(0.3 - 2) \times 10^7$
Microbalance	3–20	0.000001	$(0.3 - 2) \times 10^7$
Ultramicrobalance	2	0.0000001	2×10^7
Precision Balances	**Capacity**	**Readability**	**Resolution**
Industrial precision scale	30–6,000 kg	0.0001–0.1 kg	$(0.1 - 6) \times 10^5$
Precision balance	100–30,000 g	0.001–1 g	$(0.3 - 2) \times 10^5$

Source: This classification scheme and list of characteristics are from M. Kochsiek, *Glossary of Weighing Terms: A Practical Guide to the Terminology of Weighing*, Mettler-Toledo, Switzerland, 1998; and W. E. Kupper, "Laboratory Balances," In *Analytical Instrumentation Handbook*, 2nd ed., G. W. Ewing, Ed., Marcel Dekker, New York, 1997, Chapter 2.

TABLE 3.2   Good Laboratory Practices for Using a Balance

Balance Selection	The balance should have a capacity and readability that are adequate for the sample size that is being measured. In any mass determination it is best to select the balance that will give the most significant figures for the measurement at hand.
Balance Location	The balance should be kept away from any part of a room that might have air currents or that is near a source of cold or heat (for example, by doors, windows, hot plates, heating or cooling vents, and areas of heavy traffic through the laboratory). The balance should be kept on a rigid, sturdy surface that is not affected by the presence of the operator or by vibrations due to nearby machinery, doors, or elevators. It should be kept away from and not share a common circuit with any equipment that might cause erratic power fluctuations, such as electrical motors.
Balance Care and Maintenance	The balance should be properly leveled before it is used for any mass measurements. In addition, it should be calibrated at its final site of use before it is used to process any samples. The balance and its surrounding area should always be kept clean and free from any dust or spilt chemicals. For electronic balances, be sure to tare the device back to zero after you have removed a weighed item from the sample pan. For analytical balances, keep the doors to the weighing area closed except when you are placing items onto the sample pan or removing them from it.
Sample Handling	For accurate mass measurements, samples that are volatile or that might adsorb water or carbon dioxide from the air should be weighed in a closed container. The sample should be at the same temperature as the balance and its surroundings, if possible, to avoid the production of local air convection around the sample pan. Normal air humidity should be kept in the laboratory to minimize the presence of static electricity in samples, which can lead to erratic mass measurements. Avoid touching the measured object or its weighing container with your hands, because fingerprints and the moisture they adsorb can add a detectable mass to these items. Always place an object on the middle of the sample pan for the most accurate results. Do not place a liquid sample that contains a magnetic stir bar onto a balance, because the magnet will create additional force acting on the sample that can lead to errors in the weighing procedure.
Recording Mass Measurements	The measured mass values should be recorded *directly* into the laboratory notebook, using the full number of significant figures that are provided by the balance. An indication should be made in the notebook as to whether any buoyancy correction was made. Such a correction is usually required if the final mass needs to be within 0.1% of its true value. If a buoyancy correction is made, the density of the sample and the surrounding air should both be recorded in the notebook. Determining the air density will also require that you determine and record the barometric pressure, temperature, and relative humidity of the air during the time of the mass measurement.

Direct weighing can be conducted on an electronic balance by using a feature known as *taring*. Taring involves placing an empty weighing container onto the balance and resetting the balance's display (by pressing a "tare," "zero," or "rezero" button) so that it reads zero when the container is present.[4] When a chemical is placed into this container, the display can then be used as a direct reading of the amount of added substance. After the object and container have been removed, the "tare" button can be pressed again to provide a reading of zero when nothing is present on the balance. This last step is important when you are working with balances that have automatic calibration, because failure to reset the tare value might affect the calibration process.

The mass of a chemical can also be determined through an approach known as *weight by difference*. In this procedure, a sample's mass is calculated by taking the difference between the mass of the sample plus its container and the mass of the container alone. The best results are obtained when the sample is weighed in the same container that will be used for its final study or preparation. If the sample must be transferred from its weighing container to another vessel, the container should be weighed again after the sample has been removed. The difference in mass for the final container and the container plus the sample is then used to determine the mass of the substance that was transferred. This method is more accurate than using only the initial weight of the container in that some sample may have remained behind after the transfer.

Buoyancy Corrections.   Up to this point we have assumed the mass that is displayed by a balance is equal to the mass of the sample being weighed. This assumption is valid if the buoyancy term $[1 - (d_{air}/d_{ref})]/ [1 - (d_{air}/d_{obj})]$ in Equation 3.4 is approximately equal to one and can be ignored. But when do we need to consider buoyancy effects? We can answer this question by looking more closely at what affects the ratio $[1 - (d_{air}/d_{ref})]/[1 - (d_{air}/d_{obj})]$. First, there is the density of the surrounding medium, which appears in both the top and bottom of this ratio as the term d_{air}. For air, this density will have an average value of $1.2 \times 10^{-3} g/cm^3$. A second factor to consider is the density of the reference weight (d_{ref}). By international convention, such weights usually have a density of 8.0 g/cm^3 and are made of stainless steel.[4,6] The third factor is the density of the object being weighed (d_{obj}). This density can have a wide range of values and will depend on the type of sample we are measuring.

An illustration of how the true and measured masses of an object will differ at various sample densities is shown in Figure 3.6. Samples with lower densities than the reference weight (8.0 g/cm^3 in this example) will give apparent masses that are lower than their true values, while samples with densities greater than the reference weight will have high apparent masses. The size of this error will depend on how different the sample and reference weights are in their densities. Sample materials with densities of 2−15 g/cm^3, as occur in many solids, have an error of less than 0.01% in the measured mass, as indicated by Figure 3.6. For samples with densities of 0.8−2.0 g/cm^3, which includes many liquids, this error is around 0.1–0.2%.

A useful guideline to remember is that buoyancy effects should be considered whenever you want to measure a mass with *four* or *more* significant figures (that is, when errors of less than 0.1–0.2% become important). The process of adjusting for these effects is known as making a **buoyancy correction**. This correction can be accomplished by using a modified version of Equation 3.4, in which the mass of the reference weight is replaced by the equivalent apparent mass that is read from the balance's display ($m_{display}$).

$$m_{obj} = m_{display} \cdot \frac{[1 - (d_{air}/d_{ref})]}{[1 - (d_{air}/d_{obj})]} \qquad (3.6)$$

Equation 3.6 shows that you can obtain the correct measured mass of an object if you know (1) the apparent mass given on the balance's display, (2) the density of the sample, (3) the density of the reference weight used to calibrate the balance, and (4) the density of the surrounding medium.

| **EXERCISE 3.1** | **Correcting for Buoyancy Effects** |

A weighing container is placed onto a balance and the display is tared so that it reads zero. Calcium carbonate is then placed into this container, giving a displayed mass of 10.0150 g. It is desired in this case to know the true mass of the calcium carbonate to five significant figures. The balance has previously been calibrated with 8.0 g/cm^3 stainless steel weights and is operated in a laboratory that has an air density of 1.2×10^{-3} g/cm^3. It is known that the density of pure calcium carbonate (calcite) is 2.710 g/cm^3. What is the true mass of the calcium carbonate sample?

SOLUTION

The true mass can be obtained by substituting the measured mass and the densities of the sample, air and reference weights into Equation 3.6.

$$m_{obj} = m_{display} \cdot \frac{[1 - (d_{air}/d_{ref})]}{[1 - (d_{air}/d_{obj})]}$$

$$= (10.0150 \text{ g})$$

$$\cdot \frac{[1 - (0.0012 \text{ g/cm}^3)/(8.0 \text{ g/cm}^3)]}{[1 - (0.0012 \text{ g/cm}^3)/(2.710 \text{ g/cm}^3)]}$$

$$m_{obj} = 10.0179 \text{ g} = \mathbf{10.018 \text{ g}}$$

In checking our answer by dimensional analysis, we get a final answer with the same units as the measured mass (grams). Also, we see that the measured and actual masses differ by less than 0.03%, which agrees with the type of error we would expect from Figure 3.6.

You may have noticed in the preceding exercise that we did not consider the effects of buoyancy on the weighing container itself. This simplification was possible because we used the tare feature of the balance to subtract the container's weight from the weight of the container plus calcium carbonate. If the taring and mass measurements are carried out in a reasonably short period of time (so that the density of the air is unchanged), the buoyancy of the container should remain constant, thus allowing the container's weight to be canceled out when we tare the balance. The same assumption would have been true if we had used weight by difference to determine the mass of the calcium carbonate.

To help you make a correction like the one in the last exercise, the densities of many common chemicals can be found in references like the *CRC Handbook of Chemistry and Physics* or *Lange's Handbook of Chemistry*.[16,17] It is also important that you know the exact density of the weights used to calibrate your balance. Although 8.0 g/cm^3 stainless steel is usually employed for this purpose, other types of calibration weights may be used in some cases. Another factor to keep in mind is that 1.2×10^{-3} g/cm^3 is only an average value for the density for air. The actual air density for your laboratory will depend on the barometric pressure, temperature, and relative humidity and can vary by as much as 3% from this average value.[7] Using the actual density of air in a buoyancy correction is especially crucial when *five* or *more* significant figures are needed in a mass measurement.

3.3 VOLUME MEASUREMENTS

Another important property of matter is **volume**. *Volume* can be defined as the amount of space that is occupied by a three-dimensional object. For solid materials, volume

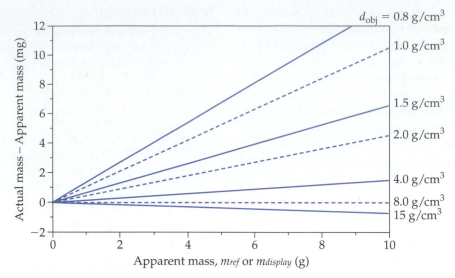

FIGURE 3.6 Difference in the actual and measured masses for samples or objects of various densities. These samples are being compared to a reference weight with a density of 8.0 g/cm^3 and are assumed to be surrounded by air with a density of 1.2×10^{-3} g/cm^3.

can be calculated from the object's height, width, and length, or by measuring the volume of liquid that is displaced by the solid. For a liquid, volume is determined by measuring the amount of space the liquid occupies in a container. Volume measurements have been made since ancient times in areas such as architecture, cooking, and commerce. These measurements also play a key role in the preparation of samples and reagents for chemical analysis, the subject that we will focus on in this chapter. Volume measurements can also be utilized directly for measuring chemical content. We will examine this last topic later in this book when we discuss a technique known as a "titration."

3.3A The Determination of Volume

Volume versus Mass. Although both volume and mass are related to the size of an object, volume has several advantages when it comes to describing materials. For instance, the volume of a sample is easier to visualize than its mass. Volumes are more convenient to measure for liquids, where all that is required is to place the liquid in a suitably marked container. A disadvantage of using volumes is that a sample's volume, unlike its mass, can vary with temperature and pressure. The mass of an object can also usually be measured with a higher precision than its volume.

The base unit of volume in the SI system is the *cubic meter* (m^3), but this unit is relatively large and not particularly convenient for use in routine chemical testing. Chemists instead often use the *liter* (L), which is now defined in the SI system as being equal to 1 cubic decimeter (or 1000 cm^3). The volume and mass of a material are related to each other through that substance's **density**. Density refers to the mass (*m*) per unit volume (*V*) of a material and is often represented by the

symbols *d* or ρ, where $d = m/V$. We have already used this relationship in the previous section when we discussed corrections for buoyancy effects. Like volume, the density of an object changes with pressure and temperature. However, the density of a material does have the same value regardless of the material's actual size, which makes density more useful than volume or mass as a means for chemical identification.

Analytical Volume Measurements. Many people are familiar with common pieces of laboratory glassware such as the beaker, Erlenmeyer flask, test tube, and graduated cylinder. These pieces of glassware are designed for the heating, mixing, and handling of solutions, but are not usually designed for the accurate determination of volumes. Even a good-quality graduated cylinder will provide a volume measurement that is accurate to within only 1% of its true value. In modern chemical laboratories more accurate volume measurements must be made on a routine basis. This higher level of accuracy is obtained by using devices that are especially designed for volume measurments, such as volumetric flasks and volumetric pipets (some of which can determine a volume to an accuracy of 0.025%).

One unique feature of these volume-measuring devices is the materials from which they are constructed. Berzelius and other scientists of his time made their own glassware out of conventional *soda-lime glass*, as is obtained by combining sand (SiO$_2$) with limestone (CaCO$_3$) and sodium carbonate (Na$_2$CO$_3$).[18] The most common glass you will find in today's laboratory is *borosilicate glass*, which contains a significant amount of boron oxide (B$_2$O$_3$) and a lower percentage of sodium oxide (Na$_2$O) and other oxides than ordinary glass. Borosilicate glass is more resistant than ordinary glass to strong acids or bases and has a third the change in size and volume with temperature.[1,18] These properties make borosilicate glass better for constructing devices for accurate volume measurements.

One problem with any type of glass is that it will eventually lose some of its mass when exposed to acids or bases over long periods of time. This effect causes the interior of the glass container to be "etched." This process will change the actual volume of the container and makes the container easier to break. Another disadvantage of glass is it may contain trace levels of metal ions that can enter and contaminate solutions. This contamination can be a problem when you are preparing samples and reagents for the analysis of trace-level metals. Because of these limitations, some analytical laboratories use special plastic containers that are made out of Teflon, polymethylpentene, or polypropylene. These materials offer good resistance to most chemical reagents and contain only trace amounts of metals. Their main disadvantage is they melt at much lower temperatures than glass, limiting the range of conditions over which they can be employed.

3.3B Types of Volumetric Equipment

There are a large number of volume-measuring devices used for analytical measurements. These devices include *volumetric flasks*, *volumetric pipets*, *burets*, *micropipets*, and *syringes*.

Volumetric Flasks. A **volumetric flask** is a device that is used to prepare solutions and to dilute them to a specific volume (generally 1–2000 mL). The general shape of a volumetric flask is shown in Figure 3.7 and consists of a long upper neck plus a round, flat-bottomed lower region for mixing and holding solutions. The top of the neck contains an opening where a stopper can be placed for mixing the flask's contents. There is also a line etched on the neck, which indicates where the *meniscus* (or curved upper surface) of the solution should be located when the volume of liquid in the flask is equal to its stated volume. Most volumetric flasks have the abbreviation "TC" on their side, indicating that they are designed "To Contain" the stated volume of liquid.

On a volumetric flask you will usually see a letter such as "A" or "B," indicating whether it is *Class A or Class B glassware*. The properties needed for a volumetric flask to be designated as Class A glassware are given in Table 3.3. Both Class A and Class B volumetric flasks provide much better volume measurements than routine glassware, but Class A flasks have only half of the maximum errors than those for Class B flasks. Class B flasks are less expensive than Class A flasks and are often fine for use in teaching or in general purpose work. Class A flasks, however, are the devices of choice whenever high-quality volume measurements are desired during solution preparation.

The procedure for using a volumetric flask is fairly simple. You begin by making sure the flask is clean and clear of any cracks or other defects. You then put into the flask a small amount of solvent and the solid or liquid you would like to place into solution. Next, you swirl the contents until all of the added material has dissolved. More solvent is then added (not yet totally filling the flask) and the swirling is repeated. While the flask is sitting on a firm flat surface, any solvent needed to fill the flask to its calibrated mark is carefully added. As you near the calibration line, it is recommended that single drops of the solvent be added with a small pipet (*not* a squirt bottle, which is harder to use for this purpose). When the bottom of the solvent's meniscus lies directly on the line, a stopper is firmly placed on the flask and the flask is repeatedly mixed by inverting it for several minutes (for example, inverting 10 times over the course of 4 to 5 minutes). This procedure ensures the flask's contents will have a uniform composition.

Volumetric Pipets. Another important device for measuring liquids is a **volumetric pipet** (also known as a *transfer pipet*). This type of pipet, shown in Figure 3.8, is designed to measure and deliver a single specific volume of liquid to a separate container, such as a volumetric flask. Volumetric pipets are used to handle volumes that range from 0.5 mL to 100 mL and are employed when volume measurements are needed that are reliable to within a few hundredths of a milliliter.

Like volumetric flasks, a volumetric pipet contains a label that gives the volume and temperature at which the pipet was calibrated. There is also a mark around the pipet's neck that indicates where this calibrated volume occurs, as well as a label indicating whether the pipet is a Class A or Class B device. Class A pipets have the

TABLE 3.3 Characteristics of Class A Volumetric Flasks[*]

Type of Flask (mL)	Maximum Allowable Error (mL)
1	± 0.01
2	± 0.015
5	± 0.02
10	± 0.02
25	± 0.03
50	± 0.05
60	± 0.05
100	± 0.08
110	± 0.08
200	± 0.10
250	± 0.12
500	± 0.20
1000	± 0.30
2000	± 0.50

[*]The properties contained in this table are those specified by the American Society for Testing Materials. The term "tolerance" is often used in place of "maximum allowable error" when describing properties of these devices.

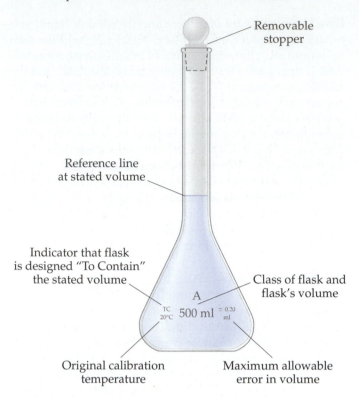

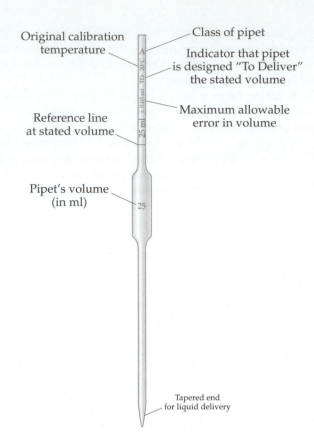

FIGURE 3.7   The general design of a volumetric flask. A line is etched on the neck of this flask to indicate the point at which the meniscus of the solution should be located when the volume of liquid in the flask is equal to the flask's stated volume. The size of this volume and the temperature at which the flask was calibrated are given near the bottom of the flask along with the symbol "TC," which means that the flask is designed "To Contain" the stated amount of liquid.

FIGURE 3.8   The general design of a volumetric pipet. A line is etched on the neck of the pipet to indicate the point at which the meniscus of a liquid should be located when its volume is equal to the pipet's stated volume. The size of this volume and the temperature at which the pipet was calibrated are given on the side along with the symbol "TD," which means that the pipet is designed "To Deliver" the stated amount of liquid without any added force or pressure.

characteristics listed in Table 3.4, while Class B pipets have maximum allowable errors that are twice these levels. One important difference from volumetric flasks is that volumetric pipets are designed "To Deliver," as represented by the symbol "TD" on their sides. This symbol means these pipets will provide the indicated volume when their contents are allowed to drain (*without* blowing or any forced delivery) into another container.

When you are using a volumetric pipet, you first need to inspect the pipet to make sure it is clean and free from any cracks or chips, especially at the end where the solvent is delivered. If the pipet is in good shape, you should then rinse its interior by using a rubber or plastic bulb or related device to draw in a small amount of the liquid you would like to measure. After you have carefully swirled this liquid within the pipet (including the region past the calibration mark), discard the rinse liquid. It is wise to carry out this rinsing step at least twice to ensure there is no dust or chemicals in the pipet from previous work. Once the pipet has been rinsed, draw a fresh portion of the desired liquid into the pipet and up past the calibration mark. The bulb or device used to draw up the liquid is then removed from the flat upper end of the pipet and this end is quickly sealed by placing the tip of your

finger onto it. Next, you gently wipe the other end of the pipet with a tissue to remove any excess liquid from the outside. The tip is then touched to the side of a waste container and the pipet is allowed to slowly drain until the bottom of the liquid's meniscus is at the calibrated mark. The pipet is ready to be moved to the final desired container, into which its liquid contents are allowed to drain freely while the tip of the pipet is kept in contact with the container's wall. At the end of this step, keep the pipet's tip against the container for a few extra seconds to make sure that draining of the liquid is complete.

It is important to *never* blow out the final contents of a volumetric pipet or to use any force other than gravity to cause liquid to flow out of such a device. It is also important to always use a rubber or plastic bulb or related device to draw liquids into the pipet and to *never* draw in a liquid by using your mouth. Finally, after you have finished using a pipet you should rinse it with water or a cleaning solution to avoid any buildup of materials that may clog the pipet or contaminate its interior.

Burets and Other Volumetric Devices.   Figure 3.9 shows some other volume-measuring devices that are often found in analytical laboratories. One of these

TABLE 3.4   Characteristics of Class A Volumetric Pipets*

Type of Pipet (mL)	Maximum Allowable Error (mL)	Minimum Flow Time (s)
0.5	± 0.006	5
1	± 0.006	10
2	± 0.006	10
3	± 0.01	10
4	± 0.01	10
5	± 0.01	15
6	± 0.02	15
7	± 0.02	15
8	± 0.02	15
9	± 0.02	15
10	± 0.02	15
15	± 0.03	25
20	± 0.03	25
25	± 0.03	25
30	± 0.05	25
40	± 0.05	25
50	± 0.05	25
75	± 0.08	30
100	± 0.08	30

*These properties are those specified by the American Society for Testing Materials. The term "tolerance" is often used in place of "maximum allowable error" when describing the properties of these devices. The "minimum flow time" is the shortest amount of time that is acceptable for all the liquid in a pipet to drain into another container.

devices is the **buret**, which is used to accurately measure and deliver variable amounts of a liquid. A buret consists of a graduated glass tube with an opening at the top for the addition of a liquid and a stopcock at the bottom for the precise delivery of this liquid into another container. A buret is used in the method of titrations, where accurate volume measurements of a reagent solution and a known reaction of this reagent with a sample are used to measure an analyte's concentration. Burets are available as both Class A and Class B devices, with common burets holding from 10 mL to 100 mL of liquid. The scale on the side of the buret varies with the device's size, with the Class A 10 mL burets having 0.05 mL divisions and maximum allowable errors of ± 0.02 mL, and 100 mL burets having ± 0.20 mL divisions and maximum errors of 0.10 mL. There are also special designs for working with smaller or larger volumes of liquids. Further information on burets and their use is provided in Chapter 12.

Another type of volumetric device is a *Mohr pipet* (or *measuring pipet*). This type of pipet has many marks on its side that allow it to measure and deliver a variety of liquid volumes within its calibrated range. Like a volumetric pipet, a Mohr pipet is designed "To Deliver" liquids through the process of natural draining, without blowing or any forced delivery. Mohr pipets have maximum volumes of 0.1–25 mL and calibrated marks at 0.1, 0.01, or 0.001 mL intervals. Mohr pipets are not as accurate as volumetric pipets, but they are more convenient when measuring a variety of volumes or when using volumes that cannot be delivered by standard volumetric pipets.

Two other types of volumetric devices are the *serological pipet* and the *Ostwald–Folin pipet*. These devices are useful when you are working with small liquid volumes or when you want *all* of a measured liquid to be delivered to a container. These pipets are similar in appearance to the Mohr pipet and volumetric pipet, respectively. An important difference is that serological pipets and Ostwald–Folin pipets are designed "*To Deliver/Blow Out*," which means they deliver the indicated volume only when the last bit of their contents is blown out with a pipet bulb.

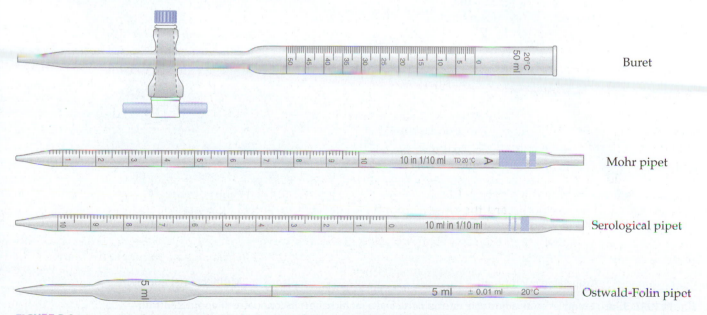

FIGURE 3.9   Examples of a buret, Mohr pipet, serological pipet, and Ostwald–Folin pipet.

Because of the different ways these pipets are operated, it is essential that you always be familiar with the type of pipet you are using during an experiment.

For handling very small liquid volumes, a *micropipet* (or *pipetter*) can be employed. These devices come with volume capacities often ranging from 0.1 μL to 5000 μL and have typical allowable errors of ± 0.5–2.0%. A micropipet uses disposable tips that can easily be replaced between samples or changed to deliver liquids over different volume ranges. Micropipets are convenient when you are dealing with small or precious samples and when errors of a few percent are acceptable. Some micropipets are operated manually, while others are electronically controlled or even equipped with several tips, allowing up to 8–12 samples to be measured and delivered at once (see Figure 3.10).

A *syringe* is a volumetric device that consists of a graduated glass or plastic barrel that holds the sample of interest. An open needle allows the sample to enter or leave the barrel, while a plunger is used to push out and dispense this sample. Syringes come with volume capacities of 0.5–500 μL or larger and, like micropipets, are used to measure and deliver small volume samples. Unlike micropipets, syringes can work with either gases or liquids. One application of syringes is their use for injecting samples into instruments for chemical measurements, as we will see when we discuss the methods of gas chromatography and liquid chromatography in Chapters 21 and 22.

3.3C Recommended Procedures for Volume Measurements

Selection and Use of Volumetric Devices. Four factors to consider when you are choosing a volumetric device are (1) the general goal of the volume measurement, (2) the volume or range of volumes to be measured, (3) the degree of reliability needed for the measurement, and (4) the number of measurements that are to be made. If you need to deliver a well-defined 5 mL volume of a liquid, a volumetric pipet would be a good choice. However, if you want to measure many different volumes (for instance, in the 3–6 mL range), a Mohr pipet might be better. If volumes less than a milliliter are to be measured, a syringe or micropipet would be preferred. Like mass measurements, volume determinations require appropriate care to provide good results. The procedures to employ when you are using volumetric flasks and volumetric pipets were discussed in Section 3.3B. Similar procedures for syringes and micropipets can be obtained from the manufacturers of these devices. Table 3.5 gives several general rules to follow when using a volumetric device.

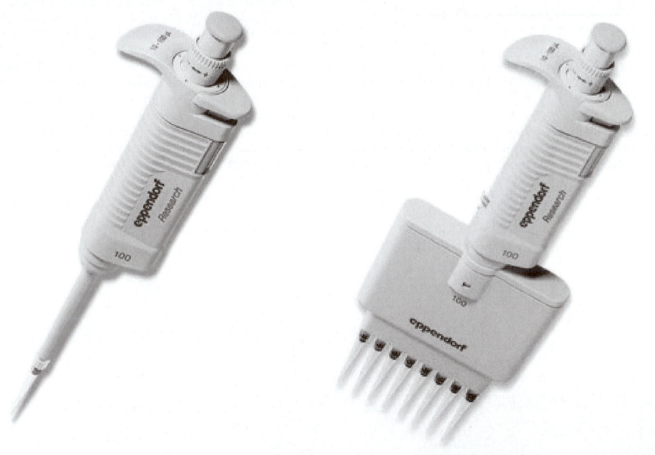

Adjustable micropipet   Multichannel micropipet

FIGURE 3.10 Examples of an adjustable micropipette and a multichannel micropipette. (Reproduced with permission and courtesy of Eppendorf.)

TABLE 3.5	**Good Laboratory Practices for Using Volumetric Glassware**
Selection of Glassware	Be familiar with the properties of your volumetric equipment, such as whether you are working with a Class A or Class B device and whether it is designed "To Deliver," "To Contain," or "To Deliver/With Blowout." Make a note of the temperature at which the glassware was calibrated and whether this will differ from the temperatures you will be using.
Glassware Condition	Always inspect your glassware before you use it. Do not use any glassware that has cracks or chips. Dirt or grease within glassware can block the openings and slow down the drainage of liquids. Dirty glassware can also place undesired substances into the measured liquid and take up part of the container's volume, causing the measured volume to be less than what is stated on the glassware.
Glassware Cleaning	Clean volumetric glassware by using a nonabrasive detergent, an acidic solution of dichromate, or an acidic alcohol solution. Follow this with several rinses of distilled or deionized water. Immediately before use, rinse your glassware with the liquid it will be used to measure or deliver.
Glassware Calibration	All volumetric equipment should be calibrated before its first use or if it is to be employed at a temperature different from that at which it was originally calibrated. Select a device with a volume and level of reliability that matches what is needed in your analytical method.
Equipment Handling	When you place a liquid into volumetric glassware, allow the liquid to settle for a short period of time before you estimate its volume. Make sure there are no air bubbles trapped in the device, which can lead to a low volume measurement. There should also be no undissolved materials remaining in your solution, because these can create inaccurate volume estimates and clog certain types of glassware (such as volumetric pipets).
Recording Results	For equipment that has reference marks for volume measurements, be sure to look at the level of your liquid and the calibrated marks at eye level. Record the measured volumes immediately in your laboratory notebook. Also record the type of device used (for example, a Class A 100.00 mL volumetric flask calibrated at 20°C) and the temperature during this measurement.

One aspect of correctly using volumetric glassware involves knowing how to properly read the level of a liquid in such devices. For volumetric flasks and pipets, the bottom of the meniscus of the liquid should be at the top of the calibration mark when viewed on a horizontal surface at eye level (see Figure 3.11). You can tell when you are looking at the meniscus properly when the calibration marks on both sides of the flask or pipet overlap one another. If you were to instead look at the liquid above or below this level,

it would give an apparent volume reading that is too high or low (an effect known as *parallax error*).[1] You can make the meniscus easier to see by placing a piece of paper, preferably with a dark color, behind the glassware. Also, when you are using a piece of glassware that has many calibrated marks on it (such as a Mohr pipet or serological pipet), remember to estimate the extent to which the liquid level occurs between the calibrated marks, thus providing one additional significant figure in your measurement.

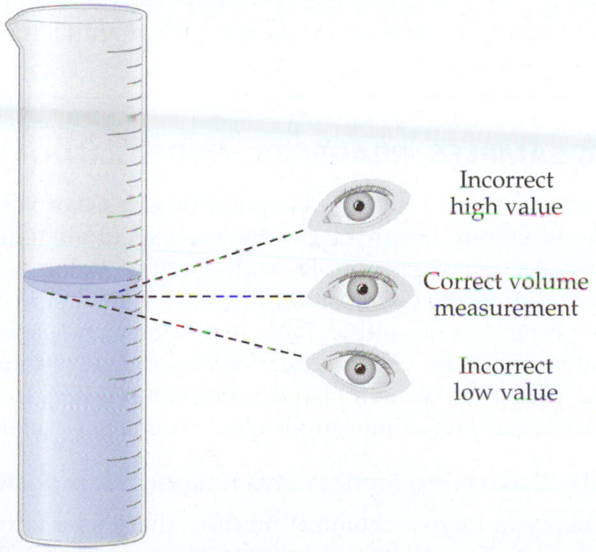

Calibration of Volumetric Devices. It is a good idea to calibrate your volumetric equipment from time to time. This is especially true when you receive a new volumetric device or are using volumetric glassware at a temperature other than the one used for its original calibration because glassware will expand or contract with a change in temperature. Table 3.6 indicates that a 1000.00 mL volumetric flask made from borosilicate glass will have a change in volume of approximately 0.01 mL (or 0.001%) for every 1°C change in temperature.[1] This relative change applies to any device made from borosilicate glass. For instance, a Class A 250.00 mL volumetric flask calibrated at 20°C would have an expected volume of 0.99950 · 250.00 mL = 249.88 mL at 15°C.

One way you can determine the true volume of a piece of glassware is to simply use it to measure a sample of distilled water. You then measure the mass of water that is contained by the device and, from the known density of water at various temperatures (see Table 3.7), you can calculate the volume of water that was present. It is important to correct for buoyancy

FIGURE 3.11 Parallax error and the correct approach for reading liquid volumes on a calibrated piece of glassware. The point where each line intersects the outer wall of the glassware is the point at which the meniscus will appear to be located.

TABLE 3.6 Change in Volume with Temperature for Borosilicate Glassware[*]

Temperature (°C)	Change from Volume at 20°C
10	$0.99990 \cdot V_{20°C}$
15	$0.99995 \cdot V_{20°C}$
20	$1.00000 \cdot V_{20°C}$
25	$1.00005 \cdot V_{20°C}$
30	$1.00010 \cdot V_{20°C}$

[*]These results were calculated based on data provided in H. Diehl, *Quantitative Analysis: Elementary Principles and Practice*, Oakland Street Science Press, Ames, IA, 1970. The term $V_{20°C}$ refers to the volume of the glassware at 20°C.

effects during this process (as described in Section 3.2C), because you will generally need to use mass measurements with four or more significant figures. An example of this calibration process is given in the next exercise.

TABLE 3.7 Density of Water at Various Temperatures*

Temperature (°C)	Density (g/cm³)	Buoyancy Correction
10	**0.999 702 6**	$m_{display} \cdot$ **1.001 052**
11	0.999 608 4	$m_{display} \cdot 1.001\ 052$
12	0.999 500 4	$m_{display} \cdot 1.001\ 052$
13	0.999 380 1	$m_{display} \cdot 1.001\ 052$
14	0.999 247 4	$m_{display} \cdot 1.001\ 052$
15	**0.999 102 6**	$m_{display} \cdot$ **1.001 052**
16	0.998 946 0	$m_{display} \cdot 1.001\ 052$
17	0.998 777 9	$m_{display} \cdot 1.001\ 053$
18	0.998 598 6	$m_{display} \cdot 1.001\ 053$
19	0.998 408 2	$m_{display} \cdot 1.001\ 053$
20	**0.998 207 1**	$m_{display} \cdot$ **1.001 053**
21	0.997 995 5	$m_{display} \cdot 1.001\ 054$
22	0.997 773 5	$m_{display} \cdot 1.001\ 054$
23	0.997 541 5	$m_{display} \cdot 1.001\ 054$
24	0.997 299 5	$m_{display} \cdot 1.001\ 054$
25	**0.997 047 9**	$m_{display} \cdot$ **1.001 055**
26	0.996 786 7	$m_{display} \cdot 1.001\ 055$
27	0.996 516 2	$m_{display} \cdot 1.001\ 055$
28	0.996 236 5	$m_{display} \cdot 1.001\ 056$
29	0.995 947 8	$m_{display} \cdot 1.001\ 056$
30	**0.995 650 2**	$m_{display} \cdot$ **1.001 056**

* All of the densities shown for pure air-free water are at a pressure of 101.325 kPa (1 atmosphere). The buoyancy corrections assume that the air density is 1.20×10^{-3} g/cm³ and the density of the reference weight is 8.00 g/cm³.

EXERCISE 3.2 **Calibration of a Volumetric Flask**

A 200.00 mL volumetric flask is placed on an electronic balance, the display is set to zero, and the flask is filled to the mark with distilled water at 25°C. The displayed mass for the water in this flask is 199.2094 g. The density of the surrounding air is 1.20×10^{-3} g/cm³ and the balance has been calibrated using a 8.00 g/cm³ reference weight. What is the true volume of the flask at 25°C?

SOLUTION

Because we need a final mass for water with at least four significant figures, the displayed mass of the water must first be corrected for buoyancy effects. The density of water and the corresponding buoyancy correction at 25°C can be found in Table 3.7. (*Note:* The same result is obtained by using Equation 3.6 and the densities of the water, air, and reference weight.)

$$m = 199.2094 \text{ g} \cdot (1.001055)$$
$$= 199.4196 \text{ g}$$

Now that we know the mass of water in the flask, the flask's volume can be determined by dividing this mass by the density of water, giving the true volume of the flask at 25°C.

$$V = \frac{199.4196 \text{ g}}{0.9970479 \text{ g/cm}^3}$$
$$V = 200.0100 \text{ cm}^3 = \mathbf{200.01 \text{ mL}}$$

In this case we rounded the final answer to be consistent with the expected error in using the flask ($\pm$ 0.10 mL, as listed in Table 3.3) and we converted the units from cm³ to mL. You could also use this result to estimate the flask's volume at other temperatures. As an example, the calculated volume at 20°C would be 200.01 mL/1.00005 = 200.00 mL, which is the volume you would expect if 20°C was the temperature used by the manufacturer to calibrate this flask.

3.4 SAMPLES, REAGENTS, AND SOLUTIONS

Most analytical techniques begin with a mass or volume measurement. For instance, the analysis of air requires the collection of gas samples with specified volumes, and determining the composition of steel might first involve weighing a steel sample. Let's now look at how masses and volumes are used to describe the content of samples and reagents. We will also discuss some issues to consider when preparing samples and reagents for analysis.

3.4A Describing Sample and Reagent Composition

When you have a chemical mixture that has a uniform distribution for all its components, this mixture is called a **solution**. The most abundant component of the solution (or the component that is used to dissolve and contain the other chemicals) is known as the **solvent**. All other substances in the mixture are called **solutes**. As an example, when a small amount of sodium chloride is dissolved in water, this salt dissociates to form sodium ions and chloride ions (the solutes) in water (the solvent).

True solutions have the same composition throughout, so their overall content will be identical for any reasonably sized portion of the solution. As a result, we can describe the content of a solution by using a **concentration**. Concentration can be defined as the amount of a substance that is present within a given volume or mass of solution.[19] We will now examine several ways for reporting the concentration of a solution, as well as how we might describe the composition of non-uniform mixtures of chemicals.

Weight and Volume Ratios.   The easiest approach for describing the composition of any chemical mixture is to simply use the masses or volumes of the various components that are present in the mixture. One way this can be done is by using a **weight-per-weight** ratio (**w/w**). This ratio is calculated by dividing the mass of the analyte or substance of interest by the total mass of the mixture. A major component of a mixture is often described by using a *percent weight-per-weight* (*% w/w*), which is found by multiplying the weight-per-weight ratio by 100.

$$\% \, w/w = 100 \cdot \frac{\text{Mass of chemical}}{\text{Mass of mixture}} \quad (3.7)$$

For example, the body of a 120 pound person contains about 72 pounds of water, so the percent weight-per-weight of water would be $100 \cdot (72 \text{ lb})/(120 \text{ lb})$, or 60% w/w. It is important when calculating this value to have the same units for the weights or masses of both the chemical and its mixture so that the answer will be expressed as a true fraction or percent.

EXERCISE 3.3   **Calculating Percent Weight-per-Weight**

The element cerium was discovered by J. J. Berzelius and other scientists in 1803. One source of cerium is the mineral bastnasite-(Ce), which has the formula $Ce(CO_3)F$. If you have one mole of pure bastnasite-(Ce), what would be the % w/w of cerium in this ore?

SOLUTION

From the given chemical formula, the formula weight for bastnasite-(Ce) is 219.12 g/mol. We also know that each mol of bastnasite-(Ce) contains one mol of cerium (atomic weight = 140.12 g/mol), so the amount of cerium in one mole of pure bastnasite-(Ce) would be as follows.

$$\% \text{ Cerium (w/w)} = 100 \cdot \frac{\text{Mass of cerium}}{\text{Mass of bastnatite-(Ce)}}$$

$$= 100 \cdot \frac{(1 \text{ mol} \cdot 140.12 \text{ g/mol})}{(1 \text{ mol} \cdot 219.12 \text{ g/mol})}$$

$$\% \text{ Cerium (w/w)} = \mathbf{63.947\%}$$

An identical % w/w value for cerium would be obtained if we were working with 0.5 mol or 2.0 mol of this ore because the *relative* content of cerium in this material will still be the same.

When you are working with minor or trace components of a mixture, other multiplying factors besides 100 can be used with weight-per-weight ratios. Using 1000 in place of 100 would give a result in *parts-per-thousand*, which is sometimes represented by the symbol $\%_{00}$. If even lower amounts of a substance are present, multiplying factors like a million (10^6), billion (10^9), or trillion (10^{12}) might be used. This process would provide results given in units of *parts-per-million* (*ppm*), *parts-per-billion* (*ppb*), or *parts-per-trillion* (*ppt*), respectively. This idea can be illustrated using rare earth metals like lanthanum that can be found in bastnasite. If a 100 g sample of bastnasite-(Ce) contained 2 mg lanthanum, the relative amount of lanthanum would be $10^6 \cdot (2 \times 10^{-3} \text{ g La})/(100 \text{ g Bastnasite-(Ce)}) = 20 \text{ ppm La}$, or 20 parts-per million. Note that the use of ppm, ppb and ppt is mostly a matter of convenience in that it allows us to describe chemical compositions in a simpler manner, such as saying "20 ppm La" instead of "0.000020 g La/g sample."

A second way of describing chemical composition is to use a **volume-per-volume (v/v)** ratio. This type of ratio is employed when you are working with mixtures of liquids or gases, for which volumes are easier to measure than masses. These ratios are often expressed as % v/v (see Equation 3.8), but can also be given as parts-per-thousand, ppm, ppb, or ppt.

$$\% \, v/v = 100 \cdot \frac{\text{Volume of chemical}}{\text{Volume of mixture}} \quad (3.8)$$

One common application of volume-per-volume ratios is to describe mixtures of alcohols with water. For instance, a container labeled "25% methanol (v/v)" should contain 25 mL of methanol for every 100 mL of solution. However, this label does *not* mean the solution contains 25 mL of methanol for every 75 mL of solvent, because the volumes of the components in a liquid mixture are not strictly additive. You can get around this problem by simply stating the amount of each component that was placed into the solution. Thus, a "25:75 (or 1:3) solution of methanol in water" would be a better way of describing a solution that is formed by adding 25 mL of methanol to 75 mL of water.

The **weight-per-volume (w/v)** ratio is another means of stating chemical content. This ratio is calculated by determining the mass of a chemical that is present in the total volume of a mixture, and is often used in describing solutions that contain dissolved solids as solutes.

$$w/v = \frac{\text{Mass of chemical}}{\text{Volume of mixture}} \quad (3.9)$$

An illustration would be a 4.0 g/L solution of iron(III) chloride in water, which could be prepared by placing 0.40 grams of solid $FeCl_3$ into water and diluting the entire solution to a final volume of 100.0 mL. The advantage of using weight-per-volume ratios in this situation is it combines the convenience of volume measurements for liquids with the ease of mass measurements for solids.

In addition to being expressed as a ratio of mass and volume units (like g/L or mg/L), weight-per-volume ratios are sometimes given as a percent or in related terms like ppm, ppb, and ppt. These units are typically employed when weight-per-volume ratios are used to describe dilute solutions of chemicals in water. The basis for this approach is that the density of water at room temperature is approximately 1.0 g/mL, meaning that one milliliter of water weighs about 1 gram. Because of this, the mass of water (in grams) can be substituted for its volume (in mL).

EXERCISE 3.4   Working with Weight-per-Volume Ratios

Figure 3.12 shows a warning sign posted near a well found to have high levels of nitrate. The amount of nitrate–nitrogen found in this water was 27.5 mg/L. What would this value be if it were given in parts-per-million?

SOLUTION

This is a dilute solution of water that is probably at or near room temperature, so we know the density will be about 1.0 g/mL. As a result, we can rewrite 27.5 mg/L as shown below.

Nitrate−nitrogen (w/v) =

$$\frac{(27.5 \text{ mg nitrate−nitrogen}) \cdot (1 \text{ g}/10^3 \text{ mg})}{(1.000 \text{ L solution}) \cdot (1.0 \text{ g/mL solution}) \cdot (10^3 \text{ mL/L})}$$

$$= \frac{(27.5 \text{ g nitrate−nitrogen})}{1.000 \text{ g solution} \cdot 10^6}$$

Nitrate−nitrogen (w/v) = 27.5 ppm

We can see from the last exercise that a concentration of 1 ppm (w/v) in a dilute aqueous solution is approximately equal to 1 mg/L (or 1 μg/mL). Similarly, a 1 ppb (w/v) aqueous solution is approximately the same as 1 μg/L (or 1 ng/mL), and 1 ppt is roughly equal to 1 ng/L (or 1 pg/mL). It needs to be emphasized that these relationships are valid *only* if the final solution has a density of 1.0 g/mL. If this is not the situation, the weight-per-volume ratio should not be given as a percent or related fraction. Instead, this ratio should be written by using the mass and volume units of the solute and solution, such as g/mL.

Molality and Molarity.   Weight and volume ratios are valuable when dealing with the masses or volumes of chemicals. To understand how chemicals might react

with each other, it is even more valuable to know the actual *number* of a given type of molecule, atom, or ion that might be present in solution. In Chapter 2, we learned that the SI unit for the number of any substance is the mole, which is equal to 6.02×10^{23}. Although this number may seem rather large, it is convenient for describing the chemical content of many materials, which often contain solutes or compounds that approach or even exceed a mole in quantity.

To relate the number of moles of a particular molecule to its mass, a chemist uses that substance's **molar mass**. The molar mass for any substance is defined as the number of grams that are contained in one mole of that substance. For molecular compounds, the molar mass is commonly referred to as the *molecular weight* (MW), while for ionic compounds and elements the molar mass is also called the *formula weight* or *atomic weight* (or *formula mass* and *atomic mass*), respectively.

One concentration unit that measures the amount of solute in moles is **molality** (represented by the symbol *m*). Molality is equal to the number of moles of a solute per kilogram of solvent.

$$m = \frac{\text{Moles of solute}}{\text{Kilograms of solvent}} \quad (3.10)$$

As an example, 0.025 mol of iron(II) chloride in 0.500 kg water would give a solution that has a concentration of 0.025 mol/0.500 kg = 0.050 *m* iron(II) chloride, or a 0.050 *molal* solution. It is important to note here that the mass given in the bottom of Equation 3.10 is for the solvent and not for the total final solution. Because molality is based on a ratio of masses, it is an important unit to use when changes in temperature, and thus changes in volume, are expected during an analysis. Molality is also useful in describing the relative amount of *both* the solute and solvent in a solution.

A disadvantage of using molality is that the total volume of a solution is much easier to measure than the mass of the solvent. This is especially true when you are using a volumetric flask to prepare solutions. Using units of molality also does not directly tell what the total volume of the solution will be. This can be a big problem if volume measurements are to be used during solution preparation or handling. It is more common in routine laboratory work to use a related unit known as **molarity** (*M*). Molarity is defined as the number of moles (or the *number of gram molecular weights*) of a substance that is present in each liter of solution.

$$M = \frac{\text{Moles of solute}}{\text{Liters of solution}} \quad (3.11)$$

To illustrate this relationship, a solution with a total volume of 500 mL that contains 1.00 g glucose (or 5.56×10^{-3} mol) would have a glucose concentration of 5.56×10^{-3} mol/0.500 L = 0.0111 *M*. This mixture could also be referred to as a 0.0111 *molar* solution.

Nebraska Department of Roads

NOTICE

In accordance with the Nebraska Department of Health *Regulations Governing Public Water Supply Systems*, the consumers of the

Chappell Eastbound Rest Area

public water supply systems are hereby notified that the system is in violation of the established drinking water standard of "10.0" mg/l for nitrate-nitrogen. The nitrate concentration as determined by the Nebraska Department of Health Laboratory from samples collected on:

27.5 mg/l

Nitrate contamination above "10.0" mg/l in drinking water has been associated with the incidence of infant methemoglobinemia or "blue baby syndrome." Symptoms occur when nitrate is reduced to nitrate in the infant internal tract. The nitrate, in turn, reacts with the blood to reduce the capacity of oxygen to reach to body cells. Infants under six months of age are at a greater risk of developing methemoglobinemia. Reports of methemoglobinemia in older children and adults from ingestion of nitrate have not been reported.

Water from the

Chappell Eastbound Rest Area

system should not be used in the preparation of infant formula or as a source of drinking water for infants six months of age or less, nursing mothers or pregnant women. Please note that the boiling water *will not* lower the nitrate concentration.

For additional information regarding this notice, interested persons may contact:

Nebraska Department of Roads   (308) 254-4712

FIGURE 3.12   Sign posted in western Nebraska that alerts travelers to the hazards of drinking water from a local well contaminated with nitrate.

EXERCISE 3.5	Using Molarity and Molality to Describe a Solution

A 2.500 g portion of sodium hydroxide (NaOH) is placed into a 250.00 mL volumetric flask and diluted to the mark with water. It is later found that the added mass of water was 497.2 g. What is the final concentration of the NaOH in units of molarity and molality?

SOLUTION

The molarity is found by dividing the moles of added NaOH by the total volume of the final solution. The moles of NaOH present in 2.500 g can be found by using a molar mass of 40.00 g/mol for NaOH, which gives (2.500 g NaOH)/(40.00 g NaOH/mol NaOH) = 0.6250 mol. The molar concentration would be (0.6250 mol NaOH/500.00 mL) · (1000 mL/1 L) = 1.250 M. The molality of the solution is obtained by dividing the moles of NaOH by the kilograms of added solvent. The result is a solution concentration of (0.6250 mol NaOH/497.2 g water) · (1000 g water/1 kg water) = 1.257 m NaOH.

If a solute dissociates into ions or produces several forms when it is placed in solution, the number of moles of this solute that is used in Equation 3.11 would be based on the *number of gram formula weights* instead of gram molecular weights. This approach is then used to describe the total amount of solute that was placed into solution, giving a concentration in units of moles per liter of solution and referred to as the *formality (F)*. For instance, placing 60.05 g acetic acid (1.000 mol) into 1.000 L of an aqueous solution would give a 1.000 *formal* solution of acetic acid, even though the acetic acid could be present in two forms, acetic acid (HAc) and the acetate ion (Ac$^-$). We can also describe the concentrations of these individual forms by using molarity, but this can lead to some confusion because formality and molarity have the same net units (mol/L). Because of this, some authorities recommend using molarity to describe the concentrations of both molecular *and* ionic solutes, a practice that will be followed in this book. As part of this process, we will use the term **analytical concentration (C)** instead of formality to refer to the total concentration of a substance in solution and brackets to represent the concentration of an individual form for the substance (for example, [HAc] or [Ac$^-$] to describe the individual concentrations of acetic acid and the acetate ion).

When you are dealing with a relatively dilute aqueous solution at room temperature (for instance, 0.01 *M* NaCl or 0.01 *m* NaCl), the molality and molarity of the solution will have about the same numerical value even though they have different units (mol NaCl/L solution versus mol NaCl/kg solvent). This situation occurs because the density of water and of the solution will both be approximately 1.0 g/mL, which means 1 kg of water will roughly be equivalent to 1 L of solution. This relationship will not be true if you have a different solution density, as would happen when you have a nonaqueous solvent or a solution with a moderate-to-high concentration of solutes. For concentrated solutions, molality is the preferred unit because it provides temperature-independent values. When working with dilute solutions, like those found in many samples for chemical analysis, molarity is the preferred unit because of the greater ease with which a solution's volume can be measured versus a solvent's mass. For this reason, molarity will be the main concentration unit used in this text.

Other Units.   There are other measures of chemical content that you will also encounter from time-to-time in books, scientific articles, or chemical catalogs. One such case is when you must describe the amount of a chemical on a surface. For instance, the amount of iron oxide (rust) on the surface of an iron bar could be expressed as a "surface concentration" with units of moles per square meter. Other units for chemical contents depend on the ability of a substance to take part in a particular reaction, like the rate at which an enzyme solution catalyzes the formation of a product. Another example is the use of radioactivity to describe chemical content, such as the measurement of carbon-14 decay for determining the age of an organic sample.

Another measure of chemical reactivity is the unit of *normality* (represented by the symbol *N*). Normality describes the amount of a chemical that is available for a specific type of reaction, as is accomplished by using the *equivalents* of the chemical per liter of solution. The number of equivalents you have of a chemical will depend on that chemical's structure and the type of reaction in which it is going to be employed. In an acid–base reaction (see Chapter 8), an equivalent is given by the moles of a chemical needed to produce or consume one unit of titratable hydrogen ions. For a reduction–oxidation reaction (see Chapter 10), an equivalent is related to the moles of a chemical that are needed to produce or consume one unit of electrons. Although chemical equivalents are still employed to describe reactions in analytical chemistry, the use of normality as a unit of concentration is now discouraged.[19] Thus, this unit will not be used in this book except to illustrate how it may be converted to other units of concentration.

3.4B Solution Preparation

Chemical Purity.   When you are preparing a solution for an analytical method, one factor that must be considered is the purity of the chemicals that will be placed into your samples and reagents. Ideally, you do not want to have compounds in the reagents that may interfere with detection of the analyte and cause an inaccurate result to be obtained. You also do not want to accidentally add the same chemicals as those you want to determine, which will create a false "positive" result. For these and other reasons, you should always use high-purity chemicals in any analytical method.

To help in this selection, commercial chemicals are often classified according to their purity, as shown in Table 3.8. Routine work that requires good but not exceptionally high quality chemicals can be performed with materials that are "Technical Grade" or "Laboratory Grade." For general analytical work, substances with higher purity can be obtained by using chemicals that meet requirements set by the American Chemical Society or by agencies that regulate analytical laboratories. There are also chemicals that have been prepared to meet the needs of particular methods. Examples are "HPLC Grade," "Trace Metal Grade," and "Biotechnology Grade" chemicals.

In some cases, chemicals with special properties may be needed. This situation occurs when a solution is to be used as a reagent in a titration, where the actual concentration of this solution is first determined by reacting it with a compound known as a *primary standard*. A primary standard is a pure substance that is stable during storage, can be weighed accurately, and undergoes a known reaction with the solution it is used to characterize. The reagent solution that is characterized by this process is then referred to as a *secondary standard*. A further discussion of primary and secondary standards for titrations can be found in Chapters 12, 13, and 15.

It is important when selecting chemicals to also consider the purity of your solvents. High-purity water is particularly crucial because many of the samples and reagents in analytical chemistry are aqueous solutions. Table 3.9 shows some contaminants that can be found in ordinary water. To remove these contaminants, the water should be purified before it is used to prepare samples or

TABLE 3.8   Common Grades of Commercially Available Chemicals

Type of Chemical	Meaning of Grade	Common Uses
Certified ACS Grade	Reagent chemicals that meet or exceed specifications by the American Chemical Society (ACS)	Various analytical applications
USP, BP, EP, NF, or FCC Grade	Reagent chemicals that meet or exceed specifications made by the U.S. Pharmacopeia (USP), the British Pharmacopeia (BP), the European Pharmacopeia (EP), the National Formulary (NF), or the Food Chemicals Codex (FCC)	Food and drug laboratories, biological testing
Technical or Laboratory Grade	Chemicals of reasonable purity for cases where no official standards exist for quality or impurity levels	Manufacturing and general laboratory use
Biotechnology Grade	Chemicals and solvents that have been purified and prepared for use in biotechnology	Molecular biology, electrophoresis assays, DNA/RNA or peptide sequencing and synthesis
HPLC Grade	Chemicals that have been purified and prepared for use in high-performance liquid chromatography (HPLC)	Preparation of reagents and samples for HPLC
Trace Metal Grade	Chemicals prepared to have low levels of trace metals	Preparation of reagents and samples for trace-metal analysis

reagents. A common approach for water purification is to use *distillation*. This is a relatively inexpensive method in which the water is heated to boiling, with the steam then being condensed and used in a purified form known as *distilled water*. Distillation is one of the oldest methods for water treatment and is good for obtaining water that is free of particulates, dissolved solids, microorganisms, and pyrogens. (*Note:* a "pyrogen" is a substance that produces a fever.) Distillation can also reduce the amount of some dissolved organic compounds, but does not help much in removing dissolved gases.[20]

Some laboratories take distilled water and treat it further by employing a second method, such as

deionization. The method of deionization uses cartridges that take the cations or anions in water and exchange these for hydrogen ions (H^+) and hydroxide ions (OH^-), which will then react to form more water. The purified water that is obtained by this method is called *deionized water* (or *DI water*). This approach is good at removing ions and dissolved gases like carbon dioxide (which is present in water as carbonic acid), making it a nice complement to distillation as a means for water treatment.[20] Systems for preparing DI water also often contain cartridges with activated carbon to remove organic compounds, as well as submicron filters to remove bacteria and microbes.

Aliquots and Dilutions.   Let's next examine some terms that are used to describe the preparation of solutions. One of these is the term **stock solution**, which is a reagent used to make other less concentrated solutions for an assay. The advantage of making a stock solution is that it allows the use of a large mass of solute, which will be easier to handle and measure than smaller quantities. Stock solutions are also easier to store than more dilute solutions because they occupy less space for the same amount of solute.

When part of a stock solution or sample is used to prepare a second less concentrated solution, the portion that is taken from the original solution or sample is known as an **aliquot**. An aliquot is often drawn and measured by using a pipet, micropipette, or syringe. If more solvent is then added to the aliquot, this process is known as a **dilution**. For instance, suppose a 10.00 mL aliquot of a stock solution were placed into a 50.00 mL volumetric flask, and this flask was then filled to the mark with water. The result would be a new solution in which the contents of the original solution had undergone a fivefold dilution. A dilution step is used to adjust

TABLE 3.9   Types of Contaminants Found in Water

Type of Contaminant	Examples
Dissolved inorganic solids	Calcium ions, magnesium ions, chloride, fluoride, iron(II) and iron(III) ions, silicates, phosphates, and nitrates
Dissolved inorganic gases	Carbon dioxide, oxygen
Dissolved organics	Pesticides and herbicides, decayed plant and animal matter, gasoline, alcohols, chloramines
Particulates (Particulate matter)	Sand, silt, clay, colloidal particles, and debris from pipes
Microorganisms	Bacteria, algae, amoebae, protozoa, diatoms and rotifers
Pyrogens	Cell wall fragments and lipopolysaccharides from bacteria

Source: The information in this table was obtained from *A Guide to Laboratory Water Purification*, Labconco, Kansas City, MO, 1998.

the concentration of samples and reagents to place them in an appropriate range for a chemical analysis.

When you are taking aliquots and conducting a dilution, it is important to keep track of the volumes you are using in each step. This practice is necessary so that you can relate the concentration of your final solution back to the contents of its initial sample or solution. As is shown in Figure 3.13, the moles of each chemical in the aliquot will be the same as the moles present in the diluted solution (assuming the same chemicals are not in the solvent or other added components). We can use this fact to relate the initial and final concentrations of the analyte as shown below.

$$(\text{mol solute})_{\text{Dilution}} = (\text{mol solute})_{\text{Aliquot}} \quad (3.12)$$

or

$$M_{\text{Dilution}} \cdot V_{\text{Dilution}} = M_{\text{Aliquot}} \cdot V_{\text{Aliquot}} \quad (3.13)$$

In these equations, M_{Aliquot} and M_{Dilution} are the molar concentrations of the solute in the aliquot and diluted sample, and V_{Aliquot} and V_{Dilution} are the volumes of the aliquot and dilution in liters. Similar equations can be written for situations in which you are working with concentrations that are in units of molality and weight or volume ratios. In the following exercise you will see how these relationships can be used to calculate what solute concentration will be present in a diluted solution. You can also use these equations to determine the aliquot size and extent of dilution that are needed to prepare a reagent from a particular stock solution or sample.

EXERCISE 3.6 Preparing a Diluted Solution

A diluted solution of sodium chloride is to be prepared from a 0.1000 M stock solution. A 10.00 mL aliquot of the stock solution is obtained with a 10.00 mL volumetric pipet and placed into a 50.00 mL volumetric flask. The contents of this flask are then diluted to the mark with water. What is the final concentration of sodium chloride in the diluted solution?

SOLUTION

The aliquot is the only source of sodium chloride in the solution, so the moles of NaCl in the diluted solution will be equal to the moles of NaCl that were in the 10.00 mL aliquot, or (0.1000 mol/L) (0.01000 L) = 1.000×10^{-3} mol. When we place this amount of sodium chloride into a total volume of 50.00 mL water (10.00 mL of which comes from the aliquot), the final concentration of NaCl is $(1.000 \times 10^{-3}$ mol$)/(0.05000$ L$) = 0.02000$ M. The same result is obtained by rearranging Equation 3.13 into the form shown below.

$$M_{\text{Dilution}} = M_{\text{Aliquot}}(V_{\text{Aliquot}}/V_{\text{Dilution}})$$

$$= 0.1000 \ M \ \text{NaCl} \cdot (0.01000 \ \text{L}/0.05000 \ \text{L})$$

$$M_{\text{Dilution}} = \mathbf{0.02000 \ \textit{M} \ NaCl}$$

Because this expression uses a ratio of volumes $(V_{\text{Aliquot}}/V_{\text{Dilution}})$, we would have arrived at the same final answer by keeping the original units of milliliters, where 0.01000L/0.05000 L = 10.00 mL/50.00 mL (or a 1:5 dilution of the original solution).

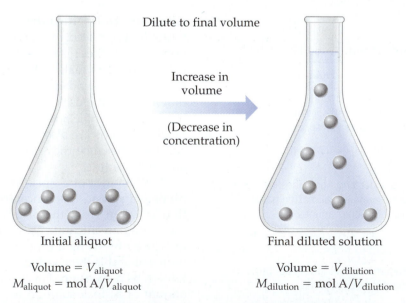

Dilute to final volume

Increase in volume

(Decrease in concentration)

Initial aliquot

Final diluted solution

Volume = V_{aliquot}
M_{aliquot} = mol A/V_{aliquot}

Volume = V_{dilution}
M_{dilution} = mol A/V_{dilution}

FIGURE 3.13 The relationship between the amount of a chemical that is present in an aliquot and the final concentration of this chemical in a diluted solution. In this example, the total moles of analyte (A) is the same in both the initial aliquot and the final diluted solution. However, the concentration of this analyte is lower in the final diluted solution because this diluted solution has a larger volume than the original aliquot.

Temperature Effects.   Another factor to consider when you are preparing solutions is the temperature at which you are making and using your samples and reagents. The volume and concentration of a solution can change as the temperature changes (see Figure 3.14). This change can result in significant errors if overlooked. When you prepare a solution, the total moles of each added chemical should be constant, provided these chemicals do not degrade or react to form other substances. Thus, if you know the volume (V_1) and molar concentration (M_1) of the solution at its original temperature and you measure the volume (V_2) of the same solution at its final temperature, the new concentration (M_2) at the final temperature can be determined by using Equation 3.14.

$$M_1 \cdot V_1 = M_2 \cdot V_2 \qquad (3.14)$$

Imagine that you had a 0.1000 M solution of sodium chloride in water that was prepared at 20°C in a 500.00 mL volumetric flask. If you look at this solution later when the temperature is 25°C, you will find that the volume has increased by 0.58 mL. This change in volume will affect the molar concentration of sodium chloride. This new concentration can be found by using Equation 3.14, which gives $M_{25\,°C} = (0.1000\ M) \cdot (0.50000\ L)/(0.50058\ L) = 0.09988\ M$.

If you do not know the volume of your final solution, but do know the densities of the original and final solutions (d_1 and d_2), the new solution volume can be found by using Equation 3.15.

$$d_1 \cdot V_1 = d_2 \cdot V_2 \qquad (3.15)$$

You can then use the value obtained for V_2 with Equation 3.14 to determine the molarity of the final solution. An example of this process is given in Exercise 3.7. If you look more closely at Equations 3.14 and 3.15 you will find that the volume units on each side cancel out to give moles = moles or mass = mass. In other words, both expressions are based on the fact that the total mass or moles of solute will be constant in the solution.

EXERCISE 3.7	Effects of Temperature on Concentration

A 1.0000 × 10^{-3} M solution of hydrochloric acid in water is prepared in a 1000.00 mL volumetric flask at 20°C. This solution is set aside for later use. The temperature of this solution is later found to be 25°C. It is known that the density was 0.998232 g/mL at 20°C and that the same solution has a density of 0.997074 g/mL at 25°C. Determine the concentration of hydrochloric acid at 25°C.

SOLUTION

We don't know the exact volume of the final solution but we do know its density, so we first need to use Equation 3.15 to find the new volume.

$$d_1 \cdot V_1 = d_2 \cdot V_2$$
$$(0.998232\ g/mL) \cdot (1000.00\ mL)$$
$$= (0.997074\ g/mL) \cdot V_2$$
$$V_2 = 1001.16\ mL$$

We can then use this volume with Equation 3.14 to calculate the molar concentration at 25°C.

$$M_1 \cdot V_1 = M_2 \cdot V_2$$
$$M_2 \cdot (1001.16\ mL) = (1.0000 \times 10^{-3}\ M) \cdot (1000.00\ mL)$$
$$\mathbf{M_2 = 0.9988 \times 10^{-3}\ M}$$

Thus, because the volume of solution has increased from 20°C to 25°C, we see a slight *decrease* in the molar concentration of hydrochloric acid.

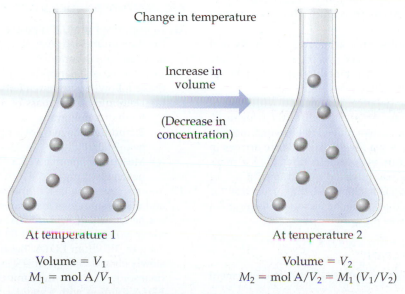

Change in temperature

Increase in volume

(Decrease in concentration)

At temperature 1

Volume = V_1
$M_1 = $ mol A$/V_1$

At temperature 2

Volume = V_2
$M_2 = $ mol A$/V_2 = M_1\ (V_1/V_2)$

FIGURE 3.14   Change in the volume and molar concentration of a solution with temperature. The total moles of analyte (A) is the same in the solution at both temperatures. However, the concentration of this analyte is lower in the solution on the right because this solution has a larger volume than the solution on the left.

Key Words

Other Terms

Questions

THE DETERMINATION OF MASS

1. Define what is meant by the terms "mass" and "weight." How do these terms differ? Which of these is preferred for use in scientific measurements?
2. Explain what is meant by "weighing" a sample? What is the goal of this process?
3. Describe what is meant by "buoyancy." How does buoyancy affect the measurement of weight? How does it affect the measurement of mass?
4. Write equations that show how the forces of gravity and buoyancy act on an object when it is on a balance. Show how these equations can be combined to relate the mass of the object to the mass of a reference weight.
5. A single-pan electronic balance is calibrated with a set of standard weights in Los Angeles, California (elevation, 340 ft above sea level), and then moved to Denver, Colorado (elevation, 5280 ft above sea level). The balance arrives in good working order but is found to now give a mass reading that is off by a few hundred milligrams for a 1 kg weight. After the balance is recalibrated, the correct result is obtained. What do you think was the reason for the original error after the balance was moved?

TYPES OF LABORATORY BALANCES

6. What is a "balance" and how is it used for the measurement of weight or mass?
7. What is a "mechanical balance"? Explain how this type of balance operates.
8. What is an "electronic balance"? Describe how this type of balance works.

9. What is a "quartz crystal microbalance"? How is this device constructed and how is it used to provide a mass measurement?
10. Describe how the mass of individual chemical species can be examined by using a mass spectrometer. How are the results of this measurement usually displayed?

RECOMMENDED PROCEDURES FOR MASS MEASUREMENTS

11. Define "resolution," "capacity," and "readability." What are the maximum load, readability, and resolution for the balances in your laboratory?
12. Which type of balances in Table 3.1 would you use for each of the following measurements.
    (a) Determining the mass of a 150 g sample to the nearest tenth of a milligram
    (b) Examining the mass of a 1.00 kg chemical product to the nearest 0.01 g
    (c) Measuring the mass of a 100 mg protein sample to the nearest 0.01 mg
13. You are given the task of measuring the amount of calcium and magnesium ions in a water sample by using ethylenediamine tetraacetic acid (EDTA) as a reagent. To perform this assay, you must prepare 1.00 L of a reagent that contains 7.4 g of disodium EDTA. Because disodium EDTA tends to slowly absorb water as it is weighed, the concentration of its final solution is later determined by reacting a portion of the EDTA solution with a standard solution of calcium carbonate, which is prepared to contain 0.35 g calcium carbonate in 250 mL water. Your laboratory has two electronic balances that you could use in this method. The first is a precision balance with a capacity of 200 g and a readability of 0.001 g. The

other is an analytical balance with a capacity of 80 g and readability of 0.0001 g. Which balance would you use to weigh the disodium EDTA and which would you use for the calcium carbonate?

14. Describe how "direct weighing" and "weight by difference" are performed. What are the advantages and disadvantages of these methods?

15. What is "taring"? How is taring used on laboratory balances?

16. What does it mean when you make a "buoyancy correction" in a mass measurement? When is such a correction important, and how is this type of correction performed?

17. Potassium hydrogen o-phthalate ($KHC_8H_4O_4$, or KHP) is a chemical used as a primary standard for determining the exact concentration of a base in a reagent. A sample of KHP is placed into a weighing container and gives a mass reading of 10.4194 g. The KHP is then transferred to a volumetric flask and the weighing container is measured again, now giving a mass of 5.3052 g. It is known that the density of KHP is 1.636 g/cm^3. If the density of the surrounding air was $1.2 \times 10^{-3} g/cm^3$ and that the balance was calibrated with a 8.0 g/cm^3 reference weight, what actual mass of KHP was placed into the volumetric flask?

18. The amount of iron in an ore sample can be determined by dissolving the ore sample and precipitating the resulting iron ions as hydrated ferric oxide, $Fe_2O_3 \cdot x\ H_2O$. The water in this precipitate is driven off by heating this material with a flame, creating solid Fe_2O_3.

$$Fe_2O_3 \cdot x\ H_2O \xrightarrow{\Delta} Fe_2O_3 + x\ H_2O \qquad (3.16)$$

The mass of Fe_2O_3 is then measured and used to calculate the amount of iron in the original sample. An ore sample with a known mass of 9.85 g is examined by this method and produces an amount of Fe_2O_3 that has an apparent mass of 0.3369 g. The density of pure Fe_2O_3 is known to be 5.25 g/cm^3, and it is known that only Fe_2O_3 was in the measured product. If air density and types of reference weights were the same as in the previous problem, what was the true mass of Fe_2O_3 that was formed from the sample? What would be the amount of iron in the original sample when given as a percent weight-per-weight?

19. Although reference weights with a density of 8.0 g/cm^3 are now used to calibrate most balances, all of these weights, in turn, are ultimately compared against a platinum–iridium cylinder with a density of 21.5 g/cm^3 that is used as the international definition of the kilogram. Suppose that a 1 kg stainless steel weight is compared to a copy of this 1 kg platinum–iridium cylinder. If the density of the surrounding air is 1.2×10^{-3} g/cm^3, what is the size of the buoyancy correction that must be used to adjust the apparent mass of the stainless steel weight to its true value?

THE DETERMINATION OF VOLUME

20. How is the volume of an object related to the object's mass? What are the advantages and disadvantages of using volume in place of mass to describe a material?

21. How do the requirements for volumetric glassware, such as for volumetric flasks and pipets, differ from those for more routine glassware, like Erlenmeyer flasks and graduated cylinders?

22. What is "borosilicate glass"? How does borosilicate glass differ from ordinary soda-lime glass? What properties of borosilicate glass make it valuable for use in glassware for analytical volume measurements?

23. What are some materials besides glass that are used to make volumetric devices? What are the advantages and disadvantages of these other materials versus glass?

TYPES OF VOLUMETRIC EQUIPMENT

24. State the function for each of the following devices and describe how the design of each device helps it to perform this function.
    (a) Volumetric flask
    (b) Volumetric pipet
    (c) Buret
    (d) Micropipet
    (e) Syringe
    (f) Serological pipet
    (g) Ostwald–Folin pipet
    (h) Mohr pipet

25. What is meant by "Class A glassware" and "Class B glassware"? Which type of glassware is preferred for analytical volume measurements?

26. What is the meaning of the terms "To Deliver," "To Contain," and "To Deliver/Blow Out"? What devices are associated with each of these labels?

RECOMMENDED PROCEDURES FOR VOLUME MEASUREMENTS

27. List four factors that should be considered when you are selecting a volumetric device.

28. What types of volumetric devices could be used for each of the following applications?
    (a) Transferring 10.00 mL of a solution from a volumetric flask to a separate container
    (b) Measuring a 250 μL solution that contains a DNA sample
    (c) Repeated dispensing of 2.0 mL portions of a reagent to a series of test tubes
    (d) Measuring a 0.2 mL blood sample from a newborn baby
    (e) Measuring a 2.0 mL gas sample
    (f) Delivery of various volumes of a 0.100 M NaOH solution for the titration of an acid

29. What are some general procedures that you should follow when examining, handling, and cleaning a volumetric device?

30. What is parallax error? What steps can be taken to minimize this type of error when you are using a volumetric device?

31. Explain why it is important to calibrate volumetric devices. How does temperature affect this calibration?

32. A 50.00 mL volumetric flask is filled at 30°C with deionized water. The mass of water found to be in the flask at this temperature (as read directly from the balance) is 49.7380 g.
    (a) What is the true internal volume of the flask at 30°C? (*Note:* You may assume a density for the surrounding air of 0.0012 g/cm^3 and that the balance has been calibrated with a 8.0 g/cm^3 reference weight.)
    (b) What is the true volume of the volumetric flask at 20°C?

33. A pipette is set to a volume of 250 μL and used at 25°C to deliver a sample of deionized water to a small beaker on a tared balance. The mass of water that is delivered to the beaker (uncorrected for buoyancy effects) is 0.2509 g.
    (a) What is the true volume of water that was delivered by the pipette under these conditions?
    (b) If the volume that is delivered by the pipette remains the same, what mass of water would be delivered by this pipet at 20°C?

DESCRIBING SAMPLE AND REAGENT COMPOSITION

34. Define the terms "solution," "solute," "solvent," and "concentration." Use a reagent that contains 0.10 M NaOH in water to illustrate each of these terms.

35. Antifreeze is a solution that is prepared by combining the liquid ethylene glycol ($C_2H_6O_2$) with water. The freezing point of the antifreeze will depend on the relative amount of each chemical in the solution. If 10.0 kg of ethylene glycol liquid is combined with 5.0 L of water (density = 1.00 g/mL), which chemical in this mixture is the solvent and which is the solute?

36. Define each of the following measures of chemical content. State how these units differ and describe the general types of situations in which each is employed.
    (a) Weight-per-weight
    (b) Volume-per-volume
    (c) Weight-per-volume

37. Calculate the content or concentration for each substance in the following mixtures.
    (a) A solution that contains 250 mL of acetonitrile and 500 mL of methanol diluted with water to a total volume of 2.00 L
    (b) A 15.2 g steel sample that contains 10.69 g iron, 2.67 g chromium, 1.22 g nickel, 0.306 g manganese, 0.153 g silicon, and 0.122 g carbon
    (c) A 5.00 L river-water sample that contains 25 g of dissolved solids

38. Determine the content or concentration of the analyte in the following samples.
    (a) A gaseous mixture for the measurement of oxygen that contains 20 mL oxygen in a total volume of 3.5 L
    (b) A 2.00 mL sample of blood that contains 12.5 μg of a drug
    (c) A 5.00 g sample of coal that contains 4.15 g carbon

39. Define each of the following terms and state how they are used to describe the content of chemical solutions and mixtures.
    (a) Parts-per-thousand
    (b) Parts-per-million
    (c) Parts-per-billion
    (d) Parts-per-trillion
    (e) Percent

40. Calculate the content for each stated chemical in the following mixtures. Give the results in percent, parts-per-thousand, ppm, ppb, or ppt.
    (a) 0.010 g Cu^{2+} in a 2.0 L aqueous solution
    (b) 6.2×10^{-3} g Be^{+2} in a 750 mL aqueous solution
    (c) 255 mg $NaIO_3$ in a 1.5 L aqueous solution

41. According to the U.S. EPA, the herbicide atrazine cannot appear in drinking water at levels above 3 μg/L. At this concentration, what is the maximum allowable mass of atrazine that can be present in a glass of water (volume, roughly 240 mL)?

42. What is meant by the "molar mass" of a chemical? Explain how molar mass is related to the terms "molecular weight," "formula weight," or "atomic mass."

43. Define what is meant by the terms "molarity" and "molality." What are the advantages and disadvantages to using each of these units?

44. Calculate the molarity of the following solutions.
    (a) 49.73 g H_2SO_4 dissolved in 500.0 mL of solution
    (b) 4.739 g $RuCl_3$ dissolved in 1.000 L of solution
    (c) 5.035 g $FeCl_3$ dissolved in 250.00 mL of solution
    (d) 27.74 g $C_{12}H_{22}O_{11}$ dissolved in 750.0 mL of solution

45. A solution is prepared by adding 5.84 g of formaldehyde (CH_2O) to 100.0 g water. The final volume of this solution is 104.0 mL. Calculate the molarity and molality of the formaldehyde in this solution.

46. The density of a 10.0% (w/w) solution of NaOH is 1.109 g/cm³. Calculate both the molar and molal concentration of NaOH in this solution.

47. What is meant when the unit of "formality" is used to describe the concentration of a chemical? How is this related to the "analytical concentration" of that chemical?

48. A 500.00 mL aqueous solution of acetic acid is found to contain 0.00538 mol of acetic acid and 0.00321 mol of its conjugate base, acetate. What are the concentrations for the acetic acid and acetate in units of formality or molarity? What is the analytical concentration of acetic acid plus acetate in this solution?

49. A 25.00 mL portion of the solution in Problem 48 is diluted with water to a total volume of 100.00 mL, and the pH is adjusted to a predetermined value. The new solution is found to contain 0.000322 mol acetic acid and 0.000108 mol acetate. What are the individual concentrations of acetate and acetic acid in this diluted solution? What is the analytical concentration of acetic acid plus acetate in this solution?

50. Explain why the molarity and molality of a dilute aqueous solution are approximately the same at room temperature. Why do differences occur in these values when you work at higher concentrations at other temperatures and when using solvents other than water?

51. Describe each of the following measures of chemical content. In what types of situations might you find each of these measures of content employed?
    (a) Surface concentration (mol/m²)
    (b) Normality (eq/L)

SOLUTION PREPARATION

52. State the meaning for each of the following terms as related to chemical purity. Which of these grades would be found in most routine analytical laboratories?
    (a) Certified ACS Grade
    (b) Technical Grade
    (c) USP Grade
    (d) Trace Metal Grade
    (e) HPLC Grade
    (f) Biotechnology Grade

53. In titrations, what is meant by a "primary standard" and a "secondary standard"?

54. What types of impurities are often found in water? Give a specific example for each type of impurity.

55. Explain how distillation and deionization are used to purify water. What are the advantages and disadvantages of each approach? Which method is used in your laboratory?

56. Define the terms "stock solution," "aliquot," and "dilution" and state the role played by each during the preparation of a solution.

57. Describe how you would prepare the following solutions.
    (a) 100.00 mL of 1.00 M NaCl in water, beginning with solid sodium chloride
    (b) 250 mL of 1.0 M Na_2SO_4 in water, beginning with a 2.5 M sodium sulfate solution
    (c) 250 mL of 0.500 M HCl in water, beginning with 12 M HCl

58. Nitric acid (HNO_3) is commercially available as a 72% (w/w) solution (density, 1.42 g/cm³). How many milliliters of this reagent are needed to prepare 2.00 L of a 1.00 M HNO_3 solution?

59. A 1.00 mL urine sample is removed from a collection container and placed into a test tube with 19.0 mL water, giving a final volume of 20.0 mL. If the chemical creatinine has a concentration of 8.5 mM in the original sample, what will be its concentration in the final diluted solution?

60. A 25.00 g portion of 1.435 M NaOH is placed into a 250.00 mL volumetric flask and diluted to the mark with deionized water. A 25.00 mL aliquot of this stock solution is removed and placed into a 500.00 mL volumetric flask. This solution is then mixed and diluted to the mark with more deionized water. What is the molar concentration of NaOH in the final solution?

61. Explain how a change in temperature can affect the concentration of a sample.

62. A pharmaceutical chemist wishes to examine the binding of a drug to its target molecules. To do this, he carefully prepares a 50.0 μM solution of the drug in an aqueous buffer. This solution is made at room temperature (25°C) but is to be used at temperatures that range from 4°C to 45°C. Assuming that the drug's solution has essentially the same density as water (1.00000 g/cm^3 at 4°C and 0.99025 g/cm^3 at 45°C), how much will the drug's concentration change in going from 25°C to the other temperatures that will be used in this study?

63. A biochemist prepares a standard solution that contains 25.0 mg/mL of the protein bovine serum albumin (BSA) in water. This solution is prepared at 30°C and placed into a −20°C freezer for storage. When it is time to use this standard, it is removed from the freezer, thawed, remixed, and allowed to warm before being used in the assay. If this solution is only at 10°C when it is used, what will be the actual concentration of the protein?

CHALLENGE PROBLEMS

64. The method that was allegedly used by Archimedes for the examining the crown of King Hiero (see Chapter 1) involved placing both the crown and an equal mass of pure gold into water and examining the amount of water that each displaced. If you have a crown of pure gold (density, 19.3 g/cm^3) that has a mass of 1000 g, what volume of water would be displaced at 25°C by this crown? If the density of pure silver is 10.5 g/cm^3, what volume of water will be displaced by a 1000 g crown that contains 80% (w/w) gold and 20% silver?

65. The value of g, the local gravitational acceleration constant, changes on Earth as you move to different altitudes or change your distance north or south of the equator. The way in which g varies with the position on the Earth is described by Equation 3.17,

$$g = 9.80632 - 0.02586 \cdot \cos(2\,v) + 0.00003 \cdot \cos(4\,v) - 0.00000293 \cdot h \qquad (3.17)$$

where v is the location above or below the equator (in degrees) and h is the height above or below sea level (in meters).[4]
(a) Obtain the approximate altitude and latitude of the city in which your laboratory is located. What is the value of g for your location? Ignoring buoyancy effects, how much would the weight of a 1.000 kg object change in going from the equator at sea level to your laboratory?
(b) Imagine that you are given the task of moving a balance from the ground floor of a building to one that is five stories higher, an increase in height of approximately 20 m.

How much would this move affect the value of g around the balance? If this change were to go uncorrected, what effect would the move have on measurements that are performed on the balance?

66. The plots given in Figure 3.6 were obtained by using the relationship shown in Equation 3.18 between the density of a sample and the measured versus actual mass,

$$(m_{obj} - m_{display})$$
$$= m_{display} \cdot \frac{[(0.0012 \text{ g/cm}^3)/d_{obj}] - 0.00015}{1 - [(0.0012 \text{ g/cm}^3)/d_{obj}]} \qquad (3.18)$$

where m_{obj} is the actual mass of the object being measured, $m_{display}$ is its apparent mass, and d_{obj} is the density of the object.
(a) Demonstrate how this equation can be derived from Equation 3.6. What assumptions were made in obtaining this relationship?
(b) Based on Equation 3.18, create a spreadsheet that can be used to generate graphs like the ones shown in Figure 3.6 for each of the following substances: cork (density, 0.2 g/cm^3), gasoline (0.7 g/cm^3), silicate-based rock (3.0 g/cm^3), and platinum (21.4 g/cm^3). Which of these materials would be expected to have the largest buoyancy effect? Which would have the smallest buoyancy effect?

67. Although a value of 0.0012 g/cm^3 is often used as the density for air in buoyancy calculations, a more exact air density (in units g/cm^3) can be found by using Equation 3.19,

$$d_{air} = 0.0012929 \cdot \frac{(273.13 \text{ K})}{T} \cdot \frac{(P - 0.3787 \cdot h)}{760} \qquad (3.19)$$

where T is the absolute temperature (in K), and P is the barometric pressure (in mm Hg). The factor h is the vapor pressure of water (in mm Hg), which is a measure of the air's relative humidity. This last term can be determined through measurement or by using the dew point of the air.[16]
(a) Using Equation 3.19, estimate the density of air at a temperature of 28°C, a barometric pressure of 745 mm Hg, and a vapor pressure of 11.99 mm Hg (corresponding to a dew point of 14°C).
(b) Under the same conditions as in Part (a), determine what the actual mass would be for a 10.000 g sample that has a density of 0.89 g/cm^3 and is being compared to a reference with a density of 8.00 g/cm^3. How do your results compare to what is obtained when 0.0012 g/cm^3 is used as the density of air?

68. Mercury is sometimes used in place of water to calibrate volumetric equipment, especially when small volumes of liquids are being measured. Look up the physical and chemical properties of mercury in its material safety data sheet (MSDS) and the *CRC Handbook of Chemistry and Physics*.[16] Based on this information, what advantages do you think there might be to using mercury for calibrating volumetric devices? What disadvantages might there be to such a procedure?

69. Equation 3.14 was used earlier to show how an adjustment can be made for temperature effects when you are dealing with molar concentrations. What type of relationship would you use to make a similar correction for solutions that have

their contents expressed as weight-per-volume ratios (w/v)? Would this same type of correction be needed when working with units of molality, % w/w or % v/v? Explain your answer.

70. It is sometimes necessary during an analysis to convert from one unit of concentration to another. Show how you would convert between each of the following pairs of concentration units. (*Note:* Some of these conversions can be found in sources such as the *CRC Handbook of Chemistry & Physics*.)[16] State what additional information would be needed for each of these conversions. Confirm your approach by using dimensional analysis.
   (a) Converting a concentration in g/L to molarity
   (b) Converting a concentration in molarity to molality
   (c) Converting a concentration in mg/L to ppm (w/w)
   (d) Converting a concentration in % (w/w) to g/L

71. A unique feature of mass spectrometry is that it cannot only provide the mass of molecules but can give information on the isotopic composition of these molecules. One result of mass spectrometry's ability to discriminate between isotopes is that it can give multiple "molecular ion" peaks for a single chemical. Examples of these isotope peaks are shown in Figure 3.15.

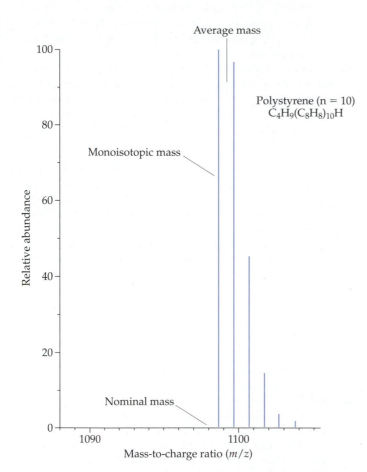

FIGURE 3.15 The isotope pattern for the molecular ion(s) of polystyrene, illustrating the differences between the nominal mass, monoisotopic mass and average mass for this chemical. (Adapted with permission from J. Yergey, D. Heller, G. Hansen, R. J. Cotter, and C. Fenselau, "Isotopic Distributions in Mass Spectra of Large Molecules," *Analytical Chemistry*, 55 (1983) 353–356.)

(a) The presence of isotope peaks makes it necessary to distinguish between several ways of describing the molecular weight of a compound in mass spectrometry. Using Reference 19 as a guide, define the terms "average mass," "nominal mass," and "monoisotopic mass" as used in Figure 3.15.

(b) Chlorpheneramine ($C_{16}H_{19}ClN_2$) is a drug found in many over-the-counter cold medicines. Based on the isotopic masses and abundances given in Reference 16, calculate the nominal mass and monoisotopic mass for this drug, with the latter value being based on the most abundant isotopes for each element in this compound. How do these values compare with the molecular mass that is found when you use the atomic masses listed in the periodic table? Explain the reason for any differences in these values.

TOPICS FOR DISCUSSION AND REPORTS

72. Visit a local analytical laboratory and discuss how mass and volume measurements are used at that facility. Obtain information on the types of mass- and volume-measuring devices that are employed in that laboratory and on the procedures that are followed in the use of this equipment. Also, learn about the types of samples or reagents that are being measured with these devices and any special precautions that are followed during these determinations.

73. Using Reference 12 and other resources, obtain further information on the quartz crystal microbalance. Discuss how a quartz crystal microbalance works. Give examples of some analytical applications for this device.

74. There are several other methods that can be used to purify laboratory water besides distillation and deionization. A list of some of these alternative techniques is provided below.[18] Obtain information of one or more of these methods and describe how they work. How do these methods compare to deionization and distillation in the removal of unwanted substances from water?
   (a) Reverse osmosis
   (b) Activated carbon filtration
   (c) Ultrafiltration
   (d) Microporous filtration
   (e) Ultraviolet oxidation
   (f) Electrodialysis

75. Locate a recent research article that used the method of atomic force microscopy (AFM). Discuss how AFM was used in this article and describe the types of information that it provided.

76. Atomic force microscopy is part of a larger family of techniques known collectively as *scanning probe microscopy (SPM)*. SPM is a type of microscopy that uses a physical probe to scan a surface and form an image of a sample. Below is a list of several types of scanning probe microscopy.[9,10,20,21,22] Obtain information on one of these methods from the Internet, a book, or a review article. Write a report that describes how this method works. Include in your report some examples of applications for this method in the area of chemical analysis
   (a) Scanning tunneling microscopy
   (b) Scanning electrochemical microscopy
   (c) Force modulation microscopy
   (d) Magnetic force microscopy

References

1. H. Diehl, *Quantitative Analysis: Elementary Principles and Practice*, Oakland Street Science Press, Ames, IA, 1970.
2. F. Szabadvary, *History of Analytical Chemistry*, Pergamon Press, New York, 1966.
3. "Baron Jöns Jakob Berzelius," *Columbia Electronic Encyclopedia*, Columbia University Press, New York, 2000.
4. M. Kochsiek, *Glossary of Weighing Terms: A Practical Guide to the Terminology of Weighing*, Mettler-Toledo, Switzerland, 1998.
5. B. Kisch, *Scales and Weights: A Historical Outline*, Yale University Press, New Haven, CT, 1966.
6. M. W. Hinds and G. Chapman, "Mass Traceability for Analytical Measurements," *Analytical Chemistry*, 68 (1996) 35A–39A.
7. W. E. Kupper, "Laboratory Balances," In: *Analytical Instrumentation Handbook*, 2nd ed., G. W. Ewing, Ed., Marcel Dekker, New York, 1997, Chapter. 2.
8. G. Binnig, C.F. Quate, and C. Gerber, "Atomic Force Microscope," *Physical Review Letters*, 56 (1986) 930–933.
9. E. Meyer, H. J. Hug, and R. Bennewitz, *Scanning Probe Microscopy: The Lab on a Tip*, Springer, New York, 2003.
10. P. Carlo Braga and D. Ricci, Eds., *Atomic Force Microscopy: Biomedical Methods and Applications*, Humana Press, Totowa, NJ, 2003.
11. N. Singer, "The Quiet Revolution in Analytical Balance Technology," *Chemistry*, Springer, New York, (2000) 14–16.
12. C. Henry, "Measuring the Masses: Quartz Crystal Microbalances," *Analytical Chemistry*, 68 (1996) 625A–628A.
13. M. L Gross, "Mass Spectrometry," In *Instrumental Analysis*, 2nd ed., G. D. Christian and J. E. O'Reilly, Eds., Allyn & Bacon, Boston, MA, 1986, Chapter 16.
14. D. A. Skoog, F. J. Holler, and T. A. Nieman, *Principles of Instrumental Analysis*, 5th ed., Saunders, Philadelphia, PA, 1998, Chapter 20.
15. D. O. Sparkman, *Mass Spectrometry Desk Reference*, Global View, Pittsburgh, PA, 2000.
16. D. R. Lide, Ed. *CRC Handbook of Chemistry and Physics*, 83rd Ed., CRC Press, Boca Raton, FL, 2002.
17. J. A. Dean, *Lange's Handbook of Chemistry*, 15th ed., McGraw-Hill, New York, 1999.
18. "Glass," *Encyclopedia Britannica*, Encyclopedia Britannica, Inc., Chicago, IL, 1999.
19. J. Inczedy, T. Lengyel, and A. M. Ure, *Compendium of Analytical Nomenclature*, 3rd ed., Blackwell Science, Malden, MA, 1997.
18. *A Guide to Laboratory Water Purification*, Labconco, Kansas City, MO, 1998.
19. J. Yergey, D. Heller, G. Hansen, R. J. Cotter, and C. Fenselau, "Isotopic Distributions in Mass Spectra of Large Molecules," *Analytical Chemistry*, 55 (1983) 353–356.
20. A. I. Kingon, P. M. Vilarinho, and Y. Rosenwaks, *Scanning Probe Microscopy: Characterization, Nanofabrication and Device Application of Functional Materials*, Kluwer Academic, Norwell, MA, 2005.
21. R. Wiesendanger, Ed., *Scanning Probe Microscopy: Analytical Methods*, Springer, New York, 1998.
22. D. Bonnell, Ed., *Scanning Probe Microscopy and Spectroscopy: Theory, Techniques and Applications*, 2nd ed., Wiley, New York, 2001.

Chapter 4

Making Decisions with Data

Chapter Outline

4.1 INTRODUCTION: TAKE ME OUT TO THE BALL GAME?

News Release—December 24, 2006: *Major-league baseball's drug cheats are the subject of an investigation and face tougher penalties and public exposure when they are caught using steroids. Congress has ratcheted up prison sentences for convicted steroid dealers, and federal drug agents are showing a new willingness to cooperate with sports officials in tracking down athletes who use banned drugs. And in virtually every high school in the country, young athletes are taught about the health risks associated with the use of performance-enhancing drugs. According to experts, those are some of the lasting impacts of the BALCO steroids scandal, the ongoing federal investigation that revealed the use of banned performance-enhancing drugs by some of the greatest athletes of the era.*[1]

Recent investigations into the use of steroids in professional baseball have lead to an increased awareness of the presence and problem of performance-enhancing drugs in sports. This issue has been around since the original Olympic games in ancient Greece, where some athletes chewed on opium-soaked rags to better their performance. This problem has become an even greater issue in modern times. Drug use by athletes not only occurs in professional sports and the Olympics but is even found in college competition and at high school events.[2–5] One concern is the actual or perceived benefit that these substances may provide athletes. Another, more immediate concern is the danger these drugs may pose to an athlete's life and health.[3]

Analytical chemistry played a key role in exposing the use of new steroids in the BALCO scandal (see Figure 4.1).[4] Chemical analysis is also routinely used to detect banned substances at the Olympics and other sports events. The first phase of this process is a fast screening assay, in which a qualitative test is conducted to determine whether a particular drug or group of substances is present above a preset level. If the result of the screening assay is negative, the athlete has passed the drug test. A positive result, however, is followed by a second more selective and quantitative method to confirm the results. If the drug is still detected, appropriate actions are then taken against the athlete.[3–5]

It is important in this process for the analyst to be fully aware of the reliability of their results because this information will be used to make decisions that could determine the winner of an event or the recipient of a gold medal. Similar decisions with data must be made daily in other areas, including the determination of food quality and the interpretation of clinical tests. This process requires knowledge of the possible errors that can occur in an assay and the assay's reproducibility. In this chapter we look at the general types of errors that might be found in analytical methods, learn how to describe the reliability of results, and discuss techniques for comparing experimental data to other values. With these tools in place, you will then be able to make better decisions based on your measurements.

4.1A Types of Laboratory Errors

A fact that must always be considered when you are performing experiments is that *all* physical and chemical measurements have some degree of uncertainty. There are two types of errors that can occur in a measurement. The first type is a **systematic error**, which is represented by a *constant bias* between your results and the true answer. For instance, a systematic error would result when you are making a solution with an improperly calibrated volumetric flask. Systematic errors are produced by a consistent problem in a scientist's technique, an instrument, or a

procedure. Although these errors can occur in any measurement, it is possible to eliminate them through good laboratory practices, as discussed in Chapter 2. One way to accomplish this goal is to have a well-kept notebook that can be used to note unusual trends in experiments and to identify sources of errors. The proper training of laboratory workers and the regular maintenance of equipment will also help you to avoid systematic errors.

A second source of experimental uncertainty is **random error**. This type of uncertainty is the result of *random variations* in experimental data. Random errors are present in all measurements. These errors are produced by such things as variations in instrumental readings or experimental conditions that cannot be controlled. For instance, fluctuations due to static electricity or air currents can cause random errors when you are using a balance. Although no measurement is totally free from random errors, it is possible to reduce the size of these errors through the proper design of experiments and the correct choice of methods for comparing data. We discuss this topic in Section 4.5 when we will look at various techniques for comparing experimental results.

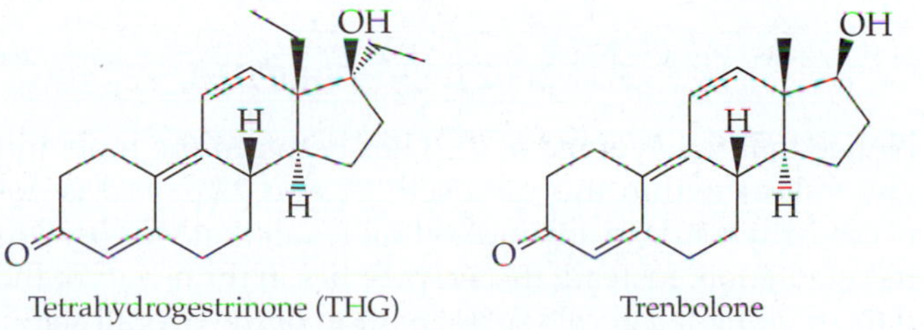

4.1B Accuracy and Precision

Now that we know what types of errors can be present during an experiment, let's consider how we might describe the effects of these errors. Two terms used for this purpose are *accuracy* and *precision*. **Accuracy** is used in science to describe the difference between an experimental result and its true value, while **precision** refers to the variation in results that are obtained under similar conditions. To illustrate the difference between these two terms, imagine that we are watching a competition between four archers at the Olympics (see Figure 4.2). The first archer hits the target with all of his or her arrows, but the arrows are spread over a wide area and are far from the center. We would say that this archer has poor accuracy (bad aim) as well as poor precision (poor consistency). The second participant has an average aim closer to the center of the target (good accuracy), but still has arrows spread across the target (poor precision). The third archer is more consistent in aim, but has a bow that causes all the arrows to hit left of the center (poor accuracy, good precision). The fourth archer has all arrows that hit near the center, giving both good accuracy and good precision.

Two terms that are used to describe accuracy are *absolute error* and *relative error*. The **absolute error (e)** of an experimental result (*x*) is found by calculating the difference between this result and its true value (μ).[6]

$$e = x - \mu \tag{4.1}$$

If you used a balance to measure a sample with a known mass of 10.0000 g and the balance gave a reading of 9.9995 g, the absolute error would be $(9.9995\ \text{g} - 10.0000\ \text{g}) = -0.0005\ \text{g}$.

Tetrahydrogestrinone (THG)

Trenbolone

FIGURE 4.1 The structure of two anabolic steroids, tetrahydrogestrinone (THG) and trenbolone. Both of these substances are now banned for human use by the U.S. Food and Drug Administration. The identification of THG as a new designer steroid that was being used by some athletes was a key development in the BALCO steroid scandal.[4]

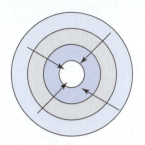

Archer 1
poor accuracy, poor precision

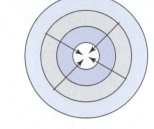

Archer 3
poor accuracy, good precision

Archer 2
good accuracy, poor precision

Archer 4
good accuracy, good precision

FIGURE 4.2 Illustration of the difference between accuracy and precision, using an archery contest as an example.

The **relative error (e_r)** is calculated by finding the difference between the true and measured values and dividing this difference by the true answer.[6]

$$e_r = \frac{x - \mu}{\mu} \qquad (4.2)$$

Relative error can be expressed as either a fraction or a related value, such as by using a percent or parts-per-million (ppm). As an illustration, the percent relative error in the mass of our 10 gram sample would be $100 \cdot (9.9995 \text{ g} - 10.0000 \text{ g})/(10.0000 \text{ g}) = -0.005\%$.

The goal in any analytical measurement is to have good accuracy and good precision. The precision will be directly related to the random errors, but the accuracy of individual results will be affected by both systematic *and* random errors. For instance, even though archer number 2 in Figure 4.2 had a good overall aim, the accuracy of each individual shot varied. This difference was caused by random errors. In the next section we will learn how to minimize the effects of random errors in measurements by increasing the size of a data set to obtain a more reliable estimate of the true result.

4.2 DESCRIBING EXPERIMENTAL RESULTS

We have seen how both systematic and random errors can affect the accuracy of a measured result. The effect due to systematic error can be eliminated if we are careful in how we perform our analysis, but random errors can never be totally eliminated. It is possible, however, to minimize the effects of random errors by carefully designing an experiment and properly treating your

data. This process requires knowing how to best describe and report the results of an experiment.

4.2A Determining the Most Representative Value

Because random errors will be present in any measurement, we can expect to see some variations even in repeated results that are generated using the same sample. We must then determine how to best represent our final answer. This task is often accomplished by using the **arithmetic mean** ($\bar{x}$, also known as the *mean* or *average*), which is calculated as shown below.[6]

$$\bar{x} = \frac{(x_1 + x_2 + \cdots x_n)}{n} = \frac{\Sigma\,(x_i)}{n} \qquad (4.3)$$

In this equation, n is the *number of observations* (or the total number of measured values within a group of results), and x_i represents each of the individual values in the set (x_1, x_2, and so on). The symbol "Σ" represents the fact that we are adding the values of x_1 through x_n. An example of how you would determine the mean for a set of experimental data is given in the following exercise.

EXERCISE 4.1   Determining the Mean for a Group of Results

Erythropoieten (EPO) is a naturally occurring hormone that increases the ability of blood to carry oxygen. Recombinant EPO has been used by athletes and is now banned from events such as the Olympics.[3] To develop an assay for this hormone, a chemist collected blood samples from eight healthy people who were not receiving any recombinant EPO. The EPO concentrations measured for these individuals were 9.1, 26.4, 32.1, 15.8, 23.7, 20.5, 13.0, and 27.6 International Units/liter (IU/L). What was the average concentration of EPO in these samples?

SOLUTION

The average concentration can be found by using Equation 4.3 (where each of the numbers in the numerator has units of IU/L).

$$\bar{x} =$$
$$\frac{(9.1 + 26.4 + 32.1 + 15.8 + 23.7 + 20.5 + 13.0 + 27.6)}{8}$$
$$\therefore \quad \bar{x} = 21.0\underline{2} \text{ IU/L} = \mathbf{21.0\ IU/L}$$

Notice that the average of 21.0 lies between the high and low values within this data set (9.1 and 27.6) and is not necessarily equal to any individual result that was used in its calculation. Instead, the average lies in the center of the data, making it a good representation of the overall set.

Throughout this book we will use the symbol $\bar{x}$ to refer to the *experimental mean* of a data set and μ to represent the *true mean*. As shown in Figure 4.3, it is necessary to make this distinction because random errors

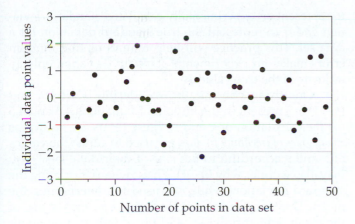

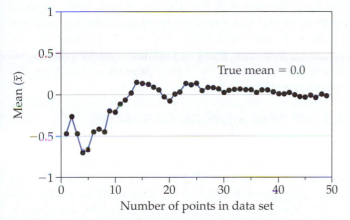

FIGURE 4.3   Effect of increasing the number of points in a data set (*n*) on the calculated experimental mean ($\bar{x}$). The results were obtained using a computer simulation in which values were selected randomly according to a normal distribution model. This plot shows that the experimental mean approaches a constant value (the true mean, μ), as the number of averaged data points is increased.

cause $\bar{x}$ to be only an approximation of μ, especially when we are dealing with small sets of numbers. It is only when we have a large group of numbers, which causes random errors to cancel out, that the experimental mean approaches the true mean in its value.

4.2B Reporting the Variation in a Group of Results

We now have a way for determining a good representative value for a data set. But how do we describe the variation of results within that set? One way to do this is to use the **range (R_x)**. The range is found by taking the difference between the largest and smallest values (x_{high} and x_{low}).[6]

$$R_x = x_{high} - x_{low} \qquad (4.4)$$

To illustrate this idea, suppose you had a set of four urine samples in which you measured their specific gravity (a method used to check for any tampering with drug samples). If these samples gave specific gravities of 1.025, 1.028, 1.032, and 1.035, the range for this set would be (1.035–1.025), or 0.010. One advantage of using a range to describe a data set is that it is easy to calculate. However, the range tends to increase as your number of data points increases (see Figure 4.4), which gives random errors a greater chance to produce extreme values. This feature makes it difficult to use ranges for comparing data sets, especially if they contain different amounts of data.

A more consistent way of describing the variation in a group of results is to use the **standard deviation (*s*)**. A standard deviation is calculated by using Equation 4.5, where each value in the set (x_1 through x_n for *n* values) is compared to the mean ($\bar{x}$) for that same group of numbers.

$$s = \sqrt{\frac{\Sigma\,(x_i - \bar{x})^2}{n - 1}} \qquad (4.5)$$

Two parameters related to the standard deviation are the *variance (V)* and the *relative standard deviation (RSD)*. The relative standard deviation is also called the *coefficient of variation* (or *CV*) and is found by taking the standard deviation for a group of results and dividing it by the mean for the same data, where RSD = $s/\bar{x}$ or RSD (%) = $100 \cdot (s/\bar{x})$. The variance is simply equal to the

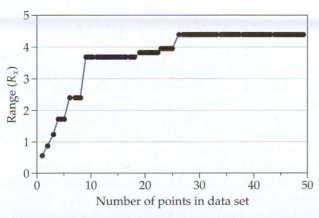

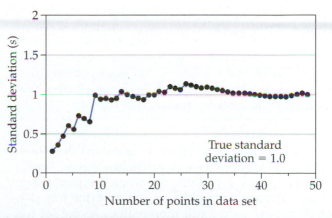

FIGURE 4.4   Effect of increasing the number of points in a data set (*n*) on the calculated experimental standard deviation (*s*) and range (R_x) for a data set. These results were obtained using the same values as given in Figure 4.3. This plot shows that the experimental standard deviation approaches a constant value (the true standard deviation, σ) as the number of data points is increased, but that the range continues to increase as *n* is increased.

square of the standard deviation, or $V = (s)^2$, and is important in describing the propagation of experimental errors, as we will see in the next section of this chapter.

EXERCISE 4.2   Describing the Variation in a Group of Numbers

What are the range, standard deviation, and RSD for the EPO concentrations in Exercise 4.1?

SOLUTION

The highest and lowest values in this data set are 32.1 and 9.1 IU/L, so the range would be $R_x = (32.1 \text{ IU/L} - 9.1 \text{ IU/L}) = 23.0 \text{ IU/L}$. The mean in Exercise 4.1 was found to be 21.0$\underline{2}$ IU/L, which can be used with Equation 4.5 to calculate the standard deviation.

$$s = \left[\frac{(9.1 - 21.0\underline{2})^2 + (26.4 - 21.0\underline{2})^2 +}{(8 - 1)} \right.$$

$$\frac{(32.1 - 21.0\underline{2})^2 + (15.8 - 21.0\underline{2})^2 + (23.7 - 21.0\underline{2})^2 +}{(8 - 1)}$$

$$\left. \frac{(20.5 - 21.0\underline{2})^2 + (13.0 - 21.0\underline{2})^2 + (27.6 - 21.0\underline{2})^2}{(8 - 1)} \right]^{1/2}$$

$$= 7.8\underline{9} \text{ IU/L} = \textbf{7.9 IU/L}$$

The relative standard deviation is then found by dividing s by $\bar{x}$.

$$\text{RSD (\%)} = 100 \cdot \frac{7.8\underline{9} \text{ IU/L}}{21.0\underline{2} \text{ IU/L}} = 37.\underline{5}\% = \textbf{38\%}$$

Notice that the standard deviation will always have the same units as $\bar{x}$. However, the RSD does not have any units because it is the ratio of s and $\bar{x}$. This is the reason why RSD is usually expressed as a fraction or as a percent (%).

Another thing to remember when you are calculating the value of s or RSD is to *not* round off any numbers in your calculation until you have reached the final answer. Instead, use guard digits, as represented by the underlined digits in the preceding equations, to help indicate how many significant figures are present in each number until you have reached the end of the calculation. This approach avoids the introduction of rounding errors into the answer. We will follow this practice throughout this chapter as we use standard deviations, means, and related values to compare and analyze data.

The physical meaning of a "standard deviation" might be a bit harder to visualize than the "range," but the standard deviation is easier to use when comparing large data sets because it approaches a constant (the true standard deviation) as we add more values to the set (see Figure 4.4). Throughout this book we will use the symbol s

to represent an experimentally calculated standard deviation and σ to represent the true standard deviation for a data set. This practice will help remind us that random errors make our experimental standard deviation only an estimate of the true value σ.

One thing you may be wondering is why n and $(n-1)$ are present in the bottom of Equations 4.3 and 4.5 for the calculation of $\bar{x}$ and s. These terms are known as the *degrees of freedom* (f) and are used to adjust the values of $\bar{x}$ and s according to the size of their data sets.[6] You may also wonder why $(n-1)$ is used when calculating the standard deviation, while n is used for determining the mean. This is because the calculation of s makes use of the mean, which provides additional information about the data. The fact that the data set is now better defined (that is, it has lost one degree of freedom because we know $\bar{x}$) is indicated by using $(n-1)$ instead of n in Equation 4.5. The more calculated values you use to determine a number, the fewer degrees of freedom you will have in the result.

4.3 THE PROPAGATION OF ERRORS

Now that we know how to describe errors and data sets, let's consider how the overall precision of an experimental result will depend on the random errors that occur in each individual step used to obtain this result. We can look at this dependence by employing **error propagation**, which is a method that can be used to predict the precision of an experimental value and to help identify the major contributions to random errors in an analysis.

4.3A Addition and Subtraction

One case where the propagation of errors may occur is when we use calculations that involve addition or subtraction. This situation shows up frequently in experiments, such as when we weigh samples, measure volumes, or calculate a molar mass. To describe the propagation of errors during these processes, we will represent our final result by the symbol "y" and the error in this result by its standard deviation, s_y. Also, we will represent the values we are adding or subtracting by the terms "a," "b," and "c," each having its own individual standard deviation (or error) of s_a, s_b, or s_c. As a further simplification, we'll assume that all of the errors in a, b, and c are independent of each other and represent only random variations (in other words, we've already eliminated all systematic errors). Under these conditions, it can be shown that the final error in y is related to the random errors in a, b, and c through Equation 4.6 (shown in Table 4.1 and derived in Appendix A). Thus, if you can estimate the random errors in each of these initial values, you can estimate the random error in the final result. An example of this process is given in the next exercise.

TABLE 4.1   General Formulas for Error Propagation in Common Mathematical Operations*

Type of Operation	Example	Relation of Error in Result (s_y) to Original Errors (s_a, s_b & s_c)	
Addition and Subtraction	$y = a + b - c$	$s_y = \sqrt{(s_a)^2 + (s_b)^2 + (s_c)^2}$	(4.6)
Multiplication and Division	$y = a \cdot b/c$	$(s_y/y) = \sqrt{(s_a/a)^2 + (s_b/b)^2 + (s_c/c)^2}$	(4.7)
Logarithms	$y = \log(a)$	$s_y = 0.434 \cdot (s_a/a)$	(4.8)
	$y = \ln(a)$	$s_y = (s_a/a)$	(4.9)
Antilogarithms	$y = 10^a$	$(s_y/y) = 2.303 \cdot s_a$	(4.10)
	$y = e^a$	$(s_y/y) = s_a$	(4.11)
Exponentiation	$y = a^x$	$(s_y/y) = x (s_a/a)$	(4.12)

* The derivation of these equations is provided in the Appendix A. In these operations, y represents the calculated result, while a, b, and c are the variables used within the calculation. The value of x in Equation 4.12 represents a known constant.

EXERCISE 4.3	Error Propagation During Addition or Subtraction

A technician wishes to prepare a solution of ephedrine for the measurement of this drug in urine samples. A weighing container is placed onto a balance, tared, and ephedrine is added until a reading of 37.5 mg is obtained. This sample is transferred to a volumetric flask and the container is reweighed, giving a displayed mass of 0.3 mg. If the precision of the balance is ±0.05 mg, what is the expected precision for the mass of ephedrine in the flask?

SOLUTION

This measurement was based on *weight by difference* (see Chapter 3), where the difference in the original and final masses of the weighing container gives the mass of transferred drug.

$$m_{\text{ephedrine}} = m_{\text{original}} - m_{\text{final}}$$

$$= 37.5 \text{ mg} - 0.3 \text{ mg} = 37.2 \text{ mg}$$

Because this calculation is based on subtraction, a modified version of Equation 4.6 can be used to determine the standard deviation for the final result.

$$s_{\text{ephidrine}} = \sqrt{(s_{\text{original}})^2 + (s_{\text{final}})^2}$$

$$= \sqrt{(0.05 \text{ mg})^2 + (0.05 \text{ mg})^2}$$

$$\therefore \quad s_{\text{ephidrine}} = 0.07\underline{1} \text{ mg} = \mathbf{0.07 \text{ mg}}$$

Thus, the amount of transferred ephinephrine was 37.2 mg ± 0.07 mg, where the second value represents a range of plus-or-minus one standard deviation in the measured mass.

Another way we could have written the preceding expression is as $(s_{\text{ephidrine}})^2 = (s_{\text{original}})^2 + (s_{\text{final}})^2$. This form shows us that the variance for the mass of ephinephrine, or $(s_{\text{ephidrine}})^2$, will be equal to the sum of the variances for the original and final masses that were measured, $(s_{\text{original}})^2$ and $(s_{\text{final}})^2$. Thus, although in this chapter we will focus on the use of standard deviations for error propagation, you should keep in mind that we are really looking at how the variance $(s)^2$ changes during this process. This is the reason why many of the relationships in Table 4.1 have terms that contain $(s)^2$.

In the preceding exercise, the random error in the calculated result, as represented by a standard deviation of ±0.07 mg, was greater than the random error in any of the numbers used to determine this result (±0.05 mg). This situation occurs whenever you have random errors that accumulate during a calculation or experimental process. This effect is why you should try to minimize such errors in each step of a laboratory procedure.

Another thing we can see from the previous problem is that the *absolute size* of the error we obtain after addition or subtraction (s_y) will be determined by the *absolute size* of the errors in the numbers used for the calculation (s_a, s_b, and s_c). This relationship is the basis for the rule we reviewed in Chapter 2 (see summary in Appendix A) for estimating the number of significant figures during addition or subtraction. In Exercise 4.3 this rule would also give an answer with one significant figure to the right of the decimal point (37.2 mg). The main difference in these two approaches (error propagation versus significant figures) is error propagation gives a more complete description of the random errors. If we had used only significant figures, we would have assumed our mass of 37.2 mg had an uncertainty of ±0.1 mg. However, through error propagation we know this result actually has a precision of ±0.07 mg. This may seem like a small difference, but it can become important when you are using this value in other calculations or are trying to optimize the precision of a measurement.

4.3B Multiplication and Division

Like addition and subtraction, it is possible to determine how random errors will be carried from one number to the next during multiplication or division. This information might be useful when you are diluting a sample or determining its concentration by using mass and volume ratios. If we again assume that only independent, random errors are present, Equation 4.7 shows how the final error in a calculated result will depend on the errors in all numbers that are used to get this value.

| **EXERCISE 4.4** | **Error Propagation in Multiplication and Division** |

The ephedrine sample from Exercise 4.3 is placed into a 250.00 mL volumetric flask and diluted to the mark with deionized water. If the random error in volume of this flask is ±0.10 mL, what is the expected precision for the concentration of ephedrine (in mg/mL) in the final solution?

SOLUTION

The concentration (C) of ephedrine is to be determined by using a mass and volume ratio ($C = m/V$), so the precision of the final concentration will be given by the following formula.

$$(s_C/C) = \sqrt{(s_m/m)^2 + (s_V/V)^2}$$

We know the values for m (37.2 mg), s_m (0.071 mg), V (250.00 mL), and s_V (0.10 mL). We can also calculate the concentration of ephedrine, where $C = (37.2\ \text{mg}/250.00\ \text{mL}) = 0.1488\ \text{mg/mL}$. The precision of the calculated concentration can be found by placing these values into the preceding relationship.

$$s_C/(0.1488\ \text{mg/mL})$$
$$= \sqrt{(0.071\ \text{mg}/37.2\ \text{mg})^2 + (0.10\ \text{mL}/250.00\ \text{mL})^2}$$
$$s_C = (0.1488\ \text{mg/mL})$$
$$\cdot \sqrt{(0.071\ \text{mg}/37.2\ \text{mg})^2 + (0.10\ \text{mL}/250.00\ \text{mL})^2}$$
$$\therefore\ s_C = 0.00029\ \text{mg/mL} = \mathbf{0.0003\ mg/mL}$$

From this result, we can say that the concentration of ephedrine was 0.1488 ± 0.0003 mg/mL ($\pm 1\ s$). Another thing we can see is that the guard digit we had retained in the value of 0.1488 for this calculation actually ended up being retained when we determined the precision of this value by error propagation. This is another reason why the use of guard digits and rounding off only at the final answer is recommended in such calculations.

Equation 4.7 indicates that multiplication and division will both give a result in which the *relative size* of the final error (s_y/y) will depend on the *relative size* of the errors in the numbers used in the calculation (s_a/a, s_b/b, and s_c/c). This is equivalent to the rule given in Appendix A for determining significant figures during multiplication or division, which would have given an answer in Exercise 4.4 with three significant figures (0.149 mg/mL). With error propagation we were able to gain one additional digit in our answer (0.1488 mg/mL). This difference

arose because the significant figure rules tend to overestimate random errors, showing why it is preferable to employ error propagation when you are carrying out calculations that make use of experimental results.

4.3C Logarithms, Antilogarithms, and Exponents

Error propagation can also be utilized with other types of operations, such as finding a logarithm or antilogarithm, or working with a value that has an exponent. The expressions used for error propagation during such calculations are given by Equations 4.8–4.12 in Table 4.1.

| **EXERCISE 4.5** | **Error Propagation with Logarithms and Antilogarithms** |

One test performed with athlete urine samples is a measurement of the urine's pH. This measurement ensures that such samples have not been altered to prevent the detection of banned substances. As we will see in Chapter 8, the pH is approximately equal to the negative logarithm of the hydrogen ion activity in a sample (a_{H^+}), where $\text{pH} = -\log a_{H^+}$ or $a_{H^+} = 10^{-\text{pH}}$.

   a. If a standard solution has a hydrogen ion activity of $4.0\ (\pm 0.2) \times 10^{-8}\ M$, what will be the pH and precision of this pH value?
   b. If the pH of a urine sample is determined to be 6.00 ± 0.05, what will be the value and precision of the hydrogen ion activity in the sample?

SOLUTION

   a. The pH of this solution can be obtained by using $\text{pH} = -\log a_{H^+}$, which gives a pH of $-\log(4.0 \times 10^{-8}\ M)$, or 7.40. This calculation involves taking the logarithm of a number, so we will use Equation 4.8 to determine how random errors in the hydrogen ion activity will affect the error in the pH. The result is a final pH of **7.40 ± 0.02**.

$$s_{\text{pH}} = 0.434 \cdot (s_{aH^+}/a_{H^+})$$
$$= 0.434 \cdot (0.2 \times 10^{-8}\ M/4.0 \times 10^{-8}\ M)$$
$$\therefore\ s_{\text{pH}} = \mathbf{0.02}$$

A similar process can be used to find the precision of a natural logarithm, or $\ln(a)$, through the use of Equation 4.9 in Table 4.1.

   b. The hydrogen ion activity in a pH 6.00 solution would be $a_{H^+} = 10^{-6.0} = 1.0 \times 10^{-6}\ M$. The precision expected for this value can be found by utilizing Equation 4.10.

$$(s_{aH^+}/a_{H^+}) = 2.303\ s_{\text{pH}}$$
$$s_{aH^+} = 2.303 \cdot (a_{H^+} \cdot s_{\text{pH}}) = 2.303 \cdot (1.0 \times 10^{-6}\ M) \cdot (0.05)$$
$$\therefore\ s_{aH^+} = \mathbf{0.1 \times 10^{-6}\ M}$$

Our final answer would be a hydrogen ion activity of $1.0\,(\pm 0.1)\times 10^{-6}\,M$. The same process can be used with Equation 4.11 to find the precision for a natural antilogarithm, e^a.

Equations 4.8 and 4.9 indicate that the *absolute size* of the error for a logarithm (s_y) will be determined by the *relative size* of the error in the number used to calculate the logarithm (s_a/a). This relationship is roughly equivalent to the significant figure rule in Appendix A, which states that the digits to the right of the decimal in a logarithm (the mantissa) should have the same number of significant figures as the original number. Both approaches give two significant figures in the pH found in the preceding exercise, because the original hydrogen-ion activity also had two significant figures.

In Equations 4.10 and 4.11, just the opposite situation occurs. Here the *relative size* of the error for an antilogarithm (s_y/y) will be determined by the *absolute size* of the error in the number used to calculate the antilogarithm (s_a). This relationship is what led to the rule in Appendix A that the significant figures for antilogarithms will have the same number of digits as those to the right of the decimal in the original log. This also fits with the results obtained in the last exercise, in which the antilog of pH 6.00 gave a hydrogen-ion activity with two significant figures ($1.0\times 10^{-6}\,M$).

4.3D Mixed Calculations

Although it is useful to know how errors propagate during simple operations like addition or multiplication, there are many situations in which several types of operations are employed to obtain a result. It is possible to generate formulas like those in Table 4.1 for various combinations of calculations. An easier, alternative approach is to separate these operations into a series of steps that consist of only addition or subtraction, multiplication or division, and so on. During each step, the equations in Table 4.1 can be used to examine the random errors that are carried through that particular operation. The result can then be used in the next step until the final answer is obtained. This is similar to the approach we used in Chapter 2 to determine the number of significant figures in a result that involved a series of operations.

The first calculations that should be examined in this procedure are those that appear in parentheses or brackets. If no brackets or parentheses are present, then carry out any calculations that involve logarithms, antilogarithms, or exponents. This is followed by steps that involve multiplication or division, and then by those that use addition or subtraction. An example of this process is given in Exercise 4.6 concerning the use of a best-fit line to describe experimental data.

EXERCISE 4.6 **Error Propagation in Mixed Calculations**

A method for the detection of morphine is used to generate a calibration curve in which the assay response (y) is plotted versus morphine concentration (x, in mg/L). This gives a straight line with a slope (m) of 0.253 and a y-intercept (b) of 0.010, where $y = mx + b$. The slope of this line has a standard deviation of ± 0.009, and the standard deviation of the intercept is ± 0.007. If the sample from an athlete gives a response of 0.541 ± 0.015 in this method, what is the concentration of morphine in the sample and estimated precision of this concentration?

SOLUTION

The concentration of morphine in the unknown sample can be found by rearranging the calibration line equation and solving for x.

$$x = \frac{y - b}{m} = \frac{(0.541 - 0.010)}{0.253} = 2.099\ \text{mg/L}$$

To determine the error in x, we can conduct this calculation in two steps: (1) the subtraction of b from y, and (2) division of the difference by the slope, m. The result of the first step, which we will call x_1, has an error that can be calculated by using Equation 4.6.

$$x_1 = (y - b) = 0.531$$
$$s_{x1} = \sqrt{(s_y)^2 + (s_b)^2}$$
$$= \sqrt{(0.015)^2 + (0.007)^2} = 0.0166$$

The error in the second step, which gives our final answer, can be found by using Equation 4.7.

$$(s_x/x) = \sqrt{(s_{x1}/x_1)^2 + (s_m/m)^2}$$
$$s_x = (2.099\ \text{mg/L}) \cdot \sqrt{(0.0166/0.531)^2 + (0.009/0.253)^2}$$
$$= 0.099 = \mathbf{0.10\ mg/L}$$

Thus, the detected concentration of morphine in the athlete's sample was $\mathbf{2.10 \pm 0.10\ mg/L}$.

Notice in this last exercise that we were careful to include guard digits as we determined x_1 and s_{x1} in the first step of our calculation. This is critical in a multistep process to avoid introducing rounding errors and will provide the most significant figures in our final result.

4.4 SAMPLE DISTRIBUTIONS AND CONFIDENCE INTERVALS

Up to this point, we have looked at how to describe data sets and at how random errors propagate through calculations that use these values. We still, however, need to consider what is actually represented by the "average" or "standard deviation" when we are describing a group of numbers. In this section we will see how this can be accomplished by using tools like the normal distribution, the standard deviation of the mean, and confidence intervals.

4.4A Describing the Variation in Large Data Sets

We now know that random variations will be present whenever we make a measurement. If we were to make the same measurement many times and plot the number of times we obtain any particular value, we would get a result similar to the plot in Figure 4.5. In this graph, the x-axis gives the range of values we have measured for our sample, and the y-axis shows the number of times each of these values was obtained. If we make enough measurements and we have an equal chance of getting either high or low variations in a result, the plot we get will have a "bell shape" with the center occurring at the average of our data set. This is known as a **normal distribution** or a *Gaussian distribution*. Normal distribution curves are commonly used in science to represent the variation in measurements because this type of distribution can be directly related to the process of random error propagation. This feature also makes the normal distribution useful in comparing results and in estimating the reliability of a measurement.

The shape of the normal distribution curve in Figure 4.5 can be described by the following equation, which relates the experimental value of x to the probability of measuring this value (y).

$$y = \frac{1}{\sigma\sqrt{2\pi}} \cdot e^{-\frac{1}{2}[(x-\mu)^2/(\sigma^2)]} \qquad (4.13)$$

Along with x and y, two other factors that appear in this equation are (1) the average of the data set (μ), which gives the central point for the distribution, and (2) the standard deviation of the data set (σ), which describes the width of this curve. These are the same parameters we used earlier to describe small data sets; however, we are now dealing with a large group of numbers so the average and standard deviation are well known. This is why μ and σ are now used in place of $\bar{x}$ and s.

Figure 4.6 shows how a change in either the mean or standard deviation will affect the shape of a normal distribution curve. As we increase or decrease the average, the entire curve shifts to higher or lower x values. As we increase or decrease the standard deviation, the curve becomes broader or narrower, respectively. For chemical measurements it is always desirable to have a small standard deviation, because this will represent a group of precise results. This, in turn, makes it easier to compare the averages of data sets to see if they represent similar or different values.

If our results follow a normal distribution, we can use the average and standard deviation for the data set to

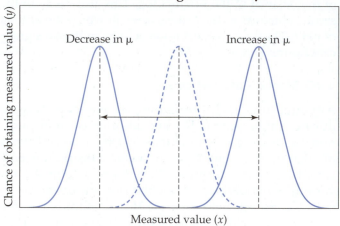

Effect of change in mean (μ)

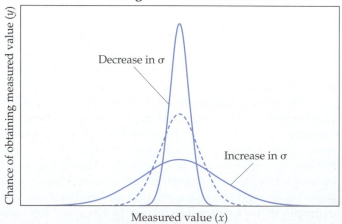

Effect of change in standard deviation (σ)

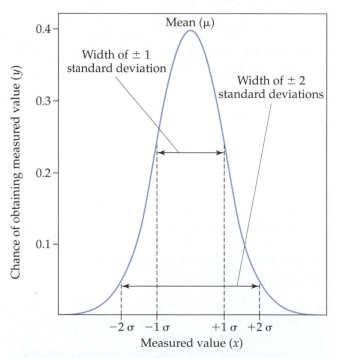

FIGURE 4.5 The normal distribution curve. The mean (μ) represents the center of this curve, while the standard deviation (σ) is a measure of the width of the curve. In a normal distribution, a width of exactly ± 1 standard deviation occurs at the *inflection points*, where a tangent line would touch the curve.

FIGURE 4.6 Effects of changing the true mean (μ) or true standard deviation (σ) on the shape and position of a normal distribution. The total area under each of the curves is constant in this set of examples.

determine what fraction of our results will fall between any two measured values. Table 4.2 shows what fraction of results (as represented by the area under the normal distribution) will occur between the mean and a value of x. This is accomplished by using the term z, where $z = (x - \mu)/\sigma$, which describes the difference between x and μ in terms of the number of standard deviations that separate these two values.

When we use a normal distribution to describe chemical measurements, there are two ranges that are particularly valuable to remember. The first is a range of one standard deviation above or below the mean ($\mu \pm 1\,\sigma$). According to Table 4.2, this range corresponds to a relative area of $2 \cdot (0.3413) = 0.6826$, or 68.26% of the results in a normal distribution, or roughly two thirds of all its values. The other range to remember is that of two standard deviations above or below the mean ($\mu \pm 2\,\sigma$), which represents a relative area of $2 \cdot (0.4772) = 0.9544$ in the normal distribution curve, or 95.44% of all results (approximately 95%, or 19 out of 20 values). These ranges are both frequently used to describe data sets and to compare experimental results.

4.4B Describing the Variation in Small Data Sets

One problem we have noted for a small set of numbers is that the experimental values for x and s are only estimates of the true average and standard deviation, μ and σ. Thus, we must always consider how precisely we know $\bar{x}$ and s when we use these values to describe experimental data.

Standard Deviation of the Mean. In the same way that we use s to describe the variation within a data set, we can employ a related value ($s_{\bar{x}}$) to describe the precision of our experimental average ($\bar{x}$). This new value, known as the **standard deviation of the mean**, is determined by using the standard deviation of the entire data set (s) and the number of data points in this set (n).[6]

$$s_{\bar{x}} = \frac{s}{\sqrt{n}} \qquad (4.14)$$

For instance, suppose you had a series of three measurements with an average of 12.0 and a standard deviation of 0.5. The standard deviation of the mean would then be $s_{\bar{x}} = 0.5/\sqrt{3} = 0.3$ for the three points. Notice that the size of $s_{\bar{x}}$ is always less than or equal to s, because n must be greater than or equal to one. In addition, whenever you report a standard deviation for a mean you should state the number of points in your data set, which acknowledges the fact that the size of $s_{\bar{x}}$ depends on n.

We saw earlier that as the number of values in a data set increases, the standard deviation for that entire set (s) will approach a constant value (σ). However, if we look at how $s_{\bar{x}}$ changes with an increase in n, we find that its value becomes smaller. This effect is illustrated in Figure 4.7 and occurs because the precision of the experimental average becomes better as we acquire more data, making $\bar{x}$ a more reliable estimate of the true average. Although using more measurements will provide a better estimate of the mean result for a sample by decreasing $s_{\bar{x}}$, acquiring more data will also increase the time, effort, and sample required to make the measurements. As a compromise between effort and reproducibility, three to five measurements are recommended for most analyses.

Confidence Intervals. It is common in science to describe the variation in experimental numbers by using a range of values. As an example, we could report a result by giving the mean plus or minus two standard deviations of the mean ($\bar{x} \pm 2\,s_{\bar{x}}$). In this approach, the range of values that follows our mean ($\pm 2\,s_{\bar{x}}$) is called the *confidence limit*, and the mean plus this range ($\bar{x} \pm 2\,s_{\bar{x}}$) is known as the

TABLE 4.2 Areas Under a Normal Distribution Curve at Various Distances from the Mean*

Distance from Mean (z)	Relative Area from Mean to z
0.0	0.0000
0.2	0.0793
0.4	0.1554
0.6	0.2258
0.8	0.2881
1.0	0.3413
1.2	0.3849
1.4	0.4032
1.6	0.4192
1.8	0.4641
2.0	0.4772
2.2	0.4861
2.4	0.4981
2.6	0.4953
2.8	0.4974
3.0	0.4987

* The mean of the normal distribution is represented by the point $z = 0$, where z is equal to the number of standard deviations that separate any point of interest from this central location. A more complete version of this table can be found in the *CRC Handbook of Chemistry and Physics*[7] (D. R. Lide, Ed., 83rd ed. CRC Press, Boca Raton, FL., 2002).

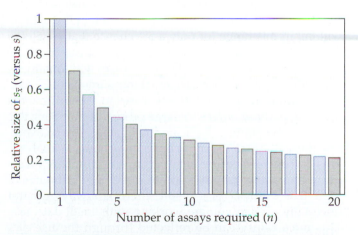

FIGURE 4.7 Decrease in the standard deviation of the mean ($s_{\bar{x}}$) as the number of individual values in a data set (n) is increased. These results were found by using the formula $s_x/s = 1/\sqrt{n}$, where s is the standard deviation for entire data set.

TABLE 4.3 Student's *t*-Values at Various Degrees of Freedom*

Degrees of Freedom (*f*)	Confidence Level		
	90%	95%	99%
1	6.31	12.7	63.7
2	2.92	4.30	9.92
3	2.35	3.18	5.84
4	2.13	2.78	4.60
5	2.02	2.57	4.03
6	1.94	2.45	3.71
7	1.90	2.36	3.50
8	1.86	2.31	3.36
9	1.83	2.26	3.25
10	1.81	2.23	3.17
11	1.80	2.20	3.11
12	1.78	2.18	3.06
13	1.77	2.16	3.01
14	1.76	2.14	2.98
15	1.75	2.13	2.95
16	1.75	2.12	2.95
17	1.74	2.11	2.92
18	1.73	2.10	2.88
19	1.73	2.09	2.86
20	1.72	2.09	2.85
∞	1.64	1.96	2.58

* All values shown in this table are for a *two-tailed* Student's *t*-test, in which we are considering the fact that our experimental value may be either higher *or* lower than the model. A more complete listing of Student's *t*-values can be found in the *CRC Handbook of Tables for Probability and Statistics*[12] (W. H. Beyer, Ed., 2nd ed., The Chemical Rubber Co., Cleveland, OH, 1968) or other sources.

confidence interval (or *C.I.*).[6] When you are reporting a confidence interval, the number placed in front of $s_{\bar{x}}$ helps specify the degree of certainty that the experimenter has in the result. As an example, we know in a normal distribution that a range of ±2 standard deviations means there is roughly a 95% chance that any given value in the data set will fall in this range (or only a 5% chance that it will fall outside this range). The same general approach can be used to give a range that in 95% of all cases will contain the true value for a set of experimental measurements.

Although it is relatively easy to determine the meaning of these ranges for large groups of numbers, this task becomes more complicated with small sets of experimental values, as we often have in a chemical analysis. This is the case because small groups of numbers give a mean and standard deviation that are only estimates of their true values. As a result, there is always greater uncertainty when you are working with small data sets. This uncertainty can be reflected through the use of a larger confidence interval and a correction factor known as the **Student's *t*-value** (*t*) (see Box 4.1). The Student's *t*-value can be used to express the confidence interval for

either a group of results (based on *s*) or for the mean (based on $s_{\bar{x}}$) by using the following expressions.[6,8]

Confidence interval for a group of results:
$$\text{C.I.} = \bar{x} \pm t \cdot s \qquad (4.15)$$
Confidence interval for a mean result:
$$\text{C.I.} = \bar{x} \pm t \cdot s_{\bar{x}}$$
$$= \bar{x} \pm t \cdot s/\sqrt{n} \qquad (4.16)$$

In this last equation, a reminder has been added for you that the value of $s_{\bar{x}}$ is found by simply calculating the ratio $s/\sqrt{n}$ when you are writing the confidence interval for an experimental mean.

Table 4.3 gives the values for *t* you would use with Equations 4.15 and 4.16 in writing a confidence interval. The Student's *t*-value chosen for this purpose will depend on the number of points (*n*) in your data set, as represented in Table 4.3 by the degrees of freedom (*f*), where $f = n - 1$. As you add more data points, the value of *t* decreases and approaches a constant. This reflects the fact that the experimental mean and standard deviations are becoming better estimates of their true values. Another factor that will determine the selected value of *t* is the degree of certainty you would like to have that the true answer falls within your calculated confidence interval. This degree of certainty is known as the **confidence level**. As you go to higher confidence levels, the size of *t* increases to provide a greater chance for the true value falling within the stated range.

EXERCISE 4.7 **Confidence Intervals and Confidence Limits**

Probenecid is a drug used by some athletes to prevent the excretion of other substances into urine, thus lowering their detectable concentrations. A scientist makes three measurements of a urine sample known to contain probenecid. The scientist gets a mean result of 11.8 µg/L and a standard deviation for the entire set of results of 0.2 µg/L. What is the 95% confidence interval for this mean?

SOLUTION

In this example we are looking at the mean rather than a population of results, so we first need to find $s_{\bar{x}}$. Using Equation 4.14, we get $s_{\bar{x}} = 0.2/\sqrt{3} = 0.12$ µg/L. Next, we need to look at Table 4.3 to see what value we should use for *t* when we have three measurements ($3 - 1 = 2$ degrees of freedom) and use a 95% confidence level. This process gives a Student's *t*-value of 4.30. Next, we place these numbers into Equation 4.16 to get the confidence interval for the mean.

$$\text{C.I.} = 11.8 \pm (4.30 \cdot 0.12 \text{ µg/L})$$
$$\therefore \text{ 95\% C.I.} = 11.8 \pm 0.5 \text{ µg/L} \quad (\text{at } n = 3)$$

We can now say with 95% confidence that the result will be between $(11.8 - 0.5 \text{ µg/L}) = 11.3$ µg/L and $(11.8 + 0.5 \text{ µg/L}) = 12.3$ µg/L. As the preceding answer indicates, you should always state the number of data points and the confidence level when you are reporting a confidence interval.

BOX 4.1
Who Was "Student"?

There have been many times throughout history in which a need for better methods of characterizing materials has led to the creation of new techniques for analyzing and describing these samples. Sometimes these new approaches come from research labs, while other times they come from an industrial setting or as the result of a combined effort between industry and research labs. An example of this last case is the Student's *t*-test, which was developed by a chemist and mathematician named William S. Gossett (see Figure 4.8).

Gossett began work as a young man in 1899 at the Guinness Brewery in Dublin, Ireland. At that time there was great interest in this company in finding ways of relating the properties of their raw materials and manufacturing conditions to the quality of their final product. Gossett was one of several staff scientists assigned to study this relationship. As part of his work, he found that he needed a way of describing the statistical distributions for small groups of samples. This issue had not been explored before, so his company arranged for Gossett to spend some time with a mathematics professor named Karl Pearson at the University College in London.

During these studies, Gossett came up with a new type of data distribution that was related to a normal distribution through a "*t*" value. Gossett studied this distribution and calculated the value of *t* for various sample sizes and degrees of freedom. After presenting his results to his company, he was allowed to publish his work provided that he did not use his real name or any actual data from the firm.[9,10] Working under the name of "Student," Gossett submitted his results to the journal *Biometrika*, which published them in 1908.[11] He continued to write and submit papers under the name Student for over 30 years, remaining anonymous to most of the world during this time.[9,10]

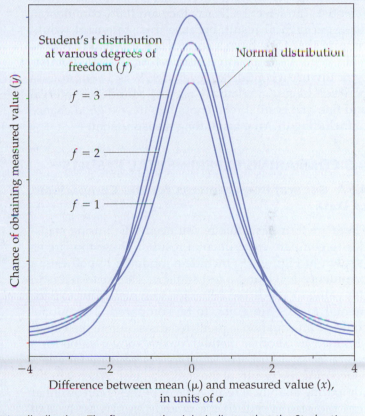

FIGURE 4.8   William Sealy Gosset (1876–1937) and the Student's *t*-distribution. The figure on the right indicates that the Student's *t*-distribution is often much broader than a normal distribution (even though these curves have the same areas), especially when working with small data sets that have low degrees of freedom. However, the Student's *t*-distribution becomes narrower and approaches the normal distribution as the number of points in a data set and degrees of freedom are both increased. The Student's *t*-values that are given in Table 4.3 represent the ranges of this distribution that contain a given percentage of all values in the population. For instance, a Student's *t*-distribution for a data set with two degrees of freedom will have 95% of its values occurring over a range that spans from –4.30 to +4.30 standard deviations about the mean. In a data set with three degrees of freedom, 95% of the values in a Student's *t*-distribution will be in a range that spans from –3.18 to +3.18 standard deviations about the mean.

Ideally, we would like the confidence interval for an experimental value to be small. For instance, a small confidence interval makes it easier to decide if two experimental values are the same or different. We can obtain a narrower confidence interval for a mean result, as given by Equation 4.16, by increasing the number of data points that we use to calculate the mean. We saw earlier with Equation 4.14 that using more data points will lower the estimated value of $s_{\bar{x}}$, which in turn will give a small confidence interval. Using more data points will also improve our estimate of the overall standard deviation s for a group of results, but the value of s approaches a constant instead of a smaller value as we increase n. The only way we can obtain a smaller confidence interval for a population of results (as described by Equation 4.15 and using s) is to improve the method of our measurement to get more precise results.

Selection of the confidence level can have a big impact on the range of results we obtain for the confidence interval. As an example, a confidence level of 90% will always give a smaller confidence interval than a confidence level of 99%. This effect occurs because as we increase the level of confidence we have that a result will fall within the given confidence interval, we also must expand the range of numbers that we use for this interval. This concept is reflected in Table 4.3 by the fact that the Student's t-value increases in size as you move from a low to high confidence level. This is also a reason why scientists can never really say they are 100% confident in an experimental result, because this statement would require that they have a confidence interval that goes from negative to positive infinity! Analytical chemists typically use a confidence level of 95% as a compromise between having a relatively narrow confidence interval and one that is still broad enough to have a good chance of including the true result for a measurement.

4.5 COMPARING EXPERIMENTAL RESULTS

4.5A General Requirements for the Comparison of Data

There are four basic things you need when using statistics to compare experimental results. These four items (shown in Figure 4.9) include a model, a hypothesis, a confidence level, and a test statistic. The *model* refers to the value or predicted behavior to which your experimental results are going to be compared. This model could be an equation, a predicted distribution, the values obtained by another method, or the known value for a reference standard. The *hypothesis* is your initial guess concerning the results of the statistical test. When you are comparing analytical results, the hypothesis can be either that the results will fit the model (known as the *null hypothesis*) or that they won't fit the model (the *alternate hypothesis*). The *confidence level* represents the degree of certainty you wish to have in your comparison. We saw earlier that all scientific results have some degree of uncertainty in them because of random errors. The confidence level helps us estimate the extent of this uncertainty and to avoid reaching any unreasonable conclusions about our data.

The last part of a statistical method is the *test statistic*. This is a numerical value that is calculated from your data to use in the comparison. One common example of a test statistic is the Student's t-value, which can be used to

The model

What is my result being compared to?

The hypothesis

What is the expected result for this comparison?

The confidence level

What degree of certainty do I want for this comparison?

The test statistic

How will I compare my result and model?

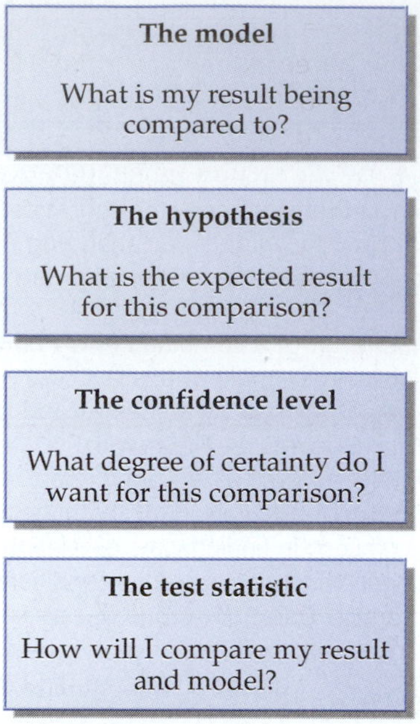

FIGURE 4.9 The four key components needed for any statistical comparison of experimental data.

compare an experimental mean to another number. In this process, the test statistic that is calculated for your result is then compared to a *critical value* (such as those given in Table 4.3), which represents the largest value for the test statistic that you would expect for the given number of data points and selected confidence level. We will see a specific example of this process in the next section when we will learn how to compare an experimental mean with a known reference value.

4.5B Comparing an Experimental Result with a Reference Value

When you work in an analytical laboratory you will often have to compare an experimental result with a known reference value. For instance, this type of comparison might be needed when you are determining the accuracy of a new method. If the reference value is known exactly (or at least has a much better precision than the experimental result), we can use this value to represent the true "mean" for the sample, μ. To compare this value to the average result measured for the sample ($\bar{x}$), we can use the Student's t-value as the test statistic. The result is a statistical method known as the **Student's t-test**.

If you begin by assuming that your reference value and experimental result are the same (the null hypothesis), you can test this assumption by calculating a Student's t-value, as shown below,[8,13]

$$t = \frac{|\bar{x} - \mu|}{s_{\bar{x}}} \qquad (4.17)$$

where $\bar{x}$ is your mean experimental result, μ is the reference value, and $s_{\bar{x}}$ is the standard deviation of the experimental mean, as given by $s_{\bar{x}} = s/\sqrt{n}$. The lines shown on either side of "$\bar{x} - \mu$" in this equation indicate we are looking at only the absolute, or positive, value of this difference. The terms $\bar{x}$, μ, and $s_{\bar{x}}$ should all have the same units, so putting these into Equation 4.17 should give a value for t that has no units. By looking more closely at this ratio, you can see that t simply represents the number of standard deviations that separate $\bar{x}$ and μ. Thus, a large Student's t-value means that $\bar{x}$ and μ are very different and probably represent different numbers.

Once we have calculated t for our data, we need to compare this result to a critical value (t_c), as obtained from Table 4.3. The t_c value we select will be determined by the number of data points that were used to find our experimental mean (as represented by the degrees of freedom, $f = n - 1$) and the confidence level we have chosen for our comparison. If we find that $t \leq t_c$, we can say that $\bar{x}$ and μ are not significantly different at the given confidence level.

| EXERCISE 4.8 | Comparing an Experimental Result and a Reference Value |

Action is taken against Olympic athletes if their urine is found to contain caffeine concentrations above 12.00 μg/mL. A sample from one athlete gives a mean caffeine concentration of 12.16 μg/mL for five measurements (range, 12.00 to 12.28 μg/mL) with the standard deviation for this mean being 0.07 μg/mL. The coach of the athlete argues that this result is statistically the same as the 12.00 μg/mL cutoff. Are these two values equivalent at the 95% confidence level?

SOLUTION

Our model in this example is the 12.00 μg/mL value, and we are told to work at the 95% confidence level. To see if our mean and reference value are the same (the underlying hypothesis), we can place these values into Equation 4.17 to calculate a Student's t-value.

$$t = \frac{|12.16\ \mu g/mL - 12.0\ \mu g/mL|}{0.07\ \mu g/mL} = \mathbf{2.29}$$

Next, we need to look up the critical Student's t-value in Table 4.3 at the 95% confidence level and for $f = (5 - 1) = 4$ degrees of freedom. This process gives a value for t_c of 2.78, which is greater than our experimental t value of 2.29. Thus, we can say with 95% confidence that the amount of caffeine in the sample from the athlete is not significantly different from the allowed cutoff level.

It is important when you are using a statistical test that you decide on the confidence interval to be employed *before* you perform the test. This is considered to be a part of good laboratory practices and helps you avoid placing any personal bias in the final results of your test. The correct choice of a confidence interval can also help you minimize the possibility of making an incorrect conclusion based on a statistical test (see Box 4.2).

4.5C Comparing Two or More Experimental Results

Individual Results. Another situation often encountered in chemical analysis is when we need to compare two experimental results. To illustrate this, suppose we have mean results for two samples ($\bar{x}_1$ and $\bar{x}_2$) that have been measured by the same method or by two methods with similar precision. In this situation, the model would be one of the two mean results and the hypothesis we would be testing is that these two results represent the same number.

The test statistic we use for this situation is again the Student's t-value, but we now need to modify this approach to allow for the fact that both our experimental result and model have some uncertainty in their values. Thus, instead of using the standard deviations for either of these means, we use a pooled standard deviation (s_{pool}) that reflects the variation in both results.[14]

$$s_{pool} = \sqrt{\frac{(n_1 - 1) \cdot (s_1)^2 + (n_2 - 1) \cdot (s_2)^2}{(n_1 + n_2 - 2)}} \qquad (4.18)$$

In this equation, s_1 and s_2 are the estimated standard deviations for the two data sets, and n_1 and n_2 are the number of points in each of these sets. You can think of s_{pool} as a weighted average of the individual standard deviations for the two groups of results. Also, just as we can use s_1 or s_2 to find the standard deviations of the means $\bar{x}_1$ and $\bar{x}_2$, we can use s_{pool} to get a standard deviation for the pooled mean ($s_{\bar{x}pool}$), which is found by using Equation 4.19.

$$s_{\bar{x}pool} = \frac{s_{pool}}{\sqrt{(n_1 \cdot n_2)/(n_1 + n_2)}} \qquad (4.19)$$

One way to view $s_{\bar{x}pool}$ is as a measure of the uncertainty in the difference between $\bar{x}_1$ and $\bar{x}_2$.

If $\bar{x}_1$ and $\bar{x}_2$ represent the same value, their difference ($\bar{x}_1 - \bar{x}_2$) should fall within a reasonably small number of standard deviations for this difference. This means we can compare ($\bar{x}_1 - \bar{x}_2$) directly to $s_{\bar{x}pool}$ and use their ratio to give a Student's t-value for this comparison.

$$t = \frac{|\bar{x}_1 - \bar{x}_2|}{s_{\bar{x}pool}} \qquad (4.20)$$

A small value for this ratio would indicate our two results are close together and probably represent the same number. Once we have calculated this Student's t-value, we must compare it to a critical value (t_c) from Table 4.3, as obtained for our selected confidence level and using $f = (n_1 + n_2 - 2)$ as the degrees of freedom. If t is less than or equal to t_c, we can say that $\bar{x}_1$ and $\bar{x}_2$ represent the same value at our selected confidence level.

BOX 4.2
Selecting a Confidence Level

Your choice of a confidence interval can be quite important in determining both the outcome and accuracy of a statistical test. For instance, in Exercise 4.8 we concluded that the measured and allowed concentrations of caffeine were not significantly different because t was less than t_c at the 95% confidence level. If we had selected a slightly lower confidence level of 90%, this would have given a smaller critical Student's t-value ($t_c = 2.13$ for 4 degrees of freedom) and created a result where t was now greater than t_c, indicating that the measured and allowed amounts of caffeine were different. At this point you are probably wondering "Why does this difference occur?" and "Why did we originally use the 95% confidence level for this comparison?"

The reason for these two different outcomes is that changing the confidence level for the test also changed the allowable errors in the results. There are actually two types of errors that are always possible during a statistical test. The first type of error (known as a *type 1 error* or "alpha error") occurs when you conclude that the model and experimental value are not the same when they really are equivalent.[8] Figure 4.10 shows that this situation occurs when random errors cause the difference in these values to fall outside the confidence interval being used for their comparison. The probability of a type 1 error (when stated as a percent) is equal to 100 minus the percent confidence level that is selected for a statistical test. As an example, the use of a 95% confidence level in Exercise 4.8 means there was only a $100 - 95 = 5\%$ chance that random errors would cause a value in the given data set to fall outside of the allowed confidence interval. When we use a lower confidence level of 90%, the chance of having a type 1 error is raised to $100 - 90 = 10\%$, making it more likely that we will incorrectly conclude that the model and experimental value are different. As you increase the confidence level, you always reduce the chance that a type 1 error will occur.

The second type of error that can be produced during a statistical test is a *type 2 error* (or "beta error"). A type 2 error occurs when you conclude that a result is the same as your model, but the result is actually part of an entirely different data distribution.[8] In Figure 4.10 the chance of a type 2 error occurring is given by the area of overlap between the distribution for the tested result and the distribution for the model. The use of a higher confidence level will increase the probability that a type 2 error will be present because this means we are using a broader range to describe our experimental value. As a result, your selection of a confidence level for a statistical test will always represent a compromise between the size of the type 1 and type 2 errors that will be present. In analytical chemistry, a 95% confidence level usually works well for this purpose.

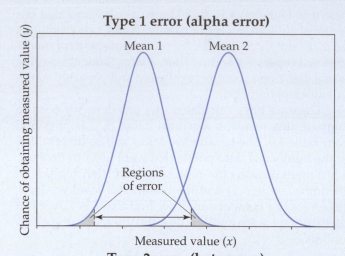

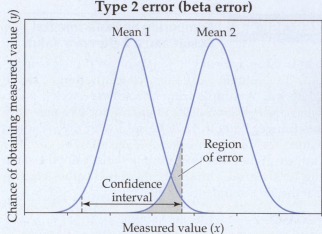

FIGURE 4.10 An illustration of type 1 and type 2 errors in the comparison of two mean values. In this example, the type 1 error is the chance that a result belonging to the distribution of Result 1 will fall outside of the range used in the comparison. The probability of this type of error is equal to the fraction of the distribution for Result 1 that falls outside of this range, and can be calculated directly from the confidence level that is used in comparing it to Result 2. The type 2 error is equal to the overlap in the distribution for Result 2, with the range being used for Result 1 in the comparison. The size of this type 2 error will depend on the difference between the mean values for Results 1 and 2 and the widths of their corresponding distributions.

EXERCISE 4.9   Comparing Two Mean Results

Human chorionic gonadotropin (hCG) is a naturally occurring substance that has been abused by some athletes because of its ability to stimulate testosterone production. Two laboratories that conduct drug testing are to be evaluated for their ability to measure this hormone by using the same sample and analysis method. The first laboratory obtains a mean hCG level of 2.99 IU/L ($n_1 = 4$) with a standard deviation of 0.06 IU/L, while the second laboratory reports a mean level of 3.13 IU/L ($n_2 = 5$) with a standard deviation of 0.08 IU/L. Are these mean results the same at the 95% confidence level?

SOLUTION

If we assume the standard deviations for the two means are approximately the same, the first step in solving this problem is to use Equation 4.18 to get the pooled standard deviation.

$$s_{pool} = \sqrt{\frac{(4-1)\cdot(0.06\ IU/L)^2 + (5-1)\cdot(0.08\ IU/L)^2}{(4+5-2)}}$$

$$= \sqrt{\frac{0.036\ (IU/L)^2}{7}} = 0.072\ IU/L$$

Next, we can use s_{pool}, n_1, and n_2 to determine the standard deviation of our pooled mean ($s_{\bar{x}pool}$).

$$s_{\bar{x}pool} = \frac{0.072\ IU/L}{\sqrt{(4\cdot5)/(4+5)}} = 0.048\ IU/L$$

We are now ready to calculate the Student's t-value for our results by using Equation 4.20.

$$t = \frac{|2.99\ IU/L - 3.13\ IU/L|}{0.048\ IU/L} = 2.9$$

The degrees of freedom in this case is $f = (4+5-2) = 7$, and we are told to work at the 95% confidence level, so the critical t_c value from Table 4.3 for this situation would be 2.36. When we compare our experimental and critical values, we find that t is greater than t_c (2.9 > 2.36). This means the results from the two laboratories are significantly different at the 95% confidence level.

Groups of Values. Another comparison that is often made is when two sets of the identical samples are analyzed by different methods. If these methods have similar precision, we can compare their results by using a procedure known as the **paired Student's t-test**.[8,14,15] To set up this test, we begin by making a list of the results obtained by both methods for each sample, as demonstrated in Table 4.4. The difference between

each set of results is then calculated (as represented by d_i), and the average of these differences ($\bar{d}$) is found, where $\bar{d} = \Sigma(d_i)/n$. It is the size of this average difference ($\bar{d}$) that is eventually employed in our comparison of the two methods.

To determine whether the differences in the two sets of results are significant, we also need to find the standard deviation for these differences (s_d), which is calculated as shown below.

$$s_d = \sqrt{\frac{\Sigma(d_i - \bar{d})^2}{(n-1)}} \quad (4.21)$$

We can then calculate the standard deviation in the average difference ($s_{\bar{d}}$).

$$s_{\bar{d}} = \frac{s_d}{\sqrt{n}} \quad (4.22)$$

If the differences in the results for the two methods occur because of only random variations, the average difference in these results should be similar in size to $s_{\bar{d}}$. Based on this reasoning, the Student's t-value to be used for this comparison is calculated according to Equation 4.23.

$$t = \frac{|\bar{d}|}{s_{\bar{d}}} \quad (4.23)$$

Once we have this t-value, we again need to compare it to a critical value from Table 4.3, as given for our desired confidence level and $f = n - 1$ degrees of freedom, where n now represents the number of data-point pairs we are comparing. If we find that $t \leq t_c$, we can say that the two methods produce statistically identical values at the given confidence level. Although this is a somewhat long procedure, it is quite valuable in comparing methods.

EXERCISE 4.10 Using the Paired Student's t-Test

Corticosteroids can legitimately be used by athletes for the relief of inflammation and pain, but the injection or inhalation of these compounds is allowed only when needed for a medical condition. A new technique for the measurement of corticosteroids in urine is to be compared with a previous method. Both approaches have similar precision and are used to analyze a series of identical samples. The new method gives mean results of 2.53, 5.19, 3.60, 6.42, and 7.08 $\mu mol/L$ for five separate samples, while another established method gives mean results of 2.68, 5.03, 3.79, 6.51, and 7.24 $\mu mol/L$ for the same samples. Are the results of these two methods equivalent at the 95% confidence level?

SOLUTION

In this example we are comparing several samples and have methods with comparable precision, so we can use a paired Student's t-test. To do this, the mean results obtained for all the samples are first listed side by side, as illustrated in Table 4.4. We then calculate the difference

TABLE 4.4 An Example of a Paired Student's t-Test

Sample Number	Mean Results ($\mu mol/L$)		Difference in Results ($\mu mol/L$) $d_i = x_{Method\ 1} - x_{Method\ 2}$
	Method 1	Method 2	
1	2.53	2.68	−0.15
2	5.19	5.03	0.16
3	3.60	3.79	−0.19
4	6.42	6.51	−0.09
5	7.08	7.24	−0.16
			$\bar{d} = (\Sigma d_i)/n$ = −0.086 $\mu mol/L$

between each pair of results, which gives an average difference between the two methods of $-0.086\ \mu mol/L$. Next, we use Equations 4.21 and 4.22 to find s_d and $s_{\bar{d}}$.

$$s_d = \left[\frac{(-0.15 - (-0.086))^2 + (0.16 - (-0.086))^2 +}{(5 - 1)} \right.$$

$$\frac{(-0.19 - (-0.086))^2 + (-0.19 - (-0.086))^2 +}{(5 - 1)}$$

$$\left. \frac{(-0.16 - (-0.086))^2}{(5 - 1)} \right]^{1/2}$$

$$= \sqrt{\frac{(0.081\ \mu mol/L)^2}{4}} = 0.14\ \mu mol/L$$

$$s_{\bar{d}} = \frac{0.14\ \mu mol/L}{\sqrt{5}} = 0.063\ \mu mol/L$$

We are now ready to calculate our Student's t-value by using Equation 4.23.

$$t = \frac{|-0.086\ \mu mol/L|}{0.063\ \mu mol/L} = \mathbf{1.4}$$

According to Table 4.3, the critical value for our test at $f = (n - 1) = 4$ degrees of freedom and at the 95% confidence level would be 2.78. Because $1.4\ (t) \leq 2.78\ (t_c)$, we can say the results of our two methods are the same at the 95% confidence level.

An even more complex situation than the last example is one in which we must compare two mean results obtained by methods with very different precision. Although there is no rigorous statistical approach available for this case, there are some empirical methods that can be used.[8,14] This task can also be accomplished by using a *correlation chart*, which we will learn about in Chapter 5.

4.5D Comparing the Variation in Results

Another situation you might encounter is when you need to compare the precision of two results or methods. This comparison is accomplished by using the **F-test**.[8,14,15] The model in this test is the method or result with the smallest standard deviation (s_1), and the hypothesis we are testing is the belief that this model is the same as the standard deviation for the other method or result (s_2). This comparison is made by looking at the ratio of the variances for these two values, as indicated in Equation 4.24, which gives us a test statistic known as the *F-value*.

$$F = \frac{(s_2)^2}{(s_1)^2} \qquad \text{(where } s_2 \geq s_1) \qquad (4.24)$$

Because we have selected s_1 to be the smaller of the two standard deviations, the value we calculate for F should always be greater than or equal to one. As F becomes larger, this represents a greater likelihood that s_1 and s_2 represent different numbers.

After we have calculated the value of F for our data, we need to compare this to an appropriate critical value (F_c) at our desired confidence level and degrees of freedom. A list of such values is given in Table 4.5. The degrees

---

TABLE 4.5 Critical F-Test Values (F_c) at the 95% Confidence Level*

Degrees of Freedom in Numerator (s_2)	Degrees of Freedom in Denominator (s_1)									
	1	**2**	**3**	**4**	**5**	**6**	**7**	**8**	**9**	**10**
1	161	18.5	10.1	7.71	6.61	5.99	5.59	5.32	5.12	4.96
2	199	19.0	9.55	6.94	5.79	5.14	4.74	4.46	4.26	4.10
3	216	19.2	9.28	6.59	5.41	4.76	4.35	4.07	3.86	3.71
4	225	19.2	9.12	6.39	5.19	4.53	4.12	3.84	3.63	3.48
5	230	19.3	9.01	6.26	5.05	4.39	3.97	3.69	3.48	3.33
6	234	19.3	8.94	6.16	4.95	4.28	3.87	3.58	3.37	3.22
7	237	19.4	8.89	6.09	4.88	4.21	3.79	3.50	3.29	3.14
8	239	19.4	8.84	6.04	4.82	4.15	3.73	3.44	3.23	3.07
9	241	19.4	8.81	6.00	4.77	4.10	3.68	3.39	3.18	3.02
10	242	19.4	8.78	5.96	4.74	4.06	3.64	3.35	3.14	2.98

* All values shown in this table are for a *one-tailed* F-test, in which we are seeing if s_2 is greater than or equal to s_1 (the model). A two-tailed F-test would be needed if the more general question is asked as to whether s_2 and s_1 are equivalent values. A more complete list of F-test values can be obtained from *Data Analysis in the Chemical Sciences*[13] (VCH, New York, 1993) or related sources.

of freedom in this table are found by using $f_1 = (n_1 - 1)$ and $f_2 = (n_2 - 1)$, where n_1 and n_2 are the number of points for data sets one and two, respectively. If we find that $F \leq F_c$, the precision of these methods is considered to be the same at the selected confidence level.

EXERCISE 4.11	Comparing the Precision of Two Methods

It is known that the two methods in Exercise 4.10 have standard deviations of 0.09 and 0.16 μmol/L (for $n_1 = n_2 = 5$) at a corticosteroid concentration of 5.0 μmol/L. Can we say the precision of the second method is not significantly greater than the first method at the 95% confidence level?

SOLUTION

In making this comparison, we would set s_2 equal to 0.16 and s_1 equal to 0.09, so that $s_2 \geq s_1$. We can then substitute these numbers into Equation 4.24 to calculate a value for F.

$$F = \frac{(0.16 \ \mu mol/L)^2}{(0.09 \ \mu mol/L)^2} = 3.2$$

In looking at Table 4.5, we see that the critical value at the 95% confidence level and for our particular degrees of freedom, where $f_1 = (n_1 - 1) = 4$ and $f_2 = (n_2 - 1) = 4$, is 6.39. Because $F \leq F_c$, we can say with 95% confidence that the precision of the second method is not significantly greater than that of the first method.

4.6 DETECTING OUTLIERS

There will always be some small variation present whenever you perform repeated measurements on a sample. Occasionally, you will find a data point that is quite different from others that are obtained under supposedly identical conditions. If this is due to a problem with the experiment, an experienced chemist will often be able to identify such a situation and to appropriately separate this data point from the rest of the results. However, there are occasions when even the best laboratory worker will find a data point that just doesn't seem to fit the trend observed for other results. The term **outlier** is used to describe such data points.[8,14]

4.6A General Strategy in Handling Outliers

Chemists and other scientists must constantly be watching for outliers. One reason for this is that outliers may indicate the presence of unanticipated errors or changes during an analysis. This situation might occur when you are learning how to use a new technique or are performing research in an area that has never been fully explored. As part of this process, it is helpful to have a plan in place as to how you will handle data points that appear to be outliers. One simple thing you can do is recheck your results to make sure that all data were correctly recorded

and that no errors were made in calculating or plotting the results. You should also determine if there were any differences in the experimental conditions for the outlier versus other data. This process is easiest if you have a well-maintained laboratory notebook that contains all pertinent observations and conditions for the experiment. A good familiarity with the techniques and samples you are using also helps in this process.

Another valuable piece of information that can be used in outlier detection is the precision of your analysis method. This is easy to do if you have a large number of data points for a sample, because you can then see if the point in question has a greater than expected difference from the mean. For example, 95% of the results in a normal distribution should be within two standard deviations of the mean. This type of comparison is more difficult to conduct for small data sets because you then have only an estimate of the precision for the data. This can be overcome by using one of two statistical methods, the Q-test and T_n-test. Remember, however, that such tests are to be used *only* for identifying outliers and *not* as the sole means for justifying their removal. A thorough knowledge of your methods and conditions is always the best tool when you are deciding whether a point should be kept in a data set.

4.6B Statistical Tests for Outliers

Q-Test. The first method we will look at for outlier testing is the **Q-test** (also known as *Dixon's test*).[16,17] This test works by taking the absolute difference between a suspected outlier's value and the nearest value in the rest of the data set, with this difference then being compared to the total range of values in the set. If the difference between the suspected point and its nearest neighbor is greater than a critical fraction of the total range, the suspected value can be said to represent a true outlier.

To perform this test, you begin by ranking the results in the data set from the lowest to highest value. For the sake of argument, let's call our suspected outlier x_o and its nearest neighbor x_n. We also need to identify the highest value (x_{high}) and lowest value (x_{low}) in the data set to give the range. Because x_o is always at one end of this range, it will be the same as either the high or low value. We then calculate the following ratio (Q).

$$Q = \frac{|x_o - x_n|}{x_{high} - x_{low}} \tag{4.25}$$

Next, we compare our calculated value for Q to a critical test value (Q_c), as given in Table 4.6. As we saw for other statistical tests, this critical value will depend on the total number of results in our data set and on the confidence level we wish to use in determining whether x_o is a true outlier. If we find that $Q > Q_c$, the suspected data point can be called an outlier and considered for rejection.

TABLE 4.6 Critical Values (Q_c) for the Q-Test*

Number of Values in Data Set	Values for Q_c at Various Confidence Levels		
	Confidence Level = 90%	95%	99%
3	0.941	0.970	0.994
4	0.765	0.829	0.926
5	**0.642**	**0.710**	**0.821**
6	0.560	0.625	0.740
7	0.507	0.568	0.680
8	0.468	0.526	0.634
9	0.437	0.493	0.598
10	**0.412**	**0.466**	**0.568**
11	0.392	0.444	0.542
12	0.376	0.426	0.522
13	0.361	0.410	0.503
14	0.349	0.396	0.488
15	**0.338**	**0.384**	**0.475**
16	0.329	0.374	0.463
17	0.320	0.365	0.452
18	0.313	0.356	0.442
19	0.306	0.349	0.433
20	**0.300**	**0.342**	**0.425**

* These values are for a *two-tailed* Q-test and are recommended for general outlier testing, where the outlier may be either a high *or* low value in a data set. A more complete list of Q_c values, including those for a one-tailed Q-test, can be found in D. B. Rorabacher, "Statistical Treatment for Rejection of Deviant Values: Critical Values of Dixon's 'Q' Parameter and Related Subrange Ratios at the 95% Confidence Level," *Analytical Chemistry*, 63 (1991) 139–146.

EXERCISE 4.12   **Outlier Detection by the *Q*-Test**

A urine sample containing a known amount of markers for marijuana is sent to several drug testing laboratories to evaluate their ability to monitor this analyte. These laboratories report the following concentrations: Lab 1—55.3 μg/L, Lab 2—57.8 μg/L, Lab 3—54.0 μg/L, Lab 4—68.1 μg/L, and Lab 5—58.7 μg/L. Use the **Q-test** to determine whether any of these results can be considered an outlier at the 95% confidence level.

SOLUTION

The low and high values in this group are 54.0 and 68.1 μg/L. The result of 68.1 μg/L is also the most likely outlier, because it is the furthest from its nearest data point (58.7 μg/L). When we place these numbers into Equation 4.25, we get the following Q-value.

$$Q = \frac{|68.1\ \mu g/L - 58.7\ \mu g/L|}{68.1\ \mu g/L - 54.0\ \mu g/L}$$

$$= \frac{|9.4\ \mu g/L|}{14.1\ \mu g/L} = \mathbf{0.67}$$

From Table 4.6, we find that the critical value for this situation is 0.710, as listed for $n = 5$ at the 95% confidence level. Our calculated Q-value is less than 0.710, so the point at 68.1 μg/L cannot be called an outlier at the given confidence level.

One appealing feature of the Q-test is that it involves only simple calculations, but there are also several disadvantages to this test.[8,18] First, it does not make use of all the available information, because it employs only three values (x_n, x_{high}, and x_{low}) from the data set. A second problem is this test is often misused. This misuse usually occurs because the critical values in Table 4.6 assume that only a *single* possible outlier is present in the data set. Thus, it is not correct to use the values in this table to eliminate more than one point from the same group of results.

T_n-Test.   A second method for detecting outliers is the T_n-test.[19] In this method, we calculate the difference between the overall mean of our data set ($\bar{x}$) and the suspected outlier (x_o). We then divide the absolute value of this difference by the standard deviation for the entire data set (s), which gives a ratio called T_n.

$$T_n = \frac{|x_o - \bar{x}|}{s} \quad (4.26)$$

In preparing for this test, you need to include x_o as part of the data set when you calculate both $\bar{x}$ and s. As a result, this point cannot be considered an outlier until *after* we have performed the test.[19]

Equation 4.26 indicates that T_n is equal to the number of standard deviations that separate x_o and $\bar{x}$. A large value for T_n will represent a large difference between x_o and $\bar{x}$, as expected if x_o is an outlier, while a small T_n-value will represent only normal variations due to random errors within the data set. The cutoff where x_o can be said to be an outlier is given by a critical value T_n^*, as obtained from Table 4.7. The use of this test is illustrated in the following exercise.

EXERCISE 4.13   **Detection of Outliers by the T_n-Test**

In Exercise 4.12 can any of the listed results be called outliers when using the T_n-test at the 95% confidence level?

SOLUTION

The mean and standard deviation for this group of numbers are 58.8 and 5.5 μg/L. We also know from the last exercise that the most likely outlier is the value at 68.1 μg/L. When we place these values into Equation 4.26, we get the following result for T_n.

$$T_n = \frac{|68.1\ \mu g/L - 58.8\ \mu g/L|}{5.5\ \mu g/L} = \mathbf{1.7}$$

From Table 4.7, the critical value for five data points at the 95% confidence level is 1.715. When we compare this to T_n, we find that $T_n \leq T_n^*$. Thus, once again the suspect value cannot be considered an outlier at the 95% confidence level.

TABLE 4.7   Critical Values (T_n^*) for the T_n-Test*

Number of Values in Data Set	Values for T_n^* at Various Confidence Levels		
	Confidence Level = 90%	95%	99%
3	1.153	1.155	1.155
4	1.463	1.481	1.496
5	1.672	1.715	1.764
6	1.822	1.887	1.973
7	1.938	2.020	2.139
8	2.032	2.126	2.274
9	2.110	2.215	2.387
10	2.176	2.290	2.482
11	2.234	2.355	2.564
12	2.285	2.412	2.636
13	2.331	2.462	2.699
14	2.371	2.507	2.755
15	2.409	2.549	2.806
16	2.443	2.585	2.852
17	2.475	2.620	2.894
18	2.504	2.651	2.932
19	2.532	2.681	2.968
20	2.557	2.709	3.001

* These values are based on data given in "Standard Practice for Dealing with Outlying Observations" by the American Society for Testing and Materials[19] and are for a *two-tailed* T_n-test, in which an outlier may be either a high *or* low value in a data set.

In the last two exercises the Q-test and T_n-test both gave the same conclusion for our data sets, however, these approaches can give different results in other cases. This situation can occur when you are looking at outliers that are near the critical values for these methods. It is for this reason that you should choose the method you will rely upon for outlier detection *before* you perform the outlier test. The main advantage of the T_n-test is it involves all the values in a data set through its use of $\bar{x}$ and s. This feature makes the T_n-test more robust than the Q-test for outlier detection. However, the T_n-test still assumes there is only one outlier present at most, making it invalid to use this method to reject more than one point at a time from a data set (see References 16-18 for alternative approaches that can be used in such a situation).

4.7 FITTING EXPERIMENTAL RESULTS

Our final investigation in this chapter is to see how we can fit an equation or line to a set of results. This procedure is often required when we are preparing a calibration curve or are comparing experimental results to a predicted response. There are many types of equations used in chemical analysis, but the most common is the one for a straight line. Let's now see how we can

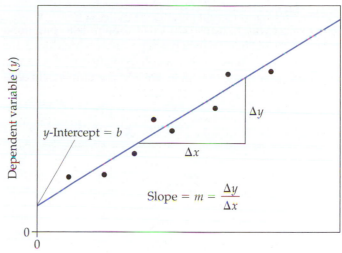

FIGURE 4.11   Parameters used to describe a best-fine line and linear plot.

determine the best-fit line for a set of data by using a process known as **linear regression**.

4.7A Linear Regression

Linear regression involves taking a set of (x,y) values (where y is the "dependent variable" and x is the "independent variable") and fitting these to an equation with the following form,

$$y_{i,\text{calc}} = m\,x_i + b \tag{4.27}$$

where m is the **slope** (representing the change in y vs. x), b is the line's **intercept** on the y-axis, x_i is a given x value in the data set, and $y_{i,\text{calc}}$ is the response predicted at x_i by the best-fit line. The result is a relationship like the one shown in Figure 4.11.

It is possible to obtain the best estimates for m and b by using the method of *least-squares analysis*, as described in Appendix A. This method gives a series of equations (summarized in Table 4.8) that allow the slope and intercept for the best-fit line to be calculated for a particular data set based on the number of points in the data set (n) and the values for each (x,y) pair.[6,8,14] Although these relationships can be used in manual calculations (as shown in the following exercise), they are also the approach used by calculators and computers to obtain best-fit lines. Further information is given at the end of this chapter on how you can create a computer spreadsheet to perform such calculations.

EXERCISE 4.14   Determining the Best-Fit Parameters for a Line

Standards that contain the drug oxymorphone are analyzed and give a calibration curve that appears to follow a straight line. The peak heights measured by liquid chromatography for standards with oxymorphone concentrations of 100, 200, 300, 400, and 500 ng/mL have relative values of 161, 342, 543, 765, and 899, respectively. Determine the best-fit slope and intercept for this line.

SOLUTION

The easiest way to approach this problem is to prepare a table (as shown here) which has separate columns for each x and y pair, as well as for the calculated values of x_i^2 and x_iy_i. After we have made this table, we add the numbers in each column to get Σx_i, Σy_i, $\Sigma x_i y_i$, and Σx_i^2.

x = Drug Concentration	y = Peak Height	$x_i y_i$	x_i^2
100	161	16,100	10,000
200	342	68,400	40,000
300	543	162,900	90,000
400	765	306,000	160,000
500	899	449,500	250,000

$\Sigma x_i = 1,500$ $\Sigma y_i = 2,710$ $\Sigma x_iy_i = 1,002,900$ $\Sigma x_i^2 = 550,000$

After we have these sums, we can place them into Equation 4.28 to get the best-fit slope (m).

$$m = \frac{[n(\Sigma x_i y_i) - (\Sigma x_i)(\Sigma y_i)]}{[n(\Sigma x_i^2) - (\Sigma x_i)^2]}$$

$$= \frac{[5(1,002,900) - (1,500)(2,710)]}{[5(550,000) - (1,500)^2]}$$

$\therefore$ **$m = 1.90$** (Note: The units here would be mL/ng.)

Similarly, we can use these sums with Equation 4.29 to get the best-fit intercept (b).

$$b = \frac{[(\Sigma y_i)(\Sigma x_i^2) - (\Sigma x_i y_i)(\Sigma x_i)]}{[n(\Sigma x_i^2) - (\Sigma x_i)^2]}$$

$$= \frac{[(2,710)(550,000) - (1,002,900)(1,500)]}{[5(550,000) - (1,500)^2]}$$

$\therefore$ **$b = -28$** (Note: This value would have relative peak height units.)

Thus, the best-fit line to our data set is $y = 1.90 \cdot x + (-28)$. Similar calculations based on Equations 4.31 and 4.32 give standard deviations for this slope and intercept of ± 0.08 mL/ng and ± 27.

Whenever you are using the equations in Table 4.8, you should be aware that these equations only give valid results if your data set meets three basic requirements. First, it is assumed that the independent variable (x) is known exactly and that only the dependent variable (y) has any significant random errors. Second, it is assumed that the variation within the y values is truly random and has a consistent type of distribution. And third, it is assumed that the variability (or standard deviation) for y is the same throughout the entire range of x values being fit to the line. Although these assumptions are often true, there are many cases in which they are not. One example

TABLE 4.8   Formulas for Determining the Best-Fit Parameters for a Straight Line ($y_{i,\text{calc}} = m x_i + b$)

Slope (m):
$$m = \frac{[n(\Sigma x_iy_i) - (\Sigma x_i)(\Sigma y_i)]}{[n(\Sigma x_i^2) - (\Sigma x_i)^2]} \quad (4.28)$$

Intercept (b):
$$b = \frac{[(\Sigma y_i)(\Sigma x_i^2) - (\Sigma x_iy_i)(\Sigma x_i)]}{[n(\Sigma x_i^2) - (\Sigma x_i)^2]} \quad (4.29)$$

Standard deviation of all y values (s_Y):
$$s_Y = [\Sigma(y_i - m x_i - b)^2/(n-2)]^{1/2} \quad (4.30)$$

Standard deviation of the slope (s_m):
$$s_m = (n/[n(\Sigma x_i^2) - (\Sigma x_i)^2])^{1/2}(s_Y) \quad (4.31)$$

Standard deviation of the intercept (s_b):
$$s_b = ((\Sigma x_i^2)/[n(\Sigma x_i^2) - (\Sigma x_i)^2])^{1/2}(s_Y) \quad (4.32)$$

Correlation coefficient (r):
$$r = s_{xy}/\sqrt{s_{xx}s_{yy}} \quad (4.33)$$

where:
$$s_{xx} = (\Sigma x_i^2) - [(\Sigma x_i)^2/n] \quad (4.34)$$
$$s_{yy} = (\Sigma y_i^2) - [(\Sigma y_i)^2/n] \quad (4.35)$$
$$s_{xy} = (\Sigma x_iy_i) - [(\Sigma x_i)(\Sigma y_i)/n] \quad (4.36)$$

is when you plot the experimental results of one method versus another, where both x and y would contain random errors. More information on how you can deal with such situations can be found in References 8 and 14.

4.7B Testing the Goodness of a Fit

Once you have a best-fit line, it is essential to check this fit to make sure the line really does give a good description of the data. This process is known as determining the "goodness of fit" of the line. Two tools that help in this process are a *correlation coefficient* and a *residual plot*.

Correlation Coefficients.   The **correlation coefficient** (r) is a number you can calculate for a best-fit line to indicate how well it describes your data. The correlation coefficient is found by using Equations 4.33–4.36 in Table 4.8 and will give a value between –1 and 1. A closely related term is the *coefficient of determination* (r^2), which is equal to the square of the correlation coefficient and has a value between 0 and 1.[8,14] Figure 4.12 shows how the correlation coefficient changes with the size of a data set and the agreement of this set to a best-fit line. In such plots, a value of r equal to 1 or –1 (or $|r| = 1$) represents perfect agreement between the data points and best-fit line, while a correlation coefficient of zero represents a

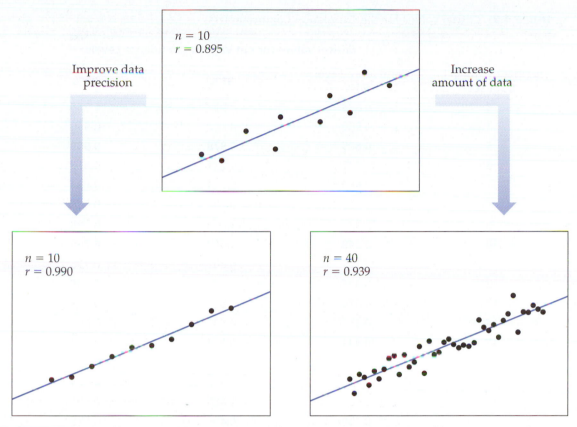

FIGURE 4.12 Change in the correlation coefficient (r) as the number of values in a data set is increased, or as the variability in the data set is increased.

random relationship between the line and the data. A positive value for r means y and x are changing in the same direction (for example, the value of y increases as x increases), while a negative value for r indicates y and x are changing in opposite directions.

Due to the presence of random errors, a correlation coefficient that is exactly equal to 0, –1, or 1 is almost never obtained. Instead, you usually obtain some value that lies between these extremes. This means you must then look at the value of r and determine whether it represents a real fit between x and y. As you might have guessed, statistical tests can help us in making this decision, as shown by the critical values in Table 4.9. To illustrate how you would use this table, let's imagine that we have a set of six (x,y) points with a correlation coefficient of 0.965 when compared to a best-fit line. Table 4.9 shows at the 99% confidence level that a minimum correlation coefficient of 0.917 would be expected for a true linear relationship with six data points. Our actual correlation coefficient of 0.965 is larger than this value, so we can say there is a greater than 99% chance that the best-fit line represents a true fit to our data.

| **EXERCISE 4.15** | **Determining the Correlation Coefficient for a Best-Fit Line** |

What is the correlation coefficient for the best-fit line in Exercise 4.14? What is the probability that this line represents a real trend between the x and y values in the data set?

SOLUTION

The correlation coefficient for this data can be calculated using Equation 4.33. This, in turn, requires that we first use Equations 4.34–4.36 to find s_{xy}, s_{xx}, and s_{yy}. The values of $\Sigma\,x_i^2$, $\Sigma\,x_i$, $\Sigma\,x_i\,y_i$, and $\Sigma\,y_i$ in these equations can be obtained from the table we made earlier in Exercise 4.14. The value of $\Sigma\,y_i^2$ can also be found by using such a table, giving $\Sigma\,y_i^2 = 1{,}831{,}160$. When we place these numbers into Equations 4.33–4.35, we get the following results.

$$s_{xx} = (\Sigma\,x_i^2) - [(\Sigma x_i)^2/n]$$
$$= (550{,}000) - [(1{,}500)^2/5] = 100{,}000$$
$$s_{yy} = (\Sigma\,y_i^2) - [(\Sigma y_i)^2/n]$$
$$= (1{,}831{,}160) - [(2{,}710)^2/5] = 362{,}340$$
$$s_{xy} = (\Sigma\,x_i\,y_i) - [(\Sigma\,x_i)(\Sigma\,y_i)/n]$$
$$= (1{,}002{,}900) - [(1{,}500)(2{,}710)/5] = 189{,}900$$

We then get the correlation coefficient by placing these numbers into Equation 4.33.

$$r = s_{xy}/\sqrt{s_{xx}\,s_{yy}}$$
$$= (189{,}900)/\sqrt{(100{,}000)\cdot(362{,}340)}$$
$$\therefore\ r = 0.998 \quad (\textit{Note: This is a unitless number.})$$

When we compare this result with the values in Table 4.9, we see there is a greater than 99% chance that a correlation coefficient of 0.998 for five data points represents a true trend for this best-fit line.

TABLE 4.9 Critical Values for the Correlation Coefficient (r)*

Number of Points in Data Set	Critical Values for r at Various Confidence Levels		
	Confidence Level = 90%	95%	99%
3	0.988	0.997	1.000
4	0.900	0.950	0.990
5	**0.805**	**0.878**	**0.959**
6	0.729	0.811	0.917
7	0.669	0.754	0.874
8	0.622	0.707	0.834
9	0.582	0.666	0.798
10	**0.549**	**0.632**	**0.765**
11	0.521	0.602	0.735
12	0.497	0.576	0.708
13	0.476	0.553	0.684
14	0.458	0.532	0.661
15	**0.441**	**0.514**	**0.641**
16	0.426	0.497	0.623
17	0.412	0.482	0.606
18	0.400	0.468	0.590
19	0.389	0.456	0.575
20	**0.378**	**0.444**	**0.561**

*These values are for a *two-tailed* comparison and were obtained from R. L. Anderson, *Practical Statistics for Analytical Chemists*, Van Nostrand Reinhold, New York, NY, 1987. The critical values shown here for n points in a linear relationship are for $(n-2)$ degrees of freedom.

Residual Plots. Although the correlation coefficient gives some indication as to how a line fits a set of data, this value should not be used alone in determining the goodness of fit. There are many cases where you can get a good correlation coefficient even if your data set does not really fit your line. An example is given in Figure 4.13. However, there is a way of detecting and avoiding this problem by using a tool known as a **residual plot**.[8,14]

A residual plot is prepared by plotting the difference, or *residual*, between each experimental value for the dependent variable (y_i) and the value predicted by the best-fit line $(y_{i,\text{calc}})$. This plot also often includes a reference line that shows where $(y_i - y_{i,\text{calc}}) = 0$, the result you would get if there was perfect agreement between the data and best-fit line. If the best-fit line is a good description of your data, the residual plot should only have a random distribution of points above and below the line at $(y_i - y_{i,\text{calc}}) = 0$. If the best-fit line does not describe the data, a trend in the residual points should instead appear, signaling that an alternative fit is needed. This approach cannot only test the fit of a straight line to data but can be used for any other type of line you wish to compare to your results.

Original plots

Residual plots

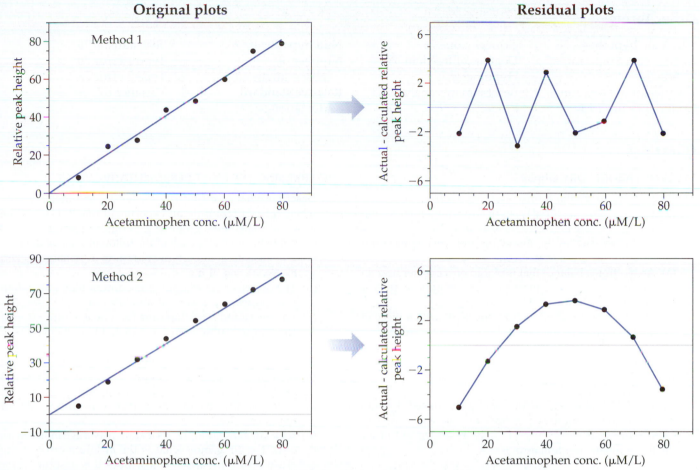

FIGURE 4.13   Examples of residual plots for linear and nonlinear calibration curves when both are compared to best-fit straight lines. The original curves shown for methods 1 and 2 have the same correlation coefficients and number of data points, but the residual plots indicate that only method 1 has a true fit to a linear relationship. The nonrandom pattern seen in the residual plot for method 2 indicates that some other type of equation should be used to describe the response of this technique.

Key Words

Absolute error *65*

Accuracy *65*

Average (mean) *66*

Confidence interval *74*

Confidence level *74*

Correlation coefficient *84*

Error propagation *68*

F-test *80*

Intercept *83*

Linear regression *83*

Normal distribution *72*

Outlier *81*

Paired Student's t-test *79*

Precision *65*

Q-test *81*

Random error *65*

Range *67*

Relative error *66*

Residual plot *86*

Slope *83*

Standard deviation *67*

Standard deviation of the
   mean *73*

Student's t-test *76*

Student's t-value *74*

Systematic error *65*

T_n-test *82*

Other Terms

Questions

TYPES OF LABORATORY ERRORS

1. What is meant by the terms "systematic error" and "random error"? Why are these types of errors important to consider when you are making a measurement?
2. State whether each of the following problems represents a systematic error or random error.
   (a) An electronic balance that is placed on a counter near an area with frequent air movement and vibrations, such as in a high-traffic area in the laboratory
   (b) The use of a 10.00 mL volumetric pipet with a piece of dust lodged in its interior
   (c) The use of a reference weight density of 8.0 g/cm^3 to make a buoyancy correction when a balance has actually been calibrated with a reference weight that has a density of 7.8 g/cm^3
3. Explain why systematic errors can be eliminated through the use of good laboratory practices. Discuss why good laboratory practices can be used to only minimize, but never totally eliminate, random errors.

ACCURACY AND PRECISION

4. What is meant by "accuracy" in a scientific measurement? What is meant by "precision"? How is each of these affected by systematic error and random error?
5. Define the terms "absolute error," "relative error," and "percent relative error." How is each of these terms used in analytical chemistry?
6. The content of ATP (adenosine triphosphate) in a tissue sample is known to be 122 μmol/mL. A new assay for ATP gives the following values for separate analyses of this tissue: 117, 119, 111, 115, and 120 μmol/mL. Calculate the absolute error and relative error for each of these results.
7. A student is asked by an instructor to deliver three samples of water from a Class A 25 mL volumetric pipet. The water delivered by this student is found by mass measurements to have volumes of 25.12, 25.15, and 25.13 mL. If this pipet has a calibrated volume of 25.02 mL, determine the absolute error and relative error for each of these results.
8. A second student is asked by the same instructor as in Problem 7 to deliver water using a volumetric pipet with a calibrated volume of 24.99 mL. Mass measurements of water samples that are delivered from this pipet indicate that this student has delivered volumes of 24.96 mL, 25.01 mL, and 25.04 mL.
   (a) Calculate the absolute error and relative error for each of these results.
   (b) Compare these results with those for the first student in Problem 7. Which student had the most accurate results? Which student had the most precise results?

DETERMINING THE MOST REPRESENTATIVE VALUE

9. Describe what is meant by the term "arithmetic mean" when you are reporting an experiment result. How is the mean calculated for a group of experimental results?
10. Calculate the mean for each of the following sets of data.
   (a) The measured amount of sodium in a graham cracker: 87, 89, 90, and 91 mg
   (b) The detected wavelength of maximum light absorption in a spectrum: 278.8, 279.0, and 279.1 nm
   (c) The measured retention time for a peak eluting from a chromatographic column: 135.2, 134.8, 135.4, 134.2, and 135.0 s
11. A chemist obtains the following results for the average molar mass of a polymer sample: 2.32×10^5, 2.19×10^5, 2.15×10^5, 2.11×10^5, and 2.27×10^5 g/mol.
   (a) What is the mean result for this entire set of measurements?
   (b) If the reference sample has a known average molar mass of 2.21×10^5 g/mol, what is the absolute error of the mean result in Part (a)? How does this absolute error compare with the absolute error that is obtained for the individual values in this data set?
   (c) Calculate the mean and absolute error that would be obtained using only the first two, three, or four measurements in this data set. How do these results compare to those obtained when using all values in the data set?
12. What is the difference between the "experimental mean" and the "true mean" for a data set? How does the number of data points in a set affect each of these two values?

REPORTING THE VARIATION IN A GROUP OF RESULTS

13. Define each of the following terms and explain how they are calculated or found for a set of experimental results.
   (a) Range
   (b) Standard deviation
   (c) Relative standard deviation
   (d) Variance
14. How are the range and standard deviation affected by the number of points in a data set? What are the advantages and disadvantages of using the range or standard deviation in describing data sets?
15. Find the range and standard deviation for each of the following groups of numbers. Compare the ranges and standard deviations that you obtain for each group.
   (a) The measured half-life for a chemical reaction: 32.8, 34.1, 33.7, 32.9, and 33.5 min
   (b) The amount of tin found in a metal sample: 0.21, 0.24, 0.19, and 0.23% (w/w)
   (c) The measured concentration of an HCl solution: 0.01005, 0.01018, and 0.00998 M

16. Calculate the relative standard deviation and variance for each of the data sets in Problem 15. Compare your results with the ranges and standard deviations that were found earlier for these same data sets. Discuss how each of these values reflects the precision of these results.

17. The data shown in the table below were obtained for replicate measurements of a control sample to be used to follow the long-term performance of a clinical instrument for blood glucose testing. Determine the mean, range, standard deviation, and relative standard deviation for these data.

Run Number	[Glucose] (mg/dL)
1	98
2	100
3	103
4	94
5	105
6	88
7	105
8	99
9	112
10	95
11	100
12	93
13	109
14	105
15	99
16	97
17	96
18	97
19	99
20	102

18. An analysis to determine the percent weight of copper in an ore sample gives the following results: 16.54%, 16.30%, 16.64%, 16.67%, 16.70%, and 16.49% (w/w).
(a) What are the mean, range, standard deviation, and relative standard deviation for this set of data?
(b) Calculate the range and standard deviation for this data set using only the first two, three, four, or five of the listed values. How do these results compare with range and standard deviation that you get when using all of the numbers in the data set?
(c) How do the range and standard deviation in Part (b) change as more data points are used to determine these values? Explain any trends that you observe.

19. Why is it necessary to distinguish between the experimental standard deviation and true standard deviation for a data set? How does the number of data points in a set affect each of these two values?

20. What is meant by the "degrees of freedom" when you are calculating a value from experimental data? What are the degrees of freedom when you are determining the mean or the standard deviation for a set of numbers?

THE PROPAGATION OF ERRORS

21. What is "error propagation"? Why is this important to consider in chemical analysis?

22. Describe the equations or general approaches that are used to carry out error propagation during each of the following mathematical operations.
(a) Addition
(b) Subtraction
(c) Multiplication
(d) Division

23. Determine the result and precision of the result for each of the following calculations by using the propagation of errors. Each number in parentheses represents ± 1 standard deviation of the preceding value. Report all final answers using the appropriate number of significant figures.
(a) $0.121 (\pm 0.009) + 2.93 (\pm 0.04)$
(b) $9.23 (\pm 0.03) + 4.21 (\pm 0.02) - 3.26 (\pm 0.06)$
(c) $91.3 (\pm 1.0) \cdot 40.3 (\pm 0.2) \cdot 21.1 (\pm 0.2)$
(d) $185 (\pm 1) \cdot 3.2 (\pm 0.3)/9.1 (\pm 0.1)$
(e) $1 + 6.4 (\pm 0.2) + 36.2 (\pm 0.3)$
(f) $7.53 (\pm 0.1) \times 10^5 \cdot 2.9 (\pm 0.1) \cdot \pi$

24. Describe the equations or general approaches that are used to carry out error propagation during each of the following mathematical operations.
(a) Logarithms
(b) Antilogarithms
(c) Exponents

25. Find the result and precision for each of the following operations by using the propagation of errors. Each number in parentheses represents ± 1 standard deviation of the preceding value. Report all final answers using the appropriate number of significant figures.
(a) $\log[2.0164 (\pm 0.0008)]$
(b) $\text{antilog}[-3.22 (\pm 0.02)]$
(c) $10^{2.384 (\pm 0.011)}$
(d) $2 \cdot \log[7.05 (\pm 0.02)]$
(e) $e^{-1.68 (\pm 0.02)}$
(f) $\ln[12.6 (\pm 0.2)]$

26. What would be the estimated precision for each of the calculations in Problems 23 and 25 if you had used only the number of significant figures as a guide in reporting the final values, as based on the rules given in Appendix A? How do these results compare to those obtained using error propagation? Explain any similarities or differences that you observe in these results.

27. Describe how error propagation can be performed for calculations that involve more than one mathematical operation.

28. Find the result and precision for each of the following operations using the propagation of errors. Each number in parentheses represents ± 1 standard deviation of the preceding value. Report all final answers using the appropriate number of significant figures.
(a) $[4.97 (\pm 0.05) - 1.86 (\pm 0.01)]/21.1 (\pm 0.2)$
(b) $[1.89 (\pm 0.03) \times 10^3 + 2.30 (\pm 0.06) \times 10^3 - 9.8 (\pm 0.2) \times 10^2] \cdot 5.80 (\pm 0.06) \times 10^{-2}$
(c) $6 \cdot 1.00794 (\pm 0.00007) + 2 \cdot 12.0107 (\pm 0.0008)$
(d) $10^{[7.40 (\pm 0.02) - 3.12 (\pm 0.01)]}$
(e) $2 \cdot \log[0.107 (\pm 0.002)/0.158 (\pm 0.003)]$
(f) $(3/2) \cdot e^{-[2.85 (\pm 0.03)/0.103 (\pm 0.002)]}$

29. Estimate the precision of the answers for the calculations in Problem 28 using only the number of significant figures as a guide in reporting the final values, as based on the rules given in Appendix A. How do these results compare to those obtained using error propagation? Discuss any similarities or differences that you observe in these results.

30. The concentration of sodium hydroxide in an aqueous solution is to be found by using an acid–base titration in which a solution containing a known concentration of hydrochloric acid is used as the titrant. The analytical concentration of sodium hydroxide (C_{NaOH}) in the original sample can be found as follows.

$$C_{NaOH} = (C_{HCl} \cdot V_{HCl})/V_{NaOH}$$

where C_{HCl} is the concentration of hydrochloric acid in the titrant, V_{HCl} is the volume of this titrant that is required to react with the ammonia, and V_{NaOH} is the original volume of the sodium hydroxide solution. The titration of a 30.00 ($\pm$ 0.03) mL solution of sodium hydroxide requires 24.37 ($\pm$ 0.04) mL of 0.0783 ($\pm$ 0.0003) M HCl for this analysis. What was the concentration of sodium hydroxide in the original solution? What is the standard deviation of this concentration?

31. The molar mass of some compounds can be determined using the ideal gas law. This is done by weighing a sample of the compound and converting it into a gas in a container with a known volume. The pressure inside the vessel is then measured at a known temperature, and the molar mass of the compound is calculated by using the following formulas.

$$P V = n R T \qquad\qquad MW = m/n$$

where P is measured pressure, V is the volume of the vessel, n is the moles of compound that is being examined, T is the absolute temperature, R is the ideal gas law constant, m is the mass of the substance being studied, and MW is the molar mass of this compound. This experiment is carried out with 0.0500 ($\pm$ 0.0002) g of a compound in a container with a volume of 1.000 ($\pm$ 0.005) L. The pressure after the compound is vaporized is 0.492 ($\pm$ 0.009) J/L when the temperature is 298.2 ($\pm$ 0.5) K. It is also known that R has a value of 8.31451 ($\pm$ 0.00007) J/(mol·K). From these data, what is the expected molar mass and precision of this mass for the compound being studied?

32. Which factor in the preceding problem made the largest contribution to the precision of the final molar mass? Which factor made the smallest contribution to this precision? Based on this information, how would you redesign this experiment to improve the precision of your final result?

33. The change in total standard free energy (ΔG^o) for an oxidation-reduction reaction (as shown below for the reaction of A_{red} with B_{ox}) can be determined under standard conditions by using the following formula that relates ΔG^o to the standard half-cell reduction potentials E_A^o and E_B^o.

$a\,A_{red} + b\,B_{ox} \rightleftharpoons a\,A_{ox} + b\,B_{red} \quad \Delta G = -a\,b\,F\,(E_B^o - E_A^o)$

where a and b are the moles of A and B involved in the reduction–oxidation reaction and F is Faraday's constant. An example of such a reaction is the combination of Fe^{2+} with Ce^{4+},

$$Fe^{2+} + Ce^{4+} \rightleftharpoons Fe^{3+} + Ce^{3+}$$

which is described under standard conditions by the following parameters: $E_A^o = E_{Fe3+/2+}^o = +0.771$ (±0.005) V vs. NHE; $E_B^o = E_{Ce4+/3+}^o = +1.44$ (±0.02) V vs. NHE; $F = 96485.309$ (±0.029) C/mol (or J/(V·mol)).   Calculate

the value of ΔG^o for the Fe^{2+}/Ce^{4+} reaction. Report your answer using the correct number of significant figures.

DESCRIBING THE VARIATION IN LARGE AND SMALL DATA SETS

34. What is a "normal distribution"? Why is this type of distribution useful in describing analytical measurements?

35. What are the main factors that determine the shape of a normal distribution? How does the shape of this distribution change as each of these factors is altered?

36. What relative area in a normal distribution falls within a range of $\mu \pm 1\sigma$? What relative area falls within a range of $\mu \pm 2\sigma$? Why are these two ranges useful to remember when comparing data sets?

37. What is meant by the "standard deviation of the mean"? How does this parameter differ from the standard deviation for a group of results? How are these two parameters related?

38. A new colorimetric method for calcium gives the following results for replicate analyses of a single serum sample: 9.2, 10.5, 9.7, 11.5, 11.6, 9.3, 10.1, 11.2, and 10.8 mg/dL. What are the mean and standard deviation of the mean for this group of values?

39. A chemist wishes to measure the iron in steel with a standard deviation of the mean that is $\pm1\%$ or less. If five replicate assays of a steel sample give a standard deviation for the mean result of $\pm5\%$, how many total assays will be needed to obtain a mean with a precision that is $\pm1\%$?

40. What is a "confidence interval"? What is a "confidence limit"? How are these terms used to describe analytical measurements?

41. What is the Student's t-value? What is a "confidence level"? How are these used to determine the confidence intervals and confidence limits for experimental data?

42. A new herbicide is measured in water samples by using gas chromatography. The following results are obtained for the replicate analysis of a single sample: 3.01 μg/L, 2.92 μg/L, 3.18 μg/L, 3.07 μg/L, and 2.84 μg/L. What are the mean result and the 90% confidence interval for the average herbicide concentration in this sample?

43. The standardization of a sodium hydroxide solution gives the following measurements for the concentration of this solution: 0.0980, 0.1000, 0.0992, 0.0997, and 0.0993 M.
    (a) What are the mean, standard deviation, and standard deviation of the mean for these results?
    (b) Calculate and compare the confidence intervals for the mean at the following confidence levels: 90%, 95%, and 99%. What trend do you observe in these confidence intervals as the confidence level is changed?

44. Explain why using a larger number of values in a data set will allow you to obtain a narrower confidence interval for the mean result of the data set.

45. Why is the proper selection of a confidence level important when you are describing experimental results? Why is a confidence level of 95% often used by analytical chemists?

GENERAL REQUIREMENTS FOR THE COMPARISON OF DATA

46. Define each of the following terms and explain how they are used during the comparison of data.
    (a) Model
    (b) Hypothesis
    (c) Confidence level
    (d) Test statistic

47. What is meant by the term "critical value" in a statistical test? How is this parameter used in such a test?

COMPARING AN EXPERIMENTAL RESULT WITH A REFERENCE VALUE OR OTHER EXPERIMENTAL RESULTS

48. What is a "Student's t-test"? How is this test performed when you are comparing an experimental result with a reference value? How is this test performed when you are comparing two individual experimental results?

49. The ratio of gallium isotopes 69 and 71 ($^{69}Ga/^{71}Ga$) is to be measured by mass spectrometry. The results obtained for eight analyses of a reference sample are 1.52660, 1.52974, 1.52592, 1.52804, 1.52685, 1.52793, 1.53210, and 1.52698. This reference sample is known to have an actual $^{69}Ga/^{71}Ga$ ratio of 1.52810. Are the results of this method the same as the true value at the 95% confidence level?

50. A food chemist wishes to validate a new method that will be used to measure the vitamin C content of food. A reference orange sample is obtained that has a known vitamin C content of 53.2 mg/100 g, or 0.0532% (w/w). Several replicate measurements of this sample by the new method give estimated vitamin C contents of 0.0482%, 0.0471%, 0.0510%, and 0.0495%. Is the mean result of the new method the same as the known content of the reference sample if these values are compared at the 90% confidence level?

51. The Ca content of a powdered mineral sample was analyzed four times by two methods believed to have similar precision. The results are shown below. Are the mean values for these two methods the same at the 90% confidence level?

Percent Ca (w/w)	
Method 1:	0.0271, 0.0282, 0.0279, 0.0271
Method 2:	0.0271, 0.0268, 0.0263, 0.0274

52. An instructor gives two students portions of the same ore sample to examine by gravimetric analysis. Student number 1 gets the following results for the iron content for this sample: 35.1, 33.8, and 34.5 mg/g. Student number 2 obtains values of 32.4, 33.1, and 32.7 mg/g. Are the mean results obtained by these two students the same at the 95% confidence level?

53. Measurements of the amino acid tryptophan in a food sample are performed by gas chromatography (GC) and high-performance liquid chromatography (HPLC), with the following results.

Tryptophan (μM)		
Analysis Number	GC Method	HPLC Method
1	24.2	23.4
2	24.8	23.9
3	24.4	23.5
4	24.6	24.0
5	24.5	23.6
6	24.3	23.7
7	24.5	
8	24.6	

Assuming that the precision of these methods is the same, determine whether their results are equivalent at the 95% confidence level.

54. What is a "paired Student's t-test"? How is this test performed? In what situations might this test be used?

55. The GC and HPLC methods used in Problem 53 are used to examine a set of five different samples that contain various tryptophan concentrations. The mean tryptophan concentrations obtained by GC for these five samples are 12.8, 35.2, 25.1, 15.8, and 31.2 μM. The mean tryptophan concentrations obtained by HPLC for the same five samples are 12.1, 34.7, 25.2, 15.9, and 29.8 μM, respectively. Determine whether these two methods are giving statistically identical results for these samples at the 95% confidence level.

56. A regulatory agency sends six samples to two food-testing laboratories. The following average moisture contents are reported back to the agency by the two laboratories. Are these two sets of results equivalent at the 95% confidence level?

Moisture Content (% w/w)		
Sample Number	Lab Number 1	Lab Number 2
1	10.2	11.0
2	15.3	16.4
3	21.0	21.5
4	13.3	14.1
5	27.8	29.3
6	30.5	32.2

COMPARING THE VARIATION IN RESULTS

57. What is the "F-test"? How is this test performed? In what situations might it be used to examine experimental results?

58. A new employee in a laboratory is asked to measure the protein content of a test sample. This employee obtains values of 75.1, 73.8, 76.9, 70.1, and 74.2 μg/mL for replicate assays. Another, more experienced employee obtains values of 74.8, 75.2, 76.3, 75.1, and 76.8 μ/mL for the same sample.
   (a) Are the mean results obtained by these two employees the same at the 95% confidence level?
   (b) Which of these employees obtained a result with a larger degree of variation? Is this level of precision comparable at the 95% confidence level to that obtained by the other employee?

59. It was assumed in Problem 53 that the GC and HPLC methods had similar precision. Which of these methods had the larger degree of variation? Is this level of precision comparable at the 95% confidence level to that obtained by the other method?

60. A graduate student wishes to improve the precision of a new drug assay. This student finds that one method of sample preparation gives a measured concentration of $1.025 (\pm 0.0020) \times 10^{-6}$ M for five replicate measurements of the drug, where the value in parentheses represents one standard deviation of the overall population of results. A second method of sample preparation gives a drug concentration of $1.017 (\pm 0.011) \times 10^{-6}$ M for five replicate measurements of the same sample. Was there any significant improvement in the precision in going from one approach to the other when these results are compared at the 95% confidence level?

DETECTING OUTLIERS

61. What is meant by the term "outlier"? What strategy should be used in handling potential outliers?

62. Explain how the Q-test can be used for outlier detection. What are the advantages and disadvantages of this method?

63. Describe how the T_n-test can be used for outlier detection. What are the advantages and disadvantages of this technique when compared to the Q-test?

64. A student in a quantitative analysis laboratory is to determine the concentration of a sodium hydroxide solution. The values that are obtained by this student are as follows: 0.0210, 0.0212, 0.0208, 0.0225, and 0.0250 M. Can any of these results be rejected by the Q- test at the 95% confidence level?

65. An analysis of the percent nickel in an ore sample gives the following results for five replicate measurements: 16.54, 16.64, 16.30, 16.67, and 16.70 % Ni (w/w). Can any of these data points be rejected by the T_n-test at the 90% level? Can any of these data points be rejected by the Q-test at the same confidence level? Discuss your findings.

66. State why each of the following situations does not follow good laboratory practices.
    (a) After finding with the Q-test that one point can be removed from a data set, a student uses the Q-test to examine and remove a second data point from the same data set.
    (b) A student performs the T_n-test on a data set, but does not use the suspected outlier to calculate either the mean or standard deviation values for this test.
    (c) A student selects a confidence level based on whether it will allow a particular data point in a set to be rejected by the Q-test.

FITTING EXPERIMENTAL RESULTS

67. What is the purpose of "linear regression"? Why is linear regression useful in analytical chemistry?

68. Define what is meant by the "slope" and "intercept" for a line. Show how these terms are used to provide a general equation for straight line.

69. A plot of absorbance versus concentration gives the following values for samples that contain the protein myoglobin.

Absorbance	Concentration ($\mu g/L$)
0.002	0.0
0.062	10.0
0.125	20.0
0.198	30.0
0.244	40.0

A visual examination of these data suggests these values follow a linear relationship. Calculate the slope and intercept for the best-fit line of these data.

70. The amount of Ca^{2+} in a sample is measured using a calcium ion-selective electrode. The following potentials (E) are obtained for standards containing known Ca^{2+} levels, where $pCa = -\log[Ca^{2+}]$. It is predicted that the response should follow a linear relationship, which appears to be true when a plot of these data is made. Find the slope and intercept of the best-fit line for these values.

pCa	E (mV)
5.00	−53.8
4.00	−27.7
3.00	+2.7
2.00	+31.9
1.00	+65.1

71. What are the standard deviations for the slope and intercept of the best-fit line in Problem 70? According to theory, the response obtained should have a slope of −29.6 (or −59.2/2) mV per unit change in pCa at 25°C. Does this predicted value agree with the slope found in Problem 70? Explain your answer.

72. What is a "correlation coefficient"? Why is a correlation coefficient useful in linear regression? How is this parameter related to the "coefficient of determination"?

73. What is the correlation coefficient for the best-fit line in Problem 69?

74. What is the correlation coefficient for the best-fit line in Problem 70? What is the probability that this line represents a true fit to the data?

75. What is a "residual plot"? How can a residual plot be used to evaluate a best-fit line?

76. Prepare a residual plot using the data and best-fit line from Problem 70. Does this plot support the assumption made in Problem 70 that the data set follows a linear relationship? Explain your answer.

77. A colorimetric method using thiocyanate (SCN^-) as a reagent is used for the determination of Fe^{3+} in water. The following results are obtained for 10 iron standards.

Concentration Fe^{3+} (mg/L)	Absorbance
10	0.205
20	0.396
30	0.612
40	0.816
50	0.987
60	1.160
70	1.251
80	1.385
90	1.489
100	1.565

(a) Prepare a plot of "Absorbance" versus "Concentration Fe^{3+}" for these results. Determine the best-fit lines for this assay using (1) the entire range of standards, and (2) just those standards that contain iron concentrations of 10–50 mg/L. Compare the slopes, intercepts, and correlation coefficients for these two best-fit lines. What can you conclude from this comparison?
(b) Prepare residual plots for the two best-fit lines from Part (a) for the 10–50 mg/L standards and for the entire set of iron standards. What can you conclude by comparing these residual plots?

CHALLENGE PROBLEMS

78. The total serum concentration of testosterone in adult males follows a normal distribution that has an average value of 22.6 nmol/L and a standard deviation of 6.0 nmol/L. If a screening test for synthetic testosterone has a lower cutoff of 40.0 nmol/L, what percent of normal adult males will have a concentration above this value? (*Hint:* See Table 4.2.)

79. The density of sediments in rivers and lakes can vary greatly from one region to the next. In one particular area, the density of 25 sediment samples was determined. The average density of the samples was 2.8 g/mL and the standard deviation of this data set was 0.8 g/mL. Assuming that

a normal distribution is present, what is the probability that a new sample from this same area will have a density between 2.6 and 3.0 g/mL?

80. The variance (V) for a set of results is often calculated by using Equation 4.37.

$$V = \frac{(\sum x_i^2) - (\sum x_i)^2/n}{n - 1} \qquad (4.37)$$

   (a) Using the fact that $V = (s)^2$, show that Equation 4.37 is equivalent to Equation 4.5.
   (b) Use *both* Equations 4.5 and Equation 4.37 along with the relationship $V = (s)^2$ to find the standard deviation and variance for the following set of numbers: 0.0998, 0.1000, 0.0992, 0.0997, and 0.0993. How do the results of Equations 4.5 and 4.37 compare in terms of the values they provide for s and V?

81. A student wishes to examine his or her technique in an acid–base titration for measuring the concentration of ammonia in an aqueous solution.
   (a) The following equations relate the concentration and volume of the ammonia solution (C_{NH3} and V_{NH3}) to the concentration and volume of HCl (C_{HCl} and V_{HCl}) that will be needed to titrate this ammonia solution.

$$(C_{NH_3})(V_{NH_3}) = (C_{HCl})(V_{HCl}) \quad or \quad C_{NH_3} = \frac{(C_{HCl})(V_{HCl})}{V_{NH_3}}$$

   If the student uses an ammonia solution with a volume of 30.00 (± 0.03) mL and the titration of this solution requires 24.37 (± 0.04) mL of 0.0783 (± 0.003) M HCl, what is the expected concentration of ammonia in the original solution? What is the estimated standard deviation for this concentration based on error propagation?
   (b) The same student performs a series of replicate measurements using identical solutions and conditions to those described in Part (a). These measurements provide the following values for the ammonia concentration: 4.25×10^{-3} M, 4.14×10^{-3} M, 4.20×10^{-3} M, and 4.09×10^{-3} M. What are the mean and standard deviation for this set of results?
   (c) Compare the results from Parts (a) and (b). What can you conclude about the accuracy and precision of the student who is performing the titration? Justify your answer.

82. It is common for scientists to perform linear regression by using a calculator or computer program that automatically calculates the slope and intercept for a given collection of (x,y) values. One way this task can be accomplished is by creating a computer spreadsheet to perform such operations. If you are unfamiliar with how to use a spreadsheet program, information on this topic can be found in Appendix C. To illustrate how you can develop a spreadsheet for linear regression, we will use the example given in Exercise 4.14 for the analysis of oxymorphone in urine samples.
   (a) Recall in this example that peak heights of 161, 342, 543, 765, and 899 relative units were measured at oxymorphone concentrations of 100, 200, 300, 400, and 500 ng/mL. Begin your spreadsheet by creating two columns that are labeled "Drug concentration (ng/mL)" and "Peak height." These will be used to represent the x and y values in your data set. Under the column of x values, enter each of your concentrations. Under the column of y values, enter the corresponding response at each concentration. After you have entered these values, use the spreadsheet to create a graph in which the peak heights are represented by the y-axis and the concentrations by the x-axis. As you prepare this plot, follow the good laboratory practices for making graphs that were given in Chapter 2.
   (b) Near the concentration and peak-height columns, add to the spreadsheet a new set of columns that are labeled "x squared" and "x times y." Under each of these new columns, use the spreadsheet to calculate and enter the values x^2 and $x \cdot y$ for each (x,y) pair. This process should give you a table similar to the one used in Exercise 4.14. At the bottom of each column, use the spreadsheet to determine the sum of all entries in that column. This will give you the values of $\sum x_i$, $\sum y_i$, $\sum x_i y_i$, and $\sum x_i^2$. At this point, compare these numbers to those determined in Exercise 4.14 to ensure that all columns and operations in your spreadsheet have been set up correctly.
   (c) In a separate part of the spreadsheet, set up an area in which the numbers you have obtained for $\sum x_i$, $\sum y_i$, $\sum x_i y_i$, and $\sum x_i^2$ are used with Equation 4.28 to get the best-fit slope (m). Also use these values with Equation 4.29 to get the best-fit intercept (b). How do these numbers compare to those found in Exercise 4.14?
   (d) Make a column near the original (x,y) data and label it "Predicted y." In this column, use the best-fit slope and intercept that you found in Part (c) to determine the best-fit value for y at each x value. Add these results to the graph you made in Part (a) by making a line that connects the predicted y values. Compare the fit of this line to the actual y values. Does your data follow a linear relationship? Explain your answer.

83. Modify the spreadsheet developed in the previous problem to provide data on the goodness of fit between a set of (x,y) values and their best-fit line. This includes the addition of calculations to determine the correlation coefficient for the best-fit line and the addition of a residual plot to compare the actual y values to those that are predicted by the best-fit line.
   (a) To determine the correlation coefficient, begin by inserting a new column that calculates the square of each y value in your data set. Label this column "y squared." At the bottom of this column, insert a formula that determines the sum of all the y squared values, which will give the term $\sum y_i^2$. Use this along with the numbers you have already determined in your spreadsheet for $\sum x_i^2$, $\sum x_i$, $\sum x_i y_i$, and $\sum y_i$ to calculate s_{xy}, s_{xx}, and s_{yy} through the use of Equations 4.34–4.36. Finally, set up an area in your spreadsheet that uses Equation 4.33 and these numbers to give the correlation coefficient for your best-fit line. How do your results compare to those in Exercise 4.15? How does the value of this correlation coefficient change as you alter the y values in your original set of numbers?
   (b) To include a residual plot in your spreadsheet, begin by inserting a column that calculates the difference between the initial y values and those that are predicted by your best-fit line. Label this new column "y – Predicted y." This new set of numbers gives the residuals that will be used to examine the agreement between your data and best-fit line. Now make a plot in which the x values are on the lower axis and the corresponding values for "y – Predicted y" are on the left axis. Also include a horizontal line where "y – Predicted y" is equal to zero as a reference. When you have completed your graph, examine the results. Do the (x,y) values still show a reasonable fit to a straight line? Justify your answer.

84. We saw earlier in this chapter that the selection of a confidence interval is related to the size of type 1 and type 2 errors that can occur when using experimental data. Type 1 and 2 errors are

particularly important to consider in clinical studies. As an example, imagine that we conducted a large-scale study of a new analytical method for blood glucose measurements and found that normal individuals gave a result of 100 ± 15 (1 s) mg/dL, with the individual points in this set following a normal distribution. In addition, when the same method was used for individuals with diabetes, it gave results that followed a normal distribution with a mean glucose level of 180 ± 30 (1 s) mg/dL. Based on this study, it was then recommended that the "normal range" for glucose concentrations be defined as 75 to 125 mg/dL. From these results, what is the probability that a normal individual will give a glucose result *outside* of this "normal range" (giving a type 1 error)? What is the probability that a diabetic patient will have a result that falls within the normal range (giving a type 2 error)?

TOPICS FOR DISCUSSION AND REPORTS

85. Obtain further information on athletic drug testing from References 1–5 or other resources. Report on the general approach that is used in this testing and on the special factors that must be considered during the collection and analysis of its samples. What drugs are examined in this process? What is the importance of each drug in terms of how it affects athletic performance or drug detection? What negative effects can such agents have on the body?

86. The Apollo missions to the Moon in the 1960s and 1970s brought many lunar samples back to the Earth for an analysis of their chemical and physical properties. Because of the unique nature and limited amount of material in these samples, it was important to use good laboratory practices during all steps of their analysis.[20] Read about the approaches that were used in this work and report on the comparison and analysis of data that were performed in this study. Include in your discussion such items as the determination of mean values, the detection of possible outliers, and the comparison of results from different laboratories or different methods.

87. One field in which laboratory errors are of ongoing concern is in the analysis of clinical samples. A recent study estimated the relative risk of errors and outliers in several common clinical tests, like those used to measure cholesterol and blood glucose levels.[21] Obtain a copy of this report and discuss its results. From this report, what can you conclude about the importance and occurrence of analytical errors in clinical laboratories?

88. In this chapter, we have looked at several techniques for comparing the mean values or variation in two numbers, using methods like the Student's t-test or F-test. But there are other tests that can be used to make additional types of comparisons or to compare larger groups of numbers. Some examples of these alternative tests are listed below. Using References 8, 13, and 14 or other resources, get more information on one or more of these methods. Discuss how these methods are performed and describe the types of situations in which each is employed.
    a. Duncan's test
    b. Cochran's test
    c. Analysis of variance (ANOVA)
    d. χ^2 (chi-square) test

References

1. M. Fainaru and L. Williams, "Steroids Scandal—The Balco Legacy," *San Francisco Chronicle*, December 24, 2006.
2. R. Kazlauskas and G. Trout, "Drugs in Sports: Analytical Trends", *Therapeutic Drug Monitoring*, 22 (2000) 103–109.
3. W. Clarke, W. Klein, and D. Palmer-Toy, "Athletic Drug Testing," *Therapeutic Drug Monitoring and Toxicology*, 21 (2000) 221–230.
4. M. Fainaru-Wada and L. Williams, *Game of Shadows*, Gotham Books, New York, 2006.
5. G. Zorpette, "All Doped Up—and Going for The Gold," *Scientific American*, 282 (2000) 20–22.
6. J. Inczedy, T. Lengyel, and A. M. Ure, *Compendium of Analytical Nomenclature*, 3rd ed., Blackwell Science, Malden, MA, 1997.
7. D. R. Lide, Ed. *CRC Handbook of Chemistry and Physics*, 83rd ed., CRC Press, Boca Raton, FL, 2002.
8. P. C. Meier and R. E. Zund, *Statistical Methods in Analytical Chemistry*, Wiley, New York, 1993.
9. R. L. Plackett and G. A. Barnard, Eds. *"Student": A Statistical Biography of William Sealy Gosset*, Clarendon Press, Oxford, 1990.
10. C. B. Read, "William Sealy Gosset," In *Leading Personalities in Statistical Sciences: From the Seventeenth Century to the Present*, N. L. Johnson and S. Kotz, Eds., Wiley, New York, 1997, pp. 327–329.
11. Student, "The Probable Error of a Mean," *Biometrika*, 6 (1908) 1–25.
12. W. H. Beyer, Ed., *CRC Handbook of Tables for Probability and Statistics*, 2nd Ed., The Chemical Rubber Co., Cleveland, OH, 1968.
13. R. C. Graham, *Data Analysis in the Chemical Sciences*, VCH, New York, 1993.
14. R. L. Anderson, *Practical Statistics for Analytical Chemists*, Van Nostrand Reinhold, New York, 1987.
15. W. J. Youden, *Experimentation and Measurement*, NIST Special Publication 672, U.S. Department of Commerce, Washington, DC, 1991.
16. W. J. Dixon, "Analysis of Extreme Values," *Annals of Mathematics and Statistics*, 21 (1950) 488–506.
17. R. B. Dean and W. J. Dixon, "Simplified Statistics for Small Numbers of Observations," *Analytical Chemistry*, 23 (1951) 636–638.
18. D. B. Rorabacher, "Statistical Treatment for Rejection of Deviant Values: Critical Values of Dixon's 'Q' Parameter and Related Subrange Ratios at the 95% Confidence Level," *Analytical Chemistry*, 63 (1991) 139–146.
19. "Standard Practice for Dealing with Outlying Observations," E 178-94, *Annual Book of ASTM Standards*, American Society for Testing and Materials, Philadelphia, PA, 1994, pp. 91–107.
20. G. H. Morrison, "Evaluation of Lunar Elemental Analyses," *Analytical Chemistry*, 43 (1971) 22A–31A.
21. D. L. Witte, S. A. VanNess, D. S. Angstadt, and B. J. Pennel, "Errors, Mistakes, Blunders, Outliers, or Unacceptable Results: How Many?" *Clinical Chemistry*, 43 (1997) 1352–1356.

Chapter 5

Characterization and Selection of Analytical Methods

Chapter Outline

5.1 INTRODUCTION: THE VINLAND MAP

In October 1965 scholars from Yale University announced they had found a map believed to be one of the earliest known depictions of the New World.[1] An inscription on this map (shown in Figure 5.1) claimed it had been prepared around 1440. This map included a drawing of Europe, Greenland, parts of Africa and Asia, and an island to the west labeled "Vinland." This last region had a remarkable resemblance to the coast of North America which, according to tradition, had been reached by the Norseman Leif Ericson around 1000 A.D. This map, now known as the Vinland map, has been the subject of controversy ever since its discovery.[2–6]

Several analytical methods have been used to determine the authenticity of this map. One research group using radiocarbon dating found that the parchment of the map matched its alleged age.[4] A second group reported that the elemental composition of various regions on the map was comparable to other documents from the same era.[5] Other scientists who used microscopy to examine single particles taken from the map's ink and parchment found a man-made form of titanium oxide (anatase) that was not available until 1917 (see Box 5.1).[3,6] Thus, some methods supported the authenticity of the map while others indicated it was a fraud.

The case of the Vinland map illustrates the importance of considering both the sample and the measurement technique when you are selecting a procedure for chemical analysis. Table 5.1 contains several questions you should ask during this process. Sample-related questions deal with issues like the chemical or physical nature of the analyte and matrix, such as whether the analyte is in an appropriate form for study. Related questions concern the

TABLE 5.1   Questions to Consider When Choosing a Chemical Analysis Method

Sample-Related Questions

What type of information is needed about the sample?

What types of analytes and samples are to be examined?

What range of analyte concentrations or properties must be measured?

If multiple forms of the analyte are present, which forms are to be measured?

What other substances in the sample might interfere in the method?

What type of sample preparation will be required?

How many samples will have to be analyzed?

Are there any special safety factors that must be considered for the samples?

Method-Related Questions

What type of information will the method provide?

Is the method capable of working with the desired samples and analytes?

What type of accuracy and precision are required?

What is the expected detection range for the method?

How selective and robust will the method be?

How fast will the method be able to analyze samples?

How many samples will the method be able to process?

What level of training or experience is required for the method?

What equipment is needed and what costs will be associated with this technique?

Will there be any special safety precautions that must be followed in this method?

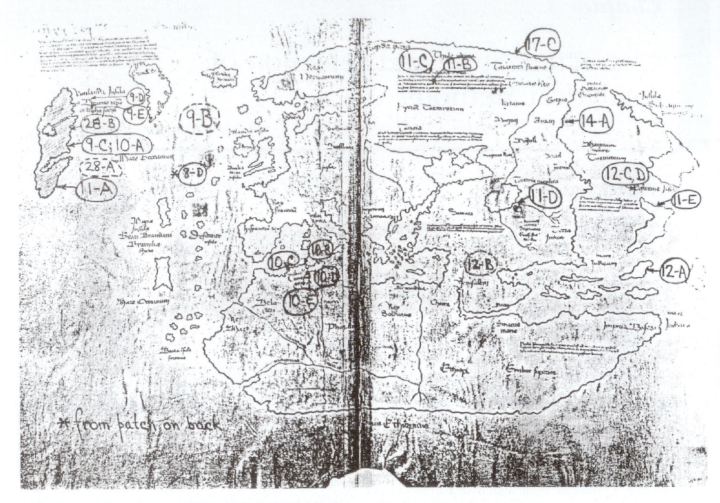

FIGURE 5.1   The Vinland map. The circled numbers and letters represent various regions that were sampled to obtain particles from the map's surface. (Reproduced with permission from W. C. McCrone, "The Vinland Map," *Analytical Chemistry*, 60 [1988] 1009–1018.)

amount of analyte that is to be measured and whether other interfering chemicals are present. In the case of the Vinland map, the sample was a potentially valuable material with a heterogeneous composition, requiring selective approaches that could deal with such an artifact without destroying it.

Method-related questions you should ask before conducting a chemical analysis include "Is this technique capable of working with the desired sample?" and "Does this method have sufficient accuracy and precision for examining the analyte?" In addition, you should think about whether the method is capable of working with the expected range of analyte concentrations and whether it is sufficiently fast, selective, and stable for your application. For the Vinland map, this included the need for a nondestructive method that could work with a nonuniform sample. The ability to discriminate between different forms of the analyte (an approach known as *speciation*, as used with the Vinland map to detect anatase) was also important. In this chapter we consider various ways of characterizing analytical methods to help answer these and related questions as we develop new procedures for chemical measurements.

5.2 METHOD CHARACTERIZATION AND VALIDATION

To help in the selection of an analytical method, you need to be familiar with the properties that are used to characterize such techniques. These properties are known as the method's *figures of merit* and include things like accuracy, precision, and the usable assay range. Many common analysis methods have well-established figures of merit. However, if you are developing a new procedure, the figures of merit will have to be determined before you can use the technique for routine analysis. The process of characterizing an analytical technique and proving that it will fulfill its intended purpose is known as **method validation**.[9,10] We now look more closely at this process and learn how to characterize and validate analytical methods.

5.2A Accuracy and Precision

Accuracy and precision are often the two most important factors to consider when choosing an analytical method. We learned in Chapter 4 that accuracy refers to the degree to which an experimental result approaches the true

BOX 5.1
A Closer Look at Small Samples

The analysis of sample particles that are less than approximately 1 mg in size often requires special analytical techniques and sample handing methods. One powerful approach for learning about the composition of these small samples is *optical microscopy* (see Figure 5.2). This is a method in which a microscope is used to visually examine a sample. Optical microscopes have been used for many years in biology to examine cells and microorganisms, but they can also be used for chemical analysis. This type of analysis is made possible by the fact that many small chemical particles have a relatively pure composition and a distinct appearance. Thus, just as it is easy to identify bacteria or pollen grains under a microscope, it is also possible to use a microscope to identify and study small chemical particles.[3,7,8]

The use of an optical microscope to examine particles taken from the Vinland map is one of many cases in which this approach has been employed.[3] Another common application of microscopes in chemical analysis is in the identification of asbestos fibers.[7,8] To the unaided eye, a sample of asbestos may look like many other types of fibers. As is shown in Figure 5.2, however, it is possible with a microscope to easily identify the presence of asbestos in a sample.

Many chemical and physical properties of small samples can be learned by using traditional microscopes or more specialized microscopes that use polarized light or heated stages. Properties that can be examined by microscopy include a particle's shape, size, color, homogeneity, refractive index, optical activity, melting point, freezing pattern, chemical reactions, and polymorphism. Chemical reactions, such as those used to stain bacteria, can be used to help identify particles. It is also possible to combine an optical microscope with other techniques of chemical analysis. A common example is the use of an optical microscope with fluorescence detection. Other microscopes can be used to conduct infrared spectroscopy on small-scale samples.

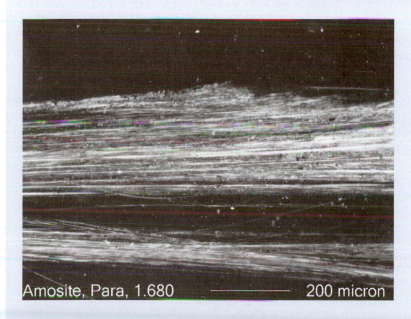

Amosite, Para, 1.680   200 micron

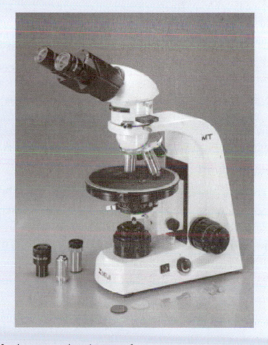

FIGURE 5.2   A polarized light/phase contrast microscope used for the detection of asbestos and an image of brown asbestos, or amosite. (The image of the MT6820 microscope is reproduced with permission from Meiji Techno; the asbestos image is reproduced with permission from MicrolabNW.)

value, while precision refers to the reproducibility of the results. In chemical analysis it is always desirable to have an accurate and highly precise technique so that the results will provide a good representation of the sample. Thus, accuracy and precision are usually the first properties examined during the method-validation process.

Spiked Recovery and Correlation Studies.   One way the accuracy of a technique can be determined is to examine reference samples with known properties or amounts of analyte.[9,10] The absolute or relative error is then calculated by comparing the measured result for the reference sample with the known value. This is relatively easy to do if you are working with a common type of sample or an analyte for which you can obtain a *certified reference material* (CRM). A CRM is a material that has documented values for its chemical content or physical properties. Such materials can be obtained from various suppliers or agencies like the U.S. National Institute of Standards and Technology (NIST).

A **spiked recovery study** can be used to evaluate the accuracy of an analytical method when you are working with an analyte or sample for which no good reference materials are available.[9,10] This type of study is conducted by taking a typical sample and "spiking" a portion of it with a known amount of the analyte. The amounts of analyte in the original and spiked samples are then measured. The difference in these values is then compared to the amount added to the spiked sample. This process allows the **percent recovery** to be calculated, which is a measure of the method's accuracy.

$$\text{Percent recovery} = 100 \cdot \frac{\text{(Change in the amount of measured analyte)}}{\text{(Amount of analyte spiked into the sample)}} \quad (5.1)$$

The closer this recovery is to 100%, the more accurate the measurement will be. Ideally, the percent recovery for a method should be determined at several analyte concentrations because the recovery may vary as you change the amount of analyte in a sample.

EXERCISE 5.1   Determining the Percent Recovery for an Analyte

Serum known to have an insulin concentration of 11.2×10^{-6} IU/mL is spiked by placing 20.0×10^{-6} IU of insulin into 2.00 mL of this sample. When this spiked sample is analyzed, the measured concentration of insulin is 20.5×10^{-6} IU/mL. What is the percent recovery for insulin in this assay?

SOLUTION

The addition of 20.0×10^{-6} IU insulin to a 2.00 mL sample will increase the insulin concentration by $(20.0 \times 10^{-6}$ IU$/2.00$ mL$) = 10.0 \times 10^{-6}$ IU/mL. The measured change in insulin concentration was $(20.5 \times 10^{-6}$ IU/mL $- 11.2 \times 10^{-6}$ IU/mL$) = 9.3 \times 10^{-6}$ IU/mL. The calculated percent recovery for insulin in this method would then be as follows.

$$\therefore \text{ Percent Recovery} = 100 \cdot \frac{9.3 \times 10^{-6} \text{ IU/mL}}{10.0 \times 10^{-6} \text{ IU/mL}} = 93\%$$

This result indicates that the method gave a slightly lower-than-expected value. If low recovery is noted for many other samples, it is a sign that a systematic error may be present in the method.

Another way accuracy can be evaluated is by taking a group of several samples and measuring the amount of analyte in each by using both your method and a second, more established technique.[7,8] This approach is known as a *correlation study*. One way you can compare the results of these two methods is to use a paired Student's *t*-test, as described in Chapter 4. A second approach is to use a **correlation chart**.[11] This is a graph in which the results of your new method are plotted versus those of the reference technique. If the two techniques have identical or similar results, their correlation chart should give a linear relationship with a slope near one and an intercept near zero (see Figure 5.3). If there is a systematic error in one of the methods, this will cause a deviation from the expected behavior.

Precision Measurements.   The precision of an analytical technique can be examined by analyzing several portions of the same reference sample. The absolute or relative standard deviation of this group of results is then calculated, as discussed in Chapter 4, and used as an index of the method's precision. This calculation should be done with several samples that contain different amounts of the analyte, because the precision of a method can show a large change with analyte content. This effect is illustrated in Figure 5.4 by using a diagram known as a **precision plot**.[11,12] The purpose of a precision plot is to give a visual representation of how the precision of a method changes as the measured property or amount of an analyte is varied. In the example shown in Figure 5.4, the assay's absolute standard deviation increases with the concentration of analyte, while the relative standard deviation first decreases and then increases. This behavior is seen in many analytical methods and indicates the need to measure the precision over a wide range of conditions.

Just as there are many factors that can influence the performance of an analytical method, there are also many ways of describing how these factors affect the technique's reproducibility. A summary of terms that are used

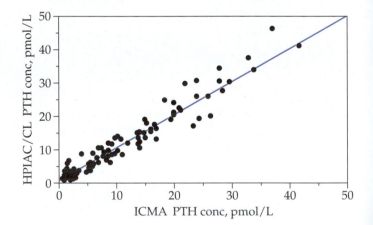

FIGURE 5.3   An example of a correlation chart, based on the analysis of parathyroid hormone (PTH) in various samples by both high-performance immunoaffinity chromatography with chemiluminescence detection (HPIAC/CL) and an immunochemiluminometric assay (ICMA). The solid line shows the best-fit that was obtained for 130 plasma samples analyzed by both methods. The best-fit line in this case had a slope of 0.99 ($\pm$ 0.04) and an intercept of 0.7 ($\pm$ 0.6) pmol/L. From this graph and this best-fit line, it can be seen that the methods represented on the *y*-axis and *x*-axis gave comparable results. (Reproduced with permission from D. S. Hage, B. Taylor, and P. C. Kao, "Intact Parathyroid Hormone: Performance and Clinical Utility of an Automated Assay Based on High-Performance Immunoaffinity Chromatography with Chemiluminescence Detection," *Clinical Chemistry*, 38 [1992] 1494–1500.)

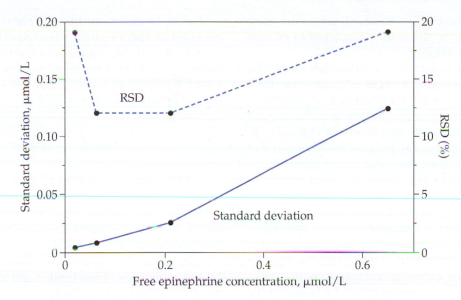

FIGURE 5.4   A precision plot showing the standard deviation and relative standard deviation (RSD) obtained in an assay for the measurement of free epinephrine in urine. This graph indicates the range of analyte concentrations that can be examined with a desired level of precision, thus helping determine the usable range of the assay and its upper and lower limits of detection. (Generated with data obtained from J. Wassell, P. Reed, J. Kane, and C. Weinkove, "Freedom from Drug Interference in New Immunoassays for Urinary Catecholamines and Metanephrines," *Clinical Chemistry*, 45 [1999] 2216–2223.)

for this purpose is given in Table 5.2.[9–11] All these items are reported using an absolute or relative standard deviation, but they include different variables. One of the most common of these is the *within-run precision*, which is a measure of how the analysis of a single sample can fluctuate during one analysis session, or "run."

EXERCISE 5.2   Estimating Within-Run Precision

Three measurements of the titanium content in a narrow region of the Vinland map gave surface concentrations of 1.9, 2.7 and 2.3 ng/cm^2 in back-to-back measurements of this sample.[5] What was the within-run precision for this analysis?

SOLUTION

The within-run precision is found by determining the standard deviation (s) or relative standard deviation (RSD) of the data, as described in Chapter 4, along with a calculated mean of 2.3 ng/cm^2 for this set of results.

$$\therefore s = \sqrt{\frac{(2.3 - 1.9)^2 + (2.3 - 2.7)^2 + (2.3 - 2.3)^2}{(3 - 1)}}$$

$$= \textbf{0.4 ng/cm}^2$$

$$\therefore \ \textbf{RSD} \ (\%) = 100 \cdot \frac{0.4 \ \text{ng/cm}^2}{2.3 \ \text{ng/cm}^2} = \textbf{17\%}$$

Thus, the within-run precision was ± 0.4 ng/cm^2 or $\pm 17\%$ at a mean titanium level of 2.3 ng/cm^2.

5.2B Assay Response

Another key characteristic of an analytical method is the way its signal (or **response**) changes with the amount of analyte or property that is being measured. The response is described through the use of parameters such as the limit of detection, range, and sensitivity or selectivity.

Limits of Detection.   The lowest or highest amount of analyte that can be detected by an analytical method is described by using a **limit of detection (LOD)**.[9–11] The *upper limit of detection* is the largest amount of analyte that

TABLE 5.2   Types of Precision Used to Characterize Analytical Methods

Type of Precision	Property Measured
Within-run precision	Variation obtained for the same sample during a single run of an assay
Within-day precision	Variation obtained during a single day, usually over the course of several runs
Day-to-day precision	Variation obtained over several days
Interoperator precision	Variation obtained with a single method and sample, but by different analysts
Interlaboratory precision	Variation obtained with a single method and sample, but by different laboratories

can reliably be measured by a method, while the *lower limit of detection* refers to the smallest amount of analyte that can be reliably measured.

The lower limit of detection is often determined by comparing the variation in a method's response for a *blank sample* (that is, a sample that contains no analyte) to the response seen for materials known to contain small amounts of the analyte. This comparison is made by calculating a **signal-to-noise ratio (S/N)**, as illustrated in Figure 5.5. In this process, the random variation in the blank signal is referred to as the *noise*, which is usually measured by using the peak-to-trough variation or standard deviation of the background response. Similarly, the net change in response measured between this background response and the response for a sample known to contain the analyte is called the *signal*. The S/N is determined by dividing the value for the signal by the noise. The larger this ratio is, the easier it will be to detect the analyte in a sample.

A S/N of two or three (S/N = 2:1 or 3:1) is often considered the smallest value for this index that will give a measurable signal for an analyte.[9,10] Other S/N values, however, can also be used to determine the lower limit of detection. For example, many drug-testing laboratories use a S/N of 3.3:1, as shown in the following relationship.[13]

$$LOD = 3.3\,(s_b/m) \tag{5.2}$$

This equation makes it possible to estimate the lower limit of detection by using the best-fit line for a linear calibration curve. The value of s_b in Equation 5.2 is the standard deviation of the blank response or the standard deviation of the intercept for a calibration curve (representing the "noise"), while m is the slope of the response near the lower limit of detection (see Chapter 4 for a review of how s_b and m can be determined for a best-fit line).

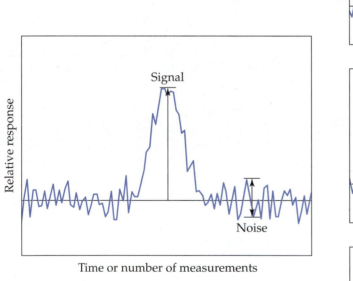

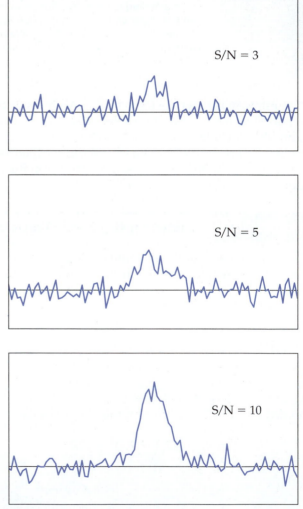

FIGURE 5.5 The determination of a signal-to-noise ratio (S/N). The signal in this example is given by the response measured above the average background in the absence of any analyte. The noise is given by the random variation in the background signal alone, as measured by using either the standard deviation of the background signal or the peak-to-trough variation in the background. Once the S/N value has been determined, this provides a measure of how many times larger the signal is than the background noise. This helps the analyst to determine at what point a reliable signal can truly be said to be present in a measurement.

A similar parameter used to describe the response of a method is the **limit of quantitation (LOQ)**. The LOQ represents the smallest or largest amount of analyte that can be measured within a given range of accuracy and/or precision. To illustrate this concept, let's look again at the precision plot in Figure 5.4. If our goal is to make measurements with a precision better than ±15%, then 0.05 and 0.4 μmol/L are the lower and upper LOQs for this assay. Even though we could detect lower or higher amounts, these results would not be useful because they do not have the required precision of at least ±15% that is needed in this particular example.

Besides using a precision plot, you can determine a method's LOQ by using a signal-to-noise ratio. This determination is performed in the same way as described for the LOD, except that a S/N of 10:1 is now used as the cutoff value.[13]

$$LOQ = 10 \, (s_b/m) \qquad (5.3)$$

Some practice in determining the lower LOD and LOQ for an assay is provided in the next exercise.

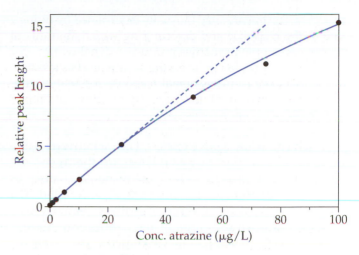

FIGURE 5.6 Calibration curve for atrazine measured in water samples by high-performance liquid chromatography. Atrazine is a herbicide used throughout the world for the control of broadleaf weeds in crops. The dashed line shows the best-fit line that was obtained by this method at atrazine concentrations below 25 μg/L. The slope and intercept of this best-fit line are given in Exercise 5.3. The solid line shows the actual response that was obtained at all observed concentrations. (Reproduced with permission from D. H. Thomas, M. Beck-Westermeyer, and D. S. Hage, "Determination of Atrazine in Water Using Tandem High-Performance Immunoaffinity Chromatography and Reversed-Phase Liquid Chromatography," *Analytical Chemistry*, 66 [1994] 3828–3829.)

EXERCISE 5.3	**Determining the Limits of Detection and Quantitation for a Method**

A new method for determining the herbicide atrazine in water results in a linear calibration curve at low-to-moderate levels of this analyte (see Figure 5.6). The best-fit line for this curve is $y = 0.207 \, (\pm 0.001) \cdot x + 0.008 \, (\pm 0.006)$, where y is the measured response, x is the concentration of atrazine, and the values in parentheses are the standard deviations of the best-fit slope and intercept. From these data, estimate the lower limit of detection and lower limit of quantitation for this assay.

SOLUTION

The value of s_b in this case is ±0.006 (the standard deviation of the intercept) and m is 0.207 (the best-fit slope). Placing this information into Equations 5.2 and 5.3 allows us to calculate the values of LOD and LOQ.

$LOD = 3.3 \, (s_b/m)$ $LOQ = 10 \, (s_b/m)$

$\quad = 3.3 \, (0.006)/(0.207)$ $\quad = 10 \, (0.006)/(0.207)$

$\quad = 0.09\underline{6} = \textbf{0.1} \, \boldsymbol{\mu}\textbf{g/L}$ $\quad = 0.2\underline{9} = \textbf{0.3} \, \boldsymbol{\mu}\textbf{g/L}$

We can see here that the lower limit of quantitation is larger than the limit of detection. This occurs because stricter criteria (S/N = 10 vs. S/N = 3.3) were used in estimating LOQ compared to LOD.

Assay Range and Linearity. The "range" of an analytical method is closely related to the limit of detection. The range refers to the entire set of analyte concentrations or sample properties that fall within the desired levels of precision and accuracy for an analytical method.[9,10] The largest possible range that can be used with an analytical technique extends from the lower limit of detection to the upper limit of detection (see Figure 5.7) and is known as the **dynamic range**. For instance, the assay in Figure 5.6 has a dynamic range that goes from the lower limit of detection (0.1 μg/L) up to at least 100 μg/L, the largest amount of atrazine that gives a measurable change in response.

The dynamic range can be divided into smaller regions that represent specific types of responses. One of these regions is the **linear range**, which is the portion of a method's range that gives a linear dependence between the response and the amount of an analyte or the measured

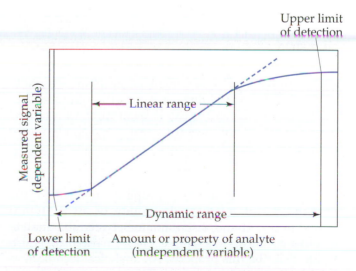

FIGURE 5.7 General relationship between the dynamic range, linear range, and limits of detection for an analytical method.

property. The linear range is usually determined by fitting a line to the response and determining which amounts of analyte or sample properties produce a signal within ±5% or 10% of this line. In Figure 5.6 the best-fit line has less than a 5% difference from the actual response for analyte concentrations between 0.1 and 32 µg/L, which would represent the linear range for this assay.

Sensitivity and Specificity. The **sensitivity** of a method is another factor related to the response. Sensitivity is a measure of how the response changes as the amount of analyte or sample property is varied. The most common way of describing the sensitivity of a method is to use the slope of the calibration curve.[10] This approach gives a parameter known as the *calibration sensitivity*. For instance, a method that follows a linear response ($y = mx + b$) will have a calibration sensitivity equal to the best-fit slope m. If the assay's response is not linear, the slope and calibration sensitivity will be different at each point in the nonlinear region of the calibration curve.

Although some people use the terms "lower limit of detection" and "sensitivity" interchangeably, these refer to quite different things. The lower limit of detection refers to the smallest amount of analyte that can be examined by a method, while the sensitivity refers to the smallest *change* in the amount of analyte that can be detected. As a result, a method with a steep slope in its calibration curve will have a high sensitivity, because this makes it possible to see small differences in an analyte's concentration. A method with low sensitivity would have a small slope in its calibration curve, which would make it more difficult to discriminate between two samples that contain similar amounts of the analyte.

detect a wide range of compounds are often referred to as *general* or *universal methods*.

The specificity of a technique is determined by comparing the method's response for the analyte to its response for other compounds that may be present in the sample. One way you can evaluate the specificity of a method is to use an **interference plot** (Figure 5.8). This is a graph in which the apparent amount of analyte that is measured by a method is plotted versus the amount of a second substance that has been added to the sample. Such a graph makes it possible to determine whether the added agent will create any problems in the assay. To illustrate this, Figure 5.8 indicates that the measurement of bilirubin by Method 1 will have a 10–85% error when 0.06–6.4 g/L hemoglobin is present in the sample, while Method 2 will have a much smaller error under these same conditions.

Another approach for describing the specificity of an analytical method is to use a *selectivity coefficient*.[14] The selectivity coefficient is a ratio that compares the signals that are produced by one substance versus another in the same method. As an example, suppose we have a method that has a linear response for compounds A and B. The selectivity coefficient of this method for chemical B versus A ($k_{B,A}$) would be equal to the ratio of the calibration slopes for these two agents, or $k_{B,A} = (m_B/m_A)$. This concept can be illustrated by using the data in Figure 5.8. This figure shows that Method 1 has a signal for hemoglobin at a concentration of 0.25 g/L that is roughly equal to the signal produced by 8.0×10^{-3} g/L bilirubin. Thus, Method 1 has a selectivity factor for bilirubin versus hemoglobin that is given by the ratio $(0.25 \text{ g/L})/(8.0 \times 10^{-3} \text{ g/L}) = 31$, which

---

EXERCISE 5.4	**Determining the Sensitivity of a Method**

What is the calibration sensitivity in Figure 5.6 for a sample containing 20 µg/L atrazine? How will this sensitivity be affected as the concentration of atrazine is increased?

SOLUTION

A concentration of 20 µg/L is in the linear range of this assay, so the calibration sensitivity at this point is equal to the slope of the best-fit line, or $m = 0.207$ L/µg. The calibration sensitivity will become lower at concentrations above 32 µg/L as we leave the linear range and the slope begins to decrease at higher analyte concentrations.

Specificity is another property related to the response[9,10] and refers to the ability of an analytical method to detect and discriminate between the analyte and other chemicals in a sample. The term *selectivity* is also used to describe this property. An analytical technique is said to be "specific" or "selective" if it responds to a single analyte or a small group of chemicals. Procedures that

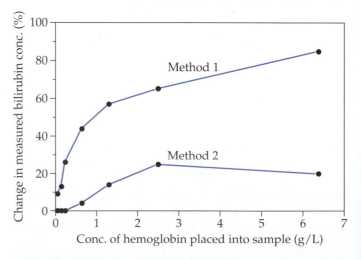

FIGURE 5.8 The effects of having hemoglobin in a serum sample on the measurement of bilirubin by two different analytical methods. Bilirubin is a normal breakdown product of hemoglobin and, like hemoglobin, has the ability to strongly absorb visible light. (Generated based on data from M. G. Scott, B. W. C. Lau, and J. W. H. Ladenson, "Improved Total Bilirubin Method for the Olympus AU 5000 that Decreases Interferences by Hemolysis or Azotemia," *Clinical Chemistry*, 34 [1988] 1921.)

means this method is over 30 times better at detecting bilirubin than hemoglobin in such samples.

5.2C Other Properties of Analytical Methods

An additional property that should be characterized for an assay is the speed at which the method can process samples. One factor to consider is the *overall analysis time* of the method, which is the time needed for all steps in the preparation and analysis of a sample. A related factor is *sample throughput*. This term refers to the number of samples that can be processed by a method in a given period of time. When describing this property, a technique with "high throughput" is one that can examine many samples in a short period, while a "low throughput" approach is one that can measure only a small number of samples in the same amount of time.

The types of samples that can be examined by a method also must be considered. Some methods require that a chemical be in the gas phase before it can be measured, while others may require a liquid or solid. Sometimes the amount of available sample is too small or the concentration too low for analysis by a particular method. If this occurs, either an alternative approach must be found or the properties of the analyte and matrix must be adjusted to meet the requirements of the method, as we will discuss in Section 5.4.

Other items you should consider during method selection are the cost, availability, and ease-of-use of a technique. The total cost should include the funds required to purchase and maintain any equipment needed for the analysis, as well as the cost of labor and supplies that are required by the method. Availability and ease of use are both related to cost because it will often cost less to use a method for which you already have the necessary equipment and trained personnel. *Robustness*, which refers to the ability of a technique to provide a consistent response when small variations are made in its experimental conditions, is another important characteristic of an analytical method.[9,10] These variations might include small changes in the temperature, reaction times, or reagent composition. It is always desirable to have a highly robust method in that fewer errors will be introduced into the results when such variations do occur.

5.3 QUALITY CONTROL

When you are working in an analytical laboratory, you not only need to know how to use various methods, but you must be able to determine when an assay is not working properly. The process of monitoring the routine performance of an analytical method is known as **quality control**.[15]

5.3A General Requirements for Quality Control

There are three items used in any quality-control process. The first of these is a **control material** (or "control"). This is a substance you will analyze periodically by an analytical method to determine whether the procedure is working in a consistent manner. There are several potential sources of control materials, including government agencies, national associations, and commercial suppliers. Control materials can also be made within your own laboratory, as might be done when you are measuring new types of chemicals or samples.

There are several features that should be present in any control material. First, the control material must contain the analyte of interest, preferably at levels similar to those in other samples that will be analyzed. Second, the control material should have a matrix similar to that of the other samples to avoid any differences due to interference effects. Third, a sufficient quantity of the control material should be available for long-term use and it should be possible to store this material for extended periods of time. Although some control materials might be stable under ordinary laboratory conditions, many may require a dry storage area or the use of refrigeration, freeze-drying, or special lighting to prevent their degradation.

The second part of a quality control program is a **control chart**. This is a graph that uses the results obtained with a control material to follow the performance of an analytical method over time (see Figure 5.9). This chart can indicate when a change has occurred in the systematic or random errors of a method by comparing the result for the control material to the range of results that would normally be expected for the same material. If a new result is outside of this range, the analytical method should be examined to ensure it is working properly before any other samples are tested.

Error assessment is the third part of a quality-control program. This is the process of identifying all sources of errors that can occur in an analytical method and determining how to correct these errors. Several tools can be used to aid in error assessment. For instance, a flow chart can show the individual steps in a measurement and be used to identify possible problems that might occur in each step. Another valuable tool might be a list of all major sources of errors that can appear in a method, along with a description of how each error will affect the results of the measurement. The proper training of laboratory workers further aids in error assessment, because this makes it easier for these workers to identify and correct problems as they arise.

5.3B Preparing and Using Control Charts

Let's now look more closely at one of the key components of a quality-control program, the **control chart**. The first step in making a control chart is to obtain a control material with the analyte in the proper matrix and at appropriate concentrations. Next, this control material is analyzed many times (typically, 20–30) by the method to be evaluated, using identical conditions to those that will be utilized for testing samples. You then calculate the mean and standard deviation of the results for this control material, which are then used to construct a control chart for the analytical technique.

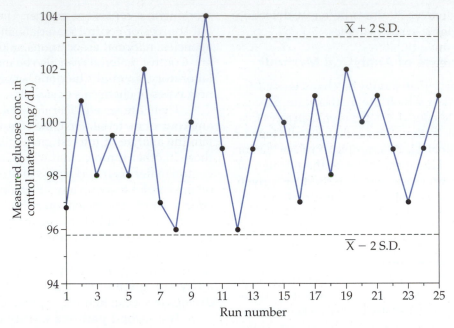

FIGURE 5.9 An example of a Levey–Jennings control chart.

Figure 5.9 shows a plot that is known as a *Levey–Jennings chart* or *individual chart*.[10,15] This chart is used in situations where just one measurement per analyte is performed on a sample. The *y*-axis in this plot represents the additional measurements that are later made of the control material, and the *x*-axis shows the time or order of these measurements. The *y*-axis also contains some reference lines based on the original measurements of the control material, which show the range over which any new values for this material should appear. In this particular plot, these lines occur at the mean and at two standard deviations above or below the mean for the control material's original results. As we learned in Chapter 4, 95% of any new determinations made for this same material should appear in this given range if there are no changes in the analytical method. If a new result is found to be outside this range (or "out of control"), corrective action may be required before other samples can be measured or examined by the method.

EXERCISE 5.5	Developing and Using a Control Chart

Analysis of the cholesterol in a control serum sample gives the following results: 187, 198, 191, 189, 194, 197, 191, 199, 195, 186, 192, 194, 196, 193, 200, 192, 190, 195, 188, and 191 mg/dL. These data are used to construct a Levey–Jennings chart for this method. A new measurement of the same control material gives a value of 200 mg/dL. Is this new value "in control" or "out of control" when using an allowable range of $\bar{x} \pm 2\,s$? Is this technique operating at a level that is satisfactory for sample analysis?

SOLUTION

Based on the 20 values that were originally measured for the control material, we get a mean result of 193 mg/dL and a standard deviation of 4 mg/dL. Thus, if we were to

make one additional measurement for the same material, there is a 95% probability that the new value will fall between $(\bar{x} - 2\,s) = 185$ and $(\bar{x} + 2\,s) = 201$ mg/dL (the reference lines for the control chart). The new value of 200 mg/dL is within this range. Thus, this result is "in control" and the method can be used for other samples.

Because it makes use of only a single measurement for each analyte, a Levey–Jennings chart is commonly employed in clinical laboratories, where the amount of available sample is often limited and a fast turnaround time is required. Alternative types of control charts can be found in other settings. One example is a *Shewhart chart*, which is often used in industrial laboratories and in situations where enough time and material are available to perform several measurements on every sample.[11,16] With this extra information, it is possible to determine a mean result, range, and standard deviation for each sample and to prepare separate control charts for these various parameters. These separate plots make it easier to identify and differentiate between any systematic or random errors that might later appear in the assay. For instance, a change in systematic errors should affect the mean result in a Shewhart chart, but have little effect on the standard deviation or range. However, these other parameters will be affected by a change in the technique's precision, representing a shift in random errors. This information can be valuable when you are trying to locate and fix a problem in an analysis method.

5.4 SAMPLE COLLECTION AND PREPARATION

Along with the selection and characterization of your measurement technique, it is also essential that your sample be properly selected and prepared for study.[17–20]

Failure to do so will give results that are an incorrect representation of the original material. To avoid this situation, let's examine some guidelines to follow when you are getting a sample ready for analysis.

5.4A Sample Collection

The type of material you want to examine will usually determine how difficult it is to obtain a representative sample. If you are working with a liquid or gas, collecting a sample may simply involve obtaining a sufficient volume or mass for testing. However, if the material is a group of solid particles or a nonuniform substance, you must consider how the analytes are distributed in this material in order to obtain a representative portion for analysis. The specific approach you use to acquire a sample is known as the *sampling plan*.[17] An example of such a plan is shown in Figure 5.10. As this figure shows, a sampling plan should include all steps involved in the selection, withdrawal, preservation, transportation, and preparation of materials to be studied in a laboratory. The particular approach you use for a sampling plan will depend on the physical state of the material, the sample size that is needed, and the extent to which the original material is uniform in composition.

Failure to obtain a truly representative sample will lead to a **sampling error**. This is an error that is created when you use only a portion of a nonuniform substance for analysis. The size of the sampling error will depend on the number of "particles" in your collected sample and the fraction of the total material that you are analyzing.[18,20] As you use more of the original material for a measurement, the sampling error will decrease, but using more material also requires a larger sample size. Thus, it is more common for analytical chemists to use only a small portion of material for analysis. This approach can still lead to small sampling errors if you are examining a large number of particles, as often occurs in the measurement of molecules or ions.

Table 5.3 lists several approaches that can be used to obtain a sample of a material.[17–20] For example, a *random sample* is acquired by arbitrarily taking part of a material and testing this portion to obtain information on the material's entire contents. This approach works well for a homogeneous material; however, it is more difficult to use with nonuniform substances. An alternative approach is to collect a *representative sample*, which requires a well-defined sampling plan that allows you to acquire one or more samples that reflect the overall composition and properties of the original material. Obtaining a representative sample involves more work than random sampling, but provides a

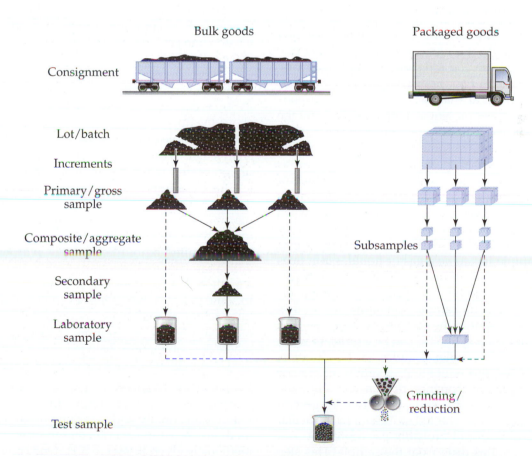

FIGURE 5.10 Sampling plan for the measurement of raw materials being used in an industrial process. After the sampling operations in this plan have been carried out, the test sample would then be taken to a laboratory where it could be weighed, placed in a solution, treated, and divided into aliquots prior to analysis. (Adapted with permission from R. E. Majors, "Nomenclature for Sampling in Analytical Chemistry," *LC·GC*, 10 [1992] 500–506.)

TABLE 5.3   General Types of Analytical Samples

Sample Type	Definition	Example
Random sample	A sample selected so that any portion of the overall material has an equal chance of being chosen	Random selection of a container from a manufacturing line for quality-control monitoring
Representative sample	A sample chosen to represent the overall properties of the material being studied	Use of a portion from a beef sample for the analysis of its nutritional content, after the sample has been processed in a blender
Selective sample	A sample selected to have certain characteristics	A preparation of soil particles that are greater than 1 mm in size
Sequential sample	Samples obtained one after another to monitor the changes in a material over time	Periodic collection of blood from a patient to monitor changes in the level of a therapeutic drug
Stratified sample	A sample taken from a specific region of a material	Collection of water from different points below the surface of a lake
Umpire sample	A sample collected and handled in an agreed upon manner for the purpose of settling a dispute	The collection and handling of urine samples for drug testing at athletic events

Source: This list is based on L. H. Keith (Ed.), *Principles of Environmental Sampling*, American Chemical Society, Washington, DC, 1996.

better picture of a material, especially when it has a heterogeneous composition. More information on techniques for obtaining specific types of samples can be found in References 18–21.

5.4B   Sample Preparation

Goals of Sample Preparation.   After you have selected and acquired a sample, it is often necessary to treat or prepare this sample before you can study it. There are many reasons why you might consider using sample pretreatment prior to a chemical analysis. One common reason is that the analytes are not present in a matrix suitable for the chosen method. This factor must be considered because all analytical techniques have some requirements regarding the types of samples that these methods can be used to study. Two specific examples are the separation methods of gas chromatography (GC) and liquid chromatography (LC). GC requires that chemicals be present in a gas when they are placed into the analysis system, while in LC they must be present in a liquid. This does not necessarily mean the original sample has to be a gas or liquid, but it does means the analytes must at least be transferred into such a medium before they can be examined by these methods.

A second reason for sample pretreatment is to deal with substances present in the sample that interfere with measurement of the analyte. This can be a problem when you are dealing with complex samples. In this situation, you can either use a procedure that separates the analyte from other sample components, or you can eliminate these interferences by converting them into a different form. For instance, if you wanted to measure a particular drug in blood, you would probably use a pretreatment step to first isolate this drug from the sample. This step makes it easier to detect the drug and reduces the possibility that other agents might create an error in the analysis. A third reason for sample pretreatment is to place the analyte into a chemical form that can be studied by your method. As an example, suppose you wanted to determine the total iron content of an ore sample. This iron may be present in several chemical forms (or "species"), making it difficult to analyze directly. Instead, this sample could be treated so that all of the individual types of iron are converted into a single species for measurement. One way of accomplishing this goal would be to convert all the iron into Fe^{2+}, which could then be measured by a variety of techniques.

A fourth reason for sample pretreatment is to place the analyte at a level that is suitable for detection. This situation is easiest to deal with when the analyte is present above the upper limit of detection for a method, in which case the sample can simply be diluted until the amount of analyte falls within the usable range. A more difficult situation occurs when the analyte is present at levels too low for measurement. This situation might be handled by using a large sample, removing the analyte from the sample, and then placing the analyte into a smaller volume for analysis. However, this approach will not work if you have a limited amount of sample or if you are dealing with a rare material. Another possible strategy is to select a method that can detect smaller amounts of the analyte. In some unique cases, it may even be possible to increase the amount of analyte by using a catalytic or enzymatic reaction. We saw an example of this case in Chapter 2, where the polymerase chain reaction was used to amplify the DNA in blood or hair samples for forensic testing.

Preparation Methods.   The wide range of analytes, samples, and methods found in analytical chemistry requires an equally wide selection of sample preparation methods. A list of several common sample preparation methods is given in Figure 5.11. One common feature of these methods is they all involve the use of either a physical or chemical change in the analyte or its matrix. A physical change might involve diluting a sample solution, weighing a portion of a solid for analysis, filtering a

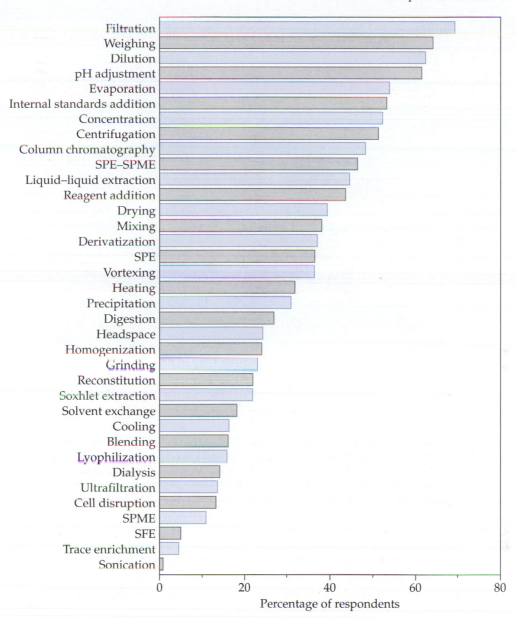

FIGURE 5.11 Common sample preparation procedures found in chemical analysis laboratories. This list is based on a survey of 467 laboratory scientists who were working with a wide variety of samples. Further information on many of the techniques listed in this figure can be found throughout this book. (*Abbreviations*: SFE: supercritical fluid extraction; SPE: solid-phase extraction; SPME: solid-phase microextraction. (Reproduced with permission from R. E. Majors, "Trends in Sample Preparation," *LC · GC*, 14 [1996] 754–766.)

precipitate from a liquid, or heating a solution to release a volatile chemical. Examples of chemical changes for sample pretreatment include the addition of reagents to make an analyte easier to detect, or the use of a reaction that converts multiple forms of an analyte into a single form.

Sample preparation often requires the use of more than one step, such as weighing, dissolving, and filtering to prepare liquid samples from solids. Many of these steps are simple to perform, but can be tedious and time-consuming when there are a large number of samples. In addition, if these steps are not carried out carefully, they can add an appreciable error to the results of the final analysis. It is for these reasons that automated systems are becoming popular for the preparation of samples. An example of a robotic system that can be used for this purpose is shown in Figure 5.12. These systems allow laboratory workers to then perform other tasks and tend to reduce random errors during sample preparation, leading to more precise results.

Recommended Precautions. Although some sample pretreatment is required in most analytical methods, there are several problems that can occur if this pretreatment is not performed properly. For instance, a portion of the analyte might be lost during this process, as could occur during analyte transfer from the original sample to

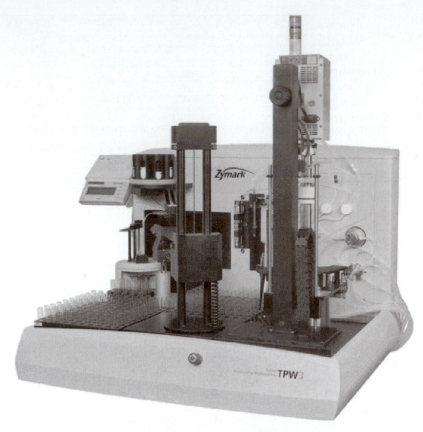

FIGURE 5.12   A robotic workstation for the preparation and analysis of drug tablets. This programmable unit is capable of performing many common sample preparation steps and some methods of chemical analysis on drug samples. (This image of the Zymark TPW3 system is reproduced with permission from Sotax Corporation.)

a new matrix. The loss of analyte might also be caused by degradation during a pretreatment step or by incomplete conversion of the analyte into a new form for measurement. This problem of analyte loss can be detected by using recovery studies (as discussed in Section 5.2A), in which the amount of analyte that remains after a pretreatment step is compared to the amount that was initially present. As an example, if you used a pretreatment step on a standard known to contain 25 mg/L of an analyte, but only measured 20 mg/L after the procedure, the percent recovery of analyte in this step would be $100 \cdot (20/25) = 80\%$.

To help correct for this loss of analyte, a *fixed* amount of another chemical can be added to the sample to act as an **internal standard**. An internal standard is a substance not present in the original sample, but that has similar properties to the analyte and the ability to be detected separately from this compound. Because the analyte and internal standard should have the same behavior during pretreatment steps, the loss of one will be reflected by the loss of the other. Thus, by measuring the final amount of both of these chemicals and taking the ratio of their signals, it is possible to correct for any changes in the amount of analyte that occurred during the sample preparation process (as illustrated in Figure 5.13). This approach can

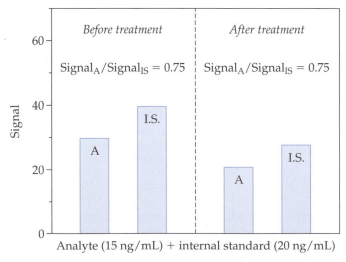

FIGURE 5.13   Principle behind the use of an internal standard. In this example, a fixed amount of an internal standard (I.S.) is added to a sample that contains the analyte (A). If the analyte and internal standard have similar chemical and physical properties, they should have similar recoveries and behavior during any sample pretreatment steps. This means that the ratio of their two signals will remain the same even if the absolute amount of each is decreased. Thus, by using the ratio of their signals rather than only the analyte's signal for quantitation, a more robust and reproducible approach is obtained for determining the amount of the analyte in samples.

also be used to adjust for variations during sample injection and analysis in methods such as atomic emission spectroscopy and gas chromatography or liquid chromatography (see Chapters 19, 21, and 22).

A second problem that can occur during sample pretreatment is when new substances are introduced to the matrix that may interfere with analyte detection. This problem is often encountered during the determination of trace metals in that even small amounts of metal contaminants in the reagents or sample containers can give artificially high results. It is possible to detect this problem by performing interference studies (as described in Section 5.2B) or by testing the specificity of the method to determine which substances may cause difficulties in the technique.

Another thing to remember when you use sample pretreatment is that this will always result in the loss of some information about the sample. Removing the analyte from its initial matrix not only simplifies the mixture to be tested but prevents you from gaining data on other sample components. Similarly, converting the analyte into a new form can prevent you from learning about its original composition. This result may be acceptable in many types of analysis. If you are not aware of such effects, however, this loss of information can lead to an improper interpretation of your final data.

Key Words

Control chart *103*	Internal standard *108*	Method validation *96*	Sensitivity *102*
Control material *103*	Limit of detection	Percent recovery *98*	Signal-to-noise ratio
Correlation chart *98*	(LOD) *99*	Precision plot *98*	(S/N) *100*
Dynamic range *101*	Limit of quantitation	Quality control *103*	Specificity *102*
Error assessment *103*	(LOQ) *101*	Response *99*	Spiked recovery study *97*
Interference plot *102*	Linear range *101*	Sampling error *105*	

Other Terms

Blank sample *100*	Interlaboratory precision *99*	Random sample *105*	Signal *100*
Calibration sensitivity *102*	Interoperator precision *99*	Representative sample *105*	Speciation *96*
Certified reference	Levey–Jennings chart *104*	Robustness *103*	Universal (general)
material *97*	Lower limit of	Sample throughput *103*	method *102*
Correlation study *98*	detection *100*	Sampling plan *105*	Upper limit of detection *99*
Day-to-day precision *99*	Noise *100*	Selectivity *102*	Within-day precision *99*
Figures of merit *96*	Optical microscopy *97*	Selectivity coefficient *102*	Within-run precision *99*
	Overall analysis time *103*	Shewhart chart *104*	

Questions

INTRODUCTION: METHOD CHARACTERIZATION AND SELECTION

1. What questions regarding the sample should be asked when you are selecting an analytical method? How do these questions affect the design of the analysis procedure?
2. What questions regarding the analysis technique should be addressed when you are designing a procedure for a chemical analysis? How do these questions affect the selection of the measurement method?
3. List several specific factors that should be considered about the sample and the method when choosing a chemical analysis technique for each of the following situations.
   (a) The measurement of oxygen and carbon dioxide in blood during a surgical procedure
   (b) The routine measurement of acidity in soil samples
   (c) Determination of a trace pesticide in a batch of vegetables that are being sent to market
   (d) Confirmation of the structure of a newly synthesized drug

4. Mercury can be found in biological samples as part of many different chemical species. An environmental scientist would like to study how mercury affects a certain type of fish in a contaminated lake. What are some specific questions that this scientist should ask in choosing a chemical analysis method for this research?
5. What is meant by the term "figures of merit" in analytical chemistry? What are some specific examples of figures of merit?
6. What is "method validation"? Explain why method validation is important when performing a chemical analysis.

SPIKED RECOVERY AND CORRELATION STUDIES

7. What are some general approaches that are used to examine the accuracy of an analytical method? In what kinds of circumstances are each of these approaches employed?
8. What is a "certified reference material"? Why is this type of material important in method validation?

9. A drug-testing laboratory wishes to market an analysis method it has developed for monitoring the drug cyclosporine A in samples from organ-transplant patients. Describe how this laboratory might evaluate the accuracy of this method using a certified reference material.

10. A certified reference material of bituminous coal is analyzed by a laboratory and found to have a mean aluminum content of 0.874% (w/w). The reported value for this same material is 0.855%. What is the absolute error and the relative error for the method that was used in this analysis?

11. How is a spiked recovery study performed? How is this type of study used during method validation?

12. A pharmaceutical chemist prepares a 1.0 mL blank serum sample that has been spiked with 23.0×10^{-9} g of digitoxin, a drug given for the treatment of irregular heartbeats. This sample is prepared and analyzed by a method that has been previously calibrated with aqueous digitoxin standards. The final result for the spiked sample is a measured digitoxin concentration of 22.7 μg/L. What is the percent recovery for digitoxin?

13. The recovery of a pesticide from a 5.0 ppb aqueous solution is examined for a solid-phase extraction method to be used for the preparation of groundwater samples. The recoveries found for five measurements of the same sample are 98.7, 101.2, 97.0, 99.1, and 97.6%. Based on these results, are any systematic errors detectable in the extraction method at the 95% confidence level?

14. Describe how a paired Student's t-test can be used during method validation.

15. The following data were obtained by a food company that wishes to adopt a new procedure for measuring the protein content of its products. Use a paired Student's t-test to determine if the new and old analysis procedures are giving the same results at the 95% confidence level. (*Note*: You may assume that the precision is approximately the same for these two methods.)

Sample Number	Measured Protein Content (% w/w)	
	Old Analysis Method	New Analysis Method
1	10.1	10.5
2	25.8	26.9
3	15.3	15.6
4	30.0	32.9
5	6.2	7.8
6	21.7	23.1
7	34.2	36.9
8	40.1	44.8

16. What is a correlation chart? How is a correlation chart prepared, and how is it used during method validation?

17. The following results were obtained by methods based on liquid chromatography (LC) and gas chromatography (GC) for measuring nicotine in smokers. Prepare a correlation chart that compares the results of the LC method versus the GC method, with the GC data being represented by the x-axis. From this graph, what can you conclude about the agreement between these two methods?

Concentration of Nicotine in Serum (μg/L)		
Sample Number	GC	LC
1	2.5	2.8
2	3.2	2.9
3	7.1	7.5
4	10.2	11.5
5	12.9	11.8
6	14.4	16.0
7	16.1	16.3
8	22.7	23.5
9	29.6	30.2
10	33.8	35.9
11	37.4	40.2
12	39.1	41.0
13	42.5	44.3
14	46.7	49.5

18. Use a correlation chart to compare the two methods that were employed in Problem 15 for measuring the total protein content of food samples. From this graph, what can you conclude about the correlation of these two methods? How does this conclusion compare with the result that was obtained in Problem 15 when using a paired Student's t-test?

PRECISION MEASUREMENTS

19. What are some ways in which the precision of an analytical method can be evaluated? What measures of precision from Chapter 4 are often used during this evaluation process?

20. A new method is developed for measuring oncofetal antigen CEA as a marker for colorectal cancer. This method gives the following results for a reference sample of CEA: 15.0, 15.5, 15.3, 14.4, and 16.1 mg/mL. What is the precision of this method, as based on the standard deviation and relative standard deviation for this sample?

21. The same method that was described in Problem 20 is used to examine a second reference sample that contains a higher concentration of CEA and gives the following results: 35.3, 36.8, 37.5, 34.0, and 35.9 ng/mL. What is the precision of the method for this second sample in terms of the standard deviation and relative standard deviation? How do these results compare with those found in Problem 20?

22. What is a precision plot and how is it prepared? Why is this type of plot useful in describing the performance of an analytical method?

23. The following data were obtained during the evaluation of a method for the analysis of acetaminophen in drug preparations. The values given in the table for the peak heights include the mean and standard deviation found for five replicate measurements at each concentration. Prepare a precision plot for this method. How do the standard deviation and relative standard deviation of the response change as the drug's concentration is increased?

Acetaminophen Concentration (mg/L)	Peak Height ($\bar{x} \pm 1\,s$)
5	150 ± 43
10	310 ± 32
20	445 ± 20
30	589 ± 15
40	763 ± 28
50	895 ± 58

24. A pharmaceutical chemist wishes to use the method in the last problem for the routine measurement of acetoaminophen in drug tablets. It is expected that samples made from these tablets will contain acetaminophen concentrations in the range of 15 to 35 mg/L. Based on the results in Problem 23, what range of relative standard deviation values would be expected over this concentration range for these samples? What approximate range of sample concentrations could be examined by this method and provide a relative standard deviation that is at or below 5%?

25. Define each of the following terms and state how they are used to describe the precision of an analytical method.
    (a) Within-run precision
    (b) Day-to-day precision
    (c) Within-day precision
    (d) Interoperator precision
    (e) Interlaboratory precision

26. Using the list in Table 5.2, state the type of precision that should be measured in each of the following situations.
    (a) A supervisor in a quality-control laboratory wishes to compare the degree of reproducibility that is obtained by several employees who take turns in performing an assay for the cholesterol content of food.
    (b) A field scientist wishes to determine the reproducibility of a portable instrument as it is used at a remote testing site over the course of several days.
    (c) A clinical laboratory wishes to determine the extent of variation that occurs for a control sample that is measured several times during the analysis of a large batch of serum samples.

LIMITS OF DETECTION

27. Describe what is meant by a "signal-to-noise ratio." How is a signal-to-noise ratio used in analytical chemistry?
28. What is the difference between a "limit of detection" and a "limit of quantitation"? How are each of these factors determined for an analytical method?
29. Explain how the limit of detection or limit of quantitation can be estimated using a precision plot.
30. A toxicologist has developed a new technique for analyzing DNA adducts that could be used to follow the exposure of humans to carcinogens in the environment. This method gives a linear relationship up to an adduct concentration of roughly 100 ng/mL, with a slope over this range of 1250 ± 50 mL/ng and an intercept of 22 ± 10. Estimate the lower limit of detection and lower limit of quantitation for this assay.
31. A method for measuring monoclonal antibodies gives a linear response for a plot of absorbance versus concentration (in units of μg/mL). The equation that describes the best-fit

line for this plot is $y = 0.0252\,(\pm 0.004)x + 0.004\,(\pm 0.003)$. What is the approximate lower limit of detection for monoclonal antibodies in this method? What is the lower limit of quantitation for this method?

ASSAY RANGE AND LINEARITY

32. Define the terms "linear range" and "dynamic range" as they are used to describe analytical techniques.
33. The following calibration data were obtained by a food chemist who used atomic absorption spectroscopy to determine the iron content of milk and baby formulas. Based on a best-fit line over 0.00–7.50 mg/L, what is the approximate linear range for this method? What is its dynamic range? In each case, state how you defined or determined these ranges.

Iron Concentration (mg/L)	Measured Absorbance
0.00	0.002
1.00	0.023
2.00	0.048
5.00	0.118
7.50	0.165
10.00	0.215
12.50	0.254
15.00	0.282
17.50	0.335
20.00	0.346

34. A new employee at the same laboratory as in Problem 33 assumes that the linear response that is obtained for samples that contain between 0.00 and 7.50 mg/L iron can also be used for samples that contain higher iron concentrations. Based on the data given in Problem 33, how large an error would be produced in the measured iron concentration if this employee were to use this linear range with a sample that gave a measured absorbance of 0.258? How large an error would occur for a sample with an absorbance of 0.340?

SENSITIVITY AND SPECIFICITY

35. What is the difference between "sensitivity" and "limit of detection" when these are used to characterize a method for chemical analysis?
36. A student writes a laboratory report in which it is stated that the sensitivity of an aluminum assay is 0.03% (w/w) because this is the smallest amount of aluminum that could be measured with a precision of ± 2%. What is wrong with this student's statement?
37. What is the calibration sensitivity for the method in Problem 31? Describe how you determined this value.
38. What is the calibration sensitivity for the method in Problem 33 at an iron concentration of 7.50 mg/L? How would this value change as you move to lower and higher concentrations in the calibration curve for this technique?
39. What does "specificity" mean in analytical chemistry? What is the difference between a specific method and a general technique?
40. What is an "interference plot"? How is an interference plot used in method validation?

41. A geochemist using atomic emission spectroscopy wishes to examine the effect that different levels of potassium will have on samples to be measured for their sodium content. Describe how an interference plot could be used to provide this information.

42. It is known that the presence of small amounts of aromatic compounds in water can affect the background signal of a fluorescence detector. A food chemist wishes to obtain a low background signal with water in order to use fluorescence to measure low concentrations of certain water-soluble vitamins. Describe how an interference plot could be used to help determine the maximum concentration of aromatic compounds that can be tolerated in this assay.

43. What is a "selectivity coefficient," and how is this used to describe the specificity of analytical methods?

OTHER PROPERTIES OF ANALYTICAL METHODS

44. Define each of the following terms as related to methods of chemical analysis. Explain how each of these parameters can be measured.
(a) Overall analysis time
(b) Sample throughput
(c) Robustness

45. A gas chromatographic (GC) method for the analysis of cocaine and its metabolites in urine requires about 10 min for extraction of an analyte from a sample, 10 min for its derivatization, and 15 min for its injection and separation on a GC column. What is the overall analysis time for each sample?

46. It is possible to modify the gas chromatographic (GC) method in the last problem so that up to 10 samples can be extracted and derivatized simultaneously, with each of these samples then being injected one at a time onto a single GC column. What is the maximum sample throughput for this method? How would this sample throughput change if two separate GC systems were operated at the same time and used to process these samples? What would be the overall analysis time for each sample when this second GC system is used?

QUALITY CONTROL

47. What is "quality control"? Why is quality control important when performing a chemical analysis?

48. What are the three parts of a quality-control program? What is the purpose of each part?

49. What is a "Levey–Jennings chart"? How is this type of chart used?

50. A clinical laboratory wishes to monitor the performance of an instrument that measures the total concentration of the hormone thyroxine in serum samples. Twenty measurements of the same serum control material give the following concentrations for thyroxine: 8.9, 9.5, 9.2, 8.8, 9.1, 9.0, 8.7, 9.1, 9.3, 9.2, 8.9, 9.0, 9.4, 9.1, 8.7, 9.2, 9.6, 8.9, 9.2, and 9.1 μg/dL. Prepare a Levey–Jennings control chart for this method, using a scale of 1–14 days on the x-axis.

51. The same control material as used in the preceding problem is analyzed on a daily basis to examine the routine behavior of the method in Problem 50. The values that are measured over the course of the next two weeks are as shown in the following table.

Day Number	Control Value (μg/dL)
1	8.9
2	9.0
3	9.2
4	9.3
5	8.8
6	9.4
7	9.7
8	9.1
9	8.9
10	9.3
11	8.6
12	9.2
13	9.0
14	9.1

Plot the preceding data on the control chart made in Problem 50 and compare these data to the allowed control limits. Which of these results are acceptable and which are "out of control"? What action would you take if you did get a result that was out of control?

52. What is a "Shewhart chart"? How does this differ from a Levey–Jennings chart? How is a Shewhart chart used?

SAMPLE COLLECTION AND PREPARATION

53. What is a "sampling plan"? What items need to be considered when developing a sampling plan?

54. What is "sampling error"? What factors determine the size of the sampling error in a chemical analysis?

55. What is a "random sample"? What is a "representative sample"? How are each of these used?

56. What are some reasons why pretreatment of a sample may be needed before an analysis?

57. Compare the sample preparation procedures that are listed in Figure 5.11 with those performed in your laboratory.

58. What are some possible problems that can occur during sample pretreatment? How can these problems be detected or avoided?

CHALLENGE PROBLEMS

59. The various properties of analytical methods (such as accuracy, precision, limit of detection, sensitivity and speed) are often related.[22–24] For instance, developing a method with higher accuracy and precision might require an increase in the cost of the assay or the need for a greater number of steps for pretreatment and analysis. State what kind of relationship might be expected between the following parameters. Explain the reasons for your answers.
(a) Effect on the lower limit of detection as the assay precision is improved
(b) Effect of an increase in sensitivity on the assay range
(c) Effect of a decrease in sensitivity on method robustness
(d) Effect of an increase in precision on the ability to determine the accuracy of a method

60. Obtain a copy of an SRM catalog from the U.S. National Institute of Standards and Technology (NIST) or visit this agency's Web site. Using this information, identify at least one specific reference material that might be used in each of the following situations.
    (a) Use of the fire assay to determine the purity of gold bullion
    (b) Analysis of the elements aluminum, cadmium, chromium, iron, manganese, nickel, and vanadium in soil samples
    (c) Measurement of creatinine in human serum or urine
    (d) Analysis of metals in auto catalysts

TOPICS FOR DISCUSSION AND REPORTS

61. The example of the Vinland map that was used at the beginning of this chapter is just one of many cases in which chemical analysis has been used to either confirm or disprove the true identity of a rare object, a work of art, or a precious document. Additional examples include studies that have been conducted on the Shroud of Turin, the Bust of Nefertiti, and the Salamander letter, among others.[25–29] Obtain information on one of these other cases and report on how analytical chemistry was used to study and characterize the given object.

62. The use of laboratory robots in the preparation of samples is just one of the many applications that are emerging for such devices within the analytical laboratory.[30,31] Identify and discuss some other applications in which robotics might prove valuable for chemical measurements.

63. Below are several routine or newer sample pretreatment methods that are used before chemical analysis. Obtain information on the function of each pretreatment step (for instance, analyte concentration or purification), describe how the method works, and give an example of an analytical technique that might employ this pretreatment step.
    (a) Soxhlet extraction
    (b) Solid-phase extraction
    (c) Microdialysis probe
    (d) Microwave digestion

References

1. R. A. Skelton, T. E. Marston, and G. D. Painter, *The Vinland Map and the Tartar Relation*, Yale University Press, New Haven, 1995.
2. E. Eakin, "Was 'Old' Map Faked to Tweak the Nazis?" *New York Times*, Sept. 14, 2002, p. 4.
3. W. C. McCrone, "The Vinland Map," *Analytical Chemistry*, 60 (1988) 1009–1018.
4. D. J. Donahue, J. S. Olin, and G. Harbottle, "Determination of the Radiocarbon Age of the Parchment of the Vinland Map," *Radiocarbon*, 44 (2002) 45–52.
5. T. A. Cahill, R. N. Schwab, B. H. Kusko, R. A. Eldred, G. Moller, D. Dutschke, D. L. Wick, and A. S. Pooley, "The Vinland Map, Revisited: New Compositional Evidence on Its Inks and Parchment," *Analytical Chemistry*, 59 (1987) 829–833.
6. K. L. Brown and R. J. H. Clark, "Analysis of Pigmentary Materials on the Vinland Map and Tartar Relation by Raman Microprobe Spectroscopy," *Analytical Chemistry*, 74 (2002) 3658–3661.
7. W. C. McCrone, *Asbestos Identification*, McCrone Research Institute, Chicago, IL, 1987.
8. W. C. McCrone, "Why Use the Polarized Light Microscope?" *American Laboratory*, 4 (1992) 17–21.
9. J. M. Green, "A Practical Guide to Analytical Method Validation," *Analytical Chemistry*, 68 (1996) 305A–309A.
10. M. E. Swartz and I. S. Krull, *Analytical Method Development and Validation*, Marcel Dekker, New York, 1997.
11. N. W. Tietz, Ed., *Textbook of Clinical Chemistry*, Saunders, Chicago, IL, 1986.
12. W. Horwitz, "Evaluation of Analytical Methods Used for Regulation of Foods and Drugs," *Analytical Chemistry*, 54 (1982) 67A–76A.
13. *Validation of Analytical Procedures, Terms and Definitions*, International Conference on Harmonization of Technical Requirements for Registration of Pharmaceuticals in Human Use, Geneva, Switzerland, March 1995, ICH-Q2A.
14. D. A. Skoog, F. J. Holler, and T. A. Nieman, *Principles of Instrumental Analysis*, 5th ed., Saunders, Chicago, IL, 1998, Appendix 1.
15. S. Levey and E. R. Jennings, "The Use of Control Charts in the Clinical Laboratory," *American Journal of Clinical Pathology*, 20 (1950) 1059–1066.
16. W. A. Shewhart, *Economic Control of Quality of the Manufactured Product*, Van Nostrand, New York, 1931.
17. R. E. Majors, "Nomenclature for Sampling in Analytical Chemistry," *LC · GC*, 10 (1992) 500–506.
18. C.A. Bicking, "Principles and Methods of Sampling," In *Treatise on Analytical Chemistry: Part I—Theory and Practice*, 2nd ed., Vol. 1, I. M. Kolthoff and P. J. Elving, Eds., Wiley, New York, 1978, Chapter 6.
19. M. Stoeppler, *Sampling and Sampling Preparation: Practical Guide for Analytical Chemists*, Springer-Verlag, New York, 1997.
20. P. Gy, *Sampling for Analytical Purposes*, Wiley, New York, 1998.
21. L. H. Keith, Ed., *Principles of Environmental Sampling*, American Chemical Society, Washington, DC, 1996.
22. M. Valcarcel and A. Rios, "The Heirarchy and Relationships of Analytical Properties," *Analytical Chemistry*, 65 (1993) 781A–787A.
23. P. C. Meier and R. E. Zund, *Statistical Methods in Analytical Chemistry*, Wiley, New York, 1993, Chapter 2.
24. R. L. Anderson, *Practical Statistics for Analytical Chemists*, Van Nostrand Reinhold, New York, 1987, Chapter 6.
25. R. Hedges, "Relic, Icon or Hoax? Carbon Dating the Turin Shroud," *Nature*, 385 (1997) 310.
26. A. A. Cantu, "Analytical Methods for Detecting Fraudulent Documents," *Analytical Chemistry*, 63 (1991) 847A–854A.
27. H. G. Wiedemann and G. Bayer, "The Bust of Nefertiti," *Analytical Chemistry*, 54 (1982) 619A–628A.

28. R. J. H. Clark and P. J. Gibbs, "Raman Microscopy of a 13th-Century Illuminated Text," *Analytical Chemistry*, 70 (1998) 99A–104A.

29. A. Burnstock, "Chemistry Beneath the Surface of Old Master Paintings," *Chemistry & Industry*, Sept. 21 (1992) 692–695.

30. A. Newman, "Running on Automatic: Laboratory Robotics and Workstations," *Analytical Chemistry*, 69 (1997) 255A–259A.

31. R. A. Felder, J. C. Boyd, K. S. Margrey, W. Holman, J. Roberts, and J. Savory, "Robots in Health Care," *Analytical Chemistry*, 63 (1991) 741A–747A.

Chapter 6

Chemical Activity and Chemical Equilibrium

Chapter Outline

6.1 INTRODUCTION: "AND THE LONG-RANGE FORECAST IS..."

In December 1997, delegates from over 150 nations met in Kyoto, Japan, to discuss the issue of global warming.[1] Of particular interest was the theory that an increase in the temperature of the Earth is occurring as result of a rise in carbon dioxide and other gases that can trap heat from sunlight. Part of the carbon dioxide entering the atmosphere is from natural sources and some comes from human activities, such as the burning of fossil fuels.[2–4] As a result of the 1997 meeting in Kyoto, a treaty was drafted calling for a dramatic reduction in human-related CO_2 production. The treaty took effect in February 2005,[5] but it has been a subject of continuing controversy related to how CO_2 emissions might be decreased and the expense of accomplishing this goal. There is also disagreement on how global warming will be affected by a change in emissions or how such a change might affect the Earth's climate.[3–9]

Analytical chemists have helped in this discussion by providing measurements of the past and present levels of CO_2 in the atmosphere (see Figure 6.1). These measurements are made at stations located throughout the world that have monitored the concentration of CO_2 in air for many decades. Carbon dioxide levels in the past are estimated by examining the content of this gas in ice core samples that are acquired at sites such as those located in Antarctica. These latter samples are useful because the ice at these sites never melts, giving an ongoing record of the CO_2 content in air. The results of these measurements are then compared to estimates of the average global temperature, allowing the effect of CO_2 on global warming to be examined.[3]

Based on such measurements, scientists have used computer models that incorporate data on past CO_2 levels and global temperatures to predict how Earth's climate will change in the future. This is an extremely complex problem in which scientists must consider how CO_2 is produced and circulated in the environment, along with the way in which sunlight interacts with the Earth, the degree of rainfall, and the circulation of water in the oceans.[9] Just as computer simulations can be used to help predict what the climate on Earth may look like in the future, we can use calculations on a smaller scale to develop or optimize a chemical analysis method, as well as to describe the composition of a sample or reagent. This type of work helps us understand the chemical reactions that are important for anyone who wishes to use such methods. In this chapter, we review several mathematical tools for describing the reactivity of chemicals and the extent to which chemicals react with other substances. We then move on in the next few chapters to see how these tools can be applied to specific reactions for the design and use of analytical methods.

6.1A Types of Chemical Reactions and Transitions

There are many types of reactions or transitions that a chemical may take part in within a sample or with a reagent. These processes are usually organized into

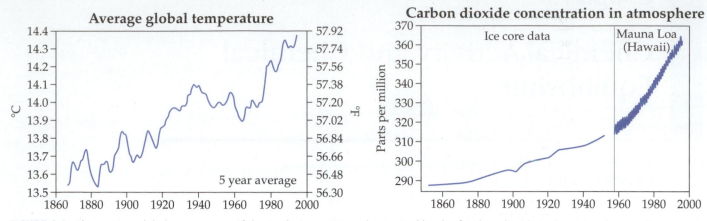

FIGURE 6.1 The average global temperature of the Earth since 1860, and measured levels of carbon dioxide in the atmosphere as determined from either ice core samples (before 1955) or direct measurements made at the Mauna Loa volcano in Hawaii (after 1955). The ice core samples were collected at sites like those in Antarctica (the South Pole station) or in Russia (the Siple station), where the ice within a given core layer is crushed to release the gas that is held in air bubbles in the ice. This released gas is then examined to measure its carbon dioxide content. (Based on data and graphs from NASA/Goddard Institute for Space Studies; the Office of Science and Technology Policy; and A. Neftel, E. Moor, H. Oeschger and B. Stauffer, "Evidence from Polar Ice Cores for the Increase in Atmospheric Carbon Dioxide in the Past Two Centuries," *Nature*, 315 (1985) 45–47.)

categories that have certain common features. Some of the more common types of reactions and transitions that are encountered in chemical analysis are shown in Table 6.1.

A *precipitation reaction* is one important process that is employed in analytical chemistry. This reaction takes place when the combination of two or more soluble chemicals leads to the creation of an insoluble substance, or *precipitate*. In the example shown in Table 6.1, carbonate ions and calcium ions in water form insoluble calcium carbonate, a solid that will drop out of the solution. Precipitation and its applications in chemical analysis are discussed in Chapters 7, 11, and 13.

Another type of reaction that is often used in chemical analysis is an *acid–base reaction,* which can be defined in the classical sense as the transfer of a hydrogen ion from one compound (an acid) to another (a base)[10,11] An example is the transfer of a hydrogen ion from carbonic acid (H_2CO_3, formed when CO_2 combines with water) to water to give bicarbonate (HCO_3^-). This type of reaction is important whenever an acid or base is to be measured or controlled in a sample, such as in buffer preparation or in acid–base titrations (Chapters 8 and 12).

Complex formation is another type of reaction that is important in chemical analysis. This reaction involves the formation of a reversible complex between two or more chemicals, such as a complex based on a coordinate covalent bond or on noncovalent interactions.[12] Many of these reactions involve the binding of a chemical that acts as an electron-pair donor (e.g., hemoglobin) with an electron-pair acceptor (O_2). However, other interactions can also be present, such as ionic forces and hydrogen bonds. In Chapters 9 and 13, we will see many examples of such reactions and discuss their use in chemical analysis (e.g., complexation titrations and immunoassays).

A fourth type of reaction that we will examine is an *oxidation–reduction reaction.*[11] An oxidation–reduction reaction occurs when there is an exchange of electrons between chemicals, in which one chemical has a net gain of electrons (or is *reduced*), while another has a net loss of electrons (or is *oxidized*). As shown in Table 6.1, the burning of fuels like octane and other carbon-containing substances is an example of an oxidation–reduction reaction; another example is respiration, in which food is "burned" by our bodies to generate energy. We will learn about the basic principles of oxidation–reduction reactions in Chapter 10 and see how such reactions are used in analytical methods in Chapters 14–16.

A fifth set of important chemical processes are those that involve only a change in the *environment* of a chemical. One example is a *phase transition*, in which there is a change in the physical but not the chemical nature of a compound.[10] A few examples of phase transitions are the conversion of solid carbon dioxide into CO_2 gas (a process known as *sublimation*), the boiling of a liquid to form a gas, and the melting of a solid to give a liquid. A change in environment also occurs when we create a solution of one chemical in another (giving a *solubility equilibrium*),[12] or when we distribute a chemical between two or more separate chemical phases (a *distribution equilibrium*).[13] We will learn more about these processes later in this book when we discuss chemical solubility (Chapter 7) and the use of phase transitions and distribution equilibria in chemical separation methods (Chapters 20–23).

6.1B Describing Chemical Reactions

Regardless of the type of chemical process we are studying, there are always certain questions we can ask about this process and how it takes place. Two such questions include "What amount of a chemical is

TABLE 6.1   Common Types of Chemical Reactions and Processes

Type of Reaction or Process	Examples
Precipitation reaction	Reaction of carbonate with Ca^{2+} to form solid calcium carbonate ($CaCO_3$), the main chemical in limestone
	$Ca^{2+} + CO_3{}^{2-} \rightleftharpoons CaCO_3(s)$
Acid–base reaction	Dissociation of carbonic acid (H_2CO_3) in water to give bicarbonate ($HCO_3{}^-$)
	$H_2CO_3 + H_2O \rightleftharpoons HCO_3{}^- + H_3O^+$
Complex formation	Binding of oxygen to hemoglobin (Hb) in blood
	$Hb + O_2 \rightleftharpoons Hb{-}O_2$
Oxidation–reduction reaction	Combustion of octane (C_8H_{18}) during the burning of this compound in fuel
	$2\,C_8H_{18} + 25\,O_2 \rightarrow 16\,CO_2 + 18\,H_2O$
Phase transition	Sublimation of dry ice to give carbon dioxide gas
	$CO_2(s) \rightleftharpoons CO_2(g)$
Solubility equilibrium	Dissolving of carbon dioxide gas in water
	$CO_2(air) \rightleftharpoons CO_2(aq)$

required in the reaction or process?" and "How much of this chemical is actually present?" We discussed these issues in Chapter 2 when we examined ways of describing the chemical content of samples by using molarity, weight-per-weight, and related units. But there are other questions about chemical processes that we have not yet addressed, such as "How does the ability of a chemical to react change as we alter its surroundings?" This question is addressed by using *chemical activity*, which refers to the actual amount of energy available from a chemical under the conditions used for a reaction or phase transition. As we will see later, the activity is closely related to the total amount of a chemical in a sample.

A second set of questions we can ask is "How far might the chemical process proceed?" or "How much energy will be given off or required by this process?" These questions are answered by using *chemical thermodynamics*,[12–15] which is the field of chemistry concerned with the changes in energy that take place during chemical reactions or phase transitions and the overall extent to which such processes can occur. To describe these properties, we will pay particular attention to a factor known as an *equilibrium constant*,[12] which we will learn about in Section 6.3.

A third question that can be asked is "How fast does a chemical process occur?" This issue is dealt with by using *chemical kinetics*, the field of chemistry that is concerned with the rates of chemical processes.[12,16,17] This information helps us determine the time required for a given process, which is related to the mechanism by which the reaction or transition takes place. An example of analysis that is based on chemical kinetics is given in Box 6.1. Later in this textbook we will see other examples of how the rate of a reaction can be used for chemical measurements.

6.2 CHEMICAL ACTIVITY

Before we discuss chemical reactions and transitions, we need to think about how we can describe the amount of each reactant or product in these processes. For example, if we were to place HCl into water, this chemical would dissociate to produce hydrogen ions and chloride ions. Although this process strongly favors the formation of such ions, to say this process is independent of concentration or that all these ions act independently of one another is just an approximation. Using this approximation can be a problem when we are dealing with a concentrated solution, because the total amount of the dissolved substances (especially ions) can alter the reactivity of chemicals and make their behavior different from what might be expected based on their total concentrations.

It is especially important to be aware of this relationship in analytical chemistry, where we often measure the apparent reactivity of a chemical with a reagent or method, but would really like to know the total concentration of the chemical. To help us describe this effect, we will use the term "chemical activity" (or simply "activity"). In this section we look at the definition of activity, see how it is related to the concentration of a chemical, and learn how to estimate or control activity in chemical measurements.

6.2A What Is Chemical Activity?

Definition of Chemical Activity. The **chemical activity** (*a*) is defined by Equation 6.1 as being related to the difference in energy μ for a chemical in a particular sample to the value $\mu°$ for the same chemical in its standard state.[12]

$$a = e^{(\mu - \mu°)/(RT)} \tag{6.1}$$

BOX 6.1
Carbon-14 Dating

Along with thinking about chemical equilibrium and the extent to which a reaction might proceed, analytical chemists must also consider how quickly a reaction might occur. *Chemical kinetics* is the area of chemistry that is concerned with the speed of chemical reactions. An excellent example of the importance of kinetics in analytical chemistry is the method of *carbon-14 dating*.[18–20] This technique is commonly used in archaeology, geology, atmospheric science, and medicine for estimating the age of carbon-based materials. This method was developed in 1947 by a group of scientists led by U.S. Chemist William F. Libby, who won the 1960 Nobel Prize in chemistry for this work.[18]

Carbon-14 dating is based on the fact that there are three main isotopes of carbon in nature: carbon-12 and carbon-13, which are both stable, and carbon-14, which is radioactive and slowly converts into nitrogen-14 plus a high-energy electron known as a beta particle. Carbon-14 is unstable but is continually being replenished by the bombardment of nitrogen with neutrons, which are formed by cosmic radiation interacting with atoms in the atmosphere. This process results in the production of additional carbon-14 and causes a relatively consistent amount of carbon-14 to be present in air. This carbon-14 then enters other regions of Earth as part of the carbon dioxide in air goes into the oceans, is placed into sediments, or is incorporated into the food chain through photosynthesis.

The way in which this carbon-14 can be used for determining the age of a plant or animal sample is shown in Figure 6.2. While the plant or animal is living it is continually taking up carbon in food or eliminating it in waste, giving the organism an amount of carbon-14 that is in balance with the surrounding environment. Once the plant or animal has died, no new carbon is being added. The amount of carbon-14 in this sample (as well as in anything that is made from the plant or animal material) begins to decrease over time as the carbon-14 undergoes radioactive decay to form nitrogen-14. The time for half the carbon-14 to undergo this process (or its "half-life") is 5730 (±40) years. If it is assumed the original ratio of carbon-14 to carbon-12 is known, the date the plant or animal died can be estimated by comparing this ratio to the measured amount of carbon-14 vs. carbon-12 in the sample. This analysis is often made by burning a small portion of the sample to convert the carbon into carbon dioxide and then collecting and measuring the carbon-14 and total amount of carbon dioxide generated by the sample. The result is then used with the half-life of carbon-14 to determine the amount of time that has elapsed since the plant or animal that gave rise to the material in the original sample was alive.

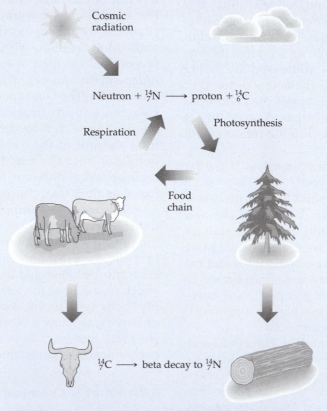

FIGURE 6.2 The production and decay of carbon-14 that forms the basis for carbon-14 dating. This figure shows how carbon-14 is formed in the atmosphere, followed by the uptake of this carbon-14 by plants and animals in the form of carbon dioxide. After the plant or animal dies, no more new carbon-14 enters this material and the remaining carbon-14 undergoes decay to form nitrogen-14.

In this equation, R is the ideal gas law constant and T is the absolute temperature of our system. This equation compares the *chemical potential* (μ) of the chemical (a measure of the energy available in one mole of this material)[12] to the *standard chemical potential* ($\mu°$) for the same material in some *standard state*. This comparison is needed so that we can judge whether the chemical is more or less reactive as we change its environment from a standard state, which is a pure form of the chemical (for example, a pure solid or pure liquid) or a solution that contains a well-defined concentration of this chemical (for example, an exactly 1 M solution). Table 6.2 lists the standard states that are commonly used for chemicals in this type of comparison.

Equation 6.1 indicates that the activity a of a chemical will be a number with no units and that simply reflects how much energy this chemical contains versus its standard state. Another way of looking at this equation is to say the activity and apparent reactivity of a chemical will change as the chemical's environment is changed. If a chemical is in its standard state, such as when we are working with the pure form of a solid or liquid, the value of μ will be the same as $\mu°$ and we will get an activity from Equation 6.1 of $a = 1.0$. If we are working with a dilute solution, the chemical potential will generally be much less than it is for the pure form of the same compound ($\mu < \mu°$), giving a value for a that is somewhere between 0 (the lowest possible value) and 1.0. There are also some situations in which a chemical can have more energy than its standard state. In this case, the chemical potential of the substance is greater than its standard state ($\mu > \mu°$) and gives a value for a greater than 1.0.

To illustrate this idea, let's see how the activity of HCl changes as we vary its total concentration in water (see Figure 6.3). At very low concentrations (that is, [HCl] < 0.01 M), the activities and concentrations for the hydrogen ions and chloride ions that are produced from HCl have similar values, differing by less than 10%. Under these conditions we have a solution that contains a large amount of solvent and only a small amount of dissolved ions. The result is a situation in which essentially all the ions are fully surrounded by the solvent and act independently from one another. Even at low

HCl concentrations, however, there is a small difference between the activity of HCl and its total concentration. This difference is observed because the ions formed from HCl do have some influence over each other's behavior. For instance, the hydrogen ions and chloride ions can attract each other, while neighboring ions with like charges will repel each other (see Figure 6.4). This process becomes greater as we go to higher concentrations and it causes the activity to be lower than would be expected based on the total concentrations for H^+ and Cl^-.

As we move to even higher concentrations, other interactions also begin to alter the activity of the HCl solution. For instance, H^+Cl^- ion pairs may form, which will behave differently than individual H^+ or Cl^- ions. There also will be changes in the repulsion or attraction of these ions as they appear at levels high enough to alter the properties of the solution. In addition, the presence of large amounts of dissolved chemicals can affect the behavior of the solvent. In the case of the HCl solutions in Figure 6.3, the water used as the solvent will interact with ionic substances like H^+ and Cl^-. This

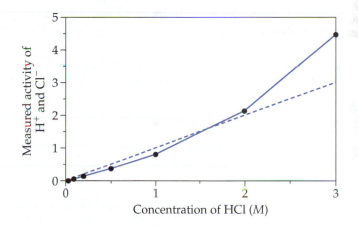

FIGURE 6.3 Change in the measured activity of H^+ and Cl^- in water as the total concentration of HCl is varied. The solid line and data points show the change in the measured activities. The dashed line is included as a reference and shows where the values of concentration and activity are equal. (This graph was generated using data from D.G. Peters, J.M. Hayes, and G.M. Hieftje, *Chemical Separations and Measurements: Theory and Practice of Analytical Chemistry*, Saunders, Philadelphia, PA, 1974, p. 46.)

TABLE 6.2   Standard States of Various Substances

Type of Substance	Standard State (where $a = 1$)
Solid	Pure form of the solid[a]
Liquid	Pure form of the liquid
Gas	Pure form of gas at 1 bar[b]
Solution of a dissolved chemical	One molar (1 M) solution of the chemical

[a]For solids with more than one form, the standard state is the most stable form of the solid.

[b]It is necessary to define the pressure and temperature of a gas in the standard state because the volume, and thus amount of gas per unit volume, will depend on both these factors. A pressure of 1 bar (formerly 1 atmosphere) and a temperature of 0°C (273.15 K) are often used for this purpose, although other temperatures can also be chosen. This combination of conditions is commonly referred to as the *standard temperature and pressure*, or *STP*.

Low ion concentration

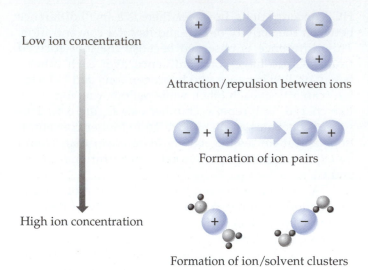

Attraction/repulsion between ions

Formation of ion pairs

High ion concentration

Formation of ion/solvent clusters

FIGURE 6.4   Examples of nonideal effects that can occur in a solution containing ions.

interaction forms large species like H_3O^+, $H_5O_2^+$, or $H_9O_4^+$, reducing the number of ordinary water molecules and increasing the activity of all solutes dissolved in the solution.

Activity Coefficients.   From Figure 6.3 we can see that even though the activity and concentration of a chemical are two different things, these two terms are closely related. We can describe this relationship by using an **activity coefficient (γ)**, as shown in Equation 6.2.[11,12,21]

$$a = \gamma \, (c/c^\circ) \qquad (6.2)$$

In this equation, a refers to the activity of our chemical, c is the concentration of the chemical under the conditions we are examining, and c° is the concentration of the same chemical under some reference conditions (for instance, a concentration of exactly 1.00 M, when we are using molarity to describe the chemical content). Sometimes the subscript "c" is added to the activity coefficient in Equation 6.2 (written as "γ_c") to indicate that this is a "concentration"-based value. Similar relationships can be used to relate the activity of a chemical to its molality or other measures of chemical content.[12] Like activity (a), the activity coefficient for a chemical is a number with no units. This is the case because the concentration c in Equation 6.2 is divided by c°, allowing us to eliminate the units that are present in both these terms. Because c° is chosen to be equal to one, it is often not shown and Equation 6.2 is replaced with the simpler relationship $a = \gamma \cdot c$. Although c° will not be shown in the rest of this book, keep in mind that it is still present when using dimensional analysis to check the units in activity calculations, as illustrated in the following exercise.

A sample of seawater has activities for bicarbonate and carbonate ions of 9.75×10^{-4} and 4.7×10^{-6}, respectively, at 25°C and one atmosphere pressure.[7] The concentrations of these same chemicals are 0.00238 M for HCO_3^- and 0.000269 M for CO_3^{2-}. What are the activity coefficients for these ions?

SOLUTION

The activity coefficient for bicarbonate can be determined from Equation 6.2, where the values of a and c are given and the value of c° is assigned a value of exactly 1.000 M.

$$a = \gamma \, (c/c^\circ)$$
$$9.75 \times 10^{-4} = \gamma \, (0.00238 \, M/1.000 \, M)$$
$$\therefore \, \gamma = \mathbf{0.410}$$

Using the same approach, the activity coefficient for carbonate is found to be **0.017**. In both cases the activity coefficients are much less than one, which reflects the fact that the activities for bicarbonate and carbonate were lower than the total concentrations of these ions. Although these are both dilute solutes, this difference in activity versus concentration is caused by the presence of other ions (e.g., Na^+ and Cl^-) that affect the overall properties of this solution.

As we can see from the last exercise, both the activity and activity coefficient for a solute will be affected by the presence of other substances in a solution. When we are dealing with very dilute solutions, there are few interactions that take place between individual solutes, making the activities and concentrations essentially the same and giving activity coefficients close to one. As we move to more concentrated solutions, more solute–solvent and solute–solute interactions take place and γ will usually be quite different from one. This effect is important to consider when work is being performed with concentrated solutions, reagents, or samples. We will come back to this problem later when we consider various strategies for dealing with the differences between activity and concentration in analytical measurements.

Ionic Strength.   We have already seen two main effects that cause chemical activities to be different from chemical concentration: solute–solvent interactions and solute–solute interactions. Solute–solvent interactions tend to occur only at moderate to high concentrations and are often (but not always) negligible for many of the samples examined in analytical chemistry. Solute–solute interactions are a bigger problem because they occur at even reasonably low analyte levels. To help us identify and control this effect, it is necessary to have some way of describing such interactions. This is particularly true for ions in that their charges can give them an influence over other charged substances over relatively long distances.

In predicting the extent of these interactions, we need to consider the charge and concentration of every type of ion in the solution. This task is accomplished by determining the **ionic strength (I)** of the overall solution,[12]

$$I = \tfrac{1}{2}(c_1 z_1^2 + c_2 z_2^2 + \cdots + c_n z_n^2)$$

or

$$I = \tfrac{1}{2}\Sigma\,(c_i z_i^2) \qquad (6.3)$$

where c_i is the concentration of a particular type of ion in solution, and z_i is the charge on that ion (for instance, "+1" for H^+, "–1" for Cl^-, and "+2" for Cu^{2+}). From Equation 6.3, we get a concentration-based ionic strength, which is sometimes written as "I_c."[12] The fact that the z-terms are all squared gives a positive $c_i z_i^2$ value for all ions, regardless of whether the ions are negative or positive. The result is that the value of I will always be positive. These squared terms also ensure that multiply-charged ions have a greater impact on chemical activities than singly charged ions.

| **EXERCISE 6.2** | **The Ionic Strength of Sea Water** |

A chemist wishes to mimic the effects of sea water on a reaction by preparing a solution that contains 0.500 M sodium chloride (NaCl) and 0.050 M magnesium chloride (MgCl$_2$). If these salts completely dissolve and no other substances are present, what is the ionic strength of this mixture?

SOLUTION

To find the ionic strength, we simply need to place the individual concentrations and charges for each of the ions in our solution (Na^+, Cl^-, and Mg^{2+}) in Equation 6.3. In this case, the total concentration for Cl^- is 0.600 M, which is the sum of what is produced from NaCl (giving 0.500 M Cl^-) and MgCl$_2$ (which gives $2 \cdot 0.050 = 0.100\ M\ Cl^-$).

$$I = \tfrac{1}{2}[(0.500\ M\ Na^+)(+1)^2 +$$
$$(0.600\ M\ Cl^-)(-1)^2 +$$
$$(0.050\ M\ Mg^{2+})(+2)^2] = \mathbf{0.650\ M}$$

Notice that the ionic strength is larger than the concentration of any single type of ion in our solution, even for the ion with the highest concentration (Cl^- in this case). This result occurs because I is a measure of the overall influence of all ions on chemical activity.

Although ionic strength can be used to describe the composition of a reagent or sample, it differs from a chemical concentration in several ways. For instance, Equation 6.3 indicates that the value of I depends only on the charge and concentration of the ions in a solution,

and (at least to a first approximation) not on the specific types of ions present. As a result, a solution of 0.10 M NaCl (dissociating into Na^+ and Cl^- ions) will have the same ionic strength as a 0.10 M solution of HNO$_3$ (forming H^+ and NO_3^-) when both chemicals completely dissociate in water. Also, the concentrations of nonionic solution components are not considered in calculating the ionic strength, because such substances generally interact only with solute or solvent molecules very close to them and do not usually have any appreciable effect on the activity of other chemicals in a solution.

6.2B Chemical Activity in Analytical Methods

Because one of the most common applications of analytical chemistry is the measurement of chemicals, it is important to consider how the difference between chemical activity and concentration will affect these measurements. In such an analysis we generally have a series of standards and reagents that contain known amounts or concentrations of the chemicals we wish to use or study. However, when we use these reagents or samples, the results we obtain will often be based on the activity of the analyte in these samples. This fact means the conversion from activity to concentration is often an inherent part of an analysis. We will now look at several approaches for dealing with this issue, including methods for estimating or controlling chemical activity.

Estimating Activity. We would ideally like for a chemical analysis to directly measure the activity of an analyte and to relate this activity back to the amount of analyte in our sample. Many of the techniques we use in analytical chemistry do give a response that is related to chemical activity (for example, gravimetric analysis, titrations, chromatography, and electrochemical methods). Unfortunately, this direct approach is complicated by the fact that any experimental measure of activity gives a weighted average for the activities of both negatively and positively charged ions in solution (because one is always present with the other). This weighted average is represented by a term known as the *mean activity coefficient* ($\gamma_\pm$).[12] The following equation shows how activities of individual ions (γ_A and γ_B for ions A^{n+} and B^{m-}) are related to the mean activity coefficient for a strong electrolyte A_mB_n (that is, a chemical where one mole dissociates to produce m moles of A^{n+} and n moles of B^{m-}).[22]

$$(\gamma_\pm)^{m+n} = (\gamma_A)^m \cdot (\gamma_B)^n \qquad (6.4)$$

We have already seen one use of this value in Figure 6.3, where the measured activity for HCl was actually the mean activity of both H^+ and Cl^-, as described by the mean activity coefficient $(\gamma_{\pm HCl})^2 = (\gamma_{H+})^1 \cdot (\gamma_{Cl-})^1$.

If it is necessary to know the activity of a chemical and we do not have the time or ability to measure this value, another option is to use one of several equations that have been developed to estimate activity coefficients. Although this approach is not as reliable as using a measured value, it usually provides a reasonably good approximation for dilute solutions. The most famous relationship used for this purpose is the **extended Debye–Hückel equation**.[11,12,22]

$$\log(\gamma) = \frac{-A \cdot z^2 \cdot \sqrt{I}}{1 + a \cdot B \cdot \sqrt{I}} \qquad (6.5)$$

This equation, derived by Peter Debye and Erich Hückel in 1923 (see Figure 6.5),[23,24] relates the activity coefficient for an ion to the ionic strength of its solution (I), the charge on the ion (z), and three adjustable parameters: a, A, and B. The first of these adjustable parameters is an ion-size term a, which represents the closest distance the ion can approach another ion. Table 6.3 gives values for this term for many common inorganic ions in water. Appendix B has a similar table for organic ions. The other two adjustable parameters in Equation 6.5 are A and B,

which represent the effects of temperature and solvent on the activity coefficient.[22]

In water at room temperature (25°C, the condition often used during chemical analysis), the value of A is approximately 0.51 and B (when a is given in picometers) is roughly 3.28×10^{-3}, or $1/(305)$. Putting these values into Equation 6.5 gives the following version of the extended Debye–Hückel equation.[22]

In Water at 25°C: $\qquad \log(\gamma) = \dfrac{-0.51 \cdot z^2 \cdot \sqrt{I}}{1 + (a \cdot \sqrt{I})/305} \qquad (6.6)$

If we are working with an even more dilute solution, Equation 6.6 can be further simplified to give an expression known as the *Debye–Hückel limiting law (DHLL)*, where $\log(\gamma) = -0.51 \cdot z^2 \cdot \sqrt{I}$. This latter equation is much more limited in use than Equations 6.5 and 6.6, so we will instead use these more complete equations throughout the rest of this chapter. (*Note*: In some texts you will see only the term $1 + \sqrt{I}$ in the denominator of Equation 6.6; this assumes you are working with an ion

Peter Debye

Erich Hückel

FIGURE 6.5 Peter Debye (1884–1966) and Erich Hückel (1896–1980). Debye was a Dutch physicist who studied the behavior of ions in solution. Hückel was born in Germany and was Debye's assistant in Zürich, where in 1923 they together developed their famous equation for describing the activity coefficients of ions in solution. In earlier work with Swiss scientist Paul Scherrer, Debye showed that the powders of crystalline solids could be examined with X rays to determine the chemical structures of these solids, a technique now known as powder X-ray diffraction. Debye won the 1936 Nobel Prize in chemistry for his work, and moved to the United States near the beginning of World War II, continuing his studies at Cornell University. The unit of measure for a dipole moment, the *debye* (D), is named in his honor. After working for Debye, Hückel entered the area of quantum mechanics and briefly worked with Neils Bohr. Hückel later became a professor of physics in Marburg, Germany, until his retirement.

TABLE 6.3 Estimated Individual Activity Coefficients for Inorganic Ions in Water at 25°C[*]

Type of Ion	Ion Size Parameter a (pm)	Activity Coefficient at Ionic strength I (M)							
		I = 0.0005	0.001	0.002	0.005	0.01	0.02	0.05	0.10
Charge = +1 or –1									
H^+	**900**	0.976[a]	0.967	0.955	0.934	0.913	0.889	0.854	0.825
Li^+	**600**	0.975	0.966	0.953	0.930	0.907	0.878	0.833	0.795
Na^+, ClO_2^-, IO_3^-, HCO_3^-, $H_2PO_4^-$, HSO_3^-, $H_2AsO_4^-$, $[Co(NH_3)_4(NO_2)_2]^+$	**400–450[b]**	0.975	0.965	0.952	0.928	0.902	0.870	0.817	0.773
OH^-, F^-, SCN^-, OCN^-, HS^-, ClO_3^-, ClO_4^-, BrO_3^-, IO_4^-, MnO_4^-	**350**	0.975	0.965	0.951	0.926	0.900	0.867	0.811	0.762
K^+, Cl^-, Br^-, I^-, CN^-, NO_2^-, NO_3^-	**300**	0.975	0.965	0.951	0.925	0.899	0.864	0.806	0.753
Rb^+, Cs^+, NH_4^+, Tl^+, Ag^+	**250**	0.975	0.965	0.951	0.924	0.897	0.862	0.801	0.745
Charge = +2 or –2									
Mg^{2+}, Be^{2+}	**800**	0.906	0.872	0.829	0.756	0.689	0.616	0.516	0.444
Ca^{2+}, Cu^{2+}, Zn^{2+}, Sn^{2+}, Mn^{2+}, Fe^{2+}, Ni^{2+}, Co^{2+}	**600**	0.904	0.870	0.824	0.747	0.675	0.595	0.482	0.400
Sr^{2+}, Ba^{2+}, Cd^{2+}, Hg^{2+}, Ra^{2+}, S^{2-}, $S_2O_4^{2-}$, WO_4^{2-}	**500**	0.904	0.868	0.822	0.743	0.668	0.583	0.464	0.376
Pb^{2+}, CO_3^{2-}, SO_3^{2-}, MoO_4^{2-}, $[Co(NH_3)_5Cl]^{2+}$, $[Fe(CN)_5NO]^{2-}$	**450**	0.903	0.868	0.821	0.740	0.664	0.577	0.454	0.363
CrO_4^{2-}, Hg_2^{2+}, HPO_4^{2-}, SO_4^{2-}, $S_2O_3^{2-}$, $S_2O_6^{2-}$, $S_2O_8^{2-}$, SeO_4^{2-}	**400**	0.903	0.867	0.820	0.738	0.660	0.571	0.444	0.350
Charge = +3 or –3									
Al^{3+}, Ce^{3+}, Cr^{3+}, Fe^{3+}, In^{3+}, La^{3+}, Nd^{3+}, Pr^{3+}, Sc^{3+}, Sm^{3+}, Y^{3+}	**900**	0.801	0.737	0.659	0.539	0.442	0.348	0.241	0.178
$[Co(NH_3)_6]^{3+}$, $[Co(NH_3)_5H_2O]^{3+}$, $[Cr(NH_3)_6]^{3+}$, $[Fe(CN)_6]^{3-}$, PO_4^{3-}	**400**	0.795	0.726	0.640	0.505	0.393	0.283	0.161	0.094
Charge = +4 or –4									
Ce^{4+}, Sn^{4+}, Th^{4+}, Zr^{4+}	**1100**	0.678	0.587	0.485	0.347	0.251	0.172	0.098	0.062
$[Fe(CN)_6]^{4-}$	**500**	0.667	0.568	0.457	0.304	0.199	0.116	0.046	0.020

[*] This table is based on data provided in J. Kielland, "Individual Activity Coefficients of Ions in Aqueous Solutions," *Journal of the American Chemical Society*, 59 (1937) 1675–1678.

[a] The last number to the right and underlined in each activity coefficient is a guard digit (see discussion of guard digits in Chapter 2).

[b] The activities coefficients given for a ± 1 charge and a = 400–450 are the averages of the values obtained at a = 400 and a = 450.

with a size term a of roughly 300 pm, making $a/305$ equal to about 1.0.)

One way we can use the extended Debye–Hückel equation is to predict how the activity coefficient for an ion will change with the ionic strength of its solution.

An example is shown in Figure 6.6, where the actual activity coefficients observed for HCl are compared to those calculated with Equation 6.6. The way these values were obtained is illustrated in the following exercise.

EXERCISE 6.3 Estimating Activity Coefficients for Ions

Using Equation 6.6, what activity coefficients would be expected for H^+ and Cl^- at 25°C in an aqueous HCl solution with an ionic strength of 0.010 M? What is the mean activity coefficient predicted for HCl and how does this compare to the measured values in Figure 6.3?

SOLUTION

In Table 6.3 the ion size parameters for H^+ and Cl^- are given as being 900 pm and 300 pm. By placing these values in Equation 6.6 along with the given ionic strength of 0.010 M, we get the following estimates for the activity coefficients for the individual H^+ and Cl^- ions.

For H^+ at $I = 0.010$ M:

$$\log(\gamma_{H^+}) = \frac{-0.51 \cdot (+1)^2 \cdot \sqrt{0.010 \ M}}{1 + (900 \ \text{pm}) \cdot \sqrt{0.010 \ M}/(305 \ \text{pm})}$$

$$\therefore \ \gamma_{H^+} = 0.91\underline{3} = \mathbf{0.91}$$

For Cl^- at $I = 0.010$ M:

$$\log(\gamma_{Cl^-}) = \frac{-0.51 \cdot (-1)^2 \cdot \sqrt{0.010 \ M}}{1 + (300 \ \text{pm}) \cdot \sqrt{0.010 \ M}/(305 \ \text{pm})}$$

$$\therefore \ \gamma_{Cl^-} = 0.89\underline{9} = \mathbf{0.90}$$

This same answer could have been reached by interpolating between the activity coefficients given for these ions in Table 6.3 at similar ionic strengths.

Now that we have estimated the activity coefficients for our separate ions, we can combine these values using Equation 6.4 to get the mean activity coefficient for HCl.

$$(\gamma_{\pm HCl})^{1+1} = (\gamma_{H^+})^1 \cdot (\gamma_{Cl^-})^1$$
$$(\gamma_{\pm HCl})^2 = (0.91\underline{3})^1 \cdot (0.899)^1$$
$$\therefore \ \gamma_{\pm HCl} = 0.90\underline{6} = \mathbf{0.91}$$

In looking back to Figure 6.3, we can see that this calculated number agrees fairly closely with the expected value at $I = 0.010$ M for HCl in water.

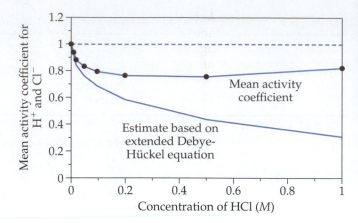

FIGURE 6.6 Actual and predicted mean activity coefficients for HCl based on the extended Debye–Hückel equation. (The experimental results are based on data from D.G. Peters, J.M. Hayes, and G.M. Hieftje, *Chemical Separations and Measurements: Theory and Practice of Analytical Chemistry*, Saunders, Philadelphia, PA, 1974, p. 46.)

of neutral compounds as the ionic strength is increased. This effect can be viewed as an increase in the activity of a neutral agent, causing it to have a higher effective concentration and come out of the solution more easily. Such an effect can be represented by the following equation,

$$\log(\gamma) = k \cdot I \qquad (6.7)$$

where γ is the activity coefficient for the neutral compound, I is the ionic strength of the solution containing this compound, and k is a constant for the compound known as its *salting coefficient*.[22] The value of k is between 0.01 and 0.10 for most noncharged compounds, which means these compounds will have an increase in activity from 1.00 to 1.02–1.26 as we increase I from 0 to 1.0 M. For low ionic strengths ($I < 0.10$ M), this effect causes less than a 2.5% change in activity and is usually insignificant. Thus, the activity coefficient for a neutral compound in such solutions is approximately equal to one. However, the salting coefficient does need to be considered when working at higher ionic strengths.

One problem with the extended Debye–Hückel equation is that it only gives good agreement with experimental activity coefficients up to an ionic strength of about 0.10 M (for instance, see Figure 6.6). This difference occurs because the extended Debye–Hückel equation only considers how the activity coefficients for ions are affected by simple attraction and repulsion between neighboring ions. At higher ionic strengths, other effects like ion pair and ion/solvent cluster formation also become significant. This situation requires more advanced equations to estimate activity coefficients (see the problems at the end of this chapter).

Although ions are the main types of chemicals affected by changes in ionic strength, noncharged compounds are also affected to a small extent by these changes. This phenomenon, known as the *salting-out effect*, is usually seen as a slight decrease in the solubility

EXERCISE 6.4 Estimating the Activity Coefficient for a Neutral Compound

The salting coefficient for H_2CO_3 in seawater is about 0.075. If a seawater sample has an ionic strength of 0.70 M, what is the activity coefficient for H_2CO_3?

SOLUTION

We can estimate the activity coefficient for H_2CO_3 by placing $k = 0.075$ and $I = 0.70$ M into Equation 6.7 and solving for γ.

$$\log(\gamma_{H_2CO_3}) = (0.075) \cdot (0.70 \ M) \ \therefore \ \gamma_{H_2CO_3} = \mathbf{1.13}$$

Under these conditions, H_2CO_3 will have an activity that is 13% higher than would occur if this same chemical were present in water without added salts.

Factors That Affect Activity. Besides allowing us to estimate activity coefficients, we can use relationships like Equations 6.5 through 6.7 to help us see what factors are most important in affecting the activity of chemicals. We already know that the ionic strength of a solution is important in determining the activity coefficients for both charged and neutral compounds. As a result, we should always try to match the matrix of a sample to that of our standards so that they will have similar ionic strengths. One way this can be done is by adding a fixed excess of a salt to all of our samples and standards to avoid any major changes in the ionic strength due to changes in the concentration of the analyte or the sample composition. For instance, a change in Ca^{2+} concentration from 0 to $10 \, \mu M$ for samples prepared in a $0.10 \, M$ NaCl solution will have an insignificant effect on the ionic strength during the measurement of Ca^{2+} levels.

The type of solvent used to prepare a solution is also important in determining chemical activity. One way the solvent can affect activity coefficients is through changes that occur in the *dielectric constant* (ε), which will alter both the A and B terms in the Debye–Hückel equation. The dielectric constant can be thought of as a measure of the degree to which a solvent or material will allow an electrostatic force from one charged body (such as an ion) to affect another.[15] The dielectric constant can also be used as a rough indicator of a chemical's "polarity." In general, it will be easier for dissolved ions to influence each other through attraction or repulsion when these ions are present in a polar solvent like water (which has a high dielectric constant of $\varepsilon = 78.54$ at 25°C) versus less polar solvents like methanol (CH_3OH, $\varepsilon = 32.63$) or ethanol (CH_3CH_2OH, $\varepsilon = 24.30$).

Equations 6.5–6.7 also show how the properties of a solute will determine the extent to which a change in ionic strength will affect its activity. The charge (z) on a solute is one item that appears in the Debye–Hückel equation. If we compare solutes with similar sizes, those with the largest charge will also be affected the most by changes in ionic strength and have the greatest difference between their chemical activity and concentration. The effective size of the solute is also important to consider. If we compare the calculated activity coefficients in Table 6.3 for hydrated ions with the same charge but different values for a, we see that in going from Rb^+ ($a = 250$ pm) to H^+ ($a = 900$ pm), the activity coefficient at $I = 0.10 \, M$ changes from 0.75 to 0.83. However, this effect is much smaller than what is seen when increasing the charge on an ion and is often ignored when working at low ionic strengths.

You may have noticed that the ion size parameters in Table 6.3 do not follow the trend that might be expected based on the periodic chart. In moving down the left column of the periodic table from H^+ to Li^+, Na^+, and K^+, you would initially expect these ions to increase in size because they are adding greater numbers of electrons, protons, and neutrons. This trend is actually just the opposite of what you see when comparing the size parameters in Table 6.3. The reason for this difference is

that the ion size parameter in the Debye–Hückel equation is actually a measure of the size of an ion plus the shell of solvent that surrounds it. When we are using water as the solvent, this shell is called the *hydration layer*, and the resulting size of the ion plus this solvent is known as the *hydrated radius*. As shown in Figure 6.7, small ions like Li^+ (and also H^+) have a small region of concentrated charge that tends to attract a large amount of solvent and form a large hydrated radius. Larger ions like K^+ and Na^+ have a less concentrated charge, causing them to have smaller hydration layers, even though the ion itself is much larger.

Although most analytical chemists do not think about chemical activity as they go about their daily work, they do deal with this issue on a routine basis through the methods and reagents they use for chemical measurements. In most analytical techniques, this issue is handled by generating a calibration curve, in which the measured response (related to chemical activity) is plotted against the concentration or content of standards that contain known amounts of the chemical of interest. If the analysis method has been designed so that samples are treated in the same way as these standards, and therefore have similar relationships between their activity and content, this calibration curve can be used to determine the amount of the analyte in unknown samples. One example of this approach is the graph we saw in Figure 6.3, in which the measured activity for HCl was plotted against its total concentration in solution. Based on this graph, we could measure the activities of other HCl solutions and relate these activities back to the concentrations of HCl in these samples.

6.3 CHEMICAL EQUILIBRIUM

So far in this chapter we have discussed various types of chemical processes and have examined the concept of chemical activity. In this section we will learn about yet another factor, the equilibrium constant, which can be used to help us describe reactions used in analytical techniques.

6.3A What Is a Chemical Equilibrium?

To understand what is meant by a "chemical equilibrium," we can go back to our opening example of CO_2 production and uptake within the ocean. Carbon dioxide and related compounds in the ocean (such as carbonate ions, CO_3^{2-}) come from several sources, two of which are the air above the ocean and materials on the bottom of the ocean. To simplify this picture, we can focus on the events taking place near the ocean floor. On the ocean floor is calcium-containing debris from dead single-celled sea life. One compound found here is calcium carbonate, $CaCO_3$. As solid calcium carbonate comes into contact with water, it dissociates to a small extent to form calcium ions and carbonate ions dissolved in water.

$$CaCO_3(s) \; \longrightarrow \; Ca^{2+} + CO_3^{2-} \qquad (6.8)$$

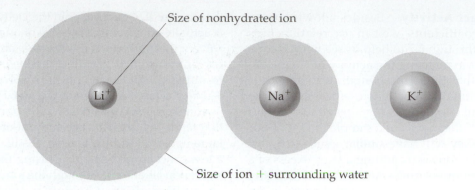

Size of nonhydrated ion

Size of ion + surrounding water

FIGURE 6.7 A comparison of the effective diameter of Li^+, Na^+, and K^+ in water versus the size of these same ions in the absence of water.

At the same time, some of the calcium ions in the surrounding water combine with carbonate ions to form new calcium carbonate that settles back on to the ocean floor.

$$Ca^{2+} + CO_3^{2-} \longrightarrow CaCO_3(s) \qquad (6.9)$$

Both processes happen at the same time but proceed in opposite directions, resulting in an ongoing process. To describe the overall reaction, we can combine these processes and use a double arrow $\rightleftarrows$ to indicate that the reaction is proceeding in both the forward and reverse directions.

$$CaCO_3(s) \rightleftarrows Ca^{2+} + CO_3^{2-} \qquad (6.10)$$

The same type of reaction can take place in a self-contained system, such as a beaker of water containing solid calcium carbonate. Given enough time, this reaction will reach a point where the rate at which calcium carbonate dissolves exactly equals the rate at which new calcium carbonate is formed. This situation, which occurs when the forward and reverse rates of a reaction are equal, is known as a **chemical equilibrium** and is represented by the symbol " $\rightleftharpoons$ ".[11,21]

At Equilibrium: $CaCO_3(s) \rightleftharpoons Ca^{2+} + CO_3^{2-}$ (6.11)

A chemical equilibrium is important in analytical chemistry because it represents the furthest extent to which an overall reaction will proceed. Knowledge of this feature can be quite useful when you are preparing reagents or designing a method to obtain the maximum amount of a product for measurement.

Definition of an Equilibrium Constant. When a reaction has reached the point of equilibrium, the overall chemical activity of each product and reactant will be a constant. One way of describing this situation is to look at the relative amount of each reactant and product that exists under these conditions. This is accomplished by using a ratio known as an **equilibrium constant**, which is represented by the general symbol K.[12,21]

To illustrate this idea, suppose we have a reaction that involves the combination of m moles of reactant A with n moles of reactant B to give r moles of product C and s moles of product D.

$$mA + nB \rightleftharpoons rC + sD \qquad (6.12)$$

If this system is at equilibrium (as is implied by the symbol $\rightleftharpoons$), the equilibrium constant for this reaction is given by the following ratio of chemical activities (a_A, a_B, a_C, and a_D) for the reactants and products, in which each activity is raised to a power equal to the stoichiometric amount of its corresponding chemical in the reaction (that is, a_A is raised to the power m, a_B is raised to the power n, and so on).

$$K^\circ = \frac{(a_C)^r (a_D)^s}{(a_A)^m (a_B)^n} \qquad (6.13)$$

Because we are using only activities in this particular case, the result is a special ratio known as a *thermodynamic equilibrium constant*, K°.[12,22] For example, we would use the following ratio to give the thermodynamic equilibrium constant for the calcium carbonate reaction in Equation 16.11.

$$K^\circ = \frac{(a_{Ca^{2+}}) (a_{CO_3^{2-}})}{(a_{CaCO_3})} \qquad (6.14)$$

It is important when we are writing an equilibrium constant expression to use a *balanced* chemical reaction, as we have done for the example in Equations 6.11 and 6.14. This practice is essential because each reaction or product in the reaction will have a term in the equilibrium constant, and the number of moles of these reactants or products that take part in the reaction (that is, the reaction **stoichiometry**)[11,21] will determine the powers that appear on these terms.

One way we can use an equilibrium constant is to determine the activities of all products and reactants in a reaction at equilibrium, as we will see later in this chapter. If we know the final activities of our reactants and products, we can also calculate the equilibrium constant for a reaction. This approach is illustrated in the next exercise. In addition, once we have calculated or measured an equilibrium constant, we can use it to predict how the same reaction will behave when we combine other amounts of products and reactants.

EXERCISE 6.5   Calculating the Value of an Equilibrium Constant

A beaker sitting at room temperature (25°C) contains an aqueous solution of calcium ions and carbonate ions in direct contact with solid calcium carbonate. The activities of the calcium ions and carbonate ions are both found to be 7.0×10^{-5}, and it is known that the activity of the solid calcium carbonate is equal to 1.0. If these chemicals are at equilibrium, what is the equilibrium constant for this reaction?

SOLUTION

The reaction taking place here is the same as given in Equation 6.11.

$$CaCO_3(s) \rightleftharpoons Ca^{2+} + CO_3{}^{2-}$$

To solve this problem, we simply need to place the activities for each of our chemicals into the equilibrium constant expression for this reaction, as given in Equation 6.14.

$$K° = \frac{(7.0 \times 10^{-5})(7.0 \times 10^{-5})}{1} = 4.9 \times 10^{-9}$$

Notice there are no units given for the value of $K°$ because it is based on chemical activities, which are also numbers with no units. Later we will see how equilibrium constants can also be written in terms of product and reactant concentrations.

Using Equilibrium Constants.   One way we can use an equilibrium constant is to determine the actual extent to which a reaction will proceed toward the formation of products. A large equilibrium constant will be obtained for a reaction in which product formation is favored, while a small value for K will occur if the reaction creates only a small amount of products (or favors the presence of the reactants). As an example, in Exercise 6.5 the equilibrium constant for dissolving solid calcium carbonate and forming calcium and carbonate ions was found to be 4.9×10^{-9}. This result tells us this reaction highly favors the formation of calcium carbonate from calcium and carbonate ions or of solid calcium carbonate staying in its present form.

Another way we can use an equilibrium constant is to obtain information on the change in energy that accompanies a reaction. We can do this because an equilibrium constant is directly related to the change in standard Gibbs free energy ($\Delta G°$) that occurs as we go from the reactants to products. The actual relationship between $K°$ and $\Delta G°$ is as follows.

$$\Delta G° = -RT \ln K° \qquad (6.15)$$

where T is the absolute temperature at which the reaction is occurring and R is the ideal gas law constant.[11,22] The relationship in Equation 6.15 is the basis for the way in which we write equilibrium constant expressions, like those shown in Equations 6.13 and 6.14. Further information on this relationship can be found in the problems at the end of this chapter.

EXERCISE 6.6   Relationship Between an Equilibrium Constant and the Energy of a Reaction

What is the change in standard Gibbs free energy for the reaction we considered in Exercise 6.5?

SOLUTION

In Exercise 6.5 we determined that the thermodynamic equilibrium constant for the dissolving of calcium carbonate in water was 4.9×10^{-9} at 25°C (or 298 K). We can then place these numbers into Equation 6.15 along with a value for R of 8.314 J/(mol · K).

$$\Delta G° = -R\,T \ln K°$$
$$= -(8.314\ \text{J/mol} \cdot \text{K}) \cdot (298\ \text{K}) \cdot \ln(4.9 \times 10^{-9})$$
$$\therefore \Delta G° = \mathbf{4.74 \times 10^4\ \text{J/mol or } 47.4\ \text{kJ/mol}}$$

Figure 6.8 shows how the values of $K°$ and $\Delta G°$ are related at room temperature. As we can see, a value of $K°$ that is greater than 1.0 will result in a negative value for $\Delta G°$, which means such a reaction will release energy as it goes from reactants to products. A reaction with an equilibrium constant less than 1.0 will give a positive value for $\Delta G°$, indicating it requires added energy to go from the desired reactants to products. Furthermore, as $K°$ becomes much smaller or larger than 1.0, the amount of free energy given off or taken up by the reaction will increase.

We can also use equilibrium constants to predict the direction in which a reaction will tend to proceed when we start with any given amount of reactants or products. It is often possible for a chemist to generally predict which direction a reaction might proceed by using **Le Châtelier's principle**.[11,21] This principle states that when a change or "stress" is placed on a system at equilibrium (such as a change in reactant or product concentrations), the system will respond to partially relieve this stress (for instance, by creating more products or reactants). We can also examine this shift in a more quantitative manner by putting the activities (or concentrations) of these chemicals into the same type of ratio that we use to give the value for $K°$ at equilibrium, but now using the *nonequilibrium conditions*. When we do this, the ratio we calculate is known as the **reaction quotient (Q)**.[12,22] If this reaction quotient is greater than $K°$, it means we have too much product and the reaction will proceed to form reactants

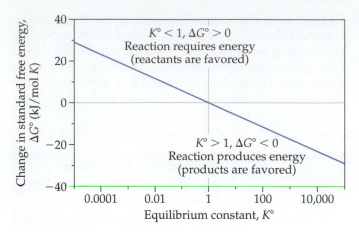

FIGURE 6.8 Relationship between the thermodynamic equilibrium constant ($K°$), the associated change in standard free energy ($\Delta G°$) for a reaction, and temperature. This plot was prepared using Equation 6.15.

until an equilibrium is reached. In this situation the "products," which are the chemicals written on the right side of the equation, are really acting as the "reactants" that give rise to other chemicals. If Q is less than $K°$, there is an excess of the reactants and more products will be formed until the reaction reaches equilibrium. As we will see later, this approach can also be used to predict the amount of each reactant and product that will be present at equilibrium.

EXERCISE 6.7	Predicting the Direction of a Chemical Reaction

A small amount of solid calcium carbonate is placed into water at room temperature along with calcium and carbonate ions that have initial concentrations of $1.0 \times 10^{-4}\ M\ Ca^{2+}$ and $1.0 \times 10^{-4}\ M\ CO_3{}^{2-}$. If we assume that the activities of all our dissolved chemicals are approximately the same as their concentrations, will this mixture react to form more calcium and carbonate ions or more calcium carbonate?

SOLUTION

The reaction we are considering for this mixture is the same as the one examined in Exercises 6.5 and 6.6, which has an equilibrium constant of 4.9×10^{-9} at 25°C. To solve this problem we simply need to use our starting activities to calculate Q and compare this value to $K°$.

Under Nonequilibrium Conditions:

$$Q = \frac{(a_{Ca^{2+}})\,(a_{CO_3{}^{2-}})}{(a_{CaCO_3})}$$

$$= \frac{(1.0 \times 10^{-4})(1.0 \times 10^{-4})}{1}$$

$$\therefore Q = 1.0 \times 10^{-8}$$

The value of Q is greater than $K°$ for this reaction ($1.0 \times 10^{-8} > 4.9 \times 10^{-9}$), so there is more product initially present than would be expected at equilibrium. Thus, for equilibrium to be established in this system some of the calcium ions and carbonate ions must combine to form calcium carbonate, which will lower the value of Q until it equals $K°$.

Concentration-Based Equilibrium Constants. Up to this point we have used chemical activities to describe the amount of each reactant and product at equilibrium. As we learned earlier, an equilibrium constant expressed this way is called a *thermodynamic equilibrium constant* ($K°$). Although these constants do depend on temperature and pressure, they are independent of ionic strength and other effects that could cause deviations from the expected behavior of ideal solutions. This feature makes these values popular for use in reference tables that provide equilibrium constants for chemical reactions. However, as we saw in Exercise 6.7, it is often convenient to use concentrations when calculating equilibrium constants or reaction quotients. For example, an equilibrium constant for the general reaction in Equation 6.12 could be written as shown below.

$$K = \frac{[C]^r\,[D]^s}{[A]^m\,[B]^n} \tag{6.16}$$

This type of relationship gives a value we will call a *concentration-dependent equilibrium constant.*[22] In this book we will sometimes use "apparent" units based on molarity in writing this type of equilibrium constant for solution-phase chemicals, but other concentration units can also be used (such as molality or mole fractions). When dealing with gas-phase chemicals in reactions, units such as partial pressures can also be employed. For any solid substance that takes part in a reaction, it is customary to use chemical activities instead of concentrations to describe this chemical in an equilibrium constant expression. We do this for solids because, as we saw in Table 6.2, the chemical activity for a solid is defined as a standard state and has an assigned chemical activity of exactly one. This is a useful feature when working with calculations and reactions that involve such materials.

Concentration-dependent equilibrium constants are convenient to use when such values have been determined under the same set of temperature, pressure, and solution conditions that we will actually be using in an analysis or experiment. It is important to keep in mind, though, that concentration-dependent equilibrium constants will depend on the ionic strength and type of mixture being examined, because concentrations for individual species are also affected by these factors. A concentration-dependent equilibrium constant (K) is only approximately equal to the thermodynamic equilibrium constant ($K°$) for a reaction when we are working with dilute solutions (that is, when chemical activities and concentrations are similar in value). This feature can create a problem if we want to study a reaction at higher chemical concentrations and are using thermodynamic equilibrium constants provided in tables. To tackle this problem, we need to know exactly how K and $K°$ are related to one another. This relationship can be found by going back to Equation 6.2 (which we will represent here as $a = \gamma c$), which shows that the activity for a chemical is related to its concentration through its activity coefficient. Based on this equation, we can show that the concentration-dependent and thermodynamic equilibrium constants are related as follows.

$$K° = \frac{(\gamma_C [C])^r (\gamma_D [D])^s}{(\gamma_A [A])^m (\gamma_B [B])^n} \quad (6.17)$$

$$K° = K \cdot \frac{(\gamma_C)^r (\gamma_D)^s}{(\gamma_A)^m (\gamma_B)^n} \quad (6.18)$$

An illustration of the differences between $K°$ and K is given in Table 6.4 for the dissociation of calcium carbonate into calcium ions and carbonate ions. As this table shows, the values for $K°$ and K are quite close at low ionic strengths, but become very different for even moderately concentrated solutions. The size of this difference will depend on such things as the ionic strength, the charge on each ion, and its effective size in the solvent. It will also depend on the number of ions involved in the reaction and whether these ions are present as both products and reactants, because these items will affect the value of the activity coefficient ratio in Equation 6.18. The more this ratio deviates from one, the greater $K°$ and K will differ under a given set of reaction conditions.

Equation 6.18 indicates that if we know the thermodynamic value $K°$ for a reaction and can estimate the activity coefficients for each product and reactant at a given ionic strength, we can also estimate K for the reaction under these conditions. This process is what was done in determining the values of K shown in Table 6.4. We can also use the relationship in Equation 6.18 in the opposite manner to go from a concentration-dependent equilibrium constant to a thermodynamic equilibrium constant by using the ionic strength of the solution to estimate the activity coefficients of the products and reactants. The following exercise illustrates this idea.

EXERCISE 6.8	Converting Between $K°$ and K

Write an expression that gives the concentration-dependent equilibrium constant for the dissociation of calcium carbonate into calcium ions and carbonate ions. What is the expected value for K if this reaction is conducted in a solution where $\gamma_\pm$ for Ca^{2+} and CO_3^{2-} is 0.80?

SOLUTION

The value of K for this reaction is given by the following equation.

$$K = \frac{[Ca^{2+}][CO_3^{2-}]}{1}$$

where an activity of one is given for the solid calcium carbonate in the denominator. By replacing activities with concentrations in the thermodynamic equilibrium expression we wrote for this same reaction in Equation 6.14, we can show that $K°$ and K for this reaction are related as follows.

$$K° = \frac{(\gamma_{Ca^{2+}})[Ca^{2+}](\gamma_{CO_3^{2-}})[CO_3^{2-}]}{1}$$

or

$$K° = K \cdot (\gamma_{Ca^{2+}})(\gamma_{CO_3^{2-}})$$

Using Equation 6.4, we estimated $\gamma_{Ca^{2+}}$ and $\gamma_{CO_3^{2-}}$ by using the fact that $(\gamma_\pm)^2 = (0.80)^2 = (\gamma_{Ca^{2+}})^1(\gamma_{CO_3^{2-}})^1$, giving a value of 0.894 for $\gamma_{Ca^{2+}}$ and $\gamma_{CO_3^{2-}}$. We also know the value for $K°$ in this reaction (4.9×10^{-9} at 25°C), so we can use the preceding relationship to obtain the expected size of K under the given reaction conditions.

$$K° = K \cdot (\gamma_{Ca^{2+}})(\gamma_{CO_3^{2-}})$$
$$4.9 \times 10^{-9} = K \cdot (0.894)(0.894)$$
$$\therefore K = 6.13 \times 10^{-9} = \mathbf{6.1 \times 10^{-9}}$$

We have already learned that thermodynamic equilibrium constants have no units because they are based on activities. A related question is "What are the units on a concentration-based equilibrium constant?" It is actually acceptable to write a concentration-dependent equilibrium constant either with or without units, depending on how such a constant is being used. For instance, the value of K in Exercise 6.8 for the reaction of Ca^{2+} with CO_3^{2-} should have apparent units of molarity raised to the second power (M^2) based on the terms that appear in the numerator and denominator of this equilibrium constant expression. (*Note:* These units will differ for other reactions and types of equilibrium constants.) It is convenient to show these units when we are using an equilibrium constant to calculate the amount of a chemical in a given reaction, which can help when we use dimensional analysis to see if we have obtained a reasonable answer.

There are also times when it is useful to "eliminate" the units on a concentration-dependent equilibrium constant. This type of situation occurred in Equation 6.18 when we related the values of $K°$ to K. In this case neither $K°$ nor activity coefficients have units, so Equation 6.18 would imply that K also does not have units. However, what really took place was that each concentration term in the expression for K was being divided by a reference concentration with a value of exactly 1.00 (e.g., a value of 1 M for a concentration that is expressed in molarity). This is the same approach we discussed in Equation 6.2 for relating activity to chemical concentration and acts as

TABLE 6.4 Thermodynamic and Concentration-Based Equilibrium Constants ($K°$ and K) for the Reaction $H_2CO_3 \rightleftharpoons HCO_3^- + H^+$ in Water at 25°C

| Ionic Strength | Activity Coefficients[a] | | | Activity Coefficient Ratio $(\gamma_{H^+})(\gamma_{HCO_3^-})/\gamma_{H_2CO_3}$ | Equilibrium Constants[b] | |
	$\gamma_{H_2CO_3}$	γ_{H^+}	$\gamma_{HCO_3^-}$		$K°$	K
0.000	1.000	1.000	1.000	1.000	4.5×10^{-7}	4.5×10^{-7}
0.001	1.000	0.967	0.965	0.933	4.5×10^{-7}	4.8×10^{-7}
0.005	1.001	0.934	0.928	0.865	4.5×10^{-7}	5.2×10^{-7}
0.010	1.002	0.913	0.902	0.822	4.5×10^{-7}	5.5×10^{-7}
0.050	1.009	0.854	0.817	0.691	4.5×10^{-7}	6.5×10^{-7}
0.100	1.017	0.825	0.773	0.627	4.5×10^{-7}	7.1×10^{-7}

[a]The activity coefficients for H^+ and HCO_3^- were taken from Table 6.3. The activity coefficients for H_2CO_3 were calculated using Equation 6.7 with a salting coefficient of $k = 0.075$.

[b]The value of K was determined by using the relationship $K = K°/$(Activity Coefficient Ratio), as obtained by rearranging Equation 6.18.

a conversion factor to remove any units from K during the calculation. To remind us of this feature, whenever any units are given later in this book for a concentration-dependent equilibrium constant K, we will refer to these as "apparent" units.

6.3B Solving Chemical Equilibrium Problems

We learned earlier that many types of reactions are used in analytical chemistry. We also learned that an understanding of the fundamental basis for these reactions is important in helping us to use and predict how these reactions will perform in such methods. Being able to predict what reaction conditions will be needed for a particular analysis often involves using calculations that involve chemical equilibria. For instance, we might use information on the expected concentration range for an analyte to determine the optimum composition of a reagent that should be used for measuring an analyte. Let's now discuss some questions and approaches that can be used in dealing with such calculations.

General Strategy. In solving any type of chemical problem, it is important to first ask several questions about the reaction. A summary of these questions is given in Table 6.5. To illustrate this process, suppose we have a reaction that involves the dissociation of carbonic acid in water, where we are told to prepare a 0.00500 M solution of carbonic acid in water at 25°C. We will now see how we can estimate the concentration of hydrogen ions in this solution.

The first question that needs to be asked in this problem is *"Which reactions are the most important to the problem at hand?"* To answer this question, we need to consider both the main reaction and any side reactions that might take place in the system. In our example, the main reaction

is the ionization of carbonic acid to give bicarbonate and hydrogen ions, as shown in Equation 6.19.

$$H_2CO_3 \rightleftharpoons HCO_3^- + H^+ \qquad (6.19)$$

However, there are other reactions that can also produce H^+, our product of interest. One is the further dissociation of the bicarbonate to form more hydrogen ions (Equation 6.20); the second is the dissociation of water to produce H^+ and hydroxide ions (Equation 6.21).

$$HCO_3^- \rightleftharpoons CO_3^{2-} + H^+ \qquad (6.20)$$

$$H_2O \rightleftharpoons H^+ + OH^- \qquad (6.21)$$

Another possible reaction that might be important is the combination of dissolved carbon dioxide with water to form more H_2CO_3, giving what is known as a *hydration reaction*.

$$CO_2 + H_2O \rightleftharpoons H_2CO_3 \qquad (6.22)$$

Although it would be important to consider this latter reaction for an open system, as would be done when modeling global warming, in our particular example we will assume we are working with a closed system in which the concentration ratio of CO_2 and H_2CO_3 is constant. This assumption, in turn, makes it possible for us to focus on the solution-phase reactions in Equations 6.19–6.21 for the remainder of this chapter.

Once we have identified the reactions we need to consider, the next question to ask is *"What is already known about the reaction?"* In answering this question, we need to think about both the reactions and conditions in the system we are studying. With regard to our system, we know the

---

TABLE 6.5   Questions to Consider When Solving Chemical Equilibrium Problems

Which reactions are the most important to the problem at hand?

What is the main reaction of interest?

Are there any side reactions that must be considered?

What is already known about the reaction?

How much of each reactant or product is there at the beginning and/or end of the reaction?

How are the reactants and products related to each other (reaction stoichiometry)?

What other chemicals are present that might affect this reaction?

What is known about the amounts of these other chemicals?

What are the conditions for the reaction (temperature, solvent, etc.)?

What is the equilibrium constant for the reaction?

Are there any simplifying assumptions that I can make about these reactions?

What degree of accuracy and complexity will be required in the calculation?

Can concentrations be used instead of chemical activities?

Can any reactions be ignored or their effects treated using constants?

Are there any reactants or products whose concentrations will not change significantly during the reaction?

What equations can I use to describe my reaction?

What are the equilibrium expressions for this reaction?

What are the mass balance equations for this reaction?

What is the charge balance equation for the system?

Are there enough equations to obtain an answer?

What mathematical approach should be used to solve these equations?

Can a simple linear equation or quadratic equation be used?

Is it necessary to use successive approximations?

Will a spreadsheet or other computer-based tool be required?

Does my final answer make sense?

Are all of the concentrations (or activities) that were obtained of a reasonable size?

Do the results provide an adequate solution to each equation that was used to describe the reaction?

---

total amount of carbonic acid originally placed in solution was 0.00500 M. We also know the reaction is taking place in water at 25°C. It is further implied that no other substances are present in significant amounts, such as other sources of carbonate or bicarbonate. In addition, we know how the chemicals in our solution will react and the stoichiometry between the reactants and products. Finally, it is helpful to see what is known about the equilibrium constants for these reactions. In this case, the values of K for the reactions in Equations 6.19–6.21 in water and at 25°C can be found from the literature to be 4.5×10^{-7}, 4.7×10^{-11}, and 1×10^{-14}, respectively.

Another question to now ask is *"Are there any simplifying assumptions that can be made about the system?"* A few simplifications can make an equilibrium problem much easier to solve, but the number of assumptions and simplifications that are made will also affect the accuracy of the final answer. In this case, we have already assumed that we are working with a closed system, which allows us to ignore the effects of the reaction in Equation 6.22. We might also assume that the ionic strength of our reaction mixture is low enough to allow the use of concentrations instead of chemical activities and use our known value for $K°$ in place of K. This second approximation is a good one in this case because (as shown in Table 6.4) even if all of the carbonic acid dissociated to form bicarbonate and hydrogen ions, the resulting ionic strength of 0.00500 M would give a difference of only 15% between the values for K and $K°$. This approximation would not be valid, however, if we needed results in which greater accuracy was required for our calculations.

We can now begin to set up the equations we need to solve our problem. We also need to see how many variables must be considered and identify at least as many equations as this number of variables to allow us to get a solution. If we assume that chemical activities and concentrations are approximately the same (or that K and $K°$ are roughly equal) and if we do not make any further assumptions about this system, then we have five variables to deal with in our problem. These variables include the hydrogen ion concentration, the main factor of interest. In solving for the hydrogen ion concentration, we must also obtain the concentrations for H_2CO_3, HCO_3^-, CO_3^{2-}, and OH^-, because these will all affect the value of $[H^+]$ through the reactions taking place in the system. The last remaining component in these reactions is water (see Equation 6.21), but we can safely assume its activity is one because it is the solvent and is in a large excess versus the other chemicals.

Fortunately, we already have three of the five equations we need to get at these five variables. These three equations are the equilibrium expressions for the three reactions in Equations 6.19–6.21, which are as follows when written in terms of concentration-based equilibrium constants.

For Reaction 6.19: $K_1 = \dfrac{[HCO_3^-][H^+]}{[H_2CO_3]} \approx 4.5 \times 10^{-7}$ (6.23)

For Reaction 6.20: $K_2 = \dfrac{[CO_3^{2-}][H^+]}{[HCO_3^-]} \approx 4.7 \times 10^{-11}$ (6.24)

For Reaction 6.21: $K_w = \dfrac{[H^+][OH^-]}{1} \approx 1.0 \times 10^{-14}$ (6.25)

If we also wish to include apparent units along with these equilibrium constants for use in later calculations, these units would be M for K_1 and K_2 and M^2 for K_w. At this point we have three unique equations that describe our reactions, but we still have five unknown concentrations. This means we must find two more equations to describe our system. The way we get these two other equations is by using the methods of mass balance and charge balance, as we will discuss in the following section.

Mass Balance and Charge Balance. One way that we can get additional equations to help us solve an equilibrium problem is to use *mass balance*. Mass balance is simply an application of the law of conservation of mass, which means that we cannot create or destroy matter as a result of an ordinary chemical reaction. One useful aspect of this concept when we are solving an equilibrium problem is that the total mass (or moles) of each element we place into a system must be equal to the sum of the masses (or moles) of all the individual forms of that element after equilibrium has been reached.

Mass balance is a particularly helpful tool when we are dealing with a set of reactions that involve many different forms of the same chemical. An example would be the reaction of carbonic acid to form bicarbonate and then carbonate ions, as is shown in Equations 6.19–6.21. In this particular case we often do not know the concentrations of each form of carbonate, but we do know the total amount of carbonic acid we initially placed into solution. We can use this information to write a **mass balance equation**, which is an equation that shows how the total concentration of a chemical is related to the concentrations of its various species. The mass balance equation for carbonate in our particular example would be as follows.

$$C_{Carbonate} = [H_2CO_3] + [HCO_3^-] + [CO_3^{2-}] \quad (6.26)$$

In this equation the term $C_{Carbonate}$ is also known as the **analytical concentration** for carbonate. The analytical concentration for a chemical is equal to the total concentration of that chemical in a solution, regardless of the final form or number of species for the chemical. In writing Equation 6.26 for carbonate, we have not introduced any new variables because we know the total concentration of carbonic acid placed into the initial solution, 0.00500 M. You may have also noticed that this relationship is written in terms of concentrations rather than chemical activities. The reason for this format is that we are now looking at the content of this chemical, rather than its reactivity. Thus, mass balance equations are always written in terms of mass, moles, or concentration rather than chemical activity.

There are several key things to remember when you are writing a mass balance equation. First, make sure you are considering all significant reactions that might involve substances in your mass balance equation. If you are not sure whether or not a particular reaction should be included, it is best to consider it until you learn more about your system and have determined whether the reaction can be safely ignored when you later make assumptions that can simplify your calculations. Second, it is important when you are writing a mass balance equation to always use *balanced* chemical reactions. This practice will provide you with the correct stoichiometry for each reactant and product, which, in turn, will ensure that your mass balance equation properly relates the amount of each form of your given chemical to that of its other possible forms.

Another resource we can use in helping solve equilibrium problems is a **charge balance equation.** Charge balance is an approach for solving chemical equations that makes use of the fact that the sum of all positive and negative charges in a closed system should be zero. For this situation to be true, the concentrations of all cations and anions must be balanced so their ionic charges cancel to give a net neutral charge to their solution. A simple example of a charge balance equation can be obtained by looking at water. As we have noted, even pure water will contain some charged species because water will dissociate to form hydrogen ions and hydroxide ions. However, we also know that the number of positive and negative charges must be equal so that the overall system has a net neutral charge. Pure water can have only two charged species present (H^+ and OH^-), which means we must have equal concentrations of these ions ($[H^+] = [OH^-]$) to obtain this net neutral charge.

In the case of pure water, the charge balance equation is the same as the mass balance equation (i.e., each mole of water that dissociates forms one mole of H^+ and one mole of OH^-, or $[H^+] = [OH^-]$). However, these equations are not necessarily the same for more complex solutions. For instance, if we were to also add sodium chloride to water (giving Na^+ and Cl^- ions), the mass balance equation between the hydrogen and hydroxide ions would still be $[H^+] = [OH^-]$, but the new charge balance equation would now be $[Na^+] + [H^+] = [OH^-] + [Cl^-]$.

Equation 6.27 gives a general relationship that can be used to write the charge balance equation for any chemical system.

$$(1)\ [C_1{}^{1+}] + (2)\ [C_2{}^{2+}] + \cdots + (n)\ [C_n{}^{n+}] =$$
$$(1)[A_1{}^{1-}] + \cdots + (2)\ [A_2{}^{2-}] + \cdots (n)\ [A_n{}^{n-} \cdots]\quad (6.27)$$

In this expression, $[C_1{}^{1+}]$, $[C_2{}^{2+}]$ and so on, refer to the concentrations of all cations in the solution, with absolute values for their charges of +1, +2, etc. Similar terms are used to represent the concentrations of all anions ($[A_1{}^{1-}]$, $[A_2{}^{2-}]$, …) and their charges. Thus, for a solution of sodium chloride in water, the charge balance would be $(1)\ [Na^+] + (1)\ [H^+] = (1)\ [OH^-] + (1)\ [Cl^-]$. Additional practice with charge balance equations is given for you in the next exercise.

EXERCISE 6.9	Writing a Charge Balance Expression

Write a charge balance expression for the combined reactions in Equations 6.19–6.21 concerning the production of hydrogen ions from carbonic acid and water. How does this expression compare to the mass balance equation for all chemicals that produce hydrogen ions in this system?

SOLUTION

There is only one type of cation in this system (H^+), but there are several different anions ($HCO_3{}^-$, $CO_3{}^{2-}$, and OH^-). When we put these terms into Equation 6.27, we get the following result.

$$[H^+] = [HCO_3{}^-] + 2\,[CO_3{}^{2-}] + [OH^-]\quad (6.28)$$

In this specific case, the charge balance equation is also the mass balance equation for hydrogen ions, because there is only one type of cation present in these reactions. However, if there were additional types of cations (such as Na^+ from sodium chloride), the charge balance and mass balance equations would not be the same for any of these ions. This situation would give us an additional equation that could be used to solve an equilibrium problem that involves such ions.

Other Tools. At this point we now have enough equations to solve our carbonic acid problem. These equations include three equilibrium constant expressions

(Equations 6.23–6.25), one mass balance expression (Equation 6.26) and one charge balance equation (Equation 6.28). The next question we must address is "*What mathematical approach should be used to solve this problem?*" In this book, we will see several ways in which a final answer to such a problem can be obtained. We will also see that the particular tool we choose to solve a problem will depend on the complexity of this problem and the number of variables that must be considered.

The simplest chemical problem to solve is one that involves only a single reaction and one or two variables. An example is the reaction and calculations shown in Table 6.6, where we start with a known total amount of carbonic acid, and the dissociation of this acid is assumed to be the only source of hydrogen ions. All that is now needed to solve this problem is to take the concentration terms from Table 6.6 and place them into the equilibrium expression for the reaction. (*Note:* Units have been included here on K_1 for use in dimensional analysis.)

$$K_1 = \frac{[HCO_3{}^-][H^+]}{[H_2CO_3]}$$

$$4.5 \times 10^{-7}\ M = \frac{(x)\,(x)}{(0.00500\ M - x)}\quad (6.29)$$

where $x = [HCO_3{}^-] = [H^+]$. Next, we can rearrange this equation so that all terms containing our unknown (x) appear on the same side.

$$0 = x^2 + (4.5 \times 10^{-7}\ M)\,x - (4.5 \times 10^{-7}\ M)\cdot$$
$$(0.00500\ M)\quad (6.30)$$

This relationship is now written as a **quadratic equation**,[11] which is an equation where the highest-order term for "x" is x^2. The specific quadratic equation just shown is written in the general form

$$0 = Ax^2 + Bx + C\quad (6.31)$$

where A, B, and C are the constants that appear in front of the x^2, x, and constant terms of this equation. For instance, in Equation 6.30 the value of A is 1, B is $4.5 \times 10^{-7}\ M$ and C is $-(4.5 \times 10^{-7}\ M)(0.00500\ M)$. It is useful to rearrange a quadratic equation into this form because the values of A, B, and C can then be used

---

TABLE 6.6   Use of Mass Balance in Solving a Simple Equilibrium Problem*

Reaction	H_2CO_3	⇌	$HCO_3{}^-$	+	H^+
Initial concentrations:	0.0050 M		0 M		0 M
Change in concentrations:	−x		x		x
Equilibrium concentrations:	0.0050 − x		x		x

*This type of table is also known as an *ICE table* because it lists the "initial," "change," and "equilibrium" concentrations for each reactant and product.

to solve for x by using the **quadratic formula**, as given below.[11]

$$x = \frac{-B \pm \sqrt{B^2 - 4AC}}{2A} \quad (6.32)$$

The "$\pm$" symbol in this equation means that there will always be two answers, or "roots," for x, one obtained when you use $-B + \sqrt{B^2 - 4AC}$ in the numerator and the other when you use $-B - \sqrt{B^2 - 4AC}$. In most chemical problems, however, only one of these roots will give a realistic answer to the problem (i.e., a value that is within the range of concentrations or values that may actually occur in your system).

EXERCISE 6.10   Using the Quadratic Equation

Based on the quadratic equation, what are the possible values for x in Equation 6.30? Based on these results, what are the approximate concentrations of carbonic acid, bicarbonate, and hydrogen ions that should be present at equilibrium for the system in Table 6.6?

SOLUTION
We have already seen that the values of A, B, and C in our particular example are $A = 1$, $B = 4.5 \times 10^{-7}\ M$, and $C = -(4.5 \times 10^{-7}\ M)(0.00500\ M) = -2.25 \times 10^{-9}\ M^2$. When we substitute these values into the quadratic equation, we get the two possible answers for x.

$$\textit{First x value} = \frac{-(4.5 \times 10^{-7}\ M)\ +}{2(1)}$$

$$\frac{\sqrt{(-4.5 \times 10^{-7}\ M)^2 - 4(1)(-2.25 \times 10^{-9}\ M^2)}}{2(1)}$$

$$= 4.7\underline{2} \times 10^{-5}\ M$$

$$\textit{Second x value} = \frac{-(4.5 \times 10^{-7}\ M)\ -}{2(1)}$$

$$\frac{\sqrt{(-4.5 \times 10^{-7}\ M)^2 - 4(1)(-2.25 \times 10^{-9}\ M^2)}}{2(1)}$$

$$= -4.7\underline{7} \times 10^{-5}\ M$$

Although both of the values we obtained for x in the preceding equations make sense mathematically, from a chemical viewpoint only $x = 4.7\underline{2} \times 10^{-5}\ M$ is reasonable, because we cannot have a negative value for $x = [HCO_3^-]$ or $[H^+]$. When we use this result with Table 6.6, we get the following final answers for our equilibrium concentrations when we round to the correct number of significant figures.

At Equilibrium: $[H_2CO_3] = 0.00500 - x = \mathbf{0.00495\ M}$

$$[HCO_3^-] = x = \mathbf{4.7 \times 10^{-5}\ M}$$

$$[H^+] = x = \mathbf{4.7 \times 10^{-5}\ M}$$

In checking these results, you can see that all of the final concentrations are within a reasonable range (in this case, between 0 and 0.00500 M, because the reaction has a 1:1 stoichiometry between each reactant and product). Also, if we plug these concentrations back into the equilibrium expression for this reaction, we get back the expected equilibrium constant of $4.5 \times 10^{-7}\ M$.

It is possible on occasion that you may get an answer of $x = 0$ when you are using the quadratic equation to solve for the concentration or content of a chemical in an equilibrium. If this situation happens, the answer does not mean that there is none of the chemical present. You cannot have an equilibrium without having some of *both* the reactants and products present, even if they are present in only a small amount. What an answer of $x = 0$ in this case does mean is that the estimated amount of the chemical is quite small compared to other concentrations and terms in your calculation and cannot be distinguished from a value of zero with the number of digits that are being used to express this answer. There are several ways around this problem. First, see if you can use more digits (whether they are significant figures or guard digits) in your calculation; this will help to avoid problems due to rounding errors. Second, you can use the new information that you have (that the value of "x" is small compared to other values in your equations) to go back and simplify your equations. This simplification should then lead to a usable answer to your calculation.

If a reaction system is too complex to solve directly by using the quadratic formula, another approach must be used. As an example, suppose that we did not ignore the production of H^+ from water or from the dissociation of bicarbonate, as was done in Exercise 6.10. Instead of working with only one reaction, as illustrated in Table 6.6, we now must consider all of these reactions by combining the equilibrium expressions in Equations 6.19–6.21 with the mass balance and charge balance expressions derived in Equations 6.26 and 6.28. The result is the following relationship between $[H^+]$ and the other known factors (K_1, K_2, and $C_{Carbonate}$) that describe this system.

$$[H^+] = C_{Carbonate} \cdot \frac{K_1[H^+] + 2K_1K_2}{([H^+]^2 + K_1[H^+] + K_1K_2)} +$$

$$\frac{1.0 \times 10^{-14}M^2}{[H^+]} \quad (6.33)$$

We will see in the next chapter exactly how this type of combined equation was obtained. For now, we will focus on how we can solve for the value of $[H^+]$. The quadratic equation cannot be used because we will not be able to get this expression into a form needed for this approach to work. One alternative approach that can be used in this case is to use the "solver function" found on many modern calculators. Another approach is to use a technique known as **successive approximations**.[22]

To perform successive approximations, begin by placing into your equation a rough estimate of what the concentration for the reagent or product of interest might be. In the carbonic acid example, suppose we guess that one-fifth of

TABLE 6.7   Example of an Equilibrium Problem Using Successive Approximations

Estimated [H$^+$], M	Calculated [H$^+$], M	Strategy for the Next Step[a]
0.001000	0.000002	Calculated value is much higher than estimate; use a smaller estimate
0.000100	0.000022	Calculated value is still much higher than estimate; decrease estimate further
0.000050	0.000045	Calculated value is slightly higher than estimate; decrease estimate slightly
0.000045	0.000050	Calculated value is slightly smaller than estimate; increase estimate slightly
0.000047	0.000047	Calculated value and estimate are the same—You have your answer!

[a]Another strategy would be to use each calculated value as the new estimated value. This approach often works well, but in this particular example this method takes a much larger number of cycles to converge on the final answer.

our 0.00500 M carbonic acid dissociates to give hydrogen ions, or [H$^+$] = 0.0010 M. As shown in Table 6.7, we then place this value into the right-hand side of Equation 6.33 and calculate the value of [H$^+$] on the left, which instead gives us a result of 0.000002 M. Because our initial guess and this calculated value for [H$^+$] are not the same, we then make a new estimate, place it into the right-hand side of Equation 6.33 and repeat the process. Each time we do this process, the value we calculate for [H$^+$] should hopefully get closer to our estimate until they approach the same value. At this point, the overall equation is balanced and we have arrived at our final answer, 4.7×10^{-5} M. Although this approach can take several cycles to get to a final answer, it does have the advantage of being able to work with complex equations that cannot be solved by other methods.

A third option for solving chemical equilibrium problems is to use computers to find possible answers for each calculation. This approach is used in modeling extremely complex systems, like when scientists use computers to study the fate of carbon dioxide in the atmosphere and to predict the consequences of changes in carbon dioxide levels. This approach involves the use of a program that is specially designed to solve multiple equations, such as the built-in "polynomial" or "solver" feature of a calculator. Another way computers can be employed in equilibrium calculations is by using a spreadsheet program to describe the reactions of chemical processes of interest.[22] An example of this method is shown in Figure 6.9, where a spreadsheet was used to determine how the concentrations of H$_2$CO$_3$, HCO$_3^-$, CO$_3^{2-}$, and

Concentrations for H$_2$CO$_3$ and related species in 0.100 M carbonic acid at various values for [H$^+$]

[H$^+$], M	K_{a1}	K_{a2}	[H$_2$CO$_3$], M	[HCO$_3^-$], M	[CO$_3^{2-}$], M
1.00E+00	4.45E−07	4.69E−11	1.00E−01	4.45E−08	2.09E−18
1.00E−01	4.45E−07	4.69E−11	1.00E−01	4.45E−07	2.09E−16
1.00E−02	4.45E−07	4.69E−11	1.00E−01	4.45E−06	2.09E−14
1.00E−03	4.45E−07	4.69E−11	1.00E−01	4.45E−05	2.09E−12
1.00E−04	4.45E−07	4.69E−11	9.96E−02	4.43E−04	2.08E−10
1.00E−05	4.45E−07	4.69E−11	9.57E−02	4.26E−03	2.00E−08
1.00E−06	4.45E−07	4.69E−11	6.92E−02	3.08E−02	1.44E−06
1.00E−07	4.45E−07	4.69E−11	1.83E−02	8.16E−02	3.83E−05
1.00E−08	4.45E−07	4.69E−11	2.19E−03	9.74E−02	4.57E−04
1.00E−09	4.45E−07	4.69E−11	2.14E−04	9.53E−02	4.47E−03
1.00E−10	4.45E−07	4.69E−11	1.53E−05	6.81E−02	3.19E−02
1.00E−11	4.45E−07	4.69E−11	3.95E−07	1.76E−02	8.24E−02
1.00E−12	4.45E−07	4.69E−11	4.69E−09	2.09E−03	9.79E−02
1.00E−13	4.45E−07	4.69E−11	4.78E−11	2.13E−04	9.98E−02
1.00E−14	4.45E−07	4.69E−11	4.79E−13	2.13E−05	1.00E−01

Formulas:

[H$_2$CO$_3$] = $(0.100*[H^+]^2)/([H^+]^2 + K_{a1}*[H^+] + K_{a1}*K_{a2})$

[HCO$_3^-$] = $(0.100*K_{a1}*[H^+])/([H^+]^2 + K_{a1}*[H^+] + K_{a1}*K_{a2})$

[CO$_3^{2-}$] = $(0.100*K_{a1}*K_{a2})/([H^+]^2 + K_{a1}*[H^+] + K_{a1}*K_{a2})$

FIGURE 6.9 Spreadsheet results predicting the change in concentration of carbonic acid and related species in an aqueous solution prepared with an initial concentration of 0.10 M carbonic acid and containing various fixed concentrations of hydrogen ions. The terms K_{a1} and K_{a2} are the equilibrium constants for the release of the first and second hydrogen ions from carbonic acid (for more on acid–base reactions, see Chapter 8).

OH^- will change as the concentration of H^+ is fixed at various values in a solution with an analytical carbonate concentration of 0.010 M. We look at this type of system in the next chapter, and later in this book we see several other examples of how spreadsheets can be used in chemical calculations.

Key Words

Activity coefficient *120*	Equilibrium constant *126*	Le Châtelier's principle *127*	Reaction quotient *127*
Analytical concentration *132*	Extended Debye–Hückel	Mass balance equation *132*	Stoichiometry *119*
Charge balance equation *132*	equation *122*	Quadratic equation *133*	Successive
Chemical activity *117*	Ionic strength *121*	Quadratic formula *134*	approximations *134*
Chemical equilibrium *126*			

Other Terms

Chemical kinetics *117*	Debye–Hückel limiting	Mean activity coefficient *121*	Standard state *119*
Chemical potential *119*	law *122*	Salting-out effect *124*	Thermodynamic equilib-
Chemical	Dielectric constant *125*	Salting coefficient *124*	rium constant *126*
thermodynamics *117*	Hydration layer *125*	Standard chemical	
Concentration-dependent	Hydrated radius *125*	potential *119*	
equilibrium constant *117*	Mass balance *132*		

Questions

TYPES OF CHEMICAL REACTIONS/TRANSITIONS AND DESCRIBING CHEMICAL REACTIONS

1. What types of chemical reactions are commonly used in analytical chemistry? What are some examples of their applications?
2. What are some general factors to consider when describing a chemical reaction?
3. What is meant by "chemical thermodynamics" and "chemical kinetics"? Why are these terms important in describing chemical reactions?

WHAT IS CHEMICAL ACTIVITY?

4. Define "chemical potential" and "chemical activity." How are these terms related?
5. What is meant when we say a chemical is in its "standard state"? What is the activity of a chemical when it is present in its standard state?
6. What are the standard states for each of the following samples?
   (a) Oxygen gas
   (b) Sodium chloride crystals
   (c) Methanol as a solvent
   (d) Sodium chloride dissolved in water
   (e) Methanol dissolved in water
   (f) Helium as a trace component of air
7. What are some effects that can cause the activity for a chemical to differ from the activity for its standard state? Illustrate these effects using a solution of NaCl in water as an example.
8. Explain why water is often assigned an activity of 1.0 for an aqueous solution. Under what circumstances might the activity of water in this solution not be equal to 1.0?

9. What is an "activity coefficient"? What are the units on an activity coefficient?
10. A 0.10 molal solution of $AgNO_3$ in water that completely dissolves is found to produce an activity of 0.0734 for Ag^+ and NO_3^-. What are the activity coefficients for each of these ions?
11. If the K^+ and I^- ions in a completely dissolved sample of 0.20 molal KI each have an activity of 0.155, what are the activity coefficients for these ions?
12. Why is it important to consider the differences between chemical activity and concentration when using or creating an analytical method? What problems might arise if these differences are not considered?
13. Sodium perchlorate, $NaClO_4$, is a salt that has a high solubility in water. What is the ionic strength of an aqueous solution prepared by adding 0.20 g of this salt to 1.00 L water?
14. What is the ionic strength of a 0.100 M solution of NaCl in water? What is the ionic strength of a 0.100 M solution of Na_2SO_4 in water? Compare these ionic strengths and explain any differences in their values.
15. Determine the ionic strength for each of the following mixtures. In each case, assume that all of the salts completely dissolve and dissociate into their respective ions.
    (a) 0.10 M NaCl plus 0.20 M KI
    (b) 0.050 M $MgSO_4$ plus 0.050 M Na_2SO_4
    (c) 0.050 g KBr plus 0.100 g KCl in 1.00 L water

CHEMICAL ACTIVITY IN ANALYTICAL METHODS

16. What is the difference between a "mean activity coefficient" and a "single-ion activity coefficient"? How are these two items related to one another?

17. A solution of NaCl produces a mean activity coefficient of 0.85. If it is assumed the Na^+ and Cl^- ions have the same individual activity coefficients, what are the values of these activity coefficients?

18. A solution of K_2SO_4 has a mean activity coefficient of 0.75. What are the individual activity coefficients for K^+ and SO_4^{2-} in this solution?

19. What is the extended Debye–Hückel equation? What information is required when using this equation to estimate ion activity coefficients?

20. What form of the extended Debye–Hückel equation is used when working with a solution in water at 25°C? Under what conditions is it no longer suitable to use the extended Debye–Hückel equation?

21. Use the extended Debye–Hückel equation to estimate each of the following activity coefficients. Assume in each case that the given chemical completely dissolves to form the listed ions.
    (a) Activity coefficients for H^+ and NO_3^- in a 0.0050 M HNO_3 solution
    (b) Activity coefficients for K^+ and OH^- in a 0.020 M KOH solution
    (c) Activity coefficients for Ba^{2+} and Cl^- in a 0.010 M $BaCl_2$ solution

22. What are the expected mean activity coefficients for the solutions in Problem 21?

23. What is the Debye–Hückel limiting law (DHLL)? Use this equation to recalculate the activity coefficients in Problem 21. How do these results compare to those obtained with the extended Debye–Hückel equation?

24. How is the activity of a neutral compound altered by a change in ionic strength? How does this compare to the effect of ionic strength on activity for an ionic substance?

25. What is the "salting-out effect"? What is a "salting coefficient"?

26. Acetic acid has a salting coefficient of 0.066 in the presence of NaCl in water. What is the approximate activity coefficient for acetic acid in a 1.0 M NaCl solution under acidic conditions (where most of the acetic acid is in its original neutral form)?

27. A chemist wishes to have the activity of a neutral compound be within 1% of its concentration. If this compound has a salting coefficient of 0.15 in the presence of KNO_3, what is the maximum ionic strength that can be present in this solution for such a condition to be met?

28. Describe the following effects on activity coefficients based on the extended Debye–Hückel equation.
    (a) Charge effects
    (b) Size effects
    (c) Ionic strength effects

29. Explain how the use of a known excess of a salt such as sodium chloride could be used to maintain a constant ionic strength in a sample or reagent.

30. What is a "hydrated radius"? How is this value related to the ion size parameters used for the extended Debye–Hückel equation?

31. Explain how a calibration curve can be used by an analytical chemist to deal with changes in chemical activity.

WHAT IS A CHEMICAL EQUILIBRIUM?

32. What is meant by "chemical equilibrium"? Why is chemical equilibrium important to consider in analytical chemistry?

33. What is an "equilibrium constant" and how is this used to describe a chemical equilibrium?

34. Write equilibrium constant expressions (in terms of activity) for each of the following reactions.
    (a) $H_2SO_4 + H_2O \rightleftharpoons H_3O^+ + HSO_4^-$
    (b) $Zn^{2+} + NH_3 \rightleftharpoons Zn(NH_3)^{2+}$
    (c) $PbCl_2(s) \rightleftharpoons Pb^{2+} + 2\,Cl^-$

35. Write equilibrium constant expressions in terms of concentrations for each of the reactions in Problem 34. (*Note*: when dealing with a solid like $PbCl_2$, continue to use the activity of the given chemical in such an expression.)

36. A chemist is studying the acid dissociation of formic acid in water, as represented by the following net reaction: $HCOOH \rightleftharpoons H^+ + HCOO^-$. It is determined at 25°C that the activities of HCOOH, H^+, and $HCOO^-$ at equilibrium are 2.0×10^{-4}, 1.0×10^{-4}, and 3.6×10^{-4}. What is the equilibrium constant for this reaction under these conditions?

37. The reaction of Cd^{2+} with S^{2-} to form solid CdS is found to give a saturated solution that has activities of 8.4×10^{-14} for Cd^{2+} and S^{2-} and 1.00 for CdS. What is the equilibrium constant for this reaction?

38. How is the equilibrium constant for a reaction related to the change in standard Gibbs free energy for that reaction?

39. The reaction of aspartic acid with Ca^{2+} at 25°C has an equilibrium constant of 40. What is the change in standard Gibbs free energy for this reaction?

40. The binding of a drug with a protein is found to have an equilibrium constant of 2.3×10^5 at 37°C. What is the change in standard Gibbs free energy for this reaction?

41. What is a "reaction quotient"? How is a reaction quotient similar to an equilibrium constant? How is it different?

42. When solid silver chromate, $Ag_2CrO_4(s)$, is placed into water, some of this solid will dissolve to form Ag^+ and CrO_4^{2-} ions according to the following reaction.

$$Ag_2CrO_4(s) \rightleftharpoons 2\,Ag^+ + CrO_4^{2-}$$

This reaction has a known equilibrium constant of 2.4×10^{-12} at 25°C. If solid silver chromate is placed into a solution that contains 1.3×10^{-4} M Ag^+ and 6.3×10^{-5} M CrO_4^{2-}, what will be the reaction quotient for this process? Based on this value, in what direction will this reaction proceed as it approaches equilibrium (i.e., will it move to form more solid or to create more dissolved ions)?

43. The following reaction has an equilibrium constant of 7.1×10^2 at 25°C.

$$H_2(g) + I_2(g) \rightleftharpoons 2\,HI(g)$$

Predict the direction this reaction will shift (toward the left or right) when beginning with each of the following sets of reagents and products. Assume in each case that the chemical activities are approximately the same as the concentrations for these listed chemicals.
    (a) $[H_2] = 0.81\ M$ $[I_2] = 0.44\ M$ $[HI] = 0.58\ M$
    (b) $[H_2] = 0.078\ M$ $[I_2] = 0.033\ M$ $[HI] = 1.35\ M$
    (c) $[H_2] = 0.034\ M$ $[I_2] = 0.035\ M$ $[HI] = 1.50\ M$

SOLVING CHEMICAL EQUILIBRIUM PROBLEMS

44. What questions should you consider when setting up and solving a chemical equilibrium problem?

45. What is a "mass balance equation"? What is an "analytical concentration"? How can these tools be used to help solve a chemical equilibrium problem?

46. When phosphoric acid (H_3PO_4) is dissolved in water it can undergo a series of acid–base reactions to produce $H_2PO_4^-$, HPO_4^{2-}, and PO_4^{3-}. Write the mass balance equation for all species related to phosphoric acid in such a solution.

47. A 0.010 M solution of Cu^{2+} combines with a 0.030 M solution of NH_3 to form several complex ions of copper that have formulas ranging from $Cu(NH_3)^{2+}$ to $Cu(NH_3)_4^{2+}$. What are the mass balance expressions for copper and ammonia in this solution?

48. What is a "charge balance equation"? How can this be used to help solve a chemical equilibrium problem?

49. A solution is prepared by adding 0.05 M $CaCl_2$, 0.10 M NaCl, and 0.10 M $MgCl_2$ to water. If all of these salts completely dissolve in water, what is the charge balance equation for this solution?

50. Placing hydrogen fluoride (HF) in water results in the partial conversion of this chemical into hydrogen ions (H^+) and fluoride ions (F^-). At the same time, some of the water itself will form hydrogen ions and hydroxide ions (OH^-). Write the charge balance equation for this system.

51. What is the "quadratic formula"? It what type of situation is this formula a useful tool in solving a chemical equilibrium problem?

52. Use the quadratic formula to solve for x (reporting both possible answers) in each of the following equations.
    (a) $0 = 8x^2 + 3x + 0.10$
    (b) $8.3 \times 10^2 x^2 = 1.5 \times 10^2 - x$
    (c) $3.4x = 1.5x^2 - 2.0x + 0.8$
    (d) $2.8 \times 10^1 x^3 = 9.1 \times 10^2 x^2 + 4.5x$

53. Use the quadratic formula to solve the unknown concentration in each of the following equations.
    (a) $0 = [H^+]^2 + (2.5 \times 10^{-4})[H^+] - (3.0 \times 10^{-6})(1.0 \times 10^{-3})$ where $x = [H^+]$
    (b) $(1.0 \times 10^{-4})/[OH^-] = 5.5 \times 10^{-4} + [OH^-]$ where $x = [OH^-]$

54. What is the "method of successive approximations"? Describe how this method can be used to help solve chemical equilibrium problems.

55. Solve each of the following using the method of successive approximations.
    (a) $0 = 2x^3 + 4x^2 + x + 0.5$
    (b) $(2.5 \times 10^{-3} + x)/x = 1.0 \times 10^{-4} x^2$
    (c) $9.5x^3 = 0.25x^2 - 1.37 x + 5.3$
    (d) $3.0 \times 10^3 x^2 = 4.1 \times 10^2 x^3 + 18.0$

56. Use the method of successive approximations to solve the unknown concentration for each of the equations given in Problem 53. How do these results compare to those obtained when using the quadratic formula?

CHALLENGE PROBLEMS

57. The Davies' equation (shown below) is an empirically modified version of the extended Debye–Hückel equation that can be used up to ionic strengths of approximately 0.2 M.[22]

$$\log(\gamma) = z^2 \cdot [0.15 \cdot I - \frac{0.51 \cdot \sqrt{I}}{1 + \sqrt{I}}]$$

$$= z^2 \cdot I' \quad \text{where} \quad I' = [0.15 \cdot I - \frac{0.51 \cdot \sqrt{I}}{1 + \sqrt{I}}] \quad (6.34)$$

(a) Use Equation 6.34 to estimate the individual ion activity coefficients for and in solutions of NaCl in water that range in concentration from 0 to 0.2 M.

(b) Make a plot of the values obtained in Part (a) and compare them to the results obtained with the extended Debye–Hückel equation. Under which conditions do these two equations give similar results? Under which conditions do they differ? Which approach is more accurate at moderate-to-high ionic strengths?

58. Another approach that is sometimes used to estimate the size of $K°$ from K or K from $K°$ is to take the logarithm of both sides and combine the resulting expression with the simplified Debye-Hückel expression in Equation 6.35.[19]

$$\log(K°) = \log(K) + [rAz_C^2 + sAz_D^2 + nAz_B^2]$$
$$\cdot (-0.51 \cdot \sqrt{I})/(1 + \sqrt{I}) \quad (6.35)$$

Use this equation to calculate the values for the concentration-dependent equilibrium constant in Table 6.4. When are the results similar? When are they different? Explain any differences you observe.

59. When chloride ions are added to a solution of silver ions the result is the formation of solid silver chloride, followed later by the formation of several soluble complexes as a large excess of chloride is added.

$$Ag^+ + Cl^- \rightleftharpoons AgCl(s)$$

$$AgCl(s) + Cl^- \rightleftharpoons AgCl_2^-$$

$$AgCl_2^- + Cl^- \rightleftharpoons AgCl_3^{2-}$$

$$AgCl_3^{2-} + Cl^- \rightleftharpoons AgCl_4^{3-}$$

Using moles as your unit of chemical content, write mass balance equations for both the silver- and chloride-containing compounds in this mixture.

60. The following formula can be used to estimate the age of an object that contains plant or animal material when this material is examined carbon-14 dating.

$$t = t_{1/2} \cdot -\frac{\ln(N_t/N_0)}{0.693} \quad (6.36)$$

In the equation, t is the estimated age of the sample, $t_{1/2}$ is the half-life of carbon-14, N_t is the relative amount of carbon-14 versus carbon-12 in the sample being studied, and N_0 is the relative amount of carbon-14 versus carbon-12 in a living plant or animal. The half-life of carbon-14 is 5730 years, and the relative amount of carbon-14 vs. carbon-12 in a living plant or animal is known to be approximately 1.3×10^{-12}.

(a) An ancient basket made from plant reeds is analyzed by carbon-14 dating and found to contain a ratio for carbon-14 vs. carbon-12 that is equal to 0.42×10^{-12}. What is the approximate age of this sample?

(b) The particular analysis method that is used in Part (a) can determine dates as far back as 60,000 years. This range is set by the limit of detection for carbon-14 vs. carbon-12 in the sample. What ratio of carbon-14 vs. carbon-12 will be present for a 60,000-year-old sample?

TOPICS FOR DISCUSSION AND REPORTS

61. Contact a meteorologist at a local television station, radio station, or newspaper. Discuss with him or her how chemical and physical measurements (such as pressure, humidity, and so on) are performed and used in weather forecasting. Write a report that discusses your findings.

62. There are many other general types of reactions besides those that were listed in Table 6.1. One example we saw briefly in this chapter was a hydration reaction, in which water will combine with a chemical to form a new "hydrated form." Below is a list of other classes of reactions. Obtain more information on one or more of these reactions and discuss how they occur. Also, try to find an example of where your reaction is used in a chemical analysis method.
    (a) Photochemical reaction
    (b) Enzymatic reaction
    (c) Ionization reaction (gas phase)
    (d) Polymerization

63. The field of paleoclimatology—the study of past climates—makes use of many techniques to look at how the composition of the atmosphere and the global temperature have changed over time. This field uses radiometric methods such as carbon 14 dating and the careful analysis of the fossil or chemical content of samples collected from the oceans, polar ice caps, or other regions of the Earth. Write a report that describes one analytical tool that is used in this field. Describe the type of information that can be gathered by the technique. Also discuss the advantages and limitations of the method.

64. Several precautions must be taken and assumptions made for carbon-14 dating to provide an accurate estimate of age for a sample. For instance, the ratio of carbon-14/carbon-12 does vary with different environments, such as air vs. the ocean. Obtain more information on carbon-14 dating and discuss the types of errors that can be present in this method. How do these errors affect the reliability of the method, and what steps must be taken to avoid or minimize these errors?

65. The use of computers in chemistry has given rise to many recent advances in our study of chemical interactions. Examine recent issues of journals such as *Scientific American*, *Science*, or *Nature* and locate an article in which computers played a major role in understanding a chemical reaction. Report on your findings.

66. One area of chemistry that frequently uses computers is research into the structures of proteins and other large biological molecules. Some researchers have used the Internet and volunteered time on many home computers to help in these efforts. Obtain information on this type of research effort. Describe, in general, how this method works and the types of chemical problems that have been examined by this approach.

67. Use Reference 25 or other sources to learn about the history behind the concept of ionic strength. Discuss how this concept was developed, and describe how it has been used over the years to characterize and study chemical reactions.

68. Learn more about the original development of the Debye–Hückel equation.[26] Describe how this equation originated and state how it has been used by chemists to help them understand chemical reactions.

References

1. P. Passell, "Global Warming Plan Would Make Emissions a Commodity," *New York Times*, October 24, 1997, p. D1.
2. Frederick K. Lutgens and Edward J. Tarbuck, *The Atmosphere: An Introduction to Meteorology*, 10th Ed., Pearson Prentice Hall, Upper Saddle River, NJ, 2007, Chapter 2.
3. *Climate Change: State of Knowledge*, Executive Office of the President, Office of Science and Technology Policy, Washington, DC, 1997. (Available on-line at http://www.usgcrp.gov/usgcrp/Library/CC-StateOfKnowledge1997.pdf)
4. C. Suplee, "Unlocking the Climate Puzzle," *National Geographic*, 1998 (193) 38–71.
5. M. Landler, "Mixed Feelings as Kyoto Pact Takes Effect," *New York Times*, February 16, 2005, p. C1.
6. A.C. Revkin, "New Warnings of Climate Change," *New York Times*, January 20, 2007, p. A7.
7. J. Haley, Ed., *Global Warming: Opposing Views*, Greenhaven Press, San Diego, CA, 2002.
8. S.G. Philander, *Is the Temperature Rising? The Uncertain Science of Global Warming*, Princeton University Press, Princeton, NJ, 1998.
9. A.C. Revkin, "Computers Add Sophistication, but Don't Resolve Climate Debate," *New York Times*, August 31, 2004, p. F3.
10. *IUPAC Compendium of Chemical Terminology*, Electronic version, http://goldbook.iupac.org
11. *The New Encyclopaedia Britannica*, 15th Ed., Encyclopaedia Britannica, Inc., Chicago, IL, 2002.
12. J. Inczedy, T. Lengyel, and A. M. Ure, *Compendium of Analytical Nomenclature*, 3rd ed., Blackwell Science, Malden, MA, 1997.
13. H. M. N. H. Irving, H. Freiser, and T. S. West, *Compendium of Analytical Nomenclature: Definitive Rules—1977*, Pergamon Press, New York, 1977.
14. B.E. Smith, *Basic Chemical Thermodynamics*, Oxford University Press, Oxford, 2004.
15. B.J. Ott and J. Buerio-Goates, *Chemical Thermodynamics—Principles and Applications*, Academic Press, New York, 2000.
16. J.H. Espenson, *Chemical Kinetics and Reaction Mechanisms*, McGraw-Hill, New York, 1981.
17. M.R. Wright, *Introduction to Chemical Kinetics*, Wiley, Hoboken, NJ, 2004.
18. E.C. Anderson, W.F. Libby, S. Weinhouse, A.F. Reid, A.D. Kirshenbau, and A.V. Grosse, "Radiocarbon from Cosmic Radiation," *Science*, 105 (1947) 576–577.
19. J.R. Arnold and W.F. Libby, "Age Determinations by Radiocarbon Content: Checks with Samples of Known Age," *Science*, 110 (1949) 678–680.
20. W.F. Libby, "Accuracy of Radiocarbon Dates," *Science*, 140 (1963) 278–280.

21. D. R. Lide, Ed., *CRC Handbook of Chemistry and Physics*, 83rd ed., CRC Press, Boca Raton, FL, 2002.

22. H. Frieser, *Concepts & Calculations in Analytical Chemistry: A Spreadsheet Approach*, CRC Press, Boca Raton, FL, 1992.

23. P. Debye and E. Hückel, "The Theory of Electrolytes. I. Lowering of Freezing Point and Related Phenomena," *Physikalische Zeitshrift*, 24 (1923) 185–206.

24. P. Debye and E. Hückel, "The Theory of Electrolytes. II," *Physikalische Zeitshrift*, 24 (1923) 305–325.

25. M.E. Sastre de Vicente, "The Concept of Ionic Strength Eighty Years After Its Introduction in Chemistry," *Journal of Chemical Education*, 81 (2004) 750–753.

26. B. Naiman, "The Debye-Hückel Theory and its Application in the Teaching of Quantitative Analysis," *Journal of Chemical Education*, 26 (1949) 280–282.

Chapter 7

Chemical Solubility and Precipitation

7.1 INTRODUCTION: FIGHTING STOMACH CANCER

Stomach cancer is the second highest cause of cancer-related deaths in the world. Fortunately, this disease can easily be treated if it is detected at an early stage. *X-ray imaging* is one tool that is often used by doctors to detect problems in the body.[1,2] To perform this method, a patient is exposed to a small dose of X rays. Some of these X rays are absorbed by regions of the body that contain elements with high atomic numbers, such as the calcium and phosphorus in bone. X rays are not normally absorbed in any significant amount by the stomach and other "soft tissues," which mainly contain low-atomic-number elements like carbon, nitrogen, hydrogen, and oxygen. This limitation can be overcome, however, by first placing into the stomach or tissue a substance that can effectively absorb X rays. This added substance is known as a *radiocontrast agent*.

Barium (atomic number 56) is used as radiocontrast agent when doctors wish to examine the stomach and digestive tract (see Figure 7.1). This agent is given to the patient to drink as a suspension of barium sulfate ($BaSO_4$). Once $BaSO_4$ has entered the digestive tract, the barium in this chemical allows X rays to be used in imaging the stomach for the detection of cancer. $BaSO_4$ is a salt that is only slightly soluble in water, even at the low pH that is present in the stomach.[1,2] This low solubility is important because it prevents the ingested barium from leaving the stomach and entering the body as barium ions, which can be toxic at high doses.

The use of $BaSO_4$ in X-ray imaging is just one example of an analytical method that relies on the ability to use or form an insoluble chemical substance. In this particular example, it is important for the analyst to understand what factors will cause $BaSO_4$ to dissolve or remain as suspension of solid $BaSO_4$. This requires that the analyst have a good working knowledge of chemical solubility and precipitation. In this chapter, we examine the topics of solubility and precipitation and see how these processes are used in chemical analysis. We will also look at the reactions that are involved as a chemical dissolves or precipitates and see how we can use equilibrium constants to describe and predict the extent of these processes.

7.1A What Is Solubility?

The term **solubility** refers to the maximum concentration or amount of a chemical that can be placed into a solvent to form a stable solution.[3,4] Let's think about this process by using a suspension of $BaSO_4$ in water as an example. We already know $BaSO_4$ is only slightly soluble in water. This means that if we add enough of this salt to water we will eventually reach a point where some of the $BaSO_4$ is dissolved and the rest is a solid in contact with this solution. This situation occurs at room temperature when just over 1 mg $BaSO_4$ has been added to one liter of water. Although some chemicals can dissolve as intact molecules, ionic substances like $BaSO_4$ tend to dissolve

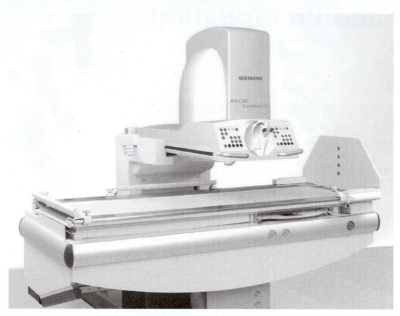

X-ray imaging system

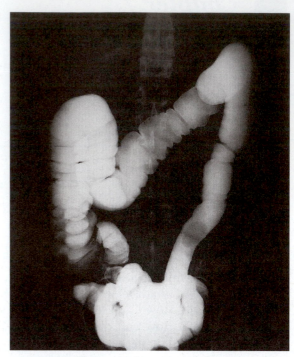

X-ray image in the presence of BaSO$_4$

FIGURE 7.1   The use of X rays to obtain an image of a patient after a suspension of BaSO$_4$ in water has been placed into the digestive tract, allowing soft tissues in this part of the body to be examined. The X rays are passed through the patient and detected on the other side. The result is an image that shows the relative intensity of X rays that have passed through a given region of the body. This result is usually used by doctors as a negative image, in which areas of the body that transmit few X rays (such as bones or the digestive tract when it contains barium sulfate) appear lighter than the surrounding regions.[2] (The image on the left shows a modern X-ray imaging system and is reproduced with permission from Siemens; the image on the right is reproduced with permission from the American Society of Radiologic Technologists.)

by forming ions. This process, shown below, is known as *dissociation*.

$$BaSO_4(s) \rightleftarrows Ba^{2+} + SO_4{}^{2-} \qquad (7.1)$$

For instance, BaSO$_4$ undergoes dissociation in water as it dissolves to form soluble Ba^{2+} and SO$_4{}^{2-}$ ions. If these ions are present at sufficiently high concentrations, they can recombine to form solid BaSO$_4$. It is this balance between the dissociation of BaSO$_4$ and its reformation that determines the overall solubility of this chemical in water.

Regardless of whether we are dealing with BaSO$_4$ or some other chemical, the observed solubility will depend on such things as the chemical's structure and the solvent in which it is being dissolved. The extent of this solubility can vary widely even between chemicals with similar formulas. For instance, the sulfate salts of strontium, calcium, and magnesium (giving SrSO$_4$, CaSO$_4$, and MgSO$_4$) are 100 to 290,000 times more soluble than BaSO$_4$ in water, even though all of these sulfate salts contain alkaline earth metals.[5] The solubility of a chemical will also depend on temperature and pressure. This is illustrated by the 1.5-fold increase in solubility that occurs for BaSO$_4$ in water between 18 and 50°C.[4] Another factor that will affect the solubility is the presence of any

side reactions that involve the dissolving chemical. We examine this last topic in more detail in Section 7.3C of this chapter.

7.1B What Is Precipitation?

A closely related term to solubility is **precipitation**. Precipitation is a process that occurs when a portion of a dissolved chemical leaves the solution to form a solid.[3] This process can be illustrated by considering what would happen when various amounts of BaSO$_4$ are placed in water. First, there is the situation in which BaSO$_4$ is added in an amount that is below its maximum solubility limit (less than 1.15 mg of BaSO$_4$ in 1 L of water at room temperature). The result, shown in Figure 7.2, is an *unsaturated solution*, in which the final concentration of the added solute can be directly determined from the total amount that was added to the solution. If we keep adding more BaSO$_4$, we will eventually create a situation in which no more can dissolve. This gives a *saturated solution*, which will have a dissolved concentration for the solute that is equal to its maximum solubility in a solution at equilibrium.[3,4,6]

A third possible situation is a *supersaturated solution*. This situation occurs when the concentration of a dissolved chemical is temporarily greater than its maximum

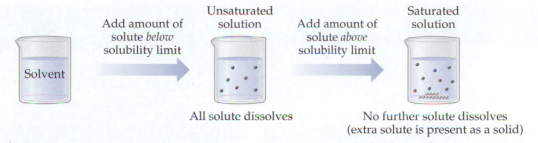

FIGURE 7.2   Formation of unsaturated and saturated solutions.

solubility at equilibrium.[3,6] This type of solution can be created by first producing a saturated or nearly saturated solution of a chemical under conditions in which it has a higher solubility than its final desired solution. Figure 7.3 shows one way this can be accomplished, in which a chemical is initially dissolved at a high temperature (which tends to increase solubility) and is then brought to a lower temperature (which tends to decrease solubility). This creates an unstable situation in which the dissolved chemical is now present at a concentration above its solubility limit. To fix this situation, some of the solute will leave the solution to reestablish a stable system. If this undissolved material is a solid, it is called a **precipitate**.

The particular form that a solid precipitate takes will depend in part on the rate at which this precipitate is created. If this solid is formed slowly, the resulting solid will be relatively pure and have a high degree of order in its structure. This type of precipitate is called a *crystal* (see Figure 7.4),[2,4,6] which is created through a special type of precipitation known as *crystallization*. Crystallization is a valuable way of purifying chemicals. The careful control of precipitation to obtain high purity crystals is also important in the use of these crystals to study chemical structure by a method known as *X-ray crystallography*, as discussed in Box 7.1.

If precipitation is performed quickly, this often produces small solid particles that have less order than crystals. However, these smaller particles can still be useful in isolating and analyzing the precipitating chemical. This is particularly true if these particles are large enough to allow them to be easily removed from the solution by filtration or centrifugation. The formation of smaller particles (with sizes between 1 nm and 1 μm) gives rise to a *colloidal dispersion*, or "colloid."[2-4] Although there are many useful substances that are colloids (such as milk, butter, paint, and ink), the presence of a colloid is usually not desirable in an analysis method if the goal is to collect this precipitate for its characterization and measurement.

7.1C Why Are Solubility and Precipitation Important in Chemical Analysis?

There are many reasons why it is important to consider the solubility or precipitation of a chemical during an analysis. First, a chemical must be at least partially soluble in a sample if we are to detect or measure it in this sample. In addition, any chemicals we wish to place into a solution as reagents must be able to dissolve in a reproducible fashion and at a sufficient concentration for our purpose. There are also many analytical methods that make use of the ability of a chemical to precipitate from a solution. One example is the method of gravimetry (see Chapter 11), where mass measurements of a pure precipitate are used to measure an analyte that is contained in this precipitate. A related method is a precipitation titration (Chapter 13), in which the amount of analyte is determined by measuring the volume of a reagent that is required to completely precipitate an analyte from a solution.

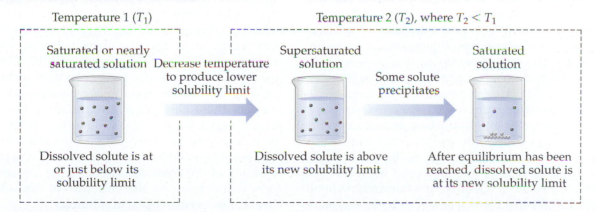

FIGURE 7.3   Formation of a supersaturated solution by using the difference in chemical solubility that occurs as temperature is varied.

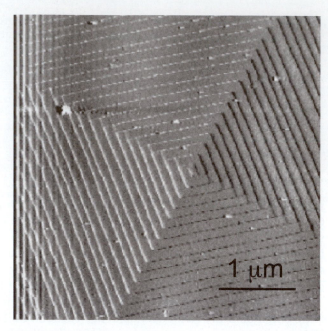

FIGURE 7.4 Crystals of barium sulfate. The large crystal is a naturally occurring mineral form of barium sulfate ($BaSO_4$) known as "barite."[2] The image on the right is a crystal of $BaSO_4$ that was grown in a laboratory and shows the high degree of order that can occur in such a solid. (The image on the right is reproduced with permission from S.R. Higgins, Wright State University; the image on the left is reproduced with permission from J. Crowley, Crystal Mine.)

Sometimes a difference in solubility for a chemical between two liquids or two different chemical environments can be used to isolate or analyze a chemical. This occurs in an *extraction* (Chapter 20), in which the components in a sample are allowed to distribute between two mutually insoluble liquids or chemical phases to help remove one or more chemicals from others in the sample. The technique of *chromatography*, which is discussed in Chapters 20–22, is another method that can make use of differences in solubility to separate and analyze various chemicals in a sample.

7.2 CHEMICAL SOLUBILITY

7.2A What Determines Chemical Solubility?

We learned earlier that the solubility of a chemical depends on the nature of the chemical and the solvent. In general, the process of placing one chemical into another involves changing **intermolecular forces**. Intermolecular forces are non-covalent, electrostatic interactions that cause separate but neighboring molecules or chemical species to attract or repel one another.[10,11] Some common types of intermolecular forces are shown in Figure 7.6. An *ionic interaction* is one example; this is when an electrostatic attraction occurs between two ions with opposite charges or repulsion occurs between two ions with the same type of charge.[4] *Hydrogen bonding* is another type of intermolecular interaction, in which a hydrogen atom is shared in a noncovalent bond between molecules that contain atoms such as nitrogen or oxygen. An example is the hydrogen bond that can form between neighboring molecules of water or ammonia.

A third type of intermolecular force is a *dipole–dipole interaction*. This interaction takes place between two chemicals that have permanent dipole moments. A permanent dipole moment occurs in a chemical when it has a nonsymmetrical arrangement of electrons, giving it regions with a slight excess of negative charge (δ^-) or positive charge (δ^+).[2,4] If two molecules with permanent dipoles are aligned properly, these regions can cause the two molecules to be attracted to each other. Both chemicals with and without permanent dipoles can also interact through *dispersion forces* (also known as *London forces* or *Van der Waals forces*).[2] Dispersion forces are similar to dipole–dipole interactions but occur when the movement of electrons in one chemical creates a temporary dipole moment, which induces another temporary but complementary dipole in a neighboring chemical.[10,11]

The formation of a solution involves breaking the interactions between individual molecules or ions in the solid that is being dissolved. This also requires disrupting some of the intermolecular forces between neighboring solvent molecules to make room for the dissolving chemical. New intermolecular forces can then form between the dissolved solute and the surrounding solvent. The stronger these new forces are, the more favorable the process will be for the solid to dissolve. The more similar the solvent and the dissolving chemical are in terms of their possible intermolecular interactions, the greater the solubility of the chemical will be in the solvent. This, in turn, will depend on the structure of the solvent and chemical we wish to place in solution.

Let's consider a few examples to illustrate this idea. For instance, water is a solvent which is a "polar"

BOX 7.1
X-Ray Crystallography

Crystals are not only good sources of pure chemicals but they can also provide important information on the structure of a chemical. This information can be obtained by a method known as *X-ray crystallography* (see Figure 7.5). This technique makes use of the fact that X rays will be diffracted by the regular arrangement of atoms in a crystal (see Chapter 17 for a more detailed discussion of how diffraction occurs). This produces an image known as a *diffraction pattern*, which will depend on the spacing and distances of the atoms in the crystal. This pattern is then recorded and analyzed to obtain information on how these atoms are arranged, providing information on the three-dimensional structure of the chemical that makes up this crystal.[2,7–9]

X-ray crystallography is often performed on single crystals of a chemical. This approach is known as *single-crystal diffraction* and can be used to provide detailed information on chemicals ranging from small molecules and ions to large biological substances like proteins and DNA. For instance, the first three-dimensional structure of a protein (myoglobin) was determined by M. Perutz and J.C. Kendrew using this method, leading to their being awarded the Nobel Prize in Chemistry in 1958.[2] X-ray crystallography with single crystals is still one of the main approaches used by chemists and biochemists to examine the structures of chemicals and biochemicals. However, single-crystal diffraction does require the use of crystals that are both pure and of a proper size for analysis. This need makes knowledge of the factors that affect the purity and preparation of crystals essential in obtaining good-quality samples for determining the structure of a chemical.

If a sufficiently large and pure crystal of a chemical is not available, it is also possible to perform X-ray crystallography by using a powder that is prepared from small crystals of a desired chemical. This method is known as *powder diffraction*. Using a powder in X-ray crystallography instead of large single crystals makes this method easier to perform but provides less information on the structure of a chemical. However, the pattern that is generated by this method can often be valuable in determining the identity, purity, crystal size, and texture of the powdered material that is being analyzed. This information makes powder diffraction useful in fields such as materials chemistry and synthetic chemistry.

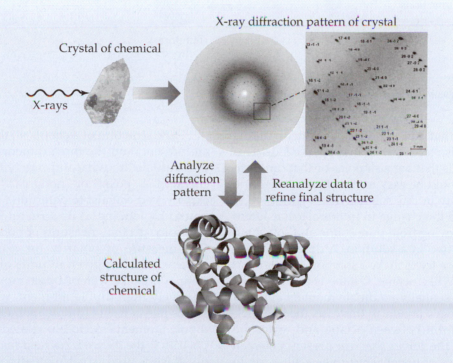

FIGURE 7.5 The general process by which X-ray crystallography is used to determine the structure of a chemical. This process begins by passing X rays through a pure crystal of the chemical. As these X rays pass through the crystal, the spacing and arrangement of atoms in the crystal will create a unique diffraction pattern. This pattern consists of a set of spots that represent regions of relatively intense X-ray radiation. The positions of these spots are then analyzed to obtain information on the arrangement of the atoms in the crystal. This arrangement of atoms makes it possible to find the three-dimensional structure of the chemical that contains these atoms. This analysis often requires many stages of refinement until a chemical structure with a satisfactory level of resolution is obtained. The crystal, diffraction pattern, and structure that are shown in this example are for myoglobin. (The diffraction pattern for myoglobin is reproduced with permission by the AAAS from F. Schotte, M. Lim, T.A. Jackson, A.V. Smirnov, J. Soman, J.S. Olson, G.N. Phillips Jr., M. Wulff, and P.A. Anfinrud, *Science*, 300 (2003) 1944–1947.)

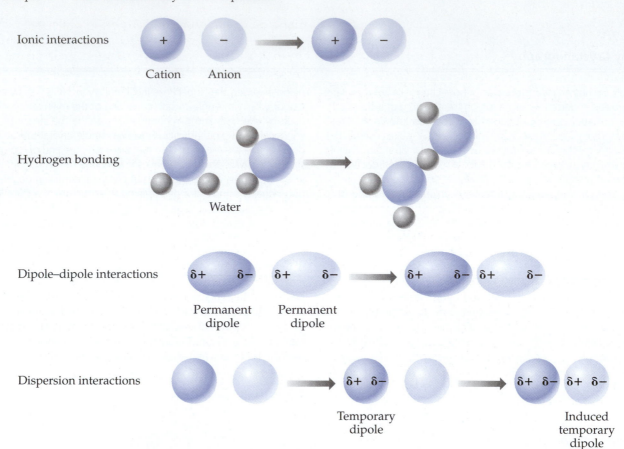

FIGURE 7.6   Examples of intermolecular forces.

chemical (or one that has a high permanent dipole moment) that can also form strong hydrogen bonds. This means that chemicals that are also polar or that can form hydrogen bonds should be easy to dissolve in water and have a high solubility in this solvent. This type of situation helps minimize the change in intermolecular forces that occurs when moving from pure solute and solvent to a mixture of these two in a solution. A different result occurs, however, if we try to make a solution in water that contains a substance like octane. Octane is a "nonpolar" chemical that has essentially no dipole moment and that does not form hydrogen bonds. In this case, the intermolecular forces that form between octane and water are weak compared to the forces that are present between individual molecules of water. This gives a situation in which only a small amount of octane will go into water and results in a low solubility for octane in this solvent.

These general observations can be summarized by the phrase "like dissolves like." This simply means that substances that are similar in their polarity and intermolecular forces will tend to dissolve well in each other. As a result, compounds that are polar, ionic, or can form hydrogen bonds (such as glucose and sodium chloride) will usually dissolve well in a polar solvent like water. Nonpolar compounds like octane will dissolve best in nonpolar solvents (e.g., benzene).

The structure of a chemical, the types of atoms that are present, and the three-dimensional arrangement of these atoms are what ultimately determine whether a chemical is "polar" or "nonpolar." This often allows an experienced chemist to determine the polarity of a compound by looking at its structure. However, there are also some general measures of polarity that can be used. One measure of polarity for a chemical is its *dipole moment*, which is given in units of a debye, or D. Polar compounds like water (dipole moment = 1.85 D) have a relatively high dipole moment, while a nonpolar substance like benzene (dipole moment = 0.00 D) has a low dipole moment.[4] A closely related measure of chemical polarity is the *dielectric constant* (as discussed in Chapter 6), which is used to calculate dipole moments. It is also possible to rank chemicals based on their relative polarity by comparing their ability to dissolve in water versus *n*-octanol (a relatively nonpolar solvent). This can be accomplished as shown in Figure 7.7 by using the *octanol–water partition ratio* (K_{ow}).[12] A low K_{ow} value is obtained for chemicals that are polar and easily dissolve in water, while a high K_{ow} value is obtained for nonpolar compounds that dissolve better in octanol. This information is often used by scientists to predict the solubility of a chemical in various matrices, such as biological tissues or soil and water.

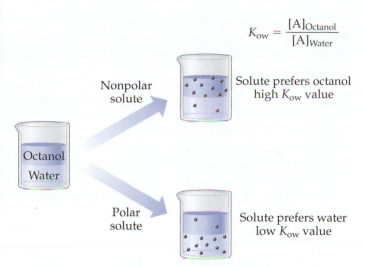

$$K_{ow} = \frac{[A]_{Octanol}}{[A]_{Water}}$$

Nonpolar solute — Solute prefers octanol high K_{ow} value

Octanol Water

Polar solute — Solute prefers water low K_{ow} value

FIGURE 7.7 A method for determining the octanol–water partition ratio (K_{ow}) for a chemical. (*Note:* This ratio is also sometimes represented by the terms P_{ow} or P.) In this method, the analyte of interest (A) is combined with known volumes of octanol and water, two mutually immiscible solvents. Some of the analyte will enter each of these solvents according to its solubility in water and octanol. Once equilibrium has been reached, the concentration of analyte in both the octanol and water layers is determined. The ratio of these concentrations is then used to determine the value of K_{ow}. Polar solutes will tend to dissolve better in water than in octanol, giving a low K_{ow} value. Nonpolar solutes will dissolve better in octanol than water and give a high K_{ow}. The equation shown for K_{ow} in this figure is based on the concentrations of a given species of analyte (A) in the octanol and water solution. In practice, however, the total concentration of analyte (regardless of its form) in each of the two phases is often used in determining an octanol–water partition ratio, where $K_{ow} = C_{A, Octanol}/C_{A, Water}$.

EXERCISE 7.1 Using P_{ow} to Estimate Chemical Polarity and Solubility

A toxicologist finds that the pesticide DDT has a K_{ow} value of approximately 10^5. What does this value tell you about the polarity of DDT? From this information, would you expect DDT to be more soluble in water or in a nonpolar environment such as fatty tissues in animals and humans?

DDT
(Dichloro-diphenyl-trichloroethane)

SOLUTION

The large K_{ow} value for DDT indicates that it is much more soluble in a nonpolar solvent like octanol versus water. This indicates that DDT is a nonpolar compound. This agrees with environmental and toxicological measurements, which have shown that this pesticide tends to accumulate in the fatty tissues of animals and to travel up

the food chain as one animal eats another that has been exposed to DDT. Although DDT is an extremely effective insecticide, its possible environmental effects led to its ban in the United States and other countries.

Another way we can think about solubility is in terms of the overall change in energy that occurs when we place a solute into a solvent. We now know that a change in intermolecular forces will take place when we dissolve a solute in a solvent. But a change in entropy, or the "order" of the system, occurs as well. Placing a solute into a solvent to produce a solution will give rise to a less ordered system, which means that the change in entropy for mixing these two chemicals will always be favorable. However, the process of breaking and forming new intermolecular forces will require the input of energy. It is this balance between the change in energy due to entropy and the change in intermolecular forces that determines the overall change in energy as we mix a solute and solvent. Solutes and solvents that are similar in their polarity and interactions will produce the smallest change in energy due to intermolecular forces and will give rise to the greatest solubility.[11,13]

7.2B How Can We Describe Chemical Solubility?

Molecular Solids. The solubility of a substance can be expressed in a variety of units of chemical content, but this is mostly commonly done in terms of molarity (M) or the mass of solute per a given volume of solvent (for instance, 1.15 mg $BaSO_4$ per 1 L water). It is also possible to write a reaction that shows the relationship between a chemical in its solid form and this chemical when it is dissolved in a solution. For a molecular compound like glucose ($C_6H_{12}O_6$), the equilibrium between the solid and dissolved forms can be described by the following reaction.

$$Glucose(s) \rightleftharpoons Glucose(aq) \qquad (7.2)$$

In this case, we are changing the surrounding environment (from a solid to a solution), but we are not changing the structure of our chemical. This type of process is sometimes called a *solubility equilibrium*,[14] and it can be described by using either a thermodynamic equilibrium constant ($K°$, based on activity) or a concentration-based equilibrium constant (K, based on molarity or other concentration units for the chemical that has been dissolved in the solution).

$$K° = \frac{a_{Glucose(aq)}}{a_{Glucose(s)}} = a_{Glucose(aq)} \quad (\text{where } a_{Glucose(s)} = 1)$$

$$(7.3)$$

$$K = \frac{[Glucose]}{a_{Glucose(s)}} = [Glucose] \qquad (7.4)$$

These equilibrium constants are known as **solubility constants.**[15] As just shown, we can simplify these equilibrium expressions by using the fact that the solid form of glucose will have an activity of exactly one, because this solid represents a standard state of glucose (see Chapter 6 for our discussion of activity and *standard states*). This simplification means the value of $K°$ for a molecular species like glucose will be exactly equal to the activity of the dissolved molecule at equilibrium. The value of K will describe the concentration of this dissolved compound, with apparent units of "M." (*Note*: Recall that the value for an equilibrium constant K can also be written more properly without any units, as discussed in Chapter 6.) Both $K°$ and K for this system are directly related to the solubility (S) of glucose in the solution.

EXERCISE 7.2   **Using Solubility Constants for a Molecular Solid**

A physician wishes to give a high-concentration solution of glucose to a patient to drink as part of research in developing a new test for diabetes. The protocol to be used in this test calls for a 1.00 M solution of glucose in water. The solubility of glucose (molar mass, 180.16 g/mol) is approximately 91 g per 100 mL water at 25°C. Is a 1.00 M solution of glucose above, below, or at this solubility limit?

SOLUTION

A 100.00 mL portion of a 1.00 M solution of glucose in water would contain (0.10000 L)(1.00 mol glucose/L)(180.16 g/mol glucose) = $18.0\underline{16}$ = **18.0 g/L glucose**. Thus, a 1.00 M solution of glucose should be well below the solubility limit and all of added glucose should dissolve under these conditions. (*Note*: In this case, the desired concentration of glucose was low enough to make it safe to ignore any changes in volume that occurred when the glucose was added to water. However, this may not be the case when examining the concentration of more concentrated solutions.)

Any type of solid can be classified as "highly soluble," "slightly soluble," or "insoluble" in a given solvent. These same terms are used in tables of chemical and physical data to describe solubility, but these tables often do not have rigid definitions for these terms. In this book, we will use a solubility of 0.5 M or greater to refer to a chemical that is "highly soluble." A solution of glucose in water would be one example of this type of system. Similarly, we will use the phrases "slightly soluble" and "insoluble" when discussing chemicals that have solubilities between 0.05 and 0.5 M and below 0.001 M, respectively. However, we should keep in mind that these concentration limits are arbitrary and that the use of terms like "soluble" and "insoluble" will depend on our particular application for a chemical and its solution.

One assumption we made in the preceding exercise was that the molecular form of glucose was the only species for this chemical that was present in solution. However, this is not the case for all molecular solids. For instance, glucose actually has two forms (α-glucose and β-glucose) that are in equilibrium with each other and that have slightly different solubilities. Other solids might react with the solvent or react with other chemicals in the solution to give additional species that must be considered when describing solubility. We will see in Section 7.3C how the presence of these side reactions can be used to our advantage to alter and control the solubility of a chemical. The previous exercise also assumed that the final solution was at equilibrium. It should be kept in mind, however, that even higher amounts of a substance such as glucose can sometimes be dissolved on a temporary basis by forming a "supersaturated solution."

Ionic Solids.   A similar equilibrium to what we saw for a molecular substance like glucose will occur when we dissolve part of an ionic solid like barium sulfate in water.

$$BaSO_4(s) \rightleftharpoons Ba^{2+} + SO_4{}^{2-} \qquad (7.5)$$

However, this reaction now differs from the one we wrote earlier for a molecular compound like glucose in that the form, and not just the environment, of the chemical changes as we go from the ionic solid to dissolved ions.

Just as we saw for glucose, we can describe the process in Equation 7.5 for barium sulfate using either a thermodynamic equilibrium constant ($K°_{sp}$) or concentration-dependent value (K_{sp}).

$$
\begin{aligned}
K°_{sp} &= \frac{(a_{Ba^{2+}})(a_{SO_4{}^{2-}})}{a_{BaSO_4}} \\
&= (a_{Ba^{2+}})(a_{SO_4{}^{2-}}) \qquad (\text{where } a_{BaSO_4} = 1)
\end{aligned}
$$
$$(7.6)$$

$$K_{sp} = \frac{[Ba^{2+}][SO_4{}^{2-}]}{a_{BaSO_4}} = [Ba^{2+}][SO_4{}^{2-}] \qquad (7.7)$$

The subscript "sp" is used with equilibrium constants for the dissolving/precipitation of an ionic solid because such a constant is often called a **solubility product.**[3,4,6,14,15] The term "product" in this name refers to the fact that these equilibrium constants are equal to some multiple of the concentrations or activities of the individual ions that are formed as the ionic solid dissolves in the solvent (e.g., "$[Ba^{2+}][SO_4{}^{2-}]$" is the *ion product* in Equation 7.7). You may also have noticed that the activity for original solid ($BaSO_4$, in this case) again has a value of one and is usually not shown in the expressions written for $K°_{sp}$ or K_{sp}. This practice is commonly followed in writing solubility products and ion products and will be the approach used in the remainder of this book.

EXERCISE 7.3 | **Writing Expressions for Solubility Products**

Write the solubility product expressions for each of the following reactions in water. What is the ion product in each of these expressions?

a. K°_{sp} for $AgBr(s) \rightleftharpoons Ag^+ + Br^-$
b. K°_{sp} for $CaF_2(s) \rightleftharpoons Ca^{2+} + 2F^-$
c. K_{sp} for $Fe(OH)_3(s) \rightleftharpoons Fe^{3+} + 3OH^-$
d. K_{sp} for $Mg(NH_4)PO_4(s) \rightleftharpoons Mg^{2+} + NH_4^+ + PO_4^{3-}$

SOLUTION

The following solubility product expressions are obtained, where the activities for pure solids are assigned activities of exactly one and are not shown: (a) $K^\circ_{sp, AgBr} = (a_{Ag^+})(a_{Br^-})$; (b) $K^\circ_{sp, CaF_2} = (a_{Ca^{2+}})(a_{F^-})^2$; (c) $K_{sp, Fe(OH)_3} = [Fe^{3+}][OH^-]^3$; (d) $K_{sp, Mg(NH_4)PO_4} = [Mg^{2+}][NH_4^+][PO_4^{3-}]$. The ion products for the expressions in (a) through (d) would be $(a_{Ag^+})(a_{Br^-})$, $(a_{Ca^{2+}})(a_{F^-})^2$, $[Fe^{3+}][OH^-]^3$, and $[Mg^{2+}][NH_4^+][PO_4^{3-}]$, respectively.

A list of K_{sp} values for common salts used in chemical analysis is given in Table 7.1. One way we can use K_{sp} is to predict the solubility of an ionic substance. For instance, barium sulfate is known to have a K_{sp} in water equal to 1.08×10^{-10} at 25°C (with apparent units of M^2 for a concentration-based value for K_{sp}). If barium sulfate is the only source of Ba^{2+} and SO_4^{2-} in this solution, where these ions are produced in a 1:1 ratio (i.e., $[Ba^{2+}] = [SO_4^{2-}]$), we can use the K_{sp} expression for barium sulfate to determine the maximum concentration of

these ions that will be in solution at equilibrium. (*Note*: If desired, apparent units of M^2 can be used for this K_{sp} for the sake of dimensional analysis.)

$$K_{sp} = [Ba^{2+}][SO_4^{2-}] \quad \text{and}$$
$$[Ba^{2+}] = [SO_4^{2-}] \Rightarrow 1.08 \times 10^{-10} M^2 = [Ba^{2+}]^2$$
$$\therefore [Ba^{2+}] = 1.04 \times 10^{-5} M \tag{7.8}$$

This concentration of Ba^{2+} is far too low to do any harm to humans, so the insoluble $BaSO_4$ that is given to a patient for an X ray will pass through the body without dissolving appreciably and entering the blood stream. But during the time this solid spends in the stomach and intestines, it will allow X rays to be used to view the structures of these organs, hopefully leading to the detection of any abnormalities.

EXERCISE 7.4 | **Estimating the Solubility of an Ionic Solid**

Although the solubility of ionic solids can be calculated from their solubility products, the degree of this solubility will depend on both the K_{sp} value for this ionic solid and the stoichiometry of the ions that make up this solid. To illustrate this, let's consider $ZnCO_3$ ($K_{sp} = 1.46 \times 10^{-10}$) and Ag_2CO_3 ($K_{sp} = 8.46 \times 10^{-12}$). What would be the expected solubilities of these solids in pure water if no side reactions or other sources for the ions in these solids are present? Which of these compounds is more soluble in water? Explain your results.

SOLUTION

We can begin solving this problem by first writing the reactions for the dissolution of these ionic solids and their solubility constant expressions (in which apparent units

TABLE 7.1   Solubility Products for Ionic Substances with Low Solubilities in Water[*]

	Substance	Dissociation Reaction	Solubility Product, K_{sp}
1:1 Salts[a]	Barium sulfate	$BaSO_4(s) \rightleftharpoons Ba^{2+} + SO_4^{2-}$	$[Ba^{2+}][SO_4^{2-}] = 1.08 \times 10^{-10}$
	Calcium sulfate	$CaSO_4(s) \rightleftharpoons Ca^{2+} + SO_4^{2-}$	$[Ca^{2+}][SO_4^{2-}] = 4.93 \times 10^{-5}$
	Copper(II) sulfide	$CuS(s) \rightleftharpoons Cu^{2+} + S^{2-}$	$[Cu^{2+}][S^{2-}] = 6.3 \times 10^{-36}$
	Silver chloride	$AgCl(s) \rightleftharpoons Ag^+ + Cl^-$	$[Ag^+][Cl^-] = 1.77 \times 10^{-10}$
	Silver bromide	$AgBr(s) \rightleftharpoons Ag^+ + Br^-$	$[Ag^+][Br^-] = 5.35 \times 10^{-13}$
	Silver iodide	$AgI(s) \rightleftharpoons Ag^+ + Cl^-$	$[Ag^+][Cl^-] = 8.52 \times 10^{-17}$
1:2 Salts	Barium fluoride	$Ba(F)_2(s) \rightleftharpoons Ba^{2+} + 2F^-$	$[Ba^{2+}][F^-]^2 = 1.84 \times 10^{-7}$
	Calcium fluoride	$CaF_2(s) \rightleftharpoons Ca^{2+} + 2F^-$	$[Ca^{2+}][F^-]^2 = 5.3 \times 10^{-9}$
	Iron(II) hydroxide	$Fe(OH)_2(s) \rightleftharpoons Fe^{2+} + 2OH^-$	$[Fe^{2+}][OH^-]^2 = 4.87 \times 10^{-17}$
1:3 Salts	Iron(III) hydroxide	$Fe(OH)_3(s) \rightleftharpoons Fe^{3+} + 3OH^-$	$[Fe^{3+}][OH^-]^3 = 2.79 \times 10^{-39}$

[*]These K_{sp} values were obtained from J.A. Dean, Ed., *Lange's Handbook of Chemistry*, 15th Ed., McGraw-Hill, New York, 1999. The listed values were acquired at temperatures between 18 and 25°C.

[a]Terms such as "1:1 salt" and "1:2 salt" refer to the stoichiometric ratio of the cation and anion that make up the salt. For instance, iron(III) hydroxide is a 1:3 salt because it contains the cation Fe^{3+} and the anion OH^- in a 1:3 ratio.

of M^2 and M^3 can be used for $K_{sp, ZnCO_3}$ and K_{sp, Ag_2CO_3}, respectively).

$$ZnCO_3(s) \rightleftharpoons Zn^{2+} + CO_3^{2-}$$

$$K_{sp, ZnCO_3} = [Zn^{2+}][CO_3^{2-}] = 1.46 \times 10^{-10}$$

$$Ag_2CO_3(s) \rightleftharpoons 2\,Ag^+ + CO_3^{2-}$$

$$K_{sp, Ag_2CO_3} = [Ag^+]^2[CO_3^{2-}] = 8.46 \times 10^{-12}$$

Because we are assuming there are no side reactions or other sources of ions, we can say for a 1:1 salt like $ZnCO_3$ that the final concentration of Zn^{2+} must equal the final concentration of CO_3^{2-} that is produced from this salt, or $[Zn^{2+}] = [CO_3^{2-}]$. This means we can substitute the value of $[Zn^{2+}]$ for $[CO_3^{2-}]$ in the preceding equation for $K_{sp, ZnCO_3}$, giving the following result.

$$K_{sp, ZnCO_3} = [Zn^{2+}]^2 = 1.46 \times 10^{-10}\,M^2$$

$$\text{or } [Zn^{2+}] = [CO_3^{2-}] = 1.21 \times 10^{-5}\,M$$

This means that the solubility of $ZnCO_3$ under these conditions (as given by the final value of either $[CO_3^{2-}]$ or $[Zn^{2+}]$) is **$1.21 \times 10^{-5}\,M$**.

Similarly, as Ag_2CO_3 dissolves, it will form two moles of Ag^+ for each mole of CO_3^{2-} that is produced. If there are no other sources for these ions, we can say that $[CO_3^{2-}] = \frac{1}{2}[Ag^+]$ in the final solution. We can then use this information along with the solubility expression for Ag_2CO_3 to obtain these values for $[Ag^+]$ and $[CO_3^{2-}]$.

$$K_{sp, Ag_2CO_3} = 8.46 \times 10^{-12} = [Ag^+]^2\,(\tfrac{1}{2}[Ag^+])$$

$$= \tfrac{1}{2}[Ag^+]^3$$

$$\therefore\ [Ag^+] = 2\,(8.46 \times 10^{-12}\,M^3)^{1/3} = 4.07\underline{5} \times 10^{-4}\,M$$

$$= \mathbf{4.08 \times 10^{-4}\,M}$$

$$[CO_3^{2-}] = \tfrac{1}{2}[Ag^-] = 2.03\underline{7} \times 10^{-4}\,M$$

$$= \mathbf{2.04 \times 10^{-4}\,M}$$

The solubility of Ag_2CO_3 in this second situation would be given as $1.02 \times 10^{-4}\,M$ (based on the final concentration of CO_3^{2-}, where one mole of CO_3^{2-} is formed for every mole of Ag_2CO_3 that dissolves). Thus, the solubility of Ag_2CO_3 is greater than that for $ZnCO_3$ even though Ag_2CO_3 has a lower K_{sp}.

In this last problem we used the concentrations of the dissolved ions for our equilibrium calculation rather than their activities. This approach was reasonable in this situation because the final concentrations of these ions (and the ionic strength of the resulting solution) were so low that the activity coefficients for these ions should have been near one. We will continue with this practice throughout this chapter as we deal with low solubility substances. However, we should always keep in mind

that a better approach is to use chemical activities if we are dealing with a solution that has a high concentration of our desired ions or a moderate-to-high ionic strength.

Some ionic solids like $NaCl$ that have a high solubility in water are generally regarded as totally soluble in this solvent. Other ionic solids that fall in this category include the chloride, bromide, and iodide salts of all cations (except those involving Ag^+, Pb^{2+}, and Hg_2^{2+}), such as $NaCl$, KI, and $CaBr_2$. Other ionic substances that are highly soluble in water are those based on ammonium cations and acetate or perchlorate anions (with the exception of $KClO_4$). The sulfate salts of Group IA elements, Mg^{2+}, and Fe^{2+} (e.g., sodium sulfate and magnesium sulfate) and nitrate salts are also highly soluble in water.

When dealing with aqueous solutions that involve any of these highly soluble ionic solids, it is often assumed that these solids will totally dissolve and dissociate into their corresponding cations and anions. We should keep in mind, though, that just because an ionic solid has a high solubility in water does not mean that the compound will dissolve readily in other solvents. For instance, sodium chloride can easily dissolve in water, but is only slightly soluble in ethanol. Also, it is possible to get a saturated solution in water for even a highly soluble salt. For $NaCl$ this occurs when 357 g of this salt have been added to 1 L of water at 273 K.

Most ionic solids can be placed in much smaller quantities in water and will be only either slightly soluble or insoluble in this solvent. Examples of ionic substances that fall into this category are given in Table 7.1. These include such chemicals as barium sulfate, silver chloride, and calcium fluoride. For these substances we must use solubility products to determine how much of their solid form will actually go into a solution. We have already seen some examples of these calculations in the last exercise and earlier in this section. Other examples dealing with more complicated situations will be given later in this chapter. We will also see in Chapters 11 and 13 how such calculations can be used to describe the behavior of analytical methods that make use of precipitation reactions (e.g., gravimetry and a precipitation titration).

Mixtures of Liquids.   Up to this point, we have discussed the solubility of molecular and ionic solids in water, but the concept of solubility can be expanded to also include other types of chemical mixtures. Let's first look at the dissolving of one liquid in another. For instance, we would expect that liquids like ethanol that are polar and capable of forming hydrogen bonds would dissolve well in water, because this would make them similar to water in some of their properties. On the other hand, a liquid that is very nonpolar and cannot form hydrogen bonds, such as benzene or carbon tetrachloride, would be very different from water and not dissolve well in this solvent.

There are several terms used to describe how well one liquid will mix with another (see Figure 7.8). One such word is *miscible*, which describes two liquids that

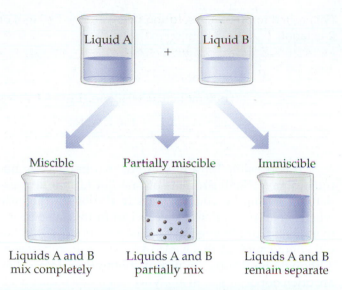

Miscible · Partially miscible · Immiscible

Liquids A and B
mix completely

Liquids A and B
partially mix

Liquids A and B
remain separate

FIGURE 7.8 Miscible, immiscible, and partially miscible liquids. The small circles in the central diagram represent the small amounts of A and B that mix with each other.

can form a stable solution when mixed in any proportion.[4] For example, Table 7.2 indicates that methanol and ethanol are miscible with water in all proportions. The solubility of many other alcohols in water is also high, but decreases as their carbon chain becomes longer and they become more nonpolar in their behavior.

The opposite case occurs when two liquids are *immiscible*, or do not dissolve to any appreciable extent in one another.[4] The presence of two immiscible liquids will result in the formation of two distinct liquid layers, with the lower density liquid laying on top of the other. A mixture of ether and water is one example of this because these liquids are immiscible and will form two separate layers when they are combined. This mutual insolubility is an important feature that is used in some analytical methods, such as liquid–liquid extractions (see Chapter 20).

If we were to look more closely at the layers in a water-ether mixture, we would actually find that the upper ether layer is a saturated solution of water in ether and that the lower water layer is a saturated solution of ether in water. However, although neither layer is pure, there is only a trace amount of water in the ether and ether in the water because of the low solubility for these liquids in each other. For some combinations of liquids this mixing can be much greater and result in a difference in volume for the final mixture versus the total volumes of the two combined liquids. Liquids that give this type of behavior are often said to be *partially miscible*.[4]

The solubility of one liquid in another is determined by the same types of intermolecular interactions that affect the ability of a solid to dissolve in a liquid. These interactions include hydrogen bonding, ionic interactions, dipole–dipole interactions, and dispersion forces.[11,13] Again, the more similar two liquids are in their intermolecular interactions, the greater they will be able to dissolve within each other. This explains why two polar liquids like methanol and water are totally miscible, while nonpolar liquids like octanol and benzene are immiscible in water. Qualitative information on the degree to which various common liquids are miscible can be easily obtained experimentally or by looking in sources like the *CRC Handbook of Chemistry and Physics*.[4]

We can also describe quantitatively how one liquid dissolves in another by using a similar approach to the one we have already used for solutions of molecular solids. For instance, if we were to mix water with hexane (C_6H_{14}, a solvent that is only slightly miscible with water), we could describe the transfer of the hexane from its pure liquid state to its dissolved state by the following process.

$$\text{Hexane(l)} \rightleftharpoons \text{Hexane (aq)} \qquad (7.9)$$

TABLE 7.2   Solubilities of Various Alcohols in Water at 20°C

Type of Alcohol		Solubility in Water	
Name	*Structure*	*(g/L)*	*(M)*
Methanol	CH_3OH	Fully miscible	Fully miscible
Ethanol	CH_3CH_2OH	Fully miscible	Fully miscible
n-Propanol	$CH_3CH_2CH_2OH$	Fully miscible	Fully miscible
n-Butanol	$CH_3CH_2CH_2CH_2OH$	79	1.07
n-Pentanol	$CH_3CH_2CH_2CH_2CH_2OH$	27	0.31
n-Hexanol	$CH_3CH_2CH_2CH_2CH_2CH_2OH$	5.9	0.057
n-Heptanol	$CH_3CH_2CH_2CH_2CH_2CH_2CH_2OH$	0.9	0.008
n-Octanol	$CH_3CH_2CH_2CH_2CH_2CH_2CH_2CH_2OH$	0.54	0.004

Source: The information in this table was obtained from D. R. Lide, Ed., *CRC Handbook of Chemistry and Physics*, 83rd ed., CRC Press, Boca Raton, FL, 2002; and J.A. Dean, Ed., *Lange's Handbook of Chemistry*, 15th ed., McGraw-Hill, New York, 1999.

This is another example of a solubility equilibrium,[14] where we are changing the surrounding environment of a chemical (from a liquid to a solution), but we are not changing the structure of the chemical. The thermodynamic and concentration-based equilibrium constants for this process are given in Equations 7.10 and 7.11.

$$K° = \frac{a_{Hexane(aq)}}{a_{Hexane(l)}} = a_{Hexane(aq)}$$

$$(\text{where } a_{Hexane(l)} = 1) \qquad (7.10)$$

$$K = \frac{[Hexane]}{a_{Hexane(l)}} = [Hexane] \qquad (7.11)$$

The *solubility constants* ($K°$ and K) that we obtained here for the dissolving of liquid hexane in water are similar to those that saw earlier in Equations 7.3 and 7.4 when solid glucose was dissolved in water. As we also saw for a molecular solid like glucose, we can simplify these equilibrium expressions by using the fact that hexane as a pure liquid has an activity of exactly one, because this represents a standard state of this chemical. The result is that the solubility constant for a dissolved liquid, with no side reactions, will also be equal to the solubility of this dissolved liquid when expressed in terms of activity (in the case of $K°$) or as a molar concentration (the apparent units when using K).

EXERCISE 7.5   **Using Solubility Constants to Prepare a Liquid Mixture**

A chemist wishes to prepare a solution that contains 1.00 mol of 2-octanol in water. 2-Octanol is relatively nonpolar compound with a molar mass of 130.23 g/mol and a density of 0.8193 g/mL. If the solubility of 2-octanol in water is 0.96 mL/L at 25°C, what volume of water will be required at this temperature to prepare the desired 2-octanol solution?

SOLUTION

One way we can solve this problem is by first using the density and molar mass of 2-octanol to describe its solubility in water in units of molarity.

S (in M) =

$$\frac{(\text{Density, in g/mL})(S, \text{ in mL/L water})}{(\text{MW, in g/mol})}$$

$$= [(0.96 \text{ mL 2-octanol/L H}_2\text{O})$$

$$\cdot (0.8193 \text{ g 2-octanol/mL})]/(130.23 \text{ g/mol 2-octanol})$$

$$= 0.006040 \underline{0} \ M$$

where MW is the molecular weight or molar mass. We can next use this solubility to find the volume of water

(V) needed to prepare a solution that contains 1.00 mol of 2-octanol.

$$V = (\text{mol 2-octanol})/(S, \text{ in g/mol})$$

$$= (1.00 \text{ mol})/(0.006040\underline{0} \text{ mol/L})$$

$$= 1.656\underline{6} \times 10^2 \text{ L} = \textbf{166 L water}$$

It is possible for a mixture of two immiscible liquids to have small droplets of one liquid that are suspended in another liquid. This is similar to the case where small precipitate particles can form a suspension in a solvent, producing a colloid (see Section 7.1B). A liquid colloid such as this is also known as an *emulsion*.[2,3] Although emulsions are useful chemical systems, they are often not desirable in analytical methods if the goal is to separate two liquids or to use these separate liquids to interact with other chemicals.

Another situation in which the combination of two liquids does not give a simple mixture is if the two liquids react with one another. A common example is when a liquid that is an acid (such as acetic acid) is combined with a liquid that can act as a base (e.g., water that can act as either an acid or base). This reaction gives a solution that now perhaps contains some of the original liquids plus their products. Such a situation will require the use of side reactions and chemical equilibria to describe the final content of the mixture. We will not deal with the subject of acid–base reactions at this time, but will come back to this topic in Chapter 8.

Dissolved Gases.   Gases can also dissolve in liquids and form saturated solutions. For instance, the presence of dissolved oxygen in water is essential to animals such as fish that live in lakes and rivers. For a similar reason, the measurement of dissolved oxygen is a useful measure of water quality and pollution. The difference between a gas as a solute and a solute that was a solid or liquid is the activity of the gas is measured by its pressure. This means the activity of pure gas is not necessarily equal to 1.00 as it is for pure solid or liquid solutes, but will now depend on the pressure of the gas. In fact, the only time the activity of a pure gas will be equal to one is at a pressure of one bar and a temperature of 273 K (standard temperature and pressure conditions).

Henry's law is used to describe the solubility of a gas in a liquid,[2,4] where C_{solute} is the saturated concentration of the gas in a particular liquid, P_{solute} is the partial pressure of the gas in equilibrium with the solution, and K_H is a proportionality factor known as *Henry's law constant*.

$$C_{solute} = K_H P_{solute} \qquad (7.12)$$

This relationship shows that the saturated concentration for a dissolved gas will increase directly with its partial pressure. Henry's law constants are known for several gases that

TABLE 7.3	Henry's Law Constants for Several Gases in Water at 25°C
Gas	K_H (mol/L · bar)[a]
O_2	1.26×10^{-3}
N_2	6.40×10^{-4}
CO_2	3.34×10^{-2}
H_2	7.80×10^{-4}
CH_4	1.32×10^{-4}

[a] These values are based on data provided in S.E. Manahan, *Environmental Chemistry*, 6th ed., Lewis Publishers, Boca Raton, FL, 1994, p. 117.

can dissolve in water. These values depend strongly on temperature and the type of solvent that is being used. Some typical values for these constants are shown in Table 7.3. Based on these values, it is possible to estimate the saturated concentration of a gas in a liquid if we know its partial pressure and Henry's law constant. An example of this process is given in the following exercise.

EXERCISE 7.6	**Estimating the Solubility of Oxygen in Water**

What concentration of oxygen will dissolve in water at 25°C if the external partial pressure for the oxygen is 240 torr? Give your answer in mg/L, or parts-per-million (ppm).

SOLUTION

To solve this problem, we first need to calculate the partial pressure in units of bar. When we use the appropriate conversion factors, we get (240 torr)(1 bar/750.06 torr) = 0.320 bar. Next, we know from Table 7.3 that the value of K_H for oxygen at 25°C in water is 1.26×10^{-3} mol/L·bar. When we place all of this information into Henry's Law, we get the following concentration for oxygen in water.

$$(1.26 \times 10^{-3} \text{ mol/L} \cdot \text{atm})(0.320 \text{ bar}) =$$

$$\mathbf{4.03 \times 10^{-4} \textit{ M} \text{ oxygen}}$$

The corresponding concentration in ppm would then be found to be as follows.

$$C_{solute} = (4.03 \times 10^{-4} \text{ mol/L})(32.00 \text{ g/mol})(1000 \text{ mg/g})$$

$$= 12.9 \text{ mg/L or approximately } \mathbf{12.9 \text{ ppm oxygen}}$$

These results indicate that the concentration of oxygen in water is relatively low under these conditions. This is not too surprising because molecular oxygen is nonpolar and does not form hydrogen bonds, so it would interact poorly with water.

We saw earlier how reactions can take place between a solvent and for some dissolved solids and liquids. It is also possible to have a chemical reaction take place when certain gases dissolve in some liquids. We saw an example of this in Chapter 6, where dissolved carbon dioxide and water could undergo an acid–base reaction to form carbonic acid (H_2CO_3). This reaction will affect the total amount of the gas that will dissolve, because part of this gas is undergoing a side reaction to form a second soluble chemical. Such a process can often be useful in trapping a gas or converting it into a more measurable form. For instance, we will see in Chapter 14 how CO_2-measuring electrodes can be developed based on the reaction of dissolved CO_2 with water.

7.2C How Can We Determine the Solubility of a Chemical?

Solubility is quite often one of the very first factors that is examined for a new chemical. This solubility might be expressed in a quantitative fashion (e.g., the grams of a chemical that will dissolve in 100 mL of a particular solvent) or in a qualitative manner by using terms such as "slightly soluble," "very soluble," and so on. Such information gives important clues as to the types of solutions and samples that are compatible with the chemical and the approaches that might be used for handling and purifying this substance.

Most common chemicals should have at least some information on solubility that can be obtained from the literature. As an example, the *CRC Handbook and Chemistry and Physics*[4] and *Lange's Handbook of Chemistry*[5] contain extensive lists on the general solubility of many organic and inorganic chemicals in water, acidic or basic solutions, and organic solvents. The International Union of Pure and Applied Chemistry also has a large series of books devoted to the solubilities of particular classes of substances (e.g., halides, hydroxides, antibiotics, and so on).[16] This type of information is also sometimes provided for a chemical in its materials safety data sheet (see Appendix A for an example).

If we are working with a new or less common chemical and wish to know more about its solubility, there are many ways to obtain such information.[17] This is often first done in a qualitative manner by mixing together a particular amount of the desired solute and solvent. For a mixture of a solid and a liquid, this process is followed by an examination of the resulting mixture to see if there is a uniform mixture or any observable solid that remains in contact with the solution. The presence of any solid in the final solution would indicate that a saturated solution has been obtained. This would also be indicated by the presence of any cloudiness or scattering of light in the solution, which is a sign that a colloidal suspension is present.

A similar approach would be used for examining a mixture of two liquids. In this case, a clear solution with no visible layers or cloudiness would represent a miscible mixture. The presence of a distinct boundary between the liquid phases in the mixture would indicate that a saturated solution is present, and a change in volume for this mixture versus the total volume of the combined liquids would indicate that these liquids are partially miscible. If any cloudiness is present, then an emulsion has formed.

A more quantitative approach that works with chemicals that have high solubility is to add a known mass of the solute to a known amount of solvent, stir at a known temperature until equilibrium is attained, and filter off any undissolved solute for measurement. The difference between the amount of added solute and nondissolved solute should give the amount that has dissolved. An alternative for nonvolatile solutes would be to evaporate off the solvent in the final solution and then weighing the mass of solute that is recovered from a known volume of the original solution.

Measuring the amount of a low-solubility compound that enters a solution is a bit more complicated. This usually requires a technique that directly determines the amount of this chemical in its final solution. The maximum solubility of such a compound is measured by examining the amount that stays in solution at equilibrium as increasing amounts of the solute are added to the solution. A common example is a solubility study in which a pharmaceutical chemist examines the ability of a drug or agent such as $BaSO_4$ to dissolve or remain in a solid form when it has been swallowed and entered the stomach.

7.3 CHEMICAL PRECIPITATION

7.3A The Process of Precipitation

Precipitate Formation. The simple-looking chemical reaction for the formation of precipitates doesn't begin to describe the series of events actually occurring in the formation of a precipitate. We know from our earlier discussion that precipitation begins when a supersaturated solution is present for a chemical. The first step in precipitation is **nucleation**, in which small particles (or "nuclei") of the precipitating chemical are formed.[6] There is some debate about how large these initial nuclei are, but for some ionic solids it has been estimated that this requires clusters of only eight to twelve ions.[18,19]

These nuclei act as centers onto which more of the desired chemical can accumulate to form even larger particles. The slow addition of more molecules or ions to these nuclei gives rise to a competing process known as **crystal growth**. This, in turn, can produce a pure precipitate or even a crystal of the precipitating chemical. However, if precipitation occurs too quickly, there will be many nuclei formed and little growth. The result in this second situation is a large number of small precipitate particles and perhaps even an undesirable colloid dispersion. As a result, the way in which precipitation is

performed often dictates the purity of a precipitate and how easy this precipitate is to handle for use in an analytical method.

Problems During Precipitation. There are several ways in which impurities can be introduced into crystals during the precipitation process (see Figure 7.9). For instance, the formation of a colloid will give a solution that contains a large number of small precipitate particles. Besides being difficult to filter and remove from the solution, these particles also have a large surface area. This causes these particles to stick together (or "coagulate") due to the electrostatic attractions of oppositely charged ions on neighboring particles. One result of this is that some solvent or molecular impurities become trapped inside of the precipitate. This type of trapping is called **occlusion** (or, more specifically, "molecular occlusion").[6] Occlusion can also occur due to the rapid growth of a precipitate around the solvent and its contents. This process can lead to large errors when precipitates are used for quantitative chemical analysis if the impurities are not later removed.[20]

Additional sources of precipitate contamination involve other ions that are present in the starting solution. For instance, it is possible that the precipitate may include some ions that are different from the desired cation or anion in the precipitate, but that have similar sizes and charges. This effect is known as **inclusion**. A good example is when Pb^{2+} ions (radius, 1.20 Å) become trapped in a precipitate of $BaSO_4$ (radius of Ba^{2+}, 1.35 Å) if the solution also contains some Pb^{2+} ions, even if the solubility of the salt between Pb^{2+} and sulfate ($PbSO_4$) is not exceeded. In addition, other impurities (including both ions and the solvent) can stick to the surface of a crystal. This process is known as **adsorption**.[4] In this case, the impurity is said to be "adsorbed" rather than "absorbed," because something that is adsorbed is stuck to the surface of a solid, whereas something that is absorbed moves into the interior of the solid.

The presence of ions and other impurities (through occlusion, inclusion, or adsorption) in the structure of a growing crystal or precipitate gives an effect known as **coprecipitation**. This effect leads to the presence of contaminant ions or solvent in a precipitant even though these ions or other chemicals are not present at levels above their own solubility limits. It is also sometimes possible for impurities to collect on a precipitate after it has been formed but is still standing in its original solution (an effect known as *postprecipitation*).

There are occasions when analytical chemists use precipitation only to remove a material from a solution. In this case the purity of the final precipitate may not be of great concern. There are times, however, when a highly pure precipitate with a known composition is desired. We will see several examples of this in Chapter 11 when we discuss the quantitative method of *gravimetric analysis*. We will also see in Chapter 11 how our

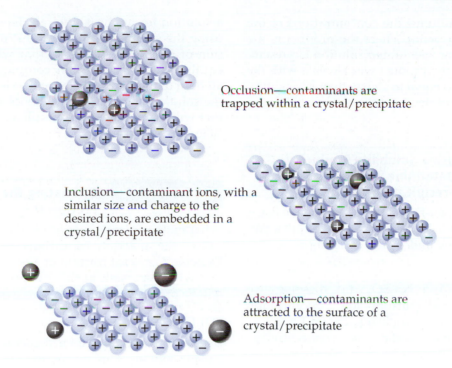

Occlusion—contaminants are trapped within a crystal/precipitate

Inclusion—contaminant ions, with a similar size and charge to the desired ions, are embedded in a crystal/precipitate

Adsorption—contaminants are attracted to the surface of a crystal/precipitate

FIGURE 7.9   Three ways in which impurities can be introduced into crystals during the precipitation process: occlusion, inclusion, and adsorption.

knowledge of the precipitation process and its problems can be used to minimize the effects of these problems through the use of such methods as *digestion* (or *Ostwald ripening*), washing of the precipitates, and *precipitation from a homogeneous solution*.

In general, when we wish to prepare as pure a precipitate as possible, we should work with a dilute solution. This minimizes the extent of supersaturation that is obtained during the precipitation process. In addition, any precipitating agent that is used should be added slowly with stirring. This minimizes any high local concentrations of the precipitating agent and helps promote growth of precipitate particles over the formation of new particles. The solution that is being used for precipitation should also be kept hot or warm while forming the precipitate, as well as for some time after completing the addition of any precipitating agent. This will cause the small, imperfect precipitate particles to dissolve and the released ions to add onto larger, better-defined particles.

Techniques that are used to improve the purity of precipitates usually involve **reprecipitation**. In this approach, the original precipitate (which may contain impurities) is filtered away from the rest of the original solution and is redissolved in pure solvent. This process allows the release of impurities that were trapped in the precipitate. A precipitating agent is then added again to reform the precipitate; however, there will now be much smaller amounts of impurities present than were in the original solution. This should, in turn, result in a purer precipitate. The main disadvantage of this procedure is

that some of the desired material in the precipitate will also be lost during the reprecipitation process.

7.3B Using Solubility Products to Examine Precipitation

Determining Whether Precipitation Will Occur.   As we saw earlier, the solubility product can be used to predict the solubility of an ionic substance in a solution. However, we can also use K_{sp} values, and their associated ion products, to predict the extent to which an ionic substance will precipitate from a solution. For instance, if barium ions and sulfate ions are added to a solution in the form of the soluble salts $BaCl_2$ and Na_2SO_4, these ions will combine to form some solid $BaSO_4$.

$$BaCl_2(aq) + Na_2SO_4(aq) \rightleftharpoons$$
$$BaSO_4(s) + 2\,NaCl(aq) \qquad (7.13)$$

Regardless of the source for the barium or sulfate, these ions will react to form insoluble $BaSO_4$ until the system has reached equilibrium. The amount of barium sulfate that forms can be estimated in the following way. First, we can calculate the initial concentration of each ion that will be present after mixing and placing these values into the ion product for the salt from its solubility product expression, which in this case is $[Ba^{2+}][SO_4^{2-}]$. We can then compare the resulting value of the ion product to the value for K_{sp}. If the ion product is larger than K_{sp}, a precipitate will form until the concentrations of the ions have reached the point where the product is the same as K_{sp}, indicating a saturated solution is present. On the other hand, if the product we calculate with the initial concentrations is equal to or less than K_{sp}, no precipitation would be expected to occur.

EXERCISE 7.7   **Using Solubility Products to Determine Whether Precipitation Will Occur**

A chemist adds 50 mL of a 4.0×10^{-4} M solution of $BaCl_2$ in water to 200 mL of 1.0×10^{-5} M Na_2SO_4 in water. Will $BaSO_4$ form a precipitate under these conditions?

SOLUTION

If all of the $BaCl_2$ and Na_2SO_4 did dissolve, we would have an initial barium ion concentration of $[Ba^{2+}] = (4.0 \times 10^{-4}$ $M)(50$ mL$/250$ mL$) = 8.0 \times 10^{-5}$ M and an initial sulfate ion concentration of $[SO_4{}^{2-}] = (1.0 \times 10^{-5}$ $M)(200$ mL$/250$ mL$) = 8.0 \times 10^{-6}$ M. Under these conditions, the expected ion product would be as follows.

$$[Ba^{2+}][SO_4{}^{2-}] = (8.0 \times 10^{-5}\ M)(8.0 \times 10^{-6}\ M)$$
$$= 6.4 \times 10^{-10}\ M^2$$

This value is greater than the solubility product for barium sulfate, where $K_{sp} = 1.08 \times 10^{-10}$, so precipitation will occur until the concentrations have decreased sufficiently to make the ion product match K_{sp}.

The procedure we used in the preceding exercise to determine whether precipitation will occur is really the same approach that we discussed in Chapter 6 for comparing the reaction quotient (Q) for a reaction with its equilibrium constant. In this case, the reaction quotient is our ion product. If we find that the value of Q (the ion product) is greater than K_{sp}, then we will have precipitation from our solution until equilibrium is reached. If our calculated ion product is less than the solubility product ($Q < K_{sp}$), then no precipitation will occur and all of our ionic solid will remain in solution. And finally, if the calculated ion product is exactly the same as the solubility product ($Q = K_{sp}$), our system is at equilibrium and no further change in the net amount of dissolved solid or precipitate will occur. At this point we have now reached equilibrium and have a thermodynamically stable system.

Estimating the Extent of Precipitation.   Another way K_{sp} values and ion products can be used is to determine how much precipitate will be formed from a solution. This can be useful if our goal is to use precipitation to remove a sufficient amount of an analyte from

a solution for analysis. We can determine this by first using the solubility product to estimate the concentration of our ionic substance that will be dissolved at equilibrium. We can then compare this to the total amount of the same substance that is initially present in the solution and use the difference to find the amount that will precipitate. An example of this is given in the following exercise.

EXERCISE 7.8   **Estimating the Mass of a Precipitate**

What mass of $BaSO_4$ would be expected to precipitate under the reaction conditions that were given in Exercise 7.7? What fraction of all the barium in this system will be present in this precipitate? What fraction will be present in solution?

SOLUTION

Barium ions and sulfate ions react in a 1:1 ratio to form barium sulfate. This means the concentration of both these ions will decrease by the same amount, which we will call x. At equilibrium (i.e., when precipitation has finished) the remaining concentrations of these ions in the solution will be $[Ba^{2+}] = (8.0 \times 10^{-5}$ $M - x)$ and $[SO_4{}^{2-}] = (8.0 \times 10^{-6}$ $M - x)$. When we place this information into the solubility product expression for barium sulfate, we get the result shown below.

$$K_{sp} = 1.08 \times 10^{-10}$$
$$= (8.0 \times 10^{-5}\ M - x)(8.0 \times 10^{-6}\ M - x)$$

or

$$0 = x^2 - (8.8 \times 10^{-5}\ M)x + 5.3\underline{2} \times 10^{-10}\ M^2$$

We can now solve for x in this equation by using the quadratic formula (see Chapter 6), where A = 1, B = -8.8×10^{-5} M, and C = $5.3\underline{2} \times 10^{-10}$ M^2 (when using apparent units of M^2 for K_{sp}). This gives us a final answer where $x = 6.5 \times 10^{-6}$ M. The other possible answer given by the quadratic formula, $x = 8.1 \times 10^{-5}$ M, is not possible because this requires more barium ion than the total added concentration of 8.0×10^{-5} M. When a concentration of $x = 6.5 \times 10^{-6}$ M is combined with the known volume of solution and the molar mass for barium sulfate, we find that $1.6\underline{2} \times 10^{-6}$ mol or 3.8×10^{-4} g $BaSO_4$ would be expected in the precipitate.

The total mass of barium in the original solution was $(0.050$ L$)(4.0 \times 10^{-4}$ mol/L$)(233.4$ g/mol$) = 0.04668$ g, and we know now the total mass of $BaSO_4$ in the precipitate will be 3.8×10^{-4} g. This means the total fraction of barium in the precipitate is $100 (3.8 \times 10^{-4}$ g$)/(0.04668$ g$) = 0.81\%$. Thus, $(100.00 - 0.81) = 99.19\%$ of the barium will remain in solution as Ba^{2+} under these conditions.

One way we can use this last type of calculation is to determine what reagent and reaction conditions must be used to give a desired degree of precipitation. As an example, in Chapter 11 we will discuss the method of gravimetry, in which the precipitation of more than 99.99% of the analyte is usually desired. We now know how to calculate whether we will obtain this degree of precipitation under a given set of conditions. A way we can alter this degree of precipitation is by changing the amounts of our precipitating ions (Ba^{2+} and SO_4^{2-} in our current example) that are added to the original solution. The temperature or type of solvent we are using can also be used to control the extent of precipitation. Another route that can be employed is to add other chemicals or side reactions to the system that can affect the precipitation of our desired analyte. Examples of this last approach are discussed in the next section.

7.3C Effects of Other Chemicals and Reactions on Precipitation

In the previous examples of precipitation we have been looking at a relatively simple system (the precipitation of $BaSO_4$) where we have only one major reaction and we have only single sources for the barium ions and sulfate ions. However, real chemical systems are usually more complex than this. For instance, we might have more than one source of barium ions or sulfate ions in our original solution. We also might have more than one reaction that affects the amount of precipitate that occurs when these ions combine to form barium sulfate. We will now look at both situations and see how they affect the precipitation process.

The Common Ion Effect. The first case we will consider is when there is more than one source for some of the ions that form an ionic solid. For instance, suppose we wish to determine the concentration of Ba^{2+} that will be present in an aqueous solution saturated with solid $BaSO_4$. The reaction involved in this case is shown below.

$$BaSO_4(s) \rightleftharpoons Ba^{2+}(aq) + SO_4^{2-}(aq) \quad (7.14)$$

If $BaSO_4$ is the only source of Ba^{2+} and SO_4^{2-} in this solution, these ions will be produced in a 1:1 ratio as this ionic solid dissolves in the water. This means that an equal concentration of Ba^{2+} and SO_4^{2-} will be present at equilibrium in such a solution. Putting this information into the K_{sp} expression for this ionic solid then gives the following relationship for this situation at equilibrium.

$$\text{If } [Ba^{2+}] = [SO_4^{2-}]: \quad K_{sp} = [Ba^{2+}][SO_4^{2-}]$$
$$= [Ba^{2+}]^2$$
$$= [SO_4^{2-}]^2 \quad (7.15)$$

When we look up the value of K_{sp} for $BaSO_4$, we find this is equal to 1.08×10^{-10} (with apparent units of M^2

according to the ion product in Equation 7.15). When we use this value along with Equation 7.15, this gives $[Ba^{2+}] = [SO_4^{2-}] = (K_{sp})^{1/2} = (1.08 \times 10^{-10} M^2)^{1/2} = 1.04 \times 10^{-5} M$.

If we were to redo this experiment, but now ask for the barium ion concentration when solid $BaSO_4$ is dissolved in a solution containing 0.010 M K_2SO_4, we would now have two sources of sulfate ions. This gives us a different answer for the barium ion concentration.

$$K_{sp} = [Ba^{2+}][SO_4^{2-}] = [Ba^{2+}](0.010\ M)$$

or

$$[Ba^{2+}] = K_{sp}/(0.010\ M) = 1.08 \times 10^{-8}\ M \quad (7.16)$$

There are two things to notice about this calculation. First, the actual sulfate concentration will be $(0.010\ M + [Ba^{2+}])$, but the additional sulfate gained by dissolving $BaSO_4$ is so small as to be insignificant when compared to that from K_2SO_4. Thus, we can safely ignore $BaSO_4$ as a source of sulfate in this calculation. Second, notice that the barium ion concentration in a solution containing sulfate ions from multiple sources is much lower than when just $BaSO_4$ is dissolved in water. This lower solubility is due to the **common ion effect**,[15] which occurs when the presence of additional sources for one or more of the ions in a dissolved salt will result in a lower solubility for that salt.

EXERCISE 7.9 **The Common Ion Effect**

What concentration of calcium ion will be present at equilibrium in a solution that is saturated with CaF_2 and contains 0.0100 M NaF? How does this calcium ion concentration compare to the concentration that would be obtained if no NaF were present?

SOLUTION

The reactions and equilibrium expressions to consider in this situation are the dissolving of NaF, which can be considered to go completely into solution (making $[F^-] = 0.0100\ M$), and the dissolution of CaF_2, which goes only slightly into solution.

$$CaF_2(s) \rightleftharpoons Ca^{2+}(aq) + 2\ F^-(aq)$$
$$K_{sp} = [Ca^{2+}][F^-]^2 = 5.3 \times 10^{-9} \quad (7.17)$$

If we assume the amount of F^- produced from CaF_2 is much lower than the amount present from NaF, we can then calculate the amount of Ca^{2+} that would be present at equilibrium (in which apparent units of M^3 can be included for K_{sp} based on the ion product in Equation 7.17).

$$(5.3 \times 10^{-9}\ M^3) = [Ca^{2+}](0.01\ M)^2$$
$$[Ca^{2+}] = (5.3 \times 10^{-9}\ M^3)/(0.01\ M)^2 = 5.3 \times 10^{-5}\ M$$

In comparison, the amount of Ca^{2+} that would have been dissolved in the absence of NaF (making CaF_2 now the only source of F^-) would have been as follows.

$$K_{sp} = [Ca^{2+}][F^-]^2 = [Ca^{2+}](2[Ca^{2+}])^2 = 4[Ca^{2+}]^3$$

$$[Ca^{2+}] = (5.3 \times 10^{-9} \, M^3/4)^{1/3} = 1.1 \times 10^{-3} \, M$$

Thus, the amount of calcium fluoride that goes into solution in the absence of NaF is almost twenty times higher than it is in the presence of 0.0100 M NaF.

Dealing with Side Reactions. One of the complicating factors in describing a precipitation reaction is that there are frequently one or more side reactions that can affect this process. This might involve any of the types of reactions we discussed in Chapter 6, such as acid–base, complexation, or oxidation–reduction reactions. To illustrate this, we will consider how the solubility of barium sulfate in water changes at a low pH, such as is found in the stomach. This is of interest because at lower pH values an acid–base reaction can take place between sulfate ions and hydrogen ions to give HSO_4^-, as shown below.

$$SO_4^{2-} + H^+ \rightleftharpoons HSO_4^-$$

$$K = \frac{[HSO_4^-]}{[SO_4^{2-}][H^+]} = 1.0 \times 10^2 \quad (7.18)$$

It is also known that the concentration of hydrogen ions in stomach fluids typically ranges from $3.2 \times 10^{-2} \, M$ to $3.2 \times 10^{-3} \, M$ (giving a pH of 1.5 to 2.5). Under these conditions, it is possible to determine that only about 24% of all sulfate-related species in solution will be present in the form SO_4^{2-}. (*Note:* We will see how to find this value in Chapter 8.) The other 76% of the dissolved sulfate will be in the form HSO_4^-, which does not react with barium ions to form a precipitate. We can describe this relationship by using Equation 7.19, where $C_{SO_4^{2-}}$ is the total concentration of all dissolved sulfate species and $\alpha_{SO_4^{2-}}$ is the fraction of sulfate that is actually present as SO_4^{2-}.

$$[SO_4^{2-}] = \alpha_{SO_4^{2-}} C_{SO_4^{2-}}$$

$$\approx (0.24) \, C_{SO_4^{2-}} \text{ at pH 1.5} \quad (7.19)$$

Because a solution with a low pH will cause less sulfate to be present as SO_4^{2-}, this would be expected to increase the solubility of barium sulfate. We can determine the extent of this change by using the same general

approach shown earlier in Equation 7.8, but now using $[Ba^{2+}] = C_{SO_4^{2-}}$ in place of $[Ba^{2+}] = [SO_4^{2-}]$ to allow for the fact that sulfate is not present in a single form.

$$K_{sp} = [Ba^{2+}][SO_4^{2-}] \quad \text{and} \quad [Ba^{2+}] = C_{SO_4^{2-}} \quad (7.20)$$

If we now substitute Equation 7.19 and the known solubility product for barium sulfate ($K_{sp} = 1.08 \times 10^{-10}$) into the K_{sp} expression in Equation 7.20, we can calculate the following value for $[Ba^{2+}]$.

$$1.08 \times 10^{-10} = [Ba^{2+}](0.24) \, C_{SO_4^{2-}}$$

$$= [Ba^{2+}](0.24)[Ba^{2+}]$$

$$= (0.24)[Ba^{2+}]^2$$

$$\therefore [Ba^{2+}] = 2.1 \times 10^{-5} \, M \text{ at pH 1.5} \quad (7.21)$$

The result is a twofold increase in solubility versus the result we obtained in Equation 7.8, when the side reaction of sulfate with hydrogen ions was assumed to be negligible. This level of solubility is still low enough to keep most barium sulfate in a solid form in the stomach, but this calculation does illustrate how side reactions can affect the solubility of a chemical.

EXERCISE 7.10   **Effect of Acid–Base Reactions on the Solubility of Fe(OH)₃**

An iron(III) ion (Fe^{3+}) can react with hydroxide ions to produce the insoluble compound $Fe(OH)_3$, as shown below.

$$Fe(OH)_3 \rightleftharpoons Fe^{3+} + 3 \, OH^-$$

$$K_{sp} = [Fe^{3+}][OH^-]^3 \quad (7.22)$$

This process will depend on the pH and concentration of hydrogen ions in the solution, because the concentration of H^+ will affect the concentration of hydroxide ions through the following acid–base reaction (as discussed in detail in Chapter 8, where K_w has apparent units of M^2).

$$H_2O \rightleftharpoons H^+ + OH^-$$

$$K_w = [H^+][OH^-] \approx 1.0 \times 10^{-14} \quad (7.23)$$

What will be the concentration of $[Fe^{3+}]$ in a solution that contains $Fe(OH)_3$ in the presence of a pH 10.00 buffer ($[H^+] = 1.0 \times 10^{-10} \, M$)? What will $[Fe^{3+}]$ be in a pH 4.00 buffer ($[H^+] = 1.0 \times 10^{-4} \, M$)?

SOLUTION

Table 7.2 states that the K_{sp} for $Fe(OH)_3$ is 1.27×10^{-39} (with apparent units of M^4). We can also determine the concentration of hydroxide ions in water at any pH by rearranging Equation 7.23 into the form $[OH^-] = (1.0 \times 10^{-14})/[H^+]$. This means at pH 10.00, the hydroxide concentration would be $[OH^-] = (1.0 \times 10^{-14} M^2)/(1.0 \times 10^{-10} M) = 1.0 \times 10^{-4} M$. Next, we can place this hydroxide concentration and the K_{sp} for $Fe(OH)_3$ into Equation 7.22 to solve for the saturated concentration of Fe^{3+} at pH 10.00.

At pH 10.00:

$$[Fe^{3+}] = K_{sp}/[OH^-]^3$$
$$= (1.27 \times 10^{-39} M^4)/(1.0 \times 10^{-4} M)^3$$
$$= 1.3 \times 10^{-27} M$$

This estimated concentration for $[Fe^{3+}]$ is remarkably low, representing a solution that contains an average of less than one Fe^{3+} ion per thousand liters of water!

The same process can be used to find the expected concentration of Fe^{3+} in water at pH 4.00. In this case, the hydroxide concentration would be $[OH^-] = (1.0 \times 10^{-14} M^2)/(1.0 \times 10^{-4} M) = 1.0 \times 10^{-10} M$. Placing this value

and the K_{sp} for $Fe(OH)_3$ into Equation 7.22 now gives the result shown below.

At pH 4.00:

$$[Fe^{3+}] = (1.27 \times 10^{-39} M^4)/(1.0 \times 10^{-10} M)^3$$
$$= 1.3 \times 10^{-9} M$$

Although this result is much higher than the concentration we estimated for Fe^{3+} at pH 10.00, this low value indicates that $Fe(OH)_3$ is still quite insoluble even in a relatively acidic solution.

Both of the previous examples involved acid–base reactions as side processes that affected the solubility of an ionic solid. There are many other types of side reactions that can also affect chemical solubility. Complex formation is another common example (see Chapter 9). For instance, the solubility of the ionic solid $Ni(OH)_2$ is strongly influenced by chemicals such as ammonia that can form complexes with Ni^{2+}. This complex formation can, in turn, affect the solubility of $Ni(OH)_2$ in water by altering the amount of Ni^{2+} that is available for reacting with hydroxide ions to form this solid. Further examples of other linked chemical reactions will be given in Chapters 8–10.

Key Words

Adsorption *152*	Inclusion *154*	Precipitate *143*	Solubility constant *148*
Common ion effect *157*	Intermolecular forces *144*	Precipitation *142*	Solubility product *148*
Coprecipitation *154*	Nucleation *154*	Reprecipitation *155*	
Crystal growth *154*	Occlusion *154*	Solubility *141*	

Other Terms

Colloidal dispersion (colloid) *143*	Dispersion forces *144*	Ionic interaction *144*	Saturated solution *142*
Crystal *143*	Dissociation *142*	Miscible *150*	Single crystal diffraction *145*
Crystallization *143*	Emulsion *152*	Octanol–water partition ratio *146*	Solubility equilibrium *147*
Diffraction pattern *145*	Henry's law *152*	Partially miscible *151*	Supersaturated solution *142*
Dipole moment *146*	Henry's law constant *152*	Postprecipitation *154*	Unsaturated solution *142*
Dipole–dipole interactions *144*	Hydrogen bonding *144*	Powder diffraction *145*	X-ray crystallography *143*
	Immiscible *151*	Radiocontrast agent *141*	X-ray imaging *141*
	Ion product *148*		

Questions

WHAT IS SOLUBILITY?

1. What is meant by the "solubility" of a chemical?
2. How do ionic substances tend to dissolve into a solution? How does this differ from the way that molecular compounds like glucose dissolve?

3. What are some general factors that can affect the solubility of a chemical?

WHAT IS PRECIPITATION?

4. Define the term "precipitation." What general conditions can lead to the precipitation of a chemical?

5. Define each of the following terms. How are these terms related to the solubility of a chemical and its ability to undergo precipitation?
   (a) Unsaturated solution
   (b) Saturated solution
   (c) Supersaturated solution
6. What is a "precipitate"? What is a "crystal"? What is a "colloid"? How are these terms related?

WHY ARE SOLUBILITY AND PRECIPITATION IMPORTANT IN CHEMICAL ANALYSIS?

7. Why is it important to consider solubility during a chemical analysis? Give two examples of situations in which this should be considered.
8. How can precipitation be used during a chemical analysis? Give two examples.
9. Figure 7.10 shows a portion of a scheme for the qualitative identification of inorganic ions such as Al^{3+}, Cr^{3+}, and Fe^{3+} in aqueous samples. Describe how precipitation is being used in this scheme for the identification of these substances.

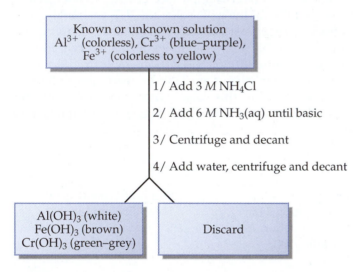

FIGURE 7.10 A portion of a scheme for the qualitative analysis of Al^{3+}, Cr^{3+}, and Fe^{3+} in an aqueous sample. The 3 M NH_4Cl added in Step 1 is to help buffer the solution and prevent the pH from increasing too greatly when ammonia is added in Step 2. The addition of NH_3 in Step 2 will lead to an increase in pH, which raises the concentration of hydroxide ions in the solution. The precipitate that is formed in Step 3 is isolated by centrifugation, followed by decanting of the liquid that is in contact with this solid. In Step 4, water is added to the precipitate, followed by another centrifugation step to give a distinct solid layer. (Adapted with permission from J. Carr, D. Kinnan, and C. McLaughlin, *Chemistry 110 Laboratory Manual, University of Nebraska-Lincoln*, Hayden-McNeil Publishing, Plymouth, MI, 2007.)

WHAT DETERMINES CHEMICAL SOLUBILITY?

10. Describe what is meant by "intermolecular forces." Explain how intermolecular forces can affect chemical solubility.
11. Define each of the following terms and describe how they occur.
    (a) Ionic interaction
    (b) Hydrogen bonding
    (c) Dipole–dipole interaction
    (d) Dispersion forces

12. State what is meant by the phrase "like dissolves like." Describe how this phrase is related to intermolecular interactions and chemical solubility.
13. What is meant by the phrase "polar chemical"? What is a "nonpolar" chemical? How does the polarity of a chemical affect its intermolecular forces with other chemicals?
14. What is a "dipole moment"? What is an "octanol–water partition ratio"? How are each of these items used to describe chemical polarity?
15. Below are given the structures and dipole moments (shown in parentheses) for several solvents that are found in chemistry laboratories.

Acetone (2.88 D)

Benzene (0.00 D)

Chloroform (1.04 D)   Dimethylsulfoxide (3.96 D)

   (a) Use the above information to rank these solvents based on their polarities. Which of these solvents do you expect to mix the best with water? Which would be expected to mix best with octane?
   (b) Look at the structures shown for these solvents. From your observations, what types of intermolecular forces do you think might be involved in determining the solubilities of these chemicals and their ability to dissolve other compounds?
16. (a) Use octanol–water partition ratios to rank the following chemicals based on their polarity.

Acetic acid (log K_{ow} = –0.17)

Cyclohexane (log K_{ow} = 3.44)

Propane (log K_{ow} = 2.33)   *n*-Propanol (log K_{ow} = 0.25).

   (b) Which of these chemicals are the most soluble in water? Which of these chemicals are the most soluble in *n*-octanol? Explain the reasons for your answer.

(c) Based on the structures of these chemicals, determine what intermolecular forces might be involved in determining their solubilities. What trends can you see in comparing this list of intermolecular forces with the K_{ow} values for these compounds?

17. (a) Discuss why both intermolecular forces and the change in entropy due to mixing are important in determining the solubility of one chemical in another.

(b) Explain how your answer to Part (a) fits with the equation shown below,

$$\Delta G^{\circ}_{mix} = \Delta H^{\circ}_{mix} - T\Delta S^{\circ}_{mix} \qquad (7.24)$$

where T is the absolute temperature, ΔG°_{mix} is the overall change in standard free energy due to the mixing of two chemicals, ΔH°_{mix} is the change in standard enthalpy due to mixing, and ΔS°_{mix} is the change in standard entropy due to mixing.

(c) Based on Equation 7.24, would you expect most chemicals to become more soluble or less soluble as the temperature is increased? Explain your answer.

HOW CAN WE DESCRIBE CHEMICAL SOLUBILITY FOR MOLECULAR SOLIDS?

18. What is a "solubility equilibrium"? Explain why the process by which a molecular solid dissolves in a solution is an example of a solubility equilibrium.

19. What is a "solubility constant"? How is this used to describe the solubility of a chemical?

20. Write "reactions" similar to Equation 7.2 to describe the processes by which the following molecular solids will dissolve in the given solvents.
(a) Naphthalene in benzene
(b) Vitamin C in water
(c) Iodine in water
(d) Iodine in CCl$_4$

21. Write the expressions that give the solubility constants K and K° for each of the reactions in Problem 20.

22. Explain why the solubility constants K and K° are often equal to the solubility (S) for a molecular solid. In what situations would this not be true?

23. A 5.00 g sample of benzoic acid ($C_7H_6O_2$) is dispersed in 250.0 mL of water. This mixture is allowed to equilibrate, and any undissolved benzoic acid is removed by filtration. If 4.15 g of solid benzoic acid is found in the filtrate, what is the solubility (in M) of benzoic acid in water?

24. A spoonful of solid iodine is placed into 250.0 mL water and allowed to reach equilibrium. Most of the iodine does not dissolve. After separating the solution from the undissolved iodine, it is found that 0.0781 g of iodine remains in the solution. What is the solubility of iodine (in units of both g/L and M) in water under these conditions?

HOW CAN WE DESCRIBE CHEMICAL SOLUBILITY FOR IONIC SOLIDS?

25. What is a solubility product? How is this used to describe the solubility of a chemical?

26. Write the solubility product expressions for each of the following salts. Express your answers in terms of both K and K°.
(a) AgBr
(b) SrF$_2$
(c) Ca$_3$(PO$_4$)$_2$
(d) Mg(NH$_4$)PO$_4$

27. What are some general ways in which solubility products can be used to describe the solubility or precipitation of a chemical?

28. Zinc sulfide (ZnS) is often used in the construction of analytical devices that measure or produce infrared radiation. A chemist wishes to use this ionic solid to prepare a new device for spectroscopic measurements.
(a) What is the expected solubility of ZnS in pure water at 25°C?
(b) Cadmium sulfide (CdS) is sometimes mixed with ZnS to produce some optical devices. What is the solubility of CdS in pure water at 25°C?

29. Rank each of the following groups of chemicals based on their solubility in pure water at 25°C. Use calculations to support your answers (*Note*: You may assume that no other significant side reactions are present.)
(a) AgBr, AgCl, and AgI
(b) CaSO$_4$ and Calcium oxalate hydrate
(c) CaSO$_4$ and BaSO$_4$
(d) Al(OH)$_3$ and Ca(OH)$_2$
(e) Fe(OH)$_2$ and Fe(OH)$_3$

30. Give at least five examples of salts that are often regarded as being "highly soluble" in water. Give five examples of salts that are only slightly soluble or insoluble in water.

HOW CAN WE DESCRIBE CHEMICAL SOLUBILITY FOR LIQUID MIXTURES?

31. Define each of the following terms as related to liquid mixture.
(a) Miscible
(b) Partially miscible
(c) Immiscible

32. Describe the role that intermolecular interactions play in determining whether two liquids will mix with one another.

33. Write reactions to describe the processes by which each of the following liquids will dissolve in the given solvents.
(a) CH$_3$COOH in water
(b) HOCH$_2$CH$_2$OH in ethanol
(c) Br$_2$ in CCl$_4$

34. Write the expressions that give the solubility constants K and K° for each of the reactions in Problem 33. Explain why K and K° are often equal to the solubility for one liquid in another.

35. An environmental chemist wishes to prepare a standard sample that contains benzene (C_6H_6) in water. Pure benzene has a density of 0.8786 g/mL at 25°C. At the same temperature, the solubility of benzene in water is 1.79 g/L.
(a) What is the maximum concentration of benzene (in units of molarity) that can be placed into water to form a stable solution at 25°C?
(b) What volume of pure benzene should be placed into water to give a solution with a final benzene concentration of 1.79 g/L and a total volume of 1.00 L?

36. The solubility of chloroform (CHCl$_3$) in water is 1.0 mL in 200 mL water at 25° C. The density of pure chloroform is 1.484 g/mL. What is the solubility of chloroform in units of M, parts per million, and parts per billion?

37. The gasoline additive MTBE (C$_4$H$_9$OCH$_3$) has a solubility of 42 g/L in water at 25°C. What is the solubility of MTBE when expressed in units of molarity? If 0.50 mol MTBE is released into the environment, how much water would be required to completely dissolve all of this chemical?

38. What is an "emulsion"? Explain how an emulsion forms.

39. Give one example of how a side reaction can affect the solubility of one liquid in another.

HOW CAN WE DESCRIBE CHEMICAL SOLUBILITY FOR DISSOLVED GASES?

40. What is Henry's law? How is this used to describe the solubility of a gas in a liquid?

41. What concentration of oxygen will dissolve in water at 20°C and at an atmospheric pressure of 600 torr?

42. What is the expected solubility of dissolved O_2, N_2, H_2, and CO_2 in air-saturated water at a pressure of 1.0 atm and 25°C?

43. Describe how side reactions may affect the solubility of a gas in a liquid. Give one specific example of this effect.

HOW CAN WE DETERMINE THE SOLUBILITY OF A CHEMICAL?

44. Describe how solubility can be studied on a qualitative basis for solids dissolved in liquids and liquid mixtures.

45. What are some books and other resources that can be used to provide information on chemical solubility?

46. Describe how the solubility of a solid or a liquid in a liquid can be examined in a qualitative fashion.

47. What are three strategies that can be used to determine the amount of a solid that will dissolve in a liquid? In what types of situations would each of these strategies be employed?

THE PROCESS OF PRECIPITATION

48. What are the general steps involved in the formation of a chemical precipitate?

49. Define the following terms and explain why they are important in precipitation.
    (a) Nucleation
    (b) Crystal growth

50. Explain why a supersaturated solution is needed for precipitate formation.

51. Why is the rate of precipitation important in determining the type of precipitate that is formed?

52. Define each of the following terms and explain why they are important to consider during the use of precipitation.
    (a) Occlusion
    (b) Inclusion
    (c) Adsorption

53. What is "coprecipitation"? Why is this effect important to consider during the use of precipitation for quantitative analysis?

54. In the method of *gathering*, a precipitate of one chemical (called a "gathering agent") is used to collect a small amount of another trace chemical in a solution. Explain why coprecipitation would be important in such a method.

55. What is "postprecipitation"? How does this differ from coprecipitation? How are these two processes similar?

56. Two identical portions of an aqueous solution containing aluminum ion and a small amount of Fe^{3+} are precipitated by raising the pH of these solutions to form $Al(OH)_3$. In one case, the precipitate is filtered, dried, and found to have a mass of 0.2543 g. In the other case, the precipitate is filtered, redissolved with dilute HCl, and reprecipitated, dried, and found to have a mass of 0.2487 g. Explain the difference in mass of these two precipitates.

57. Two students both perform an analysis of Ni^{2+} by precipitating it with the agent dimethylglyoxime (dmg). The first student works at room temperature and quickly adds the total required amount of dmg, taking about 30 s to form the nickel–dmg precipitate. The second student adds a solution of dmg a drop at a time and uses a warm sample solution that is later allowed to gradually cool to room temperature; this student takes about 30 min to do this step. The second student is later able to quickly filter his precipitate and finishes the lab on time, while the first student has difficulty in recovering her precipitate and must repeat the experiment later. What is the reason for first student's problem with this experiment?

58. What is "reprecipitation"? Why does this help create purer precipitates? What is a disadvantage to this method?

USING SOLUBILITY PRODUCTS TO EXAMINE PRECIPITATION

59. Describe how solubility products can be used to determine whether precipitation will occur in a particular solution.

60. Determine whether a precipitate will form for each of the following mixtures. If a precipitate does form, give the formula for this precipitate and estimate the mass of precipitate that will be produced.
    (a) An aqueous mixture containing 50.0 mL of 0.025 M $Pb(NO_3)_2$ plus 100 mL of 0.010 M Na_2SO_4
    (b) An aqueous mixture containing 50.0 mL of 0.0080 M $Pb(NO_3)_2$ plus 200 mL of 0.0050 M NaCl

61. Determine the concentrations for the following solutions in which the given solid has been allowed to reach equilibrium with water at 25°C.
    (a) $[Ag^+]$ in a solution where solid AgI is equilibrated with pure water
    (b) $[Pb^{2+}]$ in a solution where solid $PbCl_2$ is equilibrated with pure water
    (c) $[OH^-]$ in a solution where solid $Ca(OH)_2$ is equilibrated with pure water

62. Describe how solubility products can be used to estimate the extent of precipitation that will occur from a given solution.

63. What mass of $PbCl_2$ would be expected to precipitate if 50.0 mL of a 0.050 M solution of $Pb(NO_3)_2$ is mixed with 50.0 mL of 0.100 M HCl?

64. A student wishes to precipitate Fe^{3+} as $Fe(OH)_3$ from a 500 mL aqueous solution that initially contains 1.50 mM Fe^{3+}.
    (a) If the student adjusts the pH of this solution to 5.00, what mass of $Fe(OH)_3$ will precipitate?
    (b) What mass of Fe^{3+} will remain in solution after the pH has been adjusted to 5.00?
    (c) What fraction of the original Fe^{3+} will remain in solution after the $Fe(OH)_3$ has been precipitated at pH 5.00?

65. A person with cystic fibrosis typically contains 60–200 mM Cl^- in samples of their perspiration. The chloride in a 1.0 mL aliquot of such a sample is to be measured by combining it with 1.0 mL of a reagent containing 0.0250 M $AgNO_3$.
    (a) What mass of AgCl would be expected to precipitate if the sample originally contained 200 mM Cl^-?
    (b) What mass of Cl^- will remain in solution after this precipitation has occurred?
    (c) What percent of Cl^- in the original sample will have been precipitated during this analysis?

66. What are some general approaches that can be used to alter the extent of precipitation that occurs for a chemical?

EFFECTS OF OTHER CHEMICALS AND REACTIONS ON PRECIPITATION

67. What is the "common ion effect"? How can this affect the solubility of an ionic solid?

68. What concentration of silver ions would be expected in an aqueous solution that is saturated with solid AgBr and that contains no other added ions? How would this concentration

change if solid AgBr were instead dissolved in an aqueous solution containing 0.010 M KBr?

69. A student wishes to dissolve $Cd(IO_3)_2$ in an aqueous solution that already contains KIO_3.
    (a) What is the maximum amount of $Cd(IO_3)_2$ that can dissolve without precipitation in pure water at 25°C?
    (b) What is the maximum amount of $Cd(IO_3)_2$ that can dissolve without precipitation in 0.01 M KIO_3 at 25°C?

70. An analyst wants to dissolve CuI in an aqueous solution of KI.
    (a) What is the maximum amount of CuI that will dissolve in pure water at 25°C?
    (b) What will the solubility of CuI be if it is placed into a 0.050 M aqueous solution of KI at 25°C?

71. A 100 mL portion of 0.020 M $AgNO_3$ in water is added to 100 mL of an aqueous mixture of 0.010 M NaCl and 0.010 M KBr.
    (a) What will the solubility of Cl^- be in this solution after it has reached equilibrium?
    (b) What will the final concentration of Br^- be in this solution after it has reached equilibrium?
    (c) What would have been the final concentration of Cl^- if no KBr had originally been present?
    (d) What would have been the final concentration of Br^- if no NaCl had been originally present?

72. Describe how side reactions can affect the solubility of an ionic solid. Give one specific example of this effect.

73. It was shown in Exercise 7.10 that the solubility of $Fe(OH)_3$ will vary with the acidity of a solution.
    (a) A chemist wishes to prepare a standard aqueous solution that has a concentration of Fe^{3+} equal to 1.0×10^{-5} M. This solution is to be prepared by allowing water at a set pH to reach equilibrium with solid $Fe(OH)_3$. Given the fact that pH $\approx -\log [H^+]$, what pH should be used for the aqueous solution to give this desired concentration of $[Fe^{3+}]$?
    (b) A sample of water in contact with solid $Fe(OH)_3$ is found to have a pH of 7.5. If no other dissolved solids are present in this water, what is the expected concentration of Fe^{3+} in this sample?

74. "Milk of magnesia" is actually a suspension of solid magnesium hydroxide, $Mg(OH)_2$, in water. Assuming that this suspension is at equilibrium, use the solubility product of magnesium hydroxide to estimate the pH of this mixture.

CHALLENGE PROBLEMS

75. Look up the structures of the following chemicals. Based on these structures, determine whether each of these chemicals would be classified as "polar" or "nonpolar." Which of these compounds would you predict to be the most soluble in water? Which would you expect to be the most soluble in benzene?
    (a) Carbon tetrachloride
    (b) Formic acid
    (c) Tetrahydrofuran
    (d) Naphthalene
    (e) Cholesterol ester
    (f) Saccharin

76. (a) Use the data in Table 7.1 to prepare a plot of water solubility versus the number of carbon atoms that appear in the following alcohols: methanol, ethanol, n-propanol, n-butanol, n-pentanol, n-hexanol, n-heptanol, and n-octanol.

(b) What trend do you observe in the plot that was made in Part (a)? What do you think is the reason for this trend?

77. An analytical chemist wishes to select several pairs of immiscible solvents for use in the technique of liquid–liquid extraction. Based on information in such sources as the *CRC Handbook of Chemistry and Physics*[4] or *Lange's Handbook of Chemistry*,[5] which of the following pairs of solvents would work in this method? Which would not work?
    (a) Water and acetone
    (b) Water and ether
    (c) CCl_4 and benzene

78. Another chemist wishes to test various combinations of miscible solvents for dissolving a given sample. Using the same types of sources as listed in Problem 76, determine which of the following solvent pairs will form miscible systems.
    (a) Ethanol and water
    (b) $CHCl_3$ and CH_2Cl_2
    (c) Hexane and methanol

79. The toxicity limit of barium ion is 19.2 mg $BaCl_2 \cdot 2H_2O$ per kg of body weight. The human stomach has a volume of approximately 1.0 L and has a typical pH of 2.0 due to the presence of hydrochloric acid.
    (a) Estimate the concentration of Ba^{2+} that can be present in the stomach when the fluids in the stomach are saturated with $BaSO_4$. (*Note:* Assume no common ion effects.)
    (b) How does this maximum concentration of barium ions in the stomach compare with the upper toxicity limit for this ion?

80. The precipitation of CaF_2 (see Exercise 7.9) is affected by the acidity of its solution. This occurs because fluoride ions can react with water, as shown by the reaction below.

$$F^- + H_2O \rightleftharpoons HF + OH^-$$

$$K = \frac{[HF][OH^-]}{[F^-](a_{H_2O} = 1)} = 1.5 \times 10^{-11} \quad (7.25)$$

It can also be shown that the fraction of fluoride that is present in solution as F^- (α_{F^-}) can be found by using the following formula,

$$\alpha_{F^-} = \frac{K_w}{K_w + K \cdot 10^{-pH}} \quad (7.26)$$

where K_w is the autoprotolysis constant for water ($K_w = 10^{-14.00}$ at 25°C).
    (a) Based on the above acid–base reaction, what solubility would be expected for CaF_2 in water at pH 2.00? How does this value compare to the solubility that was obtained in Exercise 7.9 for a solution of pure CaF_2? What does this comparison tell you about the importance of pH on the solubility of calcium fluoride?
    (b) What would the expected solubility be for CaF_2 at pH 2.00 in a solution that is saturated with this solid, but which also contains 0.0100 M NaF? How does this result compare to that estimated in Exercise 7.9 for a solution in contact with CaF_2 and containing 0.0100 M NaF? What does this result tell you about the combined effect of pH and the common ion effect on the solubility of calcium fluoride?
    (c) Use a spreadsheet to prepare a plot of the predicted solubility of CaF_2 versus pH for both pure water and a 0.0100 M NaF solution. What trends do you see in these results? Explain these trends based on Equation 7.26 and the results in Parts (a) and (b).

81. Using Henry's law, explain how bubbling nitrogen gas through an aqueous solution can be used to remove dissolved oxygen from this solution.

82. The solubility of AgCl and $BaSO_4$ in various concentrations of potassium nitrate are as follows.[21]

Concentration KNO_3 (M)	Solubility AgCl (M)	Solubility $BaSO_4$ (M)
0.000	1.278×10^{-5}	0.96×10^{-5}
0.001	1.325×10^{-5}	1.16×10^{-5}
0.005	1.385×10^{-5}	1.42×10^{-5}
0.010	1.427×10^{-5}	2.35×10^{-5}

Explain why the solubility of these compounds increases at higher KNO_3 concentrations and why the impact on $BaSO_4$ is so much greater than the impact on AgCl.

TOPICS FOR DISCUSSION AND REPORTS

83. X-ray imaging is just one of many techniques that are used in modern medicine to allow physicians to examine the various regions of the body. Some examples of other imaging methods are given below. Write a report on one of these methods. Include in your report a description of the chemical and physical principles behind this method. Also state what types of information your method can provide about the body.
    (a) Magnetic resonance imaging (MRI)
    (b) Computed axial tomography (CAT scan)
    (c) Ultrasound imaging
    (d) Positron emission tomography (PET scan)

84. Barium sulfate is just one example of a radiocontrast agent that is used for X-ray imaging. Chemicals that contain iodine are also commonly used as radiocontrast agents. Obtain more information on specific iodine-containing chemicals that are used for this purpose and describe the chemical and physical properties that make them useful for this application. Give some specific examples that show how these chemicals can be used to provide images of the body.

85. Locate a recent scientific article that used the method of X-ray crystallography. Discuss how this method was used in the article. What types of information did this technique provide?

86. The solubility of a chemical like DDT is important in determining such factors as its "bioavailability" and "bioaccumulation" in living systems. Locate more information on each of these topics. Describe why these factors are important in the analysis of chemicals in living systems.

87. There are several other types of intermolecular forces besides those that are shown in Figure 7.6. Describe each of the following forces and state what general types of chemicals will give rise to each of these forces.
    (a) Dipole-induced dipole forces
    (b) Ion-dipole forces
    (c) Ion-induced dipole forces

88. The rate of growth of a crystal will determine the purity and final degree of order that is obtained in this material. Obtain more information from sources such as Reference 21 on the factors that determine and effect crystal growth. Write a report describing this process. Explain why the control of this process is important in such areas as X-ray crystallography and gravimetric analysis.

89. Reference 13 discusses how intermolecular forces and changes in entropy affect the ability of two liquids to be miscible, partially immiscible, or immiscible. Obtain more information on this process and state how the overall change in energy for the system is affected by these factors. Under what circumstances will two liquids be miscible based on this change in energy? Under what circumstances will they be partially miscible or immiscible?

References

1. MayoClinic.com, "Stomach Cancer," www.mayoclinic.com/health/stomach-cancer

2. *The New Encyclopaedia Britannica*, 15th ed., Encyclopaedia Britannica, Inc., Chicago, IL, 2002.

3. *IUPAC Compendium of Chemical Terminology*, Electronic version, http://goldbook.iupac.org

4. D. R. Lide, Ed., *CRC Handbook of Chemistry and Physics*, 83rd ed., CRC Press, Boca Raton, FL, 2002.

5. John A. Dean, Ed., *Lange's Handbook of Chemistry*, 15th ed., McGraw-Hill, New York, 1999.

6. G. Maludziska, Ed., *Dictionary of Analytical Chemistry*, Elsevier, Amsterdam, 1990.

7. J. Drenth, *Principles of Protein X-Ray Crystallography*, Springer-Verlag, New York, 1999.

8. J.P. Glusker, M. Lewis, and M. Rossi, *Crystal Structure Analysis for Chemists and Biologists*, VCH Publishers, New York, 1994.

9. G. Rhodes, *Crystallography Made Crystal Clear*, Academic Press, New York, 2000.

10. I. Kaplan, *Intermolecular Interactions*, Wiley, New York, 2006.

11. B.L. Karger, L.R. Snyder, and C. Horvath, *An Introduction to Separation Science*, Wiley, New York, 1973.

12. J. Sangster, *Octanol–Water Partition Coefficients—Fundamentals and Physical Chemistry*, Wiley, New York, 1997.

13. S.R. Logan, "The Behavior of a Pair of Partially Miscible Liquids," *Journal of Chemical Education*, 75 (1998) 339–342.

14. J. Inczedy, T. Lengyel, and A. M. Ure, *Compendium of Analytical Nomenclature*, 3rd ed., Blackwell Science, Malden, MA, 1997.

15. H. Frieser, *Concepts & Calculations in Analytical Chemistry: A Spreadsheet Approach*, CRC Press, Boca Raton, FL, 1992.

16. *IUPAC-NIST Solubility Database (NIST Standard Reference Database 106)*, National Institute of Standards and Technology, Gaithersburg, MD, 2003.

17. G.T. Hefter and R.P.T. Tomkins, Ed., *The Experimental Determination of Solubilities*, Wiley, Chichester, UK, 2003.

18. I.M. Kolthoff and B. van't Riet, "Formation and Aging of Precipitates. XLVI. Precipitation of Lead Sulfate at Room Temperature," *Journal of Physical Chemistry*, 63 (1959) 817–823.

19. R.W. Ramette, *Chemical Equilibrium and Analysis*, Addison-Wesley, Reading, MA, 1981.

20. I.M. Kolthoff, E.B. Sandell, E.J. Meehan, and S. Bruckenstein, *Quantitative Chemical Analysis*, Macmillan, New York, 1969.

21. J.M. Garcia-Ruiz, "Arcade Games for Teaching Crystal Growth," *Journal of Chemical Education*, 76 (1999) 499–501.

Chapter 8

Acid–Base Reactions

Chapter Outline

8.1 INTRODUCTION: RAIN, RAIN GO AWAY

News Release—October 16, 2006: *The U.S. Environmental Protection Agency released its Acid Rain Program 2005 Progress Report today, marking the 11th year of one of the most widely regarded and successful environmental programs in U.S. history. Since 1995, the program has significantly reduced acid deposition in the United States by decreasing sulfur dioxide (SO_2) and nitrogen oxides (NO_x) emissions. Due to rigorous emissions monitoring and allowance tracking, overall compliance with the Acid Rain Program has been consistently high—nearly 100 percent...In 2005, SO_2 emissions from electric power plants were more than 5.5 million tons below 1990 levels. NO_x emissions were down by about 3 million tons below 1990 levels. The program's emission cuts have reduced acid deposition and improved water quality in U.S. lakes and streams. The emission reductions to date also have resulted in reduced formation of particles, improved air quality and human health related benefits.*[1]

The topic of acid rain has been of interest since the mid-1800s, when it was first observed in England near the beginning of the Industrial Revolution. Acid rain has been of particular concern over the last few decades as increased emission of gases like SO_2 and NO_x have entered the atmosphere as a result of the burning of coal and other fossil fuels. This process is illustrated in Figure 8.1. Once these gases are in the atmosphere, they can react with water and other substances to form sulfuric acid (H_2SO_4) and nitric acid (HNO_3). The presence of these acids lowers the pH (a measure of hydrogen ion content) of rain. When this pH reaches a value of 5.0 or less, the result is known as "acid rain." Rain with a pH below this value is of concern because it can lead to damage in plants, animals, rocks and soil, and man-made structures.[2–4]

The negative effects of acid rain have lead to a worldwide effort to study this phenomenon and to minimize its effects. In the United States, the Environmental Protection Agency and U.S. Geological Survey have been closely monitoring the emission of gases that can lead to acid rain and has been regularly measuring the pH of rain throughout the United States, as shown in Figure 8.2. It now appears that this effort, in combination with various programs aimed at reducing acid-rain-producing emissions, is helping to reduce the size of this problem.[1,3]

Terms like "acid" and "base" are important not only in describing the environmental effects of chemicals but

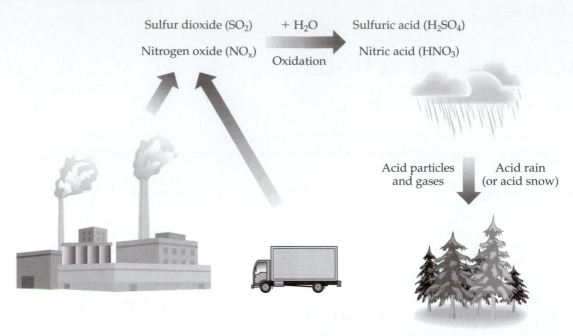

FIGURE 8.1 Processes involved in the formation of acid rain. This process begins with the emission of sulfur dioxide or nitrogen oxides into air through man-made or natural processes. These gases can later react with other agents in the atmosphere to produce sulfuric acid or nitric acid, respectively. There are two ways that these acids can reach the ground. The first is by these gases reaching the surface directly or as part of particles that fall to the surface, giving what is known as *dry deposition*. Sulfuric acid and nitric acid in air can also combine with water and fall to the earth in rain or snow. This second route is what people normally think of as "acid rain," but it is more accurately referred to as *wet deposition*.

are also essential to consider in the analysis of samples. In fact, many chemicals found in industrial, biological, and environmental samples have acid or base properties. A few examples are strong acids like hydrochloric acid, strong bases like sodium hydroxide, and weak acids or weak bases like amino acids, fatty acids, and nitrogen bases. Thus, it should be no surprise that acid–base reactions are important in the analytical laboratory. In this chapter, we examine the general properties of acids and bases and learn how to describe their reactions. We also see how we can use these reactions and related calculations to estimate the extent of an acid–base reaction and use this information in a chemical analysis.

8.1A What Is an Acid or a Base?

One way we can define the terms "acid" or "base" is by using an approach proposed in 1884 by Swedish chemist Svante Arrhenius (as part of his doctoral dissertation).[4,5] In this *Arrhenius model*, an *acid* is a chemical that results in an increase in hydrogen ions in aqueous solution, while a base is a chemical that results in an increase in hydroxide ions.[6] For example, the nitric acid (HNO$_3$) found in acid rain is an Arrhenius acid because it produces hydrogen ions in water.

$$HNO_3 \rightleftarrows H^+ + NO_3^-$$ (8.1)

The sulfuric acid (H$_2$SO$_4$) that can appear in rain would also be an acid by this definition, because it too will

produce hydrogen ions in water (in this case, involving up to two sequential reactions).

$$H_2SO_4 \rightleftarrows H^+ + HSO_4^-$$ (8.2)

$$HSO_4^- \rightleftarrows H^+ + SO_4^{2-}$$ (8.3)

In this same model, sodium hydroxide (NaOH, a chemical often used to react with and quantitatively measure acids) would be defined as a *base* because it produces hydroxide ions in water.

$$NaOH \rightleftarrows Na^+ + OH^-$$ (8.4)

Although the Arrhenius model for acids and bases is simple to visualize, it has a major limitation in that hydrogen ions do not exist by themselves in solution, as is implied by the reactions we have written in Equations 8.1–8.3. Instead, hydrogen ions are surrounded by several solvent molecules. Some of the structures produced by hydrogen ions in water are shown in Figure 8.3. We can view the formation of these ions as the transfer of a hydrogen ion to water, which acts as the base. This process makes water (the solvent) extremely important in determining the ability of a compound within that solvent to give up or accept hydrogen ions. Because the Arrhenius model does not provide a means for describing this solvent effect, this model is not a completely

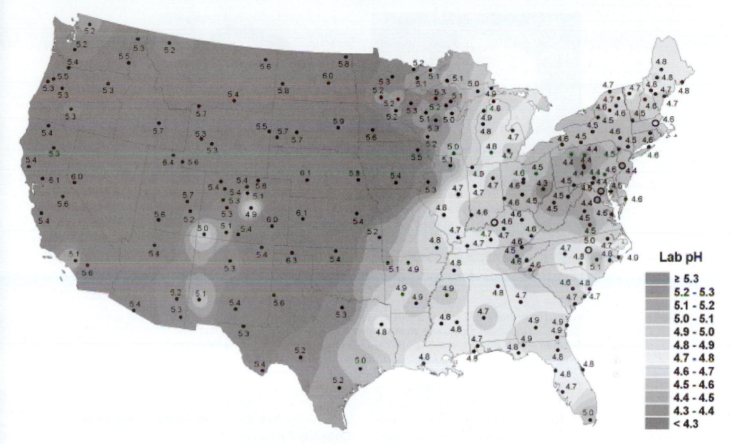

FIGURE 8.2 Activity of hydrogen ions in rain, as represented by pH measurements made in 2007 at field laboratories located throughout the United States. (Reproduced with permission from the National Atmospheric Deposition Program, *National Atmospheric Deposition Program 2007 Annual Summary*, NADP Report 2008-01, Illinois State Water Survey, University of Illinois at Urbana-Champaign, Champaign, IL, 2008.)

accurate way of describing what actually happens when an acid and base react in a solution.

A second and more useful approach for describing acids and bases is the *Brønsted–Lowry model*. This is named after Danish chemist Johannes N. Brønsted and English chemist Thomas M. Lowry, who independently proposed this model in 1923 (see Figure 8.4).[4,7,8] This model defines a **Brønsted–Lowry acid** as a chemical that donates a proton (or hydrogen ion) to another chemical. Similarly, a **Brønsted–Lowry base** is a chemical that can accept a proton from another chemical (an acid).[6] Using this model, let's look again at what happens when nitric acid is put in water.

$$HNO_3 + H_2O \rightleftharpoons H_3O^+ + NO_3^- \qquad (8.5)$$

In the **Brønsted–Lowry model** the HNO_3 in Equation 8.5 is still acting as an acid (a hydrogen ion donor), but water can now be seen to be the base (a hydrogen ion acceptor). We can similarly extend this model to the reaction in Equation 8.2, which allows us to identify H_2SO_4 as an acid (a hydrogen ion donor) and water as a base (a hydrogen ion acceptor). (*Note*: HSO_4^- in Equation 8.3 would be classified as an acid because it

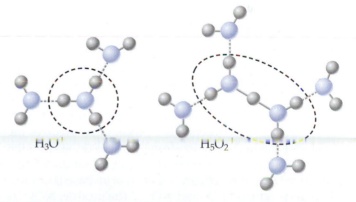

FIGURE 8.3 Two structures for a "hydrogen ion" (H^+) in water: the hydronium ion (H_3O^+) and a proton that is shared between two water molecules ($H_5O_2^+$). Each of these structures is shown within the dashed ovals. The water molecules outside of these ovals represent the first hydration shell about each structure. The smaller balls are hydrogen atoms or a hydrated proton (which is located at or near the center of each structure) and the larger balls are oxygen atoms. The short solid lines are covalent bonds and the dashed lines between atoms represent hydrogen bonds. Clusters involving even a larger number of water molecules are also possible to give other structures for H^+ in water. (Reproduced with permission from S. Borman, "Revisiting the Hydrated Proton," *Chemical & Engineering News*, July 4, 2005, 26–27.)

Thomas M. Lowry

Johannes N. Brønsted

FIGURE 8.4 Thomas Martin Lowry (1874–1936) and Johannes Nicolaus Brønsted (1879–1947). These two scientists independently proposed a new view of acids and bases in 1923, which is now known as the *Brønsted–Lowry model*. Lowry was an English chemist who was the first professor of chemistry in a London medical school. He later became the first person to hold a chair of physical chemistry at Cambridge University. Brønsted was a Danish chemist who was a professor of inorganic and physical chemistry in Copenhagen.

donates a hydrogen ion to water; the same chemical in Equation 8.2 is viewed as a base because it could, at least to a small extent, receive a hydrogen ion from H_3O^+ to form H_2SO_4). According to the same model, the hydroxide ion that is produced from NaOH in Equation 8.4 would be defined as a base because it can accept a hydrogen ion, such as in the reaction of OH^- with H_3O^+.

$$OH^- + H_3O^+ \rightleftarrows 2\,H_2O \qquad (8.6)$$

We can see from this comparison that a chemical that is an acid or base according to the Arrhenius model will also be an acid or base under the Brønsted–Lowry model.

Equation 8.5 (and also Equation 8.6) is an example of a complete **acid–base reaction**.[4] In the Brønsted–Lowry model, an acid–base reaction is a process that involves the transfer of the hydrogen ion from the acid (HNO_3) to the base (H_2O). If you look more closely at Equation 8.5 you will see that there is actually a second acid–base pair present in the products H_3O^+ and NO_3^-. The product NO_3^- is called the **conjugate base** of nitric acid because it can accept a hydrogen ion from H_3O^+ to again form HNO_3, and H_3O^+ is the **conjugate acid** of water because it can donate a hydrogen ion to NO_3^- to give back a molecule of water. All acid–base reactions in the Brønsted–Lowry model have an acid and a base on each side of the reaction, with one pair being the conjugate acid and conjugate base of the other.[4,9] In the remainder of this chapter we will continue to use the Brønsted–Lowry model as we describe acids, bases and their reactions. In Chapter 9 we will discuss a more general view of acids and bases (the *Lewis model*) that can be used with an even broader range of reactions.

8.1B Why Are Acids and Bases Important in Chemical Analysis?

You will find acids and bases in many types of reactions and samples. For instance, many of the most common industrial chemicals are acids or bases (see Chapter 11). Acids and bases also occur in many common samples, such as those taken from food, biological systems, and environmental samples. This fact is useful in that we can use the acid or base properties of these samples to help us analyze them or prepare them for analysis.

An acid–base titration is one method in which acid–base reactions are used directly for chemical analysis (see Chapter 12). This technique involves the reaction of a known amount of a reagent acid or base with an unknown amount of a base or acid in a sample. By determining the amount of the reagent that is required to complete this reaction, it is possible to calculate the moles and concentration of the acid or base in the original sample. One example of this approach is the use of an acid–base titration to determine the concentration of nitric acid or sulfuric acid in rain water.

Acid–base reactions are also often used to prepare samples for analysis. In one example, changing the pH of a sample might be used to control the extent to which an acidic or basic analyte will transfer from water to an organic liquid during the method of liquid–liquid extraction. Adjusting the pH of a solution can also be used to alter other reactions, such as precipitations or complex formation processes that involve weak acids or bases. An example of how an acid–base reaction can affect another reaction was given in Chapter 7, where we examined how a change in hydrogen ion concentration can affect the solubility of barium sulfate. The dependence of many reactions on hydrogen ion concentration is the reason why acid rain has so many undesirable effects on natural systems and man-made structures. This dependence can also be used by chemists to control and adjust the extent of such reactions during a chemical analysis.

8.2 DESCRIBING ACIDS AND BASES

8.2A Strong and Weak Acids

Not all acids have the same degree of reactivity. For instance, a spill of concentrated hydrochloric acid may cause a great deal of damage to your clothes and skin, but a small amount of vinegar (a dilute solution of acetic acid) will cause little or no damage. One way we can rank such acids is based on their strength. A "strong acid" can be defined as an acid that undergoes essentially complete dissociation to form hydrogen ions (a process called *acid dissociation*) in a given solvent. A "weak acid" is an acid that only partially dissociates to form hydrogen ions. Examples of both strong acids and weak acids are given in Table 8.1. Nitric acid and hydrochloric acid are classified as strong acids in water because they almost completely dissociate and transfer a hydrogen ion to water to form H_3O^+. The same is true for sulfuric acid, as it dissociates and transfers its first hydrogen ion to water. However, the product of the reaction between sulfuric acid and water,

HSO_4^-, is a moderately weak acid because it does not completely dissociate. Acetic acid is an even weaker acid that typically transfers only a small fraction of its hydrogen ions (usually less than 1%) to water.

Another way of describing an acid's strength is to use a special equilibrium constant known as the **acid dissociation constant**, or *acidity constant* (K_a).[10,11] Suppose we have the general reaction

$$AH + B \rightleftharpoons A + BH \qquad (8.7)$$

where AH is an acid, B is a base and A and BH are their products. The acid dissociation constant for acid A in this reaction is written as shown in Equation 8.8, where a_{AH}, a_B, a_A, and a_{BH} are the chemical activities of the reactants and products at equilibrium.

$$K_a^\circ = \frac{(a_A)(a_{BH})}{(a_{AH})(a_B)} \qquad (8.8)$$

In Chapter 6 we learned that we can also write a concentration-based equilibrium constant for such a reaction (K_a),

$$K_a = \frac{[A][BH]}{[AH][B]} \text{ where } K_a^\circ = K_a \cdot \frac{(\gamma_A)(\gamma_{BH})}{(\gamma_{AH})(\gamma_B)} \qquad (8.9)$$

TABLE 8.1   Examples of Strong and Weak Acids in Water

Strong Acids[a]				
Chloric acid:	$HClO_3 \rightleftharpoons ClO_3^- + H^+$		Hydrobromic acid:	$HBr \rightleftharpoons Br^- + H^+$
Hydrochloric acid:	$HCl \rightleftharpoons Cl^- + H^+$		Hydroiodic acid:	$HI \rightleftharpoons I^- + H^+$
Nitric acid:	$HNO_3 \rightleftharpoons NO_3^- + H^+$		Perchloric acid:	$HClO_4 \rightleftharpoons ClO_4^- + H^+$
Sulfuric acid:[b]	$H_2SO_4 \rightleftharpoons HSO_4^- + H^+$			
	$HSO_4^- \rightleftharpoons SO_4^{2-} + H^+$			

Weak Acids		**Acid Dissociation Constant, K_a (25°C)[c]**
Acetic acid:	$CH_3COOH \rightleftharpoons CH_3COO^- + H^+$	1.75×10^{-5}
Benzoic acid:	$C_6H_5COOH \rightleftharpoons C_6H_5COO^- + H^+$	6.28×10^{-5}
Boric acid:	$B(OH)_2OH \rightleftharpoons B(OH)_2O^- + H^+$	$5.79 \times 10^{-10} (K_{a1})$
Carbonic acid:	$H_2CO_3 \rightleftharpoons HCO_3^- + H^+$	$4.46 \times 10^{-7} (K_{a1})$
	$HCO_3^- \rightleftharpoons CO_3^{2-} + H^+$	$4.69 \times 10^{-11} (K_{a2})$
Hydrofluoric acid:	$HF \rightleftharpoons F^- + H^+$	6.8×10^{-4}
Phosphoric acid:	$H_3PO_4 \rightleftharpoons H_2PO_4^- + H^+$	$7.11 \times 10^{-3} (K_{a1})$
	$H_2PO_4^- \rightleftharpoons HPO_4^{2-} + H^+$	$6.34 \times 10^{-8} (K_{a2})$
	$HPO_4^{2-} \rightleftharpoons PO_4^{3-} + H^+$	$4.22 \times 10^{-13} (K_{a3})$

[a]The K_a values for these acids are greater than or near to $10^{1.74} = 55.5$, the approximate K_a for H_3O^+.

[b]Sulfuric acid is a strong acid in water only for the loss of its first hydrogen ion, a process for which K_{a1} is roughly $10^{1.99} = 98$. The loss of the second hydrogen ion involves the dissociation of HSO_4^-, for which $K_{a2} = 1.03 \times 10^{-2}$.

[c]The K_a values in this table were obtained from NIST Database 46.

where γ_A, γ_{BH}, γ_{AH}, and γ_B are the activity coefficients that relate the activities of the reactants and products to their concentrations. Some specific examples of these equilibrium constants are given below for the acid–base reaction of nitric acid with water (see reaction in Equation 8.5).

$$K^\circ_a = \frac{(a_{H_3O^+})(a_{NO_3^-})}{(a_{HNO_3})(a_{H_2O})} \quad \text{or} \quad K_a = \frac{[H_3O^+][NO_3^-]}{[HNO_3]} \quad (8.10)$$

Notice in these equations that because water is the solvent it will have an activity of one for a dilute solution (see the section on standard states in Chapter 6). Thus, rather than placing the concentration of water into the expression for K_a this activity of one is maintained and no concentration term for water is given. It is for this reason that the solvent is not generally shown in the equilibrium expression for an acid–base, even when the solvent is acting as the acid or base. The product that results from the solvent (H_3O^+) is still shown because it is a solute and does not have an activity of one. Although this product is often written simply as a hydrogen ion (H^+), a practice followed in later portions of this book, this approach does tend to minimize the importance of water in reacting with the acid.

Some typical acid dissociation constants are included in Table 8.1. Additional examples of these constants for other acids are listed in Appendix B or can be found in other sources like the *CRC Handbook of Chemistry & Physics*,[6] as well as many others.[12–14] Strong acids in water (like HCl and nitric acid) have acid dissociation constants that are greater than $10^{1.74} = 55.5$, which is the approximate K_a for H_3O^+ (the "acid" form of water).[14] This large K_a value means that the acid–base equilibrium for these strong acids will favor their transfer of hydrogen ions to water. On the other hand, weak acids in water (such as acetic acid) have acid dissociation constants that are much smaller than the K_a for H_3O^+ and usually smaller than 1.0. This feature means that acid dissociation constants can be used as a direct means to compare the relative strength of acids and to determine whether they will be strong or weak in a given solvent. Like any equilibrium constant, the value for K_a is often expressed without units (see Chapter 6), but a concentration-dependent acid association constant can also be written with an apparent unit of M when it is defined as shown in Equation 8.10.

8.2B Strong and Weak Bases

Like acids, bases can also be classified as "strong" or "weak." We saw earlier that one example of a strong base is sodium hydroxide, NaOH. When NaOH is placed into water, it almost completely ionizes to give sodium ions and the hydroxide ion, OH^-. Other examples of strong bases in water are given in Table 8.2. Ammonia (NH_3) is an example of a weak base, because when ammonia is placed in water only a portion of this base will react to form hydroxide ions.

Just as the strength of an acid was determined by its ability to donate a hydrogen ion to another chemical, the relative strength of a base is determined by its ability to accept a hydrogen ion. We can describe the strength of a base and its ability to accept a hydrogen ion from an

TABLE 8.2 Examples of Strong and Weak Bases in Water

Strong Bases[a]

Barium hydroxide:	$Ba(OH)_2 \rightleftharpoons Ba^{2+} + 2\,OH^-$	Calcium hydroxide:	$Ca(OH)_2 \rightleftharpoons Ca^{2+} + 2\,OH^-$
Cesium hydroxide:	$CsOH \rightleftharpoons Cs^+ + OH^-$	Lithium hydroxide:	$LiOH \rightleftharpoons Li^+ + OH^-$
Potassium hydroxide:	$KOH \rightleftharpoons K^+ + OH^-$	Rubidium hydroxide:	$RbOH \rightleftharpoons Rb^+ + OH^-$
Sodium hydroxide:	$NaOH \rightleftharpoons Na^+ + OH^-$	Strontium hydroxide:	$Sr(OH)_2 \rightleftharpoons Sr^{2+} + 2\,OH^-$

Weak Bases		**Base Ionization Constant, K_b (25°C)[b]**
Ammonia:	$NH_3 + \mathbf{H^+} \rightleftharpoons \mathbf{NH_4}^+$	1.78×10^{-5}
Dimethylamine:	$(CH_3)_2NH + \mathbf{H^+} \rightleftharpoons (CH_3)_2\mathbf{NH_2}^+$	6.01×10^{-4}
Ethylamine:	$C_2H_5NH_2 + \mathbf{H^+} \rightleftharpoons C_2H_5\mathbf{NH_3}^+$	4.76×10^{-4}
Methylamine:	$CH_3NH_2 + \mathbf{H^+} \rightleftharpoons CH_3\mathbf{NH_3}^+$	4.47×10^{-4}
Pyridine:	$C_5H_5N + \mathbf{H^+} \rightleftharpoons C_5H_5\mathbf{NH}^+$	1.6×10^{-9}
Trimethylamine:	$(CH_3)_3N + \mathbf{H^+} \rightleftharpoons (CH_3)_3\mathbf{NH}^+$	6.35×10^{-5}

[a]These chemicals are considered strong bases because they readily dissociate to produce hydroxide ions in water. The hydroxide ions can then act as hydrogen ion acceptors through the following reaction: $OH^- + H^+ \rightleftharpoons H_2O$

[b]The K_b values in this table were calculated based on data obtained from NIST Database 46.

TABLE 8.3   Examples of Amphiprotic Solvents and Autoprotolysis Reactions

Solvent	Autoprotolysis Reaction	K_{auto} (25°C)
Formic acid	$2\ HCOOH \rightleftharpoons HCOOH_2^+ + HCOO^-$	6.3×10^{-7}
Acetic acid	$2\ CH_3COOH \rightleftharpoons CH_3COOH_2^+ + CH_3COO^-$	3.2×10^{-15}
Methanol	$2\ CH_3OH \rightleftharpoons CH_3OH_2^+ + CH_3O^-$	2.0×10^{-17}
Ethanol	$2\ CH_3CH_2OH \rightleftharpoons CH_3CH_2OH_2^+ + CH_3CH_2O^-$	7.9×10^{-20}

acid by using an equilibrium constant known as a **base ionization constant** (also called the *basicity constant* or *protonation constant*).[10,11] This concept is illustrated below for the reaction of ammonia with water, where K_b° is the thermodynamic base ionization constant for this reaction and K_b is the concentration-dependent base ionization constant.

$$K_b^\circ = \frac{(a_{NH_4^+})(a_{OH^-})}{(a_{NH_3})(a_{H_2O})} \quad or \quad K_b = \frac{[NH_4^+][OH^-]}{[NH_3]} \quad (8.11)$$

In these equations, water is the solvent and has an activity of one, which is used in place of the concentration of water in the expression for K_b. Thus, as we saw for acid dissociation constants, water (or the solvent) is not usually shown in a K_b expression when it is acting as an acid. Although K_b° is written without units, the concentration-dependent value of K_b can be written with or without units, with an apparent unit of M being used when this constant is written as shown in Equation 8.11.

8.2C The Acid and Base Properties of Water

Autoprotolysis. In the previous acid–base reactions we have looked at, we have noted that water can act as either an acid or base. This idea is illustrated by the following reaction, in which two molecules of water react (one as an acid and one as a base) to form a hydronium ion (H_3O^+) and hydroxide ion (OH^-). (*Note*: This process is just the reverse of the reaction we saw in Equation 8.6.)

$$2\ H_2O \rightleftharpoons H_3O^+ + OH^- \quad (8.12)$$

Substances like water that can act as either an acid or a base are said to be *amphiprotic*, which means they can either donate or accept hydrogen ions.[11] Other chemicals besides water which are amphiprotic are listed in Table 8.3. One example is HSO_4^-, which acted as a base for the reaction in Equation 8.2, but acted as an acid in Equation 8.3. If an amphiprotic chemical is present in a high concentration, it can sometimes react with itself in an acid–base reaction. This type of reaction is particularly common when the amphiprotic chemical is a solvent, as occurs in Equation 8.12 for water. This special type of acid–base reaction in which the same chemical is both the acid and base is known as **autoprotolysis**.

The autoprotolysis of water is an important process that occurs in any aqueous solution. We can describe this process by using an equilibrium constant known as the **autoprotolysis constant** of water, or K_w.[10,11]

$$K_w^\circ = \frac{(a_{H_3O^+})(a_{OH^-})}{(a_{H_2O})^2} \quad or \quad K_w = [H_3O^+][OH^-] \quad (8.13)$$

Notice in these equilibrium-constant expressions that the activity of water is again taken to be one (because water is the solvent). As a result, this unit activity is kept in the denominator of K_w (even if it is not shown) and no other concentration term for water is needed. If you wish to give units for the concentration-dependent value of K_w, such as when you use this constant to calculate the concentrations of hydrogen ions or hydroxide ions, Equation 8.13 indicates that apparent units of M^2 can be employed with this value.

The values for K_w at several temperatures are given in Table 8.4. At 25°C this equilibrium constant is 1.01×10^{-14} (or a rounded value of 1.0×10^{-14}), a number that is frequently used in calculations involving acid–base reactions in water. One important application of K_w° and K_w is they allow the activity (or concentration) of H_3O^+ to be determined if the activity (or concentration) of OH^- is known for an aqueous solution. The reverse is also true, where the activity and concentration of OH^- in water can be used to calculate the activity or concentration of H_3O^+. In *pure* water, the only source of $[H_3O^+]$ and $[OH^-]$ is water itself. Thus, based on mass balance the concentrations of H_3O^+ and OH^- (which are formed from water in a 1:1 ratio) will be equal under these conditions. These concentrations, however, will not usually be equal if other substances with acid or base properties are present in the water.

EXERCISE 8.1   Using K_w to Determine $[H_3O^+]$ and $[OH^-]$

A chemist prepares a $1.5 \times 10^{-2}\ M$ solution of hydrochloric acid in water at 25°C. If the HCl completely dissociates and there is no other significant source of hydrogen ions, what will the concentrations of H_3O^+ and OH^- be in this solution? What will these concentrations be if the same solution is increased in temperature to 60°C?

SOLUTION

If we assume that all of the HCl is dissociated and there is no other significant source of hydrogen ions, the value for $[H_3O^+]$ will be the same as the concentration of dissolved HCl, or $[H_3O^+] = 1.5 \times 10^{-2}\ M$. Using Equation 8.13 and the fact that $K_w = 1.0 \times 10^{-14}$ (with apparent units of M^2) at 25°C, the concentration of hydroxide ions can then be found.

$$1.0 \times 10^{-14}\ M^2 = (1.5 \times 10^{-2}\ M)[OH^-]$$

$$\therefore [OH^-] = 6.7 \times 10^{-13}\ M \text{ at } 25°C$$

If the temperature of this solution is raised to 60°C, the value for $[H_3O^+]$ is still equal to the concentration of dissolved HCl, $1.5 \times 10^{-2}\ M$. However, we must now use $K_w = 9.71 \times 10^{-14}$ as listed for 60°C in Table 8.4. In this case, Equation 8.13 gives the following result.

$$9.71 \times 10^{-14}\ M^2 = (1.5 \times 10^{-2}\ M)[OH^-]$$

$$\therefore [OH^-] = 6.5 \times 10^{-12}\ M \text{ at } 60°C$$

The Leveling Effect.   Now that we know something about the acid and base properties of water, let's again examine the reactions of water with other acids and bases. First, we need to look back at the acid–base reaction of nitric acid with water, as shown earlier in Equation 8.5.

$$HNO_3 + H_2O \rightleftharpoons H_3O^+ + NO_3^- \tag{8.5}$$

TABLE 8.4   Autoprotolysis Constant for Water (K_w)*

Temperature (°C)	K_w	$pK_w = -\log(K_w)$
0	1.14×10^{-15}	14.944
5	1.85×10^{-15}	14.734
10	2.92×10^{-15}	14.535
15	4.51×10^{-15}	14.346
20	6.81×10^{-15}	14.167
25	1.01×10^{-14}	13.997 ($\approx$ 14.00)[a]
30	1.469×10^{-14}	13.833
35	2.09×10^{-14}	13.680
40	2.92×10^{-14}	13.535
45	4.02×10^{-14}	13.396
50	5.47×10^{-14}	13.262
55	7.29×10^{-14}	13.137
60	9.62×10^{-14}	13.017

*These results were obtained from D. R. Lide, Ed. *CRC Handbook of Chemistry and Physics*, 83rd ed., CRC Press, Boca Raton, FL, 2002.

[a]The value of pK_w is equal to 14.0000 (or $K_w = 1.000 \times 10^{-14}$) at 24°C.

On both sides of this reaction we have an acid (HNO_3 or H_3O^+) and a base (H_2O or NO_3^-). We also know this reaction goes almost totally toward the right, because HNO_3 has essentially complete dissociation. So why is this reaction so highly favored? The answer is that water is a *much* stronger base than NO_3^-, so the reaction proceeds mostly toward the right. The same is true for the acid-base reactions of HCL, HBr, HI, or $HClO_4$ with water. Because all these acids give essentially a complete transfer of hydrogen ions to water, they appear to have equal strength even though they really do not. Instead, what has happened is that water has reacted with each of these acids to produce a common product, H_3O^+.

The ability of water to equalize the strength of strong acids is known as the **leveling effect**. This effect means that the strongest acid that can exist in any appreciable amount in water is the product of these reactions, H_3O^+. The same effect occurs when a strong base such as NaOH is placed into water, but the product is now OH^-, the strongest base that can exist in any significant amount in water. This phenomenon occurs whenever a base stronger than hydroxide ion is put into water. For instance, placing the strong base sodium oxide (Na_2O) into water will lead to the quantitative production of hydroxide ions, which then act as a strong base in place of the Na_2O.

$$Na_2O + H_2O \rightarrow 2\,Na^+ + OH^- \tag{8.14}$$

This effect is also a reason why acid dissociation and base ionization constants are often not listed for strong acids and strong bases. What is really happening with these chemicals is they are reacting with water to create essentially quantitative production of either H_3O^+ (in the case of the strong acids) or OH^- (in the case of the strong bases).

Even though we cannot differentiate between strong acids and bases in water, we can rank them according to their strength if we use solvents other than water. For example, a solvent that is a weaker base than water can discriminate between HCl and $HClO_4$. This idea is illustrated below for the case where these acids are placed in acetic acid (CH_3COOH), an acidic solvent.[15]

$$HClO_4 + CH_3COOH \rightleftharpoons ClO_4^- + CH_3COOH_2^+$$
$$K = 1.3 \times 10^{-5} \quad (8.15)$$

$$HCl + CH_3COOH \rightleftharpoons Cl^- + CH_3COOH_2^+$$
$$K = 2.8 \times 10^{-9} \quad (8.16)$$

We can now see from the equilibrium constants for these reactions that $HClO_4$ is a better hydrogen ion donor (and therefore a stronger acid) than HCl. A similar comparison between strong bases can be made by using a solvent that is a weaker acid than water, such as pyridine or ethylenediamine.

Relationship of K_a to K_b.   Now that we have learned how to describe the strengths of acids and bases, let's see how this information can be used to describe conjugate bases and conjugate acids. First, given what we know about the relative strengths of nitric acid and acetic acid, what might we conclude about the strengths of their conjugate bases, nitrate, and acetate? Nitric acid is a strong acid in water, so it will almost completely dissociate in this solvent to form H_3O^+. Another way of thinking about this is to view the nitrate ion as such a weak base that very little of it goes back to form nitric acid. In the case of acetic acid, which is a weak acid in water, most of this compound stays in its original form. When water (on the reactant side) and acetate ion (on the product side) both act as bases to accept hydrogen ions, the acetate ion wins because it is a stronger base than water. From this discussion we can see that strong acids have conjugate bases that are weaker bases than water and weak acids have somewhat stronger conjugate bases. By the same argument, strong bases have conjugate acids that are weaker acids than water, while weak bases have stronger conjugate acids.

It is also possible to show the relationship between an acid and its conjugate base in a more quantitative fashion. This relationship is as follows for a dilute solution,

$$K_w = K_a \cdot K_b \qquad (8.17)$$

where K_w is the autoprotolysis constant for water, K_a is the acid dissociation constant for the acid, and K_b is the base ionization constant for its conjugate base. Equation 8.17 provides a way for us to determine the value for K_a for the conjugate acid in water if we know K_b for the parent base, or it allows us to calculate K_b for a base if we know K_a for its conjugate acid.

EXERCISE 8.2	**Determining K_b from K_a**

Acetic acid has a K_a of approximately 1.75×10^{-5} in water at 25°C. What is the value for K_b for the conjugate base, acetate (CH_3COO^-), under these conditions?

SOLUTION

The value of K_b for acetate can be found by placing the known values of K_a for acetic acid and K_w for water into Equation. 8.17. The result is shown below.

$$1.01 \times 10^{-14} = 1.75 \times 10^{-5} \cdot K_b \quad \therefore \; \mathbf{K_b = 5.77 \times 10^{-10}}$$

The same process can be used to find K_b for the conjugate base of any other acid in water, as long as we know K_a for the acid and an appropriate value of K_w for our solution.

You may be wondering at this point about the origin of Equation 8.17. This equation simply reflects the fact that an acid in an aqueous solution will react with water to form H_3O^+, while a base in an aqueous solution will react with water to form OH^-. This concept can be demonstrated through the following reactions and equilibrium expressions for a weak acid (HA) and its conjugate base (A^-) in water.

$$HA + H_2O \rightleftharpoons A^- + H_3O^+ \qquad K_a^\circ = \frac{(a_{A^-})(a_{H_3O^+})}{(a_{HA})(a_{H_2O})}$$

$$A^- + H_2O \rightleftharpoons HA + OH^- \qquad K_b^\circ = \frac{(a_{HA})(a_{OH^-})}{(a_{A^-})(a_{H_2O})} \qquad (8.18)$$

If we multiply together the equilibrium expressions for these two reactions ($K_a^\circ \cdot K_b^\circ$) and cancel out common terms in their numerators and denominators, we get the result shown below.

$$K_a^\circ \cdot K_b^\circ = \frac{(a_{A^-})(a_{H_3O^+})}{(a_{HA})(a_{H_2O})} \cdot \frac{(a_{HA})(a_{OH^-})}{(a_{A^-})(a_{H_2O})} \qquad (8.19)$$

$$\therefore \; K_a^\circ \cdot K_b^\circ = (a_{H_3O^+})(a_{OH^-}) = K_w^\circ$$

$$(\text{assuming } a_{H_2O} = 1) \qquad (8.20)$$

or

$$K_a \cdot K_b = [H_3O^+][OH^-] = K_w \qquad (8.21)$$

When we set the term $K_a^\circ \cdot K_b^\circ$ equal to K_w° (and $K_a \cdot K_b$ equal to K_w), we are assuming that the activity of water is one. This assumption is a good one if we have a dilute or only moderately concentrated solution of the acid or base in water. This type of situation is often encountered in analytical chemistry and will be the focus of the remainder of this chapter. However, Equations 8.20 and 8.21 will not be valid if you are working with a highly concentrated acid or base.

8.3 THE ACID OR BASE PROPERTIES OF A SOLUTION

8.3A What Is pH?

A Description of pH. The content of hydrogen ions (or rather H_3O^+) in a sample like rain water can be expressed in terms of activity or concentration, but this value can span over an enormously wide range. For this reason it is often more convenient to report and compare H^+ levels by using a logarithmic scale. Chemists often accomplish this by using an operator called "p," which is used to describe the negative logarithm of a given number (e.g., $pX = -\log(X)$). The use of this operator with hydrogen ion concentrations gives rise to a term known as **pH**. In this text, we will use the following **notational definition for pH**,

$$pH = -\log(a_{H^+}) \approx -\log([H^+]) \qquad (8.22)$$

where a_{H^+} and $[H^+]$ are the activity or concentration of "hydrogen ions" in the sample.[10,11] To illustrate this definition, let's use the pH of less than 5.0 that is often used to define acid rain. We can now see from Equation 8.22 that this pH range means that acid rain has a hydrogen ion concentration of greater than $1 \times 10^{-5} \, M$, because $-\log(1 \times 10^{-5}) = 5.0$. The "definition" given for pH in Equation 8.22 was first used in 1909 by S.P.L. Sørensen.[4,16] Although we will use this relationship to calculate and use pH values throughout the remainder of this text, you should always keep in mind for water that it is really the activity or concentration of hydronium ions (H_3O^+) that is being described by these pH values.

Equation 8.22 is valuable when you are carrying out chemical calculations that involve hydrogen ions, but this same relationship is not a practical definition for pH. This is the case because Equation 8.22 describes pH based on the activity of a single ion (H^+), which is a problem because this activity cannot be measured independently from the counter ions (e.g., OH^-) that must also be present to maintain charge balance in the system (see Chapter 6). A practical consequence of this fact is that any pH measurement must be performed by a method that has been calibrated with reference solutions having accepted pH values.[4,10,11,17]

The same type of operation that is used to obtain pH from a_{H^+} or $[H^+]$ can be used to describe other numbers that have a broad range of values. For instance, the term "pOH" can be used to represent the term $-\log(a_{OH^-})$ or pK_w could be used in place of $-\log(K_w)$. The general equations for converting between these parameters are shown in Figure 8.5. It is often convenient to work with pH and related values like pK_w and pOH when we are trying to describe how these terms are connected to each other. We saw an example of this in Equation 8.13, in which K_w was related to $[H_3O^+]$ (or "$[H^+]$") and $[OH^-]$ through $K_w = [H^+][OH^-]$. Another way of writing this relationship is to take the logarithm of both sides of this equation.

$$pK_w = pH + pOH \qquad (8.23)$$

When Equation 8.23 is combined with $K_w \approx 1.0 \times 10^{-14}$ at 25°C (or $pK_a = 14.00$), we get the following result.

$$\text{At 25°C } 14.00 = pH + pOH \qquad (8.24)$$

We now have two different but equally valid ways (Equations 8.21 and 8.23–8.24) for describing the relationship of hydrogen ion and hydroxide ion concentrations in water. Equations 8.23 and 8.24 are more convenient to use when we already know the pH for an aqueous solution.

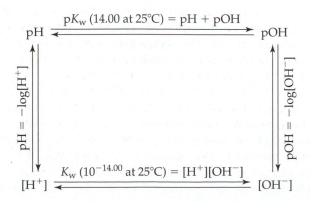

FIGURE 8.5 General relationship between pH, pOH, $[H^+]$, and $[OH^-]$, where K_w is the autoprotolysis constant for water.

Some examples of how we might use these relationships are given in the next exercise.

EXERCISE 8.3 **Calculating pH and Related Values**

Determine each of the following values.

a. pH for a river water sample where $[H^+] = 3.2 \times 10^{-4}\ M$
b. pK_a for acetic acid ($K_a = 1.75 \times 10^{-5}$)
c. pH for an industrial waste sample in which $[OH^-] = 3.1 \times 10^{-5}\ M$

SOLUTION

a. $pH = -\log(3.2 \times 10^{-4}\ M) = \mathbf{3.49}$

c. $pK_a = -\log(1.75 \times 10^{-5}) = \mathbf{4.757}$

c. $pOH = -\log(3.1 \times 10^{-5}\ M) = 4.51$ and $pH = 14.00 - 4.51 = \mathbf{9.49}$ (*Note*: See Appendix A for a reminder of how to determine the number of significant figures when you are using logarithmic values.)

In the preceding exercise, you will notice that we did not give any units for pH or pOH, even though our original values for $[H^+]$ and $[OH^-]$ had units of molarity. This is because logarithmic terms like pH and pOH (as well as pK_a) have no units. As we saw in Chapter 6 for activity coefficients, this is possible because there is a hidden factor present during each of these logarithmic conversions that makes it possible to cancel out any units for the original number. If we have a hydrogen concentration expressed in units of molarity, the more complete form of Equation 8.22 would be $pH \approx -\log([H^+]/\{1\ M\})$, where the number in the denominator is exactly equal to one and has the same units as the numerator.[10,11] The same thing occurred when we found the value of pOH in Exercise 8.3.

The pH Scale. According to Equation 8.24, the sum of the pH and pOH values for an aqueous solution should equal approximately 14.00 at 25°C. This relationship is the basis for the **pH scale** that is often used with water and aqueous samples (see Figure 8.6). In using this scale, an aqueous solution with a pH less than 7.0 is said to be *acidic* because it contains a greater amount of hydrogen ions than hydroxide ions. An aqueous solution with a pH greater than 7.0 is called *basic* (or alkaline), because it contains a greater amount of hydroxide ions.[4] An aqueous solution that is at or near a pH of 7.0 is said to be *neutral*, because it has approximately the same amounts of hydrogen ions and hydroxide ions, making the solution neither acidic or basic.

The pH values for some common substances are included in Figure 8.6. Pure water with nothing dissolved in it should have a neutral pH of exactly 7.0 on this scale. Normal tap water and stream water can actually have slightly higher or lower values (pH 6–8) due to the effects

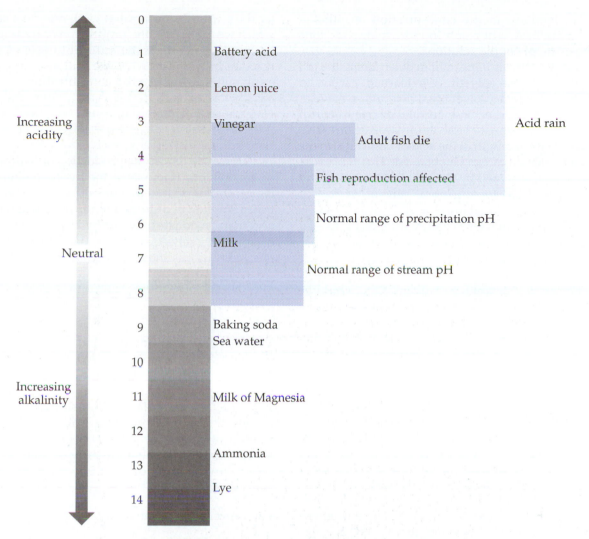

FIGURE 8.6   The pH scale in water, including the range of values seen in common samples and acid rain. (Reproduced from Environment Canada's Freshwater Web site, www.ns.ec.gc.ca, with permission from the Department.)

of dissolved solids and gases (such as CO_2) that can affect hydrogen and hydroxide ion levels. Normal rain has a pH between 6.5 (slightly acidic) down to 5.0, and acid rain is defined as having a pH below 5.0. As we move to lower pH values and higher hydrogen ion concentrations, we find substances that contain weak acids (like vinegar, which is a source of acetic acid) and strong acids (like the sulfuric acid in batteries). At the other end of the pH scale we have an increase in pH and lower hydrogen ion concentrations, or higher hydroxide ion levels, as we move to solutions of weak bases such as ammonia and solutions of strong bases such as lye (NaOH).

Because pH is a logarithmic value, even a small change in pH can represent a significant change in hydrogen ion concentration. A one-unit change in pH will correspond to a ten-fold change in hydrogen ion concentration. This is one reason why a decrease in the pH of rain is of such concern. For example, Figure 8.6 indicates that fish reproduction will be affected by a drop in the pH of stream water from 7.0 to 5.0 (the same upper pH limit used to define acid rain).[18] This

drop in pH is the same as a 100-fold increase in hydrogen ion activity and, due to the many reactions in biology that are affected by pH, can lead to damage in living organisms.[2–4]

8.3B Factors That Affect pH

There are several factors that affect the pH of a solution. The first factor is the type of acid or base that is present in the solution. In general, a strong acid like nitric acid will produce a lower pH solution than a weak acid like acetic acid at the same analytical concentration because the strong acid will react to a greater extent to form hydrogen ions. Similarly, a strong base like NaOH will produce a higher pH than a weak base like ammonia at the same analytical concentration because the strong base will create more hydroxide ions. A second factor that affects the pH is the concentration of the acid or base. In the case of acid rain, as we increase the concentration of nitric acid or sulfuric acid the amount of hydrogen ions that is produced will increase and the pH will decrease. In the same manner, an increase in the concentration of either a strong or weak base will result in more hydroxide ions and increase the pH of the solution.

A third item that is important in determining the pH of a solution is the solvent. We learned earlier in Section 8.2C that it is the solvent that determines the relative strength of an acid or base and the degree to which such substances can donate or accept hydrogen ions. This affects not only the strength of the acid or base but also the range of pH values that can be obtained. This is because the pH scale is determined by the autoprotolysis constant for the solvent (K_{auto} for a general solvent, or K_w for water). In water, K_w is equal to 1.0×10^{-14} at 25°C, so the pH scale in water at this temperature extends up to $-\log(K_w) = 14.00$. If we were to instead use deuterium oxide (D_2O) as the solvent, K_{auto} at 25°C would be 1.1×10^{-15} and the pH scale would extend up to $-\log(K_{auto}) = 14.96$. If ethanol or ammonia were the solvent, K_{auto} would be 7.9×10^{-20} or 2.0×10^{-28} and the pH scale would be represented by a range that went up to 19.1 or 27.7.

Because temperature can affect a chemical equilibrium, it will affect acid–base reactions and the pH and hydrogen ion activity that result from these reactions. One way temperature can alter these processes is by changing the values for K_a or K_b. As an example, the pK_a for phosphoric acid changes from 2.15 to 2.28 when going from 25° to 50°C. This means this acid will transfer more hydrogen ions to water as we increase the temperature over this range. Another item affected by temperature is the autoprotolysis constant of the solvent. For water, K_w has a change of about ten-fold between 25° and 60°C. This change will affect the pH scale slightly (going from a maximum of 13.997 to 13.017, according to Table 8.4) and will alter the degree to which water can act as an acid or base.

8.4 ESTIMATING THE pH OF SIMPLE ACID–BASE SOLUTIONS

Determining the pH of a sample or solution is one of the most common measurements made in analytical chemistry. This type of measurement can be accomplished in many ways. Two examples are the use of a pH electrode (see Chapters 12 and 14) or reagents known as acid–base indicators that change color with pH (discussed in Chapters 12 and 18). It is also possible to use equilibrium calculations to estimate the pH that would be expected for a given solution if we know the amounts and types of chemicals this solution will contain. This ability can allow us to plan how we will make a reagent for a chemical measurement or to predict how sample components may behave during pretreatment or analysis.

8.4A Monoprotic Strong Acids and Bases

Concentrated Solutions. One situation in which you may want to estimate the pH of a solution is when you are preparing a relatively concentrated, aqueous solution of a strong acid. During the remainder of this chap-

ter the phrase "concentrated" will refer to a solution for which we can ignore the production of hydrogen ions or hydroxide ions from the autoprotolysis of water (the reaction in Equation 8.12). We will also only be considering solutions where the concentration of the acid or base is low enough that we can assume the activity of water is approximately one, as is required when using the expression $K_a \cdot K_b = [H_3O^+][OH^-] = K_w$ in Equation 8.21.

If we can ignore the production of hydrogen ions or hydroxide ions from water, the hydrogen ion (or H_3O^+) concentration that is produced by a relatively concentrated solution of a strong **monoprotic acid** (HA) (i.e., an acid that can donate only one hydrogen ion to a base) will be approximately equal to the analytical concentration of the acid (C_{HA}), as shown in Equation 8.25.

For a strong monoprotic acid:

$$[H^+] \approx C_{HA} \text{ for } C_{HA} > 10^{-6}\,M \qquad (8.25)$$

This is the relationship that we used earlier in Exercise 8.1 when we assumed that a solution of $1.5 \times 10^{-2}\,M$ nitric acid totally dissociated in water to gave a final hydrogen ion concentration of $1.5 \times 10^{-2}\,M$. A similar relationship can be written for a relatively concentrated solution of a strong **monoprotic base** (B), which will react with an acid to accept essentially one mole of hydrogen ions (or form one mole of hydroxide ions in water) per mole of the original base.

For a strong monoprotic base:

$$[OH^-] \approx C_B \text{ for } C_B > 10^{-6}M \qquad (8.26)$$

Equations 8.25 and 8.26 do have some limitations. First, they both apply only to strong acids and strong bases with concentrations that are at or above roughly $10^{-6}\,M$. This limitation is due to the autoprotolysis of water, which becomes an important source of hydrogen ions and hydroxide ions at lower concentrations. This effect will be examined further in the next section of this chapter. Second, there is also an upper concentration limit for Equations 8.25 and 8.26. This second limitation occurs because we are assuming in these equations that the concentration of hydrogen ions is approximately the same as the activity of hydrogen ions. This assumption can lead to errors when working with solutions that have moderate or high ionic strengths, for which the concentration and activity of H^+ can be quite different. The size of this error will also be examined later in this chapter.

Dilute Solutions. When working at very low acid–base concentrations, we must consider the effects of autoprotolysis, even when we are working with strong acids or strong bases. To illustrate this concept, let's estimate what the pH would be for a $1 \times 10^{-8}\,M$ solution of nitric acid in water. If we determined this pH by using Equation 8.25 and the analytical concentration of nitric acid, we would get a calculated value of pH = $-\log(1 \times 10^{-8}\,M)$ = 8.0. This result cannot be correct, however, because it would mean that placing a strong acid into water actually made a basic solution! What actually occurs when we make a $1 \times 10^{-8}\,M$ solution of nitric acid is we get a pH that is slightly below 7.0. The reason for this difference is that using Equation 8.25 for a strong acid (or Equation 8.26 for a strong base) gives an incomplete description of our system.

To get a more realistic answer for the pH of $1 \times 10^{-8}\,M$ nitric acid, we need to consider both nitric acid and water as sources of hydrogen ions. This can be done by using the following mass balance relationship.

$$[H^+] = [H^+]_{HNO_3} + [H^+]_{Water} \qquad (8.27)$$

In this equation, $[H^+]$ represents the total concentration of all hydrogen ions in the solution, while $[H^+]_{HNO_3}$ or $[H^+]_{Water}$ are the hydrogen ion concentrations that are produced by the acid dissociation of nitric acid or water, respectively. We can now rewrite Equation 8.27 by making the following substitutions into this formula.

Substitution	Change to Equation 8.27
(1) $[H^+]_{HNO_3} = C_{HA}$	$\Rightarrow [H^+] = C_{HA} + [H^+]_{Water}$
(2) $[H^+]_{Water} = [OH^-]_{Water}$	$\Rightarrow [H^+] = C_{HA} + [OH^-]_{Water}$
(3) $[OH^-]_{Water} = [OH^-]$	$\Rightarrow [H^+] = C_{HA} + [OH^-]$
(4) $[OH^-] = K_w/[H^+]$	$\Rightarrow [H^+] = C_{HA} + K_w/[H^+]$

The first of these substitutions assumes that the nitric acid will completely dissociate to form one mole of H^+ per mole of nitric acid, which means $[H^+]_{HNO_3}$ should be equal to the analytical concentration of the added nitric acid. In the second substitution, we use the fact that each mole of H^+ that is formed from water will be accompanied by the production of one mole of OH^- from water. Third, we know that water is the only source of hydroxide ions in this system, which allows us to say $[OH^-]_{Water} = [OH^-]$, the total hydroxide ion concentration. And fourth, Equation 8.21 shows that the total concentration of hydrogen ions in an aqueous solution will be related to the total concentration of hydroxide ions, where $K_w = [H^+][OH^-]$ or $[OH^-] = K_w/[H^+]$. When we substitute these relationships into Equation 8.27 and rearrange this expression, we get the result shown below.

For a dilute, strong monoprotic acid:

$$[H^+] = C_{HA} + K_w/[H^+]$$

$$\text{or } 0 = [H^+]^2 - C_{HA}[H^+] - K_w \qquad (8.28)$$

This result gives us a quadratic formula that can be used to solve for $[H^+]$ and pH, because pH = $-\log([H^+])$, if we know the values for C_{HA} and K_w. Although we derived this equation for a solution of nitric acid, it can also be used for solutions of any other strong monoprotic acid in water. In addition, the approach by which we derived this equation is important because it illustrates the type of thought process that should be followed in dealing with this type of equilibrium problem.

We can now determine the possible values for $[H^+]$ by using the quadratic formula (as described in Chapter 6), where $x = [H^+]$, A = 1, B = $-C_{HA}$ ($-1.0 \times 10^{-8}\,M$, in this case) and C = $-K_w$ (or $-1.0 \times 10^{-14}\,M^2$, when including apparent units for K_w). The final answer that is obtained is $[H^+] = 1.05 \times 10^{-7}\,M$, or pH = $-\log(1.05 \times 10^{-7})$ = 6.98. This pH of 6.98 for a dilute solution of nitric acid now makes sense. In this situation most hydrogen ions in the solution are contributed by the autoprotolysis of water, but the addition of nitric acid, even in a small amount, is enough to give a pH that is slightly acidic. As shown in the following exercise, the same approach can be used to determine how the pH of a solution will be affected by adding a small amount of a strong base.

EXERCISE 8.4   Estimating the pH for a Dilute Solution of a Strong Base

Derive a similar expression to Equation 8.28 for a dilute solution of a strong base. Use this equation to predict the pH that would be expected for a $1 \times 10^{-8}\,M$ solution of NaOH in water.

SOLUTION

This case is similar to the situation we had with dilute nitric acid, except we now have two possible sources of hydroxide ions (NaOH and water) and only one source of hydrogen ions (water). The total concentration of hydroxide ions in this solution, which we will simply represent as $[OH^-]$, can be described by the mass balance relationship shown below.

$$[OH^-] = [OH^-]_{NaOH} + [OH^-]_{Water} \qquad (8.29)$$

We can now make a similar set of substitutions to those we used for the earlier case of a strong acid.

Substitution	Change to Equation 8.29
(1) $[OH^-]_{NaOH} = C_B$	$\Rightarrow [OH^-] = C_B + [OH^-]_{Water}$
(2) $[OH^-]_{Water} = [H^+]_{Water}$	$\Rightarrow [OH^-] = C_B + [H^+]_{Water}$
(3) $[H^+]_{Water} = [H^+]$	$\Rightarrow [OH^-] = C_B + [H^+]$
(4) $[OH^-] = K_w/[H^+]$	$\Rightarrow K_w/[H^+] = C_B + [H^+]$

The first substitution can be made because we know the concentration of hydroxide produced from the sodium hydroxide ($[OH^-]_{NaOH}$) will be approximately equal to the analytical concentration of NaOH (C_B). Second, we know one H^+ will be formed for each OH^- that is produced from water, or $[H^+]_{Water} = [OH^-]_{Water}$. In the third substitution, water is the only source of H^+ in this solution, so we can say that $[H^+]_{Water}$ is equal to the total hydrogen in concentration, $[H^+]$. And finally, the fourth substitution is based on the fact that K_w is related to the total concentrations of H^+ and OH^- by $K_w = [H^+][OH^-]$, or $K_w/[H^+] = [OH^-]$. When we make all of these changes to Equation 8.29, we get the following equivalent expressions that work for a solution of any strong monoprotic base in water.

For a dilute, strong monoprotic base:

$$K_w/[H^+] = C_B + [H^+]$$

or
$$0 = [H^+]^2 + C_B[H^+] - K_w \qquad (8.30)$$

Equation 8.30 can now be used to estimate the pH for a 1×10^{-8} M solution of NaOH in water. We can do this by using the quadratic formula to solve for $[H^+]$ in Equation 8.30, where $x = [H^+]$, $A = 1$, $B = C_B (=1.0 \times 10^{-8}$ M), and $C = -K_w$ (or $- 1.0 \times 10^{-14}$ M^2). The final answer we get for $[H^+]$ is 9.51×10^{-8} M, or pH $= -\log(9.5\underline{1} \times 10^{-8}) = $ **7.02**. As expected, this pH is just slightly higher than 7.0 for the dilute solution of our strong base.

In all of our previous calculations in this section, we have estimated the pH by using hydrogen ion concentrations, where pH $\approx -\log[H^+]$. We also know that the pH is actually related to the activity of the hydrogen ions and that the activity and concentration of an ion are

generally equal only if we have a solution with a low ionic strength. As a result, the use of $[H^+]$ in place of hydrogen ion activity (a_{H^+}) to calculate the pH will give rise to a systematic error. The size of this error is shown in Table 8.5 for rain water that contains various concentrations of a monoprotic strong acid. In this example, there is not a problem if we have solutions with total acid concentrations and ionic strengths below 1.0×10^{-3} M, but at higher ionic strengths there are differences between the actual and estimated pH. At an acid concentration of 1.0×10^{-3} M and pH around 3.0, this difference is around 0.02 pH units. This difference rises to 0.04 pH units around a pH of 2.0 and total acid concentration of 1.0×10^{-2} M, and it further increases to 0.08 pH units at a pH near 1.0 and a total acid concentration of 1.0×10^{-1} M. The same types of errors occur for solutions of strong monoprotic bases. Thus, we do need to carefully consider these effects and the concentration of the strong acid or base when deciding upon the degree of accuracy and number of significant figures that we would like to use in calculating and reporting the pH of a strong acid or strong base solution.

8.4B Monoprotic Weak Acids and Bases

Determining the pH of a solution that contains a weak acid and base is similar to the approach used for a strong acid or base in that we must determine whether the effects of water dissociation are important, and we must consider how this pH will be affected by the analytical concentration of the acid or base. With weak acid and bases, this process is complicated by the fact we can no longer assume these chemicals are totally reacting with water to produce hydrogen ions or hydroxide ions. Let's see how we can handle this situation for a simple solution of a

TABLE 8.5 Effect of Concentration on the Estimated and Actual pH of Water Containing a Monoprotic Strong Acid (HA)

Concentration of HA, M[a]	Ionic Strength, M[b]	Activity Coefficient for H^+ (γ_{H^+})[c]	Estimated pH — pH $= -\log([H^+])$	Actual pH — pH $= -\log(a_{H^+}) = -\log(\gamma_{H^+} \cdot [H^+])$
1.0×10^{-6}	1.0×10^{-6}	0.99\underline{9}	6.00	6.00\underline{1} = 6.00
1.0×10^{-5}	1.0×10^{-5}	0.99\underline{6}	5.00	5.00\underline{2} = 5.00
1.0×10^{-4}	1.0×10^{-4}	0.98\underline{9}	4.00	4.00\underline{5} = 4.00
1.0×10^{-3}	1.0×10^{-3}	0.96\underline{7}	3.00	3.01\underline{5} = 3.02
1.0×10^{-2}	1.0×10^{-2}	0.91\underline{3}	2.00	2.03\underline{9} = 2.04
1.0×10^{-1}	1.0×10^{-1}	0.82\underline{5}	1.00	1.08\underline{3} = 1.08

[a]It is assumed in this case that the strong acid completely dissociates and that there are no other significant sources of hydrogen ions or A^-, making $[H^+] = [A^-] = C_{HA}$.

[b]Because this example uses a monoprotic strong acid and dissociation of the acid is the only major source of cations or anions, the ionic strength is the same as the original concentration of the acid, since $I = \frac{1}{2}\{[H^+](+1)^2 + [A^-](-1)^2\} = \frac{1}{2}\{C_{HA}(+1)^2 + C_{HA}(-1)^2\} = C_{HA}$. However, the ionic strength would not be equal to C_{HA} if a polyprotic system was not present or if other important sources of cations or anions were present.

[c]The activity coefficient for H^+ at each given ionic strength was found by using the extended Debye–Hückel equation, as described in Chapter 6.

weak acid or weak base. We will also look at what can occur as we create mixtures of conjugate acids and bases.

Simple Solutions. The easiest situation for which we can determine the pH of a weak acid or base solution is when we add only the acid or base to a solution and the concentration of this acid or base is high enough to allow the effects of water autoprotolysis to be ignored. We will begin with the case in which we have a relatively concentrated solution that is prepared by adding a monoprotic weak acid like acetic acid to water. We can further simplify this case if the acetic acid is the only chemical that has been added to this solution. It is possible to solve for the pH of this solution if we start by writing the reaction and acid dissociation expression for our weak acid (HA).

$$HA \rightleftharpoons H^+ + A^- \qquad K_a = \frac{[H^+][A^-]}{[HA]} \qquad (8.31)$$

We can now make the following substitutions to relate the terms in Equation 8.31 to constants for this system or to $[H^+]$.

Substitution	Change to Equation 8.31
(1) $[HA] = C_{HA} - [A^-]$	$\Rightarrow K_a = \dfrac{[H^+][A^-]}{C_{HA} - [A^-]}$
(2) $[A^-] = [H^+]$	$\Rightarrow K_a = \dfrac{[H^+][H^+]}{C_{HA} - [H^+]}$

The first of these substitutions is based on the mass balance equation for all species that are related to the weak acid, where the total concentration of this acid (C_{HA}) is equal to the sum of the concentrations of its acid and base forms ($[HA]$ and $[A^-]$), or $C_{HA} = [HA] + [A^-]$. The second substitution makes use of the fact that we have a concentrated solution and started with only HA being added to the solution. According to the reaction in Equation 8.31, this fact means that every mole of A^- that forms will be accompanied by the production of one mole of H^+ (where we are assuming that the production of H^+ from water is negligible for this solution). When we make these substitutions and rearrange, we get the following equivalent expressions.

For a weak monoprotic acid: $K_a = \dfrac{[H^+]^2}{C_{HA} - [H^+]}$

or $\qquad 0 = [H^+]^2 + K_a[H^+] - K_a C_{HA} \qquad (8.32)$

If we know K_a and C_{HA} for our weak acid, we can use the quadratic equation to solve for $[H^+]$ in Equation 8.32, where $x = [H^+]$, A = 1, B = K_a, and C = $-K_a C_{HA}$. If we prepared a solution that had a total concentration of 0.10 M acetic acid ($K_a = 1.75 \times 10^{-5}$), for example, the estimated

value of $[H^+]$ from Equation 8.32 would be 1.3×10^{-3} M, or pH = $-\log(1.31 \times 10^{-3}) = 2.88$.

EXERCISE 8.5 The pH of a Weak Acid Solution

Hydrofluoric acid (HF) is a weak monoprotic acid that is formed when sulfuric acid reacts with the mineral fluorite (CaF_2). An environmental chemist wants to the study the effects of dilute HF solutions on other minerals. To do this, this chemist prepares a 0.010 M solution of HF in water. What is the expected pH of this solution at 25°C?

SOLUTION

The total analytical concentration (C_{HA}) of HF is known to be 0.010 M and the reported K_a for this acid is 6.8×10^{-4} at 25°C, as given in Table 8.1. We can use this information with Equation 8.32 to estimate the hydrogen ion concentration and the pH of the HF solution.

$$K_a = \frac{[H^+]^2}{C_{HA} - [H^+]}$$

$$\Rightarrow 6.8 \times 10^{-4} M = \frac{[H^+]^2}{1.0 \times 10^{-2} M - [H^+]}$$

or

$$0 = [H^+]^2 + (6.8 \times 10^{-4} M)[H^+] -$$
$$(6.8 \times 10^{-4} M)(1.0 \times 10^{-2} M)$$

We can solve for $[H^+]$ in this case by using the quadratic formula, where $x = [H^+]$, A = 1, B = 6.8×10^{-4} M and C = $-(6.8 \times 10^{-4} M)(1.0 \times 10^{-2} M)$. This approach gives a final answer of $[H^+] = 2.29 \times 10^{-3}$ M, or pH = **2.64**.

The process for determining the pH of a nondilute solution for a weak base is similar to the one we just used for a weak acid, but now we make use of K_b instead of K_a in describing the equilibrium in this system. We also need to use the fact that $K_w = [H^+][OH^-]$ in water to arrive at a final equation that can be solved for $[H^+]$, as given below.

For a weak monoprotic base:

$$K_b = \frac{[OH^-]^2}{C_B - [OH^-]} = \frac{(K_w/[H^+])^2}{C_B - K_w/[H^+]}$$

or

$$0 = K_b C_B[H^+]^2 + K_b K_w[H^+] - K_w^2 \qquad (8.33)$$

In these calculations it is important to remember that a systematic error is introduced whenever we use $[H^+]$ in place of hydrogen ion activity to estimate the pH. As we saw in Table 8.5, $pH = -\log[H^+]$ is a reasonable approximation for dilute solutions, but it can lead to problems as the ionic strength increases. The size of this error at various ionic strengths is actually the same for weak acids and bases as it is for strong acids and bases. If the acid or base is the only source of ions in our solution, a much higher concentration of a weak acid or base is needed to reach any given ionic strength because these have less dissociation in water than strong acids and bases. Thus, the use of $[H^+]$ in place of activity for pH calculations tends to be an even better approximation for weak acid–base solutions than it is for strong acids and bases.

Another source of errors in these pH calculations can occur if we incorrectly assume that water autoprotolysis is an insignificant source of H^+ or OH^- compared to our weak acid or weak base. We can use the same guideline that we used for strong acids and bases (a minimum acid or base concentration of $10^{-6}\ M$ to ignore autoprotolysis). Because weak acids and bases do not totally dissociate, however, the effects of autoprotolysis can sometimes be important at higher concentrations. This is especially true if the value of K_a or K_b for the acid or base is 10^{-6} or smaller. Some equations that can be used to calculate the pH in this situation can be found in the problems at the end of this chapter.

Mixtures of Conjugate Acids and Bases. Until now the solutions we have been examining have had pH values that were determined by the amount of acid or base we placed into a solution and the K_a or K_b values for these compounds. However, what do we do if we want to adjust the pH to some other value? This adjustment in pH can be accomplished by putting into the solution a known amount of the conjugate acid or base for our system. If we want to alter the pH of a solution containing acetic acid, we could add some acetate (the conjugate base of acetic acid). If we want to alter the pH of an ammonia solution, we could add NH_4^+ (the conjugate acid of ammonia).

We can look at this type of solution by using a monoprotic weak acid/conjugate base mixture as an example. If we ignore the dissociation of water, the main acid–base reaction will be as follows.

$$HA \rightleftharpoons H^+ + A^- \qquad K_a = \frac{[H^+][A^-]}{[HA]} \qquad (8.34)$$

We can now use some other equations and make substitutions into Equation 8.34 to help us solve for $[H^+]$. In this case, we know the total added concentrations of the acid and its conjugate base, which we will represent by the terms C_{HA} and C_{A^-}. We will assume that these concentrations are high enough that water can be ignored as the source of hydrogen ions in this solution. By looking at the reaction in Equation 8.34, we can see that there will be one mole of HA consumed for every mole of H^+ that is produced. Due to this relationship, if we originally have a total concentration for HA of C_{HA}, the remaining value for $[HA]$ at equilibrium will be $C_{HA} - [H^+]$. This gives us our first substitution that can be made into Equation 8.34.

Substitution	Change to Equation 8.34
(1) $[HA] = C_{HA} - [H^+]$ $\Rightarrow$	$K_a = \dfrac{[H^+][A^-]}{C_{HA} - [H^+]}$
(2) $[A^-] = (C_{A^-} + [H^+])$ $\Rightarrow$	$K_a = \dfrac{[H^+](C_{A^-} + [H^+])}{C_{HA} - [H^+]}$

We also know that each mole of hydrogen ions that is produced from HA will lead to the formation of one mole of A^-. A result of this relationship is that that the concentration of A^- at equilibrium will be given by $[A^-] = C_{A^-} + [H^+]$, which is used to make a second substitution. Equation 8.35 shows the final result that is obtained.

For a mixture of a weak monoprotic acid and its conjugate base:

$$K_a = \frac{[H^+](C_{A^-} + [H^+])}{C_{HA} - [H^+]}$$

or

$$0 = [H^+]^2 + (C_{A^-} + K_a)[H^+] - K_a C_{HA} \qquad (8.35)$$

We now have an equation in terms of $[H^+]$ that we can solve by the quadratic formula, where $x = [H^+]$, $A = 1$, $B = (C_{A^-} + K_a)$, and $C = -K_a C_{HA}$. Although we could derive a similar equation for a mixture of a weak monoprotic base with its conjugate acid, Equation 8.35 can also be used in this situation if we let the original concentration of the base be C_{A^-} and the original concentration of the conjugate acid be C_{HA}. This process requires that we have the value of K_a for the conjugate acid, as can be obtained by using Equation 8.17 (which states that $K_w = K_a \cdot K_b$ in water, or $K_a = K_w/K_b$).

EXERCISE 8.6 **Determining the pH of an Acid–Conjugate Base Mixture**

An aqueous solution is prepared that initially contains $1.5 \times 10^{-3}\ M$ acetic acid and $0.50 \times 10^{-3}\ M$ acetate. What is the expected pH of this mixture?

SOLUTION

The original concentrations of acetic acid and acetate are given (C_{HA} and C_{A^-}) and we can look up the acid dissociation constant for this reaction ($K_a = 1.75 \times 10^{-5}\ M$). If we ignore the acid–base dissociation of water (a good assumption, given the relatively high concentrations of the acetic acid and acetate), this information can be placed into Equation 8.35 to obtain the pH for the solution.

$$K_a = \frac{[H^+](C_{A^-} + [H^+])}{C_{HA} - [H^+]}$$

$$\Rightarrow 1.75 \times 10^{-5}\ M = \frac{[H^+](0.50 \times 10^{-3}M + [H^+])}{1.5 \times 10^{-3}\ M - [H^+]}$$

or

$$0 = [H^+]^2 + (0.50 \times 10^{-3}\ M + 1.75 \times 10^{-5}\ M)[H^+] -$$

$$(1.75 \times 10^{-5}\ M)(1.5 \times 10^{-3}\ M)$$

When we use the quadratic formula with this expression, we get $[H^+] = 4.6\underline{5} \times 10^{-5}\ M$, or pH = **4.33**.

If we had instead prepared a solution with the same total concentration ($2.0 \times 10^{-3}\ M$) of either acetic acid or acetate, the result would have been a pH of **3.75** for the acetic acid solution (determined using Equation 8.32) or a pH of **8.03** for the acetate solution (calculated using Equation 8.33). It is important to notice that the pH of the acid–conjugate base mixture falls between the pH values expected for the acid or conjugate base alone at the same total concentration. This feature will become important in the next section as we see how such mixtures can be used to control and adjust the pH of a solution.

8.5 BUFFERS AND POLYPROTIC ACID–BASE SYSTEMS

8.5A Buffer Solutions

What Is a Buffer?   We saw in the last exercise how it is possible to control the pH of a solution by using mixtures of acids and their conjugate bases. This type of mixture is known as a **buffer solution** (or "buffer") and results in a solution that will tend to keep the same pH, even when small amounts of additional acid, base, or water are added to it.[4] This property makes buffers important tools in controlling pH during chemical experiments and analytical measurements.

To illustrate why we might use a buffer solution, we can go back to the mixture of acetic acid and acetate that we considered in Exercise 8.6. In this exercise, we had determined that a solution that contained an initial concentration of $1.5 \times 10^{-3}\ M$ acetic acid and $0.50 \times 10^{-3}\ M$ acetate would have a final pH of 4.33. If we look closely at this answer we can see that the concentration of hydrogen ions produced by the mixture ($4.65 \times 10^{-5}\ M$) was small compared to the original concentrations of acetic acid and acetate. This result means the final concentrations of acetic acid and acetate are approximately the same as their initial concentrations. Furthermore, if we were to add a small amount of additional acetic acid or acetate to this mixture, the solution will still have about the same pH. In other words, we have created a solution with a pH that is relatively immune to small changes in its composition.

Almost any weak acid–conjugate base mixture can be used as a buffer solution (the water/OH$^-$ system being one exception). These buffers will differ, however, in terms of the pH range over which they will control the pH of a solution. Some common buffers are shown in Table 8.6 (also see Box 8.1 on "Preparing Buffers"). The center of the optimum pH range for a buffer occurs at the point where there are equal concentrations of the acid and conjugate base. This

TABLE 8.6   Buffers Used as Primary Standards*

Type of Buffer	Contents	Buffer Concentration	Density (g/mL at 25°C)	pH (at 25°C)	Change in pH with Temperature (dpH/dT)	Buffer Index (dB/dpH)
Tartrate	Saturated solution of potassium tartrate ($KHC_4H_4O_6$)	0.0341 m tartrate or 0.034 M tartrate	1.0036	**3.557**	−0.0014 pH units/°C	0.027 eq/pH unit
Phthalate	Solution containing 10.12 g/L potassium phthalate($KHC_8H_4O_4$)	0.05 m phthalate or 0.04958 M phthalate	1.0017	**4.008**	+0.0012 pH units/°C	0.016 eq/pH unit
Phosphate D	Solution containing 3.39 g/L KH_2PO_4 + 3.53 g/L Na_2HPO_4	0.050 m phosphate[a] or 0.04980 M phosphate	1.0028	**6.865**	−0.0028 pH units/°C	0.029 eq/pH unit
Phosphate E	Solution containing 1.179 g/L KH_2PO_4 + 4.30 g/L Na_2HPO_4	0.039125 m phosphate[b] or 0.038985 M phosphate	1.0020	**7.413**	−0.0028 pH units/°C	0.016 eq/pH unit
Borate	Solution containing 3.80 g/L sodium borate decahydrate ($Na_2BO_4 \cdot 10\ H_2O$)	0.010 m borate or 0.009971 M borate	0.9996	**9.180**	−0.0082 pH units/°C	0.020 eq/pH unit

*The information in this table was obtained from J. Bates, *Research of the National Bureau of Standards*, 66A, (1962), 179, and R.M. C. Dawson *et al.*, 3rd ed., Clarendon Press, Oxford (1986), p. 421.

[a]The concentration of the individual phosphate species in Phosphate D buffer is 0.025 m or 0.02490 M for both $H_2PO_4^-$ and HPO_4^{2-}.

[b]The concentration of the individual phosphate species in Phosphate E buffer is 0.008685 m or 0.008665 M for $H_2PO_4^-$ and 0.03043 m or 0.03032 M for HPO_4^{2-}.

BOX 8.1
Preparing Buffers

Buffer solutions are important in many analytical methods. The appropriate choice of a buffer will be determined by several factors. The first of these factors is the pH at which we wish to perform an analysis. The desired buffer capacity, concentration of the buffer, and allowable changes in pH with temperature are other items to consider. Sometimes additional features are desired for the buffer. For instance, a volatile buffer is useful for methods that involve converting liquid-phase samples into a vapor (such as some types of mass spectrometry). Other analytical methods might require buffers to have such properties as a small background signal (like low absorbance of light or low fluorescence) and little or no interactions with analytes such as metal ions.

There are several ways to prepare a buffer. In some cases there are specific recipes available for making a buffer that will have a particular pH and known composition. Some examples of this are given in Table 8.6 for primary standard buffers that are recommended by the National Institutes of Standards and Technology. Many analytical facilities also have standard operating procedures that describe how to prepare buffers that are commonly used in their laboratory.

One way of preparing a buffer solution is to use the K_a values for the system to predict how much of the given weak acid and its conjugate base must be combined to produce the desired pH. This is the approach that was illustrated in Section 8.5A. However, you will often find with this approach that the actual pH of the solution is slightly different from the one that you calculated. This will occur if you are using an activity-based value for K_a from a reference table and have not accounted for how the activity coefficients of hydrogen ions and the acid or conjugate base in your buffer will vary with ionic strength. A correction for this can be performed as described in Chapter 6. Preparing a buffer at a temperature different from that at which the K_a value was determined will produce a similar systematic error. If a difference in temperature of only a few degrees is present, a correction for this error can be made by obtaining literature information on how K_a for the given buffer system is expected to vary with temperature (see the right-hand column in Table 8.6 as an example). For a larger difference in temperature, it is advisable to measure or obtain a K_a value

that is at the same temperature as that which will be used to prepare your buffer.

Even if the actual and expected pH for a buffer differ, a small adjustment in the pH of this buffer can easily be made by carefully adding a small volume of a relatively concentrated strong acid such as HCl (to lower the pH) or strong base such as NaOH (to raise the pH). Adding these solutions will also increase the volume of the solution. Thus, when you are working with a volumetric flask, this adjustment in pH should be made just before you add more solvent to bring the total volume of the buffer to the mark on the flask. It is important when using strong acids and bases to take care in handling these solutions and to avoid adding a large amount of either to your buffer. Adding large amounts of these adjusting solutions will add other ions to the buffer, raising the ionic strength and making it difficult to determine the exact composition of your final solution. It is for this reason that preparing a buffer by simply adding a strong acid or strong base to a solution that initially contains only a weak acid or base (rather than a mixture of the two) is usually not recommended.

Yet another way that a buffer can be prepared is to begin with two solutions: one containing only the weak acid and the other containing only its conjugate base, but with both having the same total molar concentration of these agents. These solutions can then be combined in various ratios until the desired pH is obtained. No additional ions are added to this system, because only the acid and its conjugate base (along with their associated counter ions) are present in these solutions. This approach also gives a buffer where the total concentration of the acid and conjugate base is always the same. Such a method avoids the use of strong acids and strong bases, because it is possible to adjust the pH of the mixture by simply adding more of the weak acid solution (to lower the pH) or more of the conjugate base (to raise the pH). An estimate of the volumes of the two solutions that are needed for this process can be made based on Equation 8.37. Highly detailed calculations looking at the effects of ionic strength and temperature are not as crucial in this particular approach as in the previous methods because of the ease with which the pH can be readjusted to a desired value by adding more of the acid or conjugate base solutions.

---

situation gives a pH equal to the pK_a for the weak acid. In addition, the pH range over which an acid and conjugate base mixture will act as a good buffer will extend both above and below this central point (in this case, given as the pH range over which the ratio of the acid-to-conjugate base concentrations varies from 10/1 to 1/10).

The Henderson–Hasselbalch Equation. Another way we can estimate the pH of many buffer solutions is to rearrange the acid dissociation constant expression for the conversion of a weak acid (HA) into its conjugate base (A⁻) so that this expression is now in terms of pH.

This task can be done by taking the logarithm of the acid dissociation constant expression, as shown below.

At equilibrium:

$$K_a = \frac{[H^+][A^-]}{[HA]} \Rightarrow \log(K_a) = \log(H^+) + \log(\frac{[A^-]}{[HA]})$$

$$\Rightarrow -\log(H^+) = -\log(K_a) + \log(\frac{[A^-]}{[HA]})$$

$$\therefore pH = pK_a + \log(\frac{[A^-]}{[HA]}) \tag{8.36}$$

The final expression shown in Equation 8.36 is known as the **Henderson–Hasselbalch equation**. This expression is named after U.S. and Danish biochemists Lawrence J. Henderson (1878–1942) and Karl A. Hasselbalch (1874–1962), whose work led to this equation as they tried to understand how buffers maintain the pH of biological systems.[4,19–23]

The Henderson–Hasselbalch equation is useful because it allows us to see how a change in the ratio $[A^-]/[HA]$ will affect the pH for a buffer solution. This equation also confirms several of our previous observations. First, it shows that the pH of our solution will equal the pK_a for our weak acid when the concentrations of the weak acid and conjugate base are equal, making $[A^-]/[HA] = 1$ and $\log([A^-]/[HA]) = 0$. This relationship also predicts that the pH range over which this mixture acts as a buffer will extend an equal distance both above and below the pK_a, as long as other competing acid–base processes are not present. For example, a pH that is 1.0 units below the pK_a for the weak acid will occur when $\log([A^-]/[HA]) = -1.0$ or $[A^-]/[HA] = 0.1$. A pH that is 1.0 units above the pK_a will occur when $\log([A^-]/[HA]) = +1.0$ or $[A^-]/[HA] = 10$. In general, for every tenfold change in the ratio $[A^-]/[HA]$, there will be a change of one unit in the pH.

EXERCISE 8.7   Using the Henderson–Hasselbalch Equation

An aqueous buffer for analyzing the various ions in rain is prepared that has equilibrium concentrations of $5.0 \times 10^{-5} M$ boric acid and $2.0 \times 10^{-5} M$ borate (the conjugate base of boric acid). Using the Henderson–Hasselbalch equation, estimate the pH of this solution at 25°C.

SOLUTION
According to Table 8.1, K_{a1} for boric acid at 25°C is 5.79×10^{-10}, or $pK_{a1} = 9.237$. If we ignore the effects of water dissociation, we can determine the pH of this solution by placing the equilibrium concentrations for boric acid (HA) and borate (A^-) into Equation 8.36.

$$pH = pK_a + \log\left(\frac{[A^-]}{[HA]}\right) = 9.237 + \log\left(\frac{2.0 \times 10^{-5} M}{5.0 \times 10^{-5} M}\right)$$

$$\therefore pH = 9.237 + -0.40 = \mathbf{8.84}$$

In this case, we would have obtained essentially the same result when using the exact approach in Equation 8.35, because the concentration of hydrogen ions produced or consumed in reaching equilibrium in this solution is small compared to the concentrations of our acid or conjugate base.

Although the Henderson–Hasselbalch equation can be quite useful when dealing with buffers, it is important to remember that there are several assumptions made in this equation.[21,22] One limitation of this equation is it ignores the effects of water as a source of hydrogen ions, which means the Henderson–Hasselbalch equation only applies to nondilute buffers. It is also essential to remember that this equation only describes the weak acid and its conjugate base when they are *at equilibrium*. This last fact needs to be emphasized because the Henderson–Hasselbalch equation is often used by scientists to determine how much weak acid and conjugate base should be combined to produce a buffer with a given pH. Although the relative concentrations of these two species may be about the same as their final equilibrium concentrations when the desired pH is near the pK_a for the weak acid, this will not be the case if the pH is significantly different from the pK_a. Thus, we must always check to make sure that the assumptions behind the Henderson–Hasselbalch equation hold true for a given solution when using this expression.

If these underlying assumptions are true, the Henderson–Hasselbalch equation can be used to give a first estimate of the relative amount of acid versus conjugate base that must be combined to produce a given pH for our buffer. This is accomplished by rearranging this equation as follows.

$$\log\left(\frac{[A^-]}{[HA]}\right) = pH - pK_a \quad \text{or} \quad \frac{[A^-]}{[HA]} = 10^{(pH-pK_a)} \quad (8.37)$$

If we wanted to prepare a buffer with boric acid and borate that had a final pH of 8.50, the ratio of $[A^-]/[HA]$ needed at equilibrium to obtain this pH would be $[A^-]/[HA] = 10^{(8.50-9.24)} = 0.12$. If we also know the total concentration of HA plus A^- that we want to have present in solution, we can determine the actual concentrations of HA and A^- needed at equilibrium to provide the given pH. In our example of a boric acid/borate buffer, let's say we want a pH 8.50 buffer that has an analytical concentration (C_{HA}) for all boric acid–borate species of $0.10\ M$, where $0.10\ M = [HA] + [A^-]$. By using the fact that $[A^-]/[HA] = 0.12$ at pH 8.50, the equilibrium concentration of boric acid in this solution would be approximately given by $0.10\ M = [HA] + 0.12\ [HA]$, or $[HA] = 0.089\ M$, and the equilibrium concentration of borate would be $0.10\ M = 0.089\ M + [A^-]$, or $[A^-] = 0.011\ M$. If we had instead solved this problem by using a modified version of Equation 8.35 (now solving for C_{HA} and C_{A^-} instead of $[H^+]$, as given in the problems at the end of this chapter), we would have found that the amount of boric acid and borate that was actually needed was $C_{HA} = 0.085\ M$ and $C_{A^-} = 0.015\ M$.

Buffer Capacity. Although the central pH at which a buffer will work depends mainly on the pK_a of the acid in this buffer, the pH range over which this buffer works will also depend on the concentrations of the acid and conjugate base. Higher concentrations lead to a slightly broader pH range that can be used with the buffer and lower concentrations result in a smaller pH range.

The ability of a buffer to protect a sample or solution against large changes in pH upon the addition of an acid or base is described by a term known as the **buffer capacity**.[11,24,25] The buffer capacity is defined as the moles of strong acid or strong base that must be added per liter of buffer to produce a pH change of 1.0 units in the buffer. As is shown in the following exercise, the buffer capacity can be found by using the same equations we have used for calculating the pH of a buffer, but with the goal now being to see how much acid or base must be added to give a change in pH of 1.0.

EXERCISE 8.8 | **Estimating Buffer Capacity**

Soil that contains a large amount of limestone (calcium carbonate, $CaCO_3$) has the ability to buffer the effects of acid rain. What is the buffer capacity when nitric acid is added to a 1.000 L carbonate buffer that has original concentrations of 0.100 M for both bicarbonate and its conjugate base, carbonate? How many moles of nitric acid would be needed to lower the pH by one unit for a 0.200 L portion of this buffer at 25°C?

SOLUTION

Table 8.1 gives the K_a for bicarbonate at 25°C as 4.69×10^{-11}, or $pK_a = 10.329$. We initially have $(0.100\ M)(1.000\ L) = 1.00 \times 10^{-1}$ mol of both bicarbonate and carbonate in the buffer. Because [HA] = [A⁻] in this original solution, we further know that the initial pH must be equal to the pK_a for this buffer (or pH = 10.329). For every x moles of nitric acid we add, we will convert one mole of carbonate into bicarbonate. We also know the maximum allowed drop in pH is 1.0 units, taking our solution from pH 10.329 to pH 9.329. This information can be used with the Henderson–Hasselbalch equation to estimate how many moles of HNO_3 will be required to give this pH change.

$$pH = 9.329 = 10.329$$
$$+ \log\left(\frac{0.100\ \text{mol A}^- - x\ \text{mol HNO}_3}{0.100\ \text{mol HA} + x\ \text{mol HNO}_3}\right)$$

$$10^{(9.329 - 10.329)} = 10^{(-1.000)}$$

$$= 0.100 = \frac{0.100\ \text{mol A}^- - x\ \text{mol HNO}_3}{0.100\ \text{mol HA} + x\ \text{mol HNO}_3}$$

$$0.100 \cdot (0.100\ \text{mol HA} + x\ \text{mol HNO}_3)$$
$$= 0.100\ \text{mol A}^- - x\ \text{mol HNO}_3$$

$$\therefore x = \textbf{0.125 mol HNO}_3 \textbf{ (for 1.00 L)}$$

For a 0.200 L portion of the same buffer, we would need $(0.125\ \text{mol HNO}_3)(0.200\ L/1.00\ L) = \textbf{0.025 mol HNO}_3$ to

---

give the same pH change. If we had started with a higher buffer concentration, more HNO_3 would be needed to give the same change in pH, giving a higher buffer capacity. Similarly, a lower concentration buffer would have been able to handle only smaller amounts of acid and would have had a smaller buffer capacity.

A closely related term to the buffer capacity is the *buffer index* (β).[9,24] The buffer index is a more general term that is equal to the moles of a strong acid or strong base that are required to produce a given change in pH per unit volume. The buffer index is represented by Equation 8.38,

$$\text{Buffer index } (\beta) = -\frac{d\text{A}}{d\text{pH}} = \frac{d\text{B}}{d\text{pH}} \qquad (8.38)$$

where $-d\text{A}/d\text{pH}$ and $d\text{B}/d\text{pH}$ represent the change in the amount of acid or base that must be added to give a particular change in pH.[9,24] Figure 8.7 shows how the buffer index changes with pH for a typical monoprotic acid buffer, such as a mixture of acetic acid and acetate. One thing we can see from this figure is that the buffer index is largest at the point where the pH = pK_a for the buffer system. This maximum reflects the fact that a buffer will be most resistant to the addition of an acid or base at this pH. The buffer index also increases as we go beyond the working range of the buffer and to very acidic or basic

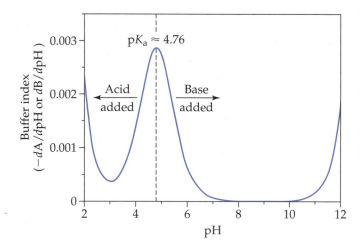

FIGURE 8.7 Change in the buffer index (–dA/dpH, or dB/dpH) vs. pH for an acetic acid/acetate buffer. The pK_a for acetic acid is included for reference. This plot was prepared using a total initial concentration of 0.10 M for all acetate-related species and an initial buffer volume of 0.050 L, to which is added various volumes of a 0.10 M solution containing a strong acid or strong base (such as HCl or NaOH). The values calculated here for dA and dB in the buffer index are based on the change in moles for the added strong acid or strong base; however, the same general plot (but with a different scale on the y-axis) would be obtained if dA or dB were instead determined by using the change in volume or concentration of an added strong acid or strong base. The rise in the buffer index at low or high pH values is due to the presence of excess strong acid or strong base in these regions. A value of $pK_a \approx 4.76$ was used in constructing this plot; the actual pK_a value for acetic acid is 4.757 at 25°C.

pH values. This increase is a reflection of the high concentrations of hydrogen ions that are present in water at a low pH and the high concentrations of hydroxide ions that are present at a high pH. Even if no other buffer components are present, these ions can be produced from water itself to now resist changes in pH as a small amount of acid or base is added to these highly acidic or basic solutions.

8.5B Polyprotic Acid–Base Systems

What Is a Polyprotic System?   Up to this point we have considered only acids and bases that can donate or accept a single hydrogen ion, but it is also possible for an acid or base to be able to give up or accept more than one hydrogen ion. Substances that can act in this manner are called **polyprotic acids** or **polyprotic bases**. An example is carbonic acid, H_2CO_3, which is produced when acid rain reacts with limestone or marble (see Figure 8.8). Although carbonic acid can undergo a dehydration reaction to release water and carbon dioxide, it is also an acid that contains two hydrogens that can be donated to a base. The result is a series of acid–base reactions that can produce a mixture of carbonic acid plus its monobasic form (bicarbonate, HCO_3^-) and its dibasic form (carbonate, CO_3^{2-}).

$$2\,H^+ + CaCO_3 \rightleftharpoons H_2CO_3 + Ca^{2+}$$

$$\Updownarrow$$

$$H_2O + CO_{2(g)}$$

FIGURE 8.8   The effect of acid rain on marble antefixes located on the roof of the Philadelphia Merchants' Exchange building (built, 1832). Acid rain has caused the originally sharp edges of these structures to gradually become rounded and less distinct. The reactions shown are those that occur between the nitric acid in acid rain and calcium carbonate ($CaCO_3$), which is present in marble and limestone. (The image of the statue is reproduced with permission from the USGS and are by E. McGee.)

Polyprotic acids and bases are quite common in nature. These can be small inorganic and organic compounds like carbonic acid and citric acid, or they can be biomolecules that range in size from individual amino acids to larger agents like peptides and proteins. One feature all of these compounds have in common is that their individual acid or base sites can be described using the same types of acid–base reactions that we use for monoprotic acids and bases. However, these compounds do differ from monoprotic acids and bases in that we now have two or more sites that can act as an acid or base, with each site having its own K_a or K_b value. This idea is illustrated in Table 8.1 for carbonic acid, where the loss of the first hydrogen occurs with a K_a of 4.46×10^{-7} and the loss of the second hydrogen has a K_a of 4.69×10^{-11}. Notice that as we lose each additional hydrogen ion, the K_a value decreases. This trend indicates that it becomes more difficult for the acid to donate hydrogen ions as fewer of these become available.

Fractions of Species.   When we place a polyprotic acid or base into solution, the presence of several possible acid dissociation or base ionization steps will also produce a mixture of many acid or base forms that are created from our original agent. When a chemical exists in multiple forms, we learned earlier that these various forms are referred to as *chemical species*.

To help predict the behavior of a polyprotic acid or base, it is useful to determine the fraction of this compound that will be present in each of its various possible species at a given pH or other set of conditions. We can illustrate this process by using a monoprotic acid as an example. First, we know from Section 8.4B that this type of compound will be present at equilibrium in either a protonated form (HA) or deprotonated form (A^-), with the ratio of these forms varying with pH. Second, we can describe the relative fractions (α_{HA} and α_{A^-}) of our original compound that is in these two forms, as shown below, where C_{HA} is the analytical concentration of our compound,

$$\text{Fraction of HA: } \alpha_{HA} = \frac{[HA]}{C_{HA}} = \frac{[HA]}{[HA] + [A^-]} \quad (8.39)$$

$$\text{Fraction of } A^-: \alpha_{A^-} = \frac{[A^-]}{C_{HA}} = \frac{[A^-]}{[HA] + [A^-]} \quad (8.40)$$

where $1 = \alpha_{HA} + \alpha_{A^-}$. Third, we can simplify these expressions by dividing the numerators and denominators by [HA] and using the fact that $[A^-]/[HA] = K_a/[H^+]$ from the acid dissociation expression for a monoprotic acid. This process allows us to rewrite these equations in the following forms.

$$\alpha_{HA} = \frac{1}{1 + K_a/[H^+]} \quad \text{or} \quad \alpha_{HA} = \frac{[H^+]}{[H^+] + K_a} \quad (8.41)$$

$$\alpha_{A^-} = \frac{K_a/[H^+]}{1 + K_a/[H^+]} \quad \text{or} \quad \alpha_{A^-} = \frac{K_a}{[H^+] + K_a} \quad (8.42)$$

In this book, this type of relationship will be referred to as a *fraction of species equation*, because it can be used to show how the fraction of a chemical species will change as a given parameter is varied for a chemical system. For example, Equations 8.41 and 8.42 indicate that the relative fraction of HA and A^- will depend only on the hydrogen ion concentration (or pH) and the acid dissociation constant for HA. In addition, neither of these fractions (α_{HA} and α_{A^-}) have any units because they are simply the ratio of the chemical concentration of one species (HA or A^-) versus the concentration of all other related species (HA and A^- together) in this same system.

This same idea can be extended to polyprotic acids and bases, as is illustrated in Table 8.7 for carbonic acid. Although it can be relatively tedious to derive equations like those in Table 8.7 from scratch, this process can be greatly simplified by using the fact that the final expressions fall into a well-defined pattern. For instance, Table 8.7 shows that the denominator will have $n+1$ terms for a compound that can lose up to n hydrogen ions (giving three terms for carbonic acid, which can lose up to two H^+). You can also see that each of these terms in the

denominator represents one of the possible species that will be formed by the acid or base. In the case of carbonic acid, the most acidic of these compounds (H_2CO_3) is represented by the term $[H^+]^2$ on the left and the most basic of these compounds (CO_3^{2-}) is given by the term $K_{a1} K_{a2}$ on the right. As we go from left to right, each term has a decrease in the exponent on $[H^+]$ by one unit, and gains one more K_a term beginning with K_{a1} and moving on to $K_{a1}K_{a2}$. Knowing this pattern, it is easy to write the denominator for the fraction of species equation for any compound that is related to a polyprotic acid. Also, once you have this denominator, you simply have to place each of its individual terms in the numerator to get the fraction of species equations for all the desired components of your system.

EXERCISE 8.9   Writing the Fraction of Species Equations for a Polyprotic Acid

Write equations for the fractions of all species that form when phosphoric acid is placed in water.

SOLUTION

Phosphoric acid is a triprotic acid with three hydrogen ions it can donate to a base. This property gives phosphoric acid four possible acid–base forms that can exist in

---

TABLE 8.7   Derivation of Fraction of Species Equations for Carbonic Acid

Fraction of H_2CO_3 $(\alpha_{H_2CO_3})$ $= \dfrac{[H_2CO_3]}{C_{Carbonate}} = \dfrac{[H_2CO_3]}{[H_2CO_3] + [HCO_3^-] + [CO_3^{2-}]}$

$$= \frac{1}{1 + [HCO_3^-]/[H_2CO_3] + [CO_3^{2-}]/[H_2CO_3]}$$

$$\therefore \quad \alpha_{H_2CO_3} = \frac{[H^+]^2}{[H^+]^2 + K_{a1}[H^+] + K_{a1} K_{a2}}$$

Fraction of HCO_3^- $(\alpha_{HCO_3^-})$ $= \dfrac{[HCO_3^-]}{C_{Carbonate}} = \dfrac{[HCO_3^-]}{[H_2CO_3] + [HCO_3^-] + [CO_3^{2-}]}$

$$= \frac{[HCO_3^-]/[H_2CO_3]}{1 + [HCO_3^-]/[H_2CO_3] + [CO_3^{2-}]/[H_2CO_3]}$$

$$\therefore \quad \alpha_{HCO_3^-} = \frac{K_{a1}[H^+]}{[H^+]^2 + K_{a1}[H^+] + K_{a1} K_{a2}}$$

Fraction of CO_3^{2-} $(\alpha_{HCO_3^{-2}})$ $= = \dfrac{[CO_3^{2-}]}{C_{Cabonate}} = \dfrac{[CO_3^{2-}]}{[H_2CO_3] + [HCO_3^-] + [CO_3^{2-}]}$

$$= \frac{[HCO_3^-]/[H_2CO_3]}{1 + [HCO_3^-]/[H_2CO_3] + [CO_3^{2-}]/[H_2CO_3]}$$

$$\therefore \quad \alpha_{CO_3^{2-}} = \frac{K_{a1} K_{a2}}{[H^+]^2 + K_{a1}[H^+] + K_{a1} K_{a2}}$$

where: $[HCO_3]/[H_2CO_3^-] = K_{a1}/[H^+]$

$$[CO_3^{2-}]/[H_2CO_3] = ([CO_3^{2-}]/[HCO_3^-]) \cdot ([HCO_3^-]/[H_2CO_3])$$

$$= (K_{a1}/[H^+])(K_{a2}/[H^+]) = K_{a1}K_{a2}/[H^+]^2$$

solution. Using the patterns we saw for carbonic acid, the fractions of species equations for phosphoric acid and its various forms would be as follows.

Fraction of H_3PO_4: $\alpha_{H_3PO_4} =$

$$\frac{[H^+]^3}{[H^+]^3 + K_a[H^+]^2 + K_{a1}K_{a2}[H^+] + K_{a1}K_{a2}K_{a3}}$$

Fraction of $H_2PO_4^-$: $\alpha_{H_2PO_4^-} =$

$$\frac{K_{a1}[H^+]^2}{[H^+]^3 + K_{a1}[H^+]^2 + K_{a1}K_{a2}[H^+] + K_{a1}K_{a2}K_{a3}}$$

Fraction of HPO_4^{2-}: $\alpha_{HPO_4^{2-}} =$

$$\frac{K_{a1}K_{a2}[H^+]}{[H^+]^3 + K_{a1}[H^+]^2 + K_{a1}K_{a2}[H^+] + K_{a1}K_{a2}K_{a3}}$$

Fraction of PO_4^{3-}: $\alpha_{PO_4^{3-}} =$

$$\frac{K_{a1}K_{a2}K_{a3}}{[H^+]^3 + K_{a1}[H^+]^2 + K_{a1}K_{a2}[H^+] + K_{a1}K_{a2}K_{a3}}$$

where $1 = \alpha_{H_3PO_4} + \alpha_{H_2PO_4^-} + \alpha_{HPO_4^{2-}} + \alpha_{PO_4^{3-}}$. The last equation in this series is a reminder that the total fraction of all species related to phosphoric acid must be equal to one (a type of mass balance equation in which the amount of these chemical species is expressed in terms of a fraction rather than a concentration).

A useful application of the fraction of species equations for a polyprotic acid is it allows us to see how the relative amounts of the products change with pH. This goal can be accomplished by using a *fraction of species plot*, as illustrated in Figure 8.9 for carbonic acid. This diagram contains a definite pattern in terms of which species dominates at a given pH. This pattern becomes clear when you place the pK_a values for carbonic acid on this plot. What this diagram shows is that carbonic acid is the main species at pH values below pK_{a1}. Furthermore, HCO_3^- is the main species at pH values between pK_{a1} and pK_{a2}, and CO_3^{2-} is the main species at a pH above pK_{a2}. This is the same pattern (moving from the most acidic species to the most basic as pH is increased) that is seen for any polyprotic acid that has a relatively large difference in its pK_a values. Knowing this trend will become important later when we discuss how to determine the pH of a polyprotic acid solution.

Another thing you may have wondered is what to do if you have a polyprotic base instead of an acid. The answer is that the same procedure is used. The only item to keep in mind when beginning with K_b values is to use these to first determine the K_a values for the corresponding conjugate acids. You can then treat this system in the same manner as a polyprotic weak acid.

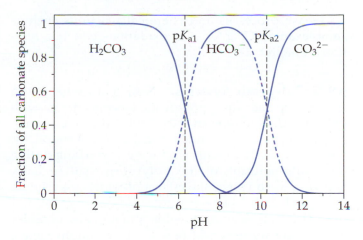

FIGURE 8.9   Fraction of species plot for carbonic acid, bicarbonate, and carbonate in water.

Fraction of species plots and relationships like those in Equations 8.41 and 8.42 are valuable in dealing with acid–base processes that occur with other side reactions. We saw an example of this application in Chapter 7 when we looked at the effect of pH on the solubility of barium sulfate. One item that was of interest in this earlier problem was how a change in pH would affect the fraction of dissolved sulfate that was present as SO_4^{2-}, or $\alpha_{SO_4^{2-}}$. As shown in Table 8.1, this involves the acid dissociation of H_2SO_4, a polyprotic acid that can lose one hydrogen ion to form HSO_4^- (an essentially complete reaction with a large value for K_{a1}), followed by the loss of a second hydrogen ion to give SO_4^{2-} ($K_{a2} = 1.03 \times 10^{-2}$). We now know that the value of $\alpha_{SO_4^{2-}}$ at a given pH can be found by using the following fraction of species equation for SO_4^{2-}.

$$\alpha_{SO_4^{2-}} = \frac{K_{a1}K_{a2}}{[H^+]^2 + K_{a1}[H^+] + K_{a1}K_{a2}}$$

$$\approx \frac{K_{a2}}{[H^+] + K_{a2}}$$

$$\text{(for } K_{a1} \gg K_{a2}) \tag{8.43}$$

If we place into these expressions $K_{a1} = 98$ and $K_{a2} = 1.03 \times 10^{-2}$ (as obtained from the footnote of Table 8.1) along with $[H^+] = 3.2 \times 10^{-2}\,M$, we get a value of 0.24 for $\alpha_{SO_4^{2-}}$ at pH 1.5, the same number we used in Chapter 7 to help estimate the solubility of barium sulfate at this pH.

Expressions like those in Equations 8.41–8.43, Table 8.7, and Exercise 8.9 can be utilized in examining the combined effects of acid–base reactions and other processes as long as the acid–base reaction is at *equilibrium*. Fortunately, this is often true because many acid–base processes have fast reaction rates. We will later see other examples of how we can use fraction of species equations to solve problems involving acid–base processes and chemical reactions such as complexation (Chapter 9) or oxidation–reduction reactions (Chapter 10).

pH in Polyprotic Systems.   Now that we have seen how the species for a polyprotic system change with pH, let's see how we can estimate the pH of a solution containing such compounds. For the sake of simplicity, we will initially assume that we are working with a nondilute solution of the polyprotic acid or base, allowing us to ignore the effects of water dissociation on the pH.

There are several possible situations to consider even under these conditions. First, we might have a solution where we begin with only the most acidic form of our agent. In the case of carbonic acid, this situation would occur if we added only H_2CO_3 to water. Once in water, this acid can donate hydrogen ions to water to first form HCO_3^- and then CO_3^{2-}, but (as shown in Figure 8.9) the second of these two processes (the formation of CO_3^{2-}) does not occur at any significant levels in the pH range where carbonic acid (our starting form) is the dominant species. Thus, it is safe in this case to consider only the formation of bicarbonate from carbonic acid and use this reaction alone to estimate the hydrogen ion concentration of our solution. In other words, we can treat the carbonic acid as if it is a weak monoprotic acid in this situation. This is accomplished by using the approach described in Section 8.4A.

A second but similar situation occurs when we prepare a solution that initially has only the most basic form of a polyprotic acid–base system. This situation occurs in Figure 8.9 when we add only the carbonate to water. As the carbonate accepts hydrogen ions from water, it will form a small amount of bicarbonate, and an insignificant amount of carbonic acid. As a result we can determine the pH of this solution by treating carbonate as a weak monoprotic base, again using the techniques we learned in Section 8.4A.

A third situation that we can have with a polyprotic acid–base system is when we prepare a solution that contains a mixture of two closely related acid–base forms. This situation occurs if we make a solution that contains carbonic acid and its conjugate base bicarbonate, which are linked through the first K_a value for carbonic acid. The same type of situation occurs when we mix bicarbonate and its conjugate base, carbonate. If we are dealing with a relatively concentrated solution, allowing us to ignore the effects of water dissociation, we can find the pH of these mixtures using the same approach that we learned earlier for mixtures of monoprotic acids and their conjugate bases. As an example, the pH for a mixture of carbonic acid and bicarbonate would be determined by using Equation 8.35, where the value of C_{HA} would be the analytical concentration of carbonic acid and C_{A^-} is the analytical concentration of bicarbonate and K_a is equal to K_{a1}.

What is the expected pH for solutions that contain the following initial compositions? (*Note:* You can assume in each case that the autoprotolysis of water can be ignored as a source of hydrogen ions.)

a. 1.0×10^{-5} M carbonic acid in water
b. 1.0×10^{-6} M sodium carbonate in water
c. 1.0×10^{-5} M sodium bicarbonate plus 1.0×10^{-6} M sodium carbonate

SOLUTION

a. The pH of this solution can be determined by using Equation 8.32 for a monoprotic acid, where $C_{HA} = 1.0 \times 10^{-5}$ M, and K_a is equal to $K_{a1} = 4.46 \times 10^{-7}$ M for the carbonate system. By using the quadratic formula with Equation 8.32, we can solve for $[H^+]$. This process gives $[H^+] = 1.90 \times 10^{-6}$ M, or **pH = 5.72**.

b. In this situation we are beginning only with a solution of the most basic component of our polyprotic system, carbonate. We can find the pH by using Equation 8.33 and treating this situation as a solution of a simple monoprotic base. In doing so, we let $K_b = K_w/K_{a2} = (1.0 \times 10^{-14})/(4.69 \times 10^{-11}) = 2.13 \times 10^{-4}$ and $C_B = 1.0 \times 10^{-6}$ M. We can then solve for $[H^+]$ in $0 = K_b C_B [H^+]^2 + K_b K_w [H^+] - K_w^2$ by employing the quadratic formula, which gives $[H^+] = 4.7 \times 10^{-11}$ M, or **pH = 10.33**.

c. The pH of this mixture can be found by using Equation 8.35 and treating this system as a mixture of a monoprotic acid (bicarbonate, in this case) and its conjugate base (carbonate). In using this equation, the values for C_{HA} and C_{A^-} would be 1.0×10^{-5} M and 1.0×10^{-6} M, and K_a would be given by $K_{a2} = 4.69 \times 10^{-11}$. When we put this information into Equation 8.35, we get the following,

$$0 = [H^+]^2 + (1.0 \times 10^{-6} M + 4.69 \times 10^{-11} M) \cdot [H^+] - (4.69 \times 10^{-11} M)(1.0 \times 10^{-5} M)$$

which can be solved with the quadratic formula to give $[H^+] = 4.7 \times 10^{-10}$ M, or **pH = 9.32**.

Another possible situation for a polyprotic system is a solution that originally contains one of the intermediate forms of the polyprotic system. For carbonic acid this situation would occur if we made a solution by adding only sodium bicarbonate to water. We now have two ways in which this compound can react, because it can either donate a hydrogen ion to water to form carbonate or accept a hydrogen ion from water to form carbonic acid. As we saw for water, this ability of bicarbonate or any other intermediate form of a polyprotic acid–base system makes such a compound amphiprotic, or able to act as either an acid or base. The plot in Figure 8.9 indicates that both reactions are equally likely to occur, so we must consider both in estimating the pH of our solution.

Although this is a more complex situation than those we have already discussed for polyprotic acid–base systems, it is still possible for us to estimate the pH of such a solution. To obtain this estimate, we must first begin by writing all of the reactions that involve our intermediate form, HA^-, when we place it as a sodium salt into water. (*Note*: This requires the reasonable assumption that all of the sodium salt completely dissolves to form sodium ions and HA^-.)

$$NaHA \rightleftharpoons Na^+ + HA^- \qquad (8.44)$$

$$HA^- \rightleftharpoons H^+ + A^{2-} \qquad K_{a2} = \frac{[H^+][A^{2-}]}{[HA^-]} \qquad (8.45)$$

$$HA^- + H_2O \rightleftharpoons H_2A + OH^-$$

$$K_{b2} = \frac{[H_2A][OH^-]}{[HA^-]} \qquad (8.46)$$

To obtain an additional equation to describe this system, we can write the charge balance equation given below (see Chapter 6 for a review of charge balance equations).

$$[H^+] + [Na^+] = [HA^-] + [OH^-] + 2[A^{2-}] \qquad (8.47)$$

But we also know that the concentration of sodium ions must be equal to the analytical concentration of HA^- (C_{HA^-}) that has been added to the solution. We also know the amount of HA^- remaining can be determined by subtracting the concentrations of A^{2-} and H_2A from this analytical concentration, or $[HA^-] = C_{HA^-} - [H_2A] - [A^{2-}]$. This fact allows us to obtain a simpler form for our equation.

$$[H^+] + C_{HA^-} = (C_{HA^-} - [H_2A] - [A^{2-}]) +$$

$$[OH^-] + 2[A^{2-}]$$

or

$$0 = [H^+] + [H_2A] - [OH^-] - [A^{2-}] \qquad (8.48)$$

We can now make some further substitutions to reduce the number of unknown terms in this equation. First, we can rearrange the equilibrium expression in Equation 8.45 to get $[A^{2-}] = K_{a2}[HA^-]/[H^+]$ and use the fact that $[OH^-] = K_w/[H^+]$. Second, we can convert the equilibrium expression in Equation 8.46 to give $[H_2A] = K_{b2}[HA^-]/[OH^-]$ or the equivalent relationship $[H_2A] = [HA^-][H^+]/K_{a1}$. When all of these substitutions are placed into Equation 8.48, we get the following result which is now written in terms of only equilibrium constants and the values for $[HA^-]$ and $[H^+]$.

$$0 = [H^+] + \frac{[HA^-][H^+]}{K_{a1}} - \frac{K_w}{[H^+]} - \frac{K_{a2}[HA^-]}{[H^+]} \qquad (8.49)$$

To determine the pH of this solution, we next need to rewrite Equation 8.49 in a form that might allow us to solve for $[H^+]$. This is accomplished by multiplying both sides of Equation 8.49 by $[H^+]$ and rearranging to solve for the concentration of hydrogen ions. Taking the square root of both sides of this last equation leads to the result shown in Equation 8.50.

$$[H^+] = \sqrt{\frac{K_{a1}(K_w + K_{a2}[HA^-])}{K_{a1} + [HA^-]}} \qquad (8.50)$$

At this point there are two ways that we can find the pH of this solution. The first is an exact approach in which we use a fraction of species equation to give $[HA^-] = \alpha_{HA^-} C_{HA^-} = \{K_{a1}[H^+]/([H^+]^2 + K_{a1}[H^+] + K_{a1}K_{a2})\}C_{HA^-}$. Although we can get an answer this way, the resulting equation is relatively complex and will require the use of successive approximations or computers to get at the final pH value. The second, and much simpler, approach is to use the fact that a solution containing only one initial form of a weak acid or base will usually have only a small amount of this species convert into other acid–base forms. This feature means we can often assume the concentration of HA^- in our final solution is approximately the same as its analytical concentration, or

Operation	Change made to Equation 8.49
(1) Multiply both sides by $[H^+]$	$\Rightarrow 0 = [H^+]^2 + \dfrac{[HA^-][H^+]^2}{K_{a1}} - \dfrac{K_w[H^+]}{[H^+]} - \dfrac{K_{a2}[HA^-]}{[H^+]}$
(2) Place all terms with $[H^+]^2$ on same side	$\Rightarrow [H^+]^2 + \dfrac{[HA^-][H^+]^2}{K_{a1}} = K_w + K_{a2}[HA^-]$
(3) Combine all terms containing $[H^+]^2$	$\Rightarrow [H^+]^2 \left(\dfrac{K_{a1} + [HA^-]}{K_{a1}}\right) = K_w + K_{a2}[HA^-]$
(4) Divide both sides by $(K_{a1} + [HA^-])/K_{a1}$	$\Rightarrow [H^+]^2 = \left\{\dfrac{K_{a1}(K_w + K_{a2}[HA^-])}{K_{a1} + [HA^-]}\right\}$

$[HA^-] \approx C_{HA^-}$. When we make this simplification, Equation 8.50 converts to the form shown below, which is now relatively easy to solve.

$$[H^+] \approx \sqrt{\frac{K_{a1}(K_w + K_{a2}C_{HA^-})}{K_{a1} + C_{HA^-}}} \qquad (8.51)$$

The following exercise illustrates how this relationship can be used to estimate the pH for such a system.

EXERCISE 8.11   **Determining the pH for a Sodium Bicarbonate Solution**

What is the expected pH for a solution prepared by mixing 1.0×10^{-3} M sodium bicarbonate in water if you assume the final concentration of bicarbonate is roughly equal to its analytical concentration?

SOLUTION

The pH for this solution can be estimated by using Equation 8.51, where $C_{HA^-} = 1.00 \times 10^{-3}$ M, $K_{a1} = 4.46 \times 10^{-7}$ and $K_{a2} = 4.69 \times 10^{-11}$ (where both K_{a1} and K_{a2} would have M as the apparent unit). The result is as follows.

$$[H^+] \approx \sqrt{\frac{K_{a1}(K_{a2}C_{HA^-} + K_w)}{K_{a1} + C_{HA^-}}} \approx$$

$$\left[\frac{(4.46 \times 10^{-7} M)(1.00 \times 10^{-14} M + 4.69 \times 10^{-11} M}{4.46 \times 10^{-7} M + 1.00 \times 10^{-3} M}\right.$$

$$\left.\frac{\cdot 1.00 \times 10^{-3} M)}{4.46 \times 10^{-7} M + 1.00 \times 10^{-3} M}\right]^{1/2}$$

$$\therefore \quad [H^+] = 5.03 \times 10^{-9} \text{ M} \quad \text{or} \quad \textbf{pH} = \textbf{8.30}$$

A more convenient form of Equation 8.51 can be obtained with a few simplifications. These simplifications can be made because many weak acid–base solutions will have some terms in Equation 8.51 that are much larger than others. Such a situation often occurs when $K_{a2}C_{HA^-} \gg K_w$ and $C_{HA^-} \gg K_{a1}$. Under these conditions, the right-hand term of Equation 8.51 reduces to $\sqrt{K_{a1}K_{a2}}$. If we then take the logarithm of both sides of this equation and substitute in pH = $-\log[H^+]$, $pK_{a1} = -\log(K_{a1})$ and $pK_{a2} = -\log(K_{a2})$, we get Equation 8.52.

$$pH \approx \frac{pK_{a1} + pK_{a2}}{2} \qquad (8.52)$$

This relationship is often useful in obtaining a first estimate of the pH for a solution made up of an amphiprotic compound like carbonate. For instance, the pH that would have been given by this equation in the preceding exercise would be 8.35, which is close to the value

we estimated using Equation 8.51. One benefit of Equation 8.52 is it is independent of the analytical concentration for a compound. However, this will only be true if the analytical concentration fits the simplifications we made in deriving Equation 8.52 (namely, $C_{HA^-} \approx [HA^-]$, $C_{HA^-} \gg K_w$ and $C_{HA^-} \gg K_{a1}$).

8.5C Zwitterions

The polyprotic acid–base systems we have looked at so far have all begun with a neutrally charged compound at a low pH, such as H_2CO_3 or H_3PO_4, and have moved to form species with more negative charges as we have increased the pH. An alternative situation that you might encounter is a polyprotic system where both acidic and basic groups are present on the same species. Some examples of such compounds are the amino acids, which have the following general structure.

All amino acids contain at least one acidic region (the terminal carboxylic acid group, —COOH) and at least one region that can act as a base (the terminal amine group, —NH₂). Many amino acids also have additional acidic or basic groups in their side chains, as occurs in lysine, arginine, and glutamic acid (see Table 8.8). Although amino acids are often drawn in a "neutral" form, as shown previously, the presence of the acidic and basic groups in amino acids does give amino acids regions that contain local charges even when the overall charge of amino acid is zero. A chemical that possesses a net charge of zero but contains groups with an equal number of negative and positive charges is known as a **zwitterion**.[26] Other zwitterions besides amino acids include peptides and proteins, which are both made from amino acids.

The presence of both basic and acidic groups means amino acids can have several species present at various pH values. An illustration of this idea is shown in Figure 8.10,

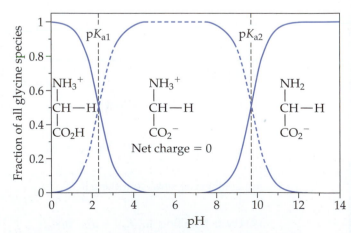

FIGURE 8.10   Fraction of species plot for glycine in water.

TABLE 8.8 Amino Acids and Their Acid Dissociation Constants

Amino Acid	Symbols[a]		Structure	pK_a (at 25°C)[b]		
				C-Terminus	N-Terminus	Side Chain
Alanine	A	Ala	NH₃⁺ | CH—CH₃ | CO₂H	2.34	9.87	
Arginine	R	Arg	NH₃⁺ | CH—CH₂CH₂CH₂NHC=NH₂⁺ / NH₂ | CO₂H	1.82	8.99	12.1[c,d]
Asparagine	N	Asn	NH₃⁺ O | || CH—CH₂CNH₂ | CO₂H	2.16[c]	8.73[c]	
Aspartic acid	D	Asp	NH₃⁺ | CH—CH₂CO₂H | CO₂H	1.99	10.00	3.90
Cysteine	C	Cys	NH₃⁺ | CH—CH₂SH | CO₂H	1.7[d]	10.74	8.36
Glutamic acid	E	Glu	NH₃⁺ | CH—CH₂CH₂CO₂H | CO₂H	2.16	9.96	4.30
Glutamine	Q	Gln	NH₃⁺ O | || CH—CH₂CH₂CNH₂ | CO₂H	2.19[c]	9.00[c]	
Glycine	G	Gly	NH₃⁺ | CH₂ (R=H) | CO₂H	2.35	9.78	
Histidine	H	His	NH₃⁺ | CH—CH₂—⟨NH / N⁺H⟩ | CO₂H	1.6[d]	9.28	5.97
Isoleucine	I	Ile	NH₃⁺ | CH—CH(CH₃)CH₂CH₃ | CO₂H	2.32	9.76	

Continued

TABLE 8.8 **Amino Acids and Their Acid Dissociation Constants** Continued

Amino Acid	Symbols[a]		Structure	pK_a(at 25°C)[b]		
				C-Terminus	N-Terminus	Side Chain
Leucine	L	Leu	NH_3^+ / CHCH_2CH(CH_3)_2 / CO_2H	2.33	9.74	
Lysine	K	Lys	NH_3^+ / CHCH_2CH_2CH_2CH_2NH_3^+ / CO_2H	1.77[d]	9.07	10.82
Methionine	M	Met	NH_3^+ / CHCH_2CH_2SCH_3 / CO_2H	2.18[c]	9.08[c]	
Phenylalanine	F	Phe	NH_3^+ / CHCH_2— / CO_2H	2.20	9.31	
Proline	P	Pro	—CO_2H / N / $^+H_2$	1.95	10.64	
Serine	S	Ser	NH_3^+ / CHCH_2OH / CO_2H	2.19	9.21	
Threonine	T	Thr	NH_3^+ / CHCHOHCH_3 / CO_2H	2.09	9.10	
Tryptophan	W	Trp	NH_3^+ / CHCH_2— / CO_2H	2.37[c]	9.33[c]	
Tyrosine	Y	Tyr	NH_3^+ / CHCH_2—OH / CO_2H	2.24[c]	9.19[c]	10.47

TABLE 8.8   Amino Acids and Their Acid Dissociation Constants   Continued

Amino Acid	Symbols[a]		Structure	C-Terminus	N-Terminus	Side Chain
				pK_a(at 25°C)[b]		
Valine	V	Val	NH_3^+ | $CHCH(CH_3)_2$ | CO_2H	2.29	9.72	

[a]These symbols are often used when writing the sequences of amino acids in peptides and proteins. The single letter symbols are currently recommended for this purpose; however, the three letter symbols are more commonly found in older resources.

[b]These pKa values were obtained from NIST Standard Reference Database 46—*NIST Critically Selected Stability Constants for Metal Complexes Database, v. 8.0*, NIST, Gaithersburg, MD, 2004.

[c]These values are for an ionic strength of 0.1. All other listed pK_a values are for an ionic strength of 0.0.

[d]These values are listed in NIST database 46 as having questionable validity.

which shows how a change in pH alters the relative fractions of the three acid–base forms of glycine, the simplest amino acid. Notice in this case that the form present at a low pH actually has a charge of "+1," which is due to the hydrogen that has been added to the basic amine group of glycine at a low pH. As the pH is increased, the carboxylic acid group of glycine loses its hydrogen. The result is that one region on glycine has a +1 charge and a second region has a −1 charge, giving a net charge of zero. At an even higher pH, the hydrogen at the amine end is lost, giving glycine an overall charge of −1. A similar pattern, but with the possibility of more species and charge states, is seen for the other amino acids. Even more complex patterns exist for peptides and proteins, which are important structural and functional components of biological systems. This relationship between pH and charge is one reason why the control of pH (or the lack of control in the case of acid rain) is so essential for plants, animals, and other living organisms.

If you look closer at Figure 8.10, you will see that there is a point on this graph in which the average charge on all glycines will be exactly equal to zero. This special pH is known as the **isoelectric point (pI)**.[26] The pI for a zwitterion occurs in the pH range where the form with zero charge (which we will call HA) is the dominant form and where the relative amounts of the +1 and −1 forms (H_2A^+ and A^-) have identical concentrations. We can find the exact value for this pH if we know the acid–base reactions and acid dissociation constants involved in the conversion of HA to A^- and H_2A^+. For instance, by rearranging the equilibrium expressions for these reactions, we can show that $[A^-] = K_{a2} [HA]/[H^+]$ and $[H_2A^+] = [HA][H^+]/K_{a1}$. We also know at the isoelectric point that $[H_2A] = [A^-]$, or $[HA][H^+]/K_{a1} = K_{a2} [HA]/[H^+]$. When [HA] is cancelled from both sides of this last equation and the logarithm of this equation is taken, we get the following result.

Isoelectric point for a zwitterion:

$$pI = \frac{pK_{a1} + pK_{a2}}{2} \qquad (8.53)$$

You may notice that Equation 8.53 is quite similar to the one given in Equation 8.52 for the pH of a solution containing an intermediate form of a polyprotic acid–base system. This similarity occurs because both equations describe the pH (in this case, the point where pH = pI) for a solution of an intermediate acid–base form. The same type of relationship can be derived to determine the isoelectric point for a zwitterion with more than two acid–base groups. In this situation, the pK_a values that are used in Equation 8.53 are those for the processes that have the "neutral" form of the zwitterion as either the conjugate acid or as the base. For instance, the pK_a values that would be used here for glutamic acid would be 2.16 and 4.30, because these are the values that represent the acid–base reactions that form and consume the form of this amino acid with a net charge of zero.

> **EXERCISE 8.12   Determining the Isoelectric Point for an Amino Acid**
>
> An environmental chemist studying the biological effects of acid rain wishes to prepare a solution of alanine with a pH that is exactly equal to the pI for this amino acid. At what pH should this solution be prepared?
>
> **SOLUTION**
>
> The form of alanine that has a net charge of zero occurs when its carboxyl group has lost a hydrogen ion (giving this group a charge of −1) and the N-terminal region has gained a hydrogen ion, giving it a local charge of +1. The pK_a values for these two processes are 2.35 and 9.87, as obtained from $K_{a1} = 4.47 \times 10^{-3}$ and $K_{a2} = 1.35 \times 10^{-10}$. When we place these values into Equation 8.53, we get the following isoelectric point.
>
> $$pI = \frac{pK_{a1} + pK_{a2}}{2} = \frac{2.35 + 9.87}{2} = \mathbf{6.11}$$
>
> Thus, a pH of 6.11 should be used to prepare a solution that gives pH = pI for alanine.

TABLE 8.9   Examples of "Good" Buffers

Acronym	Full Name and Structure[a]	pK_a (at 25°C)[b]
ACES	N-2-Acetamido-2-aminoethanesulfonic acid	6.85
	$H_2NCCH_2\overset{+}{N}H_2CH_2CH_2SO_3^-$ (with C=O above first C)	
ADA	N-2-Acetamidoiminodiacetic acid	6.84
	$H_2NCCH_2\overset{+}{N}H_2$ with two CH_2CO^- branches (each C=O)	
BICINE	N-Bis(2-hydroxyethyl)glycine	8.33
	$(HOCH_2CH_2)_2\overset{+}{C}NH_2CH_2CO^-$ (C=O)	
HEPES	N-2-(Hydroxyethyl)piperazine-N'-2-ethanesulfonic acid	7.56
	$HOCH_2CH_2\overset{+}{N}H \quad \overset{+}{H}NCH_2CH_2SO_3^-$ (piperazine ring)	
HEPPS	N-2-(Hydroxyethyl)piperazine-N'-3-propanesulfonic acid	7.96
	$HOCH_2CH_2N \quad \overset{+}{H}NCH_2CH_2CH_2SO_3^-$ (piperazine ring)	
MES	2-(N-Morpholino)ethanesulfonic acid	6.27
	$O \quad \overset{+}{H}NCH_2CH_2SO_3^-$ (morpholine ring)	
MOPS	3-(N-Morpholino)propanesulfonic acid	7.18
	$O \quad \overset{+}{H}NCH_2CH_2CH_2SO_3^-$ (morpholine ring)	
PIPES	Piperazine-N,N'-bis(2-ethanesulfonic acid)	7.14
	$^-O_3SCH_2CH_2N \quad \overset{+}{H}NCH_2CH_2SO_3^-$ (piperazine ring)	
TAPS	N-Tris(hydroxymethyl)methyl-3-aminopropane sulfonic acid	8.55
	$(HOCH_2)_3C\overset{+}{N}H_2CH_2CH_2CH_2SO_3^-$	
TES	N-Tris(hydroxymethyl)methyl-2-aminoethane sulfonic acid	7.55
	$(HOCH_2)_3C\overset{+}{N}H_2CH_2CH_2SO_3^-$	
TRICINE	N-Tris(hydroxymethyl)methylglycine	8.14
	$(HOCH_2)_3C\overset{+}{N}H_2CH_2CO^-$ (C=O)	

[a]The structure shown for each compound is the acid form that is used as part of the buffer system at a neutral pH. The hydrogen that is lost to form the conjugate base is given in boldface.

[b]The listed pK_a values for the hydrogens are in boldface. These Good buffers each have more than one acidic group and pK_a value. For instance, pK_a for the carboxylic acid group in TRICINE is 2.02, and the pK_a is 1.59 for the carboxylic acid groups in ADA.

Knowing the pI value for an amino acid, peptide, protein, or other type of zwitterion can be extremely useful in identifying and isolating these chemicals. We will see an example of this application in Chapter 23 when we discuss how *isoelectric focusing* can be used to separate biological compounds according to their isoelectric points. The isoelectric point of a zwitterion is also important in determining the solubility of such a compound, because proteins and other biological agents tend to have their lowest solubility at a pH that is equal to their pI value.

Zwitterions are important not only as analytes but also as reagents. For instance, a group of zwitterionic chemicals known as *Good buffers* are often used in work with biological samples (see Table 8.9). These buffers are named after Norman E. Good,[27] who first proposed the use of such agents in biological research. These agents were selected by Good and his coworkers based on their good solubility in water, their ability to buffer in the pH range of 6–8 (typical conditions in biological systems), their high stability and low cost, and their minimal salt effects. The buffers in Table 8.9 also tend to have low absorbance of light at wavelengths above 230 nm, a feature that gives them a low background signal when used in many assays based on spectroscopy (see Chapter 18).

Key Words

Acid (Brønsted–Lowry model) *167*	Base (Brønsted–Lowry model) *167*	Henderson–Hasselbalch equation *183*	pH Scale *174*
Acid–base reaction *168*	Base ionization constant *171*	Isoelectric point *193*	pK_a *174*
Acid dissociation constant *169*	Buffer capacity *184*	K_w (autolysis constant for water) *171*	pK_w *174*
Autoprotolysis *171*	Buffer solution *181*		Polyprotic acid *185*
Autoprotolysis constant *171*	Conjugate acid *168*	Leveling effect *172*	Polyprotic base *185*
	Conjugate base *168*	Monoprotic acid *176*	pH (Notational definition) *173*
		Monoprotic base *176*	Zwitterion *190*

Other Terms

Acid (Arrhenius model) *166*	Amphiprotic *171*	Buffer index *181*	Good buffers *194*
Acid dissociation *169*	Base (Arrhenius model) *166*	Fraction of species equation *186*	Neutral solution (in water) *174*
Acidic solution (in water) *174*	Basic solution (in water) *174*	Fraction of species plot *187*	
Acidity constant *169*	Basicity constant *171*		
	Brønsted–Lowry model *167*		

Questions

WHAT IS AN ACID OR BASE AND WHY ARE THESE IMPORTANT IN CHEMICAL ANALYSIS?

1. What is the definition of an "acid" or "base" according to the Arrhenius model? What are some examples of Arrhenius acids or bases?
2. What is the definition of an "acid" or "base" according to the Brønsted–Lowry model? What is an acid–base reaction, as defined by this same model? What is meant by the terms "conjugate acid" or "conjugate base"?
3. Two chemists in the same lab both prepare a solution of ammonia in water. The first chemist labels her solution as "Aqueous ammonia (NH_3)" while the second labels his container "Ammonium hydroxide (NH_4OH)." These same two chemists write in their laboratory notebooks the following reactions to describe the acid–base properties of their solutions.

Chemist 1: $NH_3 + H_2O \rightleftharpoons NH_4^+ + OH^-$
Chemist 2: $NH_4OH \rightleftharpoons NH_4^+ + OH^-$

Which of these chemists is using the Arrhenius model of acids and bases to describe this solution? Which chemist is using the Brønsted–Lowry model of acids and bases? Explain your answers.

4. What are some applications of acids or bases in chemical analysis? Give some examples.

STRONG AND WEAK ACIDS

5. What is meant by a "strong acid"? What is a "weak acid"? What are some examples of each in water?
6. What is an "acid dissociation constant"? How is the value of this constant related to the strength of an acid?
7. Write the $K_a°$ expressions for each of the following reactions.
   (a) $HOCN + H_2O \rightleftharpoons OCN^- + H_3O^+$
   (b) $C_2H_5COOH + H_2O \rightleftharpoons C_2H_5COO^- + H_3O^+$
   (c) $SH^- + H_2O \rightleftharpoons S^{2-} + H_3O^+$
   (d) $2 CH_3COOH \rightleftharpoons CH_3COOH_2^+ + CH_3COO^-$
8. Write the K_a expressions for the reactions in Problem 7. Show how K_a is related to the $K_a°$ values for the same reactions.
9. Rank the following inorganic acids based on their relative strength in water. What trends with regard to chemical structure and acid strength do you notice in each of these examples?
   (a) Phosphoric acid, dihydrogen phosphate ($H_2PO_4^-$), and monohydrogen phosphate (HPO_4^{2-})
   (b) Hypobromous acid, hypochlorous acid, and hypoiodous acid
   (c) Hypophosphorus acid, phosphoric acid, and phosphorus acid

10. Rank the following organic acids based on their relative strength in water. What trends with regard to chemical structure and acid strength do you notice in each of these examples?
    (a) Acetic acid, bromoacetic acid, and chloroacetic acid
    (b) 2-Nitrophenol, 3-nitrophenol, 4-nitrophenol, and phenol
    (c) Acetic acid, butanoic acid, formic acid, and propanoic acid

STRONG AND WEAK BASES

11. What is meant by a "strong base"? What is a "weak base"? Give some examples of each in water.
12. What is a "base association constant"? How is the value of this constant related to the strength of a base?
13. Write the expressions for the thermodynamic K_b values of the following reactions.
    (a) $RNH_2 + H_2O \rightleftharpoons RNH_3^+ + OH^-$
    (b) $ClO^- + H_2O \rightleftharpoons HClO + OH^-$
    (c) $RCOO^- + H_2O \rightleftharpoons RCOOH + OH^-$
    (d) $2 CH_3OH \rightleftharpoons CH_3OH_2^+ + CH_3O^-$
14. Write the expressions for the concentration-based base ionization constants for each of the reactions in Problem 13. Show how these concentration-based base ionization constants are related mathematically to the thermodynamic base ionization constants for the same reactions.
15. Rank the following bases based on their relative strength in water. What trends with regard to chemical structure and acid strength do you notice in each of these examples?
    (a) Ammonia, ethylamine, and methylamine
    (b) Phosphate, dihydrogen phosphate ($H_2PO_4^-$), and monohydrogen phosphate (HPO_4^{2-})
    (c) Pyridine, piperidine, 2,2'-bipyridine

THE ACID AND BASE PROPERTIES OF WATER

16. What is the definition of an "amphiprotic compound"? Why is water considered to be amphiprotic?
17. What is "autoprotolysis"? Write the reaction for this process in water. Write the reaction for the autoprotolysis of methanol (CH_3OH).
18. What is an "autoprotolysis constant"? Write expressions for both the thermodynamic and concentration-dependent autoprotolysis constants for water.
19. Determine the following values for aqueous solutions.
    (a) The concentration of OH^- in a solution containing $2.7 \times 10^{-4}\ M\ H^+$ at 25°C
    (b) The concentration of H^+ in a solution containing $5.1 \times 10^{-5}\ M\ OH^-$ at 25°C
    (c) The concentrations of both H^+ and OH^- in pure water at 10°C
20. Fish and other aquatic organisms are sensitive to the acid concentration in their surroundings. Lake trout can flourish when the hydrogen ion concentration in a lake is $1 \times 10^{-5}\ M$, but cannot survive if the hydrogen ion concentration rises above $1 \times 10^{-4}\ M$. Determine whether lake trout would be able to survive in (a) a solution that contains 0.365 g of nitric acid dissolved in 1000 L of water, and (b) a solution that contains 36.5 g of nitric acid dissolved in 1000 L of water.
21. What is the "leveling effect"? How does the leveling effect occur in water? Can this effect occur in other solvents?
22. Explain how changing the solvent can be used to tell the difference in strength between two strong acids. Give an example of a solvent that can be used for this purpose.

What type of solvent would you use to tell the difference in strength between two strong bases?
23. Give an equation that shows how K_a and K_b are related to each other for a given acid–base reaction. Explain why this relationship depends on the solvent in which the reaction is occurring.
24. Write the chemical formulas for the conjugate bases of the following acids. Find the K_b values for each of these conjugate bases in water at 25°C.
    (a) Formic acid, HCOOH
    (b) Phenol, C_6H_5OH
    (c) Chromic acid, H_2CrO_4
    (d) Hydrogen chromate, $HCrO_4^-$
25. Write the formulas for the conjugate acids of the following bases. Calculate the K_a value for each of these conjugate acids in water at 25°C.
    (a) Ammonia, NH_3
    (b) Pyridine, C_5H_5N
    (c) Phosphate, PO_4^{3-}
    (d) Hydrogen phosphate, HPO_4^{2-}
26. What assumptions are made in deriving the relationship $K_w = K_a \cdot K_b$? Under what conditions will this relationship no longer be valid?

pH AND THE pH SCALE

27. What is the common notational definition of pH? What are the advantages and limitations of this definition?
28. Mussels can survive in waterways with a pH of 6.8. They cannot survive at pH 5.2. What is the approximate hydrogen ion concentration for each of these samples?
29. The pH of human blood usually ranges from 7.2 to 7.6, with pH 7.4 being the most common value. What is the range of hydrogen ion activities that can occur in blood? What hydrogen ion activity would be found in most blood samples?
30. Find each of the following values. Use the correct number of significant figures in your answers.
    (a) pH for $[H^+] = 6.3 \times 10^{-6}\ M$
    (b) pK_b for $K_b = 8.24 \times 10^{-8}$
    (c) pOH for $[H^+] = 2.15 \times 10^{-3}\ M$
    (d) pCl for $[Cl^-] = 1.5 \times 10^{-4}\ M$
31. Find each of the following values, using the correct number of significant figures in your answers.
    (a) $[H^+]$ for pH = 9.3
    (b) K_a for $pK_a = 2.22$
    (c) $[OH^-]$ for pH = 11.20
    (d) $[Ca^+]$ for pCa = 5.7
32. Explain what is meant by the relationship $pK_w = pH + pOH$. Describe how this expression is obtained from Equation 8.21.
33. Discuss why the relationship $14.00 = pH + pOH$ is only strictly true at or near a temperature of 25°C. How would you change this equation to make it correct if you were working at 20°C? How would you change it for work performed at 10°C?
34. What pH or pH range corresponds to an "acidic" solution in water? What pH or pH range corresponds to a "basic" solution or a "neutral" solution?
35. Explain how each of the items below can affect the pH of a solution.
    (a) An increase in the concentration of an acid
    (b) An increase in the concentration of a base
    (c) Changing the solvent from water to acetic acid

36. Explain how a change in temperature can affect the pH of a solution.

MONOPROTIC STRONG ACIDS AND BASES

37. What is a "monoprotic acid"? What is a "monoprotic base"? Give one example of each.
38. Estimate the pH for each of the following solutions in water at 25°C. State and justify any assumptions you made in obtaining your answers.
    (a) 0.030 M HCl
    (b) 0.0060 M KOH
    (c) 0.50 g/L HNO$_3$
    (d) 0.2 g/L NaOH
39. Potassium hydroxide is a strong base that is commonly used for the analysis of acids in the method of an acid–base titration. What is the expected pH at 25°C for a 0.0600 M KOH solution that is to be used as a reagent in a titration? If this base reacts in a 1:1 ratio with HCl, what concentration of this acid must be present in a 20 mL HCl sample that is exactly neutralized by 25 mL of this KOH solution?
40. A laboratory worker purchases a bottle of concentrated hydrochloric acid (concentration, 12.0 M). A 10.00 mL aliquot of this concentrated acid is placed into a 250.00 mL volumetric flask. This acid is then carefully combined with water and diluted to the mark of the volumetric flask. What is the final concentration of HCl in this solution? What is the approximate pH that is expected for this solution at 25°C?
41. A 25.00 mL portion of a 0.100 M HCl solution in water is combined with 10.00 mL of a 0.0500 M solution of NaOH. If the H$^+$ and OH$^-$ ions from these two chemicals combine in a 1:1 ratio to form water, what will be the pH of the final solution at 25°C?
42. What limitations are there in using the total concentration of a strong acid or strong base to estimate the pH? Explain the reasons for each of these limitations.
43. Estimate the pH for each of the following solutions in water at 25°C. State and justify any assumptions you made in obtaining your answers.
    (a) 3.00 × 10^{-8} M KOH
    (b) 6.10 × 10^{-7} M HNO$_3$
    (c) 1.0 μg/L HBr
    (d) 0.2 μg/L NaOH
44. A student takes a 1.00 mL aliquot of a 5.00 × 10^{-4} M HCl solution and dilutes it to the mark with water in a 1000.00 mL volumetric flask.
    (a) What is the expected pH for the final solution at 25°C?
    (b) What would be the expected pH if this solution were later used at a temperature of 20°C?
45. In what ways does the ionic strength of a solution affect its pH? Under what conditions are these effects most significant? What types of errors can result in pH calculations if ionic strengths are not considered but are present?
46. Explain why different formulas are often used to estimate the pH for a solution of a weak acid or base versus a solution of a strong acid or base. (*Note*: Compare Equations 8.25 and 8.26 to Equations 8.32 and 8.33.) What are the reasons for these differences?
47. Estimate the pH for the following solutions in water at 25°C, using the given analytical concentrations.
    (a) 0.0500 M benzoic acid
    (b) 1.75 × 10^{-4} M formic acid
    (c) 20 g/L 2-nitrobenzoic acid
    (d) 150 mg/L chloroacetic acid

48. Estimate the pH for the following solutions in water at 25°C, using the given analytical concentrations.
    (a) 0.0040 M ammonia
    (b) 6.50 × 10^{-5} M pyridine
    (c) 7.25 × 10^{-4} M cyanate
    (d) 9.5 × 10^{-3} M methylamine
49. An aqueous solution of hydrofluoric acid (HF) is to be prepared for the analysis of silica (SiO$_2$) in ore samples. This process involves the reaction of silica with hydrofluoric acid, which produces silicon tetrafluoride (SiF$_4$) as a gas. To prepare for this analysis, a student makes an aqueous solution with an analytical concentration for hydrofluoric acid of 0.80 M. If no other significant sources of H$^+$ or F$^-$ are present, what approximate pH would be expected for this solution at 25°C? (*Note*: You may ignore ionic strength effects to simplify this problem.)
50. Estimate the pH for the following solutions in water at 25°C, using the given analytical concentrations.
    (a) 2.5 × 10^{-3} M phenol
    (b) 8.1 × 10^{-4} M hypochlorous acid
    (c) 1.0 mg/L cyclohexamine
    (d) 200 mg/L hydrogen thiocyanate
51. Lactic acid is produced in our muscle cells when we work too strenuously to maintain aerobic respiration. Lactic acid buildup lowers the pH in the muscle tissues from about 7.0 to 6.5, at which point enzymes required for an anaerobic energy supply no longer function and the muscles do not work adequately. What concentration of lactic acid must be present in water to create a pH of 6.5 at 25°C?
52. The amount of ascorbic acid (vitamin C) in clinical samples is to be determined by a colorimetric assay.
    (a) To prepare for this analysis, a laboratory worker prepares a standard solution that contains 10 mg of ascorbic acid that is dissolved in 100.00 mL of water. What is the expected pH of this standard at 25°C?
    (b) A second standard is prepared that contains 10 mg of ascorbate (the conjugate base of ascorbic acid) in 100.00 mL of water. What is the expected pH of this second standard at 25°C?
53. The typical concentration of ammonia in human plasma is approximately 9–30 μM. If two standards are prepared that contain 9 or 30 μM ammonia dissolved in water, what would be the approximate pH for each of these standard solutions?
54. Describe how an increase in ionic strength can affect the pH of a weak acid or weak base solution. How does this relationship compare to the effects of ionic strength on solutions containing strong acids or strong bases?
55. A chemist prepares a standard solution of benzoic acid (molar mass, 122.13 g/mol) by dissolving 5.0 μg of this chemical in 1.00 L water. Although, the chemical is an acid, an estimate made using Equation 8.32 suggests that this solution will give a pH of 7.38 at 25°C. What error was made in estimating this value? What must be done to fix this problem?
56. Estimate the pH for the following mixtures in water at 25°C, using the provided original concentrations of each component.
    (a) 1.0 × 10^{-3} M acetic acid and 2.5 × 10^{-4} M acetate
    (b) 4.6 × 10^{-4} M benzoic acid and 5.0 × 10^{-4} M benzoate
    (c) 6.9 × 10^{-5} M ammonia and 1.5 × 10^{-4} M ammonium

57. The amount of *o*-nitrophenol ($C_6H_5NO_3$, a weak acid) is to be determined in environmental samples by using an acid–base titration (a method we will discuss in Chapter 11).
    (a) A chemist prepares for this analysis by placing in 100.00 mL of water a 0.250 g sample known to contain about 50% (w/w) *o*-nitrophenol. If no other acids or bases are present in this sample, what is the approximate pH that is expected for this solution at 25°C?
    (b) Various volumes of a 0.0100 *M* solution of NaOH are next combined with this sample as the pH of the resulting mixture is measured. If the OH^- ions from NaOH react in a 1:1 ratio with *o*-nitrophenol to form its conjugate base, how many moles of conjugate base (*o*-nitrophenolate) will be produced after 25.00, 50.00, or 100.00 mL of the NaOH solution has been added?
    (c) What is the pH that would be predicted for each of the final mixtures in Part (b)? How does the value of pH change as the amount of added NaOH is increased?
58. Like the example given in Problem 57, an acid–base titration can also be used to determine the content of 4-aminopyridine ($C_5H_6N_2$, a weak base) in a solution.
    (a) A chemist dissolves a 0.2500 g portion of 4-aminopyridine in 250.00 mL of water. If a pure sample of 4-aminopyridine was used, what is the expected pH for this solution at 25°C?
    (b) Various volumes of a 0.01500 *M* solution of HCl are combined with this sample as the pH of the resulting mixture is measured. If the H^+ ions from HCl react in a 1:1 ratio with 4-aminopyridine to form its conjugate acid, how many moles of this conjugate acid will be produced after 50.00 or 100.00 mL of the HCl solution has been added?
    (c) What is the pH that would be predicted for each of the final mixtures in Part (b)? How does the value of pH change as the amount of added HCl is increased? How does this behavior compare to that seen in Problem 57?

BUFFER SOLUTIONS

59. What is a buffer solution? Give two examples of buffers.
60. What is the Henderson–Hasselbalch equation? What are the assumptions made in the use of this equation?
61. A biochemist wishes to use phosphoric acid and dihydrogen phosphate to make a buffer solution that has a final pH of 3.0. If the total concentration of all phosphate species in this solution must be equal to 0.050 *M*, what concentration of phosphoric acid and dihydrogen phosphate must be present at equilibrium in this solution?
62. A chemist determines that an aqueous sample containing a mixture of phenol and phenolate has equilibrium concentrations of these two chemicals equal to 2.3×10^{-6} *M* and 1.9×10^{-5} *M*. If this analysis was performed at 25°C and the sample contained no other major acids or bases, what was the pH of the sample?
63. An analytical chemist prepares an acetate buffer by combining 5.0×10^{-5} *M* acetic acid with 7.5×10^{-5} *M* sodium acetate in water at 25°C. What is the expected pH of the final solution?
64. A service laboratory has several chemicals available for the preparation of buffer solutions: acetic acid, barbituric acid, ACES, diethanolamine, MES, and tricine. Which of these chemicals would you suggest using when making buffers for work at the following pH values? Explain the reason for each of your choices.
    (a) pH 5.0
    (b) pH 6.0

(c) pH 7.0
(d) pH 8.0

65. A biochemist wishes to prepare various phosphate buffers by mixing together two 0.20 *M* solutions that contain either sodium dihydrogen phosphate (NaH_2PO_4) or disodium hydrogen phosphate (NaH_2PO_4).
    (a) In what ratio (v/v) should these solutions be combined to give a buffer with a pH of 7.0 at 25°C? What ratio should be used to give a pH of 6.5 or 7.5?
    (b) What will the total concentration of all phosphate species be in each of these solutions? What will be the approximate fraction of $H_2PO_4^-$ or HPO_4^{2-} in each of these buffers?
    (c) If a total volume of 1.00 L is needed for a pH 7.2 buffer solution, what volumes of the original two NaH_2PO_4 and Na_2HPO_4 solutions will be needed to prepare this buffer? What mass of the solids $Na_2HPO_4 \cdot 2H_2O$ and $NaH_2PO_4 \cdot 2H_2O$ will be needed to prepare these solutions?
66. A research chemist wishes to use a buffer based on boric acid, $B(OH)_3$.
    (a) A total of 10.68 g of boric acid is dissolved in 250.00 mL of water. What is the expected pH at 25°C for this solution?
    (b) A 100.00 mL portion of the solution from Part (a) is combined with 50.00 mL of 0.5000 *M* NaOH and brought up to a total volume of 250.00 mL in water. If the OH^- from NaOH reacts in a 1:1 ratio with boric acid to form the conjugate base of this compound, what will the approximate pH (at 25°C) be for the buffer solution that is formed by this mixture?
67. What is meant by "buffer capacity"? Why is the buffer capacity important to consider when you are selecting or preparing a buffer solution?
68. A buffer based on pyridine has a pH of 5.00 and a total concentration for all pyridine species of 0.050 *M*. What is the buffer capacity when sodium hydroxide is added to 1.000 L of this buffer?
69. A chemist prepares 1.000 L of a buffer solution that contains 0.10 *M* malic acid and 0.10 *M* malate. The initial pH of this buffer is 3.40. What is the buffer capacity for this solution?
70. What is a "buffer index"? How does the buffer index differ from the buffer capacity?
71. Why does the buffer index curve in Figure 8.7 have a maximum at pH 4.76? At what approximate pH would the maximum of this curve appear if you were to use a buffer based on imidazole? Explain your answer.

POLYPROTIC ACIDS AND BASES

72. What is a "polyprotic acid" or "polyprotic base"? Give one example of each.
73. What is a "fraction of species equation"? How can this type of equation be used to describe a solution containing a polyprotic acid or base?
74. Write the fraction of species equations for all of the acid–base forms of the following compounds.
    (a) Hydrogen sulfide
    (b) Succinic acid
    (c) Citric acid
    (d) Lysine
75. A food chemist prepares a solution that contains a total concentration of 0.20 *M* piperazine and that has a pH of 6.00. What are the approximate fractions for all piperazine-related species in this solution at 25°C?

76. A student is studying binding by the amino acid tryptophan to the protein albumin. The binding depends on the pH of the solution. Determine the relative fraction of each tryptophan species that will be present in a solution made at pH 7.4. What is the principal species for tryptophan at this pH?

77. Ammonia is often used as a secondary ligand to help control complexation reactions between metals and various binding agents. The ability of ammonia to take part in these reactions depends on the relative amount of this chemical that is actually present in the form NH_3 (vs. NH_4^+) at a given pH.
    (a) At 25°C, what will be the fraction of the ammonia present in the form NH_3 in a pH 11.0 aqueous solution?
    (b) What will the concentration of NH_3 be at pH 11.0 if the total concentration of ammonia (in the absence of any metal ions) is 0.025 M?
    (c) What will be the fraction of NH_3 in the same solution as in Part (a), but at a pH of 7.0? At which pH will there be the greater amount of NH_3 that is available for binding to metal ions?

78. One way of describing the relative strength of acids is by comparing the fraction of these chemicals that dissociate when placed into a solution. Use the fraction of species equations to find the approximate fraction of dissociation for each of the following acids at pH 2.00.
    (a) HCl
    (b) HNO_3
    (c) Acetic acid
    (d) Phosphoric acid

79. What are four general situations that can occur when you are trying to determine the pH of a polyprotic system? What strategy do you use to find the pH in each of these situations?

80. What is meant by the term "amphiprotic" compound? Give two specific examples of amphiprotic compounds.

81. Estimate the pH for each of the following solutions. State and justify any assumptions you make in obtaining your answers.
    (a) 2.50×10^{-3} M H_3PO_4
    (b) 4.3×10^{-4} M PO_4^{3-}
    (c) 5.0×10^{-5} M $H_2PO_4^-$
    (d) 5.0×10^{-5} M HPO_4^{2-}

82. Estimate the pH for each of the following solutions. State and justify any assumptions you make in obtaining your answers.
    (a) 5.0×10^{-5} M Chromic acid
    (b) 5.0×10^{-5} M Chromate
    (c) 2.5×10^{-5} M H_2CrO_4 and 2.5×10^{-5} M $HCrO_4^-$

83. What assumptions are made in Equation 8.51? What assumptions are made in Equation 8.52? What are the advantages and disadvantages of using each of these equations?

ZWITTERIONS

84. What is the definition of a "zwitterions"? Explain why amino acids are zwitterions.

85. What is an "isoelectric point"? Why is this parameter important?

88. Estimate the isoelectric point for the following amino acids in water at 25°C.
    (a) Asparagine
    (b) Isoleucine
    (c) Methionine
    (d) Proline

89. What pH should be used to prepare a 0.15 M solution of leucine that has a pH that is equal to the isoelectric point for this amino acid?

90. Estimate the pI values for each of the following amino acids. Explain how you obtained these answers.
    (a) Glutamic acid
    (b) Histidine
    (c) Cysteine
    (d) Tyrosine

CHALLENGE PROBLEMS

91. Use the *CRC Handbook of Chemistry and Physics* or other references to locate the following values.
    (a) Picric acid, K_a (25°(C)
    (b) Itaconic acid, K_a (25°(C)
    (c) Acetic acid, pK_a (10°(C)
    (d) Allantoin, pK_a (25°(C)
    (e) Ammonia, K_b (10°(C)
    (f) *n*-Dodecaneamine, pK_b (25°(C)

92. Obtain information from the Internet or material safety data sheets on the chemical and physical hazards that are associated with one or more of the following strong acids or bases.
    (a) Hydrochloric acid
    (b) Sodium hydroxide
    (c) Nitric acid
    (d) Perchloric acid

93. Using propagation of errors, estimate the maximum relative error that can be present in calculated values for $[H^+]$ if a pH is desired that is accurate to within 0.01 units. What maximum relative error in $[H^+]$ is needed if the pH must be accurate to within 0.02 units? What uncertainty in $[H^+]$ would be present if the pH were measured to accuracy of 0.005 units?

94. Prove that $K_{auto} = K_a \cdot K_b$ for an acid–base reaction in any amphiprotic solvent.

95. Deuterium oxide (D_2O) is often used in place of H_2O as a solvent when using the method of proton nuclear magnetic resonance spectroscopy (^1H-NMR) because, unlike ordinary water, D_2O does not give a background signal in this method. Like water, deuterium oxide can undergo autoprotolysis, with the products now being D_3O^+ and OD^-.
    (a) Write the reaction and equilibrium expression for the autoprotolysis of D_2O.
    (b) A small sample of pure (glacial) acetic acid is dissolved in D_2O. Write the acid–base reaction for acetic acid with this solvent.
    (c) It was stated earlier that pK_w for D_2O is 14.955 at 25°C. What would be the expected concentrations of D_3O^+ and OD^- in a pure sample of deuterium oxide at 25°C?
    (d) If $pD = -\log([D_3O^+])$, what would be the value of pD for a liquid sample of pure D_2O that is held at a temperature of 25°C?

96. Prepare a spreadsheet that uses Equations 8.25 and 8.28 to determine the minimum analytical concentration for a strong acid at which the acid dissociation of water can be ignored during the calculation of pH. Determine this answer for the case where the pH value must be accurate to within 0.01, 0.02, or 0.05 pH units. How do your results compare to the general cutoff of $> 10^{-6}$ M that was given in Equation 8.25?

97. Use a spreadsheet to prepare plots for each of the following chemicals that show how the fractions of each of their acid or base forms change with pH. Determine the pH conditions under which each form is the principal species.

    (a) Formic acid
    (b) Citric acid
    (c) Isoleucine
    (d) Cysteine

98. A pharmaceutical chemist prepares a pH 7.40 buffer that contains a total concentration of 0.067 M NaH_2PO_4 and Na_2HPO_4.
    (a) What is the estimated ionic strength of this solution? What assumptions, if any, did you make in obtaining this answer?
    (b) To what extent would the ionic strength of this solution cause K_a to differ from $K_a°$ for the given buffer system?
    (c) What concentrations of $H_2PO_4^-$ and HPO_4^{2-} would you predict to be present in this buffer at pH 7.4 if the ionic strength were not considered? How many moles for each of these species would you use to prepare this buffer if ionic strength effects were ignored?
    (d) Repeat the calculations in Part (c), but with ionic strength now being considered in these calculations. How do your new results compare to those found in Part (c)?

99. Derive the relationship that is shown in Equation 8.33 for a weak monoprotic base. What assumptions were made during this derivation?

100. Another type of buffer that is often used in pharmaceutical studies is phosphate-buffered saline (PBS). One recipe for this buffer has the following composition, which is then adjusted to pH 7.4 after it has been prepared.

---

137 mM NaCl
2.7 mM KCl
4.3 mM Na_2HPO_4
1.4 mM KH_2PO_4

    (a) What is the ionic strength of this buffer solution? Which of the reagents just listed make the largest contribution to this ionic strength?
    (b) To what extent would the ionic strength cause K_a to differ from $K_a°$ for this buffer system?
    (c) What pH would be expected for this buffer if the ionic strength was not considered? How does this compare to the final adjusted pH of 7.4? How different are these values? What is the corresponding difference in terms of hydrogen ion concentration?

101. For a dilute weak acid, we can get an equation in terms of $[H^+]$ by using the same approach described in Section 8.4A for a dilute strong acid. The derivation begins with the following mass balance equation, which shows that we have two important sources of hydrogen ions: water and the weak acid (HA).

$$[H^+] = [H^+]_{HA} + [H^+]_{Water} \qquad (8.54)$$

The end result is the set of expressions that are shown below.

For a dilute, weak monoprotic acid:

$$[H^+] = C_{HA} \cdot \frac{K_a}{[H^+] + K_a} + \frac{K_w}{[H^+]}$$

or

$$0 = [H^+]^3 + K_a [H^+]^2 - (K_w + K_a C_{HA}[H^+]) - K_a K_w \quad (8.55)$$

    (a) Identify each term in Equation 8.55. Show how this expression can be derived from Equation 8.54. Explain the reasons behind each step in your derivation.
    (b) A student takes a 0.10 M solution of acetic acid and dilutes it with water to an analytical concentration of $5.0 \times 10^{-7} M$. What is the expected pH for this solution?
    (c) How does the pH obtained in Part (b) compare to the result that is predicted if the effects of water dissociation are ignored?

102. Prepare a spreadsheet that uses both Equations 8.32 and 8.55 to determine what pH would be obtained at a given analytical concentration for a weak monoprotic acid.
    (a) Use your spreadsheet to prepare plots of pH versus analytical concentration for weak acids that have K_a values of 10^{-4}, 10^{-6}, 10^{-8}, or 10^{-10}.
    (b) From the plot in Part (a), determine the minimum analytical concentration of weak acid that must be present at each of the given K_a values to produce calculated pH values for Equation 8.32 that are within 0.01, 0.02, or 0.05 pH units of those predicted by Equation 8.55.
    (c) How do the minimum analytical concentrations that you found in Part (b) compare to the results that were obtained for a strong acid in Problem 53? Explain any observed differences in these results.

103. If the pH and total concentration (C_T) of acid and conjugate base are known for a buffer, Equation 8.35 can be rearranged as follows.

$$C_{A^-} = \frac{K_a C_T - K_a [H^+] - [H^+]^2}{K_a + [H^+]}$$

$$\text{and } C_{HA} = C_T - C_{A^-} \qquad (8.56)$$

where C_{HA} is the total concentration of acid that was originally added to the solution and C_{A^-} is the total concentration of conjugate base that was initially placed into the solution.
    (a) Show how the relationships in Equation 8.56 can be obtained from Equation 8.35.
    (b) Prepare a spreadsheet in which both Equation 8.56 and the Henderson–Hasselbalch equation are used to estimate the pH for a mixture of a weak acid and its conjugate base at several different total concentrations of these two components. Prepare this spreadsheet using K_a values of 10^{-4}, 10^{-6}, and 10^{-8}. How do the results of the two approaches differ in your plots? Explain any differences that you observe.

104. The buffer index for a relatively concentrated monoprotic acid–base system, can be approximated by using Equation 8.57.[24]

$$\beta = 2.3 \cdot \frac{K_a [H^+]}{(K_a + [H^+])^2} \cdot C_{HA} \qquad (8.57)$$

    (a) Prepare a spreadsheet in which you use Equation 8.57 to plot the buffer index versus pH for a weak acid with a K_a value of 10^{-6} and a total analytical concentration of $10^{-2} M$. Compare your results to those shown in Figure 8.7. What similarities or differences do you see?
    (b) Prepare similar plots of buffer index versus pH for weak acids with total concentrations of $10^{-2} M$ and K_a

values of 10^{-4}, 10^{-6}, or 10^{-8}. How are these plots affected as you vary the K_a value of the weak acid?

(c) Use your spreadsheet to prepare plots of the buffer index versus pH for a weak acid with a K_a value of 10^{-4} and total analytical concentrations of 10^{-2}, 10^{-3}, 10^{-4}, or 10^{-5} M. What changes in these plots do you see as you vary C_{HA}? What do your observations tell you about the effect of buffer concentration on the buffer index?

105. A value that is often quite similar to the isoelectric point for a zwitterion is its *isoionic point*.[26] This is the pH that arises when only the neutral form of zwitterionic compound (HA) is placed into a solution. The isoionic point of a zwitterion can be found by using the same equation we have already derived for a solution for an intermediate form of a polyprotic acid–base system, where we are now specifically looking at the intermediate form HA.

Isoionic point for HA: $\quad [H^+] \cdot \sqrt{\dfrac{K_{a1}(K_{a2}C_{HA} + K_w)}{K_{a1} + C_{HA}}}$ $\quad$ (8.58)

(a) Estimate the isoionic points for the following solutions in water at 25°C.
   1.5×10^{-4} M valine
   2.0×10^{-5} M glutamic acid
   5.5×10^{-3} M lysine
   2.5×10^{-4} M histidine
(b) What are the isoelectric values for the compounds in Part (a)?
(c) How do the isoelectic points in Part (b) compare with the isoionic points from Part (a)? Explain any similarities or differences that you observe in these values.

TOPICS FOR DISCUSSION AND REPORTS

106. Locate information from the Internet on the effects of acid rain in your area. How do these effects compare to other areas of the United States or the world? What analytical techniques were used to monitor these effects?

107. The U.S. Environmental Protection Agency and U.S. Geological Survey are two agencies that routinely monitor various chemicals that are present in our environment. Write a report on one of these agencies and on the areas in which it performs chemical analysis.

108. The way in which pH should be defined has been the subject of much discussion over the years. Obtain more information from sources such as References 10, 11, and 17 on how pH is defined for various areas of work and research. Report on your findings.

109. Contact a clinical or environmental laboratory in your area and ask them how they use pH measurements in their work. Write a report on your findings.

110. Along with contributing to acid–base theory, Svante Arrhenius made many other contributions to our modern picture of chemical reactions. Obtain more information on Arrhenius and his work. Describe how his studies have contributed to analytical chemistry.

111. The Henderson–Hasselbalch equation has played an important role in the past in helping scientists study common acid–base processes.[21–23] Learn more about the history behind this equation and write a report on how it has affected the description and use of acid–base reactions in chemistry and other fields.

112. The structure of the hydrogen ion or "hydrated proton" in water has long been a subject of scientific research. Using sources like References 28–30, describe some other possible structures that a hydrogen ion can have with water. What analytical methods have been used to study this issue? What have been some of the difficulties encountered in obtaining methods for such work? What types of information have been obtained on the structures of hydrogen ions in water?

113. Contact a local research or service laboratory and ask them how buffers are utilized in their work. Include in your report a description of the types of buffer solutions that they commonly use and the reasons why they work with these particular buffers.

References

1. J. Millett, "Acid Rain Program Shows Continued Success and High Compliance, EPA Reports," U.S. Environmental Protection Agency, October 16, 2006.
2. W.B. Grant, "Acid Rain and Deposition," *Handbook of Weather, Climate and Water*, 1 (2003) 269–284.
3. U.S. Environmental Protection Agency, "*Acid Rain Program: 2005 Progress Report*," 2006.
4. *The New Encyclopaedia Britannica*, 15th ed., Encyclopaedia Britannica, Inc., Chicago, IL, 2002.
5. S.A. Arrhenius, *Investigations on the Galvanic Conductivity of Electrolytes*, Doctoral Dissertation, Swedish Academy of Sciences, Stockholm, Sweden, 1884.
6. D. R. Lide, Ed., *CRC Handbook of Chemistry and Physics*, 83rd ed., CRC Press, Boca Raton, FL, 2002.
7. J.N. Brønsted, "Some Remarks on the Concept of Acids and Bases," *Recueil des Travaux Chimiques des Pays-Bas*, 42 (1923) 718–728.
8. T.M. Lowry, "The Uniqueness of Hydrogen," *Chemistry and Industry*, 42 (1923) 43–47.
9. H. Frieser, *Concepts & Calculations in Analytical Chemistry: A Spreadsheet Approach*, CRC Press, Boca Raton, FL, 1992.
10. J. Inczedy, T. Lengyel, and A.M. Ure, *Compendium of Analytical Nomenclature*, 3rd ed., Blackwell Science, Malden, MA, 1997.
11. H.M.N.H. Irving, H. Freiser, and T.S. West, *Compendium of Analytical Nomenclature: Definitive Rules–1977*, Pergamon Press, New York, 1977.
12. John A. Dean, Ed., *Lange's Handbook of Chemistry*, 15th ed., McGraw-Hill, New York, 1999.
13. A.E. Martell and R.M. Smith, *Critical Stability Constants*, Plenum Press, New York, 1974.
14. E.P. Serjeant and B. Dempsey, Ed., *Ionization Constants of Organic Acids in Solution*, Pergamon Press, Oxford, UK, 1979.
15. J.S. Fritz, *Titrations in Nonaqueous Solvents*, Allyn and Bacon, Boston, MD, 1973.
16. S.P.L. Sörenson, "Enzyme Studies II. The Measurement and Meaning of Hydrogen Ion Concentration in Enzymatic Processes," *Biochemische Zeitschrift*, 21 (1909) 131–200.

17. H.B. Kristensen, A. Salomon, and G. Kokholm, "International pH Scales and Certification of pH," *Analytical Chemistry*, 63 (1991) 885A–891A.

18. A. LaBastille, "Acid Rain—How Great a Menace?" *National Geographic*, 170 (1981) 652–681.

19. L.J. Henderson, "Concerning the Relationship Between the Strength of Acids and Their Capacity to Preserve Neutrality," *American Journal of Physiology*, 21 (1908) 173–179.

20. K.A. Hasselbalch, "The Calculation of the Hydrogen Number of the Blood from the Free and Bound Carbon Dioxide of the Same and the Binding of Oxygen by the Blood as a Function of the Hydrogen Number," *Biochemische Zeitschrift*, 78 (1916) 112–144.

21. H.N. Po and N.M. Senozan, "Henderson-Hasselbalch Equation: Its History and Limitations," *Journal of Chemical Education*, 78 (2001) 1499–1503.

22. R. de Levie, "Henderson-Hasselbalch Equation: Its History and Limitations," *Journal of Chemical Education*, 80 (2003) 146.

23. J.W. Severinghaus, P. Astrup, and J.F. Murray, "Blood Gas Analysis and Critical Care Medicine," *American Journal of Respiratory and Critical Care Medicine*, 157 (1998) S114–S122.

24. V. Chiriac and G. Balea, "Buffer Index and Buffer Capacity for a Simple Buffer Solution," *Journal of Chemical Education*, 74 (1997) 937–939.

25. E.T. Urbansky and M.R. Schock, "Understanding, Deriving and Computing Buffer Capacity," *Journal of Chemical Education*, 77 (2000) 1640.

26. *IUPAC Compendium of Chemical Terminology*, Electronic version, http://goldbook.iupac.org.

27. N.E. Good, G.D. Winget, W. Winter, T.N. Connolly, S. Izawa, and R.M.M. Singh, "Hydrogen Ion Buffers for Biological Research," *Biochemistry*, 5 (1966) 467–477.

28. Timothy S. Zwier, "The Structure of Protonated Water Clusters," www.sciencexpress.org/29 April 2004/Page 1/10.1126/science.1098129

29. J.M. Headrick, E. C. Diken, R.S. Walters, N.I. Hammer, R.A. Christie, J. Cui, E.M. Myshakin, M.A. Duncan, M.A. Johnson, and K.D. Jordan, "Spectral Signatures of Hydrated Proton Vibrations in Water Clusters," *Science*, 308 (2005) 1765–1769.

30. S. Borman, "Revisiting the Hydrated Proton," *Chemical & Engineering News*, July 4, 2005, pp. 26–27.

Chapter 9

Complex Formation

Chapter Outline

9.1 INTRODUCTION: WHAT'S IN MY MAYO?

The food industry makes use of analytical chemistry on a daily basis to examine the content of their products. The importance of chemical analysis in this area can be easily seen by looking at the large amount of information that is given on a food label. A typical required label on food shows the amounts of protein, fat, carbohydrates, sodium, and cholesterol that are present, all of which is determined at some point by chemical analysis. Sometimes other ingredients are also listed on these labels. For instance, the label in Figure 9.1 for a jar of mayonnaise lists "calcium disodium EDTA" as an ingredient. EDTA is a chemical that binds tightly to many metal ions, such as Fe^{3+}. The presence of free Fe^{3+} ions in mayonnaise is undesirable because these ions can lead to the oxidation of unsaturated fats. This process, in turn, creates an unpleasant taste and rancid smell. However, if EDTA is present it will bind to Fe^{3+} and prevent this metal ion from being used in the oxidation of fats. It is for this reason that EDTA is often added as a preservative in products that contain fats and oils, such as mayonnaise and salad dressing.[1] The analysis of EDTA in these foods is thus a key part of the quality-control process by which the manufacturer ensures that their product meets required standards before this product can be sold to consumers.

Agents such as EDTA are also important reagents in chemical analysis methods. For example, EDTA binds to Fe^{3+} and many other metal ions through a process known as *complex formation*.[2,3] This ability of EDTA makes this chemical a popular reagent for the measurement of these ions in methods such as a complexometric titration (see Chapter 13). In this chapter, we learn about agents like EDTA that can take part in complex formation. We will also see how we can describe complex formation reactions. To do this, we will again make use of chemical equilibria and equilibrium constants as tools to help us understand and control these processes. In later chapters we will see how this information can be used in analytical methods that are based on complex formation reactions.

9.1A What Is Complex Formation?

Complex formation is a reaction in which there is reversible binding between two or more distinct chemical species, such as EDTA and Fe^{3+}. The product of this reaction is called a **complex**.[4,5] There are many types of complex-formation reactions. The first type of complex formation we will examine in this chapter is that involving metal ions, as is illustrated by the binding of EDTA with Fe^{3+} (Sections 9.2 and 9.3). The second type of complex formation we will examine is that which occurs between various types of biological or organic molecules (Section 9.4). Examples of this latter group of reactions include the binding of an antibody with a foreign agent or the binding

Nutrition Facts

Serving Size 1 Tbsp (13g)
Servings Per Container 60

Amount Per Serving

Calories 90	Calories from Fat 90

	% Daily Value*
Total Fat 10g	15%
Saturated Fat 1.5g	8%
Trans Fat 0g	
Polyunsaturated Fat 6g	
Monounsaturated Fat 2.5g	
Cholesterol 5mg	2%
Sodium 90mg	4%
Total Carbohydrate 0g	0%
Proteins 0g	

Not a significant source of Dietary Fiber, Sugars, Vitamin A, Vitamin C, Calcium and Iron.

*Percent Daily Values are based on a 2,000 calorie diet.

INGREDIENTS: SOYBEAN OIL, WATER, WHOLE EGGS AND EGG YOLKS, VINEGAR, SALT , SUGAR , LEMON JUICE, CALCIUM DISODIUM EDTA (USED TO PROTECT QUALITY), NATURAL FLAVORS.

Calcium disodium EDTA

FIGURE 9.1 Mayonnaise and a close up of the label for this product that lists "calcium disodium EDTA" as one of the ingredients. (Photo by B.R. Hage.)

of two strands of DNA, as used in DNA testing and the polymerase chain reaction (see Chapter 2).

9.1B What Are Some Analytical Applications of Complex Formation?

Complex formation is used in many ways in analytical chemistry. One way complex formation can be utilized is to measure the amount of an analyte. In the case of EDTA and metal ions such as Fe^{3+}, this approach often involves the use of a titration (see Chapter 13). Binding agents like EDTA can also be employed in other formats for detection. An example would be a reagent that forms a colored product when it forms a complex with an analyte (see Chapter 18).

A second way complex formation can be used in chemical analysis is to control the effective amount of an analyte that is available for other reactions. This effect is illustrated by the addition of EDTA to mayonnaise for binding Fe^{3+}, a process that prevents this metal ion from taking part in the oxidation of fats. This same idea is often employed in the analysis of metal ions, where complex formation can prevent certain ions from interfering in the detection of the desired analyte (a method known as "masking," as discussed in Chapter 13).

A third way complex formation can be used in analytical methods is as a tool for the separation of chemicals. For instance, complexing agents can be attached to a solid support for binding and separating their target

compounds from other substances in a sample. This approach is found in some types of extractions and in liquid chromatography, methods that will be discussed in Chapters 20–22. Complex formation can also lead to a change in the physical properties of an analyte, such as its apparent size and charge. This effect can also be helpful in chemical separations, as we will see when we discuss the method of electrophoresis (Chapter 23).

9.2 SIMPLE METAL–LIGAND COMPLEXES

The first type of complex formation we will consider is that which occurs between a metal ion and a simple binding agent like ammonia or a chloride ion. Chemists have known for over 100 years that Cl^- can combine with Ag^+ to form insoluble AgCl (an example of precipitation, as we learned in Chapter 7). However, these same chemists did not understand why Cl^- would not precipitate Ag^+ as easily if the chloride ions were first placed into a solution that contained a metal ion like Co^{3+}. These chemists were also puzzled by the fact that a solution of ammonia and copper ions gave a soluble deep-blue product, where it was possible to recover intact ammonia from this solution if it was later heated (see Figure 9.2).

These observations were explained in 1893 by a Swiss chemist named Alfred Werner,[6–9] who suggested that these reactions were based on a special type of chemical bond. He proposed that these reactions were due to

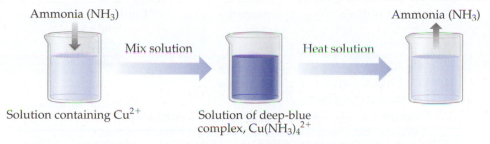

FIGURE 9.2   The reaction of Cu^{2+} with ammonia in water to form a blue-colored complex, $Cu(NH_3)_4{}^{2+}$. This complex forms through a reversible reaction. Ammonia can later be released from this complex and removed from the solution through heating, which leaves behind Cu^{2+} ions.

the reversible formation of a complex between two chemical species, such as Co^{3+} and Cl^- or ammonia and copper ions. He further believed that this complex formed through the sharing of an electron pair by one chemical (Cl^- or ammonia) with a metal ion (Co^{3+} or Cu^{2+}). The type of bond that forms when one chemical shares a pair of electrons with a metal ion is called a *coordinate bond* (also known as a "coordinate covalent bond" or "dative bond").[5] The chemical that shares its electron pair with the metal ion is known as the **ligand** (from the Latin word *ligare*, which means "to bind" or "to tie"). The product of this reaction is called a **metal–ligand complex** or *metal coordination complex*.[10]

9.2A What Is a Metal–Ligand Complex?

Formation of Metal–Ligand Complexes.   The formation of a metal–ligand complex is really a special type of acid–base reaction. In Chapter 8 we learned about the Brønsted–Lowry model of acids and bases, in which an acid is a proton donor and a base is a proton acceptor. A more general model that also covers the formation of metal–ligand complexes was proposed in 1923 by American chemist Gilbert N. Lewis.[9,11,12] In this broader definition, a **Lewis acid** is a chemical that can accept a pair of electrons from another substance, and a **Lewis base** is a chemical that can donate a pair of electrons to another substance.[5,10,13]

Figure 9.3 illustrates how the Lewis acid–base model can be applied to the formation of a metal–ligand complex. In this example, ammonia furnishes a pair of electrons to the outer orbitals of a Cu^{2+} ion, forming a complex between ammonia and this metal ion. Because the Cu^{2+} ion is accepting electrons from ammonia, the Cu^{2+} ion is acting as a Lewis acid during this reaction and ammonia is acting as a Lewis base. This is a reversible process, which is why the ammonia can be released from this complex through heating (see Figure 9.2). It is also possible for this reaction to continue through additional steps where more than one ammonia molecule combines with a Cu^{2+} ion. For example, the deep blue color of Cu^{2+} plus ammonia solutions is due to the addition of four ammonia molecules per Cu^{2+}, giving the complex $Cu(NH_3)_4{}^{2+}$.

The pair of electrons on ammonia that allow its bonding to Cu^{2+} also take part in the binding of ammonia to a hydrogen ion, in which ammonia acts as a Brønsted–Lowry base (a proton acceptor). In fact, any chemical that is an acid or base in the Brønsted–Lowry model is also an acid or base in the Lewis model. This statement is true because the transfer of a proton (or hydrogen ion) in a Brønsted–Lowry acid–base reaction requires that the proton be added onto an electron pair on the base, making the base act as an "electron-pair donor." You should keep in mind, though, that not all Lewis acids and bases are also Brønsted–Lowry acids and bases, because many Lewis acid–base reactions do not involve the transfer of protons.

The Lewis acid–base reaction in Figure 9.3 is a simplified view of the reaction that is actually occurring between Cu^{2+} and ammonia. This particular figure implies there is nothing bound to the Cu^{2+} ions at the beginning of the reaction. However, what is actually present at the start of this reaction is a complex between Cu^{2+} and water, in which water acts as a ligand to form the complex $Cu(H_2O)_6{}^{2+}$. A similar complex occurs when other metal ions are dissolved in aqueous solutions.

The presence of the complex $Cu(H_2O)_6{}^{2+}$ means that the reaction of aqueous Cu^{2+} with ammonia actually requires the substitution of a relatively strong Lewis base (NH_3) for a weaker Lewis base (H_2O) on the copper ion.

Addition of one NH_3 to Cu^{2+}:

$$Cu(H_2O)_6{}^{2+} + NH_3 \rightleftharpoons Cu(NH_3)(H_2O)_5{}^{2+} + H_2O \quad (9.1)$$

If even more ammonia is added to this solution, a total of six ammonia molecules can combine with each Cu^{2+} ion, with the first four ammonia molecules being added at low ammonia concentrations and the last two being added at only very high concentrations.[14]

Overall addition of six NH_3 to Cu^{2+}:

$$Cu(H_2O)_6{}^{2+} + 6\,NH_3 \rightleftharpoons Cu(NH_3)_6{}^{2+} + 6\,H_2O \quad (9.2)$$

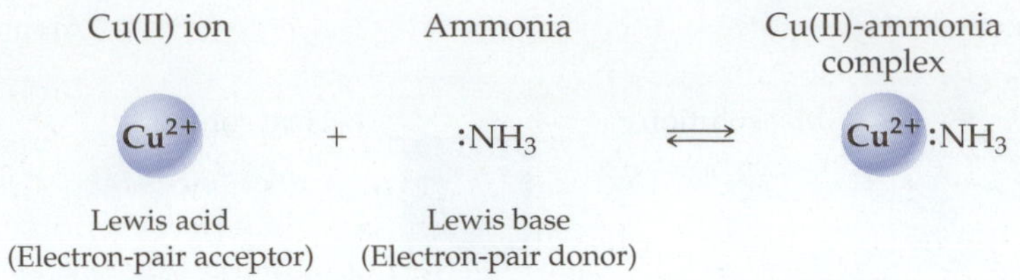

Cu(II) ion   Ammonia   Cu(II)-ammonia complex

$$Cu^{2+} \quad + \quad :NH_3 \quad \rightleftharpoons \quad Cu^{2+}:NH_3$$

Lewis acid   Lewis base
(Electron-pair acceptor)   (Electron-pair donor)

FIGURE 9.3 Example of a complex between a metal ion (Cu^{2+}) and monodentate ligand (NH_3). In this reaction, the metal ion acts as a Lewis acid (electron-pair acceptor) and the ligand acts as a Lewis base (electron-pair donor). A more complete view of this reaction is given in Equation 9.1, in which water is included as part of the reactants and products.

We can now see that each of these reactions has two sets of Lewis acids and bases—one set on the reactant side and one set on the product side. In Equation 9.1, $Cu(H_2O)_6^{2+}$ is the Lewis acid and NH_3 is the Lewis base on the reactant side, while $Cu(NH_3)(H_2O)_5^{2+}$ is the Lewis acid and H_2O is the Lewis base on the product side. The relative extent of this reaction will be determined by the change in energy that occurs in going between these two acid–base pairs.

Describing Metal–Ligand Complexes. There are many chemicals besides ammonia that can act as ligands for metal ions. One example we have noted is water. An anion like chloride can also act as a ligand, as occurs when Cl^- binds with Co^{3+} to produce a soluble deep-blue complex. What all of these ligands have in common is they have a pair of electrons that they can donate to an empty orbital on a metal ion. A chemical like ammonia, water, or Cl^- that can donate only one pair of electrons to a metal ion is called a **monodentate ligand** or *simple ligand*, where the term monodentate means "one-toothed."

Examples of several monodentate ligands are given in Table 9.1. All of these ligands can interact with a metal ion by sharing a single pair of electrons with this ion. For neutral ligands like NH_3 and H_2O,

this binding occurs through the nonbonding electrons on nitrogen or oxygen atoms. It is also possible for a negatively charged ligand like Cl^- or OH^- to share a pair of electrons with a metal ion, where one of these electrons is also responsible for the negative charge on the ligand.

It can be time-consuming to always show all of the water molecules that are involved in metal–ligand complex formation when we are working in water. Thus, these complex formation reactions are often shown in a simplified form with no water bound to the metal ion. When we write Equations 9.1 and 9.2 in this matter, we get the following results.

Addition of one NH_3 to Cu^{2+}:

$$Cu^{2+} + NH_3 \rightleftharpoons Cu(NH_3)^{2+} \qquad (9.3)$$

Overall addition of six NH_3 to Cu^{2+}:

$$Cu^{2+} + 6\,NH_3 \rightleftharpoons Cu(NH_3)_6^{2+} \qquad (9.4)$$

This type of chemical shorthand, which we will follow in the remainder of this chapter, is similar to the strategy we used in Chapter 8 to describe Brønsted–Lowry acid–base reactions in water. Even though the water is not shown, however, you should keep in mind that this

TABLE 9.1 Examples of Monodentate Ligands	
Type of Ligand	**Metal Ions that Bind to Ligand**
Ammonia (NH_3)	Cd^{2+}, Co^{2+}, Cu^{2+}, Hg^{2+}, Ni^{2+}, Sn^{2+}, Zn^{2+}
Bromide (Br^-)	Cd^{2+}, Co^{2+}, Cu^{2+}, Hg^{2+}, Ni^{2+}, Pb^{2+}, Sn^{2+}, Zn^{2+}
Chloride (Cl^-)	Ag^+, Cd^{2+}, Co^{2+}, Cu^{2+}, Fe^{3+}, Hg^{2+}, Mn^{2+}, Ni^{2+}, Pb^{2+}, Sn^{2+}, Zn^{2+}
Cyanide (CN^-)	Cd^{2+}, Fe^{2+}, Fe^{3+}, Hg^{2+}, Ni^{2+}, Zn^{2+}
Fluoride (F^-)	Ba^{2+}, Be^{2+}, Ca^{2+}, Fe^{2+}, Fe^{3+}, Mg^{2+}, Mn^{2+}
Hydroxide (OH^-)	Ba^{2+}, Be^{2+}, Ca^{2+}, Cd^{2+}, Co^{2+}, Cu^{2+}, Fe^{2+}, Hg^{2+}, Mg^{2+}, Mn^{2+}, Ni^{2+}, Pb^{2+}, Sr^{2+}, Zn^{2+}
Iodide (I^-)	Cd^{2+}, Co^{2+}, Hg^{2+}, Ni^{2+}, Pb^{2+}, Sn^{2+}, Zn^{2+}
Thiocyanate (SCN^-)	Be^{2+}, Cd^{2+}, Co^{2+}, Cu^{2+}, Fe^{2+}, Hg^{2+}, Mn^{2+}, Ni^{2+}, Pb^{2+}, Sn^{2+}, Zn^{2+}

solvent can still play an active role during metal–ligand complex formation. Because water can act as a Lewis base, the presence of this solvent will affect the extent to which a metal ion can react with a ligand like ammonia. The stronger a ligand like ammonia is as a Lewis base, the easier it will be for this ligand to replace water molecules that are bound to the metal ion.

If you compare the list of ligands and metal ions in Table 9.1 with the precipitation reactions we discussed in Chapter 7, you will find some overlap. A good example is the reaction of Ag^+ with Cl^-, which leads to the insoluble precipitate AgCl when these two agents combine in a 1:1 ratio. If we add even more Cl^- to such a system, some of this precipitate can actually redissolve and form soluble complexes like $AgCl_2^-$, $AgCl_3^{2-}$, and $AgCl_4^{3-}$. A similar thing happens in the combination of OH^- with metal ions, leading to either a soluble complex (such as $AlOH^{2+}$ when OH^- reacts in a 1:1 with Al^{3+}) or an insoluble precipitate (like the 1:3 salt $Al(OH)_3$, depending on the starting concentrations of these reagents. (*Note*: The addition of a fourth hydroxide in this case can lead to another soluble species, $Al(OH)_4^-$.) For now we will focus on only complex formation, but in Section 9.3D we will learn how to deal with situations like these in which more than one type of reaction can occur at the same time for a ligand or metal ion.

9.2B Formation Constants for Metal–Ligand Complexes

It is important to remember that a metal–ligand complex is created through a reversible reaction in which the complex is usually in equilibrium with the metal ion and unbound ligand. There may also be more than one ligand that can add to the same metal ion. The result is a series of stepwise reactions for addition of the ligand, with each step having its own equilibrium constant. This idea is illustrated in Table 9.2 for the reaction of Ni^{2+} in water with ammonia, which occurs through a process that is similar to the reaction of Cu^{2+} with ammonia.

The original form of Ni^{2+} in this case is one nickel ion surrounded by six molecules of water, or $Ni(H_2O)_6^{2+}$. As the nickel combines with ammonia, one mole of water is displaced for every mole of ammonia that is added. The result is a series of six sequential reactions in which each water molecule surrounding the Ni^{2+} can be displaced by NH_3. Table 9.2 shows the six equilibrium expressions for this system, where the equilibrium constant for each step during this complex formation process is known as a **formation constant** or *stability constant*.[2,3,5,10] A number of other formation constants are listed in Appendix B. Other sources of this information include databases that are available from the International Union of Pure and Applied Chemistry (IUPAC) and the National Institutes

TABLE 9.2   Equilibrium Expressions for the Reaction of Ni^{2+} and Related Nickel Species with Ammonia in Water*

Reaction	Equilibrium Expression	K_f (at 25°C)[a,b]
Addition of first NH_3 to Ni^{2+}: $Ni^{2+} + NH_3 \rightleftharpoons Ni(NH_3)^{2+}$	$K_{f1} = \dfrac{[Ni(NH_3)^{2+}]}{[Ni^{2+}][NH_3]}$	52$\underline{5}$
Addition of second NH_3: $Ni(NH_3)^{2+} + NH_3 \rightleftharpoons Ni(NH_3)_2^{2+}$	$K_{f2} = \dfrac{[Ni(NH_3)_2^{2+}]}{[Ni(NH_3)^{2+}][NH_3]}$	14$\underline{5}$
Addition of third NH_3: $Ni(NH_3)_2^{2+} + NH_3 \rightleftharpoons Ni(NH_3)_3^{2+}$	$K_{f3} = \dfrac{[Ni(NH_3)_3^{2+}]}{[Ni(NH_3)_2^{2+}][NH_3]}$	45.$\underline{7}$
Addition of fourth NH_3: $Ni(NH_3)_3^{2+} + NH_3 \rightleftharpoons Ni(NH_3)_4^{2+}$	$K_{f4} = \dfrac{[Ni(NH_3)_4^{2+}]}{[Ni(NH_3)_3^{2+}][NH_3]}$	13.5
Addition of fifth NH_3: $Ni(NH_3)_4^{2+} + NH_3 \rightleftharpoons Ni(NH_3)_5^{2+}$	$K_{f5} = \dfrac{[Ni(NH_3)_5^{2+}]}{[Ni(NH_3)_4^{2+}][NH_3]}$	4.5$\underline{7}$
Addition of sixth NH_3: $Ni(NH_3)_5^{2+} + NH_3 \rightleftharpoons Ni(NH_3)_6^{2+}$	$K_{f6} = \dfrac{[Ni(NH_3)_6^{2+}]}{[Ni(NH_3)_5^{2+}][NH_3]}$	0.93$\underline{3}$

* All of the K_f values given in this table and Tables 9.4 and 9.5 are written as concentration-based equilibrium constants. Similar expressions based on activities would be used to give the thermodynamic equilibrium constants for these reactions (see Chapter 6).

a The K_f values were calculated based on overall formation constants reported at an ionic strength of 0.0 M in NIST Standard Reference Database 46—NIST *Critically Selected Stability Constants for Metal Complexes Database, vol. 8.0*, NIST, Gaithersburg, MD, 2004.

b The **underlined numbers** represent guard digits (see Chapter 2). When these K_f values are used in calculations, they should be treated as having two significant digits in their values.

of Standards and Technology (NIST),[15,16] along with texts such *Critical Stability Constants* and *Lange's Handbook of Chemistry*.[17,18]

For the addition of one ammonia to Ni^{2+} (or rather $Ni(H_2O)_6{}^{2+}$), the thermodynamic formation constant ($K_f{}^\circ$) would be written as follows,

$$K_f^\circ = \frac{\left(a_{Ni(H_2O)_5\,(NH_3)^{2+}}\right)\left(a_{H_2O}\right)}{\left(a_{Ni(H_2O)_6^{2+}}\right)\left(a_{NH_3}\right)}$$

$$\text{or } K_f^\circ = \frac{\left(a_{Ni(NH_3)^{2+}}\right)}{\left(a_{Ni^{2+}}\right)\left(a_{NH_3}\right)} \tag{9.5}$$

where the equation on the right is the simplified form we will use in this chapter. (*Note*: Remember $a_{H_2O} = 1$ when we are working with an aqueous solution, which is why this term is not shown in the above simplified form just given.) The concentration-based formation constant (K_f) for the same reaction would be written as shown in Equation 9.6,

$$K_f = \frac{\left[Ni(NH_3)^{2+}\right]}{\left[Ni^{2+}\right]\left[NH_3\right]}$$

$$\text{where } K_f^\circ = K_f \frac{\left(\gamma_{Ni(NH_3)^{2+}}\right)}{\left(\gamma_{Ni^{2+}}\right)\left(\gamma_{NH_3}\right)} \tag{9.6}$$

where $\gamma_{Ni(NH_3)^{2+}}$, $\gamma_{Ni^{2+}}$, and γ_{NH_3} are the activity coefficients for the reactants and products during this complex formation reaction.[6] Similar equilibrium expressions can be written for the addition of every other ammonia to Ni^{2+}, with each of these reactions having its own formation constant. If you need to include units with these formation constants for use in calculations, the appropriate apparent unit to use would be M^{-1}.

Table 9.2 shows that the concentration-based formation constants for the reaction of the NH_3 with Ni^{2+} in water (K_{f1} through K_{f6}) have values that decrease from 525 to 0.93 at 25°C. The size of these formation constants is a direct measure of how effectively the ligand will combine with the metal ion. The decrease in formation constant as each additional NH_3 is added to the nickel ions is an effect that happens in nearly all complexation reactions that have multiple steps for adding a ligand. This decrease is an indication that it becomes harder and harder for the metal ion (Ni^{2+}, in this case) to combine with more of the ligand (NH_3) as greater amounts of ligand are complexed by the metal ion. This is the same reason why it is relatively easy to add four ammonia molecules to Cu^{2+} to form $Cu(NH_3)_4{}^{2+}$, while the addition of a fifth or sixth

ammonia molecule to a requires very high concentrations of ammonia.

9.2C Predicting the Distribution of Metal–Ligand Complexes

We can see from Table 9.2 that the reaction of a metal ion with a simple ligand like NH_3 can lead to a complex mixture of products. In the case of Ni^{2+} and NH_3 in water, each complex will contain six ligands, but will have different combinations of water and ammonia (see Box 9.1 for a further discussion of this process). If we know a few things about these reactions, we can predict the distribution of the complexes and relative amount of each complex that will be present under a given set of conditions. This can be accomplished by using an approach similar to the one we used in Chapter 8 for describing the fraction of species for polyprotic acids.

We can describe the fraction of species for a series of metal–ligand complexes by first writing a mass balance expression for all of the metal-containing species in solution. This gives the following mass balance equation for the system in Table 9.2, where C_{Ni} is the analytical concentration of all nickel species in the solution.

$$C_{Ni} = [Ni^{2+}] + [Ni(NH_3)^{2+}] + [Ni(NH_3)_2{}^{2+}] +$$

$$[Ni(NH_3)_3{}^{2+}] + [Ni(NH_3)_4{}^{2+}]$$

$$+ [Ni(NH_3)_5{}^{2+}] + [Ni(NH_3)_6{}^{2+}] \tag{9.9}$$

This mass balance equation can be used with the equilibrium expressions in Table 9.2 to obtain equations for the fraction of each nickel species that will be present at a given concentration ammonia (see Table 9.4). To accomplish this goal, we first use the fact that the fraction of any nickel species, as represented by the symbol "α", is equal to the concentration of that species at equilibrium divided by the analytical concentration of all soluble nickel species, or $[Ni^{2+}]/C_{Ni}$. The resulting fraction of species equation for Ni^{2+} would be written as shown below.

$$\alpha_{Ni^{2+}} = [Ni^{2+}]\Big/\Big([Ni^{2+}] + [Ni(NH_3)^{2+}] +$$

$$[Ni(NH_3)_2{}^{2+}] + [Ni(NH_3)_3{}^{2+}] +$$

$$[Ni(NH_3)_4{}^{2+}] + [Ni(NH_3)_5{}^{2+}] +$$

$$[Ni(NH_3)_6{}^{2+}]\Big) \tag{9.10}$$

BOX 9.1

A Closer Look at Metal–Ligand Complex Formation

The rate of a reaction is an important to consider in many analytical methods. Some metal–ligand complexes seem to form almost instantaneously in water, but others form slowly. The speed of this reaction depends on how rapidly a water molecule can be lost from the metal ion to make way for another ligand. The loss of one water molecule will not lead to a new complex, however, unless the ligand is quite close when this loss occurs. If this is not the case, then another water molecule will take the place of the first and no net change in the complex will be observed.

We can describe this process by looking closer at the *coordination sphere* of the complex, which includes the metal ion and its surrounding ligands. In order for a ligand to be close enough for complex formation, it must be present in the layer of molecules that is immediately next to this complex. For instance, imagine that the species $\{M(H_2O)_6^{2+}\}L$ represents a metal ion that has formed a complex with only water molecules and that has ligand (L) present next to this complex. To simplify matters, let's also suppose that this species is in equilibrium with a form of the metal ion that is not associated with L, or $M(H_2O)_6^{2+}$. We can describe this equilibrium by using the following *outer sphere association constant*, K_{os}, as shown below.[19]

$$M(H_2O)_6^{2+} + L \rightleftharpoons \{M(H_2O)_6^{2+}\}L$$

$$K_{os} = \frac{\left[\{M(H_2O)_6^{2+}\}L\right]}{\left[M(H_2O)_6^{2+}\right][L]} \tag{9.7}$$

The size of this equilibrium constant will depend on the charge of the metal ion (z_M), the charge of the ligand (z_L), and the distance between these agents, which is usually 0.4 nm. Some typical values for this equilibrium constant are $K_{os} = 0.16$ when the product ($z_M z_L$) is zero, $K_{os} = 1.8$ when $z_M z_L) = -2$, $K_{os} = 23$ when ($z_M z_L$) = -4, and $K_{os} = 280$ when ($z_M z_L$) = -6.[20]

The rate of complex formation between a metal ion and monodentate ligand will be determined by K_{os}, and the rate constant for the loss of water (k_{-H_2O}) from the metal ion, where the product of these two terms ($k_{-H_2O} K_{os}$) is equal to the formation rate constant for the complex (k_f).

$$\text{Rate of complex formation} = k_{-H_2O} \left[\{M(H_2O)_6^{2+}\}L\right]$$

$$= k_{-H_2O} K_{os} [M(H_2O)_6][L]$$

$$= k_f [M(H_2O)_6][L] \tag{9.8}$$

Some typical values for k_{-H_2O} are given in Table 9.3. As an example, the reaction of Ni^{2+} in water with one NH_3 would have a formation rate constant that is described by the relationship $k_f = k_{-H_2O} K_{os} = (2.7 \times 10^4 \text{ s}^{-1})(0.16 \text{ } M^{-1})$, or $4.3 \times 10^3 \text{ } M^{-1} \text{ s}^{-1}$. The rate constants for the loss of water from metal ions that are already partially substituted by other ligands may differ significantly from the values shown in Table 9.3.

Table 9.3 Rate Constants for the Loss of Water (k_{-H_2O}) from Metal Ions*

Metal Ion	k_{-H_2O} (s^{-1})
Mg^{2+}	5.3×10^5
Ca^{2+}	5×10^8
Fe^{2+}	3.2×10^6
Fe^{3+}	1.5×10^2
Co^{2+}	2.4×10^6
Co^{3+}	$\ll 10^{-3}$
Ni^{2+}	2.7×10^4
Cu^{2+}	5×10^9
Zn^{2+}	7×10^7

*These values were obtained from D.W. Margerum, G.R. Kayley, D.C. Weatherburn, and G.K. Pagenkopf, "Kinetics and Mechanisms of Complex Formation and Ligand Exchange," In *Coordination Chemistry*, Vol. 2, A. E. Martell (Ed.), ACS Publications, Washington, DC, 1978.

TABLE 9.4   Fraction of Species Equations for the Complexes of Ni^{2+} and Related Nickel Ions with Ammonia in Water

Fraction of Ni^{2+}:

$$\alpha_{Ni^{2+}} = \frac{[Ni^{2+}]}{C_{Ni}} = \frac{1}{1 + \beta_1[NH_3] + \beta_2[NH_3]^2 + \beta_3[NH_3]^2 + \beta_4[NH_3]^4 + \beta_5[NH_3]^5 + \beta_6[NH_3]^6}$$

Fraction of $Ni(NH_3)^{2+}$:

$$\alpha_{Ni(NH_3)^{2+}} = \frac{[Ni(NH_3)^{2+}]}{C_{Ni}} = \frac{\beta_1[NH_3]}{1 + \beta_1[NH_3] + \beta_2[NH_3]^2 + \beta_3[NH_3]^3 + \beta_4[NH_3]^4 + \beta_5[NH_3]^5 + \beta_6[NH_3]^6}$$

Fraction of $Ni(NH_3)_2^{2+}$:

$$\alpha_{Ni(NH_3)_2^{2+}} = \frac{[Ni(NH_3)_2^{2+}]}{C_{Ni}} = \frac{\beta_2[NH_3]^2}{1 + \beta_1[NH_3] + \beta_2[NH_3]^2 + \beta_3[NH_3]^2 + \beta_4[NH_3]^4 + \beta_5[NH_3]^5 + \beta_6[NH_3]^6}$$

Fraction of $Ni(NH_3)_3^{2+}$:

$$\alpha_{Ni(NH_3)_3^{2+}} = \frac{[Ni(NH_3)_3^{2+}]}{C_{Ni}} = \frac{\beta_3[NH_3]^3}{1 + \beta_1[NH_3] + \beta_2[NH_3]^2 + \beta_3[NH_3]^3 + \beta_4[NH_3]^4 + \beta_5[NH_3]^5 + \beta_6[NH_3]^6}$$

Fraction of $Ni(NH_3)_4^{2+}$:

$$\alpha_{Ni(NH_3)_4^{2+}} = \frac{[Ni(NH_3)_4^{2+}]}{C_{Ni}} = \frac{\beta_4[NH_3]^4}{1 + \beta_1[NH_3] + \beta_2[NH_3]^2 + \beta_3[NH_3]^3 + \beta_4[NH_3]^4 + \beta_5[NH_3]^5 + \beta_6[NH_3]^6}$$

Fraction of $Ni(NH_3)_5^{2+}$:

$$\alpha_{Ni(NH_3)_5^{2+}} = \frac{[Ni(NH_3)_5^{2+}]}{C_{Ni}} = \frac{\beta_5[NH_3]^5}{1 + \beta_1[NH_3] + \beta_2[NH_3]^2 + \beta_3[NH_3]^3 + \beta_4[NH_3]^4 + \beta_5[NH_3]^5 + \beta_6[NH_3]^6}$$

Fraction of $Ni(NH_3)_6^{2+}$:

$$\alpha_{Ni(NH_3)_6^{2+}} = \frac{[Ni(NH_3)_6^{2+}]}{C_{Ni}} = \frac{\beta_6[NH_3]^6}{1 + \beta_1[NH_3] + \beta_2[NH_3]^2 + \beta_3[NH_3]^3 + \beta_4[NH_3]^4 + \beta_5[NH_3]^5 + \beta_6[NH_3]^6}$$

Second, we need to simplify this type of relationship by relating each of the concentrations in the denominator to the known concentration of ammonia, $[NH_3]$, and the formation constants for the system. We can show by rearranging the expression for K_{f1} in Table 9.2 that $[Ni(NH_3)^{2+}] = K_{f1}[Ni^{2+}][NH_3]$. Similarly, we can show that $[Ni(NH_3)_2^{2+}] = K_{f2}[Ni(NH_3)^{2+}][NH_3]$, and so on. This process gives the initial equilibrium relationships that are listed in Table 9.5.

At this point we still have too many unknown terms in these equations. One example is the term $[Ni(NH_3)^{2+}]$ that appears on the right-hand side of the initial equilibrium relationship for $[Ni(NH_3)_2^{2+}]$. The same problem occurs for all of the initial relationships we have written in Table 9.5 other than the first one for $[Ni(NH_3)^{2+}]$, which is expressed only in terms of K_{f1}, $[NH_3]$, and $[Ni^{2+}]$ (the last of these terms being our desired unknown). To get around this problem, we can substitute the upper equilibrium relationships in Table 9.5 into the lower ones. The first of these equations gives the relatively simple relationship $[Ni(NH_3)^{2+}] = K_{f1}[Ni^{2+}][NH_3]$. This relationship means we can use $K_{f1}[Ni^{2+}][NH_3]$ in place of $[Ni(NH_3)^{2+}]$ in the next equation for $[Ni(NH_3)^{2+}]$, providing the alternative equation $[Ni(NH_3)_2^{2+}] = K_{f1}K_{f2}[Ni^{2+}][NH_3]^2$. If we continue this substitution process through Table 9.5, a new set of equations is obtained for all nickel species in terms of only $[Ni^{2+}]$, $[NH_3]$, and the formation constants for this system.

A well-defined pattern appears in the equations we obtain during this substitution process. The first pattern that appears as we go down Table 9.5 from the first addition of ammonia to the sixth is that the power on the $[NH_3]$ term increases for each term by one, going from $[NH_3]$ to $[NH_3]^6$. This pattern reflects the fact that more and more molecules of NH_3 are combining with the nickel ions as they form these higher-order complexes. Another pattern is that the number of formation constants that are being multiplied together is increasing from K_{f1} for the first Ni^{2+} complex with ammonia ($Ni(NH_3)^{2+}$) to $K_{f1}K_{f2}K_{f3}K_{f4}K_{f5}K_{f6}$ for the last of these complexes ($Ni(NH_3)_6^{2+}$).

Equations like these can be further simplified by replacing each product of formation constants with a single term known as an **overall formation constant (β)**, *cumulative formation constant*.[2,3,5,10] This term is defined as follows for n stepwise reactions between a metal (M) and ligand (L) leading to the complex $M(L)_n$ and in a system that is at equilibrium.

For net reaction
$$M + n\,L \rightleftharpoons M(L)_n \qquad \beta_n = K_{f1}K_{f2}\cdots K_{fn} \quad (9.11)$$

Based on this definition, we can use β_6 in place of $K_{f1}K_{f2}K_{f3}K_{f4}K_{f5}K_{f6}$, and so on. This substitution gives rise to the alternative relationships given on the right of Table 9.5 for the Ni^{2+} and ammonia system. In this type of multistep process (known as a *stepwise reaction*),[5] the equilibrium constants K_{f1}, K_{f2}, and so on, for the individual reactions are referred to as **stepwise formation constants**.[2,3,10]

TABLE 9.5 Derivation of Fraction of Species Equations for Complexes of Ni^{2+} and Related Nickel Ions with Ammonia in Water

Reaction	Initial Equilibrium Relationship		Alternative Equilibrium Relationships	
Addition on first NH_3 to Ni^{2+}:	$[Ni(NH_3)^{2+}] = K_{f1}[Ni^{2+}][NH_3]$	$=$	$K_{f1}[Ni^{2+}][NH_3]$ $=$	$\beta_1[Ni^{2+}][NH_3]$
Addition of second NH_3:	$[Ni(NH_3)_2{}^{2+}] = K_{f2}[Ni(NH_3)^{2+}][NH_3]$	$=$	$K_{f1}K_{f2}[Ni^{2+}][NH_3]^2$ $=$	$\beta_2[Ni^{2+}][NH_3]$
Addition of third NH_3:	$[Ni(NH_3)_3{}^{2+}] = K_{f3}[Ni(NH_3)_2{}^{2+}][NH_3]$	$=$	$K_{f1}K_{f2}K_{f3}[Ni^{2+}][NH_3]^3$ $=$	$\beta_3[Ni^{2+}][NH_3]$
Addition of fourth NH_3:	$[Ni(NH_3)_4{}^{2+}] = K_{f4}[Ni(NH_3)_3{}^{2+}][NH_3]$	$=$	$K_{f1}K_{f2}K_{f3}K_{f4}[Ni^{2+}][NH_3]^4$ $=$	$\beta_4[Ni^{2+}][NH_3]$
Addition of fifth NH_3:	$[Ni(NH_3)_5{}^{2+}] = K_{f5}[Ni(NH_3)_4{}^{2+}][NH_3]$	$=$	$K_{f1}K_{f2}K_{f3}K_{f4}K_{f5}[Ni^{2+}][NH_3]^5$ $=$	$\beta_5[Ni^{2+}][NH_3]$
Addition of sixth NH_3:	$[Ni(NH_3)_6{}^{2+}] = K_{f6}[Ni(NH_3)_5{}^{2+}][NH_3]$	$=$	$K_{f1}K_{f2}K_{f3}K_{f4}K_{f5}K_{f6}[Ni^{2+}][NH_3]^6$ $=$	$\beta_6[Ni^{2+}][NH_3]$

Overall formation constants at 25°C: $\beta_1 = 5.25 \times 10^2$; $\beta_2 = 7.59 \times 10^4$; $\beta_3 = 3.47 \times 10^6$; $\beta_4 = 4.68 \times 10^7$; $\beta_5 = 2.14 \times 10^8$; $\beta_6 = 2.00 \times 10^8$.

Now that we have equations for all of the nickel species in this series of reactions, we can use them to obtain the fraction for any of these compounds under a given set of reaction conditions. To illustrate this, let's place each of the equilibrium relationships from Table 9.4 back into our expression for $\alpha_{Ni^{2+}}$ in Equation 9.10.

$$\alpha_{Ni)^+} = [Ni^{2+}] / \Big([Ni^{2+}] + \beta_1[Ni^{2+}][NH_3] +$$

$$\beta_2[Ni^{2+}][NH_3]^2 + \beta_3[Ni^{2+}][NH_3]^3 +$$

$$\beta_4[Ni^{2+}][NH_3]^4 + \beta_5[Ni^{2+}][NH_3]^5$$

$$+ \beta_6[Ni^{2+}][NH_3]^6 \Big) \qquad (9.12)$$

The last issue we must address is the presence of $[Ni^{2+}]$, an unknown concentration that appears in both the numerator and denominator of Equation 9.12. This problem is easily handled by dividing both the top and bottom of this expression by the term $[Ni^{2+}]$. The result is the final relationship shown for $\alpha_{Ni^{2+}}$ in Table 9.4, which is now written only in terms of the ligand concentration and the overall formation constants (β_1 through β_6). The same process can be used to obtain the fraction of species equations for any of the other complexes of Ni^{2+} with ammonia. We can now use these equations to predict how much of these species will be present at equilibrium under a given set of conditions.

EXERCISE 9.1 **Concentration of Ni^{2+} in the Presence of Excess Ammonia**

Ammonia is often used to complex with Ni^{2+} and help control the precipitation of this metal ion by the reagent dimethylglyoxime. An aqueous solution is prepared at 25°C that contains an analytical concentration for Ni^{2+} of 1.00×10^{-4} M. This solution also contains 1.0 M ammonia at a high pH. (a) If no other significant reactions are present in this solution other than the combination of Ni^{2+} with ammonia, what fraction of all nickel species will be present as Ni^{2+} at equilibrium? (b) What will be the molar concentration of Ni^{2+} in this solution? What will be the concentration of Ni^{2+} when expressed as the term "pNi", where pNi $= -\log([Ni^{2+}])$?

SOLUTION

(a) We can solve the first part of this problem by using the equation given for $\alpha_{Ni^{2+}}$ in Table 9.4. To get a result for $\alpha_{Ni^{2+}}$ with this equation, we also must have the values for $[NH_3]$ and β_1 through β_6. We are given the starting concentration of ammonia (1.00 M), which is much larger than the total amount of Ni^{2+} present (a 10^4-fold excess). Thus, we can safely assume that the total ammonia concentration after our reaction is approximately the same as its initial value, or $[NH_3] \approx 1.00$ M. It is stated that we are working at a high pH, so we can ignore the conversion of NH_3 to $NH_4{}^+$. We can also obtain the values of β_1 through β_6 from Table 9.5. In this example, the values of

β_1 through β_6 are shown for the sake of dimensional analysis as having apparent units of M^{-1} through M^{-6}, although this calculation could also be conducted without including any units for these constants. When we place these values into the expression for $\alpha_{Ni^{2+}}$ we get the following result.

$$\alpha_{Ni^{2+}} = 1 / \{1 + (5.2\underline{5} \times 10^2 \, M^{-1})(1.00 \, M) +$$
$$(7.5\underline{9} \times 10^4 \, M^{-2})(1.00 \, M)^2 + (3.4\underline{7} \times 10^6 \, M^{-3})$$
$$(1.00 \, M)^3 + (4.6\underline{8} \times 10^7 \, M^{-4})(1.00 \, M)^4 +$$
$$(2.1\underline{4} \times 10^8 \, M^{-5})(1.00 \, M)^5 +$$
$$(2.0\underline{0} \times 10^8 \, M^{-6})(1.00 \, M)^6\}$$
$$\therefore \alpha_{Ni^{2+}} = 2.15 \times 10^{-9} = \mathbf{2.2 \times 10^{-9}}$$

(b) We know the analytical concentration of all nickel species in solution (C_{Ni}), which means we can find $[Ni^{2+}]$ by rearranging the relationship between $\alpha_{Ni^{2+}}$, C_{Ni}, and $[Ni^{2+}]$ in Table 9.4.

$$\left(Ni^{2+}\right) = \alpha_{Ni^{2+}} C_{Ni}$$
$$= (2.1\underline{5} \times 10^{-9})(1.00 \times 10^{-4} \, M)$$
$$= \mathbf{2.2 \times 10^{-13} \, M}$$

The corresponding value of pNi is $-\log(2.2 \times 10^{-13})$ = **12.66**. These results show that of the total nickel ions placed into this solution, only $2.2 \times 10^{-13} \, M$ will be present as Ni^{2+} after it has reached an equilibrium with ammonia. This calculation demonstrates the ability of complex formation to play an important role in controlling the amount of a free metal ion that will be available for other chemical reactions or analysis.

Modern spreadsheets have made these types of calculations relatively easy to perform.[10] This can be accomplished by placing the appropriate fraction of species equations into the spreadsheet and determining their values while varying as $[H^+]$ or another parameter. This information can then be used to determine the relative amount of the various chemical species that will be present under the given set of conditions. For example, Figure 9.4 shows that several types of nickel–ammonia complexes are present at most ammonia concentrations. This type of situation is common for complexes that form between metal ions and many monodentate ligands.

It is important to notice in Table 9.4 that the equations for fraction of species follow a well-defined pattern. As an example, we have one term in the denominator for each type of species that can be formed from the original metal ion. The free metal ion is represented by the "1" and the addition of each ligand to this metal ion is represented by each new term as we move toward the right of this denominator. Thus, all you have to do in writing the fraction of species equations for a simple metal–ligand reaction is to first put down the denominator and then determine which of its terms is the one for your particular

species of interest.[6] You then put this term into the numerator to get your final answer. The next exercise will give you some practice with this approach.

EXERCISE 9.2   Using Fraction of Species Equations for Metal–Ligand Complexes

In a scheme for the qualitative identification of silver, mercury, and lead ions in water, these ions will all form precipitates in the presence of a suitable concentration of Cl^-. However, only silver chloride will dissolve when this precipitate is later placed into a solution containing ammonia. The reactions that take place in this case between Ag^+ and ammonia are shown below.

$$Ag^+ + NH_3 \rightleftharpoons Ag(NH_3)^+$$
$$Ag(NH_3)^+ + NH_3 \rightleftharpoons Ag(NH_3)_2{}^+$$

Write a series of equations that describe the fraction of all soluble silver species that will be present at equilibrium when Ag^+ reacts with ammonia to form $AgNH_3{}^+$ and $Ag(NH_3)_2{}^+$.

SOLUTION

There are three possible forms for soluble silver species in this system, Ag^+, $AgNH_3{}^+$, and $Ag(NH_3)_2{}^+$. The reactions just given can be represented by the formation constants K_{f1} and K_{f2} or the overall formation constants β_1 and β_2, where $\beta_1 = K_{f1}$ and $\beta_2 = K_{f1}K_{f2}$. This will give us a fraction of species equations with three terms in the denominator, or $1 + \beta_1[NH_3] + \beta_2[NH_3]^2$. We also know that the first term must represent the free metal ion (Ag^+), the second term represents $AgNH_3{}^+$, and the last term represents $Ag(NH_3)_2{}^+$. We then obtain the following fraction of species equations.

$$\alpha_{Ag} = \frac{1}{1 + \beta_1\,[NH_3] + \beta_2\,[NH_3]^2}$$

$$\alpha_{AgNH_3} = \frac{\beta_1\,[NH_3]}{1 + \beta_1\,[NH_3] + \beta_2\,[NH_3]^2}$$

$$\alpha_{Ag(NH_3)^2} = \frac{\beta_2\,[NH_3]^2}{1 + \beta_1\,[NH_3] + \beta_2\,[NH_3]^2}$$

This is the same answer we would have obtained if we had gone through the more lengthy process of deriving these expressions from the mass balance equation and equilibrium expressions for this system, as was illustrated earlier for the reaction of nickel ions with ammonia in Tables 9.4 and 9.5.

It is important to keep a few things in mind when you are using the fraction of species equations like those we have just derived for Ag^+ and Ni^{2+} in the presence of ammonia. First, these particular equations assume that

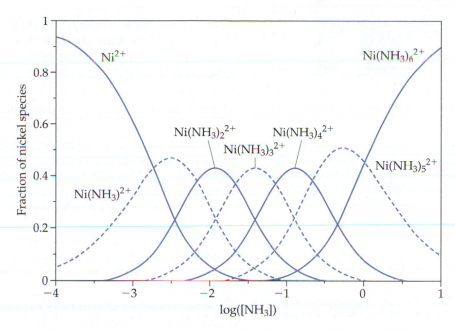

FIGURE 9.4 Distribution of various nickel species in water in the presence of ammonia as a ligand. Theses species are created through the complex formation reactions that are shown in Table 9.2. These results in this plot assume that essentially all of the ammonia is present as NH_3 or that there is a known concentration of ammonia in this form. It is also assumed that no other significant reactions are occurring that might affect the concentrations of Ni^{2+}, ammonia, or the nickel–ammonia complexes.

ammonia is the only type of ligand that is competing with water for the formation of complexes with these metals. If this were not the case, we would have to go back to the more lengthy approach of using mass balance equations and equilibrium expressions to examine the combined effects of these other ligands on the fraction of all metal-related species in solution. Second, we have assumed that we know or can estimate the concentration of free ammonia at equilibrium, because it is only the noncomplexed form of this ligand that is included in the term $[NH_3]$. The fact that ammonia is a weak base means that some of this ligand may also be present as its conjugate acid (NH_4^+) if the pH is sufficiently low in our sample. If we know the pH, we can determine what fraction of the noncomplexed ammonia will be present in both its base and conjugate acid forms. We can then adjust for the effect of this side reaction in our fraction of species equations. We will come back to this idea later in Section 9.3D when we discuss how to deal with side reactions during complex formation.

9.3 COMPLEXES OF CHELATING AGENTS WITH METAL IONS

9.3A What Is a Chelating Agent?

Besides having complex formation based on a monodentate ligand, it is also possible to form a complex in which the ligand has more than one binding site for a metal ion.

This second type of ligand is known as a **chelating agent**.[4] This type of binding requires that the ligand contain two or more atoms that have an unshared pair of electrons (e.g., two nitrogen or oxygen atoms), with these two atoms being separated by at least two or three -CH_2- groups. The -CH_2- groups allow both atoms with unshared electrons to reach the metal ion and bind to it at the same time, producing a stable ring-type structure that typically has five or six atoms within the ring.[4] One common chelating agent is *ethylenediamine* (see Figure 9.5).[9] This agent has the formula $H_2NCH_2CH_2NH_2$, where two nitrogen atoms are separated by a -CH_2CH_2- spacing group. These characteristics allow both ends of ethylenediamine to bind to the same metal ion and give rise to a structure that contains a five-member ring.

Chelating agents can be divided into various subcategories based on how many binding sites they possess for a metal ion. Ethylenediamine is an example of a *bidentate ligand*, or one that has two binding sites for a metal ion. Other chelating agents may have three, four, or even more sites for binding metals, making them *tridendate*, *tetradentate*, or *polydentate ligands*. All of these ligands produce a ring-type structure when they form a complex with a metal ion. The type of complex that forms between a metal ion and a chelating agent is known as a **chelate**.[4,5,9] The term "chelate" comes from the Greek word *chele* for the "claw" of lobsters and crabs, a reference to the way in which a chelating agent wraps around and binds to a metal ion.[21]

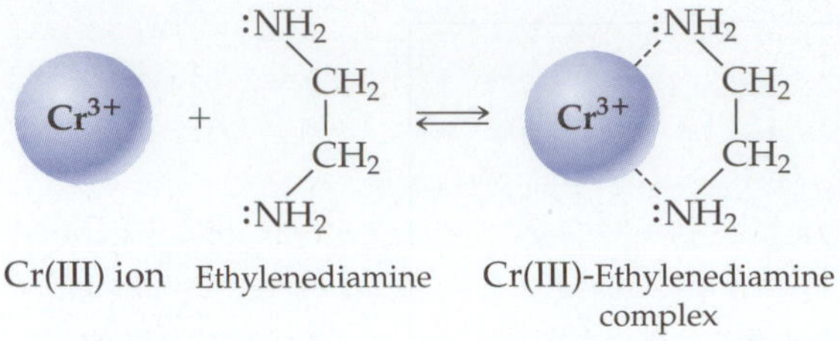

Cr(III) ion   Ethylenediamine   Cr(III)-Ethylenediamine
complex

FIGURE 9.5 A complex formed between a metal ion (Cr^{3+}) and ethylenediamine (en), a bidentate chelating agent. The water molecules that are also involved in forming a complex with Cr^{3+} during this process are not shown for the sake of simplicity.

9.3B The Chelate Effect

Chelating agents are valuable for complexing metal ions because their formation constants are usually much larger than those for monodentate ligands. This can be illustrated by looking at the formation constants for ethylenediamine (abbreviated here as "en") with Ni^{2+} at 25°C.[16]

$$Ni^{2+} + \rightleftharpoons Ni(en)^{2+} \qquad K_{f1} = 2.1 \times 10^7 \qquad (9.13)$$

$$Ni(en^{2+}) + \rightleftharpoons Ni(en)_2^{2+} \qquad K_{f2} = 1.5 \times 10^6 \qquad (9.14)$$

$$Ni(en)_2^{2+} + \rightleftharpoons Ni(en)_3^{2+} \qquad K_{f3} = 1.3 \times 10^4 \qquad (9.15)$$

In comparison, the first three formation constants listed in Table 9.2 for the reaction between Ni^{2+} and ammonia (a simple ligand that also binds through a nitrogen atom), are only 520, 140, and 46. There is also a large difference in the overall formation constant between the highest-order complex that can be formed between Ni^{2+} and ethylenediamine, and ammonia. Up to three molecules of ethylenediamine can react with one Ni^{2+} ion to

form the complex $Ni(en)_3^{2+}$. This process has the overall formation constant β_3, which is equal to $K_{f1} K_{f2} K_{f3} = (2.1 \times 10^7)(1.5 \times 10^6)(1.3 \times 10^4) = 4.1 \times 10^{17}$. In comparison, even the addition of six ammonia molecules to Ni^{2+} has an overall formation constant (β_6) of only 2.0×10^8. The tendency of chelating agents to give more stable complexes with metal ions and provide larger overall formation constants than monodentate ligands is known as the **chelate effect**.

The chelate effect is useful in chemical analysis for several reasons. First, the presence of larger formation constants makes it easier to obtain quantitative reactions between metal ions and chelating agents than between metal ions and monodentate ligands. We will discuss this effect more in Chapter 13 when we consider the method of complexometric titrations. The presence of multiple binding sites on a chelating agent means there will also be fewer species that must be considered for a metal ion and its complexes than when we are working with a monodentate ligand. The ideal situation for a chemical analysis is the use of a chelating agent that has a large formation constant and that reacts only in a 1:1 ratio with the metal ion, such as occurs when working with EDTA.

9.3C Ethylenediaminetetraacetic Acid

Structure of EDTA. EDTA is one of the most commonly used chelating agents. The term "EDTA" is an abbreviation for the chemical name **ethylenediaminetetraacetic acid**. We saw earlier in this chapter how EDTA is commonly used in foods and other products as a preservative and metal binding agent. In fact, this ligand was originally synthesized in the mid-1930s with the purpose of building a chemical agent that could form strong and stable complexes with many metal ions. This reagent was later tested in 1945 for use in the analysis of metal ions, leading to a

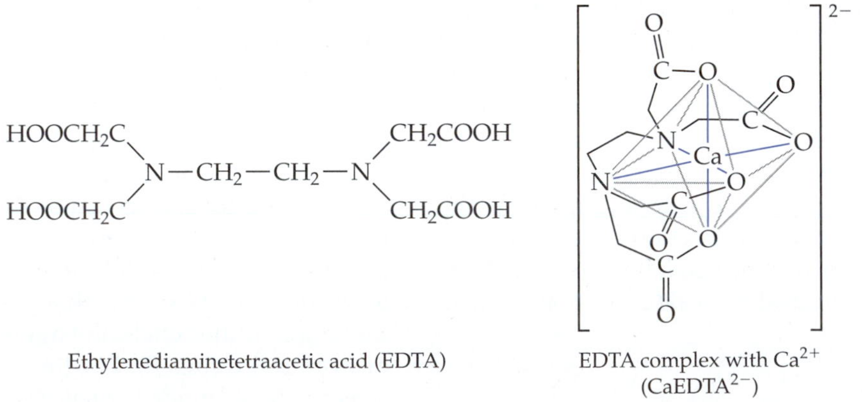

Ethylenediaminetetraacetic acid (EDTA)   EDTA complex with Ca^{2+}
($CaEDTA^{2-}$)

FIGURE 9.6 The structure of EDTA and its complex with a metal ion (Ca^{2+}). EDTA is sold as both a tetraprotic acid (H_4EDTA, molar mass = 292.24 g/mol) and as a disodium salt ($Na_2H_2EDTA \cdot 2\ H_2O$, molar mass = 372.24 g/mol). The tetraprotic form is only very slightly soluble in water, so a base such as NaOH or KOH is often added to this to make up reagent solutions with reasonable concentrations. The disodium form is considerably more soluble, but gives a solution pH of about 6 when it dissolves and is also best dissolved by adding a strong base.

group of methods that are still used to this day for the measurement of such ions.[9,22]

The structure of EDTA is shown in Figure 9.6. This structure has six possible locations where EDTA can bind to a metal ion: two nitrogen atoms and four carboxylate groups. This means that a single molecule of EDTA can form up to six coordinate bonds with the same metal ion. The result is a structure with several five-membered rings, creating a highly stable 1:1 complex.

The general reaction between EDTA and a metal ion (M^{n+}) can be written as follows, where K_f is the formation constant for this complex and $EDTA^{4-}$ is the tetrabasic form of EDTA (that is the form that is generally viewed as binding to metal ions).

$$M^{n+} + EDTA^{4-} \rightleftharpoons M(EDTA)^{n-4}$$

$$K_f = \frac{\left(M(EDTA)^{n-4}\right)}{\left(M^{n+}\right)\left(EDTA^{4-}\right)} \qquad (9.16)$$

The stability of EDTA-metal ion complexes is reflected in the large formation constants for these complexes (Table 9.6), where these constants are often listed as logarithms due to the large range of values that they span.[16] For instance, the complex that forms between Ca^{2+} and $EDTA^{4-}$ has a formation constant of $10^{10.65} = 4.47 \times 10^{10}$ at 25°C, while the formation constant between Fe^{3+} and $EDTA^{4-}$ at

the same temperature is $10^{25.1} = 1.3 \times 10^{25}$. This strong binding, and the ability to react with many metal ions in a 1:1 ratio, has made EDTA useful as a binding agent for metal ions in applications that range from its use as a food additive and cleaning agent to its use in the treatment of heavy-metal poisoning. These same properties have made this chelating agent an invaluable tool for chemical analysis.[22]

Acid–Base Properties of EDTA. A closer look at the structure of EDTA in Figure 9.6 indicates that all of its metal-binding sites can also act as weak acids or weak bases. Both the two nitrogens and four carboxylic acid groups on EDTA must be in their nonprotonated forms to bind to metal ions. This situation occurs because these are the forms in which a pair of electrons will be available for creating a coordinate bond with the metal ion. This feature is the reason why the tetrabasic form of this chelating agent ($EDTA^{4-}$) is written in Equation 9.16 as the form of EDTA that is responsible for binding to metal ions.

To determine how strongly metal ions will bind to a given solution of EDTA, we must consider the pH of our solution and the relative amount of this chelating agent that will be present in the form $EDTA^{4-}$. The pK_a values for the six acid–base sites on EDTA at 25°C are listed in Table 9.7. Using the same approach that was described in Chapter 8 for other polyprotic acids, we can use these values to determine the relative fraction of

TABLE 9.6 Formation Constants for Complexes Formed Between EDTA and Metal Ions at 25°C*

Metal Ion	log $K_{f,MEDTA}$	Metal Ion	log $K_{f,MEDTA}$
Ag^+	7.20	Lu^{3+}	19.74
Al^{3+}	16.4	Mg^{2+}	8.79
Ba^{2+}	7.88	Mn^{2+}	13.89
Ca^{2+}	10.65	Na^+	1.86
Cd^{2+}	16.5	Ni^{2+}	18.4
Co^{2+}	16.45	Pb^{2+}	18.0
Cu^{2+}	18.78	Sc^{3+}	23.1[a]
Fe^{2+}	14.30	Sm^{3+}	17.06
Fe^{3+}	25.1	Sr^{2+}	8.72
Ga^{3+}	(21.7)	Th^{4+}	23.2
Hg^{2+}	21.5	VO^{2+}	18.7
In^{3+}	25.0	Y^{3+}	18.08
La^{3+}	15.36	Zn^{2+}	16.5

*The formation constants given are for an ionic strength of 0.10 M and were obtained from NIST Standard Reference Database 46—*NIST Critically Selected Stability Constants for Metal Complexes Database*, vol. 8.0, NIST, Gaithersburg, MD, 2004. Additional formation constants for EDTA and related species with metal ions can be found in Appendix B. Values in parentheses are listed in NIST Database 46 as having questionable validity.

[a]The value for Sc^{3+} was measured at 20°C.

TABLE 9.7 pK$_a$ Values for EDTA*

Acid–Base Reaction	K$_a$ (at 25°C)^a	pK$_a$ = −log(K$_a$)
$H_6EDTA^{2+} \rightleftharpoons H^+ + H_5EDTA^+$	1.0×10^0	$pK_{a1} = 0.0^{b,c}$
$H_5EDTA^+ \rightleftharpoons H^+ + H_4EDTA$	3.2×10^{-2}	$pK_{a2} = 1.5^c$
$H_4EDTA \rightleftharpoons H^+ + H_3EDTA^-$	1.02×10^{-2}	$pK_{a3} = 1.99$
$H_3EDTA^- \rightleftharpoons H^+ + H_2EDTA^{2-}$	2.14×10^{-3}	$pK_{a4} = 2.67$
$H_2EDTA^- \rightleftharpoons H^+ + HEDTA^{2-}$	6.92×10^{-7}	$pK_{a5} = 6.16$
$HEDTA^{3-} \rightleftharpoons H^+ + EDTA^{4-}$	6.46×10^{-11}	$pK_{a6} = 10.19^d$

*The listed pK$_a$ values were obtained from NIST Standard Reference Database 46—*NIST Critically Selected Stability Constants for Metal Complexes Database, vol. 8.0*, NIST, Gaithersburg, MD, 2004.

^{a}The underlined numbers represent guard digits. When these K$_a$ values are used in calculations they should generally be treated as having two significant digits in their values; the exceptions are K$_{a1}$ and K$_{a2}$ for EDTA, which each have only one significant digit.

^{b}This pK$_a$ value is for an ionic strength of 1.0 M. All other listed values are for an ionic strength of 0.10 M.

^{c}These values are listed in NIST database 46 as having questionable validity.

^{d}This value was determined using K$^+$ as a background electrolyte. A pK$_{a6}$ of 9.52 has been reported when Na$^+$ is a background electrolyte.

each form of EDTA at various pH values. Figure 9.7 shows the fraction of species plot that is obtained when we do this, where EDTA^{4-} becomes the principle form only at high pH values.

The fraction of EDTA that is present as EDTA^{4-} at equilibrium can be calculated at any pH by using the following formula, which can be derived for polyprotic acids as discussed in Chapter 8.

$$\alpha_{EDTA^{4-}} = K_{a1}K_{a2}K_{a3}K_{a4}K_{a5}K_{a6} \Big/ \Big([H^+]^6 + K_{a1}[H^+]^5$$

$$+ K_{a1}K_{a2}[H^+]^4 + K_{a1}K_{a2}K_{a3}[H^+]^3 +$$

$$K_{a1}K_{a2}K_{a3}K_{a4}[H^+]^2 +$$

$$K_{a1}K_{a2}K_{a3}K_{a4}K_{a5}[H^+] +$$

$$K_{a1}K_{a2}K_{a3}K_{a4}K_{a5}K_{a6} \Big) \qquad (9.17)$$

Table 9.8 shows the values for EDTA^{4-} that are obtained with this formula at several pH values. Equation 9.17 can also be used directly to calculate the fraction of EDTA^{4-} at a given pH, as illustrated in the following exercise.

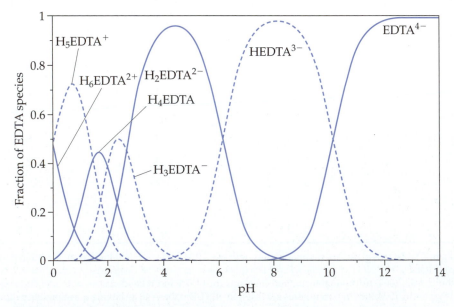

FIGURE 9.7 The distribution of the fractions of various acid–base forms of EDTA as a function of pH. These results are based on the K$_a$ values listed in Table 9.7.

EXERCISE 9.3   The Effect of pH on EDTA

Many soft drinks contain EDTA as a chelating agent. One popular soft drink has a pH of 3.22. What fraction of the EDTA in this product will be present as $EDTA^{4-}$ at equilibrium? What is the main form of EDTA at this pH?

SOLUTION

The fraction of $EDTA^{4-}$ present at pH 3.22 can be determined by placing into Equation 9.17 the K_a values from Table 9.8 and $[H^+] = 10^{-3.22} = 6.03 \times 10^{-4} M$, based on the given pH of 3.22. (*Note: Apparent units of M are given in this example for K_a for the sake of dimensional analysis, but it should be remembered that, as discussed in Chapter 8, K_a values can also be correctly expressed without any units.*)

$$\alpha_{EDTA^{4-}} = \{(1.0 \times 10^0 M)(3.2 \times 10^{-2} M)(1.02 \times 10^{-2} M)$$

$$(2.14 \times 10^{-3} M)(6.92 \times 10^{-7} M)$$

$$(6.46 \times 10^{-11} M)\} / \{(6.03 \times 10^{-4} M)^6 +$$

$$(1.0 \times 10^0 M)(6.03 \times 10^{-4} M)^5 +$$

$$(1.0 \times 10^0 M)(3.2 \times 10^{-2} M)$$

$$(6.03 \times 10^{-4} M)^4 + (1.0 \times 10^0 M)$$

$$(3.2 \times 10^{-2} M)(1.02 \times 10^{-2} M)$$

$$(6.03 \times 10^{-4} M)^3 + (10 \times 10^0 M)$$

$$(3.2 \times 10^{-2} M)(1.02 \times 10^{-2} M)$$

$$(2.14 \times 10^{-3} M)(6.03 \times 10^{-4} M)^2$$

$$+ (1.0 \times 10^0 M)(3.2 \times 10^{-2} M)$$

$$(1.02 \times 10^{-2} M)(2.14 \times 10^{-3} M)$$

$$(6.92 \times 10^{-7} M)(6.03 \times 10^{-4} M)$$

$$+ (1.0 \times 10^0 M)(3.2 \times 10^{-2} M)$$

$$(1.02 \times 10^{-2} M)(2.14 \times 10^{-3} M)$$

$$(6.92 \times 10^{-7} M)(6.46 \times 10^{-11} M)\}$$

$$\therefore \quad \alpha_{EDTA^{4-}} = 9.47 \times 10^{-11} = \mathbf{9.5 \times 10^{-11}}$$

This result indicates that less than 0.1 parts-per-billion of the EDTA is in the tetrabasic form at pH 3.22. The principal form is actually the dibasic form H_2EDTA^{2-}, which makes up about 77% of the EDTA under these pH conditions (see Figure 9.7).

9.3D Dealing with Side Reactions

Conditional Formation Constants. Many ligands and chelating agents like EDTA also have acid–base properties. The preceding exercise demonstrated that pH can have a large effect on the active fraction of such binding agents. This effect makes it important to consider how a change in pH affects the overall binding of these ligands to metal ions. It is possible to account for this effect by using a **conditional formation constant (K'_f)**, or *effective stability constant*.[2] A conditional formation constant is an equilibrium constant that describes the complex formation under a given set of reaction conditions.

A good illustration of where you might use a conditional formation constant is when you are considering the effect of pH on the ability of EDTA to complex with a metal ion. We already know that EDTA can exist in many forms through acid–base reactions, with only the $EDTA^{4-}$ form showing any significant binding for most metal ions. We also know that the fraction of this particular from ($\alpha_{EDTA^{4-}}$) can be calculated from Equation 9.17, which requires knowledge of only the pH of our solution. If we know the total concentration of EDTA that is present in this solution (C_{EDTA}), we can find the concentration of $EDTA^{4-}$ by using the relationships in Equation 9.18.

$$\alpha_{EDTA^{4-}} = \frac{[EDTA^{4-}]}{C_{EDTA}}$$

$$\text{or } [EDTA^{4-}] = \alpha_{EDTA^{4-}} C_{EDTA} \quad (9.18)$$

We can then substitute $\alpha_{EDTA^{4-}} C_{EDTA}$ in for $[EDTA^{4-}]$ in Equation 9.16, and rearrange this expression in terms of a conditional formation constant K'_f that will depend on the pH of the system.

$$K'_f = K_f \alpha_{EDTA^{4-}} = \frac{[M(EDTA)^{n-4}]}{[M^{n+}]C_{EDTA}} \quad (9.19)$$

Another benefit of using Equation 9.19 is it provides an equilibrium expression based on the total concentration of EDTA, a quantity that is usually known when we are using this chelating agent.

EXERCISE 9.4   Using a Conditional Formation Constant

EDTA is often added to tubes for the collection of whole blood. The EDTA is used to bind Ca^{2+} in the blood, which prevents the clotting process from occurring (see Figure 9.8). A clinical chemist wishes to test one of these tubes by using two samples with a pH of 7.0 or 8.0 (solutions that bracket the expected pH of 7.4 for blood). What will be the conditional formation constant for the complex of calcium with EDTA in such samples at 25°C?

SOLUTION

We know from Table 9.6 that K_f for CaEDTA is 4.47×10^{10} at 25°C. In addition, Table 9.8 gives the fraction of EDTA^{4-} at 25°C as being $5.6\underline{4} \times 10^{-4}$ at pH 7.0 and $6.3\underline{3} \times 10^{-3}$ at pH 8.0. We can then use these values with Equation 9.19 to get following values for K'_f.

At pH 7.0: $K'_{f,\text{CaEDTA}} = (4.4\underline{7} \times 10^{10}) \cdot (5.6\underline{4} \times 10^{-4})$

$= 2.5\underline{2} \times 10^7 = \mathbf{2.5 \times 10^7}$

At pH 8.0: $K'_{f,\text{CaEDTA}} = (4.4\underline{7} \times 10^{10}) \cdot (6.3\underline{3} \times 10^{-3})$

$= 2.8\underline{3} \times 10^8 = \mathbf{2.8 \times 10^8}$

These results indicate that the conditional formation constant for Ca^{2+} with EDTA in blood will be between 2.5×10^7 and 2.8×10^8. If we were to instead use Equation 9.17 to find $\alpha_{\text{EDTA}^{4-}}$ at pH 7.4, this would give $\alpha_{\text{EDTA}^{4-}} = 1.5\underline{0} \times 10^{-3}$ and $K'_{f,\text{CaEDTA}} = (4.4\underline{7} \times 10^{10})(1.5\underline{0} \times 10^{-3}) = \mathbf{6.7 \times 10^7}$.

The use of conditional formation constants to adjust for the effects of pH on acid–base reactions is not limited to EDTA, but can also be used for other ligands that act as weak acids or bases. As an example, the effect of ammonia's

TABLE 9.8 Calculated Fraction of EDTA^{4-} as a Function of pH at 25°C*

pH	Fraction of EDTA as EDTA^{4-} $(\alpha_{Y^{4-}})^a$
0	$1.5\underline{4} \times 10^{-23}$
1	$2.1\underline{5} \times 10^{-18}$
2	$3.8\underline{2} \times 10^{-14}$
3	$2.9\underline{5} \times 10^{-11}$
4	$4.2\underline{4} \times 10^{-9}$
5	$4.1\underline{6} \times 10^{-7}$
6	$2.6\underline{4} \times 10^{-5}$
7	$5.6\underline{4} \times 10^{-4}$
8	$6.3\underline{3} \times 10^{-3}$
9	$6.0\underline{6} \times 10^{-2}$
10	$0.39\underline{2}$
11	$0.86\underline{6}$
12	$0.98\underline{5}$
13	$0.99\underline{8}$
14	1.00

*These results were calculated using the K_a values given in Table 9.9.

^{a}The underlined numbers represent guard digits. When these fractions of EDTA^{4-} are used in calculations they should be treated as having two significant digits in their values.

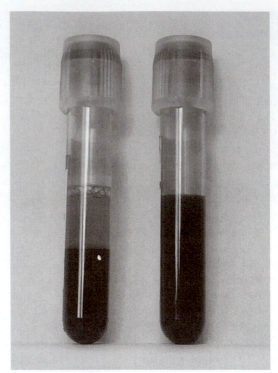

Blood without EDTA added (a)   Blood with EDTA added (b)

FIGURE 9.8 The use of EDTA in tubes for collecting whole blood samples in clinical laboratories. The presence of free Ca^{2+} ions in blood is required for clotting to occur. (a) The first tube contains no EDTA, so Ca^{2+} ions in its blood sample were available for clotting. This caused the blood to separate into its clotted components (on the bottom) and its nonclotting liquid components (on the top and referred to as "serum"). (b) This tube on the right contains EDTA, which was used to bind the Ca^{2+} in blood and to prevent the clotting process from occurring. Both serum and whole-blood samples such as these are useful in clinical testing. (Reproduced with permission under the GNU Free Documentation License).

conversion into its conjugate acid NH$_4$$^+$ can be examined through this same approach by using the total concentration of this ligand (C_{NH_3}) and α_{NH_3}, the fraction of the acid–base species for the ligand that is present as NH$_3$ (the form that has a free pair of electrons and that can bind to metal ions). These terms can then be used to modify an equilibrium expression like that of Equation 9.6 to give a new expression that makes use of a conditional formation constant that is dependent on pH.

$$K'_f = K_f \cdot \alpha_{\text{NH}_3} = \frac{[\text{Ni}(\text{NH}_3)^{2+}]}{[\text{Ni}^{2+}]C_{\text{NH}_3}}$$

$$\text{where} \quad \alpha_{\text{NH}_3} = \frac{K_{a,\text{NH}_4^+}}{[\text{H}^+] + K_{a,\text{NH}_4^+}} \quad (9.20)$$

This same technique can be used to obtain conditional formation constants for any other ligand that also takes part in acid–base reactions by using the appropriate fraction of species equation for this ligand (see Chapter 8 for a review on how to write such an equation).

Predicting the Effects of Side Reactions. Conditional formation constants can be highly valuable in predicting how the extent of a reaction will change with a given value of pH. This idea is illustrated in Figure 9.9 for the binding of Ca^{2+} with EDTA, as described by the conditional formation constant we derived in Equation 9.19. We can see from this figure that the size of K_f' increases with pH until it approaches its maximum value around a pH of 11. We can also use this type of plot to determine what pH range must be employed to obtain a particular range of conditional formation constants. An example of this is shown in Figure 9.9 by the horizontal line, which indicates the pH range over which K_f' exceeds a minimum value of 10^8, a condition that is often desired in titrations that make use of EDTA as a reagent.[23]

A conditional formation constant can also be used to examine the effects of more than one type of side reaction. This can be illustrated by going back to the binding of Ca^{2+} with EDTA. Although Equation 9.19 gives a useful estimate of how pH will affect this reaction, there are some additional side reactions it does not consider. One such reaction takes place at a high pH when Ca^{2+} reacts with OH^- (a ligand now competing with $EDTA^{4-}$) to form the complex $CaOH^+$.[16]

$$Ca^{2+} + OH^- \rightleftharpoons CaOH^+ \quad K_{f,\,CaOH^+} = 2.0 \times 10^1 \quad (9.21)$$

Another side reaction that can occur is the simultaneous binding of $EDTA^4$ with both Ca^{2+} and a hydrogen ion.[16]

$$CaEDTA^{2-} + H^+ \rightleftharpoons CaHEDTA^-$$

$$K_{f,\,CaHEDTA^-} = 1.3 \times 10^3 \quad (9.22)$$

Although this can quickly lead to a complex series of reactions, we can again deal with the overall effect of pH on these various processes by using a conditional formation constant.

To consider all of these reactions, we can write the equilibrium expression for the binding of $EDTA^{4-}$ with Ca^{2+} as given below,

$$K_{f,\,CaEDTA^{2-}} \cdot \frac{[CaEDTA^{2-}]}{[Ca^{2+}][EDTA^{4-}]} =$$

$$\frac{(\alpha_{CaEDTA^{2-}}\, C_{CaEDTA})}{(\alpha_{Ca^{2+}}\, C_{Ca})(\alpha_{EDTA^{4-}}\, C_{EDTA})} \quad (9.23)$$

where $\alpha_{EDTA^{4-}}$ and C_{EDTA} are defined the same as they were in Equation 9.19. The new terms we have used in place of $[CaEDTA^{2-}]$ and $[Ca^{2+}]$ in the right-hand relationship are described by the following mass balance expressions.

$$C_{Ca} = [Ca^{2+}] + [CaOH^+] \quad (9.24)$$

$$C_{CaEDTA} = [CaEDTA^{2-}] + [CaHEDTA^-] \quad (9.25)$$

These mass balance expressions, in turn, can be combined with the equilibrium expressions in Equations 9.21 and 9.22 to get the following formulas for $\alpha_{Ca^{2+}}$ and $\alpha_{CaEDTA^{2-}}$ that are related to the hydrogen ion concentration (or pH) of the system.

$$\alpha_{Ca^{2+}} = \frac{1}{1 + K_{CaOH^+}\left(OH^-\right)} \quad (9.26)$$

$$\alpha_{CaEDTA^{2-}} = \frac{1}{1 + K_{CaHEDTA^-}\left(H^+\right)} \quad (9.27)$$

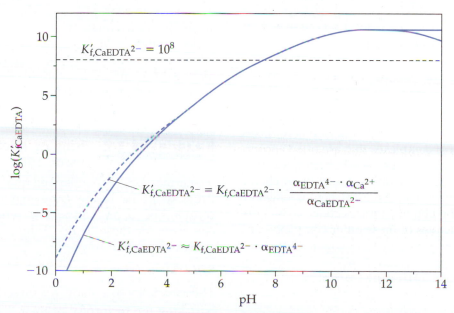

FIGURE 9.9 The overall effect of pH on the conditional formation constant of EDTA for Ca^{2+}. The solid line was calculated based on Equation 9.19 and only considers the effect of pH on the acid–base reactions of EDTA. The dashed line shows the results predicted by Equation 9.28, which also considers the effects of pH on the complexation of Ca^{2+} to OH^- and the simultaneous binding of $EDTA^{4-}$ with Ca^{2+} and H^+. The horizontal line shows the point at which a conditional formation constant would be equal to 10^8.

If you wish, you can further simplify Equation 9.26 by substituting in $K_w/[H^+]$ for $[OH^-]$, as based on the relationship $K_w = [H^+][OH^-]$ for water (see Chapter 8).

We can now rearrange Equation 9.23 to get the following conditional formation constant,

$$K'_{f, CaEDTA^{2-}} = K_{f, CaEDTA^{2-}} \cdot \frac{\alpha_{EDTA^{4-}} \cdot \alpha_{Ca^{2+}}}{\alpha_{CaEDTA^{2-}}}$$

$$= \frac{C_{CaEDTA}}{C_{Ca} \cdot C_{EDTA}} \tag{9.28}$$

where the values of $\alpha_{EDTA^{4-}}$, $\alpha_{Ca^{2+}}$ and $\alpha_{CaEDTA^{2-}}$ can be calculated based on only the equilibrium constants for our reactions and the pH (which, in turn, gives $[H^+]$ and $[OH^-]$). It is now possible to get a more detailed picture of how $K'_{f, CaEDTA^{2-}}$ really changes with pH by looking at the combined effects of all these reactions. The results are included as a dotted line in Figure 9.9. This graph indicates that the first plot we made based on Equation 9.19 gave a good description of calcium binding to EDTA over a broad pH range. There are some differences, however, between this first plot and the more detailed one that is based on Equation 9.28. First, the new plot indicates that there will be slightly stronger binding than we originally expected at pH below 4. This is due to the additional reaction in Equation 9.22, which helps promote the overall binding between Ca^{2+} and EDTA. Second, our new plot shows weaker than expected binding at a pH above 12. This deviation is due to the formation $Ca(OH)^+$, as shown in Equation 9.21, which prevents Ca^{2+} from binding to EDTA. This type of information can be important in optimizing the use of binding agent like EDTA, especially in cases where a relatively small change in the degree of binding may have a big effect on the selectivity of this reagent in a chemical analysis.

9.4 OTHER TYPES OF COMPLEXES

Up to this point, we have focused on complexes between metal ions and ligands that were based on Lewis acid–base reactions and the formation of coordination complexes. There are other types of complexes that are often encountered in analytical chemistry. These other complexes are still formed through the reversible interaction of chemicals, however, they often involve noncovalent interactions instead of coordination bonds. In this section we will discuss the general principles of these other complexes and consider a few of their applications in chemical analysis.

9.4A A General Description of Complex Formation

We can expand our view of complex formation to include processes besides Lewis acid–base reactions by using the following general equilibrium expression,

$$A + L \rightleftharpoons A{-}L \qquad K_f = \frac{[A{-}L]}{[A][L]} \tag{9.29}$$

where A is analyte and L is the ligand that binds to the analyte. As we saw for metal–ligand complex formation, we can describe the preceding reaction in terms of a formation constant, K_f. However, we are no longer limiting the analyte to being only a metal ion, nor must this bind to the ligand through a coordinate bond.

When we view complex formation in this broader sense, there are actually many types of interactions that can lead to the formation of a stable analyte–ligand complex. For instance, hydrogen bonds between the analyte and ligand can be one force that holds these two chemicals together. Other possible forces that can help stabilize the resulting complex include dipole–dipole interactions, dispersion forces, hydrogen bonding, and ionic interactions (see Chapter 7 for a review of these noncovalent interactions). The fit of a particular analyte with a ligand will also determine whether the analyte–ligand complex is stable. Although many of these forces and interactions are by themselves weak (with the exception of ionic interactions), when all these forces occur at the same time this can result in a large formation constant.

When we are referring to complexes other than those between metal ions and their ligands, the term **association constant (K_A)** or *affinity* is sometimes used in place of the formation constant (K_f). This is particularly true if the ligand or analyte are biological compounds. A closely related term is the **dissociation constant (K_D)**,[5] which is equal to the reciprocal of K_A.

$$K_D = \frac{1}{K_A} = \frac{[A][L]}{[A{-}L]} \tag{9.30}$$

The value of K_A is often reported with units of inverse concentration, while K_D is reported with units of concentration. As an example, an analyte and ligand that have an association constant of 10^7 (usually written with apparent units of M^{-1}) will have a corresponding dissociation constant of 10^{-7} M, or 0.1 μM. As you move to a more stable complex, the value of K_D will decrease, while the value of K_A value will increase.

EXERCISE 9.5 **Using Association and Dissociation Constants**

A pharmaceutical chemist determines that two potential drugs will bind to the same enzyme with dissociation constants of 1.2 pM and 8.5 nM. What are the association constants for these drugs? Which drug forms a strong complex with the enzyme?

SOLUTION

The association constants for these two drugs are found by taking the reciprocal of their dissociation constants. This gives K_A values and apparent units of $1/(1.2 \times 10^{-12}\,M) = 8.3 \times 10^{11}\,M^{-1}$ and $1/(8.5 \times 10^{-9}\,M) = 1.2 \times 10^8\,M^{-1}$, respectively. The drug that forms the most stable complex with the enzyme will be the one with the lowest K_D and highest K_A values. This occurs in this case for the drug that has $K_D = 1.2\ pM$ and $K_A = 8.3 \times 10^{11}\ M^{-1}$.

9.4B Examples of Alternative Complexes

You may have guessed from the last exercise that one area in which complex formation is often found is in biological systems. This can include metal–ligand complexes or it might involve the formation of other types of complexes. Examples of this second group include things such as the complex that is formed between the two strands of DNA and the binding of an enzyme with its substrate. Both of these cases involve a relatively large ligand (an enzyme or strand of DNA) that interacts through one or more forces with a given target (a molecule that binds to the enzyme or a strand of DNA that is complementary to the first strand).

It is through this type of complex formation that many of the reactions in our bodies are controlled and conducted on a regular basis. By making use of these same reactions, it is also possible to design analytical methods that can be used to detect and measure specific agents. One common example is the use of antibodies for analyte detection. An **antibody** is a protein produced by the body's immune system that has the ability to specifically bind to a foreign agent, such as a bacterial cell, virus, or protein from another organism.[11] The basic structure of an antibody is shown in Figure 9.10. This molecule has a typical molar mass of 150,000–160,000 g/mol, making it much larger than any of the ligands we discussed in previous parts of this chapter. However, it has the same ability to form reversible complexes with analytes through two identical binding sites that are located at the upper ends of its structure.[24]

Because antibodies can be produced to a wide range of foreign substances, they can also be used as reagents in assays for a variety of chemicals. Any analytical method that uses an antibody as a reagent is known as an **immunoassay**.[5,25,26] This name comes from the term *immunoglobulin*, which is another name used for an antibody.[5,9] There are types of assays that are performed every day through the use of antibodies. These include kits sold by drug stores for pregnancy testing, tests used for food testing labs to detect bacteria, and many of the assays that are run by hospitals to monitor drugs given to patients. Two common ways in which antibodies are used are in the *competitive binding immunoassay and sandwich immunoassay* (see Box 9.2).

There are many other ligands besides antibodies that can be used in analytical methods. Another, smaller example is a *cyclodextrin* (see Figure 9.12). A cyclodextrin is a cyclic polymer of glucose that is formed by certain types of bacteria.[5,28] The result is a truncated, cone-shaped structure that has a nonpolar interior and polar upper and lower edges that are ringed by alcohol groups. This shape and arrangement of groups is useful in that some small organic compounds can enter into the nonpolar interior and form a relatively strong complex with a cyclodextrin. The size of the formation constant for this complex will depend on a number of factors, including the fit of the compound into the cyclodextrin cavity and its ability to form hydrogen bonds or other interactions with the alcohol groups on the cyclodextrin. We will come back to cyclodextrins and discuss some of their analytical applications when we later examine the methods of liquid chromatography and electrophoresis (Chapters 22 and 23).

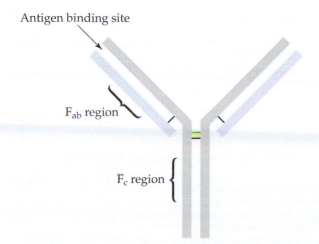

FIGURE 9.10 The structure of a typical antibody. This particular figure shows immunoglobulin G (or IgG), the type of antibody that is most common in blood. An IgG-class antibody has a Y-shaped structure with a diameter of approximately 8–10 nm (or 80–100 Å). The lower part of this structure is the same from one antibody to the next and is known as the F_c *region*. The two upper arms of the antibody are identical and each contains a binding site for a foreign agent, or *antigen*. These two upper arms are known as the F_{ab} *regions*. A change in the amino acid sequence of these F_{ab} regions is what allows our bodies to produce antibodies with binding sites that are specific for different antigens.

Antigen binding site

F_{ab} region

F_c region

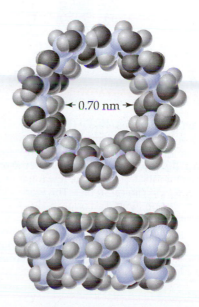

0.70 nm

FIGURE 9.12 The structure of β-cyclodextrin. (Reproduced with permission from M. Chaplin.)

BOX 9.2
Immunoassays

There are many ways in which antibodies can be used in methods for chemical analysis. Two common ways in which these are used are in a competitive binding immunoassay or in a sandwich immunoassay.[25,26] A *competitive binding immunoassay* (shown in Figure 9.11) involves the incubation of analyte in the sample with a fixed amount of a labeled analyte analog (containing an easily measured tag) and a limited amount of antibodies that bind to both the native analyte and labeled analog. Because there is only a limited amount of antibodies present, the analyte and labeled molecules must compete for binding sites on these antibodies. After this competition has been allowed to take place, the compounds that are bound to the antibodies are separated from those that remain free in solution. The amount of the labeled analog present in either the bound or free fraction is then measured. In the absence of any sample analyte, the largest amount of labeled analyte in the bound fraction will be observed. As the amount of sample analyte increases, the level of the bound label will also decrease, giving an indirect measure of the amount of analyte in the sample. This method was originally discovered in the 1959 by American scientists Rosalyn Yalow and Solomon Berson, who used radioisotopes as the labels for this method.[27]

A *sandwich immunoassay* instead involves the use of two different types of antibodies that each bind to the analyte of interest.[25,26] The first of these two antibodies is attached to a solid-phase support and is used for extraction of the analyte from samples. The second antibody contains an easily measured tag and is added in solution to the analyte either before or after this extraction; this second antibody serves to place a label onto the analyte, allowing the amount of analyte on the support to be measured. An important advantage of a sandwich immunoassay is it produces a signal for the bound label that is directly proportional to the amount of analyte. The fact that two types of antibodies are used gives a sandwich immunoassay much higher selectivity than a competitive binding immunoassay. The main disadvantage of a sandwich immunoassay is it can only be used for analytes large enough to bind simultaneously to two antibodies.

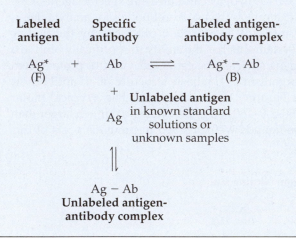

Labeled antigen		Specific antibody		Labeled antigen-antibody complex
Ag* (F)	+	Ab	$\rightleftharpoons$	Ag* − Ab (B)

+

Unlabeled antigen
Ag   in known standard
solutions or
unknown samples

$\Updownarrow$

Ag − Ab
Unlabeled antigen-antibody complex

Ab = antibody   Ag = antigen   Ag* = labeled antigen

F = free fraction of labeled antigen

B = bound fraction of labeled antigen

FIGURE 9.11 Rosalyn S. Yalow (1921–Present), recipient of the 1977 Nobel Prize in physiology and medicine for her development of the radioimmunoassay (RIA). A type of competitive binding immunoassay, RIA is shown in the figure on the right. Dr. Yalow was born in New York City and received her Ph.D. in physics in 1945. She later went to work in the area of nuclear physics and radioisotopes at the Bronx Veterans Administration Hospital. It was here that she began studying how radioisotopes could be used with antibodies in clinical studies. This work, in turn, lead to the development of the RIA by Yalow and Solomon A. Berson (who died prior to 1977). Yalow and Berson initially used the RIA in 1959 to measure the concentration of insulin in the blood of diabetic patients, but this method was soon adapted for hundreds of other applications. The RIA and its more general form of the competitive binding immunoassay are still used to this day in clinical and biochemical laboratories throughout the world.[11] (The photo is reproduced from USIA; the figure is reproduced with permission © The Nobel Foundation 1977 and was used in Dr. Yalow's Nobel lecture.)

Key Words

Other Terms

Questions

WHAT IS COMPLEX FORMATION AND WHAT ARE SOME ANALYTICAL APPLICATIONS OF COMPLEX FORMATION?

1. What is "complex formation"? What are two examples of reactions that involve complex formation?

2. Explain why EDTA is added to mayonnaise. Why is this an example of complex formation?

3. What are three ways in which complex formation is used in chemical analysis?

WHAT IS A METAL–LIGAND COMPLEX?

4. What contribution did Alfred Werner make to the understanding of complex formation reactions?

5. Define each of the following terms, as used in describing complex formation.
   (a) Coordinate bond
   (b) Ligand
   (c) Metal–ligand complex

6. What is a "Lewis acid" or a "Lewis base"? How do these differ from Brönsted acids and bases?

7. Explain why Cu^{2+} is considered to be a Lewis acid when it forms a metal–ligand complex with ammonia.

8. Identify the Lewis acid(s) and the Lewis base(s) in each of the following reactions.

   (a) $OH^- + Mg^{2+} \rightleftharpoons MgOH^-$

   (b) $AgCl_3^{2-} + Cl^- \rightleftharpoons AgCl_4^{3-}$

   (c) $Fe^{3+} + EDTA^{4-} \rightleftharpoons FeEDTA^-$

   (d) $NH_3 + CH_3COOH \rightleftharpoons NH_4^+ + CH_3COO^-$

9. Describe why water is considered to be a ligand for many metal ions.

10. Explain why all Lewis acids and bases are also Brønsted–Lowry acids and bases, but not all Brønsted–Lowry acids and bases are Lewis acids and bases.

11. Define the term "mondentate ligand." Give three specific examples of monodentate ligands.

12. Explain why the two following reactions are equivalent if both are performed in water. What are the advantages of using the first type of expression? What are the advantages of using the second?

Reaction 1:
$$Ni(H_2O)_4(NH_3)_2{}^{2+} + NH_3 \rightleftharpoons Ni(H_2O)_3(NH_3)_3{}^{2+} + H_2O$$

Reaction 2 :
$$Ni(NH_3)_2{}^{2+} + NH_3 \rightleftharpoons Ni(NH_3)_3{}^{2+}$$

13. Discuss why a reagent such as Cl^- or OH^- can sometimes be used as a precipitating agent and other times as a complexing agent for the same metal ion.

FORMATION CONSTANTS FOR METAL–LIGAND COMPLEXES

14. What is a formation constant? Explain why the reaction of a metal ion with a monodentate ligand can have more than one formation constant.

15. Write the concentration-dependent equilibrium expressions for K_f for each of the following reactions.

   (a) $Ba^{2+} + OH^- \rightleftharpoons BaOH^+$

   (b) $Cu^{2+} + 2\,NH_3 \rightleftharpoons Cu(NH_3)_2{}^{2+}$

   (c) $Ni^{2+} + 4\,CN^- \rightleftharpoons Ni(CN)_4{}^{2-}$

   (d) $Fe^{3+} + 6\,F^- \rightleftharpoons FeF_6{}^{3-}$

16. For each of the reactions in Problem 15, write the equilibrium expressions for K_f in terms of chemical activities. Using activity coefficients, show how the activity- and concentration-based values of these formation constants are related to each other.

17. Explain why the value of a formation constant tends to decrease as more ligands are added to a metal ion.

PREDICTING THE DISTRIBUTION OF METAL–LIGAND COMPLEXES

18. Explain why a single metal ion such as Ni^{2+} or Cu^{2+} often has a mixture of many different complexes when it is combined with a ligand like NH_3.

19. What is an "overall formation constant"? How does this differ from a "stepwise formation constant"?

20. Cyanide can act as a ligand for Cd^{2+} to form 1:1 through 1:4 complexes. The stepwise formation constants for these reactions are $K_{f1} = 1.02 \times 10^6$, $K_{f2} = 1.3 \times 10^5$, $K_{f3} = 3.4 \times 10^4$, and $K_{f4} = 1.9 \times 10^2$. Calculate the overall formation constants for these reactions.

21. A reference book lists the following overall formation constants for the combination of bromide with Pb^{2+}: $\log(\beta_1) = 1.77$, $\log(\beta_2) = 2.6$, $\log(\beta_3) = 3.0$, and $\log(\beta_4) = 2.3$. What are the stepwise formation constants for these reactions?

22. What will be the concentration of Ni^{2+} in a solution that is made at 25°C and a high pH by adding 1.0 mL of $0.250\ M\ Ni(NO_3)_2$ to 100 mL of $0.058\ M$ ammonia? You can assume that all of the $Ni(NO_3)_2$ originally dissolves to produce Ni^{2+} and NO_3^- and that the value of $[NH_3]$ at equilibrium is approximately equal to the analytical concentration of ammonia.

23. A textbook lists formation constants for the creation of 1:1 through 1:4 complexes between Cl^- and Hg^{2+}.
    (a) Write a mass balance equation for soluble forms of Hg^{2+} (the metal ion and all of its complexes with chloride) in such a solution.
    (b) Write an equation for the fraction of all soluble mercury(II) species that exists as Hg^{2+} in an aqueous solution that contains chloride ions. Your final equation should be written only in terms of the overall formation constants for the system and the concentration of free chloride ions in the solution.
    (c) Use the equation you derived in Part (b) to calculate the fraction of all soluble mercury(II) species that will be present as Hg^{2+} at 25°C in a solution where $[Cl^-] = 0.050\ M$. Under these same conditions, what will be the value of pHg if pHg $= -\log([Hg^{2+}])$?

24. It was stated earlier in this chapter that, at a sufficiently high ligand concentration, between one and six molecules of ammonia can combine with Cu^{2+} ions.
    (a) Write a mass balance equation for the various forms of Cu^{2+} and its complexes with ammonia in such a system.
    (b) Write two separate equations that describe the fraction of all soluble copper(II) species that will exist as Cu^{2+} or $Cu(NH_3)_4^{2+}$ in an aqueous solution of ammonia. The final form of these equations should be expressed in terms of only the overall formation constants for the system and the concentration of free ammonia in solution.
    (c) Use the equations you derived in Part (b) to calculate the fractions of all soluble copper(II) species that will be present as Cu^{2+} and $Cu(NH_3)_4^{2+}$ in a solution where $[NH_3] = 0.10\ M$.

25. Thiocyanate (SCN^-) and Zn^{2+} are known to react in water to form the soluble complexes $Zn(SCN)^+$ and $Zn(SCN)_2$. Write fraction of species equations for Zn^{2+} and each of its complexes with SCN^-. The final form of these equations should be expressed in terms of only the overall formation constants for the system and the concentration of free thiocyanate in solution.

26. Stannous fluoride (SnF_2) is sometimes used in toothpaste and other products as an agent to prevent tooth decay. Fluoride ions can form the following soluble complexes with Sn^{2+} ions: SnF^+, SnF_2, and SnF_3^-. Write fraction of species equations for Sn^{2+} and each of these complexes involving F^-. The final form of these equations should be expressed in terms of only the overall formation constants for the system and $[F^-]$.

CHELATING AGENTS

27. What is a "chelating agent"? How does this differ from a monodentate ligand?

28. Explain why ethylenediamine is an example of a chelating agent.

29. What type of structure is produced when ethylenediamine binds to a metal ion?

30. Define each of the following terms.
    (a) Bidentate ligand
    (b) Tridentate ligand
    (c) Tetradentate ligand
    (d) Polydentate ligand

31. What is a "chelate"? Describe the general process by which a chelate is formed.

THE CHELATE EFFECT

32. An inorganic chemist finds that when two molecules of ethylenediamine bind to Co^{2+} the overall formation constant is 4.2×10^{10} at 25°C. This binding involves four nitrogen atoms in the molecules of ethylenediamine. However, Co^{2+} has an overall formation constant of only 2.0×10^5 at the same temperature when it binds to four molecules of ammonia (another nitrogen-based ligand). Explain why ethylenediamine has a strong complex with Co^{2+} than ammonia under these conditions.

33. What is the "chelate effect"? Why is this effect believed to occur?

34. Explain why the chelate effect can be useful in a chemical analysis.

ETHYLENEDIAMINETETRAACETIC ACID

35. What is EDTA? Explain why EDTA is an example of a chelating agent.

36. One of the earliest uses of EDTA was as a binding agent for toxic metals like Pb^{2+}. Write the reaction for this process and give the formation constant for this reaction. Discuss the properties of EDTA that allow it to form a strong complex with Pb^{2+}.

37. A service laboratory routinely determines "water hardness" by measuring the amount of Ca^{2+} and Mg^{2+} in water samples by using EDTA as a reagent.
    (a) Write the reactions for the binding of EDTA with each of these metal ions.
    (b) What are the formation constants for the complexes that are formed between these ions and EDTA?
    (c) Approximately what volume of $0.0132\ M$ EDTA would be necessary to complex all the calcium in a standard solution that is prepared by dissolving 0.5764 g calcium carbonate in water?

38. What mass of $Na_2H_2EDTA \cdot 2H_2O$ must be used to prepare 500 mL of $0.0200\ M$ EDTA? What would be the molarity of an EDTA solution prepared by placing 7.50 g of Na_2H_2EDTA into water to make 750 mL of solution?

39. Based on the structure of EDTA, explain why this binding agent can act as a Brønsted–Lowry acid or base. What regions in EDTA's structure give rise to these acid–base properties. What regions of EDTA give rise to its ability to act as a ligand for metal ions?

40. A manufacturer of shampoo wishes to include EDTA as an additive in a product that will have a "neutral" pH of 7.5 and 25°C.
    (a) What fraction of EDTA will be present at equilibrium as $EDTA^{4-}$ at this pH?
    (b) What will be the principal form of EDTA in this product?

41. An analytical chemist prepares a reagent that contains EDTA in a solution that is buffered at pH 9.5 and kept at 25°C.
    (a) What fraction of the EDTA will be present as its tetrabasic form in this solution?
    (b) What will be the principal form of EDTA in this reagent?

42. What fraction of EDTA will be present in the $EDTA^{4-}$ form at 25°C and pH 5.0 or pH 9.0? How will an increase in pH from 5.0 to 9.0 be expected to change the binding of a metal ion to EDTA?

43. An environmental scientist notices that the degree of binding between EDTA and Hg^{2+} increases almost 15,000-fold as the pH is raised from 6.0 to 10.0 in a water sample that contains Hg^{2+} ions. What is the cause of this effect?

DEALING WITH SIDE REACTIONS

44. What is a "conditional formation constant"? Why is this often useful when describing complex formation?

45. Explain how a correction for the acid–base properties of EDTA can be made by using a conditional formation constant for the reaction of EDTA with metal ions.

46. What will be the conditional formation constants for each of the following reactions at 25°C in an aqueous solution? Assume that the only significant side reactions are the acid–base reactions of EDTA.
    (a) Cu^{2+} + EDTA at pH 4.0
    (b) Ni^{2+} + EDTA at pH 9.0
    (c) Zn^{2+} + EDTA at pH 8.0
    (d) Mg^{2+} + EDTA at pH 3.0

47. Calculate the conditional formation constants for each of the following complexes in water at 25°C. You may assume that the acid–base reactions of EDTA are the only significant side reactions occurring in the solution.
    (a) $LaEDTA^-$ at pH 7.5
    (b) $ScEDTA^-$ at pH 4.2
    (c) $ThEDTA$ at pH 5.8
    (d) $CdEDTA^{2-}$ at pH 6.5

48. Discuss how conditional formation constants can be used with ligands other than EDTA that also have acid–base properties. Illustrate your answer using ammonia as an example.

49. What will be the conditional formation constant for the complex $Ni(NH_3)^{2+}$ if this complex is allowed to form in water at 25°C and a pH of 5.0? How does this result differ from the conditional formation constant that would be obtained at pH 10.0? Which of these values is closest to the true formation constant for $Ni(NH_3)^{2+}$?

50. A chemist wishes to examine the effect of pH on the ability of ammonia to complex with Co^{2+}. These experiments are to be performed at pH values of 4.0, 6.0, 8.0, and 10.0.
    (a) Write expressions for the conditional formation constants for the following complexes between Co^{2+} ions and ammonia: $Co(NH_3)^{2+}$, $Co(NH_3)_2^{2+}$, $Co(NH_3)_3^{2+}$, and $Co(NH_3)_4^{2+}$.
    (b) What fraction of the various acid–base forms of ammonia will be present as NH_3 at the given pH values?
    (c) What will be the values of the conditional formation constants for all of the complexes in Part (a) when these complexes are formed at each of the given pH values and in water at 25°C?

51. Explain why the conditional formation constant for the complex of Ca^{2+} with EDTA has a maximum at approximately pH 10–12. How does this result differ from the case where only the acid–base properties of EDTA are considered when accounting for possible side reactions?

A GENERAL DESCRIPTION OF COMPLEX FORMATION

52. Write a general reaction and equilibrium expression that can be used to describe the formation of a 1:1 complex between any type of analyte and ligand.

53. If complexes other than those between metal ions and ligands are considered, what interactions besides Lewis acid–base reactions can lead to the formation of such complexes?

54. What is an "association constant"? How is this related to the formation constant for a complex?

55. What is a "dissociation constant"? How is this related to the association constant and formation constant for a complex?

56. A biochemist determines that a receptor binds to a hormone with a dissociation constant of 6.3×10^{-10} M. What is the association constant for the resulting receptor–hormone complex?

57. An analytical chemist determines that an antibody binds to an analyte and related contaminant with association constants of 2.7×10^{10} M^{-1} and 4.0×10^6 M^{-1}, respectively. What are the dissociation constants for these two compounds? Which of these compounds binds more tightly to the antibody?

EXAMPLES OF ALTERNATIVE COMPLEXES

58. What is an "antibody"? Describe the general structure and physical properties of an antibody.

59. How is an antibody similar to the binding agents that were discussed earlier in this chapter? How is an antibody different from these other binding agents?

60. What is an "immunoassay"? Describe two different types of immunoassays.

61. What is a "cyclodextrin"? What is the structure of this binding agent?

62. Describe how a cyclodextrin can form a complex with another compound.

CHALLENGE PROBLEMS

63. Use sources such as References 15–18 to locate the formation constants for each of the following metal–ligand complexes or chelates.
    (a) EDTA + Ca^{2+} (37°C)
    (b) Ammonia + Mg^{2+} (25°C)
    (c) 4-Sulfonic acid + Cu^{2+} (25°C)
    (d) Nitroso-2-naphthol + Zn^{2+} (25°C)
    (e) Thiourea + Pb^{2+} (1–6 ligands at 25°C)
    (f) Glycine + Ni^{2+} (1–3 ligands at 25°C)

64. Locate a method in your laboratory that involves the use of a solution that contains EDTA. Write a standard operating procedure for the preparation of this solution and its standardization.

65. Obtain a copy of the material safety data sheet for EDTA. State how this chemical should be stored and describe any chemical and physical hazards that are associated with EDTA.

66. Reference 16 lists the following values of $\log(K_f)$ for EDTA and Mg^{2+} at 25°C and various ionic strengths: $\log(K_f) = 8.79$ at $I = 0.10\ M$, 8.67 at $I = 0.50\ M$, and 8.61 at $I = 1.00\ M$. Explain this trend based on your knowledge of activity coefficients and their dependence on ionic strength.

67. Obtain the values for the formation constants for the 1:1 complexes that form between Ni^{2+} ions when using Cl^-, CN^-, NH_3, or F^- as the ligand. Also obtain the pK_b values for each of these ligands. Use a spreadsheet to prepare a plot of K_{f1} for the nickel complexes versus the pK_b for each of these ligands. What trend do you observe in this plot? What do you think is the basis for this trend?

68. If EDTA is already complexed with one metal ion, it is possible that this may be displaced by a different metal ion with stronger binding to EDTA. For instance, when EDTA is used as a preservative in food, it is a calcium–EDTA complex that is actually added. When this ligand binds to Fe^{3+} in the food to prevent spoiling, this metal ion must displace the Ca^{2+} that is already bound to the EDTA. We can represent this competition by the following reaction and equilibrium expression.

$$Fe^{3+} + CaEDTA^{2-} \rightleftharpoons FeEDTA^- + Ca^{2+}$$

$$K = \frac{[FeEDTA^-][Ca^{2+}]}{[Fe^{3+}][CaEDTA^{2-}]} \qquad (9.31)$$

(a) The equilibrium constant for the preceding reaction can be calculated by using the formation constants for $CaEDTA^{2-}$ and $FeEDTA^-$. What is the value of this equilibrium constant at 25°C?

(b) What is the change in total standard free energy ($\Delta G°$) for the reaction in Equation 9.31 at 25°C? How does this change in free energy compared to that for the separate formation of $CaEDTA^{2-}$ and $FeEDTA^-$ at 25°C?

69. Suppose that an aqueous solution containing 0.00268 mol of EDTA at pH 5.0 is combined with 0.00100 mol Cu^{2+} and 0.00185 mol Ni^{2+}, and that this mixture is then brought to a final volume of 500 mL. Using a spreadsheet, calculate the concentrations of Cu^{2+}, Ni^{2+}, $NiEDTA^{2-}$, $CuEDTA^{2-}$, and all forms of EDTA that will present at 25°C when this mixture is at an equilibrium.

70. Construct a spreadsheet that allows you to plot the fraction of Ni^{2+} and its various complexes with ammonia. Include in this spreadsheet columns in which you can change both the total concentration of ammonia (C_{NH_3}) and the pH of the solution, which will affect the fraction of the acid–base forms of ammonia that are present as NH_3.

(a) Use this spreadsheet to prepare a plot where it is assumed that all of the ammonia is present in the form NH_3 (as would occur at a high pH). This plot should include lines that show the fractions of Ni^{2+} and the complexes $Ni(NH_3)^{2+}$ through $Ni(NH_3)_6^{2+}$ when these are plotted versus $\log(C_{NH_3})$ values that range from -4 to 1. Compare your plot with the one that is shown in Figure 9.4.

(b) Prepare two other plots using the same total concentrations of ammonia as in Part (a), but where the pH is now 6.0 or 8.0. How does the appearance of these new plots compare to the one from Part (a) or Figure 9.4?

71. The formation constants at 25°C for EDTA with Mg^{2+}, Ca^{2+}, Sr^{2+}, Ba^{2+}, and Ra^{2+} are $10^{8.79}$, $10^{10.65}$, $10^{8.72}$, $10^{7.88}$, and approximately $10^{7.0}$, respectively.[16] Compare this trend to the position of these elements in the periodic chart. What trend do you notice in these values?

72. The observed pK_{a6} for EDTA at 25°C is 10.19 when measured in 0.10 M KCl, but 9.52 when measured in 0.10 M NaCl.[16] If it is assumed that EDTA does not complex with K^+, show how this information can be used to find the formation constant for the complex of Na^+ with $EDTA^{4-}$.

73. Ethylenediamine monoacetic acid (EDMA) is a chelating agent related to EDTA and that binds to a variety of metal ions. However, EDMA differs from EDTA in that it only has three instead of six locations in its structure that can undergo acid–base reactions.

(a) Write fraction of species equations for each of the four possible acid–base forms of EDMA.

(b) The pK_a values for EDMA are 2.15, 6.65, and 10.15. Using this information, make a plot that shows how the fraction of each acid–base species of EDMA changes with pH.

(c) Over what pH range will the most basic form of EDMA be the principal main species? How does this range compare to the pH range over which the tetrabasic form of EDTA is the main form of EDTA?

74. Under appropriate pH conditions, Cu^{2+} and its complexes with EDTA can have similar side reactions to those shown in Section 9.3D for Ca^{2+} and EDTA. For instance, Cu^{2+} can react with hydroxide ions, and $EDTA^{4-}$ can bind to both Cu^{2+} and H^+, as illustrated in Equations 9.32–9.33 using K_f values that have been measured at 25°C.[16]

$$Cu^{2+} + OH^- \rightleftharpoons CaOH^+$$

$$K_{f,\ CaOH^+} = 3.2 \times 10^6 \qquad (9.32)$$

$$CuEDTA^{2-} + H^+ \rightleftharpoons CuHEDTA^-$$

$$K_{f,\ CaHEDTA^-} = 1.3 \times 10^3 \qquad (9.33)$$

(a) Derive fraction of species equations similar to Equations 9.26 and 9.27 for the binding of Cu^{2+} with EDTA.

(b) Use the equations from Part (a) to produce a plot that shows the effects of pH on the conditional formation constant for Cu^{2+} with EDTA. Based on this plot, what pH range would be expected to give the strongest binding between Cu^{2+} and EDTA?

(c) Another side reaction that can occur during the reaction of $EDTA^{4-}$ with Cu^{2+} is a combination of the resulting complex with hydroxide ions, as shown in Equation 9.34.[16]

$$CuEDTA^{2-} + OH^- \rightleftharpoons Cu(OH)EDTA^{3-}$$

$$K_{Cu(OH)EDTA^{3-}} = 3.2 \times 10^{-2} \qquad (9.34)$$

Modify the expressions you derived in Part (a) to also include this side reaction and prepare a plot similar to that made in Part (b). How does this additional side reaction affect the conditional formation constant for Cu^{2+} with EDTA as the pH is varied?

75. If k_{-H_2O} for Cr^{3+} is $5.8 \times 10^{-7}\ s^{-1}$, what rate constant would be expected for the following reaction? Show how you obtained your answer.

$$Cr^{3+} + NH_3 \rightarrow Cr(NH_3)^{3+} \qquad (9.35)$$

76. It was mentioned earlier in this chapter that Cl^- cannot only combine with Ag^+ to form the precipitate $AgCl$ but can also result in soluble complexes such as $AgCl_2^-$ and $AgCl_3^-$. Table 9.9 shows how the solubility of $AgCl$ changes as the concentration of chloride ions is varied.[29,30]

TABLE 9.9 Solubility of AgCl at Various Concentrations of Soluble Chloride Ions

Solution Concentration of Cl⁻ (*M*)	Solubility of AgCl (*M*)
5.38×10^{-5}	5.37×10^{-6}
5.92×10^{-5}	3.31×10^{-6}
1.12×10^{-4}	2.04×10^{-6}
2.08×10^{-4}	1.66×10^{-6}
3.44×10^{-4}	1.02×10^{-6}
5.51×10^{-4}	6.92×10^{-7}
9.66×10^{-4}	6.92×10^{-7}
1.10×10^{-3}	5.25×10^{-7}
1.27×10^{-3}	6.02×10^{-7}
1.59×10^{-3}	5.62×10^{-7}
2.75×10^{-3}	4.90×10^{-7}
5.50×10^{-3}	5.75×10^{-7}
1.10×10^{-2}	6.60×10^{-7}
2.75×10^{-2}	1.10×10^{-6}
5.50×10^{-2}	1.95×10^{-6}
1.10×10^{-1}	3.80×10^{-6}

(a) Make a plot of the solubility of $AgCl$ versus the concentration of Cl^- in solution. Make a similar plot using expected solubility of $AgCl$ in the absence of any complexation, as given by the term $K_{sp, AgCl}/[Cl^-]$.

(b) What similarities or difference do you see when comparing the two plots in Part (a)? What are the reasons for these similarities or differences? What type of reaction between Ag^+ and Cl^- would you expect to be most important at low chloride concentrations. What type of reaction would be more important at high chloride concentrations?

(c) Prepare a plot of the solubility of $AgCl$ versus the term $1/([Cl^-]\gamma_\pm^2)$ at chloride concentrations below 2.75×10^{-3} M (the chloride concentration producing the minimum solubility for $AgCl$ in Table 9.9). (*Note:* $\gamma_\pm$ is the mean activity coefficient for Ag^+ and Cl^-, as discussed in Chapter 6.) What type of behavior do you observe from this plot? What information can you obtain from the slope and intercept of this plot?

(d) Prepare a plot of the solubility of $AgCl$ versus $[Cl^-]$ at high chloride concentrations (i.e., those above 2.75×10^{-3} M in Table 9.9). What type of behavior do you observe from this graph? What information can you obtain from the slope and intercept of this plot?

77. It is possible to use a ligand or chelating agent like EDTA to form a solution known as a *metal buffer* or *pM buffer*) that is used to help maintain a consistent concentration for the free form of a metal ion.[3]

(a) Using an approach similar to that described in Chapter 8 for the Henderson–Hasselbalch equation, show that it is possible to convert Equation 9.16 into the following form, where $pM = -\log([M^{n+}])$.

$$pM = \log K_f + \log\left(\frac{[EDTA^{4-}]}{[MEDTA^{n-4}]}\right) \quad (9.36)$$

(b) Derive an expression similar to that shown in Equation 9.36, but now beginning with Equation 9.19 and using the conditional formation constant K'_f instead of K_f.

(c) A biochemist wishes to create a metal buffer for calcium ions. To do this, he prepares a mixture of 0.10 M EDTA (prepared from H_4EDTA) and 0.10 M CaEDTA. This solution is then adjusted to and maintained at a pH of 7.0. What is the value of pCa (where $pCa = -\log([Ca^{2+}])$ for this solution?

(d) How does the value of pCa change in Part (c) if the biochemist instead uses a pH 7.0 solution that contains 0.05 M EDTA and 0.15 M CaEDTA? What is pCa if the biochemist uses a pH 7.0 solution that contains 0.15 M EDTA and 0.05 M CaEDTA? What do your results tell you about the metal buffering ability of this solution?

78. The amount of insulin in a sample of human serum is to be determined by a competitive binding immunoassay (an RIA, as discussed in Box 9.2). This procedure involves mixing 100 μL of the sample with a small fixed amount of anti-insulin antibodies plus a fixed amount of ^{125}I-labeled insulin. The total level of radioactivity for this added, labeled insulin is 1500 counts per minute (cpm). After this mixture has been allowed to incubate for 1 day, the antibody-bound fractions of the insulin and ^{125}I-labeled insulin are removed from the mixture and measured. The following results are obtained for the insulin sample and a series of standards that contain known amounts of insulin.

Concentration of Insulin (μUnits/mL)	Radioactivity of Bound Fraction (cpm)
0	850
5	825
10	790
20	750
40	525
80	300
160	105
Unknown sample	450

(a) One way of preparing a calibration curve for this type of assay is to plot the ratio B/T (where B is the radioactivity measured in the bound fraction and T is the total radioactivity initially added to the sample) versus the concentration of analyte in the measured standards. Prepare this type of plot using the above data in the table. What type of response do you observe? What concentration does this curve give for the unknown sample?

(b) A second way of preparing a calibration curve for this type of assay is to plot the ratio B/T versus the base-10 logarithm of the concentration of analyte in the measured standards. Prepare this type of plot using the data in the table and determine the concentration of insulin in the unknown sample. What type of response do you observe for this second plot? What advantages or disadvantages do you think this second plot might offer versus the plot that was prepared in Part (a)?

79. The amount of parathyrin (a hormone with a molar mass of 9500 g/mol) is to be determined in human plasma by using a sandwich immunoassay. This method involves first mixing 100 μL of plasma or standards with 200 μL of a buffer solution containing a fixed amount of labeled but soluble antibodies that can bind to parathyrin. A small bead is then added to each of these solutions, which contains immobilized antibodies that will also bind parathyrin. After the bead and solution have been allowed to incubate for 24 h, the bead is removed from the solution and washed to remove any nonbound sample components or excess reagents. The amount of labeled antibodies that remain on the bead is then measured. This assay gives the following results for an unknown plasma sample and a series of standards that contain known amounts of parathyrin.

Concentration of Parathyrin (pg/mL)	Signal Due to Bound, Labeled Antibodies
0	100
5	850
10	1,610
20	3,000
40	6,150
80	11,500
160	22,500
Unknown sample	7,500

(a) A calibration curve is prepared for this assay by plotting the signal due to the bound, labeled antibodies versus the concentration of parathyrin in the standards. Prepare this type of plot using the data in the table. What response do you observe? What concentration does this curve give for parathyrin in the unknown sample?

(b) Compare the plot that you prepared in Part (a) with the plots that were generated for the competitive binding immunoassay in Problem 78. Based on these plots, what advantages do you think a sandwich immunoassay will have in analyte detection? What are some possible limitations of this assay?

TOPICS FOR DISCUSSION AND REPORTS

80. Obtain more information from Ref. 1 and other articles on EDTA and other preservatives that are used in food. Describe the function of each type of preservative and the type of chemical reaction (for example, acid–base reaction,

complex formation, and so on) that these preservatives use to maintain food freshness.

81. Analytical chemistry is important to the food industry as a tool to learn about the content of food and to assure that this food is of good quality and has a known content. Find information on how the fat, protein, and carbohydrate content of food is commonly measured. Write a report on this topic and discuss it with other students.

82. The U.S. Food and Drug Administration (FDA) and U.S. Department of Agriculture (USDA) have played important roles in the past in dealing with problems in our food and drug supply (for example, see Reference. 31). Obtain information on a recent issue dealing with contamination or other problems with a specific type of food or drug in the United States What role did the FDA or USDA play in resolving this problem? What role did analytical methods play in dealing with this issue?

83. In 2008, approximately 300,000 infants in China became sick and six died as a result of the unauthorized addition of melamine to milk products. Melamine is a nitrogen-rich compound that had been added to these products by some manufacturers to increase the apparent protein content. Several individuals involved in this scandal were given death sentences or life imprisonment. This event also led to a call for greater efforts and better analytical methods to detect such contaminants in food.[32] Obtain more information on efforts that have been made by China and other governments in this area and write a report on this topic.

84. Contact scientists at a local laboratory and discuss with them how they use EDTA or other complexing agents. Write a report on your findings.

85. Locate a recent research article from *Analytical Chemistry* or some other journal dealing with chemical analysis that made use of a chelating agent (such as EDTA) or a biological complexing agent (such as antibodies or cyclodextrins). Discuss how the chelating agent or biological agent was used in the article. How did this application make use of the ability of the chelating agent or biological agent to form complexes with other chemicals?

86. One of the earliest assays that used antibodies for testing was the ABO blood typing scheme that we still use today. Obtain information on this method and describe how antibodies (usually present in this assay in the form of "antisera") are used in this approach.

87. There are many chemicals besides ammonia, EDTA, antibodies, and cyclodextrins that can form complexes. Obtain more information on one of the following biological agents and learn about how it is used as a binding agent for chemical analysis.
    (a) Crown ethers
    (b) Steptavidin or avidin
    (c) Aptamers
    (d) Concanavalin A or wheat germ agglutinin

References

1. "What's That Stuff?," *Chemical & Engineering News*, November 11, 2002, p. 40.
2. J. Inczedy, T. Lengyel, and A.M. Ure, *Compendium of Analytical Nomenclature*, 3rd ed., Blackwell Science, Malden, MA, 1997.
3. H.M.N.H. Irving, H. Freiser, and T.S. West, *Compendium of Analytical Nomenclature: Definitive Rules–1977*, Pergamon Press, New York, 1977

4. G. Maludziska, Ed., *Dictionary of Analytical Chemistry*, Elsevier, Amsterdam, 1990.

5. *IUPAC Compendium of Chemical Terminology*, Electronic version, http://goldbook.iupac.org

6. A. Werner, "Contribution to the Theory of Affinity and Valence," *Zeitschrift für Anorganische Chemie*, 3 (1893) 267–330.

7. G.B. Kauffman, Ed., *Coordination Chemistry–A Century of Progress, Oxford University Press*, New York, 1997.

8. G.B. Kauffman, *Classics in Coordination Chemistry. Part I: The Selected Papers of Alfred Werner*, Dover Publications, New York, 1968.

9. *The New Encyclopaedia Britannica*, 15th Ed. Encyclopaedia Britannica, Inc., Chicago, IL, 2002.

10. H. Frieser, *Concepts & Calculations in Analytical Chemistry: A Spreadsheet Approach*, CRC Press, Boca Raton, FL, 1992.

11. G.N. Lewis, *Valence and Structure of Atoms and Molecules*, Dover Publications, New York, 1966 (originally published in 1923).

12. D.A. Davenport, "Gilbert Newton Lewis: Report of the Symposium," *Journal of Chemical Education*, 61 (1984) 2–21.

13. D. R. Lide, Ed., *CRC Handbook of Chemistry and Physics*, 83rd ed., CRC Press, Boca Raton, FL, 2002.

14. L.G. Sillen and A.E. Martell, *Stability Constants of Metal–Ion Complexes, Special Publication 25, Supplement No. 1*, Chemical Society, London, 1971.

15. IUPAC, *Stability Constants* Database, IUPAC/Academic Software, Otley, UK, 1993.

16. (*NIST Standard Reference Database 46*) *NIST Critically Selected Stability Constants for Metal Complexes Database, vol. 8.0*, NIST, Gaithersburg, MD, 2004.

17. A.E. Martell and R.M. Smith, *Critical Stability Constants*, Vols. 1–5, Plenum Press, New York, 1974.

18. John A. Dean, Ed., *Lange's Handbook of Chemistry*, 15th Ed., McGraw-Hill, New York, 1999.

19. R.M. Fuoss, "Ionic Association. III. The Equilibrium Between Ion Pairs and Free Ions," *Journal of the American Chemical Society*, 80 (1958) 5059–5061.

20. D.W. Margerum, G.R. Kayley, D.C. Weatherburn, and G.K. Pagenkopf, "Kinetics and Mechanisms of Complex Formation and Ligand Exchange." In *Coordination Chemistry*, Vol. 2, A. E. Martell., Ed. ACS Publications, Washington, DC, 1978, pp. 1–220.

21. G.T. Morgan and H.D.K. Drew, "Residual Affinity and Coordination. II. Acetylacetones of Selenium and Tellurium," *Journal of the Chemical Society*, 117 (1920) 1456–1465.

22. H.A. Laitinen and G.W. Ewing, *A History of Analytical Chemistry*, American Chemical Society/Maple Press, York, PA, 1977.

23. C.N. Reilley and R.W. Schmid, "Chelometric Titrations with Potentiometric End Point Detection," *Analytical Chemistry*, 30 (1958) 947–953.

24. C.A. Janeway and P. Travers, *Immunobiology: The Immune System in Health and Disease*, Current Biology Ltd., London, 1996.

25. C.P. Price and D.J. Newman, Eds., *Principles and Practice of Immunoassay*, 2nd ed., Macmillan, London, 1997.

26. J.E. Butler, Ed., *Immunochemistry of Solid-Phase Immunoassay*, CRC Press, Boca Raton, FL, 1991.

27. R.S. Yalow and S.A. Berson, "Immunoassay of Endogenous Plasma Insulin in Man," *Journal of Clinical Investigation*, 39 (1960) 1157–1175.

28. J. Szejtle, "Introduction and General Overview of Cyclodextrin Chemistry," *Chemical Reviews*, 98 (1998) 1743–1753.

29. R. Ramette "Solubility and Equilibria of Silver Chloride," *Journal of Chemical Education*, 37 (1960) 348.

30. J.H. Jonte et al. "The Solubility of Silver Chloride and the Formation of Complexes in Chloride Solution," *Journal of the American Chemical Society*, 74 (1952) 2052.

31. A.R. Newman, "The Great Fruit Scares of 1989," *Analytical Chemistry*, 61 (1989) 861A–863A.

32. S.L Rovner, "Silver Lining in Melamine Crisis," *Chemical & Engineering News*, 87(21) (2009) 36–38.

Chapter 10

Oxidation–Reduction Reactions

Chapter Outline

10.1 INTRODUCTION: SAVING THE *ARIZONA*

On December 7, 1941, an air strike was made by the Japanese on the U.S. fleet at Pearl Harbor in Hawaii. This event led to the entry of the United States into World War II. The battleship USS *Arizona*, which was located at Pearl Harbor, was struck and sunk during the attack. A total of 1177 sailors and marines died as a result, representing the largest loss of life for a single ship in the history of the U.S. Navy.[1] A memorial operated by the U.S. National Park Service is now located at the site of this sunken ship (see Figure 10.1). There is, however, growing interest in how quickly the *Arizona* might be deteriorating as it sits submerged in seawater. Part of the reason for this concern is a desire to maintain this monument. There are also almost half a million gallons of fuel oil still trapped in the ship, which would pollute the surrounding water if any new holes developed in the Arizona's hull.[2–3]

Various analytical techniques are now being used to study the deterioration of the USS *Arizona*, with the goal of controlling this degradation and preserving this ship. Of particular interest is the extent to which corrosion is occurring in the steel hull.[1–4] Corrosion is a type of *oxidation–reduction reaction*. Other examples of oxidation–reduction reactions are photosynthesis, combustion, and the reactions that are used in batteries to generate electricity. In this chapter, we learn about oxidation–reduction reactions and see how we can describe these processes. We also see how we can determine whether a given oxidation–reduction reaction might occur and learn how to predict the extent to which this reaction will take place. Later in this book we will use this information to describe and help us understand various analytical methods that rely on oxidation–reduction reactions.

10.1A What Are Oxidation–Reduction Reactions?

Oxidation is a process in which a chemical loses one or more electrons, becoming "oxidized." **Reduction** is a process in which a chemical gains one or more electrons, causing this chemical to become "reduced."[5,6] Oxidation must always take place when there is reduction, and reduction is always present during oxidation, so a reaction that involves both these processes is called an **oxidation–reduction reaction** (or *redox reaction*).[6,7] Oxidation–reduction reactions are also known as *electrochemical reactions*, which can be defined as reactions that involve the exchange of electrons between chemicals. The study of electrochemical reactions and their applications is a field known as *electrochemistry*.[6]

To illustrate an oxidation–reduction reaction, let's look closer at what occurs during corrosion of the steel hull in the USS *Arizona* (see Figure 10.2). The steel used in this hull contains mainly solid iron, along with small

USS *Arizona*

USS *Arizona* memorial

FIGURE 10.1   The USS *Arizona* when it was still in service (shown around 1916 by New York City) and the USS *Arizona* Memorial at Pearl Harbor. (Reproduced with permission of the U.S. National Park Service.)

amounts of carbon, phosphorus, sulfur, and silica that make up less than 0.34% (w/w) of the steel. When this steel sits in seawater, some of the iron at its surface will give up electrons to form Fe^{2+} ions that will enter the water, as represented by the process in Equation 10.1. (*Note*: We can also represent these Fe^{2+} ions in water by using the term "Fe^{2+}(aq)" however, the subscript (aq) will not be used in the remainder of this chapter in the interest of simplicity, because Fe^{2+} and other ions will be assumed to be present in water unless otherwise indicated.)

$$Fe(s) \rightleftharpoons Fe^{2+} + 2\,e^- \text{ or } 2\,Fe(s) \rightleftharpoons 2\,Fe^{2+} + 4\,e^- \quad (10.1)$$

This reaction is an example of an oxidation process because each atom of Fe(s) loses two electrons and moves from an oxidation number of 0 to +2 as it forms Fe^{2+}. The electrons that are lost are conducted through the steel to another location where they can take part in reduction. For example, these electrons might be acquired by dissolved oxygen near the surface of the hull to form water, as shown in Equation 10.2.

$$O_2(g) + 4\,H^+ + 4\,e^- \rightleftharpoons 2\,H_2O \quad (10.2)$$

This second process involves reduction because electrons are gained by oxygen as its oxidation number goes from zero in O_2 to a value of –2 in water. The hydrogen does not change its oxidation number during this reaction when it changes from H^+ to part of a water molecule.

Although we have written the corrosion process in Equations 10.1 and 10.2 by using steps, it is important to remember that these oxidation and reduction events occur at the same time and will depend on one another

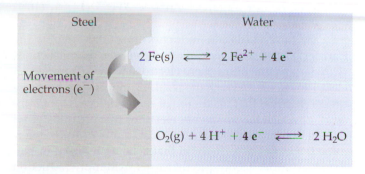

FIGURE 10.2   An example of a possible oxidation–reduction reaction that can lead to corrosion of steel in water, such as is occurring in the hull of the USS *Arizona*.

for their exchange of electrons. Another way we could represent this overall reaction is shown in Equation 10.3.

Oxidation: $\qquad 2\,Fe(s) \rightleftarrows 2\,Fe^{2+} + 4\,e^-$

Reduction: $\qquad O_2(g) + 4\,H^+ + 4\,e^- \rightleftarrows 2\,H_2O$

Net Reaction:

$$2\,Fe(s) + O_2(g) + 4\,H^+ \rightleftarrows 2\,Fe^{2+} + 2\,H_2O \qquad (10.3)$$

In this case, both the oxidation and reduction processes have been combined to give a single balanced equation. The electrons are no longer shown directly in this overall reaction, but they are still being transferred as the Fe(s) is oxidized to Fe^{2+} and as oxygen is being reduced to give water. The result for the USS *Arizona* is that part of the Fe(s) at the surface of the steel hull forms soluble Fe^{2+} and is lost to water, slowly creating pits and holes in this hull (see Figure 10.3).

10.1B How Are Oxidation–Reduction Reactions Used in Analytical Chemistry?

Oxidation–reduction reactions not only occur in corrosion and other processes, but they can be used as tools for chemical analysis. One important application of these reactions is in a "redox titration" (see Chapter 14), which is a method that can be used to measure the concentration of an analyte that can undergo either oxidation or reduction in the presence of a particular reagent. Oxidation–reduction reactions are also often used to pretreat samples prior to a chemical analysis. For instance, Fe^{3+} can be converted to Fe^{2+} for complexation with 1,10-phenanthroline, which gives a colored complex that allows the measurement of iron in a sample. This reduction step can be accomplished by first reacting Fe^{3+} with zinc metal, as shown in Equation 10.4, during which the zinc is oxidized to Zn^{2+}.

$$2\,Fe^{3+} + Zn(s) \rightleftarrows Zn^{2+} + 2\,Fe^{2+} \qquad (10.4)$$

Because oxidation–reduction reactions involve the transfer of electrons, we can often combine these reactions with devices that allow these electrons to travel through an electrical circuit in which the resulting current can be measured. A method that makes use of such a device is "coulometry," in which the number of electrons that are transferred during an oxidation–reduction reaction is used to determine the moles of a chemical that is being reduced or oxidized. Another technique that uses oxidation–reduction reactions along with an external electrical circuit is "voltammetry," in which the change in current of a chemical system is examined as changes are made in the electrical potential applied to this system. The principles of these methods and their applications are described in Chapter 16.

Corrosion and pitting at surface of hull

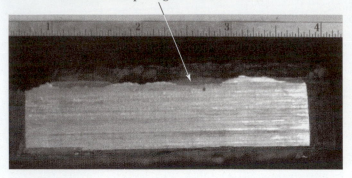

FIGURE 10.3 A cross section of a steel hull sample from the USS *Arizona*, taken at a depth of 5.9 m from the port side. The pits at the top of this sample are due to corrosion of the hull in the presence of seawater. (Reproduced with permission from D.L. Johnson et al., "Corrosion of Steel Shipwreck in the Marine Environment: USS Arizona—Part 1," *Materials Performance*, 45 (2006) 40–44.)

10.2 GENERAL PRINCIPLES OF OXIDATION–REDUCTION REACTIONS

10.2A Describing Oxidation–Reduction Reactions

We can use the following general equations to represent an oxidation–reduction reaction like the one that takes place during corrosion.

Oxidation: $\qquad Red' \rightleftarrows Ox' + ne^- \qquad (10.5)$

Reduction: $\qquad Ox + ne^- \rightleftarrows Red \qquad (10.6)$

Net Reaction: $\qquad Ox + Red' \rightleftarrows Red + Ox' \qquad (10.7)$

In the net reaction, Ox is the chemical being reduced and Red' is the chemical that is oxidized as n electrons are transferred from Red' to Ox. Red and Ox' are the products of this reaction, where Red is the reduced form of Ox and Ox' is the oxidized form of Red'. During the corrosion of iron by water in Equation 10.3, Red' and Ox' would be Fe(s) and Fe^{2+}, while Ox and Red would be $O_2(g)$ and H_2O. It is often necessary to include other chemical species to balance an oxidation–reduction reaction, such as the H^+ that appears in Equation 10.3 during the corrosion of iron.

Every oxidation–reduction reaction has two related pairs of chemicals (Ox/Red and Ox'/Red') that each form a *redox couple* or *redox system*.[6] A redox couple consists of a pair of two different oxidized and reduced forms of the same element in an oxidation–reduction reaction. For instance, Fe(s) and Fe^{2+} represent one redox couple for the reaction in Equation 10.3. The chemicals $O_2(g)$ and H_2O represent another redox couple in this reaction because they show the oxidized and reduced forms of oxygen during the corrosion process.

10.2B Identifying Oxidation–Reduction Reactions

Using Oxidation Numbers. All oxidation–reduction reactions can be identified by the fact that they involve the oxidation of one chemical and the reduction of another. This oxidation and reduction process is accompanied by a change in the *oxidation number* (or *oxidation state*) for some of the elements in these chemicals. An oxidation number is the charge that an element in a chemical would have if this element existed as a solitary ion but still possessed the same number of electrons that it has in the chemical. Table 10.1 summarizes some rules that can be used to determine the oxidation number for each element in a molecule or ion.

One way we can use Table 10.1 is to determine if any elements in a reaction change in their oxidation number. If there is a change in oxidation number, this reaction is then an oxidation–reduction process. For instance, the corrosion process in Equation 10.3 involves a change in the oxidation number of iron from 0 in its elemental state, $Fe(s)$, to +2 in the Fe^{2+} ion. The reduction of oxygen during this same reaction involves a change in the oxidation number of oxygen from zero in O_2 to an oxidation number of –2 for oxygen in water. During this process hydrogen has an oxidation number of +1 in both H^+ and H_2O. Thus, because some elements are being reduced and others are oxidized, we have confirmed that the corrosion process in Equation 10.3 is an example of an oxidation–reduction process.

EXERCISE 10.1	**Oxidation Numbers and Combustion**

The burning of an organic compound in the presence of oxygen is a process known as *combustion*. An example of combustion is the burning of octane (C_8H_{18}) in fuel oil, such as is trapped aboard the USS *Arizona*, to give carbon dioxide and water.

$$2\,C_8H_{18}(l) + 25\,O_2(g) \rightarrow 16\,CO_2(g) + 18\,H_2O(g) \quad (10.8)$$

Determine which elements are being oxidized or reduced during this reaction. Use your results to explain why combustion is an example of an oxidation–reduction reaction.

SOLUTION

The rules in Table 10.1 indicate hydrogen will have an oxidation number of +1 on both sides of this reaction, so this element is not being oxidized or reduced. Oxygen is being reduced from its elemental form in O_2 (where oxygen has an oxidation number of zero) to an oxidation number of –2 in CO_2. At the same time, carbon is oxidized as it goes from an average oxidation number of $-18/8 = -2.25$ in octane to an oxidation number of +4 in CO_2. This makes combustion a true example of an oxidation–reduction reaction.

TABLE 10.1   Rules for Assigning Oxidation Numbers*

1. **The oxidation number of a free, uncombined element is zero.**

   Examples: The oxidation number for the iron atoms in $Fe(s)$ is zero. The oxidation number for oxygen atoms in $O_2(g)$ is also zero because oxygen in this chemical is not combined with any other element.

2. **The oxidation number for a simple monatomic ion is the charge on that ion.**

   Examples: The oxidation number for the iron in Fe^{2+} is +2, and the oxidation number for hydrogen in H^+ is +1.

3. **When hydrogen is combined with other elements, the hydrogen usually has an oxidation number of +1. An exception occurs when hydrogen is bonded to a metal to form a metal hydride, in which case the hydrogen has an oxidation number of –1.**

   Examples: The oxidation number of the hydrogen in H_2O is +1. The oxidation number of the hydrogen in lithium hydride (LiH) is –1.

4. **When oxygen is combined with other elements, the oxygen usually has an oxidation number of –2. An exception occurs when oxygen is present in a peroxide, in which case the oxygen has an oxidation number of –1.**

   Examples: The oxidation number of the oxygen in H_2O is –2. The oxidation number of the oxygen in hydrogen peroxide (H_2O_2) is –1.

5. **The sum of the oxidation numbers for all atoms in a neutral chemical must be zero.**

   Examples: In water (H_2O), the oxidation number of the hydrogen is +1, and the oxidation number of oxygen is –2, giving a sum for these oxidation numbers of $2(+1) + 1(-2) = 0$. In hydrogen peroxide, the oxidation number of hydrogen is +1 and the oxidation number of oxygen is –1, giving a sum for these oxidation numbers of $2(+1) + 2(-1) = 0$.

6. **In a polyatomic ion, the net charge on the ion is equal to the sum of the oxidation numbers for all of the atoms in that ion.**

   Example: For the hydroxide ion (OH^-), the oxygen has an oxidation number of –2, and the hydrogen has an oxidation number of +1, giving a sum for these oxidation numbers of $(-2) + (+1) = -1$, or the charge on this ion.

---

*These rules were obtained from the *IUPAC Compendium of Chemical Terminology*, Electronic version, http://goldbook.iupac.org

Using Half-Reactions. Another way of identifying many common oxidation–reduction processes is by using **half-reactions**. A half-reaction is a chemical "reaction" that is written to show electrons among either the products or reactants.[8,9] Some examples of half-reactions are those that occur during the corrosion of steel in water.

Oxidation Half-Reaction (Equation 10.1):

$$2\,Fe(s) \rightleftharpoons 2\,Fe^{2+} + 4\,e^-$$

Reduction Half-Reaction (Equation 10.2):

$$O_2(g) + 4\,H^+ + 4\,e^- \rightleftarrows 2\,H_2O$$

As this example indicates, there are always two half-reactions that make up every overall oxidation–reduction reaction. First, there is an *oxidation half-reaction*, in which electrons are one of the products. Second, there is a simultaneous *reduction half-reaction*, in which electrons are one of the reactants. When we combine balanced oxidation and reduction half-reactions, the electrons on the product and reactant sides should cancel and give the overall observed reaction, as we saw when combining Equations 10.1 and 10.2 to get Equation 10.3 for the corrosion of iron.

Half-reactions are, in a sense, an artificial way of looking at an oxidation–reduction process, because they do not occur by themselves, but always take place in pairs—an oxidation process plus a reduction process. In addition, the electrons that are shown in half-reactions do not actually exist in a solution, but are instead transferred directly to another chemical. (*Note*: In some cases, this other "chemical" may be an electrode.) However, half-reactions are quite useful to chemists in describing what is occurring during an oxidation–reduction process.

A list of common half-reactions is shown Table 10.2 where, according to current convention, they are shown as *reduction processes*. Additional examples of half-reactions can be found in Appendix B, the *CRC Handbook of Chemistry & Physics*,[10–14] and *Lange's Handbook of Chemistry*,[11] among other sources.[12–14] You can also balance and obtain new half-reactions by using methods that are described in Appendix A. To obtain the oxidation form of a half-reaction, all you have to do is reverse the corresponding reduction half-reaction. We will see later how half-reactions can be used to compare the ability of chemicals to be oxidized or reduced. These half-reactions also show the forms that a given element might take as it undergoes oxidation or reduction and can be used to identify an oxidation–reduction reaction.

EXERCISE 10.2	Using Half-Reactions to Examine Oxidation–Reduction Processes

A chemist wishes to examine the iron content of a ship hull by dissolving a small sample of this hull and converting all of the iron in this sample into Fe^{2+} ions. These

---

TABLE 10.2 Examples of Reduction Half-Reactions

Half-Reaction	Standard Potential E^0 (V) vs. SHE[a]
Strong Oxidizing Agent	*Weak Reducing Agent*
$MnO_4^- + 8\,H^+ + 5\,e^- \rightleftarrows Mn^{2+} + 4\,H_2O$	+1.51
$Cr_2O_7^{2-} + 14\,H^+ + 6\,e^- \rightleftarrows 2\,Cr^{3+} + 7\,H_2O$	+1.36
$O_2(g) + 4\,H^+ + 4\,e^- \rightleftarrows 2\,H_2O$	+1.23
$Ag^+ + e^- \rightleftarrows Ag(s)$	+0.80
$Fe^{3+} + e^- \rightleftarrows Fe^{2+}$	+0.77
$O_2(g) + 2\,H_2O + 4\,e^- \rightleftarrows 4\,OH^-$	+0.40
$Cu^{2+} + 2\,e^- \rightleftarrows Cu(s)$	+0.34
$2H^+ + 2\,e^- \rightleftarrows H_2(g)$	0.00 (Reference Reaction)
$Ni^{2+} + 2\,e^- \rightleftarrows Ni(s)$	−0.26
$Fe^{2+} + 2\,e^- \rightleftarrows Fe(s)$	−0.44
$Zn^{2+} + 2\,e^- \rightleftarrows Zn(s)$	−0.76
$2\,H_2O + 2\,e^- \rightleftarrows H_2(g) + 2\,OH^-$	−0.83
$Al^{3+} + 3\,e^- \rightleftarrows Al(s)$	−1.68
$Mg^{2+} + 2\,e^- \rightleftarrows Mg(s)$	−2.36
$Ca^{2+} + 2\,e^- \rightleftarrows Ca(s)$	−2.84
Weak Oxidizing Agent	*Strong Reducing Agent*

[a]The standard potentials are all shown for the reduction half-reaction at 25°C vs. the standard hydrogen electrode. The chemicals listed in boldface include the element that is being oxidized or reduced in the given half-reaction. The abbreviation SHE stands for the **standard hydrogen electrode** (see Section 10.3B). The listed values were obtained from J. A. Dean, Ed., *Lange's Handbook of Chemistry*, 15th ed., McGraw-Hill, New York, 1999, and have been rounded to nearest 0.01 V.

ions are then reacted with permanganate (MnO_4^-) during a titration that involves the following reaction.

$$5\,Fe^{2+} + MnO_4^- + 8\,H^+ \rightleftharpoons 5\,Fe^{3+} +$$
$$Mn^{2+} + 4\,H_2O \qquad (10.9)$$

Use the information in Table 10.2 to help determine which chemicals are undergoing oxidation and reduction in this reaction and to write the half-reactions for these processes.

SOLUTION

We can see from the list of half-reactions in Table 10.2 that both Fe^{2+} and MnO_4^- change oxidation states in this reaction. The half-reactions are given below, where the first is written as an oxidation step to match Fe^{2+} and Fe^{3+} with the sides in which they appear in Equation 10.9.

Oxidation Half-Reaction:

$$Fe^{2+} \rightleftharpoons Fe^{3+} + e^- \text{ (or } 5\,Fe^{2+} \rightleftharpoons 5\,Fe^{3+} + 5\,e^-) \quad (10.10)$$

Reduction Half-Reaction:

$$MnO_4^- + 8\,H^+ + 5\,e^- \rightleftharpoons Mn^{2+} + 4\,H_2O \quad (10.11)$$

The redox couple for the first half-reaction is Fe^{2+} and Fe^{3+}, which is the same answer we would get if we were to look only at oxidation numbers. The redox couple for the second half-reaction is MnO_4^- and Mn^{2+} because manganese is undergoing a change in oxidation number during this reaction. (*Note*: The elements in H^+ and H_2O are not undergoing any change in oxidation number.)

10.2C Predicting the Extent of Oxidation–Reduction Reactions

Using Equilibrium Constants. One way we can describe the ability of one chemical to oxidize or reduce another is to use an equilibrium constant for this reaction. For instance, the general oxidation–reduction reaction in Equation 10.7 would have the following thermodynamic equilibrium constant.

$$Ox + Red' \rightleftharpoons Red + Ox'$$

$$K° = \frac{(a_{Red})(a_{Ox'})}{(a_{Ox})(a_{Red'})} \qquad (10.12)$$

We can also write a concentration-based equilibrium constant for this reaction (K), where γ_{Ox}, $\gamma_{Red'}$, γ_{Red}, and $\gamma_{Ox'}$ are the activity coefficients for the given reactants and products.

$$K = \frac{[Red][Ox']}{[Ox][Red']} \quad \text{and} \quad K° = K \cdot \frac{(\gamma_{Red})(\gamma_{Ox'})}{(\gamma_{Ox})(\gamma_{Red'})} \quad (10.13)$$

As a specific example, the expressions for $K°$ and K for the corrosion reaction in Equation 10.3 would be written as follows.

$$K° = \frac{(a_{Fe^{2+}})^2\,(a_{H_2O})^2}{(a_{O_2})(a_{Fe})^2(a_{H^+})^4} \quad \text{or} \quad K = \frac{[Fe^{2+}]^2}{P_{O_2}[H^+]^4} \quad (10.14)$$

In this case, an activity of one is used in the expression for K for the values of a_{Fe} and a_{H_2O}, because both the Fe(s) and water are present in their standard states (the iron as a solid and water as the solvent—see Chapter 6). It is also necessary in the expression for K to use a partial pressure instead of a concentration for O_2, because the oxygen is present as a dissolved gas.

EXERCISE 10.3	Equilibrium Constants for Oxidation–Reduction Reactions

Write an expression for the thermodynamic and concentration-dependent equilibrium constants for the titration reaction in Equation 10.9, in which water is being used as the solvent.

SOLUTION

The expression for $K°$ and K that are obtained for this titration reaction are as follows.

$$K° = \frac{(a_{Fe^{3+}})^5(a_{Mn^{2+}})(a_{H_2O})^4}{(a_{Fe^{2+}})^5(a_{MnO_4^-})(a_{H^+})^8}$$

$$\text{or} \quad K = \frac{[Fe^{3+}]^5[Mn^{2+}]}{[Fe^{2+}]^5[MnO_4^-][H^+]^8} \qquad (10.15)$$

No concentration term is shown for water in the expression for K because water is the solvent and thus has an activity of one. All other species in this reaction are dissolved solutes and can be described in the expression for K by using their molar concentrations.

In Chapter 6 we learned that a reaction will tend to proceed from reactants to products if the equilibrium constant for this reaction is large (i.e., much greater than one). The fact that the corrosion of iron proceeds spontaneously indicates that the equilibrium constant for this process is greater than one. A large equilibrium constant is also desired when using the reaction in Equation 10.9 as the basis for a chemical analysis because a large equilibrium constant will ensure that essentially all of the added MnO_4^- will react with Fe^{2+} in the sample. The equilibrium constant for the corrosion reaction is 8.2×10^{112} and the equilibrium constant for the titration of Fe^{2+} by permanganate is 1.4×10^{62}, which are both extremely large values. We will see how to calculate these values in the next section.

It is important to remember that the concentration or activity ratios that appear in expressions like Equations 10.12–10.14 will be equal to $K°$ or K only if the system is truly at equilibrium. However, many oxidation–reductions are used under conditions that are not at equilibrium. A good example is the use of oxidation–reduction reactions to produce energy in a battery. This energy output can only occur up until the system has reached equilibrium, at which point the battery will be "dead." Under nonequilibrium conditions, the ratio of the activities or concentrations for all species in the reaction is said to be equal to the reaction quotient Q, which can be compared to the value of $K°$ or K to determine the direction in which the reaction will proceed as it approaches equilibrium (see Chapter 6).

Using Standard Potentials. If you look closely at Table 10.2, you will see that the reduction half-reactions in this list are ranked according to their "standard potentials." The term *potential* in this case refers to an **electric potential**, which is the work required per unit charge to move a charged particle (such as an electron) from one point to another.[7,10] An electric potential has units of a volt (V), where $1 \text{ V} = 1 \text{ J/A} \cdot \text{s}$ in the SI system (see Chapter 2). The list in Table 10.2 gives the **standard electrode potential** ($E°$) for various half-reactions, which is the electric potential expected for a half-reaction under standard conditions when compared to the following reference half-reaction.

Reference Half-Reaction:

$$2 \text{ H}^+ + 2 \text{ e}^- \rightleftharpoons \text{H}_2(\text{g}) \qquad E° = 0.000 \text{ V exactly} \quad (10.16)$$

The conditions that are used in determining the value of $E°$ include a temperature of 25°C, a concentration of exactly 1 M for all dissolved chemicals, and a pressure of 1 bar for all gases that are reactants or products in the half-reaction. We will learn in Section 10.4 how the potential for a half-reaction can be determined when utilizing other conditions. Regardless of the temperature that is being used, the standard potential of the reference half-reaction for the reduction of H^+ to produce $\text{H}_2(\text{g})$ (Equation 10.16) is always assigned a value of exactly 0.000 V.

Standard electrode potentials can be used to compare the relative ability of chemicals to undergo oxidation or reduction. A large positive potential for a reduction half-reaction in Table 10.2 (as obtained for MnO_4^- or $\text{Cr}_2\text{O}_7^{2-}$) indicates that the chemical being reduced will easily take up electrons. This same property makes this chemical a good *oxidizing agent* (also called an *oxidant* or *oxidizer*)[6] because it can take electrons from other chemicals and cause them to be oxidized. A large negative potential in Table 10.2 indicates that the corresponding reduction process is not favorable, or that the reverse oxidation reaction is more likely to occur. This situation is present for the reduction of Mg^{2+} and Ca^{2+}. However, the products of these same half-reactions (Mg or Ca metal) should be easy to oxidize and cause other chemicals to be reduced, making these products act as good *reducing agents* or *reductants*.[6]

The standard potentials for half-reactions can also be used to determine the potential difference under standard conditions for an oxidation–reduction reaction ($E°_{\text{Net}}$). We can do this by taking the standard potential for the chemical that is being reduced ($E°_{\text{Reduction}}$) and subtracting from it the standard potential for the chemical that is being oxidized ($E°_{\text{Oxidation}}$).[8]

Under Standard Conditions:

$$E°_{\text{Net}} = E°_{\text{Reduction}} - E°_{\text{Oxidation}} \quad (10.17)$$

This value of $E°_{\text{Net}}$ gives the difference in electric potential, or work per unit charge, that is produced under standard conditions as electrons are transferred from the chemical that is being oxidized to the chemical that is being reduced. Figure 10.4 illustrates this idea for the corrosion of steel in water, in which $E°_{\text{Net}}$ for Equation 10.3 can be found by taking the standard reduction potential for the half-reaction in Equation 10.1 and subtracting from this the standard reduction potential for the half-reaction in Equation 10.2. The result is $E°_{\text{Net}} = E°_{\text{Reduction}} - E°_{\text{Oxidation}} = 1.23 - (-0.44) = 1.67 \text{ V}$, where the positive value obtained for $E°_{\text{Net}}$ indicates that this corrosion reaction will occur spontaneously under standard conditions.

The standard potential is related, but not equal, to the change in free energy that occurs during an oxidation–reduction reaction. The relationship between $E°_{\text{Net}}$ and the total change in free energy for an oxidation–reduction reaction under standard conditions ($\Delta G°$) is given below,

$$\Delta G° = -nFE°_{\text{Net}} \qquad \text{or} \qquad E°_{\text{Net}} = -\frac{\Delta G°}{nF} \quad (10.18)$$

where n is the number of electrons involved in the oxidation–reduction reaction and F is the **Faraday constant**.[6,8] The Faraday constant is equal to the charge that is present in one mole of electrons. This charge is given in units of the coulomb (C, where $1 \text{ C} = 1 \text{ A} \cdot \text{s}$ in the SI system), resulting in a value for the Faraday constant of $9.6485 \times 10^4 \text{ C/mol}$. By using Equation 10.18, it is possible to convert between $\Delta G°$ and $E°_{\text{Net}}$ for an oxidation–reduction reaction.

EXERCISE 10.4 **Estimating the Standard Potential for an Oxidation–Reduction Reaction**

What is the value of $E°_{\text{Net}}$ for the titration of Fe^{2+} with permanganate in Equation 10.9? What is the value of $\Delta G°$ for this reaction? Would this reaction be expected to occur spontaneously as written when working under standard conditions?

SOLUTION

This titration involves the oxidation of Fe^{2+} to Fe^{3+} (Equation 10.10) and the corresponding reduction of MnO_4^- to Mn^{2+} (Equation 10.11). The standard potentials from Table 10.2 for the corresponding reduction half-reactions are +0.77 V and +1.51 V, respectively. Placing these values into Equation 10.17 gives the following value for E°_{Net}.

$$E^{\circ}_{Net} = E^{\circ}_{Reduction} - E^{\circ}_{Oxidation}$$

$$= 1.51 \text{ V} - (0.77 \text{ V}) = \textbf{0.74 V}$$

The value of ΔG° for this reaction is obtained by using Equation 10.18 and the fact that $n = 5$ for this oxidation–reduction reaction (see the balanced half-reactions in Equations 10.10 and 10.11).

$$\Delta G^{\circ} = -nFE^{\circ}_{Net} = -(5)(9.6485 \times 10^4 \text{ C/mol})(0.74 \text{ V})$$

$$= -(5)(9.6485 \times 10^4 \text{ A} \cdot \text{s/mol})(0.74 \text{ J/A} \cdot \text{s})$$

$$= -3.57 \times 10^5 \text{ J/mol}$$

$$= -3.6 \times 10^5 \text{ J/mol}$$

$$(\text{or } -3.6 \times 10^2 \text{ kJ/mol})$$

The large positive value we have obtained for E°_{Net} and the large negative value found for ΔG° both tell us that this titration reaction is highly favored to occur under standard conditions.

Because the standard potential is related to ΔG°, it can also be used to estimate the equilibrium constant for an oxidation–reduction reaction. If we know the number of electrons n that are being transferred during a balanced oxidation–reduction reaction, the following formulas make it possible to convert between the standard potential for the net reaction (E°_{Net}) and the thermodynamic equilibrium constant (K°) at 25°C.

$$E^{\circ}_{Net} = \frac{(0.05916 \text{ V})}{n} \log(K^{\circ})$$

$$\text{or} \quad K^{\circ} = 10^{(nE^{\circ}_{Net})/(0.05916 \text{ V})} \qquad (10.19)$$

These relationships indicate that an oxidation–reduction reaction with a large equilibrium constant will have a large, positive value for E°_{Net}. Similarly, an oxidation–reduction reaction with a small equilibrium constant ($K^{\circ} < 1$, representing a reaction that proceeds in a direction opposite to that in which it is written), will have a large, negative value for E°_{Net}.

EXERCISE 10.5 **Relating a Standard Potential to an Equilibrium Constant**

Calculate the value of E°_{Net} for the corrosion process in Equation 10.3. What is the expected equilibrium constant for this reaction under standard conditions?

SOLUTION

The value of E°_{Net} can be found by using Equation 10.17 to combine the standard reduction potentials for the half-reactions in Equations 10.1 and 10.2, where $E^{\circ}_{Reduction} = 1.23$ V and $E^{\circ}_{Oxidation} = -0.44$ V from the values that are listed in Table 10.2. This gives the following net standard potential for the corrosion process in Equation 10.3.

$$E^{\circ}_{Net} = E^{\circ}_{Reduction} - E^{\circ}_{Oxidation}$$

$$= 1.23 \text{ V} - (-0.44 \text{ V}) = \textbf{1.67 V}$$

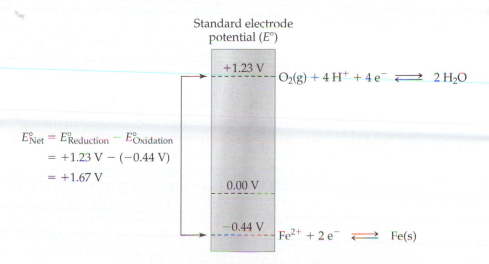

FIGURE 10.4 Calculation of the net potential difference under standard conditions (E°_{Net}) that occurs during the corrosion of iron in water in the presence of dissolved oxygen. Although both half-reactions are written in this diagram as reduction processes, the half-reaction with the lower standard electrode potential will occur as an oxidation reaction, giving the result shown in Equation 10.3. In this type of diagram, electrons would be given up by the oxidation half-reaction (i.e., the half-reaction with the lower E°_{Net} value) and be consumed by the reduction half-reaction (the half-reaction with the higher E°_{Net} value). This effect is the same as is illustrated by the half-reactions in Equations 10.1 and 10.2, as they are combined to give the overall reaction in Equation 10.3.

We can use Equation 10.19 to obtain $K°$ for this reaction by using the calculated value for $E°_{Net}$ and $n = 4$ for the number of electrons being transferred in this oxidation–reduction reaction.

$$K° = 10^{(nE°_{Net})/(0.05916\ V)} = 10^{(4)(1.67\ V)/(0.05916\ V)}$$

$$\therefore K° = 8.2 \times 10^{112}$$

We can again see from the large value obtained for $K°$ that this corrosion process is highly favored, at least thermodynamically, to occur under standard conditions.

The previous two exercises have illustrated how the value $E°_{Net}$ can be used to predict whether a given oxidation–reduction reaction is likely to occur in the way in which it is written. This feature is important because many times you may not know in advance in which direction such a reaction might proceed. For example, if we had instead written the titration reaction in Equation 10.9 in the opposite direction, with Fe^{3+} reacting with Mn^{2+} to form Fe^{2+} and MnO_4^-, we would have obtained a negative value for the overall standard potential ($E°_{Net} = 0.77\ V - 1.51\ V = -0.74\ V$). This negative value would have simply indicated that the reaction would not occur spontaneously as written, but would actually proceed in the opposite direction. You should also keep in mind that although the values of the standard potential $E°_{Net}$ and equilibrium constant $K°$ indicate the overall extent to which a reaction might eventually occur, they do not tell you anything about the *rate* at which this reaction might occur. This is fortunate in the case of corrosion of a ship's hull, because the rate of this reaction can be quite slow, even though we now know that this process is highly favored in terms of its values for $E°_{Net}$ and $K°$.

10.3 ELECTROCHEMICAL CELLS

A valuable feature of oxidation–reduction reactions is that it is often possible to separate their oxidation and reduction processes and carry them out in separate locations. This feature allows us to measure or control the flow of electrons between the oxidizing agent and reducing agent. All of this is made possible by using a device known as an *electrochemical cell*.

10.3A Describing Electrochemical Cells

General Concepts. An **electrochemical cell** is a device in which the oxidation and reduction processes of an oxidation–reduction reaction occur in different locations, with electrons flowing from one location to the other through an external circuit.[6] A common example of such a device is a *Daniell cell* (see Figure 10.5).[15,16] This cell makes use of the ability of Cu^{2+} ions in an aqueous solution to be reduced by zinc metal and form copper metal

plus Zn^{2+} ions. The half-reactions and overall reaction that take place in this cell are shown below.

Oxidation Half-Reaction:

$$Zn(s) \rightleftarrows Zn^{2+} + 2\,e^- \qquad (10.20)$$

Reduction Half-Reaction:

$$\underline{Cu^{2+} + 2\,e^- \rightleftarrows Cu(s)} \qquad (10.21)$$

Overall Reaction:

$$Zn(s) + Cu^{2+} \rightleftarrows Zn^{2+} + Cu(s) \qquad (10.22)$$

The standard reduction potential for the conversion of Cu^{2+} to $Cu(s)$ in Equation 10.21 is $+0.34\ V$ and the standard reduction potential for the conversion of Zn^{2+} to $Zn(s)$ (the reverse process of that shown in Equation 10.20) is $-0.76\ V$. This means that the overall reaction in Equation 10.22 has a standard potential of $E°_{Net} = (+0.34\ V) - (-0.76\ V) = +1.10\ V$, indicating that the reaction written in Equation 10.22 highly favors the products on the right. We can confirm this experimentally by simply adding zinc metal to an aqueous solution containing Cu^{2+} ions under standard conditions. When we do this, the zinc metal will be oxidized to form Zn^{2+}, while the copper ions will form reddish copper metal that will precipitate from solution.

When we combine zinc metal and Cu^{2+} together in an aqueous solution, these two reagents will exchange electrons directly with one another as they are oxidized and reduced. It is possible, however, to separate these reagents into two separate areas and instead have them exchange electrons through an electrical circuit. This is the idea behind the electrochemical cell in Figure 10.5. The individual half-reactions that are occurring in this cell are exactly the same as those shown in Equations 10.20 and 10.21 for the reaction of Cu^{2+} ions with zinc metal. What is different in these two situations is the means by which electrons are transferred from the zinc to the Cu^{2+} ions. In Figure 10.5, the electrons that become available due to the oxidation of zinc metal must travel through an external circuit to reach copper ions in a separate solution, producing a current. This design makes it possible to use this flow of electrons (such as in a battery) or to measure this current when studying what is happening in the electrochemical cell.

The pieces of copper and zinc metal in Figure 10.5 are called *electrodes*. An **electrode** is a conducting material at which one of the half-reactions in an electrochemical cell is taking place.[6] In particular, it is at the *surface* of each electrode where electrons are being transferred from chemicals in the solution to the internal circuit. You may have noticed in Figure 10.5 that the zinc electrode is labeled as the *anode* and the copper electrode is called the *cathode*. The **anode** in an electrochemical cell is the electrode at which a chemical is

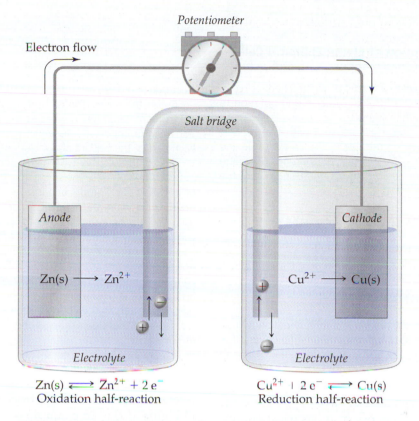

Potentiometer

Electron flow

Salt bridge

Anode

Cathode

Zn(s) $\longrightarrow$ Zn^{2+}

Cu^{2+} $\longrightarrow$ Cu(s)

Electrolyte

Electrolyte

Zn(s) $\rightleftharpoons$ Zn^{2+} + 2 e$^-$
Oxidation half-reaction

Cu^{2+} + 2 e$^-$ $\rightleftharpoons$ Cu(s)
Reduction half-reaction

FIGURE 10.5 The general design of a Daniell cell. This type of electrochemical cell is named after British chemist John Frederic Daniell (1790–1845), who invented this cell in 1836 to produce electricity for telegraphs. The *potentiometer* that is shown in this diagram (also known as a *voltmeter*) is a device used to measure the potential difference between the two electrodes in this cell.

being oxidized.[5–7] The anode for the cell in Figure 10.5 is the zinc electrode because this is where Zn(s) is oxidized to Zn^{2+}. The **cathode** is the electrode at which a chemical is being reduced.[5–7] In Figure 10.5, Cu^{2+} is reduced to Cu(s) at the copper electrode, making this electrode the cathode.

The two electrodes of an electrochemical cell are connected through a wire and circuit by which electrons can flow from the oxidation half-reaction to the reduction half-reaction. The solutions and metals that are involved in these two half-reactions are isolated from each other by placing them in separate regions of the cell. However, it is still necessary to have some contact between the two solutions so that current can flow. This current is carried by the movement of ions in the solution that surrounds each electrode, with this solution of ions being referred to as an *electrolyte*.[6,7] The term *half-cell* is often used to describe the combination of an individual electrode and its electrolyte plus the associated chemicals that are needed for the half-reaction occurring at this electrode.[8] Because there are always two half-reactions and electrodes in any cell, there will also be two half-cells present. A *salt bridge* is another feature that may be present in the cell, this is a device that allows current to flow between two electrodes while preventing the mixing of their electrolytes.[6] The movement of ions through both the electrolyte and salt bridge is neces-

sary to provide a flow of current through the entire electrochemical cell and to maintain a balance of charge in this device as chemicals are oxidized or reduced at the electrodes (Box 10-1).

Figure 10.6 shows an example of an electrochemical cell that might be used to study the effects of corrosion on a material such as the steel hull from a ship like the USS *Arizona*. The general design of the cell is similar to the one we saw for the cell in Figure 10.5. The anode in this new cell is the electrode to be studied. The cathode is made of an inert material such as platinum or gold, which will allow oxidation to occur as dissolved oxygen combines with hydrogen ions and electrons supplied by this electrode to form water. The electrolyte in this case might be a solution such as seawater that will supply any ions that are needed for charge to be carried between the two electrodes and the salt bridge. This particular cell also includes a source of oxygen gas that can bubble through the electrolyte near the cathode to help mimic the desired reduction and corrosion reactions (see Figure 10.2).

Types of Electrochemical Cells. All electrochemical cells can be divided into two general types: (1) a galvanic cell, or (2) an electrolytic cell. A **galvanic cell** (or *voltaic cell*) is an electrochemical cell in which the oxidation–reduction reaction occurs spontaneously,

BOX 10.1

A Shorthand Description of Electrochemical Cells

Rather than using complete diagrams of electrochemical cells, chemists often use a standard set of symbols and notation to describe these cells to other scientists. For example, suppose we construct a cell like that in Figure 10.5 with a 0.0125 M solution of $ZnSO_4$ in water on the side of the Zn electrode and a 0.0125 M solution of $CuSO_4$ in water on the side containing the Cu electrode. This cell could be described using one of the forms shown below.

$$Zn(s) \mid Zn^{2+} \text{ (aq, 0.0125 } M) \parallel Cu^{2+} \text{ (aq, 0.0125 } M) \mid Cu(s)$$

or

$$Zn(s) \mid ZnSO_4 \text{ (aq, 0.0125 } M) \parallel CuSO_4 \text{ (aq, 0.0125 } M) \mid Cu(s)$$

This notation always has the anode and its half-cell shown on the left, beginning with the electrode material (Zn(s)) and followed by the chemicals or solution that are in contact with this electrode (a 0.0125 M aqueous solution of Zn^{2+} prepared from $ZnSO_4$, in this case). The cathode is shown on the right, with electrode material being listed at the far right. The concentrations of the important dissolved ions, molecules, or gases in each part of the cell are also included.

The lines in this type of notation represent boundaries between two phases. The single line (|) between Zn(s) and Zn^{2+} or between Cu^{2+} and Cu(s) in the cell shown represents the interface between the solid electrode and its surrounding solution, and it is here that a potential is produced between these components of each half-cell. The two lines (||) in the middle of this notation represent the salt bridge, which has two boundaries—one for each side of the salt bridge where it comes in contact with the two half-cells. These boundaries can also produce a small difference in potential when a current is passed through the cell and as different ion concentrations are created on either side of the boundaries. This last effect can be important to consider when using a cell to make chemical measurements (see Chapter 14 for more details).

resulting in a flow of electrons.[6,7] This type of cell is used to convert chemical energy into electrical energy, as would be present in a battery (where the term *battery* refers to a collection of electrochemical cells). The device in Figure 10.5 is an example of a galvanic cell because it

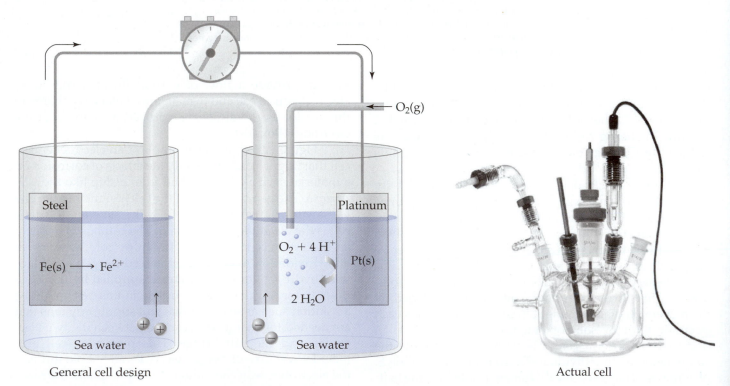

General cell design   Actual cell

FIGURE 10.6 Design of an electrochemical cell that could be used to study the corrosion of steel in the presence of oxygen and water (left) and an actual electrochemical cell that is designed to study the corrosion of metals (right). The real electrochemical cell shown on the right has a water jacket for temperature control, with the inlet and outlet ports for water being present on the side of this cell; there is also a port on the upper left for the introduction of oxygen or other gases. The material to be studied for corrosion makes up the center electrode, with a reference electrode and bridge being present to its right. This particular cell also has a third, counterelectrode on the left—a feature that is present in many modern cells, as we will learn in Chapter 14. (The image on the right is reproduced with permission from Gamry Instruments.)

involves an oxidation–reduction reaction that will occur spontaneously to provide a flow of electrons.

In an **electrolytic cell**, an external power source is used to apply an electric current and cause a particular oxidation–reduction reaction to occur. (*Note*: The flow of current and the reaction that it creates in an electrolytic cell is sometimes referred to as *electrolysis*.)[6,7] An electrolytic cell is the type of cell that is present when you are using an external power supply to recharge a battery. Another example of an electrolytic cell is in *electroplating*, in which an external power supply is used to cause metal ions in a solution to be reduced to form a solid metal coating on a surface (see Figure 10.7). One application of this process is to create galvanized steel, where a coating of zinc is placed onto steel to prevent corrosion. A similar approach is used to place a chrome, nickel, or silver coating on a surface. Electrolytic cells are also used on an industrial scale to produce aluminum, chlorine, calcium, and sodium, among other chemicals.[17]

10.3B Predicting the Behavior of Electrochemical Cells

Standard Cell Potentials. We saw in Section 10.2C how it was possible to use standard potentials to predict the extent to which an oxidation–reduction reaction might occur. We can use the same approach to estimate the standard potential for an electrochemical cell. This can be accomplished by knowing that a cell potential (in the absence of any significant current flow) is simply the *difference* in potentials for the two half-reactions occurring in the cell.

A special type of potential that is important to determine for an electrochemical system is the **standard cell potential (E°_{Cell})**. This is the potential that develops between an anode and a cathode when all of the components in an electrochemical cell are in their standard states (i.e., all chemicals in the oxidation–reduction reaction have an activity of one).[8] The standard cell potential for a given oxidation–reduction system can be estimated by using Equation 10.23,

Under Standard Conditions:

$$E^\circ_{Cell} = E^\circ_{Cathode} - E^\circ_{Anode} \qquad (10.23)$$

where $E^\circ_{Cathode}$ and E°_{Anode} are the standard electrode potentials for the half-reactions that are occurring at the cathode and anode, respectively. As an example, the values of $E^\circ_{Cathode}$ and E°_{Anode} for the cell in Figure 10.5 are +0.34 V and –0.76 V, which represent the standard electrode potentials for the reduction of Cu^{2+} to $Cu(s)$ and Zn^{2+} to $Zn(s)$. Using these values in Equation 10.23 gives an estimated standard cell potential of $E^\circ_{Cell} = (+0.34 \text{ V}) - (-0.76 \text{ V}) = +1.10 \text{ V}$, which is the potential we would obtain under standard conditions when directly combining Cu^{2+} with $Zn(s)$. The same approach can be used to determine the standard cell potential that would be expected for any other combination of half-reactions in an electrochemical cell.

EXERCISE 10.6	Estimating a Standard Cell Potential

A chemist wishes to construct an electrochemical cell that is based on the following half-reactions and overall reaction.

Oxidation Half-Reaction:

$$Ni(s) \underset{\leftarrow}{\rightarrow} Ni^{2+} + 2\,e^- \qquad (10.24)$$

Reduction Half-Reaction:

$$2\,Ag^+ + 2\,e^- \underset{\leftarrow}{\rightarrow} 2\,Ag(s)$$

$$\underline{(\text{or } Ag^+ + e^- \underset{\leftarrow}{\rightarrow} Ag(s))} \qquad (10.25)$$

Net Reaction:

$$2\,Ag^+ + Ni(s) \underset{\leftarrow}{\rightarrow} 2\,Ag(s) + Ni^{2+} \qquad (10.26)$$

What is the estimated standard potential for this electrochemical cell? Will the reactions in this cell proceed as written under standard reaction conditions?

SOLUTION

If the half-reactions in this electrochemical cell proceed as indicated, then the process taking place at the cathode would be the reduction of Ag^+ to $Ag(s)$ and the process at the anode would be the oxidation of $Ni(s)$ to Ni^{2+}. We can use this information along with the standard reduction potentials in Table 10.2 to determine that $E^\circ_{Cathode} = +0.80 \text{ V}$ (for the $Ag^+/Ag(s)$ redox couple) and

FIGURE 10.7 An example of an electroplating center in industry. Each tank in this image represents an electrochemical cell that has the material to be coated as one of the electrodes, where metal ions from the surrounding solution are reduced and coated onto this material. (Reproduced with permission from SGR Global.)

$E°_{\text{Anode}} = -0.24$ V (for the $Ni^{2+}/Ni(s)$ redox couple). When we place these values into Equation 10.23, we get the following standard cell potential.

$$E°_{\text{Cell}} = E°_{\text{Cathode}} - E°_{\text{Anode}}$$
$$= 0.80 \text{ V} - (-0.24 \text{ V}) = +1.03 \text{ V}$$

The fact that we obtained a positive value for $E°_{\text{Cell}}$ indicates that the net reaction will occur spontaneously as we have written it, making this system act as a galvanic cell.

Although Equation 10.23 is often useful in estimating the standard potential that is expected for an electrochemical cell, there are several reasons why the actual measured potential of a cell may be different from this value. For instance, we may not be working under standard conditions and the activities of our reactants and products may not be equal to one—a situation we will examine further in Section 10.4. The measured potential may also be affected by the fact that we have placed additional components in our system when constructing a device that separates the two half-reactions. A common example of this occurs when the presence of a salt bridge leads to the formation of a "junction potential" at each boundary between the electrolyte and salt bridge, although the proper construction of the salt bridge can minimize this effect. There will also be some resistance to the flow of ions through the electrolyte between the electrodes and salt bridge when current is passing through the cell, creating a change in potential due to an effect known as *IR drop*.[18] Even though we will not

explore these last two topics further in this chapter, we will come back to the issue of junction potential in Chapter 14 where we see how we can minimize this effect during the use of potential measurements for chemical analysis.

The Standard Hydrogen Electrode. We have already learned that the standard potential for an oxidation–reduction reaction or for an electrochemical cell is always based on a difference between the potentials of two electrodes or half-reactions. This comparison is necessary because it is not possible to directly measure the standard potential of any electrode or any half-reaction without referencing this to some other electrode or half-reaction. To help in this process, we also learned in Section 10.2C that the half-reaction involving reduction of hydrogen ions to form hydrogen gas is assigned a reference value of exactly 0.000 V for determining the standard potentials of all other half-reactions. This half-reaction is used as the basis for a reference electrode, known as the **standard hydrogen electrode** (or **SHE**), that can be used to determine the standard potential for other electrodes.

The basic design of a standard hydrogen electrode is shown in Figure 10.8. This design consists of an inert platinum electrode that is in contact with an aqueous solution that contains hydrogen ions with activity of $1\ M$ and bubbling hydrogen gas at a pressure of 1 bar.[5,6] A standard hydrogen electrode is used as the anode when it is placed in an electrochemical cell, with the components for the half-reaction to be examined used as the cathode. The standard electrode potential for the standard hydrogen electrode is automatically assigned a value of exactly 0.000 V at any temperature. This means that the standard

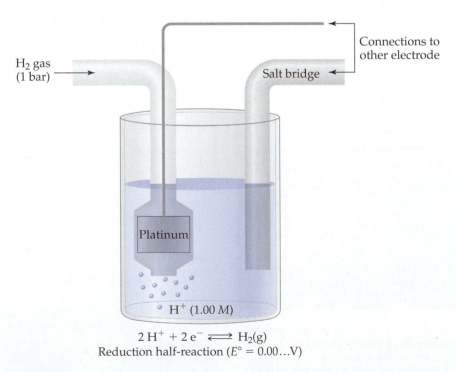

$$2\,H^+ + 2\,e^- \rightleftharpoons H_2(g)$$
Reduction half-reaction ($E° = 0.00...V$)

FIGURE 10.8 General design of a standard hydrogen electrode (SHE).

potential for an electrochemical cell that contains a SHE can be estimated by using Equation 10.27.

When using a SHE:

$$E_{Cell}^{\circ} = E_{Cathode}^{\circ} - E_{SHE}^{\circ}$$

$$= E_{Cathode}^{\circ} - (0.000\ V)$$

$$\therefore E_{Cell}^{\circ} = E_{Cathode}^{\circ} \qquad (10.27)$$

The result is that the value of E_{Cell}° that is measured for such an electrochemical cell will automatically be equal to the standard potential for the half-reaction taking place at the cathode.

In the past, the standard hydrogen electrode has been used to help determine the standard potentials for many of the half-reactions listed in Table 10.2 and in Appendix B. The standard potentials for even more systems were determined by using this group of half-reactions as secondary references. This has given the standard hydrogen electrode an important historical role in helping chemists compare and study half-reactions. Unfortunately, the standard hydrogen electrode is also inconvenient to use and is not found in most modern laboratories. In Chapter 14, we will see some other reference electrodes that are now used in place of the SHE, such as the saturated calomel electrode and the silver–silver chloride electrode.

10.4 THE NERNST EQUATION

Up to this point we have only considered cells and half-reactions that are examined under standard reaction conditions (1 bar pressure, 1 mol/L activity for each solution-phase species, and so on). It is much more common to have an electrochemical cell or oxidation–reduction reaction that is occurring under other nonstandard conditions. This situation can be examined by using a relationship known as the **Nernst equation** (see Box 10.2).

10.4A Working with the Nernst Equation

General Principles.   The Nernst equation is an expression used for a *reversible* half-reaction to relate the reduction potential under nonstandard conditions (E) to the activities of the reactants and products and to the half-reaction's standard reduction potential (E°). For the general half-reaction in Equation 10.28, the Nernst equation can be written as shown in Equation 10.32.

General Half-Reaction: $Ox + ne^{-} \rightleftharpoons Red \qquad (10.28)$

Nernst Equation: $E = E^{\circ} - \dfrac{RT}{nF} \ln\left(\dfrac{a_{Red}}{a_{Ox}}\right) \qquad (10.32)$

In Equation 10.32, a_{Ox} is the activity of the species being oxidized (Ox), a_{Red} is the activity of the species being reduced (Red), R is the gas-law constant, T is the temperature (in kelvin), F is the Faraday constant, and n is the number of electrons involved in the half-reaction.[6]

If we are working specifically at 25°C, we can combine the known values for R, T, and F ($R = 8.314\ J/(K \cdot mol)$, $T = 298.15\ K$, and $F = 96{,}485\ C/mol$) to get a single constant. When we combine this new constant with a conversion factor of 2.303 in going from a natural logarithm to a base-ten logarithm, we get the following form of the Nernst equation.

Nernst Equation at 25°C:

$$E = E^{\circ} - \frac{0.05916\ V}{n} \log\left(\frac{a_{Red}}{a_{Ox}}\right) \qquad (10.33)$$

Both Equations 10.32 and 10.33 show us that the potential for a reversible half-reaction will depend on its standard potential, the number of electrons involved in the half-reaction, and the activities of the chemicals that are involved in this half-reaction.

There are other, equivalent ways of writing the Nernst equation, however, we will use the forms in either Equation 10.32 or 10.33 throughout this text. These equations apply to a half-reaction that is written either as a reduction or oxidation process, but are given in a way that always gives the expected potential when this half-reaction is written as a *reduction* process (such as listed in Table 10.2). This approach will make it more convenient for us to later use these equations with expressions like Equations 10.17 and 10.23 for calculating E_{Cell} and E_{Net}.

Simple Half-Reactions.   As a simple example of how you might write a Nernst equation for a half-reaction, let's first look at the conversion of Fe(s) to Fe^{2+} that occurs during the corrosion of iron in steel. (*Note:* The reduction form of this process is also shown for reference.)

Oxidation Half-Reaction (Equation 10.1):

$$Fe(s) \rightleftharpoons Fe^{2+} + 2\ e^{-}$$

Reduction Form of Half-Reaction:

$$Fe^{2+} + 2\ e^{-} \rightleftharpoons Fe(s)$$

The Nernst equation for this half-reaction at 25°C can be written as shown below, where we make use of the fact that $E^{\circ}_{Fe^{2+}/Fe(s)} = -0.44\ V$ and $a_{Fe(s)} = 1$ (because Fe(s) is in a standard state).

Nernst Equation at 25°C:

$$E_{Fe^{2+}/Fe(s)} = E^{\circ}_{Fe^{2+}/Fe(s)} - \frac{0.05916\ V}{2} \log\left(\frac{a_{Fe(s)}}{a_{Fe^{2+}}}\right)$$

$$\therefore E_{Fe^{2+}/Fe(s)} = (-0.44\ V)$$

$$- \frac{0.05916\ V}{2} \log\left(\frac{1}{a_{Fe^{2+}}}\right) \qquad (10.34)$$

Once you have written the Nernst equation for a half-reaction, you can calculate the potential that would be expected for that half-reaction under various conditions. We will come back to this concept in Section 10.4B when we learn how to use the Nernst equation to estimate the potential for a cell or oxidation–reduction reaction under nonstandard conditions.

BOX 10.2

A Closer Look at the Nernst Equation

The Nernst equation is probably the most famous equation in the field of electrochemistry. This equation was first developed by Walther H. Nernst, a German chemist who helped found the field of physical chemistry (see Figure 10.9).[18,19] Besides his work in the area of electrochemistry and his derivation of the "Nernst equation," Nernst performed research in the areas of thermodynamics, photochemistry, and solid-state chemistry. He received the 1920 Nobel Prize in chemistry for his discovery of the third law of thermodynamics.

There are various ways of deriving the Nernst equation.[8,20,21] An approach based on thermodynamics is shown in Figure 10.9. This approach begins by writing an expression for the total change in free energy (ΔG) that is expected for the reduction of Ox to produce Red (Equation 10.29). This change in free energy is related to the

total change in free energy of this reaction under standard conditions ($\Delta G°$) and the reaction quotient for the reduction process (as given by the ratio of activities, a_{Red}/a_{Ox}. Nernst also realized that the total change in free energy for this reaction was related to the change in electric potential created during this reaction. This relationship is given in Equation 10.30 by the formula $\Delta G° = -nFE°$ under standard conditions and the more general formula $\Delta G = -nFE$.

The next step is to use these two sets of relationships to produce a combined expression. This gives a form of Equation 10.29 that is now expressed using $-nFE°$ instead of $\Delta G°$ and $-nFE$ in place of ΔG (see Equation 10.31). Both sides of this equation are then divided by the term $-nF$. The final result is the general form of the Nernst equation that is given in Equation 10.32.

Walther H. Nernst

$$Ox + n\,e^- \;\rightleftharpoons\; Red \tag{10.28}$$

Initial equations:

$$\Delta G = \Delta G° + R\,T \ln\left(\frac{a_{Red}}{a_{Ox}}\right) \tag{10.29}$$

$$\Delta G° = -n\,F\,E° \qquad \Delta G = -n\,F\,E \tag{10.30}$$

Combined expression:

$$-(n\,F\,E) = -(n\,F\,E°) + RT \ln\left(\frac{a_{Red}}{a_{Ox}}\right) \tag{10.31}$$

$$E = E° - \frac{R\,T}{n\,F}\ln\left(\frac{a_{Red}}{a_{Ox}}\right) \tag{10.32}$$

FIGURE 10.9   (a) Walther Hermann Nernst (1864–1941), and (b) a derivation of the Nernst equation. The derivation is discussed in Box 10.2.

EXERCISE 10.7   Writing a Nernst Equation for a Simple Half-Reaction

Write the Nernst equation at 25°C for the reduction of Cu^{2+} to $Cu(s)$ in the cell in Figure 10.5.

SOLUTION

The $Cu^{2+}/Cu(s)$ redox couple has the following reduction half-reaction and Nernst equation expression at 25°C.

Reduction Half-Reaction (Equation 10.21):

$$Cu^{2+} + 2\,e^- \rightleftharpoons Cu(s)$$

Nernst Equation at 25°C:

$$E_{Cu^{2+}/Cu(s)} = E°_{Cu^{2+}/Cu(s)} - \frac{0.05916\ \text{V}}{2}\log\left(\frac{a_{Cu(s)}}{a_{Cu^{2+}}}\right)$$

$$\therefore \quad E_{Cu^{2+}/Cu(s)} = (+0.34\ \text{V}) -$$

$$\frac{0.05916\ \text{V}}{2}\log\left(\frac{1}{a_{Cu^{2+}}}\right) \tag{10.35}$$

The final form of this equation includes the known standard potential for this half-reaction ($E°_{Cu^{2+}/Cu} = +0.34$ V from Table 10.2) and the fact that the activity of $Cu(s)$ is equal to 1.0, because this form represents a standard state for copper. The same approach could be used to write a Nernst equation for the $Zn^{2+}/Zn(s)$ redox couple that is present for the cell in Figure 10.5.

Complex Half-Reactions. There are many situations in which there is more than one reactant or product in a half-reaction, not including the electrons. In this situation, an activity term for each product will appear in the numerator of the logarithmic portion of the Nernst equation (when written in the form shown in Equation 10.32 or 10.33) and an activity term for each reactant will appear in the denominator. Also, each of these activities will be raised to a power that is identical to the number of moles that appear before that product or reactant in the balanced half-reaction. This approach follows the same rules that we have used throughout this book in writing expressions for equilibrium constants and reaction quotients. To illustrate this idea, let's write the Nernst equation at 25°C for the half-reaction that describes the reduction of O_2 under acidic conditions to form water.

Reduction Half-Reaction (Equation 10.2):

$$O_2(g) + 4\,H^+ + 4\,e^- \rightleftharpoons 2\,H_2O$$

Nernst Equation at 25°C:

$$E_{O_2/H_2O} = E^\circ_{O_2/H_2O} - \frac{0.05916\text{ V}}{n}\log\left(\frac{(a_{H_2O})^2}{(a_{O_2})(a_{H^+})^4}\right)$$

$$\therefore\ E_{O_2/H_2O} = (+1.23\text{ V})$$

$$-\frac{0.05916\,V}{4}\log\left(\frac{1}{(a_{O_2})(a_{H^+})^4}\right) \quad (10.36)$$

In this last equation we substituted in $n = 4$ and $E^\circ = 1.23$ V for the O_2/H_2O redox couple (see Table 10.2), along with the fact that $a_{H_2O} = 1$, because water is the solvent (a standard state). This approach can also be used for other half-reactions that have multiple reactants or products.

EXERCISE 10.8   Writing a Nernst Equation for a Complex Half-Reaction

Write a Nernst equation at 25°C for the half-reaction shown in Equation 10.11 for permanganate, as used for the titration of Fe^{2+} from an iron or steel sample.

SOLUTION

The MnO_4^-/Mn^{2+} redox couple has the following reduction half-reaction and Nernst equation expression at 25°C.

Reduction Half-Reaction (Equation 10.11):

$$MnO_4^- + 8\,H^+ + 5\,e^- \rightleftharpoons Mn^{2+} + 4\,H_2O$$

Nernst Equation at 25°C:

$$E_{MnO_4^-/Mn^{2+}} = E^\circ_{MnO_4^-/Mn^{2+}}$$

$$-\frac{0.05916\text{ V}}{5}\log\left(\frac{(a_{Mn^{2+}})(a_{H_2O})^4}{(a_{MnO_4^-})(a_{H^+})^8}\right)$$

$$\therefore\ E_{MnO_4^-/Mn^{2+}} = (+1.51\text{ V})$$

$$-\frac{0.05916\text{ V}}{5}\log\left(\frac{a_{Mn^{2+}}}{(a_{MnO_4^-})(a_{H^+})^8}\right) \quad (10.37)$$

The final form of this equation includes the known standard reduction potential for this half-reaction ($E^\circ_{MnO_4^-/Mn^{2+}} = +1.51$ V from Table 10.2) and the fact that the activity of a_{H_2O} is equal to 1.0, because water is the solvent for this reaction and is in its standard state.

10.4B Calculating Potentials for Oxidation–Reduction Reactions

General Approach.   A large benefit of using the Nernst equation is it allows us to predict the overall potential that is expected for a reversible oxidation–reduction reaction or an electrochemical cell under nonstandard conditions. This can be accomplished by utilizing the Nernst equation to find the potential expected for pertinent half-reactions under the given reaction conditions and then using these values to find the overall potential for the system. This second step is accomplished by using the following forms of Equations 10.17 and 10.23, which can now be used under either standard or nonstandard conditions.

Under Standard or Nonstandard Conditions:

$$E_{Net} = E_{Reduction} - E_{Oxidation} \quad (10.38)$$

$$E_{Cell} = E_{Cathode} - E_{Anode} \quad (10.39)$$

You can see from these two equations that we can still relate the overall potential values of E_{Net} to $E_{Reduction}$ and $E_{Oxidation}$ (for an oxidation–reduction reaction) or E_{Cell} to $E_{Cathode}$ and E_{Anode} (for an electrochemical cell) as long as we first use the Nernst equation to find the potentials for the two half-reactions that are present in this system.

The use of the Nernst equation to find a cell potential can be demonstrated with the cell in Figure 10.5. The Nernst equations at 25°C are shown below for the two half-reactions in this cell.

Reduction Half-Reaction (Equation 10.21):

$$Cu^{2+} + 2\,e^- \rightleftharpoons Cu(s)$$

$$E_{Cu^{2+}/Cu(s)} = (+0.34\text{ V}) - \frac{0.05916\text{ V}}{2}\log\left(\frac{1}{a_{Cu^{2+}}}\right) \quad (10.35)$$

Oxidation Half-Reaction (Equation 10.20):

$$Zn(s) \rightleftarrows Zn^{2+} + 2\,e^-$$

$$E_{Zn^{2+}/Zn(s)} = (-0.76\ V) - \frac{0.05916\ V}{2}\ \log\left(\frac{1}{a_{Zn^{2+}}}\right)\ (10.40)$$

We can now put in the activities for all of the reactants and products in these half-reactions and see how this will affect their potentials. The overall cell potential under these conditions is obtained by placing these results into Equation 10.39.

$$E_{Cell} = E_{Cathode} - E_{Anode}$$

$$= E_{Cu^{2+}/Cu(s)} - E_{Zn^{2+}/Zn(s)}\ (10.41)$$

The next exercise will provide you with some practice in performing these calculations.

EXERCISE 10.9 Calculating the Potential of an Oxidation–Reduction Reaction

A cell is constructed with the copper electrode initially in contact with an aqueous solution containing 0.0125 M $CuSO_4$ ($a_{Cu^{2+}} = 0.0606$, $\gamma_{Cu^{2+}} = 0.485$ M), and the zinc electrode in contact with an aqueous solution containing 0.0125 M $ZnSO_4$ ($a_{Zn^{2+}} = 0.0606$, $\gamma_{Zn^{2+}} = 0.485$). This cell has an original potential of approximately +1.10 V at 25°C. After being operated for some time, the concentration of Cu^{2+} at the cathode decreases to 0.0025 M ($a_{Cu^{2+}} = 0.00169$, $\gamma_{Cu^{2+}} = 0.675$) and the concentration of Zn^{2+} by the anode increases to 0.020 M ($a_{Zn^{2+}} = 0.00852$, $\gamma_{Zn^{2+}} = 0.426$). What is the estimated potential for this cell under these new conditions?

SOLUTION

To solve this problem, we first substitute in the activities for Cu^{2+} and Zn^{2+} into the Nernst equations for their two corresponding half-reactions.

$$E_{Cu^{2+}/Cu(s)} = (+0.34\ V) - \frac{0.05916\ V}{2}\ \log\left(\frac{1}{0.00169}\right)$$

$$= 0.34\ V - (0.0820\ V) = 0.258\ V$$

$$E_{Zn^{2+}/Zn(s)} = (-0.76\ V) - \frac{0.05916\ V}{2}\ \log\left(\frac{1}{0.00852}\right)$$

$$= -0.76\ V - (0.0612\ V) = -0.821\ V$$

Next, we place these two half-cell potentials into Equation 10.41 to find the overall cell potential.

$$E_{Cell} = E_{Cu^{2+}/Cu(s)} - E_{Zn^{2+}/Zn(s)}$$

$$= 0.258\ V - (-0.821\ V) = +1.079 = +1.08\ V$$

The result is a slightly lower potential than was present for the initial cell (+1.10 V). If we were to continue to operate this cell, the value of E_{Cell} will eventually approach 0.00 V as equilibrium is established and no more net energy is gained from the cell.

Another approach that we might have used in the last exercise is to combine the Nernst equations for our two half-reactions directly with Equation 10.41, with this overall expression then being used to find E_{Net}. This approach is illustrated in Table 10.3 for a general oxidation–reduction reaction. The final expression for E_{Net} that is obtained by this method (see Equation 10.44 or 10.45) is now given in terms of E_{Net}° and the reaction quotient (Q) for the overall balanced oxidation–reduction reaction. The value of n in this equation is the number of electrons required in each half-reaction (Equations 10.5 and 10.6) to get the net reaction to balance with no electrons being shown. Regardless of whether you use this approach or the one illustrated in the previous exercise, you should obtain the same calculated value for E_{Net}.

A useful feature of the final general equation in Table 10.3 for E_{Net} is it indicates that there are a few special situations that can occur when you are using an electrochemical cell. The first special situation appears when the reaction quotient is exactly equal to 1.00, which causes Equation 10.44 to take on the following form.

Special Case 1—Reaction Quotient $Q = 1$:

$$E_{Net} = E_{Net}^\circ - \frac{RT}{nF}\ \ln(1)$$

$$\therefore\ E_{Net} = E_{Net}^\circ - 0 \ \text{or}\ E_{Net} = E_{Net}^\circ\ (10.46)$$

This situation was present in the last exercise when we began with a cell that had identical concentrations for Zn^{2+} and Cu^{2+}, causing Q to be approximately equal to 1.00. (*Note*: Q may have differed a little from 1.00 in this case because we used concentrations instead of activities in the Nernst equation.) The result was a value for E_{Net} that was roughly equal to E_{Net}°.

A second special situation occurs when an electrochemical cell is at equilibrium. In this case, there will be no more net energy production by the cell, as indicated by an E_{Net} value of 0.00 V. At equilibrium, the reaction quotient will also be equal to the equilibrium constant for a reaction ($Q = K^\circ$), which causes Equation 10.44 to take the form shown below.

Special Case 2—Reaction is at Equilibrium:

$$0 = E_{Net}^\circ - \frac{RT}{nF}\ \ln(K^\circ)$$

$$\therefore\ E_{Net}^\circ = \frac{RT}{nF}\ \ln(K^\circ) \ \text{or}\ K^\circ = e^{(nFE_{Net}^\circ/RT)}\ (10.47)$$

TABLE 10.3   General Equation for Calculating a Cell Potential

General Reaction:

Oxidation Half-Reaction:	$Red' \rightleftarrows Ox' + ne^-$	(10.5)
Reduction Half-Reaction:	$Ox + ne^- \rightleftarrows Red$	(10.6)
Net Reaction:	$Ox + Red' \rightleftarrows Red + Ox'$	(10.7)

Derivation of General Equation for Calculating Cell Potential:

$$E_{Net} = E_{Reduction} - E_{Oxidation} \tag{10.38}$$

$$= \left\{ E^{\circ}_{Red/Ox} - \frac{RT}{nF} \ln\left(\frac{a_{Red}}{a_{Ox}}\right) \right\} - \left\{ E^{\circ}_{Red'/Ox'} - \frac{RT}{nF} \ln\left(\frac{a_{Red'}}{a_{Ox'}}\right) \right\} \tag{10.42}$$

$$= \underbrace{\left(E^{\circ}_{Red/Ox} - E^{\circ}_{Red'/Ox'} \right)}_{E^{\circ}_{Net}} - \frac{RT}{nF} \underbrace{\left\{ \ln\left(\frac{a_{Red}\, a_{Ox'}}{a_{Ox}\, a_{Red'}}\right) \right\}}_{\text{Reaction Quotient, } Q}$$

Final Equation for Calculating Cell Potential:

General Equation:

$$E_{Net} = E^{\circ}_{Net} - \frac{RT}{nF} \ln(Q) \tag{10.44}$$

Equation at 25° C:

$$E_{Net} = E^{\circ}_{Net} - \frac{0.05916\ V}{n} \log(Q) \tag{10.45}$$

This result gives a direct relationship between E°_{Net} and K°. If we further modify Equation 10.47 for work at 25°C and with base-ten logarithms, we get Equation 10.19 in Section 10.2C.

Using Concentration versus Activity. Because we often know the concentration of a chemical rather than its activity, it is typically more convenient to use concentrations rather than activities with the Nernst equation. This approach does introduce a systematic error into the calculated potential, as demonstrated by Equation 10.51 in Table 10.4. This equation indicates that the error will depend on the activity coefficients for each of the reactants and products in the oxidation–reduction process we are examining. These activity coefficients will, in turn, depend on such things as the ionic strength and temperature of the system (see Chapter 2).

If we are working with a dilute chemical and a low ionic strength, these activity coefficients will be near one, which will make the error term ($E_{Net} - E_{Net,Conc}$) in Equation 10.51 close to zero. For example, the potential that we would have obtained in Exercise 10.9 when using the concentrations of Zn^{2+} and Cu^{2+} would have given an error of only +0.01 V versus the result obtained with activities. Under these conditions the use of concentrations in place of activities in the Nernst equation generally results in only a small error in the calculated potential. These conditions can also allow us to use the Nernst equation and a measured potential to determine the concentration of an oxidized or reduced chemical in a sample (see Chapters 14 and 15). You should keep in mind, however, that larger errors can be produced by

this practice when you are dealing with a charged analyte that has a moderate-to-high concentration or that is present in a solution with a reasonably high ionic strength. Under these conditions chemical activities must be used with the Nernst equation to provide accurate results.

10.4C Effects of the Sample Matrix and Side Reactions

pH Effects. Many of the half-reactions that are shown in Table 10.2 and in Appendix B will be affected by a change in pH. This occurs if either H^+ or OH^- appears as one of the reactants or products in the half-reaction. Let's go back to our example of the corrosion of the steel in seawater to examine this effect.

Oxidation Half-Reaction (Equation 10.1):

$$2\ Fe(s) \rightleftarrows 2\ Fe^{2+} + 4\ e^-$$

Reduction Half-Reaction (Equation 10.2):

$$O_2\ (g) + 4\ H^+ + 4\ e^- \rightleftarrows 2\ H_2O$$

The oxidation of Fe(s) to Fe^{2+} does not involve H^+ or OH^- directly, so the potential for this particular half-reaction will not itself change with pH. However, the reduction of O_2 to H_2O does involve H^+ and will depend on pH. This can be further demonstrated by rearranging the Nernst equation for this half-reaction into the following form, in which we first separate out the activity term due to H^+ and convert from H^+ activity to pH.

TABLE 10.4 Effect of Using Concentration to Calculate a Cell Potential

Net Reaction:
$$Ox + Red' \rightleftarrows Red + Ox' \tag{10.7}$$

Calculated Cell Potential when Using Activities (E_{Net}):

$$E_{Net} = E_{Net}^{\circ} - \frac{RT}{nF} \ln\left(\frac{a_{Red}a_{Ox'}}{a_{Ox}a_{Red'}}\right) \tag{10.43}$$

$$E_{Net} = E_{Net}^{\circ} - \frac{RT}{nF}\ln\left(\frac{(\gamma_{Red})(C_{Red})(\gamma_{Ox'})(C_{Ox'})}{(\gamma_{Ox})(C_{Ox})(\gamma_{Red'})(C_{Red'})}\right) \tag{10.48}$$

$$E_{Net} = E_{Net}^{\circ} - \frac{RT}{nF}\ln\left(\frac{(\gamma_{Red})(\gamma_{Ox'})}{(\gamma_{Ox})(\gamma_{Red'})}\right) - \frac{RT}{nF}\ln\left(\frac{(C_{Red})(C_{Ox'})}{(C_{Ox})(C_{Red'})}\right) \tag{10.49}$$

Calculated Cell Potential when Using Concentrations ($E_{Net,Conc}$):

$$E_{Net,Conc} = E_{Net}^{\circ} - \frac{RT}{nF}\ln\left(\frac{(C_{Red})(C_{Ox'})}{(C_{Ox})(C_{Red'})}\right) \tag{10.50}$$

Error when Using Concentrations Instead of Activities ($E_{Net} - E_{Net,Conc}$):

$$(E_{Net} - E_{Net,Conc}) = -\frac{RT}{nF}\ln\left(\frac{(\gamma_{Red})(\gamma_{Ox'})}{(\gamma_{Ox})(\gamma_{Red'})}\right) \tag{10.51}$$

Original Expression (Equation 10.36):

$$E_{O_2/H_2O} = (+1.23 \text{ V}) - \frac{0.05916 \text{ V}}{4} \log\left(\frac{1}{(a_{O_2})(a_{H^+})^4}\right)$$

Step 1—Separate Out a_{H^+} Term:

$$\Rightarrow E_{O_2/H_2O} = (+1.23 \text{ V}) - \frac{0.05916 \text{ V}}{4} \log\left(\frac{1}{(a_{H^+})^4}\right)$$

$$- \frac{0.05916 \text{ V}}{4} \cdot \log\left(\frac{1}{a_{O_2}}\right)$$

Step 2—Substitute in $\log(1/(a_{H^+})^4) = -4\log(a_{H^+})$ and $pH = -\log(a_{H^+})$:

$$\Rightarrow E_{O_2/H_2O} = (+1.23 \text{ V}) - \frac{0.05916 \text{ V}}{4} \cdot 4 \cdot pH$$

$$- \frac{(0.05916 \text{ V})}{4} \log\left(\frac{1}{a_{O_2}}\right) \tag{10.52}$$

This new equation indicates that a change in pH of one unit will lead to a change in the value of E_{O_2/H_2O} by $4 \cdot (0.05916 \text{ V})/4 = 0.05916 \text{ V}$, where a decrease in pH will create an increase in E_{O_2/H_2O} to a more positive value. This result should make sense to you based on the reaction in Equation 10.2, because a decrease in pH will mean that there is a higher activity of H^+ (a reactant in Equation 10.2). This higher activity will, in turn, make the reduction of O_2 more likely to occur and create a more positive value for E_{O_2/H_2O}. A similar approach can be used to predict the effect of pH on other half-reactions in water that involve H^+ or OH^- as reactants or products.

EXERCISE 10.10 **Effect of pH on a Reduction Half-Reaction**

The reduction of dichromate ($Cr_2O_7^{2-}$) to Cr^{3+} in water can be represented by the following half-reaction and Nernst Equation when working at 25°C. What is the expected effect of an increase in pH on the potential for this half-reaction?

Reduction Half-Reaction:

$$Cr_2O_7^{2-} + 14 H^+ + 6 e^- \rightleftarrows 2 Cr^{3+} + 7 H_2O$$

$$E_{Cr_2O_7^{2-}/Cr^{3+}} = E_{Cr_2O_7^{2-}/Cr^{3+}}^{\circ}$$

$$- \frac{0.05916 V}{6} \log\left(\frac{(a_{Cr^{3+}})^2}{(a_{Cr_2O_7^{2-}})(a_{H^+})^{14}}\right) \tag{10.53}$$

SOLUTION

We can use the same approach demonstrated for the O_2/H_2O redox couple to convert Equation 10.53 into a form that is directly related to pH.

Step 1—Separate Out a_{H^+} Term:

$$\Rightarrow E_{Cr_2O_7^{2-}/Cr^{3+}} = E_{Cr_2O_7^{2-}/Cr^{3+}}^{\circ}$$

$$- \frac{0.05916 V}{6} \log\left(\frac{1}{(a_{H^+})^{14}}\right) - \frac{0.05916 V}{6} \log\left(\frac{(a_{Cr^{3+}})^2}{a_{Cr_2O_7^{2-}}}\right)$$

Step 2—Substitute in $\log(1/(a_{H^+})^{14}) = -14\log(a_{H^+})$ and $pH = -\log(a_{H^+})$:

$$\Rightarrow E_{Cr_2O_7^{2-}/Cr^{3+}} = E^\circ_{Cr_2O_7^{2-}/Cr^{3+}} - \frac{0.05916\ V}{6}\cdot 14 \cdot pH$$

$$- \frac{0.05916\ V}{6}\log\left(\frac{(a_{Cr^{3+}})^2}{a_{Cr_2O_7^{2-}}}\right) \qquad (10.54)$$

This result shows that each change of one pH unit will cause a change of $14 \cdot (0.05916\ V)/6 = \mathbf{0.138\ V}$ in the measured potential for this half-reaction, where a decrease in pH (or an increase in a_{H^+}) leads to a more positive value for $E_{Cr_2O_7^{2-}/Cr^{3+}}$.

The last two examples indicate that pH can have a large effect on the potentials for some half-reactions. This fact makes it important to control the pH for these reactions if you wish to use them to study or analyze species other than H^+ or OH^-. This effect makes the use of acid–base buffers (as discussed in Chapter 8) essential for many oxidation–reduction reactions. Another way we can use this effect is as a means for determining pH through a potential measurement. We will come back to this idea in Chapter 14 when we examine the method of potentiometry.

Dealing with Side Reactions. The pH of a solution is not the only thing that can affect the potential expected for an oxidation–reduction reaction. The presence of side reactions can also affect this potential if they alter the activities of any of the reactants or products that are involved in the oxidation–reduction reaction. An example that can occur during the corrosion of steel in seawater is the reaction of Fe^{2+} with carbonate (CO_3^{2-}) to form insoluble $FeCO_3$ (see Figure 10.10). This reaction occurs as is shown in Equation 10.55, and is described by the solubility product (K_{sp}) for $FeCO_3(s)$ (see Chapter 7 for a review of solubility products).

$$Fe^{2+} + CO_3^{2-} \rightleftarrows FeCO_3(s)$$

$$K_{sp,FeCO_3} = (a_{Fe^{2+}})(a_{CO_3^{2-}}) = 3.13 \times 10^{-11}\,(\text{at }25°C)\ (10.55)$$

If the product of activities for Fe^{2+} and CO_3^{2-} exceeds the K_{sp} value given in Equation 10.55, these two ions will combine to form $FeCO_3(s)$ as a precipitate. This process will lower the activity of Fe^{2+}, which will then affect the potential for the $Fe^{2+}/Fe(s)$ redox couple.

We can look directly at the change in this potential by combining the equilibrium expression for the side reaction with the Nernst equation for the half-reaction of interest. For example, we can do this by using the fact that $a_{Fe^{2+}} = (K_{sp,FeCO_3})/(a_{CO_3^{2-}})$ from the solubility product expression in Equation 10.55 and substituting this in

FIGURE 10.10 Formation of insoluble deposits of iron-related species found on the hull of the USS *Arizona*. Some of these deposits are due to formation of siderite ($FeCO_3$) and magnetite (Fe_3O_4). Another mineral found in these deposits is aragonite ($CaCO_3$).

(This figure is reproduced with permission from D.L. Johnson, B.M. Wilson, J.D. Carr, M.A. Russell, L.E. Murphy, and D.L. Colin, "Corrosion of Steel Shipwrecks in the Marine Environment: USS *Arizona*—Part 1," *Materials Performance*, 45 (2006) 40–44.)

for $a_{Fe^{2+}}$ in the Nernst equation at 25°C for the $Fe^{2+}/Fe(s)$ redox couple (see Equation 10.34), as shown below.

$$E_{Fe^{2+}/Fe(s)} = (-0.44\ V)$$

$$- \frac{0.05916\ V}{2}\log\left(\frac{a_{CO_3^{2-}}}{K_{sp,FeCO_3}}\right) \qquad (10.56)$$

What we have done in creating this modified Nernst equation is equivalent to combining the half-reaction for the reduction of Fe^{2+} to $Fe(s)$ with the solubility reaction for $FeCO_3(s)$.

Reduction Half-Reaction: $Fe^{2+} + 2\,e^- \rightleftarrows Fe(s)$

Solubility Reaction: $\underline{FeCO_3(s) \rightleftarrows Fe^{2+} + CO_3^{2-}}$

Combined Half-Reaction:

$$FeCO_3(s) + 2\,e^- \rightleftarrows Fe(s) + CO_3^{2-} \qquad (10.57)$$

If you look closely at the oxidation numbers for the elements in this half-reaction, you will still find that this process involves the reduction of Fe^{2+} to $Fe(s)$. This means that using the Nernst equation for either the new combined half-reaction (Equation 10.57) or for the $Fe^{2+}/Fe(s)$ redox couple (Equation 10.34) should provide us with the same calculated half-reaction potential. However, the new combined half-reaction we obtained also allows us to examine how a change in the concentration of carbonate in the solution might affect the potential for this system.

EXERCISE 10.11 Effect of Side Reactions on the Nernst Equation

A sample of steel from one section of the USS *Arizona* is found to have a deposit of solid $FeCO_3$ that is in contact with water that has an activity of 1.0×10^{-4} for Fe^{2+} and 3.13×10^{-7} for $CO_3{}^{2-}$. (*Note:* Most of the carbonate in water at a neutral pH is present as bicarbonate.) What potential is predicted by the original Nernst expression (Equation 10.34) for the $Fe^{2+}/Fe(s)$ redox couple under these conditions? What potential is predicted by the modified Nernst expression in Equation 10.56 under these conditions? Explain your results.

SOLUTION

We get the following potential for this system when using the original Nernst equation for the $Fe^{2+}/Fe(s)$ redox couple in Equation 10.34.

$$E_{Fe^{2+}/Fe(s)} = (-0.44 \text{ V}) - \frac{0.05916 \text{ V}}{2} \log\left(\frac{1}{1.0 \times 10^{-4}}\right)$$

$$= -0.558 \text{ V} = \mathbf{-0.56 \text{ V}}$$

The result obtained when using the modified Nernst expression is given in Equation 10.56.

$$E_{Fe^{2+}/Fe(s)} = (-0.44 \text{ V}) - \frac{0.05916 \text{ V}}{2} \log\left(\frac{3.13 \times 10^{-7}}{3.13 \times 10^{-11}}\right)$$

$$= -0.558 \text{ V} = \mathbf{-0.56 \text{ V}}$$

We can now see that both of these Nernst equations give exactly the same result for the calculated half-reaction potential. This is because we simply substituted in $a_{Fe^{2+}} = (K_{sp,FeCO_3})/(a_{CO_3{}^{2-}})$ by using the solubility product expression for $FeCO_3(s)$.

Another way we could have dealt with this problem is by writing a Nernst equation for the combined reaction that is given in Equation 10.58,

$$E_{FeCO_3(s)/Fe(s)} =$$

$$\underbrace{(-0.44 \text{ V}) - \frac{0.05916 \text{ V}}{2} \log\left(\frac{1}{K_{sp,FeCO_3}}\right)}_{E^{\circ}_{FeCO_3(s)/Fe(s)} = -0.75 \text{ V}}$$

$$- \frac{0.05916 \text{ V}}{2} \log\left(a_{CO_3{}^{2-}}\right) \quad (10.58)$$

where the Nernst equation for this combined reaction, the value of $E^{\circ}_{FeCO_3(s)/Fe(s)}$ is simply the sum of $E^{\circ}_{Fe^{2+}/Fe(s)}$ and the log term for $K_{sp,FeCO_3}$. The expression in Equation 10.58 will also result in a calculated potential of -0.56 V under the given reaction conditions in this exercise.

Combined reactions work well with systems that have only a few side reactions. However, for other systems there might be many side reactions and half-reactions that occur simultaneously. The corrosion of steel in seawater is one such example. In this type of situation it is often useful to prepare a plot of potential versus pH that shows which principal form of a particular element (in this case, the iron in steel) is expected under a set of reaction conditions. The resulting plot of potential versus pH is known as a *Pourbaix diagram*.[8,22–24] Figure 10.11 shows an example of such a plot for a sample of iron in water (using a 1 *M* concentration for all soluble forms of iron, in this case). Each region of this plot represents conditions in which a different form of iron is the principal species. For instance, the main form of iron that would be expected at a potential below –0.5 V and a pH of 5.0 would be Fe(s), the form that is present in steel. If the pH is kept at 5.0, but the potential of the system is raised just above –0.5 V, the principal form of iron will now be Fe^{2+}. If the potential is kept around –0.5 V, but the pH is raised to 8, the main form of iron will be $Fe(OH)_2$. This information can help a chemist to decide which

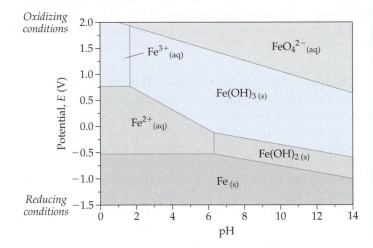

FIGURE 10.11 A Pourbaix diagram (or a *potential/pH diagram*) that shows the principal forms of iron that are present under various potential and pH conditions in an aqueous environment.[22,23] This type of graph is named after Marcel Pourbaix (1904–1998), a Russian-born chemist who first used such a plot.[23,24] Each line in this particular Pourbaix diagram indicates the point at which the principal species of iron changes from one form to another at 25°C, with a concentration of 1 *M* being used for all soluble forms of iron. Each line in this plot indicates the point at which the two forms of iron on either side of the line are present in equal amounts. The vertical lines are based on a Nernst equation that is independent of pH, while the diagonal lines do have a dependence on pH. Regions with identical shading represent the same oxidation state of iron (i.e., Fe^{2+} and $Fe(OH)_2$ both contain iron with an oxidation state of +2). A similar plot for iron in the presence of a carbonate solution can be made that includes a region in which $FeCO_3(s)$ is the principal species. A graph like this will sometimes use pE instead of *E* on the *y*-axis, where $pE = E/(0.05916 \text{ V})$.

TABLE 10.5 Formal Potentials for the Reduction of Fe^{3+} to Fe^{2+} in Water

Reduction Half-Reaction: $Fe^{3+} + e^- \rightleftarrows Fe^{2+}$

Solvent	Formal Potential, $E^{\circ\prime}$ (vs. SHE)[a]
Water	+0.77 (E°)
1.0 M HCl	+0.70
0.67 M H_2SO_4	+0.67
0.3 M H_3PO_4	+0.44

[a]These values were obtained from J. A. Dean, Ed., *Lange's Handbook of Chemistry*, 15th ed., McGraw-Hill, New York, 1999.

reactions are important to consider under a given set of potential and pH conditions.

Formal Potentials. There are times in which we can not accurately predict all of the side reactions that might be taking place in a sample during a potential measurement or in which we wish to use a particular set of conditions for measuring a potential. A common way of dealing with both of these situations is to use a **formal potential**, or *conditional potential*. A formal potential (represented as $E^{\circ\prime}$) is similar to a standard potential in that it represents the expected potential for a given redox couple when the activities of the species undergoing oxidation or reduction are exactly 1.0. However, a formal potential is instead reported for a specific type of solution or electrolyte in which the oxidation–reaction is being examined.[6]

A common way in which formal potentials are utilized is to describe an oxidation–reduction process at a specific pH. For instance, suppose that a chemist wishes to study the oxidation of a ship's steel hull specifically at pH 7.00. At this pH, the Nernst equation we derived for the reduction of O_2 to H_2O in Equation 10.52 can be written as shown in the following expression, where the

pH term is now a constant because we are working specifically at pH 7.00.

Formal Potential for Equation 10.52 at pH 7.00:

$$E_{O_2/H_2O} = (+1.23 \text{ V}) - \underbrace{\frac{0.05916 \text{ V}}{4} \cdot 4 \cdot (7.00)}_{E^{\circ\prime}{}_{O_2/H_2O} = -0.81\underline{5}\text{V}}$$

$$- \frac{0.05916 \text{ V}}{4} \log\left(\frac{1}{a_{O_2}}\right) \qquad (10.59)$$

We can now combine this constant pH term with the standard reduction potential to obtain a new constant that represents the standard formal potential for this half-reaction at pH 7.00. The same approach can be used to correct for ionic strength effects by combining the activity coefficient term with the standard potential in Equation 10.49 (see Table 10.4). The result is a formal potential that allows the use of concentrations with the Nernst equation for a particular type of solution or sample.

Another situation in which formal potentials are useful is in when multiple side reactions can occur for a redox couple. A good example of this situation occurs during the reduction of Fe^{3+} to Fe^{2+} in water. This half-reaction has a standard electrode potential of +0.77 V, but under acidic conditions the potential for this system will be affected by the type of acid that is present (see Table 10.5). For instance, Fe^{3+} can form several complexes with chloride (such as $FeCl^{2+}$, $FeCl_2{}^+$, and so on) in the presence of HCl. Similar side-reactions can occur with other strong acids and will change the activity of Fe^{3+}, affecting the observed potential for its reduction to Fe^{2+}. The use of formal potentials in place of standard potentials can be an aid in allowing the extent of these effects to be described in a relatively simple manner.

Key Words

Other Terms

Questions

WHAT ARE OXIDATION–REDUCTION REACTIONS AND HOW ARE THEY USED IN ANALYTICAL CHEMISTRY?

1. Define the terms "oxidation" and "reduction." Explain why these two processes always occur simultaneously.
2. Describe what is meant by the term "oxidation–reduction reaction." Explain why reactions are also sometimes called "electrochemical reactions."
3. Explain why corrosion is an example of an oxidation–reduction reaction.
4. Describe at least three ways in which oxidation–reduction reactions might be used as part of a chemical analysis.
5. Hydroxylamine (NH_2OH) can be used in a pretreatment step to convert Fe^{3+} into Fe^{2+} prior to performing a spectrophotometric assay on a sample to measure its iron content. This pretreatment step takes place according to the following reaction.

$$4\,Fe^{3+} + 2\,NH_2OH \rightleftarrows 4\,Fe^{2+} + N_2O + 4\,H^+ + H_2O$$

Explain why this step is an example of an oxidation–reduction reaction.

DESCRIBING OXIDATION–REDUCTION REACTIONS

6. What is meant by the term "redox couple"? How is this term used to describe oxidation–reduction reactions?
7. Identify the redox couples in each of the following oxidation–reduction reactions.
   (a) $Zn + 2\,Ag^+ \rightleftarrows 2\,Ag + Zn^{2+}$
   (b) $Pb + PbO_2 + 4\,H^+ \rightleftarrows 2\,Pb^{2+} + 2\,H_2O$
   (c) $I_2 + 2\,S_2O_3^{2-} \rightleftarrows 2\,I^- + S_4O_6^{2-}$
8. Identify the redox couples in each of the following half-reactions.
   (a) $Ru^{3+} + e^- \rightleftarrows Ru^{2+}$
   (b) $Ni(OH)_3 + e^- \rightleftarrows Ni(OH)_2 + OH^-$
   (c) $2\,HOCl + 2\,H^+ + 2\,e^- \rightleftarrows Cl_2 + 2\,H_2O$
   (d) $H_2S \rightleftarrows S + 2\,H^+ + 2\,e^-$
9. In what ways are oxidation–reduction reactions similar to acid–base reactions (see Chapter 8). How are these two types of reactions different?

IDENTIFYING OXIDATION–REDUCTION REACTIONS USING OXIDATION NUMBERS

10. What is an "oxidation number"? What are the rules for determining the oxidation number of an atom in a chemical?
11. Determine the oxidation number of each atom in the following chemicals.
    (a) Cl_2
    (b) $Au(s)$
    (c) CaO
    (d) K_2SO_4
    (e) Fe_3O_4
    (f) H_2O_2
    (g) XeF_4
    (h) $(NH_4)_2Cr_2O_7$
    (i) CH_3OH
12. Determine the oxidation number of each element in the following chemicals.
    (a) Cl^-
    (b) Ca^{2+}

(c) H_3O^+
(d) MnO_4^-
(e) NH_4^+
(f) IO_3^-
(g) $Cr_2O_7^{2-}$
(h) HPO_4^{2-}
(i) AsO_4^{3-}

13. Explain how oxidation numbers can be used to identify an oxidation–reduction reaction.
14. Assign oxidation numbers to each atom in each of the following equations and identify which elements get reduced and which get oxidized.
    (a) $Fe_2O_3(s) + 2\,Al(s) \rightleftarrows Al_2O_3(s) + 2\,Fe(s)$
    (b) $2\,Co(OH)_3(s) + Sn(s) + OH^-(aq) \rightleftarrows$
        $2\,Co(OH)_2(s) + HSnO_2^-(aq) + H_2O(l)$
    (c) $2\,H_2O_2 \rightleftarrows 2\,H_2O + O_2$
    (d) $F_2 + 2\,Br^- \rightleftarrows Br_2 + 2\,F^-$
15. Determine whether or not each of the following equations represents an oxidation–reduction reaction. If it is an oxidation–reduction reaction, explain which atoms are undergoing oxidation or reduction.
    (a) $[Ag(NH_3)_2]^+ + Cl^- + 2\,H^+ \rightleftarrows AgCl + 2\,NH_4^+$
    (b) $Cl_2(g) + 2\,KI(aq) \rightleftarrows I_2(g) + 2\,KCl(aq)$
    (c) $BaCl_2(aq) + H_2SO_4(aq) \rightleftarrows BaSO_4(s) + 2\,HCl(aq)$
    (d) $Zn(s) + CuSO_4(aq) \rightleftarrows ZnSO_4(aq) + Cu(s)$

IDENTIFYING OXIDATION–REDUCTION REACTIONS USING HALF-REACTIONS

16. Define what is meant by the term "half-reaction." What is the difference between an "oxidation half-reaction" and a "reduction half-reaction"?
17. Explain why half-reactions are an "artificial" but useful way of looking at oxidation–reduction processes.
18. Dichromate ($Cr_2O_7^{2-}$) is often used as a reagent to titrate Fe^{2+}. This titration involves the following oxidation–reduction reaction.

$$Cr_2O_7^{2-} + 6\,Fe^{2+} + 14\,H^+ \rightleftarrows 2\,Cr^{3+} + 6\,Fe^{3+} + 7\,H_2O$$

Determine which chemicals are undergoing oxidation or reduction in this reaction and write the half-reactions for these processes.

19. Examine each of the oxidation–reduction reactions in Problem 15 by using half-reactions. Write the half-reactions for each of these processes. Show how these half-reactions can be combined to give the overall oxidation–reduction reactions that are shown in Problem 15. Also determine how many electrons are being transferred in each of these reactions.
20. Write the half-reactions for each of the following processes. Identify which chemicals are being oxidized or reduced based on these half-reactions and determine how many electrons are being transferred in each of these reactions.
    (a) $Sn^{2+} + 2\,HgCl_2 \rightleftarrows Sn^{4+} + Hg_2Cl_2 + 2\,Cl^-$
    (b) $I_2 + 2\,S_2O_3^{2-} \rightleftarrows 2\,I^- + S_4O_6^{2-}$
    (c) $H_2O_2 + 2\,Fe^{2+} + 2\,H^+ \rightleftarrows 2\,H_2O + 2\,Fe^{3+}$
    (d) $5\,Br^- + BrO_3^- + 6\,H^+ \rightleftarrows 3\,Br_2(aq) + 3\,H_2O$

USING EQUILIBRIUM CONSTANTS TO DESCRIBE OXIDATION–REDUCTION REACTIONS

21. Write an expression for the thermodynamic equilibrium constant ($K°$) for each of the oxidation–reduction reactions in Problem 20.

22. Write an expression for the concentration-dependent equilibrium constant (K) for the oxidation–reduction reactions in Problem 20. Show how K is related to the $K°$ expressions for these same reactions.

23. Write expressions for the concentration-dependent reaction quotient (Q) for the oxidation–reduction reactions in Problem 20. Explain how these expressions are different from those given for the concentration-dependent equilibrium constant in Problem 22.

24. A solution is prepared that originally contains two solutes, Ox and Red. Both of these solutes have an initial concentration of 1.0 M, but begin to undergo the following oxidation-reduction once they are combined.

$$Ox + 2\,Red \rightleftarrows Red' + 2\,Ox'$$

When the solution has reached equilibrium at 25°C, the concentration of Red is measured and found to be 0.00450 M. If it is known that no other significant side reactions are present, what are the concentrations of the other reactants and products at equilibrium? What is the value of the equilibrium constant at 25°C for this reaction?

25. It was indicated earlier that the equilibrium constant for the reaction of permanganate with Fe^{2+} is 1.4×10^{62} at 25°C. A mixture of Fe^{2+} (originally at a concentration of 0.0500 M) and a stoichiometric amount of permanganate that is buffered at pH 2.50 is allowed to react and reach equilibrium. It is found that the final concentration of Fe^{3+} is now the same as the original concentration of Fe^{2+} and that Mn^{2+} now has a concentration that is 20.00% that of Fe^{3+}. In theory, what will be the concentration of permanganate in this solution at equilibrium?

USING STANDARD POTENTIALS TO DESCRIBE OXIDATION–REDUCTION REACTIONS

26. Define what is meant by a "standard electrode potential." What are the conditions used to measure such a potential?

27. What is the reference half-reaction that is used for determining a standard electrode potential? What does it mean if a given half-reaction has a greater value for $E°$ than this reference half-reaction? What does it mean if a half-reaction has a lower value for $E°$ than this reference half-reaction?

28. Rank the following chemicals in terms of their ability to undergo reduction: Cr^{3+}, Cl_2, Zn^{2+}, Na^+, and O_3. Which of these chemicals would be the strongest oxidizing agents?

29. Rank the following chemicals in terms of their ability to undergo oxidation: $Cr(s)$, Cl^-, $Zn(s)$, $Na(s)$, and O_2. Which of these chemicals would be the strongest reducing agents?

30. Give an equation that allows the standard potential for an oxidation–reduction reaction to be calculated by using the standard potentials for the half-reactions that make up this overall reaction.

31. Calculate the standard potential for the following oxidation–reduction reactions. Rank these reactions in terms of their tendency to occur under standard reaction conditions.

    (a) $Cd(s) + Ni^{2+} \rightleftarrows Ni(s) + Cd^{2+}$

    (b) $Zn^{2+} + Cu(s) \rightleftarrows Cu^{2+} + Zn(s)$

    (c) $Sn^{2+} + Fe(s) \rightleftarrows Sn(s) + Fe^{2+}$

32. Silver ions in an aqueous solution will react with bismuth as shown below.

$$3\,Ag^+ + Bi(s) \rightleftarrows Bi^{3+} + 3\,Ag(s)$$

Write the two half-reactions and calculate the standard potential for this reaction.

33. How is the standard potential for an oxidation–reduction reaction related to the total change in energy for this same reaction under standard reactions conditions?

34. What is the "Faraday constant"? What is the typical value of this constant?

35. Calculate the standard potential and change in free energy under standard conditions for the reaction that takes place when $Zn(s)$ is converted to Zn^{2+} as it is used to reduce Fe^{3+} to Fe^{2+} in the presence of 1.0 M HCl.

36. Calculate the value of the potential and change in free energy under standard conditions for the following oxidation–reduction reaction.

$$Cl_2\,(aq) + 2\,I^- \rightleftarrows 2\,Cl^- + I_2(aq)$$

37. Give an equation that relates the standard potential for an oxidation–reduction reaction to the equilibrium constant for this same reaction. Using this equation, demonstrate why a large positive value for $E°$ would represent a favorable spontaneous reaction, while a large negative value for $E°$ represents a unfavorable nonspontaneous reaction.

38. What is the thermodynamic equilibrium constant at 25°C for the oxidation–reduction reaction that takes place in a Daniell cell (see Figure 10.5)? Explain how you obtained your answer.

39. Iron(III) tris-phenanthroline ($Fe(phen)_3^{3+}$) can undergo an oxidation–reduction reaction with $IrCl_6^{3-}$ to form $IrCl_6^{2-}$ and $Fe(phen)_3^{2+}$. The equilibrium constant for this reaction at 25°C is found by a spectrophotometric method to be 100. What is the expected potential for this oxidation–reduction reaction under these conditions?

40. A 1.00 g sample of zinc metal is placed into a 50 mL aqueous solution of 0.050 M $CuCl_2$.

    (a) Write the oxidation–reduction reaction that will occur between these chemicals. What is the equilibrium constant for this reaction at 25°C?

    (b) If this system is allowed to reach equilibrium 25°C, what will be the final expected concentrations of all the reactants and products?

41. A solution is prepared in which 0.0400 mol each of Ce^{4+}, Ce^{3+}, Fe^{2+}, and Fe^{3+} is dissolved in a 2.00 L aqueous solution of 1.0 M HCl.

    (a) Write the oxidation–reduction reaction that will occur between these chemicals. What is the equilibrium constant for this reaction at 25°C?

    (b) If this solution is allowed to reach equilibrium 25°C, what are the final expected concentrations of all the reactants and products?

DESCRIBING ELECTROCHEMICAL CELLS

42. What is an "electrochemical cell"? Explain why an electrochemical cell is a valuable tool in the use and study of oxidation–reduction reactions.

43. Describe how the cell in Figure 10.5, or the Daniell cell, works. What are the half-reactions that are used in this cell? How does the design of this particular cell make it possible to separate the components of these two half-reactions?

44. Define each of the following terms and the function of each in an electrochemical cell.
    (a) Electrode
    (b) Anode
    (c) Cathode
    (d) Salt bridge
    (e) Electrolyte
45. What is a "galvanic cell"? What is an "electrolytic cell"? How do these two types of cell differ?
46. Explain how it is possible for the same set of oxidation–reduction reactions to sometimes be used in the form as a galvanic cell and other times as an electrolytic cell. Give one specific example.

PREDICTING THE BEHAVIOR OF AN ELECTROCHEMICAL CELL

47. Define the term "standard cell potential." How do you calculate this value?
48. Calculate the standard cell potentials for the following reactions.
    (a) $2 Ce^{4+} + Zn(s) \rightleftarrows Zn^{2+} + 2 Ce^{3+}$
    (b) $MnO_4^- + 5 Fe^{2+} + 8 H^+ \rightleftarrows Mn^{2+} + 5 Fe^{3+} + 4 H_2O$
    (c) $Cr_2O_7^{2-} + 6 VO^{2+} + 2 H^+ \rightleftarrows 2 Cr^{3+} + 6 VO_2^+ + H_2O$
    (d) $2 Pu^{3+} + 3 Mg(s) \rightleftarrows 3 Mg^{2+} + 2 Pu(s)$
49. Write the two half-reactions for a cell in which one electrode is uranium metal in contact with a solution of U^{3+} (prepared by dissolving UCl_3 in water) and the other electrode is a platinum wire in contact with a solution that contains a mixture of V^{2+} and V^{3+} (prepared by dissolving VCl_2 and VCl_3 in water). Calculate the standard cell potential that is expected for this system.
50. Explain why the use of only standard potentials for half-reactions when calculating a standard cell potential can give a different value than the true standard potential for a cell.
51. Explain why the potential for an electrochemical cell always involves a comparison between the potentials that are present for two electrodes or half-reactions.
52. What is a "standard hydrogen electrode"? How is this type of electrode used in the study and characterization of oxidation–reduction reactions?
53. The heading of the column of reduction potentials that are listed in Table 10.2 indicates that all of these potentials were measured versus a standard hydrogen electrode. Use Equation 10.27 to explain why this type of comparison is useful in generating such a table.
54. An electrochemical cell is constructed in which a calomel electrode (see list of half-reactions in Appendix B) is used as the cathode and a standard hydrogen electrode is used as anode. What is the standard cell potential for this system?
55. A standard hydrogen electrode is used as the anode in an electrochemical cell that has the following overall oxidation–reduction reaction.

    $$Cu^{2+} + H_2 \rightleftarrows Cu(s) + 2 H^+$$

    (a) Write the half-reactions that are taking place at the cathode and anode in this cell.
    (b) What is the standard cell potential for this system?

WORKING WITH THE NERNST EQUATION

56. What is the "Nernst equation"? How is this equation used in describing oxidation–reduction reactions?
57. Write a general form of the Nernst equation for the reduction of Ox to Red through the addition of n electrons. What is one common form of this equation when it is used specifically at 25°C?
58. Some books use the following form of the Nernst equation for the general half-reaction.

    $$E = E° + \frac{RT}{nF} \ln\left(\frac{a_{Ox}}{a_{Red}}\right)$$

    (a) Show that the preceding form of the Nernst equation is mathematically equivalent to the one that was given earlier in Equation 10.32.
    (b) Derive a form of the preceding Nernst equation that could be used at 25°C and with base-10 logarithms rather than natural logarithms.
59. Write a general form of Nernst equation that can be used at any temperature for each of the following reduction half-reactions.
    (a) $Cu^{2+} + e^- \rightleftarrows Cu^+$
    (b) $Hg^{2+} + 2 e^- \rightleftarrows Hg$
    (c) $Tl^+ + e^- \rightleftarrows Tl$
    (d) $Sn^{4+} + 2 e^- \rightleftarrows Sn^{2+}$
60. Write a Nernst equation that could be used specifically at 25°C for each of the half-reactions in the previous problem.
61. When we obtained Equation 10.3 earlier in this chapter, we assumed that the two following half-reactions were equivalent for the oxidation of Fe(s) to Fe^{2+}.

    $$2 Fe(s) \rightleftarrows 2 Fe^{2+} + 4 e^- \quad or \quad Fe(s) \rightleftarrows Fe^{2+} + 2 e^-$$

    Write Nernst equations for the reduction form of each of these half-reactions. Compare these two Nernst equations to see if they will provide calculated potentials for these half-reactions that are mathematically equivalent.
62. Write a general form of Nernst equation that can be used at any temperature for each of the following reduction half-reactions.
    (a) $AgCl + e^- \rightleftarrows Ag + Cl^-$
    (b) $2 H_2O + 2 e^- \rightleftarrows H_2 + 2 OH^-$
    (c) $NO_3^- + 3 H^+ + 2 e^- \rightleftarrows HNO_2 + H_2O$
    (d) $2 HOCl + 2 H^+ + 2 e^- \rightleftarrows Cl_2 + 2 H_2O$
63. Write a Nernst equation that could be used specifically at 25°C for each of the half-reactions in the previous problem.
64. Write Nernst equations for the two half-reactions that make up the following oxidation–reduction reaction.

    $$H_3PO_4 + H_3AsO_3 \rightleftarrows H_3AsO_4 + H_3PO_3$$

CALCULATING POTENTIALS FOR OXIDATION–REDUCTION REACTIONS

65. Describe a general approach by which the Nernst equation can be used to find the net potential of an oxidation–reduction reaction or electrochemical cell.

66. What is the expected potential of a Daniell cell in which $a_{Zn^{2+}} = 0.340$ and $a_{Cu^{2+}} = 0.135$? Show how you obtained your answer.

67. Identify the two half-reactions for the following oxidation–reduction process.

$$Br_2(aq) + H_3AsO_3 + H_2O \rightleftarrows 2\,Br^- + H_3AsO_4 + 2\,H^+$$

   (a) Write the Nernst equations for the half-reactions that are involved in this process.
   (b) Calculate the potential that would be expected for this overall reaction at 25°C and under the following reaction conditions: pH 2.00, $[Br_2] = 0.00500\ M$, $[Br^-] = 0.244\ M$, $[H_3AsO_3] = 0.128\ M$, and $[H_3AsO_4] = 0.00367\ M$. (*Note:* You may assume that the activities and concentrations of these chemicals are approximately the same in this example.)

68. Write Nernst equations for the two half-reactions that are involved in the oxidation–reduction process in Problem 32. Find the net potential that would be expected for this overall reaction at 25°C if $[Ag^+] = 0.0100\ M$ and the $[Bi^{3+}] = 0.0100\ M$, assuming that the activities and concentrations of these chemicals are approximately the same.

69. An electrochemical cell is constructed in which one side of the cell contains a nickel electrode that is in contact with an aqueous solution of $0.00040\ M$ Ni^{2+}. The other side of the cell contains an electrode made of cobalt metal that is in contact with an aqueous solution of $0.050\ M$ Co^{2+}.
   (a) What are the two half-reactions for this cell?
   (b) Which electrode would you expect to be the anode and which would you expect to be the cathode based on the standard reduction potentials of their corresponding half-reactions?
   (c) What is the estimated potential for this cell at 25°C based on the concentrations of the listed species?
   (d) Which electrode is the cathode and which is the anode in this cell? Explain your answer.

70. An electrochemical cell is constructed in which one electrode is a piece of copper metal that is immersed in an aqueous solution containing $0.00500\ M$ $Cu(NO_3)_2$ and the other electrode is a piece of silver metal that is immersed in an aqueous solution containing $0.235\ M$ $AgNO_3$.
   (a) What are the expected half-reactions for this cell?
   (b) What is the approximate potential at 25°C based on the concentrations of the given chemicals in this cell?
   (c) Which electrode is the anode and which is the cathode in this cell?

71. A *concentration cell* is a special type of electrochemical cell in which the same redox couples are used in both the anode and cathode, but with concentrations of the species at these two electrodes being different.[6] For instance, suppose we have an electrochemical cell in which both electrodes are made of copper metal, but where one electrode is immersed in an aqueous solution containing Cu^{2+} at an activity of 0.236 and the other electrode is in an aqueous solution containing Cu^{2+} at an activity of 0.875. Write the half-reactions that occur at the cathode and anode and estimate the potential of this cell at 25°C. Explain why a difference in potential is observed between these two electrodes.

72. Explain why using concentrations instead of chemical activities can lead to errors when you calculate potentials with the Nernst equation.

73. Derive an equation that shows the error that will be introduced in Problem 68 if concentrations are used instead of chemical activities in calculating the cell potential. Do you expect this error to be significant if $I = 0.07\ M$?

74. Derive an equation that shows the error that will be introduced in Problem 70 if concentrations are used instead of chemical activities in calculating the cell potential. Do you expect this error to be significant in this particular case? Explain your answer.

75. Suppose that an electrochemical cell is established in which one electrode is platinum in contact with a solution containing $[Fe^{2+}] = 0.00235\ M$ and $[Fe^{3+}] = 0.00764\ M$ in $0.10\ M$ H^+, and the other is also platinum in $0.01\ M$ H^+ and $[Mn^{2+}] = 0.0439\ M$ and $[MnO_4^-] = 0.0764\ M$.
   (a) Predict the potential for this cell at 25°C.
   (b) Calculate the error that was introduced in this calculated potential by using concentration rather than chemical activity if both solutions in this cell have an ionic strength of approximately $0.10\ M$. (*Note:* See Chapter 6 if you need a review of how an activity coefficient can be estimated for an ion in water.)

EFFECTS OF THE SAMPLE MATRIX AND SIDE REACTIONS

76. Which of the following oxidation–reduction reactions or half-reactions will be directly affected by a change in pH?
   (a) $BiOCl + 2\,H^+ + 3\,e^- \rightleftarrows Bi + H_2O + Cl^-$
   (b) $I_3^- + 2\,S_2O_3^{2-} \rightleftarrows 3\,I^- + S_4O_6^{2-}$
   (c) $CrO_4^{2-} + 4\,H_2O + 3\,e^- \rightleftarrows Cr(OH)_3 + 5\,OH^-$
   (d) $2\,Mn^{2+} + 5\,S_2O_8^{2-} + 8\,H_2O \rightleftarrows 10\,SO_4^{2-} + 2\,MnO_4^- + 16\,H^+$

77. The reduction of dichromate ($Cr_2O_7^{2-}$) to Cr^{3+} in water can be represented by the following half-reaction and Nernst Equation under basic conditions.

$$Cr_2O_7^{2+} + 7\,H_2O + 6\,e^- \rightleftarrows 2\,Cr^{3+} + 14\,OH^-$$

$$E_{Cr_2O_7^{2-}/Cr^{3+}} = E^\circ_{Cr_2O_7^{2-}/Cr^{3+}}$$
$$- \frac{0.05916\ V}{6}\log\left(\frac{(a_{Cr^{3+}})^2(a_{OH^-})^{14}}{(a_{Cr_2O_7^{2-}})}\right)$$

What is the expected effect of an increase in pH on the potential for this half-reaction? How do these results compare with those obtained for Equation 10.54?

78. A chemist wishes to use a solution of dichromate/chromate at pH 7.00 for a redox titration. Using Equation 10.54, estimate the half-reaction potential for the reduction of $Cr_2O_7^{2-}$ to Cr^{3+} at pH 7.00. Also calculate the potential for this process by using the Nernst equation given in Problem 76. How do the results of these two equations compare?

79. Estimate the potential at 25°C of an electrochemical cell in which one electrode is a piece of platinum held in a pH 2.00 aqueous solution of $0.0015\ M$ $KMnO_4$ and $0.023\ M$ Mn^{2+}, while the other electrode is a piece of platinum held in a aqueous solution that contains

0.025 M Sn^{4+} and 0.014 M Sn^{2+}. What is the estimated potential for this same cell at pH 4.00?

80. Explain how it is possible to use a modified form of a Nernst equation to deal with a side reaction that involves the oxidizing agent or reducing agent in a half-reaction.

81. A silver wire coated with AgCl is immersed in an aqueous solution containing 0.0025 M NaCl. This wire and solution are then used as an electrode in a cell that also contains a standard hydrogen electrode.
    (a) What are the half-reactions for this process? What are the Nernst equations for these half-reactions?
    (b) What side reactions might occur that will affect the half-reactions in this system?
    (c) What is the expected potential of this system at 25°C?

82. A sample of solid bismuth oxychloride (BiOCl) is to be reduced by zinc metal in the presence of 0.00500 M HCl and 0.0200 M Zn^{2+}.
    (a) What are the half-reactions for this process? What are the Nernst equations for these half-reactions?
    (b) What side reactions might occur that will affect the half-reactions in this system?
    (c) What is the expected potential of this system at 25°C?

83. What is a "Pourbaix diagram"? Explain why this type of diagram is useful when dealing with oxidation–reduction systems that involve many simultaneous reactions.

84. Use the Pourbaix diagram given in Figure 10.10 to determine the principal form of iron that will be present under each of the following conditions.
    (a) pH 4.0 and a potential of 0.0 V
    (b) pH 10 and a potential of –0.6 V
    (c) pH 1.0 and a potential of +1.5 V
    (d) pH 8.0 and a potential of +1.5 V

85. What is a "formal potential"? How does this differ from a standard potential?

86. An electrochemical cell is constructed in which both electrodes are made of platinum, but in which one electrode is in contact with an aqueous solution that has $[Fe^{2+}] = 0.0765$ M and $[Fe^{3+}] = 0.176$ M, while the other electrode is in contact with an aqueous solution that contains $[Ce^{4+}] = 0.0376$ M and $[Ce^{3+}] = 0.0987$ M.
    (a) What is the estimated potential for this cell at 25°C? Which electrode is the cathode and which is the anode in this cell?
    (b) What is the estimated potential for this cell at 25°C when the aqueous solutions in both sides of the cell also contain 1.0 M HCl? Compare this result to the potential that was estimated in Part (a).
    (c) What is the estimated potential for this cell at 25°C when the aqueous solutions in both sides of the cell contain 1.0 M H_2SO_4 instead of 1.0 M HCl?

87. Biochemists usually do their experiments in solutions that are at or near pH 7.0. Many compounds of biological interest that undergo oxidation–reduction reactions can also act as acids or bases. For instance, the reduction of pyruvate ($C_3H_3O_3^{-}$) to lactate ($C_3H_5O_3^{-}$) occurs according to the following half-reaction and has a standard potential of +0.224 V.

$$C_3H_3O_3^{-} + 2\,H^{+} + 2\,e^{-} \rightleftarrows C_3H_5O_3^{-}$$

    (a) Write a Nernst equation for this half-reaction at 25°C.
    (b) Rearrange the equation you obtained in Part (a) to separate out all terms that are directly hydrogen ion

activity. Write the modified form of the Nernst equation that you obtain.
    (c) The formal potential for this system at pH 7.00 can be obtained by combining the terms for E° and hydrogen ion activity that you obtained in Part (b). Calculate the value of this formal potential. How does it compare to the value for E°?

CHALLENGE PROBLEMS

88. Appendix A describes various approaches for balanced oxidation–reduction reactions. Use one of these methods to balance the following reactions.
    (a) $Ag^{+} + Cu \rightleftarrows Cu^{2+} + Ag$
    (b) $Sn^{4+} + V^{2+} \rightleftarrows V^{3+} + Sn^{2+}$
    (c) $Br_2 + S_2O_3^{2-} \rightleftarrows Br^{-} + S_4O_6^{2-}$
    (d) $H_2O_2 + CrO_4^{2-} \rightleftarrows O_2 + Cr^{3+}$

89. Use the methods described in Appendix A to balance each of the following oxidation–reduction reactions.
    (a) $MnO_4^{-} + Zn(s) \rightleftarrows Mn^{2+} + Zn^{2+}$ (acidic conditions)
    (b) Fe^{2+} reacting with Au^{+} in water
    (c) $Cr_2O_7^{2-} + S_2O_3^{2-} \rightleftarrows Cr^{3+} + S_4O_6^{2-}$
        (basic conditions)
    (d) In(s) reacting with Sn^{2+} in water
    (e) $VO_3^{-} + Zn(s) \rightleftarrows VO^{2-} + Zn^{2+}$ (acidic conditions)
    (f) IO_3^{-} reacting with I^{-} in water (acidic conditions)

90. When we use the method described in Section 10.2B to calculate the oxidation number for carbon in an organic compound, we obtain an average value for all carbon atoms in the molecule. A different method will allow calculation of the oxidation number for each carbon in a compound. To do this, start counting the oxidation number of a carbon atom at zero and increase by one for every bond that carbon has to an atom more electronegative than carbon (such as oxygen, nitrogen, or halogen) and decrease by one for every bond to an atom less electronegative than carbon (which is usually hydrogen).
    (a) Use this approach to find the oxidation numbers for the two terminal carbons and six internal carbons of n-octane, $CH_3(CH_2)_6CH_3$. What is the average oxidation number for the carbon atoms in this n-octane?
    (b) Calculate the oxidation number of each carbon in acetic acid, CH₃COOH. What is the average oxidation number for the carbon atoms in acetic acid?

91. Given the fact that the net potential for an oxidation–reduction reaction is exactly 0.00 at equilibrium, use the Nernst equation to derive the relationship in Equation 10.19 that is give between K° and E_{Net}°.

92. Use the *CRC Handbook of Chemistry and Physics*[10] or other references[11–14] to locate standard reduction potentials or formal potentials at 25°C for the following half-reactions.
    (a) $2\,D^{+} + e^{-} \rightleftarrows D_2$ (E°)
    (b) $Ir_2O_3 + 3\,H_2O + 6\,e^{-} \rightleftarrows 2\,Ir + 6\,OH^{-}$ (E°)
    (c) $Np^{4+} + e^{-} \rightleftarrows Np^{3+}$ ($E^{\circ\prime}$ in 1 M $HClO_4$)
    (d) $BrO^{-} + H_2O + 2\,e^{-} \rightleftarrows Br^{-} + 2\,OH^{-}$
        (E° in 1 M NaOH)

93. The following list gives several common oxidizing agents and reducing agents that are found or used in chemical laboratories. Obtain information from the Internet or

material safety data sheets on the chemical and physical hazards that are associated with these agents.
(a) Sodium borohydride
(b) Hydrogen peroxide
(c) Ozone
(d) Periodic acid

94. Use cell notation (see Box 10.1) to describe Problems 79, 81, and 86. (*Note:* Assume that a salt bridge like the one in Figure 10.5 is present in each of these cells.)

95. Compare the general form of the Nernst equation with that of the Henderson–Hasselbalch equation (see Chapter 8). What similarities are present in these two equations? What differences are there in these equations and in their underlying assumptions?

96. Ozone (O_3) will react with bromide to make bromine and oxygen (O_2) in water.
(a) Prepare a spreadsheet in which you calculate and plot the potential for this reaction at 25°C and pH values ranging from 0.00 to 10.00.
(b) Prepare a second plot in which you calculate and plot the equilibrium constant for this reaction at 25°C and pH values ranging from 0.00 to 10.00.
(c) Based on the plots in Part (a) and (b), determine whether this reaction is more favored at an acidic pH or basic pH. Compare this result with what would be expected based on the pH dependence of the Nernst equations for the two half-reactions that are involved in this process.

97. Hydrogen peroxide disproportionates to form water and molecular oxygen according to the following reaction.

$$2\,H_2O_2 \rightleftarrows 2\,H_2O + O_2$$

(a) What are the half-reactions that are involved in this process? What are the Nernst equations for these half-reactions?
(b) Prepare a spreadsheet in which you calculate and plot the potential and equilibrium constant for this reaction at 25°C and pH values ranging from 0.00 to 10.00. What trends do you observed in these plots?

98. Write the reactions that are occurring at the various boundaries within the Pourbaix diagram that is shown in Figure 10.10. Write an equilibrium expression for each solubility reaction in this plot and a Nernst equation for each oxidation–reduction reaction. Use these equations to explain why some of the boundaries in this plot change in position with pH.

99. It is sometimes useful to combine two different half-reactions to devise a third, new half-reaction. Suppose we know the standard potentials for the following reduction half-reactions, involving the one-electron reduction of Fe^{3+} to Fe^{2+}, followed by the two electron reduction of Fe^{2+} to Fe(s).

$$
\begin{aligned}
Fe^{3+} + 1\,e^- &\rightleftarrows Fe^{2+} \quad & E_1^\circ &= +0.77\text{ V (at 25°C)} \\
Fe^{2+} + 2\,e^- &\rightleftarrows Fe_{(s)} \quad & E_2^\circ &= -0.41\text{ V (at 25°C)} \quad (10.60) \\
\hline
Fe^{3+} + 3\,e^- &\rightleftarrows Fe_{(s)} \quad & E_3^\circ &= ?\text{ V}
\end{aligned}
$$

With the information provided in first two of these half-reactions, it is also possible to calculate the standard potential that would be expected for the new half-cell reaction that involves the direct reduction of Fe^{3+} to Fe(s) by a three-electron process. To calculate the standard potential for a new half-reaction that is obtained through

the combination of others, we must carefully account for the number of electrons that are being transferred in each of the old and new half-reactions. This is necessary to appropriately weigh the thermodynamic contributions for each half-reaction that is being combined to give the final overall half-reaction. We can make this adjustment by using Equation 10.61,

$$n_1 E_1^\circ + n_2 E_2^\circ = (n_1 + n_2)\,E_3^\circ \qquad (10.61)$$

where E° and n represent the standard potentials and number of electrons being transferred in each of these half-reactions, and the subscripts 1 and 2 represent the two half-reactions that are being combined to give a third half-reaction.
(a) What is the expected standard potential for the half-reaction in Equation 10.61? How does this standard potential compare to those for the half-reactions that were combined to obtain this new half-reaction?
(b) The standard potential for the reduction of Cu^{2+} to Cu(s) is +0.34 V at 25°C, and the standard potential for reduction of Cu^+ to Cu(s) is +0.52 V at 25°C. What is the expected standard potential at 25°C for the reduction of Cu^{2+} to Cu^+?

100. Another way Equation 10.61 can be used is to obtain the standard potential for a new half-reaction that can be combined with a known half-reaction to give a third half-reaction that is also known. For example, the standard potentials at 25°C for the reduction of O_2 to water and for the reduction of hydrogen peroxide (H_2O_2) to water have been previously determined, as shown below.

$$
\begin{aligned}
O_2(g) + 2\,H^+ + 2\,e^- &\rightleftarrows H_2O_2 \quad & E^\circ &= ?\text{ V} \\
H_2O_2 + 2\,H^+ + 2\,e^- &\rightleftarrows 2\,H_2O \quad & E^\circ &= +1.78\text{ V (at 25°C)} \\
\hline
O_2(g) + 4\,H^+ + 4\,e^- &\rightleftarrows 2\,H_2O \quad & E^\circ &= +1.23\text{ V (at 25°C)}
\end{aligned}
$$

Use this information to estimate the standard potential at 25°C for the reduction of oxygen to hydrogen peroxide. Explain how you obtained your final answer.

TOPICS FOR DISCUSSION AND REPORTS

101. The field of electrochemistry deals with the study and use of oxidation–reduction reactions. There are many scientists besides Walther Nernst who have contributed to this field. Obtain more information on one of the following individuals and describe how their work contributed to our current understanding of oxidation–reduction reactions or the applications of these reactions.
(a) Alessandro Volta
(b) Michael Faraday
(c) Luigi Galvani
(d) John Daniell

102. There are various analytical methods that are now being used to study and preserve ships like the USS Arizona. Obtain more information on one or more of these methods and describe how they are used in the preservation of sunken ships.[1-4]

103. There are many types of oxidation–reduction reactions that can be used to create batteries. Obtain more information on one of the following batteries and describe how they work. Include in your description the oxidation and

reduction half-reactions that are used within this battery and some applications for the battery.
(a) Mercury battery
(b) Zinc–carbon dry cell-battery
(c) Lead–acid battery
(d) Lithium battery

104. The use of hydrogen as a fuel for automobiles has been a topic of great interest in recent years. Obtain more information on how hydrogen can be produced from water and on how this hydrogen can be used as a fuel. Write a report that shows how the production and use of hydrogen as a fuel involve oxidation–reduction reactions.

References

1. M.A. Russell, D.L. Conlin, L.E. Murphy, D.L. Johnson, B.M. Wilson, and J.D. Carr, "A Minimum-Impact Method for Measuring Corrosion Rate of Steel-Hulled Shipwrecks in Seawater," *The International Journal of Nautical Archaelology*, 35 (2006) 310–318.

2. D.L. Johnson, B.M. Wilson, J.D. Carr, M.A. Russell, L.E. Murphy, and D.L. Colin, "Corrosion of Steel Shipwrecks in the Marine Environment: USS *Arizona*—Part 1," *Materials Performance*, 45 (2006) 40–44.

3. D.L. Johnson, B.M. Wilson, J.D. Carr, M.A. Russell, L.E. Murphy, and D.L. Colin, "Corrosion of Steel Shipwrecks in the Marine Environment: USS *Arizona*—Part 2," *Materials Performance*, 45 (2006) 54–57.

4. C.H. Arnaud, "Saving Shipwrecks," *Chemical & Engineering News*, 85 (2007) 45–47.

5. *IUPAC Compendium of Chemical Terminology*, Electronic version, http://goldbook.iupac.org

6. G. Maludziska, Ed., *Dictionary of Analytical Chemistry*, Elsevier, Amsterdam, 1990.

7. J. Inczedy, T. Lengyel, A.M. Ure, *Compendium of Analytical Nomenclature*, 3rd Ed., Blackwell Science, Malden, MA, 1997.

8. H. Frieser, *Concepts & Calculations in Analytical Chemistry: A Spreadsheet Approach*, CRC Press, Boca Raton, FL, 1992.

9. A.J. Bard and L.R. Faulkner, *Electrochemical Methods: Fundamentals and Applications*, 2nd ed., Wiley, New York, 2004.

10. D.R. Lide, Ed., *CRC Handbook of Chemistry and Physics*, 83rd ed., CRC Press, Boca Raton, FL, 2002.

11. John A. Dean, Ed., *Lange's Handbook of Chemistry*, 15th ed., McGraw-Hill, New York, 1999.

12. S.G. Bratsch, "Standard Electrode Potentials and Temperature Coefficients in Water at 298.15 K," *Journal of Physical and Chemical Reference Data*, 18 (1989) 1–21.

13. A.J. Bard, R. Parsons, and J. Jordan, *Standard Potentials in Aqueous Solution*, Marcel Dekker, New York, 1985.

14. G. Milazzo and S. Caroli, *Tables of Standard Electrode Potentials*, Wiley, New York, 1978.

15. J.F. Daniell, "On Voltaic Combinations," *Philosophical Transactions*, 1836.

16. D.I. Davies, "John Frederic Daniell 1791–1845," *Chemistry in Britain*, 26 (1990) 946–947, 949, 960.

17. *The New Encyclopaedia Britannica*, 15th ed., Encyclopaedia Britannica, Inc., Chicago, IL, 2002.

18. D.A. Skoog, F.J. Holler, and S.R. Crouch, *Principles of Instrumental Analysis*, 6th ed., Brooks/Cole, Pacific Grove, CA, 2006.

19. D.K. Barkan, *Walther Nernst and the Transition to Modern Physical Science*, Cambridge University Press, New York 1998.

20. A.S. Feiner and A.J. McEvoy, "The Nernst Equation," *Journal of Chemical Education*, 71 (1994) 493–494.

21. L. Meites, "A 'Derivation' of the Nernst Equation for Elementary Quantitative Analysis," *Journal of Chemical Education*, 29 (1952) 142–143.

22. D.A. Jones, *Principles and Prevention of Corrosion*, 2nd ed., Prentice Hall, Upper Saddle River, NJ, 1996.

23. M. Pourbaix, *Atlas of Electrochemical Equilibria in Aqueous Solutions*, 2nd ed., National Association of Corrosion Engineers, Houston, TX, 1974.

24. A. Napoli and L. Pogliani, "Potential-pH Diagrams," *Education in Chemistry*, 34 (1997) 51–52.

Chapter 11

Gravimetric Analysis

Chapter Outline

11.1 INTRODUCTION: FIXING THE PERIODIC TABLE

The periodic table is probably the most valuable and widely used tool in chemistry. This table (originally developed in 1872 by Dimitri Mendeleev) provides a wealth of information on chemical and physical properties of the elements.[1,2] One important number in this table is the atomic mass of each element. This value is employed on almost a daily basis by today's chemists, who routinely use this information to determine the formula weights of chemicals and the amount of a chemical that must be used to prepare a particular sample or reagent.

The reliability of the atomic masses listed in the periodic table is often taken for granted. However, the accuracy of these values was of great concern in the late 1800s and early 1900s.[3] This topic was of particular interest to an American chemist named T.W. Richards (see Figure 11.1).[4–7] Richards used various chemical reactions and his knowledge of chemical equilibrium to determine the atomic masses of elements by using the technique of *gravimetric analysis*.[3–9] He compared the atomic masses of silver and chloride by dissolving a known mass of pure silver and precipitating the resulting Ag^+ ions with Cl^-, giving a weighable and pure sample of solid silver chloride (AgCl). He then converted this solid into silver oxide (Ag_2O) and found the atomic mass of silver by using an atomic mass of exactly 16.00000 for oxygen (the atomic mass reference at the time of his studies). The same approach also gave the atomic mass of chloride once the value for silver had been determined.

By using gravimetric analysis, Richards and his students were able to determine the atomic masses of 55 elements. His results were considered to be the most reliable values of their time and are still remarkably close to modern values for atomic masses.[3–5] Gravimetric analysis is still an important method for chemical measurements. In this chapter, we will learn how to perform a gravimetric analysis and look at several examples of how this technique is used in modern chemical measurements.

11.1A What Is Gravimetric Analysis?

A **gravimetric analysis** (also known as *gravimetry*) is an analytical method that uses only measurements of mass and information on reaction stoichiometry to determine the amount of an analyte in a sample. A good example of such a method can be seen in the work by T.W. Richards when he determined the atomic mass of silver (see Figure 11.2). In this case, silver ions that had been placed into an aqueous solution were reacted with chloride ions to form an insoluble precipitate of silver chloride (AgCl). We learned in Chapter 7 that this reaction has a known

T. W. Richards

Element	Previous atomic mass	Richards' result	Present value
Hydrogen	1.002	1.0082	(1.0079)
Copper	63.3	63.57	(63.55)
Barium	137.0	137.37	(137.33)
Strontium	87.5	87.62	(87.62)
Zinc	65.0	65.37	(65.41)
Magnesium	24.2	24.32	(24.31)
Cobalt	59.1	58.97	(58.93)
Nickel	58.5	58.68	(58.69)
Iron	56.00	55.85	(55.85)
Uranium	240.2	238.4	(238.0)
Rubidium	85.5	85.42	(85.47)
Sodium	23.05	22.995	(22.990)
Chlorine	35.45	35.458	(35.453)
Bromine	79.95	79.917	(79.904)
Potassium	39.14	39.095	(39.098)
Nitrogen	14.04	14.008	(14.007)
Sulfur	32.06	32.07	(32.07)
Silver	107.93	107.88	(107.87)
Lithium	7.03	6.94	(6.94)
Calcium	40.00	40.07	(40.08)
Carbon	12.0	12.005	(12.011)
Aluminum	27.1	26.96	(26.98)
Gallium	69.9	69.716	(69.723)
Cesium	132.9	132.81	(132.91)

FIGURE 11.1 Theodore William Richards (1868–1928) and a list of atomic masses that were determined by Richards and his students. Richards was a professor at Harvard University who believed that the atomic mass was a key property in understanding the properties of the elements. Of his over 300 publications, more than half were devoted to the determination of atomic mass and in improving methods for its measurement. He received the 1914 Nobel Prize in Chemistry for this work, making Richards the first American to receive this award. Among his other research was a paper in 1914 in which he and Max E. Lambert gave some of the first experimental evidence for the existence of isotopes, based on their careful measurements of the apparent atomic mass of radioactive lead from various sources.[7] (The data in the table were obtained from A. J. Ihde, "Theodore William Richards and the Atomic Weight Problem," *Science*, 164 (1969) 647–651.)

stoichiometry and a small solubility product (favoring the formation of AgCl). Richards used this reaction to find the atomic mass of silver by measuring the amount of AgCl that was produced from a sample that contained a *known* initial mass of silver. Modern chemists use this same approach for the quantitative analysis of silver by using the known atomic mass of silver and the measured mass of AgCl to determine the amount of silver that is present in a sample. Both types of experiments make use of the known stoichiometry of this reaction, in which one mole of Ag^+ reacts with one mole of Cl^- for every mole of AgCl that is formed.

The combination of silver ions with chloride ions to form AgCl is one of hundreds of reactions that can be used in gravimetric analysis. Although this approach is simple in theory, obtaining accurate results by gravimetric analysis requires great attention to detail in order to avoid serious errors. T.W. Richards made use of good laboratory technique and his knowledge of chemical reactions to overcome errors that were made by others and to greatly improve the accuracy of atomic mass values.[4,5] Later in this chapter we will see what types of methods and steps can be used to help minimize errors and provide accurate measurements during a gravimetric analysis for the measurement of chemicals in unknown samples.

11.1B How Is Gravimetric Analysis Used in Analytical Chemistry?

There are several ways in which you might conduct a gravimetric analysis, but all involve a measurement of mass. Table 11.1 shows various strategies that are used for gravimetric analysis. Most traditional gravimetric methods determine the mass of an analyte by measuring the mass of a precipitate or related solid that contains this analyte in a known ratio. This is the type of analysis that we focus on in most of this chapter. This method usually involves adding a precipitating agent (or **precipitant**) to form a weighable solid whose mass can be used to calculate the mass of analyte that is present. In the closely related area of precipitation titrations (see Chapter 13), the same type of reaction is used, but it is now the volume of the added reagent that is measured rather than the mass of the resulting solid.

Two other types of gravimetric analysis methods are combustion analysis and thermogravimetric analysis, which we examine in Section 11.3. These are both methods in which mass measurements are combined with the formation or loss of volatile chemicals. In the case of combustion analysis, a sample is burned to release gases such as carbon dioxide and water, which are then collected and weighed to determine the carbon and hydrogen

Determination of atomic mass of silver by T.W. Richards

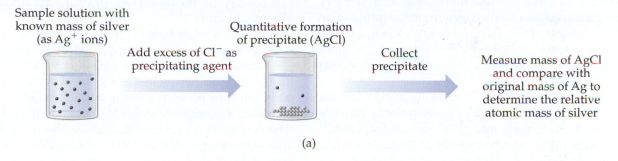

(a)

Determination of silver content in a sample

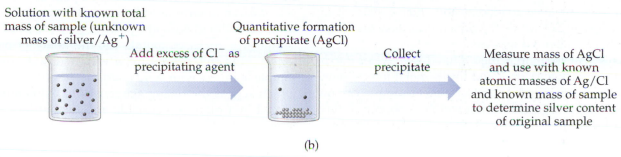

(b)

FIGURE 11.2   Use of gravimetric analysis and the precipitation of silver chloride by T. W. Richards to determine the atomic mass of silver (a), and use of gravimetric analysis by a modern chemist to determine the amount of silver in a sample (b).

content of the original sample.[10] In a thermogravimetric analysis, a sample is heated in a controlled fashion and its change in mass is measured with temperature as it releases volatile components or reacts with gases in its surrounding atmosphere.[11] In both cases, mass measurements and reaction stoichiometry are used to provide information on the chemical content of a sample.

A big advantage of gravimetric analysis is that determining the mass of a substance is one of the most accurate measurements that can be made. Many of the gravimetric methods that we discuss in this chapter can be carried out with errors of less than 0.2%, even when they are performed by relatively untrained students. A traditional gravimetric analysis is also inexpensive to conduct and requires only a minimal amount of equipment, such as a high-quality laboratory balance and perhaps a

drying oven. Although this approach can be tedious when used for a large number of samples, it is a relatively simple method for obtaining highly accurate and precise results when working with only a few samples.

An inherent requirement of gravimetric analysis is that the sample being examined must contain enough analyte to give a weighable mass. To meet this requirement in a traditional gravimetric analysis, the final precipitate or solid that is weighed must have a mass of greater than 0.10 g to provide a relative error of less than 0.2%. Gravimetric analysis is used mainly in modern laboratories when high accuracy is absolutely essential and time is of little concern. For instance, the National Institute of Standards and Technology uses gravimetric analysis as a "gold standard technique," or reference method, to evaluate the accuracy of other analytical techniques.[8,9]

11.2 PERFORMING A TRADITIONAL GRAVIMETRIC ANALYSIS

11.2A General Strategies and Methods

The steps that are involved in a traditional gravimetric analysis, based on the precipitation and formation of a weighable solid, are the same general steps used by T.W. Richards to measure the atomic masses of many elements. After a sample has been acquired, the first step is to convert it into a form that can be used in a gravimetric analysis. Traditional gravimetric methods require that the analyte be placed into a solution in a form that is both soluble and that can be precipitated. In the case of a metal or mineral-based sample, the sample must be

TABLE 11.1 General Types of Gravimetric Methods	
Type of Method	**Example**
Measurement of the mass for a solid that is formed through precipitation of an analyte from a solution	Traditional gravimetric analysis
Measurement of the gain in mass due to the collection of a chemical from a sample	Combustion analysis
Measurement of the change in mass of a sample as the temperature is varied	Thermogravimetric analysis

converted into metal ions that will be soluble in an aqueous solution. To produce these metal ions, some samples must be dissolved in a fairly concentrated solution of an acid (for example, HCl, HNO_3, H_2SO_4, $HClO_4$, or HF). Even more extreme conditions are necessary for other samples, requiring pretreatment methods such as wet ashing, dry ashing, or fusion.

The term *ashing*, when used in chemical analysis, refers to the pretreatment by dry or wet methods that convert metals in the sample into metal ions in a solution.[12] **Wet ashing** is a method often used with organic samples that involves adding a weighed amount of sample to a concentrated acid and heating it to the boiling point of the acid, usually in a porcelain dish.[13,14] This procedure is done in such a way as to oxidize any organic material so that it is lost as CO_2 while the mineral components of the sample remain behind and dissolve in the acid as metal ions. This acid solution is then diluted with water into a volumetric flask and analyzed for its metal-ion content. For example, the iron content of a meat sample can be determined gravimetrically if a weighed portion of this meat is boiled in a mixture of nitric acid and perchloric acid. This digestion reaction converts all of the organic material in the sample into CO_2, H_2O, and NO_2 gas, but leaves behind an acidic solution that contains iron ions and ions of any other metals that might have been present in the sample. This solution can then be analyzed to determine the original iron or metal content of the sample.

Dry ashing is a method of sample preparation in which a weighed portion of a sample is heated to a red-hot temperature in a porcelain dish that is open to the air. This procedure burns off any organic material and leaves nonvolatile metal oxides behind in the dish.[14] These metal oxides are usually soluble in a dilute solution of hydrochloric acid. For instance, if you were to use dry ashing on a sample that contains a protein-like hemoglobin (which contains the elements carbon, hydrogen, nitrogen, sulfur, and iron), the reaction of oxygen with this protein in the presence of heat would give CO_2, H_2O, NO_2, SO_2, and Fe_2O_3. In this situation all products except the iron oxide are gases at elevated temperatures, leaving only solid Fe_2O_3 in the crucible for later analysis.

A special method of sample preparation is the use of hydrofluoric acid (HF) to remove silica (SiO_2)-related materials from rocks and minerals.[13,14] This procedure would be used prior to a gravimetric analysis like the one that is utilized to measure aluminum in kaolinite (a clay used to make bricks and chinaware). Kaolinite clay has the general formula $Al_2(OH)_4Si_2O_5$ and undergoes the following reaction when it is dissolved in HF.

$$Al_2(OH)_4Si_2O_5(s) + 14\ HF \rightarrow 9\ H_2O +$$

$$2\ SiF_4(g) + 2\ AlF_3 \qquad (11.1)$$

The silicon tetrafluoride (SiF_4) that is formed by this reaction is lost as a gas, which leaves an HF solution of aluminum that is dissolved in the form AlF_3, which can readily be measured by gravimetric analysis or other techniques.

A closely related use of HF is in the analysis of silica in rocks and ores, which makes use of the fact that the product SiF_4 is volatile. In this case, the sample is heated in HF, with the resulting SiF_4 being lost as a gas. The measured loss in mass can then be related to the amount of SiO_2 that was present in the original sample.[13] Great care must be taken when using HF for such work, because this reagent can be quite damaging to skin and other tissues. Also, because all glass and porcelain contain silica, any sample pretreatment using HF to dissolve silica-related materials must be carried out in a metal container such as a platinum crucible.

Fusion (a term used here to refer to a method that involves melting) is another way to pretreat and dissolve samples of rocks or metals for the analysis of metal ions.[13–16] In this procedure, a weighed and powdered sample is mixed with solid sodium carbonate (Na_2CO_3), which acts as the fusing agent, or *flux*. This mixture is placed in a platinum crucible and heated to a red-hot temperature. Under these conditions, sodium carbonate decomposes to form CO_2 and sodium oxide (Na_2O).

$$Na_2CO_3(s) \rightarrow Na_2O(s) + CO_2(g) \qquad (11.2)$$

Molten sodium oxide is extremely basic and will dissolve rocks and silica-based materials by reacting with SiO_2 to form the water-soluble salt Na_2SiO_3.

$$Na_2O(molten) + SiO_2(s) \rightarrow Na_2SiO_3(s) \qquad (11.3)$$

After this fusion process is complete and everything has been cooled, the remaining material (which contains Na_2SiO_3) is dissolved in a dilute solution of hydrochloric acid and analyzed for its metal-ion content. Other reagents that can also be used to pretreat samples by fusion include potassium or sodium pyrosulfate, sodium peroxide, and potassium or sodium hydroxide. Further details on these and other sample preparation methods can be found in References 13–15.

11.2B Filtering Precipitates

After a precipitate has been formed from a solution, it is necessary to isolate this precipitate so that we can measure its mass. This step involves **filtration**, which is a process by which a filter is used to physically separate a solid material from a liquid.[16] The original mixture of the solid precipitate and its surrounding liquid is called a "slurry," and the liquid that is in contact with the precipitate is known as the *supernate* or "supernatant liquid." The **filter** that is used in this method is a porous structure that forms a barrier to the solids but that allows liquid to pass through. The liquid that passes through this filter is called the *filtrate*.[16] Filtration is often a critical step in a gravimetric analysis, because a high recovery of the precipitate is needed to provide accurate results when its mass is

We also learned in Chapter 7 that peptization occurs as a result of the adsorption of ions onto the surface of small particles of a precipitate. For instance, if we were precipitating silver chloride by adding an excess of sodium chloride to a solution with silver ions, a large number of excess chloride ions would be adsorbed onto the surface of the AgCl particles, imparting a negative charge to each particle. This negatively charged particle will then attract cations from the surrounding solution, such as Na^+, which will be loosely attracted to the surface of the particle. Because these adsorbed ions will give a greater than expected mass for our precipitate, we need to remove these ions before our measurement. If we simply use water to wash a silver chloride precipitate that contains adsorbed chloride and sodium ions, the loosely bound sodium ions will be removed, but the more tightly held chloride ions will remain. This type of washing will cause AgCl to form many small particles with negative charges that repel each other, creating a colloidal suspension of fine particles that is extremely difficult to filter.

To avoid this problem, we need to wash the precipitate with a solution that can replace its adsorbed ions (Cl^- and Na^+ in our particular example) with other ions that can later be removed by other means. This washing is often done by using a solution that contains an ionic compound that can replace such ions but has ions that can be removed by a process such as heating. For example, the wash solution for silver chloride is usually a very dilute solution of nitric acid (typically 0.5 mL of HNO_3 per 200 mL of distilled water). This wash solution keeps AgCl particles from becoming charged and the nitric acid and water can later be easily removed by evaporation as the precipitate is dried.

Ostwald Ripening. Another technique for obtaining precipitates with both large and pure particles is **Ostwald ripening**.[16] This technique involves heating a precipitate in its original solution to a temperature that is near the boiling point of the solution. This heating is done for an hour or so and alters the solubility product (K_{sp}) for the precipitate so that more can dissolve in the solution. Because solubility is a dynamic equilibrium, a solid forms at the same rate as dissolution occurs. The result is a situation in which the small precipitate particles (which have the greatest surface area per mass) tend to dissolve and release ions. At the same time, larger particles tend to take on some of these released ions and grow in size, making these particles easier to filter and remove from the solution. After this process has been allowed to occur and larger crystals have been obtained, the solution is gradually cooled and filtered for collection of the precipitate.

Ostwald ripening helps increase the purity of the final precipitate. Part of this increase in purity is because of the slow dissolving and reformation of the precipitate particles, which makes it possible for some of the entrapped impurities to be released from the original

particles. The larger particles that are obtained also have less surface area than the original collection of small particles. As we saw in the last section, the outer surface of these particles can be bound with additional ions that must later be removed. If this surface area is decreased by working with larger particles, there will be fewer of these adsorbed ions to be removed or to cause problems in our analysis.

11.3 EXAMPLES OF GRAVIMETRIC METHODS

The use of precipitation for gravimetric analysis has long been a common and relatively simple approach for measuring both inorganic and organic substances. There are literally hundreds of gravimetric methods that have been described for chemical analysis, with a few of the more common of these techniques being listed in Table 11.4 (see References 13–15 for other examples). In this section, we look at several specific gravimetric methods and see how knowledge of chemical reactions can be used to help design, optimize, and properly perform such assays.

11.3A Precipitation of Silver with Chloride

The first method we will consider is the gravimetric analysis of silver by reacting silver ions in water with an excess of chloride ions to form a precipitate of silver chloride, as shown below.

$$Ag^+ + Cl^- \text{ (excess)} \rightleftarrows AgCl(s) \qquad (11.9)$$

In this method, we must first be sure that we are using conditions that will allow essentially all the silver ions to be precipitated as AgCl. We also need this precipitate to be as pure as possible, because we will be assuming in our analysis that it contains only AgCl. If either of these conditions is not met, systematic errors will be present in the analysis. If some of the Ag^+ remains in solution, our final estimated value for the silver content of the sample will be too low. If the precipitate contains impurities besides AgCl, the estimated silver content of the sample will be too high. Thus, we need to select reaction conditions that will allow us to avoid or minimize both of these problems.

It is impossible to precipitate all the silver ions, however, as some must remain as Ag^+ in solution to maintain an equilibrium in the solution and satisfy the solubility product. We do, however, still need to precipitate such a high fraction of the analyte that the nonprecipitated part is too small to notice in the weighing step. An analytical balance, which would be used for such a method, can measure a mass to the nearest 0.0001 g. This means the nonprecipitated silver ions must have a mass less than 0.00005 g and any impurities in the AgCl precipitate must also have a mass less than 0.00005 g. We can determine that if we are capable of weighing to the nearest 0.0001 grams and we want an answer accurate to the nearest parts-per-thousand,

TABLE 11.4 Examples of Common Methods for Traditional Gravimetric Analysis*

Analyte	Precipitating Agent	Precipitate and Final Weighed Product[a]
Analytes in the Form of Cations		
Aluminum (Al^{3+})	8-Hydroxyquinoline (HC_9H_6ON)	$Al(C_9H_6ON)_3$
Barium (Ba^{2+})	Sulfate (SO_4^{2-})	$BaSO_4$
Calcium (Ca^{2+})	Oxalic acid ($H_2C_2O_4$)	$CaC_2O_4 \cdot H_2O \rightarrow CaCO_3$ or CaO
Cobalt ($Co^{2+} \rightarrow Co^{3+}$)[b]	1-Nitroso-2-naphthol ($HC_{10}H_6O_2N$)	$Co(C_{10}H_6O_2N)_3 \rightarrow CoSO_4$
Copper ($Cu^{2+} \rightarrow Cu^+$)[c]	Thiocyanate (SCN^-)	$CuSCN$
Iron (Fe^{3+})	Hydroxide (OH^-)	$Fe(OH)_3 \rightarrow Fe_2O_3$
Lead (Pb^{2+})	Sulfate (SO_4^{2-})	$PbSO_4$
Nickel (Ni^{2+})	Dimethylglyoxime ($HC_4H_7O_2N_2$)	$Ni(C_4H_7O_2N_2)_2$
Silver (Ag^+)	Chloride (Cl^-)	$AgCl$
Tin (Sn^{4+})	Cupferron ($NH_4C_6H_5O_2N_2$)	$Sn(C_6H_5O_2N_2)_4 \rightarrow SnO_2$
Analytes in the Form of Anions		
Bromide (Br^-)	Silver (Ag^+)	$AgBr$
Chloride (Cl^-)	Silver (Ag^+)	$AgCl$
Iodide (I^-)	Silver (Ag^+)	AgI
Phosphate (PO_4^{3-})	Magnesium (Mg^{2+}) in $NH_3(aq)$	$Mg(NH_4)PO_4 \cdot 6 H_2O \rightarrow Mg_2(P_2O_7)$
Sulfate (SO_4^{2-})	Barium (Ba^{2+})	$BaSO_4$
Thiocyanate (SCN^-)	Copper (Cu^+)	$CuSCN$

*Further details on these methods can be found in sources such as References 13–15.

[a]For those methods in which the precipitated form and final weighed form differ, the final weighed form is usually obtained by ignition of the precipitated form. The only exception in this list is the conversion of $Co(C_{10}H_6O_2N)_3$ to $CoSO_4$ through a reduction of Co^{3+} to Co^{2+} and reaction with H_2SO_4.

[b]The precipitation of cobalt ions by 1-nitroso-2-naphtholate works best when Co^{2+} has first been oxidized to Co^{3+}; the precipitate shown is for the product formed with Co^{3+}.

[c]The precipitation of copper ions by thiocyanate is formed after the Cu^{2+} ions in a sample have been reduced to Cu^+, which then forms a precipitate with SCN^-.

the weighed precipitate must have a mass of greater than $(0.0001 \text{ g})(1000) = 0.10$ g.

Many of the steps in this analysis will require the general operations we discussed in Section 11.2. Examples of such steps include preparing and drying the sample, filtering the precipitate, and weighing the final product. There are, however, a few special features that should be pointed out. For instance, concentrated nitric acid is added to the dissolved sample to prevent the precipitation of silver oxide that would occur at a pH over approximately 7.3, as shown below.

$$2 Ag^+ + 2 OH^- \rightarrow Ag_2O(s) + H_2O \qquad (11.10)$$

The wash water is also made slightly acidic with nitric acid to prevent peptization of the precipitate.

It is best to conduct the precipitation of AgCl in dim light, because silver chloride reacts with light to form silver metal and a chlorine molecule, as indicated in Equation 11.11.

$$2 AgCl(s) \xrightarrow{\text{light}} 2 Ag(s) + Cl_2(g) \qquad (11.11)$$

This is the same reaction that forms an image in photography when you are using black-and-white film. However, this process will lead to errors when we try to determine the amount of silver in our sample based on the mass of our product (now a mixture of Ag and AgCl).

Another way we might introduce errors into this analysis is by using too large an excess of Cl^- during the precipitation. Precipitation using an initial excess chloride concentration of 1×10^{-3} M will allow quantitative precipitation of silver ions from solutions that contain up to 1×10^{-7} M Ag^+. If we use even higher chloride concentrations, there is a danger of having the silver react with Cl^- to form complexes such as $AgCl_2^-$ and

$AgCl_3{}^{2-}$ that are now soluble in water and do not precipitate from solution.

EXERCISE 11.3	**Analysis of Silver in Ore by Gravimetric Analysis**

When using the method shown in Table 11.4, what percent of silver must have been present in an ore if a 10.4784 g sample of this ore is found to form 0.1763 g AgCl?

SOLUTION

The mass of silver in the final precipitate is obtained by using the mass of this precipitate and its known formula.

$$\text{Mass Ag} = (0.1763 \text{ g AgCl}) \cdot \frac{1 \text{ mol AgCl}}{143.321 \text{ g AgCl}} \cdot$$

$$\frac{1 \text{ mol Ag}}{1 \text{ mol AgCl}} \cdot \frac{107.868 \text{ g Ag}}{1 \text{ mol Ag}}$$

$$= 0.1327 \text{ g Ag}$$

From this result, we can determine that the percent of silver in the original ore sample was $100 \cdot (0.1327 \text{ g Ag})/(10.4784 \text{ g ore}) = \textbf{1.266\% Ag}$.

To help with calculations in a gravimetric analysis, a **gravimetric factor** is sometimes used. This is a conversion factor that can be utilized to multiply the measured mass of the precipitate to obtain the mass of the desired analyte.[13,22] In the last exercise, the gravimetric factor for determining the amount of silver in the AgCl precipitate was (1 mol AgCl/143.321 g AgCl)(1 mol Ag/mol AgCl) (107.868 g Ag/mol Ag) = 0.7526 g Ag/g AgCl. If we had simply used this conversion factor instead of the longer term shown in our dimensional analysis, we would have arrived at the same answer because (0.1763 g AgCl)(0.7526 g Ag/g AgCl) = 0.1327 g Ag. The value of the gravimetric factor will depend on the analyte you are examining and the formula for the precipitate that contains this analyte. For instance, the gravimetric factor for chloride when using a AgCl precipitate would instead be as follows.

$$\frac{35.453 \text{ g Cl}}{\text{mol Cl}} \cdot \frac{1 \text{ mol Cl}}{1 \text{ mol AgCl}} \cdot \frac{1 \text{ mol AgCl}}{143.321 \text{ g AgCl}} =$$

$$0.2474 \text{ g Cl/g AgCl}$$

It is relatively easy to calculate the value of a gravimetric factor once you know the reactions that are present during a gravimetric analysis. Many of these values can also be obtained in references such as the *CRC Handbook of Chemistry and Physics* or *Lange's Handbook of Chemistry* for a wide range of gravimetric methods.[22,23] It is important when using a gravimetric factor to include,

if possible, a greater number of significant figures in this factor than will be present in any of the experimental masses employed in calculating the result of the analysis. This rule helps avoid the introduction of rounding errors in the final result (as discussed in Chapter 2).

11.3B Precipitation of Iron with Hydroxide

The second type of gravimetric analysis we will examine is a method for the analysis of iron. This analysis is based on the fact that iron in the form of Fe^{3+} forms an insoluble precipitate with hydroxide. This precipitate is often called iron(III) hydroxide, or $Fe(OH)_3$, and forms at a pH higher than about 5. The solubility product for this precipitate is 1.6×10^{-39}, but its composition is actually a mixture of $Fe(OH)_3$ and $FeO(OH)$, which can be represented by the formula $Fe_2O_3 \cdot xH_2O$.

Unfortunately, there are many problems associated with this analysis. The resulting precipitate is gelatinous, difficult to filter, and coprecipitates many substances that might be present in the solution from which iron is precipitated. Also, the precipitate becomes colloidal when the ionic strength of the supernatant solution is low. In addition, many samples contain other ions such as Al^{3+} and Cr^{3+}, which can also form insoluble hydroxide precipitates. Yet another complicating factor is the fact that iron is frequently in solution at least partially as Fe^{2+}, which does not precipitate with hydroxide until a much higher pH than is needed for Fe^{3+}.

Several steps can be used in a gravimetric method to overcome these problems. The samples can be heated with a mild oxidizing agent such as bromine water or nitric acid to convert the Fe^{2+} to Fe^{3+}. Also, the pH is raised to promote precipitation by adding ammonia instead of NaOH, because ammonia forms a complex ion with some metal ions (such as Zn^{2+} and Cu^{2+}) that will help prevent their coprecipitation as hydroxides. Furthermore, ammonia is highly volatile and can easily be removed from the final sample so that it does not add to the mass of the final iron(III) oxide precipitate. This ammonia is added slowly and with stirring during the precipitation process to prevent high local concentrations of hydroxide ions that may cause a poor quality precipitate to form.

After the precipitation step in the analysis, the sample is filtered through paper and the liquid is discarded. The precipitate is then redissolved by adding HCl to lower the pH. The dissolved Fe^{3+} ions are then allowed to precipitate again by adding fresh ammonia. This second precipitation step helps to further lower the amount of metal ions besides Fe^{3+} that precipitate. This final precipitate is then filtered using paper, with as much liquid and as little precipitate as possible initially being transferred to the filter paper. This precipitate is washed several times with a 1% solution of ammonium nitrate. Finally, the remaining precipitate is washed onto the filter paper with more of the ammonium nitrate wash solution.

The gelatinous iron(III) hydroxide precipitate present at this point is next ignited to convert it into the more well-defined form of iron(III) oxide, Fe_2O_3. To carry out this ignition, the filter paper containing the damp iron(III) hydroxide is put into a porcelain crucible. The contents are then gently heated to volatilize any remaining ammonium nitrate and to dry and burn away the paper. The precipitate is next ignited at the full power of the Meker burner to fully dehydrate the precipitate to form Fe_2O_3. Sometimes part of the iron(III) is reduced to iron(II) by hot carbon monoxide that is formed when burning off the paper. To prevent this reaction, the ignited precipitate is treated with a few drops of nitric acid, after which the precipitate is reignited, cooled in a desiccator, and weighed.

EXERCISE 11.4   **Determination of Iron by Gravimetric Analysis**

A newly-discovered iron ore was examined to determine the percent iron in the ore. Three samples were obtained with masses of 5.408, 4.768, and 4.209 g, respectively. After treatment, these samples gave rise to 0.3785, 0.3348, and 0.2957 g of ignited precipitate. What is the average percent iron in these samples? What is the gravimetric factor for this analysis?

SOLUTION

The mass and % Fe in the first sample can be determined as shown below.

Sample 1: Mass iron $= (0.3785 \text{ g Fe}_2\text{O}_3) \cdot \dfrac{1 \text{ mol Fe}_2\text{O}_3}{159.70 \text{ g Fe}_2\text{O}_3} \cdot$

$$\dfrac{2 \text{ mol Fe}}{\text{mol Fe}_2\text{O}_3} \cdot \dfrac{55.85 \text{ g Fe}}{\text{mol Fe}} = 0.2647 \text{ g Fe}$$

% Fe $= 100 \cdot (0.2647 \text{ g Fe}/5.408 \text{ g ore}) =$

4.895% Fe (w/w)

The same calculation for the second and third ore samples gives % Fe values of 4.911% and 4.914%, with an average for all three samples equal to 4.907% Fe. The gravimetric factor for this analysis is shown below.

$(1 \text{ mol Fe}_2\text{O}_3/159.70 \text{ g F}_2\text{O}_3) \cdot (2 \text{ mol Fe}/\text{mol Fe}_2\text{O}_3) \cdot$

$(55.85 \text{ g Fe}/\text{mol Fe}) = $ **0.6994 g Fe/g Fe$_2$O$_3$**

11.3C Precipitation of Nickel with Dimethylglyoxime

The third example of a gravimetric method that we will consider is a homogeneous precipitation method for nickel. Until now, we have only discussed the precipitation of metal ions by using inorganic agents like Cl^- and OH^-. This is the approach used for most of the analyses

listed earlier in Table 11.4. Attempts also have been made since the early years of the twentieth century to develop specific organic-based precipitating agents for each of the elements. One case in which these attempts were particularly successful was in the use of *dimethylglyoxime* (or *dmg*) for the precipitation and analysis of nickel ions.[13–15,24]

Nickel ions precipitate in the presence of dmg from a neutral or weakly basic ammonia solution, where two dmg anions combine with Ni^{2+} to form a dark-red precipitate (see Figure 11.10). Each molecule of dmg loses one hydrogen ion and the two molecules of dmg form a square planar, neutral complex with nickel ion. Each dmg molecule also forms hydrogen bonds with the other molecule of dmg to add to the stability of this complex. This precipitate is then filtered on a sintered glass crucible, dried at 110°C, and weighed. This precipitate is very difficult to filter if appropriate precautions are not taken. The precipitate that initially forms is in very small crystals that sometimes seem to creep up the sides or even out the top of the beaker in which the precipitate has been made. This is a situation where precipitation from homogeneous solution can be useful. In this case, dimethylglyoxime is formed by the slow reaction of biacetyl with hydroxylamine, as illustrated in Equation 11.12.

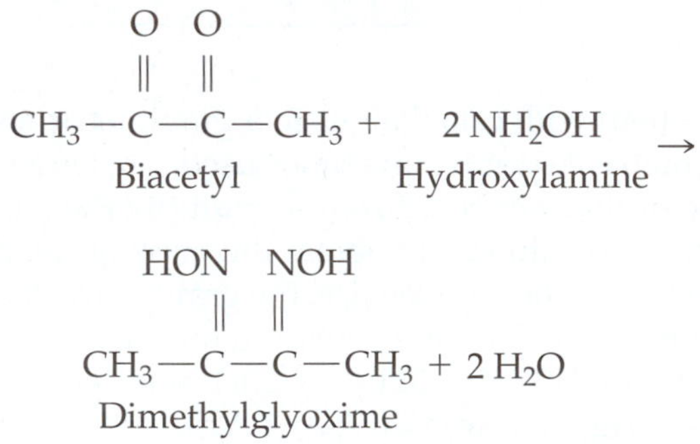

$$(11.12)$$

As dimethylglyoxime is formed by this reaction, it will combine with Ni^{2+} and precipitate this metal ion from solution as $Ni(dmg)_2$.

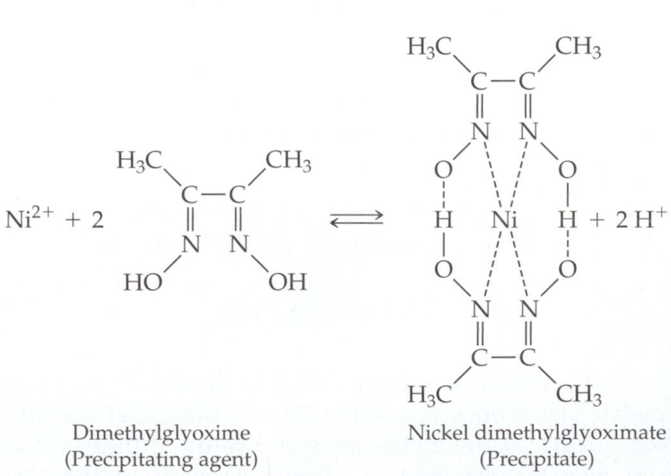

Dimethylglyoxime
(Precipitating agent)

Nickel dimethylglyoximate
(Precipitate)

FIGURE 11.10   The structure of dimethylglyoxime, or dmg, and its reaction with Ni^{2+} to form a precipitate (nickel dimethylglyoximate, or $Ni(dmg)_2$).

EXERCISE 11.5	Using Homogeneous Precipitation for the Gravimetric Analysis of Nickel

If we are using an excess of biacetyl, how many grams of hydroxylammonium chloride ($NH_2OH \cdot HCl$, 69.50 g/mol), are necessary to have a 20% excess of dmg present during the precipitation of 0.0687 g of Ni^{2+} from a homogeneous solution?

SOLUTION

Assuming we have converted all nickel in the sample into Ni^{2+} ions, the amount of these ions will be as follows.

$$(0.0687 \text{ g Ni}) \cdot \frac{1 \text{ mol Ni}}{58.69 \text{ g Ni}} \cdot \frac{1 \text{ mol Ni}^{2+}}{1 \text{ mol Ni}} =$$

$$1.17 \times 10^{-3} \text{ mol Ni}^{2+}$$

Because each Ni^{2+} ion will react with two molecules of dmg, to precipitate these nickel ions we will need at least $(1.17 \times 10^{-3} \text{ mol Ni}^{2+}) \cdot (2 \text{ mol dmg/mol Ni}^{2+}) = 2.34 \times 10^{-3}$ mol dmg. If we want an extra 20% of dmg (or 1.2 times what we actually need), we will require $(1.20) \cdot (2.34 \times 10^{-3} \text{ mol dmg}) = 2.81 \times 10^{-3}$ mol dmg for this analysis.

We know from Equation 11.12 that the homogeneous production of dmg requires 2 mol of hydroxylamine and 1 mol biacetyl for each mol of dmg that is produced. However, we are also told that an excess of biacetyl is present, so hydroxylamine is the limiting reagent that controls the amount of dmg we will produce. This fact means we will need $(2.81 \times 10^{-3} \text{ mol dmg}) \cdot (2 \text{ mol hydroxylamine/1 mol dmg}) = \mathbf{5.62 \times 10^{-3}} \textbf{ mol}$ hydroxylamine, or **0.39 g** $NH_2OH \cdot HCl$.

In addition to dmg, there are a variety of other organic reagents that can be used to precipitate metal ions.[14,15] A few examples are shown in Table 11.5. One of these other reagents is 8-hydroxyquinoline (or oxine), which is a complexing agent that we discussed in Chapter 9. In fact, all of the reagents shown in Table 11.5 form a precipitate with metal ions through the creation of metal–ligand complexes. These complexes then precipitate from solution because of the large decrease in solubility that occurs as metal ions form a much larger and neutral complex with one of these reagents. Although all of these reagents will react with more than one type of metal ion (even dmg, which can precipitate Ni^{2+} or Pd^{2+}), the type of metal ion that is precipitated can be controlled by altering such factors as the pH of the solution or the concentration of a secondary ligand such as ethylenediamine tetraacetic acid (EDTA) that is added to bind and mask other metal ions.[14,15]

11.3D Combustion Analysis

Another past application of gravimetric methods has been their use in **combustion analysis**. Combustion analysis is a method in which a sample is burned to measure the relative amount of carbon, hydrogen, and other elements in a sample.[10] Gravimetric analysis can then be used to determine the mass of the released gases and use this information along with the original mass of the sample to determine this sample's original composition. This approach differs from the traditional methods we discussed in the previous section in that we are no longer examining the mass of a precipitate from a solution. Instead, we are examining the masses of certain gases (such as carbon dioxide and water) that are formed as we burn our sample.

Combustion analysis is often carried out by burning a known mass of sample in the presence of an excess of oxygen. Under these conditions, all of the carbon in the sample should combine with the oxygen to form carbon dioxide, and all the hydrogen should react with oxygen to form water. This process can be represented by the following general reaction, where $C_xH_yO_z$ is used as a general formula for a typical organic compound that makes up our sample.

$$C_xH_yO_z + O_2(g) \text{ (excess)} \rightarrow x \, CO_2(g) +$$

$$y/2 \, H_2O(g) \quad (11.13)$$

As we burn the sample, the mixture of gases that is generated is allowed to pass through a series of previously weighed cartridges that can each react with or adsorb one of these gases. An example of such a device is shown in Figure 11.11. A cartridge containing magnesium perchlorate $Mg(ClO_4)_2$ can be used to absorb the water in this gas mixture; P_4O_{10} is also often used for this purpose.

$$Mg(ClO_4)_2(s) + 2 \, H_2O(g) \rightarrow Mg(ClO_4)_2 \cdot 2 \, H_2O(s) \quad (11.14)$$

The remaining gases are then passed through a cartridge containing *ascarite*, which is a material made up of sodium hydroxide adsorbed onto asbestos.[16] Equation 11.15 shows how the NaOH in this cartridge can react with CO_2 to form sodium carbonate and water, which both stay in the cartridge and add to its mass.

$$2 \, NaOH(s) + CO_2(g) \rightarrow Na_2CO_3(s) + H_2O(l) \quad (11.15)$$

Once the sample has been completely burned and all of its emitted gases have been collected, the cartridges with the adsorbed gases are weighed. According to Equation 11.14, the increase in mass of the magnesium perchlorate cartridge should be equal to the mass of water that was formed during sample combustion. The increase in mass for the ascarite cartridge will give the mass of carbon dioxide that was formed during combustion. These values can be used to give the mass or moles of H (from the water result) and C (from the carbon dioxide measurement), allowing the carbon/hydrogen content of the sample to be determined. If it is known that only oxygen makes up the rest of the sample, the oxygen content of this sample can also be determined.

TABLE 11.5 Examples of Organic Precipitating Agents

Name and Structure	Precipitated Analytes[a]
Cupferron	Ce^{4+}, Fe^{3+}, Ga^{3+}, Nb^{5+}, Sn^{4+}, Ta^{5+}, Ti^{4+}, VO_2^+, Zr^{4+}
Dimethylglyoxime	Ni^{2+}, Pd^{2+}
8-Hydroxyquinoline (Oxine)	Al^{3+}, Bi^{3+}, Cd^{2+}, Cu^{2+}, Fe^{3+}, Ga^{3+}, Mg^{2+}, Ni^{2+}, Pb^{2+}, Th^{4+}, UO_2^{2+}, WO_2^{2+}, Zn^{2+}, Zr^{4+}
1-Nitroso-2-naphthol	Co^{2+}, Fe^{3+}, Pd^{2+}, Zr^{4+}

[a]This is a partial list of analytes that can be quantitatively precipitated. A more complete list, and the conditions needed to cause this precipitation, can be found in J. Bassett, R. C. Denney, G. H. Jeffery, and J. Mendham, *Vogel's Textbook of Quantitative Inorganic Analysis*, 4th ed., Longman, New York, 1978.

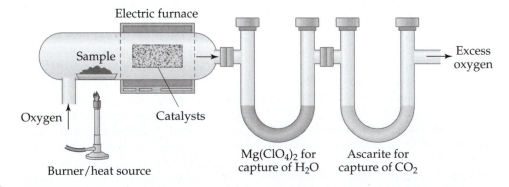

FIGURE 11.11 A general system for performing combustion analysis for hydrogen and carbon based on gravimetric measurements of the water and carbon dioxide that are released when burning a sample in the presence of excess oxygen. If it is known that carbon, hydrogen, and oxygen are the only elements present in the original sample, the oxygen content of this sample can be calculated by taking the original weight of the sample and subtracting the amount of hydrogen and carbon that was found to be present in this same sample.

Determining C/H Content of a Sample by Combustion Analysis

A new organic material is to be examined by combustion analysis. Burning 0.09303 g of this sample in the presence of excess oxygen gives an increase in mass of 0.04256 g on a magnesium perchlorate cartridge and a 0.1387 g increase in mass on an ascarite cartridge. If we know carbon, hydrogen, and oxygen are the only elements present, what is the percent C, H, and O in the sample?

SOLUTION

The mass of carbon and percent carbon in the sample can be determined from the change in mass of the ascarite cartridge, which is due to the adsorption of CO_2.

Mass C in sample =

(0.1387 g CO_2 adsorbed by cartridge) ·

$(12.01 \text{ g C}/44.01 \text{ g CO}_2)$

= 0.03785 g C

Percent C (w/w) = 100 · (0.03785 g C)/(0.09303 g sample)

= **40.68% C (w/w)**

The mass of hydrogen in the sample can be found by using the change in mass for the magnesium perchlorate cartridge, which is a result of the adsorption of H_2O.

Mass H in sample =

(0.04256 g H_2O adsorbed by cartridge) ·

$(2.017 \text{ g H}/18.017 \text{ g H}_2O)$

= 0.004764 g H

Percent H (w/w) =

100 · (0.004764 g H)/(0.09303 g sample) = **5.12% H (w/w)**

Finally, we are told that carbon, hydrogen, and oxygen are the only elements in this sample, which means that all of the remaining sample mass must be due to oxygen.

Mass O in sample =

0.09303 g sample − 0.03785 g C − 0.004764 g H

= 0.05042 g O

Percent O (w/w) =

100 · (0.05042 g O)/(0.09303 g sample) = **54.19% O (w/w)**

Gravimetric analysis has long played an important role in the past for determining the results of a combustion analysis. However, there are now other, more automated measurement techniques that can also be used with combustion analysis, as discussed in Box 11.1.

11.3E Thermogravimetric Analysis

Another special method that makes use of mass measurements in chemical analysis is **thermogravimetric analysis** (also known as *TGA* or *thermogravimetry*). Thermogravimetric analysis is a technique in which the mass of a sample is measured as the temperature of the sample is varied.[11,16] An instrument for performing this type of study is known as a *thermobalance* (see Figure 11.13). This instrument includes a high-quality analytical balance along with a furnace for heating the sample in a controlled fashion. A computer system is also required in this system to control and monitor the temperature of the sample, as well as to measure the mass of the sample as its temperature is varied.[11,30,31]

Thermogravimetric analysis is valuable in examining how a sample changes with temperature. This information is represented by a graph called a *thermogravimetric curve*,[11] in which the measured mass of the sample is plotted versus temperature (see example in Figure 11.14). As is shown in this example, the mass of a sample will often decrease as the sample is heated. This decrease in mass occurs as water, carbon dioxide, or other gaseous components are lost from the sample. It is possible on some occasions for the sample to gain weight with temperature if it reacts with a gas in the surrounding atmosphere to create a product with a higher mass. The types of reactions that lead to these changes in mass are related to the temperatures at which the changes in mass are occurring. Also, the change in mass with each transition gives important quantitative information on the composition of the sample. These properties make graphs like Figure 11.14 valuable as both qualitative and quantitative tools for sample analysis.

Using Thermogravimetric Analysis

A pure sample of calcium oxalate monohydrate ($CaC_2O_4 \cdot H_2O$) has an expected molar mass of 146.112 g/mol. If a chemist heats 2.5100 g of supposedly pure calcium oxalate monohydrate in a TGA system from 100°C to 300°C, what is the expected change in mass for this sample? What will be the mass of this sample after it is further heated to 600°C?

SOLUTION

A change from 100°C to 300°C would involve the release of water from the sample to form calcium oxalate, as shown in Figure 11.14. Because there is one mole of water released per mole of calcium oxalate monohydrate, the mass of water released and the change in mass for the sample would be found as follows.

Mass released H_2O =

$[(2.510 \text{ g } CaC_2O_4 \cdot H_2O)/(146.112 \text{ g/mol } CaC_2O_4 \cdot H_2O)] \cdot$
$[(1 \text{ mol } H_2O)/(1 \text{ mol } CaC_2O_4 \cdot H_2O)][18.015 \text{ g/mol } H_2O]$

$= \mathbf{0.309 \text{ g } H_2O}$

This loss of water would correspond to a decrease in mass of $100 \cdot (0.309 \text{ g})/(2.510 \text{ g}) = 12.3\%$ versus the mass of the original sample. A further increase in temperature to 600°C would result in the release of one mole of carbon monoxide per mole of the original calcium oxalate monohydrate. Using the same approach as used for the loss of water, the change in mass due to the loss of carbon monoxide would be 0.481 g, or a decrease in mass of 19.2% versus the mass of the original sample.

BOX 11.1

Combustion Analysis, Then and Now

Combustion analysis has been an important tool in chemical analysis for over 200 years. This approach was the main way in which new organic chemicals were characterized and identified up through the time of World War II. Although many other newer analytical methods are also now used by synthetic chemists (such as mass spectrometry and nuclear magnetic resonance spectroscopy), combustion analysis is still often used to confirm the empirical formulas of new organic compounds.[25]

Combustion analysis was first developed by the French chemist Antoine Lavoisier in the late 1780s, who used it to examine the content of organic oils. The apparatus that he used for this analysis is shown in Figure 11.12(a).[26] Although the general principles of this analysis were the same then as those used now for combustion analysis, Lavoisier's approach required more than 50 g of sample, several operators, and expensive equipment. Around 1831, German chemist Justus von Liebig devised an approach that allowed combustion analysis to be conducted with only 0.5 g of sample (1/100 the amount needed by Lavoisier) and that could be performed by a single worker on inexpensive equipment.[27] A century later, Austrian physician and chemist Fritz Pregl further improved combustion analysis so that the amount of required sample was reduced by another 100-fold to only 3–5 mg.[28] This

last development, for which Pregl won the 1923 Nobel Prize in chemistry, made it much easier for chemists to use combustion and gravimetric analysis for newly discovered compounds and revolutionized the field of organic chemistry.[29]

The method of gravimetric analysis that was used by Pregl is still considered the standard method for determining the carbon and hydrogen content of chemical samples. However, there are now many other techniques that can accomplish this same goal without using gravimetric analysis. For instance, rather than collecting and weighing the carbon dioxide, water, and other gases that are produced during sample combustion, these gases can be separated and measured by employing gas chromatography (a technique discussed in Chapter 21) or without a separation by using infrared spectroscopy (see Chapter 18). This type of instrument is called a *CHN analyzer* because it is generally designed to determine the carbon, hydrogen, and nitrogen content of a sample. An example of one such instrument is shown in Figure 11.12(b). Advantages of these devices are that they can be easily automated and require even smaller amounts of sample than gravimetric methods. Similar instruments are available for determining the content of almost any other element that is found in organic samples.[11]

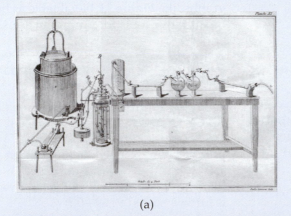

(a)

(b)

FIGURE 11.12 An example of an early combustion analysis system used by Antoine Lavoisier in 1789 that made use of gravimetric measurements (a), and a modern combustion analysis that uses gas chromatography to separate and analyze the gases that result from the combustion of a sample (b). (The figure in (a) is from a book by Lavoisier entitled *Traité Élémentaire de Chimie*, considered by some to be the first modern chemistry textbook; the figure in (b) is used with permission from Perkin-Elmer and is a Model 2400 Series II CHNS/O system.)

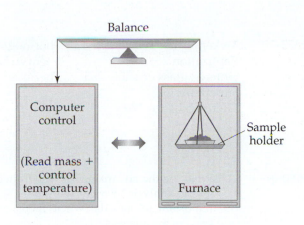

FIGURE 11.13 A possible design for an instrument for performing thermogravimetric analysis (TGA) and an example of a commercial TGA system. The general design shown on the left is based on a null-point balance, which is used in many types of TGA instruments.[11] (The image on the right is reproduced with permission from Shimadzu.)

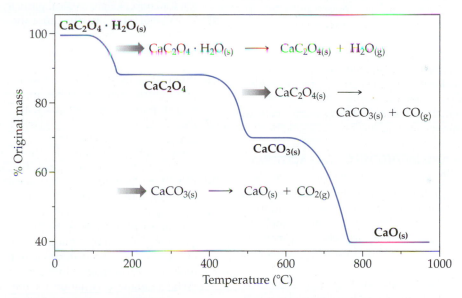

FIGURE 11.14 A typical thermogravimetric curve obtained when heating a sample of calcium oxalate monohydrate ($CaC_2O_4 \cdot H_2O$). Several well-defined plateaus in the measured mass occur as this sample is slowly heated between 100°C and 900°C. First, the calcium oxalate monohydrate forms calcium oxalate (CaC_2O_4) through the loss of water vapor, which then gives a plateau with a steady mass reading beginning at 182°C. Second, the calcium oxalate forms calcium carbonate ($CaCO_3$) through the loss of carbon monoxide, which leads to a steady mass reading that begins at 503°C. Third, the calcium carbonate forms calcium oxide (CaO) through the loss of carbon dioxide, which leads to a steady mass reading between 759°C and 1020°C. (This plot is based on data from D. Dollimore, *Analytical Instrumentation Handbook*, 2nd ed., G.W. Ewing, Ed., Marcel Dekker, New York, 1997, Chapter 17.)

Key Words

Other Terms

Ascarite 273	Dimethylglyoxime 272	Meker burner 264	Thermogravimetric
Ashing 262	Filtrate 262	Peptization 268	curve 275
Ashless paper 263	Flux 262	Sintered glass crucible 265	Thermogravimetry 275
Aspirator 265	Gravimetry 259	Supernate 262	
CHN analyzer 276	Hygroscopic 266	Thermobalance 275	

Questions

WHAT IS GRAVIMETRIC ANALYSIS AND HOW IS IT USED IN ANALYTICAL CHEMISTRY?

1. Define the term "gravimetric analysis." What is actually measured in this approach? How is this measurement used to determine the amount of a particular analyte in the sample?

2. Explain how T.W. Richards made use of gravimetric analysis to determine the atomic masses for silver and other elements. How does this differ from the way in which modern chemists typically use gravimetric analysis? How is the approach that was utilized by Richards similar to those that are used by modern chemists for gravimetric analysis?

3. Describe three general ways in which measuring a mass or a change in mass can be used for a gravimetric analysis.

4. What is a "precipitant"? How is a precipitant used in many methods based on traditional gravimetric analysis?

5. List two general examples of gravimetric methods that involve the formation or loss of volatile chemicals.

6. What are some important advantages of gravimetric analysis? What are some requirements or limitations of this method?

GENERAL STRATEGIES AND METHODS FOR A TRADITIONAL GRAVIMETRIC ANALYSIS

7. Explain why sample pretreatment is an important early step in many types of gravimetric analysis. In what form must an analyte be present for a traditional gravimetric analysis?

8. What is meant by the term "ashing" in a chemical analysis? Why is ashing often needed when examining metals by a traditional gravimetric analysis?

9. Define each of the following terms and explain how they can be used to pretreat a sample prior to a chemical analysis.
   (a) Wet ashing
   (b) Dry ashing
   (c) Fusion

10. Which of the pretreatment methods in Problem 9 would you use to prepare each of the following samples for analysis? Explain each of your choices.
   (a) Determination of the iron content in a chocolate bar
   (b) Determination of the amount of nickel in a silicate ore
   (c) Measurement of the amount of mercury in a tuna sample

FILTERING PRECIPITATES

11. What is "filtration"? Explain why filtration is often an important part of a gravimetric analysis.

12. Define each of the following terms. Use these terms to describe the filtration of AgCl from a mixture that is prepared by combining aqueous solutions of $AgNO_3$ and NaCl.
   (a) Slurry
   (b) Supernate
   (c) Filtrate

13. What are the advantages of using filter paper for a gravimetric analysis? Explain why pore size or "porosity" is important to consider when choosing filter paper for a gravimetric analysis.

14. Which type of filter paper in Table 11.2 would you use for each of the following precipitates? Give the reasons behind each of your choices.
   (a) Silica (typical particle size, 25-40 μm)
   (b) Ammonium phosphomolybdate (typical particle size, 20 μm)
   (c) Calcium oxalate (typical particle size, 15 μm)

15. The use of ripening can cause the average size of barium sulfate particles to increase from 3 μm to 8 μm. Explain why this ripening will make it easier to isolate this precipitate by using filter papers like those listed in Table 11.2.

16. What is "ashless paper"? How is ashless paper used in a gravimetric analysis? What are some desirable properties for this paper in such an analysis?

17. Describe how you would fold and use a piece of filter paper as part of a gravimetric analysis.

18. What is the process of "ignition" as used in gravimetric analysis? What is the purpose of ignition in such a method?

19. What is a "Meker burner"? How is this type of burner used in a traditional gravimetric analysis?

20. A method for the gravimetric determination of either magnesium or phosphate is the precipitation of the triple salt, $Mg(NH_4)PO_4 \cdot 6 H_2O$ (magnesium ammonium phosphate hexahydrate). This material is not easily weighed, so it is ignited to form magnesium pyrophosphate, or $Mg_2(P_2O_7)$. Write a balanced equation for this ignition process. (*Note:* You may assume an excess of oxygen is available in the surrounding air.)

21. The reaction of Sn^{4+} with cupferron produces a precipitate with the formula $Sn(C_6H_5O_2N_2)_4$. This precipitate is collected on filter paper and converted to SnO_2 by ignition prior to weighing. Write a balanced equation for the ignition step, assuming that an excess of oxygen is present in the surrounding air during this process.

22. Describe how a porous crucible can be used to collect a precipitate during a traditional gravimetric analysis.

23. Describe each of the following devices and explain how they might be used in a traditional gravimetric analysis: (a) aspirator and (b) sintered glass crucible.

DRYING AND WEIGHING PRECIPITATES

24. Why is it necessary for the sample and final precipitate to be dry before taking their mass in a gravimetric analysis? What methods are typically used to dry these materials?

25. What is a "desiccator"? What is the function of this device in an analytical laboratory?

26. What is a "desiccant"? Give three specific examples of chemicals that are desiccants.
27. What is meant when a chemical is said to be "hygroscopic"? Explain why a precipitant (or associated reagent) that is hygroscopic might cause a problem during a traditional gravimetric analysis.
28. What types of balance are most commonly used in a gravimetric analysis? What properties of these balances make them useful in in such a method?
29. To how many decimal places must the mass of an iron precipitate be known to give an uncertainty of no more than 2 parts-per-thousand (ppt) if the sample mass is 0.1253 g and the sample contains 1.87% (w/w) iron?
30. If a solid sample contains 2.50% iron, what is the smallest sample size that can give an answer with an error smaller than 2 parts-per-thousand (ppt) if an analytical balance is used to weigh the sample and solid Fe_2O_3 that is formed from iron?

METHODS FOR OBTAINING HIGH-QUALITY PRECIPITATES

31. Describe what is meant by the phrase "precipitation from a homogeneous solution." What advantages does this technique have over more traditional precipitation methods?
32. Discuss how each of the following reagents can be produced for use in precipitation from a homogeneous solution. Show the corresponding reactions that are involved in each of these processes.
    (a) Phosphate
    (b) Sulfate
    (c) Oxalate
33. Aluminum in the form Al^{3+} can be precipitated as $Al(OH)_3$ by adding ammonia to an aluminum solution, but the precipitate is usually gelatinous and difficult to filter. Precipitation from a homogeneous solution gives a more filterable precipitate. The reaction utilized for this process is the hydrolysis of urea.

$$H_2NCONH_2 + H_2O \rightarrow 2\,NH_3 + CO_2(g) \qquad (11.16)$$

Suppose this reaction takes place in a 100 mL solution with an initial Al^{3+} concentration of 1.5×10^{-3} M. How many moles of urea must be hydrolyzed to supply enough hydroxide to stoichiometrically combine with Al^{3+} and form the precipitate $Al(OH)_3$? How many moles of hydroxide will be formed if 0.500 g of urea are hydrolyzed and the final pH of the solution is 9.5, the pK_a of ammonium (NH_4^+)?
34. Precipitation of nickel dimethylglyoxime can be performed through the homogeneous formation of dimethylglyoxime (dmg) from biacetyl (2,3-diketobutane) and hydroxylamine, as shown below.

$$(11.17)$$

How much biacetyl and hydroxylamine are necessary to form enough dimethylglyoxime to precipitate 0.0587 g of Ni^{2+} if a twofold excess of hydroxylamine is necessary to cause the reaction in Equation 11.17 to go essentially to completion?

35. Explain why peptization can be a problem in a gravimetric analysis.
36. What are some techniques that can be used to avoid peptization during a gravimetric analysis?
37. What is "Ostwald ripening"? What occurs during this process?
38. Explain how Ostwald ripening can be used to improve the accuracy of a gravimetric analysis.

EXAMPLES OF TRADITIONAL GRAVIMETRIC METHODS

39. What is a "gravimetric factor"? Why is such a factor often used in a gravimetric analysis?
40. Calculate the gravimetric factors for each of the following methods shown in Table 11.4.
    (a) Analysis of barium using sulfate as a precipitating agent
    (b) Analysis of copper using thiocyanate as a precipitating agent
    (c) Analysis of tin using cupferron as a precipitating agent
    (d) Analysis of phosphate using magnesium ions in an aqueous solution of ammonia for precipitation
41. The precipitation of calcium with oxalic acid can be used to produce a final product in the form of either $CaCO_3$ or CaO. Calculate the gravimetric factor when using each of these final products.
42. Write the reactions that are involved in the gravimetric analysis of silver by its precipitation with chloride. What are some additional, undesirable reactions that must be considered during this analysis? What steps can be taken to prevent or minimize errors due to these other reactions?
43. A sample of silver metal is suspected of being contaminated with zinc. A 0.2365 g portion of this metal is dissolved in nitric acid and the silver ions precipitated with excess aqueous NaCl. After filtering, drying, and other steps, the mass of the precipitate is found to be 0.2865 g. What was the purity of the original silver metal?
44. What volume of 0.15 M HCl must be added to 50.0 mL of 0.035 M $AgNO_3$ to provide a twofold excess of chloride for the precipitation of Ag^+?
45. Suppose that you precipitate a mixture of AgCl and AgBr from a 0.3654 g sample of NaCl plus NaBr and find the combined mass of these silver halides to be 0.8783 g. What was the composition of NaCl and NaBr in the original sample?
46. The same reactions as are used for the determination of Ag^+ by precipitation with Cl^- can also be used to determine Cl^- by its precipitation with excess Ag^+.
    (a) Write the reactions for this assay. How would the conditions for these reactions be similar to those used in the precipitation of Ag^+ with Cl^-? How would they be different?
    (b) In this type of assay, calculate the salinity (or content of NaCl) of a brackish water sample if the precipitation of Cl^- from 100 mL of this sample in the presence of excess silver nitrate produces 3.295 g AgCl. Express your final answer in terms of both molarity and in grams of NaCl per liter of water.
47. A sample containing only NaCl and NaBr has a mass of 0.8764 g. This sample is placed into a solution and some of the resulting ions are precipitated with excess silver, giving a mixture of solid AgCl and AgBr. This precipitate has a mass of 1.8758 g. What were the amounts of NaCl and NaBr in the original sample?
48. Write the reactions involved in the precipitation of Fe^{3+} when using hydroxide as a precipitating agent. Be sure to

consider all reactions involved in this analysis. What are some possible sources of errors in this method? What are some ways that the effects of these errors can be minimized?

49. What mass of Fe_2O_3 would you expect from a gravimetric analysis of a sample that contains 0.257 mol of iron?

50. Describe how nickel can be analyzed with dimethygly-oxime through homogenous precipitation. Write the reactions involved in this method and state the purpose for each reaction.

51. What is dimethylglyoxime? How is dimethylglyoxime used for the analysis of nickel? What is the reaction for this process?

52. Discuss how a reagent like dimethylglyoxime or 8-hydroxyquinoline can cause the precipitation of a metal ion.

53. What is the concentration of Ni^{2+} (in units of molarity) in a solution if 10.00 mL of this solution gives rise to 0.0658 g of nickel dimethylglyoxime. (*Note:* For this example you can assume that the reaction of Ni^{2+} with dimethylglyoxime is nearly 100% complete.)

54. Aluminum can be determined gravimetrically by precipitation with 8-hydroxyquinoline (C_9H_7NO, or 8-quinolinolate). Three molecules of 8-hydroxyquinoline each lose a hydrogen ion on the OH group and form an insoluble complex with an aluminum ion, as shown below.

$$Al^{3+} + 3\ HC_9H_6NO \rightarrow Al(C_9H_6ON)_3\ (s) + 3\ H^+ \quad (11.18)$$

   (a) Calculate the value of the gravimetric factor for the determination of aluminum by using the precipitate that is created through this method.
   (b) Use the gravimetric factor from part (a) to calculate the mass of aluminum that is present in a 0.1653 g precipitate of $Al(C_9H_6ON)_3$. (*Note:* You can assume the precipitate is essentially 100% pure.)

55. A U.S. "nickel" coin that weighs 4.945 g is dissolved in nitric acid and treated with an excess of dimethylglyoxime. The precipitate has a mass of 6.0797 g. What percent of the original coin actually consisted of the element nickel?

COMBUSTION ANALYSIS

56. Describe what is meant by a "combustion analysis." What is the purpose of such an analysis? How is the use of combustion analysis with gravimetry similar to a precipitation-based gravimetric analysis? How is it different?

57. What is the purpose of the ascarite and magnesium perchlorate cartridges in a system for combustion analysis? Write the reactions that take place in each of these cartridges.

58. A 0.0537 g portion of a pure organic compound is burned in the presence of excess oxygen. The CO_2 is trapped and found to weigh 0.1362 g. The amount of water that is captured from the burned sample is found to weigh 0.04953 g. No other products are formed.
   (a) What are the masses of carbon, hydrogen, and oxygen in this sample?
   (b) What is the empirical formula for this organic compound?

59. Calculate the expected mass of CO_2 and H_2O when 0.1753 g of cholesterol ($C_{27}H_{46}O$) is burned in the presence of excess oxygen.

60. What is a "CHN analyzer"? What are some techniques that can be used in such a device to conduct a chemical analysis?

THERMOGRAVIMETRIC ANALYSIS

61. What is "thermogravimetric analysis"? How is this technique performed?

62. What is a "thermobalance"? What are some important components that make up this type of device?

63. What is a "thermogravimetric curve"? What information can be obtained about a sample by using a thermogravimetric curve?

64. A sample of iron oxide contains a mixture of Fe_2O_3 and $FeO(OH)$. It is known that $FeO(OH)$ loses water upon heating to 200°C according to the reaction shown below, while Fe_2O_3 has no change in mass under the same temperature conditions.

$$2\ FeO(OH)(s) \rightarrow Fe_2O_3(s) + H_2O(g) \quad (11.19)$$

A 0.2564 g portion of this mixture is subjected to thermogravimetric analysis and found to have a weight loss of 0.0097 g at 300°C. Find the percentage of both Fe_2O_3 and $FeO(OH)$ in the original sample.

65. A sample is a mixture of ammonium carbonate, sodium carbonate, and sodium chloride. Thermogravimetric analysis indicates that a 0.0965 g portion of this sample loses 0.0574 g over a temperature range of 50–75°C and then loses another 0.0124 g at about 800°C. What is the composition of the original sample?

CHALLENGE PROBLEMS

66. Write standard operating procedures for each of the following methods.
   (a) Proper use of filter paper for a gravimetric analysis
   (b) Proper use of a desiccator for chemical storage
   (c) Drying a sample or sample container for a gravimetric analysis

67. Locate material safety data sheets for each of the following reagents that might be used as part of a gravimetric analysis. What physical or chemical hazards are associated with each of these chemicals? What precautions should be followed in their use?
   (a) Perchloric acid (used for wet ashing)
   (b) Hydrofluoric acid (used to pretreat and analyze silica-related materials)
   (c) Sodium carbonate (used in sodium carbonate fusion)

68. Obtain from your instructor or the literature a detailed method for the gravimetric analysis of silver or iron. Explain the reasons for each step in this procedure.

69. Suppose that a crystal of AgCl is a cube having a mass of 0.0500 μg and is suspended in a solution with an excess of chloride ions.
   (a) Calculate how many silver ions and how many chloride ions are part of such a crystal.
   (b) Estimate the number of chloride ions that are on the surface of the crystal and compare this value with the number within the body of the crystal.

70. A silver sample is dissolved in an aqueous solution and the resulting silver ions are measured by using gravimetric analysis, with chloride acting as the precipitating agent. A total of 0.1543 g AgCl is collected after filtering this precipitate from a 150 mL solution that originally contained 0.020 M NaCl.
   (a) How many moles and grams of silver ions would be expected to remain in solution after the AgCl precipitate

has been collected? (*Hint:* Consider the solubility product for silver chloride.)

(b) How large an error would these silver ions produce in the use of this gravimetric analysis to determine the total silver content of the original sample?

71. How large an error would be expected in a gravimetric determination of chloride if this analysis is performed by precipitating the chloride as AgCl from a sample that also contains 0.010 mol NaBr for every mol of chloride ions?

72. Use the *CRC Handbook of Chemistry and Physics* or another resource to locate the gravimetric factors for the following analyses. Compare these values with those that you calculate when using the atomic or molar masses for the given analytes and products.
   (a) Determination of aluminum through the measurement of aluminum oxide
   (b) Determination of bromide through the measurement of silver bromide
   (c) Determination of nickel through the measurement of nickel dimethylglyoxime

73. T.W. Richards was puzzled when he found that the overall atomic mass of lead from radioactive samples varied with the source of such samples. We now know this effect is caused by the presence of different mixtures of lead isotopes, which were unknown prior to Richards' time. Suppose that an excess of dissolved lead in an aqueous solution is reacted with a solution that contains 0.1273 g of Na_2SO_4. How accurately must you be able to measure the mass of $PbSO_4$ if you wish to calculate an atomic mass for lead that has an error of only ± 0.01 g/mol?

74. Locate the structures for the following reagents and describe how each can be used for the gravimetric analysis of metal ions.[15,24]
   (a) Nitron
   (b) Neocupferron
   (c) Salicylaldehyde oxime
   (d) Tetraphenylarsonium chloride
   (e) Ethylenediamine
   (f) Pyrogallol

75. Biacetyl is a natural constituent of butter and is used in butter-flavored oil for microwave popcorn. Describe how you might measure the amount of biacetyl in butter-flavored oil by making use of a modified nickel dimethylglyoxime gravimetric analysis. (*Hint:* See Figure 11.10 and Equation 11.12.)

76. In studying radioactive fission, very small amounts of radioactive ^{141}Ba must be converted into $BaSO_4$ to remove the barium from other fission products. If a sample contains 0.0000024 g of this barium isotope in a 1.50 L solution, what fraction of this analyte will be precipitated when adding 0.10 mol H_2SO_4? How would this analysis be improved if before adding the sulfuric acid, 0.020 mol of nonradioactive barium chloride were added? (*Note:* In this second situation, the added barium is called a "cold carrier" and acts as a gathering agent for the radioactive barium.)

77. Calcium can be determined through a gravimetric analysis by precipitation of calcium ions in the form of their hydrated oxalate, $CaC_2O_4 \cdot H_2O$, followed by ignition to form $CaCO_3$ or CaO. This precipitation must be performed from a neutral solution so that the oxalate is present as a dianion instead of its protonated form. Use the solubility product of calcium oxalate and the pK_a values for oxalic acid to calculate the percent of calcium ions that will be pre-cipitated from a 100.00 mL solution that initially contains 0.00300 mol of Ca^{2+} and 0.00800 mol oxalate at pH 5.00. How high must the pH be in this analysis to achieve 99.99% precipitation for the calcium ions?

TOPICS FOR DISCUSSION AND REPORTS

83. A report was published in 1997 that examined the changes that have occurred in the atomic masses over the past century.[3] Obtain a copy of this article and report on how the accepted masses for several elements have changed over this period of time. Describe the types of errors that were present in earlier values and indicate how the size of such errors has changed over time.

84. There are many other individuals besides T.W. Richards who have also contributed to the development of gravimetric methods. One example is J.J. Berzelius, who we learned about in Chapter 3. Below are several other individuals who have contributed to this field. Find more information on one or more of the following people and describe the type of work they performed in gravimetric analysis and analytical chemistry.[6]
   (a) Karl Fresenius
   (b) William Hildebrand
   (c) William Ostwald
   (d) Fritz Pregl
   (e) Justus von Liebig
   (f) Torbern Bergman

85. Using resources such as References 13–15, locate a procedure for performing a gravimetric analysis in one or more of the following situations. Describe the reactions that take place in this method and state the purpose for each step in the procedure.
   (a) Determination of ammonium in a sample by using sodium tetraphenylboronate
   (b) Measurement of the percent carbonate in a limestone sample
   (c) Determination of silicon in a solid mineral sample
   (d) Determination of zinc in an ore sample

86. The quality of labware was for many years a limiting factor in the characterization of chemicals by gravimetric analysis and other classic methods. The crucible is one device that has been particularly critical in such work. A recent study was conducted on special crucibles from the Hesse region of Germany that were highly sought after by early chemists for their superior strength and resistance to high temperatures and chemical deterioration.[32] Describe the findings of this study. Compare the properties of the Hessian crucibles with those listed by the manufacturers of modern crucibles that are used in analytical laboratories.

87. Thermogravimetric analysis is not the only method in which the properties of a sample are monitored as the temperature is changed. Other examples are given below. Locate more information on one of these methods.[11,30,31] Write a report that states how the method works and that describes the basic equipment that is used to perform the method. Discuss the types of information that are provided by the method and list a few applications.
   (a) Differential thermal analysis (DTA)
   (b) Differential scanning calorimetry (DSC)
   (c) Thermomechanical analysis (TMA)
   (d) Evolved gas analysis (EGA)

References

1. R. T. Sanderson, *Chemical Periodicity*, Reinhold, New York, 1960.
2. N. N. Greenwood and A. Earnshaw, *Chemistry of the Elements, 2nd ed.*, Butterworth, UK, 1997.
3. T. B. Coplen and H. S. Peiser, "History of the Recommended Atomic-Weight Values from 1882 to 1997: A Comparison of the Differences from Current Values to the Estimated Uncertainties of Earlier Values," *Pure and Applied Chemistry*, 70 (1998) 237–257.
4. J. B. Conant, "Theodore William Richards and the Periodic Table," *Science*, 168 (1970) 425–428.
5. A. J. Ihde, "Theodore William Richards and the Atomic Weight Problem," *Science*, 164 (1969) 647–651.
6. F. Szabadvary, *History of Analytical Chemistry*, Pergamon Press, New York, 1966.
7. T. W. Richards and M. E. Lembert, "The Atomic Weight of Lead of Radioactive Origin," *Journal of the American Chemical Society*, 36 (1914) 1329–1344.
8. C. M. Beck II, "A Brief History of Inorganic Classical Analysis." In *The Australian Chemistry Resource Book*, Vol. 10, Charles Sturt University Bathurst, New South Wales, Australia, 2000.
9. C. M. Beck II, "Classical Analysis: A Look at the Past, Present, and Future," *Analytical Chemistry*, 66 (1994) 224A–239A.
10. T. S. Ma, "Organic Elemental Analysis." In *Analytical Instrumentation Handbook*, 2nd ed., G. W. Ewing, Ed., Marcel Dekker, New York, 1997, Chapter 3.
11. D. Dollimore, "Thermoanalytical Instrumentation and Applications." In *Analytical Instrumentation Handbook*, 2nd ed., G.W. Ewing, Ed., Marcel Dekker, New York, 1997, Chapter 17.
12. *IUPAC Compendium of Chemical Terminology*, Electronic version, http://goldbook.iupac.org
13. H. Diehl, *Quantitative Analysis: Elementary Principles and Practice*, 2nd ed., Oakland Street Science Press, Ames, IA, 1974.
14. K. Kodama, *Methods of Quantitative Inorganic Analysis: An Encyclopedia of Gravimetric, Titrimetric and Colorimetric Methods*, Interscience, New York, 1963.
15. J. Bassett, R. C. Denney, G. H. Jeffery, and J. Mendham, *Vogel's Textbook of Quantitative Inorganic Analysis*, 4th ed., Longman, New York, 1978.
16. G. Maludzinska, Ed., *Dictionary of Analytical Chemistry*, Elsevier, Amsterdam, The Netherlands, 1990.
17. *The American Heritage College Dictionary*, 4th ed., Houghton Mifflin, Boston, MA, 2004.
18. D. J. Pietrzyk and C. W. Fank, *Analytical Chemistry*, 2nd ed., Elsevier, Amsterdam, The Netherlands, 1979.
19. P. F. S. Cartwright, E. J. Newman, and D. Woodburn, "Precipitation from Homogeneous Solution," *Analyst*, 92 (1967) 663–679.
20. L. Gordon, M. L. Salutsky, and H. H. Willard, *Precipitation from Homogeneous Solution*, Wiley, New York, 1959.
21. M. L. Salutsky and W. R. Grace, "Precipitates: Their Formation, Properties, and Purity." In I.M. Koltoff and P.J. Elving, Editors. *Treatise on Analytical Chemistry, Part 1, Theory and Practice*, Vol. 1, Interscience, New York, 1959, Chapter 18.
22. D. R. Lide, Ed., *CRC Handbook of Chemistry and Physics*, 83rd ed., CRC Press, Boca Raton, FL, 2002.
23. J. A. Dean, *Lange's Handbook of Chemistry*, 15th ed., McGraw-Hill, New York, 1999.
24. M. Windholz, Ed., *The Merck Index*, 10th ed., Merck, Rahway, NJ, 1983.
25. F. L. Holmes, "Elementary Analysis and the Origins of Physiological Chemistry," *Isis*, 54 (1963) 50–81.
26. A. Lavoisier, *Traité Élémentaire de Chimie*, Vol. 2, Chapter VII, 1789.
27. J. Liebig, "Ueber einen neuen Apparat zur Analyse organischer Korper," *Annalen der Physik und Chemie*, 21 (1831) 1–43.
28. F. Pregl, *Die Quantitative Microanalyse*, J. Springer, Berlin, 1917.
29. *The Nobel Prize Internet Archive*, www.almaz.com/nobel
30. M. E. Brown, *Introduction to Thermal Analysis: Techniques and Applications*, 2nd ed., Springer-Verlag, New York, 2001.
31. P. Gabbott, Ed., *Principles and Applications of Thermal Analysis*, Blackwell Publishing, Malden, MA, 2007.
32. I. Amato, "Crucible Secrets," *Chemical & Engineering News*, Aug. 14 (2006) 56.

Chapter 12

Acid–Base Titrations

Chapter Outline

12.1 INTRODUCTION: RISE OF THE TITRATIONS

The Industrial Revolution was a major turning point in human history. It began in Europe during the late 1700s when manual labor was replaced in some industries by machines. Part of this change was created by the use of steam power to drive the machines, which lead to an increase in production speed and a decrease in cost. The creation of better transportation networks helped increase trading and the distribution of products. In addition, advances were made in many areas of science, including chemistry. These advances resulted in better methods for manufacturing steel, dyes for textiles, paints, and other products.[1,2]

Many of the earliest chemicals manufactured during the Industrial Revolution were acids and bases. Sulfuric acid is one example. This acid is still the chemical produced in the largest quantity by the United States and many other countries. Many other acids and bases are also produced by industry on a large scale, such as nitric acid, hydrochloric acid, phosphoric acid, ammonia, sodium hydroxide, and sodium carbonate (see Figure 12.1).[3] After these chemicals are produced, they are used in making various products. This process relies on the availability of reliable analytical methods to test the purity of these acids and bases. For instance, during the Industrial Revolution this need led to the establishment of analytical laboratories in many factories and to the creation of better methods for determining the concentration of acids or bases in raw materials and products.[2]

One important type of analysis method that arose as part of this process was the *acid–base titration*.[2] This technique is still commonly used to measure the purity and concentration of acids and bases. In this chapter, we will learn about the basic principles of an acid–base titration and see how this method is carried out. We will also see how equations for describing an acid–base equilibrium (see Chapter 8) can be used to predict the behavior of an acid–base titration. Finally, we will learn how to use this information to develop and optimize acid–base titrations for chemical analysis.

12.1A What Is an Acid–Base Titration?

A **titration**, or "titrimetric analysis," is a procedure in which the quantity of an analyte in a sample is determined by adding a known quantity of a reagent that reacts completely with the analyte in a well-defined manner.[1,4,5] The reagent that is combined with the analyte in this method is known as the **titrant**.[1,4,5] A common way of performing a titration is shown in Figure 12.2. This approach involves the use of a buret to carefully deliver a known volume of titrant to a sample. During this process, the titrant/sample mixture is examined for a change in color, pH, or other measurable property that can be used to signal when the analyte has been completely consumed by the titrant. An **acid–base titration** is a special type of titration in which the reaction of an acid with a base is used for measuring an analyte.[4,5] For example, if the analyte is an acid such as hydrochloric acid, the titrant would be a base like sodium hydroxide. Similarly, if the analyte is a base such as sodium hydroxide, the titrant would be an acid like hydrochloric acid.

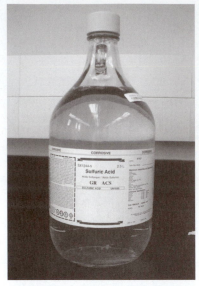

Industrial production of acids and bases in the U.S.	
Rank - chemical	Annual production
1 – Sulfuric acid	39.6×10^9 kg
4 – Phosphoric acid	16.2×10^9 kg
5 – Ammonia	15.0×10^9 kg
8 – Sodium hydroxide	11.0×10^9 kg
9 – Sodium carbonate	10.2×10^9 kg
11 – Nitric acid	8.0×10^9 kg
16 – Hydrochloric acid	4.3×10^9 kg

FIGURE 12.1 A bottle of sulfuric acid and typical production levels in the U.S. of common acids and bases. The rank shown by each acid or base is their relative position in the top 20 chemicals that are produced by the U.S. chemical industry, based on information from 2000.[3] This list does not include minerals that can be used without processing (such as salt or sulfur) and petroleum feedstocks.

A plot of the measured response versus the amount of added titrant during a titration is called a **titration curve** (see example in Figure 12.2).[4,5] The point in this curve at which exactly enough titrant has been added to react with all of the analyte is known as the **equivalence point**.[1,4,5] Once we have determined the amount of titrant that must be added to reach the equivalence point, we can determine the amount of analyte that was present in the original sample by using the known stoichiometry for the reaction of the titrant with the analyte. Besides having a well-characterized reaction between the analyte and titrant, it is also desirable in a titration for this reaction to be fast and to have a large equilibrium constant. These properties help to ensure that the titrant will quickly and completely combine with the analyte as the titrant is added to the sample.

There are various ways in which the progress of a titration can be followed. Both visual methods and instrumental methods are discussed later in this chapter (for example, the use of a colored indicator or pH measurements). Although it is always desirable to get the most accurate estimate possible of the equivalence point in a titration, the method for detecting this point may have some systematic error. This experimental estimate of the equivalence point is referred to as the **end point**, and the difference in the amount of titrant that is needed to reach the true equivalence versus the end point is called the *titration error*.[4,5] The absolute value of this error (in this case, in units of volume for the titrant solution) is given by Equation 12.1,

$$\text{Titration error} = V_{T,\text{End Pt}} - V_E \qquad (12.1)$$

where $V_{T,\text{End Pt}}$ is the volume of titrant needed to reach the end point and V_E is the volume of titrant needed to reach the true equivalence point. One topic we will examine later in this chapter is how good laboratory practices can be used to minimize this titration error and allow for accurate sample measurements.

The general type of titration that is illustrated in Figure 12.2 relies on the use of volume measurements to describe how much titrant was added to the sample. This particular approach is an example of a **volumetric analysis**, in which volume measurements are used for characterizing a sample.[5] When volume measurements are used during a titration, the resulting method is sometimes called a *volumetric titration*. A volumetric titration is the approach that will be emphasized in this chapter and book as we discuss "titrations." The titrant in a volumetric analysis is delivered by using a *buret* (see Chapter 3), which is a piece of volumetric glassware that makes it possible to both deliver and accurately measure the titrant as it is placed into the sample. It is also possible to perform a titration by measuring the mass of titrant that has been added to a sample. This second approach is known as a *gravimetric titration*, or "weight titration." Although a gravimetric titration tends to be more difficult to carry out than a volumetric titration, it can be useful in situations that require highly accurate and precise measurements of an analyte.[6]

12.1B How Are Acid–Base Titrations Used in Analytical Chemistry?

Applications of Titrations. There are many advantages to using acid–base titrations and related titration methods (see Chapters 13 and 15). Titrations are usually inexpensive to conduct and require only simple, standard laboratory equipment. Titrations are also capable of providing both good accuracy and precision during a chemical analysis. Acid–base reactions are especially well

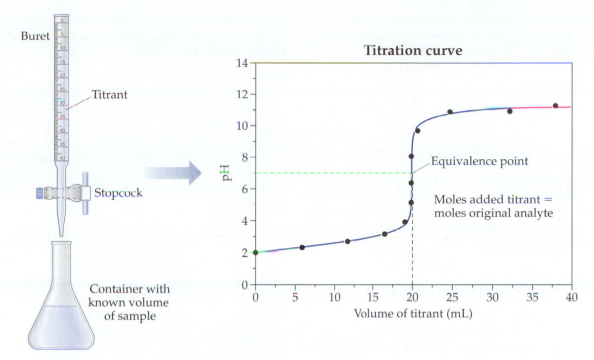

Buret

Titrant

Stopcock

Container with
known volume
of sample

Titration curve

Equivalence point

Moles added titrant =
moles original analyte

pH

Volume of titrant (mL)

FIGURE 12.2 A general example of the equipment used in an acid–base titration, and a typical titration curve. The curve shown is a typical titration curve for the titration of a strong monoprotic acid (such as HCl) with a strong base (such as NaOH). The dashed lines in this graph indicate the volume of titrant that is needed to reach the equivalence point and show the pH at this point in the titration. A titration in which the measured response makes use of a logarithmic expression of concentration or activity (such as pH) and is known as a *logarithmic titration curve*.[4] A logarithmic titration curve often has the curved behavior that is shown in this example. It is also possible to have a measured response such as absorbance that is directly related to an analyte's concentration or activity; this second type of plot is known as a *linear titration curve* (see examples under "spectrophotometric titrations" in Chapter 18).

suited for titrations, because these reactions tend to have large equilibrium constants and fast reaction rates. These properties make it possible to quickly obtain essentially stoichiometric reactions between the analyte and titrant. A disadvantage of a traditional acid–base titration is it usually needs at least 0.01 mol of an analyte for measurement. Also, many acid–base titrations are often manual methods that can be time-consuming to perform for a large number of samples. It is, however, possible to automate these titrations by using a device like that shown in Figure 12.3.

The most common application for a titration is to determine the amount of an analyte that is in a sample. This measurement is made by using the known reaction between the titrant and analyte and by determining the amount of titrant that is required to reach the equivalence point for a sample that contains this analyte. Figure 12.2 shows a typical titration curve that would be obtained for the analysis of a strong monoprotic acid like HCl when using a strong monoprotic base such as NaOH as the titrant. In this situation, H^+ from HCl reacts in a 1:1 ratio with OH^- from NaOH. The moles of analyte ($n_{Analyte}$) in the original sample (with a total molar concentration of $C_{Analyte}$ and an original volume of $V_{Analyte}$) should then be equal to the moles of titrant ($n_{Titrant}$) that are required to reach the equivalence point of this titration. The moles of titrant needed to reach the equivalence point can, in turn, be determined by using

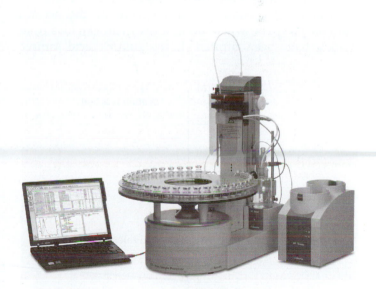

FIGURE 12.3 A system that is designed for use in automating an acid–base titration. This type of device, known as an *autotitrator*, is capable of precisely delivering various amounts of titrant to a sample (held in the beaker on the right) while an electrode is used to measure the pH of the sample/ titrant mixture. Similar devices that make use of other means for measuring a response in the sample/titrant mixture can be used to automate additional types of titrations. (Reproduced with permission and courtesy of Metrohm USA.)

the measured volume of titrant that is needed to reach the equivalence point ($V_{Titrant}$) and the known molar concentration of the titrant ($C_{Titrant}$), as shown below.

At the equivalence point for titration of a monoprotic acid with a monoprotic base:

$$n_{Analyte} = n_{Titrant}$$

or

$$C_{Analyte} V_{Analyte} = C_{Titrant} V_{Titrant} \qquad (12.2)$$

For example, suppose that we find that a 10.00 mL sample of HCl requires the addition of 20.00 mL of 0.005000 M NaOH to reach the equivalence point in an acid–base titration. We can determine the concentration of HCl in the sample by using Equation 12.2, as shown below.

$$C_{HCl} \cdot (0.01000 \text{ L HCl})$$
$$= (0.005000 \ M \text{ NaOH}) \cdot (0.02000 \text{ L NaOH})$$
$$\therefore \quad C_{HCl} = 0.01000 \ M \text{ HCl}$$

The same general approach can be used to find the concentration of a monoprotic base that is titrated with a monoprotic acid or to find the moles of a monoprotic titrant that are needed to reach each equivalence point for a polyprotic acid or base.

EXERCISE 12.1   Using a Titration to Determining the Concentration of an Acid

A chemist working in an industrial analytical laboratory titrates a 10.00 mL portion of a sulfuric acid sample with a $5.000 \times 10^{-3}\ M$ solution of sodium hydroxide. The titration curve obtained is shown in Figure 12.4, in which the measured end point actually occurs at the second equivalence point where both of the hydrogen ions from H_2SO_4 have been titrated. If the sulfuric acid sample requires 40.00 mL of the NaOH titrant to reach the second equivalence point, what is the concentration of the original sample of sulfuric acid?

SOLUTION

This problem involves the complete reaction of a diprotic acid (H_2SO_4) with a monoprotic base (NaOH). We can find the concentration of sulfuric acid in this case by using the relationship shown below, in which the "2" that appears before "[H_2SO_4]" is used because we are doing this calculation for the second equivalence point in which both the hydrogen ions of H_2SO_4 have been titrated with NaOH.

$$2\,[H_2SO_4]\, V_{H_2SO_4} = [NaOH]\, V_{NaOH}$$

$$[H_2SO_4] = \frac{[NaOH]\, V_{NaOH}}{2\, V_{H_2SO_4}} = \frac{(0.005000\ M) \cdot (0.04000\ L)}{2 \cdot (0.01000\ L)}$$

$$\therefore [H_2SO_4] = \mathbf{0.01000\ M}$$

Another way we could have expressed the concentration of sulfuric acid is by using units of normality (see Appendix A). The normality (N) of an acid is related to the number of moles (or equivalents) of hydroxide ion that are required to neutralize a liter of the acid's solution. In the same manner, the normality of a base is the number of moles (or equivalents) of hydrogen ions that are needed to neutralize a liter of the base in its solution. Sulfuric acid is a diprotic acid, so the normality of its solution will be double the value of its molarity, or (2 equivalents/mol) $\cdot$ 0.01000 M = 0.02000 N H_2SO_4. However, because NaOH reacts with only one mole of hydrogen ions per mole in the titration of H_2SO_4 (or any other acid in water), the $5.000 \times 10^{-3}\ M$ NaOH solution will also have a concentration of $5.000 \times 10^{-3}\ N$.

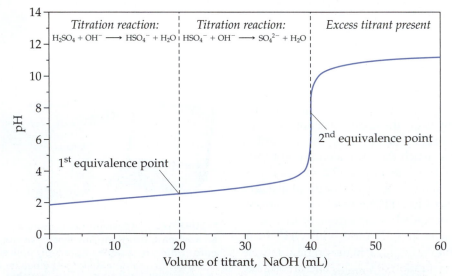

FIGURE 12.4   Titration curve obtained for a 10.00 mL sample of sulfuric acid (H_2SO_4) when using 0.005000 M sodium hydroxide as the titrant. The first equivalence point in this titration is not readily detectable in this curve due to the relatively high K_a value for the conjugate base of sulfuric acid (HSO_4^-), but the second equivalence point can be observed and provides a usable end point for this titration.

A second, related application for titrations is to determine the purity of a sample. In this case, we know the molar mass of the analyte but not its actual mass in the sample, even though we do know the total sample mass. For instance, suppose a pharmaceutical chemist wishes to determine the purity of a benzoic acid sample (C_6H_5COOH, molar mass = 122.12 g/mol) by titrating 0.4370 g of this sample with NaOH. If 25.57 mL of 0.08653 M NaOH is needed to reach the equivalence point for this sample, the mass of benzoic acid in the sample can be found by starting with the relationship in Equation 12.2, as shown in the following.

$$n_{Analyte} = C_{Analyte} V_{Analyte} = C_{Titrant} V_{Titrant}$$

$$= (0.08653 \text{ mol/L NaOH})(0.02557 \text{ L NaOH})$$

$$\cdot \frac{1 \text{ mol Benzoic acid}}{1 \text{ mol NaOH}}$$

$$= 0.002213 \text{ mol Benzoic acid}$$

$$\text{mass Benzoic acid} = (0.002213 \text{ mol Benzoic acid})$$
$$(122.12 \text{ g/mol Benzoic acid})$$
$$= 0.2702 \text{ g Benzoic acid}$$

This result means that the purity of the benzoic sample was $100 \cdot (0.2702 \text{ g benzoic acid})/(0.4370 \text{ g sample}) = 61.83\%$ (w/w). It should be kept in mind, however, that this calculation does assume that no sample components other than the analyte are capable of reacting with the titrant.

A third general way a titration can be used is to determine some fundamental properties of the analyte, such as its molar mass. This application requires a pure sample of the analyte. If it is assumed that the analyte and titrant react in a 1:1 ratio, the moles of titrant required to reach the equivalence point for a solution of this analyte, as determined from its concentration ($M_{Titrant}$) and added volume ($V_{Titrant}$), can be combined with the known mass of the analyte ($m_{Analyte}$) and concentration of titrant to get the analyte's molar mass ($MW_{Analyte}$).

At the Equivalence Point:

$$n_{Analyte} = n_{Titrant}$$

or

$$m_{Analyte}/MW_{Analyte} = M_{Titrant} V_{Titrant}$$

$$\therefore \quad MW_{Analyte} = m_{Analyte}/(M_{Titrant} V_{Titrant}) \quad (12.3)$$

As an example, if 0.4370 g of a pure acid requires 41.38 mL of 0.08653 M NaOH to reach the equivalence point, the molar mass of this acid would be found as shown in the following.

$$MW_{Analyte} = \frac{0.4370 \text{ g}}{(0.08653 \text{ mol/L})(0.04138 \text{ L})} = 122.0 \text{ g/mol}$$

Notice in this calculation that the solution volume must be expressed in liters to obtain the proper units for the final answer. If the acid were polyprotic, the apparent molar mass (or "equivalent weight") determined by this approach will be equal to the true molar mass divided by the number of hydrogen ions or hydroxide ions from the titrant that react with each molecule of the analyte. This means a diprotic acid like sulfuric acid will give an equivalent weight that is half of its true molar mass if a pure sample of this acid is analyzed by a titration like the one in Figure 12.4.

A titration can also provide fundamental information on the acid–base properties of the analyte. The position of the end point and shape of an acid–base titration curve like that in Figure 12.2 can tell us whether the analyte is monoprotic or polyprotic and whether the analyte is a strong or weak acid or base. In some cases, we can also use such a curve to determine the acid dissociation constant (K_a) or base ionization constant (K_b) for an analyte that is an acid or base. We will see more on how this information can be obtained in Section 12.C.

General Types of Titrations. There are several ways in which an acid–base titration or any other type of titration can be performed. The technique that was used in Figure 12.2 and Figure 12.4 is known as a **direct titration**. In this approach, the amount of analyte is determined by combining it directly with the titrant while the appearance of the end point for this titration is monitored.[5] A direct titration is the simplest type of titration, but it does require an end point that is sharp and easy to detect. Ideally, a direct titration also requires that the reaction between the analyte and titrant be rapid and have a large equilibrium constant, thus providing a relatively fast method that can accurately determine the amount of analyte that was present in the sample.

There are alternative methods that can be employed if the reaction between the analyte and titrant is slow or does not have a suitably large equilibrium constant. One example is a **back titration**. In this technique, an excess of a known amount of a reagent is added to react with an analyte in a sample. The amount of reagent that remains after this process is then determined by a titration.[5,6] The difference between the original amount of reagent and the amount that is titrated is then used to determine how much analyte was present in the sample.

A back titration makes up part of an important analytical technique known as the **Kjeldahl method**, which is used to measure the nitrogen content in organic samples (see Box 12.1).[5,6] In this method, the nitrogen in organic compounds such as proteins is converted to ammonium ions, which are then neutralized with a large excess of concentrated NaOH. This neutralization process converts the ammonium ions into ammonia, which is distilled from the sample as a gas. The gaseous ammonia is then redissolved and collected in a solution that has a known amount of HCl, which is in excess of the expected amount of generated ammonia. After the ammonia has dissolved in this solution and reacted with HCl, the remaining HCl is then measured by carrying out a back titration with NaOH. The amount of HCl that is measured is then compared to the original concentration of HCl. This information provides the moles of ammonia

BOX 12.1
The Kjeldahl Method

The Kjeldahl method is perhaps the most common of all acid–base titrations in modern industrial laboratories. This method was first described by Danish chemist Johan Kjeldahl in 1883 as a means for determining the nitrogen content in foods, fertilizer, and other samples (see Figure 12.5).[1,6]

This technique involves several steps. First, the sample is digested with sulfuric acid, which causes all of the nitrogen in an organic-based sample to convert to ammonium ions. Second, the pH of this digested sample is raised by carefully adding sodium hydroxide. This change to a basic pH causes the ammonium ions to covert to ammonia. Next, this ammonia is removed from the sample through distillation and is collected by in a separate container that contains a measured volume of a standard acid solution, which reacts with the ammonia and causes it to convert back into ammonium ions. The amount of acid that remains in this solution is then determined through a back titration with a strong base. The difference in the concentration of the acid solution before and after the sample distillation is then used to determine how much ammonia was captured, thereby giving the amount of nitrogen that was present in the original sample.[1,5–7]

The Kjeldahl method remains an extremely important technique for examining the quality of materials that are produced and used in the food industry and in agriculture. Many industrial laboratories that work in these areas perform this method on a routine basis. Although this method can be labor intensive, these laboratories use special distillation and collection units that allow many samples to be processed at the same time, such as the one shown in Figure 12.5. These laboratories can also use devices like the one in Figure 12.3 to help automate the titration of these samples.[8]

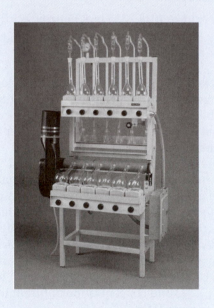

Johan Kjeldahl

FIGURE 12.5 Johan Kjeldahl (1849–1900) and a commercial distillation system for use in the Kjeldahl method for organic nitrogen analysis. The flasks in the front of the given apparatus are used for sample digestion and those on top are ready for distillation into a set of flasks at the rear that will receive the distilled ammonia and redissolve it in an acidic solution. (The image of the distillation system is used with permission from Labconco.)

that were collected and, thus, the amount of nitrogen that was present in the original sample.

EXERCISE 12.2   Performing a Kjeldahl Analysis

A manufacturer of baked goods wishes to find the relative amount of protein (measured as % nitrogen) that is present in a shipment of flour. A chemist weighs out 4.0030 g of the flour and examines it by the Kjeldahl method. After this sample has been digested, the ammonia it produces is trapped in a flask that originally contains 50.00 mL of 0.1450 M HCl. It is later found that this HCl solution with the trapped ammonia gives an end point at 17.54 mL when titrated with 0.1290 M NaOH. What is the percent (w/w) nitrogen in the flour?

SOLUTION

The amount of HCl that is originally in the receiving flask is (0.05000 L)(0.1450 mol/L) = 0.007250 mol. The amount of HCl that remains after trapping ammonia from the sample is equal to the moles of NaOH that are needed to back titrate the final solution.

mol remaining HCl = (0.01754 L NaOH) ·

$$(0.1290 \text{ mol/L NaOH}) \cdot \frac{1 \text{ mol HCl}}{1 \text{ mol NaOH}}$$

$$= 0.002263 \text{ mol}$$

The amount of ammonia that has been collected in the HCl solution will be equal to the difference between the initial and remaining amounts of HCl.

Amount NH$_3$ = [(0.007250 mol initial HCl) −

$$(0.002263 \text{ mol remaining HCl})] \cdot \frac{1 \text{ mol NH}_3}{1 \text{ mol HCl}}$$

$$= 0.004987 \text{ mol NH}_3$$

This amount of ammonia can be used to give the mass of nitrogen in the original sample.

Mass of nitrogen in sample =

$$(0.004987 \text{ mol NH}_3) \cdot \frac{1 \text{ mol N}}{1 \text{ mol NH}_3} \cdot \frac{14.0067 \text{ g}}{\text{mol N}}$$

$$= 0.06985 \text{ g N}$$

We can now calculate the % nitrogen in the sample by using the following relationship.

$$\% \text{ nitrogen (w/w)} = 100 \cdot \frac{0.06985 \text{ g N}}{4.0030 \text{ g flour}}$$

$$= \mathbf{1.745\% \text{ nitrogen (w/w)}}$$

A typical protein is roughly 17.5% nitrogen, so this batch of flour contains about 100 · (1.745 g N/100 g flour)/(17.5 g N /100 g protein) = 9.97% = 10.0% (w/w) protein.

12.2 PERFORMING AN ACID–BASE TITRATION

12.2A Preparing Titrant and Sample Solutions

Standardizing Titrants. To carry out a successful titration, it is necessary to know the concentration of the titrant. Determining the concentration of this solution is known as *standardization*[5] and provides a solution with an accurately known concentration that is referred to as a **standard solution**.[4] The most common strong acid used in acid–base titrations is hydrochloric acid and the most common strong base is sodium hydroxide. However, it is not possible to simply weigh out either HCl or NaOH, dissolve these in water, and accurately know the molarity of their resulting solution. The problem with a commercial preparation of HCl is that it might contain a range of HCl concentrations (approximately 36–37% HCl in water, or roughly 12 M HCl). NaOH is sold as pellets that are only about 95% pure, with the other 5% consisting of water and sodium carbonate. Thus, neither HCl nor

NaOH in their common commercial forms is of a high enough purity or made with sufficient reproducibility to give more than two significant figures in concentration when they are weighed and placed into a solution.

To overcome this problem, the concentration of a HCl or NaOH solution can be standardized by titrating it with another chemical that does have a known level of high purity. The acid or base that is used to determine the concentration of NaOH or HCl in this situation is a "primary standard." As we learned in Chapter 3, a primary standard is a substance of known high purity that can be dissolved to prepare a solution with a well-known concentration (a *primary standard solution*).[4,5] This primary standard solution is then used to determine the purity of another chemical for use as a reagent (referred to as a "secondary standard") or the concentration of this chemical in a solution (giving a *secondary standard solution*).[4,5] Potassium hydrogen phthalate is a primary standard acid used for titrating strong bases like NaOH, and sodium carbonate is a common primary standard base used for titrating strong acids like HCl. Some important characteristics of these and other primary standards for acid–base titrations are listed in Table 12.1.

There are several properties that should be present before a chemical can be used as a primary standard. First, this chemical must be capable of undergoing the desired type of reaction (for instance, an acid–base reaction). Second, the reaction of this compound should be fast and have no side products. Third, this chemical should be readily available in a highly pure form, allowing it to be used in accurate mass measurements. Fourth, this chemical must not absorb appreciable amounts of water or carbon dioxide, making it possible to accurately determine its mass even when it is exposed to air. All of the primary standards in Table 12.1 meet these criteria.

| EXERCISE 12.3 | Standardizing a Sodium Hydroxide Solution |

A chemist in an industrial laboratory would like to standardize a solution of NaOH for use in the titration of weak acids. What is the concentration of NaOH if a 42.83 mL aliquot of this solution is needed to titrate 0.1765 g of the primary standard potassium hydrogen phthalate (KHP)?

SOLUTION

At the equivalence point of this titration, the moles of added NaOH will be equal to the moles of KHP in the original sample. From this, we can write the following relationship based on Equation 12.3 between the concentration and volume of the NaOH solution and the amount of KHP that has been titrated.

$$M_{\text{Titrant}} V_{\text{Titrant}} = m_{\text{Analyte}} / MW_{\text{Analyte}}$$

$$[\text{NaOH}] V_{\text{NaOH}} = \frac{\text{Mass KHP}}{\text{Molar mass KHP (in g/mol)}}$$

It is now possible to rearrange this equation and to solve for the concentration of NaOH.

$$[NaOH] = \frac{0.1765 \text{ g KHP}}{(204.221 \text{ g/mol KHP})(0.04283 \text{ L})} =$$

0.02043 M NaOH

We can see from this calculation that this standardization leads to a measured concentration for NaOH that has up to four significant figures if good technique is used during the titration and if both the solution volumes and mass of KHP have also been determined to at least four significant figures.

Effects of Titrant and Sample Concentration. The general effect of changing the concentration of a sample or titrant in an acid–base titration is shown in Figure 12.6. This particular example is for the titration of a strong acid with a strong base, but the overall effects are similar to those that occur in other types of acid–base titrations. (*Note:* In the case of a titration of a base with a strong acid, the pH would instead begin at a high value and decrease as more titrant is added.)

Let's first look at how a change in concentration for the titrant will affect an acid–base titration, as illustrated in Figure 12.6(a). Because we are using different concentrations for the titrant but the moles of titrant that is needed remains the same, the equivalence point will shift to a lower volume of titrant when the titrant's concentration is increased. Similarly, the equivalence point will shift to a higher volume of titrant when the concentration of the titrant is decreased. The size of this shift can be predicted by using Equation 12.2. The concentration of the titrant also has some effect on the

relative degree to which the pH changes just before and after the equivalence point. To achieve maximum accuracy in a titration when using a buret (see Section 12.2B), the concentration of the titrant should be selected so that the equivalence point will occur when the total volume of added titrant is approximately 80% of the buret's capacity.

The analyte's concentration will also have a large effect on the results obtained for an acid–base titration. Figure 12.6(b) shows an example for the titration of a strong acid with a strong base where the concentration of the titrant is also varied to be exactly half the concentration of the sample, thereby giving the same volume for the equivalence point in these titrations. You can see in this example that as the sample becomes more dilute, the change in pH near the equivalence point becomes shallower and the strong acid more difficult to detect. Eventually, this change is so small that an acid–base titration is no longer a feasible option to measure this analyte. For strong acids and bases the analyte concentrations that are usually examined by an acid–base titration are in the range of 0.1 M to 0.0001 M, with a concentration around 0.010 M being most common.

The concentration of the analyte has a similar effect when a solution is being titrated that contains a weak acid or weak base, but the relative strength of the weak acid or weak base must also now be considered. An analyte that is a weak acid with a $pK_a = -\log(K_a)$ of approximately 2.0 or lower will act about the same as a strong acid in a titration, and a weak base with a $pK_b = -\log(K_b)$ of 2.0 or lower will act about the same as a strong base in an acid–base titration (see Figure 12.7). However, as the pK_a of the weak acid (or pK_b of the weak base) is increased, the change in pH near the equivalence point becomes smaller and the end point is more difficult to

---

TABLE 12.1 Primary Standards for Acid–Base Titrations in Water*

Primary Standard	Formula Weight	Notes on Preparation and Use
Acids for Standardizing Solutions of Bases		
Potassium hydrogen phthalate, o-C_6H_4(COOK)(COOH)	204.221 g/mol	Dry at 105°C (<135°C). Use to titrate a strong base with phenolphthalein as the indicator
Potassium hydrogen iodate, $KH(IO_3)_2$	389.912 g/mol	Use to titrate a base with any indicator suitable for an end point between pH 5 and 9
Sulfamic acid, H_3NSO_3	97.095 g/mol	Use to titrate a base with any indicator suitable for an end point between pH 5 and 9
Bases for Standardizing Solutions of Acids		
Tris(hydroxymethyl)aminomethane, $(HOCH_3)_3CNH_2$	121.137 g/mol	Dry at 100–103°C (< 110°C). When used in a titration to standardize a strong acid solution, the end point will be approximately between pH 4.5 and 5
Sodium carbonate, Na_2CO_3	105.989 g/mol	Obtain as reagent-grade chemical and heat for 1 h at 255–256°C, followed by cooling in a desiccator

*This information was obtained from J. A. Dean, *Lange's Handbook of Chemistry*, 15th Ed., McGraw-Hill, New York, 1999.

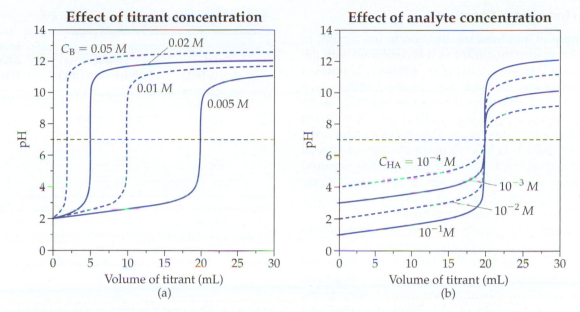

FIGURE 12.6 The effects of (a) titrant concentration and (b) analyte concentration on the titration curve that is obtained for a sample that contains a strong monoprotic acid (such as HCl) when titrated with a solution containing a strong base (NaOH). The concentrations shown in these graphs are the total concentrations of the acid or base in their original solutions. In (b), the concentration of the base (C_B) was varied to give the same equivalence point by using a titrant concentration that was always equal to half the concentration of the acid that was the analyte ($C_B = 0.5\, C_A$). The concentration of the acid in the original sample was kept constant at 0.01000 M in (a), while the value of C_B in the titrant was varied.

detect. The result is that even for a 0.010 M solution of analyte it is not practical to examine a weak acid with a pK_a of 10.0 or greater in water or, similarly, a weak base that has a pK_b of 10.0 or greater in water. This means it is important to consider both the acid–base strength of the analyte and its expected concentration when you are determining whether it will be feasible to measure this analyte by an acid–base titration.

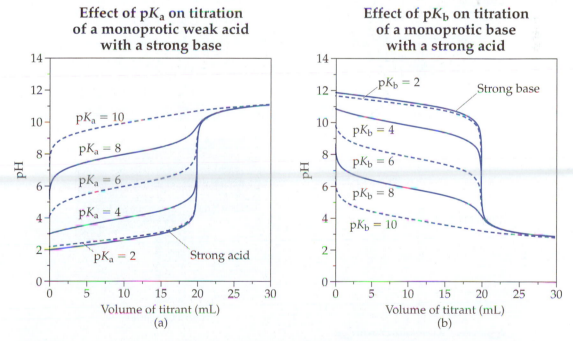

FIGURE 12.7 The effects of (a) the relative strength of a weak acid, as represented by pK_a, during a titration with a strong base, and (b) the relative strength of a weak base, as represented by pK_b, during a titration with a strong acid. The concentration of the weak acid or the weak base in the original sample was 0.01000 M and the sample volume was 10.00 mL. The concentration of the titrant in these calculations was 0.005000 M.

12.2B Performing a Titration

It is important in a volumetric titration to use a buret to deliver and measure the titrant as it is combined with the sample. As we learned in Chapter 3, a buret is graduated glass tube that has an open top for the addition of a titrant, numerous markings for volume readings on its side, and a stopcock at the bottom through which titrant is delivered to a sample. Figure 12.8 shows a picture of a typical buret, which is held in place by using a special clamp and a ringstand. Table 12.2 shows the required properties for Class A burets. Of these various sizes, burets with total capacities of 25 or 50 mL are the ones that are most commonly used in titrations.

   To perform an accurate titration, it is necessary to follow the standard operating procedures listed in Chapter 3 for the proper use of volumetric glassware. For example, it is essential that you always work with a clean buret to prevent droplets of titrant from sticking to the walls of the buret as this titrant is being delivered to a sample container. There must also be no air bubbles anywhere in the buret after it has been filled with the titrant. Failure to remove all air bubbles will affect both the precision and accuracy of the readings as the titrant is delivered to the sample container.

   Some practice is also required to properly read the volume of liquid in a buret. During a volumetric titration, two readings are required for each addition

TABLE 12.2   Characteristics of Class A Burets*

Total Capacity (mL)	Smallest Subdivision (mL)	Maximum Allowable Error (mL)
5	0.01	± 0.01
10	0.02 or 0.05	± 0.02
25	0.10	± 0.03
50	0.10	± 0.05
100	0.20	± 0.10

*These properties are those specified by the American Society for Testing Materials. The term "smallest subdivision" refers to the smallest interval in volume that appears in the markings on the side of the buret. The term "tolerance" is often used in place of "maximum allowable error" when describing the properties of these devices.

of titrant. The first reading is made before the desired volume of titrant is added, and the second reading is made after it is added. The difference in these two volume readings is used to give the volume of titrant that was actually delivered. Both readings must be made in the same fashion, with the meniscus of the liquid in the buret being exactly at eye level. Failure to make both readings at the same eye level will lead to "parallax error" (see Chapter 3). Because the amount of titrant in

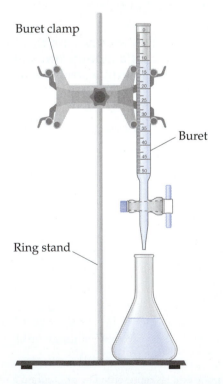

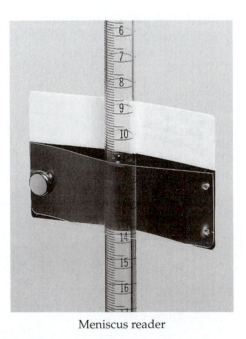

Meniscus reader

FIGURE 12.8   A typical experimental arrangement for the use of a buret and an example of a device that can be used to help read the meniscus level of the titrant in a buret. (*Note:* See Section 3.3C in Chapter 3 for a closer view of a meniscus and a discussion of parallax error.) (The figure on the left is reproduced with permission from Wikipedia. The figure on the right is reproduced with permission from Fisher Scientific)

the buret changes during the titration, you must adjust the level of your eye when viewing the liquid level to minimize this error.

Another problem with reading a buret is that the curved top of a liquid column (the "meniscus") acts like a lens and can give the appearance of at least two curved surfaces at the top region of the liquid. This is another reason why it is important to have your eyes level with this meniscus when determining the level of titrant in a buret. Many chemists place behind the buret a white card or a white card with a black region to make it easier to make a volume reading (see Figure 12.8).

When reading a volume from a buret, it is *always* important to estimate the relative distance that each volume is between its nearest markings (see our discussion on the use of an analog scale in Chapter 2). This will allow you to obtain an extra significant figure in your volume measurement. A 25 or 50 mL buret has a mark every 0.1 mL, which allows an experienced chemist to estimate the volume of solution in this buret to the nearest 0.01 mL. When carrying out a titration, carefully read and record the initial volume as well as all later volumes of titrant as they are added to the sample.

The titrant can often be added in a relatively large volume (in increments on the scale of mL) during the initial stages of the titration, but this added volume should be reduced to single drops or even a fraction of a drop as the end point is approached. You can tell when this should be done by looking for a small amount of temporary color change in the mixture when a visual indicator is used to signal the end point (as described in the next section) or if the measured pH of this mixture begins to rise more steeply. As you get really close to the end point, add the titrant in single drops or even as partial drops to hit the end point more accurately. When you think you are at the end point, allow the level of liquid in the buret to stabilize and read it carefully. Record the volume reading and then proceed to add another half drop to observe if any further change occurs that signals you have indeed reached and gone on beyond the end point.

12.2C Determining the End Point

pH Measurements and Acid–Base Indicators.   There are many ways of determining the end point for an acid–base titration. One common approach is to simply follow the course of a titration by measuring the pH of the titrant/sample mixture as more titrant is added. This type of measurement is carried out by placing a pH electrode into this mixture and measuring the pH after the sample has been mixed with each new addition of titrant (for example, see the automated system shown earlier in Figure 12.3). A graph of pH versus the volume of added titrant is then made and used to find the end point of the titration. Several examples of these plots are shown in Figures 12.2 and 12.4. It is important while you are collecting these pH measurements to carefully monitor them

and watch how fast they are increasing with each addition of titrant. The pH for an acid–base titration will show the sharpest rise at or near the equivalence point, so it is particularly important that you add only small volumes of the titrant in this range to allow for an accurate estimate to be obtained for the end of the titration.

Another way the course of an acid–base titration can be followed is by using an **acid–base indicator**. An acid–base indicator is a chemical, or mixture of chemicals, that changes color over a known range of pH. If this pH range for the color transition includes the pH of the equivalence point, the indicator can provide a quick and easy way of finding the end point of an acid–base titration.[1,4,5] All that is needed is for you to add a small amount of the indicator to the sample and to watch for the appropriate change in color as titrant is added to this sample.

An acid–base indicator is simply a chemical that is part of a weak acid–base system in which the acid form has a different color than its conjugate base. As a result, the color of this indicator will change with pH, and particularly when the pH changes in the region near the pK_a value for the indicator. The most widely used indicator in acid–base titrations is *phenolphthalein.*

$$pK_a = 9.4$$

$$\cdots = O + 2\,H^+$$

Acid form (H_2In)
Colorless

Base form (In^{2-})
Red/pink color

Phenolphthalein is colorless in acid and red in base, with this change in color occurring between approximately pH 8.0 and 10.0. This property makes phenolphthalein useful for the titration of a strong acid or moderately strong acid with a strong base, for which the equivalence point will be at pH 7.0 or only a slightly higher value.[6,9,10]

Table 12.3 lists several acid–base indicators along with their K_a values and pH ranges of color change (also see colored insert in the center of this book). Indicators typically change colors over about a range of 2.0 pH units that is centered around their pK_a. When the pH is equal to the pK_a for the indicator, 50% of the indicator will be in its acid form and 50% will be in the basic form. Typically, one can see a color change when 10% of the material is different from the rest, which will occur when the pH is one unit below or above this pK_a value. A good visual indicator should have an intense color so that only a small amount must be added to be

TABLE 12.3 Examples of Indicators for Acid–Base Titrations in Water[*]

Indicator	pK_a (25°C)	pH Range of Color Change	Color Change (Acid form → Base form)
Cresol purple	1.51	1.2–2.8	Red → Yellow
Thymol blue	1.65	1.2–2.8	Red → Yellow
Tropeolin OO	2.0	1.3–3.2	Red → Yellow
Methyl orange	3.40	3.1–4.4	Red → Yellow
2,6-Dinitrophenol	3.69	2.4–4.0	Colorless → Yellow
2,4-Dinitrophenol	3.90	2.5–4.3	Colorless → Yellow
Methyl yellow	3.3	2.9–4.0	Red → Yellow
Methyl orange	3.40	3.1–4.4	Red → Orange
Bromocresol blue[a]	3.85	3.0–4.6	Yellow → Blue violet
Bromocresol green[a]	4.68	4.0–5.6	Yellow → Blue
Methyl red	4.95	4.4–6.2	Red → Yellow
Chlorophenol red	6.0	5.4–6.8	Yellow → Red
Bromocresol purple[a]	6.3	5.2–6.8	Yellow → Purple
p-Nitrophenol	7.15	5.3–7.6	Colorless → Yellow
Bromothymol blue[a]	7.1	6.2–7.6	Yellow → Blue
Phenol red	7.9	6.4–8.0	Yellow → Red
Neutral red	7.4	6.8–8.0	Red → Yellow
m-Nitrophenol	8.3	6.4–8.8	Colorless → Yellow
Cresol red[b]	8.2	7.2–8.8	Yellow → Red
Cresol purple	8.32	7.6–9.2	Yellow → Purple
Thymol blue	8.9	8.0–9.6	Yellow → Blue
Phenolphthalein	9.4	8.0–9.6	Colorless → Red
Thymolphthalein	10.0	9.4–10.6	Colorless → Blue
Alizarin yellow R	11.16	10.1–12.0	Yellow → Violet

[*] The data in this table were obtained from J. A. Dean, *Lange's Handbook of Chemistry*, 15th ed., McGraw-Hill, New York, 1999.
[a] Some sources refer to these indicators as bromcresol blue, bromcresol green, bromcresol purple, and bromthymol blue.
[b] Cresol red also has a color change from red to yellow that occurs between pH 0.2 and 1.8.

noticeable to the eye. Because the indicator is part of an acid–base system, it will react with a base or acid titrant to convert to its conjugate form. This occurs when the pH of the sample/titrant mixture approaches the pK_a of the indicator, which should be chosen so that the observed end point is as near as possible to the expected pH for the equivalence point in the titration. For instance, the titration shown in Figure 12.9 has an equivalence point that occurs at 20.00 mL titrant and pH 8.14. Of the three indicators that are shown in this figure, cresol red has a pK_a value that is closest in value to the pH at the equivalence point (pK_a = 8.22), followed by phenol red (pK_a = 7.9) and phenolphthalein (pK_a = 9.4). The pH range for the color transition of

cresol red also gives the best match with the expected pH for the equivalence point. Thus, of these three indicators, cresol red would be expected to give the most accurate end point and lowest titration error for this analysis.

Graphical Methods. If you have difficulty in locating the end point by pH measurements, one type of graph that may help is a derivative plot. This type of plot makes use of the pH versus volume measurements that are obtained during a regular acid–base titration. However, we now use this information to calculate and plot the slope (or first derivative, $d\text{pH}/dV_{\text{Titrant}}$) of our

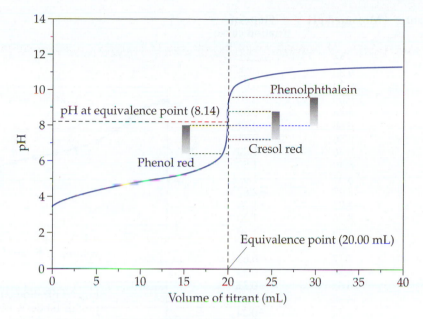

FIGURE 12.9 Use of various indicators for the analysis of a 10.00 mL solution of 0.01000 M acetic acid by titrating this solution with 0.005000 M NaOH. The pH range over which each of these indicators undergoes a color change is from pH 6.4 to 8.0 for phenol red (19.55–19.99 mL titrant), pH 7.2 to 8.8 for cresol red (19.93–20.04 mL), and pH 8 to 9.6 to phenolphthalein (19.99–20.04 mL).

data versus the added volume of titrant. An example of this type of plot is shown in Figure 12.10 for a situation in which the "alkalinity" of a water sample is being determined by looking at its content of carbonate and bicarbonate ions. The change in pH at the first equivalence point in this curve (due to the titration of carbonate) is too small to provide a suitably sharp end point for determination by a normal plot of pH versus titrant volume or through the use of an acid–base indicator. When a plot of the slope is made using the given data, it is now possible to locate this end point as well as the second end point (due to the titration of bicarbonate). These end points are represented by the two sharp peaks that appear in the plot of the slope. The top of each of these peaks, where the slope goes through a maximum and changes from an increasing to a decreasing value, is called an *inflection point*. If the analyte is a weak acid or a weak base, a second inflection point will also be present at the minimum slope in the middle of the titration curve, as occurs between titrant volumes of 5 and 6 mL in Figure 12.10 during the titration of bicarbonate.

A second type of graph that can also be used to locate and provide an accurate estimate of the equivalence point is a **Gran plot**.[5,11,12] This graph is constructed by plotting a special function of pH versus the volume of titrant to give a linear response for the titration curve. For the titration of a weak acid, this type of graph is made by plotting $V_B \, 10^{-pH}$ versus V_B as based on Equation 12.4 (see derivation in Table 12.4), where

V_E is the volume of titrant needed to reach the equivalence point.

Gran Plot for a Weak Acid Titrated with a Strong Base:

$$V_B \cdot 10^{-pH} = K_a{}^\circ \cdot \frac{(\gamma_{HA})}{(\gamma_{A^-})} \cdot (V_E - V_B) \qquad (12.4)$$

A similar relationship and plot can be used for the titration of a weak base by a strong base.

Gran Plot for a Weak Base Titrated with a Strong Acid:

$$V_A \cdot 10^{pH} = \frac{(\gamma_B)}{K_a{}^\circ (\gamma_{BH^+})} \cdot (V_E - V_A) \qquad (12.5)$$

An example of a Gran plot for a weak acid is given in Figure 12.11. According to Equation 12.4, the x-intercept for the linear portion of this plot should give the value for V_B that is equal to V_E. The acid dissociation constant for the system can be obtained from the slope, which will be equal to $K_a{}^\circ \cdot (\gamma_{HA})/(\gamma_{A^-})$.

A nice feature of a Gran plot is that it allows more than just one or a few data points to be used to determine the position of the equivalence point. It is also not necessary to acquire data directly in the region of the equivalence point to determine this value. These are important advantages when compared to end-point detection methods that are based on acid–base indicators or the use of a traditional titration curve and results in a more precise and robust technique for locating the equivalence point.

Volume of titrant (mL)	Measured pH	Slope of titration curve
0.00	9.35	———
0.20	9.30	−0.25
0.50	9.20	−0.33
1.00	9.00	−0.40
1.25	8.86	−0.56
1.50	8.68	−0.72
1.66	8.56	−0.75
1.76	8.46	−1.00
1.80	8.40	−1.50
1.83	8.35	−1.67
1.86	8.27	−2.67
1.90	8.22	−1.25
1.95	8.16	−1.20
2.05	8.05	−1.10
2.50	7.80	−0.56
3.50	7.17	−0.63
4.50	6.84	−0.33
5.50	6.63	−0.21
6.50	6.38	−0.25
7.50	6.14	−0.24
8.50	5.85	−0.29
9.50	5.39	−0.46
10.00	4.85	−1.08
10.08	4.68	−2.13
10.11	4.60	−2.67
10.14	4.50	−3.33
10.17	4.43	−2.33
10.20	4.40	−1.00
10.25	4.35	−1.00
10.35	4.24	−1.10
10.50	4.12	−0.80
10.70	3.99	−0.65

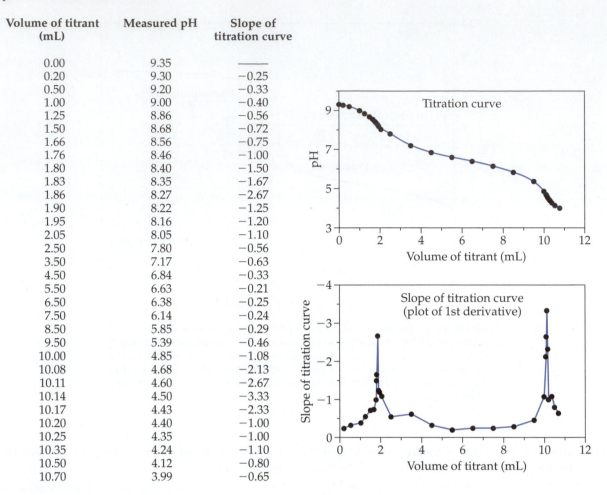

FIGURE 12.10 A titration curve and a plot of the slope (also known as the "first derivative") of this titration curve. The example used in this case is the titration of a natural water sample with a strong acid to determine the carbonate and bicarbonate content of this sample as a measure of the water's alkalinity. The titration of carbonate is represented by the first end point. The second end point represents the titration of bicarbonate in the original sample plus bicarbonate that was produced during the titration of carbonate when reaching the first end point. (This spreadsheet and plot are based on data obtained from the U.S. Geological Survey.)

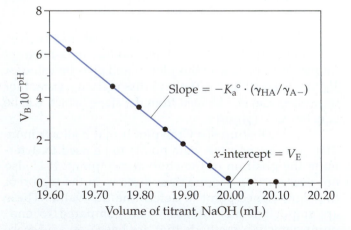

FIGURE 12.11 Example of a Gran plot for the titration of a weak acid with a strong base. This particular example is for the titration of a 10.00 mL aliquot of 0.01000 M acetic acid with 0.005000 M NaOH. The x-intercept of the best-fit line provides an estimate of V_E, the volume of titrant that is needed to reach the equivalence point.

12.3 PREDICTING AND OPTIMIZING ACID–BASE TITRATIONS

12.3A Describing Acid–Base Titrations

Now that we have discussed the basic concept of an acid–base titration, let's look at this method more closely to see how we can describe and optimize this technique. First, we need to think more about the reaction that is actually occurring during the titration. For a titration in water that involves a strong acid and strong base, the overall reaction is the neutralization of a hydronium ion with a hydroxide ion, as often represented by the reaction of H^+ with OH^-.

$$H_3O^+ + OH^- \rightleftharpoons 2\,H_2O$$

$$(\text{or} \quad H^+ + OH^- \rightleftharpoons H_2O) \tag{12.6}$$

We learned in Chapter 8 that the equilibrium constant for this process is quite large and lies heavily to the right side of the reaction, with a value of $1/K_w = 1.0 \times 10^{14}$ at

TABLE 12.4   Gran Equation for a Weak Acid (HA) Titrated with a Strong Base (OH–)

Titration Reaction:

$$HA + OH^- \rightarrow A^- + H_2O$$

$K_a°$ Expression for Weak Acid:

$$HA \rightleftharpoons H^+ + A^-$$

$$K_a° = \frac{(a_{H^+})(a_{A^-})}{(a_{HA})} = \frac{(\gamma_{H^+})[H^+](\gamma_{A^-})[A^-]}{(\gamma_{HA})[HA]}$$

Mass Balance Equations:

$$\text{Mol } A^- \text{ produced} = \text{Mol } OH^- \text{ added} \quad \Rightarrow \quad [A^-] = \frac{C_B V_B}{V_A + V_B}$$

$$\text{Mol } HA \text{ remaining} = \text{Mol original } HA - \text{Mol } OH^- \text{ added} \quad \Rightarrow \quad [HA] = \frac{C_A V_A - C_B V_B}{V_A + V_B}$$

Substitute Mass Balance Equations into $K_a°$ Expression; Eliminate Common Terms:

$$\Rightarrow K_a° = \frac{(\gamma_{H^+})[H^+](\gamma_{A^-})(C_B V_B)/(V_A + V_B)}{(\gamma_{HA})(C_A V_A - C_B V_B)/(V_A + V_B)} = \frac{(\gamma_{H^+})[H^+](\gamma_{A^-})(C_B V_B)}{(\gamma_{HA})(C_A V_A - C_B V_B)}$$

Rearrange to Solve in Terms of $V_B(\gamma_{H^+})[H^+]$; Substitute in $10^{-pH} = (\gamma_{H^+})[H^+]$:

$$\Rightarrow \quad V_B \cdot (\gamma_{H^+})[H^+] = K_a° \cdot \frac{(\gamma_{HA})(C_A V_A - C_B V_B)}{(\gamma_{A^-})C_B}$$

$$\Rightarrow \quad V_B \cdot 10^{-pH} = K_a° \cdot \frac{(\gamma_{HA})(C_A V_A - C_B V_B)}{(\gamma_{A^-})C_B}$$

Substitute in $V_E = (C_A V_A)/C_B$, where V_E is the Titrant Volume at the Equivalence Point:

$$\therefore \quad V_B \cdot 10^{-pH} = K_a° \cdot \frac{(\gamma_{HA})}{(\gamma_{A^-})} \cdot (V_E - V_B) \qquad (12.4)$$

25°C. This reaction is also rapid, with the hydronium ion reacting with a hydroxide ion essentially as fast as these two ions diffuse together in solution. As a result, this reaction appears to occur almost instantaneously as the titrant is added to the sample.

If a titration involves either a weak acid or weak base, the K_a or K_b values for these substances must be considered. To illustrate this, let's examine the reaction of the weak acid HF with the strong base NaOH, as represented by the following net reaction and equilibrium constant,

$$HF + OH^- \rightleftharpoons H_2O + F^- \qquad K = \frac{(1)[F^-]}{[HF][OH^-]} \qquad (12.7)$$

where the activity of water in the expression for K is given a value of $a_{H_2O} = 1$ because water is the solvent for this reaction and is in its standard state. To obtain a value for this equilibrium constant, we can multiply both the top and bottom of Equation 12.7 by $[H^+]$ and reorganize our terms to get the following expression.

$$K = \frac{[H^+][F^-]}{[HF][OH^-][H^+]}$$

$$= \frac{[H^+][F^-]}{[HF]} \cdot \frac{1}{[OH^-][H^+]}$$

$$= K_{a,HF}/K_w \qquad (12.8)$$

This new expression makes it possible for us to calculate the equilibrium constant for this titration reaction by using the known values for $K_{a,HF}$ (6.8×10^{-4}) and K_w (1.0×10^{-14}), giving $K = (6.8 \times 10^{-4})/(1.0 \times 10^{-14}) = 6.8 \times 10^{10}$.

The equilibrium constant we calculated for the titration of HF with NaOH is relatively large, but is much smaller than the value of 1.0×10^{14} that would be present for the titration of a strong acid with a strong base in water. This leads us to the question "How large must K be for an acid–base titration in water to be practical?" We saw earlier in Figure 12.6 that an acid must have a K_a value greater than 1.0×10^{-10} (or pK_a less than 10.0) to give a titration curve that has an easy-to-detect end point when using a strong base in water as the titrant. Based on this cutoff value for K_a, we get an overall equilibrium constant (K) of at least $K_a/K_w = (1.0 \times 10^{-10})/(1.0 \times 10^{-14}) = 1.0 \times 10^4$ that is needed for such a titration. As we see in the following exercise, the same minimum value for K is required for the titration of a base by a strong acid in water.

Determining the Equilibrium Constant for the Titration of a Weak Base

A chemist working at a fertilizer production plant wishes to titrate ammonia by using HCl as the titrant. What is the equilibrium constant that would be expected for this titration reaction?

SOLUTION

The overall reaction and concentration-based equilibrium constant for the titration are shown below, where an activity of one is used in the equilibrium expression for water (the solvent).

$$H_3O^+ + NH_3 \rightleftharpoons NH_4^+ + H_2O$$

$$K = \frac{[NH_4^+] \, a_{H_2O}}{[NH_3][H_3O^+]} = \frac{[NH_4^+]}{[NH_3][H_3O^+]}$$

We can rewrite this equilibrium expression in terms of the K_b value for ammonia and K_w by multiplying the top and bottom of this expression by $[OH^-]$.

$$K = \frac{[NH_4^+][OH^-]}{[NH_3][H_3O^+][OH^-]}$$

$$= \frac{[NH_4^+][OH^-]}{[NH_3]} \cdot \frac{1}{[OH^-][H_3O^+]} = K_{b,NH_3}/K_w$$

If we now use this new expression along with the known values for K_{b,NH_3} (1.75×10^{-5}) and K_w, we get an equilibrium constant for this titration reaction of $K = (1.75 \times 10^{-5})/(1.0 \times 10^{-14}) = \mathbf{1.75 \times 10^9}$. This result is smaller than the maximum value of $K = 1/K_w = 1.0 \times 10^{14}$ that would be expected for a strong base titrated with a strong acid like HCl in water. However, an equilibrium constant of 1.75×10^9 is still much larger than the minimum value $K = K_b/K_w = (1.0 \times 10^{-10})/(1.0 \times 10^{-14}) = 1.0 \times 10^4$ needed to make this titration feasible, based on a lower cutoff for K_b of 1.0×10^{-10} (or a pK_b less than 10.0), as noted earlier in Figure 12.6.

12.3B Titration Curves for Strong Acids and Bases

In the case of an acid–base titration, the titration curve is usually a graph of pH for the sample/titrant mixture versus the volume of titrant that has been added. When titrating an acid with a base, it should come as no surprise that the pH will increase as the base is added to the acid. Determining the exact extent of this change is important to consider if we are to determine the best conditions for conducting the titration and for detecting the equivalence point. This can be accomplished by using the tools we learned in Chapter 7 for describing acid–base reactions. We will first see how such methods can be used

to predict the shape of a titration curve for the combination of a strong acid and strong base and will then move on to look at titrations of weak acids and bases.

Titration of a Strong Acid. The analysis of strong acids like HCl or HNO_3 is important in industry for determining the purity and concentration of these reagents. A titration of these strong acids is generally carried out by using a strong base as the titrant. The general shape of the titration curve that will be obtained in water is shown in Figure 12.12. This titration curve can be divided into four distinct regions based on how much titrant has been added to the sample: (1) the original sample, (2) the middle of the titration in which only some of the analyte has been titrated, (3) the equivalence point, and (4) the region beyond the equivalence point in which excess titrant is present. We will see later that these same four regions are also present in any other type of acid–base titration.

The *first region* for the titration in Figure 12.12 occurs before the titration even begins (a point we will refer to as "0% Titration"). This is simply a solution of a strong acid, so the pH in this situation will be given by the hydrogen ion content of the original sample. If the total concentration of the acid is greater than 10^{-6} M (allowing us to ignore contributions due to water, as discussed in Chapter 8), the concentration of H^+ will be equal to the total concentration of the acid (C_{HA}).

Original Sample (0% Titration): $\quad [H^+] = C_{HA}$ (12.9)

Once we use this equation to get $[H^+]$, we can calculate the pH by using the approximate relationship $pH \approx -\log[H^+]$. (*Note:* Remember that this last equation is an approximation because pH is actually based on activity rather than concentration.)

The *second region* present in an acid–base titration occurs when some titrant has been added to the sample but we have not reached the equivalence point. For the titration of a strong acid with a strong base in water, each mole of added base produces one mole of OH^-, which then reacts with one mole of the strong acid (or, more specifically, with H_3O^+/H^+ ions that are produced by the strong acid in water). If we can ignore the pH contribution of water in this reaction (as occurs if $C_{HA} > 10^{-6}$ M), the moles of remaining H^+ ions from the acid at any point in this region will be equal to the initial moles of acid in the sample ($C_{HA}V_{HA}$) minus the moles of added base ($C_B V_B$), where C_B is the total concentration of the base in the titrant, V_{HA} is the volume of the original sample, and V_B is the volume of the added titrant. We can then find the concentration of these hydrogen ions by dividing this difference in moles by the total volume of the sample/titrant mixture ($V_{HA} + V_B$), as shown below.

Middle of Titration (1–99% Complete):

$$[H^+] = \frac{C_{HA}V_{HA} - C_B V_B}{V_{HA} + V_B}$$ (12.10)

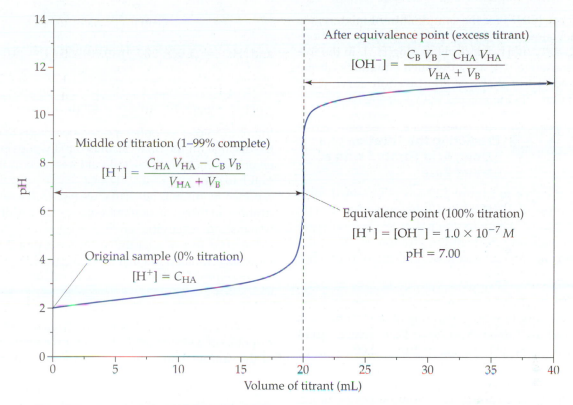

FIGURE 12.12   Equations for predicting the response for the titration of a strong monoprotic acid with a strong base in water. These equations assume that the ionic strength is sufficiently low to allow concentrations to be used in place of chemical activities. The effects of water dissociation are also not considered except for the equation that is given at the equivalence point. All terms used in these equations are defined in the text.

The pH of the sample/titrant mixture is then estimated by using this calculated value for $[H^+]$. It is important to remember when you are doing this calculation to use proper units to obtain the proper dimensions in the final answer. In this case, it is recommended that C_{HA} and C_B be given in units of molarity and that V_{HA} and V_B be expressed in units of liters, even if the original volumes are given in milliliters.

The *third region* in a titration curve occurs at the equivalence point. At this point, exactly enough strong base in Figure 12.12 has been added to react with all of the strong acid that was originally present. Under these conditions the only source of hydrogen ions or hydroxide ions is now the water that is used as the solvent for this titration. To determine the pH at this point, we can simply use the K_w expression for water ($K_w = [H^+][OH^-] = 1.0 \times 10^{-14}$ at 25°C) and the fact that one hydrogen ion is formed for each hydroxide ion if there is no other source of either ion in the solution. When we put this information together, we get the following result.

Equivalence Point (100% Titration):

$$[H^+] = [OH^-] = \sqrt{K_w} = 1.0 \times 10^{-7}\ M$$

or

$$pH \approx -\log(1.0 \times 10^{-7}) = 7.00 \qquad (12.11)$$

As this result indicates, the pH for the titration of a strong acid with a strong base will always be equal to 7.00 at the equivalence point. This number is useful to keep in mind when we are planning such a titration and determining when the equivalence point has been reached or are selecting an acid–base indicator for detecting the end point of a titration (see Section 12.2C).

The *fourth region* in a titration curve occurs after the equivalence point. In this region, all of the original analyte has been consumed and we are now simply adding excess titrant to the sample. In the case of a strong acid that is being titrated with a strong base, the moles of excess base that have been added after the equivalence point will be equal to the difference in the original moles of acid and added base, or $(C_B V_B) - (C_{HA} V_{HA})$. We also know the total volume of the sample/titrant mixture $(V_{HA} + V_B)$, which allows us to calculate the concentration of hydroxide ions that will be present for any mixture of titrant and sample after the equivalence point.

After Equivalence Point (Excess Titrant):

$$[OH^-] = \frac{C_B V_B - C_{HA} V_{HA}}{V_{HA} + V_B} \qquad (12.12)$$

This equation is based on the same general approach that we used in Chapter 8 to determine the hydroxide ion concentration of a strong base in water.

Once we know [OH⁻] for the sample/titrant mixture, we can find the value for [H⁺] and pH by using the relationships $[H^+] = K_w/[OH^-]$ and $pH \approx -\log[H^+]$. In the following exercise we will see how such equations can be used to predict the shape of the titration curve for a strong acid that is being reacted with a strong base.

EXERCISE 12.5 | **Predicting the Titration of a Strong Acid Titrated with a Strong Base**

The titration curve in Figure 12.12 was obtained for a 10.00 mL sample of 0.01000 M HCl that was titrated with 0.005000 M NaOH. Estimate the pH at the beginning of the titration, half-way through the titration, at the equivalence point, and after an excess of 10.00 mL of titrant has been added to the sample.

SOLUTION

The pH of the sample at the beginning of the titration can be calculated by using Equation 12.9, which gives $[H^+] = 0.010\ M$ and $pH \approx -\log(0.010) = \mathbf{2.00}$.

To find the pH half-way through the titration, we first need to calculate what volume of titrant is needed to reach the equivalence point. This volume can be determined by using Equation 12.2 and the known stoichiometry for the reaction between HCl and NaOH.

$$\text{mol HCl} = \text{mol NaOH} \quad \text{or} \quad C_{HCl}V_{HCl} = C_{NaOH}V_{NaOH}$$

By rearranging this relationship we get $V_{NaOH} = C_{HCl}$ $V_{HCl}/C_{NaOH} = (0.005000\ M)(0.02500\ L)/\ (0.01000\ M) = 0.02000$ L or 20.00 mL. This means that the volume needed to go half-way through the titration will by (20.00 mL)/2 = 10.00 mL. The pH at a point half-way through the titration can then be determined by using Equation 12.10.

$$[H^+] = \frac{(0.01000\ M\ \text{HCl})(0.01000\ L\ \text{HCl}) -}{0.01000\ L\ \text{HCl} + 0.01000\ L\ \text{NaOH}}$$

$$\frac{(0.005000\ M\ \text{NaOH})(0.01000\ L\ \text{NaOH})}{0.01000\ L\ \text{HCl} + 0.01000\ L\ \text{NaOH}}$$

$$\therefore [H^+] = 0.002500\ M \text{ or } pH \approx -\log(0.002500) = \mathbf{2.60}$$

Because this is a titration of a strong acid with a strong base, we also know from Equation 12.11 that **pH = 7.00** at the equivalence point for this titration.

If 20.00 mL of titrant is required to reach the equivalence point, an additional 10.00 mL of titrant will result in the total addition of 30.00 mL of titrant. The pH at this point can be determined using Equation 12.12.

$$[OH^-] = \frac{(0.005000\ M\ \text{NaOH})(0.03000\ L\ \text{NaOH})}{0.03000\ L\ \text{NaOH} + 0.01000\ L\ \text{HCl}} -$$

$$\frac{(0.01000\ M\ \text{HCl})(0.01000\ L\ \text{HCl})}{0.03000\ L\ \text{NaOH} + 0.01000\ L\ \text{HCl}}$$

$$\therefore [OH^-] = 0.001250\ M$$

and $[H^+] = (1.0 \times 10^{-14})/(0.001250\ M) = 8.0 \times 10^{-12} M$

or

$$pH \approx -\log(8.0 \times 10^{-12}) = \mathbf{11.10}$$

In this example, we only reported two digits to the right of the decimal for the final calculated pH values even though all of the concentrations and volumes used in this calculation had four significant figures. The reason the pH was reported in this manner is that we have made some simplifications in these calculations (such as the use of concentration instead of activities and ignoring water's contribution to the pH before or after the equivalence point), which provides an estimated pH that is often only accurate to within two significant digits in this type of calculation.

We can see that it is relatively easy to estimate the response of a strong acid titration if we know the volume of the sample, know the concentration of the titrant, and have a ballpark value for the expected concentration of the sample. This process also makes it possible to determine the volumes of titrant we will have to use for the analysis and the response that would be expected at each of these volumes. If desired, we can perform such calculations manually and use them to prepare a plot of the expected curve to help in the optimization of the method. It is important, however, to note that there are some limitations to the use of Equations 12.9–12.12 for this purpose. In Section 12.3D we examine a more general approach using spreadsheets and a single equation to estimate the pH throughout all four regions of the titration curve.

Titration of a Strong Base. A strong base like NaOH is also important in both an industrial and routine laboratory setting. This analysis is typically conducted by titrating the strong base with a strong acid and results in a titration curve like the one in Figure 12.13. The pH at different regions in this curve can be estimated by using a similar approach to that used for the titration of a strong acid with a strong base. The pH of the original sample can be determined by using the approach given in Chapter 8 for solution of a strong base ($C_B > 10^{-7} M$), as represented by Equation 12.13.

Original Sample (0% Titration): $\quad [OH^-] = C_B$ (12.13)

We can then determine the values of [H⁺] and pH for this sample by using $[H^+] = K_w/[OH^-]$ and $pH \approx -\log[H^+]$.

The next region of the titration curve is when we have added some acid but have not added enough to reach the equivalence point. In this region, one mole of hydroxide ion from the original base will be removed for each mole of hydrogen ion that is added in the titrant. We can find the concentration of the remaining hydroxide ions by calculating the difference in the initial moles of base versus the moles of added acid ($C_B V_B - C_{HA} V_{HA}$)

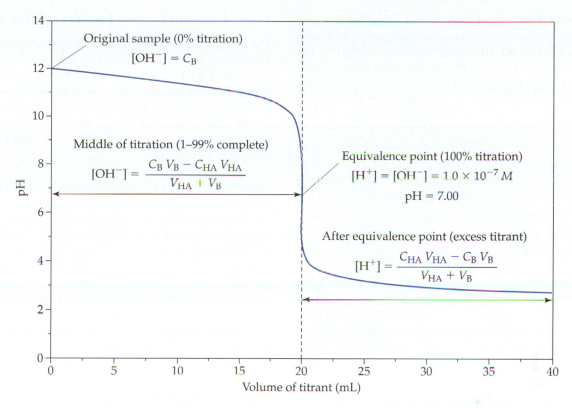

FIGURE 12.13 Equations for predicting the response for the titration of a strong monoprotic base with a strong acid in water. The assumptions made in these equations are the same as made in Figure 12.11 for the titration of a strong acid with a strong base. All terms used in these equations are defined in the text.

and dividing this by the total volume of the titrant and sample ($V_{HA} + V_B$).

Middle of Titration (1−99% Titration):

$$[OH^-] = \frac{(C_BV_B - C_{HA}V_{HA})}{(V_{HA} + V_B)} \qquad (12.14)$$

The resulting value for $[OH^-]$ is used to find $[H^+]$ through $[H^+] = K_w/[OH^-]$, which is then used to calculate the pH of the sample/titrant mixture.

The titration of a strong base with a strong acid is also similar to the titration of a strong acid with a strong base with regards to the pH at the equivalence point. In this situation, the original base has been consumed and no additional strong acid has been added, so water is the only remaining source of H^+ and OH^-. This gives the following result when we use the K_w expression for water and the fact that one hydrogen ion is formed for each hydroxide ion from water:

Equivalence Point (100% Titration):

$$[H^+] = [OH^-] = 1.0 \times 10^{-7}\,M$$

or

$$pH \approx -\log(1.0 \times 10^{-7}) = 7.00 \qquad (12.15)$$

This is exactly the same result we obtained in Equation 12.11 for a strong acid titration, where the equivalence point also had a pH of 7.0.

After the equivalence point in this titration, the base from the sample has now all been neutralized and we are simply adding extra titrant. To determine the pH of this mixture, we have to calculate the moles of excess acid that are present (an amount equal to $C_{HA}V_{HA} - C_BV_B$) and divide this by the total volume of the sample/titrant mixture ($V_{HA} + V_B$).

After Equivalence Point (> 101% Titration):

$$[H^+] = \frac{(C_{HA}V_{HA} - C_BV_B)}{(V_{HA} + V_B)} \qquad (12.16)$$

Once we have $[H^+]$, we can determine the pH by using $pH \approx -\log[H^+]$. This is essentially the same approach we used in Chapter 8 to determine the pH for a simple solution of a strong acid.

EXERCISE 12.6   **Predicting the Titration of a Strong Base with a Strong Acid**

The titration that is shown in Figure 12.13 is just the opposite of that illustrated for the acid–base titration in Figure 12.12 in that we are now titrating a 10.00 mL sample of 0.01000 *M* NaOH with 0.005000 *M* HCl. Estimate the pH at the beginning of this titration, after 10.00 mL of titrant has been added, at the equivalence point, and after an excess of 10.00 mL of titrant has been added to the sample. How do these results compare with those for the titration of HCl in Figure 12.12?

SOLUTION

The pH of the NaOH sample at the beginning of this titration can be found by using Equation 12.13, where $[OH^-] = C_B = 0.005000\ M$. This results in a calculated value of $[H^+] = K_w/[OH^-] = 1.0 \times 10^{-14}/(0.005000\ M) = 1.0 \times 10^{-12}\ M$ or pH = **12.00** at the beginning of the titration. We also know from Equation 12.15 that at the equivalence point the pH will be equal to **7.00**.

To find the pH when 10.00 mL of titrant has been added, we need to determine what volume of titrant is needed to reach the equivalence point. This volume can be determined by using Equation 12.2, which gives $V_{HCl} = C_{NaOH}V_{NaOH}/C_{HCl} = (0.005000\ M)(0.01000\ L)/(0.01000\ M) = 0.02000\ L$ or 20.00 mL. This means that the addition of 10.00 mL titrant represent a point at 50% titration. The pH at this point (after the addition of 10.00 mL titrant) can then be found by using Equation 12.14.

$$[OH^-] = \frac{(0.01000\ M\ NaOH)(0.01000\ L\ NaOH)}{0.01000\ L\ HCl + 0.01000\ L\ NaOH} -$$

$$\frac{(0.005000\ M\ HCl)(0.01000\ L\ HCl)}{0.01000\ L\ HCl + 0.01000\ L\ NaOH}$$

$$\therefore [OH^-] = 0.002500\ M \text{ and}$$

$$[H^+] = (1.00 \times 10^{-14})/(0.002500\ M) = 4.00 \times 10^{-12}\ M$$

or

$$pH \approx -\log(4.00 \times 10^{-12}) = \mathbf{11.40}$$

We now know that a volume of 20.00 mL titrant represents the equivalence point, so an excess of 10.00 mL titrant would be equal to a total of 30.00 mL of titrant that has been added to the sample. Equation 12.16 can be used to find the pH under these conditions (at 10.00 mL of excess titrant).

$$[H^+] = \frac{(0.005000\ M\ HCl)(0.03000\ L\ HCl)}{0.03000\ L\ HCl + 0.01000\ L\ NaOH} -$$

$$\frac{(0.01000\ M\ NaOH)(0.01000\ L\ NaOH)}{0.03000\ L\ HCl + 0.01000\ L\ NaOH}$$

$$\therefore [H^+] = 0.001250\ M \text{ or } pH \approx -\log(1.0 \times 10^{-7}) = \mathbf{2.90}$$

All of these values show good agreement with the pH that is seen at the same locations in the titration curve in Figure 12.13. Doing these calculations at strategic points during a titration can provide a good estimate of the shape of the overall titration curve that will be obtained.

12.3C Titration Curves for Weak Acids and Bases

The next titrations we will consider are those that involve samples of weak acids or weak bases. There are many more weak acids and bases than strong acids and bases, so this is a relatively common situation encountered in acid–base titrations. To perform such titrations, either a strong base is used as the titrant (in the case where the analyte is a weak acid) or the titrant is a strong acid (if the analyte is a weak base).

Titration of a Weak Monoprotic Acid. The titration of a weak acid by the addition of a strong base results in a titration curve like the one shown in Figure 12.14. This particular example is for the titration of acetic acid, with NaOH being used as the titrant. Along with being an important component of many food products (e.g., vinegar and pickling solutions), acetic acid is also an important reagent and solvent used for the preparation of many other chemical compounds. We learned in Chapter 8 that the strength of a weak acid like acetic acid is described by its acid dissociation constant (K_a), where $K_a = [H^+][A^-]/[HA]$. We also saw in Chapter 8 how this expression can be used to obtain the hydrogen ion concentration for a solution of a weak monoprotic acid. This type of calculation gives us the initial pH for a weak acid sample before we begin its titration.

Original Sample (0% Titration): $K_a = \dfrac{[H^+]^2}{C_{HA} - [H^+]}$

$$\text{or}\quad 0 = [H^+]^2 + K_a[H^+] - K_aC_{HA} \tag{12.17}$$

Thus, if we have an estimate of the total concentration for the weak acid (C_{HA}) and we know the acid dissociation constant for this acid, we can use Equation 12.17 along with the quadratic formula to solve for $[H^+]$ and the pH of the original sample.

As we begin to titrate this sample with a strong base, we will convert some of our original weak acid (HA) into its conjugate base (A^-). If we assume that each mole of added base converts one mole of HA into A^- (an assumption generally valid over the region of 5–95% titration), we can determine the pH of the sample/titrant mixture by using the Henderson–Hasselbalch equation, as shown in Equation 12.18.

Middle of Titration (5−95% Titration):

$$pH = pK_a + \log\left(\frac{[A^-]}{[HA]}\right)\quad \text{or}$$

$$pH \approx pK_a + \log\left(\frac{C_BV_B}{C_{HA}V_{HA} - C_BV_B}\right) \tag{12.18}$$

When using this equation to describe the titration of a weak acid, the ratio $[A^-]/[HA]$ is approximately equal

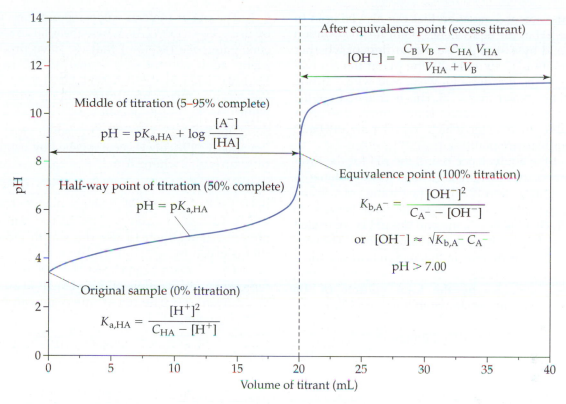

FIGURE 12.14   Equations for predicting the response for the titration of a weak monoprotic acid (HA) with a strong base in water. The assumptions made in these equations are the same as are made in Figure 12.11 for the titration of a strong acid with a strong base. All terms used in these equations are defined in the text.

to $[(C_B \cdot V_B)/(V_B + V_{HA})]/[(C_{HA} \cdot V_{HA} - C_B \cdot V_B)/(V_B + V_{HA})]$. This term can be further simplified to the ratio $(C_B \cdot V_B)/(C_{HA} \cdot V_{HA} - C_B \cdot V_B)$, because the term $(V_B + V_{HA})$ that appears in both the numerator and denominator will cancel out. Thus, all you need to know when using Equation 12.18 to describe a weak-acid titration curve is the pK_a of the weak acid and the concentrations and volumes of the acid and base solutions. A special case for this equation occurs at the half-way point for the titration, at which the concentrations of the weak acid and its conjugate base will be equal and the term $\log([A^-]/[HA])$ in Equation 12.18 will be zero. Under these conditions the pH of the sample/titrant mixture will be the same as the pK_a for the weak acid. This is useful to know in that it gives us one easy point (at 50% titration) to estimate the pH when predicting the shape of a titration curve for a weak acid. This also means we can use the pH at 50% titration for a weak acid as a means for determining the pK_a. This part of the titration curve for a weak acid is often called the *buffer region* because only a small change in pH is produced with a large change in the volume of added titrant.

At the equivalence point of a weak acid titration, we have added enough base in the titrant to exactly neutralize the weak acid in the original sample. This means we have converted essentially all of our original weak acid into its conjugate base (A^-). The pH at this point can be

found by using the approach described in Chapter 8 for a solution of a weak base, as shown below.

Equivalence Point (100% Titration): $K_b = \dfrac{[OH^-]^2}{C_{A^-} - [OH^-]}$

or $0 = [OH^-]^2 + K_b[OH^-] - K_b C_{A^-}$ (12.19)

In this case, the total concentration of conjugate base that we have produced from the original acid will be determined by the original moles of the weak acid that were in the sample and the total volume of the sample plus titrant at the equivalence point, or $C_{A^-} = (C_{HA} V_{HA})/(V_B + V_{HA})$. We can use this concentration along with the known value of K_{b,A^-} for this conjugate base (where $K_{b,A^-} = K_w/K_{a,HA}$) to solve for $[OH^-]$ in Equation 12.19. This is accomplished by using the quadratic formula, where $x = [OH^-]$, $A = 1$, $B = K_{b,A^-}$ and $C = -K_{b,A^-} C_{A^-}$. We can then use the value that we obtain for $[OH^-]$ to calculate $[H^+]$ and the pH.

It is almost always true during the titration of a weak acid with a strong base like hydroxide that the concentration of the resulting conjugate base will be much greater than the hydroxide concentration that results from the dissociation of water. In this situation, Equation 12.19 can be simplified to the following form.

$$K_{b,A^-} \approx \frac{[OH^-]^2}{C_{A^-}} \quad \text{or} \quad [OH^-] \approx \sqrt{K_{b,A^-} C_{A^-}} \quad (12.20)$$

This gives us a second equation that can also be used to estimate the pH at the equivalence point for the titration of a weak acid by a strong base. Regardless of whether you use Equation 12.19 or 12.20, the pH at the equivalence point for the titration of a weak acid with a strong base should be greater than 7.00. This is true because the product of a weak acid titration is now the acid's conjugate base rather than water, which creates a slightly basic pH at the equivalence point.

Beyond the equivalence point, the pH for the titration of a weak acid is calculated in an identical manner to that described in Section 12.3B for the titration of a strong acid. These two situations are now the same because in both cases the acid is now gone and the pH is controlled by the excess of strong base that is being added. The concentration of hydroxide ions in this region is again calculated by finding the moles of excess base and dividing this number of moles by the total volume of the sample/titrant mixture, as shown in Equation 12.21.

After Equivalence Point (> 101% Titration):

$$[OH^-] = \frac{C_B V_B - C_{HA} V_{HA}}{V_{HA} + V_B} \quad (12.21)$$

After we have obtained the concentration of hydroxide, we can determine the value for $[H^+]$ and pH by using the relationships $[H^+] = K_w/[OH^-]$ and $pH \approx -\log[H^+]$, in the same manner as illustrated in Section 12.3B for the titration of a strong acid with a strong base.

EXERCISE 12.7 **Predicting the Titration of a Weak Acid with a Strong Base**

The titration curve given in Figure 12.14 was obtained for a sample that contains $1.0 \times 10^{-2} M$ acetic acid ($K_a = 1.7 \times 10^{-5}$ at 25°C) that is titrated with 0.005000 M NaOH. Determine the pH that would be expected during this titration for the original sample, at the half-way point of the titration, at the equivalence point, and after an excess of 10.00 mL titrant has been added to the sample.

SOLUTION

The pH of the sample at the beginning of the titration can be calculated by using Equation 12.17, where $0 = [H^+]^2 + K_a[H^+] - K_a C_{HA}$. We can use the quadratic formula to solve for $[H^+]$ in this equation by letting $x = [H^+]$, $A = 1$, $B = 1.7 \times 10^{-5}$, and $C = -(1.7 \times 10^{-5})(1.0 \times 10^{-2} M)$. This process gives a final answer for $[H^+]$ of $4.04 \times 10^{-4} M$, or pH = **3.39**.

We know that the half-way point of this titration will occur when $pH = pK_a$, giving $pH = -\log(1.7 \times 10^{-5}) =$ **4.77** under these conditions. The volume of titrant that is needed to reach the equivalence for this titration is $V_{NaOH} = C_{HA} V_{HA}/C_{NaOH} = (0.00500\ M)(0.01000\ L)/(0.005000\ M) = 0.02000$ L or 20.00 mL. Thus, the half-way point of this titration will occur when the volume of added titrant is $(0.50)(20.00\ \text{mL}) = 10.00$ mL.

The pH at the equivalence point can be estimated by using Equation 12.19, where the total concentration of conjugate base (acetate) that we have produced from acetic acid will be $C_{A^-} = (0.01000\ M\ HA)(0.01000\ L\ HA)/(0.01000\ L\ HA + 0.02000\ L\ NaOH) = 0.00333\ M$. We can use this concentration along with the known value for K_{b,A^-} of acetate ($K_w/K_{a,HA} = 1.0 \times 10^{-14}/6.8 \times 10^{-4} = 1.47 \times 10^{-11}$) to solve for $[OH^-]$ in Equation 12.19. This is accomplished by using the quadratic formula, where $A = 1$, $B = K_{b,A^-} = 1.47 \times 10^{-11}$, and $C = -K_{b,A^-} C_{A^-} = -4.9 \times 10^{-14}$. This gives us $[OH^-] = 2.21 \times 10^{-7}\ M$, which means that $[H^+] = 4.52 \times 10^{-8}\ M$ or pH = **7.35**. The same result is obtained if we instead use the simplified expression that is given in Equation 12.20.

The addition of an excess of 10.00 mL of titrant will represent a total volume of titrant of 10.00 mL + 20.00 mL in this example. The pH at this point can be found by using Equation 12.21.

$$[OH^-] = \frac{(5.00 \times 10^{-3}\ M\ NaOH)(0.03000\ L\ NaOH) -}{0.03000\ L\ NaOH + 0.01000\ L\ HA}$$

$$\frac{(1.00 \times 10^{-2}\ M\ HA)(0.01000\ L\ HA)}{0.03000\ L\ NaOH + 0.01000\ L\ HA}$$

$$\therefore [OH^-] = 0.00125\ M$$

$$\text{and } [H^+] = (1.0 \times 10^{-14})/(0.00125\ M)$$

$$= 8.00 \times 10^{-12}\ M$$

or pH = **11.10**. This last result is the same as we obtained in Exercise 12.6 for the titration of an identical concentration and volume of a strong acid with the same concentration of NaOH as the titrant.

If we compare the curves in Figures 12.12 and 12.14 when using a strong base to titrate a strong acid like HCl or a weak acid like acetic acid, we can see that the shapes of these titration curves differ in several ways. One difference is that during the early part of a strong acid titration there is little change in pH, while a weak acid titration shows a rapid initial rise in pH until it has been about 10% titrated. In addition, a strong acid titration curve shows no inflection point at the 50% point of the titration, but such an inflection does occur in the middle of the weak acid titration curve. These curves also differ in their pH at the equivalence point. For the titration of a strong acid with a strong base, the pH at the equivalence point is always 7.00, while the titration of a weak acid with a strong base will give a pH at the equivalence point that is greater than 7.00.

In all of the equations given in this section, we have ignored the role of water as a source of hydrogen ions or hydroxide ions. This is not a problem throughout most of the titration curve when you are working with the relatively concentrated solutions of weak acids that are often encountered in titrations ($> 10^{-3}$ M). However, this can be an issue when you are working with more dilute solutions. To deal with this situation, you need to consider the acid, its conjugate base, and water, as described in Chapter 8. Another more general approach for describing the pH of a titration curve under these and other conditions will be discussed in Section 12.3D.

Titration of a Weak Monoprotic Base. The next case we will consider is the titration of a weak monoprotic base with a strong acid. An example of this titration is given in Figure 12.15, in which an aqueous solution containing the weak base ammonia is being titrated with a solution of HCl. Ammonia is widely used in industry as a fertilizer and to produce other chemicals. The titration curve that is produced by ammonia again has four distinct regions. To estimate the pH of the sample or sample/titrant mixture in each of these regions, we use a process that is similar to the approach just described for a weak acid titrated with a strong base. To begin this analysis, recall that the reaction of a weak base in water can be described by the equilibrium constant K_b for the weak base, where $K_b = [HB^+][OH^-]/[B]$. We learned in Chapter 8 that we can use this relationship to determine the concentration of hydrogen ions that will form when we place a weak base in water, as shown in Equation 12.22 where C_B is the total concentration of the weak base.

Original Sample (0% Titration):

$$K_b = \frac{(K_w/[H^+])^2}{C_B - (K_w/[H^+])} \quad \text{or}$$

$$0 = K_b C_B [H^+]^2 - K_b K_w [H^+] - K_w{}^2 \quad (12.22)$$

We can solve this equation for $[H^+]$ and pH by employing the quadratic equation, in which $x = [H^+]$, $A = K_b C_B$, $B = -K_b K_w$, and $C = -K_w{}^2$.

In the middle of the titration (5–95%), we can estimate the pH of our sample/titrant mixture by using the Henderson–Hasselbalch equation. This can be accomplished by using the base ionization for B to find the value of K_{a,BH^+} for its conjugate acid, where $pK_{a,BH+} = pK_w - pK_{b,B}$.

Middle of Titration (5–95% Titration):

$$pH = pK_{a,BH+} + \log\left(\frac{[B]}{[BH^+]}\right) \quad \text{or}$$

$$pH \approx pK_{a,BH^+} + \log\left(\frac{C_B V_B - C_{HA} V_{HA}}{C_{HA} V_{HA}}\right) \quad (12.23)$$

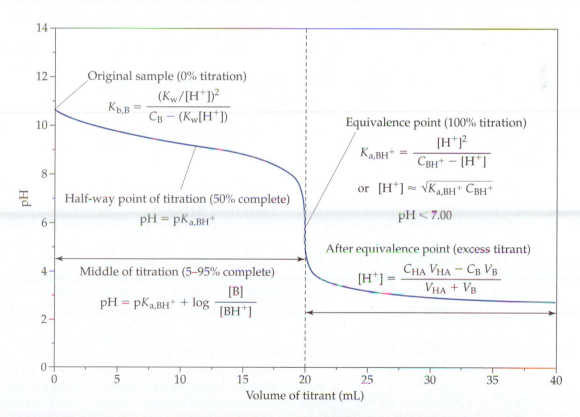

FIGURE 12.15 Equations for predicting the response for the titration of a weak monoprotic base with a strong acid in water. The assumptions made in these equations are the same as made in Figure 12.11 for the titration of a strong acid with a strong base. All terms used in these equations are defined in the text.

In this case, the second relationship that is shown in Equation 12.23 makes use of the fact each mol of added acid will lead to the formation of one mole of BH^+ from B in the central region of the titration curve. At exactly halfway through the titration, this will create a situation in which $[B]/[BH^+] = 1$ and $pH = pK_{a,BH^+}$, in a manner similar to what occurs during the titration of a weak acid with a strong base. This central part of the titration of a weak base represents a "buffer region" because the pH in this portion of the titration curve shows only a small change as more titrant is added to the sample.

At the equivalence point of a weak base titration, the amount of strong acid that has been added will be exactly equal to the moles of weak base in the original sample. In other words, we have converted essentially all the weak base into its conjugate acid. The pH of this mixture can then be determined by using the equations given in Chapter 8 for a solution of a monoprotic weak acid (represented here by BH^+).

Equivalence Point (100% Titration):

$$K_{a,BH^+} = \frac{[H^+]^2}{C_{BH^+} - [H^+]} \quad \text{or}$$

$$0 = [H^+]^2 + K_{a,BH^+}[H^+] - K_a C_{BH^+} \quad (12.24)$$

In this situation, the analytical concentration of the conjugate acid (C_{BH^+}) is equal to the moles of weak base in the original sample divided by the total volume of the sample/titrant mixture, as given by $C_{BH^+} = (C_B V_B)/(V_A + V_B)$. We can then solve for $[H^+]$ and pH in this situation by using the quadratic formula or by using the approximation $[H^+] = \sqrt{K_{a,BH^+} C_{BH^+}}$, which is sufficient for the titration of most weak bases by strong acids. The pH of this equivalence point will always be below 7.0 because the product of this titration is the acid BH^+.

After the equivalence point, the pH for a weak base titration is calculated in the same way as it is for the titration of a strong base. This outcome is a result of the fact that in both situations the original base is now gone so that the values of $[H^+]$ and pH are determined by the excess of strong acid that has been added and total volume of the sample/titrant mixture.

After Equivalence Point (> 101% Titration):

$$[H^+] = \frac{C_A V_A - C_B V_B}{V_A + V_B} \quad (12.25)$$

Once we have $[H^+]$, we can determine the pH by using $pH \approx -\log[H^+]$, the same approach employed in Section 12.3B when a strong base was titrated with strong acid.

Figure 12.15 shows the results that would be expected during the titration of 10.00 mL of 0.01000 M ammonia with 0.005000 M HCl. Estimate the pH that would be expected at the beginning of this titration, at 50% titration, at the equivalence point, and 150% titration.

SOLUTION

At the beginning of the titration, the pH of the sample can be found by using Equation 12.22 and solving with the quadratic equation, where $x = [H^+]$, $A = (1.76 \times 10^{-5})(0.01000\ M)$, $B = -(1.76 \times 10^{-5})(1.0 \times 10^{-14})$, and $C = -(1.0 \times 10^{-14})^2$. This gives a result in which $[H^+] = 2.43 \times 10^{-11}\ M$ and pH = **10.61** for the original solution of ammonia.

At 50% titration, the pH is simply equal to **9.25** (the pK_a for BH^+) because at this point the ratio $[BH^+]/[B]$ will be equal to one and $\log([BH^+]/[B])$ will equal zero according to the form of the Henderson–Hasselbalch equation that is shown in Equation 12.23.

The volume of titrant needed to reach the equivalence point will be $(0.01000\ L)(0.01000\ M)/(0.005000\ M) = 0.02000\ L$ or 20.00 mL. The pH at this point can be solved by using Equation 12.24 or the approximation $[H^+] = \sqrt{K_{a,BH^+} C_{BH^+}}$, in which $C_{BH^+} = (0.01000\ M)(0.01000\ L)/(0.01000\ L + 0.02000\ L) = 0.003333\ M$. The result at the equivalence point is $[H^+] = 1.38 \times 10^{-6}\ M$ and pH = **5.86**.

At 150% titration a total volume of $(1.5)(20.00\ \text{mL}) = 30.00$ mL titrant will have been added. We know from Exercise 12.6 that this situation would correspond to $[H^+] = 0.001250\ M$ and pH = **2.90**, which is identical to the result in Section 12.3B when the same excess of HCl was added to a sample of NaOH.

12.3D A Closer Look at Acid–Base Titrations

Polyprotic Systems. Up to this point we have mostly considered samples that contained monoprotic acids or bases, but there are many acids–base titrations in which more complex samples are encountered. A common example of this occurs when an acid–base titration is used to examine a polyprotic analyte. Phthalic acid ($HOOCC_6H_4COOH$) is one example. This chemical is the conjugate acid of potassium hydrogen phthalate, which we now know is often used as a primary standard for acid–base titrations. Phthalic acid contains two hydrogen ions with $pK_{a1} = 2.95$ and $pK_{a2} = 5.41$ and gives a titration curve like the one shown in Figure 12.16. This curve contains two equivalence points—one for the titration of each hydrogen ion in the order of their pK_a values. It is possible to detect each of the equivalence points for a polyprotic analyte if the analyte has a reasonable

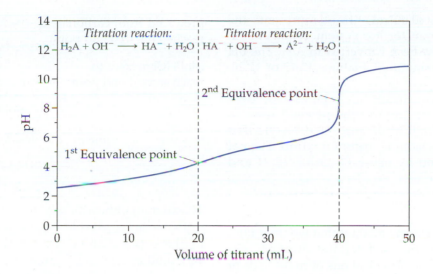

FIGURE 12.16 Titration curve obtained for a 10.00 mL sample of 0.01000 M phthalic acid, a diprotic weak acid, when using 0.005000 M sodium hydroxide as the titrant.

concentration and if the pK_a values for the analyte differ by 4.0 or more units. If the pK_a values are closer in their values, it will be difficult to detect the corresponding transitions in the titration curve.

If two or more distinct equivalence points are seen for a polyprotic analyte (with forms H_2A, HA^-, etc.), the response prior to each equivalence point will generally follow the curve expected for each individual form of the analyte. In the case of Figure 12.16, the response up to the first equivalence point can be estimated by using the equations listed earlier for the titration of the weak acid H_2A by a strong base, where $pK_a = pK_{a1}$ during this part of the titration. The response between the first and second equivalence points can also be estimated by using the expression given for the titration of a weak acid with a strong base, but with analyte now being HA^- and $pK_a = pK_{a2}$. There is, however, one difference in this system that we must also consider. This is the fact that at the first equivalence point we now have a product from the first titration (H_2A giving HA^-) that can act as either an acid (going on to form A^{2-}) or as a base (combining with H_2O to reform H_2A and OH^-). We learned in Chapter 8 that the pH of such a solution can be estimated by using the pK_a values (pK_{a1} and pK_{a2}) for the two acid–base reactions that involve this intermediate, as shown below.

At an Intermediate Equivalence Point for a Polyprotic System:

$$pH \approx \frac{pK_{a1} + pK_{a2}}{2} \tag{12.26}$$

The same approach can be used to estimate the pH at an intermediate equivalence point for any other titration that involves a polyprotic analyte.

EXERCISE 12.9 **Titration of a Polyprotic Weak Acid with a Strong Base**

The titration curve for phthalic acid in Figure 12.16 was obtained by starting with 10.00 mL of a 0.010 M solution of the pure acidic form of this chemical and titrating with 0.005000 M NaOH. Estimate the pH at the beginning of the titration and after 10.00, 20.00, 30.00, 40.00, or 50.00 mL of the titrant has been added to the sample.

SOLUTION

At the beginning of this titration we have a solution that contains only the acidic form of phthalic acid. In this case, we can use Equation 12.17 for a solution of a weak acid, which gives pH = **2.19** for the original sample.

Figure 12.16 indicates that the first equivalence point occurs at a titrant volume of 20.00, so the addition of 10.00 mL titrant will correspond to half-way to the first equivalence point. At 50% titration the pH will simply be equal to pK_{a1} for phthalic acid, or pH = **2.95**.

At the first equivalence point, all of the original phthalic acid (H_2A) will have been converted to its conjugate base (HA^-), which can then undergo further reactions with added titrant. The pH at this first equivalence point is described by Equation 12.26, giving pH = (2.95 + 5.41)/2 = **4.18**. (*Note:* A distinct change in the titration curve at this equivalence point is only slightly visible because the difference between pK_{a1} and pK_{a2} is less than 4.0.)

When 30.00 mL of titrant has been added, half of the HA^- formed from the first titration will have now been converted to A^{2-}. The pH at this point (representing 150% titration versus the original H_2A) is simply equal to the pK_a for HA^-, or pH = **5.41**.

The addition of 40.00 mL titrant represents the second equivalence point for this system. At this point, all of the HA^- has now been converted to A^{2-}. The pH can be estimated by using the Equations 12.19 or 12.20, giving pH = **8.97**.

When 50.00 mL of titrant has been added, there will be an excess of titrant and the pH will now be determined by the amount of unreacted NaOH that has been added to the sample/titrant mixture. The pH in this region is found by using Equation 12.21 and gives pH = **10.92**.

Mixed Samples. The same method that we used for a polyprotic acid could be used for the titration of a polyprotic base or even a mixture of one or more acids or bases. For instance, if a strong acid such as HCl were mixed with a weak acid we would see two distinct equivalence points in the titration curve if the difference in the pK_a values for these two acids is greater than 4.0 and if the weak acid is sufficiently weak as not to begin ionizing significantly before the strong acid is mostly titrated. In this case, the volume of titrant that is needed to reach the first equivalence point can be used to determine the concentration of the stronger of the two acids in the mixture. The difference between this volume and that needed to reach the second equivalence point will make it possible to calculate how much weak acid was present in the original sample. The same idea would also apply to a mixture of two bases with significantly different K_b values.

EXERCISE 12.10	Titrating a Mixture of Acids or Bases in Water

A 25.00 mL water sample is titrated with 0.01500 M HCl to determine the water's alkalinity by using the same general approach as shown earlier in Figure 12.10. An end point due to carbonate is obtained at 2.52 mL and a second end point due to bicarbonate is detected at 11.35 mL. What were the concentrations of carbonate and bicarbonate in the original sample?

SOLUTION
The moles of HCl needed to reach the first end point will be equal to the moles of carbonate in the original sample.

Mol HCl added at first end point =

Mol carbonate in sample

$$(0.01500 \text{ mol/L})(0.00252 \text{ L}) = 3.78 \times 10^{-5} \text{ mol}$$

The amount of HCl needed to reach the second end point will be equal to the moles of bicarbonate that were originally present in the sample plus the bicarbonate that was

produced from carbonate as the sample/titrant mixture reached the first end point.

Mol HCl needed to
reach second end point = Mol carbonate

+ mol bicarbonate in sample

$$= (0.01500 \text{ mol/L}) \cdot$$

$$(0.01135 \text{ L} - 0.00252 \text{ L})$$

$$= 1.324 \times 10^{-4} \text{ mol}$$

The amount of bicarbonate that must have been present in the original sample will be given by the difference in these two values, or $(1.324 \times 10^{-4} \text{ mol} - 3.78 \times 10^{-5} \text{ mol}) = 9.46 \times 10^{-5}$ mol bicarbonate in the sample. We can then use the sample volume to find the original concentrations of these two analytes, giving $(3.78 \times 10^{-5} \text{ mol})/(0.02500 \text{ L}) = \mathbf{1.51 \times 10^{-3}}\, \boldsymbol{M}$ **carbonate** and $(9.46 \times 10^{-5} \text{ mol})/(0.02500 \text{ L}) = \mathbf{3.78 \times 10^{-3}}\, \boldsymbol{M}$ **bicarbonate**.

Using the Fraction of Titration. The equations given in Sections 12.3B and 12.3C allow us to quickly estimate the pH at various key points in the titration curve for a monoprotic acid or base. It is also possible to use a more general equation to describe the titration curve over a broad range of conditions. This second type of equation can be produced by looking at the fraction of titration. The **fraction of titration (F)** is defined here as the ratio of the moles of titrant that have been added at any given point in the titration versus the moles of analyte that were originally present in the sample. We can describe this fraction by using the following expressions.

Titration of an Acid with a Base:

$$F = \frac{\text{Moles added base}}{\text{Moles original acid}} = \frac{C_B V_B}{C_A V_A} \qquad (12.27)$$

Titration of a Base with an Acid:

$$F = \frac{\text{Moles added acid}}{\text{Moles original base}} = \frac{C_A V_A}{C_B V_B} \qquad (12.28)$$

Both of these equations are based on the moles of titrant that are needed to reach the first equivalence point of a titration, at which $F = 1$. If the analyte is a polyprotic acid or base with multiple equivalence points, the first equivalence point will occur when $F = 1$, the second will occur when $F = 2$, and so on.

The reason the preceding definitions for F are useful is they can be combined with charge balance and mass balance equations to obtain a general expression that describes a titration. For instance, if we are titrating a solution of HCl with a solution of NaOH as the titrant, the charge balance equation (see Chapter 6) will be given by the expression shown at the top of Table 12.5. This

TABLE 12.5 Derivation of a Fraction of Titration Equation (HCl titrated with NaOH)

Titration Reaction:

$$H^+ \text{ (from HCl)} + OH^- \text{ (from NaOH)} \rightarrow H_2O$$

Charge Balance Equation:

$$[H^+] + [Na^+] = [Cl^-] + [OH^-]$$

Mass Balance Equations:

$$\text{Mol } Na^+ = \text{Mol added NaOH} \implies [Na^+] = \frac{C_B V_B}{V_A + V_B}$$

$$\text{Mol } Cl^- = \text{Mol original HCl} \implies [Cl^-] = \frac{C_A V_A}{V_A + V_B}$$

Substitute Mass Balance Equations into Charge Balance Equation:

$$\implies [H^+] + \frac{C_B V_B}{V_A + V_B} = \frac{C_A V_A}{V_A + V_B} + [OH^-]$$

Rearrange to Solve for $C_B V_B$ and Multiple Both Sides by $(V_A + V_B)$:

$$\implies C_B V_B = C_A V_A - (V_A + V_B)\left([H^+]-[OH^-]\right)$$

Combine Common Terms Containing V_A or V_B:

$$\implies C_B V_B + V_B\left([H^+]-[OH^-]\right) = C_A V_A - V_A\left([H^+]-[OH^-]\right)$$

Factor out $C_B V_B$ (on left) and $C_A V_A$ (on right):

$$\implies C_B V_B\left(1 + \frac{[H^+]-[OH^-]}{C_B}\right) = C_A V_A\left(1 - \frac{[H^+]-[OH^-]}{C_A}\right)$$

Rearrange to Obtain $(C_B V_B)/(C_A V_A)$ on Left Side of Equation:

$$\therefore \quad F = \frac{C_B V_B}{C_A V_A} = \frac{1 - \dfrac{[H^+]-[OH^-]}{C_A}}{1 + \dfrac{[H^+]-[OH^-]}{C_B}} \tag{12.29}$$

equation is based on the fact that each mole of HCl that is present in the original sample will completely dissociate to form H^+ and Cl^-. The same is true for the NaOH that is dissolved in the titration solution, which dissociates to form Na^+ and OH^-. These make up the only positive and negative ions in the system. During the titration, each mole of H^+ will react with one mole of OH^-, thereby reducing the values of both the right- and left-hand sides of the charge balance equation by equal amounts. This charge balance equation also includes the contributions of the dissociation of water, which leads to the formation of one H^+ ion for each OH^- that is produced by water.

Because HCl and NaOH are both completely dissociated in water, we can also write a mass balance equation that relates the concentrations of Cl^- and Na^+ to the amount of HCl that was originally present in the sample or the NaOH that has been added in the titrant. These mass balance equations can then be used to determine the concentration of Cl^- and Na^+ that will be present at each point in the titration (see Table 12.5). It is now

possible to substitute these mass balance equations into the charge balance expression in Table 12.5 and to rearrange this new equation to give the fraction of titration for this system, as represented by the final result in Equation 12.29. The same general approach can be used to derive expressions that describe the fraction of titration for other systems (see Table 12.6).

One common feature of all the equations in Table 12.6 is that they are written in forms that depend only on the pH (as represented by $[H^+]$ and $[OH^-]$), the original concentrations of the acid and base (C_A and C_B), and other constants for the system, such as acid dissociation constants ($K_{a,HA}$ or K_{a,BH^+}). This feature makes these equations useful if you wish to see what fraction of titration, and the volume of added titrant, that will correspond to a particular pH in an acid–base titration. As illustrated in the next exercise, an example of one situation in which you may wish to have this type of information is when you are estimating the titration error that results when you are using an acid–base indicator to follow such a titration.

TABLE 12.6 Fraction of Titration Equations

Titration of a Strong Acid with a Strong Base:

$$F = \frac{C_B V_B}{C_A V_A} = \frac{1 - \dfrac{[H^+] - [OH^-]}{C_A}}{1 + \dfrac{[H^+] - [OH^-]}{C_B}} \qquad \text{where:} \qquad [OH^-] = K_w/[H^+] \qquad (12.29)$$

Titration of a Strong Base with a Strong Acid:

$$F = \frac{C_A V_A}{C_B V_B} = \frac{1 + \dfrac{[H^+] - [OH^-]}{C_B}}{1 - \dfrac{[H^+] - [OH^-]}{C_A}} \qquad \text{where:} \qquad [OH^-] = K_w/[H^+] \qquad (12.30)$$

Titration of a Weak Monoprotic Acid (HA) with a Strong Base:

$$F = \frac{C_B V_B}{C_A V_A} = \frac{\alpha_{A^-} - \dfrac{[H^+] - [OH^-]}{C_A}}{1 + \dfrac{[H^+] - [OH^-]}{C_B}} \qquad \text{where:} \quad [OH^-] = K_w/[H^+] \qquad \alpha_{A^-} = \frac{K_{a,HA}}{[H^+] + K_{a,HA}} \quad (12.31)$$

Titration of a Weak Monoprotic Base (B) with a Strong Acid:

$$F = \frac{C_A V_A}{C_B V_B} = \frac{\alpha_{BH^+} + \dfrac{[H^+] - [OH^-]}{C_B}}{1 - \dfrac{[H^+] - [OH^-]}{C_A}} \qquad \text{where:} \quad [OH^-] = K_w/[H^+] \qquad \alpha_{BH^+} = \frac{[H^+]}{[H^+] + K_{a,BH^+}} \quad (12.32)$$

EXERCISE 12.11 | **Estimating Titration Error Using a Fraction of Titration Equation**

An industrial chemist wishes to use bromocresol purple as an indicator for the titration described in Exercise 12.7, in which a 10.00 mL sample thought to contain 1.0×10^{-2} M acetic acid is combined with 5.000×10^{-3} M NaOH in water as the titrant. What is the expected titration error if the middle of the color transition for bromocresol purple is used to signal the end point for this titration?

SOLUTION

We know from Exercise 12.7 that the volume of titrant needed to reach the equivalence point in this example is 20.00 mL and that pH at this equivalence point will be 7.35. Table 12.3 indicates that bromocresol purple will have a change in color over the pH range of 5.2–6.8, with the center of this color change occurring at about pH 6.0. The maximum expected titration error at the center of this color transition can then be found by using Equation 12.31 in Table 12.6.

$$[H^+] \approx 10^{-pH}$$

$$= 10^{-6.00} = 1.0\underline{0} \times 10^{-6} \ M$$

$$[OH^-] = K_w/[H^+]$$

$$= (1.0 \times 10^{-14} M^2)/(1.0\underline{0} \times 10^{-6} M)$$

$$= 1.0\underline{0} \times 10^{-8} M$$

$$\alpha_{A^-} = \frac{K_{a,HA}}{[H^+] + K_{a,HA}}$$

$$= \frac{1.7 \times 10^{-5} M}{1.0\underline{0} \times 10^{-6} M + 1.7 \times 10^{-5} M}$$

$$= 0.94\underline{4}$$

$$F = \frac{\alpha_{A^-} - \dfrac{[H^+] - [OH^-]}{C_A}}{1 + \dfrac{[H^+] - [OH^-]}{C_B}}$$

$$= \frac{0.944 - \dfrac{1.0\underline{0} \times 10^{-6} M - 1.00 \times 10^{-8} M}{1.0 \times 10^{-2} M}}{1 + \dfrac{1.0\underline{0} \times 10^{-6} M - 1.0\underline{0} \times 10^{-8} M}{5.000 \times 10^{-3} M}}$$

$$= 0.94\underline{37}$$

Once we know the value of both F and V_E, we can also find the value of V_B at the apparent end point (as indicated by the center of the color transition for bromocresol purple) and the resulting titration error (see Equation 12.1).

$$V_B = F \cdot V_E = (0.94\underline{37})\,(20.00\text{ mL})$$

$$= 18.\underline{87}\text{ mL}$$

$$\text{Titration error} = V_{B,\text{End Pt}} - V_E = 18.87\text{ mL} - 20.00\text{ mL}$$

$$= -\mathbf{1.13}\text{ mL}$$

This error would correspond to an estimated concentration for the acetic acid solution that is off by $100 \cdot (V_{B,\text{End Pt}} - V_E)/V_E = 100 \cdot (1.\underline{13}\text{ mL})/(20.00\text{ mL}) = 5.6\%$.

Another valuable way in which fraction of titration equations can be utilized is to predict the overall shape of a titration curve. This use is most easily performed by using a spreadsheet, such as the one shown in Figure 12.17. This spreadsheet is set up to perform the same type of calculation as used in the last exercise to find the values of F and V_{Titrant} that will correspond to a given pH. However, the spreadsheet also allows the values of F and V_{Titrant} to be found over a wide range of pH values. A plot of these values is then made with V_{Titrant} (or F) appearing on the x-axis and pH on the y-axis. This approach can be used to generate the plot for an entire titration curve or to find the pH and titrant volume that will correspond to a given point in the titration, such as the half-way point ($F = 0.50$) or the equivalence point ($F = 1.00$).

Titration of a weak monoprotic acid with a strong base (NaOH)

pK_a for weak acid:	4.76 (Acetic acid)		K_a for weak acid:	1.75E−05

Conditions used in titration:

C_A (M) =	0.01	C_B (M) =	0.005
V_A (M) =	0.01	V_E (L) =	0.02

Calculated results for titration:

pH	[H$^+$]	[OH$^-$]	Alpha A$^-$	Fraction to 1st equivalence pt.	V_B (L)	V_B (mL)
3.388	4.10E−04	2.44E−11	4.10E−02	0.0000	0.0000	0.00
3.400	3.98E−04	2.51E−11	4.21E−02	0.0021	0.0000	0.04
3.500	3.16E−04	3.16E−11	5.24E−02	0.0196	0.0004	0.39
3.600	2.51E−04	3.98E−11	6.51E−02	0.0381	0.0008	0.76
3.700	2.00E−04	5.01E−11	8.06E−02	0.0584	0.0012	1.17
3.800	1.58E−04	6.31E−11	9.94E−02	0.0810	0.0016	1.62
3.900	1.26E−04	7.94E−11	1.22E−01	0.1068	0.0021	2.14
4.000	1.00E−04	1.00E−11	1.49E−01	0.1362	0.0027	2.72
4.100	7.94E−05	1.26E−10	1.81E−01	0.1699	0.0034	3.40
4.200	6.31E−05	1.58E−10	2.17E−01	0.2082	0.0042	4.16
4.300	5.01E−05	2.00E−10	2.59E−01	0.2513	0.0050	5.03
4.400	3.98E−05	2.51E−10	3.05E−01	0.2990	0.0060	5.98
4.500	3.16E−05	3.16E−10	3.56E−01	0.3509	0.0070	7.02
4.600	2.51E−05	3.98E−10	4.11E−01	0.4061	0.0081	8.12
4.700	2.00E−05	5.01E−10	4.67E−01	0.4634	0.0093	9.27
4.800	1.58E−05	6.31E−10	5.25E−01	0.5215	0.0104	10.43
4.900	1.26E−05	7.94E−10	5.82E−01	0.5789	0.0116	11.58
5.000	1.00E−05	1.00E−09	6.36E−01	0.6341	0.0127	12.68
5.100	7.94E−06	1.26E−09	6.88E−01	0.6859	0.0137	13.72
5.200	6.31E−06	1.58E−09	7.35E−01	0.7334	0.0147	14.67
5.300	5.01E−06	2.00E−09	7.77E−01	0.7761	0.0155	15.52
5.400	3.98E−06	2.51E−09	8.15E−01	0.8136	0.0163	16.27
5.500	3.16E−06	3.16E−09	8.47E−01	0.8461	0.0169	16.92
5.600	2.51E−06	3.98E−09	8.74E−01	0.8738	0.0175	17.48
5.700	2.00E−06	5.01E−09	8.98E−01	0.8971	0.0179	17.94
5.800	1.58E−06	6.31E−09	9.17E−01	0.9165	0.0183	18.33
5.900	1.26E−06	7.94E−09	9.33E−01	0.9325	0.0187	18.65
6.000	1.00E−06	1.00E−08	9.46 E−01	0.9457	0.0189	18.91

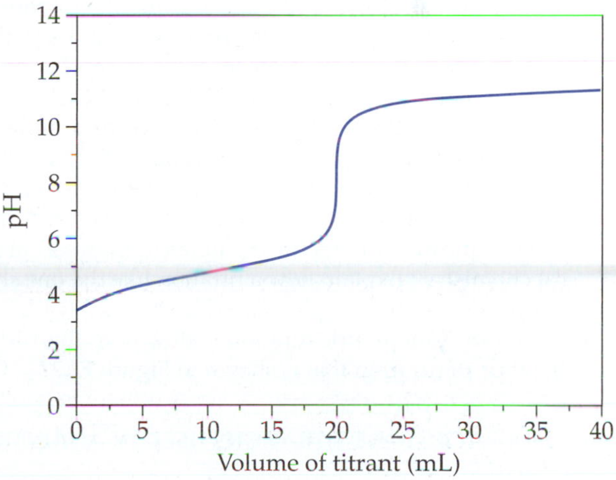

FIGURE 12.17 Example of a spreadsheet using a fraction of titration equation to predict the response of an acid–base titration. This spreadsheet is based on Equation 12.31 and is the same spreadsheet that was used to draw the titration curve for acetic acid in Figure 12.14. Only part of the actual pH values that were used in this calculation are shown in the portion of the spreadsheet that is included in this figure.

Key Words

Other Terms

Questions

WHAT IS AN ACID–BASE TITRATION?

1. What is meant by the term "titration" in analytical chemistry? What is meant by the term "acid–base titration"?
2. What is a "titrant"? Identify the titrant for the titration that is shown in Figure 12.4.
3. Describe what is meant by a "titration curve" and "equivalence point." Use the results for sulfuric acid in Figure 12.4 to discuss these terms.
4. What is the difference between an "end point" and an "equivalence point" in a titration? Explain how this difference is related to the "titration error."
5. An acid–base indicator used for the titration of a standard sample of HCl with a solution of NaOH gives an end point when 17.32 mL of titrant has been added. The actual equivalence point for this titration is known to be 18.05 mL. What is the titration error for this analysis?
6. Two students carry out a titration of 10.00 mL of 0.1261 M acetic acid with 0.05013 M NaOH using phenolphthalein as an acid–base indicator. The first student uses the first appearance of any pink color to signal the end point, which corresponds to the addition of 25.12 mL of titrant. The second student waits until the indicator turns a more intense red color and records a titrant volume of 25.17 mL as representing the end point. The true equivalence point occurs at 25.15 mL. What titration error was obtained by each student?
7. What is meant by the term "volumetric analysis" in analytical chemistry? Explain how a titration like the one shown in Figure 12.2 is an example of a volumetric analysis.
8. What is a "gravimetric titration"? How does this differ from the type of titration that is shown in Figure 12.2?

HOW ARE ACID–BASE TITRATIONS USED IN ANALYTICAL CHEMISTRY?

9. What are some advantages of using a titration for chemical measurements? What are some disadvantages of these methods?
10. Explain how a titration could be used to determine the amount of acid or base that is present in a sample. Give an equation that shows how the amount of analyte and titrant will be related to one another at the equivalence point in such a titration.
11. Acetylsalicylic acid ($C_9H_8O_4$) is a weak monoprotic acid that is commonly known as the drug "aspirin." Calculate the mass of aspirin that must have been present in a drug tablet if an end point of 42.76 mL is obtained when using 0.1354 M NaOH as a titrant.
12. A 10.00 mL portion of a household product containing ammonia gives an end point of 46.56 mL when titrated with 0.2034 M HCl. What is the concentration of ammonia in this product in units of molarity? If the overall density of the product is 0.980 g/mL, what is the amount of ammonia in terms of % (w/w)?
13. Discuss how a titration can be used to estimate the molar mass of an acid or base. Give an equation that shows how the information about the equivalence point can be used in this application.
14. An unknown sample of a chemical is shown to be acidic and quite pure; however, the identity of the chemical is not known. To help identify this chemical, its molar mass is determined by a titration. A 0.05465 g sample of this material requires 24.55 mL of 0.01265 M NaOH to reach an end point that is detected by using phenolphthalein as an acid–base indicator. What is the molar mass of the acid in this sample?
15. An unknown material is shown to be a pure base, but the identity of this base is not known. A 3.576 g portion of this material is dissolved in water and diluted to 250.0 mL. Three 25.00 mL portions of this solution are titrated separately with 0.1380 M HCl and give end points (using methyl red as an indicator) after the addition of 13.15, 13.22, and 13.19 mL of titrant.
    (a) If it is assumed that this base is monoprotic, calculate the average molar mass of this compound and the standard deviation of this molar mass.
    (b) What would the average molar mass be if it were instead assumed that the base was diprotic?
    (c) Which of the two answers in Parts (a) and (b) makes the most sense from a chemical perspective?
16. Explain how a titration can be used to determine the purity of a sample that contains an acid or base. What assumption is typically made in this type of analysis?
17. A commercial product designed to prevent cut fruit from turning brown contains a mixture of ascorbic acid ($C_6H_8O_6$) and sugar ($C_{12}H_{22}O_{11}$). Calculate the percent (w/w) of ascorbic acid (a weak monoprotic acid) in this product if a 2.0654 g portion of this product requires 34.55 mL of 0.2378 M NaOH to reach a good end point using

phenolphthalein as an indicator. (*Note:* You can assume that the sugar will not be titrated during this process.)

18. Trishydroxymethylaminomethane (THAM, $C_4H_{11}O_3N$) is a weak monoprotic base that is often used as a primary standard when examining solutions of acids. To test the purity of a new shipment of THAM, several portions of the shipment are titrated with 0.05794 *M* HCl. The following end points are observed when using methyl red as an indicator.

Mass THAM (g)	Volume Added HCl at End Point (mL)
0.1367	19.45
0.1563	22.24
0.1490	21.20

What is the average purity of the THAM when expressed in units of % (w/w)? What is the standard deviation of this purity?

19. In addition to the amount that is present, what other types of information can be learned about an analyte that is an acid or base when this is examined by means of an acid–base titration?

20. What is meant by the term "direct titration"? Give one specific example of a direct titration.

21. What is a "back titration"? How is this method performed? Give one specific example of a back titration.

22. Baking soda is a common name for sodium bicarbonate ($NaHCO_3$). A 0.2087 g portion of commercial baking soda is treated with 25.00 mL of 0.1028 *M* HCl and then back titrated to give an end point at 2.53 mL during the addition of 0.0565 *M* NaOH. What is the purity of the baking soda when expressed in units of % (w/w)?

23. Describe how the Kjeldahl method is conducted. What is the purpose of each step in this method and what types of information does this technique provide about a sample?

24. The amount of protein in a new type of cold cereal was measured using a Kjeldahl titration. The ammonia that was produced for one serving of the cereal (1.00 ounces) was distilled from the digested sample and captured in a 50.00 mL solution of 0.6000 *M* HCl. This solution was later examined in a back titration using 0.1000 *M* NaOH and gave an end point at 7.84 mL of added titrant. Calculate the percent nitrogen and percent protein in the original sample. (*Note:* Recall that proteins typically contain 17.5% nitrogen by weight.)

25. The percent nitrogen in a chemical compound that is isolated from tobacco is to be measured using the Kjeldahl method. A 0.248 g sample of this isolated compound is subjected to the Kjeldahl digestion procedure and the ammonia that is formed is distilled into 50.00 mL of 0.1234 *M* HCl. Titration of the remaining excess HCl requires 35.59 mL of 0.08736 *M* NaOH to reach the end point. What is the percent of nitrogen in the isolated compound?

PREPARING TITRANT AND SAMPLE SOLUTIONS

26. Describe why standardization is needed to conduct a successful titration. What is a "standard solution" and how is it related to this process?

27. What is meant by the term "primary standard"? What are some examples of primary standards for acid–base titrations?

28. What properties should be present in any compound that is to be used as a primary standard? Use this list of properties to explain why NaOH or HCl are not appropriate primary standards for an acid–base titration.

29. A laboratory purchases five liters of a solution that is advertised to contain 0.1000 *M* NaOH and to be free of carbonate. Workers in this laboratory routinely check such purchases by performing three titrations with the primary standard KHP. The following data were acquired following one such purchase.

Mass KHP (g)	Volume Added NaOH at End Point (mL)
0.8127	39.86
0.7549	37.03
0.8650	42.44

(a) What is the average molarity for the newly purchased NaOH solution?

(b) Is there a significant difference between the measured molarity for this solution and its advertised value of 0.1000 *M*?

30. What mass of tris(hydroxymethyl)aminomethane (Tris) would be necessary to standardize a solution made up by diluting 10.0 mL of concentrated HCl into a 250 mL volumetric flask?

31. Explain how an increase in titrant concentration will affect the general shape of an acid–base titration curve. What are some general guidelines that can be used in selecting a titrant concentration for this type of titration?

32. What general effects will a change in analyte concentration have on the shape of an acid–base titration curve? What is the typical range of analyte concentrations that is used in such a titration?

33. Explain how the strength of a weak acid or weak base will affect its ability to be examined by an acid–base titration. What range of pK_a or pK_b values will usually allow for a successful acid–base titration with these analytes?

PERFORMING A TITRATION AND DETERMINING THE END POINT

34. Discuss the role of a buret in a volumetric titration. Explain why a clean buret is important to this type of analysis.

35. Describe the proper procedure for reading the level of a titrant solution in a buret. Include in your description a discussion of parallax error and approaches for minimizing this error.

36. Explain why it is useful to follow the pH during an acid–base titration. Describe a general experimental system that can be used to measure the pH of a sample/titrant mixture during a titration.

37. Define the term "acid–base indicator." What is the role of such an indicator in an acid–base titration?

38. Discuss the general way in which an acid–base indicator gives a change in color with a change in pH. Describe how the pH range of this color change is related to the properties of the indicator.

39. What are some disadvantages of using acid–base indicators? What do you think are some of their advantages?

40. The titration of an acetic acid solution by NaOH gives an equivalence point at a pH of 8.20 after the addition of 38.65 mL titrant. One student uses bromothymol blue to find the end point of this titration, which they estimate to occur at pH 7.00 and at 38.08 mL. What is the titration error

for this analysis? Suggest one alternative acid–base indicator that the student might use to reduce the size of this error.

41. In a quantitative analysis lab, each student was asked to determine the purity of a solid KHP sample by titrating this sample with NaOH. Most students used phenolphthalein, but one student has red and green color blindness and cannot see the color change of this indicator. What is one other indicator that could be used by this student? If the true equivalence point for this titration occurs at a pH of 8.90, how will the relative size of the titration error obtained by this student compare to those for the other students in the laboratory?

42. Explain how a derivative plot can be used to help find the equivalence point in an acid–base titration.

43. The following data were collected by a student during the titration of a 10.00 mL KHP solution with 0.1034 M NaOH. Use this information to prepare a plot of pH vs. the volume of titrant. Also prepare plots that show the first derivatives of these data. Use these plots to find the end point of this titration and to determine the concentration of the KHP.

Added NaOH (mL)	pH	Added NaOH (mL)	pH
0.0	3.90	18.26	7.39
3.76	4.39	18.31	8.59
7.86	4.80	18.38	9.11
11.91	5.19	18.55	9.73
15.85	5.74	18.72	10.02
16.94	6.04	18.97	10.31
17.15	6.12	19.10	10.43
17.42	6.25	19.36	10.62
17.85	6.59	20.76	11.07
17.96	6.72	24.77	11.52
18.09	6.93		
18.19	7.16		

44. What is a Gran plot? How is this type of plot used in an acid–base titration?

45. Prepare a Gran plot using the data given in Problem 43. Use this plot to find the end point of this titration and to determine the concentration of the KHP sample and K_a value for KHP. Compare your results to those from Problem 43.

46. A Gran plot for the titration of a 25.00 mL solution of a weak acid with 0.1250 M NaOH gives a response that has the following best-fit line: $y = -(3.61 \times 10^{-4})x + 1.16 \times 10^{-5}$, where units of liters are present on both the x-axis and y-axis.
   (a) What volume of titrant is needed to reach the equivalence point for this titration?
   (b) What is the approximate value of pK_a for the weak acid that is being titrated?

DESCRIBING ACID–BASE TITRATIONS

47. Write a general reaction that can be used to describe the titration of a strong acid with a strong base in water. What is the equilibrium constant for this reaction at 25°C? Explain why the same reaction also describes the titration of a strong base with a strong acid in water.

48. Explain why the K_a for a weak acid is important to consider in determining the extent to which this analyte will undergo

a titration with a strong base. Similarly, explain why the K_b for a weak base is important to consider in determining the extent to which this analyte will undergo a titration with a strong acid.

49. A pH of 5.35 is measured after 10.75 mL of 0.1234 M NaOH has been added to a 10.00 mL sample of a weak monoprotic acid at 25°C. The end point for this titration is found to occur after 16.77 mL of the titrant has been added. What was the concentration of the weak acid in the original sample and what is its estimated acid dissociation constant under these conditions?

50. A chemist wishes to determine the feasibility of titrating the weak acid HF with the weak base ammonia. The overall reaction for this titration follows:

$$HF + NH_3 \rightleftharpoons NH_4^+ + F^-$$

   (a) Write an expression for the overall equilibrium constant for the titration, K. Show how the value of this equilibrium constant K is related to K_a and K_b for HF and NH_3.
   (b) Calculate the expected value for an aqueous solution at 25°C. How does the size of this equilibrium constant compare to that expected for the titration of HF with NaOH?

TITRATION CURVES FOR STRONG ACIDS AND BASES

51. Describe the four general regions that are present in the titration of a strong acid with a strong base. Describe how the pH of the sample or sample/titrant mixture can be estimated for each of these regions.

52. Nitric acid is used to etch copper during printmaking. To test the strength of a nitric acid solution, 5.00 mL of a commercial preparation of this acid is dissolved in water and diluted to 100.00 mL in a volumetric flask. A 25.00 mL portion of this diluted acid is titrated with 0.5365 M NaOH and gives an end point at 36.87 mL.
   (a) What is the concentration of the nitric acid in the original and undiluted commercial preparation?
   (b) What is the expected pH of the diluted nitric sample solution at the beginning of the titration? What is the expected pH at the equivalence point?
   (c) What is the expected pH during this titration after 10.00 mL, 25.00 mL, and 40.00 mL of titrant have been added?

53. A reaction gives off HCl gas, which is collected by passing this gas through 50.00 mL of distilled water. To measure the amount of HCl that is now in solution, it is titrated with 0.1000 M NaOH.
   (a) If this titration has an end point at 16.08 mL titrant, how many grams of HCl were given off in the original reaction? What was the concentration of HCl after its collection in water?
   (b) What pH should have been present for the original solution of HCl after it had been collected in the water?
   (c) What pH should have been present after 5.00, 10.00, and 20.00 mL of titrant had been added to the solution of collected HCl. What was the expected pH at the equivalence point?

54. Describe how the pH of the sample or sample/titrant mixture can be estimated for each of the four general regions during the titration of a strong base with a strong acid. How are these calculations similar to those performed for the titration of a strong acid with a strong base? How are the calculations for these two types of titrations different?

55. A 20.00 mL aqueous sample that is thought to contain a 0.08000 M of the strong base KOH is to be titrated with a 0.1000 M aqueous solution of HCl.
  (a) Estimate the pH that will be obtained during this analysis at 0% and 100% titration.
  (b) Estimate the pH at 50% titration and after a total of 20.00 mL titrant has been added to the sample.
  (c) Use the results form Parts (a) and (b) to draw a rough titration curve for this analysis.

56. Calcium oxide (CaO) is mixed with sand in making mortar. Calcium oxide reacts with water to produce calcium hydroxide $Ca(OH)_2$, which is a strong base. A 0.5654 g sample of a mortar mix is dissolved/suspended in 100.00 mL of water and is titrated with 0.2500 M HCl. An end point at 38.96 mL is found when using methyl red as the indicator.
  (a) What is the expected pH for the original sample if it is assumed that calcium hydroxide is the only basic (or acidic) chemical present at any significant concentration?
  (b) What is the pH expected for this titration after 10.00, 20.00, or 30.00 of titrant has been added?
  (c) What is the expected pH for this titration at the equivalence point or after 40.00 mL of titrant has been added?
  (d) What is the amount of CaO in the original mortar mix in terms of % (w/w)?

TITRATION CURVES FOR WEAK ACIDS AND BASES

57. Explain how the pH of the sample or sample/titrant mixture can be estimated for each of the four general regions during the titration of a weak monoprotic acid with a strong base. How are these calculations similar to those performed for the titration of a strong acid with a strong base? How are the calculations for these two types of titrations different?

58. Hydrofluoric acid is an important reagent used to remove surface oxides from silicon during the production of semiconductors and computer chips. A chemist working in the semiconductor industry wishes to determine the concentration of an HF solution by using an acid–base titration. A 25.00 mL sample thought to contain approximately 5.00×10^{-3} M HF is to be titrated with a 7.50×10^{-3} M of NaOH. What is the expected pH that will be obtained during this titration for the original sample, after 50% of the HF has been titrated and at the equivalence point of this titration? What will be the pH after 30.00 mL of titrant has been added to the HF solution?

59. A 0.3654 g portion of pure formic acid is dissolved in 50.00 mL of water and is titrated with 0.1086 M NaOH. What pH is expected during this analysis when the titration is 0%, 25%, 50%, 75%, 100%, and 110% complete?

60. A solution of a weak monoprotic acid solution requires 37.65 mL of 0.02465 M NaOH to reach the equivalence point. During this titration a pH of 5.87 is measured after the addition of 10.87 mL titrant. What is the K_a for this weak acid under these conditions?

61. Explain how the pH of the sample or sample/titrant mixture can be estimated for each of the four general regions during the titration of a weak monoprotic base with a strong acid. Compare and contrast these calculations with those that are performed for the titration of a strong base with a strong acid.

62. A 25.00 mL sample thought to contain 0.0445 M methylamine (CH_3NH_2) is titrated with a 0.07000 M solution of HCl. Calculate the pH that is expected when 0 mL, 10.00 mL, and 20.00 mL of titrant have been added as well at the equivalence point.

63. A new compound has been isolated from an unusual plant and has been shown to have properties that make this chemical act as a weak monoprotic base. A 0.0356 g pure sample of this compound is dissolved in 25.00 mL of distilled water. The pH of the water after the compound fully dissolves is found to be 11.42. This sample is then titrated with 0.01000 M HCl and gives an end point at pH 6.60 and after the addition of 22.67 mL of titrant. After the addition of 9.45 mL and 16.03 mL titrant, the pH of the sample/titrant mixture is determined to be 11.02 and 10.49, respectively. Estimate the molar mass and pK_b for this compound. If you don't get the same pK_b from all your calculations, state which value you think is the most reliable.

POLYPROTIC SYSTEMS

64. Explain how the titration curve for a polyprotic analyte differs from that of a monoprotic acid or base. What effect does the size of the pK_a or pK_b values for a polyprotic analyte have on the appearance of its titration curve?

65. Give a general equation that can be used to estimate the pH of an intermediate equivalence point for a polyprotic acid (e.g., the first equivalence point for a diprotic acid). Explain the origins of this equation. (*Hint:* See Chapter 8.)

66. During the manufacture of nylon, adipic acid ($HOOC(CH_2)_4COOH$, a diprotic weak acid) is polymerized by combining it with 1,6-diaminohexane ($H_2N(CH_2)_6NH_2$, a diprotic weak base). A polymer chemist studying this reaction wishes to analyze a sample thought to contain 0.1000 M adipic acid by titrating it with 0.1000 M NaOH.
  (a) Estimate the pH of the original sample, at both equivalence points, at the half-way point to the first equivalence point, and at the half-way point between the first and second equivalence points. Use this information to prepare a rough plot of pH versus titrant volume for this titration.
  (b) Calculate the pH that would be expected between for points that are 35.0% and 70.0% of the way to the first equivalence point in this titration, and 35% and 70% of the way between the first and second equivalence points. Compare these results to those that would be predicted by the plot you made in Part (a).

67. The same polymer chemist as in preceding problem wishes to analyze a sample thought to contain 0.1000 M 1,6-diaminohexane by titrating it with 0.1200 M HCl.
  (a) Estimate the pH of the original sample, at both equivalence points, at the half-way point to the first equivalence point, and at the half-way point between the first and second equivalence points.
  (b) Use the information from Part (a) to prepare a rough plot of pH versus titrant volume for this titration.
  (c) How does the titration curve you obtained in Part (b) for the titration of 1,6-diaminohexane compare with the titration curve that was generated for adipic acid in the previous problem? What similarities are present in these plots? What are their differences?

68. A 50.00 mL aliquot of a solution of a monoprotic weak base in water is titrated with 0.1147 M HCl. It is found that 37.58 mL of this titrant is needed to reach the equivalence point.
  (a) What was the concentration of the base in the original sample?
  (b) During this titration, a pH of 10.25 is measured for the the the sample/titrant mixture after 20.48 mL of HCl has been added to the sample. What are K_b and pK_b for this base?

MIXED SAMPLES

69. Explain how you can use the equations given earlier in this chapter for the titrations of simple samples of acids or bases to help estimate the pH at various points in a titration curve for a sample that contains a mixture of two different acids or bases.

70. Exactly 10.00 mL of a mixture of HCl (0.2356 M) and phosphoric acid (0.1075 M) is titrated with 0.1000 M NaOH. Calculate the pH of this titration at several places in the titration curve. Choose locations in the curve that are important in establishing the shape of the curve.

71. Explain how the equivalence points obtained during the titration of a mixed acid or base sample can be used to obtain information on the original composition of such a sample.

72. A 25.00 mL portion of a mixture of oxalic acid and sulfuric acid is titrated with 0.1200 M NaOH and monitored with a pH meter. Two end points are observed. The first occurs after addition of 17.44 mL and the second at a total of 21.98 mL. What were the concentrations of the two acids in the original sample?

73. A 25.00 mL solution of hypochlorous acid (HOCl) is partially reduced to form HCl. After this reduction is carried out, the resulting solution is titrated with 0.1000 M NaOH. Shifts in the response of the titration curve at two equivalence points are noted: one at a pH of roughly 3.5 (at 21.66 mL titrant) and the other at approximately pH 9.5 (at 30.17 mL titrant). What fraction of the original HOCl was reduced to HCl in this sample?

74. An industrial laboratory receives a sample that is thought to contain a mixture of sulfuric acid (H_2SO_4) and sulfurous acid (H_2SO_3). A 25.00 mL aliquot of this sample is titrated with 0.1000 M NaOH and gives two recognizable end points when the pH of the sample/titrant mixture is monitored. The first apparent end point occurs when 37.89 mL of titrant has been added (pH of sample/titrant mixture ≈ 7.0) and is a result of the titration of both the H_2SO_4 and HSO_4^- forms of sulfuric acid and the H_2SO_3 form of sulfurous acid. The second apparent end point occurs when 43.57 mL of titrant has been added (pH = 9.8) and corresponds to the titration of only the HSO_3^- form of sulfurous acid. What are the concentrations of H_2SO_4 and H_2SO_3 in the original sample?

USING THE FRACTION OF TITRATION

75. Define what is meant mathematically by the "fraction of titration." Write the equations that would be used to calculate this value during the titration of an acid with a base or a base with an acid.

76. Describe how charge balance equations and mass balance equations can be used to obtain a fraction of titration equation. Use the titration of a strong acid with a strong base as an example in your answer.

77. Using an approach similar to that shown in Table 12.5, derive Equation 12.30 that gives the fraction of titration for the titration of a strong base with a strong acid.

78. A student wishes to estimate the titration error that will be present during the titration of 25.00 mL 0.1000 M HCl with 0.1000 M NaOH if phenolphthalein is used as the indicator.
    (a) Use a fraction of the titration equation to determine the titration error that will be present if the student uses the center of the color transition for phenolphthalein to

signal the end point for this titration. What is the titration error if they use the beginning of this color transition to signal the end point.
    (b) Repeat the calculations in Part (a) using methyl red or thymolphthalein as indicators. Compare these results to those that were found in Part (a). Based on your calculations, which type of indicator to you think is best for this particular titration?

79. A 25.00 mL solution of 0.1000 M NH_3 is to be titrated with 0.1000 M HCl using methyl red as an indicator for detecting the end point.
    (a) Use a fraction of the titration equation to estimate the titrant volume that will correspond to the pH at which 50% of the methyl red is in either its acid form or base form. What titrant volumes will correspond to the pH values at which methyl red is 10% in its acid form or 90% in the form of its conjugate base? (*Hint:* This may also require use of the Henderson–Hasselbalch equation.)
    (b) b) What is the expected titration error if a chemist uses the middle of the range for the color change of methyl red to signal the end point of the titration? What is the titration error if the beginning or end of this range is used, assuming that these points correspond to a pH where 10% of the methyl red is in its acid form or 90% is in its base form, respectively?

80. Explain how fraction of titration equations can be used with spreadsheets to construct a titration curve for an acid–base titration.

CHALLENGE PROBLEMS

81. What advantages do you think there would be in using a gravimetric titration instead of a volumetric titration to determine the amount of titrant that has been added to an analyte solution? What are the possible disadvantages of such an approach? (*Hint:* See Chapters 3 and 11.)

82. Obtain more information on each of the following topics and write a standard operating procedure for the given piece of equipment or task that might be performed in an analytical laboratory.
    (a) Proper use of a buret for an acid–base titration
    (b) The use of potassium hydrogen phthalate in the standardization of a solution for use in an acid–base titration
    (c) The preparation and use of a phenolphthalein solution as an acid–base indicator

83. There are many acid–base indicators that are available besides those listed in Table 12.3. A few examples of other possible indicators are given below. Using a source such as *Lange's Handbook of Chemistry* or the *CRC Handbook of Chemistry and Physics*,[9,10] determine the pH range over which each of these indicators might be used and describe the change in color that they produce.
    (a) Methyl violet
    (b) Congo red
    (c) α-Naphtholphthalein
    (d) Thymolphthalein
    (e) Nitramine
    (f) Tropeolin O

84. One way to find the inflection points in a titration curve is to make a plot that shows the change in slope versus the change in the volume of titrant, giving a term that is also

known as the second derivative, $d(dpH/dV_{Titrant})$. In a second derivative plot the inflection points occur where the second derivative crosses the x-axis and has a value of zero.

(a) Prepare a spreadsheet to calculate the second derivative for the titration in Figure 12.10.

(b) Use you plot from Part (a) to locate the inflection points in the titration curve. Based on your plot, what volumes of titrant were needed to reach the end points in this titration?

(c) Compare the plot you prepared in Part (a) with the plot of the slope that is given in Figure 12.10. How are these plots similar in the information they provide? How are they different?

85. Derive the Gran plot equation that is given in Equation 12.5 for the analysis of a weak monoprotic base using a strong acid as the titrant. (*Hint:* See Table 12.4.)

86. Many solid dishwashing products in the past contained mixtures of $Na_3PO_4 \cdot 12H_2O$, Na_2CO_3, and $NaOH$ as the major ingredients. A 0.600 g sample of such a product was dissolved in water and titrated with 0.1234 M HCl. When the titration was begun, phenolphthalein was added as an indicator and gave a red to colorless change in color after the addition of 25.10 mL titrant. Bromocresol green was then added as a second indicator and gave a blue-to-yellow change in color at 38.45 mL. An additional 5.00 mL of HCl was added and the solution stayed yellow. The solution was next warmed and bubbled with nitrogen, followed by a back titration using 0.1456 M NaOH. A yellow-to-blue end point for bromocresol green was observed after the addition of 4.24 mL titrant, and a blue-to-purple end point was observed after a total of 12.14 mL of NaOH had been added.

(a) Write equations that show the titration reactions that are occurring during each step of this process. What do you think the purpose is of the step during which the solution is warmed and bubbled with nitrogen? What is the purpose of the back titration?

(b) Use the given information to find the mass and relative amount, in % (w/w), for each of the three main ingredients in the detergent sample. Explain how you determined each of these values.

87. In 2007 and 2008 it was discovered that some unscrupulous food providers in China had added the nitrogen-rich chemical melamine ($C_3H_6N_6$) to pet food and milk products. The alleged purpose of this additive was to fool analysts using the Kjeldahl method into believing the contaminated product had a higher protein content than was actually present. Suppose a food product that has 5.0% (w/w) protein is also spiked with 1.0% (w/w) melamine. What would be the apparent percent nitrogen and percent protein for this product as determined by the Kjeldahl method, if you assume a typical nitrogen content of 17.5% is present for the protein alone? What would be the relative error of this measured protein content when compared to the true protein content of the sample?

88. A student uses Equation 12.10 to estimate the value of $[H^+]$ at the equivalence point for the titration of HCl with NaOH and obtains and incorrect answer of $[H^+] = 0$. The same problem occurs when the student uses Equation 12.12 to estimate this value. Explain why these equations cannot be used for this particular point in the titration curve.

89. Prepare a spreadsheet like the one in Figure 12.17 to plot a titration curve for a weak acid.
Use this spreadsheet to prepare a plot for the titration of a 10.00 mL sample of 0.010 M formic acid with 0.0050 M NaOH. Compare the curve that you obtain to the one that was given in Figure 12.14 under similar conditions for acetic acid. What similarities do you see in these two curves? What differences do you observe?

90. Use the spreadsheet that you prepared in Problem 89 to examine the effect of sample concentration on the titration of a 10.00 mL sample of formic acid with 0.005000 M NaOH.
(a) How does this titration curve change as you go from a sample concentration of 0.010 M to 0.0050 M or 0.0010 M?
(b) At what concentration of formic acid is it no longer feasible to do this titration? Explain your answer.

91. Derive the fraction of titration equation that is shown in Table 12.6 for the titration of a strong base with a strong acid. Show the various steps that you took in obtaining this final equation.

92. Derive the fraction of titration equation that is shown in Table 12.6 for the titration of a weak acid with a strong base. Show the various steps that you took in obtaining this final equation.

TOPICS FOR DISCUSSION AND REPORTS

93. Locate an industrial laboratory in your area and learn how workers in this laboratory use acid–base titrations, including related techniques such as the Kjeldahl method. Write a report on your findings.

94. Litmus is a common acid–base indicator that is often used to identify acidic or basic solutions in water. Obtain information on the chemical and physical properties of litmus. Use this information to describe the ability of litmus to act as an acid–base indicator.

95. When the color change due to an acid–base indicator is not sufficiently sharp at an end point, one alternative strategy is to mix two indicators together (or an indicator plus a dye) to give a more distinct change in color at the desired pH. Some examples of mixed indicators can be found in sources such as References 9, 12, or 13. Find a few examples of mixed indicators and describe their color change and the pH over which this color occurs. Discuss how these mixed indicators work and explain why they can provide an improved end point over the use of a single indicator.

96. Below are listed several scientists who made important contributions to the method of titrations. Write a report on one of these individuals and state how they assisted in the development of this technique.
(a) Joseph Louis Gay-Lussac
(b) Karl Friedrich Mohr
(c) Claude Louis Berthollet
(d) Adolf Baeyer

97. Locate or suggest a procedure for performing an acid–base titration for one or more of the following situations. Describe the reactions that take place in this method and state the purpose for each step in the procedure.
(a) Titration of the sulfuric acid in water from mine acidic waste
(b) Determination of the purity of an aspirin tablet
(c) Measurement of the percent sulfur in a coal sample

References

1. *The New Encyclopaedia Britannica*, 15th ed., Encyclopaedia Britannica, Inc., Chicago, IL, 2002.
2. F. Szabadvary, *History of Analytical Chemistry*, Pergamon Press, New York, 1966.
3. "Facts and Figures for the Chemical Industry," *Chemical and Engineering News*, June 25, 2001, p. 79.
4. *IUPAC Compendium of Chemical Terminology*, Electronic version, http://goldbook.iupac.org
5. G. Maludzinska, Ed., *Dictionary of Analytical Chemistry*, Elsevier, Amsterdam, the Netherlands, 1990.
6. H. Diehl, *Quantitative Analysis: Elementary Principles and Practice*, 2nd ed., Oakland Street Science Press, Ames, IA, 1974.
7. T. S. Ma and R. C. Rittner, *Modern Organic Elemental Analysis*, Marcel Dekker, New York, 1979.
8. T. S. Ma, "Organic Elemental Analysis," In *Analytical Instrumentation Handbook*, 2nd ed., G. W. Ewing, Ed., Marcel Dekker, New York, 1997, Chapter 3.
9. J. A. Dean, *Lange's Handbook of Chemistry*, 15th ed., McGraw-Hill, New York, 1999.
10. D. R. Lide, Ed., *CRC Handbook of Chemistry and Physics*, 83rd ed., CRC Press, Boca Raton, FL, 2002.
11. G. Gran, "Determination of the Equivalence Point in Potentiometric Titrations, Part II," *Analyst*, 77 (1952) 661–671.
12. Kolthoff and Stenger, *Volumetric Analysis*, Vol. 1, Interscience Publishers, New York, 1942.
13. Kolthoff and Stenger, *Volumetric Analysis*, Vol. 2, Interscience Publishers, New York, 1947.

COLOR PLATE 1

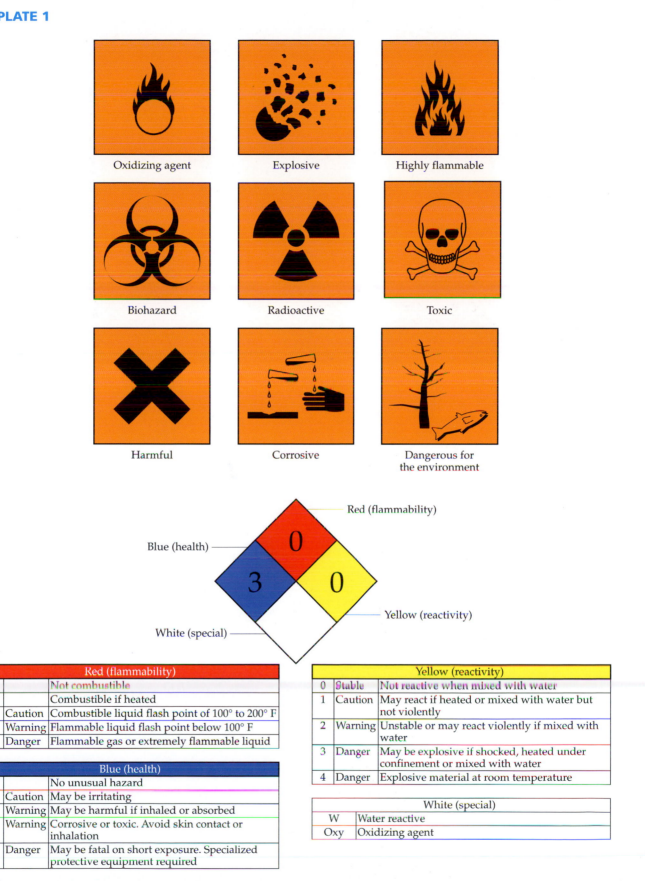

Oxidizing agent

Explosive

Highly flammable

Biohazard

Radioactive

Toxic

Harmful

Corrosive

Dangerous for
the environment

Red (flammability)

Blue (health)

White (special)

Yellow (reactivity)

Red (flammability)		
0		Not combustible
1		Combustible if heated
2	Caution	Combustible liquid flash point of 100° to 200° F
3	Warning	Flammable liquid flash point below 100° F
4	Danger	Flammable gas or extremely flammable liquid

Blue (health)		
0		No unusual hazard
1	Caution	May be irritating
2	Warning	May be harmful if inhaled or absorbed
3	Warning	Corrosive or toxic. Avoid skin contact or inhalation
4	Danger	May be fatal on short exposure. Specialized protective equipment required

Yellow (reactivity)		
0	Stable	Not reactive when mixed with water
1	Caution	May react if heated or mixed with water but not violently
2	Warning	Unstable or may react violently if mixed with water
3	Danger	May be explosive if shocked, heated under confinement or mixed with water
4	Danger	Explosive material at room temperature

White (special)	
W	Water reactive
Oxy	Oxidizing agent

(Top figure) Common symbols for chemical hazards. See Chapter 2 for more details on methods for chemical labeling and on chemical hazards. (Bottom figure) The National Fire Prevention Association (NFPA) labeling system. This particular NFPA label is for the chemical ethidium bromide. See Chapter 2 for more details on this labeling system.

Standard precipitation   Homogeneous precipitation

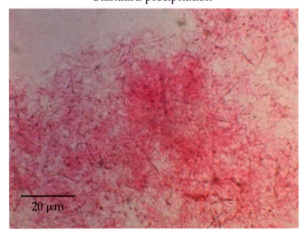
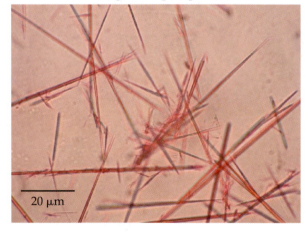

20 μm   20 μm

Precipitation of Ni^{2+} with dimethylglyoxime by using a standard precipitation method (direct addition of dimethylglyoxime to a solution of the analyte) versus precipitation from a homogeneous solution (generation of dimethylglyoxime in solution). The top images show that standard precipitation in this case produces a solid that clings to walls and floats in solution, while a solid that is easier to handle and filter is obtained when precipitating the same solid from a homogeneous solution. The microscopic view in the bottom images shows that the standard precipitation method gives extremely tiny, nearly amorphous solid particles while homogeneous precipitation gives large and well-formed needle crystals that are much easier to filter. See Chapter 11 for a further discussion of these precipitation methods.

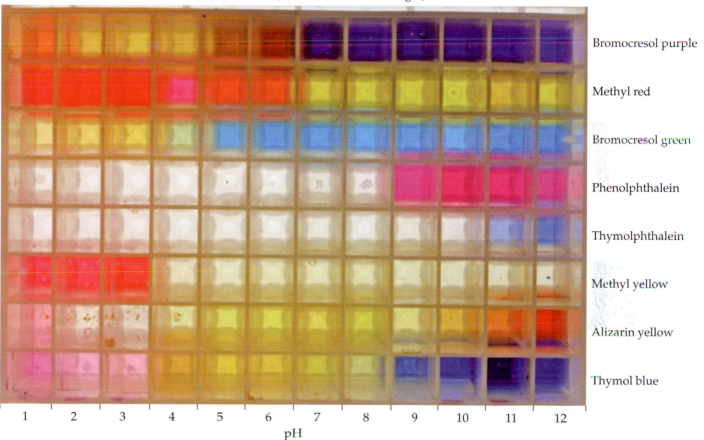

Bromocresol purple

Methyl red

Bromocresol green

Phenolphthalein

Thymolphthalein

Methyl yellow

Alizarin yellow

Thymol blue

pH

Examples of common acid-base indicators. The scale at the bottom shows the pH of each corresponding well, with the color change in indicators as a function of pH being shown across each row. See Chapter 12 for a further discussion of acid-base indicators.

COLOR PLATE 4

Complexometric indicators (under standard white light)

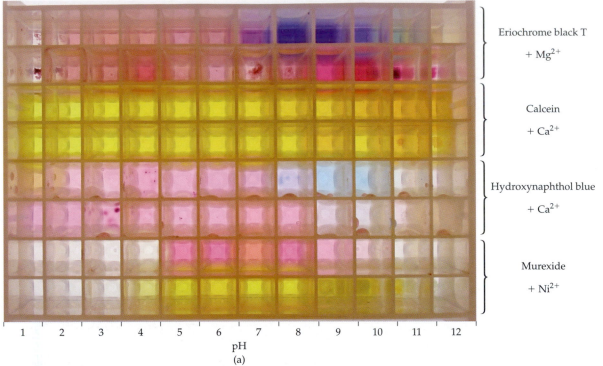

Eriochrome black T
+ Mg^{2+}

Calcein
+ Ca^{2+}

Hydroxynaphthol blue
+ Ca^{2+}

Murexide
+ Ni^{2+}

pH
(a)

Complexometric indicators (under UV light)

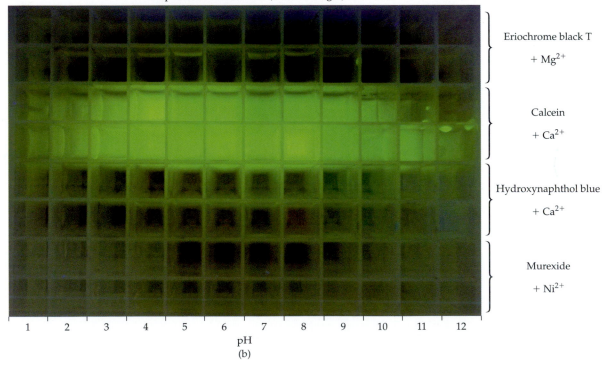

Eriochrome black T
+ Mg^{2+}

Calcein
+ Ca^{2+}

Hydroxynaphthol blue
+ Ca^{2+}

Murexide
+ Ni^{2+}

pH
(b)

Examples of common metallochromic indicators for complexometric titrations. The scale at the bottom shows the pH of each corresponding well, with the color change in indicators as a function of pH being shown across each row. Two rows are shown for each indicator, the top row showing the color of the indicator in the absence of any added metal ions and bottom row showing the color in the presence of a particular metal ion. The image in (a) was obtained under normal laboratory lighting, while the image in (b) was obtained under UV light, illustrating the ability of calcein (see rows 3 and 4) to be used as an indicator for Ca^{2+} with either visible or fluorescence detection. See Chapter 13 for a further discussion of metallochromic indicators.

COLOR PLATE 5

Volhard method

Before titration   Before the end point   After the end point

(a)

Mohr method

(b)

Fajans method

Before titration   Before the end point   After the end point

(c)

Various methods for detecting the end point in precipitation titrations in which Ag^+ is either the analyte or titrant. In (a) the Volhard method, Ag^+ is titrated with thiocyanate (SCN^-) to give solid AgSCN; Fe^{3+} is used as the indicator and reacts with excess thiocyanate after the equivalence point to form $FeSCN^{2+}$, a red soluble complex. In (b) the Mohr method, Ag^+ is the titrant for analytes such as Cl^- and the end point is detected through the reaction of the indicator chromate (CrO_4^{2-}) with Ag^+ to form silver chromate (Ag_2CrO_4), a red precipitate. In (c) the Fajans method, Ag^+ is the titrant for Cl^- and the indicator is a negatively-charged dye such as fluorescein (e.g., dichlorofluorescein in this example); this dye absorbs to the AgCl precipitate in the presence of a slight positive charge on this solid, as occurs after the end point when excess Ag^+ has been added. See Chapter 13 for a further discussion of these methods.

Common redox indicators. Each pair of beakers shows the oxidized and reduced forms of these indicators. In the case of starch, the pair of beakers shows the response obtained for the indicator in the presence of the I_3^- (left) or I^- (right). See Chapter 15 for a further discussion of redox indicators.

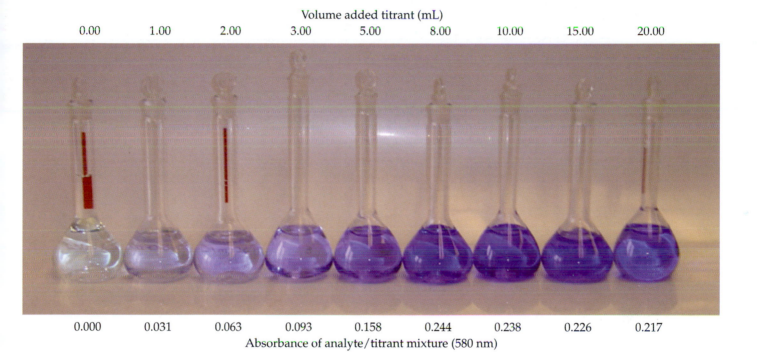

Volume added titrant (mL)

| 0.00 | 1.00 | 2.00 | 3.00 | 5.00 | 8.00 | 10.00 | 15.00 | 20.00 |

| 0.000 | 0.031 | 0.063 | 0.093 | 0.158 | 0.244 | 0.238 | 0.226 | 0.217 |

Absorbance of analyte/titrant mixture (580 nm)

An example of a spectrophotometric titration, in which a sample containing a fixed amount of Cu^{2+} is combined with various amounts of triethylenetetramine (trien) as the titrant to form a colored complex. The final concentration of Cu^{2+} in each volumetric flask was 1.43×10^{-3} M. The volumes of titrant that were added to each volumetric flask (before diluting the contents to the mark with water) are given at the top of the figure. The absorbance readings taken of the resulting analyte/titrant mixtures are given at the bottom. It was determined from these data that the end point of this titration occurred when the equivalent of 7.73 mL of trien had been added to one of these flasks (see Chapter 18 for more details on this method).

$$Fe^{2+} + 3\ \text{phen} \longrightarrow Fe(\text{phen})_3^{2+}$$

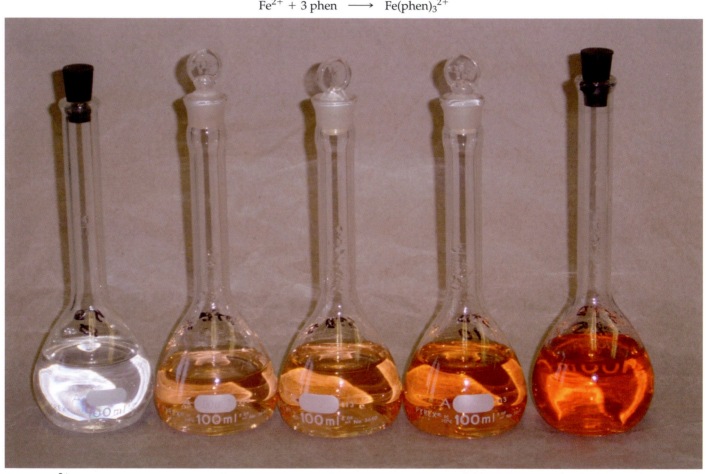

Conc. Fe²⁺:	0.000 ppm	0.359 ppm	0.599 ppm	1.198 ppm	2.995 ppm
Abs. (510 nm):	0.004	0.084	0.133	0.264	0.640

Standard solutions for the spectrophotometric determination of Fe^{2+} through its reaction with 1,10-phenanthroline (phen) to form a colored complex. The solutions in these volumetric flasks were prepared to contain an excess of 1,10-phenantholine as a complexing agent for Fe^{2+} and color-forming agent. These flasks also contained hydroxylamine (a reducing agent to prevent the air oxidation of Fe^{2+} to Fe^{3+}) and sodium acetate as a buffer. The final concentration of Fe^{2+} in each of these standards is shown at the top. The absorbance values measured at 510 nm for these standard solutions are listed at the bottom. Other examples of such assays can be found in Chapter 18.

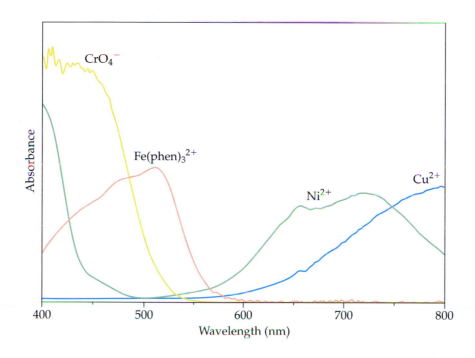

Examples of how the observed color for a chemical solution is related to the absorbance spectrum for this solution in the visible wavelength range. The containers at the bottom are 1.0 cm cuvettes that contain aqueous solutions of Ni^{2+} (green), CrO_4^{2-} (yellow), Cu^{2+} (blue) or $Fe(phen)_3^{2+}$ (orange). The corresponding absorption spectra for these same solutions are shown at the top. This comparison illustrates how the observed color of the solutions is related to the wavelengths of visible light that are not absorbed, but rather are transmitted, through the solutions. This effect is discussed in more detail in Chapter 17.

Before the end point (all added I_3^- titrant
is consumed by the analyte)

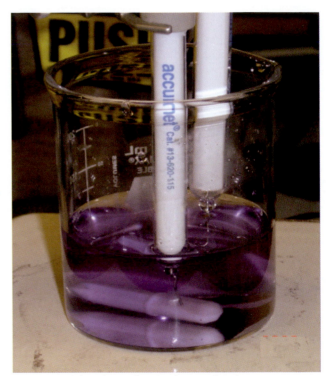

After the end point (excess I_3^- titrant
reacts with starch indicator)

Results obtained before and after the end point in a coulometric titration for ascorbic acid using the coulometric generation of I_3^- as a titrant. The end point is detected by using starch as the indicator, which forms a dark blue complex with I_3^- once excess titrant is present in the analyte/titrant mixture. This approach is described in more detail in Chapter 16.

Solution of Cu^{2+} and platinum electrode
before electrodeposition of copper

Solution and platinum electrode after
electrodeposition of copper

The electrodeposition of Cu^{2+} onto a platinum electrode acting as a cathode. The image on the left shows the original, blue solution of Cu^{2+} in water and the silver-colored electrode before it was used for electrodeposition. The image on the right demonstrates how the Cu^{2+} has now been removed from the solution as is now deposited as a coating of copper metal on the platinum electrode. This approach is described in more detail in Chapter 16.

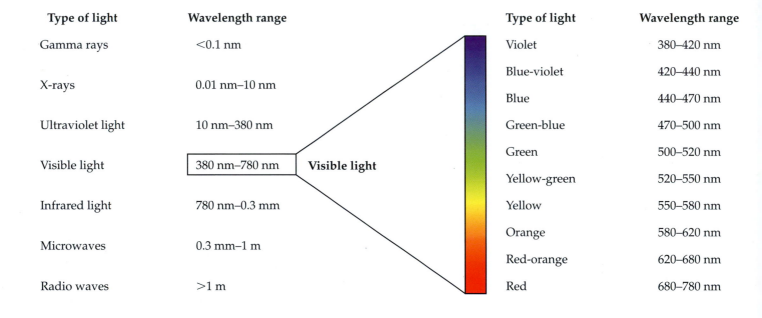

Type of light	Wavelength range		Type of light	Wavelength range
Gamma rays	<0.1 nm		Violet	380–420 nm
X-rays	0.01 nm–10 nm		Blue-violet	420–440 nm
			Blue	440–470 nm
Ultraviolet light	10 nm–380 nm		Green-blue	470–500 nm
			Green	500–520 nm
Visible light	380 nm–780 nm	Visible light	Yellow-green	520–550 nm
Infrared light	780 nm–0.3 mm		Yellow	550–580 nm
			Orange	580–620 nm
Microwaves	0.3 mm–1 m		Red-orange	620–680 nm
Radio waves	>1 m		Red	680–780 nm

Various types of electromagnetic radiation, or "light." The approximate frequencies and wavelengths are shown for each type of electromagnetic radiation, with the scale on the right showing an enlarged view of the wavelengths that make up visible light. More details on the applications of these various types of light in chemical analysis can be found in Chapters 17–19. (The wavelength ranges used in this figure are based primarily on those given in "Guide for Use of Terms in Reporting Data in Analytical Chemistry: Spectroscopy Nomenclature", *Analytical Chemistry*, 62 (1990) 91–92; the approximate ranges given for the colors of visible light vary slightly from one source to the next in the literature.)

Chapter 13

Complexometric and Precipitation Titrations

Chapter Outline

13.1 INTRODUCTION: HOW HARD IS THE WATER?

"Hard water" has been a problem for humans ever since the invention of plumbing. *Water hardness* is a measure of the total concentration of calcium and magnesium ions in water, and is expressed as milligrams of $CaCO_3$ per liter of water (see Figure 13.1). Hard water can lead to the buildup of mineral deposits in pipes and in appliances that use water on a regular basis, affecting the performance and usable lifetime of these items. This makes the measurement of water hardness an important test that is conducted by many industries and water-treatment plants.[1–3]

The degree of water hardness varies throughout the United States and depends on the geology of the area from which the water is obtained. Calcium ions get into groundwater when this water comes in contact with limestone ($CaCO_3$),

$$CaCO_3(s) + H_2O \rightarrow Ca^{2+} + HCO_3^- + OH^- \quad (13.1)$$

a process that results in the formation of bicarbonate and hydroxide ions. When this hard water is later heated or evaporated, it reforms calcium carbonate and deposits this mineral on surfaces such as the inside of a pipe or a water heater. Another problem with hard water is that the fatty acids in soap will form an insoluble precipitate with calcium ions. This leads to the formation of soap "scum" and makes it difficult to clean things with hard water.[1,2]

The measurement of calcium and magnesium ions for the determination of water hardness is just one of many examples in which specific ions in water must be analyzed. Two approaches that can be used to measure specific ions in water are a complexometric titration or a precipitation titration. We discuss both of these methods in this chapter and look at their applications.[3–5] We will also see how we can predict the behavior of these titrations during a chemical analysis.

13.1A What Is a Complexometric or Precipitation Titration?

A complexometric titration or a precipitation titration is similar to an acid–base titration in that these are often performed as volumetric methods that use a measurement of titrant volume to determine the amount of an analyte that is in a sample. However, these titration methods make use of different types of reactions. A **complexometric titration** is a titration that involves complex formation (see Chapter 9 for a review of this type of reaction).[6,7] A typical complexometric titration is shown in Figure 13.2, using the analysis of Ca^{2+} in water with ethylenediamine tetraacetic acid (EDTA) as an example. In most complexometric titrations the analyte is a Lewis acid (such as Ca^{2+}) and the complexing agent is a Lewis base (such as EDTA), which has one or more pairs of electrons that can be donated to the analyte. It is also possible for the complexing agent to be the analyte and a solution of the metal ion to act as the

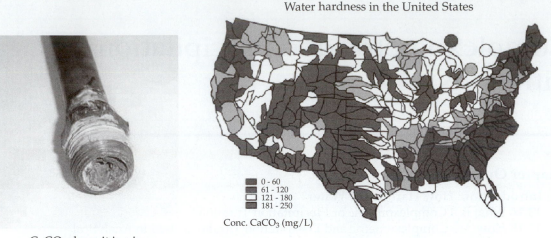

Water hardness in the United States

Conc. CaCO$_3$ (mg/L)

0 - 60
61 - 120
121 - 180
181 - 250

CaCO$_3$ deposit in pipe

FIGURE 13.1 The average hardness of water (reported here as the concentration of calcium carbonate in units of mg/L or ppm) and the effects of hard water on pipes. In general, water is considered to be "soft" if it has a hardness value less than or equal to 55 ppm, "moderately hard" if it has a hardness of 55–120 ppm, "hard" if this value is 120–250 ppm, and "very hard" if this value is greater than 250 ppm. (The map on the right shows data from 1975 and is provided courtesy of the U.S. Geological Survey.)

titrant, as might occur when we are standardizing a solution of EDTA by titrating it with a known concentration of Ca^{2+}.

A **precipitation titration** is a titration method in which the reaction of a titrant with a sample leads to an insoluble precipitate.[6,7] This method is generally carried out by adding to a sample known volumes of a solution containing a precipitating agent until no further precipitate is formed. An example of such a method is shown in Figure 13.3, in which a sample containing Ag$^+$ is titrated with a solution containing a known concentration of Cl$^-$, leading to the formation of insoluble AgCl(s). When the

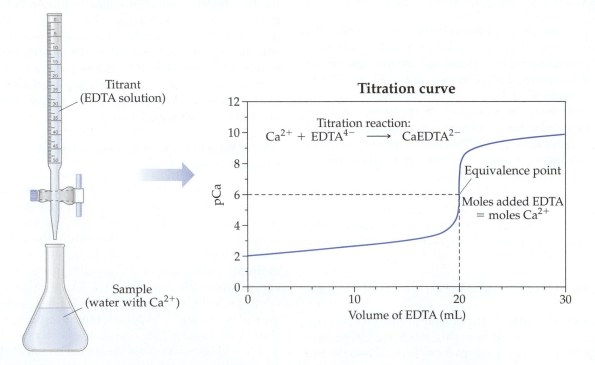

FIGURE 13.2 A typical complexometric titration, using the analysis of Ca^{2+} in water by its titration with ethylenediaminetetraacetic acid (EDTA) as an example. The dashed lines shown in the titration curve indicate the volume of the EDTA titrant that is needed to reach the equivalence point and gives the Ca^{2+} concentration at this point by using the function pCa, where pCa $= -\log([Ca^{2+}])$. The titration curve in this example shows the response expected for a 10.00 mL aliquot of a 0.01000 M Ca^{2+} solution in water at pH 10.00 that is titrated using 0.005000 M EDTA.

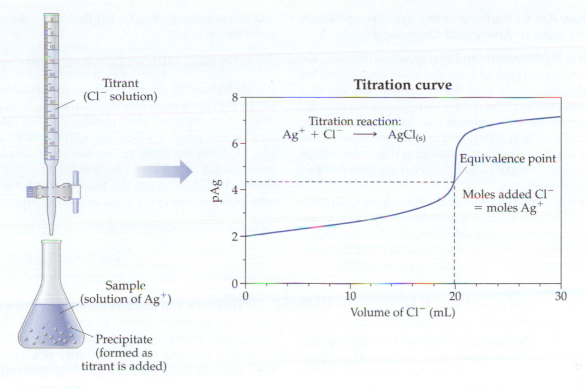

FIGURE 13.3 A typical precipitation titration, using the analysis of Ag^+ in an aqueous solution by its titration with Cl^- as an example. An insoluble precipitate ($AgCl(s)$, in this case) is formed as the titrant is added to the sample. The dashed lines shown in the titration curve indicate the volume of titrant that is needed to reach the equivalence point and gives the concentration of Ag^+ at this point by using the function pAg, where $pAg = -\log([Ag^+])$. The titration curve shown is the response expected for a 10.00 mL aliquot of a 0.01000 M aqueous solution of Ag^+ that is titrated using 0.005000 M Cl^-.

lack of further precipitation (or some related signal) indicates that the end point has been reached, the delivered volume of titrant (which has a known concentration and reaction stoichimetry with analyte) is used to determine the concentration of the analyte in the original sample. Figure 13.3 gives one application for this approach, in which it is used to measure a metal ion such as Ag^+ by adding a titrant that will precipitate this metal ion from solution in a known manner. Another common way of using this type of titration is to use a metal ion like Ag^+ as a titrant to measure a chemical that will react with this titrant to give a precipitate, as could be done to determine the concentration of Cl^- in a sample.

The titration curves for complexometric and precipitation titrations are often made by plotting a function of the analyte concentration on the y-axis versus volume of titrant that has been added on the x-axis. The concentration of analyte M (representing a metal ion, in this case) is often given on the y-axis by using a function commonly referred to as **pM**, where $pM = -\log[M^{n+}]$ for a solution of ion M^{n+} in water. As an example, the symbol pCa in Figure 13.2 represents the value of $-\log([Ca^{2+}])$ at any point along the titration curve on the left. Similarly, pAg is equal to the value $-\log([Ag^+])$ for the precipitation titration curve on the right. This is similar to the approach we followed when using pH to represent $-\log(a_{H^+}) \approx -\log([H^+])$ during an acid–base titration. The same general technique can also be used to describe the concentration of other analytes that are not metal ions (e.g., $pCl = -\log([Cl^-])$ if this analyte is being titrated with Ag^+).

The general shape of the titration curves in Figure 13.2 and Figure 13.3 is similar in many ways to the curve for a titration involving a strong acid and strong base (see Chapter 12). All of these curves have a small change in response in the early to intermediate phases of the titration, followed by a sharp change in response as the moles of titrant approach the moles of analyte in the sample. This response then gives a small change as we move into the region of the curve where an excess of titrant has been added. Many of the same terms that are used to describe acid–base titrations are also helpful in describing complexometric and precipitation titrations. For example, the point at which the moles of added titrant is exactly equal to the moles of original analyte is again called the "equivalence point," and the point at which we detect that enough titrant has been added represents the "end point." We will learn later how we can predict the shape of these titration curves to help create methods in which the end point is a good estimate of the equivalence point. To do this, we will use tools similar to those described in Chapter 12 to predict the response of acid–base titrations.

13.1B How Are Complexometric and Precipitation Titrations Used in Analytical Chemistry?

Although complexometric and precipitation titrations are used for different types of ions, the main application for both of these methods is to determine the amount of a specific ion that is present in a sample. As we saw for acid–base titrations, this goal is met by determining the amount of titrant that is required to reach the equivalence point for a sample and by combining this information with the known reaction that is occurring between this titrant and the analyte.

The most common type of complexometric titration involves the use of EDTA as a complexing agent for metal ions (see Chapter 8). A valuable feature of EDTA is that it has strong binding to many metal ions and that the product of this reaction is a 1:1 complex, as shown below.

$$M^{n+} + EDTA^{4-} \rightleftharpoons M(EDTA)^{n-4} \qquad (13.2)$$

In this reaction, the moles of a metal ion (n_M) in the sample (total molar concentration = C_M and original volume = V_M) will be equal to the moles of EDTA (n_{EDTA}) that are required to reach the equivalence point. The moles of titrant needed to reach the equivalence point can, in turn, be determined by using the measured volume of titrant added to the sample up to this point (V_{EDTA}) and the known total molar concentration of the titrant (C_{EDTA}), as shown below.

At the Equivalence Point for Titration of a Metal Ion with EDTA:

$$n_M = n_{EDTA} \qquad \text{or} \qquad C_M V_M = C_{EDTA} V_{EDTA} \qquad (13.3)$$

This relationship makes it easy to determine the concentration of the metal ion by using the volume of EDTA that is needed to reach the end point in a complexometric titration. The same concepts can also be applied to determine the concentration of an analyte through the use of a precipitation titration if the moles of titrant needed to reach the equivalence point are determined and the reaction of this titrant with the analyte has a known stoichiometry.

EXERCISE 13.1 | **Determining the Hardness of Water**

What is the "hardness" of a 250.00 mL water sample if this sample gives an equivalence point at 15.68 mL when it is titrated with 0.02578 M EDTA?

SOLUTION

The moles of EDTA that are needed to completely react with the Ca^{2+} and Mg^{2+} in the sample can be determined from the concentration of EDTA in the titrant solution

and the volume of this titrant that was added up to the equivalence point.

$$(0.02578 \text{ mol/L})(0.01568 \text{ L}) = 4.042 \times 10^{-4} \text{ mol EDTA}$$

We also know from the reaction in Equation 13.2 that EDTA will react with both Ca^{2+} and Mg^{2+} in a 1:1 ratio. This means that the total moles of EDTA added at the equivalence point will be equal to the total moles of Ca^{2+} and Mg^{2+} in the sample. We can then divide the total moles of Ca^{2+} and Mg^{2+} by the volume of the original sample (0.250 L) to get the total concentration of these ions in the sample, giving the following answer.

$$(4.042 \times 10^{-4} \text{ mol})/(0.25000 \text{ L}) = 1.61\underline{7} \times 10^{-3} \text{ } M$$

$$= \mathbf{1.62 \times 10^{-3} \text{ } M \text{ } Ca^{2+} \text{ and } Mg^{2+}}$$

Titrations can be used to determine the concentration of an analyte not only in molarity but also in other units. For instance, water hardness is usually not expressed in terms of total molarity of Ca^{2+} and Mg^{2+}, but rather as their equivalent of concentration in milligrams of $CaCO_3$ per liter water (or parts-per-million $CaCO_3$). This conversion is accomplished by assuming that all of the water hardness is due to Ca^{2+}, even though in reality both Ca^{2+} and Mg^{2+} were probably present and reacted with the EDTA. This calculation is illustrated below using data from the previous exercise.

Water hardness (in ppm $CaCO_3$)

$$= (1.617 \times 10^{-3} \text{ mol/L } Ca^{2+} \text{ and } Mg^{2+}) \cdot$$

$$\frac{1 \text{ mol } CaCO_3}{1 \text{ mol } Ca^{2+}} \cdot \frac{100.087 \text{ g}}{\text{mol } CaCO_3} \cdot \frac{10^3 \text{ mg}}{\text{g}}$$

$$= 161.7 \text{ mg/L or } 161.7 \text{ ppm } CaCO_3$$

It is this type of calculation and these units that were used to prepare the map of water hardness in the United States that was given in Figure 13.2.

13.2 PERFORMING A COMPLEXOMETRIC TITRATION

Many of the procedures that are used in complexometric titrations are similar to those described in the previous chapter for acid–base titrations. Both methods make use of burets to dispense titrant and involve the use of standardized titrant solutions. There are, however, some differences in specific aspects of these techniques, as we will see in this section.

13.2A Titrants and Standard Solutions

EDTA as a Titrant. EDTA is by far the most common titrant used in complexometric titrations.[3–5,8] A good example of this is the use of EDTA to measure Ca^{2+} and

Mg^{2+} in water to determine water hardness. EDTA rapidly forms a stable 1:1 complex with nearly all metal ions (as illustrated in Figure 13.4). The form of EDTA that is involved to the greatest extent in this complexation reaction is the tetraprotic acid form ($EDTA^{4-}$), in which each of the four carboxylate anions on EDTA and the two tertiary amine nitrogen atoms can donate pairs of electrons to a metal ion (see Chapter 9). This complexation reaction often has a large equilibrium constant whose exact value will depend on pH. This provides a reaction that is essentially quantitative at the equivalence point of the titration and that has a simple stoichiometry. These features make it relatively easy to determine the concentration of ions like Ca^{2+} and Mg^{2+} when using EDTA as a titrant.

A change in pH will affect the fraction of EDTA that is present in its $EDTA^{4-}$ form, which we can describe by using the fraction of species term $\alpha_{EDTA^{4-}}$. This change with pH will, in turn, change the conditional formation constant between EDTA and a metal ion, a topic we discussed in Chapter 9. The result of this effect is illustrated in Figure 13.5 during the use of a 0.005000 M EDTA solution to titrate 0.01000 M Ca^{2+} in 10.00 mL water, but the same trend is seen at other EDTA concentrations and in the EDTA titration of other metal ions. In this example, the conditional formation constant $K'_f = K_f\,\alpha_{EDTA^{4-}}$, where K_f is the true formation constant for EDTA with Ca^{2+}. At pH 11.0 most of the EDTA is in the form $EDTA^{4-}$, so the conditional formation constant is near the true formation constant of $\log(K_f) = 10.65$ for this particular reaction. As the pH of the sample and titrant is decreased, the fraction of EDTA that is present in the appropriate form for the titration is dramatically lowered and there is a corresponding change in the conditional formation constant for this titration. This produces a lower plateau after the equivalence point and a less sharp change in pCa in the region around the equivalence point. A value for K_f' of greater than 10^8 (or $\log(K_f') > 8$) is usually desired in the titration of a metal ion with EDTA (as discussed in Chapter 9). This condition occurs at a pH of approximately 8 or greater in this example. At a lower pH the change in response at or near

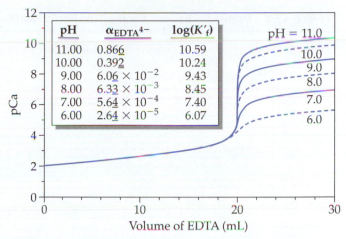

pH	$\alpha_{EDTA^{4-}}$	$\log(K'_f)$
11.00	0.866	10.59
10.00	0.392	10.24
9.00	6.06×10^{-2}	9.43
8.00	6.33×10^{-3}	8.45
7.00	5.64×10^{-4}	7.40
6.00	2.64×10^{-5}	6.07

FIGURE 13.5 Effect of pH on the relative fraction of ethylenediaminetetraacetic acid (EDTA) in the form $EDTA^{4-}$ and the resulting change in the titration curve for the complexometric titration of Ca^{2+} by EDTA. This titration curve is the response expected for a 10.00 mL aliquot of a 0.01000 M Ca^{2+} solution in water at the given pH values when using 0.005000 M EDTA as the titrant. It is assumed that no significant side reactions other than the acid–base dissociation of EDTA are present over the given pH range. The calculated values of $\log(K_f')$ were found using the relationship $K'_f = K_f\,\alpha_{EDTA^{4-}}$ (see Chapter 9, Section 9.3D), along with a value of $\log(K_f) = 10.65$ or $K_f = 4.47 \times 10^{10}$ for the binding of EDTA with Ca^{2+}. The values for $\alpha_{EDTA^{4-}}$ at each pH were obtained from Table 9.8, but could also be found by using Equation 9.17 (see Chapter 9).

the equivalence point becomes too small to allow an accurate and reproducible analysis of Ca^{2+}.

Along with a change in the distribution of the acid–base forms of EDTA, other side reactions can also occur as we go to a higher pH. An example that pertains to the titration of Ca^{2+} by EDTA is the combination of Ca^{2+} at a high pH with OH^- to form $CaOH^+$ (see Chapter 9). Other metal ions (e.g., Ni^{2+}) will combine with OH^- at a high pH to form insoluble metal hydroxides. At a lower pH, another side reaction that can take place is the simultaneous binding of $EDTA^{4-}$ with both Ca^{2+} and a hydrogen ion. The combined effect of these side reactions is they limit the pH range over which a given metal ion can be titrated with EDTA. This pH range will depend on the type of metal ion that is being titrated and on the nature of these side reactions.

We saw in Chapter 9 (Section 9.3D) how the overall effects of these side reactions can be predicted through the use of conditional formation constants. Table 13.1 shows the pH range over which a titration by EDTA will give a conditional formation constant of 10^8 or greater for a given metal ion without significant problems from such side reactions.[4,5,9] This information is not only useful in finding what pH range can be used in the EDTA titration of a given analyte, but it can also be used to control the selectivity of this titration by adjusting the pH so that a particular metal ion or group of metal ions can be measured without interference from other sample components.

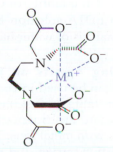

FIGURE 13.4 Structure of the 1:1 complex that is formed between a metal ion M^{n+} and $EDTA^{4-}$.

TABLE 13.1 Usable pH Ranges for the Titration of Various Metal Ions with EDTA*[a]

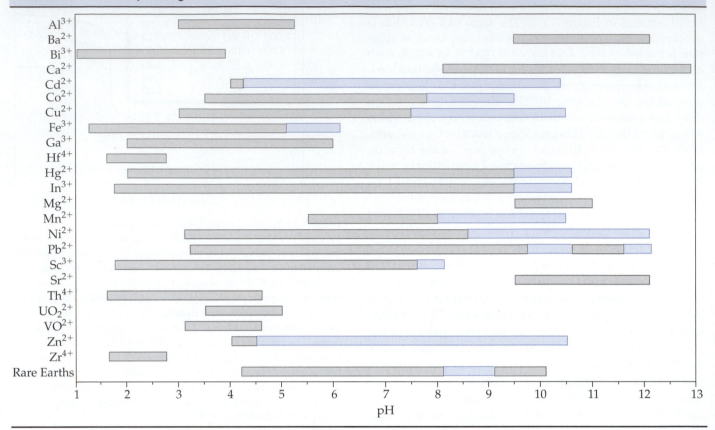

*This table is based on data from K. Ueno, *Journal of Chemical Education*, 42 (1965) 432; R. Pribil, *Applied Complexometry*, Pergamon Press, New York, 1982; and A. Ringbom, *Complexation in Analytical Chemistry*, Krieger Publishing, Huntington, NY, 1979.

[a]The light shaded regions indicate pH values at which an auxiliary agent is needed to successfully carry out the EDTA titration.

EXERCISE 13.2	Selecting a pH for an EDTA Titration

A water scientist wants to use an EDTA titration to measure the Ca^{2+} content of water with no significant interferences from Mg^{2+}. A pH of 12 is chosen for this titration. Is this an appropriate pH for such a method? What pH values could be used to titrate both Ca^{2+} and Mg^{2+}.

SOLUTION

According to Table 13.1, the overall range that can be used for the titration of Ca^{2+} with EDTA occurs at a pH of roughly 8.0 or greater. Mg^{2+} will only be titrated by EDTA if a pH of approximately 9.4 to 10.9 is used. (*Note:* At a higher pH, Mg^{2+} begins to precipitate as $Mg(OH)_2$ and only reacts slowly with EDTA.) Thus, it should be possible to obtain a selective titration of Ca^{2+} by EDTA and in the presence of Mg^{2+} at pH 12.0. Table 13.1 also indicates that the titration of both Ca^{2+} and Mg^{2+} with EDTA should be possible when using a pH of 9.4–10.9.

Standard solutions of EDTA for use in a complexometric titration can be prepared directly from commercially-available preparations of H_4EDTA or

$Na_2H_2EDTA \cdot 2H_2O$. Neither of these forms of EDTA will dissolve well in water unless a strong base like NaOH or KOH is also added. It is important to ensure that the EDTA has completely dissolved when making its solution to avoid introducing a systematic error in the calculated concentration of this titrant. It is also possible to prepare a solution of EDTA and then to standardize this solution by using it to titrate a standard solution that contains a known quantity of the desired metal ion. As an example, a standard solution prepared by dissolving a known quantity of $CaCO_3$ can be used to standardize an EDTA solution for the titration of Ca^{2+} during the analysis of water hardness.

Other Complexing Agents as Titrants. Several other titrants that can be used in complexometric titrations are given in Table 13.2. Like EDTA, all of the titrants in this list are multidentate ligands that can form a 1:1 complex with metal ions. This property gives much simpler titration curves and sharper end points than when using ligands that form higher-order complexes. In the case of a monodentate ligand like ammonia (which forms 1:1 up to 1:6 complexes with Ni^{2+}), a mixture of complexes will be present throughout most

TABLE 13.2   Common Titrants for Metal Ions in Complexometric Titrations

Common Name and Structure[a,b]	Properties/Applications
EDTA	General-purpose titrant for a broad range of metal ions
EGTA	Binds to Ca^{2+} more strongly than Mg^{2+}
DTPA	Complexes formed by DTPA with many metal ions are 1000-fold more stable than those for EDTA
CDTA	Complexes formed by CDTA with large metal ions are more stable than those for EDTA
Tetren	Binds strongly to transition metal ions but not to ions of Group IA or IIA elements

[a]Abbreviations: EDTA = ethylenediaminetetraacetic acid; EGTA = ethyleneglycol tetraacetic acid; DTPA = diethylenetriamineypentaacetic acid; CDTA = 1,2-diaminocyclohexane tetracetic acid (*trans* form shown); tetren = tetraethylenepentamine.

[b]These structures show the neutral form of each complexing agent. The most basic form for each of these complexing agents is the species that is typically used in a complexometric titration. The fraction of this species at a given pH can be calculated by using the pK_a values for these agents that are provided in Appendix B.

of the titration curve, which creates only a gradual change in pM versus titrant volume and makes it difficult to find a usable end point.

Although EDTA is a good general-purpose titrant, some complexing agents are better titrants than EDTA in certain situations. 1,2-Diaminocyclohexanetetraacetic acid (also known as DCTA or CDTA) forms stronger complexes than EDTA with some metal ions, making it easier to titrate dilute samples of these analytes. Polyamine-containing agents such as triethylenetetramine (trien) or tetraethylenepentamine (tetren) both form quite stable complexes with transition elements, but do not form complexes with alkaline earth elements like Ca^{2+} or Mg^{2+}. Ethyleneglycoldiamine tetraacetic acid (EGTA) forms a much stronger complex with Ca^{2+} than with Mg^{2+}, allowing this titrant to be used to selectively determine calcium ions in the presence of magnesium ions. As is true for EDTA, each of the titrants in Table 13.2 is part of a polyprotic acid–base system, so their ability to form complexes with metal ions will be affected by pH. Like EDTA, these agents tend to have their tightest binding to metal ions when in their most deprotonated and basic form.

All of the titrants shown in Table 13.2 are based on multidentate, organic complexing agents. There are some inorganic agents that can also be used as titrants in complexometric titrations. A good example is the use of Ag^+ as a titrant to measure CN^- through formation of the complex $Ag(CN)_2^-$ in the presence of ammonia, a process that gives a sharp end point when I^- is used as an indicator by reacting with excess Ag^+ to form solid AgI(s). Another example is the use of Hg^{2+} as a titrant for ions like Br^-, Cl^-, CN^-, and SCN^- by forming soluble 1:2 complexes such as $HgBr_2$, $HgCl_2$, $Hg(CN)_2$, and $Hg(SCN)_2$. Other examples of organic and inorganic titrants that can be used in complexometric titrations can be found in References 3–5.

13.2B Using Auxiliary Ligands and Masking Agents

Although ammonia and other monodentate ligands may not be suitable as titrants, they can still play a useful role as reagents that help control the selectivity of a complexometric titration or to reduce the effects of undesired side reactions. If an additional complexing agent is added to make the titration of an analyte easier to conduct (such as by reducing side reactions involving the analyte), this complexing agent is called an **auxiliary ligand**.[10] For example, a side reaction that must be considered during the titration of Ni^{2+} with EDTA is the combination of Ni^{2+} with OH^- to precipitate as $Ni(OH)_2$(s) at a pH above about 8.5. To prevent this side reaction, ammonia can be added as an auxiliary ligand to form soluble complexes with Ni^{2+}. This additional reaction raises the value of pNi (or lowers $[Ni^{2+}]$) before the equivalence point as ammonia complexes with some of the nickel ions and lowers the conditional formation constant between Ni^{2+} and EDTA (see Figure 13.6 for a titration conducted at pH 8.0), but it also prevents the loss of Ni^{2+} through its precipitation with hydroxide ions. The overall result is a titration for Ni^{2+} by EDTA that now can be performed up to pH 12 without problems due to undesired precipitation of the analyte.

If a ligand is added to prevent the titrant from reacting with a particular substance, this ligand is called a **masking agent**.[5,7] This type of agent is used when there may be more than one substance in a sample that might react with the titrant. A good illustration of this approach is the use of F^- as a masking agent to prevent Fe^{3+} in water from reacting with EDTA during the titration of Ca^{2+} to determine water hardness. In this case, the overall formation constant for the complex between F^- and Fe^{3+} (giving the complex FeF_6^{3-}) is so large that EDTA cannot effectively compete with the F^- for Fe^{3+}; however, there is no effect on the ability of EDTA to react with Ca^{2+}. The result is that Ca^{2+} will be titrated with no interference from Fe^{3+} in the sample.

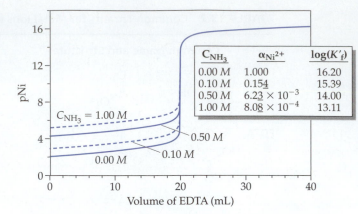

C_{NH_3}	$\alpha_{Ni^{2+}}$	$\log(K'_f)$
0.00 M	1.000	16.20
0.10 M	0.154	15.39
0.50 M	6.23×10^{-3}	14.00
1.00 M	8.08×10^{-4}	13.11

FIGURE 13.6 Effect of using ammonia as an auxiliary ligand on the titration of Ni^{2+} with EDTA at pH 8.0. The sample volume and concentration are the same as in Figure 13.7; the concentration of EDTA is also the same as in Figure 13.7. The value of $\alpha_{Ni^{2+}}$ was calculated as described in Chapter 9 and represents the fraction of all nickel species in solution that are present in the form Ni^{2+}. The value of pNi is given by $-\log([Ni^{2+}])$ and represents only the noncomplexed form of Ni^{2+} that is in solution. The value of K_f' in this case is given by the expression $K_f' = K_f (\alpha_{Ni^{2+}})(\alpha_{EDTA^{4-}})$, also according to methods described in Chapter 9 for dealing with side reactions during complex formation.

A *demasking agent* can sometimes be used to release a metal ion from a masking agent so that it can be measured again.[7] To illustrate this idea, suppose we wished to determine the concentration of both Zn^{2+} and Ca^{2+} by using EDTA, but did not want to measure Ni^{2+}. We could do this by first adding cyanide as a masking agent, which will form a strong complex with Zn^{2+} and Ni^{2+} (keeping these metal ions from reacting with EDTA), but not with calcium ions. After the Ca^{2+} has been titrated by EDTA, the Zn^{2+} can be released from its complex with cyanide by placing formaldehyde into the sample/titrant mixture. Formaldehyde (CH_2O) will combine with much of the cyanide (shown here as HCN) through the following reaction.

$$CH_2O + HCN \longrightarrow HOCH_2CN \qquad (13.4)$$

This reaction greatly lowers the concentration of free CN^- in solution, causing more cyanide to be released from the complexes it has formed in solution. Most of this CN^- is released from its complex with Zn^{2+}, allowing this metal ion to then be titrated by EDTA. Meanwhile, many of the complexes between nickel and cyanide remain intact so that Ni^{2+} is still masked even during this second titration step with EDTA.

There are hundreds of combinations of titrants, auxiliary ligands, masking agents, demasking agents, and titration conditions (such as pH) that have been proposed and developed for complexation titrations. You can find more details on these agents and on specific types of complexometric titrations in books such as *Vogel's Textbook of Quantitative Inorganic Analysis* and *Complexation in Analytical Chemistry*.[3,5] The particular set of agents and conditions that are needed for a given titration will be determined by the type of analytes and concentrations that are to be measured, as well as the presence of other substances in the sample that might interfere with the titration. Another factor that can be used to control the response of a titration is the choice of an end point detection method (see following section).

13.2C Determining the End Point

Metallochromic Indicators. The most common method for finding the end point of a complexometric titration is to use a **metallochromic indicator**.[6-8] This type of indicator has a change in its color or its fluorescence properties when it is free in solution or complexed to a metal ion. A small amount of this indicator is added at the beginning of the complexometric titration, which will place it in a state where it is bound to metal ions in the sample. As a titrant like EDTA is added to the sample and nears the equivalence point, this complexing agent will remove most of the remaining metal ions from solution and will eventually begin to compete for metal ions that are bound by the indicator. This causes the metallochromic indicator to go from its complex form to a free form, resulting in a change in color or fluorescence that signals the end point.

There are many metallochromatic indicators that are available, but only a few are in common use (see Table 13.3). *Eriochrome black T* is a metallochromic indicator that is frequently used to signal the end point during the titration of Ca^{2+} by EDTA to determine water hardness. Eriochrome black T is also an acid–base indicator. When it is not bound to a metal ion, the principle species for this indicator is a red form (H_2In^-) below about pH 6.3, a blue form (HIn^{2-}) between pH 6.3 and 11.5, and an orange form (In^{3-}) above pH 11.5. At any pH in this range, this indicator can bind to calcium and give a red complex ($CaIn^-$) that is used for end point detection. (*Note:* The reaction given below shows this process starting with the HIn^{2-} form of this indicator, which is the principal form present over the pH range used in CaEDTA titrations.)

There is a requirement that must be met for a metallochromic indicator to be useful in signaling the end point for a complexometric titration. The conditional formation constant between the metal ion and indicator must be smaller than the conditional formation constant between the same metal ion and the titrant. It is this requirement that makes it possible for the titrant to effectively compete with this indicator for a metal ion as the titration reaches the equivalence point. This competition can be described for an EDTA titration by the following general reaction, in which MIn and In have different colors. (*Note:* For simplicity, the charges and H^+ ions are not shown as part of this process.)

$$MIn + EDTA \rightleftharpoons MEDTA + In \quad (13.5)$$

We can also write a general reaction and equilibrium expression for the binding of the indicator with the metal ion,

$$M + In \rightleftharpoons MIn \quad K_{f,MIn} = \frac{[MIn]}{[M][In]} \quad \text{or}$$

$$K_{f',MIn} = \frac{[MIn]}{[M]C_{In}} \quad (13.6)$$

where $[In] = \alpha_{In} C_{In}$ and $K_{f',MIn} = \alpha_{In} K_{f,MIn}$. In the preceding equations, the conditional formation constant $K_{f',MIn}$ is used to reflect the effect of pH in determining the fraction of the indicator that exists in the proper acid–base form to react with the metal ion. The titration error can be minimized for this analysis if we have a 50:50 mixture of the MIn and In forms of the indicator present at the equivalence point.[3,6,10] This situation occurs when $[MIn] = C_{In}$ in the preceding expression for the conditional formation constant $K_{f',MIn}$, which also means that $[MIn] = [In]$ in the relationship shown for $K_{f,MIn}$. Under these conditions, the value of $K_{f',MIn}$ should be directly related to the concentration of M at the equivalence point, as shown in Equation 13.7.

$$\text{At Equivalence Point } ([MIn] = C_{In}): \; K_{f',MIn} = \frac{1}{[M]} \quad (13.7)$$

This result indicates that the titration error can be minimized by adjusting the pH so that the conditional stability constant for the metal-indicator complex is equal to the expected value for $1/[M]$ at the equivalence point.

Free indicator (H_2In^-) $+ Ca^{2+} \rightleftharpoons$ Ca^{2+} indicator complex ($CaIn^-$) $+ 2H^+$

TABLE 13.3 Examples of Metallochromic Indicators for Complexometric Titrations[*]

Indicator	Analytes (log $K_{f,MIn}$)	Initial Color		Final Color
Calmagite	Ca^{2+} (log $K_{f,CaIn}$ = 6.1) Mg^{2+} (log $K_{f,MgIn}$ = 8.1)	H_2In (Red) pK_{a1} = 8.1 HIn^- (Blue) pK_{a2} = 12.1 In^{2-} (Orange)	$\Rightarrow$	MIn (Red)
Eriochrome black T	Ca^{2+} (log $K_{f,CaIn}$ = 5.4) Mg^{2+} (log $K_{f,MgIn}$ = 7.0) Mn^{2+} (log $K_{f,MnIn}$ = 9.6) Zn^{2+} (log $K_{f,ZnIn}$ = 12.9)	H_2In (Red) pK_{a1} = 6.3 HIn^- (Blue) pK_{a2} = 11.6 In^{2-} (Orange)	$\Rightarrow$	MIn (Red)
Pyrocatechol violet	Bi^{3+} (log $K_{f,BiIn}$ = 27.1) Th^{4+} (log $K_{f,ThIn}$ = 23.1)	H_3In (Red) pK_{a1} = 7.8 H_2In^- (Yellow) pK_{a2} = 9.8 HIn^{2-} (Violet) pK_{a3} = 6.4 In^{3-} (Red - Purple)	$\Rightarrow$	MIn (Red)

[*]The constants given in this table were obtained from H. Freiser, *Concepts & Calculations in Analytical Chemistry*, CRC Press, Boca Raton, FL, 1992.

Let's see how this condition holds for the Ca-EDTA titration in Figure 13.2. In this titration [M] = 4.4 × 10^{-7} M at the equivalence point. This result means that the ideal value for $K_f'_{,MIn}$ in this situation is 1/(4.4 × 10^{-7} M) = 2.3 × 10^6 (with apparent units of M^{-1}). This condition cannot be met by using a complex between calcium and eriochrome black T, which has a formation constant well below this value (K_f = 2.5 × 10^5). However, this condition can be easily met by using a complex between this same indicator and Mg^{2+}, which has a much larger formation constant (K_f = 1.0 × 10^7). If calcium is being titrated in the absence of magnesium, this means that a small amount of Mg^{2+} should be added to form a complex with eriochrome T and give a good end point. The pH of this titration also needs to be adjusted so that both Ca^{2+} and Mg^{2+} can react quantitatively with EDTA and still give a good color change for the indicator (as obtained by using a pH around pH 10). After all of the Ca^{2+} has been titrated, the small amount of added Mg^{2+} will also be titrated by EDTA. The result is a color change and an end point that occurs just after the true equivalence point for Ca^{2+}.

Back Titrations. If the formation of the complex of a metal with the titrant is slow, it can be difficult to detect the end point or to conduct this analysis in a reasonable amount of time. This can be a problem, especially near the equivalence point, where the concentrations of the metal ion and complexing agent are small and their reaction will be particularly slow. This problem can be overcome by using a back titration (a technique discussed in Chapter 12).

A good example of the use of back titration in a complexometric method is the EDTA titration of Cr^{3+}. Formation of the 1:1 complex $CrEDTA^-$ is so slow that it can only be accomplished by adding an excess amount of EDTA, boiling the mixture for 10 to 15 minutes, cooling, and then back titrating the excess EDTA with bismuth as Bi^{3+} ions. The reaction of EDTA with Bi^{3+} is quite rapid and provides a good sharp end point. In this situation, the moles of EDTA is equal to the sum of the moles of Cr^{3+} and Bi^{3+}. Back titrations can be used in a similar manner with other complexing agents or analytes.

EXERCISE 13.3 **Using a Back Titration for a Complexation Reaction**

A sample is believed to contain Cr(III). A 25.00 mL portion of this sample is combined with 10.00 mL of a 0.0875 M solution that contains the complexing agent CDTA. This mixture is heated to boiling for several minutes to allow the CDTA to complex with the Cr(III) species. The mixture is then cooled, and the excess CDTA is measured by a back titration using 0.0258 M Bi^{3+}, requiring 4.20 mL to reach the end point. What was the original concentration of Cr(III) species in the sample?

SOLUTION

The moles of Cr(III) species in the sample can be determined by finding the difference in the total moles of CDTA that were added to the sample and the moles of remaining CDTA that were measured in the back titration with Bi^{3+}.

$$\text{mol Cr(III)} = \text{total mol CDTA} - \text{mol Bi}^{3+}$$

$$\text{used in back titration}$$

$$= (0.0875 \text{ mol/L})(0.01000 \text{ L}) -$$

$$(0.0258 \text{ mol/L})(0.00420 \text{ L})$$

$$= (8.75 \times 10^{-4} \text{ mol}) - (1.08 \times 10^{-4} \text{ mol})$$

$$= 7.67 \times 10^{-4} \text{ mol}$$

We can determine from this result and the volume of the sample that the original concentration of Cr(III) species was $(7.67 \times 10^{-4} \text{ mol})/(0.02500 \text{ L})$ = **3.07×10^{-2} M**.

Other Techniques. You will occasionally find a situation in which you want to use a complexometric titration, but there isn't any indicator that will give a satisfactory end point for your analyte and titration conditions. One way you might deal with this situation is to conduct a **displacement titration**.[3,7] This titration uses a metal ion that is the analyte to compete with and displace another metal ion from EDTA, where the second metal ion does have a suitable indicator available for its detection. This approach can be used to measure Ni^{2+} by having these nickel ions displace Mg^{2+} from EDTA. This analysis is performed by adding a sample to a solution that contains an excess of $MgEDTA^{2-}$ versus the expected Ni^{2+} content in the sample. The conditional formation constant for Ni^{2+} and EDTA is much larger than it is for Mg^{2+} and EDTA at a given pH. This means Ni^{2+} will tend to displace some of the Mg^{2+} and form $NiEDTA^{2-}$.

$$MgEDTA^{2-} + Ni^{2+} \rightarrow Mg^{2+} + NiEDTA^{2-} \quad (13.8)$$

The Mg^{2+} that has been released into solution can then be titrated with EDTA while using eriochrome black T as an indicator. The moles of Mg^{2+} that are measured in this titration will be equal to the moles of Ni^{2+} that displaced Mg^{2+} and that were in the original sample.

Another approach for extending the types of analytes that can be examined by using EDTA or other titrants is to use an **indirect titration**.[7] This is a technique that indirectly measures an analyte through the effect that this analyte has on the concentration of another chemical (such as a metal ion) that can be titrated in a solution. This approach is usually used in an EDTA titration to determine the concentration of an anion that will precipitate a metal ion that can also be readily titrated with EDTA. Sulfate ions can be determined in this way by adding them to a solution that contains a known concentration and excess of Pb^{2+}. The sulfate ions will react with some of the Pb^{2+} to form $PbSO_4(s)$. The Pb^{2+} ions that remain in solution (after the solid has been removed) can then be determined by titrating them with EDTA. This makes it possible to calculate the amount of sulfate that had been added to the lead solution by using the known original amount of Pb^{2+} and the moles that were measured by an EDTA titration after precipitation.

13.2D Predicting and Optimizing Complexometric Titrations

General Approach to Calculations. It is possible to use our knowledge of complexation reactions to predict the shape of a curve for a complexometric titration, in which we usually plot the value of pM versus the volume of added titrant. This goal can be accomplished by dividing the titration curve into four general regions: (1) the original sample, (2) the middle of the titration in which only some of the analyte has been titrated, (3) the equivalence point, and (4) the region beyond the equivalence point in which excess titrant is present. Much of the logic behind this process is the same as we used in Chapter 12 to predict the shape of the titration curve for an acid–base titration.

Figure 13.7 gives the equations that can be used to predict a titration curve for analyte M that reacts with titrant L to produce a 1:1 complex. This is the situation that occurs when EDTA is used to titrate Ca^{2+} or any other metal ion. The *first region* in a complexometric titration is represented by a sample to which no titrant has yet been added. For this solution, we can determine the value of pM by using the original concentration of our analyte M in the sample (C_M), as shown in Equation 13.9.

Original Sample (0% Titration): $pM = -\log(C_M)$ (13.9)

The *second region* in the curve for a complexometric titration occurs when some complexing agent titrant has been added, but not enough has been combined with the sample to reach the equivalence point. If we are using a 1:1 complexing agent such as EDTA, the moles of analyte that remain in solution at any point in this region will be equal to the difference in the total moles of analyte that were in the original sample (as determined from the product $C_M V_M$, where V_M is the volume of the sample that is being titrated) and the moles of

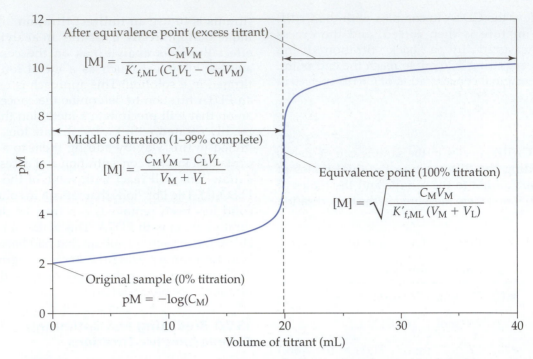

FIGURE 13.7 Equations for predicting the response for the titration of a metal ion (M) with a complexing agent (L) to form a 1:1 complex. All terms used in these equations are defined in the text.

analyte that have reacted to form a complex with the titrant (which will be equal to $C_L V_L$, where C_L is the original concentration of the titrant and V_L is the volume of titrant that has been added to the sample). We can then divide the moles of remaining analyte by the total volume of the sample/titrant mixture ($V_M + V_L$) to find the analyte's concentration in this mixture, [M].

Middle of Titration (1–99% Complete):

$$[M] = \frac{C_M V_M - C_L V_L}{V_M + V_L} \qquad (13.10)$$

The value of pM is then estimated by using this calculated value of [M] for the sample/titrant mixture.

The *third region* in a complexometric titration occurs at the equivalence point. At this point all of the original analyte has combined with the complexing agent and the only non-complexed analyte that is now present is produced from the complex as it establishes an equilibrium between the reactant and products. We can determine the value of [M] under these conditions by using the overall formation constant for the titration reaction, as defined in Equation (13.11) for the formation of a 1:1 complex between M and L.

$$M + L \rightleftharpoons ML \qquad K_{f}',_{ML} = \frac{[ML]}{[M]\, C_{L,Eq\,Pt}} \qquad (13.11)$$

At the equivalence point almost 100% of the analyte and ligand will be present in the form ML because of the large formation constant that is present during most titrations involving EDTA or other multidentate ligands. This means we can approximate the value of [ML] by setting it equal to the moles of M in the original sample divided by the total volume of the sample/titrant mixture, or

$C_{ML} = (C_M V_M)/(V_M + V_L)$. We also know that $[M] = C_{L,Eq\,Pt}$ at the equivalence point because the only way M or L can be present at this point is if some of the complex ML dissociates, producing M and L in a 1:1 ratio. If we make these substitutions into Equation 13.11 and solve for [M], we get the following result.

Equivalence Point (100% Titration):

$$[M] = \sqrt{\frac{C_M V_M}{K_{f}',_{ML}(V_M + V_L)}} \qquad (13.12)$$

The value of pM is then calculated by using this estimated value for [M] at the equivalence point.

The *fourth region* of a complexometric titration curve occurs after the equivalence point. In this region an excess of titrant is being added to the sample. The value of [M] in this region is also found by starting with the expression for the conditional formation constant in Equation 13.10. As we add more titrant to the sample, the value of [ML] can still be found by setting it equal to the moles of M in the original sample divided by the total volume of the sample/titrant mixture, or $C_{ML} = (C_M V_M)/(V_M + V_L)$. The value of C_L will be approximately equal to the moles of excess titrant divided by the total volume of the sample/titrant mixture, where $C_L = (C_L V_M - C_M V_M)/(V_M + V_L)$. If we make these substitutions into Equation 13.11, cancel common terms in the numerator and denominator, and solve for [M], we get Equation 13.13.

After Equivalence Point (Excess Titrant):

$$[M] = \frac{C_M V_M}{K_{f}',_{ML}\,(C_L V_M - C_M V_M)} \qquad (13.13)$$

As before, the value of pM is then estimated by using this calculated result for [M]. The next exercise will give you some practice with these calculations.

EXERCISE 13.4 Predicting the Behavior of a Complexometric Titration

The titration curve in Figure 13.7 was obtained for the analysis of a 10.00 mL aliquot of a 0.0100 M Ca^{2+} solution in water at pH 10.00 that was titrated using 0.0050 M EDTA. The conditional formation constant for this reaction under these conditions is 1.75×10^{10}. What values of pCa would be expected (a) at the beginning of the titration, (b) 50% of the way through this titration, (c) at the equivalence point, and (d) after an excess of 10.00 mL titrant has been added to this sample?

SOLUTION

a. At the beginning of the titration, the value of pCa can be found by using Equation 13.9 and the concentration of calcium ions in the original sample.

$$[Ca^{2+}] = 0.0100 \ M \text{ or } pCa = -\log(0.0100) = \mathbf{2.000}$$

b. The volume of titrant at exactly 50% titration can be determined by using the fact that there is a 1:1 stoichiometry for the reaction between Ca^{2+} and EDTA as they form a complex. This gives us a titrant volume at the equivalence point of $V_L = C_M V_M / C_L = (0.005000 \ M)(0.02500 \ L)/(0.01000 \ M) = 0.02000 \ L$ or 20.00 mL, or a volume of EDTA at 50% titration that is equal to $(0.5000)(20.00 \text{ mL}) = 10.00 \text{ mL}$. We can then use Equation 13.10 to determine the value of pCa when 10.00 mL of 0.0050 M EDTA has been added to a 10.00 mL sample of 0.0100 M Ca^{2+}.

$$[Ca^{2+}] =$$
$$[(0.01000 \ M)(0.01000 \ L) - (0.005000 \ M)(0.01000 \ L)] /$$
$$(0.01000 \ L + 0.01000 \ L) = 0.0025 \ M$$

or

$$pCa = -\log(0.0025) = \mathbf{2.60}$$

c. The value of pCa at the equivalence point can be found by using Equation 13.11 and the fact $K_f{'}_{,ML} = 1.75 \times 10^{10}$ (with apparent units of M^{-1}) and $V_L = 20.00$ mL at this point in the titration, as calculated in Part (b).

$$[Ca^{2+}] =$$
$$\sqrt{\frac{(0.01000 \ M)(0.01000 \ L)}{(1.75 \times 10^{10} \ M^{-1})(0.01000 \ L + 0.01000 \ L)}}$$
$$= 5.35 \times 10^{-7} \ M$$

or

$$pCa = -\log(5.35 \times 10^{-7}) = \mathbf{6.27}$$

d. After the equivalence point, we can use Equation 13.12 to find pCa. When an excess of 10.00 mL titrant

has been added (or a total titrant volume of 10.00 mL + 20.00 mL), we get the following result for pCa, again using the fact that $K_f{'}_{,ML} = 1.75 \times 10^{10}$.

$$[Ca^{2+}] =$$
$$(0.01000 \ M)(0.01000 \ L) / \{(1.75 \times 10^{10} \ M^{-1})[(0.00500 \ M)$$
$$(0.03000 \ L) - (0.01000 \ M)(0.01000 \ L)]\}$$
$$= 1.1\underline{4} \times 10^{-10} \ M$$

or

$$pCa = -\log(1.1\underline{4} \times 10^{-10}) = \mathbf{9.94}$$

Using the Fraction of Titration. One way we can use the general calculations described in the last section is to determine how the conditions for a complexometric titration might affect the response we will obtain. Calculations like these were used to generate the results in Figure 13.5 to show how the titration curve of Ca^{2+} with EDTA changes with pH. The equations listed in the last section make it possible for us to quickly determine the value of pCa at key points in this titration curve, but it is also possible to use a more general approach for these calculations that makes use of the *fraction of titration* (F). The value of F for a complexometric titration that produces a 1:1 complex between the analyte and titrant (such as the titration of Ca^{2+} or other metal ions with EDTA) is described by Equation 13.14.

Titration of M with L to produce ML:

$$F = \frac{\text{Moles added to titrant}}{\text{Moles of original analyte}} = \frac{C_L V_L}{C_M V_M} \quad (13.14)$$

A similar relationship can be written for the analysis of L using M as the titrant.

We can use the preceding definition for F, along with mass balance equations for our reaction, to obtain a general expression that describes the titration of M with L. The process, which is involved in this derivation is shown in Table 13.4, and it results in the following relationship.

Fraction of Titration Equation for M + L giving ML:

$$F = \frac{C_L V_L}{C_M V_M} = \frac{1 - \dfrac{[M]}{C_M}}{\dfrac{[M]}{C_L} + \dfrac{K_{f,ML}[M]}{1 + K_{f,ML}[M]}} \quad (13.15)$$

Notice the value of F in this expression can be determined if we know the value for [M] (or pM) we want to use in our calculation, the original concentrations of the analyte and titrant (C_M and C_L), and the formation constant for ML ($K_{f,ML}$). (*Note:* The same general relationship can be used with a conditional formation constant by replacing $K_{f,ML}$ with the appropriate value for $K'_{f,ML}$.) As a result, we can use this equation to see what fraction of titration and volume of added titrant will correspond to a particular value for pM, a practice that can be useful when you are developing and optimizing a complexometric titration.

TABLE 13.4	Fraction of Titration Equation for Titration of M with L to give ML*

Titration Reaction:

$$M + L \rightarrow ML$$

Equilibrium Expression:

$$K_f = \frac{[ML]}{[M][L]} \quad \Rightarrow \quad [ML] = K_f[M][L]$$

Mass Balance Equations:

$$C_M = [M] + [ML] = \frac{C_M V_M}{V_M + V_L}$$

$$C_L = [L] + [ML] = \frac{C_L V_L}{V_M + V_L}$$

Substitute the Equilibrium Expression into the Mass Balance Equations:

$$\Rightarrow \quad C_M = [M] + K_f[M][L] = \frac{C_M V_M}{V_M + V_L}$$

$$\Rightarrow \quad C_L = [L] + K_f[M][L] = \frac{C_L V_L}{V_M + V_L}$$

Solve Equation to Express in Terms of [L]:

$$\Rightarrow \quad [L] = \frac{C_L V_L}{V_M + V_L} \cdot \frac{1}{1 + K_f[M]}$$

Substitute Equation for [L] into Equation for C_M:

$$\Rightarrow \quad [M] + \frac{C_L V_L}{V_M + V_L} \cdot \frac{K_f[M]}{1 + K_f[M]} = \frac{C_M V_M}{V_M + V_L}$$

Mutiply Both Sides of Combined Equation by ($V_M + V_L$):

$$\Rightarrow \quad [M](V_M + V_L) + C_L V_L \cdot \frac{K_f[M]}{1 + K_f[M]} = C_M V_M$$

Combine Terms Containing V_M or V_L; Factor Out $C_L V_L$ (on left) and $C_M V_M$ (on right):

$$\Rightarrow \quad [M]V_L + C_L V_L \cdot \frac{K_f[M]}{1 + K_f[M]} = C_M V_M - [M]V_M$$

$$\Rightarrow \quad C_L V_L\left(\frac{[M]}{C_L} + \frac{K_f[M]}{1 + K_f[M]}\right) = C_M V_M\left(1 - \frac{[M]}{C_M}\right)$$

Rearrange to Obtain $(C_L V_L)/(C_M V_M)$ on the Left Side of Equation:

$$\therefore \quad F = \frac{C_L V_L}{C_M V_M} = \frac{1 - \dfrac{[M]}{C_M}}{\dfrac{[M]}{C_L} + \dfrac{K_f[M]}{1 + K_f[M]}} \tag{13.15}$$

*Terms: F = fraction of titration; C_M = total concentration of M in the original sample; V_M = volume of original sample; C_L = total concentration of L in the titrant; V_L = volume of added titrant at a given point in the titration; K_f = formation constant for the complex ML; [M] = concentration of non-complexed metal ion M in the sample/titrant mixture.

EXERCISE 13.5	Choosing an Indicator for the Titration of Ca^{2+} with EDTA

It was mentioned in Section 13.2B that a complex of eriochrome black T with Mg^{2+} can be used to give a reasonable end point during the titration of Ca^{2+} with EDTA. It was also stated that for optimum accuracy, the value of $K_f'_{,MIn}$ at this end point should be equal to $1/[Ca^{2+}]$. The conditions for the titration in Figure 13.2 give $K_f'_{MIn} = 2.75 \times 10^5$ (with apparent units of M^{-1}) when using a mixture of eriochrome black T with Mg^{2+} as the indicator at pH 10.00.[5] What is the fraction of titration at the end point for this indicator and what is the resulting titration error for the analysis that is shown in Figure 13.2?

SOLUTION

We know from the previous exercise that the equivalence point for the titration in Figure 13.2 occurs when pCa = 6.27 and $V_L = V_E$ = 20.00 mL. We next need to determine what the fraction of titration would be when pCa = 5.44 (or $[Ca^{2+}] = 1/K_f'$,MIn = 3.64×10^{-6} M), the point at which our indicator will signal that an end point has been reached. We can do this by substituting into Equation 13.15 the given value for $[Ca^{2+}]$ along with the known values for C_M, C_L, and K_f'ML (using $K'_{f,CaEDTA} = 1.74 \times 10^{10}$ M^{-1} at pH 10.0, in this case) to solve for F at this point in the titration curve.

$$F = \left[1 - \frac{(3.64 \times 10^{-6}\ M)}{(0.01000\ M)}\right] \bigg/ \left[\frac{(3.64 \times 10^{-6}\ M)}{(0.00500\ M)} + \frac{(1.74 \times 10^{10}\ M^{-1})(3.64 \times 10^{-6}\ M)}{1 + (1.74 \times 10^{10}\ M^{-1})(3.64 \times 10^{-6}\ M)}\right] = 0.9989$$

Once we know the values for both F and V_E, we can multiple these numbers to find the volume of titrant that has been added to reach this point of the titration ($V_{L,End\ Pt}$).

$$V_{L,\ End\ Pt} = F \cdot V_E = (0.9989)(20.00\ mL) = 19.98\ mL$$

We can then find the titration error, as defined below in Equation 12.1 from Chapter 12.

Titration error = $V_{L,\ End\ Pt} - V_L$

$$= 19.98\ mL - 20.00\ mL = -0.02\ mL$$

This is a small error that would give an estimated concentration for Ca^{2+} that is off by only $100 \cdot \left(V_{L,\ End\ Pt} - V_E\right)/V_E$ = $100 \cdot (0.02\ mL)/(20.00\ mL)$ = 0.1%

The fraction of titration expression given in Equation 13.15 can also be used to predict the overall shape of a complexometric titration using EDTA. Figure 13.8 shows a spreadsheet that can be used with this equation for this purpose. This example is similar to the spreadsheet that was shown in Chapter 12, in which the fraction of titration was used to calculate the response during an acid–base titration. This spreadsheet can perform the same calculation as used in the last exercise to use a particular result for pM to find the values of F and $V_{Titrant}$ at this point in the titration curve, but now performs this task over a wide range of pM values. A plot of these values is then made with $V_{Titrant}$ (or F) appearing on the x-axis and pM being used on the y-axis to provide the titration curve. This spreadsheet can also be used to see how changing the pH of the titration or how the addition of auxillary ligands will affect the response through the change that these parameters cause in the value of K'_f that is used to calculate each fraction of titration.

13.3 PERFORMING A PRECIPITATION TITRATION

13.3A Titrants and Standard Solutions

Methods Based on Silver. Most precipitation titrations that are used today are based on either the analysis of silver ions or the use of Ag^+ as a precipitating agent.[3] The earliest of these was a method developed by French chemist Joseph Louis Gay-Lussac in 1829 for determining the purity of silver in currency (see Box 13.1). This method involved dissolving the silver and converting it into Ag^+, which was then titrated with Cl^- to give insoluble silver chloride. This is the same basic approach that was used in Figure 13.3 as an example of a precipitation titration.[11] Other potential titrants that can be used for this analysis include Br^- or I^- (see comparison in Figure 13.10).

Several improvements were later made in this technique by other analysts. An example is a titration method that was created in 1874 by Jacob Volhard, which is now known as the *Volhard method*.[3,7,8] This technique involves the titration of Ag^+ with thiocyanate (SCN^-) to give solid AgSCN.

$$Ag^+ + SCN^- \rightarrow AgSCN(s) \qquad (13.16)$$

The indicator for this titration is Fe^{3+}, which reacts with the first bit of excess thiocyanate after the equivalence point to form $FeSCN^{2+}$, a red soluble complex that is used to signal the end point.

Silver ions can also be used as the titrant to detect anions like Cl^- or SCN^-. A titration method that uses Ag^+ as a titrant is also known as an *argentometric titration* (after the Latin word *argentum* for "silver").[7] There are many anions that can be measured by using Ag^+ as a titrant. For instance, the Volhard method can readily be used to determine the concentration of any anion that forms an insoluble silver salt. Some examples of these anions are bromide, iodide, arsenate, carbonate, chromate, ferricyanide, molybdate, oxalate, phosphate, sulfite, and sulfide. For instance, bromide can be analyzed by reacting it with a known excess amount of Ag^+ to form AgBr(s). The amount of Ag^+ that remains in solution is then determined by titrating it with thiocyanate, giving us another example of an "indirect titration" (see Section 13.2C).

The titration curve for a method that uses Ag^+ as the titrant does have one unusual feature in that it is often prepared by plotting on the y-axis the value of pAg (that is, the amount of soluble Ag^+ that can be present in the titrant/sample mixture at a given concentration of the analyte) rather than plotting a value that is more directly related to the concentration of the analyte. This approach, illustrated in Figure 13.11, is used because many of the methods for end point detection in an argentometric titration are based on a change in the concentration of Ag^+. We just saw one example of this in the analysis of Cl^- by the Volhard method, in which it was the presence of excess Ag^+ from the titrant that gave a color change at the end point. Other

Titration of a metal ion (M) with a complexing agent (L) - Ca²⁺ titrated with EDTA versus pH

| Log (K_f) for complex: | | 10.65 | K_f' for complex: | 1.75E+10 | Log (K_f') | |
| pH value: | | 10 | Alpha for EDTA: | 0.392 | 10.243 | |

Conditions used in titration:

| C_M (M) = | 0.01 | C_L (M) = | 0.005 | | | |
| V_M (M) = | 0.01 | V_E (L) = | 0.02 | | | |

Calculated results for titration:

pM	[M]	Fraction to 1st Eq pt	V_L (L)	V_L (mL)
2.00	1.0000E−02	0.0000E+00	0.00000	0.00
2.10	7.9433E−03	7.9451E−02	0.00159	1.59
2.20	6.3096E−03	1.6315E−01	0.00326	3.26
2.30	5.0119E−03	2.4911E−01	0.00498	4.98
2.40	3.9811E−03	3.3509E−01	0.00670	6.70
2.50	3.1623E−03	4.1886E−01	0.00838	8.38
2.60	2.5119E−03	4.9842E−01	0.00997	9.97
2.70	1.9953E−03	5.7215E−01	0.01144	11.44
2.80	1.5849E−03	6.3897E−01	0.01278	12.78
2.90	1.2589E−03	6.9829E−01	0.01397	13.97
3.00	1.0000E−03	7.5000E−01	0.01500	15.00
3.10	7.9433E−04	7.9437E−01	0.01589	15.89
3.20	6.3096E−04	8.3192E−01	0.01664	16.64
3.30	5.0119E−04	8.6334E−01	0.01727	17.27
3.40	3.9811E−04	8.8938E−01	0.01779	17.79
3.50	3.1623E−04	9.1077E−01	0.01822	18.22
3.60	2.5119E−04	9.2825E−01	0.01856	18.56
3.70	1.9953E−04	9.4244E−01	0.01885	18.85
3.80	1.5849E−04	9.5391E−01	0.01908	19.08
3.90	1.2589E−04	9.6316E−01	0.01926	19.26
4.00	1.0000E−04	9.7059E−01	0.01941	19.41
4.10	7.9433E−05	9.7654E−01	0.01953	19.53
4.20	6.3096E−05	9.8131E−01	0.01963	19.63
4.30	5.0119E−05	9.8511E−01	0.01970	19.70
4.40	3.9811E−05	9.8815E−01	0.01976	19.76
4.50	3.1623E−05	9.9057E−01	0.01981	19.81

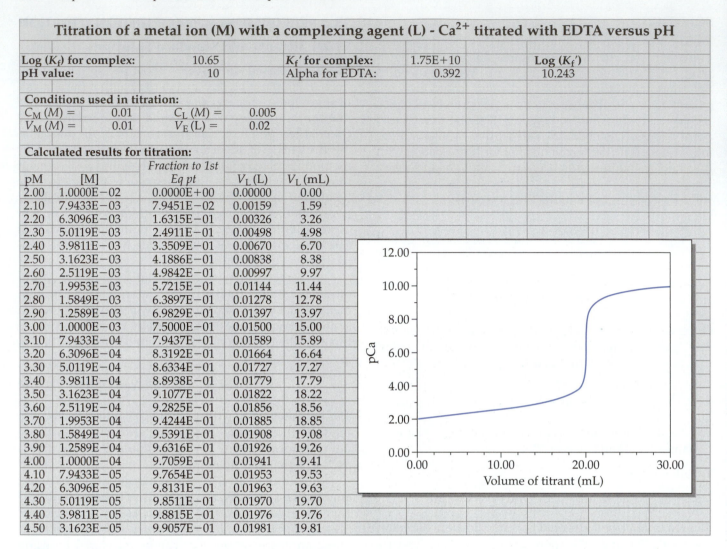

FIGURE 13.8 Example of a spreadsheet using a fraction of titration equation to predict the response of a complexometric titration. This spreadsheet is based on Equation 13.15 and is the same spreadsheet that was used to draw the titration curve for Ca²⁺ by EDTA in Figure 13.2. Only part of the actual pM (or pCa) values that were used in Figure 13.2 are shown in the portion of the spreadsheet that is included in this figure.

precipitation titrations that can also make use of Ag⁺ as a titrant and for end point detection are the *Mohr method* and *Fajans method*, as is discussed in Section 13.3B.[3,7,8,11]

To prepare a titrant for such methods, both silver nitrate (to prepare a standard solution of Ag⁺) and potassium chloride (to prepare a Cl⁻ solution) can be purchased in a very pure form and both can be used as primary standards for precipitation titrations. However, solutions containing potassium thiocyanate (for preparing a SCN⁻ solution) must be standardized by titrating them with known amounts of Ag⁺. Titrations in which silver ion is either the titrant or the analyte must not be carried out in a basic solution, because a high pH can lead to side reactions that form silver hydroxide or silver oxide as additional precipitates.

Other Precipitation Methods. There have been other precipitation reactions that have been explored in the past for use in titrations; however, only a few of these are still used in modern laboratories. One example is the use of Ba²⁺ as a titrant to determine the content of sulfate in a sample through the precipitation of BaSO₄(s). Other titration methods that make use of precipitation reactions are the analysis of I⁻ with Tl⁺ to form TlI(s), and the similar analysis of Tl⁺ by using I⁻ as a titrant.[8]

13.3B Determining the End Point

A precipitation titration is unique in that we can easily see if the titration reaction is taking place as it forms an insoluble precipitate. This feature makes it possible to visually determine the end point through several approaches. For instance, we can simply use the point in the titration at which no more precipitate is seen to form as titrant is added to the sample. This technique is sometimes called the "clear-point" method and is how Gay-Lussac and others first followed the titration of Ag⁺ with Cl⁻.[8,11]

BOX 13.1

A King with a Problem

In the year 1829 the political and economic situation in France was chaotic. The French Revolution, the Terror, the Napoleonic era, and the Bourbon restoration had all occurred within the previous few decades. In addition to this, King Charles X of France also had a drain on the nation's treasury. At that time anyone could go to any office of the French mint and buy or sell silver. To determine the price that should be paid or given for this silver, the purity of the silver was measured by using a gravimetric analysis based on cupellation, or the "fire assay" (see Chapter 1).

The existing silver analysis method was subject to many sources of errors and gave results that depended on the skill and experience of the analyst. Some of analysts at the mints tended to give results that had a high systematic error, while others gave results with a low systematic error. It therefore became very profitable for a person to buy silver at an office that obtained low results for purity and then sell the same silver at another office that gave high purity values. The

money that was made during this transaction could then be used to buy more silver at the first office, and the whole process was repeated.

To solve this problem, the king offered a prize to anyone who could develop a method for determining the purity of silver that was precise and accurate (better than 0.5 parts-per-thousand), simple, and faster than the current method (requiring no more than a few minutes). Joseph Gay-Lussac (Figure 13.9) had been working with titrations for several years and soon claimed this prize. His method involved dissolving a small portion of the silver and titrating the resulting Ag^+ solution with a standard solution of sodium chloride, giving AgCl as the precipitate. In June 1830, King Charles issued a proclamation ordering that this new method be adopted in all offices of the mint. Gay-Lussac became famous for this method and continued to extend the use of titrations to other analytes.[11]

Joseph Gay-Lussac

FIGURE 13.9 Joseph Gay-Lussac (1778–1850) and an 1827 two franc silver coin from the reign of King Charles X of France. The image of Charles X appears in the top front view of the coin. The assay developed by Gay-Lussac was used to examine the content of silver ingots, which were rectangular pieces of this metal that could be bought and sold at designated offices in France at that time. (The images of the coin are used with permission from http:/home.eckerd.edu.)

Another possibility for end point detection is to use an indicator that forms a colored product to signal the end point. This approach is used in the Volhard method by having excess SCN^- react with some Fe^{3+} (added as the indicator) to form the colored complex $FeSCN^{2+}$. Another example of this approach is the Mohr method, developed in 1855 by Fredrich Mohr. The Mohr method detects an end point in an argentometric titration by using the reaction of Ag^+ with chromate ($CrO_4{}^{2-}$) to form silver chromate (Ag_2CrO_4), a red precipitate.[3,7,8,11,12] This product forms when there is an excess of Ag^+ in solution, as occurs after the equivalence point for the titration of Cl^- by Ag^+.

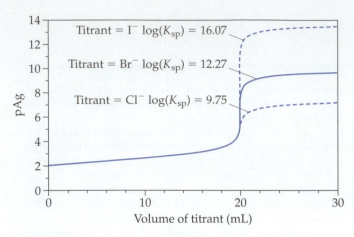

FIGURE 13.10 Titration curves for the analysis of Ag^+ in an aqueous solution by its titration with I^-, Cl^- or Br^- to form a 1:1 precipitate. These titration curves show the response expected for a 10.00 mL aliquot of a 0.01000 M aqueous solution of Ag^+ that is titrated with 0.005000 M of the precipitating agent.

This method of end point detection is often employed in the titration of Cl^- with Ag^+, but this technique for endpoint detection can also be used when titrating Ag^+ with Cl^-. In this second case, a known excess of chloride is added to a dissolved silver sample, with the excess chloride then determined by a back titration using Ag^+ as the titrant and chromate as the indicator.[12]

The Fajans method was created by Polish chemist Kasimer Fajans in 1923[13] and is also a method for the determination of Cl^- by its reaction with Ag^+. This method detects the end point by using a negatively charged dye such as fluorescein or its derivatives dichlorofluorescein or tetrabromofluorescein (see Figure 13.12).[8,11] These dyes are all weak acids that are used as indicators in the form In^{2-}. The negative charge of this form will cause these dyes to adsorb onto the surface of precipitate particles that have a slight positive charge but not on particles that have a slight negative charge. It is for this reason that such a dye is sometimes called an **adsorption indicator**.[6,7] During the titration of Cl^- with Ag^+, the precipitate will have a net negative charge before the equivalence point, when there is an excess of Cl^- in the solution and adsorbed to the particle's outer layer. However, a positive charge will be present on the precipitate particles after the equivalence point when an excess of Ag^+ has been added and some of these ions adsorb to the surface of the AgCl particles. The presence of this net positive charge also causes the indicator In^{2-} to adsorb, which changes the observed color of this precipitate from white to pink. This change is used to signal the end point of this titration.

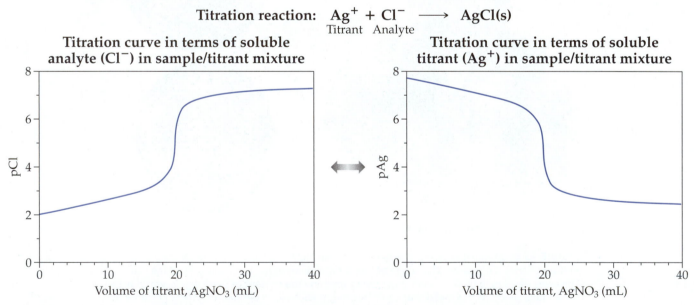

FIGURE 13.11 Titration curves for the analysis of Cl^- in an aqueous solution by its titration with Ag^+ to form AgCl(s). The curve on the left shows the response in terms of the amount of Cl^- that remains in solution, where $pCl = log([Cl^-])$. The curve on the right shows the response in terms of the amount of remaining soluble titrant by using $pAg = -log([Ag^+])$ to represent the amount of nonprecipitated titrant that can be present in solution at a given concentration of analyte. It is possible to convert between the response of these two curves in this particular titration by using the relationship $K_{sp,AgCl} = [Ag^+][Cl^-]$ or $[Ag^+] = K_{sp,AgCl}/[Cl^-]$. These titration curves show the response expected for a 10.00 mL aliquot of a 0.01000 M aqueous solution of Cl^- that is titrated with 0.005000 M Ag^+.

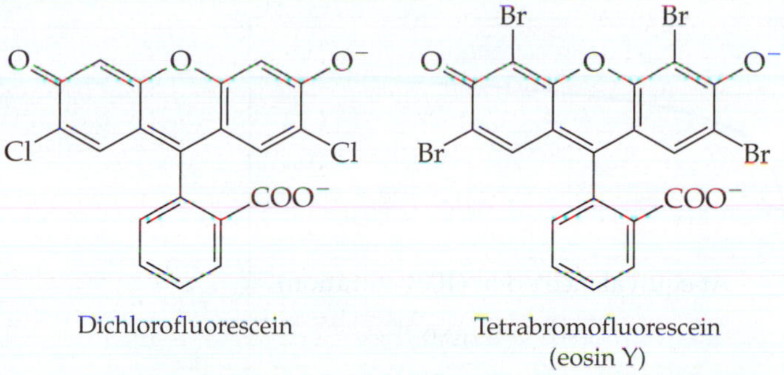

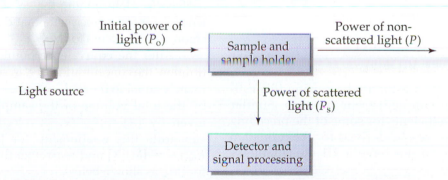

Dichlorofluorescein

Tetrabromofluorescein
(eosin Y)

FIGURE 13.12 Examples of two derivatives of fluorescein that can be used as adsorption indicators during precipitation titrations. These indicators are used in the form In^{2-}. The negative charge of this form causes these indicators to adsorb to the surface of a precipitate like AgCl when a slight positive charge is present on this precipitate. When Ag^+ is the titrant, this condition occurs after the equivalence point when excess Ag^+ is present. The adsorption of these indicators creates a change in the color of the precipitate, which signals the end point.

There are a few instrumental approaches that can be used to detect the end point for a precipitation titration. In the case of a titration that uses Ag^+ as a titrant, electrodes that give a signal related to the amount of Ag^+ in solution can be used to follow the course of the titration. This method is similar to the use of an electrode to measure the pH during an acid–base titration and makes use of a technique known as *potentiometry* (a topic discussed in Chapter 14). Another more general method for following the course of a precipitation titration is to examine the change in the *turbidity* (or "cloudiness") of the sample/titrant mixture as a precipitate is formed. The turbidity of such a mixture is a result of light being scattered by precipitate particles.[7] This light scattering

can be measured by using the technique of *nephelometry* (see Figure 13.13), in which the intensity of light that is scattered by a solution is compared to the original intensity of this light (with the scattered light measured at a right angle to the incoming light). The use of this method with a precipitation titration is sometimes called a "nephelometric titration."

13.3C Predicting and Optimizing Precipitation Titrations

General Approach to Calculations.   Just as we have seen in this chapter for complexometric titrations and in the last chapter for acid–base titrations, we can also predict the shape of a curve for a precipitation titration. The response we get will be similar to what we see for a complexometric titration, in that we will again be plotting pM versus the volume of titrant. However, the value we calculate for pM will now depend on our precipitaion reaction rather than on complex formation.

As an example of this process, let's consider the titration of Ag^+ with Cl^- (see Figure 13.14). We can again break this titration into four general regions: (1) the beginning of the titration, (2) before the equivalence point, (3) at the equivalence point, and (4) after the equivalence point. At the beginning of the titration, the value of pM (or pAg in our example) is simply obtained from the concentration of the analyte in the original sample (C_M).

Original Sample (0% Titration):

$$pM = -\log(C_M) \qquad (13.17)$$

Once we have added the titrant ($X = Cl^-$ in this case), it will react with the analyte to form a precipitate. Because

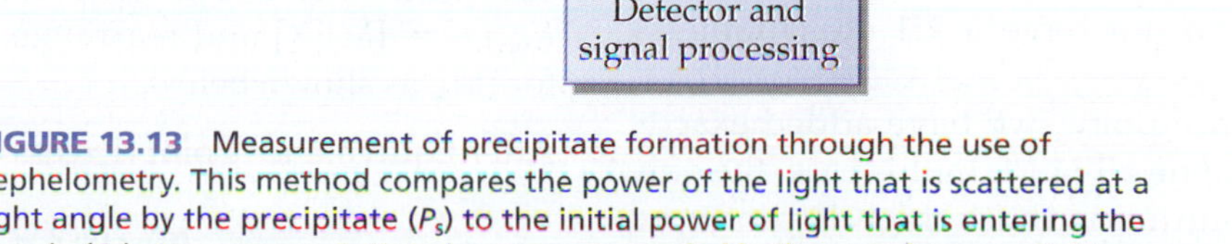

Light source

Initial power of light (P_o)

Sample and sample holder

Power of non-scattered light (P)

Power of scattered light (P_s)

Detector and signal processing

FIGURE 13.13 Measurement of precipitate formation through the use of nephelometry. This method compares the power of the light that is scattered at a right angle by the precipitate (P_s) to the initial power of light that is entering the sample (P_o). As more precipitate becomes suspended in the sample, more light is scattered and the size of P_s increases vs. P_o, causing the ratio (P_s/P_o) to increase. It is also possible to measure the decrease in the power of light that makes it through the sample without being scattered (P). This second method is known as *turbidimetry* and leads to a decrease in P and the ratio (P/P_o) as more light is scattered by the sample.

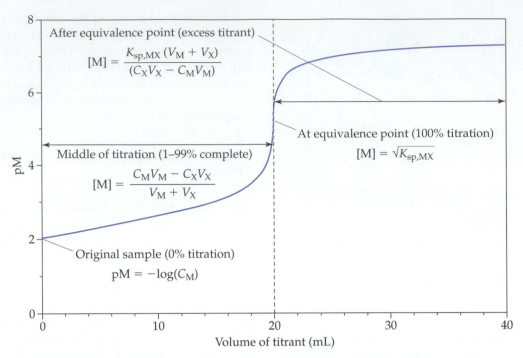

FIGURE 13.14 Equations for predicting the response for the titration of a metal ion (M) with a precipitating agent (X) to form a 1:1 precipitate. All terms used in these equations are defined in the text.

we are generally using a precipitation reaction with a low solubility product for this type of titration, we can safely assume that this reaction uses essentially all of the added titrant until the titration is about 99% complete. This means we can determine the amount of analyte that remains in solution by simply subtracting the moles of added titrant ($C_X V_X$) from the original moles of analyte ($C_M V_M$) and dividing this difference by the total volume of the sample/titrant mixture ($V_M + V_X$). We can express this relationship by using Equation 13.18 for the formation of a 1:1 precipitate.

Middle of Titration (1−99% Complete):

$$[M] = \frac{C_M V_M - C_X V_X}{V_M + V_X} \qquad (13.18)$$

The value of pM is then estimated by using this estimated [M] for the sample/titrant mixture. Similar expressions, with coefficients for some of the preceding terms other than one, can be derived for the formation of precipitates that do not have a 1:1 stoichiometry (e.g., MX_2).

At the equivalence point, we have added exactly enough of X to precipitate all of M. In this case, the only soluble M or X in solution is that which is produced from the precipitate as an equilbrium is established in the sample/titrant mixture. The extent to which M or X is formed from precipitate MX will be described by the solubility product for this reaction. For a 1:1 precipitate,

M and X will be formed in a 1:1 ratio as some of MX redissolves. This means that [M] = [X] at the equivalence point, or $K_{sp,MX} = [M][X] = [M]^2$. We can now rearrange this expression and use it to find the value of [M] at this point in the titration curve.

Equivalence Point for a 1:1 Precipitate (100% Titration):

$$[M] = \sqrt{K_{sp,MX}} \qquad (13.19)$$

Again, similar but more complex expressions can be derived for determining the value of [M] at the equivalence point for a precipitate that does not have a 1:1 stoichiometry, as occurs for MX_2.

After the equivalence point for any precipitation titration, the concentration of X in solution will be approximately equal to the moles of excess added titrant divided by the total volume of the sample/titrant mixture, as given by $[X] = (C_X V_X - C_M V_M)/(V_M + V_X)$. We can substitute this relationship for [X] in the expression $K_{sp,MX} = [M][X]$ and rearrange this expression to solve for [M], as shown below.

After Equivalence Point (Excess Titrant):

$$[M] = \frac{K_{sp,MX}(V_M + V_X)}{(C_X V_X - C_M V_M)} \qquad (13.20)$$

Examples of such calculations are provided in the next exercise.

Predicting the Behavior of a Precipitation Titration

Figure 13.14 shows the titration curve for a 10.00 mL aliquot of a 0.0100 M Ag^+ sample that is titrated with a 0.0050 M solution of AgCl. What are the expected values for pAg (a) at the beginning of the titration, (b) half-way through this titration, (c) at the equivalence point, and (d) after a 10.00 mL excess of titrant has been added to this sample?

SOLUTION

a. At the beginning of the titration, pAg can be determined by using Equation 13.17 and the Ag^+ concentration in the original sample.

$$[Ag^+] = 0.0100 \ M \quad \text{or}$$

$$pAg = -\log(0.0100) = \textbf{2.000}$$

b. The volume of titrant needed to get half-way through the titration can be calculated by using the fact that Ag^+ and Cl^- combine in a 1:1 stoichiometry during this titration to form AgCl(s). This means that the volume of added titrant at the equivalence point will be as follows.

$$V_X = \frac{C_M V_M}{C_X}$$

$$= \frac{(0.005000 \ M)(0.02500 \ L)}{0.01000 \ M}$$

$$= 0.02000 \ L \quad \text{or} \quad 20.00 \ mL$$

The volume of Cl^- solution needed to reach the half-way point of this titration will then be $(0.5000) \cdot (20.00 \ mL) = 10.00 \ mL$. This information can be used with Equation 13.17, to determine the value of pAg at this point.

$$[Ag^+] =$$

$$\frac{(0.01000 \ M)(0.01000 \ L) - (0.005000 \ M)(0.01000 \ L)}{(0.01000 \ L + 0.01000 \ L)}$$

$$= 0.0025 \ M \quad \text{or} \quad pAg = -\log(0.0025) = \textbf{2.60}$$

c. The value of pAg at the equivalence point can be found by using Equation 13.19 and the solubility product for AgCl(s), where $K_{sp,AgCl} = 1.77 \times 10^{-10}$ (with apparent units of M^2).

$$[Ag^+] = \sqrt{1.77 \times 10^{-10} \ M^2} = 1.33 \times 10^{-5} \ M$$

$$\text{or} \quad pAg = -\log(1.33 \times 10^{-5}) = \textbf{4.88}$$

d. After the equivalence point, the value of pAg can be calculated through the use of Equation 13.20 and the fact that a 10.00 mL excess of titrant past the equivalence point (at 20.00 mL) will give us a total titrant volume of 30.00 mL.

$$[Ag^+] =$$

$$\frac{(1.77 \times 10^{-10} \ M^2)(0.01000 \ L + 0.03000 \ L)}{[(0.00500 \ M)(0.03000 \ L) - (0.01000 \ M)(0.01000 \ L)]}$$

$$= 1.42 \times 10^{-7} \ M$$

$$\text{or} \quad pAg = -\log(1.42 \times 10^{-7}) = \textbf{6.85}$$

The same general process that was illustrated in the last exercise can be used to estimate a titration curve in which Cl^- is the analyte and Ag^+ is the titrant. In this case, Equations 13.17–13.20 can again be used, but with "X" now being substituted for "M" (and "M" for "X") to reflect the fact that X is now the analyte and M is the titrant. This approach makes it possible to then prepare a plot of pCl (or pX) versus the volume of titrant for such a method. Alternatively, a plot of pM versus titrant can still be used if Ag^+ is the titrant, because the end point for this method is generally related to the concentration of Ag^+ in solution (see Section 13.3A). This second type of plot is common in argentometric titrations and can be estimated by simply performing calculations based on Equations 13.17–13.20 to first obtain $[Cl^-]$ at a given point in the titration. This result is then converted to the maximum expected value for $[Ag^+]$ and pAg at the same point in the titration by using the relationships $[Ag^+] = K_{sp,AgCl}/[Cl^-]$ and $pAg = -\log([Ag^+])$.

Using the Fraction of Titration. A general equation to predict the response and fraction of a precipitation titration can be derived by using mass balance equations for both the analtye and titrant. An example of this derivation is shown in Table 13.5 for the titration of M with X to form a 1:1 precipitate MX, as is often used during an argentometric method. The final result is shown in Equation 13.21.

$$F = \frac{C_X V_X}{C_M V_M} = \frac{\left(1 - \dfrac{[M] + K_{sp}/[M]}{C_M}\right)}{\left(1 + \dfrac{[M] - K_{sp}/[M]}{C_X}\right)} \tag{13.21}$$

A similar process to that shown in Table 13.5 can be used to obtain a fraction of titration equation for the titration of X with M or for precipitation titrations that involves a precipitate with a stoichiometry besides a 1:1 combination of M with X.

TABLE 13.5 Fraction of Titration Equation for M Titrated with X, Giving MX(s)*

Titration Reaction:

$$M + X \longrightarrow MX(s)$$

Equilibrium Expression:

$$K_{sp} = [M][X] \Rightarrow [X] = \frac{K_{sp}}{[M]}$$

Mass Balance Equations:

Equation for M: Total moles M = moles M in solution + moles M in solid

$$\Rightarrow \quad C_M V_M = [M](V_M + V_X) + \text{moles MX(s)}$$

$$\Rightarrow \quad \text{moles MX(s)} = C_M V_M - [M](V_M + V_X)$$

Equation for X: Total moles X = moles X in solution + moles X in solid

$$\Rightarrow \quad C_X V_X = [X](V_M + V_X) + \text{moles MX(s)}$$

$$\Rightarrow \quad \text{moles MX(s)} = C_X V_X - [X](V_M + V_X)$$

Set Both Mass Balance Equations Equal to Each Other and Combine Common Volume Terms:

$$\Rightarrow \quad C_M V_M - [M](V_M + V_X) = C_X V_X - [X](V_M + V_X)$$

$$\Rightarrow \quad C_M V_M - [M]V_M + [X]V_M = C_X V_X + [M]V_X - [X]V_X$$

Factor out ($C_M V_M$) on the Left and ($C_X V_X$) on the Right:

$$\Rightarrow \quad C_M V_M \left(1 - \frac{[M] + [X]}{C_M} \right) = C_X V_X \left(1 + \frac{[M] - [X]}{C_X} \right)$$

Substitute in [X] = K_{sp}/[M] from Equilibrium Expression:

$$\Rightarrow \quad C_M V_M \left(1 - \frac{[M] + K_{sp}/[M]}{C_M} \right) = C_X V_X \left(1 + \frac{[M] - K_{sp}/[M]}{C_X} \right)$$

Rearrange to Obtain ($C_X V_X$)/($C_M V_M$) on Left Side of Equation:

$$\therefore \quad F = \frac{C_X V_X}{C_M V_M} = \frac{\left(1 - \dfrac{[M] + K_{sp}/[M]}{C_M} \right)}{\left(1 + \dfrac{[M] - K_{sp}/[M]}{C_X} \right)} \tag{13.21}$$

Terms: F = fraction of titration; C_M = total concentration of M in the original sample; V_M = volume of original sample; C_X = total concentration of X in the titrant; V_X = volume of added titrant at a given point in the titration; K_{sp} = solubility product for the formation of precipitate MX; [M] = concentration of noncomplexed metal ion M in the sample/titrant mixture.

Equation 13.21 and related expressions can be used to predict the response of a precipitation titration by determining the volume of titrant that would correspond to a given response. This process can be carried out by using either individual calculations or by using a spreadsheet like the one shown in Figure 13.15. This information can then, in turn, be used to determine how a specific change in reaction conditions will affect the titration. An example of this type of application is given in Figure 13.10, which shows how changing the titrant from Cl^- to I^- or Br^- will affect the analysis of Ag^+ in water when using these reagents. This example clearly indicates that the sharpness of the titration curve will be related to the solubility product (K_{sp}) for the resulting precipitate. The same type of calculation could be used to see how the titration curve will change as the concentration of the titrant and the concentration or volume of analyte are changed, along with the presence of side reactions or a variation in other reaction conditions (e.g., pH).

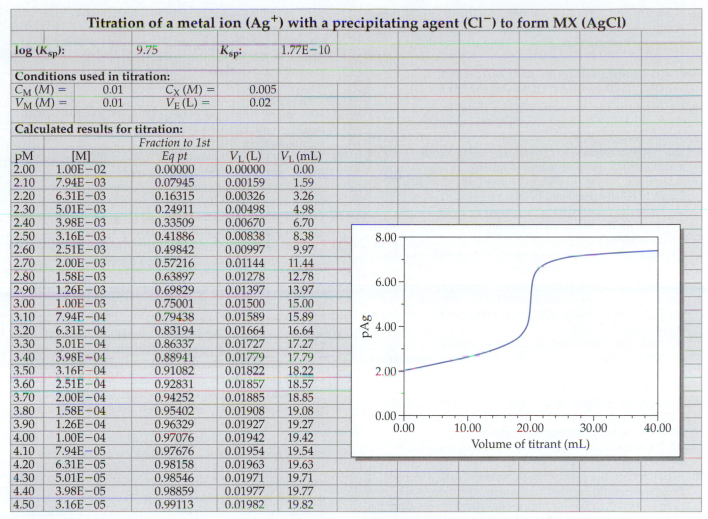

Titration of a metal ion (Ag⁺) with a precipitating agent (Cl⁻) to form MX (AgCl)				
log (K_{sp}):	9.75	K_{sp}:	1.77E−10	

Conditions used in titration:				
C_M (M) =	0.01	C_X (M) =	0.005	
V_M (M) =	0.01	V_E (L) =	0.02	

Calculated results for titration:

pM	[M]	Fraction to 1st Eq pt	V_L (L)	V_L (mL)
2.00	1.00E−02	0.00000	0.00000	0.00
2.10	7.94E−03	0.07945	0.00159	1.59
2.20	6.31E−03	0.16315	0.00326	3.26
2.30	5.01E−03	0.24911	0.00498	4.98
2.40	3.98E−03	0.33509	0.00670	6.70
2.50	3.16E−03	0.41886	0.00838	8.38
2.60	2.51E−03	0.49842	0.00997	9.97
2.70	2.00E−03	0.57216	0.01144	11.44
2.80	1.58E−03	0.63897	0.01278	12.78
2.90	1.26E−03	0.69829	0.01397	13.97
3.00	1.00E−03	0.75001	0.01500	15.00
3.10	7.94E−04	0.79438	0.01589	15.89
3.20	6.31E−04	0.83194	0.01664	16.64
3.30	5.01E−04	0.86337	0.01727	17.27
3.40	3.98E−04	0.88941	0.01779	17.79
3.50	3.16E−04	0.91082	0.01822	18.22
3.60	2.51E−04	0.92831	0.01857	18.57
3.70	2.00E−04	0.94252	0.01885	18.85
3.80	1.58E−04	0.95402	0.01908	19.08
3.90	1.26E−04	0.96329	0.01927	19.27
4.00	1.00E−04	0.97076	0.01942	19.42
4.10	7.94E−05	0.97676	0.01954	19.54
4.20	6.31E−05	0.98158	0.01963	19.63
4.30	5.01E−05	0.98546	0.01971	19.71
4.40	3.98E−05	0.98859	0.01977	19.77
4.50	3.16E−05	0.99113	0.01982	19.82

FIGURE 13.15 Example of a spreadsheet using a fraction of titration equation to predict the response of a precipitation titration between a metal ion (M) and a precipitating agent (X) to form a 1:1 precipitate. This spreadsheet is based on Equation 13.21 and is the same spreadsheet that was used to draw the titration curve for Ag⁺ by Cl⁻ in Figure 13.3. Only part of the actual pM (or pAg) values that were used in Figure 13.3 are shown in the portion of the spreadsheet that is included in this figure.

Key Words

Adsorption indicator *336*	Displacement titration *329*	Metallochromic	pM *321*
Auxiliary ligand *326*	Indirect titration *329*	indicator *327*	Precipitation titration *320*
Complexometric titration *319*	Masking agent *326*		

Other Terms

Argentometric titration *333*	Fajans method *334*	Turbidity *337*	Volhard method *333*
Demasking agent *326*	Mohr method *334*	Turbidimetry *337*	Water hardness *319*
Eriochrome black T *327*	Nephelometry *337*		

Questions

WHAT IS A COMPLEXOMETRIC OR PRECIPITATION TITRATION?

1. Describe what is meant by a "complexometric titration." Explain how the measurement of water hardness by using EDTA is an example of this type of titration.

2. What is meant by the term "precipitation titration"? How does this method differ from a complexometric titration?

3. Describe how you would plot a typical titration curve for a complexometric or precipitation titration. Define the term "pM" and state how this term is used in such a plot.

4. Compare and contrast a complexometric or precipitation titration to the titration of a strong acid with a strong base. How are these methods similar? How are they different?

HOW ARE COMPLEXOMETRIC AND PRECIPITATION TITRATIONS USED IN ANALYTICAL CHEMISTRY?

5. Explain how the concentration of a metal ion in a sample can be measured by using a complexometric titration.
6. Write a general titration reaction between EDTA and a metal ion. Use this reaction to explain why EDTA is often used as a titration in a complexometric titration.
7. A 50.00 mL portion of water requires 7.46 mL of 0.02752 M EDTA to reach a good end point.
   (a) Write the titration reaction for this analysis.
   (b) What is the water hardness for this sample in units of molarity?
   (c) What is the water hardness for this sample in units of parts-per-million $CaCO_3$?
8. A 2.589 g sample of copper ore is dissolved and the resulting Cu^{2+} ions are titrated with EDTA. An end point is reached after 7.56 mL of 0.02785 M EDTA has been added to the sample.
   (a) Write the titration reaction for this analysis.
   (b) What is the percent copper (w/w) in the original ore sample if it is known that no significant amount of other metals are present in the sample?
9. A 25.00 mL sample containing Ag^+ is titrated using a 0.01500 M solution of Cl^-. An end point is reached after 15.20 mL of the titrant has been added.
   (a) Write the titration reaction for this analysis.
   (b) What was the concentration of Ag^+ in the original sample?
10. Explain how the precipitation reaction that was used in Problem 9 could be modified for the titration of Cl^- in a sample. How would this new method be different from the one used in Problem 9? In what ways would these two methods be similar?

TITRANTS AND STANDARD SOLUTIONS FOR COMPLEXOMETRIC TITRATIONS

11. What are some properties of EDTA that make it valuable as a titrant for complexometric titrations?
12. Explain why pH is important to control during a complexometric titration that uses EDTA as a titrant.
13. Give some examples of side reactions that might occur during the titration of a metal ion with EDTA. Explain how these side reactions might affect the pH at which this titration can be performed.
14. What range of pH values could be used for each of the following titrations?
    (a) The analysis of Al^{3+} in a dissolved ore sample
    (b) The measurement of Hg^{2+} in the presence of Cd^{2+}
    (c) The analysis of rare earths in the presence of Ga^{3+} and Sc^{3+}
15. A geochemist wishes to determine the concentrations of Fe^{3+}, Al^{3+}, and Pb^{2+} that are present in a solution obtained from a dissolved mineral sample. This is to be done through a series of three titrations using EDTA and conducted at different pH values. Describe how a change in pH can be used to determine the content of Fe^{3+}, Al^{3+}, and Pb^{2+} in this sample.
16. A water scientist desires to learn both the calcium and the magnesium ion concentrations in a water sample. Two 50.00 mL

portions of the sample are titrated with 0.06500 M EDTA. The first titration is carried out at pH 10.00 and requires 18.54 mL of titrant to reach the end point. The second titration is done at pH 12.00 with KOH being used to first precipitate Mg^{2+} as $Mg(OH)_2$ and requires 13.82 mL of the titrant to reach the end point. What are the concentrations of Ca^{2+} and Mg^{2+} in the water sample?

17. How many grams of $Na_2H_2EDTA \cdot 2H_2O$ are needed to make 500 mL of a 0.0800 M solution of EDTA for use in a complexometric titration?
18. What will the concentration (in units of molarity) be for an EDTA solution that is prepared from 3.576 g of $Na_2H_2EDTA \cdot 2H_2O$ that is dissolved in water and diluted to a final volume of 500.0 mL?
19. Explain why monodentate ligands like ammonia are usually not used as titrants in complexometric titrations.
20. Describe each of the following titrants and explain why they might be used in place of EDTA for a complexometric titration.
    (a) DCTA
    (b) Trien
    (c) Tetren
    (d) EGTA
21. A 25.00 mL sample of a solution containing both Ni^{2+} and Mg^{2+} gives an end point at 17.86 mL when it is titrated with 0.04765 M EDTA. A second 25.00 mL portion of the same original sample gives an end point at 6.74 mL when it is titrated with 0.05643 M triethylenetetramine. What were the concentrations of magnesium ions and nickel ions in the original sample?
22. Copper triethylenetetramine (Cutrien) is much more intensely colored (blue) than a solution of Cu^{2+}, but a solution that contains trien alone is colorless. A 0.05000 M trien solution is added as a titrant to 25.00 mL of a copper sulfate solution. The intensity of color in the sample/titrant mixture is observed to increase until 19.0 mL of the trien solution has been added. What was the approximate concentration of Cu^{2+} in the sample?
23. Give two examples of complexometric titrations that make use of inorganic agents as complexing agents.
24. One of the few successful complexometric titrations that uses a monodentate ligand is the titration of cyanide with silver ions. Suppose that a 25.00 mL portion of a solution containing CN^- is titrated with 0.1000 M $AgNO_3$. As the titrant is added, Ag^+ and CN^- first react to form the soluble complex $Ag(CN)_2{}^-$. However, a cream-colored precipitate will occur when enough titrant has been added to also cause the $Ag(CN)_2{}^-$ to form $AgCN(s)$. This precipitate is found to appear after 18.64 mL of the silver nitrate has been added to the sample. What is the concentration of cyanide in the original sample?

USING AUXILIARY LIGANDS AND MASKING AGENTS FOR COMPLEXOMETRIC TITRATIONS

25. Describe what is meant by an "auxiliary ligand." What role can this type of ligand play during a complexometric titration?
26. Explain how ammonia can be used as an auxiliary ligand during the titration of Ni^{2+} with EDTA. How does the presence of ammonia affect the conditions that can be used for this titration? How does the presence of ammonia affect the appearance of the titration curve?
27. Calculate the conditional formation constant for the formation of nickel EDTA in the presence of 0.0100 M ammonia at pH 9.00. How do you think the titration curve shown in

Figure 13.6 would change if you were to instead use a pH of 9.00 for this titration?

28. Use Table 13.1 to determine the pH range over which an auxiliary ligand is needed when titrating each of the following analytes with EDTA.
 (a) Mn^{2+}
 (b) Hg^{2+}
 (c) Cu^{2+} plus Zn^{2+}

29. What is the purpose of a masking agent in a complexometric titration? What is the purpose of a demasking agent? Give one specific example of each.

30. Iron(III) ions can be masked by adding NaF to a solution while measuring calcium ions and magnesium ions by an EDTA titration. What fraction of the dissolved iron(III) species will be present as Fe^{3+} when these species are in equilibrium with a 0.050 M NaF solution at pH 10.00, as is often used during an EDTA titration to measure water hardness. (*Hint*: See Chapter 9.)

31. To determine both manganese ions and iron ions in the same sample, one can titrate the sum of these two analytes with EDTA and then, in a separate sample, mask the iron ions with cyanide and titrate only the manganese ions. Calculate the concentration of both elements if a 20.00 mL sample requires 18.76 mL of 0.05753 M EDTA to titrate both of these ions and requires 7.56 mL to reach an end point in the presence of cyanide.

32. The total concentration of Zn^{2+} and Ni^{2+} in a solution can be determined by their titration with EDTA. To measure only Zn^{2+}, you can add cyanide to mask both types of metal ions so that neither will react with EDTA. The Zn^{2+} can then be demasked by reacting this mixture with formaldehyde, which reacts with some of the cyanide to form $HOCH_2CN$. Under these conditions, Ni^{2+} will remain masked by cyanide. Calculate both the Zn^{2+} and Ni^{2+} concentrations in a sample if 34.75 mL of 0.05000 M EDTA are needed to titrate 25.00 mL of the original sample but only 11.47 mL of this EDTA solution are needed to titrate 25.00 mL of the sample after the addition of cyanide and formaldehyde.

DETERMINING THE END POINT FOR A COMPLEXOMETRIC TITRATION

33. What is meant by the term "metallochromic indicator"? What chemical or physical properties are needed for a chemical to be useful as a metallochromic indicator?

34. Explain how eriochrome black T can be used as a metallochromic indicator for Ca^{2+}. Discuss how this indicator also acts as a complexing agent.

35. What general requirement in terms of the conditional formation constant must be met for a metallochromic indicator to be useful in a particular titration? Explain how this requirement affects the use of eriochrome black T as an indicator during the titration of Mg^{2+} with EDTA.

36. Murexide is a commonly-used metallochromic indicator during the titration of Ni^{2+} with EDTA. This indicator is often added in the form of its calcium-murexide complex. The conditional formation constants for murexide with both Ca^{2+} and Ni^{2+} are shown in Figure 13.16. If the titration of Ni^{2+} with EDTA is conducted at pH 8, what color change in this indicator will be expected at the end point? Write a reaction that shows what happens to the form of this indicator once the end point has been passed during this titration.

37. A chemist wishes to use murexide as a metallochromic indicator during the titration of Cu^{2+} with EDTA. What change in color will be expected at pH 6, 8, or 10 at the end point during this titration if murexide is added in the form of its calcium complex? At which of these pH values would you expect to see the sharpest change in color at the end point? Explain your answer.

38. Calcein is an indicator that fluoresces when it is complexed to Ca^{2+}, but does not fluoresce in its noncomplexed form. When Ca^{2+} is titrated with EDTA at pH 12 and illuminated by an ultraviolet lamp, a calcein indicator will provide a sudden disappearance in green fluorescence at the end point. Write a reaction that describes what is happening to the calcein at the end point of this titration. Clearly indicate the role that calcein, Ca^{2+}, and EDTA all play in this process.

39. Under what conditions might a back titration be used in an analysis involving a complex formation? Give one specific example.

40. Aluminum ions react slowly with EDTA, so it is best to measure these by using a back titration. A 10.00 mL portion of an Al^{3+} solution is treated with 5.0 mL of 0.1000 M EDTA and allowed to react for 30 minutes before being back titrated with 8.08 mL of 0.02000 M $ZnCl_2$. What was the initial concentration of Al^{3+} in the sample?

41. A solution contains dissolved ions that represent both Cr(III) or Cr(VI) species. Two 50.00 mL portions of the sample are taken for an analysis. An excess of ascorbic acid is added to one of these samples to convert all of the Cr(VI) to Cr(III) species. A 25.00 mL portion of 0.0895 M CDTA is added to both samples, which represents an excess of CDTA compared to all chromium-containing ions in the sample. Both sample mixtures are heated to boiling for several minutes, cooled, and the remaining excess CDTA is titrated with 0.0458 M bismuth nitrate. The solution that was treated with ascorbic acid requires 5.75 mL of the bismuth titrant to reach the end point and the other sample requires 9.73 mL of the bismuth titrant. What was the original concentration of Cr(III) species and of Cr(VI) species in the original sample?

42. Explain how a displacement titration is used. Give one specific example for this type of titration.

43. Mercury-EDTA is a sufficiently strong oxidizing agent that can be used to analyze strong reducing agents such as hydroxylamine and hydrazine. A chemist wishes to determine the concentration of hydroxylamine in a 50.00 mL solution. This solution is combined with 10.00 mL of 0.0500 M $HgEDTA^{2-}$ and the mercury ions are reduced to elemental mercury while the hydroxylamine is oxidized to nitrous oxide. The EDTA which is released is then titrated with 7.55 mL of 0.0300 M $ZnCl_2$, and gives an end point at 7.55 mL. What was the concentration of hydroxylamine in the sample?

44. What is an "indirect titration"? Describe how this method can be used to expand the range of analytes that can be measured by using EDTA or related complexing agents.

45. The concentration of sulfate in a mixture of acids is to be determined. A 25.00 mL portion of the sample is treated with 10.00 mL of 0.0500 M $BaCl_2$. The resulting precipitate is filtered off and the supernate solution is adjusted to pH 10. This solution is then titrated with 0.0354 M EDTA and gives an end point at 7.45 mL. What was the sulfate concentration in the original sample?

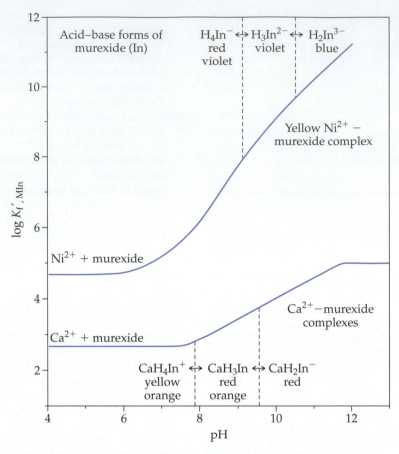

FIGURE 13.16 Value of conditional formation constant $K_f'_{MIn}$ as a function of pH for the 1:1 complexes that are formed between murexide and Ca^{2+} or Ni^{2+}. The principle acid–base forms of murexide at pH values of 4 to 12 are also shown in this graph (H_4In^-, H_3In^{2-}, and H_2In^{3-}). (Based on data from A. Ringbom, *Complexation in Analytical Chemistry*, R.E. Krieger, Huntington, NY, 1979.)

PREDICTING AND OPTIMIZING COMPLEXOMETRIC TITRATIONS

46. Describe the four general regions that are present during the titration of a metal ion with a complexing agent such as EDTA. Describe how the value of pM for the sample or sample/titrant mixture can be estimated in each of these regions.

47. A 25.00 mL sample thought to contain 0.01250 M Ca^{2+} is titrated at pH 10.0 using a 0.01050 M solution of EDTA as the titrant.
    (a) Write the titration reaction for this analysis and determine the volume of titrant that is required to reach the end point.
    (b) Estimate the value of pCa for the sample/titrant mixture at the beginning of the titration, at the equivalence point, and after 10.00 or 30.00 mL of titrant has been added.
    (c) Use the information from Part (b) to prepare a plot for the titration curve that is expected during this analysis.

48. A 50.00 mL solution prepared from dissolved copper and thought to contain 0.04500 M Cu^{2+} is to be titrated with a 0.1011 M solution of EDTA at pH 5.00.
    (a) Write the titration reaction for this analyis and determine the volume of titrant at the equivalence point.
    (b) What is the expected value of pCu at the beginning of the titration? What is the expected pCu value at the equivalence point?

(c) What is the expected pCu value during this titration after 5.00 mL, 10.00 mL, and 20.00 mL of titrant have been added?
(d) Use the information from Parts (c) and (d) to draw a titration curve for this analysis.

49. Describe how mass balance equations and an equilibrium expression can be used to obtain a fraction of titration equation for a complexometric titration. Use the titration of a metal ion with a ligand to form a 1:1 complex to illustrate your answer.

50. Use an approach similar to that in Table 13.4 to derive an equation that gives the fraction of titration for the titration of a ligand with a metal ion to form a 1:1 complex. Compare your final answer with the equation that is shown in Table 13.4 for the titration of a metal ion with a ligand.

51. A 25.00 mL water sample thought to contain 0.02500 M Ca^{2+} is to be titrated with 0.01000 M EDTA at pH 10.00. A complex of eriochrome black T with Mg^{2+} is used as the indicator during this titration. What is the fraction of titration at the end point that is signaled by this indicator? What is the resulting titration error for this analysis?

52. Use the fraction of titration equation in Table 13.4 and successive approximations to find the value of pM and corresponding volume of titrant at the beginning of the titration in Problem 48, as well at the equivalence point and half-way point of this titration. How do these results compare with

those that were obtained in Problem 48 when using Equations 13.9–13.12?

TITRANTS AND STANDARD SOLUTIONS FOR COMPLEXOMETRIC TITRATIONS

53. What is meant by the term "argentometric titration"? Give two specific examples of this approach.
54. Describe the Volhard method. How does this technique differ from the titration for silver that was developed earlier by Gay-Lussac?
55. An analysis is performed by the Volhard method in which a 50.00 mL portion of a sample thought to contain bromide is treated with 10.00 mL of 0.1065 M AgNO$_3$ and allowed to form a precipitate. The excess Ag$^+$ that remains in solution is then titrated with a 0.09875 M solution of potassium thiocyanate, giving an end point at 8.76 mL. What was the concentration of bromide in the original sample if we assume no other anions were present that could form a precipitate with Ag$^+$?
56. The Volhard method is used to determine the concentration of phosphate in a 100.00 mL aqueous sample. This sample is first mixed with 10.00 mL of 0.1000 M AgNO$_3$. The resulting silver phosphate precipitate that is formed is isolated by filtration, rinsed, and dissolved in 20.00 mL of an aqueous solution containing dilute nitric acid. The Ag$^+$ in this solution is titrated to a red end point when 12.47 mL of 0.06543 M KSCN has been added. What was the concentration of phosphate in the original sample?
57. Explain why the curve for the titration of an analyte with Ag$^+$ is often plotted with pAg being shown on the y-axis instead of a concentration term that is directly related to the analyte. Why might such a plot be used when describing a titration that is performed by the Volhard method?
58. Give two examples of precipitation titrations that make use of reactions that do not involve Ag$^+$. Write the titration reaction for each of these methods.

DETERMINING THE END POINT IN A PRECIPITATION TITRATION

59. Describe the Mohr method. Explain why in a Mohr titration Ag$^+$ must be in the titrant, while in a Volhard titration Ag$^+$ must be in the container that contains the sample.
60. What is the percent silver in a piece of metal that has a mass of 0.06978 g and requires 23.98 mL of 0.02654 M KCl to reach the end point in the Mohr method?
61. Water from Chesapeake Bay is partly fresh water from the Potomac River and partly ocean water (which will contain a significant amount of Cl$^-$). A 25.00 mL portion of this water requires 16.04 mL of 0.05647 M AgNO$_3$ to reach an end point in the Mohr method. What was the concentration of chloride ions in this sample?
62. Describe the Fajans method. How does this technique compare with the Mohr method or Volhard method?
63. What is an "adsorption indicator"? Explain how this type of indicator is used in the Fajans method.
64. The salinity of pickle juice (as represented by the concentration of Cl$^-$) is to be measured using the Fajans method. A 5.00 mL portion of this juice is transferred into an Erlenmeyer flask and diluted with distilled water. The pH is adjusted and dichlorofluorescein is added as an indicator. The solution is originally colorless but forms a white precipitate when some 0.1000 M AgNO$_3$ is added as a titrant. This precipitate continues to form and finally turns pink after

18.65 mL of the titrant has been added. What is the concentration of Cl$^-$ in the pickle juice?
65. The original formulation of Bromoseltzer contained sodium bromide. This ingredient was not used after 1975. A 5.00 g portion of a product that is supposedly Bromoseltzer is titrated with 0.1000 M AgNO$_3$ in the presence of tetrabromofluorescein. The pale cream-colored precipitate that is formed during this titration turns red (signaling the end point) after the addition of 31.85 mL of titrant. What is the weight percent of NaBr in this product?
66. Describe the method of nephelometry. Explain how this method can be used to determine the end point of a precipitation titration.
67. A student notices that shining a small laser pointer through a sample/titrant mixture can be used to help determine the end point for the titration of Ag$^+$ with Cl$^-$. Explain how this approach might be used to identify the end point for this titration. What property of the sample/titrant mixture is being examined by this approach?

PREDICTING AND OPTIMIZING PRECIPITATION TITRATIONS

68. Describe the four general regions that are present during the titration of a metal ion with a precipitating agent to create a 1:1 precipitate. Describe how the value of pM for the sample or sample/titrant mixture can be estimated in each of these regions.
69. A 50.00 mL solution thought to contain 0.0500 M Ag$^+$ is titrated by adding 0.02500 M NaCl.
    (a) Write the titration reaction for this analysis and determine the volume of titrant that is needed to reach the equivalence point.
    (b) Estimate the value of pAg for this titration at the beginning of the titration, at 50% titration, at the equivalence point, and when an excess of 10.00 mL of titrant has been added to the sample.
    (c) Use the information from Part (b) to prepare a titration curve for this method.
70. A 25.00 mL sample believed to contain 0.0500 M NaCl is titrated by adding 0.0250 M AgNO$_3$.
    (a) Write the titration reaction for this analysis and calculate the volume of titrant that will be needed to reach the equivalence point.
    (b) Estimate the value of pCl at the beginning of the titration, and after 25.00, 50.00, or 75.00 mL of titrant has been added. Use these values to make an approximate plot for the titration curve.
    (c) Use the values from Part (b) to estimate the value of pAg at each of the indicated points in the titration. Also use these values to prepare a titration curve for this analysis.
    (d) Compare the plots that were made in Parts (b) and (c). In what situations might you use the plot from Part (b)? When might you use the plot from Part (c)?
71. Describe how mass balance equations and a solubility product expression can be used to obtain a fraction of titration equation for a precipitation titration. Use the titration of a metal ion with a precipitating agent to form a 1:1 precipitate to illustrate your answer.
72. Use an approach like that in Table 13.5 to derive a fraction of titration equation for the titration of X with a metal ion M to form a 1:1 precipitate. Compare your final equation with the one that is shown in Table 13.5 for the titration of M with X. How are these two equations similar? How are they different?

73. Use Equation 13.21 and successive approximations to find the value of pAg and corresponding volume of titrant at the beginning of the titration in Problem 69, as well as at the equivalence point and half-way point of this titration. How do these results compare with those that were obtained in Problem 69 when using Equations 13.17–13.19?

CHALLENGE PROBLEMS

74. It has been said that the most rapid replacement of an old analytical method by a new one was the introduction of EDTA as a titrant in measuring water hardness. The old method, utilized prior to about 1950, involved use of a standardized soap solution. This soap solution was added with a buret to a sample and formed an insoluble precipitate with calcium ions and magnesium ions. When an excess amount of soap had been added, the soap did not precipitate and was free to form soap bubbles when shaken. Thus, the titrant and the indicator were the same substance. Compare and contrast this approach to the current EDTA titration method for measuring water hardness in terms of how the methods that are carried out and how their end point is detected. What do you think are some important advantages of the EDTA method over this previous technique?

75. Water hardness caused by the presence of $Ca(HCO_3)_2$ is called "temporary hardness" and can be removed by boiling water. Water hardness caused by the presence of $CaSO_4$ is called "permanent hardness" and cannot be removed by boiling, but can be removed by adding Na_2CO_3 to the sample. Use your knowledge of chemical reactions to explain these observations and differences.

76. The addition of a base is often needed to prepare an EDTA solution. A scientist wishes to dissolve 1.365 g of H_4EDTA in 500 mL water. What mass of KOH will be needed to dissolve this EDTA and raise the pH of its solution to pH 10.0?

77. Obtain the chemical structure for one or more of the metallochromic indicators that are listed in Table 13.3. Using this chemical structure, explain why this indicator can form a complex with certain metal ions. Also use this chemical structure to explain why this indicator is a part of a polyprotic acid–base system. Discuss how these acid–base properties would be expected to affect the ability of this indicator to form a complex with a metal ion.

78. We noticed in Chapter 12 that the titration curve for a "weak acid" (i.e., a Brønsted–Lowry acid) will look different before the equivalence point but the same after the equivalence point as the K_a of this acid is changed. However, in this chapter we saw that the titration of a metal ion with EDTA (a Lewis acid–base system) will give titration curves that look the same before the equivalence point and different after the equivalence point when the value of K_f' is changed. Explain these differences in behavior.

79. Prepare a spreadsheet like the one in Figure 13.18 to plot a titration curve for a metal ion that is titrated with EDTA. Use this spreadsheet to prepare a plot for the titration of a 10.00 mL sample of 0.010 M Mg^{2+} that is titrated at pH 10.0 with 0.0050 M EDTA. Compare the curve that you obtain to the one that was given in Figure 13.18 under similar conditions for Ca^{2+}. What similarities do you see in these two curves? What differences do you observe? Explain these differences.

80. It is possible to appreciate the demand by Charles X for a better method of silver measurements by looking at how an error in this analysis would affect the apparent value for a piece of this metal. Look up the current price of silver in a newspaper or on the Internet in which the price as listed in U.S. dollars per troy ounce. (*Note*: There are 31.103 g per troy ounce.)

(a) Suppose that one analyst consistently gives a systematic error with a mass and silver value that is low by 2%, while another gives a systematic error that is high by 2%. How much silver (in kg) would a trader have to buy and sell at these sites to make a profit of $1000?

(b) How accurate must a silver assay be to justify all of the significant figures shown in the listed price of silver? What does this tell you about modern methods for silver analysis?

81. Although the Mohr method is important from a historical point of view in the development of precipitation titrations, it is no longer commonly used in modern chemical laboratories. This is largely because of some health hazards that are associated with Cr(VI), or the chromate ion, which is a reagent in this method. Locate a material safety data sheet (MSDS) for sodium chromate and report on the chemical and physical hazards that are associated with this reagent.

82. When the Fajans method is used to titrate and measure Ag^+, it is desirable to produce small precipitate particles for AgCl that have a surface area.

(a) Explain why a precipitate with a high surface area is desired in this method.

(b) What general reaction conditions can be used to obtain this type of precipitate. (*Hint*: See the discussion on precipitate formation in Chapter 7.)

(c) How do these reaction conditions compare to those that are used to produce AgCl during a gravimetric analysis?

83. Prepare a spreadsheet like the one in Figure 13.15 to plot a titration curve for a metal ion that is titrated with an anion to form a 1:1 precipitate. Use this spreadsheet to prepare a plot for the titration of a 10.00 mL sample of 0.010 M Ag^+ that is titrated with 0.0050 M Br^-. Compare the curve that you obtain to the one that was given in Figure 13.15 for the titration of Ag^+ with Cl^-. Explain any differences that you observe in these two plots.

84. Prepare a modified spreadsheet like the one in Figure 13.15 to plot a titration curve for an anion that is titrated with a metal ion to form a 1:1 precipitate. Use this spreadsheet to prepare a plot for the titration of a 25.00 mL sample of 0.0050 M Br^- that is titrated with 0.0050 M Ag^+, in which the value of pBr is given on the y-axis. Also prepare a plot in which the value of pAg is shown on the y-axis, where $pAg = -\log([Ag^+])$ and $[Ag^+] = K_{sp}/[Br^-]$. Compare the response of these curves. Explain why you might use a plot with pAg on this y-axis when you are optimizing this type of titration.

TOPICS FOR DISCUSSION AND REPORTS

85. Obtain more information on the topic of water hardness. Write a report that describes the nature of this problem, its economic impact, and the analytical methods that are used for the analysis of water hardness.

86. Contact a local environmental or water-testing laboratory. Discuss with them how complexometric titrations are used at their facility. Discuss your findings with your class.

87. There are many other metallochromic indicators besides those that are given in Table 13.3. Obtain more information on one of the following indicators. Give the structure of the indicator, list the types of metal ions to which it binds, and describe the

change in color that is associated with this indicator. Also include a discussion of the indicator's acid–base properties.
 (a) Murexide
 (b) PAN, or 1-(2-pyridilazo)-2-naphthol
 (c) Calcein
 (d) Pyrogallol red
 (e) Methylthymol blue
 (f) Hematoxylin
88. Although there are many types of precipitation reactions that can be used for gravimetric analysis (see Chapter 11), only those reactions that involve silver halides are commonly

used in precipitation reactions. Suggest one or more reasons why the following insoluble compounds are not determined by precipitation titrations: $CaCO_3$, $PbCl_2$, CuS, and $BaCrO_4$.
89. Locate or suggest a procedure for performing a complexometric or precipitation titration for one or more of the following situations. Describe the reactions that take place in this method and state the purpose for each step in the procedure.
 (a) Measurement of the percent calcium in Portland cement
 (b) Determination of the amount of copper in aluminum foil
 (c) Assaying the phosphate content in a sample of $BiPO_4$

References

1. F. R. Spellman, *The Science of Water: Concepts and Applications*, Taylor & Francis, New York, 2007.
2. J. C. Briggs and J. F. Ficke, J.F. *Quality of Rivers of the United States, 1975 Water Year — Based on the National Stream Quality Accounting Network (NASQAN): U.S. Geological Survey Open-File Report 78-200*, U.S. Department of the Interior, Washington, DC, 1977.
3. J. Bassett, R. C. Denney, G. H. Jeffery, and J. Mendham, *Vogel's Textbook of Quantitative Inorganic Analysis, 4th ed.*, Longman, New York, 1978.
4. R. Pribil, *Applied Complexometry*, Pergamon Press, New York, 1982.
5. A. Ringbom, *Complexation in Analytical Chemistry*, Krieger Publishers, Huntington, NY, 1979.
6. J. Inczedy, T. Lengyel, and A. M. Ure, *Compendium of Analytical Nomenclature*, 3rd ed., Blackwell Science, Malden, MA, 1997.
7. G. Maludzinska, Ed. *Dictionary of Analytical Chemistry*, Elsevier, Amsterdam, the Netherlands, 1990.
8. H. Diehl, *Quantitative Analysis*, Oakland Street Science Press, Ames, IA, 1970.
9. K. Ueno, "Guide for Selecting Conditions for EDTA Titrations," *Journal of Chemical Education*, 42 (1965) 432.
10. H. Frieser, *Concepts & Calculations in Analytical Chemistry: A Spreadsheet Approach*, CRC Press, Boca Raton, FL, 1992.
11. F. Szabadvary, *History of Analytical Chemistry*, Pergamon Press, New York, 1966.
12. K. F. Mohr, *Lehrbuch der chemish-analytishen Titrirmethode*, 1877, Vieweg, Braunschweig.
13. K. Fajans, "Eine neue Methode zur Titration von Silver- und Halogenionen mit organischen Farbstoffindicatoren (nach Versuchen von O. Hassel)," *Chemiker Zeitung*, 47 (1923) 427.

Chapter 14

An Introduction to Electrochemical Analysis

Chapter Outline

14.1 INTRODUCTION: GETTING A BRIGHTER SMILE

In the 1940s it was found that children in some regions of the world had fewer cavities and less tooth decay than children from other areas. This difference was eventually found to be due to the presence of fluoride in the drinking water. Many public drinking-water systems in the United States now either contain fluoride as a natural component or have fluoride added to the water to prevent cavities. This practice has been listed by the U.S. Centers for Disease Control as one of the ten greatest public health achievements of the twentieth century[1,2] (see Figure 14.1).

It is important when fluoride is being added to drinking water (or to products such as toothpaste) to ensure that just the right amount of this chemical is present. This means water-treatment facilities must regularly measure the level of fluoride that is present in drinking water. Such a measurement is now carried out using an electrode that can selectively measure fluoride ions.[3] Prior to the development of this electrode, the measurement of fluoride in water was based on a time-consuming, colorimetric assay. However, this measurement can now be easily made on a routine basis or even in a continuous manner by measuring the electrical potential that is formed between the fluoride-selective electrode and a reference electrode in the presence of drinking water or the desired sample.

We learned in Chapter 10 that the study of electrochemical reactions and their applications is a field known as "electrochemistry." In this chapter we refer to the use of electrochemistry for the analysis of chemicals as **electrochemical analysis**. There are many types of methods that can be employed for electrochemical analysis. For instance, the approach used with the fluoride electrode is a method known as *potentiometry*. In this chapter we will learn the basic concepts and terms that are used in the field of electrochemical analysis. We will then look in more detail at the method of potentiometry, including a discussion of how pH, fluoride ion activities, or concentrations and other chemical measurements can be made by this technique. In Chapters 15 and 16, we examine other approaches for electrochemical analysis, such as redox titrations, voltammetry, and coulometry.

14.1A Units of Electrical Measurements

Before we begin to look at potentiometry and other techniques for electrochemical analysis, we first need to consider several important quantities that are measured or used in these techniques. Table 14.1 provides a summary of fundamental SI units and derived SI units that are used in electrochemical measurements.[4,5] One property that is commonly used, measured, or controlled in electrochemical methods is **charge**. The term "charge" is defined in this type of application as being equal to the integral of electrical current over time.

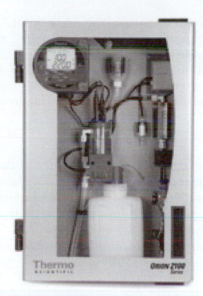

FIGURE 14.1   The analysis of fluoride as a means to monitor and control fluoridation in drinking water in public water systems in the United States. The instrument shown on the right is designed for the continuous analysis of fluoride in water by using a fluoride ion-selective electrode to measure this analyte. (The photo on the left is provided courtesy of the U.S. Centers for Disease Control; the photo on the right is provided courtesy and with permission from Thermo Fisher.)

Charge is represented by the symbol Q and is described in the SI system by using a unit known as the coulomb (C).

The smallest quantity of an elementary charge (represented by the symbol e) is the "+1" charge that is associated with a single proton or the "−1" charge that is present on a single electron.[6] The presence of a large number of charged particles can be described in terms of coulombs, where one mole of electrons is equal to a charge of 96,485 C. As noted in Chapter 10, this particular value of 96,485 C/mol is called the *Faraday constant (F)*, and exactly one mole of electrons is sometimes called one "Faraday."[6] The value of the Faraday constant can be used to determine moles of electrons (n_e)

that are needed to provide a certain charge, as shown by Equation 14.1.

$$Q = n_e F \tag{14.1}$$

Based on this equation, the charge on one electron can be written as $Q = (1 \text{ mol}/6.023 \times 10^{23}) \cdot (96,485 \text{ C/mol}) = 1.602 \times 10^{-19} \text{ C}$.

Another property that is often used or measured in electrochemical methods is current. **Current**, which is represented by the symbol I, is a measure of the amount of electrical charge that flows through a conducting medium in a given amount of time. The fundamental SI unit for current is the ampere (A, or "amp"). Many electrochemical

TABLE 14.1   SI Units and Derived SI Units for Electrochemical Measurements

Measured Quantity (symbol)	Unit (symbol)	Relationship to Other SI Units
Electric current (I)	ampere (A)	Fundamental SI unit[a]
Electric charge (Q)	coulomb (C)	$1 \text{ C} = 1 \text{ A} \cdot \text{s}$
Electric potential (E)	volt (V)	$1 \text{ V} = 1 \text{ W/A} = 1 \text{ J/A} \cdot \text{s}$
Electric resistance (R)	ohm (Ω)	$1 \text{ }\Omega = 1 \text{ V/A}$
Time (t)	second (s)	Fundamental SI unit[b]
Frequency	hertz (Hz)	$1 \text{ Hz} = 1/\text{s}$

[a]One ampere is defined as the constant current that produces a force of 2×10^{-7} newton per meter of length when maintained in two straight parallel conductors of infinite length and negligible circular cross section that are placed one meter apart in a vacuum.

[b]One second is defined as the amount of time equal to 9,192,631,770 periods of the radiation corresponding to the transition between the two hyperfine levels of the ground state of cesium-133.

techniques involve small currents, so the related units of milliamps (mA = 10^{-3} A) and microamps (μA = 10^{-6} A) are often used in these methods.[5-7]

Current is related to charge and the amount of time (t) it takes this charge to pass through a system. This relationship for a system with a constant current is as follows,

$$I = Q/t \quad \text{or} \quad Q = I \cdot t \qquad (14.2)$$

where the expression on the right is a reminder that charge is really the integral of current over time.[6-8] Based on the preceding equations, the current for an electrochemical measurement is sometimes given in units of coulombs per second (C/s), where 1 A = 1 C/s. Alternatively, it can also be said from the expressions in Equation 14.2 that 1 C = 1 A$\cdot$s.

EXERCISE 14.1   Relating Current and Charge

An electrochemical cell has a constant current of 250 μA that is allowed to flow through a wire for 220.0 s. What is the charge (in units of C) that was allowed to pass through this system? How many moles of electrons must have passed through the wire to produce this current and charge?

SOLUTION

We can first find the charge that passed through this system by using the known current and time of current flow along with the right-hand expression in Equation 14.2. (*Note*: To obtain a final answer for charge with units of "coulombs" it is necessary to make sure the current is expressed in units of amps and the time is given in seconds.)

$Q = I \cdot t$

$\quad = (250 \times 10^{-6}\,\text{A})(220.0\,\text{s}) = \textbf{5.50} \times \textbf{10}^{-2}\,\textbf{C}$

We can then determine how many moles of electrons were needed to produce this change by using Equation 14.1 and the Faraday constant.

$n_e = Q/F$

$\quad = (5.50 \times 10^{-2}\,\text{C})/(96,485\,\text{C/mol})$

$\quad = 5.70 \times 10^{-7}\,\text{mol electrons}$

The preceding exercise and Equation 14.2 indicate that time is another important parameter in many types of electrochemical measurements. In this chapter, we represent time by the symbol t and use the fundamental SI unit of the second (s) to describe this parameter. A closely related term to time is *frequency* (a term we discussed in Chapter 17), which is a measure of how many cycles of an event occur per unit of time. Frequency is expressed in the SI system in units of hertz (Hz), where 1 Hz = 1/s.

Another factor that is controlled or measured in electrochemical methods is *electrical potential* (or the "potential"). Electric potential was defined in Chapter 10 as a measure of the work that is required to bring a charge from one point to another. The difference in electrical potential between two points (E) is expressed in the unit of volts (V) in the SI system. We have already discussed the use of the differences in electrical potential to describe electrochemical cells in Chapter 10. This definition is reflected in Table 14.1 by the way in which the unit of volt is related to other SI units, where 1 volt is equal to 1 watt of power per ampere. A term that can be used in place of potential in an electrochemical cell that has no appreciable current flowing is the *electromotive force*, or "emf." In each of these situations, the electrical potential represents the driving force behind the movement of electrons through a conducting medium.[4-8]

Whenever there is an electrical potential that creates a flow of current, there will also be some resistance to this flow of current. This **resistance (R)** is expressed in a unit called the "ohm," as represented by the capital Greek letter omega (Ω, where 1 Ω = 1 V/A). The reciprocal of resistance ($1/R$) is known as the *conductance*, a value that is commonly given in units of the "mho" (Ω^{-1}, where 1 Ω^{-1} = A/V) or the siemen (S).

The potential, current, and resistance for an electrochemical system are all related through **Ohm's law**, which is shown in the following formula.

$$Ohm's\ law: \quad E = I \cdot R \qquad (14.3)$$

With Ohm's law it is possible to directly relate the potential of an electrical system to the current and resistance to current flow. This relationship makes Ohm's law valuable in finding one of these three factors if the other two parameters in Equation 14.3 are already known. Such a relationship can be quite valuable in the design and description of systems for electrochemical analysis.

EXERCISE 14.2   Using Ohm's Law

If the resistance of the wire in Exercise 14.2 was 1000. ohms, what electric potential must have been present to create a constant current of 250. μA?

SOLUTION

We know the current (250 $\times$ 10^{-6} A) and the resistance in this case, so we can use Ohm's law to also find the electric potential.

$E = I \cdot R$

$\quad = (250. \times 10^{-6}\,\text{A})(1000.\,\Omega) = \textbf{2.50} \times \textbf{10}^{-2}\,\textbf{V}$

Notice in this calculation that a value with units of volts is obtained if I is given in units of amps and R is given in ohms (Ω). We can show that this should be the case through dimensional analysis by using these units for I and R, because (1 A)(1 Ω) = (1 A)(1 V/A) = 1 V.

There are two types of current that can be used in electrical systems and in methods for electrochemical analysis. If the direction of electron movement and the current always proceed in the same direction, the current is called a *direct current* (*DC*). Most of the electrochemical methods discussed in this book make use of this type of current, but there are other methods that instead use an *alternating current* (AC).[9,10] In a system with an alternating current, the direction of the movement of electrons reverses at a regular rate. Batteries are examples of power supplies that produce a DC current, while the electricity used in most houses is based on an AC current, in which the direction of the current is alternated with a sinusoidal frequency of 60 Hz, or 60 cycles per second.

14.1B Methods for Electrochemical Analysis

There are many methods for electrochemical analysis. Table 14.2 summarizes several of the main types of electrochemical analysis techniques and lists the methods that we discuss in this book. The first of these methods is potentiometry. **Potentiometry** is a technique for electrochemical analysis that is based on the measurement of a cell potential with essentially zero current passing through the system.[7–10] This measured potential is related to the chemical composition of the two electrodes and the solutions into which they are placed. Potentiometry is the type of electrochemical method that forms the basis for use of the fluoride electrode and the pH electrode. This is also the technique we will focus on in the remainder of this chapter.

A special subcategory potentiometry is a *potentiometric titration*. This approach utilizes a potential measurement to follow the course of a titration as various amounts of titrant are combined with the analyte.[7,8] We have already seen several examples of potentiometric titrations in Chapter 12, where we discussed the use of a pH meter to follow the course of an acid–base titration. More examples are given in Chapter 15 in the use of cell potential measurements during titrations that involve oxidization–reduction reactions.

Two additional and related methods for electrochemical analysis are the techniques of amperometry and voltammetry. In **amperometry**, the current passing through an electrochemical cell is measured at a fixed potential. In **voltammetry**, the current is also measured but the potential is now varied over time.[7–10] The cell potential can be changed in a variety of ways, which creates a large number of subcategories for methods that all involve the use of voltammetry. Some examples of these subcategories that we will consider in Chapter 16 include DC voltammetry, anodic stripping voltammetry, and cyclic voltammetry.

A third type of method for electrochemical analysis is **coulometry**. This technique uses the measurement of charge for chemical analysis.[7,10] For example, the amount of current that is needed to completely reduce a particular analyte can be measured under conditions in which no other material undergoes reduction. The number of moles of electrons needed to produce this current is then calculated and used to find the moles of analyte that have been reduced. The same type of approach can be used to look at the oxidation of a chemical. A more detailed discussion of coulometry and related methods is provided in Chapter 16.

14.2 GENERAL PRINCIPLES OF POTENTIOMETRY

14.2A Cell Potentials and the Nernst Equation

Any chemical analysis that is carried out by using potentiometry will involve the measurement of a difference in potential between two electrodes in an electrochemical cell. Figure 14.2 shows a general cell that could be used in potentiometry. These components include the same basic features that we saw for other electrochemical cells in Chapter 10. This type of system includes at least two electrodes, identified here as an indicator electrode and a reference electrode, which act as the cathode and anode. Each electrode is in contact with either the sample (in the case of the "indicator electrode") or a reference solution (in the case of the "reference electrode"). There is also usually some type of salt bridge present to provide contact between these two parts of the electrochemical cell. The circuit is completed by making an electrical contact between the two electrodes, which also provides a means for measuring the difference in potential across the cell.

It is important to remember from the definition of potentiometry that measurements in this method are made under conditions in which essentially zero current

TABLE 14.2   Examples of Methods for Electrochemical Analysis

Method	Definition[a]
Potentiometry	A method in which cell potential is measured, and used for chemical analysis, under conditions giving essentially zero flow of current
Amperometry	A method in which current is measured, and used for chemical analysis, at a constant cell potential
Voltammetry	A method in which current is measured, and used for chemical analysis, as the cell potential is varied
Coulometry	A method in which charge is measured and used for chemical analysis

[a]These definitions are based on those found in J. Inczedy, T. Lengyel, and A.M. Ure, *International Union of Pure and Applied Chemistry—Compendium of Analytical Nomenclature: Definitive Rules 1997*, Blackwell Science, Malden, MA.

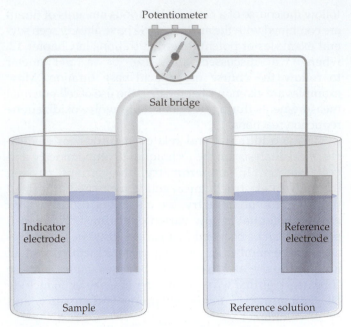

Potentiometer

Salt bridge

Indicator electrode

Reference electrode

Sample

Reference solution

FIGURE 14.2 The general components of an electrochemical cell for potentiometry. In many types of electrodes for potentiometry, several of these components are combined in the electrode design. For example, the salt bridge is often present as a porous frit on the side of the indicator electrode. In other cases, both electrodes and the salt bridge are used as part of a "combination electrode," as discussed in Section 14.3A.

is flowing through this system. This means that although an oxidation–reduction reaction may have the potential to go from reactants to products, the resistance of the electrical circuit that is used to measure this potential is high enough to prevent this reaction from occurring to any significant extent during the measurement.

We learned in Chapter 10 that the potential for each electrode in an electrochemical cell can be described in terms of the standard electrode potential for the half-reaction occurring at that electrode and the activities or concentrations of the species that are involved in this half-reaction. This relationship for a reversible half-reaction like the one in Equation 14.4 at 25°C is given by the Nernst equation in Equation 14.5 (see Chapter 10 for a more general form of the Nernst equation for work at other temperatures).

General Half-Reaction:
$$\text{Ox} + n\,\text{e}^- \rightleftarrows \text{Red} \tag{14.4}$$

Nernst Equation at 25°C:
$$E = E° - \frac{0.05916\text{ V}}{n}\log\left[\frac{a_{\text{Red}}}{a_{\text{Ox}}}\right] \tag{14.5}$$

If we use the Nernst equation to find the expected potentials at both the cathode and anode in an electrochemical cell, the difference in potential measured between these electrodes will be given by Equation 14.6 if essentially no current is present.

$$E_{\text{Cell}} = E_{\text{Cathode}} - E_{\text{Anode}} \tag{14.6}$$

The **indicator electrode** in an electrochemial cell for potentiometry is the electrode that is in contact with the sample and gives a potential related to the activity and concentration of the analyte. The reference electrode provides a fixed potential against which the potential of the indicator electrode can be measured. By convention, the reference electrode is initially assigned the role of the "anode" in an electrochemical cell used for potentiometry and the indicator electrode is assigned the role of the "cathode." This assignment means that Equation 14.7 can also be written in the following form during a measurement that is performed by potentimetry,

$$E_{\text{Cell}} = E_{\text{Ind}} - E_{\text{Ref}} \tag{14.7}$$

where E_{Ind} and E_{Ref} are now the potentials present at the indicator electrode and reference electrode, respectively, and E_{Cell} is the difference in potential that is measured between these electrodes.[7,10]

Measurements in potentiometry deal with galvanic cells, in which we are measuring the potential of a cell as it approaches equilibrium by undergoing a spontaneous oxidation–reduction reaction (see Section 10.3A in Chapter 10). This type of cell should have a value for E_{Cell} that is either positive (indicating it has not yet reached equilibrium) or zero (indicating that equilibrium is present in the cell). A cell potential that is determined by potentiometry to give a "negative" value for E_{Cell} simply means that the roles of the two electrodes are actually the opposite of those that have been assigned to them (i.e., the indicator electrode is actually the anode and reference electrode is actually the cathode), or that the overall oxidation–reduction reaction for this cell will occur in the opposite direction to that in which it is currently written.

14.2B Cell Components in Potentiometry

Reference Electrodes. The reference electrode that is used in potentiometry plays the same role that it does in any type of electrochemical cell. This purpose is to provide a reproducible, known, and constant potential against which the potential of another electrode can be measured. Although the standard hydrogen electrode (SHE) is the ultimate reference electrode against which all other potentials are compared (see Chapter 10), the SHE is far too inconvenient for general use. This problem is due to the reaction components that are needed for the SHE, as indicated by the following half-reaction for this electrode.

Standard Hydrogen Electrode:
$$2\,\text{H}^+ + 2\,\text{e}^- \rightleftarrows \text{H}_2 \quad E° = 0.000...\text{ V} \tag{14.8}$$

It was stated in Chapter 10 that a SHE must have both hydrogen ions and hydrogen gas at an activity of 1.000. This is easy to achieve for the hydrogen ions, but work with hydrogen gas is more difficult and requires that hydrogen gas at a pressure of 1 bar be present around a platinum electrode. In addition, the platinum electrode is coated with a very porous form of platinum called "platinum black" that can absorb materials besides H_2 or H^+. These other materials can lead to "poisoning" of the platinum electrode and alter its properties, creating a system that no longer gives a reproducible potential.

Two more convenient and useful reference electrodes for potentiometry are those based on the **silver/silver chloride electrode** and the **calomel electrode** (or the mercury/mercury chloride electrode, in which "calomel" is another name for mercury(I) chloride). The half-reactions for these two electrodes are given below.

Silver/silver chloride electrode:

$$AgCl + e^- \rightleftarrows Ag + Cl^- \quad E° = 0.2222 \text{ V} \quad (14.9)$$

Calomel electrode:

$$Hg_2Cl_2 + 2e^- \rightleftarrows 2Hg + 2Cl^- \quad E° = 0.268 \text{ V} \quad (14.10)$$

The general designs of these electrodes are shown in Figure 14.3 and Figure 14.4. In each of these electrodes, an insoluble chloride salt coats the free element (Ag or Hg) and both the salt and free element are immersed in a KCl solution of known concentration (often saturated KCl). A calomel electrode that contains a saturated solution of KCl is also known as a *saturated calomel electrode (SCE)*, which has a potential of 0.242 V at 25°C.[7–10]

Indicator Electrodes. There are many indicator electrodes that can be used in potentiometry. These indicator electrodes can be divided into several categories based on how their signal is related to the activity of an analyte.[7,10] Table 14.3 lists four classes of indicator electrodes that make use of a metal. One

possibility is the use of an inert metal as an electrode to oxidize or reduce another substance. This type of indicator electrode is known as a *metallic indicator electrode* and is made from a material such as platinum, palladium, or gold. An example of this type of electrode would be if we used a platinum wire as a cathode to reduce Fe^{3+} to Fe^{2+}. The half-reaction and Nernst equation for this part of the half-reaction are shown below.

Reduction Half-Reaction: $\quad Fe^{3+} + e^- \rightleftarrows Fe^{2+}$ (14.11)

Nernst Equation at 25°C:

$$E_{Fe^{3+}/Fe^{2+}} = E°_{Fe^{3+}/Fe^{2+}} - \frac{0.05916 \text{ V}}{1} \log \left[\frac{a_{Fe^{2+}}}{a_{Fe^{3+}}} \right] \quad (14.12)$$

Notice in the preceding equations that only Fe^{3+} and Fe^{2+} are shown as taking part in the half-reaction and measured response of this electrode. The reason for this is that the platinum electrode is merely acting as a source of electrons for this half-reaction but is not changed itself as part of this redox process.

A *class one electrode* ("electrode of the first kind") consists of a metal that is in contact with a solution that contains metal ions of that element. This type of electrode is used to produce a potential that is related to the activity of the metal ions in the solution-phase sample. An example of this type of electrode would be a silver wire that is immersed in a solution of silver nitrate. The potential of this electrode will depend on the activity and concentration of the Ag^+ ions in solution, as indicated by the Nernst equation for this electrode. (*Note*: An activity of 1.0 is used for Ag(s) in the Nernst equation because this represents a standard state for silver.)

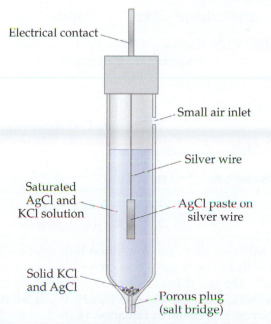

FIGURE 14.3 The general design of a silver/silver chloride (Ag/AgCl) electrode.

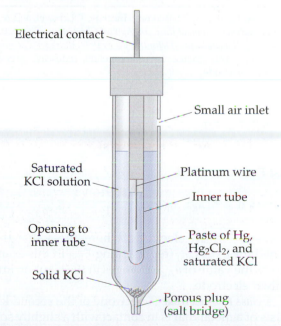

FIGURE 14.4 The general design of a saturated calomel electrode (SCE). A calomel electrode has the same basic design, but does not have solid KCl at the bottom. The calomel electrode has a lower, but fixed and known concentration of KCl instead of a saturated KCl solution.

TABLE 14.3 General Types of Indicator Electrodes Based on Metals

Type of Electrode[a]	Definition
Metallic redox indicator	An electrode made from an inert material such as platinum, palladium, or gold
Example:	A platinum electrode that serves as a site for electron exchange between Fe^{3+} and Fe^{2+}
Class one electrode	An electrode made of metal in contact with a solution that contains metal ions of the same element
Example:	A silver electrode in a solution that contains Ag^+
Class two electrode	An electrode made of metal in contact with a slightly soluble salt of that metal and in a solution containing the anion of the salt
Example:	A silver wire in contact with $AgCl(s)$ and in a solution that contains chloride ions
Class three electrode	An electrode made of metal in contact with a salt of the metal ion (or a complex of this metal ion) and a second, coupled reaction involving a similar salt (or complex) with a different metal ion
Example:	A lead wire in contact with insoluble lead oxalate, which is in contact with a solution that contains Ca^{2+} and in contact with insoluble calcium oxalate

[a]These classifications are based on J. Inczedy, T. Lengyel, and A. M. Ure, *International Union of Pure and Applied Chemistry—Compendium of Analytical Nomenclature: Definitive Rules 1997*, Blackwell Science, Malden, MA. This scheme also calls a metallic redox indicator a "class zero electrode."

Reduction Half-Reaction:

$$Ag^+ + e^- \rightleftarrows Ag(s) \quad (14.13)$$

Nernst Equation at 25°C:

$$E_{Ag^+/Ag(s)} = E°_{Ag^+/Ag(s)} - \frac{0.05916\ V}{1} \log\left[\frac{1}{a_{Ag^+}}\right] \quad (14.14)$$

This results in an electrode where the value that is obtained for $E_{Ag^+/Ag(s)}$ (representing E_{Ind} in this example) varies as the value of a_{Ag^+} changes in the solution around the silver electrode.

A *class two electrode* ("electrode of the second kind") consists of a metal that is in contact with a slightly soluble salt of that metal and that is in a solution containing the anion of this salt. An example of a class 2 electrode would be a silver wire that is in contact with $AgCl(s)$ and that is immersed in a solution that contains chloride ions (the

anion that reacts with Ag^+ to form solid AgCl). The combined half-reaction and modified Nernst equation that we get for this overall process is shown below (see Chapter 10 for a further discussion of combined half-reactions).

Reduction Half-Reaction: $\qquad Ag^+ + e^- \rightleftarrows Ag(s)$

Solubility Reaction: $\qquad AgCl(s) \rightleftarrows Ag^+ + Cl^-$

Combined Half-Reaction:

$$AgCl(s) + e^- \rightleftarrows Ag(s) + Cl^- \quad (14.15)$$

Overall Nernst Equation at 25°C:

$$E_{AgCl(s)/Ag(s)} = E°_{Ag^+/Ag(s)} -$$
$$\frac{0.05916\ V}{1} \log\left[\frac{a_{Cl^-}}{K_{sp,AgCl}}\right] \quad (14.16)$$

The result of this system is an electrode that now gives a response that is related to the activity of the anion (Cl^-) that is present in the surrounding solution.

A *class three electrode* ("electrode of the third kind") uses a metal electrode that is in contact with a salt of its metal ion (or a complex of this metal ion) and a second, coupled reaction involving a similar salt (or complex) with a different metal ion. This type of electrode is illustrated by the use of a lead wire that is in contact with insoluble lead oxalate, which in turn is in contact with a solution that contains Ca^{2+} and is in contact with insoluble calcium oxalate.

Reduction Half-Reaction:

$$Pb^{2+} + 2\ e^- \rightleftarrows Pb(s)$$

Solubility Reaction:

$$Pb(Oxalate)(s) \rightleftarrows Pb^{2+} + Oxalate^{2-}$$

Precipitation Reaction:

$$Ca^{2+} + Oxalate^{2-} \rightleftarrows Ca(Oxalate)(s)$$

Combined Half-Reaction:

$$Pb(Oxalate)(s) + Ca^{2+} + 2\ e^-$$
$$\rightleftarrows Pb(s) + Ca(Oxalate)(s) \quad (14.17)$$

Overall Nernst Equation at 25°C:

$$E_{Pb(Oxalate)(s)/Pb(s)} = E°_{Pb^{2+}/Pb(s)}$$
$$- \frac{0.05916\ V}{2}\left[\frac{K_{sp,Ca(Oxalate)}}{K_{sp,Pb(Oxalate)}\, a_{Ca^{2+}}}\right] \quad (14.18)$$

The only species in the combined half-reaction for this system that is not a solid and present in a standard state is Ca^{2+}. The result is a measured potential for the electrode that is related to the activity of Ca^{2+}, as indicated by the Nernst expression in Equation 14.18.

Along with these four classes of metal electrodes, there are several other types of indicator electrodes that can be used in potentiometry. Most of these other indicator electrodes use a thin film or membrane as a recognition element to detect a particular analyte. We will come back to this other group of indicator electrodes in Section 14.3 when we discuss the pH electrode and other ion-selective electrodes.

Salt Bridges and Junction Potentials. If two electrodes are placed into separate solutions and connected to a potentiometer, no reading can be made. This occurs because some contact must be present between the two solutions. This contact is needed to allow the flow of ions to complete the electrical circuit. However, we do not want the solution by the reference electrode to be contaminated by the sample that is in contact with the indicator electrode. It was shown in Chapter 10 that this problem can be solved by using a salt bridge to connect the two half-cells of this system and yet keep the contents of each half-cell separate.

A salt bridge can take many shapes but is often in the form of a U-shaped glass tube (see Figure 14.2). This tube is filled with agar that contains an aqueous solution of potassium chloride. Agar is a gel that will prevent mixing of solutions on either side of the salt bridge, while the solution within the agar will allow ions to travel between these solutions. When a solution in a half-cell on one side of this salt bridge begins to be depleted of negative charge, chloride ions from the salt bridge will migrate to this electrode to reestablish charge neutrality. At the same time, potassium ions in the salt bridge will move in the other direction to counter the excess of negative charge that has begun to appear at the other electrode. Potassium chloride is often used as a component of a salt bridge because K^+ and Cl^- ions have similar ionic mobilities in an aqueous environment. This feature means these ions will each be able to carry about the same amount of current in an aqueous solution and in the salt bridge. There are some situations in which salts other than KCl are used in the salt bridge. For instance, a salt

other than KCl is needed when work is being carried out with a solution that contains Ag^+, which will precipitate in the presence of Cl^-.

Although salt bridges are necessary in most electrochemical cells, they do create an additional problem when this cell is being used for potentiometry. This problem arises from the creation of a **junction potential** at each interface between the salt bridge and one of the solutions. A junction potential is present whenever two solutions or regions exist in an electrochemical cell that have different chemical compositions.[9,10] Figure 14.5 shows an example of a *liquid junction potential*, which forms between two solutions of different composition, such as when an electrode is in an aqueous solution of 0.10 *M* HCl and this solution is in contact with an aqueous solution of 0.10 *M* NaCl that is in contact with a second electrode. At the boundary between these solutions there will be movement of ions across the interface to equalize the concentrations on each side. The rate of this movement due to a difference in concentration (known as "diffusion," a process discussed in Chapter 20) will depend on the types of ions that are present on either side of the interface and their concentrations on each side.

In the example shown in Figure 14.5, chloride ions are already present at the same concentration in both solutions, so these ions will have no net movement across the boundary between these solutions. However, the concentrations of H^+ and Na^+ do differ from one side to the next, so some of these ions will tend to migrate to the other solution. Even though H^+ and Na^+ originally have equal concentrations in their respective solutions in Figure 14.5, H^+ has a much faster rate of travel in water than Na^+. This fact means H^+ will tend to travel across the solution boundary and into the NaCl solution faster than Na^+ can enter the HCl solution. The result of this initial process is that the charge at the boundary on the HCl side becomes slightly more negative than the side of the

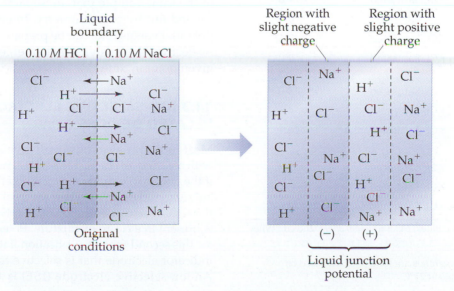

FIGURE 14.5   An example of the creation of a liquid junction potential.

boundary that faces the NaCl solution. This difference in charge represents a small change in electrical potential that creates a junction potential.

Table 14.4 provides some examples of values for liquid junction potentials. These values are typically in the range of 5 to 30 mV. A liquid junction potential can be present even between two solutions that contain the same chemical but at different concentrations, as shown in Table 14.4 for various solutions of KCl that form a boundary with a saturated KCl solution. It is important to consider liquid junction potentials in potential measurements because this factor will contribute to the overall difference in a potential that is observed for an electrochemical cell. This effect is given by the following equation,

$$E_{\text{Cell,Observed}} = E_{\text{Cell}} + E_{\text{Liq junction}} \qquad (14.19)$$

in which $E_{\text{Cell,Observed}}$ is the experimentally measured cell potential, E_{Cell} is the true potential difference between the cathode and anode, and $E_{\text{Liq junction}}$ is the contribution due to a liquid junction potential. The exact size of the liquid junction potential is often an unknown quantity in an electrochemical cell. However, the size of this potential can be minimized by using in a salt bridge a salt like KCl that has a cation and anion with similar ionic mobilities. There are also cases in which a junction potential is intentionally created and used for chemical analysis. An example of this occurs in the use of a pH electrode, as is discussed in the next section.

TABLE 14.4 Examples of Liquid Junction Potentials*

Composition of Solutions at Boundary

Solution A	Solution B	Liquid Junction Potential (mV)[a]
KCl, 0.1 *M*	KCl, Saturated	1.8
KCl, 1.0 *M*	KCl, Saturated	0.7
KCl, 4.0 *M*	KCl, Saturated	0.1
KCl, Saturated	KCl, Saturated	0.0
HCl, 0.01 *M*	KCl, Saturated	3.0
HCl, 0.1 *M*	KCl, Saturated	4.6
HCl, 4.0 *M*	KCl, Saturated	14.1
NaOH, 0.01 *M*	KCl, Saturated	2.3
NaOH, 0.1 *M*	KCl, Saturated	−0.4
NaOH, 1.0 *M*	KCl, Saturated	−8.6

*These data are from R.G. Bates, *Determination of pH*, 2nd ed., Wiley, New York, 1973.

[a]These liquid junction potentials are for a junction based on Solution A | KCl, Saturated at 25°C.

14.2C Applications of Potentiometry

There are a number of applications for potentiometry in chemical analysis. One of the most common and powerful of these applications is the use of the potential measurements to give direct information on the activity or concentration of an analyte in a sample. The most successful use of potentiometry for this type of application is in pH measurements. This success is partly a result of the availability of inexpensive and reliable equipment for making such measurements (see Box 14.1 and Figure 14.6 regarding the invention of the pH meter).[11,12] Another reason for the success of potentiometry in this area is the selectivity with which pH measurements can be made by this approach. In the next section we consider how a pH electrode works and learn the reason why it has such a high selectivity for hydrogen ions.

Direct analyte measurements are not the only application for potentiometry. A closely related application is the use of potential measurements to follow the course of a titration, as occurs in a potentiometric titration. This approach is carried out by using an appropriate reference electrode and indicator electrode to follow the progress of a titration. Examples of this technique were given in Chapter 12 regarding the use of a pH electrode and pH measurements to follow the course of an acid–base titration. More examples will be given in Chapter 15 in a discussion of redox titrations.

Potentiometry can also be combined with other methods for analyte detection. Two examples are the use of potential measurements to monitor electroactive analytes in samples that are being processed by flow-injection analysis or liquid chromatography (see Chapters 18 and 21). In both these techniques, potentiometry can be used to measure the concentration of certain analytes as they exit from a tube or column. In flow-injection analysis, the same analyte is measured in a sequence of samples that are being injected onto a flow-based system. In liquid chromatography, there are often several possible analytes in the same sample that are separated before detection is carried out by potentiometry. In either case, a graph of potential versus time can be prepared to show the amount of electroactive analytes that emerge from the system at a given point in time.[13,14]

14.3 ION-SELECTIVE ELECTRODES AND RELATED DEVICES

Most indicator electrodes will give a response or interact with a variety of chemical species. This can be an advantage if the goal is to employ a general method of analysis. There are many other times when the goal is to instead measure the activity or concentration of a particular analyte, even if it is present in a complex mixture. Potentiometry can be used for this second type of application if it is used along with an indicator electrode that is selective for the desired analyte. An **ion-selective electrode (ISE)** is an indicator electrode

BOX 14.1
Creation of the pH Meter

Early in the twentieth century most chemical analyses involved either gravimetry or titrimetry. This began to change in the 1930s with the introduction of instrumental techniques for chemical analysis. One of the key events to take place at this time was the development of the pH meter by Arnold Beckman in 1935 (Figure 14.6).

Beckman built his first pH meter to help a chemist in the California citrus industry who needed to measure the acidity of lemon juice. Beckman did this by constructing a potentiometer based on vacuum-tube amplification that would measure a potential with only a very small current. Many people had spent many years in examining the behavior of acids in solution and the concept of pH had been developed many years earlier. Beckman's contribution was to design and build an instrument that could be easily used with an ion-selective electrode to make this important measurement.

With his device, the pH of a sample could now be measured in just a few seconds and with a device that did not affect the sample in any way. This was accomplished by having the instrument make a simple measurement of electrical potential involving an ion-selective electrode for hydrogen ions and converting this measured potential into a reading of pH. The result was a new and valuable method for chemical analysis that we still use to this day.

Arnold Beckman

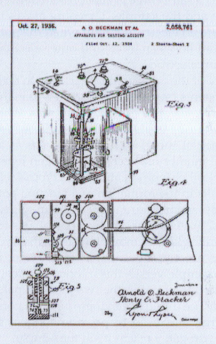

FIGURE 14.6 Arnold Beckman (1900–2004) and a drawing from his 1936 patent for the original Beckman pH meter. (The photo on the right is reproduced with permission from the Chemical Heritage Foundation; the photo on the left is reproduced with permission from the Beckman Foundation.)

that can respond to individual types of anions or cations, and is one tool that can be utilized for such a task.[3,10]

14.3A Glass Membrane Electrodes

pH Electrode. The most common type of ISE is the **pH electrode**, an indicator electrode that is selective for the detection of hydrogen ions. The most common type of pH electrode is a **glass-membrane electrode**, which is a type of indicator electrode that uses a thin glass membrane for selectively detecting the desired ion (in this case, H^+). The glass used in a typical pH electrode is based on a special mixture of lithium, barium, lanthanum, and silicon oxides (see Table 14.5). The glass membrane pH electrode was first used as part of an instrumental system for chemical analysis in the late 1930s.[12]

The design of a typical modern pH electrode is given in Figure 14.7. This design actually contains two electrodes in one, giving a device known as a *combination electrode*. Both the inner and outer electrodes in this device are Ag/AgCl electrodes. The outer part of this device contains a Ag/AgCl electrode that is surrounded by an enclosed solution saturated with AgCl and KCl. The inner part of this device has a second

TABLE 14.5 Composition of Ion-Selective Electrodes Based on Glass Membranes*

Type of Electrode	Composition of Glass	Usable Range (*M*)	Selectivity
pH Electrode	Li, Ba, La, and Si Oxides	$1-10^{-14}$	$H^+ >> Li^+, Na^+ > K^+$
Sodium Electrode	Na, Al, and Si Oxides	$1-10^{-6}$	$Ag^+>H^+>Na^+ >>Li^+,K^+,NH_4^+$
Electrode for Univalent Cations	Na, Al, and Si Oxides	$1-10^{-5}$	$K^+>NH_4^+>Na^+,H^+,Li^+$

*These data were obtained from T.S. Light, "Potentiometry: pH and Ion-Selective Electrodes." In *Analytical Instrumentation Handbook*, 2nd ed., G.W. Ewing, Ed., Marcel Dekker, New York, 1997, Chapter 18.

Ag/AgCl electrode and a saturated AgCl solution with a fixed concentration of HCl. A thin glass membrane separates the inner electrode from the sample. A porous plug that acts as a salt bridge is also present between the outer electrode and the sample. This plug makes it possible to complete an electrical circuit when the potential is to be measured between inner and outer electrodes in this combination electrode.[3,7]

The ability of the combination electrode to make pH measurements stems from the use of glass within the thin membrane that is selective for hydrogen ions. When the combination electrode is placed into an aqueous sample, the other surface of the glass membrane acts as an ion exchanger. The membrane accepts hydrogen ions more readily than any other type of cation. This type of selective interaction results in the formation of a junction potential between the glass membrane and the surrounding sample (see Section 14.2B). A similar junction potential is formed on the inside of the glass membrane where a fixed concentration of HCl is present. If the activity of hydrogen ions is different in the sample versus this interior solution, a difference will also be present in the two junction potentials that are formed. The difference in these junction potentials is then measured and used to provide a signal that is related to the hydrogen ion activity in the sample.

The relationship between the measured potential and hydrogen ion activity for a pH electrode can be described by the following equation.

$$E = K + 0.05916 \text{ (pH)}$$
$$\text{or } E = K - 0.05916 \log(a_{H^+}) \quad (14.20)$$

This equation describes the difference in junction potentials between the inside and outside of the glass membrane in the case when the potentials of the inner and outer electrodes are otherwise equal, as they are for the device shown in Figure 14.7. The term *K* in this equation is a system constant that varies from one pH electrode to the next. However, it is still possible to use Equation 14.20 and the pH electrode for pH measurements by first calibrating this system with buffers that have known pH values. This calibration should be done for each type of pH meter that is being used. This process should also ideally

involve the use of at least two reference buffers with pH values that match the range of pHs that are expected in the samples.[3,7]

The principal interferences for this type of pH electrode are alkali metal ions, such as Na^+, Li^+, and K^+ (see Table 14.5). These ions also can interact with the external surface of the glass membrane and create a junction potential that is no longer related to just the hydrogen ion activity within the sample. Although Li^+ has the largest effect on these junction potentials, this type of interference is usually called a "sodium error" because sodium salts are much more commonly present in samples than lithium salts. This error only happens when the activity of hydrogen ions is low (representing a high pH). For instance, this error can occur if NaOH is used to adjust the pH of a solution to a high value. A simple way to reduce this effect is to use KOH instead of NaOH to adjust the pH of an aqueous solution when an accurate pH reading is

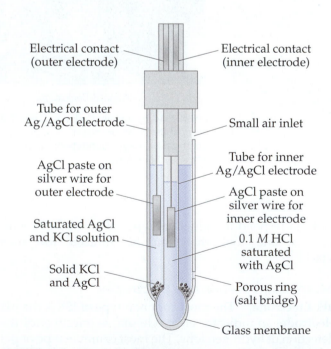

FIGURE 14.7 The design of a modern pH electrode. This is an example of a combination electrode, because both the indicator electrode and reference electrode are included with the same device.

needed. Another possible approach is to use a pH electrode that contains a glass membrane that has less interference in the presence of sodium ions.[3]

Sodium Ion-Selective Electrode. Various compositions of glass can be prepared to create ion-selective electrodes for cations besides H^+. One formulation based on a mixture of sodium, aluminum, and silicon oxides is used to make a *sodium ion–selective electrode*. This type of electrode creates a signal in the same general manner as a pH electrode. This signal is again based on a difference in the junction potentials that formed on both sides of the glass membrane in the presence of the sample on the outside of the electrode and a reference solution on the inside that contains a fixed concentration of the ion of interest.

The glass that is used in a sodium electrode still has a response that is almost 100 times greater for H^+ than for Na^+ when these ions are present at equal levels. This makes it necessary to use this type of electrode in an alkaline solution (i.e., one that has low hydrogen ion activity). A buffered solution is often added to samples and standards to control the pH (and ionic strength) for use with these electrodes and to help provide a response that is related to sodium ion concentration. The response of the sodium ion–selective electrode under these conditions will be proportional to the value of pNa, where $pNa = -\log(a_{Na^+}) \approx -\log([Na^+])$. This type of electrode will also have a strong response to silver ions. Yet another mixture of sodium, aluminum, and silicon oxides can be prepared to give an ion-selective electrode for a variety of univalent cations. This type of electrode gives a strong response for K^+, followed by a lower response for NH_4^+ and other cations.[3]

14.3B Solid-State Ion-Selective Electrodes

Other materials besides glass can be used to make ion-selective electrodes. An example is a **solid-state ion-selective electrode**, or "solid membrane electrode." This type of electrode contains a sensing element that is a crystalline material or a homogeneous pressed pellet. It is necessary in this type of electrode for the sensing element to have selective adsorption or interactions with the ion of interest. This element must also be able to conduct a small amount of current when it is used to provide a potential measurement. These are the same general requirements that are needed when using glass membranes in the pH electrode and other ion-selective electrodes.

The general design for this type of electrode is shown in Figure 14.8. This design consists of an internal reference electrode that is in contact with a reference solution containing a fixed concentration of the ion of interest. This reference solution is then in contact with the crystalline or pressed-pellet sensing element. The sensing element is also in contact with the sample solution on its outer surface. A separate reference electrode is also in contact with the sample and is used to complete the circuit for the potential measurement. The sensing element will preferentially interact with the desired ions in the solutions on both its interior and exterior surfaces. If these two solutions contain different activities of this ion, there will be different junction potentials created at these surfaces. The result is a difference in potential that is related to the activity of this ion in the sample.

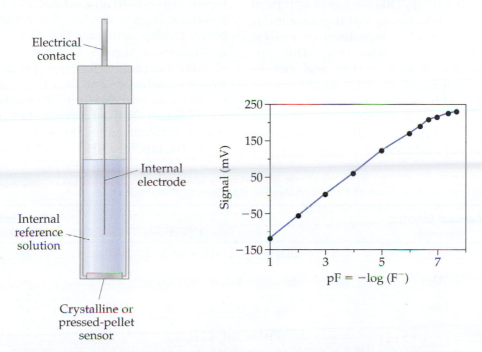

FIGURE 14.8 The general design of a solid-state ion-selective electrode (left), and an example of the response for a fluoride ion-selective electrode that is based on this design (right).

There are many types of solid-state ion-selective electrodes that have been developed. One common example is the *fluoride ion–selective electrode* that is used to measure fluoride in drinking water. In this case, the sensing element in the electrode is a pellet of lanthanum fluoride (LaF_3) that contains a trace amount of europium fluoride (EuF_2). LaF_3 is highly insoluble in water, with a K_{sp} value of only 7×10^{-17}. The surface of a crystal of LaF_3 will act as an ion exchanger for fluoride ions, as well as lanthanum ions. This feature makes such an electrode a useful tool in measuring the activity of fluoride in water samples. When LaF_3 is placed in water that contains no fluoride ions, some of this solid will dissolve according to the following solubility reaction and K_{sp} expression.

$$LaF_3(s) \rightleftharpoons La^{3+} + 3\ F^-$$

$$K_{sp} = [La^{3+}][F^-]^3 = 7 \times 10^{-17} \tag{14.21}$$

According to this K_{sp} expression, the maximum solubility of LaF_3 in water during this process will be given by $K_{sp} = [F^-]^4/3 = 7 \times 10^{-17}$, or $[F^-] = (2.1 \times 10^{-16})^{1/4} = 1.2 \times 10^{-4}\ M$. Whenever the fluoride concentration of a sample is much less than the solubility limit for LaF_3 of $1.2 \times 10^{-4}\ M$, the measured potential will begin to reflect the fluoride that has dissolved out of the electrode. At this point, the electrode becomes unresponsive to the actual fluoride content in the sample. This dissolution occurs slowly when the LaF_3 is in a compact pellet.

Figure 14.8 shows the response of a fluoride electrode when plotted as a function of pF, where $pF = -\log(a_{F^-}) \approx -\log([F^-])$. This response is typical of many ion-selective electrodes in that it gives a linear response versus the negative logarithm of analyte activity or concentration over a wide range. This type of electrode is relatively easy to use and can be employed as part of a system for continuously monitoring a sample, as is used by many water plants to monitor the fluoride content of drinking water. There are also some practical limitations to a fluoride electrode that are related to pH and sample composition. If a solution containing fluoride has a pH that is too low, the fluoride will mainly exist in solution as HF. Because the fluoride electrode responds to F^- and not HF, the pH that is used with this type of electrode should be at least two units above the pK_a of HF ($3.17 + 2 = 5.17$). At a pH greater than 10, another problem that can occur is that hydroxide ion (which has the same charge and a similar size to F^-) can also form an insoluble lanthanum salt, adsorb onto the surface of the LaF_3 crystal, and give a false high reading. If the sample contains metal ions such as Fe^{3+} or Al^{3+}, some of these metal ions can form soluble complex ions with fluoride and prevent this ion from interacting with the LaF_3 sensing element. To deal with these problems, a solution can first be added to each sample and standard to adjust the pH, control the ionic strength, and complex metal ions (e.g., Fe^{3+} and Al^{3+}) with ethylenediamine tetraacetic acid (EDTA).[3]

14.3C Compound Electrodes

Gas-Sensing Electrodes. Devices like the pH electrode are not limited to the detection of solution-phase chemicals, but can also be modified for used in other types of measurements. The modification of a pH electrode or other type of ion-selective electrode for the measurement of other analytes gives a device known as a *compound electrode*. One group of compound electrodes are those that have been modified for the analysis of certain gases. The result is known as a **gas-sensing electrode**. Some examples of gas-sensing electrodes are given in Table 14.6.

An ammonia gas-sensing electrode (illustrated in Figure 14.9) is both an electrode that can sense a gas and detect a molecular species. This device is a pH electrode covered with a membrane that allows passage of only low molecular-weight gases. This membrane is typically made of a very thin piece of Teflon or polyethylene. Between the covering membrane and the pH-sensitive glass is a small volume of an internal electrolyte solution ($0.1\ M$ KCl) that has an essentially fixed concentration of NH_4^+. When dissolved ammonia enters this solution through the membrane, the ratio of $[NH_4^+]$ to $[NH_3]$ is changed and the

TABLE 14.6 Examples of Reactions Used in Gas-Sensing Electrodes*

Chemical Entering Electrode	Reaction at Electrode	Detected Chemical
CO_2	$CO_2 + H_2O \rightleftharpoons H^+ + HCO_3^-$	H^+
SO_2	$SO_2 + H_2O \rightleftharpoons H^+ + HSO_3^-$	H^+
NH_3	$NH_3 + H_2O \rightleftharpoons NH_4^+ + OH^-$	H^+
	$H^+ + OH^- \rightleftharpoons H_2O$	
NO_2	$2\ NO_2 + H_2O \rightleftharpoons NO_3^- + NO_2^- + 2\ H^+$	H^+ or NO_3^-

*Additional examples can be found in T.S. Light, "Potentiometry: pH and Ion-Selective Electrodes." In *Analytical Instrumentation Handbook*, 2nd ed., G.W. Ewing, Ed., Marcel Dekker, New York, 1997, Chapter 18.

pH is increased, as shown by the reactions in Table 14.6. This change creates a response by the pH electrode that is related to the activity of ammonia that was in the sample. Similar electrodes can be made to respond to other basic or acidic gases such as CO_2, SO_2, and NO_2.

Enzyme Electrodes. Even more elaborate electrodes can be created by using enzymes to convert analytes into products that can be measured by potentiometry. This type of compound electrode is called an **enzyme electrode** or "enzyme substrate electrode." An example is an enzyme electrode that has been created for the measurement of urea. This electrode is constructed by immobilizing the enzyme urease onto a semipermeable membrane. Urease catalyzes the hydrolysis of urea into ammonia and carbon dioxide.

$$H_2NC(O)NH_2 + H_2O \rightleftharpoons 2\,NH_3 + CO_2 \qquad (14.22)$$

Because ammonia is so much more soluble than carbon dioxide in water, the CO_2 mainly bubbles out of solution and does not lower the pH. The ammonia, however, dissolves and goes through the membrane to raise the pH of an electrolyte solution that surrounds a pH electrode. The increase in pH is caused by a change in hydroxide ion activity and concentration, which is proportional to the amount of ammonia that has been produced by the enzyme and the original amount of urea that was in the sample. An ion-selective electrode for NH_4^+ can also

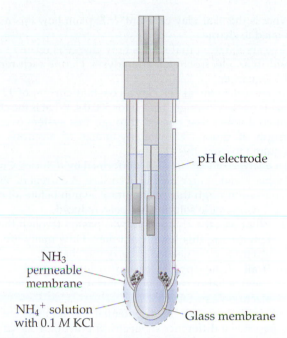

be used with such a system for the final measurement.[3] The use of other enzymes allows additional analytes to be detected by such an approach. Examples of other chemicals that can be detected through the use of enzyme electrodes include glucose, amino acids, alcohols, penicillin, and cholesterol.[3]

FIGURE 14.9 Design of a gas-sensing electrode for ammonia, based on the use of a glass-membrane pH electrode for detection.

Key Words

Amperometry 351	Enzyme electrode 361	Junction potential 355	Silver/silver chloride
Calomel electrode 353	Gas sensing electrode 360	Ohm's law 350	electrode 353
Charge 348	Glass-membrane	pH electrode 357	Solid-state ion-selective
Coulometry 351	electrode 357	Potentiometry 351	electrode 359
Current 349	Indicator electrode 352	Resistance 350	Voltammetry 351
Electrochemical analysis 348	Ion-selective electrode 356		

Other Terms

Alternating current 351	Compound electrode 360	Fluoride ion–selective	Potentiometric titration 351
Class one electrode 353	Conductance 350	electrode 360	Saturated calomel
Class three electrode 354	Direct current 351	Liquid junction	electrode 353
Class two electrode 354	Electromotive force 350	potential 355	Sodium ion–selective
Combination electrode 357		Metallic indicator	electrode 359
		electrode 353	

Questions

UNITS OF ELECTRICAL MEASUREMENTS AND METHODS FOR ELECTROCHEMICAL ANALYSIS

1. What is meant by the term "electrochemical analysis"? Give one specific example for this type of analysis.

2. Define each of the following terms and state what units are employed with each of these parameters in the SI system.
   (a) Current
   (b) Charge
   (c) Electric potential
   (d) Resistance

3. What is the "Faraday constant"? Explain how this term is related to charge.

4. Give an equation that shows how charge is related to current in an electrochemical analysis. Define each term in this equation.

5. An electrochemical cell has a constant current of 125 μA that is passed through the cell for 500.0 s. What is the charge (in coulombs) that passed through this system over this length of time? How many moles of electrons were required to carry this charge?

6. An analysis of Cu^{2+} is to be performed by reducing Cu^{2+} to copper metal at an electrode's surface. A current of 560 μA is passed through this system for 2.50 min before all of the Cu^{2+} in a sample solution has been reduced.
    (a) What was the charge that was passed through the system during this length of time? How many moles of electrons were needed to carry this charge?
    (b) If all of the applied current went to reduce Cu^{2+} to Cu(s), what mass of copper metal was deposited at the electrode's surface?

7. Define the term "electrical potential." What units are used to describe a difference in electrical potential? What is an "electromotive force"?

8. What is meant by "resistance" in an electrochemical system? What units are used to describe resistance?

9. What is "conductance"? How is conductance related to resistance? What units are used to describe conductance?

10. What is "Ohm's law"? Explain how Ohm's law can be used to examine an electrochemical system.

11. What current must be present in an electrochemical cell if the potential is 140 mV and the resistance is 4×10^{12} ohms?

12. What must the resistance be across a certain part of an electrical circuit if the current through this component is 8.5 μA when the applied potential is 59.1 mV?

13. The glass membrane in a pH electrode has a resistance of 200,000,000 ohms. What current will be passed through this membrane if the measured potential is 400 mV?

14. Explain the difference between a "direct current" and an "alternating current". Give an example of an application for each of these two types of current.

15. Draw a graph of current versus time for a 5.0 A signal that is based on a direct current and a 5.0 A signal that is based on a 60 Hz alternating current.

TYPES OF METHODS FOR ELECTROCHEMICAL ANALYSIS

16. What is meant by the term "potentiometry"? Explain why the use of a fluoride ion–selective electrode is an example of potentiometry.

17. Define the terms "amperometry" and "voltammetry." How are these methods similar? How are these methods different?

18. What is "coulometry"? Explain why the Faraday constant is often used in coulometry.

CELL POTENTIALS AND THE NERNST EQUATION

19. Describe the general parts of an electrochemical cell that are used in potentiometry. Compare the general components of this cell to those that were described in Chapter 10 for the study of oxidation–reduction reactions.

20. Why is it necessary in potentiometry to have "essentially zero flow of current"? What would happen if the current was not close to zero?

21. Discuss how the Nernst equation can be used in potentiometry.

22. What is meant by an "indicator electrode"? What is the role of this electrode in potentiometry?

23. What does it mean when a "negative" cell potential is measured in potentiometry? Explain how this type of situation can occur.

CELL COMPONENTS IN POTENTIOMETRY

24. Why is it uncommon for a standard hydrogen electrode to be used as a reference electrode in potentiometry?

25. Describe a silver/silver chloride electrode. What are the key components of this electrode and how does it work?

26. What is a "calomel electrode"? Describe how this type of electrode works.

27. What is a "saturated calomel electrode"?

28. List four types of metal indicator electrodes. Give an example of each type.

29. Determine whether each of the following electrodes is a metallic indicator electrode or a class one, two, or three electrode.
    (a) A copper wire in a solution of copper sulfate
    (b) A gold wire in a solution containing V(II) and V(III)
    (c) Mercury coated with Hg_2Cl_2 in a solution containing NaCl

30. A solution has a total concentration of iron ions of 0.0763 M in 1 M HCl. A platinum electrode that is placed in this solution gives a measured potential of 0.465 V versus SCE.
    (a) If there are no other species than Fe^{2+} and Fe^{3+} that are being detected, what is the ratio of $[Fe^{2+}]/[Fe^{3+}]$ in this solution?
    (b) What are the individual concentrations of Fe^{2+} and Fe^{3+} in this solution?

31. What is a "salt bridge"? What role does a salt bridge play in an electrochemical cell?

32. Define the term "junction potential." How does a junction potential affected the measured difference in potential for an electrochemical cell?

33. Explain how a junction potential can be formed by the presence of a salt bridge in an electrochemical cell.

34. What is a "liquid junction potential"? Give an example.

35. Sometimes both electrodes can be placed in the same solution, but this is unusual because then the redox reaction can occur in the beaker without influencing the electrodes. An example of a successful "junctionless cell" is one in which the first electrode is Ag/AgCl in a HCl solution and the other is a hydrogen electrode in the same HCl solution. What is a possible benefit of using this type of junctionless cell?

APPLICATIONS OF POTENTIOMETRY

36. What are some possible advantages of using potentiometry for chemical analysis? (*Hint*: Use the measurement of pH as an example.)

37. Calcium ion in a water sample was measured using a calcium-selective electrode. A 50 mL portion of the water showed a potential of −0.0650 V versus SCE. When a 1.0 mL portion of 0.0850 M Ca(NO$_3$)$_2$ solution was added, the potential changed to −0.0477 V. What was the original calcium concentration?

38. Explain how potentiometry can be used as part of a titration. Give a specific example of such an approach.
39. Discuss how potentiometry can be used with methods such as flow injection analysis or liquid chromatography. What do you think are some possible advantages for using these combinations of methods?

GLASS MEMBRANE ELECTRODES

40. Define the terms "ion-selective electrode" and "glass-membrane electrode." Illustrate both of these ideas using a typical pH electrode.
41. Describe how a modern pH electrode is constructed. Explain why this type of electrode is also known as a "combination electrode."
42. Explain how pH is measured by a pH electrode. What role does the glass membrane play in this process?
43. State why it is necessary to calibrate a pH electrode.
44. What is "sodium error"? Why is this type of error important to consider when using a pH electrode? What steps can be taken to minimize sodium error?
45. A pH meter reads pH = 2.50 when it is present in a dilute solution of HCl. Predict what would happen when solid NaCl is added to this solution.
46. Explain how a sodium ion–selective electrode works. How is this similar to a typical pH electrode? How are these two types of electrodes different?
47. The activity of a sodium ion as measured by a sodium-selective electrode is 0.0674 M. If the solution ionic strength is 0.0500, what is the concentration of the sodium ion in the solution?
48. A sodium-selective electrode has a response ratio of 2.00 for sodium compared to hydrogen ion. What is the lowest concentration of sodium ion that will have less than a 10% error if the measurement is made at pH 7.00?

SOLID-STATE ION-SELECTIVE ELECTRODES

49. What is a "solid-state ion-selective electrode"? How is this similar to a glass-membrane ion-selective electrode? How is it different?
50. Describe the general design of a solid-state ion-selective electrode.
51. Explain how a fluoride ion-selective electrode produces a signal that is related to fluoride activity or concentration.
52. State why it is often necessary to control the pH and ionic strength and to add EDTA to samples and standards when using a fluoride ion–selective electrode.
53. The following data represent potentials of a fluoride ion–selective electrode vs. SCE for several solutions. All solutions

Solution (M)	Potential (mV)
5.0×10^{-2}	−22.4
5.0×10^{-3}	36.8
5.0×10^{-4}	96.0
5.0×10^{-5}	155.2
Unknown 1	74.3
Unknown 2	190.6
Unknown 3	−54.3

are made up in a way to control their pH and ionic strength, and contain added EDTA. Prepare a plot of the measured potential vs. pF based on this information and determine the concentration of fluoride in each unknown solution.

54. If a solution of silver ion and copper ion is precipitated by addition of sulfide ion, a mixture of Ag_2S and CuS is formed. This material when used in jewelry is called "niello." When this is collected by filtration, dried, and pressed into a thin pellet it can be made into a membrane suitable to make an electrode responsive to silver ion or copper ion. Because copper, especially, can be in equilibrium with ligands such as EDTA to give extremely low concentrations of Cu^{2+}, such an electrode can be used to detect the endpoint of a titration of copper by EDTA.

$$Cu^{2+} + H_2EDTA^{2-} \rightleftarrows CuEDTA^{2-} + 2\,H^+ \quad (14.23)$$

The conditional formation constant for copper with EDTA is known to be greater than 10^{10} at a pH above 4.0. That means that if total copper concentration is 0.05 M and total EDTA concentration is 0.10 M, then the actual $[Cu^{2+}] < 10^{-10}$ M. The electrode can respond to such low concentrations. What is the value of pCu if the effective formation constant is 1.0×10^{14}, the total copper ion concentration is 0.050 M, and total EDTA concentration is 0.10 M.

COMPOUND ELECTRODES

55. What is meant by the term "compound electrode"? Give two general examples.
56. What is a "gas-sensing electrode"? Describe one specific example of a gas-sensing electrode.
57. Describe how a pH electrode can be modified for the detection of ammonia.
58. Using Table 14.6 as a guide, describe how the electrode in Figure 14.9 could be modified for the detection of CO_2 instead of ammonia.
59. The amount of protein in a wheat flour sample is to be measured. A 0.3476 gram portion is dissolved in concentrated sulfuric acid and heated to boiling in the presence of copper ion, which serves as a catalyst to destroy the biomolecules in the sample and convert the protein nitrogen into ammonium ion. After cooling, raising the pH with NaOH, and diluting the solution to 100 mL, an ammonia-selective electrode is used to measure the ammonium concentration to be 0.32 M. Protein is typically 16% nitrogen. Calculate the percent protein in this sample. Explain how an unscrupulous grain dealer could make the grain look more rich in protein by adding melamine ($C_3H_6N_6$).
60. A gas-sensing electrode is used to measure the concentration of carbonate in a solution that also contains other basic substances. Explain what will happen when the solution is made acidic with sulfuric acid while being monitored with a gas-sensing electrode.
61. What is an "enzyme electrode"? How is this similar to a gas-sensing electrode? How is it different?
62. Describe how a pH electrode can be made for the detection of urea through the use of enzymes.

CHALLENGE PROBLEMS

63. Suppose that both a pH and a pF electrode are used during a titration of a solution of 0.10 M HF with 0.10 M NaOH. Sketch the response of both electrodes vs. volume of NaOH.

64. Obtain more information from sources such as Reference 3 on the calomel and saturated calomel electrodes. Use this information to explain why a 0.10 M KCl calomel electrode is superior to an SCE if the electrode is to be used at different temperatures.

65. Explain why the seemingly complicated electrode of the third type described in the chapter is used to measure pCa instead of a class one electrode utilizing elemental calcium.

66. Use the solubility product K_{sp} of AgBr to calculate the expected potential for a reference electrode that is similar in design to the Ag/AgCl electrode shown in Figure 14.4, but that uses AgBr in place of AgCl.

67. Barium sulfate is about as insoluble as is AgCl. Suggest why a $BaSO_4$ reference electrode has never been seriously considered.

68. What will be the error in measured pH if the true pH is 4.56, but the measured potential of a pH electrode is 1.0 mV too high?

69. A fluoride ion–selective electrode has about a 1000-fold higher response for F^- vs. Cl^- at the same concentration. Do you think an electrode would give reliable results for measurements of fluoride in seawater? Justify your answer.

TOPICS FOR DISCUSSION AND REPORTS

70. Contact a water-treatment plant in your area that practices water fluoridation. Ask workers at this facility about the approaches they use to follow the levels of fluoride that are added to the water. Discuss your findings.

71. Gas-sensing electrodes are often used in blood-gas measurements. Talk to a worker in a hospital laboratory or a surgical room to obtain more information on the use of blood-gas measurements.

72. Liquid-membrane electrodes are another class of devices that can be used in potentiometry for the selective detection of ions. Obtain more information on this topic and write a report on a specific example of this type of electrode.

73. Ion-selective field-effect transistors (ISFET) are yet another group of sensors that can be used in potentiometry. Locate a review article or book on this topic. Describe how this type of sensor works and list some of its applications.

References

1. L. W. Ripa, "A Half-Century of Community Water Fluoridation in the United States: Review and Commentary," *Journal of Public Health Dentistry*, 53 (1993) 17–44.

2. CDC, "Ten Great Public Heath Achievements—United States, 1900–1999," *Journal of the American Medical Association*, 281 (1999) 1481.

3. T. S. Light, "Potentiometry: pH and Ion-Selective Electrodes." In *Analytical Instrumentation Handbook*, 2nd ed., G. W. Ewing, Ed., Marcel Dekker, New York, 1997.

4. B. N. Taylor, Ed., *The International System of Units (SI)*, NIST Special Publication 330, National Institute of Standards and Technology, Gaithersburg, MD, 1991.

5. *Correct SI Metric Usage*, United States Metric Association.

6. *IUPAC Compendium of Chemical Terminology*, Electronic version, http://goldbook.iupac.org

7. J. Inczedy, T. Lengyel, and A. M. Ure, *International Union of Pure and Applied Chemistry—Compendium of Analytical Nomenclature: Definitive Rules 1997*, Blackwell Science, Malden, MA, 1998.

8. G. Maludzinska Ed., *Dictionary of Analytical Chemistry*, Elsevier, Amsterdam, the Netherlands, 1990.

9. A. J. Bard and L. R. Faulkner, *Electrochemical Methods: Fundamentals and Applications*, 2nd ed., Wiley, Hoboken, NJ, 2001.

10. D. A. Skoog, F. J. Holler, and T. A. Nieman, *Principles of Instrumental Analysis*, 5th ed., Saunders, Philadelphia, PA, 1998.

11. E. Wilson, "Arnold Beckman at 100," *Chemical and Engineering News*, 78(2000) 17–20.

12. J. Poudrier and J. Moynihan, "Instrumentation Hall of Fame." In *Made to Measure: A History of Analytical Instrumentation*, J. F. Ryan, Ed., American Chemical Society, Washington, DC, 1999, pp. 10–38.

13. J. Ruzicka and E. H. Hansen, *Flow Injection Methods*, 2nd ed., Wiley, New York, 1988.

14. C. F. Poole and S. K. Poole, *Chromatography Today*, Elsevier, New York, 1991.

Chapter 15

Redox Titrations

Chapter Outline

15.1 INTRODUCTION: CHEMICAL OXYGEN DEMAND

Water in the environment often contains a wide variety of dissolved organic compounds. These organic compounds can be oxidized by O_2, which is also dissolved in the water and is replenished by contact of the water with air. If the amount of organic compounds in water is too great, as often occurs in polluted water, the transfer of O_2 from air to the water can't keep up with the rate of disappearance of oxygen as it reacts with these compounds. This situation causes the concentration of oxygen in the water to decrease dramatically and makes it difficult for fish and other aquatic organisms to survive.

A common way of describing the overall content of organic pollutants in water is to use the **chemical oxygen demand** (or **COD**).[1,2] In a COD measurement, a sample of water is reacted under acidic conditions with an excess of a strong oxidizing agent such as **dichromate** ($Cr_2O_7{}^{2-}$). This oxidizing agent will react with organic compounds in the water to form carbon dioxide, giving an oxidation–reduction reaction similar to that which occurs between these compounds and oxygen (see example in Figure 15.1). The amount of oxidizing agent that remains after this process is measured by using a back titration, which also involves an oxidation–reduction reaction. The difference in the initial and remaining amounts of oxidizing agent is then used to calculate the equivalent amount of oxygen that would have been consumed by the same organic compounds, which gives the COD value for the sample.[1,2] This approach for measuring COD is an example of an analytical method known as a *redox titration*. This chapter discusses the principles behind a redox titration and looks at how this method is performed. Approaches for predicting the response of redox titrations will also be considered. We will then examine several applications of this method for chemical analysis.

15.1A What Is a Redox Titration?

A **redox titration** (more formerly known as an *oxidation–reduction titration*) is a titration method that makes use of an oxidation–reduction reaction.[3–5] In the case of a COD analysis, the titrant is a reducing agent (Fe^{2+}) and the analyte (excess dichromate) is an oxidizing agent. A typical system that is used to perform a redox titration is shown in Figure 15.2. This system has many of the same components used in other titrations, such as a buret to deliver titrant to the sample and a means for detecting the end point. The end point in a redox titration can be detected either visually or by an instrumental method (e.g., by making potential measurements).

Like other types of titrations, a successful redox titration is based on a reaction that has a known stoichiometry

Measurement of chemical oxygen demand

Step 1: React sample with dichromate

$$C_2H_5OH + 2\,Cr_2O_7{}^{2-} + 16\,H^+ \longrightarrow 2\,CO_2 + 11\,H_2O + 4\,Cr^{3+}$$

Dichromate
(in excess)

Step 2: Back titrate remaining dichromate

$$Cr_2O_7{}^{2-} + 6\,Fe^{2+} + 14\,H^+ \longrightarrow 2\,Cr^{3+} + 6\,Fe^{3+} + 7\,H_2O$$

Titrant

Step 3: Calculate COD from amount of dichromate that reacted with the sample

Actual reaction in water with O_2:

$$C_2H_5OH + 3\,O_2 \longrightarrow 2\,CO_2 + 3\,H_2O$$

FIGURE 15.1 The determination of chemical oxygen demand (COD) for a water sample, using ethanol as an example of a simple organic compound that might be present in water. The reactions in the figure on the left show what occurs if the COD of this sample is determined by using dichromate as an oxidizing agent. The reaction in the image on the right shows what will occur if ethanol completely reacts with oxygen.

between the analyte and titrant, a large equilibrium constant, and a fast reaction rate. Equation 15.1 shows how the equilibrium constant $K°$ at 25°C can be calculated for a redox titration based on the standard potential ($E°_{Net}$) for the oxidation–reduction reaction that occurs between the analyte and titrant (see derivation in Chapter 10).

$$K° = 10^{(n\,E°_{Net})/(0.05916V)} \tag{15.1}$$

To get a large equilibrium constant in this case will require a relatively large positive value for $E°_{Net}$, which occurs if the half-reactions for the analyte and titrant have quite different standard reduction potentials. For example, the titration of dichromate with Fe^{2+} during a COD analysis has a value for $E°_{Net}$ of 0.59 V and an equilibrium constant that is more than 10^{59} at 25°C. The rates

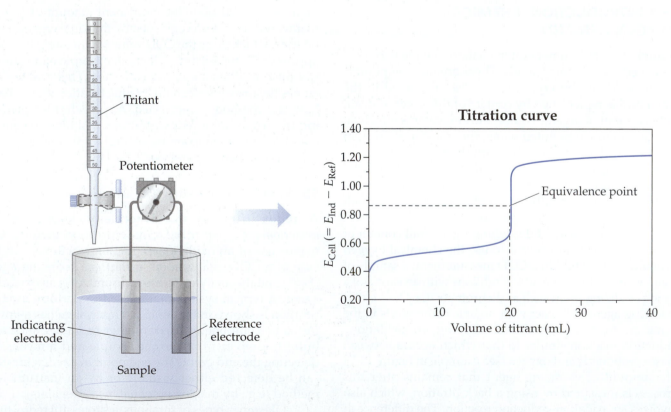

FIGURE 15.2 A typical system used to perform a redox titration that is monitored through the use of potential measurements. The titration curve that is shown here is for the analysis of a 10.00 mL portion of a 0.01000 M Fe^{2+} solution using 0.005000 M Ce^{4+} as the titrant.

of oxidation–reduction reactions are more difficult to predict, but must be fast to result in accurate end point detection. This is one reason why a back titration is used during a COD analysis; the reaction of dichromate with organic compounds in a sample may occur at a variable and often unknown rate, while the reaction of dichromate with Fe^{2+} is fast and allows for the easy detection of an end point.

15.1B How Are Redox Titrations Used in Analytical Chemistry?

Like acid–base and complexometric titrations, redox titrations are primarily used to determine the concentration of an analyte in a given mass or volume of sample. Redox titrations, however, are much more varied than acid–base or complexometric titrations in terms of the possible titrants and analytes with which they can be used. In addition, many redox titrations do not involve a 1:1 reaction between the analyte and titrant. An example is the titration of dichromate with Fe^{2+}, which reacts in a 1:6 ratio during the back titration that is performed as part of a COD measurement (see reactions in Figure 15.1). This feature often makes redox titrations more challenging to use and describe than acid–base reactions or complexometric titrations.

To describe a redox titration, we first need to write the overall balanced reaction that occurs during the titration, as is shown as follows for the *oxidation* of analyte A by titrant T.

Oxidation Half-Reaction:

$$A_{red} \rightleftharpoons A_{ox} + n_A e^-$$

$$(\text{or } n_T A_{red} \Longrightarrow n_T A_{ox} + n_T n_A e^-)$$

Reduction Half-Reaction:

$$T_{ox} + n_T e^- \rightleftharpoons T_{red}$$

$$(\text{or } n_A T_{ox} + n_T n_A e^- \Longrightarrow n_A T_{red})$$

Overall Titration Reaction:

$$n_T A_{red} + n_A T_{ox} \rightleftharpoons n_T A_{ox} + n_A T_{red} \quad (15.2)$$

In these reactions, n_A represents the number of electrons that are released during the oxidation of A_{red} to A_{ox}, and n_T represents the number of electrons that are consumed during the reduction of T_{ox} to T_{red}. We can see from this general reaction that A and T must react in the ratio n_T/n_A to properly balance the number of electrons that are being transferred during this oxidation–reduction process. One consequence of having a ratio for n_T/n_A other than 1.0, as often occurs in redox titrations, is that the resulting titration curve is not symmetrical in the region near the equivalence point. This effect is illustrated in Figure 15.3 for the titration of dichromate with Fe^{2+} during a COD analysis.

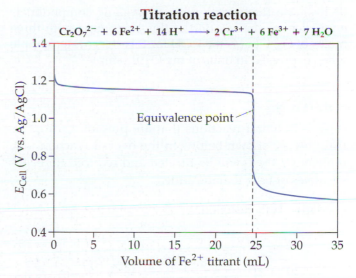

Titration reaction

$$Cr_2O_7{}^{2-} + 6\,Fe^{2+} + 14\,H^+ \longrightarrow 2\,Cr^{3+} + 6\,Fe^{3+} + 7\,H_2O$$

FIGURE 15.3 A titration curve for the analysis of excess dichromate in a 10.00 mL water sample using a 0.005000 M solution of Fe^{2+} as the titrant. This titration shows the results that would be obtained when the treated water sample contains excess dichromate at a concentration of 0.01000 M. The nonsymmetrical shape of the curve in the region of the equivalence point is mainly the result of the fact that one mole of dichromate is reduced for every six moles of Fe^{2+} that are oxidized during this titration. Another unusual feature of this analysis reaction is that each mole of dichromate produces two moles of Cr^{3+} in the titration reaction.

The stoichiometry for a balanced redox titration reaction can be used to relate the concentration of the titrant (C_T) and the volume of titrant that has been added at the equivalence point ($V_T = V_E$) to the sample volume (V_A) and concentration of analyte in the sample (C_A). The result is the following relationship for the general titration that is represented by Equation 15.2.

At the Equivalence Point for a Redox Titration.

$$n_T C_T V_T = n_A C_A V_A$$

or $\quad\quad (n_T/n_A) \cdot (\text{mol T}) = \text{mol A} \quad\quad (15.3)$

This expression now makes it possible to determine the concentration of an analyte by using the volume of titrant needed to reach the end point and the stoichiometry for the titration reaction.

EXERCISE 15.1	Measuring the Chemical Oxygen Demand for Water

A 50.00 mL sample of river water is treated with 10.00 mL of 0.2000 M $K_2Cr_2O_7$ at pH 0.0. After this mixture is heated and allowed to react, it is cooled and the remaining dichromate is titrated with 0.3000 M Fe^{2+}, giving an end point when 24.65 mL of this titrant has been added. (a) What half-reactions are involved in this titration (see Figure 15.1)? (b) How many moles of dichromate reacted with the original sample? (c) If every two moles of dichromate that reacts with an organic compound is equivalent to the reaction of this compound with three moles of O_2, what is the COD for this river-water sample when expressed in units of mg O_2/L sample?

SOLUTION

(a) The two half-reactions that are present during this titration are shown below. In the overall reaction, one mole of $Cr_2O_7^{2-}$ will be titrated and reduced for every six moles of Fe^{2+} that are added.

Oxidation Half-Reaction:

$$Fe^{2+} \rightleftarrows Fe^{3+} + e^-$$

(or $6\,Fe^{2+} \Longrightarrow 6\,Fe^{3+} + 6\,e^-$)

Reduction Half-Reaction:

$$Cr_2O_7^{2-} + 14\,H^+ + 6\,e^- \rightleftarrows 2\,Cr^{3+} + 7\,H_2O$$

Overall Titration Reaction:

$$6\,Fe^{2+} + Cr_2O_7^{2-} + 14\,H^+ \Longrightarrow$$
$$6\,Fe^{3+} + 2\,Cr^{3+} + 7\,H_2O$$

(b) Based on the overall reaction for this titration, Equation 15.3 can be written as follows.

At the Equivalence Point for the Titration of Dichromate with Fe^{2+}:

Mol titrated dichromate =

$$\left(0.3000M\ Fe^{2+}\right)(0.02465\ L) \cdot \frac{1\ \text{mol Dichromate}}{6\ \text{mol Fe}^{2+}}$$

$$= 0.0012325\ \text{mol}$$

We can now use the difference in the total added dichromate and the amount remaining before the back titration to find how much dichromate reacted with organic compounds in the water.

Mol dichromate reacted with sample =

(Mol added dichromate) − (Mol titrated dichromate)

$$= (0.2000\ M)(0.01000\ L) - (0.0012325\ \text{mol})$$

$$= 0.0007675\ \text{mol} = \mathbf{0.00768\ mol}$$

(c) The reaction of organic compounds with 0.007675 mol dichromate would be equivalent to the reaction of these same compounds with (0.0007675 mol dichromate) (3 mol O_2/2 mol dichromate) = 0.001151 mol O_2. We can now use this equivalent amount of O_2 to obtain a COD value that is expressed in units of mg O_2/L.

$$COD = (0.001151\ \text{mol O}_2) \cdot (31.999\ \text{g O}_2/\text{mol}) \cdot$$

$$(1000\ \text{mg/g})/(0.05000\ L)$$

$$= 736.6\ \text{mg/L} = \mathbf{737\ mg\ O_2/L}$$

This result is a typical value for river water, which often has a COD of 15–2000 mg O_2/L.

15.2 PERFORMING A REDOX TITRATION

15.2A Preparing Titrants and Samples

Standardizing Titrants. Because there is such a wide variety of titrants that can be used for redox titrations, there are also many methods for preparing these titrants. Examples of these methods can be found in resources such as References 6–9. Some redox titrants can be prepared directly from reagents that are primary standards, while other redox titrants must be standardized after their preparation by using additional reagents that are primary standards.

EXERCISE 15.2 **Preparing a Potassium Dichromate Solution**

Potassium dichromate ($K_2Cr_2O_7$, 294.19 g/mol) is not only used in COD measurements but can be used directly as a titrant for redox titrations. This chemical is available in a stable and pure form and is considered a primary standard. The half-reaction for the reduction of dichromate to Cr^{3+} in acidic solution is shown as follows, as might be used during a redox titration or COD analysis.

Reduction Half-Reaction:

$$Cr_2O_7^{2-} + 14\,H^+ + 6\,e^- \rightleftarrows 2\,Cr^{3+} + 7\,H_2O$$

(a) What mass of $K_2Cr_2O_7$ is necessary to make 500.0 mL of a 0.02500 M dichromate solution? (b) The *normality* of a solution for a chemical that takes part in an oxidation–reduction reaction is the number of equivalent weights of that chemical per liter, where the equivalent weight is the mass of the chemical, which requires one mole of electrons for its oxidation or reduction.[6] Based on this definition, what is the concentration of a 0.02500 M dichromate solution in units of normality?

SOLUTION

(a) The mass of potassium dichromate that is required to make a 0.02500 M solution can be determined from the molar mass of this chemical and the final desired solution volume.

$$\text{Mass K}_2Cr_2O_7 = (0.02500\ \text{mol/L}) \cdot (0.5000\ L) \cdot$$

$$(294.19\ \text{g/mol}) = \mathbf{3.677\ g}$$

(b) Because the reduction half-reaction for dichromate involves six electrons per dichromate ion, one mole of dichromate is the same as six equivalents. Thus, the concentration in units of normality for 0.02500 M dichromate will be (6 eq/mol dichromate)(0.02500 mol dichromate/L) = **0.1500 N**.

Two titrants for redox titrations that must be standardized using other chemicals are **permanganate** (MnO_4^-) and **thiosulfate** ($S_2O_3^{2-}$). Neither of these reagents can be made up directly as a usable titrant because their solid forms have insufficient purity and stability to act as primary standards. Solutions of permanganate are unstable because MnO_4^- is capable of oxidizing water. This reaction is quite slow but does mean that permanganate solutions must be standardized on the same day that they are used for a titration. A permanganate solution is often standardized by using it to titrate samples of ferrous ammonium sulfate ($Fe(NH_4)_2(SO_4)_2 \cdot 6H_2O$), arsenious oxide ($As_2O_3$), sodium oxalate ($Na_2C_2O_4$), or pure iron wire, all of which are available as primary standard materials. Thiosulfate is typically standardized by using a solution of this reagent to titrate a standard solution of potassium triiodide or potassium dichromate. Solutions containing thiosulfate are also unstable and should be standardized on the day of their use.

Sample Pretreatment Methods. It is usually desirable for all of an analyte to be in the same oxidation state at the beginning of a redox titration, but this may not be the case for the original sample. For instance, if the total amount of an analyte is to be measured through oxidation during a redox titration, this analyte must all be in the same low oxidation state before the titration begins. If some analyte is present in a higher oxidation state, this state must first be reduced through a sample pretreatment process called *prereduction*. This process must be done in a way that does not leave any reducing agent in the sample that might react with the titrant and create errors in the analysis.

One way to accomplish sample prereduction is to use a **reductor**, which is a device that consists of a column that contains an insoluble form of a reducing agent (see Figure 15.4).[7–9] Examples of these devices include a *Jones reductor* (or "zinc reductor"),[10–12] which contains amalgamated zinc, and a *Walden reductor* (or "silver reductor"), which contains silver granules.[13] Table 5.1 compares the ability of these two devices to reduce various chemicals. Zinc is more commonly used than silver for prereduction because it is a stronger reducing agent and is less expensive. Silver, however, does have some advantages over zinc, such as the fact that it will not reduce Cr^{3+} to Cr^{2+}. Other types of reductors might contain amalgams of cadmium or lead.[7]

EXERCISE 15.3   **Standardizing a Potassium Permanganate Solution**

A permanganate solution is prepared by dissolving approximately 1.6 g of $KMnO_4$ in 500.0 mL of water. This solution is then immediately standardized by using it to titrate 1.2167 g of ferrous ammonium sulfate (molar mass = 392.16 g/mol). The reaction for this titration is shown.

$$MnO_4^- + 5\,Fe^{2+} + 8\,H^+ \rightarrow Mn^{2+} + 5\,Fe^{3+} + 4\,H_2O$$

If 41.79 mL of the permanganate solution is required to reach the end point of this titration, what is the concentration of this permanganate solution?

SOLUTION

The preceeding titration reaction shows that five moles of MnO_4^- will react with one mole of Fe^{2+} from the ferrous ammonium sulfate. We can use this stoichiometry and the molar mass of ferrous ammonium sulfate to find the concentration of the permanganate solution.

Conc. MnO_4^- = (1.2167 g ferrous ammonium sulfate) · (1 mol/392.16 g)/(0.04179 L)

= **0.01478 M**

Once we have standardized this permanganate solution, we can then use it as a titrant for analytes that can be oxidized by this reagent.

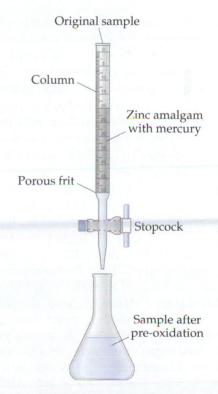

Original sample

Column

Zinc amalgam with mercury

Porous frit

Stopcock

Sample after pre-oxidation

FIGURE 15.4   A typical design of a Jones reductor for the preoxidation of analytes. This device is named after American chemist Harry Clair Jones (1865–1916), who used a tube filled with zinc particles for sample pretreatment. A decade later, work by others revealed that the reducing activity of this device could be improved by using a zinc amalgam with mercury in place of zinc particles.[10–12]

TABLE 15.1 Comparison of Common Reductors for Sample Pretreatment*

| | Reduced Product[a] | |
Chemical	Jones Reductor	Walden Reductor
Ag^+	Ag^0	Not applicable[b]
CrO_4^{2-}	Cr^{2+}	Cr^{3+}
Cu^{2+}	Cu^0	$CuCl_3^{2-}$
Fe^{3+}	Fe^{2+}	Fe^{2+}
MnO_4^-	Mn^{2+}	Mn^{2+}
MoO_4^{2-}	Mo^{3+}	Mo^{3+}
Ti^{4+}	Ti^{3+}	No reduction
UO_2^{2+}	U^{3+}/U^{4+}	U^{4+}
VO_3^-	V^{2+}	VO^{2+}

*This information was obtained from H. Diehl, *Quantitative Analysis: Elementary Principles and Practice*, 2nd ed., Oakland Street Science Press, Ames, IA, 1974.

[a]The products for the Jones reductor are shown for a reaction that is carried out in the presence of a sulfuric acid solution. The products for the Walden reductor are for a reaction that is carried out in the presence of a hydrochloric acid solution.

[b]The Walden reductor cannot be used in this case because this reductor uses silver as the reducing agent.

Prereduction can also be carried out in a solution or by using a suspension. For instance, particles of silver or a zinc amalgam might be added to a sample instead of using a Jones reductor or Walden reductor. These particles can then be removed by filtration before the pretreated sample is titrated. Reducing reagents like hydrogen sulfide (H_2S) and sulfurous acid (H_2SO_3) can be added to a sample and later removed by boiling. Stannous chloride ($SnCl_2$) is another reducing agent that can be added to a sample for pretreatment, where the remaining excess of $SnCl_2$ is later eliminated by oxidizing Sn^{2+} with $HgCl_2$ to form Sn^{4+} and insoluble Hg_2Cl_2.[7–9]

There are situations in which an analyte must be converted from a lower to higher oxidation state before it can undergo a redox titration. This type of sample pretreatment is called *preoxidation*. A number of chemicals can be used as reagents for preoxidation. Sodium bismuthate ($NaBiO_3$), lead dioxide (PbO_2), potassium periodate (KIO_4), potassium persulfate ($K_2S_2O_8$, often used with Ag^+ as a catalyst), ozone (O_3), and hydrogen peroxide (H_2O_2) can all be employed for this purpose.[7–9] Any remaining sodium bismuthate or lead dioxide can be removed by filtration. Periodate can be removed from a sample by precipitating it as $Hg_5(IO_6)_2$. Excess persulfate, ozone, and hydrogen peroxide can be eliminated by boiling, which causes these reagents to be consumed as they oxidize water to form oxygen gas.[7–9]

15.2B Finding the End Point

Potential Measurements and Gran Plots. Redox titrations are often monitored by measuring the difference in potential between a reference electrode and an indicating electrode that is present in the sample/titrant mixture. This type of titration is carried out with a system like the one in Figure 15.2 and can be used whenever we have half-reactions for the analyte and titrant that can be described by the Nernst equation.

The indicating electrode will respond to the activity of electroactive species in the sample and titrant while the reference electrode provides a constant, known potential against which the potential of the indicating electrode can be compared. The cell potential that is measured will be related to the difference in the potentials of the indicating electrode and reference electrode, as shown in Equation 15.4.

$$E_{Cell} = E_{Ind} - E_{Ref} \qquad (15.4)$$

The shape of the resulting titration curve is similar to that which is obtained for pH measurements during an acid–base titration, in which the cell potential changes only slightly during the early stages of a titration but changes dramatically in the region near the equivalence point (see Figures 15.2 and 15.4). This similarity is a result of the fact that both a cell potential and pH are related to a logarithmic function of the activity (or concentration) of their respective analytes. In the case of potential measurements, this relationship is described by the Nernst equation. For an acid–base titration, this relationship is a result of the working definition of pH, where $pH = -\log(a_{H^+})$.

Acid–base titrations and redox titrations are also similar in that data obtained through potential measurements during a redox titration can be converted into a linear form by using a Gran plot.[14] Table 15.2 shows how an equation for a Gran plot can be developed for the general redox titration reaction that was given earlier in Equation 15.2. Based on the final result that is shown in Equation 15.5, a Gran plot for a redox titration can be made by plotting the function $(V_T \cdot 10^{-n_A E_{Cell}/(0.05916 \text{ V})})$ versus the volume of added titrant (V_T).

Figure 15.5 gives an example of a Gran plot for the titration of Fe^{2+} by cerate (Ce^{4+}). According to Equation 15.5, this type of plot should provide a linear relationship for the titration before the equivalence point if the ionic strength of the sample/titrant mixture is essentially constant. The x-intercept from this plot will then be equal to the volume of titrant that is needed to reach the end point (V_E). If the ionic strength does vary during the titration, it is still possible to use a Gran plot to find V_E by employing a more limited range of titrant volumes, such as those that make up the last 10–20% of the titration prior to the end point.

Visible Detection and Indicators. Another similarity between redox titrations and acid–base titrations is that visual indicators are used in both as an additional means for end point detection. There are some situations in which the

TABLE 15.2 Gran Equation for an Analyte (A_{red}) Titrated with an Oxidizing Agent (T_{ox})

Titration Reactions:

Oxidation Half-Reaction: $\quad A_{red} \rightleftharpoons A_{ox} + n_A e^-$ (or $\quad n_T A_{red} \rightleftharpoons n_T A_{ox} + n_T n_A e^-$)

Reduction Half-Reaction: $\quad T_{Ox} + n_T e^- \rightleftharpoons T_{red}$ (or $\quad n_A T_{Ox} + n_T n_A e^- \rightleftharpoons n_A T_{red}$)

Overall Titration Reaction: $\quad n_T A_{red} + n_A T_{ox} \rightleftharpoons n_T A_{ox} + n_A T_{red}$

Nernst Equation at 25°C for Analyte:

$$E_A = E_A^\circ - \frac{0.05916\ \text{V}}{n_A} \log\left(\frac{a_{A_{red}}}{a_{A_{ox}}}\right)$$

Mass Balance Equations (Assuming Essentially Complete Reaction of A_{red} with T_{ox}):

$$\text{mol } A_{ox} \text{ produced} = (n_A / n_T)(\text{mol } T_{ox} \text{ added}) \quad \Rightarrow \quad [A_{ox}] = \frac{(n_A / n_T)\, C_T V_T}{V_A + V_T}$$

$$\text{mol } A_{ox} \text{ remaining} = \text{mol original } A_{ox} - \text{mol } T_{ox} \text{ added} \Rightarrow [A_{red}] = \frac{C_A V_A - (n_A / n_T)\, C_T V_T}{V_A + V_T}$$

Substitute Mass Balance Equations into Nernst Equation and Eliminate Common Terms:

$$E_A = E_A^\circ - \frac{0.05916\ \text{V}}{n_A} \log\left[\frac{(\gamma_{A_{red}})\,[A_{red}]}{(\gamma_{A_{ox}})\,[A_{ox}]}\right]$$

$$\rightarrow \quad E_A = E_A^\circ - \frac{0.05916\ \text{V}}{n_A} \log\left[\frac{(\gamma_{A_{red}})\,[C_A V_A - (n_A / n_T)\, C_T V_T]}{(\gamma_{A_{ox}})\,(n_A / n_T)\, C_T V_T}\right]$$

$$\Rightarrow \quad E_A = E_A^\circ - \frac{0.05916\ \text{V}}{n_A} \log\left[\frac{\gamma_{A_{red}}}{\gamma_{A_{ox}}}\right] - \underbrace{\frac{0.05916\ \text{V}}{n_A} \log\left[\frac{C_A V_A - (n_A / n_T)\, C_T V_T}{(n_A / n_T)\, C_T V_T}\right]}_{E_A^{\circ\prime}}$$

Substitute in $(n_A/n_T)(V_E C_T) = (C_A V_A)$, where V_E is the Titrant Volume at the Equivalence Point:

$$\Rightarrow \quad E_A = E_A^{\circ\prime} - \frac{0.05916\ \text{V}}{n_A} \log\left(\frac{(n_A / n_T)\, C_T V_E - (n_A /n_T)\, C_T V_T}{(n_A / n_T)\, C_T V_T}\right)$$

$$\Rightarrow \quad E_A = E_A^{\circ\prime} - \frac{0.05916\ \text{V}}{n_A} \log\left(\frac{V_E - V_T}{V_T}\right)$$

Let $E_{Cell} = E_A - E_{Ref}$ and Rearrange Combined Nernst Equation for E_{Cell}:

$$E_{Cell} = E_A - E_{Ref}$$

$$\Rightarrow \quad E_{Cell} = \left(E_A^{\circ\prime} - E_{Ref}\right) - \frac{0.05916\ \text{V}}{n_A} \log\left(\frac{V_E - V_T}{V_T}\right)$$

$$\Rightarrow \quad -\left[E_{Cell} - \left(E_A^{\circ\prime} - E_{Ref}\right)\right] \cdot \frac{n_A}{0.05916\ \text{V}} = \log\left(\frac{V_E - V_T}{V_T}\right)$$

Find the Antilogarithm of Both Sides of Equation and Rearrange:

$$\Rightarrow \quad 10^{-n_A[E_{Cell} - (E_A^{\circ\prime} - E_{Ref})]/(0.05916\ \text{V})} = \frac{V_E - V_T}{V_T}$$

$$\Rightarrow \quad V_T \cdot 10^{-n_A E_{Cell}/(0.05916\ \text{V})} \cdot 10^{-n_A (E_A^{\circ\prime} - E_{Ref})/(0.05916\ \text{V})} = V_E - V_T$$

$$\therefore \underbrace{V_T \cdot 10^{-n_A E_{Cell}/(0.05916\ \text{V})}}_{y} = \underbrace{-10^{-n_A (E_{Ref} - E_A^{\circ\prime})/(0.05916\ \text{V})}}_{m} \cdot \underbrace{V_T}_{x} + \underbrace{10^{-n_A (E_{Ref} - E_A^{\circ\prime})/(0.05916\ \text{V})} \cdot V_E}_{b} \qquad (15.5)$$

$$ y \qquad\qquad = \qquad\qquad m \qquad\qquad x \qquad + \qquad b$$

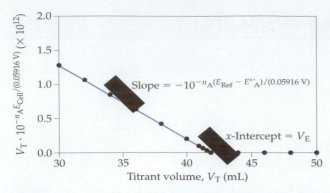

FIGURE 15.5 A typical Gran plot for a redox titration, using the analysis of a 50.00 mL sample of 0.08374 M Fe^{2+} with 0.1000 M Ce^{4+} as the titrant.

color of the titrant or sample can be used directly to monitor the progress of a redox titration. This situation occurs when we are using permanganate as a titrant. A solution of permanganate has an intense purple color, but becomes clear when it has undergone an oxidation–reduction reaction with a sample. As a result, we can find the end point of a titration that uses permanganate by noting at which point the addition of titrant first gives a permanent purple or pink color to the sample/titrant mixture if no other colored materials are present in the sample.

Unlike permanganate titrations, most other redox titrations (including the analysis of excess dichromate during a COD analysis) require that a separate indicator be added to the sample or titrant for visual detection of an end point. In an acid–base titration, a visual indicator is selected that will undergo a change in color that is near the pH of the true equivalence point. In a redox titration, the indicator is selected to undergo a change in color at a cell potential that is at or near the potential at the equivalence point. This requirement means the indicator for a redox titration must be a chemical that can also undergo an oxidation–reduction reaction. It is for this reason that such a chemical is known as a **redox indicator**.[3–5]

Many redox indicators are available and allow a change in color to be seen over a wide range of potentials (see Table 15.3). One common example of a redox indicator is *ferroin* (see structure below and Figure 15.6), which is often used for end point detection in a COD analysis.[2,6,15] Ferroin has a change in color as it undergoes a one-electron reduction from a "+3" to "+2" state. The standard reduction

TABLE 15.3 Examples of Redox Indicators*

Indicator	$E^{\circ\prime}$ (V)	Color of Indicator Oxidized form		Reduced form
Methylene blue	0.53 ($n = 2$)	Blue	↔	Colorless
Diphenylamine	0.76 ($n = 2$)	Violet	↔	Colorless
Diphenylamine-4-sulfonic acid	0.85 ($n = 2$)	Violet	↔	Colorless
Ferroin	1.06 ($n = 1$)	Blue	↔	Red
Tris(5-nitro-1, 10-phenanthroline)iron	1.25 ($n = 1$)	Blue	↔	Red

*This information was obtained from J. A. Dean, *Lange's Handbook of Chemistry*, 15th ed., McGraw Hill, New York, 1999. The formal potentials shown are for an aqueous solution at 30°C that has a pH of 0.0 or, in the case of ferroin, contains 1.0 M H_2SO_4. The values in parentheses represent the number of electrons that are involved in the reduction half-reaction for each indicator, as obtained from A. Hulanicki and S. Glab, *Pure and Applied Chemistry*, 40 (1978) 468–498.

potential for this half-reaction is +1.11 V (versus the standard hydrogen electrode [SHE]) at 25°C, and the formal potential is +1.06 V (versus SHE) near room temperature and in the presence of 1 M H_2SO_4. This means that ferroin can be used as a visual indicator for a redox titration that has an equivalence point that occurs at or near such a potential.

We can estimate the range in potentials over which a redox indicator will give a change in color by using the half-reaction for the indicator and the Nernst equation, as shown below (*Note:* Many redox indicators are also weak acids or bases, so the formal potential $E^{\circ\prime}$ is often used in place of E° in Equation 15.6 to describe the value for E_{In} that is expected at a given pH).

For a Redox Indicator at 25°C:

$$In_{ox} + n\,e^- \rightleftharpoons In_{red}$$

$$E_{In} = E_{In}^\circ - \frac{0.05916\ \text{V}}{n}\log\left[\frac{a_{In_{red}}}{a_{In_{ox}}}\right] \qquad (15.6)$$

In these equations, the oxidized and reduced forms of the indicator are represented by In_{ox} and In_{red}, where the charges on these species are omitted for the sake of simplicity. The change in color for an indicator often occurs when we go from a 10:1 ratio of the oxidized

Oxidized form (Pale blue) $+ e^- \rightleftharpoons$ Reduced form (red)

A *starch indicator* is a special type of visual indicator that is employed in redox titrations that have iodine as the titrant or the analyte.[6–9] An iodide ion (I^-) has no reaction with starch, but iodine (I_2, probably in the form of the linear I_3^- ion) can insert itself into the helical structure of starch to form a complex that has an intense blue color (see Figure 15.7). This type of indicator is sometimes described as having $E^\circ = 0.54$ V (versus SHE), which is simply the standard reduction potential for the iodine/triiodide redox couple. Starch works best as an indicator when iodine is the titrant. When iodine is the analyte, the violet starch/iodine complex forms in such a way that it is slow to dissociate. As a result, when iodine is the analyte, starch should only be added when the titration is almost at the equivalence point, as can be judged by the disappearance of the yellow color of iodine when it is dissolved in an aqueous solution.

15.3 PREDICTING AND OPTIMIZING REDOX TITRATIONS

15.3A General Approach to Calculations for Redox Titrations

We can use the Nernst equation to predict the shape of a redox titration curve if this titration involves *reversible* half-reactions. This process requires that we calculate the value of E_{Cell} that will be expected at a given volume of added titrant. In the next few sections we see how we can accomplish this for the general redox titration that was given earlier in Equation 15.2. As a specific example, we will look at the titration of Fe^{2+} with Ce^{4+}, a common oxidizing agent used for redox titrations. The following equations show the two half-reactions and overall oxidation–reduction reaction that occur during this analysis.

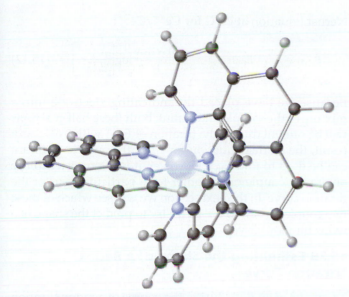

FIGURE 15.6 Structure of ferroin.

and reduced forms to a 1:10 ratio of the same forms. When we substitute these ratios in place of the activities in Equation 15.6 and combine this expression with Equation 15.4, the result is the following equation that describes the cell potentials over which the indicator will give a visible color change.

$$E_{Cell} = E_{In}^\circ \pm \left[\frac{0.05916 \text{ V}}{n} \right] - E_{Ref} \qquad (15.7)$$

When using ferroin in the presence of 1 M sulfuric acid and in a cell where a silver/silver chloride electrode is used as the reference, Equation 15.7 predicts that a color change will take place between $(+1.06 \text{ V}) + (0.05916 \text{ V}/1) - (+0.222 \text{ V}) = +0.89\underline{7}$ V and $(+1.06 \text{ V}) - (0.05916 \text{ V}/1) - (+0.222 \text{ V}) = +0.77\underline{9}$ V. In the case of a COD analysis, we can see from the titration curve in Figure 15.3 that this potential range occurs just after the equivalence point for the titration of excess dichromate by Fe^{2+}, making ferroin a suitable indicator for this method.

Oxidation Half-Reaction:	$Fe^{2+} \rightleftarrows Fe^{3+} + e^-$	(15.8)

Reduction Half-Reaction:	$Ce^{4+} + e^- \rightleftarrows Ce^{3+}$	(15.9)

Overall Titration Reaction:	$Fe^{2+} + Ce^{4+} \rightleftarrows Fe^{3+} + Ce^{3+}$	(15.10)

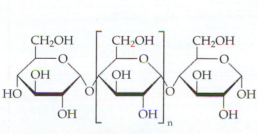

Starch (Amylose)

Starch Complex with I_3^-

FIGURE 15.7 Structure of amylose (left), a linear carbohydrate that is a major component of starch, and the colored complex (right) that this component of starch forms with the triiodide ion, I_3^-. To produce this complex, a strand of amylose forms a helix around a linear I_3^- ion. Starch also contains a branched carbohydrate known as amylopectin.

To monitor this titration, we can place an indicating electrode made of an inert material like platinum into the sample/titrant mixture. We can then use this electrode to find the difference in potential between this mixture and a reference electrode, giving E_{Cell} (Equation 15.4).

Because the reference electrode has a fixed potential, any change in E_{Cell} that is seen as we add more titrant to the sample will be due only to a change in potential for the sample/titrant and the corresponding half-reactions that are occurring in this mixture. This, in turn, means we can find the expected value for E_{Cell} if we first estimate E_{Ind} by using the Nernst equation. However, the question that then arises is "Which half-reaction do we use for this calculation?" When we mix our sample and titrant we create a system in which we have two redox couples present—one representing the oxidized and reduced forms of the analyte and the other representing the oxidized and reduced forms of the titrant. In our specific example, Fe^{2+} and Fe^{3+} would be the first redox couple and Ce^{4+} and Ce^{3+} would be the second. Each of these redox couples has its own Nernst equation to describe its half-reaction, as shown.

Nernst Equation at 25°C for Fe^{3+}/Fe^{2+}:

$$E_{Fe^{3+}/Fe^{2+}} = E^{\circ}_{Fe^{3+}/Fe^{2+}} - \frac{0.05916\ V}{1} \log\left[\frac{a_{Fe^{2+}}}{a_{Fe^{3+}}}\right] \quad (15.11)$$

Nernst Equation at 25°C for Ce^{4+}/Ce^{3+}:

$$E_{Ce^{4+}/Ce^{3+}} = E^{\circ}_{Ce^{4+}/Ce^{3+}} - \frac{0.05916\ V}{1} \log\left[\frac{a_{Ce^{3+}}}{a_{Ce^{4+}}}\right] \quad (15.12)$$

Because we have placed the indicating electrode into a mixture of the sample and titrant, both these half-reactions will be present during the titration of Fe^{2+} with Ce^{4+}. As a result, the Nernst equation for *either* the analyte or titrant can be used to find E_{Ind}. However, it is often easier to use one redox couple or the other for a particular part of the titration curve. In the next section we will see which of these two couples is best to use as we try to predict the shape of a redox titration curve.

15.3B Estimating the Shape of a Redox Titration Curve

The value of an E_{Cell} during the course of a redox titration can be estimated by dividing the titration curve into four general regions: (1) the original sample, (2) the middle of the titration in which only some of the analyte has been titrated, (3) the equivalence point, and (4) the region beyond the equivalence point in which excess titrant is present (see Figure 15.8). This process follows the same pattern described in Chapters 12 and 13 for predicting

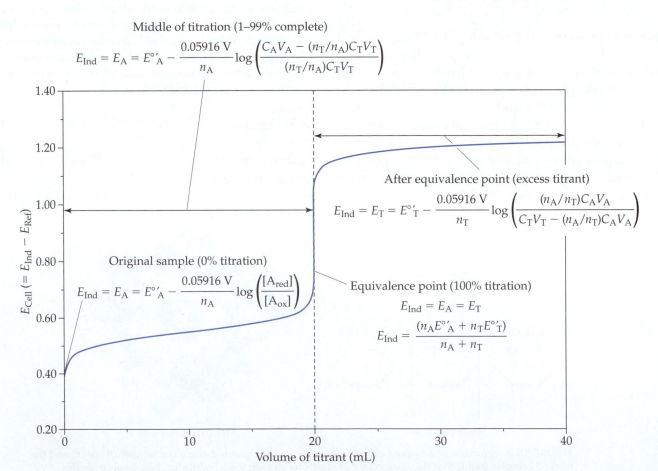

Middle of titration (1–99% complete)

$$E_{Ind} = E_A = E^{\circ'}_A - \frac{0.05916\ V}{n_A} \log\left(\frac{C_A V_A - (n_T/n_A)C_T V_T}{(n_T/n_A)C_T V_T}\right)$$

After equivalence point (excess titrant)

$$E_{Ind} = E_T = E^{\circ'}_T - \frac{0.05916\ V}{n_T} \log\left(\frac{(n_A/n_T)C_A V_A}{C_T V_T - (n_A/n_T)C_A V_A}\right)$$

Original sample (0% titration)

$$E_{Ind} = E_A = E^{\circ'}_A - \frac{0.05916\ V}{n_A} \log\left(\frac{[A_{red}]}{[A_{ox}]}\right)$$

Equivalence point (100% titration)

$$E_{Ind} = E_A = E_T$$

$$E_{Ind} = \frac{(n_A E^{\circ'}_A + n_T E^{\circ'}_T)}{n_A + n_T}$$

$E_{Cell}\ (= E_{Ind} - E_{Ref})$

Volume of titrant (mL)

FIGURE 15.8 Four general regions in a redox titration curve. The specific curve that is shown here is for the analysis of a 10.00 mL sample of 0.0100 M Fe^{2+} with 0.005000 M Ce^{4+} as the titrant.

the shape of titration curves for acid–base, complexometric and precipitation titrations, but now requires our use of the Nernst equation.

The *first region* for a redox titration occurs when we have only our original sample. Because we have not yet added any titrant, it is only possible to use the Nernst equation for the analyte's redox couple to find E_{Ind} at this point. This situation is represented by Equation 15.13.

Original Sample (0% Titration):

$$E_{Ind} = E_A \text{ and } E_A = E_A^{\circ\prime} - \frac{0.05916 \text{ V}}{n_A} \log\left[\frac{[A_{red}]}{[A_{ox}]}\right] \quad (15.13)$$

where we are using the formal potential ($E_A^{\circ\prime}$) in place of the standard reduction potential (E°) to correct for the fact that we included the concentrations of A_{red} and A_{ox} in this equation instead of their activities (see Chapter 10). At this point, you may be wondering how it is possible to use Equation 15.13 if only the original form of the analyte is present in the sample. The answer is that a small amount of the other form of the analyte must also be present if this redox couple makes up part of a truly reversible half-reaction, as is assumed by the Nernst equation. The amount of this other species (Fe^{3+}, if we are titrating Fe^{2+}) is often quite small or has an unknown value. Thus, at the beginning of a redox titration, it is usually not possible to predict what E_{Ind} will be unless further information is given on the sample conditions. In this situation, an alternative approach is to calculate the value of E_{Cell} at an early point in the titration (e.g., 1% of the way to the equivalence point) by using the method described in the next paragraph (see Equation 15.14). However, if we do have sufficient information to estimate the value of E_{Ind} by using Equation 15.13, we can obtain E_{Cell} by using the relationship $E_{Cell} = E_{Ind} - E_{Ref}$, where E_{Ref} is a known constant.

The *second region* in a redox titration curve occurs when some titrant has been added to the sample but we have not reached the equivalence point. In this region we now have significant amounts of both oxidation states of the analyte. We can also now estimate the concentrations for both forms of the analyte based on the original concentration of the analyte, the amount of titrant that has been added, and the stoichiometry for the reaction between the analyte and titrant. For example, in the general redox titration that is represented by Equation 15.2, n_T moles of product A_{ox} will be formed for every n_T moles of analyte (A_{red}) that react with n_A moles of titrant (T_{ox}). If we know this titration reaction has a large equilibrium constant, we can safely assume that essentially all of the added titrant will react with the analyte as long as the analyte is present in excess. Under these conditions, the remaining moles of analyte (A_{red}) will be equal to the initial moles of analyte ($C_A V_A$) minus the term $(n_T/n_A)(C_T V_T)$, where C_A is the original concentration of analyte, C_T is the total concentration of the titrant, V_A is the volume of the original sample, V_T is the volume of added titrant, and (n_T/n_A) is the stoichiometric ratio for

the analyte and titrant in the titration reaction. The moles of A_{ox} that are formed will then be $(n_T/n_A)(C_T V_T)$, and the concentration of the analyte or this product will be equal to the moles for each of these species divided by the total volume of the sample/titrant mixture ($V_A + V_T$). Placing this information into the Nernst equation for the analyte gives the following result.

Middle of Titration (1–99% Complete):

$$E_{Ind} = E_A \text{ and } E_A =$$
$$E_A^{\circ\prime} - \frac{0.05916 \text{ V}}{n_A} \log\left[\frac{C_A V_A - (n_T/n_A)C_T V_T}{(n_T/n_A)C_T V_T}\right] \quad (15.14)$$

The same general type of expression is obtained if the titrant is acting as a reducing agent rather than an oxidizing agent, in which case the numerator and denominator terms in the logarithm portion of Equation 15.14 are switched because A_{ox} would then be the analyte and A_{red} would be the product.

EXERCISE 15.4 **Before the Equivalence Point during the Titration of Fe^{2+} with Ce^{4+}**

The titration curve in Figure 15.8 was obtained for a 10.00 mL sample of 0.01000 M Fe^{2+} that was titrated with 0.005000 M Ce^{4+}. The original sample is believed to also contain 1.0×10^{-5} M Fe^{3+}, as created by the oxidation of Fe^{2+} to Fe^{3+} by dissolved oxygen in the sample. Estimate the value of E_{Cell} at (a) the beginning of the titration and (b) half-way through the titration if a silver/silver chloride electrode is used as the reference electrode. In addition, determine (c) the value of E_{Cell} that would be expected after only 1.00% of the initial Fe^{2+} has been titrated. How does this last value compare to the results obtained in Parts (a) and (b)?

SOLUTION

(a) The value of E_{Ind} at the beginning of this titration can be found by using the Nernst equation for the Fe^{3+}/Fe^{2+} redox couple and by assuming that the activities for Fe^{3+}/Fe^{2+} in this sample are approximately the same as the concentrations that are listed for these chemicals.

$$E_{Ind} = E_{Fe^{3+}/Fe^{2+}} \approx$$
$$E_{Fe^{3+}/Fe^{2+}}^{\circ} - \frac{0.05916 \text{ V}}{1} \log\left[\frac{[Fe^{2+}]}{[Fe^{3+}]}\right]$$
$$= (+0.77 \text{ V}) - \frac{0.05916 \text{ V}}{1} \log\left[\frac{0.01000 \text{ M}}{1.0 \times 10^{-5}\text{M}}\right]$$
$$= +0.592 \text{ V}$$

We can then use this value along with Equation 15.4 and the value of E_{Ref} for the silver/silver chloride electrode (+0.222 V) to find E_{Cell}. The answer we obtain through this process is $E_{Cell} = (+0.592 V) - (+0.222 V) = +0.37 V$ (versus Ag/AgCl).

(b) The equivalence point for this titration will occur when the volume of added titrant is $(0.01000 M Fe^{2+})(0.01000 L Fe^{2+})/(0.005000 M Ce^{4+}) = 0.02000 L Ce^{4+}$, so only halfway to this point would require $\frac{1}{2}(0.02000 L) = 0.01000 L Ce^{4+}$. The value of E_{Ind} at this halfway point in the titration can be found by using the Nernst equation for the Fe^{3+}/Fe^{2+} redox couple in the form that is given by Equation 15.14 where $(n_T/n_A) = 1$.

$$E_{Ind} = (+0.77 V) - \frac{0.05916 V}{1} \cdot$$
$$\log\left[\{(0.01000 M)(0.01000 L) - (1)(0.005000 M)(0.01000 L)\} / \{(1)(0.005000 M)(0.01000 L)\}\right]$$
$$= (+0.77 V) - (0.000 V)$$
$$= +0.770 V$$

(Note: In the case where there is a 1:1 relationship between A_{ox} and A_{red}, the value of E_{Ind} at the halfway point in a redox titration will be equal to $E°$ for the A_{ox}/A_{red} couple.) We can then use this value for E_{Ind} and E_{Ref} for the silver/silver chloride electrode to find E_{Cell}, giving an answer of $E_{Cell} = (+0.770 V) - (+0.222 V) = +0.59 V$ (versus Ag/AgCl).

(c) It was found in Part (b) that 0.02000 L of titrant was needed to reach the equivalence point, so the volume of titrant needed to titrate 1.00% of the original analyte would be $(0.01)(0.02000 L) = 0.000200 L$. The value of E_{Ind} at this point can be found by using the Nernst equation for the Fe^{3+}/Fe^{2+} redox couple, as again based on Equation 15.14 with $(n_T/n_A) = 1$.

$$E_{Ind} = (+0.77 V) - \frac{0.05916 V}{1} \cdot$$
$$\log\left[\{(0.01000 M)(0.01000 L) - (1)(0.005000 M)(0.000200 L)\} / \{(1)(0.005000 M)(0.000200 L)\}\right]$$
$$= (+0.77 V) - (0.118 V)$$
$$= +0.652 V = +0.65 V$$

The corresponding value of E_{Cell} at this point would then be $E_{Cell} = (+0.652 V) - (+0.222 V) = +0.43 V$ (versus Ag/AgCl), which is close the value of E_{Cell} found in Part (a) for the original sample. This indicates that a calculated result at 1.00% titration could have been used to describe E_{Cell} during the early stages of this titration, if we did not have additional information on how much Fe^{2+} and Fe^{3+} were initially present in the sample.

The *third region* in a redox titration curve occurs at the equivalence point. At this point, exactly enough titrant has been added to react with all of the original analyte. The only major source for either of these species in the resulting mixture is now through the equilibrium that is established by their oxidation–reduction reaction. We do not usually know the exact values for the trace amounts of analyte and titrant that are present at this point, so it is difficult to find E_{Ind} by using the separate Nernst equations for either the analyte or titrant. We can, however, use both Nernst equations at the same time to solve for E_{Ind}. To do this, we first need to set these two Nernst equations equal to E_{Ind}.

$$E_{Ind} = E_A = E_A^{°'} - \frac{0.05916 V}{n_A} \log\left(\frac{[A_{red}]}{[A_{ox}]}\right) \quad (15.15)$$

$$E_{Ind} = E_T = E_T^{°'} - \frac{0.05916 V}{n_T} \log\left(\frac{[T_{red}]}{[T_{ox}]}\right) \quad (15.16)$$

Next, we can multiply each of these equations by the number of electrons that are involved in their half-reactions (n_A or n_T) and add these two equations together.

$$n_A E_{Ind} = n_A E_A^{°'} - (0.05916 V) \log\left(\frac{[A_{red}]}{[A_{ox}]}\right)$$

$$+ n_T E_{Ind} = n_T E_T^{°'} - (0.05916 V) \log\left(\frac{[T_{red}]}{[T_{ox}]}\right)$$

$$(n_A + n_T)E_{Ind} = (n_A E_A^{°'} + n_T E_T^{°'})$$
$$- (0.05916 V)\log\left(\frac{[A_{red}][T_{red}]}{[A_{ox}][T_{ox}]}\right) \quad (15.17)$$

We also know from the stiochiometry of titration reaction that at the equivalence point it can be said that $(n_A/n_T)[A_{ox}] = [T_{red}]$ and $(n_A/n_T)[A_{red}] = [T_{ox}]$. When we substitute this information into Equation 15.17, we get the simplified expression in Equation 15.18.

$$(n_A + n_T)E_{Ind} = \left(n_A E_A^{o'} + n_T E_T^{o'}\right)$$

$$- (0.05916 \text{ V}) \log\left(\frac{[A_{red}](n_A/n_T)[T_{red}]}{[A_{ox}](n_A/n_T)[T_{ox}]}\right)$$

$$\Rightarrow (n_A + n_T) E_{Ind} = \left(n_A E_A^{o'} + n_T E_T^{o'}\right)$$

$$- (0.05916 \text{ V}) \log(1)$$

$$\Rightarrow (n_A + n_T) E_{Ind} = \left(n_A E_A^{o'} + n_T E_T^{o'}\right) \quad (15.18)$$

This last equation can now be rearranged to solve for E_{Ind} at the equivalence point.

At Equivalence Point (100% Titration):

$$E_{Ind} = \frac{\left(n_A E_A^{o'} + n_T E_T^{o'}\right)}{\left(n_A + n_T\right)} \quad (15.19)$$

Thus, the value for E_{Ind} at the equivalence point in this general redox titration is a weighted average of the standard reduction potentials for the analyte and titrant. You should note that this derivation did assume that there was a 1:1 stiochometry in the redox couples for both the analyte and titrant. If this is not the case, then additional terms that are related to either the analyte or titrant may also appear in a relationship like Equation 15.19. We will see an example of this in Section 15.4C when we discuss the use of dichromate as a reagent in redox titrations.

The *fourth region* in a redox titration curve occurs after the equivalence point. In this region, all of the original analyte has been consumed and we are simply adding excess titrant to the sample. As a result, there are now significant amounts for both oxidation states of the titrant. In addition, we can estimate the concentrations for each of these forms based on the original moles of analyte in the sample, the moles of titrant that have been added, and the stoichiometry for the reaction between the analyte and the titrant. In the case of the general titration reaction in Equation 15.2, the moles of the product for the titrant (T_{red}) will be equal to $(n_A/n_T)(C_A V_A)$, while the moles of excess titrant will given by the difference in the total moles of added titrant ($C_T V_T$) and $(n_A/n_T)(C_A V_A)$. We also know that the total volume of the sample/titrant mixture is ($V_A + V_T$), which we can use to help calculate the concentrations of the oxidized and reduced forms of the titrant. If we substitute these values into the Nernst equation for the titrant, we obtain the following result when the titrant is an oxidizing agent.

After Equivalence Point (Excess Titrant):

$$E_{Ind} = E_T \quad \text{and} \quad E_T = E_T^{o'} - \frac{0.05916 \text{ V}}{n_T} \cdot$$

$$\log\left(\frac{(n_A/n_T)C_A V_A}{C_T V_T - (n_A/n_T)C_A V_A}\right) \quad (15.20)$$

A similar expression is obtained if the titrant is a reducing agent, in which case the numerator and denominator

in the logarithm portion of Equation 15.20 would be switched from what was just shown, because T_{red} would now be the titrant and T_{ox} would be its product.

EXERCISE 15.5 **At or Beyond the Equivalence Point for the Titration of Fe^{2+} with Ce^{4+}**

What value for E_{Cell} is expected at the equivalence point for the titration in Figure 15.8? What value for E_{Cell} is expected after 30 mL of titrant have been added?

SOLUTION

According to Equation 15.19, the potential of the indicating electrode at the equivalence point and the cell potential at this point will be as follows.

$$E_{Ind} \approx [(1) + (+0.77 \text{ V}) + (1)(+1.44 \text{ V})]/2 = +1.10\underline{5} \text{ V}$$

and

$$E_{Cell} = (+1.10\underline{5} \text{ V}) - (+0.222 \text{ V}) =$$

$$+0.88 \text{ V (versus Ag/AgCl)}$$

We know from the previous exercise that the equivalence point for this titration occurs when 20.00 mL of titrant have been added. As a result, the addition of 30.00 mL titrant will be in the region of the titration curve that is after the equivalence point. The value of E_{Ind} at this point can be found by using Equation 15.20 and the Nernst equation for the redox couple of the titrant.

$$E_{Ind} \approx E_{Ce^{4+}/Ce^{3+}}^{\circ} - \frac{0.05916 \text{ V}}{1} \cdot$$

$$\log\Big(\{(1)(0.01000 M)(0.01000 \text{ L})\} / \{(0.005000 M)$$

$$(0.03000 \text{ L}) - (1)(0.01000 M)(0.01000 \text{ L})\}\Big)$$

$$= (+1.44 \text{ V}) - (0.01\underline{8} \text{ V}) = +1.42\underline{2} \text{ V}$$

We can then use this calculated value for E_{Ind} along with Equation 15.4 to find the cell potential, which gives $E_{Cell} = (+1.42\underline{2} \text{ V}) - (+0.222 \text{ V}) = +1.20 \text{ V}$ (versus Ag/AgCl).

15.3C Using the Fraction of Titration

A second way we can predict the behavior of a redox titration is to use the *fraction of titration* (F), a concept we originally discussed in Chapter 12. The value of F for the general redox titration in Equation 15.2 can be defined by using the following relationship.[5]

Titration of A with T:

$$F = \frac{n_A \text{ (moles added titrant)}}{n_T \text{ (moles original analyte)}} = \frac{n_A C_T V_T}{n_T C_A V_A} \quad (15.21)$$

We can then use this definition along with the mass balance equations for our titration to obtain a general expression that describes the titration of A with T, as shown below and in Table 15.4.

Fraction of Titration Equation for Titration of A with T:

$$F = \frac{n_A C_T V_T}{n_T C_A V_A} = \frac{1}{1 + 10^{-n_A\left[E_{Cell}+E_{Ref}-E_A^{\circ\prime}\right]/(0.05916\ V)}} \cdot \frac{10^{-n_T\left[E_{Cell}+E_{Ref}-E_T^{\circ\prime}\right]/(0.05916\ V)}}{1 + 10^{-n_T\left[E_{Cell}+E_{Ref}-E_T^{\circ\prime}\right]/(0.05916\ V)}} \quad (15.22)$$

According to this relationship, we can find the value of F at any point in this redox titration by using the value of E_{Cell} at that point along with the known standard reduction potentials or formal potentials for the analyte and titrant and the number of electrons that are involved in these half-reactions. We can then use the calculated value of F to find the volume of titrant that would be required to reach this point in the titration for a given concentration of analyte and titrant (C_A and C_T), and volume of our original sample (V_A). Figure 15.9 shows a spreadsheet that can be used with Equation 15.22 to predict the shape of a redox titration curve. This approach can also be used to see how a change in conditions will affect the titration, as is demonstrated in the following exercise.

| EXERCISE 15.6 | Choosing an Indicator for a Redox Titration |

An environmental scientist wishes to use diphenylamine-4-sulfonic acid instead of ferroin as an indicator for the titration of Fe^{2+} with Ce^{4+} in Exercise 15.5, which is to be carried out at pH 0.0. What will be the titration error if the end point occurs when $E_{Ind} = E^{\circ\prime}$ for diphenylamine? What would be the titration error if ferroin were instead used as the indicator?

SOLUTION

We know from the last exercise that the equivalence point for this titration occurs when $E_{Ind} = +1.105$ V (versus SHE) and $V_T = V_E = 20.00$ mL. We also know from Table 15.3 that $E^{\circ\prime}$ for diphenylamine-4-sulfonic acid is +0.85 V (versus SHE) at pH 0.0. This would be equivalent to a measured cell potential of $E_{Cell} = (+0.85\ V) - (+0.222\ V) = +0.628$ V (versus Ag/AgCl). We can now substitute this information into Equation 15.22 to find F at this given cell potential.

$$F = \frac{1}{1 + 10^{-1\,[0.628\ V+0.222\ V-0.77\ V]/(0.05916\ V)}} \cdot \frac{10^{-1[0.628\ V+0.222\ V-1.44\ V]/(0.05916\ V)}}{1 + 10^{-1\,[0.628\ V+0.222\ V-1.44\ V]/(0.05916\ V)}} = 0.9575$$

Next, we can use the values for F and V_E to find V_T at the end point and calculate the resulting titration error (see our earlier discussion of titration errors in Chapter 12).

$$V_L = F \cdot V_E = (0.9575)(20.00\ mL) = 19.15\ mL$$

Titration error $= V_{L,End\ Pt} - V_L =$
$$19.15\ mL - 20.00\ mL = -0.85\ mL$$

These conditions give a relatively large titration error and an estimated concentration for Fe^{2+} that is high by 4.3%. If this scientist were to instead use ferroin as the indicator ($E^{\circ\prime} = +1.06$ V versus SHE), the titration error would be only 0.24 µL and the measured concentration for Fe^{2+} would have an error of only 0.0012%.

15.4 EXAMPLES OF REDOX TITRATIONS

15.4A Titrations Involving Cerate

In Section 15.3 we saw how Ce^{4+}, or **cerate**, can be used as a titrant when performing a redox titration for Fe^{2+}. Cerate is a strong oxidizing agent that is commonly used in this and other types of redox titrations. Solutions of Ce^{4+} can be prepared by using $(NH_4)_2Ce(NO_3)_6$ as a primary standard. Cerate is a one-electron oxidizing agent whose formal potential does not change with pH, as indicated by the following half-reaction.

Reduction Half-Reaction:

$$Ce^{4+} + e^- \rightleftharpoons Ce^{3+} \quad (15.9)$$

However, pH is important to control when using Ce^{4+} as a titrant because this reagent can form an insoluble hydroxide ($Ce(OH)_4$) in a neutral or basic solution. Titrations with cerate are typically carried out in a strongly acidic solution (often containing 1 M H_2SO_4) to keep all such species in solution. In addition, the formal potential for this half-reaction can vary depending on the identity and concentration of anions that are present along with Ce^{4+} in solution (see Table 15.5).[7–9]

An aqueous solution of Ce^{4+} is orange and a solution of Ce^{3+} is yellow, but these colors are not sufficiently intense or different enough to be used directly for end point detection during a redox titration. The most common redox indicator that is used for a cerium titration is ferroin. Titrations involving Ce^{4+} also can be monitored through the use of potential measurements, as indicated in Section 15.3B.

15.4B Titrations Involving Permanganate

Permanganate is another titrant that is frequently used in redox titrations. The manganese in potassium permanganate ($KMnO_4$) is in its highest oxidation state (+7) and has a large positive E° value (+1.51 V versus SHE). These properties make potassium permanganate a good oxidizing agent that is widely used for chemical analysis as well as in organic synthesis and industrial processes.[7–9] Although potassium permanganate can be used in a COD analysis, dichromate is preferred for this purpose because it is more effective in oxidizing a wide range of organic compounds in water.

TABLE 15.4 Derivation of a Fraction of Titration Equation for A_{Red} Titrated with T_{Ox}

Titration Reactions:

Oxidation Half-Reaction:

$$A_{red} \rightleftharpoons A_{ox} + n_A e^- \quad (\text{or} \quad n_T A_{red} \rightleftharpoons n_T A_{ox} + n_T n_A e^-)$$

Reduction Half-Reaction:

$$T_{ox} + n_T e^- \rightleftharpoons T_{red} \quad (\text{or} \quad n_A T_{ox} + n_T n_A e^- \rightleftharpoons n_A T_{red})$$

Overall Titration Reaction:

$$n_T A_{red} + n_A T_{ox} \rightleftharpoons n_T A_{ox} + n_A T_{red}$$

Mass Balance Equations:

Mass Balance for A:

$$\text{Total mol A} = C_A V_A = \text{mol } A_{red} + \text{mol } A_{ox}$$
$$= \alpha_{A_{red}} (C_A V_A) + \alpha_{A_{ox}} (C_A V_A)$$

Mass Balance for T:

$$\text{Total mol T} = C_T V_T = \text{mol } T_{red} + \text{mol } T_{ox}$$
$$= \alpha_{T_{red}} (C_T V_T) + \alpha_{T_{ox}} (C_T V_T)$$

For Titration Reaction:

$$n_T (\text{mol } A_{ox}) = n_A (\text{mol } T_{red})$$

$$\Rightarrow \quad n_T \alpha_{A_{ox}} (\text{Total mol A}) = n_A \alpha_{T_{red}} (\text{Total mol T})$$

$$\Rightarrow \quad n_T \alpha_{A_{ox}} (C_A V_A) = n_A \alpha_{T_{red}} (C_T V_T)$$

$$\Rightarrow \quad \frac{n_A C_T V_T}{n_T C_A V_A} = \frac{\alpha_{A_{ox}}}{\alpha_{T_{red}}}$$

Nernst Equations for Titration:

Nernst Equation for A:

$$E_A = E_A^{o\prime} - \frac{0.05916 \text{ V}}{n_A} \log\left[\frac{[A_{red}]}{[A_{ox}]}\right]$$

$$(\text{mol } A_{red})/(\text{mol } A_{ox}) = (1 - \alpha_{Aox})/(\alpha_{Aox})$$

Nernst Equation for T:

$$E_T = E_T^{o\prime} - \frac{0.05916 \text{ V}}{n_T} \log\left[\frac{[T_{red}]}{[T_{ox}]}\right]$$

$$(\text{mol } T_{red})/(\text{mol } T_{ox}) = (\alpha_{T_{red}})/(1 - \alpha_{T_{red}})$$

Rearrange Nernst Equations to Get Alpha Expressions for A_{ox} and T_{red}:

Using Nernst Equation for A:

$$\Rightarrow \quad \left(E_A - E_A^{o\prime}\right) = -\frac{0.05916 \text{ V}}{n_A} \log\left[\frac{1 - \alpha_{A_{ox}}}{\alpha_{A_{ox}}}\right]$$

$$\Rightarrow \quad 10^{-n_A[E_A - E_A^{o\prime}]/(0.05916 \text{ V})} = \frac{1 - \alpha_{A_{ox}}}{\alpha_{A_{ox}}}$$

$$\Rightarrow \quad \alpha_{A_{ox}} \cdot 10^{-n_A[E_A - E_A^{o\prime}]/(0.05916 \text{ V})} = 1 - \alpha_{A_{ox}}$$

$$\Rightarrow \quad \alpha_{A_{ox}} = \frac{1}{1 + 10^{-n_A[E_A - E_A^{o\prime}]/(0.05916 \text{ V})}}$$

Using Nernst Equation for T:

$$\Rightarrow \quad \left(E_T - E_T^{o\prime}\right) = -\frac{0.05916 \text{ V}}{n_T} \log\left(\frac{\alpha_{T_{red}}}{1 - \alpha_{T_{red}}}\right)$$

$$\Rightarrow \quad 10^{-n_T[E_T - E_T^{o\prime}]/(0.05916 \text{ V})} = \frac{\alpha_{T_{red}}}{1 - \alpha_{T_{red}}}$$

$$\Rightarrow \quad \left(1 - \alpha_{T_{red}}\right) \cdot 10^{-n_T[E_T - E_T^{o\prime}]/(0.05916 \text{ V})} = \alpha_{T_{red}}$$

$$\Rightarrow \quad \alpha_{T_{red}} = \frac{10^{-n_T[E_T - E_T^{o\prime}]/(0.05916 \text{ V})}}{1 + 10^{-n_T[E_T - E_T^{o\prime}]/(0.05916 \text{ V})}}$$

TABLE 15.4   Derivation of a Fraction of Titration Equation for A_{Red} Titrated with T_{Ox} (continued)

Substitute into Alpha Expressions $E_{Ind} = E_T$ or E_A and $E_{Ind} = E_{Cell} + E_{Ref}$:

$$\Rightarrow \qquad \alpha_{A_{ox}} = \frac{1}{1 + 10^{-n_A[E_{Cell}+E_{Ref}-E_A^{o\prime}]/(0.05916\ V)}}$$

$$\Rightarrow \qquad \alpha_{T_{red}} = \frac{10^{-n_T[E_{Cell}+E_{Ref}-E_T^{o\prime}]/(0.05916\ V)}}{1 + 10^{-n_T[E_{Cell}+E_{Ref}-E_T^{o\prime}]/(0.05916\ V)}}$$

Substitute in Alpha Expressions into Mass Balance Equation for Titration:

$$\therefore \qquad F = \frac{n_A\ C_T\ V_T}{n_T\ C_A\ V_A} = \frac{\dfrac{1}{1 + 10^{-n_A[E_{Cell}+E_{Ref}-E_A^{o\prime}]/(0.05916\ V)}}}{\dfrac{10^{-n_T[E_{Cell}+E_{Ref}-E_A^{o\prime}]/(0.05916\ V)}}{1 + 10^{-n_T[E_{Cell}+E_{Ref}-E_A^{o\prime}]/(0.05916\ V)}}} \qquad (15.22)$$

Permanganate solutions must be standardized by titrating a portion of a primary standard such as ferrous ammonium sulfate or ferrous ethylenediammonium sulfate. As we noted in Section 15.2B, the intense purple color of permanganate allows the direct use of this titrant for the visual detection of an end point. Titrations that involve permanganate can also be followed through the use of potential measurements.

The reduced form of permanganate that is created during an oxidation–reduction reaction will depend on

Titration of a A_{red} with T_{ox}			(Titration of a Fe^{2+} with Ce^{4+} vs. Ag/AgCl reference electrode)			

$E^°_{Ared/Aox}$ (V):	0.77	n_A:	1
$E^°_{Tred/Tox}$ (V):	1.44	n_T:	1
E_{Ref} (V):	0.222		

Conditions used in titration:

C_A (M) =	0.01	C_T (M) =	0.005
V_A (M) =	0.01	V_E (L) =	0.02

Calculated results for titration:

E_{Cell}	E_{Ind}	Alpha A_{ox}	Alpha T_{red}	Fraction to 1st equivalence pt	V_T (L)	V_T (mL)
0.300	0.522	6.43E−05	1.00E+00	0.0001	0.0000	0.00
0.310	0.532	9.48E−05	1.00E+00	0.0001	0.0000	0.00
0.320	0.542	1.40E−04	1.00E+00	0.0001	0.0004	0.00
0.330	0.552	2.07E−04	1.00E+00	0.0002	0.0000	0.00
0.340	0.562	3.05E−04	1.00E+00	0.0003	0.0000	0.01
0.350	0.572	4.50E−04	1.00E+00	0.0004	0.0000	0.01
0.360	0.582	6.64E−04	1.00E+00	0.0007	0.0000	0.01
0.370	0.592	9.79E−04	1.00E+00	0.0010	0.0000	0.02
0.380	0.602	1.44E−03	1.00E+00	0.0014	0.0000	0.03
0.390	0.612	2.13E−03	1.00E+00	0.0021	0.0000	0.04
0.400	0.622	3.14E−03	1.00E+00	0.0031	0.0001	0.06
0.410	0.632	4.63E−03	1.00E+00	0.0046	0.0001	0.09
0.420	0.642	6.81E−03	1.00E+00	0.0068	0.0001	0.14
0.430	0.652	1.00E−02	1.00E+00	0.0100	0.0002	0.20
0.440	0.662	1.47E−02	1.00E+00	0.0147	0.0003	0.29
0.450	0.672	2.16E−02	1.00E+00	0.0216	0.0004	0.43
0.460	0.682	3.15E−02	1.00E+00	0.0315	0.0006	0.63
0.470	0.692	4.58E−02	1.00E+00	0.0458	0.0009	0.92
0.480	0.702	6.62E−02	1.00E+00	0.0662	0.0013	1.32
0.490	0.712	9.47E−02	1.00E+00	0.0947	0.0019	1.89
0.500	0.722	1.34E−01	1.00E+00	0.1337	0.0027	2.67
0.510	0.732	1.86E−01	1.00E+00	0.1856	0.0037	3.71
0.520	0.742	2.52E−01	1.00E+00	0.2517	0.0050	5.03
0.530	0.752	3.32E−01	1.00E+00	0.3317	0.0066	6.63
0.540	0.762	4.23E−01	1.00E+00	0.4228	0.0085	8.46
0.550	0.772	5.19E−01	1.00E+00	0.5195	0.0104	10.39
0.560	0.782	6.15E−01	1.00E+00	0.6147	0.0123	12.29
0.570	0.792	7.02E−01	1.00E+00	0.7019	0.0140	14.04

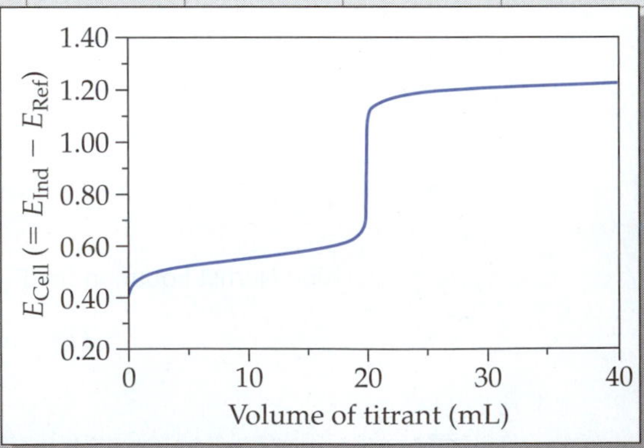

FIGURE 15.9   An example of a spreadsheet that can be used with the fraction of titration expression in Equation 15.22 to predict the shape and response of a titration curve.

TABLE 15.5 Formal Potentials for the Ce^{4+}/Ce^{3+} Redox Couple at 25°C	
Conditions	$E^{o'}_{Ce^{4+}/Ce^{3+}}$ **(V vs. SHE)***
1.0 M HClO$_4$	+1.70
1.0 M HNO$_3$	+1.61
0.5 M H$_2$SO$_4$	+1.44
1.0 M HCl	+1.28

*This information was obtained from J. A. Dean, *Lange's Handbook of Chemistry*, 15th ed., McGraw-Hill, New York, 1999.

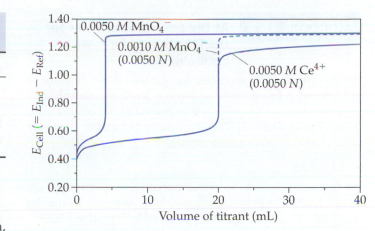

FIGURE 15.10 Comparison of titration curves for the reaction of a 10.00 mL sample of 0.01000 M Fe^{2+} using cerate or permanganate at the titrant.

the pH at which this reaction occurs. In an acidic solution, the reduced product of permanganate is the nearly colorless Mn^{2+} ion. In a solution with a neutral pH, the product will be the insoluble brown-black solid MnO_2. The two half-reactions for these processes are as follows.

Reduction Half-Reaction (Acidic pH):

$$MnO_4^- + 8 H^+ + 5 e^- \rightleftharpoons Mn^{2+} + 4 H_2O \quad (15.23)$$

Reduction Half-Reaction (Neutral pH):
$$MnO_4^- + 4 H^+ + 3 e^- \rightleftharpoons MnO_2(s) + 2 H_2O \quad (15.24)$$

Permanganate-based titrations are often carried out under strongly acidic conditions (e.g., in 1 M H$_2$SO$_4$) to avoid the formation of $MnO_2(s)$ and keep all species of interest in solution.

There are some important differences between Ce^{4+} and permanganate that will affect the calculations we use to estimate their titration curves. These differences can be illustrated by comparing the titration of Fe^{2+} by both these agents. The half-reactions and overall reaction that occur during the titration of Fe^{2+} with permanganate are shown.

Oxidation Half-Reaction:

$$Fe^{2+} \rightleftharpoons Fe^{3+} + e^- (or \quad 5 Fe^{2+} \rightleftharpoons 5 Fe^{3+} + 5 e^-) \quad (15.8)$$

Reduction Half-Reaction:

$$MnO_4^- + 8 H^+ + 5 e^- \rightleftharpoons Mn^{2+} + 4 H_2O \quad (15.23)$$

Overall Titration Reaction:

$$5 Fe^{2+} + MnO_4^- + 8 H^+ \rightleftharpoons 5 Fe^{3+} + Mn^{2+} + 4 H_2O \quad (15.24)$$

One difference between this titration and the one we looked at for Fe^{2+} with Ce^{4+} in Equations 15.8–15.10 is that the titration of Fe^{2+} with MnO_4^- is based on the transfer of five electrons, while the titration of Fe^{2+} with Ce^{4+} involves the transfer of a single electron between these agents. This means a permanganate solution with the same molar concentration (M) as a Ce^{4+} solution will require a smaller volume of titrant to reach the equivalence point for a sample of Fe^{2+} (see Figure 15.10). However,

these two analyses will require the same volume of titrant if these reagents have the same concentration in units of normality (N), which expresses the amount of these agents by using equivalents rather than moles. The shapes of these titration curves will also differ slightly due to the different numbers of electrons that are involved in their oxidation–reduction reactions.

A second difference between permanganate and cerate is that the reduction half-reaction for permanganate has a strong dependence on pH, as shown in Equation 15.25, while the reduction half-reaction for Ce^{4+} does not.

$$E_{MnO_4^-/Mn^{2+}} = E^\circ_{MnO_4^-/Mn^{2+}} -$$
$$\frac{0.05916 \text{ V}}{5} \log\left(\frac{a_{Mn^{2+}}}{(a_{MnO_4^-})(a_{H^+})^8}\right) \quad (15.25)$$

We can still use the equations in Figure 15.10 to predict the response of this titration by separating the activity of the hydrogen ion term from the other terms in the logarithm portion of Equation 15.25.

$$E_{MnO_4^-/Mn^{2+}} = E^\circ_{MnO_4^-/Mn^{2+}} -$$
$$\underbrace{\frac{0.05916 \text{ V}}{5} \log\left(\frac{1}{(a_{H^+})^8}\right)}_{E^{o'}_{MnO_4^-/Mn^{2+}}} - \frac{0.05916 \text{ V}}{5} \log\left(\frac{a_{Mn^{2+}}}{a_{MnO_4^-}}\right)$$

$$\quad (15.26)$$

This additional logarithm term can be combined with the standard reduction potential to give a formal potential ($E^{\circ'}$) that corresponds to a given pH and set of solution conditions. If we wish, we can also use a formal potential to combine the effects of pH and the use of concentrations in place of activities in Equation 15.26. With this formal potential it is then possible to work directly with the expressions described in Sections 15.3B and 15.3C to predict behavior of a titration that involves the use of permanganate as a reagent or analyte.

EXERCISE 15.7 | **At the Equivalence Point for the Titration of Fe²⁺ with Permanganate**

A titration of Fe^{2+} with permanganate is carried out at pH 1.00 by adding some sulfuric acid to the sample and titrant. What value for E_{Cell} is expected at the equivalence point of this titration if a silver/silver chloride electrode is used as the reference electrode?

SOLUTION

We first need to find the formal potential for permanganate at pH 1.00 (or $a_{H^+} = 10^{-1.00} = 0.100$). We can do this by using the information that is provided in Equation 15.26.

$$E_{MnO_4^-/Mn^{2+}} = E^\circ_{MnO_4^-/Mn^{2+}} - \frac{0.05916\ V}{5} \log\left[\frac{1}{(a_{H^+})^8}\right]$$

$$= (+1.51\ V) - \frac{0.05916\ V}{5} \log\left[\frac{1}{(0.100)^8}\right] = +1.41\underline{5}\ V$$

The value of E_{Ind} for this titration at the equivalence point can be found by using Equation 15.19.

$$E_{Ind} = \frac{(+0.77\ V) + 5(+1.415\ V)}{1 + 5} = +1.30\underline{8}\ V$$

When we use this value with E_{Ref} for the silver/silver chloride electrode, we get $E_{Cell} = (+1.30\underline{8}\ V) - (+0.222\ V) = \mathbf{+1.09\ V}$ (versus Ag/AgCl).

15.4C Titrations Involving Dichromate

Dichromate is another common reagent that is used in redox titrations. We have already seen how this chemical can be used in a COD analysis for the oxidation of organic material in water. However, dichromate can also be used directly as a titrant in a redox titration.[7–9] A standard solution of dichromate can be made directly by dissolving an appropriate amount of $K_2Cr_2O_7$ in water. The chromium in dichromate is in its highest oxidation state (+6) and has a large positive E° value (1.36 V). As we saw for permanganate, these properties make potassium dichromate a good oxidizing agent for redox titrations.

The reduction half-reaction for this dichromate is shown in Equation 15.27.

Reduction Half-Reaction:
$$Cr_2O_7^{2-} + 14\ H^+ + 6\ e^- \rightleftharpoons 2\ Cr^{3+} + 7\ H_2O \quad (15.27)$$

This reaction indicates that a redox titration that involves dichromate as the oxidizing agent will give a potential that will strongly depend on pH. This half-reaction is also somewhat unique in that two moles of the reduced species of chromium (Cr^{3+}) are produced for every one mole of the oxidized species ($Cr_2O_7^{2-}$) that undergoes a reduction.

Dichromate is used as a titrant in an acidic solution. This is because its reduction product in an acidic solution is the soluble Cr^{3+} ion, but in neutral or basic solutions its

reduction leads to the production of the insoluble green solid Cr_2O_3. In addition, under neutral or basic conditions, dichromate is converted into chromate (CrO_4^{2-}), as shown below.

$$Cr_2O_7^{2-} + 2\ OH^- \rightarrow 2\ CrO_4^{2-} + H_2O \quad (15.28)$$

For these reasons, standard solutions of potassium dichromate are usually made up in an acidic aqueous solution, often containing 1.0 M sulfuric acid. Solutions of dichromate that are made in this fashion are quite stable if kept in closed containers and can be used reliably for many months after their preparation. An additional feature to consider when using potassium dichromate is that it is listed as a carcinogen and as a hazardous substance for the environment. Thus, appropriate safety precautions and chemical handling procedures must be followed when using this reagent.

Potential measurements can be used to follow the course of a dichromate-based titration. Although potassium dichromate is orange, this color is not intense enough to allow its use as a direct means for the visual detection of an end point. The most suitable indicator for a redox titration involving dichromate is diphenylamine, which is usually obtained in a sulfonated form to increase its solubility. Upon its oxidation, diphenylamine is converted irreversibly from a colorless form to diphenylbenzidine (a product that is also colorless) and then to diphenylbenzidine violet (which has an intense purple color that can be used to signal the end point).

Dichromate is similar to permanganate as a titrant in that reduction of both these chemicals is a pH-dependent process that involves multiple electrons. This means that calculations that are performed for a dichromate titration will also follow a similar approach to what was described in the last section for dichromate. However, it is also now necessary to consider the fact that *two* moles of Cr^{3+} are produced for every *one* mole of $Cr_2O_7^{2-}$ that undergoes a reduction. This feature adds an extra level of complexity to the calculation of the measured potential at the equivalence point and after the equivalence point.

The calculation of cell potential during a dichromate-based titration can be illustrated by looking at the measurement of Fe^{2+} by such an approach. The half-reactions and overall titration in this analysis are given by the following reactions.

Oxidation Half-Reaction:
$$Fe^{2+} \rightleftharpoons Fe^{3+} + e^-$$
$$(or\quad 6\ Fe^{2+} \rightleftharpoons 6\ Fe^{3+} + 6\ e^-) \quad (15.8)$$

Reduction Half-Reaction:
$$Cr_2O_7^{2-} + 14\ H^+ + 6\ e^- \rightleftharpoons 2\ Cr^{3+} + 7\ H_2O \quad (15.27)$$

Overall Titration Reaction:
$$6\ Fe^{2+} + Cr_2O_7^{2-} + 14\ H^+ \rightleftharpoons$$
$$6\ Fe^{3+} + 2\ Cr^{3+} + 7\ H_2O \quad (15.29)$$

The value of E_{Ind} at the beginning and middle of this titration can be determined by using the Nernst equation for the Fe^{3+}/Fe^{2+} redox couple, as described in Section 15.3B. However, we need to consider both pH and Cr^{3+} when determining the value of E_{Ind} at the equivalence point and beyond when dichromate is employed as the titrant.

The effect of pH on the dichromate redox couple can be dealt with by using a formal potential, as shown.

$$E_{Cr_2O_7^{2-}/Cr^{3+}} = \underbrace{E^\circ_{Cr_2O_7^{2-}/Cr^{3+}} - \frac{0.05916 \text{ V}}{6} \log\left(\frac{1}{(a_{H^+})^{14}}\right)}_{E^{\circ\prime}_{Cr_2O_7^{2-}/Cr^{3+}}}$$

$$- \frac{0.05916 \text{ V}}{6} \log\left(\frac{(a_{Cr^{3+}})^2}{a_{Cr_2O_7^{2-}}}\right) \quad (15.30)$$

Using the same approach discussed in Section 15.3B, we can combine this formal potential with the Nernst equations for Fe^{3+} and Cr^{3+} to obtain an equation for E_{Ind} at the equivalence point.

$$E_{Ind} =$$

$$\frac{(E^{\circ\prime}_{Fe^{3+}/Fe^{2+}} + 6E^{\circ\prime}_{Cr_2O_7^{2-}/Cr^{3+}}) - (0.05916 \text{ V}) \log([Cr^{3+}])}{(1 + 6)}$$

$$(15.31)$$

Unlike the general expression given in Equation 15.19 for E_{Ind}, this relationship depends on the concentration of one of the redox products (Cr^{3+}) because dichromate and Cr^{3+} do not have a 1:1 stoichiometry in their half-reaction. Fortunately, we can estimate the value of $[Cr^{3+}]$ because essentially all of the dissolved chromium species will be present in this form at the equivalence point. This means we can get a good estimate for $[Cr^{3+}]$ from the total moles of $Cr_2O_7^{2-}$ that have been added, the total volume of the sample/titrant mixture, and the stoichiometry of the titration reaction. Similar information can be used with the formal potential and Nernst equation in Equation 15.30 to find the value of E_{Ind} after the equivalence point when dichromate is utilized as a titrant.

EXERCISE 15.8	**Titration of Fe^{2+} with Dichromate**

A 10.00 mL aqueous solution of 0.01000 M Fe^{2+} is titrated at pH 1.00 with a 0.005000 M solution of potassium dichromate. This titration is followed through potential measurements using a silver/silver chloride reference electrode. What value for E_{Cell} is expected at the equivalence point of this titration?

SOLUTION

Equation 15.31 can be used to find E_{Ind} for this titration at the equivalence point, but we first need to find $[Cr^{3+}]$ at this point. The total amount of Fe^{2+} in the original sample

was $(0.01000 \text{ } M)(0.01000 \text{ L}) = 1.000 \times 10^{-4}$ mol. We also know from the titration reaction in Equation 15.29 that six moles of Fe^{2+} react with each mole of dichromate that is added to the sample. As a result, the moles of dichromate needed to reach the equivalence point will be $(1.000 \times 10^{-4} \text{ mol})/(6 \text{ mol } Fe^{2+}/\text{mol } Cr_2O_7^-) = 1.667 \times 10^{-5}$ mol $Cr_2O_7^-$ and the volume of titrant needed to reach this point is $(1.667 \times 10^{-5} \text{ mol } Cr_2O_7^-)/(0.005000 \text{ mol/L}) = 0.003333$ L or 3.33 mL. If essentially all the dichromate has reacted with Fe^{2+}, the concentration of Cr^{3+} at the equivalence point is as follows.

$$[Cr^{3+}] = \{(1.667 \times 10^{-5} \text{ mol } Cr_2O_7^-) \cdot$$
$$(2 \text{ mol } Cr^{3+} / 1 \text{ mol } Cr_2O_7^-)\} /$$
$$(0.01000 \text{ L} + 0.003333 \text{ L}) = 2.500 \times 10^{-3} \text{ } M$$

Next, we need to consider the effect of pH on this titration by calculating the formal potential of the dichromate redox couple at pH 1.00. This can be done by using the expression for $E^{\circ\prime}_{Cr_2O_7^{2-}/Cr^{3+}}$ in Equation 15.30 along with $E^\circ_{Cr_2O_7^{2-}/Cr^{3+}} = +1.36$ V and $a_{H^+} = 10^{-pH} = 0.100$ at pH = 1.0.

$$E^{\circ\prime}_{Cr_2O_7^{2-}/Cr^{3+}} =$$

$$(+1.36 \text{ V}) - \frac{0.05916 \text{ V}}{6} \log\left(\frac{1}{(0.100)^{14}}\right) = +1.222 \text{ V}$$

We can now place our calculated values for $[Cr^{3+}]$ and $E^{\circ\prime}_{Cr_2O_7^{2-}/Cr^{3+}}$ into Equation 15.31 to solve for E_{Ind}, which then makes it possible to estimate E_{Cell}.

$$E_{Ind} = \{(+0.77 \text{ V}) + 6(+1.222 \text{ V})$$
$$- (0.05916 \text{ V}) \log (2.500 \times 10^{-3} \text{ } M)\} / (1 + 6)$$
$$= +1.179 \text{ V}$$

$$\text{or } E_{Cell} = (+1.179 \text{ V}) - (+0.222 \text{ V})$$
$$= \textbf{+0.96 V} \text{ (vs. Ag/AgCl)}$$

15.4D Titrations Involving Iodine

Another important group of redox titrations involve the use of iodine as a titrant, reagent or analyte. This type of method is called an **iodimetric titration** (or *iodimetry*).[3,4,7–9] Iodine is a milder oxidizing agent than Ce^{4+}, permanganate, or dichromate and can be used at nearly any pH. Iodine in the form I_2 has a limited solubility in water (~1.1×10^{-3} M), but this solubility can be increased through the reaction of I_2 with iodide (I^-) to form triiodide (I_3^-), which is highly soluble.

$$I_2 + I^- \rightleftharpoons I_3^- \quad (15.32)$$

Iodimetric titrations are used in a variety of ways, but frequently involve the reaction of iodine (or triiodide) with thiosulfate ($S_2O_3^{2-}$) to form iodide and tetrathionate

$(S_4O_6^{2-})$. This reaction can be written using either iodine or triiodide as a reactant, as shown below.[7-9]

$$I_2 + 2\,S_2O_3^{2-} \rightleftharpoons 2\,I^- + S_4O_6^{2-}$$

or $\qquad I_3^- + 2\,S_2O_3^{2-} \rightleftharpoons 3\,I^- + S_4O_6^{2-} \qquad$ (15.33)

An iodimetric titration is not usually followed through potential measurements, but through the use of starch as a visual indicator. As we saw in Section 15.2B, starch (especially amylose, the main component of starch in potatoes) can form a complex with a triiodide ion. This complex is a dark blue in color when triiodide binds to amylose and a less intense violet color when it involves the binding of triiodide to amylopectin, another major component of starch.[9,16]

Most iodimetric titrations involve an indirect method in which an excess of iodide is added to a sample that contains an analyte that can oxidize iodide to form iodine. The iodine that is formed is then titrated with thiosulfate, using starch as an indicator.[7] Analytes that act as oxidizing agents for iodide in this approach include hydrogen peroxide, ozone, hypochlorite, iodate, bromate, Fe^{3+}, and bromine. This method can also be used to measure permanganate, dichromate, and Ce^{4+}. During titration of the iodine (or triiodide) with thiosulfate in this analysis, the starch indicator should not be added until near the end point. If the starch is added too soon it will be exposed to a high concentration of triiodide, which will give a blue species that will protect the triiodide and decrease its rate of reaction with thiosulfate, thus leading to an error in end point detection. The starch should instead be added when the yellow color due to iodine has become faint, indicating that most of the iodine has already been titrated with thiosulfate.

EXERCISE 15.9 **Performing an Iodimetric Titration**

The concentration of hydrogen peroxide in an aqueous sample is to be determined by using an iodimetric titration. In this method, an excess of iodide is first reacted with hydrogen peroxide in the sample (see the following reaction). The iodine that is formed is then titrated with thiosulfate.

$$H_2O_2 + 2\,I^- + 2\,H^+ \rightarrow 2\,H_2O + I_2$$

A 50.00 mL portion of the sample is treated with 50.00 mL of 0.20 M KI and titrated with 0.1087 M $Na_2S_2O_3$. The end point is signaled by a starch indicator when 17.53 mL of the titrant has been added. What was the concentration of hydrogen peroxide in the sample?

SOLUTION

We first need to find how many moles of thiosulfate were added at the end point.

$$\text{mol } S_2O_3^{2-} = (0.01753\ L)(0.1087\ mol/L) = 0.001905\ mol$$

The reaction in Equation 15.33 shows that one mole of I_2 (or I_3^-) will be titrated for every two moles of $S_2O_3^{2-}$ that are needed for this titration. Thus, there was 0.5 (0.001905 mol) = 0.009525 mol I_2 (or I_3^-) in the sample after it had reacted with the excess iodide. We also can see from the reaction given between iodide and hydrogen peroxide that one mole of I_2 was formed for every mole of hydrogen peroxide that reacted with I^-. This means we also had 0.009525 mol H_2O_2 in the sample, or that the original concentration of hydrogen peroxide was $[H_2O_2]$ = (0.0095925 mol)/(0.05000 L) = $1.906 \times 10^{-2}\ M$.

There are many applications for iodimetric titrations. One important example is the **Karl Fisher method** for the analysis of water[17] (see Box 15.1). The analysis of unsaturated fats is another common use of an iodimetric titration. This process involves the addition of iodine across carbon–carbon double bonds in an unsaturated fat, as shown below.

$$I_2 + R{-}CH{=}CH{-}R \rightarrow R{-}CHI{-}CHI{-}R \qquad (15.34)$$

The moles of iodine that are consumed by this process can be determined by measuring the amount of I_2 remaining by means of its titration with thiosulfate. This result is then used to calculate the *iodine number*, which is defined as the grams of I_2 that reacted per 100 g of fat.[20,21] Table 15.6 lists some typical iodine numbers for various types of fat.

A similar method using bromine instead of iodine is employed in modern measurements of unsaturated fats

TABLE 15.6   Iodine Numbers for Various Types of Fats and Oils*

Type of Fat or Oil	Iodine Number (g I_2 consumed/100 g sample)
Coconut oil	6–10
Butterfat	26–38
Cocoa butter	33–42
Beef tallow	35–42
Chicken fat	66–72
Olive oil	79–88
Peanut oil	88–98
Corn oil	111–128
Soybean oil	122–134
Safflower oil	122–141
Sunflower seed oil	129–136
Cod-liver oil	137–166

*This information was obtained from J. A. Dean, *Lange's Handbook of Chemistry*, 15th ed., McGraw-Hill, New York, 1999.

BOX 15.1
The Karl Fischer Method

A redox titration cannot only be used to measure analytes in water, but it can also be used to measure the content of water itself in a sample. This type of measurement is used to find the amount of water in drugs, foods, and everyday household products. These measurements are performed by using a special redox titration developed by German chemist Karl Fischer in 1935.[18] This approach, known as the **Karl Fischer method**, is usually carried out in the presence of a mixture of methanol and pyridine that also contains a known concentration of SO_2 and I_2.[19]

The reactions that take place during a Karl Fischer titration are shown in Figure 15.11 when methanol is used as the solvent. First, the methanol reacts with dissolved SO_2 in the titrant through a special type of acid–base reaction known as *solvolysis*. The products of this reaction then combine with a dissolved base (pyridine) through a second acid–base reaction to form an alkyl sulfite (as represented by $BH^+SO_3R^-$ in Figure 15.11). When this titrant is added to a sample containing water, an oxidation–reduction reaction occurs in which the sulfur in $BH^+SO_3R^-$ is oxidized to form

the product $BH^+SO_4R^-$ and the iodine in I_2 is reduced to give I^- (represented by "BHI" in the presence of BH^+). This method can be performed by either adding various volumes of a titrant containing I_2 to the sample or by using an electrolytic cell to generate I_2 from I^- for the titration of water.[19]

The end point of a Karl Fischer titration is detected by looking for the point at which the addition of titrant produces a small amount of nonreacted I_2. Starch cannot be employed as an indicator in this case, because it is typically used as an aqueous solution, which would place additional water in the sample and create an error in the analysis. An alternative approach for end point detection is to look directly for the color of the Karl Fischer reagent, which will cause the sample/titrant mixture to turn from yellow to brown when a slight excess of titrant has been added. It is also possible to detect the end point by using electrochemical measurements based on biamperometry (as described in Chapter 16), which will give a response only when significant amounts of both I_2 and I^- are present in the sample/titrant mixture.[19]

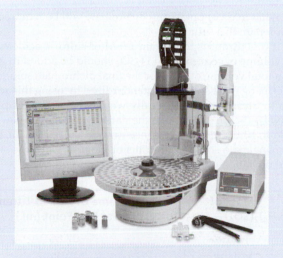

Reactions in the Karl Fischer method

1) Reaction of methanol (ROH) with SO_2

$$2\,ROH + SO_2 \rightleftharpoons RSO_3^- + ROH_2^+$$

2) Acid-base reaction with pyridine (B)

$$B + RSO_3^- + ROH_2^+ \rightleftharpoons BH^+RSO_3^- + ROH$$

3) Oxidation-Reduction reaction

$$\underset{\text{Analyte}}{H_2O} + \underbrace{I_2 + BH^+RSO_3^- + 2\,B}_{\text{Titrant}} \longrightarrow BH^+RSO_4^- + 2\,BHI$$

FIGURE 15.11 A commercial system that is used to perform an automated Karl Fischer titration and the reactions that are involved in this titration for the measurement of water.[19] The general reactant "ROH" is used in this scheme to represent methanol or any related alcohol in the Karl Fischer reaction, while "B" is used to represent a base such as pyridine that is used as part of this reaction. (The photo is reproduced with permission from Metrohm USA Inc.)

and the "iodine number." Bromine reacts with double bonds more rapidly and reliably than iodine, which gives a more accurate assessment of the degree of fat unsaturation. This analysis is performed by first reacting an excess of bromine with the sample to ensure that all the double bonds are brominated (see Equation 15.35). The remaining bromine is then reacted with an excess of iodide to form iodine (Equation 15.36).

$$Br_2 + R{-}CH{=}CH{-}R \rightarrow R{-}CHBr{-}CHBr{-}R \tag{15.35}$$

$$Br_2 + 2\,I^- \rightarrow I_2 + 2\,Br^- \tag{15.36}$$

The iodine that is formed is then titrated with thiosulfate to determine the amount of bromine that reacted with fat in the sample.[22]

Key Words

Other Terms

Questions

WHAT IS A REDOX TITRATION?

1. What is meant by the term "redox titration"? Explain why a COD analysis is an example of a method that uses a redox titration.
2. Explain how a redox titration is similar to an acid–base titration or a complexometric titration. How is a redox titration different from these other methods?
3. What requirements are there for a particular oxidation–reduction reaction to be useful as the basis for a redox titration?
4. Why is it important to have a relatively large, positive value for E°_{Net} for a successful redox titration?
5. Why is necessary to have a fast reaction rate between a titrant and analyte for a successful redox titration?
6. Explain why a COD analysis is performed by using a back titration.

HOW ARE REDOX TITRATIONS USED IN ANALYTICAL CHEMISTRY?

7. Explain why redox titrations are often more challenging to describe than acid–base titrations or a complexometric titration involving EDTA.
8. Discuss why the stoichiometry for the reaction between an analyte and titrant in a redox titration is often not 1:1.
9. Give a general equation that can be used to relate the concentrations and volumes of the analyte and titrant at the equivalence point of a redox titration. Explain the reason for all terms that appear in this equation.
10. A 50.00 mL sample of river water is treated with 25.00 mL of 0.2306 M $K_2Cr_2O_7$ at pH 0.0. After this mixture is heated and allowed to react, it is cooled and the remaining dichromate is titrated with 0.3165 M Fe^{2+}, giving an end point when 34.24 mL of this titrant has been added. How many moles of dichromate reacted with the original sample? What is the COD for this river water sample when expressed in units of mg O_2/L?
11. A 25.00 mL portion of a sample containing both Fe^{2+} and Fe^{3+} gives an end point that is detected by ferroin when 17.86 mL of 0.1234 M Ce^{4+} is used as the titrant. A second 25.00 mL portion of the same original sample is passed through a Jones reductor and is titrated with 22.54 mL of the same cerium solution. What were the concentrations of Fe^{2+} and Fe^{3+} in the sample?
12. A solution containing arsenic in the form of arsenite (AsO_3^{3-}) is titrated to a good end point with 37.55 mL of 0.05895 M $KMnO_4$ in the presence of 0.05 M H_2SO_4. How many grams of arsenic were in the original sample?

PREPARING TITRANTS AND SAMPLES

13. Give an example of a redox titrant that can be prepared directly from a primary reagent. What properties are required for the primary reagent to be useful as such a titrant? (*Hint:* See the discussion of primary standards in Chapter 12.)
14. Give two examples of redox titrants that must be standardized. What types of primary standards are used during this process?
15. A chemist wishes to prepare 2.0 L of a 0.02000 M dichromate solution.
    (a) What mass of potassium dichromate will be required to prepare this solution?
    (b) If this titrant is to contain 1.0 M of sulfuric acid, what volume of concentrated H_2SO_4 should be added to give a total volume of 2.0 L for the final dichromate solution?
16. A permanganate solution is standardized by using this solution to titrate several carefully weighed portions of pure $Fe(NH_4)_2(SO_4)_2 \cdot 6H_2O$. Use the data given in the following table to find the average concentration measured for this permanganate solution (in units of molarity) and the standard deviation of this estimated concentration.

Mass of $Fe(NH_4)_2(SO_4)_2 \cdot 6H_2O$ (g)	Volume of Added Titrant at End Point (mL)
1.3657	37.86
1.4498	40.23
1.5108	41.95

17. Explain why sample pretreatment is sometimes needed prior to conducting a redox titration.
18. What is the difference between "preoxidation" and "prereduction"? In which situations might each of these be used as part of an analytical method that involves a redox titration?
19. What is a "Jones reductor"? What is a "Walden reductor"? Explain how each of these works.
20. What general types of pretreatment would be necessary before a redox titration can be used for the following measurements?
    (a) Analysis of both the total chromium and Cr(VI) content in a sample that also contains Cr(III)
    (b) Measurement of the total manganese content in a sample that contains both $MnSO_4$ and MnO_2
    (c) Determination of both nitrite and nitrate in an aqueous solution
21. Two 25.00 mL portions of solution containing dissolved iron are titrated with 0.0254 M Ce^{4+} in 1.0 M H_2SO_4. The

first is titrated without prior treatment and requires 17.08 mL of titrant to reach an end point. The second solution is mixed with granular zinc, allowed to equilibrate, the remaining zinc is filtered off, and the iron solution titrated with 24.57 mL of the titrant. What were the concentrations of iron(II) and iron(III) in the original solution?

FINDING THE END POINT

22. Explain how potential measurements can be used to follow the course of a redox titration. Draw a simple diagram of a typical system that is used for this type of measurement.
23. What is the typical shape of a titration curve that is obtained through the use of potential measurements? How is this type of curve similar to that which is obtained during an acid–base titration when using pH measurements?
24. Explain how a Gran plot can be used to analyze data that are obtained from a redox titration. State how such a plot is prepared and describe the information that it provides.
25. The titration of a 50.00 mL aqueous sample of Fe^{2+} with $0.09553\ M\ Ce^{4+}$ gives the following cell potentials when using a platinum electrode indicating electrode and a Ag/AgCl reference electrode (in 1 M KCl). Use a Gran plot to determine the concentration of Fe^{2+} in this sample.

Volume of Added Titrant (mL)	Measured Cell Potential (V)
32.00	0.600
34.00	0.608
36.00	0.617
38.00	0.629
40.00	0.649
42.00	1.355
46.00	1.444
50.00	1.461

26. Give an example of titrant that can also be used as a visual indicator during a redox titration. What change in color is observed at the end point when using this titrant?
27. What is meant by the term "redox indicator"? Give one specific example of a redox indicator and explain how it works.
28. Explain why the expected potential at the equivalence point for a redox titration is important to consider when you are selecting a visual indicator for detecting the end point of this titration.
29. Over what potential range do you expect the indicator diphenylamine-4-sulfonic acid to change color in a titration using a 0.1000 M dichromate solution as the titrant?
30. Over what potential range do you expect tris(5-nitro-1, 10-phenanthroline) iron will change color when it is used as an indicator during the titration of Fe^{2+} with Ce^{4+}. Is this a suitable indicator for this titration?
31. Describe how starch can be used as a visual indicator for a redox titration. What particular titrant or reagent is actually detected by the starch during this process?

GENERAL APPROACH TO CALCULATIONS FOR REDOX TITRATIONS

32. State why two electrodes are needed when you are using potential measurements to follow the course of a redox titration. What is the function for each electrode in this system?

33. Explain why at least two half-reactions are actually present in the sample/titrant mixture during the course of a redox titration.
34. Explain why the Nernst equation for either of the two half-reactions mentioned in Problem 33 can be used, in theory, to estimate the potential during a redox titration.

ESTIMATING THE SHAPE OF A REDOX TITRATION CURVE

35. What are the four general regions that are present during a redox titration?
36. Explain, in general, how the cell potential can be estimated for each of the four regions in a redox titration curve.
37. Why is it sometimes difficult to estimate the cell potential at the beginning of a redox titration? Why is this more difficult that the calculation of pH at the beginning of an acid–base titration or pM at the beginning of a complexometric titration?
38. A 100.00 mL sample of $0.0200\ M\ Fe^{2+}$ in 0.5 M H_2SO_4 is titrated with a $0.05000\ M\ Ce^{4+}$ solution. The original sample is also believed to contain $1.5 \times 10^{-5}\ M\ Fe^{3+}$. The course of this titration is followed by potential measurements using a platinum indicating electrode and a silver/silver chloride reference electrode.
    (a) Write the half-reactions and overall oxidation–reduction reaction for this process. What volume of titrant will be needed to reach the equivalence point?
    (b) What cell potential is expected before any titrant is added to the sample? What is the expected potential after 1.00% of the analyte has been titrated? What is the expected cell potential after 10.00 mL of titrant has been added?
    (c) What will be the cell potential at the equivalence point? What will be the cell potential after enough titrant has been added to go 20.00 mL beyond the equivalence point?
39. A 10.00 mL aqueous sample thought to contain $0.2547\ M\ Fe^{2+}$ is to be titrated with $0.01543\ M$ $KMnO_4$ in the presence of 1.0 M H_2SO_4 using a platinum indicating electrode and a silver/silver chloride reference electrode.
    (a) Write the half-reactions and overall oxidation–reduction reaction for this process. What volume of titrant will be needed to reach the equivalence point?
    (b) Estimate the cell potential that will be obtained after the titration is 1%, 30%, or 70% complete.
    (c) Estimate the cell potential at the equivalence point of this titration and after a 30% excess of titrant has been added. (*Hint:* See the discussion on permanganate in Section 15.4B.)
    (d) Use the information from Parts (a)–(c) to plot a rough titration curve for this analysis.
40. A 25.00 mL sample containing vanadium is to be titrated with a $0.05746\ M$ dichromate solution in the presence of 1.0 M H_2SO_4 using a platinum indicating electrode and an SCE as the reference. Prior to this titration, all the vanadium in the sample is prereduced to V^{2+}, which is then oxidized to VO^{2+} by dichromate during the redox titration. The resulting pretreated sample solution is thought to contain $0.0750\ M\ V^{2+}$ at the beginning of the titration.
    (a) Write the half-reactions and overall oxidation–reduction reaction for this process. Write the half-reaction for vanadium that occurs when it is prereduced during sample pretreatment.
    (b) What volume of titrant will be needed to reach the equivalence point for this titration?
    (c) Estimate the cell potential that will be obtained after the titration is 1%, 25%, 50%, or 75% complete. Estimate the

cell potential at the equivalence point of this titration and after a 10.00 mL excess of titrant has been added. (*Hint:* See the discussion on dichromate in Section 15.4C.)

(d) Use the information from Parts (b) and (c) to plot a rough titration curve for this analysis.

USING THE FRACTION OF TITRATION

41. Describe how mass balance equations and an equilibrium expression can be used to obtain a fraction of titration equation for a redox titration.

42. Use an approach similar to that in Table 15.4 to derive an equation that gives the fraction of titration for a method in which the titrant is a reducing agent. Use the reaction between T_{red} and A_{ox} in Equation 15.2 for this derivation. Compare your result with Equation 15.22 for the titration of A_{red} with T_{ox} as an oxidixing agent.

43. Use the fraction of titration expression in Equation 15.22 to compare the titration error that will be obtained if the end point is said to occur at +0.900 V (vs. Ag/AgCl) for the titration of Fe^{2+} with 0.0050 M Ce^{4+} or MnO_4^- in Figure 15.10.

44. Use the fraction of the titration equation in Table 15.4 and successive approximations to find E_{Cell} and the corresponding volume of titrant at the equivalence points and half-way points of the titration in Problem 43.

TITRATIONS INVOLVING CERATE

45. Discuss the properties of Ce^{4+} that make it useful in redox titrations.

46. Explain why titrations that make use of Ce^{4+} are usually carried out in an acidic solution.

47. Describe how end point detection can be accomplished during a cerate-based redox titration.

48. Write the half-reactions and overall titration reaction that would be expected for each of the following analytes when they are measured by using Ce^{4+} as the titrant.
(a) Sn^{2+}
(b) Nitrite (NO_2^-)
(c) Hydroxylamine (NH_2OH)

49. What volume of 0.0543 M Ce^{4+} will be needed to reach the equivalence point for a 25.00 mL solution that contains 6.54 mg of dissolved $FeSO_4$?

50. Calculate the volume of Ce^{4+} needed to reach the end point that will be marked by a color change in ferroin when this chemical is used as a redox indicator for the titration in Problem 49. What color change will be observed? What percent of Fe^{2+} in the sample will actually be titrated at this end point?

51. A 5.00 g ore sample containing uranium is dissolved and prereduced to convert all of the uranium species to U^{4+}. This sample solution, which has a total volume of 25.00 mL, is then titrated to give UO_2^{2+} by adding 0.08755 M Ce^{4+}. The end point of this titration is detected after 31.25 mL of titrant has been added. What was the content of uranium in units of % (w/w) in the ore sample?

TITRATIONS INVOLVING PERMANGANATE

52. What properties of permanganate make it useful as a reagent for redox titrations?

53. Explain why titrations that make use of permanganate are carried out in an acidic solution. What happens if permanganate is instead used in a solution that has a neutral or basic pH?

54. Describe how end point detection can be accomplished during a permanganate-based redox titration. What feature of permanganate makes it easier to carry out end point detection in these titrations?

55. Write the half-reactions and overall titration reaction that would be expected for each of the following analytes when they are measured using permanganate as the titrant.
(a) Hydrogen peroxide (H_2O_2)
(b) Bromide (Br^-)
(c) H_3AsO_3

56. State how calculations of cell potential during a permanganate-based redox titration are different from those for a titration that uses Ce^{4+} as the titrant. In what ways are these calculations similar?

57. A 50.00 mL aqueous sample thought to contain 0.01500 M Fe^{2+} in 1.0 M H_2SO_4 is titrated with a 0.01000 M potassium permanganate solution that has also been prepared in 1.0 M H_2SO_4. The course of this titration is followed by potential measurements using a platinum indicating electrode and a silver/silver chloride reference electrode.
(a) Write the half-reactions and overall oxidation–reduction reaction for this process. What volume of titrant will be needed to reach the equivalence point?
(b) What is the expected cell potential at the half-way point of this titration?
(c) What will the cell potential be at the equivalence point?
(d) What will the cell potential be after enough of the permanganate solution has been added to go 10.00 mL beyond the equivalence point?

58. A 10.00 mL sample thought to contain 0.113 M H_3AsO_3 is to be titrated with a 0.02000 M solution of $KMnO_4$. Both the sample and titrant have been prepared in 1 M HCl, so that the products of the titration reaction are Mn^{2+} and H_3AsO_4. The course of this titration is followed by potential measurements using a platinum indicating electrode and a silver/silver chloride reference electrode.
(a) Write the half-reactions and overall oxidation–reduction reaction for this process. What volume of titrant will be needed to reach the equivalence point?
(b) What is the expected cell potential at the half-way point of this titration?
(c) What will be the cell potential at the equivalence point?
(d) What will be the cell potential after enough of the permanganate solution has been added to go 10.00 mL beyond the equivalence point?

TITRATIONS INVOLVING DICHROMATE

59. What properties of dichromate make it useful as a reagent for redox titrations? What properties of dichromate related to health or safety issues might lead to the selection of an alternative reagent for a redox titration?

60. Explain why titrations that make use of dichromate are carried out in an acidic solution. What happens if dichromate is instead used in a solution that has a neutral or basic pH?

61. Describe how end point detection can be accomplished during a dichromate-based redox titration.

62. Write the half-reactions and overall titration reaction that would be expected for each of the following analytes when they are measured by using dichromate as the titrant.
(a) Sn^{2+}
(b) Thiosulfate ($S_2O_3^{2-}$)
(c) UO^{2+} (to form UO_2^{2+})

63. State how calculations of cell potential during a dichromate-based redox titration are different from those for a titration that uses Ce^{4+} or permanganate as the titrant. In what ways are these calculations similar?

64. If you could choose between performing a titration using 0.10 M Ce^{4+}, 0.10 M MnO_4^-, or 0.10 M $Cr_2O_7^{2-}$ as the titrant, which of these reagents would require the greatest volume to reach the equivalence point when they are added to 25.00 mL of 0.10 M Fe^{2+}? Which of these titrants will have to be added in the smallest volume to reach the equivalence point of the titration?

65. A 50.00 mL aqueous sample thought to contain 0.01500 M Fe^{2+} in 1.0 M H_2SO_4 is titrated with a 0.01000 M potassium dichromate solution that has also been prepared in 1.0 M H_2SO_4. The course of this titration is followed by potential measurements using a platinum indicating electrode and a silver/silver chloride reference electrode.
    (a) Write the half-reactions and overall oxidation–reduction reaction for this process. What volume of titrant will be needed to reach the equivalence point?
    (b) What is the expected cell potential at the half-way point of this titration?
    (c) What will the cell potential be at the equivalence point?
    (d) What will the cell potential be after enough of the dichromate solution has been added to go 10.00 mL beyond the equivalence point?

66. To measure the chemical oxygen demand for a water sample, 5.00 mL of the water is treated with 5.0 mL of sulfuric acid with 0.100 M $K_2Cr_2O_7$ and heated until all the organic material has been oxidized to CO_2. The remaining dichromate is titrated with a 0.100 M Fe^{2+} solution. How much dichromate was reduced by the organic material if 17.23 mL of the Fe^{2+} solution is required to reach a good end point?

TITRATIONS INVOLVING IODINE

67. What is an "iodimetric titration"? What properties of iodine make it useful as a reagent for an iodimetric titration?

68. Explain why thiosulfate is often used as a reagent in an iodimetric titration. What is the function of reagent?

69. Describe how starch can be used for end point detection during an iodimetric titration. What is the chemical basis for this means of end point detection?

70. Explain how an iodimetric titration can be used to measure an oxidizing agent such as hydrogen peroxide. Write the reactions that are used for sample treatment and analysis in this method.

71. Write the half-reactions and overall oxidation–reduction reaction for each of the following analytes when they are measured by using an iodimetric titration.
    (a) H_3AsO_3
    (b) Hydrazine (N_2H_4), forming N_2
    (c) Formaldehyde (CH_2O), forming formic acid (HCO_2H)

72. What volume of 0.123 M $Na_2S_2O_3$ will be needed to titrate iodine formed if 100 mL of 0.00456 M ozone reacts with an excess of KI?

73. What is the concentration of Cu^{2+} in a solution if 5.00 mL of it is treated with excess iodide and the resulting iodine requires 17.84 mL of a 0.0548 M $Na_2S_2O_3$ solution to reach a starch end point?

74. What is the Karl Fischer method? What does this method analyze? Explain why this technique is an example of a redox titration and an iodimetric method?

75. A 5.00 g portion of wheat flour is dispersed in anhydrous methanol and titrated by the Karl Fischer method. An end point is observed after 7.84 mL of a titrant that will combine with approximately 5.0 mg of water per mL of titrant is added. To standardize this titrant, it is used to measure a sample known to contain 0.0525 g of water dissolved in 5.00 mL of anhydrous methanol, which requires 10.62 mL of the titrant to reach the end point. What is the concentration of the titrant and what is the percent content of water in the flour sample?

76. Discuss how an iodimetric titration can be used to find the iodine number for a sample that contains an unsaturated fat. Why is bromine often used in place of iodine for part of this analysis?

77. A 11.54 g sample of fat is treated with 25.00 mL of a 0.350 M Br_2 solution. After this mixture has been allowed to react, excess KI is added and the resulting iodine is titrated with 0.576 M $Na_2S_2O_3$. The end point of this titration occurs when 4.87 mL of the titrant has been added. What is the "iodine number" of this fat?

78. Bromate ion reacts with iodide to give iodine and bromide. Write a balanced equation for this reaction. What is the bromate concentration in 25.00 mL of a solution that is treated with an excess of KI and the resulting iodine titrated to a starch end point with 7.24 mL of a 0.0453 M sodium thiosulfate solution?

CHALLENGE PROBLEMS

79. Obtain more information on each of the following topics and write a standard operating procedure for the given task.
    (a) Preparation and handling of a potassium dichromate solution for use as a titrant.
    (b) The preparation and standardization of a potassium permanganate solution.
    (c) Preparation of a starch solution for use as a visual indicator in an iodimetric titration

80. There are many redox indicators besides those listed in Table 15.3. A few examples of other possible indicators are given below. Using a source such as *Lange's Handbook of Chemistry* or a related reference,[6,15] determine the potential range over which each of these indicators might be used and describe the change in color that will be observed.
    (a) Indigo monosulfate
    (b) Erioglaucine A
    (c) *N*-Phenylanthranilic acid
    (d) Tris(2,2'-bipyridine) reuthenium
    (e) Phenosafranine
    (f) Setopaline

81. Look up the iodine values for avocado oil, sesame oil, fish oil, and sheep fat.[6]
    (a) What is the highest iodine value you can find for these fats or oils?
    (b) What happens to the iodine number of a fat when that fat is partially hydrogenated, fully hydrogenated? What else happens to the fat during the hydrogenation process?

82. Cerium(IV) will oxidize vanadium(II) stepwise all the way to vanadium(V). Look up the half-reactions and standard potentials for these species and construct a titration curve for a 0.1000 M solution of Ce^{4+} that is used to titrate a 10.00 mL sample that contains 0.10 M V(II).

83. Permanganate in a strong base (2 M KOH) will slowly oxidize many organic compounds to carbonate ion and form the green manganate(VI) ion (MnO_4^{2-}).

(a) Write a balanced equation for this reaction.

(b) Suppose that 50.00 mL of 0.1000 M $KMnO_4$ reacts with 25.00 mL of a glycerine ($C_3H_8O_3$) solution until all the glycerine has formed carbonate. After this has come to equilibrium, the solution is acidified and the high oxidation states of manganese are titrated with 19.13 mL of 0.1235 M Fe^{2+}. How much glycerine was in the original sample?

84. Imagine that you have a sample that contains significant amounts of both Fe^{2+} and Fe^{3+}. The total concentration of both these species is 0.01000 M, with Fe^{3+} making up 30.0% of this initial concentration. A 25.00 mL portion of this sample is titrated using a 0.1000 M solution of Ce^{4+} at an acidic pH. This titration is followed through potential measurements using platinum as the indicating electrode and a silver/silver chloride reference electrode.

(a) What volume of titrant will be needed to reach the equivalence point for this titration?

(b) What will be the measured cell potential at the beginning of this titration?

(c) What will be the cell potential half-way through this titration? How does this result compare with what you would expect if there was no significant amount of Fe^{3+} in the original sample?

(d) What is the expected cell potential at the equivalence point? How does this result compare to what you would expect if there was no significant amount of Fe^{3+} in the original sample?

(e) What is the cell potential after an excess of 20.00 mL titrant has been added to the sample? How does this result compare to what you would expect if there was no significant amount of Fe^{3+} in the original sample?

85. Potassium ferrate (K_2FeO_4) has received considerable attention for use as a strong oxidizing agent. This chemical reacts very quickly with water in a neutral or acidic solution, but slowly in a base. The purity of potassium ferrate can be determined by treating it with an excess of $Cr(OH)_4^-$ in 1.0 M NaOH and titrating the resulting Cr(VI) with Fe^{2+} in an acidic solution.

(a) Write the half-reactions and overall reaction for the pretreatment step that is involved in this analysis.

(b) Write the half-reactions and the overall titration reaction for this analysis.

(c) Why do you think the sample pretreatment step and this indirect titration approach are required for the analysis of potassium ferrate?

86. In the titration of a sample containing Fe^{2+} with Ce^{4+}, the cell potential of the platinum electrode before adding any cerium is 0.234 V versus SCE. What is the ratio of Fe^{3+}/Fe^{2+} at this point in the titration? How much error will this contribute to the estimated percent iron content of the sample as determined from this titration if the end point is observed after adding 17.64 mL of 0.08765 M Ce^{4+} to the sample?

TOPICS FOR DISCUSSION AND REPORTS

87. Another approach besides a COD analysis for monitoring water pollution is to measure the oxygen needed by microorganisms to oxidize the dissolved organic material in water. This type of measurement gives a value called the *biological oxygen demand* (or *BOD*). Obtain more information on BOD and write a report that describes how this measurement is made. How does the information provided by a BOD analysis compare with the information that is obtained when finding the COD for a water sample? What are the advantages and disadvantages of using BOD versus COD measurements?

88. Locate a water-testing laboratory in your area and learn how workers in this laboratory make and use COD measurements. Discuss your findings with your class.

89. Obtain more information on the Karl Fischer method and its applications. Write a report on your findings.

90. Locate or suggest a procedure for performing a redox titration for one or more of the following situations. Describe the reactions that take place in this method and state the purpose for each step in the procedure.

(a) Determination of the uranium content in an ore sample

(b) Measurement of the amount of iodine in a sample of dried kelp. (*Note:* When dried, kelp is used as a commercial source of iodine.)

(c) Determination of the purity of sodium borohydride ($NaBH_4$), a reducing agent that is commonly used in chemical laboratories and in organic synthesis

91. A well-known chemical demonstration is the "Blue Bottle" reaction. In this demonstration a closed flask holding a colorless aqueous solution that contains glucose, potassium hydroxide, and methylene blue is shaken, causing this solution to turn blue. After a few minutes the solution becomes colorless again, but the blue color can be restored by shaking the flask. This sequence can be repeated many times. Explain what is happening during this demonstration that causes this color change.

References

1. C. N. Sawyer, P. L. McCarty, and G. F. Perkin, *Chemistry for Environmental Engineering and Science*, 5th ed., McGraw-Hill, New York, 2003.

2. L. S. Clescerl, A. E. Greenberg, and A. D. Easton, Ed., *Standard Methods for Examination of Water and Wastewater*, 20th ed., American Public Health Association, Washington, DC, 1999.

3. G. Maludziska, Ed., *Dictionary of Analytical Chemistry*, Elsevier, Amsterdam, the Netherlands, 1990.

4. J. Inczedy, T. Lengyel, and A. M. Ure, *Compendium of Analytical Nomenclature*, 3rd ed., Blackwell Science, Malden, MA, 1997.

5. H. Frieser, *Concepts & Calculations in Analytical Chemistry: A Spreadsheet Approach*, CRC Press, Boca Raton, FL, 1992.

6. J. A. Dean, Ed., *Lange's Handbook of Chemistry*, 15th ed., McGraw-Hill, New York, 1999.

7. H. Diehl, *Quantitative Analysis: Elementary Principles and Practice*, 2nd ed., Oakland Street Science Press, Ames, IA, 1974.

8. K. Kodama, *Methods of Quantitative Inorganic Analysis: An Encyclopedia of Gravimetric, Titrimetric and Colorimetric Methods*, Interscience, New York, 1963.

9. J. Bassett, R. C. Denney, G. H. Jeffery, and J. Mendham, *Vogel's Textbook of Quantitative Inorganic Analysis,* 4th ed., Longman, New York, 1978.

10. I. M. Kolthoff and P. J. Elving, *Treatise on Analytical Chemistry*, Part 1, Vol. II, Wiley, New York, 1975.

11. P. W. Shimer, "A Simplified Reductor," *American Chemical Society Journal*, 21 (1899) 723.

12. F. Szabadvary, *History of Analytical Chemistry*, Pergamon Press, New York, 1966.

13. G. H. Walden, Jr., L. P. Hammett, and S. M. Edmonds, *Journal of the American Chemical Society*, 56(1934) 350.

14. G. Gran, "Equivalence Volumes in Potentiometric Titrations," *Analytica Chimica Acta*, 206 (1988) 111.

15. A. Hulanicki and S. Glab, "Redox Indicators, Characteristics and Applications," *Pure and Applied Chemistry*, 40 (1978) 468–498.

16. R. D. Hancock and B. J. Tarbet, "The Other Double Helix: The Fascinating History of Starch," *Journal of Chemical Education*, 77 (2000) 988.

17. K. Fischer, "A New Method for the Volumetric Determination of the Water Content of Liquids and Solids," *Zeitschrift fur Analytische Chemie*, 48 (1935) 394–396.

18. S. K. MacLeod, "Moisture Determination Using Karl Fischer Titrations," *Analytical Chemistry*, 63 (1991) 557A–566A.

19. R. P. Ruiz, "Karl Fisher Titration." In *Handbook of Food Analytical Chemistry: Water, Proteins, Enzymes, Lipids and Carbohydrates*, 2005, pp 13–16.

20. *Official and Tentative Methods of Analysis*, 4th ed., Association of Official Agricultural Chemists, Washington D.C. 1935.

21. W. C. Forbes and H. A. Neville, "Wijs Iodine Numbers for Conjugated Double Bonds," *Industrial and Engineering Chemistry, Analytical Edition*, 12 (1940) 72–72.

22. H. D. DuBois and D. A. Skoog, "Determination of Bromine Addition Numbers: An Electrometric Method," *Analytical Chemistry*, 20 (1948) 624–627.

Chapter 16

Coulometry, Voltammetry, and Related Methods

16.1 INTRODUCTION: THE DEAD ZONE

The concentration of dissolved oxygen in the ocean, rivers, and lakes is important to the survival of fish and other life in these waters. Oxygen is not very soluble in water, with only $2.9 \times 10^{-4}\,M$ being present in water that is saturated with air at 20°C. This solubility decreases at higher temperatures and if oxidizable solutes such as pollutants are present. Fish utilize dissolved oxygen much as we use oxygen in air as a necessary part of our metabolism. Green plants in the water give off oxygen while living, but remove oxygen when they die and decay. The corresponding lack of sufficient oxygen in water can lead to the creation of a region in the ocean or other bodies of water that is known as a "dead zone." The concentration of dissolved oxygen in natural waters is therefore an important quantity to measure to be able to avoid the creation of such a zone by pollution and to ensure that a proper oxygen content is present for fish and wildlife to live and thrive.[1,2]

It is difficult to take water samples from rivers or the ocean and return them to a laboratory for dissolved-oxygen measurements. This difficulty arises because contact with air or a change in temperature can affect the oxygen content of the sample. As a result, it is instead necessary to make a measurement of the dissolved oxygen levels in the field while the sample of water is still in its original setting. This task is often carried out by using a dissolved-oxygen electrode that measures the reduction of O_2 in water through a method known as *voltammetry* (see Figure 16.1).

In the last two chapters we have focused on an electrochemical method known as potentiometry, in which the measurement of a potential difference is used for chemical analysis. These measurements are all made in the presence of essentially zero current. Such conditions are used in potentiometry to ensure that no appreciable amounts of any oxidation–reduction reactions are taking place in the sample during the measurement, even though a difference in potential might exist that favors such a reaction. In this chapter, we will deal with other methods of electrochemical analysis in which a measurable current does flow between two electrodes, such as in the dissolved-oxygen electrode. Important examples of these methods include electrogravimetry, coulometry, voltammetry, and amperomentry.[3–5]

16.2 ELECTROGRAVIMETRY

Electrogravimetry, or "electrodeposition," is a type of gravimetric analysis where a dissolved analyte is converted into a solid by either oxidation or reduction in such a way that the product is tightly attached to an inert electrode. The increase in mass of the electrode after the analyte has been deposited on it can then be used as a direct measure of the amount of analyte that was originally in the sample.[6,7] Copper and silver are elements that can easily be measured by electrogravimetry through the following reduction half-reactions.

$$Cu^{2+} + 2\,e^- \rightleftharpoons Cu(s) \qquad (16.1)$$

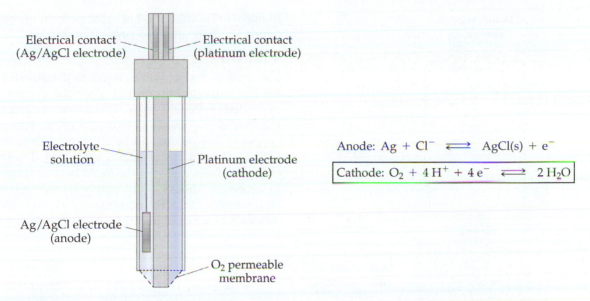

FIGURE 16.1 The general design of an electrode for measuring dissolved oxygen. This type of electrode is also known as a "Clark electrode" and is named after American chemist Leland Clark, who first developed such an electrode in the 1950s. The particular design shown here is an example of a combination electrode (a topic discussed in Chapter 14), in which both the anode and cathode are part of a single sensing device. The Ag/AgCl electrode in this device acts as the anode and reference electrode. The platinum electrode is the cathode and indicator electrode for oxygen that is able to cross the gas-permeable membrane and enter the electrolyte solution that surrounds this electrode.

$$Ag^+ + e^- \rightleftharpoons Ag(s) \qquad (16.2)$$

A necessary feature in this type of analysis is that essentially all of the analyte must be reduced and must attach to the electrode. This means the mass of any solute that is not reduced must be less than the smallest amount that can be detected by the balance, which is usually 0.0001 g.

The electrode employed in electrogravimetry is typically a piece of platinum gauze that has an area of several square centimeters, as illustrated by the system in Figure 16.2. Because reduction is occurring at this electrode, it represents the cathode in this electrochemical cell. In this situation, the cathode is electrically negative so that metal cations will be attracted to it and gain electrons as they convert from soluble ions into an insoluble metal. The other electrode, where oxidation occurs, represents the anode and is also made of platinum. The oxidation product that is created at this anode may involve the formation of oxygen gas from water, as shown in Equation 16.3, or the oxidation of some other component in the sample.

$$2\,H_2O \rightleftharpoons O_2 + 4\,H^+ + 4\,e^- \qquad (16.3)$$

Although reduction is often used in electrogravimetry for the analysis of metal ions, there are cases in which an oxidation reaction can also be employed. An example of a metal ion that can be converted to a solid form by means of oxidation is Pb^{2+}. The oxidation product of Pb^{2+} in an aqueous solution is PbO_2, which will adhere to the anode as lead is oxidized from the +2 state in Pb^{2+} to the +4 state in PbO_2. The half-reaction for this oxidation step is as follows.

$$Pb^{2+} + 2\,H_2O \rightleftharpoons PbO_2 + 4\,H^+ + 2\,e^- \qquad (16.4)$$

When just one type of metal ion is present in a solution, it is relatively easy to select a potential that can be used for electrogravimetry. However, this situation becomes more complicated when more than one type of metal ion is present. This concept can be illustrated by using a mixture of Ag^+ and Cu^{2+} as an example. The standard reduction potential for silver is +0.80 V versus a standard hydrogen electrode (SHE), and for Cu^{2+} it is +0.34 V versus SHE. These values indicate that if Ag^+ and Cu^{2+} are in the same solution, the silver ions will be more easily reduced. Thus, it is possible to have selective electrode position for Ag^+ by using a potential that is high enough to reduce silver ions, but not high enough to reduce copper ions. This approach is called **controlled potential electrolysis**, and the device used to supply the desired potential is known as a **potentiostat**.

During electrodeposition, it is desirable to have the analyte deposited on an electrode in such a way that this deposited material can easily be weighed. This is not a problem for a metal such as copper, which forms a smooth adherent layer on a platinum electrode. However, the deposition of silver from an aqueous solution can result in large crystals that adhere poorly to a platinum electrode and often fall off. A smoother, more adherent silver deposit can be obtained by instead reducing Ag^+ from a

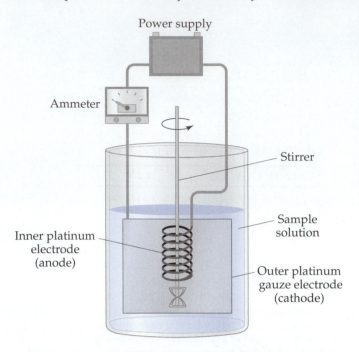

Power supply

Ammeter

Stirrer

Inner platinum electrode (anode)

Sample solution

Outer platinum gauze electrode (cathode)

FIGURE 16.2 A system for the electrogravimetric analysis of metals. This device contains two concentric platinum electrodes. The outside platinum-gauze electrode acts as the cathode and is where a metal ion is reduced and deposited as solid metal. The interior platinum electrode acts as the anode. Mechanical stirring is used to promote the movement of the metal ions from the bulk of the sample solution to the surface of the cathode for reduction.

solution that contains a complexing agent such as cyanide. The net reduction reaction that then occurs is shown in Figure 16.5.

$$Ag(CN)_2{}^- + e^- \rightleftharpoons Ag(s) + 2\,CN^- \qquad (16.5)$$

There are also cases in which analytes can be deposited at both the cathode and anode when using electrogravimetry. This situation occurs during the measurement of copper ions and lead ions in a dissolved sample of brass. In this case, copper ions are reduced and deposited as copper metal at the cathode, according to the reduction half-reaction in Equation 16.1, and lead ions are oxidized and deposited at the anode as PbO_2, according to the oxidation half-reaction in Equation 16.4. After rinsing and drying each electrode, the mass of both metals in the original sample can be determined.

EXERCISE 16.1   Using Electrogravimetry

A new penny having a mass of 2.5133 g is dissolved in nitric acid and the resulting copper ions are plated out onto a platinum cathode that has a mass of 12.0476 g. After all the copper ions have been reduced, the measured mass at the cathode is 12.1454 g. There is no increase

in mass at the anode. What is the percent of copper in the penny? Is there any lead present in the penny?

SOLUTION

The mass of copper will be equal to the difference in mass for the cathode before and after the copper has been deposited on this electrode. This difference gives a mass of (12.1554 g) − (12.0476 g) = 0.1078 g Cu. The percent of copper in this penny will be given by 100 · (0.1078 g Cu)/ (2.5133 g total mass) = **4.25% Cu (w/w)**. The lack of any change in mass at the anode indicates that there is no lead in the penny. The rest of the penny is made up of nearly all zinc, which will not be reduced or oxidized under the conditions that are used in this type of analysis.

It is important to note in electrogravimetry that 100% conversion of the soluble form of the analyte to the solid form is necessary, but 100% use of the applied current for this process is not necessarily required. If some water or other species are also oxidized or reduced, there will be no problem as long as these side reactions do not deposit any solid products on the electrodes. This is not the case in the next method we will consider, a technique known as "coulometry."

16.3 COULOMETRY

Coulometry is a technique that uses a measure of charge for chemical analysis.[6,8] In this method, the amount of an electroactive analyte can be determined based on a measurement of the total coulombs of electricity that are needed to quantitatively oxidize or reduce this analyte. For instance, the half-reaction for the reduction of Ag^+ to silver metal in Equation 16.1 indicates that one mole of electrons is needed for every mole of Ag^+ that is reduced to form $Ag(s)$. If we know the current and amount of time over which this current was applied to carry out this reduction, we can determine how much charge was required and use the Faraday constant (F, where F = a charge of 96,485 C per mole of electrons) to determine the moles of electrons that were needed to attain this current, as discussed in Chapter 14. Thus, we can use information on charge (as obtained from current and time) to measure the amount of Ag^+ that has undergone reduction. An example of this process is given in the following exercise.

EXERCISE 16.2   Analyzing Silver Using Coulometry

A constant current of 5.00 mA is allowed to flow through an electrochemical cell for 528 s as Ag^+ is reduced to silver metal from a 100.0 mL aqueous sample. If all the Ag^+ was reduced and all of the applied current was used for this reduction process, what was the original concentration of Ag^+ in the sample?

SOLUTION

We can first determine how many moles of electrons (n_e) were passed through this system by using Equations 14.1 and 14.2 from Chapter 14, in which I is the current, t is the time, and Q is the corresponding charge.

Equation 14.2: $Q = I \cdot t$

$$= (5.00 \times 10^{-3}\,A)(528\,s) = 2.640\,C$$

Equation 14.1: $Q = n_e F$

$$2.640\,C = n_e(96{,}485\,C/mol)$$

$$n_e = (2.64\,C)/(96{,}485\,C/mol)$$

$$= 2.736 \times 10^{-5}\,mol$$

The reaction in Equation 16.1 indicates each mole of electrons that is consumed will result in the reduction of one mole of silver ions if the electrons are not taking part in any other oxidation–reduction reactions. Thus, the moles of Ag^+ that were reduced will also be 2.736×10^{-5} mol. This information can then be used to find the concentration of Ag^+ in the original sample.

$$\text{Conc. } Ag^+ = (2.736 \times 10^{-5}\,mol\,Ag^+)/(0.1000\,L)$$

$$= 2.74 \times 10^{-4}\,M$$

16.3A Direct Coulometry

The example in the preceding exercise involved the use of both constant current coulometry and direct coulometry. The term **constant current coulometry** refers to the fact that the current is maintained at a constant level during the analysis, while the phrase **direct coulometry** means that the analyte itself is what is being oxidized or reduced during the coulometric analysis. In order for this type of analysis to be accurate, two requirements must be met. First, there must be *100% current efficiency*.[6–8] This term means all the electrons that are passed through the electrochemical cell must be used to oxidize or reduce the analyte. In the previous exercise it was assumed this requirement was met by stating that all the applied current was being utilized to reduce silver ions and nothing else, such as water or hydrogen ions.

A second requirement is that essentially all of the analyte must be oxidized or reduced during the coulometric analysis. It can be challenging to meet both this requirement and the need for 100% current efficiency. In the last exercise, these conditions would require that Ag^+ be the most easily reduced species in the sample, even when its concentration becomes quite low after most of the silver has been reduced. If this is not the case, any other species that can undergo a similar reduction (or oxidation) must first be removed or masked so that they do not interfere in the analysis. Of course, it is theoretically impossible to reduce *all* of the silver ions in the last example, because

chemical equilibrium requires that some be left behind. This small amount will not create a problem as long as it is insignificant compared to the total amount of analyte and still provides the desired level of accuracy for the final measurement.

The progress of the coulometric reduction of Ag^+ can be monitored by potentiometry through the use of a silver electrode (see discussion of "class one electrodes" in Chapter 14). In this case, the difference in potential between the silver electrode and a reference electrode will change as the silver ion concentration decreases (see Figure 16.3). The rapid change in potential as the reduction nears completion can be used to signal when this process should be stopped. The electrochemical system that is used to perform and follow this reduction will actually consist of four electrodes. Two of these electrodes will be used to carry out the reduction of Ag^+ by coulometry, with one electrode acting as the anode and the other as the cathode. There will also be two electrodes that will be used as the indicator electrode and reference electrode for the measurement of any remaining Ag^+ by potentiometry.

16.3B Coulometric Titrations

A **coulometric titration** is a special type of titration in which the titrant is generated by means of coulometry and in the presence of the analyte.[6,8] This method is in contrast to direct coulometry, in which electrons are used to directly reduce or oxidize the analyte. A good example of a coulometric titration is the determination of ascorbic acid, or vitamin C. Vitamin C ($C_6H_8O_6$, see structure given below) is found in many fruits and vegetables and is one of the more popular additives in food products. It is a moderately strong organic acid and a good reducing agent.

Ascorbic acid   Dehydroascorbic acid

It is difficult to measure vitamin C by direct coulometry, but this analyte can be measured by using a coulometric titration that uses iodine as an oxidizing agent. Equation 16.6 shows the reaction of vitamin C with iodine, in which the reduced product of vitamin C is dehydroascorbic acid ($C_6H_6O_6$).

$$C_6H_8O_6 + I_3^- \rightarrow C_6H_6O_6 + 2\,H^+ + 3\,I^- \qquad (16.6)$$

The volatility of I_2 makes it difficult to prepare and use in standard solutions for titrations, so an excess of I^- is also

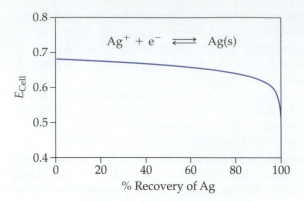

FIGURE 16.3 Expected change in the applied potential during the conversion of Ag^+ to $Ag(s)$ for analysis by electrogravimetry. This curve was calculated for an experiment conducted at 25°C and assumes that all of the current passing through the system goes toward the reduction of Ag^+ in a solution that initially contains this metal ion at a concentration of 0.010 M. The cell potential shown was calculated by utilizing the Nernst equation for this system and a standard hydrogen electrode as the reference electrode.

added to combine with I_2 and form the triiodide ion (I_3^-), which acts in Equation 16.6 as the actual titrant. A successful way of titrating something with iodine is to generate the I_2 (and I_3^-) through the oxidation of I^-. This production is carried out quantitatively between two platinum electrodes by controlling the current and time during which the oxidation of I^- is allowed to occur. The reactions that lead to the eventual formation of I_3^- during this process are given in Equations 16.7 and 16.8.

$$2\,I^- \rightleftharpoons I_2 + 2\,e^- \qquad (16.7)$$

$$I_2 + I^- \rightleftharpoons I_3^- \qquad (16.8)$$

At the cathode, the corresponding reduction half-reaction that typically occurs in the presence of an aqueous solution is the reduction of hydrogen ions from the surrounding water.

This titration is carried out by generating iodine by means of coulometry in the presence of vitamin C and by using starch as a visual indicator. When ascorbic acid is still present, the titrant reacts with it as quickly as iodine, and thus, I_3^- is formed. When no more vitamin C is present, the excess triiodide will react with starch to give the characteristic blue color that marks the end point. The time over which the current has been applied to reach the end point is then noted and used to determine how much analyte was present. An example of this type of analysis is provided in the color figures in the middle of this text. The calculations used in this approach are illustrated in the next exercise.

EXERCISE 16.3 | **Coulometric Titrations**

The vitamin C content in a 25.00 mL sample of a fruit drink is analyzed by a coulometric titration using I_3^- as the titrant and starch as the indicator. A current of 25.00 mA requires 6 min and 17 s to reach an end point during this titration. What is the concentration of vitamin C in the fruit drink (in units of g/L)?

SOLUTION

The moles of vitamin C can be found by using the fact that each mole of I_2 that is generated up to the end point will give one mole of I_3^-, which reacts with one mole of vitamin C. We can determine how many moles of I_2 were generated up to the end point by using the current and the exact amount of time this current was applied. The mass and concentration of vitamin C in the sample can then be found from this information, as shown below.

Mass of vitamin C = $(25.00 \times 10^{-3}\ C/s) \cdot$
$(377\ s) \cdot (1\ mol\ e^-/96{,}485\ C) \cdot$
$(1\ mol\ vitamin\ C/mol\ I_2) \cdot$
$(1\ mol\ I_2/2\ mol\ e^-) \cdot$
$(176\ g\ vitamin\ C/mol)$
$= 0.008596\ g\ vitamin\ C$

Conc. vitamin C = $(0.008596\ g)/(0.02500\ L)$ = **0.3438 g/L**

Another highly accurate coulometric titration is the titration of either strong or weak acids through the coulometric generation of hydroxide ions. These hydroxide ions are produced through the reduction of water, as given by the following reduction half-reaction.

$$2\,H_2O + 2\,e^- \rightleftharpoons H_2 + 2\,OH^- \qquad (16.9)$$

When carried out carefully, six significant figures can be achieved in the final answer when using this method. The end point of this acid–base titration can be detected either through the use of an acid–base indicator or by using a pH electrode. As this and the previous example demonstrate, a great advantage of coulometric titrations compared to a volumetric titration is that there is no need to prepare or keep standard solutions of a titrant. Instead, the titrant is now generated in a known amount, as needed during the analysis.

Silver coulometry was an early method used for determining the value of the Faraday constant. A weighed mass of silver was dissolved and then reduced back to silver metal by using constant current coulometry. The product of the current and time was used to

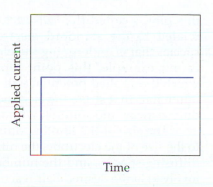
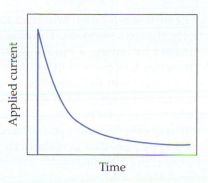

FIGURE 16.4 Typical plots of applied current versus time in constant current coulometry (left) and constant potential coulometry (right).

determine the number of coulombs that were needed to react with a known amount of silver. Because the atomic mass of silver was also known quite accurately from chemical methods, the Faraday constant could then be found by dividing the number of coulombs that were used by the moles of silver that had been reduced. This approach led to a value for the Faraday constant of 96,485 C/mol that is still used to this day, although more advanced and modern measurements have supplied additional significant figures for this value.

16.3C Constant Potential Coulometry

The potential needed for efficient oxidation or reduction in coulometry will change as the concentration of the analyte decreases. This can be illustrated through the Nernst equation.

Nernst Equation at 25°C:

$$E = E° - \frac{0.05916 \text{ V}}{n} \log\left(\frac{a_{Red}}{a_{Ox}}\right) \qquad (16.10)$$

In the case where an analyte is undergoing reduction, the ratio of $(a_{Red})/(a_{Ox})$ will increase as the analyte is reduced by coulometry. The decrease in this ratio means that a larger term is now being subtracted from $E°$, or that the electrode potential is becoming more negative. If another component of the sample can also be reduced at this lower potential, species other than the analyte will now begin to utilize some of the applied current.

A solution to this problem is to maintain the potential at a fixed value as the coulometry is carried out. This approach is known as **constant potential coulometry**. One complicating feature of this approach is that the current will now decrease as the coulometry is allowed to proceed and works with only a single species that will be decreasing in concentration over time. Under these conditions, the

charge that has been passed through the system will no longer be the simple product of current and time, but rather the integrated area of a current versus time plot (see Figure 16.4). This should not be surprising, because charge is defined as the integral of electrical current over time,[6,7] as discussed in Chapter 14.

16.4 VOLTAMMETRY AND AMPEROMETRY

Another important set of methods for electrochemical analysis are those in which current is measured as the potential is controlled. Techniques in this group include voltammetry and amperometry. **Voltammetry** is a method in which a current is measured as the potential is changed as a function of time. **Amperometry** is a method in which current is measured at a constant potential.[1] Many types of voltammetry are based on the reduction of analyte, but in some cases an oxidation process can also be employed. One example is the use of a dissolved oxygen electrode to measure O_2 in water, as we discussed at the beginning of this chapter.

16.4A Direct Current Voltammetry

The simplest kind of voltammetry is one in which the potential is gradually increased (or "ramped") from zero to a more negative value. This method is called "direct current voltammetry" or **DC voltammetry**. In order for a particular redox reaction to be studied by this method, or by any other type of voltammetry, the electroactive species of interest must be at the surface of the electrode when a potential is applied that is suitable for the desired redox reaction to occur. Of course, most species are out in solution far from the electrode surface. There are three means by which a solute ion can arrive at the electrode: convection, migration, and diffusion.[3,5]

Convection implies that the solvent is moving, which usually means the solution is being stirred. Thus, convection can be eliminated by simply using a quiet, unstirred solution. Migration occurs if the electrode and the analyte are oppositely charged. For instance, a cation will tend to migrate toward a negatively charged electrode. The effect of migration can be minimized by having a much higher concentration of nonelectroactive ions present in solution. Voltammetry is usually carried out in the presence of a high concentration of an inert salt such as KCl, which is used as a "supporting electrolyte." Diffusion is represented by the random motion of dissolved ions and solutes through the solvent. It is this process that is usually the desired mechanism of transport in voltammetry.[3,5]

An example of an analytical application of voltammetry is the use of this method to examine the concentration of cadmium ions in wastewater. The reduction half-reaction for this process is as follows.

$$Cd^{2+} + 2\,e^- \rightleftharpoons Cd(s) \qquad (16.11)$$

Cadmium ions in a homogeneous sample of the wastewater will diffuse in all directions at a variety of speeds. A small fraction will hit the electrode, where they will have a chance of being reduced. If the electrode is not sufficiently negative, the cadmium ions will simply diffuse away again and not be reduced. However, as the applied potential at the electrode is made more negative, there is a good chance that the cadmium ions will be reduced as they reach the electrode's surface. This reduction will cause current to flow through the electrode and electrochemical cell.

A typical plot of current versus applied potential that is obtained for voltammetry (giving a graph known as a **voltammogram**) is shown in Figure 16.5. The current at the plateau of such a voltammogram is called the *limiting diffusion current* (I_d) and implies that 100% of the analyte hitting the surface is undergoing an oxidation–reduction reaction. In our example, all of the cadmium ions hitting the electrode at this point are being reduced to cadmium metal, which then adheres to the electrode. The potential half-way up the wave in

this plot is called the *half-wave potential* ($E_{1/2}$) and is related to the standard electrode potential for the species that is undergoing the electrochemical reaction. In our example, this point represents a situation in which the applied potential will reduce half of the cadmium ions that strike the electrode, while the other half diffuse away into solution without being reduced.[3,5]

The size of the limiting diffusion current is related to the size of the electrode, the diffusion coefficient of the diffusing species, and the number of electrons that are involved in the desired half-reaction. The size of this current is also directly proportional to the concentration of the analyte that is being examined in the solution. Thus, both qualitative information on the identity of this species (through the half-wave potential) and quantitative information on concentration (through the limiting diffusion current) can be obtained through this analysis. The general relationship between the limiting diffusion current and analyte concentration (C) is given by Equation 16.12.

$$I_d = k \cdot C \qquad (16.12)$$

In this equation, the proportionality constant k is related to the analyte diffusion coefficient, electrode properties, number of electrons being transferred, and so on.[3,5] The exact value of this constant does not have to be known as long as it is constant during the analysis of the samples and standards. It is also necessary when using Equation 16.12 to subtract the background and charging current from the overall measured current at the plateau of the voltammogram in order to obtain the correct value for I_d.

EXERCISE 16.4   Using Voltammetry

A standard solution that contains 3.50×10^{-3} M $CdCl_2$ is examined by voltammetry and gives a limited diffusion current at the plateau of 65.3 mA. An unknown water sample that is also thought to contain cadmium ions is analyzed under identical conditions and gives a limiting diffusion current of 45.3 mA. What is the concentration of Cd^{2+} in the sample?

SOLUTION

One way of solving for the concentration of Cd^{2+} in the unknown is to use the results for the standard to find the value of k in Equation 16.12. Rearranging Equation 16.12 to solve for this constant gives $k = I_d/C = (65.3\text{ mA})/(3.50 \times 10^{-3}$ M $Cd^{2+}) = 1.86\underline{6} \times 10^4$ mA/M. We can then use this value with Equation 16.12 and the measured current for the unknown to find the concentration of Cd^{2+} in the sample.

$$I_d = k \cdot C \Rightarrow 45.3\text{ mA} = (1.86\underline{6} \times 10^4\text{ mA}/M)\,C$$
$$\therefore C = \mathbf{2.43 \times 10^{-3}\ } \boldsymbol{M}$$

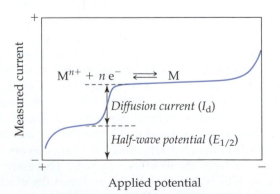

FIGURE 16.5 Examples of a general voltammogram for DC voltammetry. The size of the diffusion current (I_d) is related to the concentration of the analyte that is being reduced. The location of the half-wave potential ($E_{1/2}$) is related to the standard reduction potential for this analyte.

The same answer is obtained by using Equation 16.12 to set up a simple ratio between the measured currents for the standard and unknown sample, where $C = (3.50 \times 10^{-3} M)(45.3 \text{ mA}/65.3 \text{ mA}) = 2.43 \times 10^{-3} M$. Both approaches assume that the reduction of Cd^{2+} is the only source of the limiting diffusion current during this measurement and that all other sources of current have been accounted for during the measurement of I_d.

TABLE 16.1 Half-Wave Potentials for Cd^{2+} and Zn^{2+} versus a Saturated Calomel Electrode at 25°C*

Solution Conditions	$E_{1/2}$ for Cd^{2+} (V)	$E_{1/2}$ for Zn^{2+} (V)
1 M NaOH	−0.78	−1.53
2 M acetic acid/ammonium acetate	−0.65	−1.1
1 M KCl	−0.64	−1.00
1 M Na citrate + 0.1 M NaOH	−1.46	−1.43
1 M NH_3 + 1 M NH_4Cl	−0.81	−1.35

*Based on data from L. Meites, *Polarographic Techinques*, Interscience Publishers, New York, 1955.

There are two types of current that can be present during this measurement. The current that is created by the oxidation or reduction of the analyte or some other electroactive species is known as the *Faradaic current*. This is the current we wish to measure and relate to the concentration of an analyte that is undergoing oxidation or reduction in our sample. However, there is also a current that is produced when we first change the applied potential of an electrode. This "non-Faradaic" current is created by a charging of the electric double layer at the electrode solution interface when we change the potential and is known as the *charging current* or "double-layer current." This charging current is produced because the electrode and solution immediately in contact with this electrode act as a capacitor as the potential is changed and as an electric double layer of ions builds up at this interface. For instance, an electrode that is placed at a more negative potential will attract positively charged ions from the supporting electrolyte. This positive charge attracts negative ions in another diffusion layer until we eventually obtain the same composition of ions that we have in the bulk solution far away from the electrode. This accumulation of ions represents only a movement of charge and is not associated with an oxidation–reduction reaction. Thus, the Faradaic current we wish to measure must be greater than the charging current for us to make a useful measurement of an analyte by voltammetry (see Box 16.1 and Figure 16.6).[3,5,9]

Instead of using just two electrodes, voltammetry is usually performed using a system with three electrodes. First, there is a **working electrode** at which the reduction (or oxidation) of the analyte is carried out. This electrode is made from a material that is inert and that will not be oxidized or reduced itself as the analyte undergoes a redox reaction. For many years the working electrode was often liquid mercury, but now most modern voltammetry is done using a solid electrode that is made from a material like platinum, gold, or carbon. Second, there is a reference electrode that is used to control and set the potential of the working electrode. Enough current flows during an experiment in voltammetry that if that same current also goes through the reference electrode it would cause a chemical change within this electrode and change its potential over time. To overcome this problem, a third electrode is used that is called the **auxiliary electrode** or "counter electrode." It is this third electrode that is used

to pass current and to provide the complementary half-reaction to that taking place at the working electrode (for instance, oxidation would occur here if reduction is happening at the working electrode). It is possible in this way to provide a complete electrical circuit without running the risk of changing the properties of the reference electrode over time.[3,5]

It is important to note that the solution composition and pH can cause the half-wave potential to be different for even seemingly simple situations. Table 16.1 shows the different half-wave potentials for a few solutes. If more than one reducible species is present in solution and if their half-wave potentials differ sufficiently, separate waves for these species can be seen. For instance, Cd^{2+} and Zn^{2+} in an acetate buffer will show two well-separated waves in DC voltammetry that can be used to measure both these species.

16.4B Amperometry

In the method of voltammetry, current is measured as the applied potential is changed over time. The measurement of current when the working electrode is held at a suitable, constant potential is called *amperometry*.[6] For example, an amperometric titration is carried out by measuring the current that is associated with the reduction (or oxidation) of an electroactive solute or titrant during the course of a titration. An example of this approach is the precipitation titration of lead by using chromate as the titrant.

$$Pb^{2+} + CrO_4{}^{2-} \rightleftharpoons PbCrO_4(s) \qquad (16.13)$$

If we measure the current that is associated with the reduction of Pb^{2+} as chromate is added, we will get a curve that approaches zero current as the lead ion concentration approaches zero. A useful feature for this titration is that one does not need data specifically at the end point. Instead, extrapolation of the response before and after the equivalence point can be used to determine the end point.

BOX 16.1
Cyclic Voltammetry

There are many ways in which the potential can be varied during an experiment that uses voltammetry. One popular approach that many electrochemists use for preliminary studies is the method of *cyclic voltammetry*. This is a type of voltammetry in which the potential is scanned back and forth in a linear fashion over time. The current that is produced by a sample is then measured during this scan at an electrode under conditions in which no convection is present in the solution. The result of this type of experiment is shown as a plot of the measured current versus applied potential and is called a *cyclic voltammogram*.

A typical cyclic voltammetry experiment is shown in Figure 16.6. In this particular case, only the oxidized form of the analyte is initially present at any significant concentration in solution. The experiment begins at a starting potential that is above the expected standard reduction potential ($E°$) for the analyte. This potential is ramped in a linear manner toward a more negative value (the "forward scan") while the resulting current at the working electrode is measured. As the potential approaches $E°$ for the analyte, this chemical will be reduced and a positive "cathodic" current will be measured. This reduction continues as the potential is made even more negative, but the amount of analyte that can reach the electrode's surface soon becomes limited by diffusion and causes the current to decrease. A switch is then made and the potential is ramped back in a positive direction as part of a "reverse scan." This scan causes the analyte that has been reduced at the electrode to be reoxidized as the applied potential again approaches $E°$ for the analyte, producing a negative "anodic" current. This current eventually approaches zero as the potential is further increased and the amount of reduced analyte at the electrode is depleted.

Cyclic voltammetry can provide a variety of information on an oxidation–reduction reaction. For example, the number of waves that are observed can indicate how many oxidation–reduction events are occurring for an analyte. The location of these peaks will give the potentials at which these events occur. The difference in potential between the cathodic and anodic peaks is related to how many electrons are involved in an oxidation–reduction reaction. A comparison of the size of the peak cathodic current and peak anodic current will indicate whether the reaction is fully reversible or if the reduced and oxidized forms of the analyte have any side reactions. Changing the rate of the forward and reverse scans can also provide information on the rates of these side reactions.[5,8,9]

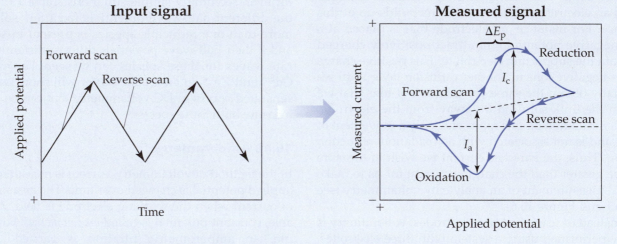

Input signal   **Measured signal**

FIGURE 16.6 A general signal input and plotted response for cyclic voltammetry. The results shown in this case are for a reversible oxidation–reduction reaction with no side reactions. The peak currents that are measured for the cathodic current (I_c) and anodic current (I_a) are related to the amount of analyte that was reduced and oxidized again during the forward and reverse scans, respectively. The difference in the peak potentials (ΔE_p) is related to the number of electrons that are involved in this oxidation–reduction process.

EXERCISE 16.5   Using Amperometry

The half-wave potential for the reduction of Pb^{2+} to $Pb(s)$ occurs at about −0.5 V versus SHE. An amperometric titration is carried out at −0.7 V versus SHE on a 50.00 mL sample containing Pb^{2+} and using chromate as the titrant. The following results were obtained when this sample was titrated with 0.0654 M Na_2CrO_4. What was the original concentration of Pb^{2+} in the sample?

Titrant Volume (mL)	Measured Current (mA)
0.00	43.7
5.00	32.6
10.00	21.5
15.00	10.4
20.00	0.0
25.00	0.0

SOLUTION

When we prepare a plot of current versus titrant volume, the current reaches a value of 0.0 mA at 19.68 mL of added titrant. At this point, we have added $(0.01968 \text{ L})(0.0654 \text{ M CrO}_4{}^{2-}) = 0.001287 \text{ mol CrO}_4{}^{2-}$. The reaction in Equation 16.13 shows that Pb^{2+} will react with $CrO_4{}^{2-}$ in a 1:1 ratio, so 0.001287 mol of Pb^{2+} must also have been present in the original sample. The concentration of Pb^{2+} would then be $[Pb^{2+}] = (0.001287 \text{ mol})/(0.05000 \text{ L}) = \mathbf{2.57 \times 10^{-2} \, M}$.

A well-known analytical method that uses an amperometric detection scheme is the "Karl Fischer method" for measuring water in a sample, a technique that was discussed in Chapter 15. The key reaction in the Karl Fischer method is carried out in a methanol solution containing pyridine. Water from the sample is the limiting reagent that is titrated with a reagent that consists of a standard solution of SO_2 dissolved in methanol with pyridine and iodine. The overall titration reaction is shown in a simplified form below and is discussed in more detail in Chapter 15.

$$H_2O + SO_2 + I_2 \rightarrow 2 \, HI + SO_3 \qquad (16.14)$$

This titration is conducted in the presence of two electrodes, each of which has a controlled potential such that current will only flow when both iodine and iodide are present in solution. The Karl Fisher reagent is added until all the water has reacted. At that point, the next drop of titrant results in the presence of excess iodine, so that now both I_2 and I^- are present in the solution. Under these conditions a significant current will now be present, signaling that the end point of the titration has been reached. Because the potential of both electrodes is controlled in this approach, this method of detection is called *biamperometry*.[6,8]

Probably the most common type of analysis that is carried out by amperometry is the measurement of dissolved oxygen (see Figure 16.1). Oxygen is an excellent oxidizing agent and can easily be reduced at an electrode. At an applied potential that is more negative than about -1.5 V versus SHE, oxygen will be reduced to water in a four-electron process, as shown in Equation 16.15. This reduction gives a diffusion current that is proportional to the dissolved oxygen concentration.

$$O_2 + 4 \, H^+ + 4 \, e^- \rightleftharpoons 2 \, H_2O \qquad (16.15)$$

Special instruments designed to measure dissolved oxygen use a gold electrode that is covered with a thin plastic membrane. The membrane allows dissolved gasses to pass through, but prevents ions or large molecules from reaching the working electrode. Dissolved oxygen electrodes are often supplied with a long connecting cable and usually include a temperature sensor. Such features allow these electrodes to be used for the direct measurement of

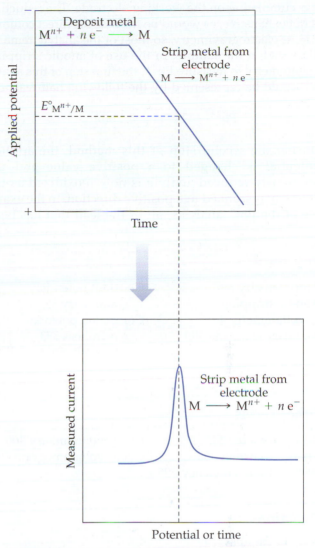

FIGURE 16.7 An illustration of the use of anodic stripping voltammetry for the reduction and later oxidation of a metal ion from a sample solution. The top graph shows how the applied potential is changed over time during this analysis. The bottom graph gives an example of the peak in current that is produced as the deposited metal is oxidized back into the form of metal ions. The size of this peak can be used with the analysis of similar standards for measuring the amount of the given metal ions in the original sample. This method can also allow several metal ions to be examined in one run if the corresponding metals have sufficient differences in their standard reduction potentials.

dissolved oxygen and temperature deep in the ocean, lakes, rivers, or wells.[6,9]

16.4C Anodic Stripping Voltammetry

Anodic stripping voltammetry is a combination of coulometry and voltammetry that is employed when measuring trace metal ions.[4,9] In this method, the working electrode is first set at a potential that is suitable for reduction of the analyte (see Figure 16.7). Reduction is allowed to occur at this potential for several minutes in a stirred solution. During this time, the reduction product is accumulating on the working electrode. The reduction is not exhaustive, as would be the case for direct coulometry or electrogravimetry, so most of the analyte remains dissolved in the sample. In the use of anodic stripping voltammetry to measure Pb^{2+}, the first step of this analysis would be represented by the following half-reaction.

$$Pb^{2+} + 2\,e^- \rightleftharpoons Pb(s) \qquad (16.18)$$

During the second step of this method, the applied potential is changed to a positive value and the previously reduced analyte is now reoxidized as the potential is scanned in a positive direction. In the example of the Pb^{2+} analysis, all of the lead that has been accumulated at the electrode during the reduction step is now oxidized at a higher potential. The current and time required for reoxidation are then determined and used to determined how many moles of lead had been placed onto the electrode during the first step. This process is carried out for both the sample and a set of standards under the same set of analysis conditions. A comparison of these results then makes it possible to determine the concentration of the analyte in the sample. Simple mixtures of two or three low concentration and electroactive analytes in the range of 10^{-8} to 10^{-10} M can be examined by this method if these species can be reduced together, but have sufficiently different oxidation potentials.

Key Words

Amperometry *397*	Constant potential	Coulometry *394*	Voltammetry *397*
Anodic stripping	coulometry *397*	DC voltammetry *397*	Voltammogram *398*
voltammetry *401*	Controlled potential	Direct coulometry *395*	Working electrode *399*
Auxiliary electrode *399*	electrolysis *393*	Electrogravimetry *392*	
Constant current	Coulometric titration *395*	Potentiostat *393*	
coulometry *395*			

Other Terms

Biamperometry *401*	Cyclic voltammetry *400*	Limiting diffusion	Faradaic current *399*
Charging current *399*	Cyclic voltammogram *400*	current *398*	Half-wave potential *398*
100% Current efficiency *395*			

Questions

ELECTROGRAVIMETRY

1. What is "electrogravimetry"? How does this approach differ for a traditional gravimetric analysis?
2. Describe how electrogravimetry could be used for the analysis of metal ions such as Cu^{2+} or Ag^+.
3. How many moles of Cu^{2+} will be reduced during electrogravimetry if a current is 5.0 mA flowing for 7 min and 36 s?
4. What mass of copper will be reduced when electrogravimetry is performed at an appropriate potential on 150 mL of a 0.0764 M solution of $CuSO_4$?
5. A 250.0 mL portion of a solution containing copper ions and lead ions is subjected to electrogravimetric analysis. The cathode original mass is 23.9854 g and the anode mass is 10.6489 g. When the analysis is complete, the two electrodes have masses of 24.5673 g and 10.9858 g, respectively. What was the concentration of copper ions and lead ions in the original solution?
6. A 4.5631 g brass sample containing only copper and zinc is analyzed by electrogravimetry. The mass of the cathode increases by 3.7618 g and the mass of the anode does not change. What is the composition of the sample?
7. A 1.2764 g portion of copper ore was dissolved in acid, filtered to give a blue colored solution, and diluted to 250 mL. The solution was then subjected to electrogravimetry. The original mass of the platinum electrode was 15.7649 g and after deposition was complete, it had a mass of 16.0467 g. What is the percent copper in the ore?
8. What is the expected precision of the silver ion concentration in a sample if a 25.00 mL pipet is used to deliver the sample for electrogravimetry if the mass of the cathode increases from 27.8645 to 28.7654 g?
9. What is meant by the phrase "controlled potential electrolysis"? Explain how this term is related to the method of electrogravimetry.
10. What is a "potentiostat"? What is the function of a potentiostat?

DIRECT COULOMETRY

11. What is "coulometry"? What is measured in this method and how is this information used for chemical analysis?
12. What is meant by the phrase "100% current efficiency"? Why is it important to have 100% current efficiency in coulometry but not in electrogravimetry?
13. Define what is meant by the terms "direct coulometry" and "constant current coulometry." Give an example of an analysis that makes use of these approaches.
14. A 25.00 mL portion of a nickel ion solution and a 25.00 mL portion of a silver ion solution each require the same amount of time to be fully reduced at the same constant current. Is it therefore correct to state that the concentrations of the two solutions are the same? Explain.

15. How many coulombs are necessary to give essentially 100% reduction of the Ag^+ that is present in a 100.0 mL solution of 0.100 M $AgNO_3$?

COULOMETRIC TITRATIONS

16. Describe how a typical coulometric titration is performed. How does this differ from a more traditional volumetric titration?

17. Analysis of a solid sample (250 mg) containing vitamin C gives an end point with starch as the indicator after 6 min and 24 s when analyzed by a constant current coulometric titration with iodine at a current of 30.00 mA. What is the percent vitamin C in the sample?

18. A forensic chemist wishes to measure the concentration of EDTA in a solution by coulometric generation of copper ions from copper metal. What is the EDTA concentration in a 100.0 mL portion of the sample if 198.5 s is needed to reach an end point when the current is 0.01000 A?

CONSTANT POTENTIAL COULOMETRY

19. Using the Nernst equation, explain why the potential changes when you are performing constant current coulometry.

20. What is "constant potential coulometry"? What are the advantages and disadvantages of this method compared to constant current coulometry?

21. The area beneath a current versus time graph of constant potential ($E = -0.320$ V) coulometric determination of nickel ions ($E° = -0.236$ V) in the presence of cadmium ions ($E° = -0.403$ V) is 458 A·s. The total solution volume is 250 mL. What can be said about the concentrations of these two ions in the solution?

DIRECT CURRENT VOLTAMMETRY

22. What is "voltammetry"? What parameter is measured in this method? What parameter is varied or controlled in this method?

23. What are three ways that a solute can arrive at the surface of an electrode during voltammetry?

24. What is DC voltammetry? How is this method performed and how can it be used in a chemical analysis?

25. Define or describe each of the following terms and explain how they are used in voltammetry.
(a) Voltammogram
(b) Limiting diffusion current
(c) Half-wave potential

26. What is the difference between a Faradaic current and a charging current? How are each of these currents created? Which of these currents can be related to the concentration of an electroactive analyte?

27. Voltammetry can distinguish different oxidation states of the same element. For instance, CrO_4^{2-} is reduced to Cr^{3+} at a potential of 1.33 V, and Cr^{3+} is reduced to Cr^{2+} at −0.41 V, and Cr^{2+} is further reduced to $Cr(s)$ at −0.91 V. A voltammogram shows a limiting diffusion current of 34.5 mA at a potential where CrO_4^{2-} is reduced, 46.0 mA at a potential where Cr^{3+} is also reduced, and finally 69.0 mA at a potential where Cr^{2+} is reduced as well. What was these composition of the original sample in terms of these three soluble oxidation states? (*Note*: For the purpose of this problem, you can assume that the reaction of CrO_4^{2-} with Cr^{2+}

to give Cr^{3+} can be ignored; however, in practice this reaction will occur fairly rapidly and make it difficult to see all three waves for the given species in a voltammogram.)

28. Calculate the concentration of hydrogen peroxide (H_2O_2) in an aerated water sample at 20°C if a voltammogram shows a first wave having a diffusion current of 43.5 mA and a second wave of 104.6 mA total diffusion current.

29. A solution containing both copper ions and silver ions was subjected to voltammetry. Two cathodic waves were seen, the first having a diffusion current of 12.4 mA and the second having a total diffusion current of 34.2 mA. Which wave corresponds to which metal? What are the relative concentrations of the two metal ions? You may assume that the two ions have equal diffusion coefficients.

30. The charging current in a voltammetric measurement is 0.065 mA and a 2.5×10^{-2} M solution of $CdCl_2$ gives a diffusion current of 56.8 mA during this measurement. What is the limit of detection for this analysis if this detection limit is equal to the concentration of $CdCl_2$, which gives a Faradaic current that is three times the charging current?

31. Explain why a three-electrode system is used during voltammetry. What is the function for each of the electrodes within such a system?

32. Why is it important to control the pH and solution composition during a measurement that is based on voltammetry?

AMPEROMETRY

33. What is "amperometry?" Explain how this can be used as a tool to perform an amperometric titration.

34. Describe the Karl Fischer method. Give the titration reaction for this process and explain how the end point is detected.

35. Explain how amperometry is used to determine the concentration of dissolved oxygen in water.

36. A dissolved-oxygen probe is lowered into a deep lake from a canoe. This probe reads 8.0 ppm at the surface and for the first 15 ft of depth, but then suddenly changes to 2.3 ppm at 20 ft and the value continues to be decrease from 2.3 to 1.3 ppm when the probe reaches the lake bottom at 70 ft. Explain these observations.

ANODIC STRIPPING VOLTAMMETRY

37. Describe the method "anodic stripping voltammetry" and state how it is used for chemical analysis.

38. In an anodic stripping measurement of cadmium, 100 mL of sample solution is subjected to electrolysis for 500 s to reduce Cd^{2+} to elemental cadmium. The cadmium that is stripped is found to require 4.0×10^{-6} coulombs of charge to be reoxidized to Cd^{2+}.
(a) What mass of cadmium was reduced in the electrolysis step?
(b) How many coulombs of charge would be required if the electrolysis instead went on for 1000?
(c) If the original solution had instead contained 4.5×10^{-8} M Cd^{2+}, what fraction of this original cadmium would have been reduced and later reoxidized?

39. The amount of lead ions in water is to be measured by using anodic stripping voltammetry. A 100 mL portion of sample solution is preelectrolyzed for exactly 10.0 min. Upon anodic stripping, the area of the peak corresponding to lead is 17.5×10^{-7} A·s. A standard lead solution (5.0×10^{-8} M) run in the same manner gave a peak area of

27.8×10^{-7} A·s. What is the concentration of lead ions in the sample? What fraction of the original lead was reduced in the preelectrolysis step?

CHALLENGE PROBLEMS

40. Why do you think it is important in electrogravimetry to use electrodes with large surface areas, although this is not important in coulometry or voltammetry?

41. Will an error be introduced in a coulometry experiment if distilled water is occasionally squirted in to rinse everything into the solution? Answer the same question if the measurement is done by voltammetry. Explain.

42. The dissolved oxygen concentration in a major river is measured at several positions upstream and downstream of a power plant that uses river water to remove excess heat. The following results are measured, all at a depth of 1.0 meter in the center of the river. Explain the difference in these measurements.

Position	Concentration of Dissolved O_2 (ppm)
500 m upstream	7.3
25 m downstream	3.0
500 m downstream	3.2
2000 m downstream	7.0

TOPICS FOR DISCUSSION AND REPORTS

43. Obtain more information from the literature or the Internet on an oxygen-sensing probe that is used in the measurement of biological oxygen demand (BOD). Explain how this probe works.

44. Find out what the dissolved oxygen and temperature requirements are for fish and other aquatic organisms in your state. Are bodies of water in your state meeting these requirements?

45. The earliest successful voltammetry was carried out using a dropping-mercury electrode (DME). This electrode had a bulb of mercury attached to a small-diameter capillary from which drops of liquid mercury fell into the sample solution every few seconds. This technique is called *polarography*. Suggest some advantages and some disadvantages of such an electrode and a few reasons that this was so widely used for many years but is seldom used today.

46. The person who originally developed the method of polarography was Jaroslav Heyrovsky. Obtain more information about the life and scientific career of Heyrovsky and write a report on what you find.

47. Read about the history of the measurement of the Faraday constant and the role that electroanalytical methods have played in these measurements. Write a report on what you find.

48. Locate an article in a scientific journal in which the method of cyclic voltammetry is used for part of the study. Describe how this approach was used in that study and state what types of information this method was used to provide.

49. There are many types of voltammetry besides those that were discussed in this chapter. Below are a few examples. Obtain more information on any one of these methods. Write a short report that describes the way in which this method is performed and types of information it can provide on an electroactive analyte.
   (a) Differential pulse polarography
   (b) Square-wave voltammetry
   (c) Hydrodynamic voltammetry

50. There has been a great deal of recent interest in the use of microelectrodes and ultramicroelectrodes in voltammetry. Get some information on this topic and discuss the advantages that these small electrodes offer in electrochemical measurements.

References

1. R. J. Diaz and R. Rosenberg, "Spreading Dead Zones and Consequences for Marine Ecosystems," *Science* 321 (2008) 926–929.

2. D. T. Sawyer, A. Sobkowiak, and J. L. Roberts, Jr., *Electrochemistry for Chemists*, 2nd ed., Wiley, New York, 1995.

3. B. H. Vassos, "Voltammetry." In *Analytical Instrumentation Handbook*, 2nd ed., G. W. Ewing, Ed., Marcel Dekker, New York, 1997, Chapter 19.

4. J. Wang, "Instrumentation for Stripping Analysis." In *Analytical Instrumentation Handbook*, 2nd ed., G. W. Ewing, Ed., Marcel Dekker, New York, 1997, Chapter 20.

5. A. J. Bard and L. R. Faulkner, *Electrochemical Methods: Fundamentals and Applications*, 2nd ed., Wiley, Hoboken, NJ, 2001.

6. J. Inczedy, T. Lengyel, and A. M. Ure, *International Union of Pure and Applied Chemistry—Compendium of Analytical Nomenclature: Definitive Rules 1997*, Blackwell Science, Malden, MA, 1998.

7. G. Maludzinska, Ed., *Dictionary of Analytical Chemistry*, Elsevier, Amsterdam, the Netherlands, 1990.

8. D. A. Skoog, F. J. Holler, and T. A. Nieman, *Principles of Instrumental Analysis*, 5th ed., Saunders, Philadelphia, PA, 1998.

9. W. R. Heineman and P. T. Kissinger, "Cyclic Voltammetry," *Journal of Chemical Education*, 60 (1983) 702–706.

Chapter 17

An Introduction to Spectroscopy

Chapter Outline

17.1 INTRODUCTION: THE VIEW FROM ABOVE

In April 1999, NASA launched *Terra*, the first in a series of satellites that are now being used to make detailed studies of the life forms, land, oceans, and atmosphere of Earth. *Terra* is the size of a small school bus and contains several instruments for examining the radiation that is reflected by or absorbed by Earth. These instruments can then be used to obtain images of Earth (see Figure 17.1) and provide detailed information on its chemical and physical composition.[1,2]

The use of *Terra* and other satellites to provide such information is known as *remote sensing*. Remote sensing can be defined as the use of an analytical instrument to examine a distant sample, as occurs when a satellite records an image using light that is being reflected from the surface of Earth. In the case of *Terra*, there are five sets of sensors that are designed to measure different types of light. This information is used by scientists to learn about the distribution of plant life on Earth and the effects of climate change on the atmosphere, land, and sea.[3,4]

Remote sensing often involves measurements of light because light can interact in many ways with matter and can quickly travel across great distances. Many laboratory instruments also utilize light for chemical or physical measurements. The use of light to obtain information on the chemical or physical properties of a sample is a technique known as *spectroscopy*. In this chapter we will learn about the basic principles of spectroscopy and see how this method can be used in chemical analysis.

Chapters 18 and 19 will then examine some more specific applications of spectroscopy, such as examining particular elements or specific types of molecules.

17.1A What Is Spectroscopy?

The term **spectroscopy** refers to the field of science that deals with the measurement and interpretation of light that is absorbed or emitted by a sample.[5] This type of analysis often involves the use of a **spectrum** (plural form, "spectra"), which is the pattern that is observed when light is separated into its various colors, or spectral bands.[5,6] Examples of some spectra are shown in Figure 17.2, which shows light that is emitted by the sun and the intensity of this light after it has passed through the atmosphere and interacted with the chemicals in air. Table 17.1 shows that there are various types of instruments and equipment used to collect such a spectrum. In this text we focus on the general type of instrument known as a **spectrometer**, which is designed to electronically measure the amount of light that occurs in a spectrum at a particular spectral band or group of bands.[5,6]

The *x*-axis of a spectrum indicates the type of light that is being measured or observed. For instance, in Figure 17.2 this axis distinguishes between the different types of light by using their "wavelengths," a term we will discuss in Section 17.2A. The *y*-axis of a spectrum shows the amount of light that is emitted by a particular source (such as the sun) or that interacts with a sample (Earth's atmosphere). A spectrum can provide both qualitative information on the chemical composition of a

Terra (EOS AM-1)

FIGURE 17.1 An artist's view of the satellite *Terra* (officially known as EOS AM-1) and the use of the Multiangle Imaging SpectroRadiometer (MISR) instrument that is aboard this satellite. *Terra* is the flagship satellite of the Earth Observing System and is located in a 100 minute polar orbit at a height of 437 miles above Earth. Each day this satellite collects about 200,000 megabytes of information on our planet. (This image is reproduced with permission from NASA's Jet Propulsion Laboratory and is by S. Suzuki and E.M. De Jong.)

source or sample through the types of light that are detected and quantitative information on this composition based on the amount of light that is detected.

EXERCISE 17.1	Using a Spectrum to Learn About a Sample

According to Figure 17.2, what wavelengths of light have the most intense emission when given off by the sun? Which wavelengths of light are taken up by (or "absorbed") to the greatest extent by Earth's atmosphere?

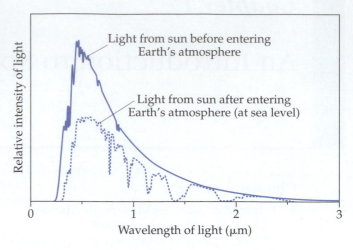

FIGURE 17.2 Spectra for light that is emitted by the sun and for sunlight that has passed through the atmosphere of Earth. The values on the upper graph with a solid line are related to the power or intensity of this light. The values on the lower graph with the dashed line show the percent of light that is transmitted at each wavelength. (These data are based on information in the American Society of Testing Materials Terrestrial Reference Spectra.)

SOLUTION

The upper graph in Figure 17.2 shows that the most intense emission of light from the sun occurs in the range of 0.5 μm (or 500 nm). From the lower graph we learn that at sea level there is an approximately equal intensity of light with wavelengths of 500 to 650 nm. (*Note*: These wavelengths are what give the sun its yellow color.) There is a decrease in intensity at several specific wavelengths after light from the sun has passed through the atmosphere. This decrease is a result of the uptake of light by gases such as water vapor, carbon dioxide, and ozone in the air and can be used to measure these chemicals in the atmosphere. The use of remote sensing to examine Earth's surface can be performed by selecting other wavelengths of light that can pass through the atmosphere, allowing this light to interact with this surface.

TABLE 17.1 General Types of Instruments Used to Perform Spectroscopy

Type of Instrument	Description[a]
Spectrometer	An instrument with an entrance slit and one or more exit slits, which makes measurements either by scanning a spectrum (point by point) or by simultaneous monitoring several positions in a spectrum; the quantity that is measured is a function of radiant power
Spectrophotometer	A spectrometer with associated equipment that is designed to furnish the ratio (or a function of the ratio) of the radiant power of two beams of light as a function of position in a spectrum
Spectrograph	An instrument with one slit and a wavelength selector that uses photography to obtain a simultaneous record of a spectrum
Spectroscope	An instrument with one slit and a wavelength selector, which forms a spectrum for visual inspection

[a]These definitions were adapted from "Guide for Use of Terms in Reporting Data: Spectroscopy Nomenclature," *Analytical Chemistry*, 62 (1990) 91–92; and from G. Maludzinska, Ed., *Dictionary of Analytical Chemistry*, Elsevier, Amsterdam, the Netherlands, 1990.

17.1B How Is Spectroscopy Used in Analytical Chemistry?

Spectroscopy is one of the most commonly used analytical tools for both qualitative and quantitative chemical analysis. One way spectroscopic methods can be classified is according to how these techniques are employed. For instance, the use of spectroscopy to identify a sample or measure chemicals in a sample is called *spectrochemical analysis*, while the use of spectroscopy to measure a spectrum is known as *spectrometry*.[5,6] Spectroscopic methods can also be subdivided according to the type of analytes they are examining or the types of light that they employ. As an example, the use of these methods to study analytes that are molecules is known as "molecular spectroscopy" (see Chapter 18), and the study of atoms or elements by these methods is known as "atomic spectroscopy" (Chapter 19).

Probably the most common way of classifying spectroscopic techniques is according to the type of radiation they employ and the way in which this radiation interacts with matter. Many examples of this type of classification scheme are given in Table 17.2. These methods not only include those that use ultraviolet or visible light, but also methods that make use of infrared light, X rays, radio waves, and microwaves, among other types of interactions.

The types of interactions that can occur with matter and this radiation vary from low energy changes involving a change in spin state (as used in nuclear magnetic resonance [NMR] spectroscopy) to electronic transitions (ultraviolet/visible-absorption spectroscopy) and transitions in core shell electrons (X-ray fluorescence). We will discuss the principles and applications for many of these methods in the next two chapters.

Spectroscopy can be used alone for chemical analysis or it can be used in combination with other analytical methods. Chapters 12–13 and 15 contained examples of

this last case in the use of visual end-point indicators for a titration. In this situation, the change in "color" (or the visually observed spectrum) for the indicator is used to help us determine when the end point has been reached in the titration. It is also sometimes possible during a titration to determine the end point by observing the change in color and spectrum for the analyte or titrant as they combine to form a product (see Chapter 18). Another way spectrometers are often used is as detectors for other analytical methods (see Chapters 22 and 23 on liquid chromatography and electrophoresis).

Color has been used since ancient times as a means for evaluating dyes and other commercial products, but the use of spectroscopy for chemical analysis is a more recent development. The area of spectroscopy began in 1672 when Sir Isaac Newton used a prism to separate a beam of white sunlight into such colors as red, orange, yellow, green, and blue (see Figure 17.3).[7,8] This ability made it possible for others to learn that the type of light that is emitted by a material is related to the sample's chemical composition. For example, in 1752 a Scotsman named Thomas Melville noticed that the addition of sea salt to a sample of alcohol produced a yellow color in a flame (an effect that we now know is due to the presence of sodium in sodium chloride).[8] In 1826, another Scotsman, William Henry Talbot, found that changing the types of salt that were added to samples created different colors in a flame.[7,8] These were the first examples of spectroscopy being used for chemical analysis. Modern chemists use spectroscopy for examining a large variety of materials and to obtain a wealth of information on the chemical composition of these materials.

17.2 THE PROPERTIES OF LIGHT

To understand spectroscopy, it is necessary to first learn about the properties of light and of how light interacts with matter. We will see in the next two chapters how

TABLE 17.2   Common Types of Spectroscopic Methods

Method[a]	Type of Radiation Employed[b]	Process Examined
NMR spectroscopy	Radio waves (λ = 100 cm to 10 m)	Change in nuclear spin
ESR spectroscopy	Radio waves (λ = 1 cm to 100 cm)	Change in electron spin
Microwave spectroscopy	Microwaves (λ = 100 μm to 1 cm)	Change in chemical rotation
Infrared spectroscopy	Infrared light (λ = 1 μm to 100 μm)	Change in chemical vibration
UV-visible spectroscopy	Ultraviolet and visible light (λ = 10 nm to 1 μm)	Change in electron distribution (outer-shell electrons)
X-ray spectroscopy	X rays (λ = 100 pm to 10 nm)	Change in electron distribution (inner-shell electrons)
Gamma ray spectroscopy	Gamma rays (λ = 10 nm to 1 μm)	Change in nuclear configuration

[a]*Abbreviations*: NMR, nuclear magnetic resonance; ESR, electron spin resonance; UV-visible, ultraviolet-visible.

[b]The wavelength regions given for each method and type of process are approximate and are based on values provided in C.N. Banwell, *Fundamentals of Molecular Spectroscopy*, 3rd ed., McGraw-Hill, New York, 1983, p. 7.

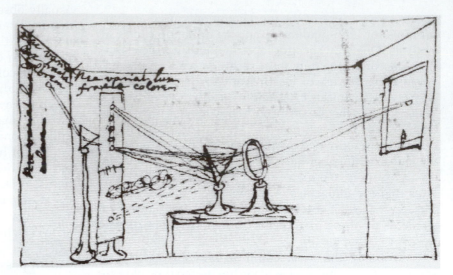

FIGURE 17.3 Sir Isaac Newton (1642–1727), and a sketch made by Newton in his notebook of an experiment in which he used a glass prism to separate a beam of sunlight into several distinct colors. In this experiment Newton first had the sunlight pass through a small slit in a window shutter, as shown on the right-hand side of the sketch. This beam of light was then passed through a lens and shown onto a prism, with the prism being used to separate the light into various color bands, which were observed on a screen. Newton also used a second prism (shown to the left of the sketch) to test whether one of the separated color bands (red, in this case) could be separated into further components. Newton called this work his "crucial experiment," or *experimentum crucis*, because he felt it would prove to even his many skeptics that white light was a mixture of many colors. (The image of Newton is from a 1689 painting by Godfrey Kneller.)

these properties and interactions can be used for the design of spectrometers and for chemical analysis.

17.2A What Is Light?

An answer to the question "What is light?" has been sought by scientists for many centuries. Modern scientists define **light** as *electromagnetic radiation*, which is a wave of energy that propagates through space with both electrical- and magnetic-field components.[5,9–11] However, there are actually two ways in which we can describe light. One of these views looks at light as having the properties of a wave, while the second view considers light to be made up of distinct particles of energy. Together these two views make up what is called the "wave–particle duality" of light. Although it may seem odd at first that we can view light as both a wave and a particle, each of these views is necessary to adequately describe how light behaves and interacts with matter.[9,10,12]

The Wave Nature of Light.   The first view of light is that of a wave of energy that moves through space. This wave can travel through a vacuum or in other transmitting media, such as air, water, or glass. Figure 17.4 provides a diagram of how light is pictured when using the wave model. In this model, a wave of light consists of an oscillating electric field that is perpendicular to an oscillating magnetic field. Like any other wave, these oscillating fields produce regular regions of high or

maximum intensity (called "crests") and regions of low or minimum intensity ("troughs"). The intensity of this wave, as measured by the height of the crests, is called the *amplitude*.[11] The idea that light acts as a wave was first suggested in 1678 by Christian Huygens, a Dutch mathematician and physicist. Our current mathematical description of light as electromagnetic radiation is the result of work by Scottish physicist James Maxwell in the middle of the nineteenth century.[9,10]

There are several properties that we can use to describe light as a wave. First, there is the velocity at which the light is traveling. Light has its fastest rate of travel in a true vacuum. This velocity is a physical constant represented by the symbol c, which in the SI system is equal to exactly 299,792,458 m/s. The velocity of light in a medium other than a vacuum is represented by the symbol "v." The ratio of these two velocities gives a parameter known as the **refractive index (n)**, also called the "index of refraction."[11]

$$\text{Definition of refractive index: } n = c/v \qquad (17.1)$$

Table 17.3 gives the refractive index of various common materials. Because the velocity of light through a medium other than a vacuum will be less than or equal to c, the result will be a refractive index for that medium that is greater than or equal to 1.000. (*Note*: Air has a typical refractive index of 1.0003, which allows us to use a rounded value of 3.00×10^8 m/s to describe the velocity of light in either a true vacuum or air.) The value of n does

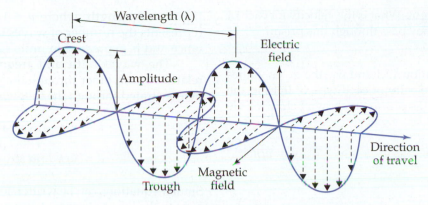

FIGURE 17.4   The wave model of light. In this model, light is viewed as being an oscillating wave with an electrical-field component (shown in this figure as moving from top to bottom) and a perpendicular magnetic-field component (shown here moving from left to right). The crest represents the maximum point of each wave, and the trough is the point at which the wave reaches its minimum. The height of the wave's crest is represented by the amplitude, and the distance between one crest and the next is known as the wavelength.

not have any units, but it does depend on the type of light that is being used for the measurement. The refractive index is highly characteristic of a material and will depend on the chemical concentration and composition of a sample. These properties make the refractive index useful in both chemical measurement and identification.

EXERCISE 17.2   **Relating the Velocity of Light to the Refractive Index**

Sir Isaac Newton separated sunlight into various colors by using a glass prism. The material used in one type of glass prism has a refractive index of approximately 1.61 for red

TABLE 17.3   Values of the Refractive Index for Various Common Materials

Type of Material or Medium		Refractive Index (n)[a]
Reference state	True vacuum	1.00000 (exact value)
Gases	Air	1.0003 (25°C, 1 atm)
	Helium	1.000036 (0°C, 1 atm)
	Carbon dioxide	1.00045 (0°C, 1 atm)
	Oxygen	1.00027 (0°C, 1 atm)
Liquids	Ethanol	1.36
	Water	1.333 (20°C)
	Sugar Solution (30%)	1.38
	Sugar Solution (80%)	1.49
Solids	Crown glass	1.52
	Diamond	2.417
	Glass	1.575 (light flint) – 1.89 (heaviest flint)
	Ice	1.309
	Polystyrene	1.55
	Quartz (fused)	1.46
	Ruby/sapphire	1.77
	Sodium chloride (salt)	1.544 (type 1) – 1.644 (type 2)
	Cellulose acetate (hard)	1.53

[a]These values for *n* were all determined using light from the sodium D line at 589 nm. The information in this table was obtained from D.R. Lide, Editor, *CRC Handbook of Chemistry and Physics*, 83rd ed., CRC Press, Boca Raton, FL, 2002.

light and 1.65 for blue light. What is the velocity for each of these types of light as they pass through this prism?

SOLUTION

We can rearrange Equation 17.1 and use the given values for n to solve for the velocity of each type of light in the prism. An example of this process is shown below for the red light.

Solving for v: $n = c/v \implies v = c/n$

Red light:

$$v = (3.00 \times 10^8 \text{ m/s})/(1.61) = \mathbf{1.86 \times 10^8 \text{ m/s}}$$

Blue light:

$$v = (3.00 \times 10^8 \text{ m/s})/(1.65) = \mathbf{1.82 \times 10^8 \text{ m/s}}$$

As this example indicates, the refractive index of a material (and the corresponding velocity of light in that material) will depend on the type of light that we are examining. It is this difference that allowed Sir Isaac Newton to use a glass prism for separating (or "dispersing") sunlight into red, blue, and other bands of color. The same effect is used in many modern analytical instruments to obtain a spectrum and to perform a chemical analysis using specific types of light.

A second property of light as a wave is the **frequency**, which is the number of waves (or "cycles") that occur in a specified amount of time.[6,11,13] The frequency of light is represented by the symbol ν (the Greek letter *nu*) and is given in units of cycles per second or hertz (Hz), where 1 Hz = 1/s. This frequency is a characteristic property of the light and is independent of its rate of travel or the medium in which the light is traveling. We will see later in this section that the frequency of light is *directly* related to its energy, where high-frequency light has a higher energy than low-frequency light.

A third and related property of light as a wave is the **wavelength**, which is the distance between any two neighboring crests in a wave.[6,11,13] The wavelength of light is represented by the symbol λ (the Greek letter *lambda*). Wavelength is measured in units of distance, such as meters, nanometers (nm), or micrometers (μm). Some older texts and resources also use the angstrom (Å) as a unit of distance to describe wavelengths, where 10 Å = 1 nm. The wavelength of light is *inversely* related to its energy, which means that light with a long wavelength will have lower energy than light that has a short wavelength. A closely related term that is directly proportional to energy is the *wave number* ($\bar{\nu}$, called "*nu* bar").[6,9] The wavenumber for any type of light is equal to the reciprocal of

light's wavelength, where $\bar{\nu} = 1/\lambda$. The wave number represents the number of waves that occur per unit distance and is expressed in units such as cm^{-1}.

The wavelength and frequency of light can be related to each other through the velocity of the light, as demonstrated through the following relationships.

$$\nu = c/\lambda \quad \text{(in a vacuum)}$$

or $\nu = v/\lambda$ (in any medium)   (17.2)

Similar equations can be written to relate the frequency of light to the wave number.

$$\nu = c\bar{\nu} \quad \text{(in a vacuum)}$$

or $\nu = v\bar{\nu}$ (in any medium)   (17.3)

These expressions can be used to find the wavelength or wave number of light that has a given frequency, or they can be used to find the frequency of light that has a known wavelength or wave number. The next exercise illustrates this process.

EXERCISE 17.3   **Calculating the Wavelength and Frequency of Light**

The most intense wavelength of light that reaches Earth from the sun has a wavelength of around 500 nm (see Figure 17.2). What is the frequency of this light in space? What is the wave number for this light?

SOLUTION

The velocity for this light in the vacuum of space will be equal to c. This information can be used with Equations 17.2 and 17.3 to find the frequency and wave number for light with a wavelength of 500 nm.

Conversion to Frequency:

$\nu = c/\lambda$

$= [(3.00 \times 10^8 \text{ m/s})(1 \text{ s/Hz})]/[(500 \text{ nm})(10^{-9} \text{ m/1 nm})]$

$\therefore \nu = \mathbf{6.00 \times 10^{14} \text{ Hz}}$

Conversion to Wave numbers:

$\bar{\nu} = 1/\lambda$

$= 1/[(500 \text{ nm})(10^{-9} \text{ m/1 nm})(10^2 \text{ cm/1 m})]$

$\therefore \bar{\nu} = \mathbf{2.00 \times 10^4 \text{ cm}^{-1}}$

It is interesting to note that if we had performed these same calculations for the light as it traveled through a glass prism, this light would have had the same frequency ($\nu = 6.00 \times 10^{14} \text{ s}^{-1}$), but the wave number and wavelength would have changed as the velocity of the light decreased. The wave number and wavelength would then return to their original values once the light had exited the prism and entered the surrounding air, which has a refractive index close to 1.000.

Figure 17.2 shows that sunlight is composed of light with many different wavelengths. Some of these wavelengths occur in a range that can be seen by the eye (the "visible range"), while others are above or below this range. Figure 17.5 shows the full range of electromagnetic radiation. Humans can see light with wavelengths that span from about 380 nm (violet light) to 780 nm (red light). There is, however, a *much* broader range of electromagnetic radiation that extends to both higher and lower wavelengths. Radiation with smaller wavelengths (and higher energies) than visible light include ultraviolet light, X rays, and gamma rays. Radiation with longer wavelengths (and lower energies) than visible light include infrared light, microwaves, and radio waves.[12,13] All of these wavelengths can be used in analytical chemistry but provide information on different chemical properties or types of matter (for instance, see Box 17.1).

The Particle Nature of Light.   A very different view of light from the wave model was first suggested by Sir Isaac Newton. Newton proposed that light was composed of small particles that moved at great speed.[7,8] Among the evidence for this "particle theory" of light

was work performed in 1905 by Albert Einstein, who found that electrons were ejected when "particles" of light hit the surface of certain materials (a phenomenon known as the *photoelectrical effect*).[11,21] The term **photon** is now used to describe these individual particles of light.[11]

The energy of a single photon of light (E_{Photon}) can be related to its frequency (ν, a wave property) by using *Planck's equation*.[11]

$$E_{Photon} = h\nu \qquad (17.4)$$

This equation is named after its discoverer, the German physicist Max Planck.[22] In this equation, the values of E_{Photon} and ν are related to each other through a proportionality term known as *Planck's constant* (h). This constant has a value of approximately 6.626×10^{-34} J·s regardless of the type of light or photons that are being examined.[11]

Planck's equation is useful in determining the energy of a photon from its frequency or the frequency of light from its energy. In addition, this equation is the source of the symbol "$h\nu$" that is often used by scientists as an abbreviation for light. Equation 17.4 can also be combined with the expressions in Equations 17.2 and 17.3 to relate the energy of a photon to the wavelength and wave number of this light.

Relationships for a True Vacuum:

$$E_{Photon} = hc/\lambda \qquad (17.5)$$

$$E_{Photon} = hc\bar{\nu} \qquad (17.6)$$

Type of light	Wavelength range		Type of light	Wavelength range
Gamma rays	<0.1 nm		Violet	380–420 nm
			Blue-violet	420–440 nm
X rays	0.01 nm–10 nm		Blue	440–470 nm
Ultraviolet light	10 nm–380 nm		Green-blue	470–500 nm
			Green	500–520 nm
Visible light	380 nm–780 nm	Visible light	Yellow-green	520–550 nm
			Yellow	550–580 nm
Infrared light	780 nm–0.3 mm		Orange	580–620 nm
			Red-orange	620–680 nm
Microwaves	0.3 mm–1 m		Red	680–780 nm
Radio waves	>1 m			

FIGURE 17.5   Various types of electromagnetic radiation, or "light." The approximate wavelengths are shown for each type of electromagnetic radiation, with the scale on the right showing an enlarged view of the wavelengths that make up visible light. (The wavelength ranges used in this figure are based primarily on those given in "Guide for Use of Terms in Reporting Data in Analytical Chemistry: Spectroscopy Nomenclature," *Analytical Chemistry*, 62 (1990) 91–92; the approximate ranges given for the colors of visible light vary slightly from one source to the next in the literature. A colored version of this figure, depicting the visible range of light, can be found in the center of this text.)

BOX 17.1

NMR: Tuning into Chemical Structure

Nuclear magnetic resonance spectroscopy (commonly known as "NMR spectroscopy") is a spectroscopic method that is valuable for determining the structure of molecules.[14,15] NMR makes use of the fact that the nuclei of some atoms possess a "spin." (*Note:* This is similar to the spin that is present for electrons.) The spin states for a given type of nucleus are of equal energy in the absence of a magnetic field, but differ by a small amount when the nuclei are placed in a magnetic field. This difference occurs in the range of energies that corresponds to electromagnetic radiation in the radio frequency (RF) range. The energy required for this transition is determined by the type of nuclei that are being examined and their local environments, including how an atom containing each nucleus is connected to other atoms. This feature results in an NMR spectrum in which a plot of the intensity of absorbed radiation versus the frequency of this radiation can be used to identify molecules and determine their structures (see Figure 17.6).

The practical use of NMR spectroscopy for chemical analysis was first reported in 1946 by American scientists Felix Bloch and Edward Purcell,[16,17] who were awarded the 1952 Nobel Prize in Physics for this work. The Nobel Prize in Chemistry was later awarded in 1991 to Swiss chemist Richard Ernst, who introduced the method of "Fourier transform NMR," and in 2002 to Swiss chemist Kurt Wüthrich for the use of NMR spectroscopy in three-dimensional studies of biological macromolecules such as proteins.[18] Modern NMR spectroscopy is now used in many fields, including organic chemistry, inorganic chemistry, analytical chemistry, and biochemistry. NMR spectroscopy is also commonly used for medical imaging under the name "magnetic resonance imaging" (MRI).[19,20]

A typical NMR spectrum is given in Figure 17.6, using a solution of ethanol dissolved in D_2O as an example. This spectrum shows the response measured for ^1H, which is found in most organic compounds and makes up 99.985% of all hydrogen in nature. The position of the peaks in this spectrum reflect the natural resonance frequency of the ^1H nucleus, as well the local environments of the ^1H nuclei in ethanol, which causes small differences in the locations of these peaks. This effect, known as "chemical shift," is why there are three groups of peaks in ethanol's ^1H NMR spectrum. These groups have relative areas of 1:2:3 and represent the -OH, $-CH_2$-, and $-CH_3$ portions of ethanol. It is possible from the splitting pattern in each group of peaks to determine how many ^1H atoms are on the neighboring carbon atoms in ethanol. The difference in energies between the individual peaks in each splitting pattern can be used to help determine which of the detected ^1H atoms are present on neighboring atoms. This information makes it possible to identify a molecule or determine its structure. A similar approach can be used with measurements based on the ^{13}C isotope of carbon (which makes up 1.11% of carbon in nature) or a combination of measurements looking at ^1H and ^{13}C (a method known as "2D NMR spectroscopy"), as well as methods using other isotopes (e.g., ^{15}N, ^{19}F, or ^{13}P). Although Figure 17.6 shows only a simple application of NMR spectroscopy, the same basic approach can be employed to study much more complex organic molecules and even large biological molecules such as proteins.[14,15]

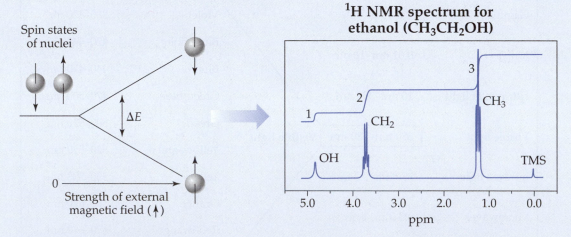

FIGURE 17.6 The effect of an external magnetic field on the difference in energy (ΔE) between the two spin states for a given type of nuclei, and an example of a typical ^1H NMR spectrum, using ethanol as an example. The positions of the peaks for the ^1H nuclei in the various portions of ethanol are shown in this spectrum along with the signal for the ^1H nuclei in trimethylsilane (TMS), which was added to the sample as a reference. The upper tracing in the ^1H NMR spectrum shows the results obtained when the detected peaks are integrated and represent the relative number of equivalent ^1H nuclei that are present at each of the indicated positions in ethanol.

Similar expressions can be written for other types of media by using the velocity v in place of c in Equations 17.5 and 17.6. The result is a series of relationships that we can use to easily convert between the energy of light and the properties such as its wavelength and frequency.

EXERCISE 17.4   **Using Planck's Equation**

a. What is the energy of a photon that has a wavelength of 500 nm, if $n = 1.00$?
b. What is the energy contained in one mole of photons with this wavelength?

SOLUTION

(a) We are told $n = 1.00$, which means the velocity of this light is approximately equal to c. We can then use the given wavelength of this light and Equation 17.5 to solve for E_{Photon}.

$$E_{Photon} = hc/\lambda$$

$$= \frac{(6.626 \times 10^{-34} \text{ J} \cdot \text{s})(3.00 \times 10^8 \text{ m/s})}{(500 \text{ nm})(10^{-9} \text{ m/nm})}$$

$$= 3.98 \times 10^{-19} \text{ J}$$

(b) We can find the energy (E_{Total}) contained in a mole of photons with a wavelength of 500 nm by taking the energy in a single photon and multiplying this value of Avogadro's number.

$$E_{Total} = E_{Photon} N_A$$

$$= (3.98 \times 10^{-19} \text{ J})(6.023 \times 10^{23} \text{ mol}^{-1})$$

$$= 2.40 \times 10^5 \text{ J or 240 kJ}$$

The same process can be used to determine the energy contained in any amount of photons. This process is particularly valuable if we wish to examine how light will interact with matter, as will be considered in the next section.

17.2B Uptake and Release of Light by Matter

There are various ways in which light can interact with matter. One way these interactions can occur is if the light is either released by or taken up by matter. For instance, the *Terra* satellite makes use of light that is released by matter in the sun, with some sensors examining how this radiation is taken up by chemicals in the atmosphere or at Earth's surface (see Figure 17.7). The same processes are used in other types of chemical analysis to study the composition of samples.

Emission of Light. The release of light by matter is referred to as **emission**.[6] The emission of light occurs when matter such as an atom, ion, or molecule goes from an excited state to a lower-energy state. This process is illustrated in Figure 17.8. For atoms and other chemical

species that emit light from the sun, this excited energy state is created by thermal energy. It is also possible to create an excited state in a chemical by other means, such as through the uptake of light (as occurs in "fluorescence") or through the input of energy from a chemical reaction (which occurs in "chemiluminescence"). (*Note*: These processes are discussed in more detail in Chapter 18.)[12] When the excited state of this chemical relaxes to a lower energy state, it must release its extra energy. One way this energy can be released is by the chemical giving off a photon of light. The photon that is released will have an energy exactly equal to the difference in energy between the initial excited state of the chemical and its final lower-energy state.

EXERCISE 17.5   **Determining the Energy and Wavelength of Emitted Light**

One application of emission in remote sensing is in the detection of forest fires, as shown on the cover of this text. The *Landsat* satellite has a sensor that detects active fires by examining their emission in the wavelength region of 1.55 μm to 1.75 μm.

a. What type of light (visible, ultraviolet, etc.) is being detected by this sensor?
b. What are the energies for single photons of light at these wavelengths?

SOLUTION

(a) According to Figure 17.5, wavelengths of 1.55 μm to 1.75 μm are in the range expected for infrared light. This type of radiation represents some of the "heat" that is being emitted by the forest fire. (b) Equation 17.5 can be used to find the energy of single photons that have the given wavelengths, if we assume that the velocity of light in air is approximately equal to c.

For 1.55 μm light:

$$E_{Photon} = hc/\lambda$$

$$= \frac{(6.626 \times 10^{-34} \text{ J} \cdot \text{s})(3.00 \times 10^8 \text{ m/s})}{(1.55 \ \mu\text{m})(1 \text{ m}/10^6 \ \mu\text{m})}$$

$$= 1.28 \times 10^{-19} \text{ J}$$

The same approach gives a quantity of energy equal to 1.14×10^{-19} J for 1.75 μm light. Both these energies are smaller than the result we found for shorter-wavelength light in the previous exercise (2.40×10^5 J at 500 nm, which represents visible light). This is the case because the processes that lead to the emission of visible light involve a larger change in energy than the changes that result in the release of infrared light. We will learn more about these processes in Chapters 18 and 19.

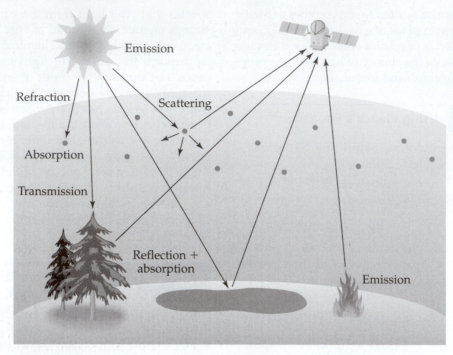

FIGURE 17.7 Interactions of sunlight with the atmosphere and Earth during remote sensing, including emission, absorption, transmission, reflection, refraction, and scattering. Emission also occurs during remote sensing as radiation is given off by the sun or by sources such as fire on Earth. This example is based on *passive remote sensing*, as performed by *Terra*, in which a satellite only makes use of radiation that originates from the sun or from Earth. It is also possible to perform *active remote sensing*, in which electromagnetic radiation (e.g., radio waves) are sent from the satellite to Earth and recorded once they have returned to the satellite.

A plot of the intensity of light that is emitted by matter at various wavelengths, frequencies, or energies is known as an *emission spectrum*.[6] One example of an emission spectrum was given earlier in Figure 17.2 for the light that is given off by the sun. We also saw in Exercise 17.1 that the wavelengths of visible light that are emitted by an object are what give that object a particular color (e.g., the yellow color of the sun). The difference in energy of a chemical as it goes from an excited state to a lower energy state is often a unique value for that chemical. This fact, in turn, means that we can use an emission spectrum and the measured energy, frequency, or wavelength of the emitted light to help us identify a chemical or material (e.g., as used to detect a forest fire). The amount of released light

will be directly related to the amount of chemical that emitted this light. As a result, the intensity of this light can be used to determine how much of this chemical is present in a sample if we compare this emission to that obtained with standard samples.

Absorption of Light. A second way in which matter can interact with light is through absorption. **Absorption** can be defined as the transfer of energy from an electromagnetic field (as possessed by light) to a chemical entity (e.g., an atom or molecule).[6] The general process that occurs during light absorption is shown in Figure 17.9. Unlike emission, we now start with a chemical species that is in a low-energy state and move this species to a

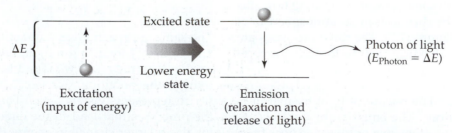

FIGURE 17.8 The general processes involved in the emission of light by matter. The value of ΔE represents the difference in energy between the excited state and lower energy state. The value of E_{Photon} represents the energy in one photon of emitted light.

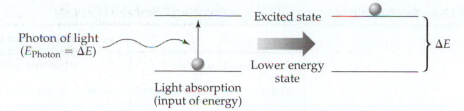

FIGURE 17.9   The general processes involved in the absorption of light by matter. The value of ΔE represents the difference in energy between the excited state and lower energy state. The value of E_{Photon} represents the energy in one photon of absorbed light.

higher energy state by having the chemical absorb a photon of light. This process again requires that the photon have an energy that is exactly equal to the difference in energy of the chemical in its original low-energy state and its final excited state.

The result of the absorption of light is that the intensity of this light after it leaves the sample will be lower than its original value at the energy or wavelength that was absorbed by the sample. The remaining light that passes through the sample is said to have undergone **transmission**,[5] which can be defined as the passage of electromagnetic radiation through matter with no change in energy taking place. The amount of light that is transmitted by a sample plus the amount that is absorbed by that sample will be equal to the total amount of light that originally entered the sample. A plot of the intensity of light that is absorbed (or transmitted) by a sample at various wavelengths, frequencies, or energies is called an *absorption spectrum*.[5,6] Figure 17.10 gives an example of this type of spectrum, as based on the uptake of light by chlorophylls *a* and *b* (i.e., the pigments that create the green color in plants and algae and that lead to the absorption of light for photosynthesis).

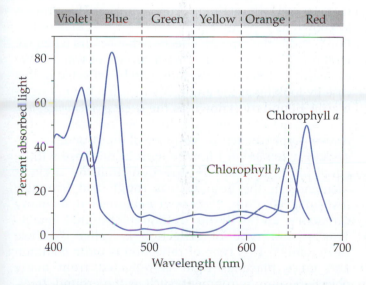

FIGURE 17.10   The absorption spectra for chlorophyll *a* and chlorophyll *b*. The scale at the top is shown as a reminder of the colors that are associated with various wavelengths of visible light.

EXERCISE 17.6   **Using an Absorption Spectrum**

The interactions of light with chlorophyll are used in remote sensing to examine the plant and algae content of the land and sea.

a. What wavelengths of light in Figure 17.10 are the most strongly absorbed by chlorophyll *a* and *b*? What wavelengths are the most easily transmitted by these pigments?

b. Which wavelengths would you select if you wished to use light absorption to measure the chlorophyll *a* and *b* content of a sample? What types of light (visible, ultraviolet, etc.) would be present at these wavelengths?

SOLUTION

(a) The strongest absorption of light for chlorophyll *a* occurs at approximately 435 nm and 660 nm. The strongest absorption of light for chlorophyll *b* takes place at roughly 460 and 635 nm. Wavelengths between 500 and 600 nm have the greatest degree of transmittance by both chlorophyll *a* and *b*. Small differences in these wavelength ranges are present for these two types of chlorophyll because of their different chemical structures, which create slight differences in their energy levels and in the types of light they can absorb.
(b) If we ignore the effects of other chemicals in the sample, the measurement of chlorophyll *a* and *b* would be best performed by using the wavelengths at which these pigments have their strongest absorption of light (435 nm or 660 nm for chlorophyll *a*, and 460 or 635 nm for chlorophyll *b*, which represent visible light).

If you look closely at Figure 17.10, you will notice that it is the light that is not absorbed by chlorophyll *a* and *b* between 500 and 600 nm that gives these pigments and plants their green/yellow color. This last effect is quite common in nature, in which the color of an absorbing object is determined by the remaining types of light that are transmitted (or reflected) by the object. For instance, the passage of white light through a blue solution of copper sulfate indicates that blue light is being transmitted while its complementary color (orange, in this case) is being absorbed. Table 17.4 shows how other

TABLE 17.4 Relationship of Absorbed Light to Observed Color*

	Absorbed light		Transmitted or reflected light and observed color[a]
	Violet	(380–420 nm)	Yellow-green
	Blue-violet	(420–440 nm)	Yellow
	Blue	(440–470 nm)	Orange
	Green-blue	(470–500 nm)	Red
	Green	(500–520 nm)	Purple
	Yellow-green	(520–550 nm)	Violet
	Yellow	(550–580 nm)	Blue-violet
	Orange	(580–620 nm)	Blue
	Red-orange	(620–680 nm)	Blue-green
	Red	(680–780 nm)	Green

White light
(all visible wavelengths)

*The listed wavelengths and color ranges are approximate and may vary in other sources.

[a]The color purple is created by a combination of red and violet. The color brown (not shown in this list) requires a combination of at least three colors, such as red, blue and yellow. An observed color of black is produced by the absence of any transmitted or reflected light in the visible range.

colors can be created through the absorption of specific types of visible light by a chemical sample. This information can be useful in estimating the wavelengths of light that are absorbed by an object based on its appearance.

17.2C Physical Interactions of Light with Matter

Besides being emitted or absorbed, it is possible for light to have other interactions with matter. In many cases, there are physical interactions that do not affect the energy or frequency of the light, but do affect such things as its velocity or direction of travel. These physical interactions are also important to consider in the use of light for chemical measurements. Examples of such interactions are refraction, reflection, some types of scattering, and diffraction.

Reflection. The process of **reflection** occurs whenever light encounters a boundary between two regions that have different refractive indices, where at least part of the light changes its direction of travel and returns to the medium in which it was originally traveling (see Figure 17.11).[12] We make use of this process every time we look at a reflection in a mirror. The use of mirrors and reflection is also employed in analytical instruments to help control how light travels within the instrument. Reflection is also used in some types of

spectroscopy to obtain information about a sample. A good example in remote sensing is when the reflection of sunlight is used by satellites to obtain information on the structure and composition of Earth's surface.

There are several types of reflection. If the boundary between two regions that causes the reflection is a flat plane, the light will be reflected in a well-defined manner and will retain its original image. This process is known as "specular reflection" (or "regular reflection") and is the type of reflection that occurs when we look in a mirror or on a smooth surface of water.[12] We can easily predict how the light will be reflected in this case by looking at the angle at which the light is striking the boundary (known as the *angle of incidence*, θ_1), as compared to a reference line (the "normal") that is drawn perpendicular to the plane of the boundary. The *angle of reflection* (θ_{1r}) for the light in this case will be equal to but opposite in direction from the angle of incidence. If the boundary is rough and irregular instead of smooth, light will be reflected in many directions and will not retain its original image. This second type of reflection is called "diffuse reflection."[12] This type of reflection is quite common and takes place as light is reflected from many objects in our environment, such as the ground, trees, and buildings.

The degree to which light will be reflected at a boundary will depend on the relative difference in the

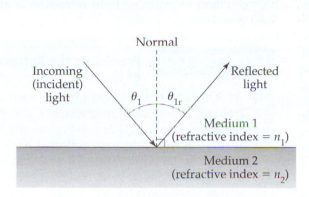

Normal

Incoming (incident) light

Reflected light

θ_1 θ_{1r}

Medium 1 (refractive index = n_1)

Medium 2 (refractive index = n_2)

FIGURE 17.11   The process of reflection. The model on the left shows the expected angle of reflection (θ_{1r}) for light that approaches a planar boundary at an incident angle of of θ_1 (versus the normal) and between two media with refractive indices of n_1 and n_2. This image on the right illustrates two types of reflection. Specular or "regular" reflection occurs for light that is reflected from still water, while diffuse reflection (involving an irregular surface) occurs for light that is reflected from the land and trees. Both types of reflection occur during remote sensing. (Photo by B.R. Hage.)

refractive indices for the two sides of the boundary. The larger this difference, the greater the fraction of the light that will be reflected. This idea is illustrated by Equation 17.7 (the *Fresnel equation*), which gives the fraction of light that will be reflected as it enters the boundary at a right angle. (*Note*: An expanded form of this expression is required for work at other angles.)[12]

$$\frac{P_R}{P_0} = \frac{(n_2 - n_1)^2}{(n_2 + n_1)^2} \qquad (17.7)$$

The symbol P in this equation represents the *radiant power* of the light (in units of watts), which is defined as the energy in a beam of light that strikes a given area per unit time. (*Note*: Although the term "intensity" is often used interchangeably with "radiant power," these two terms do have different units and refer to slightly different aspects of light.)[13] The term P_0 in Equation 17.7 represents the original or "incident" radiant power of the light, and P_R describes the radiant power of the reflected light, while P_R/P_0 is the fraction of the original light versus the reflected light. For boundaries that have only a small difference in refractive index, such as between the vacuum in space and air, the fraction of reflected light will be small and most of the light will pass through the boundary and into the new medium. If a large difference in refractive index is present, as occurs between air or glass and the silver-coated surface of a mirror, a large fraction of light will be reflected.

EXERCISE 17.7   **Working with Reflection**

Some sensors on the *Terra* satellite make use of reflection patterns to map the surface of Earth.

   a. If a beam of light passes through air ($n = 1.0003$) and strikes the smooth surface of water ($n = 1.333$) at a right angle, what fraction of this light will be reflected by the water back into the air?
   b. If this beam of light strikes the water at an angle of 65.0°, what will be the angle of reflection?

SOLUTION

(a) The relative fraction of reflected light in this case can be found by using Equation 17.7 and the given values for n_2 (the refractive index of water) and n_1 (the refractive index of air).

$$\frac{P_R}{P_0} = \frac{(1.333 - 1.0003)^2}{(1.333 + 1.0003)^2} = \mathbf{0.0203} \text{ (or 2.03\% reflection)}$$

(b) If the light is undergoing perfect regular reflection, it will be reflected at an angle of 65.0° on the other side of the normal from the incoming light. If the surface of the water is rough and diffuse reflectance instead occurs, the light will be reflected at many different angles.

Refraction.   A second way in which light can be affected by matter is through refraction. **Refraction** is a process in which the direction of travel for a beam of light is changed as it passes *through* a boundary between two media having different refractive indices.[12] We learned earlier that light will have different velocities of travel for two media with different values for the refractive index, n. This change in velocity can also affect the angle at which the light is traveling. This idea is illustrated in Figure 17.12, which compares the angle of incidence for the incoming light versus the angle of refraction for the light in the new medium. Both of these angles are again being compared to the normal (i.e., a reference line drawn perpendicular to the boundary plane).

If the value of the refractive index for the new medium is larger than that for the original medium in which the light was traveling ($n_2 > n_1$), the angle of travel for the light will be bent toward the normal. Similarly, this angle is bent away from the normal if n_2 is less than n_1. It is easy to predict the size of these angles by using a relationship known as **Snell's law** (also known as "Descartes' law"),[11,12]

Snell's Law: $n_2 \cdot \sin(\theta_2) = n_1 \cdot \sin(\theta_1)$

or $\sin(\theta_2) = \dfrac{n_1 \cdot \sin(\theta_1)}{n_2}$ (17.8)

where n_1 is the refractive index for the medium in which the light is originally traveling, n_2 is the refractive index for the medium that the light is entering, θ_1 is the angle of incidence at which the light approaches the boundary on the side of medium 1, and θ_2 is the angle at which the light is refracted after it has passed through the boundary and entered medium 2. An example of how you can use this equation to examine light refraction is given in the next exercise.

EXERCISE 17.8   Predicting the Refraction of Light

a. At what angle would light be refracted if it passed from air ($n = 1.0003$) into water ($n = 1.333$), hitting the water at an angle of 45.0° versus the normal?
b. If some of this light were reflected back to the surface of water at an angle of 45.0°, at what angle would this light be refracted as it exited the water and again entered the air?

SOLUTION

(a) When the light goes from air to water, the value of n_1 in Snell's law will be equal to the refractive index for air and n_2 will be the refractive index for water. We also know that the angle of incidence (θ_1) is 45.0°, which makes it possible for us to solve for the angle of refraction (θ_2).

$$\sin(\theta_2) = \frac{(1.0003 \cdot \sin(45.0°))}{(1.333)}$$

$$\sin(\theta_2) = 0.531 \quad \text{or} \quad \theta_2 = \mathbf{32.0°}$$

This result tells us that the beam of light is bent toward the normal, because water has a greater refractive index than air. (b) The refraction of light as it moves from the

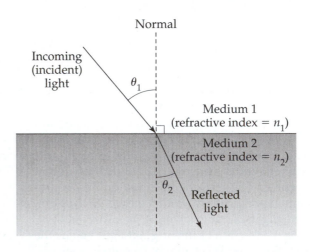

FIGURE 17.12   The process of refraction. The image on the left shows how refraction causes the image of a pencil to be bent as a result of the refraction of light as it passes between air and water. The model on the right shows the angle of refraction (θ_2) for light that strikes a planar boundary at an angle of θ_1 (vs. the normal) and that passes between two media with refractive indices of n_1 and n_2. The relationship between angles θ_1 and θ_2 is given by Snell's law (see Equation 17.8). (Photo by B.R. Hage.)

water back into the air can also be determined by using Snell's law. However, now the value of n_1 is the refractive index for water and n_2 is the refractive index of air. For the sake of comparison, the angle of incidence (θ_1) we are considering is again given as 45.0°. Under these conditions, the angle of refraction as the light moves from the water to the air would be found as follows.

$$\sin(\theta_2) = \frac{1.333 \cdot \sin(45.0°)}{(1.0003)}$$

$$\sin(\theta_2) = 0.942 \quad \text{or} \quad \theta_2 = 70.4°$$

This answer indicates that the direction of travel for light is bent away from the normal as this light moves from a medium with a higher refractive index than the medium that the light is entering.

We have seen that both refraction and reflection are created by a boundary between media with different refractive indices. This feature, in turn, means that the amount of light that is reflected will affect the amount that can be refracted, and then absorbed or transmitted by a sample. For instance, if 2% of the light is reflected as it travels through air and strikes a boundary with water, the other 98% of the light will be able to enter the water and undergo refraction. If we increase

the difference in the refractive index for our two media, the fraction of light that is reflected will increase and the amount that is refracted will decrease. It is even possible in some cases to have conditions in which 100% of the light is reflected. This situation can occur when there is a large difference in the refractive index across the boundary and when the light strikes this boundary at an appropriate angle that prevents it from crossing the boundary. This effect is employed in the design of optical fibers (see Figure 17.13) and can be valuable in controlling the travel of light within analytical instruments or to and from samples.[12]

Scattering.   The term **scattering** is used in chemistry and physics to refer to the change in travel of one particle (such as a photon) due to its collision with another particle (e.g., an atom or molecule).[12] One common type of scattering is *Rayleigh scattering*, or "small-particle scattering" (see Figure 17.14). This type of scattering occurs when photons of light are scattered by particles such as atoms or molecules that are much smaller than the light's wavelength (particle diameter $< 0.05 \lambda$). Rayleigh scattering results in light beams being redirected in a symmetrical pattern about the particles, but does not involve any change in the energy of the light. This process does depend on the wavelength of light, with short wavelengths being more effectively scattered than long wavelengths. This effect is

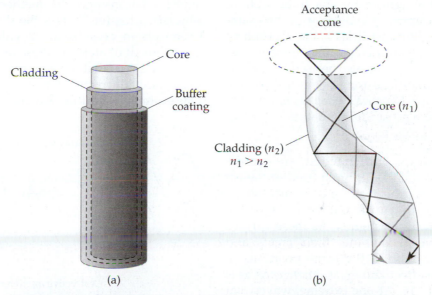

(a)   (b)

FIGURE 17.13   The general design and use of an optical fiber. An optical fiber (a) contains at least two layers with different refractive indices. In the center of an optical fiber is the "core" (medium 1), which is a thin glass cylinder through which the light travels. Surrounding the core is the "cladding" (medium 2), a material that has a refractive index less than that of the core ($n_2 < n_1$) and that is used to keep light traveling within the core. There is also often an exterior buffer coating that protects the optical fiber from water and the outside environment. If light enters the optical fiber (b) at an appropriate angle (a range of angles known as the "acceptance cone"), this light can have an angle of refraction between the core and cladding that is at or exceeds 90° (versus the normal), a condition that causes all of the light to be reflected from the cladding back into the core. This effect is known as *total internal reflection* and is what allows optical fibers to carry light over great distances and through various paths without a large loss in intensity.

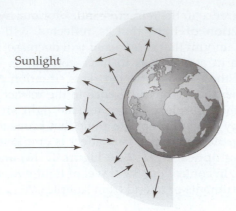

FIGURE 17.14   The scattering of sunlight by particles such as atoms and molecules in Earth's atmosphere. This process, which is based on *Rayleigh scattering*, is most effective for light with short wavelengths (such as blue light). This scattering is the reason why the sky appears to be blue during the daytime. This effect also explains why the sky at sunset is red, at which point we can better see the red light from the sun that is being transmitted through the atmosphere.

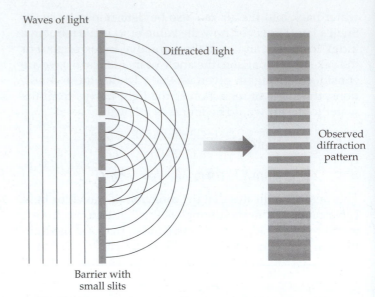

FIGURE 17.15   The production of a diffraction pattern as parallel waves of light strike a barrier with a series of small slits. The diffraction pattern is a result of constructive and destructive interference as the diffracted waves overlap. The position of the bands in this diffraction pattern will depend on the wavelength of light that is being diffracted and the spacing of the slits. A device that uses this effect to separate different wavelengths of light is called a *diffraction grating*.

what creates the blue color of the sky because blue light has a shorter wavelength than other types of visible light and so is more effectively scattered by particles in the atmosphere.[12]

Scattering is important to consider in the measurement of light absorption or emission because this process will affect the amount of light that is able to travel from the light source and sample to the detector. As an example, the scattering of light by particles in the atmosphere is one factor that must be considered during remote sensing. The process of scattering can also be used by itself as a means for detecting and measuring chemicals. We will see an example of this in Chapter 22 when we discuss the use of an "evaporative light scattering detector" for liquid chromatography.

Diffraction.   The process of **diffraction** refers to the spreading of a wave, such as light, around an object (see Figure 17.15).[12] As the wave moves around an obstacle, the distance that different portions of the wave must travel will differ slightly. As these various portions of the wave recombine, their crests and troughs may no longer be in the same positions in space. The result is an effect known as *interference*, as is illustrated in Figure 17.16. At one extreme, waves that have regions with matched crests will combine to give an overall observed amplitude that is increased (known as "constructive interference"), while regions in which the crests of some waves combine with the troughs of others will give an overall observed amplitude that is decreased ("destructive interference"). This effect becomes particularly prominent as the obstacle that is encountered by the light approaches the same size as the wavelength of the light. The result is a pattern called a *diffraction pattern* that contains regions with either constructive or destructive interference.

Diffraction is used in several ways for chemical analysis. For instance, the method of X-ray crystallography (see Chapter 7) uses the diffraction of X rays by chemicals in crystals to provide information on the arrangement of atoms in these molecules. There are also several devices that are employed in spectrometers that make use of diffraction for isolating a given type of light. One example is a diffraction grating, a simple example of

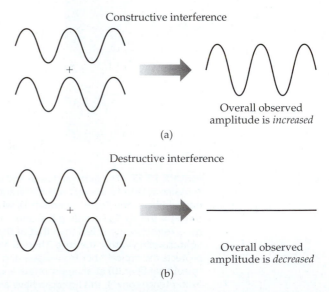

FIGURE 17.16   Examples of (a) constructive interference and (b) destructive interference for two overlapping waves. Various intermediate forms of constructive and destructive interference are also possible.

which is given by the two-slit system in Figure 17.15. This type of device gives an interference pattern in which different wavelengths of light will give regions of constructive interference at different locations around the device. It is then possible to select light from one of these regions (such as through the use of a slit) and to pass this light through a sample or onto a detector for use in an analysis.[5,12]

EXERCISE 17.9   Diffraction of X Rays by a Crystal

It is possible to predict the angles at which constructive interference will be observed for X-ray diffraction by the atoms in a crystal. This can be achieved by using the *Bragg equation*,

$$n\lambda = 2d \sin(\theta) \qquad (17.9)$$

where λ is the wavelength of the X rays passed through the crystal, d is the interplanar distance (or "lattice spacing") between atoms in the crystal, and n is the order of diffraction for the observed constructive interference band (e.g., n = 1 for first-order interference). The term θ represents the specific angle at which the X rays will strike the crystal surface and produce, on the other side of the normal and at the same angle, a diffraction band for the given order of constructive interference.[12] If X rays with a wavelength of 0.711 Å are diffracted by a crystal of sodium chloride (d = 2.820 Å), at what angle will first-order constructive interference be observed for this crystal?

SOLUTION

We can rearrange Equation 17.9, as shown next, and solve for the angle θ by substituting in the known values for n, d, and λ.

$$\sin(\theta) = \frac{n\lambda}{2d} = \frac{(1)(0.0711 \text{ nm})}{(2)(0.2820)}$$

$$\therefore \sin(\theta) = 0.126 \text{ or } \theta = 7.24°$$

The same type of calculation predicts that additional bands of constructive interferences will be seen at angles of 14.6° (n = 2), 22.2° (n = 3), and so on for higher orders of interference.

17.3 QUANTITATIVE ANALYSIS BASED ON SPECTROSCOPY

17.3A Analysis Based on Emission

General Instrumentation. Figure 17.17(a) shows a simple spectrometer that could be used to examine the emission of a sample. This type of instrument consists of the following basic components: (1) the sample and a means for exciting chemicals in this sample, (2) a wavelength selector for isolating a particular type of light for analysis, and (3) a detector for recording this light. These are the same three components that were used by Sir Isaac Newton to study the emission of light by the sun (as shown earlier in Figure 17.3). For example, in Newton's work the sun was the "sample" and the high temperature of the sun was the means for exciting chemicals and allowing them to emit light. The wavelength selector in Newton's experiment was made up of the slit in the window, the lens, and the prism that were utilized to separate a beam of sunlight into various colored bands. The detector and recording device consisted of the screen that Newton used to allow him to see the separate bands of color.

Modern instruments are a bit more complicated and sophisticated than the equipment that was used by Newton, but the three basic components shown in Figure 17.9 still make up part of any spectrometer that measures light emission. When detecting forest fires by remote sensing, the fire and its heat would be the sample and excitation source, while the wavelength selector and detector would be present on a satellite such as *Terra*. A laboratory instrument for performing chemical analysis based on emission would also have all of these components, although on a small scale. We will see several examples of these instruments in Chapter 19, when we discuss the method of atomic emission spectroscopy.

Emission and Chemical Concentration. The amount of light that is emitted by a sample will be directly related to the concentration of the atoms or molecules in the sample that are creating this emission. We can represent this relationship through the following equation,

$$P_E = kC \qquad (17.10)$$

where P_E represents the radiant power of the emitted light, C is the concentration of the species emitting this light, and k is a proportionality constant.[12] This equation can be used whether the excited state of the atom or molecule is produced by heat (as is the case for the sun) or by other methods (e.g., a chemical reaction).

Although it is necessary to correct for any background signal due to other sources of light, this type of correction can be made by preparing a calibration curve in which the emission of standard samples is used to determine the response that would be expected for the analyte in samples with a similar composition to that of the standards. We will see some specific examples of such measurements in Chapters 18 and 19 when we discuss the methods of fluorescence, chemiluminescence, and atomic emission spectroscopy.

17.3B Analysis Based on Absorption

General Instrumentation. An illustration of a basic spectrometer for absorption measurements is shown in Figure 17.17(b). This instrument contains four main

Measurement of light emission

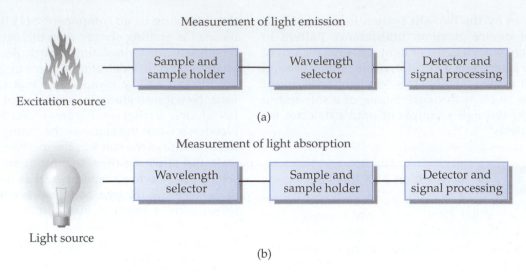

(a)

Measurement of light absorption

(b)

FIGURE 17.17 Design of a simple spectrometer for detecting (a) light emission or (b) light absorption.

components: (1) a light source, (2) a wavelength selector for isolating a particular type of light for use in the analysis, (3) the sample and a sample holder, and (4) a detector for recording the amount of light that passes through the sample.

The addition of the light source, when compared to the system for the emission spectrometer in Figure 17.17, is needed because now we are looking at how a sample absorbs or transmits light that is applied from an outside source rather than light that originates from the sample. The particular light source, wavelength selector, and detector that will be used in this system will depend on the type of light that we are examining. We will see some specific examples of these instruments later when we discuss the use of absorption measurements in molecular spectroscopy and atomic spectroscopy (Chapters 18 and 19).

Absorption and Beer's Law. To measure the amount of light that is absorbed by a sample, we must compare the original amount of light that is applied to the sample and the amount that is transmitted by the sample. If the initial radiant power of this light is represented by P_0 and the power of this light after passing through a sample (or the "transmitted radiant power") is given by P, the fraction of transmitted light can be found by using a term known as T, the **transmittance**.[6,13]

$$T = P/P_0 \qquad (17.11)$$

A closely related value is the *percent transmittance* (% T), where % $T = 100 \cdot T$.

Although T and % T can easily be measured by using a spectrometer like the one in Figure 17.17(b), these values have an inherently nonlinear relationship with the concentration of an absorbing analyte. A term that is more convenient to relate to concentration is the **absorbance (A)**, which is a measure of absorption that is found by taking the base-10 logarithm of the transmittance.[6,13]

$$A = -\log(T) = \log(P_0/P) \quad \text{or} \quad T = 10^{-A} \qquad (17.12)$$

Table 17.5 gives some typical values for the absorbance and transmittance, which helps demonstrate the logarithmic relationship between A and T. Like the transmittance, the absorbance for a sample has no units. In some older resources the absorbance is also known as the "optical density" of a sample.

The absorbance of a homogeneous sample can be related to the concentration of a dilute absorbing analyte through an expression called the "Beer–Lambert law," or simply **Beer's law** (see Box 17.2 and the derivation in Appendix A).[12,23]

$$\text{Beer's law:} \quad A = \varepsilon b C \qquad (17.13)$$

TABLE 17.5 Relationship Between Transmittance and Absorbance

Transmittance (T)	% Transmittance (% T = 100 T)	Absorbance = –log(T)
0.00010	0.010	4.00
0.0010	0.10	3.00
0.010	1.0	2.00
0.10	10	1.00
0.25	25	0.602
0.50	50	0.301
0.75	75	0.125
1.00	100	0.000

BOX 17.2

A Closer Look at Beer's Law

The equation we now call "Beer's law" or the "Beer–Lambert law" is a result of research that was performed by many scientists over the course of more than 100 years.[7,8] This process began in 1729 when a French scientist named Pierre Bouguer (1698–1758) noticed that the same relative fraction of light was absorbed for each additional unit of distance that light had to travel through matter. For instance, if half of the light was absorbed by one glass plate, then one-half of the remaining light (or one-quarter of the original amount) was absorbed by the next plate, and so on.[24] This same observation was later made by German physicist Johann Lambert (1728–1777), who in 1760 published a "law of absorption" in which he gave a mathematical description for this relationship between the amount of absorbed light and the distance of travel.[25] This same relationship is represented in our modern version of the Beer–Lambert law by the fact that the cell path length b is directly proportional to the absorbance A, where $A = -\log(P/P_0)$.

Another important development occured in 1852 when the description of light absorption was extended to solutions by the German scientist August Beer (1825–1863).[26] Beer found that the amount of light that was absorbed by an aqueous solution was proportional to the amount of an absorbing chemical that was present in that solution. (*Note:* This idea was also reported at about the same time by a French scientist named F. Bernard.)[27] We now represent this concept in the Beer–Lambert law by showing the absorbance A as being proportional to C, the concentration of the analyte. Beer also used a proportionality constant to relate the degree of absorbed light to analyte concentration.[7,8] We employ this same general approach in the modern version of the Beer–Lambert law when we use the molar absorptibity (ε) as a constant to relate the measured absorbance of a sample to an analyte's concentration and the cell path length.

Beer's law is useful for chemical analysis because it gives a linear relationship between the measured absorbance A for an analyte and the concentration of this analyte (C). This particular form of Beer's law has C as the molar concentration of the absorbing species (in units of M, or mol/L). The term b is the *path length*, or distance that the light must travel through the sample (in units of cm), and the term ε (the Greek letter *epsilon*) is the **molar absorptivity** (with units of L/mol · cm).[6] The molar absorptivity is a proportionality constant and will have a value that depends on the wavelength of light being used, the identity of the absorbing species, and the environment of this absorbing species. If units other than mol/L and cm are used for the concentration and path length, the proportionality constant in Beer's law will have units besides L/mol · cm and is then called the "absorptivity" (with recommended units of kg/m^3 or g/L) or the "extinction coefficient."[6,13]

EXERCISE 17.10   **Calculating Absorbance and Using Beer's Law**

Ozone is found not only naturally in the atmosphere but it is also used as a strong oxidizing agent in water treatment. One way to measure the ozone concentration in water is through its absorption of ultraviolet light. A solution of dissolved ozone in water gives a percent transmittance of 83.4% when measured at 258 nm and using a sample holder that has a path length of 5.00 cm. The molar absorptivity of ozone at 258 nm is known to be 2950 L/mol · cm.

a. What is the absorbance of this sample, and what is the concentration of ozone in the sample?

b. If no other absorbing species are present in the sample, what concentration of ozone would be expected to give an absorbance of 0.250 under these conditions?

SOLUTION

(a) The absorbance of this sample can be found by using Equation 17.12 and the measured transmittance, where $T = 0.834$ or % $T = 83.4\%$.

$$A = -\log(0.834) = \mathbf{0.0788}$$

If it is known that no other absorbing species is present in this sample, we can use the molar absorptivity for ozone along with Beer's law to determine the concentration of ozone.

Solving for C: $A = \varepsilon bC \Rightarrow C = A/(\varepsilon b)$

$$\therefore C = (0.0788)/[(5.00 \text{ cm})(2950 \text{ L/mol} \cdot \text{cm})]$$

$$= \mathbf{5.34 \times 10^{-6}\,M}$$

(b) The same approach as used to solve for the ozone concentration at $A = 0.0788$ can be used to estimate the absorbance that will be produced by this chemical at other concentrations if the wavelength and other conditions are kept the same.

Solving for C: $A = \varepsilon bC \Rightarrow C = A/(\varepsilon b)$

$$\therefore C = (0.250)/[(5.00 \text{ cm})(2950 \text{ L/mol} \cdot \text{cm})]$$

$$= \mathbf{1.69 \times 10^{-5}\,M}$$

This procedure can also be used to estimate the concentrations of ozone that give rise to other absorbance values, provided that these concentrations are not so large that they create deviations from the response that is predicted by Beer's law (a topic we examine in the next part of this chapter).

The preceding exercise indicates that it is sometimes possible to use Beer's law along with a measurement performed with a single standard to relate the absorbance of an analyte to its concentration. However, it is common for there to be more than one absorbing species in a solution or sample, which creates a "background" signal for the absorbance measurement. We can still use Beer's law in this situation by using several standards and making a plot of the measured absorbance for these standards versus their concentration for the analyte (see Figure 17.18). This type of graph is known as a *Beer's law plot*. This plot often has an intercept other than zero, as caused by the presence of species other than the analyte that might absorb light or cause the amount of transmitted light to be decreased (e.g., such as through light scattering).

The overall measured absorbance (A) for a sample that contains several absorbing species is the sum of the absorbance for each of these species (A_1, A_2, and so on). Equation 17.14 illustrates this relationship for two absorbing species,

$$A = A_1 + A_2$$
$$= \varepsilon_1 b C_1 + \varepsilon_2 b C_2 \qquad (17.14)$$

where ε_1 and C_1 are the molar absorptivity and concentration for species 1, and ε_2 and C_2 are the molar absorptivity and concentration for species 2. (*Note*: The path length b is the same for both species because they are in the same sample.) If, however, we prepare a series of standards in which only the concentration of the analyte (species 1) is varied, the absorbance of the other component will be constant. The result will be a calibration curve like the one in Figure 17.18, in which a nonzero

intercept is present, but that still provides a straight line that can be used to determine the concentration of an analyte in unknown samples with compositions similar to those of the standards. In next chapter we will also see how absorbance measurements made at several wavelengths can be used to determine the concentrations of multiple chemical species that have different molar absorptivity values at the given wavelengths.

Another factor to consider when we are measuring absorbance and transmittance is that several sources of random errors can limit the precision of these values. Some typical precision plots for absorbance spectrometers are shown in Figure 17.19. At a low absorbance (high transmittance) there is a loss in precision because it becomes more difficult for the spectrometer to differentiate the amount of light that is transmitted in the presence and absence of the sample (or P is close to P_0). At a high absorbance (low transmittance) there is a large difference between P and P_0, but P is now quite small, making this measurement subject to imprecision due to factors such as noise from the detector. The best precision for transmittance and absorbance measurements will occur at some intermediate value, which is indicated by the minimum in %RSD which is obtained at an absorbance of approximately 0.4 for the instrument in Figure 17.19(a). Fortunately, even simple absorbance spectrometers can be used over a relatively broad absorbance range and provide good precision (typically, A = 0.1 to 0.8). More sophisticated instruments, like the one in Figure 17.19(b) tend to provide good precision over an ever larger range of absorbance values.[12]

Limitations of Beer's Law. A big advantage of Beer's law is it can be used to relate concentration to the absorption of light for any type of light. It should be kept in mind, however, that there are several assumptions made in this equation that limit the circumstances under which it can be utilized (see Table 17.6).[12] The first assumption in Beer's law is that all of the absorbing species in the sample are acting *independently* of each other. This assumption makes Beer's law valid for only relatively

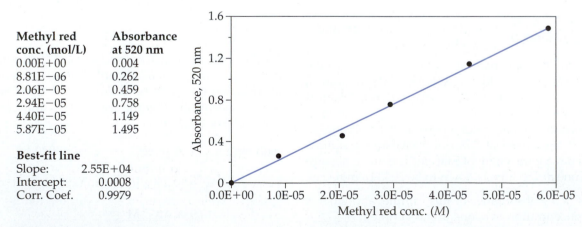

Methyl red conc. (mol/L)	Absorbance at 520 nm
0.00E+00	0.004
8.81E−06	0.262
2.06E−05	0.459
2.94E−05	0.758
4.40E−05	1.149
5.87E−05	1.495

Best-fit line
Slope:	2.55E+04
Intercept:	0.0008
Corr. Coef.	0.9979

FIGURE 17.18 A Beer's law plot and a spreadsheet to prepare such a plot, as demonstrated for the measurement of methyl red in water.

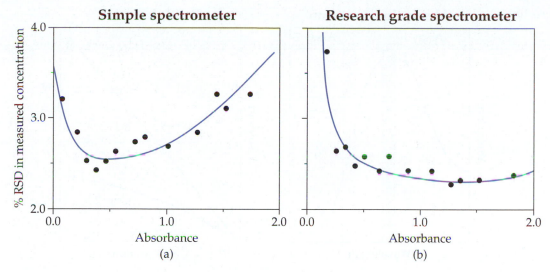

FIGURE 17.19 Precision curves for two UV-visible absorption spectometers. The curve in (a) is for a relatively simple instrument that is often used in teaching laboratories (the Spectronic 20), while the curve in (b) is for a more advanced instrument used in research laboratories (the Cary 118). The *percent relative standard deviation (% RSD)* values were calculated by using the formula % RSD = 100 (s_C/C), where C is the measured concentration and s_C is the standard deviation of this measured concentration. (These results are based on L.D. Rothman, S.R. Crouch, and J.D. Ingle, *Analytical Chemistry*, 44 (1972) 1375).

dilute samples (i.e., usually below 0.01 M). Deviations from this assumption will occur at higher concentrations as the distance between individual analytes becomes sufficiently small to allow the electric field around one analyte to affect the absorption of light by another. Similar effects occur if high concentrations of other species (particularly ions) are present in the solution (see the discussion on chemical activity in Chapter 6). The result is a lower-than-expected absorbance for high concentration samples and standards, causing a negative deviation in a Beer's law plot.

Another way in which a high analyte concentration can lead to deviations from Beer's law is through the corresponding change in refractive index. This effect occurs because the molar absorptivity is dependent on the refractive index, and the refractive index is dependent on a sample's composition and concentration. Fortunately, it is possible to correct for this effect if the refractive index is known or measured for the sample. This correction is made by using $(\varepsilon n)/(n^2 + 2)^2$ in place of ε in Beer's law, but is not usually required for analytes with concentrations below 0.01 M.[12]

A second assumption made in Beer's Law is that the absorbance of the sample is being measured using only *monochromatic light* (that is, light that contains only one wavelength). There is really no such thing as perfectly monochromatic light, but modern spectrometers come quite close in providing such light for absorbance measurements. What these spectrometers actually provide is *polychromatic light* (or light containing a mixture of two or more wavelengths), where the range of these wavelengths is described by the term $\Delta\lambda$. The effects of polychromatic light on a Beer's law plot are shown in Figure 17.20. The presence of polychromatic light will not create significant deviations from Beer's law as long as the change in the molar absorptivity for the analyte ($\Delta\varepsilon$) is small over the wavelength range $\Delta\lambda$ that is present. A change in the molar absorbitivity of less than 1% ($\Delta\varepsilon/\varepsilon < 0.01$) is generally desired. It is for this reason that absorbance measurements are often performed at or near the top of a peak in an absorption spectrum, where ε is approximately constant. If a wavelength range is used on the side of a peak or $\Delta\varepsilon/\varepsilon$ does change due to the presence of polychromatic light, the result will be a curved Beer's law plot where a lower-then-expected absorbance is measured as the analyte concentration is increased.[12]

A third assumption made in Beer's law is that all light that is passed through the sample has the same distance of travel. This assumption requires that the path length b be a constant for the measurement of a given sample and is usually met by using a square sample cell and parallel rays of light that strike the sample cell at a perpendicular to the surface to the cell. If a round sample

TABLE 17.6   Important Assumptions Made in Beer's Law

(1) All absorbing species act independently of each other

(2) The light being used for the absorbance measurement is "monochromatic"

(3) All detected rays of light that pass through the sample have the same distance of travel

(4) The concentration of absorbing species is constant throughout the path of light in the sample

(5) The light that is being used to measure absorbance is not scattered by the sample

(6) The amount of light entering the sample is not large enough to cause saturation of the absorbing species in this sample

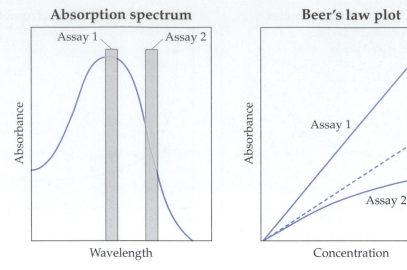

FIGURE 17.20   Effects of polychromatic light on plots made according to Beer's law.

cell is instead used (as can occur in some inexpensive instruments), curvature and negative deviations can appear at high analyte concentrations in a Beer's law plot.

There are several other inherent assumptions in Beer's law. For instance, a fourth assumption is that the concentration of the absorbing species is constant throughout the path length of light through the sample. Thus, we should make sure that we are working with a homogeneous sample when using Beer's law to relate the absorbance to analyte concentration in this sample. A fifth assumption in Beer's law is that all light entering the sample is either transmitted or absorbed. This assumption must be met if the relative difference in P versus P_0 is to be used to determine the absorbance and transmittance of light through the sample. Finally, a sixth assumption in Beer's law is that the amount of incoming light (as given by P_0) is not large enough to cause saturation of the absorbing species. This last requirement is needed to ensure that the measured value of P will increase in proportion to P_0 and that the value of P/P_0 will not be affected by the size of P. Deviations from this assumption can occur when using high-intensity light sources such as lasers (see Chapter 19).

Another situation in which curvature can occur in a Beer's law plot is if the absorbing species is involved in a chemical reaction that will alter the concentration of this particular species as its total analytical concentration is varied. This problem can appear in reactions that involve weak acids or weak bases. For instance, we learned in Chapter 8 that as the concentration of a solution of a weak acid in water is increased, the fraction of this acid in the dissociated form will decrease (i.e., the ratio of $[A^-]/[HA]$ decreases). If the molar

absorptivity is different for the two species HA and A^-, a plot of the overall measured absorbance versus the total analytical concentration can give a curved relationship (see Figure 17.21). Similar effects can occur for reactions that involve dimer formations, ion pairing, and other types of equilibria.

There are some instrumental limitations that can also cause deviations in a Beer's law plot. For instance, light that reaches the detector without going through the sample (known as *stray light*) can cause errors in the measured absorbance.[12] This stray light (which can be

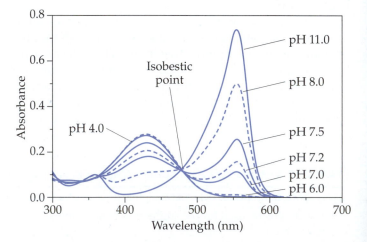

FIGURE 17.21   Change in absorbance spectrum for phenol red with a variation in pH. The observed change in these plots with pH is due to the conversion of phenol red from its acid form to its base form, which have different absorption spectra. This graph shows the location of an isobestic point at 480 nm. There are also two other isobestic points in these plots, which occur at 367 and 338 nm.

represented as having a radiant power of P_S) will add to values of both P and P_0, as shown in Equation 17.15. The net result is an observed absorbance (A_{Obs}) that is lower than the true absorbance of the sample.

Effect of Stray Light:
$$A_{Obs} = -\log\left(\frac{P + P_S}{P_0 + P_S}\right)$$

$$\approx -\log\left(\frac{P + P_S}{P_0}\right) \quad (\text{if } P_0 \gg P_S) \quad (17.15)$$

The result of this scattering is an increase in the apparent amount of light that is transmitted (or a decrease in A_{Obs}), leading to *negative* deviations in a Beer's law plot.

The reflection of light off of surfaces of the sample cell will also create deviations to Beer's law due to the instrument. In this case, the reflections will result in light that goes through the sample but that does not reach the detector. This effect will decrease the observed absorbance according to the relationship shown in Equation 17.16,

Effect of Reflection:
$$A_{Obs} = -\log\left(\frac{P - P_R}{P_0 - P_R}\right)$$

$$\approx -\log\left(\frac{P - P_R}{P_0}\right) \quad (\text{if } P_0 \gg P_R) \quad (17.16)$$

where P_R represents the power of the reflected light. This change, in turn, increases the apparent absorption of light and gives *positive* deviations in a Beer's law plot.

It is important to consider all of these possible sources of deviations when using Beer's law to perform a chemical analysis. Minimizing these effects will help to increase the linear range of a Beer's law plot. However, even if precautions are taken to reduce these effects it is important that we *always* carefully examine a Beer's law plot to determine whether this plot is or is not giving a linear response at the concentrations we wish to measure. This linearity can easily be examined by visually inspecting the Beer's law plot or by using such tools as a residual plot (as discussed in Chapter 5). If we wish to analyze a sample that has a concentration higher than this linear range, one common option is to dilute the sample until it has a concentration in this range.

Key Words

Absorbance 422	Frequency 409	Refraction 418	Spectroscopy 405
Absorption 414	Light 408	Refractive index 408	Spectrum 405
Beer's law 422	Molar absorptivity 423	Scattering 419	Transmission 415
Diffraction 420	Photon 411	Snell's law 418	Transmittance 422
Emission 413	Reflection 416	Spectrometer 405	Wavelength 409

Other Terms

Absorption spectrum 415	Emission spectrum 414	Percent transmittance 422	Remote sensing 405
Amplitude 408	Fresnel equation 417	Photoelectrical effect 411	Spectrochemical
Angle of incidence 416	Interference 420	Planck's constant 411	analysis 407
Angle of reflection 416	Monochromatic light 425	Planck's equation 411	Spectrometry 407
Beer's law plot 424	Nuclear magnetic	Polychromatic light 425	Stray light 426
Bragg equation 421	resonance	Radiant power 417	Wave number 409
Diffraction pattern 420	spectroscopy 412	Rayleigh scattering 419	
Electromagnetic	Path length 423		
radiation 408			

Questions

WHAT IS SPECTROSCOPY AND HOW IS IT USED FOR CHEMICAL ANALYSIS?

1. What is "spectroscopy"? What is a "spectrometer"? Explain how light is used by each of these two items.
2. What is "remote sensing"? What are some ways in which spectroscopy is used in remote sensing?

3. What is a "spectrum"? What information in used in plotting a spectrum? How can this information be used in chemical analysis?
4. Figure 17.22 shows a spectrum that was acquired for the complex Fe(1,10-phenanthroline)$_3{}^{2+}$, which is often used to measure the concentration of iron in water samples. At what

wavelength does this complex have its strongest absorption of light? What wavelength(s) would you use to measure this complex in absorption spectroscopy?

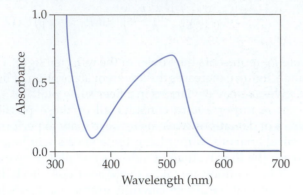

FIGURE 17.22   Spectrum for Fe(1,10-phenanthroline)$_3^{2+}$. This spectrum was acquired using a 6.26×10^{-5} M aqueous solution of this complex and using a sample cell that has a path length of 1.00 cm.

5. The following data were obtained at various wavelengths for a 3.6×10^{-4} M solution of permanganate held in a square cuvette with a path length of 1.00 cm.

Wavelength (nm)	% T	Wavelength (nm)	% T
400	89	575	50
425	92	600	82
450	83	625	86
475	60	650	88
500	27	675	93
525	15	700	96
550	29		

Plot a simple absorption spectrum for permanganate using these data. At what wavelength(s) in this spectrum does permanganate have the strongest absorption of light? At what wavelength(s) does it have the weakest absorption of light?

6. What is the difference between a "spectrochemical analysis" and "spectrometry"?

7. List two types of spectroscopy that are classified based on the types of analytes they examine. List two types of spectroscopy that are classified based on the type of light that they employ.

8. Briefly describe three different ways in which spectroscopy can be used for chemical analysis.

WHAT IS LIGHT?

9. Explain how "light" is defined by modern scientists. How does this definition differ from the common use of this term when referring to "visible light"?

10. What is meant by the "wave–particle duality" of light? Explain why this concept is essential in understanding the properties of light.

11. Define the following terms, as related to the wave model of light.
   (a) Crest
   (b) Trough
   (c) Amplitude

(d) Frequency
(e) Wavelength
(f) Wave number

12. What is the value of constant c? What is the physical meaning of this constant?

13. What is the definition of the "refractive index"? How is this index related to the velocity of light?

14. Calculate the following values.
   (a) The velocity of 589 nm light in pure water ($n = 1.333$ at 25°C)
   (b) The velocity of 250 nm light in fused quartz ($n = 1.507$ at 25°C)
   (c) The refractive index of white light in CO_2(g) ($v = 2.9965 \times 10^8$ m/s at 25°C and 1 atm)
   (d) The refractive index of 589 nm light in acetone ($v = 2.2063 \times 10^8$ m/s at 20°C)

15. A spectrometer has a path of 25 cm (roughly 1 ft) that must be traveled by a beam of light. Most of this travel occurs in air, which has an average refractive index of 1.0003 for the light. Approximately how long will it take the beam of light to cover this distance?

16. In 1999, a group of scientists used a special type of matter known as a Bose–Einstein condensate to slow a beam of light to a velocity of only 17 m/s (see Reference 8). What was the refractive index of this condensate?

17. Calculate the following values, using a velocity of c for light. In each case, indicate what type of electromagnetic radiation is present (e.g., visible light, ultraviolet light, etc.).
   (a) Frequency for 3.50 μm light
   (b) Wave number for 635 nm light
   (c) Wavelength for 2.1×10^{18} Hz radiation
   (d) Wave number for 5.5×10^{11} Hz radiation

18. A scientist wishes to study a chemical that has an energy transition that occurs at 4.5×10^{14} Hz. What is the wavelength (in nm) and wave number (in cm^{-1}) of the electromagnetic radiation that will be needed to study this transition? What general type of electromagnetic radiation (visible, ultraviolet, etc.) will be needed for this experiment?

19. An organic chemist uses both infrared and ultraviolet light to examine a newly synthesized chemical. This chemical is found to have strong absorption of light at both 2.5 μm and 400 nm. What general types of light are being absorbed at each of these wavelengths? How many photons of light at 2.5 μm are needed to give the same energy as one photon of light at 400 nm?

20. A chemist wishes to use a spectrometer that is capable of discriminating between light with a wavelength of 500 nm from light with a wavelength of 501 nm (when traveling through air at standard temperature and pressure [STP]). What is the difference in frequency for these two types of light? What is the difference in energy for a photon of light at each of these wavelengths?

21. The oxygen–oxygen bond in hydrogen peroxide contains 145 kJ/mol in energy. This bond can be broken if the hydrogen peroxide absorbs a photon that has the same amount of energy as is stored in this bond. What wavelength of light is needed to break this bond?

UPTAKE AND RELEASE OF LIGHT BY MATTER

22. Define "emission" as related to light. Describe the process of light emission by matter.

23. Sodium that is heated in a flame will emit light at 589 nm. This process is due to an electron falling from a 3p orbital to a 3s orbital.
    (a) Explain what is happening during this emission process, using the scheme in Figure 17.8 as a guide.
    (b) Light at 589 nm is the most intense emission that is observed for sodium. Use the transitions that are involved in this process to explain why this is the case.

24. Explain how the chemicals in a flame emit light, using the model in Figure 17.8 as an example.

25. What is an "emission spectrum"? How can an emission spectrum be used for chemical analysis?

26. What color and type of light would be associated with light that is emitted at each of the following wavelengths?
    (a) 475 nm
    (b) 250 nm
    (c) 675 nm
    (d) 1000 nm

27. What is meant by the terms "absorption" and "transmission" as related to light? Explain how these two terms are related.

28. How does light absorption differ from light emission?

29. What is an "absorption spectrum"? Discuss how this type of spectrum can be used for chemical analysis.

30. Based on the spectrum in Figure 17.2, over what ranges of wavelengths does sunlight have the greatest amount of absorption by Earth's atmosphere? Use this same figure to explain why the atmosphere is often said to be "transparent" to the visible range of sunlight.

31. Explain how the wavelengths of absorbed light are related to the color of a chemical.

32. What is the expected color of a solution if it strongly absorbs light from 450 to 500 nm? What is the expected color if the solution absorbs strongly at only 250–300 nm?

33. A pharmaceutical company develops a test for pregnancy in which a product is formed on a test strip that absorbs visible light and has a dark blue color. What wavelengths of visible light are probably absorbed by this colored product?

PHYSICAL INTERACTIONS OF LIGHT WITH MATTER

34. What is light "reflection"? What is needed for light reflection to occur?

35. Define the following terms.
    (a) Specular reflection
    (b) Diffuse reflection
    (c) Angle of incidence
    (d) Angle of reflection

36. What is the "Fresnell equation" and how is this equation used?

37. What factors determine the fraction of reflected light that will be reflected by a surface? Do you think this fraction of reflection will also be affected by the wavelength of light? Explain your answer.

38. Estimate the fraction of light that will be reflected in each of the following examples. In each case, you may assume that the light is striking the boundary at a right angle (90° to the plane of the boundary). (*Note*: You may assume that the light has a wavelength of 589 nm unless indicated otherwise.)
    (a) Light passing through air and striking a diamond
    (b) Light passing through water and striking ice
    (c) Light traveling from air to a surface made of hard cellulose acetate
    (d) Light passing through helium and striking a cuvette made of light flint glass

39. It is possible to estimate the amount of reflection that occurs for sunlight by Earth's atmosphere by using a model similar to that shown in Figure 17.8. In this model, we will assume there is a well-defined boundary between space and the atmosphere and that the atmosphere has an average refractive index of 1.0003 for visible light.
    (a) Calculate the fraction of reflection that would be expected for visible light at this boundary. Based on your results, do you think this reflection will play an important role during the use of sunlight for remote sensing?
    (b) The actual change from space to Earth's atmosphere is more gradual than the model in Figure 17.7, and the refractive index of air will change with altitude and other conditions. In general, how do you think these factors will affect the amount of visible light that is reflected by Earth's atmosphere?

40. What is light "refraction"? What is needed for refraction to occur?

41. What is "Snell's law"? What factors appear in this law?

42. At what angle would light be refracted if it passes from air ($n = 1.0003$) into glass ($n = 1.5171$) and hits the glass at an angle of 30° vs. the normal? How will this angle of refraction change if the light instead hits the glass at 45.0° or 60° vs. the normal?

43. Different types of glass will have different values for the refractive index of light. For instance, light with a wavelength of 589 nm has the following values of n for various types of glass: 1.51714 (ordinary crown glass), 1.52430 (borosilicate glass), and 1.65548 (dense flint glass).
    (a) Calculate the angle of refraction of 589 nm light as it passes from air into each of these glasses at an angle of 50.0° vs. the normal.
    (b) Repeat the calculation in Part (a), but now calculate the angle of refraction of 589 nm light as it passes from pure water into each glass at 50.0° vs. the normal.
    (c) Use the results in Parts (a) and (b) to rank the three types of glass based on their ability to refract 589 nm light. Which type of glass gives the largest change in direction of travel for this light? Which type of glass gives the smallest change due to refraction? Explain these results using Snell's law.

44. What is meant by light "scattering"? What is "Rayleigh scattering"?

45. Why is scattering important in chemical analysis?

46. Define the following terms. Explain how diffraction and interference are related.
    (a) Diffraction
    (b) Interference
    (c) Diffraction pattern
    (d) Constructive interference
    (e) Destructive interference

47. What are two ways in which diffraction is used for chemical analysis?

48. What is the "Bragg equation?" What terms appear in this equation?

49. A chemist wishes to use examine diffraction by a crystal of topaz that is exposed to X rays with a wavelength of 0.63 Å. If the interplanar distance of topaz is 1.356 Å, at what angle will a first-order constructive interference band be observed for this crystal?

50. Diffraction gratings are often used in place of prisms as wavelength selectors for ultraviolet-visible (UV-vis) spectroscopy. Typical gratings used in these devices have 300–2400 groves

per millimeter. If light approaches this grating at a right angle, the angle θ at which constructive interference will occur for the light is given by Equation 17.17,

$$d \sin \theta = n\lambda \qquad (17.17)$$

where λ is the wavelength of the light, d is the distance between adjacent grooves on the grating, and n is the order of diffraction for the observed constructive interference band.[12]

(a) At what angle will first-order constructive interference be observed for 600 nm light that hits at a right angle a grating that contains 1000 groves per mm?

(b) At what angle will first-order constructive interference be observed for 610 nm light when using the same grating as in Part (a)?

(c) If constructive interference bands are observed at a perpendicular distance that is 10.0 cm away from the grating, how far apart will be the first-order interference bands be at this location for the 600 and 610 nm light?

ANALYSIS BASED ON EMISSION

51. List the main components of a simple spectrometer for measuring light emission. Describe the function of each component.

52. Give a general equation that can be used to relate the amount of light emission to the chemical concentration of a sample. Describe each term that appears in this equation.

53. A series of calcium standards and two unknown calcium solutions are measured under the same conditions at 620 nm by flame emission spectroscopy. The following results are recorded. Determine the concentrations of Ca^{2+} in the unknown samples.

Concentration of Ca^{2+} (M)	Relative Measured Emission Intensity
0.000	0.0010
1.00×10^{-4}	5.02
2.00×10^{-4}	10.4
3.00×10^{-4}	16.1
5.00×10^{-4}	26.5
10.00×10^{-4}	51.3
Unknown no. 1	20.5
Unknown no. 2	57.0

54. The same system as in Problem 53 gives a signal of 127.6 for an unknown sample. (*Note*: This system is known to give a linear response up to a signal of roughly 150.) The student analyzing the sample takes a 10.00 mL aliquot, places it into a 50.00 mL volumetric flask, and fills the flask to the mark with deionized water. What approximate signal would be expected for the diluted sample? What is the concentration of Ca^{2+} in this sample?

ANALYSIS BASED ON ABSORPTION

55. List the main components of a simple spectrometer for measuring light absorption. Describe the function of each component.

56. Define "transmittance," "percent transmittance," and "absorbance." Show how these terms are related to each other.

57. Complete the following table, calculating all missing values.

Transmittance	% Transmittance	Absorbance
0.156	—	—
—	35.8	—
—	—	0.251
0.689	—	—
—	78.0	—
—	—	1.250

58. What is "Beer's law"? What terms appear in this law?

59. Calculate the molar absorptivity of an analyte if a 3.40×10^{-4} M solution of this chemical is placed in a sample cell with a path length of 5.0 cm and is found to give a value for % T of 67.4 at 450 nm.

60. Calculate the concentration of an analyte in solution placed in a 1.00 cm wide sample cell if the measured absorbance of this sample is 0.367 and the analyte has a known molar absorptivity of 6.87×10^3 L/mol · cm.

61. In the presence of a solution that has a background absorbance of 0.050, a 1.0×10^{-4} M solution of an analyte is found to give an overall absorbance of 0.350 ($b = 1.00$ cm). A similar solution that contains 3.0×10^{-4} M of the same analyte gives an overall measured absorbance of 0.900. What is the molar absorptivity of this analyte at the wavelength that is used in these experiments?

62. What is a "Beer's law plot"? How is this plot used in chemical analysis?

63. In Chapter 2, a Beer's law plot was shown for the measurement of "protein A" (see Figure 2.11). Based on this plot, what would be the expected concentration of this protein in an unknown sample that had an absorbance reading of 0.76? What would be the concentration in a sample that had a percent transmittance of 51%? Explain how you obtained your answers.

64. The following results were obtained by a clinical chemist when measuring morphine at 285 nm and using a 1.00 cm square cuvette as the sample holder.[28]

Morphine Concentration (M)	Absorbance
5.0×10^{-5}	0.229
1.0×10^{-4}	0.308
2.0×10^{-4}	0.467
5.0×10^{-4}	0.942

Prepare a Beer's law plot for these data. Use this plot to determine the concentration of morphine in an unknown sample that gives a measured absorbance of 0.615 in this assay.

65. Show how the overall measured absorbance for a sample will be affected by the presence of two or more analytes that absorb light at the same wavelength. Give an equation that describes this relationship.

66. What are some sources of random errors that can affect absorbance or transmittance measurements?

67. It was stated earlier that simple absorbance spectrometers tend to give the most reliable measurements for absorbance values between 0.1 and 0.8. Use Figure 17.19(a)

to explain this statement. Also explain why this range is not necessarily valid for all spectrometers that are used for absorbance measurements.

68. A pharmaceutical chemist wishes to determine the range at which both of the spectrometers in Figure 17.19 will provide a given level of precision for an absorbance measurement.
    (a) Over what range of absorbance values can the instrument in Figure 17.19(a) be used to provide a precision of 2.5% or better? Over what range of absorbance values will this instrument provide a precision of 3.0% or better?
    (b) Over what range of absorbance values will the instrument in Figure 17.19(b) give a precision of 2.5% or better? Over what range will the precision be 3.0% or better?

69. Explain why a Beer's law plot can show deviations from a straight line as the concentration of an analyte increases. Give two reasons why such deviations may occur.

70. Define the terms "monochromatic light" and "polychromatic light." Which type of light is assumed to be present when using Beer's law?

71. What happens to a Beer's law plot when there is an increase in the range of wavelengths that are used to make an absorbance measurement? Explain how selecting the proper wavelengths (e.g., those at the top of a peak in a spectrum vs. those on the side of a peak) can minimize this effect.

72. Based on the spectra in Figure 17.10, state how you would expect the slope and linear range to compare in Beer's law plots made for each of the following pairs of wavelengths.
    (a) 465 nm vs. 645 nm for chlorophyll b
    (b) 645 nm vs. 660 nm for chlorophyll a
    (c) 475 nm vs. 645 nm for chlorophyll b
    (d) 440 nm vs. 660 nm for chlorophyll a

73. What is "stray light"? How can stray light affect an absorbance measurement? How can stray light affect a Beer's law plot?

74. Discuss why reflection in a spectrometer can affect an absorbance measurement.

75. A sample is known to have an observed absorbance of 1.30, or % T = 5.0. What will be the apparent absorbance of this sample if stray light is present that has a radiant power equal to 1.0% of the incident radiant power ($P_S = 0.010\,P_0$)? What will the size of the error be in this observed absorbance vs. the true absorbance of the sample?

76. What will the observed absorbance of the sample in Problem 72 be if there is no stray light, but there is reflection of light at a level that is equal to 1.0% of the light's initial radiant power ($P_R = 0.010\,P_0$)? What will the size of the error in this observed absorbance be vs. the true absorbance of the sample?

CHALLENGE PROBLEMS

77. There are several different types of spectrometers used in analytical chemistry. One way these spectrometers can be classified is by the means with which they detect light. Use the Internet or resources like References 5 and 6 to determine how each of the following types of spectrometers is able to detect light.
    (a) Spectrophotometer
    (b) Spectrograph
    (c) Spectroscope

78. Locate the refractive index for each of the following materials by using the *CRC Handbook of Chemistry and Physics*[11] or other resources.
    (a) Benzene (sodium D light—589 nm, 20°C)
    (b) Rock salt (1.229 μm infrared light)

(c) 10% (w/w) aqueous solution of sucrose (sodium D light—589 nm, 20°C)
(d) Polyethylene plastic (medium density, sodium D light—589 nm light)

79. The Heisenberg uncertainty principle sets certain limits on the precision with which certain pairs of physical parameters can be determined. Two such related parameters are the lifetime of atom or molecule in its excited state and the energy that is released as this excited state goes to a lower energy state. This relationship is represented by the following equation,

$$\Delta E \cdot \Delta t \geq h/4\pi \qquad (17.18)$$

where ΔE is the uncertainty in energy of this transition, Δt is the lifetime of the excited state, and h is Planck's constant. One outcome of this relationship is that it's impossible to have truly monochromatic light due to an emission process because Equation 17.18 states that the shorter the lifetime of the excited state, the greater the uncertainty in its energy and, hence, the greater uncertainty there will be in the wavelength of light that is emitted from this excited state.
    (a) Light with a wavelength 589.3 nm is often emitted by sodium atoms. If the lifetime of these sodium atoms in their excited state is 10 ns, what is the uncertainty in the energy of this emitted light?
    (b) What will the uncertainty in the wavelength of light that is emitted by the sodium atoms in Part (a) be? From this information, determine the range of wavelengths that are actually emitted by the sodium atoms as they give off 589.3 nm light.
    (c) According to Equation 17.18, what would be required to have the emission of truly monochromatic light (i.e., what happens if $\Delta E = 0$)?

80. When light passes through a glass sample cell, it will encounter several boundaries between regions with different refractive indices. First, the light passes from air into the glass wall of the cell. This light then passes through the glass and onto the boundary between the glass and the sample. The light then travels through the sample until it reaches the other side of the glass sample cell. After reentering the glass, the light then passes onto the exterior of the sample cell, where the light goes back into the surrounding air and moves on to the detector.
    (a) At each of these boundaries a small amount of light will be lost due to reflection. Estimate the fraction of light that is reflected at each step of this process, assuming that the light is hitting each boundary at a normal angle (90° to the plane of the boundary).
    (b) What is the total fraction of light that is lost to reflection in this system? Which boundaries lead to the greatest loss of light due to reflection?

81. Molar absorptivity is not the only way we can describe the ability of a chemical to absorb light. Another closely related parameter is the *absorption cross section*, which can be calculated directly from the molar absorptivity by using unit conversions and dimensional analysis.[6,12]
    (a) What units are obtained from the value of ε when using the substitution 1 L = 1000 cm^3? What units are obtained when an additional substitution is made using Avogadro's number, where $N_A = 6.023 \times 10^{23}$ molecules/mol? Use your results to explain why the resulting value is referred to as an absorption "cross section."

(b) The absorption spectrum for benzene gives a peak with a molar absorptivity of 60,000 L/mol · cm at 184 nm.[29] What is the absorption cross section for benzene at this wavelength? How does this value compare with benzene's actual cross section of approximately 2.5×10^4 pm^2?

82. Any heated object will emit a continuum of wavelengths. This effect is known as "blackbody radiation" and is responsible for the broad range of wavelengths that are emitted by the sun.

(a) Equation 17.19 shows how the wavelength of maximum emission (λ_{max}) will change with temperature for a blackbody source,

$$\lambda_{max} = 2.879 \times 10^6/T \qquad (17.19)$$

where λ_{max} is given in units of nm, and T is the absolute temperature in units of Kelvin.[12] The sun is often modeled as a blackbody source with a temperature of 5900 K. What is the expected value for λ_{max} under these conditions? How does this value compare with the wavelength of maximum emission that was shown earlier in Figure 17.2?

(b) The amount of light that is given off at a given wavelength by a blackbody source will also depend on temperature. This relationship is described by Equation 17.20,

$$B = (2hc^2/\lambda^5)(1/e^{hc/\{\lambda kT\}} - 1) \qquad (17.20)$$

where B is the total radiance, a value that is proportional to the relative intensity of emitted light.[12] The other terms in this equation included the speed of light (c), the absolute temperature (T, in kelvin), the wavelength of light (λ, in meters), Planck's constant (h), and Boltzmann's constant (k). Use Equation 17.20 and a spreadsheet to plot the spectrum that would be expected between wavelengths of 0.1 and 3.0 μm for a blackbody source such as the sun at 5900 K. (*Note*: Be sure to use consistent SI units for all terms in Equation 17.20 when doing your calculations.) How do your results compare with the actual emission spectrum that was shown in Figure 17.2?

83. Optical fibers are an efficient way to move light from one place to another. The composition of such a fiber consists of a transparent core with a refractive index of n_1. This core is surrounded with a cladding material that has a lower refractive index (n_2, where $n_2 < n_1$). Total internal reflection will occur for light if this light strikes the core/cladding interface at an angle equal to or greater than θ, where $\sin(\theta) \geq (n_2/n_1)$. The angle $(90°-\theta)$, for the case in which θ is given in degrees, then describes the acceptance cone for the optical fiber, as shown earlier in Figure 17.13. One particular optical fiber has a core with a refractive index of 1.3334 and cladding with a refractive index of 1.4567. What range of angles make up the acceptance cone for this optical fiber?

84. The index of refraction for fused quartz, which changes at various wavelengths, is as follows: n = 1.54727 (202 nm), 1.53386 (214 nm), 1.46968 (404 nm), 1.46690 (434 nm), 1.45674 (644 nm), and 1.45640 (656 nm).

(a) A plot of refractive index vs. wavelength for a material is called a dispersion curve. Use the preceding data and a spreadsheet to prepare a dispersion curve for fused quartz.

(b) Use your spreadsheet to determine the angle at which each of the listed wavelengths will be refracted if light at these wavelengths were to pass through air and strike the surface of a fused-quartz prism at an incident angle of 45°. Make a plot that shows how this angle of refraction changes with the wavelength of light.

(c) Based on your results in Part (b), which wavelengths of light do you think will be the most easily separated based on their refraction by a fused/quartz prism? What colors of visible light will be the most easily separated by this prism?

85. The following data were obtained when using absorbance measurements to determine the amount of iron in a corrosion sample from a Civil War–era ship sunk in the ocean in the 1860s. A 0.0465 g portion of this sample was dissolved in acid and diluted to 100.00 mL. A 1.00 mL portion of this solution was then treated with 1,10-phenanthroline, hydroxylamine, and acetate buffer and diluted to a total volume of 100.00 mL. Several standard solutions were prepared from ferrous ammonium sulfate and treated in the same way. The following results were obtained after measuring the absorbances of these samples and standards at 510 nm.

Iron Concentration (mg/L)	Absorbance
0.000	0.002
0.359	0.077
1.198	0.127
2.995	0.256
5.990	1.090
8.985	1.556
Unknown sample	0.406

(a) Prepare a spreadsheet that uses these data to construct a Beer's law plot.

(b) Use your spreadsheet to examine the fit of a best-fit line to your Beer's law plot. What is the approximate linear range for this plot? What are the best-fit slope and intercept over this linear range?

(c) What is the concentration of iron in the unknown sample? What absorbance would have been expected if this sample had contained 4.15 mg/L iron? What would have been the approximate absorbance for a sample containing 15.7 mg/L iron?

86. Although a Beer's law plot is a popular way of plotting absorption data, another way of examining such data is to use a *Ringbom plot*. A Ringbom plot is prepared by plotting the transmittance of a sample vs. the concentration of an absorbing analyte in this sample.[30]

(a) Use a spreadsheet to calculate the transmittance for each sample and standard listed in Problem 85. Use these new values to prepare a Ringbom plot for these data. What type of response do you see in this plot?

(b) Use your Ringbom plot to determine the concentration of the unknown sample in Problem 85. How does your answer compare to the result that was obtained when using the Beer's law plot in Problem 85?

(c) Beginning with the definition of transmittance, derive an equation that shows how the transmittance for a sample containing a single absorbing analyte will be related to the concentration of this analyte. Use this derivation to explain the response that you saw for the Ringbom plot in Part (a).

TOPICS FOR DISCUSSION AND REPORTS

87. Locate more information on the area of remote sensing and its application. Discuss one specific example of remote sensing with your class. State how spectroscopy is being used in this application.

88. Obtain information on one of the following persons and explain who contributed to the field of spectroscopy or our current understanding of light.
    (a) James Clerk Maxwell
    (b) Heinrich Hertz
    (c) Max Planck
    (d) Robert Bunsen

89. Refractive-index measurements are often used to identify or confirm the identity of a relatively pure chemical compound. This type of measurement is performed by using an instrument known as a refractometer. Obtain more information on refractometers from the literature and manufacturers of these devices and learn about how they are used to measure refractive index. Discuss what you learn with other members of your class.

90. Visit a laboratory that has an instrument that makes use of NMR spectroscopy for chemical measurements. Describe how NMR spectroscopy is used in this laboratory and describe the types of equipment that are used to perform these studies.

91. NMR spectroscopy is also an important tool for the study of biochemical systems. Obtain a research article that used NMR spectroscopy to examine a biochemical system. Write a short report discussing how NMR spectroscopy was used in this article.

92. Obtain more information on one of the following methods. Report on how the method is performed, the types of chemicals it is used to examine, and the information that it provides on these chemicals.
    (a) Microwave absorption spectroscopy
    (b) Gamma-ray spectroscopy
    (c) Electron-spin resonance spectroscopy

93. Locate a recent research article in a journal like *Analytical Chemistry* in which optical fibers are used for chemical analysis. Describe the type of analysis that is being performed in this article and state how optical fibers were used to make this analysis possible.

References

1. National Aeronautics and Space Administration, *Terra: Flagship of the Earth Observing System*, Release No. 99-120, 1999.
2. M. Sharpe, "Focus Analyst in the Sky: Satellite-Based Remote Sensing," *Journal of Environmental Monitoring*, 2 (2000) 41N–44N.
3. T. M. Lillesand, R. W. Kiefer, and J. W. Chipman, *Remote Sensing and Image Interpretation*, 6th ed., Wiley, New York, 2007.
4. C. Elachi and J. Van Zyl, *Introduction to the Physics and Techniques of Remote Sensing*, 2nd ed., Wiley, New York, 2006.
5. G. Maludzinska, Ed., *Dictionary of Analytical Chemisty*, Elsevier, Amsterdam, the Netherlands, 1990.
6. J. Inczedy, T. Lengyel, and A. M. Ure, *Compendium of Analytical Nomenclature*, 3rd ed., Blackwell Science, Malden, MA, 1997.
7. F. Szabadvary, *History of Analytical Chemistry*, Pergamon Press, New York, 1966.
8. H. A. Laitinen and G. W. Ewing, *A History of Analytical Chemistry*, Maple Press, York, PA, 1977.
9. *The New Encyclopaedia Britannica*, 15th ed., Encyclopaedia Britannica, Chicago, IL, 2002.
10. H. D. Young, R. A. Freedman, A. L. Ford, and T. Sandlin, *University Physics*, Addison-Wesley, New York, 2007.
11. D. R. Lide, Ed., *CRC Handbook of Chemistry and Physics*, 83rd ed., CRC Press, Boca Raton, FL, 2002.
12. J. D. Ingle Jr. and S. R. Crouch, *Spectrochemical Analysis*, Prentice Hall, Upper Saddle River, NJ, 1988.
13. "Guide for Use of Terms in Reporting Data in Analytical Chemistry: Spectroscopy Nomenclature," *Analytical Chemistry*, 62 (1990) 91–92.
14. H. Geunther, *NMR Spectroscopy: Basic Principles, Concepts and Applications in Chemistry*, Wiley, New York, 1995.
15. J. W. Akitt and B. E. Mann, *NMR and Chemistry: An Introduction to Modern NMR Spectroscopy*, Thornes, London, 2000.
16. F. Bloch, W. W. Hansen, and M. Packard, "The Nuclear Induction Experiment," *Physics Review*, 70 (1946) 474–485.
17. E. M. Purcell, R. V. Pound, and N. Bloembergen, "Nuclear Magnetic Resonance Absorption by Hydrogen Gas," *Physics Review*, 70 (1946) 980–987.
18. The Nobel Prize Internet Archive, http://www.almaz.com/nobel/
19. C. Westbrook, C. K. Roth, and J. Talbot, *MRI in Practice*, Wiley, New York, 2007.
20. S. C. Bushong, *Magnetic Resonance Imaging: Physical and Biological Principles*, Mosby, Amsterdam, the Netherlands, 2003.
21. A. Einstein, "Heuristic Viewpoint on the Production and Conversion of Light," *Annalen der Physik*, 17 (1905) 132–148.
22. G. Gamow, *Thirty Years that Shook Physics: The Story of Quantum Theory*, Dover, Dover, DE, 1985.
23. L. D. Rothman, S. R. Crouch, and J. D. Ingle Jr., "Theoretical and Experimental Investigation of Factors Affecting Precision in Molecular Absorption Spectrophotometry," *Analytical Chemistry*, 47 (1975) 1226–1233.
24. P. Bougouer, *Essais d'Optique sur la Graduation de la Lumière*, Paris, 1729.
25. J. Lambert, *Photometria*, 1760.
26. A. Beer, "Bestimmung der Absorption des rothen Lichts in farbigen Fl" ussigketiten," *Annalen der Physik*, 86 (1852) 78–88.
27. D. R. Malinin and J. H. Yoe, "Development of the Laws of Colorimetry: A Historical Sketch," *Journal of Chemical Education*, 38 (1961) 129–131.
28. F. D. Snell and C. T. Snell, *Colorimetric Methods of Analysis, Including Photometric Methods*, Vol. IVAA, Van Nostrand Reinhold, New York, 1970.
29. D. A. Skoog, F. J. Holler, and S. R. Crouch, *Principles of Instrumental Analysis*, 6th ed., Brooks/Cole, Pacific Grove, CA, 2006.
30. A. Ringbom, "Accuracy of Colorimetric Determinations. Part 1," *Zeitschrift fur Analytische Chemie*, 115 (1939) 332–343.

Chapter 18

Molecular Spectroscopy

Chapter Outline

18.1 INTRODUCTION: THE GOOD, THE BAD, AND THE UGLY

Cholesterol is often viewed as being an undesirable molecule. This view is true in that high levels of cholesterol in the diet have been associated with heart disease, which is a leading cause of death in the United States and in many industrialized nations.[1–3] As a result, the screening and measurement of cholesterol levels in blood is now a routine part of medical exams in these countries. Simply measuring the total cholesterol content of blood, however, only gives a partial picture of the role this molecule plays in heart disease.[2–4]

There are actually several "forms" of cholesterol in blood. These forms consist of different particles of proteins, phospholipids, triglycerides, and cholesterol or cholesterol esters that transport cholesterol throughout the body.[4] Two important types of these cholesterol-containing particles are low-density lipoprotein (LDL) and high-density lipoprotein (HDL) (see general structure given in Figure 18.1). LDL particles carry cholesterol and cholesterol esters that have been formed in the liver and deliver these chemicals to other parts of the body. A large amount of LDL can lead to the creation of plaque in arteries, making LDL represent the "bad cholesterol" in the circulation. HDL particles act to remove excess cholesterol from the body and to take this cholesterol back to

the liver for excretion or recycling. This action has given HDL the label of "good cholesterol" because it helps prevent heart disease. The result of an imbalance in these two types of cholesterol-containing particles results in the "ugly" effects of cardiovascular disease.[2,4]

Spectroscopy is a valuable tool that is used in various ways by clinical laboratories to study and measure cholesterol, as well as many other molecules of interest in the diagnosis and treatment of disease. In this chapter, we examine a set of methods known as *molecular spectroscopy* that examine the interaction of light with an intact molecule.[5–7] Various approaches can be utilized in these measurements, including ultraviolet-visible (UV-vis) spectroscopy, infrared spectroscopy, and luminescence spectroscopy. We will also discuss in this chapter how each of these methods is carried out and learn about some typical applications and results for these measurements.

18.1A What Is Molecular Spectroscopy?

Molecular spectroscopy can be defined as the examination of the interactions of light with molecules.[5–7] Molecules can absorb light, emit light, and scatter light. All of these interactions can lead to chemical information. It was shown in Chapter 17 how absorption occurs when an analyte absorbs the energy of a photon, is raised to a higher energy level, and later emits light as the analyte

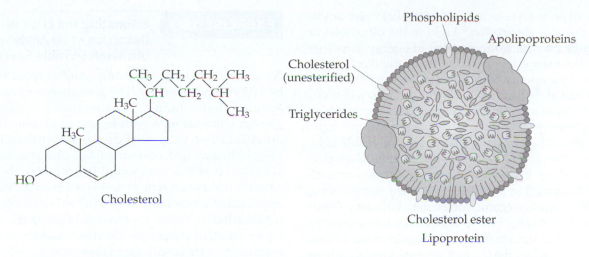

FIGURE 18.1 The structure of cholesterol and the basic structure of a lipoprotein, such as high-density lipoprotein (HDL) or low-density lipoprotein (LDL). Lipoproteins contain a nonpolar core of triglycerides and cholesterol esters that is surrounded by an outer coating of phospholipids, cholesterol, and special proteins known as "apolipoproteins." Lipoproteins are used in blood to help deliver and transport cholesterol, triglycerides, and other related agents throughout the body. (The diagram of a lipoprotein is courtesy of M. Sobansky.)

goes from the high energy level to a lower level. This absorption in molecules can involve a change in electronic energy levels, as well as a change in vibrational or rotational energy levels. This feature makes spectroscopy a valuable tool in providing both qualitative and quantitative information on molecules.

18.1B How Is Molecular Spectroscopy Used in Chemical Analysis?

Molecular spectroscopy has many uses in chemical analysis. UV-vis spectroscopy is used primarily for quantitative analysis or screening assays, while infrared spectroscopy is used primarily to identify molecular substances. Visible spectroscopy began with a technique called *colorimetry*.[6,7] In colorimetry, the analyte is combined with a reagent that will form a colored product. The color of this product is then compared to the color of standards, making it possible to determine the amount of analyte that is present in the sample or to simply see if the analyte is present above a certain level as part of a screening assay. One of the first known uses of colorimetry was by Pliny the Elder around A.D. 60 when he used an extract of gallnuts to test for the presence of iron in vinegar.[8] Colorimetry is still employed in modern chemical analysis, with common examples being the use of color-forming reactions in home kits sold for pregnancy detection or for monitoring blood cholesterol levels.[9]

In the 1830s and 1840s methods were developed to measure the concentrations of specific metal ions in slender tubes. This method was developed in 1846 by the Italian chemist Augustin Jacquelain for the analysis of Cu^{2+} in the presence of ammonia in water, giving $Cu(NH_3)_4{}^{2+}$ as a colored product.[8] The color of an unknown solution was compared with several standards by this approach to determine the concentration of the

unknown. Visual comparison colorimeters are still widely used for measurement of a wide variety of substances in field measurements (e.g., by environmental chemists) so that samples need not be returned to the laboratory for measurement. However, most quantitative measurements made by molecular spectroscopy are conducted in the laboratory using instruments that are designed to conduct more accurate and precise measurements of light absorbance or emission for chemical analysis.[7,10,11] We will see several examples of such instruments in Sections 18.2 and 18.4.

Spectroscopy can also be used to help identify chemicals and to provide information on their structure. Some information on chemical identity can be obtained by using the interactions of chemicals with ultraviolet or visible light, but infrared (IR) light is much more frequently utilized for this purpose. This is the case because an IR spectrum for a molecule often has many peaks instead of the one or two that are typically seen for the absorption of ultraviolet or visible light by molecules. In addition, the location, number, and intensities of the peaks in an IR spectrum form a useful fingerprint pattern for a molecule that can be used in identification or to determine the types of functional groups this molecule possesses.[7,10,11] The use of IR light and IR spectroscopy for this purpose is discussed in more detail in Section 18.3.

18.2 ULTRAVIOLET-VISIBLE SPECTROSCOPY

18.2A General Principles of Ultraviolet-Visible Spectroscopy

UV-visible spectroscopy (often called "UV-vis spectroscopy") is a common method for the analysis of molecules and other types of chemicals. This technique can be defined as a type of spectroscopy that is used to examine the ability of an analyte to interact with ultraviolet or visible light through absorption.[5] Light in the ultraviolet or visible range has the same amount of energy as occurs between the energy levels for some of the electrons in molecules. If the energy of this light exactly matches a difference in one of the energy levels, the electron will move from an orbital in a lower energy state to an empty orbital in a higher energy state if the molecule absorbs this light.

The absorption of both ultraviolet and visible light, especially in the range of 200–780 nm, often involves electronic transitions in molecules by π electrons or nonbonded electrons (n) as they go to an excited electron state, π^*. For this reason, organic molecules with only single bonds and σ electrons, but no π electrons or nonbonded electrons, tend to not absorb in this region of the UV-vis spectrum.[7,11] This fact means that saturated hydrocarbons can't be measured by routine UV-vis spectroscopy, because they contain only single bonds and σ electrons. However, a molecule like cholesterol that contains a number of carbon–carbon double bonds and oxygen atoms (which have nonbonding electrons) does have strong absorbance above 200 nm, as shown by the spectrum in Figure 18.2.

The portion of a molecule that has properties that allow it to absorb light is known as a *chromophore*.[5–7] A typical chromophore in an organic molecule will contain π bonds (such as those in double or triple bonds) and will often have extended conjugation (i.e., a large number of sequential double and single bonds) that tend to absorb ultraviolet or visible light efficiently. This absorption process can be described by using Beer's law, as discussed in Chapter 17. Table 18.1 shows some typical molar absorptivities for chromophores in organic molecules that absorb in the ultraviolet or visible range. The size of these molar absorptivities is sufficiently high to allow many organic compounds containing these chromophores to be measured at μM concentrations or less when using an appropriate detection wavelength and sample path length.

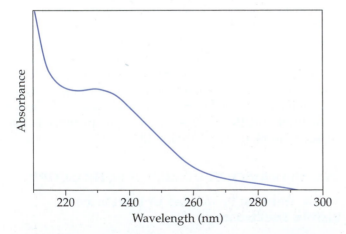

FIGURE 18.2 A typical Ultraviolet-visible (UV-vis) absorption spectrum, using the analysis of cholesterol as an example. This spectrum illustrates the broad absorption bands that are often seen with molecules for ultraviolet or visible light.

EXERCISE 18.1 | **Estimating the Limit of Detection of an Analyte in Ultraviolet-Visible Spectroscopy**

An unsaturated organic compound is to be examined by UV-vis spectroscopy. The absorbance of an aqueous solution that contains this compound is measured at 450 nm when using a path length of 1.0 cm. The smallest absorbance value that can be reliably measured by this particular instrument is 0.002 absorbance units. It is known that this compound has a molar absorptivity of 1.5×10^4 L/mol·cm at 450 nm and that there is no appreciable absorbance from any other components in its solution at this wavelength. What is the expected lower limit of detection for this analyte under these conditions if its absorbance is measured at 450 nm?

SOLUTION

We can find the lower limit of detection for the concentration of the analyte (C_{LOD}) by using Beer's law (see Equation 17.12 in Chapter 17). This can be done by using the smallest reliable absorbance measurement ($A = 0.002$), along with the known molar absorptivity for the analyte ($\varepsilon = 1.5 \times 10^4$ L/mol·cm) and the path length of light through the sample ($b = 1.0$ cm).

Beer's Law: $\quad A = \varepsilon b C$

$$(0.002) = (1.5 \times 10^4 \text{ L/mol·cm})(1.0 \text{ cm})\, C_{LOD}$$

$$\therefore \quad C_{LOD} = \frac{0.002}{(1.5 \times 10^4 \text{ L/mol·cm})(1.0 \text{ cm})}$$

$$= \mathbf{1.3 \times 10^{-7}\ M}$$

An even lower limit of detection would be expected for an analyte with a higher molar absorptivity. For instance, a molecule that has a 10-fold higher value for ε (1.5×10^5 L/mol·cm) would have a 10-fold lower limit of detection ($C_{LOD} = 1.3 \times 10^{-8}$ M).

The absorption of light in the ultraviolet or visible range is also possible for transition metal ions as they undergo electronic transitions that involve d- or f-shell electrons.[7,11] This type of absorption is seen for the ions of most lanthanide and actinide elements in the periodic table. Chemical species that can undergo charge-transfer absorption are also quite important in many assay methods because of their large molar absorptivities and ease of measurement in UV-vis spectroscopy. This type of light absorption is seen for many inorganic complexes. A few common examples include the complex that is formed between Fe^{2+} and 1,10-phenanthroline, and the complexes that are produced when Fe^{3+} reacts with thiocyanate and phenol. This type of process is also responsible for CrO_4^{2-} and MnO_4^{-} having strong absorption of visible light.[7]

TABLE 18.1   Examples of Chromophores in Ultraviolet-Visible Spectroscopy for Organic Molecules*

Chromophore	Wavelength of Absorbance Maximum (nm)	Molar Absorptivity (L/mol·cm)
C=C	182	25
	174	16,000
	170	16,500
	162	10,000
—C≡C—	172	2,500
C=C—C=C	209	25,000
(benzene ring)	255	200
	200	6,300
	180	100,000
(naphthalene)	270	5,000
	221	100,000
C=O	295	10
	185	Strong

*The data in this table are from J.D. Ingle Jr. and S.R. Crouch, *Spectrochemical Analysis*, Prentice Hall, Upper Saddle River, NJ, 1988. Futher examples of chromophores can be found in this reference.

A typical UV-vis absorption spectrum was given earlier in Figure 18.2, using cholesterol as an example. In this particular case, wavelengths up to only 300 nm are shown in the spectrum because cholesterol does not absorb at higher wavelengths between 300 and 800 nm. The x-axis for this type of spectrum is usually plotted in terms of the wavelength, which is commonly expressed in units of nanometers for ultraviolet or visible light. The y-axis is often plotted using the measured absorbance at a given wavelength. When working with analytes that are molecules, the bands in this type of spectrum tend to be quite broad. For instance, cholesterol has an absorption band that spans at least 30–40 nm from 225 nm to 260 nm in Figure 18.2. The reason for the width of these molecular absorption bands is that the electronic energy levels that are involved in these absorption processes also contain many smaller vibrational and rotational transitions. The result is many possible transitions in a molecule that have only small differences in energy, which creates the broad peak observed in the absorption spectrum. In addition, spectra that are acquired for a molecule in a solution are broad because of the rapid, repeated collisions each molecule undergoes with neighboring solvent molecules. These collisions shorten the lifetime of each vibrational/rotational state in the molecule to such a great extent that the Heisenberg uncertainty principle comes into play (see problems at end of Chapter 17). This effect causes a loss in precision for the measured changes in energy due to light absorption, and in the wavelengths of absorbed light that are observed.[7]

18.2B Instrumentation for Ultraviolet-Visible Spectroscopy

Typical System Components.   An instrument that is used to examine the absorption of light in UV-vis spectroscopy is known as a *UV-vis absorbance spectrometer*. In Chapter 17 we saw that an absorbance spectrometer has four basic components: a light source, a means for selecting a particular type of light for analysis, a sample container, and a detector. A UV-vis absorbance spectrometer has all of these components, but each of these components must be capable of working with ultraviolet or visible light.

A common light source for visible light is a *tungsten lamp*, as illustrated in Figure 18.3. In this device, a heated tungsten wire gives off a wide spectrum of light with both an intensity and wavelength maximum that depend on the temperature of the wire. This type of light emission is known as "blackbody radiation" (see problems at end of Chapter 17). Tungsten is used in this lamp because it can be heated hotter than any other metal without melting, making it possible to obtain strong-emission and high-emission wavelengths with this material. At a typical operating temperature (2000–3000 K), the wavelength of maximum emission is at about 1000 nm for a tungsten lamp, with a usable range that spans from 320 to 2500 nm.[7]

A modified form of this previous design is the *tungsten/halogen lamp*. In this device a small amount of iodine is also present inside the tungsten lamp. As the tungsten gets hot, small amounts of tungsten atoms will

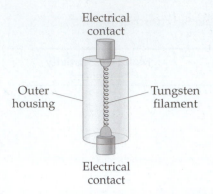

Electrical
contact

Outer
housing

Tungsten
filament

Electrical
contact

FIGURE 18.3 The general design of a tungsten lamp. A tungsten/halogen lamp has a similar design, but includes some I_2 in the chamber that surrounds the tungsten filament.

sublime off the surface and coat the interior of the lamp with a gray solid. The presence of iodine, however, will cause the gas-phase tungsten atoms to react and form tungsten(II) iodide, WI_2.

$$W + I_2 \rightarrow WI_2 \qquad (18.1)$$

WI_2 is stable at low temperatures, but if a WI_2 molecule hits the hot tungsten filament, it will decompose to put tungsten back onto the filament (the iodine atoms also recombine to form I_2 as part of this process). In this way, a more stable lamp is created that can be operated at a higher temperature (up to 3600 K) and with a greater intensity of light emission. This lamp can provide emitted light up to about 3000 nm.[7] Both the tungsten and tungsten/halogen lamps are mostly used in the visible region of the spectrum and emit almost all their radiation as infrared or visible light. As a result, a different type of light source is needed for work in the ultraviolet region in UV-vis spectroscopy.

The *hydrogen lamp* and *deuterium lamp* are two other light sources that can be used in UV-vis spectroscopy.[7,11] Both lamps consist of two inert electrodes across which a high voltage is imposed in a quartz bulb that is filled at a low pressure with either H_2 or D_2. The presence of the high voltage results in the excitation of H_2 or D_2 and their dissociation into H or D atoms plus with the emission of light. A hydrogen lamp was used by Neils Bohr as he sought to understand the nature and electronic structure of the hydrogen atom. Bohr's interest was in the several bright lines of emitted light that corresponded to electrons falling from a higher orbital into the $n = 2$ orbital. This same type of lamp provides a continuous source of radiation in the ultraviolet region by using light that is emitted when an electron falls from a nonquantized energy level into the $n = 2$ orbital of what had been a gas-phase hydrogen ion. Hydrogen and deuterium lamps provide ultraviolet radiation as a continuous band that spans from roughly 180 to 370 nm. A deuterium lamp is more common in modern laboratories because it gives more intense light emission than a hydrogen lamp.[7]

The wavelength selector (or monochromator) in an absorbance instrument often consists of a narrow slit through which light enters, a device for separating this light into its various wavelengths, and another narrow slit through which a particular portion of this light is allowed to exit and pass onto the sample and detector. Instruments for absorption measurements in UV-vis spectroscopy often use a prism or grating to separate light. The use of a prism for this purpose is demonstrated in Figure 18.4. Prisms used in the visible region are made of glass. The difference in the index of refraction of glass as a function of wavelength creates a separation of light into its various colors. The bigger that this change in angle is with wavelength, the better the separation will be for light with wavelengths in this range of the spectrum.

There are two types of gratings that can be used as monochromators. A *transmission grating* is one in which diffraction is produced by having light pass through a grating consisting of a series of small slits that create constructive and destructive interference of the light (see discussion in Chapter 17). Different wavelengths of light will have different angles at which they will produce constructive interference, making it possible to then select these wavelengths for use in absorbance measurements. A *reflection grating* is more common in modern instruments. This type of grating uses a polished and reflective surface that has a series of parallel and closely spaced steps cut into its surface, as is also illustrated in Figure 18.4. When a beam of light is reflected from this surface, a pattern of constructive and destructive interference is again created in which certain angles have constructive interference for

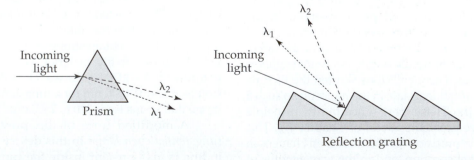

Incoming
light

Prism

λ_2

λ_1

Incoming
light

λ_1

λ_2

Reflection grating

FIGURE 18.4 Use of a prism (left) or a reflection grating (right) for the separation of light containing various wavelengths.

particular wavelengths of light. Changing the orientation and sample angle of either a transmission or reflection grating can allow the desired wavelengths of light with constructive interference to strike an exit slit and pass onto the sample and detector for use in absorbance measurements.[7,11]

The sample holder in UV-vis spectroscopy is usually a cuvette into which a sample or standard solution can be placed. This cuvette must be transparent to the wavelengths of light that will be used for the measurement and have a well-defined geometry. Cuvettes for work with light in the visible region are typically made of glass or a clear plastic. For work with ultraviolet light, cuvettes made from quartz or fused silica are required, because glass or many types of plastics will absorb such light. Many modern spectrometers use cuvettes with a square cross section and with an internal path length of 1.00 cm. However, some instruments use cylindrical test tube-looking cuvettes and/or cuvettes with longer path lengths.[7]

Many detectors that are used to monitor ultraviolet or visible light make use of the photoelectrical effect (or "photoelectric effect"), as discussed in Chapter 17, which occurs when light strikes certain substances and causes the ejection of an electron from this surface. Such an effect can be used in a phototube or photomultiplier tube to detect the light (see Figure 18.5 for an example). Both of these devices are designed so that light can enter them and strike a photoactive substance. The result is the production of electrons and a current that is proportional in size to the number of photons that have struck the photoactive surface. The absorption of a greater number of photons by the sample will result in a lower flux striking the phototube and a decreased current measurment.[7,10,11]

Single- and Dual-Beam Instruments.

Two common types of devices for UV-vis spectroscopy are "single-beam" and "double-beam" instruments. As its name implies, a **single-beam instrument** has a single path for the light to take through the instrument, as shown in Figure 18.6(a).[7,11] The light in this instrument originates from the source, a wavelength is selected by using the monochromator, this light is passed through the sample in a cuvette, and the intensity of the remaining light is measured by a detector. Because transmittance and absorbance are defined in terms of the ratio of light intensity that passes through the sample divided by the intensity of light that is entering the sample, there must be some way to measure both of these quantities in this device. In a single-beam instrument, the device is first adjusted so that it gives an output of 0% transmittance when a shutter is closed and no light reaches the sample (the "dark current"). The instrument is then set to read 100% transmittance (or $A = 0$) when a blank solution containing no analyte is in place. Next, the sample is introduced and its % transmittance (% T) or absorbance is measured. With older single-beam instruments, one must reset the dark current and 100% T current whenever wavelength is changed. With modern single-beam instruments, a computer remembers these settings during the collection of an entire spectrum. One possible problem with this process is that the intensity of the lamp or the response of the detector may change between the time the instrument is adjusted and the sample is analyzed. This type of change, if not corrected for, would give a systematic error in the final results.

A **double-beam instrument** in spectroscopy is a device in which the original beam of light is split so that half the light goes through a reference (or "blank") solution, while the other half passes through the sample.[7,11] This type of instrument, as shown in Figure 18.6(b), helps minimize errors that are caused by any drift in the lamp's intensity or in the detector's response. If either the lamp intensity or the detector response changes, this change will affect both the signal from the sample and from the blank, while the ratio of these signals will remain unchanged.

Related Devices.

Another design for a UV-vis spectrophotometer is one based on a **diode-array detector** (see Figure 18.7). This device differs from a standard spectrophotometer in that the monochromator has an entrance slit but no exit slit, which allows light of many wavelengths to enter the sample and to be detected simultaneously by an array of small diode detectors. Each diode in this array monitors a small range of wavelengths that have been separated *after* they have passed through the sample. This process allows the simultaneous measurement of an entire spectrum for a sample. These are single-beam instruments with a computer to remember the signal at all wavelengths for the dark current and the 100% T signal, as well as to monitor the light at each of the photodiodes in the array detector.[7,11–13]

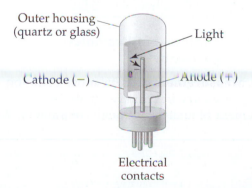

FIGURE 18.5 The basic components of a phototube. As light enters this device and strikes the cathode, electrons are given off by a photoemissive material on the cathode's surface. These electrons then travel to the more positive anode. This produces a current that is related to the intensity of light entering the phototube and striking the cathode. A photomultiplier tube has a similar design, but contains a series of intermediate electrodes between the cathode and anode that are each at a progressively more positive potential; this design results in a large number of electrons being produced for each collision of a photon with the cathode.

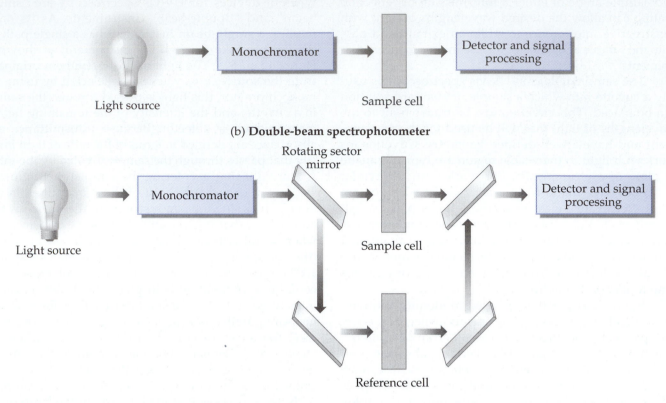

FIGURE 18.6 General design of (a) a single-beam instrument, or (b) a double-beam instrument for UV-vis spectroscopy.

Flow injection analysis (FIA) is a useful way in which UV-vis spectroscopy can be used to conduct a large number of routine analyses in a short period of time. In this approach, samples are injected sequentially into a flowing stream of a reagent, which then reacts with the contents of these samples to give a colored product.[14–15] After color development, the flowing stream passes through a spectrophotometer that is part of the FIA system (see Figure 18.8). The absorbance of the solution that results from the reaction of the analyte with the color-forming reagent is then measured, allowing the concentration of the analyte to be determined. The most desirable analyses for FIA are those in which the colored product forms rapidly, so that only a few seconds pass

between injection of the sample into the stream of reagent and the measurement of its absorbance. FIA is commonly used in laboratories where the same analytes are monitored on a regular basis in a large number of samples. A good example is the use of FIA and related flow-based devices to determine the concentration of cholesterol in blood samples in a clinical laboratory.[2,4]

18.2C Applications of Ultraviolet-Visible Spectroscopy

Direct Measurements. The most common use of spectrophotometry in chemical analysis is in the direct measurement of analytes through colorimetry. The term

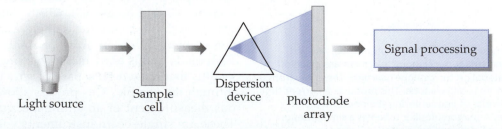

FIGURE 18.7 The general design of a diode array detector for UV-vis spectroscopy. Although this particular design shows a prism as the dispersing device for the sake of simplicity, gratings can also be used for light dispersion in this type of instrument.

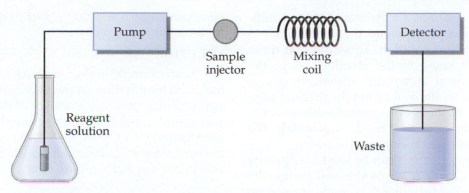

FIGURE 18.8 An example of a simple system for performing flow injection analysis (FIA). In this particular case, the sample is injected into a flowing stream of a reagent that results in the formation of a product that is monitored later at the detector. A mixing coil is present between the sample injector and detector to provide time for this reaction to occur. More advanced systems that use multiple reagents and flow streams and temperature control for the mixing chamber can also be developed based on this general design.

"colorimetry" is often used to describe the use of spectrometry in the visible region of the spectrum, where one can visually observe the color of a sample. This approach can also be use in the form of direct measurements made by instruments in the ultraviolet region. In this technique, the concentration of an analyte is determined by comparing the absorbance of an unknown concentration of a substance with the absorbance of a known concentration of the same material. Beer's law is then used to relate the measured absorbance of an analyte to its concentration.

If the molar absorptivity of the analyte is known and a sample has a known or negligible background absorbance, the concentration of the analyte can be determined directly by Beer's law. For instance, suppose the molar absorptivity of an analyte is 5.34×10^3 L/mol·cm at a particular wavelength. The same analyte gives an absorbance of 0.573 in a sample placed into a 2.00 cm sample holder, where a blank is used to correct for any background absorbance from the sample. One way the concentration of analyte in the sample can then be estimated is by rearranging Beer's law to give $C = A/(\epsilon b)$, or $C = 0.573/(5.34 \times 10^3$ L/mol·cm)(2.00 cm) $= 5.36 \times 10^{-5}$ M. Although this single-point method is simple to do, it is often subject to errors if the sample and standard are not both in the linear response region of Beer's law (see discussion of deviations from Beer's law in Chapter 17) or if there are differences in conditions, such as the solution pH or small changes in the detection wavelength, that were used to examine these solutions. A better approach is to use multiple standards and to prepare a Beer's law plot (see Figure 18.9), where these standards are measured under the same solution and wavelength conditions that will be used for the samples. The concentration of an analyte in a sample can then be determined by comparing the absorbance of this sample to the response of the plot.

Standard Addition Method. Another approach that can be used in UV-vis spectroscopy for analyte measurement is the **standard addition method**.[5-7] This is a technique that is used to determine the concentration of an

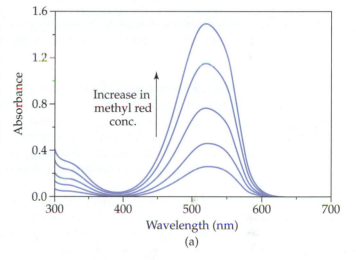

(a)

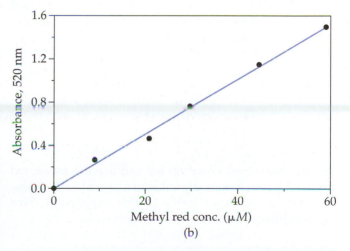

(b)

FIGURE 18.9 (a) A series of absorption spectra, and (b) a Beer's law plot for an analyte that is being examined by UV-vis spectroscopy, using methyl red as an example. These results were obtained for aqueous solutions of methyl red that were prepared in an acidic solution (pH < 4.4) and measured at 520 nm using a cuvette with a 1 cm path length. The same solutions of methyl red were used to acquire the spectra and to obtain the absorbance values that were used in the Beer's law plot.

analyte in a sample that has a matrix or solution conditions (such as pH or ionic strength) that are difficult to reproduce in a standard solution. To assure that these parameters are the same in the standard as in the unknown, the standard material is added directly to part of the sample. The signal for both the original sample and the sample that has been spiked with the standard are then measured and used to calculate the concentration of analyte in the original sample.

The following derivation can be used to illustrate how this method works. First, we need to assume that the signal we are measuring for either the sample or the standard will be proportional to the concentration of the analyte in each. We can represent this idea for absorbance measurements by the following relationships,

Absorbance of Original Sample: $\quad A_o = k\,C_o \quad$ (18.2)

Absorbance of Spiked Sample: $\quad A_{sp} = k\,C_{sp} \quad$ (18.3)

where C_o is the concentration of analyte in the original sample, C_{sp} is the total concentration of analyte in the sample that has been spiked with a known amount of the analyte, and k is a proportionality constant. (*Note:* $k = \varepsilon b$, for an absorbance measurement based on Beer's law.) We are assuming that the constant k is the same for both the original and spiked sample, which should be true if they contain the same matrix and we are working in the linear range of Beer's law. We are also assuming in this case that there is no background absorbance from the sample or its matrix.

Along with the measured absorbances of the original and spiked samples, we also know in this method (1) the original volume of the sample (V_o), (2) the volume of standard solution that we spiked into this sample (V_s), and (3) the concentration of analyte in this standard solution (C_s). This information can be combined to find the value of C_o, which is the goal of the standard addition method. To do this, we divide Equation 18.2 by Equation 18.3 to find the ratio of A_o/A_{sp} and substitute in the fact that $C_{sp} = (C_oV_o + C_sV_s)/(V_o + V_s)$.

$$\frac{A_o}{A_{sp}} = \frac{C_o\,(V_o + V_s)}{C_oV_o + C_sV_s} \qquad (18.4)$$

With this combined equation we can use the measured ratio A_o/A_{sp} to calculate the value of C_o because we know the values of all other terms in this expression. This process is illustrated in the following exercise.

A chemist performs a colorimetric assay that selectively measures iron. A 20.0 mL portion of the original sample gives an absorbance reading of 0.367 and a 20.00 mL portion of the same sample that has been spiked with 5.00 mL of a $2.00 \times 10^{-2}\,M$ iron solution gives an absorbance of 0.538. What was the concentration of iron in the original sample?

SOLUTION

This is an example of the standard addition method. We are given information on the measured absorbance, solution volumes, and concentration of the iron solution that was spiked into the sample. We simply have to place this information into Equation 18.4 and rearrange this equation to solve for C_o, which represents the concentration of iron in our original sample.

$$\frac{0.367}{0.538} = \frac{C_o\,(0.02000\text{ L} + 0.00500\text{ L})}{C_o\,(0.02000\text{ L}) + (0.005000\text{ L})(2.00 \times 10^{-2}\,M)}$$

$$\Rightarrow \quad 0.6822 = \frac{C_o\,(0.02500\text{ L})}{C_o\,(0.02000\text{ L}) + (1.00 \times 10^{-4}\text{ mol})}$$

$$\Rightarrow \quad (0.6822)[C_o\,(0.02000\text{ L}) + (1.00 \times 10^{-4}\text{ mol})]$$

$$= C_o(0.02500\text{ L})$$

$$\Rightarrow \quad (0.6822)(1.00 \times 10^{-4}\text{ mol})$$

$$= C_o(0.02500\text{ L} - 0.02000\text{ L})$$

$$\therefore \; C_o = \frac{(0.6822)(1.00 \times 10^{-4}\text{ mol})}{(0.02500\text{ L} - 0.02000\text{ L})}$$

$$= 0.01364\,M = \mathbf{0.0136\,M}$$

The previous equation shows how standard addition can be conducted when using only one spiked sample. It is also possible to use multiple spikes of a standard into the sample, with the assay response being measured for each spiked sample. The results of this assay would then be used to prepare a plot of $A_{sp}\,(V_o + V_s)$ on the y-axis and $C_s\,(V_s/V_o)$ on the x-axis (see the example in Figure 18.10). For absorbance measurements that are made in the linear region of Beer's law, the resulting plot should give a straight line where the value of the x-intercept is equal to $-C_o$, which provides the concentration of the analyte in the original sample.

Spectrophotometric Titrations. A common way in which UV-vis spectroscopy is employed in titrations is in the detection of visual indicators (see Chapters 12, 13, and 15). For example, one acid–base indicator that we have already discussed is methyl red (see spectra in Figure 18.11). Methyl red is a "two-color indicator" which is red in an acidic solution and yellow in a solution with a basic pH. Both forms of methyl red are able to strongly absorb visible light but do so at different wavelengths, thus explaining why they have different colors. As the pH is changed during the course of a titration, we alter the relative amount of methyl red that exists in the acid or base form. This alteration in the relative amount

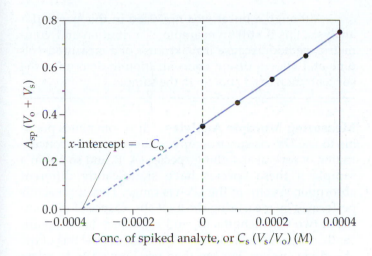

FIGURE 18.10   A plot for determining the concentration of analyte in a sample through the method of standard addition. The apparent amount of spiked analyte that corresponds to the negative value of the x-intercept is then determined, which provides a value that gives the amount of analyte in the original sample.

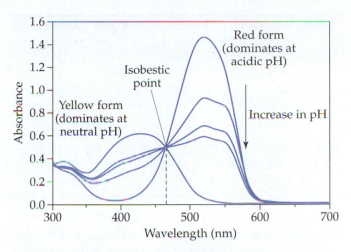

FIGURE 18.11   Absorption spectra obtained at several pH values for methyl red, a common visual indicator used for acid–base titrations. The pH range used here corresponds to the same range over which methyl red is used as an indicator, in which this chemical is red in an acidic solution (pH < 4.4) and yellow in a more basic solution (pH > 6.2). The approximate isobestic point for these acid and base forms occurs at 464 nm in these spectra.

of these two species is what leads to the observed change in color at the end point of the titration.

Besides using a visual indicator and our eyes to locate the end point of a titration, it is possible to use UV-vis spectroscopy to follow the course of a titration by measuring the absorbance due to an indicator, the analyte, the titrant, or the titration product. In the case of an indicator like methyl red, the end point could be detected by measuring the absorbance at a wavelength at which either the acid or base forms of the indicator have different molar absorptivities. Measurements at multiple wavelengths can also be used to determine the relative amounts of the different forms of the indicator at a given point during the titration (as is discussed in the next section). It is important during this type of analysis to select a wavelength at which the two forms of indicator have very different absorption spectra. This is not possible if we are using a wavelength at which the spectra for these two forms intersect (as occurs in Figure 18.11 at a wavelength of 464 nm). This point of intersection in the spectra for two absorbing species is called an *isobestic point* (see Chapter 17) and represents a place where the two forms have an identical molar absorptivity.[5,6] Although an isobestic point should not be used if we are trying to differentiate between two absorbing species, this point can be valuable if we wish to measure the *total* amount of these species or want to work at a wavelength that will not give a change in absorbance as the relative fraction of these two species is altered.

Absorption spectroscopy can also be used to follow a titration if the analyte, titrant, or product of the titration reaction has any significant absorption of visible or UV light. In this case, we can use UV-vis spectroscopy to measure the absorbance at different points along the titration and use this information to locate the end point. This approach is called a **spectrophotometric titration**.[5–7] The actual shape of the titration curve in this type of

measurement will depend on the relative size of the molar absorptivities for the analyte, titrant, and titration product (see general examples in Figure 18.12 and a specific example in the center portion of this text). However, in each case there are a series of linear regions, and their point of intersection can be used to locate the end point. The reason why these titration curves have a linear form (rather than the nonlinear response we typically saw when measuring pH or pM in Chapters 12 and 13) is that the measured quantity of absorbance is directly related to the concentration of absorbing species through Beer's law. In contrast to this, values measured for pH or pM during an acid–base or complexometric titration (as well as E_{Cell} in a redox titration) are all related to a logarthimic function of an analyte's activity or concentration.

EXERCISE 18.3   **Performing a Spectrophotometric Titration**

The following mixtures of a sample and titrant (with a concentration of 2.50×10^{-3} M) are placed with pipettes into a series of 25.00 mL volumetric flasks and diluted to the mark. The absorbance of each mixture is then measured. If the analyte and titrant are known to react in a 1:1 ratio, what volume of titrant is needed to reach the equivalence point for this titration? What was the concentration of analyte in the original sample?

SOLUTION

A titration curve can be prepared from these data by making a graph of the absorbance versus the volume of added titrant. This plot gives a linear region for the first five sample/titrant mixtures with a slope of 0.040 absorbance units per mL, followed by a region with no significant change in slope at absorbance = 0.170 for the last three sample/titrant mixtures. The intersection of

Sample Volume (mL)	Titrant Volume (mL)	Absorbance
5.00	0.00	0.000
5.00	1.00	0.040
5.00	2.00	0.080
5.00	3.00	0.120
5.00	4.00	0.160
5.00	5.00	0.170
5.00	6.00	0.170
5.00	7.00	0.170

these two linear regions occurs at 4.25 mL, which represents the volume of titrant needed to reach the end point. The concentration of the sample can then be calculated from this volume, the concentration of the added titrant, and the volume of the original sample.

Conc. analyte =

$$\frac{(2.50 \times 10^{-3} \, M \text{ titrant})(0.00425 \text{ L titrant})}{0.00500 \text{ L analyte}} \cdot$$

$$(1 \text{ mol analyte}/1 \text{ mol titrant})$$

$$= 2.12\underline{5} \times 10^{-3} \, M = \mathbf{2.12 \times 10^{-3} \, M}$$

Although a buret was not used in this particular analysis, this is still an example of a titration and volumetric method because it makes use of a measured volume and known titrant concentration to determine the concentration of an analyte in the sample.

Measuring Multiple Analytes. It is sometimes possible to use UV-vis spectroscopy to determine the concentration of several absorbing species (A, B, and so on) in a sample if these species have significantly different absorption spectra in the UV-vis range. This approach is performed by measuring the total absorbance of the sample at two wavelengths (ε_1 and ε_2), which have significantly different molar absorptivities for at least one of the absorbing species. We can then use Beer's law to relate these measured absorbance values (A_1 and A_2) to the concentrations of the two species (C_A and C_B) and the molar absorptivity of each species at the two wavelengths used in these measurements (ε_{A_1} and ε_{A_2} for species A, and ε_{B_1} and ε_{B_2} for species B).

Absorbance at Wavelength ε_1:

$$A_1 = (\varepsilon_{A_1} b C_A) + (\varepsilon_{B_1} b C_B) \tag{18.5}$$

Absorbance at Wavelength ε_2:

$$A_2 = (\varepsilon_{A_2} b C_A) + (\varepsilon_{B_2} b C_B) \tag{18.6}$$

Titration reaction: Analyte (A) + Titrant (T) $\longrightarrow$ Product (P)

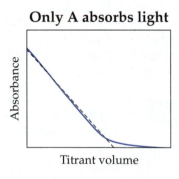

Only A absorbs light

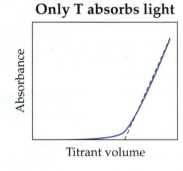

Only T absorbs light

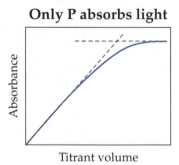

Only P absorbs light

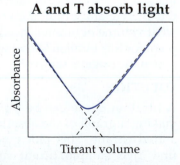

A and T absorb light

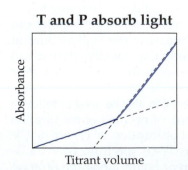

T and P absorb light

FIGURE 18.12 Example of general plots obtained for a spectrophotometric titration. These plots are for various types of analytes, titrants, and products in terms of their ability to absorb the light that is being used to follow the titration. An example of such a titration is provided in the color plates located in the center of this book.

If we measure A_1 and A_2 and know the values of b and the molar absorptivities (as obtained from the analysis of standards), we can rearrange Equations 18.5 and 18.6 to solve for both C_A and C_B. The same approach can be used to look at more than two sample components, as long as an equal or greater number of wavelengths is used for the corresponding absorbance measurements *and* these wavelengths have significantly different molar absorptivities for each of the given analytes.

EXERCISE 18.4	Spectroscopic Analysis of an Acid–Base Mixture

The two acid–base forms of methyl red present in the pH range over which it changes color are known to have the following molar absorptivities at 515 nm and 425 nm.

λ_1 = 515 nm	λ_2 = 425 nm
Acid form:	
ε_{A_1} = 2.49 × 10^4 L/mol·cm	ε_{A_2} = 2.04 × 10^3 L/mol·cm
Base form:	
ε_{B_1} = 1.49 × 10^3 L/mol·cm	ε_{B_2} = 1.06 × 10^4 L/mol·cm

A total concentration solution of $5.00 \times 10^{-5} M$ methyl red is placed into a sample and the absorbance of this mixture is measured in a cuvette with a 1.0 cm path length. An absorbance of 0.379 is measured at 515 nm and an absorbance of 0.419 is found at 425 nm. It is known from previous measurements that there are no other chemicals in the sample that absorb at these wavelengths. What are the concentrations of the acid and base forms of methyl red in this sample?

SOLUTION

This analysis is the same system represented by Equations 18.5 and 18.6, in which we have two equations and two unknowns (the concentrations of the acid and base forms of methyl red, C_A and C_B). To solve these equations, we can first rearrange Equation 18.5 to solve for C_A.

Absorbance at 515 nm (λ_1): $A_1 = (\varepsilon_{A_1} b C_A) + (\varepsilon_{B_1} b C_B)$

$$\Rightarrow \quad C_A = \frac{A_1 - (\varepsilon_{B_1} b C_B)}{(\varepsilon_{A_1} b)}$$

We can now substitute this equation for C_A into Equation 18.5 to get a combined expression that only has C_B as an unknown quantity.

Absorbance at 425 nm (λ_2): $A_2 = (\varepsilon_{A_2} b C_A) + (\varepsilon_{B_2} b C_B)$

$$\Rightarrow \quad A_2 = (\varepsilon_{A_2} b) \cdot \frac{A_1 - (\varepsilon_{B_1} b C_B)}{(\varepsilon_{A_1} b)} + (\varepsilon_{B_2} b C_B)$$

Next, we place our known values for the measured absorbances, molar absorptivities, and path length into this new equation and solve for C_B.

$$0.419 = (2.04 \times 10^3 \text{ L/mol·cm})(1.00 \text{ cm}) \cdot$$

$$\frac{0.397 - (1.49 \times 10^3 \text{ L/mol·cm})(1.00 \text{ cm}) C_B}{(2.49 \times 10^4 \text{ L/mol·cm})(1.00 \text{ cm})}$$

$$+ (1.06 \times 10^4 \text{ L/mol·cm})(1.00 \text{ cm}) C_B$$

$$\Rightarrow 0.419 = 0.0325 - (1.221 \times 10^2 \text{ L/mol}) C_B$$

$$+ (1.06 \times 10^4 \text{ L/mol}) C_B$$

$$C_B = \frac{0.419 - 0.0325}{(1.06 \times 10^4 \text{ L/mol}) - (1.221 \times 10^2 \text{ L/mol})}$$

$$= \mathbf{3.69 \times 10^{-5} \, M}$$

Placing our calculated value for C_B and our other known parameters back into Equation 18.5 then allows us to also calculate the concentration for the acid form of methyl red (C_A), or we can simply take the difference between the known total concentration of the methyl red and C_B to find C_A. Using either approach gives a concentration of $\mathbf{1.31 \times 10^{-5} \, M}$ for the acid form.

18.3 INFRARED SPECTROSCOPY

18.3A General Principles of Infrared Spectroscopy

In UV-vis spectroscopy the absorption of visible or ultraviolet light can lead to an increase in the energy of the absorbing molecules due to *electronic transitions*. Molecules can also absorb other types of radiation, but these processes may involve different types of transitions than those that make use of a change in an electronic state. If a molecule absorbs infrared light (which has a lower energy than visible or ultraviolet light), this absorption is based on a change in the energy due to vibrations or rotations that are occurring in the molecule. A spectroscopic method that uses infrared light to study or measure chemicals is called **infrared spectroscopy** (or "IR spectroscopy").[5]

The simplest kind of bond vibration involves stretching of a bond between two atoms. For example, water is often depicted as a triatomic, bent molecule with a bond angle of about 105° and bond lengths of roughly 96 pm. This model implies that a water molecule is a static object, but in reality a water molecule is always undergoing some changes in its bond angles and bond lengths (see examples of such processes in Figure 18.13). The static model merely depicts the *average* bond angle and bond length, which can have slightly higher or lower values at any given point in time. A better model is to view these bonds as acting as

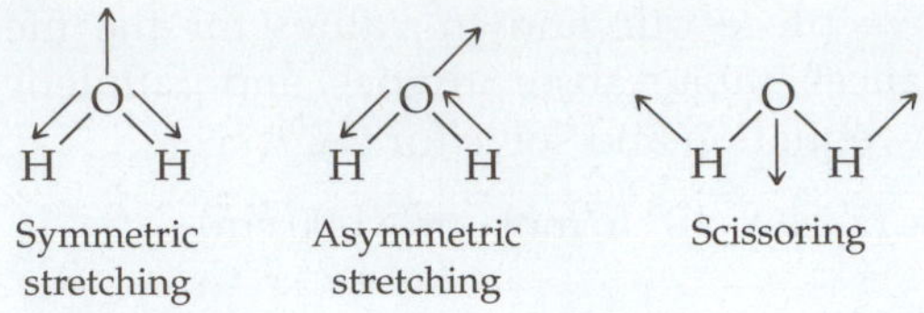

FIGURE 18.13 The vibrational modes for water. The labels given below each of these vibrational modes are terms commonly used to describe these types of vibrations. Both these and alternative types of vibrations can occur in other types of molecules.

small springs rather than fixed rods. Even at a temperature of absolute zero, molecules will vibrate as these bonds continue to contract and expand.

Energy absorption is required to cause a molecular vibration to occur more energetically, and the energy levels that describe these vibrations occur at distinct values (or are "quantized"). The differences in these vibrational energy levels are typically much smaller than the differences in energy present for electronic transitions. As a result, the energy for a photon that is needed to cause excitation in a vibration is also much smaller and is in the same range as is found for infrared light. Even smaller amounts of energy are needed to cause molecules to rotate more rapidly, corresponding to the energy present in microwave radiation. These small changes in rotational energies are often superimposed on the changes in vibrational energies that are seen in IR spectroscopy and lead to broadening of the observed absorption bands. This broadening occurs in a similar manner to the way in which vibrational and rotational transitions plus electronic transitions lead to broad absorption bands for molecules in UV-vis spectroscopy (see Section 18.2A).

Figure 18.14 shows a typical IR absorption spectrum, with cholesterol again being used as an example. We can see by comparing this spectrum with the spectrum for cholesterol in Figure 18.2 that the appearance of an infrared absorption spectrum is totally different from a UV or visible absorption spectrum. Most UV-vis absorption spectra exhibit one or two broad peaks, whereas IR spectra might have dozens of very narrow peaks. This is because only one or two electronic transitions often dominate a UV-vis spectrum, but each molecule can have many ways of undergoing vibrational transitions to produce an IR spectrum.

IR spectra are most commonly represented as a plot of % T versus wave number (in units of cm^{-1}), although some spectra are plotted in terms of % T versus wavelength (expressed in units of micrometers). This way of plotting a spectrum is in contrast to UV-vis spectra, which are usually plotted in terms of absorbance versus wavelength (in units of nanometers). The result is that an IR spectrum has a very different appearance from a typical UV-vis spectrum. The former

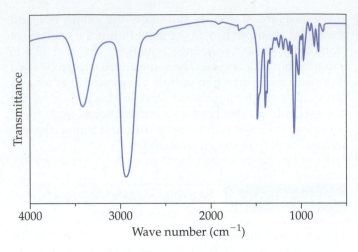

FIGURE 18.14 A typical IR absorption spectrum, using cholesterol as an example. In this type of spectrum the y-axis is often expressed in terms of transmittance or % transmittance rather than absorbance. The x-axis is often given in terms of the wave number of light instead of the light's wavelength.

typically show a large number of peaks, each of which is quite narrow compared to the entire range of wavelengths in the spectrum. In contrast to this UV-vis spectra typically have only one or a few broad peaks for each analyte (see Box 18.1).

18.3B Instrumentation for Infrared Spectroscopy

Typical System Components. IR spectroscopy is similar to UV-vis spectroscopy in that it requires a source of light, a means for separating this light into different wavelengths, a sample holder, and a detector. However, the specific instrument components in IR spectroscopy are made of different materials and often operate on different principles than the devices that are used for UV-vis spectroscopy.

The source of light in IR spectroscopy is usually an inert rod that is heated to a much lower temperature than is used for light sources in visible spectroscopy. As a result, the maximum absorbance that is now obtained through blackbody radiation will occur in the infrared region. Glass and fused silica are opaque at wavelengths greater than 2.5 μm, so the glowing source must not be in a glass bulb or in a casing that is made of these substances. The heated material is either silicon carbide (SiC, giving a device called a *globar*), or is a mixture of rare-earth oxides (producing a device known as a *Nernst glower*).[7,10–11] The general construction of such a device, the Nernst glower, is shown in Figure 18.16 (see following pages). This design includes the material that is to be heated, a heating source, and a reflector to help pass the resulting radiation in the desired direction. The light that is produced by such a device closely matches what would be expected for blackbody radiation. For a globar heated at 1300–1500 K, usable light is provided at wavelengths of 0.4–20 μm. A Nernst glower that is heated

BOX 18.1
Raman Spectroscopy

IR spectroscopy is not the only way information can be obtained on vibrational transitions in a molecule. *Raman spectroscopy* is another method that provides information on such transitions. This technique is named after Sir Chandrasekhara Vancata Raman, a scientist from India who was awarded the 1930 Nobel Prize in Physics (see Figure 18.15). Sir Raman studied an effect in which the scattering of light by molecules sometimes involved a small change in the wavelength of the light. In this effect, the difference in energy between the light from the original source and the scattered light is equal to the difference in energy of vibrational levels in a molecule. Raman spectroscopy is a method that makes use of this phenomenon (known as "Raman scattering") to study or measure chemicals.[7,10,11]

The general process of Raman scattering is illustrated in Figure 18.15. In this effect, light is scattered as it interacts with a molecule. The amount of time that passes during this interaction is around 10^{-13} s. During this time, the molecule is temporarily raised to a higher energy level called a "virtual state," and the molecule returns to a lower energy state after the light is scattered. Most of these molecules return to their original energy level, which gives the incoming light and scattered light the same wavelength (a process known as "Rayleigh scattering"). However, occasionally a molecule will also undergo a change in vibrational level during the scattering process. This change means the scattered light and incoming light will now have a difference in energy that is equal to the energy involved in the vibrational transition. This effect provides these two types of light with slightly different wavelengths.[7,10]

These changes in wavelength are quite small and can only be seen when using incoming light that is monochromatic. Sir Raman conducted his experiments using an intense mercury discharge lamp, but modern instruments for Raman spectroscopy use a laser as the light source. Raman spectroscopy is similar to IR spectroscopy in that both techniques can be used for chemical identification or measurement by using spectroscopy to examine vibrational transitions in molecules. However, the ability to use a laser as a light source and to use visible light instead of infrared light for these measurements are two important advantages of Raman spectroscopy.[7,10,11]

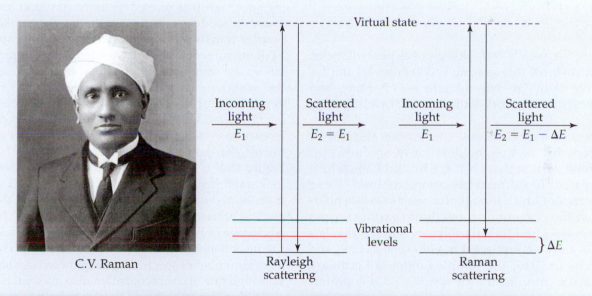

C.V. Raman

FIGURE 18.15 Sir C.V. Raman and an illustration of the types of energy transitions that occur in Rayleigh scattering vs. Raman scattering. In this example, a photon of incoming light in each type of scattering has an energy equal to E_1. In Rayleigh scattering, the scattered light has the same energy as the incoming light. In Raman scattering, the scattered light differs by a value of ΔE from the energy of the incoming light. The type of Raman scattering that is shown here, in which the molecule ends at a higher vibrational energy state and the light loses energy, is known as "Stokes scattering." It is also possible for a molecule to lose energy by ending at a lower vibrational state, in which case the scattered light gains energy; this process is known as "anti-Stokes scattering." (The photo of Sir Raman is by A. Bortzells Tryckeri and is provided courtesy of the AIP Emilio Segre Visual Archives, W.F. Meggers Gallery of Nobel Laureates.)

from 1200 to 2000 K produces adequate amounts of light at wavelengths ranging from 1 to 40 μm.[7]

A glass or quartz prism cannot be used as a part of a monochromator to separate light of different wavelengths in IR spectroscopy because of the absorbance of IR radiation by glass or quartz. Fortunately, a grating can still be used for this purpose. (*Note*: Another device that can be used for this purpose is an "interferometer," which will be discussed later in this section.) A grating in IR spectroscopy functions in the same manner as a grating that is used in UV-vis spectroscopy. However, the spacing of the lines in these gratings is different. For UV-vis spectroscopy,

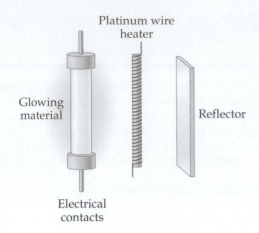

Platinum wire
heater

Glowing
material

Reflector

Electrical
contacts

FIGURE 18.16 The general design of a Nernst glower. The glowing element in this light source contains a semiconductor material that emits infrared light when a current is passed through the material and resistive heating occurs. This material must be preheated to achieve conductance, as is accomplished by using a separate platinum-wire heater. A reflector helps to collect and direct the radiation that is given off by this source in the desired direction for use.

this spacing is typically 300–2400 grooves per millimeter. In IR spectroscopy, this spacing is 300 grooves/mm for work with light at wavelengths of 2–5 μm and 100 grooves/mm for work with light at wavelengths of 5–16 μm.

Materials like glass and quartz can also not be used to construct sample holders for IR spectroscopy. Instead, ionic salts such as NaCl, KBr, and CsBr, which are transparent to infrared radiation, are used. These materials are not ideal, though, for use as sample holders because they cannot be formed into various shapes as easily as glass and they dissolve in water. If this last problem is an issue, less soluble salts such as CaF_2 and AgCl can be used for the sample holder. All ordinary solvents have complicated IR spectra, so it is preferable to measure spectra of pure substances rather than a solution, especially a dilute solution. As a result, a drop of a liquid sample is often simply put onto a flat plate of NaCl and another similar NaCl flat plate is put on top of it, clamped into place, and a spectrum taken of the resulting film. Solid samples can be mixed with dry KBr and pressed into a thin disk, which is put into the instrument for analysis.[7,11]

Finding suitable detectors for IR spectroscopy is another challenge because photons of infrared light do not have sufficient energy to dislodge an electron in a photomultiplier tube or phototube. Another problem with these latter devices is that they are surrounded with a casing made of glass or silica, which would absorb infrared light and prevent it from being detected. Instead, most conventional scanning IR instruments utilize a heat sensing detector, such as a *thermocouple*. IR radiation heats the thermocouple by causing its atoms to move more rapidly. A thermocouple is a junction of two different conductors that generates an electrical voltage that depends on the temperature difference between the ends of two wires, one of which is maintained at a constant temperature. The intensity of IR radiation falling on this detector causes warming and a change in voltage, thus making it possible to detect the radiation.[7]

Scanning Instruments and Fourier Transform Infrared Spectroscopy. One difficulty in using IR spectroscopy is that CO_2 and H_2O in air both absorb IR radiation considerably and obscure the spectrum of the desired sample. For that reason, IR spectrophotometers are often double-beam devices in which the spectrum of air is subtracted from the spectrum of the sample, leaving only the spectrum of the desired sample.

Until recently, most IR instruments were double-beam scanning instruments. That is, the wavelength was changed gradually as % T was measured and the resulting spectrum was plotted. A more common instrument found in modern laboratories is one that makes use of **Fourier transform infrared (FTIR) spectroscopy**.[7,11] An FTIR instrument allows all wavelengths of IR radiation to fall on the sample simultaneously. Instead of separating the wavelengths in time or space, the wavelength dependence of % T is gained by use of a device called an *interferometer*, which causes positive and negative interference to occur at sequential wavelengths as a moving mirror changes the path length of the light beam (see Figure 18.17).[7] The initial output of the detector doesn't look anything like a spectrum, but this direct output is transformed into a spectrum by application of the mathematical process called a "Fourier transform." The major advantage of FTIR is in the speed with which a spectrum can be obtained, typically just a few seconds. This means a large number of spectra can be gathered in a short time. This high rate of data acquisition also makes it possible to combine a large number of spectra to help to remove random fluctuations in the signal, or "noise." The more spectra that are averaged, the better the signal-to-noise ratio will become. This approach, in turn, means that a good spectrum can be achieved for a small concentration of analyte and that a lower limit of detection for measurement of the analyte can be obtained.

18.3C Applications of Infrared Spectroscopy

IR spectroscopy is most frequently employed for qualitative identification of nearly pure compounds. Because each compound gives several peaks, an IR spectrum of a mixture is very difficult to interpret. The groups of atoms we call "functional groups" have characteristic vibrational energies and characteristic IR absorption wavelengths that can be used in this process.

Table 18.2 gives an example of a correlation chart that can be used to identify functional groups in a compound from its IR spectrum. For instance, cholesterol has an OH group that has absorption bands at 3300 and 1100 cm^{-1}. Cholesterol also has a large number of C—C single bonds (giving a band at 2900 cm^{-1}), a C=C double bond (with a band at 1650 cm^{-1}), and numerous C—H bonds (with a band near 3000 cm^{-1}). A chemist can learn a great deal about the structure of a compound from its IR spectrum. One can be even more certain as to the identity of a compound if its spectrum has been included in a library of IR spectra. A match between a measured spectrum of an unknown and a library spectrum is regarded as good evidence that the unknown substance is the same as the identity of the library spectrum. Modern IR instruments frequently come with a computer that contains a library of several hundred or thousand compounds that can be searched rapidly for agreement with a measured spectrum.

TABLE 18.2 Correlation Chart for Various Types of Organic Molecules in IR Spectroscopy*

Functional Group	Bond	Wave number(s), cm^{-1}	Relative Intensity
Acyl halide	C=O	1815–1770	
Alcohol	C–O	1200–1100	Strong, 3° > 2° > 1°
	O–H	3500–3200	Strong and broad
Aldehyde	C–H	2850–2700	
	C=O	1740–1685	Strong
Alkane	sp^3 C–H	2950–2850	Strong
	C–C	1200	
Alkene	sp^2 C–H	3100–3000	Medium, sharp
	C=C	1680–1620	
cis-Alkene	C=C	730–665	
trans-Alkene	C=C	980–960	
Alkyne	C≡C	2200–2100	
	sp C–H	3300	Medium–weak, sharp
Amide	C=O	1695–1616	Strong
Amide, Amine	N–H	3500–3350	Broad (with spikes)
Aromatic	C–H	3100–3000	
Carboxylic acid	O–H	3600–2500	Strong and broad
	C=O	1725–1665	Strong
	C–O	1350–1210	Medium–strong
Ester	C=O	1750–1730	Strong
	C–O	1310–1160	Strong
Ketone	C=O	1750–1660	Strong
Nitrile	C≡N	2280–2240	
Phenol	O–H	3500–3200	Strong and broad
	C–O	1300–1180	Strong

*The information in this table was obtained from F.A. Carey, *Organic Chemistry*, 6th ed., McGraw-Hill, Boston, 2006, and L.G. Wade Jr., *Organic Chemistry*, 7th ed., Prentice Hall, New York, 2010.

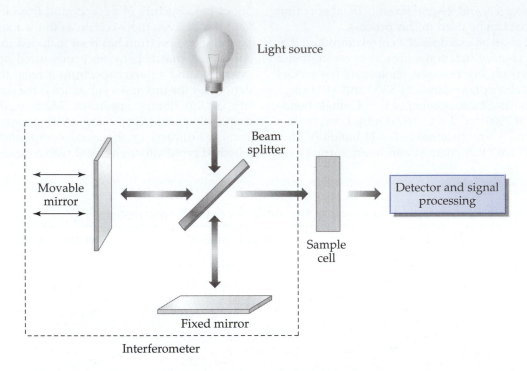

FIGURE 18.17 The general design of an instrument for FTIR spectroscopy. As one mirror is moved in this device, different wavelengths of light from the original source will have constructive interference and make it onto the sample. By moving this mirror it is possible to have different sets of wavelengths pass on to the sample. The absorption of light for each set of wavelengths and at each position of the mirror is measured and converted through the process of Fourier transform into a spectrum. A laser (not shown) is also used in this device to precisely record the position of the movable mirror. The section of the instrument that is in the dashed box and that is used for wavelength selection is known as an *interferometer*.

18.4 MOLECULAR LUMINESCENCE

18.4A General Principles of Luminescence

Luminescence is a general term that describes the emission of light from an excited-state chemical. There are three specific types of luminescence that we will consider: fluorescence, phosphorescence, and chemiluminescence. The use of these processes in spectroscopy to study molecules is sometimes known as *molecular luminescence spectroscopy*.

Fluorescence is a term used to describe light emitted by a sample after it has become electronically excited by absorbance of a photon, with the light emission being due to a "spin-allowed" transition (such as a singlet–singlet transition).[5,6] This type of process is illustrated in Figure 18.18. The emitted light in fluorescence is frequently in the visible region, while the original light absorbed by the analyte is often in the ultraviolet region but can also occur in the visible range. For low concentrations of a fluorescing compound, the relationship between fluorescence intensity and concentration is nearly linear (see Chapter 17). Fluorescence occurs rapidly, with the excited state usually lasting less than 10 nanoseconds. A method that uses fluorescence to characterize or measure chemicals is called **fluorescence spectroscopy**.[7,11]

Most molecules do not fluoresce efficiently. Instead of releasing most of the energy from their excited state in the form of light, much of this energy is lost as heat to their surroundings. Molecules that fluoresce usually have rigid structures that are often planar and have aromatic groups, as illustrated by the example if Figure 18.19. The efficiency of fluorescence by a chemical is described by using a *fluorescence quantum yield* (φ_F). This quantity is the ratio of the number of fluoresced photons that are produced divided by the number of absorbed photons. A chemical with perfect fluorescence will have all its absorbed photons lead to the emission of other photons by fluorescence, giving a maximum value for φ_F of 1.0. A chemical that absorbs light but does not undergo any fluorescence will have a value of zero for φ_F. The fluorescent quantum yield for other chemicals will be somewhere between these two limits, with compounds that have a degree of fluorescence that is appropriate for analysis typically having values for φ_F that are greater than 0.01.[5–7]

Phosphorescence also follows excitation of a molecule, but instead of immediately undergoing fluorescence, the excited electron first undergoes an intersystem crossing into a triplet state.[5–7] This means that the release of light from this excited state will now require a "spin-forbidden" transition from this triplet state to singlet state. This type of emission process is much less likely to occur and is a slower process than the singlet-to-singlet transitions that led to light emission in fluorescence. A spectroscopy technique that utilizes

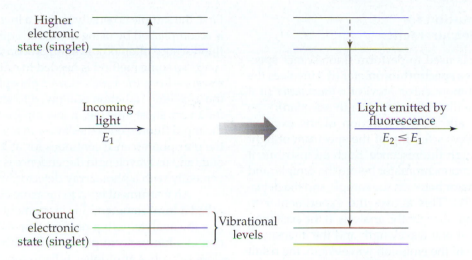

FIGURE 18.18   The basic processes that lead to light emission in fluorescence. After light has been absorbed by a molecule, some of this energy will be lost as heat through collisions as electrons move to the lowest vibrational level of the excited electronic state (a process known as "vibrational relaxation"). As the excited state of the molecule then emits light, these electrons in the excited electronic state may return to various vibrational states in the lower electronic state. The result is a series of emission wavelengths that are created when using even a single wavelength of light for excitation. Phosphorescence is a similar process, but also involves the conversion of the excited state electron from a singlet state to a triplet state before light emission takes place.

phosphorescence to characterize or measure chemicals is called **phosphorescence spectroscopy**.[7,11]

Phosphorescence is much more difficult to measure than fluorescence. A phosphorescence measurement usually requires the use of liquid nitrogen for work at low temperatures to provide a reasonable signal. The reason low temperatures are needed is that the lifetime of an excited triplet state is much greater than an excited singlet state (typically, $10^{-4} - 10^{1}$ s versus $10^{-9} - 10^{-8}$ s, respectively). This longer life time means the probability of energy loss through collisions and heat loss is also much greater in phosphorescence than in fluorescence. The use of a lower temperature for this measurement will minimize the degree of molecular motion around the analyte and make its collisions with the solvent or other sample components less likely to occur. This, in turn, increases the chance that the excited triplet state can instead give off its energy in the form of phosphorescence as it returns to the ground state.[7]

Chemiluminescence results from the emission of light by an excited state that is formed in a chemical reaction.[7,16,17] The term *bioluminescence* is sometimes also used

when the chemical reaction is of biological origin. Fireflies and glowworms are well-known examples of bioluminescence. A good example of a nonbiological chemical that can undergo chemiluminescence is luminol (5-amino-1, 4-phthalazdione), which reacts with hydrogen peroxide to form the excited molecule shown in Figure 18.20, which quickly emits a photon to give blue light. The reaction for this process is given in Figure 18.20. The timescale for this process is mainly determined by the rate of the underlying chemical reaction and varies from one chemiluminescence reaction to the next. In some cases, this process is relatively fast, such as the burst of bioluminescence that occurs in fireflies. In other cases, this process can occur over several seconds or minutes, as takes place in glow sticks and in chemiluminescent reactions that are based on luminol.[16,17]

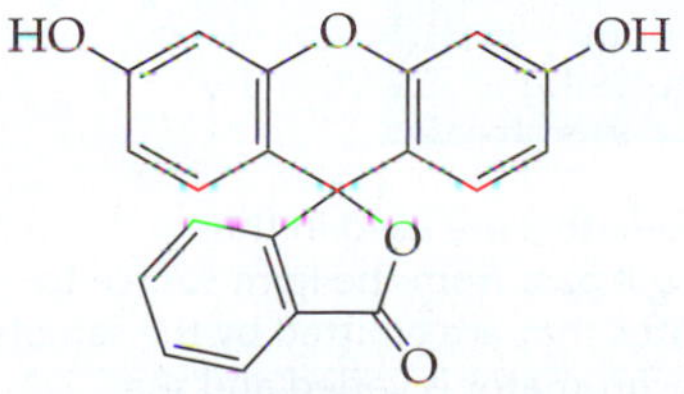

Fluorescein

FIGURE 18.19   The structure of fluorescein, a molecule that is often used as a fluorescent label in chemical assays.

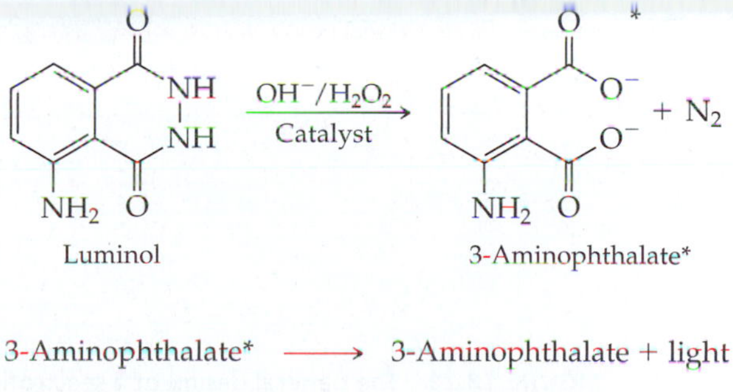

$$\text{Luminol} \xrightarrow[\text{Catalyst}]{\text{OH}^-/\text{H}_2\text{O}_2} \text{3-Aminophthalate*} + \text{N}_2$$

Luminol   3-Aminophthalate*

3-Aminophthalate* $\longrightarrow$ 3-Aminophthalate + light

FIGURE 18.20   The reactions involved in the production of chemiluminescence by luminol. The luminol is first reacted under basic conditions with the oxidizing agent H_2O_2 and in the presence of a catalyst. This reaction results in a product that is a molecule of 3-aminophthalate in an excited state. Some of this excited product releases its extra energy in the form of light.

18.4B Instrumentation for Luminescence Measurements

An instrument that is used to perform fluorescence spectroscopy is known as a **spectrofluorometer** (if it involves the use of sophisticated monochromator) or a *fluorometer* (if it makes use of simple filters for wavelength selection).[5–7] A spectrofluorometer allows the selection of the exciting wavelength and allows scanning of the spectrum of light that is emitted through fluorescence. Such an instrument has a light source, a monochromator before the sample, and another monochromator between the sample and the detector (see Figure 18.21). This allows the experimenter to choose optimum conditions for the analysis. If the excitation wavelength is fixed at one wavelength and the intensity is plotted as a function of the emission wavelength, the result is called an "emission spectrum." The other possibility is to fix the wavelength at which the fluorescence intensity is measured by varying the excitation wavelength. This results in an "excitation spectrum" and is similar to the absorption spectrum for the analyte that is undergoing fluorescence.[5,6]

The design of a fluorometer is simpler than that of a spectrofluorometer and uses only filters to select the wavelength of light that is used for excitation and that is analyzed for fluorescence. Although this design does not allow a fluorescence spectrum to be obtained, it does make it possible to measure the emission intensity at a given set of wavelengths for the quantitative analysis of a particular analyte that undergoes fluorescence at these wavelengths.[7,11]

An instrument designed for phosphorescence measurements is similar, but differs in two important respects.

First, the sample is usually kept at a low temperature, which is accomplished by using dry ice or liquid nitrogen. Second, fluorescence often occurs simultaneously with phosphorescence, so some method is needed to differentiate rapid fluorescence from the much slower phosphorescence. Usually, the excitation radiation is allowed to strike the sample only for a very short time (i.e., a few milliseconds at most). Then the rapid fluorescence dies away in a few nanoseconds and the phosphorescence continues for at least several milliseconds and its wavelength dependence is measured most conveniently with a photoarray detector.[7,11]

An instrument used to measure chemiluminescence is called a **luminometer**. This instrument includes a device to mix the analyte with a reagent that will lead to the formation of a luminescent product. The mixing device is placed close to a photomultiplier tube to measure the intensity of light given off by the excited product. A simple device merely measures this intensity, whereas a more complicated one passes the light through a monochromator to allow study of the wavelengths of the luminescence.[16]

18.4C Applications of Molecular Luminescence

Fluorescence, and to a smaller extent phosphorescence, is a valuable tool for measuring analytes at low concentrations. The more stringent requirements needed for solutes to undergo these processes also provide fluorescence and phosphorescence with greater selectivity and lower limits of detection than absorbance measurements for molecules.[7,10,11] It is also possible to use an approach based on fluorescence to examine many types of nonfluorescent

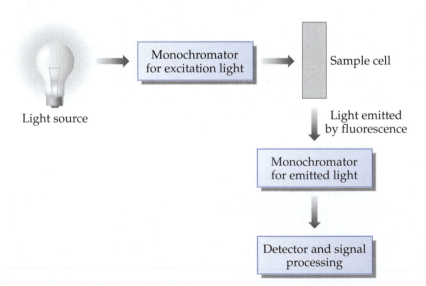

FIGURE 18.21 The general design of a spectrofluorometer. Two monochromators are used in this instrument. The first monochromator selects the wavelengths of light that will pass from the light source to the sample for excitation. The second monochromator selects the wavelengths that are emitted by the sample through fluorescence for measurement. If the wavelength in the first monochromator is varied and the wavelength setting in the second monochromator is held constant, the resulting plot of fluorescence intensity vs. wavelength is known as an "excitation spectrum." If the wavelength in the second monochromator is held constant and the wavelength setting in the second monochromator is varied, the resulting plot of fluorescence intensity vs. wavelength is known as an "emission spectrum" (or "fluorescence spectrum").

chemicals by first reacting these analytes with a reagent that converts them into a fluorescent form. A good example is the detection of amines and amino acids by fluorescence. Most amino acids that do not have significant fluorescence can be combined with the reagent o-phthaldialdehyde to yield a strongly fluorescent product. Similar reactions are available for chemicals that contain alcohol groups, aldehydes, or ketones as part of their structure.[7,11,18]

EXERCISE 18.5   **Use of Fluorescence for Chemical Analysis**

Estimate the concentration of glycine in the following unknown sample based on the fluorescence intensity that is measured for this sample and a series of standards that have each been reacted with an excess of o-phthaldialdehyde to create a fluorescent product.

Concentration of Glycine (μM)	Measured Fluorescence Intensity, I_F
0.00	0.10
0.20	3.4
0.40	6.9
1.00	17.1
2.00	34.3
5.00	83.2
10.0	152.
Sample	22.8

SOLUTION

A plot of the measured fluorescence intensity versus glycine concentration in the standards gives a linear response of the range of concentrations that were examined in this study (e.g., see Chapter 17 for the expected relationship between an emission signal and analyte concentration). The best-fit line for this graph has a slope of 15.3 μM^{-1} and an intercept of 1.77. Using this plot and fluorescence intensity obtained for the sample, the concentration of glycine in this sample is estimated to be **1.37 μM**.

Chemiluminescence can also be used as a selective means for the measurement of analytes in samples. For example, chemiluminescence can be used to measure nitric oxide (NO) in the atmosphere by first reacting NO with ozone (O_3).

$$\text{Step 1:} \quad NO + O_3 \rightarrow NO_2^* + O_2 \qquad (18.7)$$

$$\text{Step 2:} \quad NO_2^* \rightarrow NO_2 + h\nu$$

In the first step of this process, an excited state molecule of NO_2^* is produced. This excited molecule then later returns to its ground state by emitting light that can be used for its detection. Examples of many other reactions that are based on chemiluminescence, and analytical applications of such reactions, can be found in References 16 and 17.

Key Words

Chemiluminescence *451*
Diode-array detector *439*
Double-beam instrument *439*
Fluorescence *450*
Fluorescence spectroscopy *450*

Fourier transform infrared spectroscopy *448*
Infrared spectroscopy *445*
Luminometer *452*
Molecular spectroscopy *434*
Phosphorescence *450*

Phosphorescence spectroscopy *451*
Single-beam instrument *439*
Spectrofluorometer *452*
Spectrophotometric titration *443*

Standard addition method *441*
UV-visible spectroscopy *435*

Other Terms

Bioluminescence *451*
Chromophore *436*
Colorimetry *435*
Deuterium lamp *438*
Fluorescence quantum yield *450*

Fluorometer *452*
Flow injection analysis *440*
Globar *446*
Hydrogen lamp *438*
Interferometer *448*
Isobestic point *443*

Luminescence *450*
Molecular luminescence spectroscopy *450*
Nernst glower *446*
Raman spectroscopy *447*
Reflection grating *438*

Thermocouple *448*
Transmission grating *438*
Tungsten lamp *437*
Tungsten/halogen lamp *437*
UV-vis absorbance spectrometer *437*

Questions

WHAT IS MOLECULAR SPECTROSCOPY AND HOW IS IT USED IN CHEMICAL ANALYSIS?

1. Explain what is meant by the term "molecular spectroscopy."
2. What are three ways in which molecules can interact with light?

3. What types of energy levels can be involved in the study or measurement of analytes by molecular spectroscopy? List one specific type of molecular spectroscopy that makes use of each of these changes in energy levels.

4. Describe the method of "colorimetry." Give one example of an early application that was developed for chemical analysis based on colorimetry.

5. Explain how molecular spectroscopy can be used for either the measurement or identification of a molecule. List one specific type of spectroscopy that is commonly used for each of these general applications.

GENERAL PRINCIPLES OF ULTRAVIOLET-VISIBLE SPECTROSCOPY

6. What is "UV-vis spectroscopy"? Explain why ultraviolet or visible light is useful in absorbance measurements for many types of organic molecules.

7. Define the word "chromophore." What are some typical features that are found in a chromophore of an organic molecule that can absorb ultraviolet or visible light?

8. Rank the following compounds in the order that you think will be easiest to measure by UV-vis spectrometry. Explain the reasons for the order of your ranking of these chemicals.
   (a) $CH_3—CH_2—CH_2—CH_2—CH_2—CH_3$
   (b) $CH_2=CH—CH_2—CH_2—CH_2—CH_3$
   (c) $CH_3—CH=CH—CH_2—CH=CH_2$
   (d) $CH_3—CH=CH—CH=CH—CH_3$

9. Explain why Beer's law is often used for the measurement of analytes in UV-vis spectroscopy.

10. An analyte that is to be measured by UV-vis spectroscopy has a molar absorptivity of 5.6×10^5 L/mol·cm at the wavelength that is chosen for its measurement. If the smallest absorbance that can be measured is 0.002 and the sample cuvette has a path length of 1.00 cm, what is the expected lower limit of detection for this analyte?

11. A solution has an analyte concentration of 5.7×10^{-3} M that gives a transmittance of 43.6% at 480 nm and when measured in a 5.00 cm cuvette. Calculate the molar absorptivity of the analyte. What is the expected lower limit of detection for this analyte if the smallest absorbance that can be measured by the instrument is 0.001.

12. What is the upper limit of the linear range for the analyte in Problem 11 if deviations for Beer's law are found to occur at an absorbance of approximately 1.00?

INSTRUMENTATION IN ULTRAVIOLET-VISIBLE SPECTROSCOPY

13. Describe the basic components of a UV-vis absorbance spectrometer. Give the function for each of these components.

14. Explain how a tungsten lamp works, including the basis for how this lamp emits light.

15. How does a hydrogen lamp work? How does a deuterium lamp differ from a hydrogen lamp?

16. What are two types of monochromators that are used in UV-vis spectroscopy?

17. What are some requirements for a sample holder in UV-vis spectroscopy? Describe the construction of a typical cuvette that is used for this purpose.

18. List two types of detectors for light that can be used in UV-vis spectroscopy. Describe how each of these detects the light.

19. What is the difference between a "single-beam instrument" and a "double-beam instrument"? What are some advantages and disadvantages for each of these instrument designs?

20. Explain what is meant by a "diode-array detector." How does this device differ from a single beam or double-beam instrument that is used for UV-vis spectroscopy?

21. Define "flow injection analysis" and explain how UV-vis spectroscopy can be used in this method.

APPLICATIONS OF ULTRAVIOLET-VISIBLE SPECTROSCOPY

22. Explain how direct measurements of an analyte can be made by UV-vis spectroscopy. Describe the role of Beer's law in such an analysis.

23. The complex of Fe^{2+} with 1,10-phenanthroline has a molar absorptivity of about 11,000 L/mol·cm in water at 510 nm. A 20.00 mL water sample thought to contain Fe^{2+} is mixed with an excess of 1,10-phenanthroline. Acetic acid, sodium acetate, and hydroxylamine are also added to buffer the solution and assure that all the iron is reduced to Fe(II). This mixture is diluted with distilled water to a total volume of 50.00 mL, and the final solution is found to give an absorbance of 0.762 at 510 nm when using a 1.0 cm cuvette.
   (a) If it is assumed that there are no absorbing species in this solution other than the Fe^{2+} complex with 1,10-phenanthroline, what was the concentration of Fe^{2+} in the original sample?
   (b) Discuss how the presence of other species that also absorb at 510 nm would have affected the accuracy of this measurement. Would a positive error or a negative error be obtained?

24. A geochemist wishes to estimate the volume of an irregularly shaped pool of water. This task is to be accomplished by placing a known amount of highly colored dye into this solution and then measuring the absorbance by this dye after it has been in the pool. To do this, the geochemist uses 1.00 L of a dye solution that has an absorbance of 0.768 when it is measured in a 1 mm cuvette at 450 nm. After all of this solution has been placed into the pool and allowed to mixed thoroughly, a sample of the pool water is taken and found to have an absorbance at 450 nm of 0.142 in a 10.00 cm cuvette. What is the volume of the pool? What assumptions are made during this analysis?

25. Four standard solutions and an unknown sample containing the same compound give the following absorbance readings at 535 nm when using 1.00 cm cuvettes. What is the concentration of the analyte in the sample?

Solution	Analyte Concentration (*M*)	Absorbance
Standard #1	0.00	0.005
Standard #2	2.5×10^{-3}	0.085
Standard #3	5.0×10^{-3}	0.175
Standard #4	25.0×10^{-3}	0.805
Unknown sample	?	0.465

26. The content of nitrate–nitrogen in water can be measured by evaporating a known amount of water (100.00 mL) to dryness. The resulting residue is then mixed with phenoldisulfonic acid and heated until all of the solid material has dissolved. This solution is diluted to a volume of 50.00 mL with water and ammonia, giving a solution with a yellow color that is related to the amount of the nitrogen–nitrate in the original sample. Absorbance of the sample and standards that are prepared in the same

manner is measured at 410 nm using a 1.00 cm cuvette. The follow data were obtained for a series of standards and an unknown sample that were examined by this method. What was the concentration of nitrate–nitrogen in the original sample?

Mass of Nitrate–Nitrogen in Sample/Standard (mg)	Absorbance
0.00	0.000
0.10	0.103
0.25	0.257
0.50	0.515
Unknown sample	0.318

27. What is the "standard addition method"? Under what circumstances is this method typically used for analysis?
28. A sample of coffee is analyzed to determine its caffeine concentration. Two portions of this sample are prepared for analysis. The first portion contains 50.0 mL of brewed coffee, to which is added 10.0 mL of water. The second portion contains 50.0 mL of brewed coffee that has been spiked with 10.0 mL of an aqueous solution that contains 1.0×10^{-2} M caffeine. The first portion of the sample is found to give a measured absorbance of 243 units, and the second portion gives an absorbance of 387 units. What is the concentration of caffeine in the brewed coffee?
29. An aqueous sample containing Fe^{2+} is treated with 1,10-phenanthroline to form a colored complex for detection. This treated solution gives an absorbance of 0.367 when measured with a 1.00 cm long cuvette at 510 nm. Next, 5.0 mL of a 0.0200 M Fe^{2+} solution is added to 10.0 mL of the unknown sample, treated with 1,10-phenanthroline in the same fashion as the previous sample, and found to give an absorbance of 0.538 at 510 nm. Based on this information, what was the concentration of Fe^{2+} in the original unknown?
30. Explain why a change in color occurs when an acid–base indicator is used for end-point detection. Use methyl red as an example to illustrate your answer.
31. What is a "spectrophotometric titration"? What is measured in this type of titration? How is the end point detected?
32. Explain why a linear response is typically seen in a spectrophotometric titration.
33. A spectrophotometric titration that is performed has the general reaction A + T → P, where A is the analyte, T is the titrant, and P is the product of the reaction. Draw the titration curve that would be expected under each of the following conditions. Clearly indicate in each diagram the response that would be expected both before and after the equivalence point.

	ε_A	ε_T	ε_P
(a)	0	0	500
(b)	500	0	0
(c)	0	500	0
(d)	200	0	400
(e)	200	0	200

34. The following results were obtained for a spectrophotometric titration that was carried out at 600 nm to measure Cu^{2+} in a water sample by reacting Cu^{2+} with triethylenetetramine (trien) to form a colored complex, as shown below.

$$Cu^{2+} + trien \rightarrow Cu(trien)^{2+}$$

A 10.00 mL sample was used for this analysis, which was mixed with various volumes of a 0.0500 M solution of triethylenetetramine and distilled water to give a total final solution volume of 50.00 mL. The absorbance of each mixture was then determined, giving the following results.

Sample Volume (mL)	Volume Trien (mL)	Volume H_2O (mL)	Absorbance
10.00	0.0	40.0	0.005
10.00	2.0	38.0	0.217
10.00	4.0	36.0	0.428
10.00	6.0	34.0	0.643
10.00	8.0	32.0	0.750
10.00	10.0	30.0	0.750

(a) What was the concentration of Cu^{2+} in the original sample?
(b) Explain the shape of the titration curve that is obtained for this analysis. What does this curve tell you about the ability of the analyte, titrant, and product to absorb light at the wavelength that was used to follow this titration?
(c) Use the information provided to estimate the value of the molar absorptivities for the analtye, titrant, and product at 600 nm.
35. Explain how UV-vis spectroscopy can be used to examine several analytes in a sample by using absorbance measurements at multiple wavelengths. What requirements must be met for this approach to work?
36. A scientist wishes to measure three different analytes that all have significantly different spectra in the UV-vis range. How many absorbance measurements of the sample will be required for this analysis? What criteria should be used in selecting the wavelengths for these absorbance measurements?
37. What is an "isobestic point"? When should absorbance measurements at an isobestic point be avoided? When is an isobestic point useful in such measurements?
38. An unknown solution contains two absorbing species, P and Q, both of which are to be measured by UV-vis spectroscopy. Compound P has molar absorptivities of 570 L/mol · cm at 400 nm and 35 L/mol · cm at 600 nm. Compound Q has molar absorptivities of 220 L/mol · cm at 400 nm and 820 L/mol · cm at 600 nm. The unknown mixture of P and Q is placed in a 1.00 cm long cuvette and gives absorbance values of 0.436 at 400 nm and 0.644 at 600 nm. If no other absorbing species are present in this sample, what are the concentrations of P and Q?
39. The pH of the methyl red solution in Exercise 18.4 was 5.20. Based on the information given earlier in this exercise, determine the K_a value for the acid–base transition that was present under these conditions.

GENERAL PRINCIPLES OF INFRARED SPECTROSCOPY

40. What types of energy transitions in a molecule are involved in the absorption of light in IR spectroscopy? How are these energy transitions different from those that are used in UV-vis spectroscopy?
41. Describe what happens to the motions within a molecule when this molecule absorbs infrared radiation.
42. How is IR spectroscopy typically used for chemical analysis? How does the typical application of IR spectroscopy in chemical analysis differ from the applications of UV-vis spectroscopy? What are the reasons for these differences?
43. Describe the appearance of a typical IR spectrum, including the terms that are plotted on the x-axis and y-axis. How is this type of spectrum different from an absorbance spectrum that is used in UV-vis spectroscopy?

INSTRUMENTATION FOR INFRARED SPECTROSCOPY

44. How is the general design of an instrument for IR spectroscopy similar to one that is used in UV-vis spectroscopy? How are these two types of instrument different?
45. Describe a typical light source for IR spectroscopy. What are some special requirements for this type of light source?
46. What is a "Nernst glower"? What is a "globar"? How are each of these devices used in IR spectroscopy?
47. Explain why a glass or quartz prism cannot be used in IR spectroscopy.
48. How is a diffraction grating used in IR spectroscopy? How does a diffraction grating in IR spectroscopy differ from one that is used in UV-vis spectroscopy?
49. What types of materials are used for the sample holders in IR spectroscopy? What properties are desired for such materials?
50. Describe how you would prepare a liquid sample for analysis by IR spectroscopy? How would you prepare a solid sample for analysis by IR spectroscopy?
51. What problems are associated with finding a suitable detector for IR radiation? Give one example of a device that can be used in IR spectroscopy for this purpose.
52. Compare and contrast the design of a double-beam instrument in IR spectroscopy with the design of a more modern single-beam instrument.
53. What is meant by "Fourier transform infrared spectroscopy"? How is the measurement of an IR spectrum obtained in this method?
54. Describe how IR spectroscopy can be used for the identification of chemicals. What features of an IR spectrum are useful for this type of application?
55. What is a "correlation chart"? Explain how you can use this type of chart for chemical identification in IR spectroscopy.
56. A student receives an unknown organic compound that is either cyclohexane or cyclohexene. Explain how the student could use IR spectroscopy to tell which of these two compounds is present in the sample. Be as specific as possible in your answer.
57. A can of paint solvent that is found at the scene of an arson attempt is believed by a forensic laboratory to be either a mixture of hydrocarbons or acetone, $(CH_3)_2 C{=}O$. An IR spectrum for a sample of this solvent is found to show no appreciable absorbance in the region of 1700 cm^{-1}. Of the given possibilities, what is the most likely identity for this paint solvent?
58. A compound is known to have either a carbon–carbon double bond or triple bond in its structure. An IR spectrum for this chemical has a sharp peak at 2200 cm^{-1}, but nothing at 1650 cm^{-1}. Determine whether a double or triple carbon–carbon bond is present in this compound.
59. Describe how computers and libraries of spectra can be used for chemical identification in IR spectroscopy.

GENERAL PRINCIPLES OF LUMINESCENCE

60. What is "fluorescence"? Describe the general process by which light is emitted during fluorescence.
61. What features are often found in molecules that undergo fluorescence?
62. Explain why the wavelength of light fluoresced by a molecule is longer than the exciting light, but light fluoresced by an atom is the same wavelength as the exciting light.
63. How is the intensity of light that is emitted by fluorescence related to the concentration of an analyte that is undergoing fluorescence?
64. What is "phosphorescence"? How is light emitted during this process?
65. How is phosphorescence similar to fluorescence? How are these two processes different?
66. How is the intensity of light that is emitted by phosphorescence related to the concentration of the analyte that is undergoing this phosphorescence?
67. What is "chemiluminescence"? Describe how light is emitted by this process.
68. What is "bioluminescence"? Give one example of a bioluminescence process.

INSTRUMENTATION FOR LUMINESCENCE MEASUREMENTS

69. Describe the general design of a spectrofluorometer. Explain the function of each major component in this design.
70. Describe the general design of a simple fluorometer. How does this design differ from that for a spectrofluorometer?
71. What is an "excitation spectrum" in fluorescence spectroscopy? What is an "emission spectrum"?
72. What are the differences in an instrument that is used to measure phosphorescence and an instrument that is used to measure fluorescence? How are these two types of instruments similar?
73. What is a "luminometer"? Describe how this type of instrument works.

APPLICATIONS OF MOLECULAR LUMINESCENCE

74. Riboflavin emits yellow-green light through fluorescence when this molecule is excited with ultraviolet light. The following data were obtained when measuring the fluorescence intensity of this analyte in a series of standards and a sample. Estimate the concentration of riboflavin in the sample.

Concentration (*M*)	Fluorescence intensity, I_F
1.0×10^{-5}	4.0
2.0×10^{-5}	8.0
4.0×10^{-5}	16.0
8.0×10^{-5}	32.0
16×10^{-5}	58.0
$32 \times 10^{-5}{}_s$	105
64×10^{-5}	170
Unknown sample	25.8

CHALLENGE PROBLEMS

75. What is the probable color of an aqueous solution that shows a maximum molar absorptivity at (a) 500 nm, or (b) 320 nm?
76. The most abundant substances in unpolluted air are nitrogen, oxygen, argon, carbon dioxide, and water. Explain why only CO_2 and H_2O are regarded as greenhouse gases.
77. The molar absorptivity has units of $L/mol \cdot cm$. Given the fact that one liter is equal to a cubic decimeter, show how you can calculate the apparent cross section of a chromophore if you know its molar absorptivity.
78. A compound has a molar absorptivity of 15,460 $L/mol \cdot cm$ at 585 nm. The intensity of radiation at this wavelength that is incident upon a 2.40×10^{-4} M sample in a 5.0 cm cuvette is 450 lumens. What is the transmitted intensity of this light, in units of lumens?
79. The oxidation of lactic acid (LA) by NAD^+ to form pyruvic acid and NADH and H^+ is slow unless it is catalyzed by the enzyme lactic acid dehydrogenase (LDH). The overall reaction for this process is shown below.

$$\underset{\text{Lactic Acid}}{C_3H_6O_3} + NAD^+ \xrightarrow{\text{LDH}} \underset{\text{Pyruvic Acid}}{C_3H_4O_3} + NADH + H^+$$

The progress of this reaction can be monitored in a 1.0 cm cuvette at 340 nm, a wavelength at which NADH has an absorption maximum ($\varepsilon = 6,000$ $L/mol \cdot cm$), but NAD^+ and the other reactants or products have essentially no light absorption. This reaction is first order with respect to both reactants and the catalyst. Suppose that the initial conditions in this reaction include a lactic concentration of 1.0×10^{-3} M and NAD^+ concentration of 2.0×10^{-5} M for a sample with an unknown concentration of LDH. The following absorbances are measured for this system as a function of time.

Time (s)	Absorbance
0	0.000
40	0.060
80	0.090
120	0.105
400	0.120

Find the concentration of the catalyst LDH if the fixed rate law for this process is Rate = $k[LA][NAD^+][LDH]$ and the rate constant for this reaction is $k = 2.4 \times 10^6$.

80. A technique related to spectrophotometric titration is the method of continuous variation, also called Job's method. This procedure differs from the preceding in that the total concentration of both reagents remains the same instead of keeping the concentration of one species constant. The goal of Job's method is seldom to learn the concentration of an unknown, but rather to learn the stoichiometry of a color-forming reaction. An example of data for this method is given below for the reaction of U with R to form product P, which absorbs at the wavelength being used for the absorbance measurement.

Concentration of U (M)	Concentration of R (M)	Absorbance of P
0.000	0.008	0.000
0.001	0.007	0.250
0.002	0.006	0.500
0.003	0.005	0.750
0.004	0.004	0.950
0.005	0.003	0.750
0.006	0.002	0.500
0.007	0.001	0.250
0.008	0.000	0.000

Use the preceding information to prepare a plot of "Absorbance of P" vs. the concentration of U. What can you tell about the stoichiometry of the reaction between and U and R based on this plot?

TOPICS FOR DISCUSSION AND REPORTS

81. Visit a local clinical laboratory and obtain information on how methods such as UV-vis spectroscopy or fluorescence spectroscopy are used at the facility for chemical analysis.
82. Obtain a research article that uses flow injection analysis for chemical measurement. Write a report on your findings.
83. IR spectroscopy is commonly used as the basis for alcohol detection in breathalyzer devices. Obtain more information on this type of device and write a report on how it works.
84. Obtain more information on Raman spectroscopy and its use in chemical analysis. Discuss how this approach was used and the types of information it provided about the samples that were being examined.
85. Photoacoustic spectroscopy is another technique that can be used to measure and study various types of transitions in molecules. Locate some information on this method and how it is performed. Write a report on this method and describe some of its applications.
86. Use References 16 and 17 or related resources to obtain more information on the topic of chemiluminescence and bioluminescence. Describe one specific type of reaction that is based on these processes and that has been used in chemical analysis. Give an example of an application in which this type of analysis has been conducted.

References

1. W. J. Marshall, *Clinical Chemistry*, Elsevier, Amsterdam, the Netherlands, 2004.
2. C. A. Burtis, E. R. Ashwood, and D. E. Bruns, Eds., *Tietz Fundamentals of Clinical Chemistry*, 6th ed., Elsevier, Amsterdam, the Netherlands, 2007.
3. A. G. Gornall, Ed., *Applied Biochemistry of Clinical Disorders*, 2nd ed., J. P. Lippincott, New York, 1986.
4. W. Clarke and D. R. Dufour, Eds., *Contemporary Practice in Clinical Chemistry*, AACC Press, Washington, DC, 2006.
5. J. Inczedy, T. Lengyel, and A. M. Ure, *International Union of Pure and Applied Chemistry—Compendium of Analytical*

Nomenclature: Definitive Rules 1997, Blackwell Science, Malden, MA, 1998.

6. G. Maludzinska, Ed., *Dictionary of Analytical Chemistry*, Elsevier, Amsterdam, the Netherlands, 1990.

7. J. D. Ingle and S. R. Crouch, *Spectrochemical Analysis*, Prentice Hall, Upper Saddle River, NJ, 1988.

8. F. Szabadvary, *History of Analytical Chemistry*, Pergamon Press, New York, 1966.

9. J. Ross, "Home Test Measures Total Cholesterol," *The Nurse Practitioner*, 28 (2003) 52–53.

10. G. W. Ewing, Ed., *Analytical Instrumentation Handbook*, 2nd ed., Marcel Dekker, New York, 1997.

11. D. A. Skoog, F. J. Holler, and T. A. Nieman, *Principles of Instrumental Analysis*, 5th ed., Saunders, Philadelphia, 1998.

12. D. G. Jones, "Photodiode Array Detectors in UV/Vis Spectroscopy: Part I," *Analytical Chemistry*, 57 (1985) 1057A–1073A.

13. S. A. Borman, "Photodiode Array Detectors for LC," *Analytical Chemistry*, 55 (1983) 836A–842A.

14. J. Ruzicka and E. H. Hansen, *Flow Injection Methods*, 2nd ed., Wiley, New York, 1988.

15. S. D. Xoleve and I. D. McKelvie, Eds., *Advances in Flow Injection Analysis and Related Techniques*, Elsevier, Amsterdam, the Netherlands, 2008.

16. A. K. Campbell, *Chemiluminescence*, VCH Publishers, New York, 1988.

17. K. Van Dyke, Ed., *Bioluminescence and Chemiluminescence: Instruments and Applications*, CRC Press, Boca Raton, FL, 1985.

18. G. Lunn and L. C. Hellwig, *Handbook of Derivatization Reactions for HPLC*, Wiley, New York, 1998.

Chapter 19

Atomic Spectroscopy

Chapter Outline

19.1 INTRODUCTION: STAR LIGHT, STAR BRIGHT

IC 1613 is a dwarf galaxy located in the constellation Cetus. It is approximately 2.4 million light years from Earth and has been the subject of many astronomical studies.[1] Scientists learn about the composition of the stars by examining the light that is emitted from them. At the very high temperatures within a star, no chemical bonds can exist and everything is present as free atoms and atomic ions. It has been known for some time that the wavelengths of light emitted by hot atoms are characteristic of the element that is being heated. This fact has been used to not only measure samples of elements in flames but also to use the light from stars to examine their elemental composition. It is also possible to estimate the age of stars from these measurements, such as by determining the ratio of hydrogen versus helium that is present.

This approach has recently been used with the Hubble space telescope to look at the elemental composition of stars in the IC 1613 galaxy (see Figure 19.1). Light emission from three supergiant stars was measured to estimate the relative amounts of fifteen elements in these stars, including O, Na, Mg, Al, Si, Ca, Sc, Ti, Cr, Fe, Co, Ni, La, Eu, and H. The ratio of iron versus hydrogen and other elements also was determined to estimate the age of the stars. The results were found to agree nicely with modern theories of galaxy and star formation.[2]

The same general approach based on light and measurements involving atoms is used in chemical laboratories to determine the elemental composition of samples. This approach is known as *atomic spectroscopy*. In this chapter, we will examine the principles of this method and learn how it is performed. We also see how atomic spectroscopy can be used for chemical analysis and how this technique compares with molecular spectroscopy (Chapter 18).

19.1A What Is Atomic Spectroscopy?

Atomic spectroscopy refers to the measurement of the wavelength or intensity of light that is emitted or absorbed by free atoms.[3] To do this type of measurement, we first need to convert a sample into atoms. This process is automatic for helium and the other noble gases and is relatively easy for mercury, but it can be a challenge for some of the other elements. Assuming that this task can be accomplished, we have two choices of how to do our measurement. We can either examine light that is emitted by the atoms or examine light that is absorbed by these atoms. These two approaches are known as **atomic emission spectroscopy (AES)** and **atomic absorption spectroscopy (AAS)**, respectively.[3-7]

IC 1613

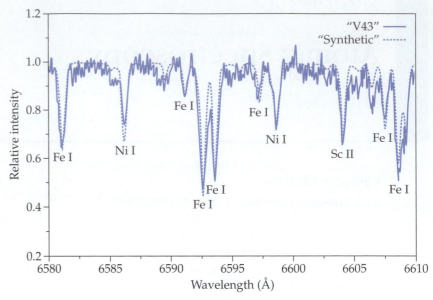

FIGURE 19.1 The irregular dwarf galaxy IC 1613, and a spectrum obtained for star V43 in this galaxy. The spectrum shows both the summation of several individual spectra (bold) and a predicted spectrum for the given elements that were detected (dashed line). (The photo of the IC 1613 galaxy is reprinted with permission and courtesy of NASA/JPL-Caltech/Spritzer Science Center; the spectrum is reproduced with permission from G. Tautvaisievne *et al.*, "First Stellar Abundances in the Dwarf Irregular Galaxy IC 1613," *The Astronomical Journal*, 134 (2007) 2318–2327.)

The requirement in atomic spectroscopy that the sample be converted to free atoms is unique compared to molecular spectroscopy, in which we desire the sample components to retain their chemical identity as molecules, polyatomic ions, or other substances. Atomic and molecular spectroscopy also differ in that atoms can undergo electronic transitions but do not exhibit any energy changes due to vibrational or rotational transitions. This feature makes the spectra that are obtained for atoms much simpler and sharper than the spectra that are observed for molecules or polyatomic species (see Chapter 18).

Atomic emission spectrometry was utilized for chemical analysis almost a century before atomic absorption spectroscopy. Shortly after Robert Bunsen had invented the "Bunsen burner," he started putting things into the flame of this device and found that different chemicals resulted in different-colored flames. Bunsen soon found that many samples gave colors dominated by abundant elements (especially sodium), which obscured the light given off by other less abundant elements. His colleague, Robert Kirchoff, suggested that they pass this light through a prism to display the colors at separate locations. By this approach they were able to observe the faint colors of emitted light from minor elements even though the more abundant elements were still present. They were also able to make a permanent record of this light on a photographic plate. The position of the lines in this photo made it possible to identify the element producing each line, and the darkness of the lines was a measure of how much of each element was present. This observation led to the use of atomic emission spectrometry for both the qualitative and quantitative analysis of elements.[6]

For over fifty years, the origin of the colored flames was unexplained. Neils Bohr in the early twentieth century demonstrated that light was emitted from atoms when one of the electrons of an atom became excited by the heat of the flame and later returned to its normal energy state, giving off the surplus energy as a photon of light. This explanation could not have been reached by Bunsen because in his day subatomic particles were unknown. It remained for Thompson, Einstein, Bohr, and others to later supply the understanding of atomic spectroscopy that allowed for an explanation of the various bands of light that are observed in this method.

19.1B How Is Atomic Spectroscopy Used in Chemical Analysis?

There are two general questions that can be addressed by atomic spectroscopy. These questions are "What is the concentration of a certain element in my sample?" and "What elements are present in my sample, and what are their concentrations?" The first of these two questions can be answered by a selective method that examines the light that results from a specific element. The technique first used by Bunsen is an example of this approach and is a method now called **flame emission spectroscopy (FES)**, which is a type of atomic emission spectroscopy.[3,4] Atomic absorption spectroscopy is another type of atomic spectroscopy that can be used to examine a specific element.

The second question is best answered by a technique that allows multiple elements to be measured simultaneously. Ideally, this approach would allow a measurement of all suitable elements without prior knowledge of what elements are present. This type of method is what Bunsen and Kirchoff used when they separated the spectrum of light emitted by a sample and used this spectrum to determine the type and amount of each element that was present.[6] An example of a method that is now used for such an analysis is inductively coupled plasma atomic emission spectroscopy, as discussed in Section 19.4B.

19.2 PRINCIPLES OF ATOMIC SPECTROSCOPY

Several factors need to be considered when we use atomic spectroscopy for chemical analysis. One of these factors is the method used for sample *atomization,* which is the process of converting a chemical compound into its constituent atoms.[3–5] A second factor to consider is the means for the detection of light that is transmitted or emitted by atoms that are produced from the sample.

19.2A Sample Atomization

In atomic spectroscopy, a high temperature is generally used to convert sample ions or molecules into atoms for analysis. A sample solution can be examined in this type of analysis by first forcing it through a narrow opening to form small droplets. These droplets are then passed along to a heating source, such as a flame. As the droplets enter the heating source several things happen quickly. First, the solvent in these droplets is removed by allowing it to evaporate or burn away. This process is called *desolvation* and leaves behind a cluster of tiny particles containing the analyte and nonvolatile materials from the sample. Second, further heating of these particles leads to *volatization,* in which the analyte enters the gas phase. The third step in the process occurs when more thermal energy from the heating sources causes all the chemical bonds in the analyte to be broken and produce atoms, a process known as *dissociation* (or "atomization").[3–5]

After these atoms have formed, some may receive additional energy from the heating source, leading to *excitation* of these atoms.[5] This step is useful if the goal is to later measure the emission of light from these excited-state atoms as they return to a lower energy state. Some of the atoms might receive too much energy and even form gas-phase ions, creating *ionization.*[5] In atomic spectroscopy this ionization is generally undesirable because it reduces the number of atoms that are available for analysis. All of these steps must happen quickly for a successful measurement to be made of atoms produced from a sample as the sample passes through the heating source. The time allowed for this process in a flame is typically less than 10^{-4} s. If the analyte does not convert into atoms within this short time, it cannot be measured before it exits the region that contains the heating source.

19.2B Sample Excitation

Elements that contain even a few electrons can have quite complicated patterns for their energy levels. As an example, a ground-state sodium atom has the electron configuration $1s^2 2s^2 2p^6 3s^1$ and has many possible transitions for its electrons to higher levels. One important excited state of sodium is produced when the absorption of energy raises the 3s electron to a 3p level, giving a sodium atom with the electronic configuration $1s^2 2s^2 2p^6 3p^1$. Typically, the lifetime of an excited atom such as this is only a few nanoseconds. As this excited atom returns to the ground state, the energy it acquired is given off as a photon of light. The energy and wavelength of this photon are characteristic of that particular transition and the type of atom that is producing the transition. In the case of a sodium atom, the difference in energy between the 3p and 3s energy levels corresponds to light with a wavelength of 589 nm, which leads to the yellow-orange color of sodium in a flame.

Two things contribute to the brightness of the emitted light that is produced by an atom in a heat source such as a flame: the concentration of atoms in the heating source and the temperature of the region around these atoms. An increase in the concentration of atoms leads to a direct increase in the intensity of light that is emitted by these atoms as they go from an excited state to the ground state. The use of a higher temperature in the heating source will cause a greater fraction of atoms to be present in an excited state, which also creates a greater intensity of light emission.

Some elements can be detected at far lower concentrations than others. This fact is related to the energy requirement that is needed to cause excitation of a ground-state atom for a particular element. If this excitation energy is low, a modest temperature can cause a significant number of atoms of that element to be excited. If the excitation energy is high, only a few atoms will be excited at a moderate temperature, producing a small emission signal. For this second type of element, a higher temperature is needed to cause more atoms to be excited and to later produce sufficient light emission to give a sufficient signal for measurement.[5,7]

19.2C Properties of Flames

The flame that was originally used by Bunsen and Kirchoff for atomic emission spectroscopy was produced by using a gaseous fuel.[7] This approach is still often employed in flame emission spectroscopy. Table 19.1 lists several fuel combinations and the typical temperatures that are achieved when using them.[5–7] A flame is a very heterogeneous chemical system. The temperatures listed in Table 19.1 are not present throughout the flame. For instance, the bottom region of a flame is much cooler. Figure 19.2 shows the general regions of a flame. The region in the center and the bottom of the flame is called the *primary combustion zone,* and it is here that combustion of the sample begins. Above this region is a relatively narrow *interzonal region* where the flame temperature reaches its maximum and a local thermal equilibrium is reached. Beyond this is the outer cone and

TABLE 19.1 Common Fuels & Oxidants Used for Flames in Atomic Spectroscopy*

Fuel	Oxidant	Maximum Temperature[a] (K)
Propane	Air	2267
Hydrogen	Air	2380
Acetylene	Air	2540
Hydrogen	Oxygen	3080
Propane	Oxygen	3094
Acetylene	N_2O (nitrous oxide)	3150
Acetylene	Oxygen	3342

*This information was obtained from C.Th.J. Alkemade and R. Herrmann, *Fundamentals of Spectroscopy*, Halsted Press, New York, 1979.

[a]The temperatures listed here are theoretical values. The actual temperatures attained experimentally are slightly lower. For comparison to these values, the temperature of an argon plasma used in inductively coupled plasma atomic emission spectroscopy can reach 6000–10,000 K.

secondary combustion zone, where oxygen from the surrounding air can lead to additional combustion.[3,5,8]

A more recent technique in atomic emission spectroscopy uses a plasma, usually involving argon, to reach much higher temperatures than can be achieved with a flame. The favored method for achieving an argon plasma is based on the use of an inductively coupled plasma. In such a device, a plasma is created when a flow of argon gas is placed within a high-frequency alternating electric field (see Section 19.4B for more details). This type of heat source can achieve temperatures of 6000–10,000 K,[8,9] but also provides a plasma with regions that are analogous to the preheating region, interconal region, and outer cone in flames.

The advantage of using a high temperature in atomic emission spectroscopy is related to the equilibrium concentration of excited atoms as a function of temperature.[5,8] This relationship is governed by the **Boltzmann distribution**, as shown in Equation 19.1,

$$\frac{N_i}{N_0} = \frac{P_i}{P_0} e^{-\Delta E/(k \cdot T)} \tag{19.1}$$

where N_i and N_0 are the number of excited atoms and ground-state atoms, P_i and P_0 are integers that describe the number of possible ways these two energy levels can come about, ΔE is the energy difference between these two states per atom, k is the Boltzmann constant (with a value of 1.38×10^{-23} J/K), and T is the absolute temperature (in kelvin). Table 19.2 gives some typical results when using the Boltzmann equation to describe sodium atoms at several temperatures, including those attained in both flames and plasma sources.

Table 19.2 indicates that even at high temperatures only a small fraction of sodium atoms are in an excited state at any one time. Still, it is clear that in comparing a temperature of 2000 K with 3000 K, as can be obtained by using different types of flames, there are almost 60 times as many excited sodium atoms at the higher temperature. This will result in a 60-fold greater intensity of light being emitted as atoms move from the higher energy level to the ground state and provide for a greater sensitivity and lower limit of detection. One

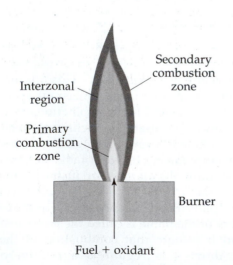

FIGURE 19.2 General regions in a flame. The role played by each of these regions is described in the text.

TABLE 19.2 Fraction of Sodium Atoms in the 3p versus 3s Orbitals at Various Temperatures

Temperature (K)	Fraction of Na in 3p versus 3s Orbitals[a]
1000	7.5×10^{-11}
2000	1.5×10^{-5}
3000	8.8×10^{-4}
6000	5.1×10^{-2}

[a]These values were calculated by using the Boltzmann equation, as demonstrated in the text. The fraction of sodium in the 3p vs. 3s orbitals is represented by the ratio (N_i/N_0) in the Boltzmann equation for this particular example.

disadvantage to using higher temperatures is that atoms that have low ionization energies can be ionized instead of just being excited. Ions can also be excited and can emit light, but will do so at different wavelengths than the corresponding atoms. The emission of light by ions can be used for chemical analysis under some conditions.

Using the Boltzmann Distribution

Lithium emits light at a wavelength of 670.8 nm when it is placed into a hydrogen/oxygen flame (temperature, 3100 K). This light emission occurs as a result of an electronic transition where $P_i = 2$ and $P_0 = 1$ for the excited state and ground state, respectively. At the given temperature of the flame, what fraction of the lithium atoms will be present in the excited state? How would this fraction change if the temperature of the flame were increased to 3300 K?

SOLUTION

We first need to find the difference in energy ΔE that is involved in this transition. This can be done by using the following relationship between ΔE and the wavelength of the emitted light, λ (see discussion in Chapter 17).

$$\Delta E = hc/\lambda$$

$$= \frac{(6.62 \times 10^{-34} \, J \cdot s)(3.00 \times 10^8 \, m/s)}{(670.8 \, nm)(1 \times 10^{-9} \, m/nm)}$$

$$= 2.96 \times 10^{-19} \, J$$

It is important to note that this result corresponds to the energy of a single emitted photon. We can now place this information into Equation 19.1, along with the given temperature and Boltzmann's constant, to calculate the fraction of lithium atoms that will be present in the excited state, as is represented by the ratio N_i/N_0.

$$\frac{N_i}{N_0} = \frac{2}{1} \cdot e^{-2.96 \times 10^{-19} \, J/[(1.38 \times 10^{-23} \, J/K)(3100 \, K)]} = \mathbf{0.0020}$$

This result indicates that 0.2% of the lithium atoms will be in the excited state at 3100 K. If we had instead used a flame temperature of 3300 K, the same type of calculation indicates that 0.3% of lithium atoms would then be in the excited state.

19.2D Analyte Measurement

There are three main techniques by which analytes can be measured through the use of atomic spectroscopy. These techniques are (1) atomic emission spectroscopy, (2) atomic absorption spectroscopy, and (3) atomic fluorescence spectroscopy. In atomic emission spectroscopy,

the intensity of emitted light at an appropriate wavelength will be directly proportional to the concentration of analyte in the flame and, therefore, will be proportional to the concentration of analyte in the solution that is entering into the flame. The proportionality constant for this relationship is a function of many variables, such as the flame temperature and the rate at which the solution is placed into the flame.

For measurements that are made by atomic absorption spectroscopy, Beer's law can be used to relate the absorbance (A) that is observed for an analyte to the detected concentration of atoms (C) that are related to this analyte, as shown in Equation 19.2.

$$A = -\log(T) = \log(P_0/P) = \varepsilon bC \qquad (19.2)$$

Just as is true for molecular spectroscopy, the values of A and C are related to each other by Equation 19.2, where ε is the molar absorptivity of the atoms at the detection wavelength, and b is the path length of the light as it passes through these atoms. Other terms that are shown in Equation 19.2 include the transmittance of the sample, the initial power of the light (P_0), and the power of light that is transmitted through the sample (P).

Some of the limitations of Beer's law that were discussed in Chapter 17 are important to consider in atomic absorption spectroscopy. The requirement that the light source be monochromatic is especially important because the absorption peak for an atom is extremely narrow. This feature means that even a narrow band of light that is isolated by a monochromator will produce a range of molar absorptivity values as it is absorbed by the atom. An approach for dealing with this problem will be described in the next section.

Atomic fluorescence spectroscopy (AFS) is similar to atomic emission spectroscopy in that both examine light that is emitted by the sample.[3–5] However, these methods differ in that the sample in atomic fluorescence spectroscopy emits light after its atoms have first been excited by absorbing a photon rather than through thermal excitation (as is used in atomic emission spectroscopy). The source of light in AFS is often a laser (see Box 19.1 and Figure 19.3). The use of this type of intense light source helps to increase the intensity of the observed fluorescence that is measured. It is not important to have an especially narrow band of wavelengths in the excitation light in this case because the transmitted light is not what is examined. What is measured instead is the light that is emitted by the sample at a 90° angle to the path of the excitation light.[5]

The energy of light from the excitation source in atomic fluorescence spectroscopy is often used to excite an atom to a higher energy level than the one from which we hope to measure fluorescence. This approach is used to avoid any interference from scattered light that originates from the light source. The intensity of the measured fluorescence will be proportional to the

BOX 19.1
Tuning into Lasers

The invention of lasers in the 1960s revolutionized many aspects of spectroscopy. The word "laser" is taken from the initials of the more formal name "**L**ight **A**mplification by **S**timulated **E**mission of **R**adiation." This type of light source is useful in spectroscopy because it provides monochromatic light that is intense and is highly directional (i.e., all of the output of the laser is going in the same direction, unlike a light bulb, which sends its light in all directions). Some lasers operate only at a fixed wavelength, but others can be tuned to operate over a fairly wide range of wavelengths. A variety of materials have been used as a "lasing" medium (see the diagram in Figure 19.3). These materials include solids such as ruby, gases such as helium and neon, dyes, and semiconductors such as gallium arsenide.[5,8,10,11]

A laser works by creating a "population inversion." This term means there are more chemical species in the excited state than in the ground state. This situation is in opposition to the behavior described by the Boltzmann equation, which predicts that normally there will always be more chemical species in the ground state than in an excited state. Population inversion is accomplished by "pumping" the lasing medium with an intense light source or an electrical discharge. Once this happens, the excited species in the lasing medium will emit light, which stimulates other excited species in the same medium to also emit. Mirrors are used to pass most of the emitted light back through the lasing medium to stimulate further emission. One of the mirrors has 100% reflectivity and the other is a partially transparent mirror that reflects about 99% of the light. The small fraction of light that is not reflected by the second mirror is allowed to exit and forms the laser beam that can be used for the desired measurement or application.[5,10,11]

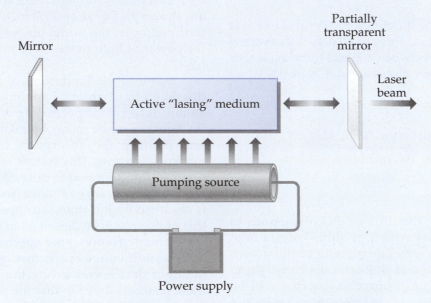

FIGURE 19.3 The general design for a typical laser. A description of how a laser works is provided in the text. Although flat mirrors are shown in this diagram, spherical mirrors are normally used to minimize the loss of light due to diffraction.[5]

concentration of the atoms that are undergoing fluorescence and will depend on the quantum efficiency of this emission process (see Chapter 18). The intensity of this light will also depend on several properties of the system. Unlike atomic absorption spectroscopy, the intensity of the light measured in atomic fluorescence spectroscopy will be proportional to the intensity of the excitation light. In addition, fluorescence depends on the atoms first absorbing a photon of light, which means that a high molar absorptivity will create a greater number of excited atoms that can later undergo fluorescence.

19.3 ATOMIC ABSORPTION SPECTROMETRY

Atomic absorption spectrometry began with the observation in the 1820s by Joseph von Fraunhofer of dark lines in the solar spectrum.[6] He observed and cataloged 576 such lines and realized that they were inherent in the light from the sun and not some instrumental artifact. Later it was shown that the wavelengths of these dark lines are the same as the bright lines of terrestrial elements when they absorb light while in a flame. This was of interest to Bohr in his study of the origin of specific colors of emitted light and of the energy levels in atoms. Our modern understanding is that the dark lines in the solar spectrum are caused by cooler atoms in the sun's atmosphere that absorb some of light being emitted by the sun. This same principle can be used for chemical analysis by utilizing flames and other heating sources to help determine the composition of certain elements in a sample.

19.3A Laminar Flow Instruments

There are several types of instruments that can be used for atomic absorption spectroscopy. One important device of this type is one which makes use of laminar flow for sample analysis. The phrase "laminar flow" describes the way in which the sample, fuel, and oxidant are fed into the region where the flame is present. This approach is in contrast to "total consumption burners" such as the Bunsen burner that are frequently used in flame emission spectroscopy, but are seldom used in atomic absorption measurements.

One type of atomic absorption instrument uses a **laminar flow burner** (Figure 19.4).[5,6] In this type of instrument the fuel and oxidant are mixed in a chamber beneath the flame. Into this chamber is fed a liquid sample. This sample is passed through a device (known as a "nebulizer") that disperses the solution to form a mist of tiny droplets. Droplets that are too large will condense onto a "spoiler," fall to the bottom of the chamber, and go through a drain to a waste container. The fuel/oxidant mixture and remaining small droplets are allowed to rise into the burner head and through a long slot into the flame. It is in the flame where sample atomization then occurs. A problem that sometimes occurs with such burners is when the flow of gas is diminished so that the flame

front moves back into the chamber where the entire mixture burns in a minor explosion, called "flashback." This typically blows the burner head into the air and can create a safety hazard. The danger of this hazard is minimized in modern instruments by attaching the burner head to restraining cables that will keep it from flying into a laboratory and at the operator. A blow-out plug at the base of the burner is also often used to avoid flashbacks.

It wasn't until about 1955 that the practical use of atomic absorption spectroscopy began in analytical laboratories. The use of atomic absorption spectroscopy for routine chemical analysis first required an adequate light source to ensure that light passing through the sample had just the right wavelength for absorption by the desired atoms. The **hollow cathode lamp** is the light source that is now commonly used for this purpose (see Figure 19.5).[3–6] A hollow cathode lamp contains a small, hollow cylindrical piece of the same metal/element that the lamp will later be used to measure in a flame. This hollow piece of material is intended for use as a cathode, the negatively charged electrode in this lamp. A voltage is applied between this cathode and a neighboring anode. This voltage difference causes ionization of some of the argon or neon gas that fills the hollow cathode lamp and results in Ar^+ or Ne^+ ions. These positive ions will then travel to the cathode and collide into the cathode at a high velocity. These collisions cause some metal atoms to be dislodged from the cathode and into the surrounding gas phase. This process is known as *sputtering*.[5,7,8] Some of the sputtered metal atoms will also receive sufficient energy from the collision to place them into an excited state. These excited atoms later return to the ground state through the emission

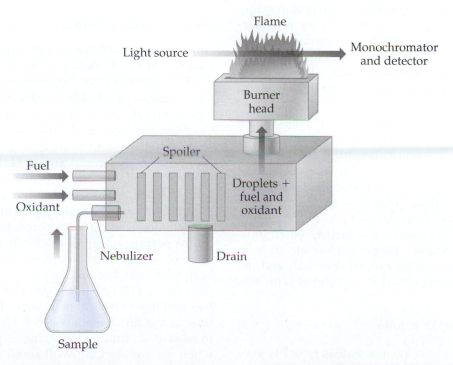

FIGURE 19.4 Design of a laminar flow burner for atomic absorption spectroscopy. The functions of the various components of this device are described in the text.

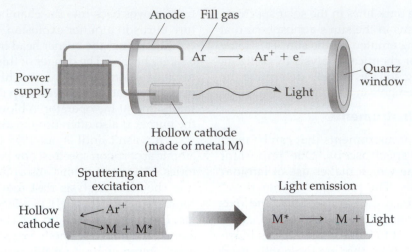

FIGURE 19.5 General design of a hollow cathode lamp (top) and the mechanism by which light is emitted by this light source (bottom).

of light. This light will have just the right energy for later exciting atoms of the same element when this light passes through a sample in a heating source such as a flame. This lamp is designed so that most of the excited atoms will return to the cathode surface when they lose their excess energy. This feature allows the atoms to be excited again later and reused for light emission.

We learned in Chapter 17 that Beer's law assumes that monochromatic light is being used for an absorption measurement. This assumption requires that the bandwidth of the incoming radiation be narrow with respect to the absorption spectrum for an analyte. This requirement is usually not a problem in molecular spectroscopy due to the broad absorption bands for molecules and polyatomic species. However, this issue can be a serious problem in atomic absorption spectroscopy because the absorption spectrum for an atom is so narrow. It is for this reason that the hollow cathode lamp is so useful in atomic absorption spectroscopy, in which the light emitted by an element in such a lamp exactly matches the light that can be absorbed by this same element in a sample.

At this point you may be wondering "What causes there to be any width to the absorption spectrum for an atom?" and "Why is this absorption band not just a single, unique wavelength instead of a very narrow band of wavelengths?" There are several factors that contribute to the width of an absorption line in atomic spectroscopy. The most fundamental factor is the *Heisenberg uncertainty principle*, as mentioned earlier, which indicates that we cannot know with perfect knowledge both the lifetime of an atom in an excited state (Δt) and the uncertainty in the energy that is associated with this excited state (ΔE).

$$\Delta E \Delta t \geq h/4\pi \qquad (19.3)$$

Because the lifetime of the excited state is typically very short in atomic spectroscopy (i.e., the value of Δt is small), the value for the uncertainty in the energy of the atom

(ΔE) will be relatively large. This results in a distribution of energies and wavelengths when we examine the absorption or emission of light by atoms.[5]

The collisions of excited atoms with other atoms in a flame can further shorten the lifetime of an atom in an excited state, which leads to a further increase in ΔE in Equation 19.3. Another factor that creates a distribution of wavelengths in the absorption or emission spectrum for an atom is the *Doppler effect*. At the high temperature of a flame, atoms are moving very rapidly in all directions. Atoms moving toward the detector will give rise to photons that appear to have a higher energy than those that are moving away from the detector, thus leading to an increase in the observed distribution of energies and wavelengths of this light.[5]

19.3B Graphite-Furnace Instruments

Another approach for sample atomization in atomic absorption spectroscopy is to use a **graphite furnace** instead of a flame (see Figure 19.6).[3,4,5,7] A graphite furnace consists of a hollow, cylindrical piece of graphite that is typically 10 cm long and about 3 cm in diameter. Graphite can carry an electric current, but also produces some resistance and heating when a current is passed through this material. Various designs of graphite furnaces are available, but they all allow a fixed amount of sample (such as a 10 µL droplet) to be introduced into the cylinder and heated by passing a current through the cylinder, leading to "resistive heating." This heating can be performed either gradually or quite quickly. The main advantage of this design is that once a sample has been atomized, the resulting atoms will stay in the furnace for several seconds instead of the fraction of a millisecond that is typical of a flame. This longer residence time in the furnace allows each atom to be excited and to relax many times. This feature is particularly valuable when we are working with small samples because the sample does not have to be continually fed into the instrument throughout the measurement. A single drop

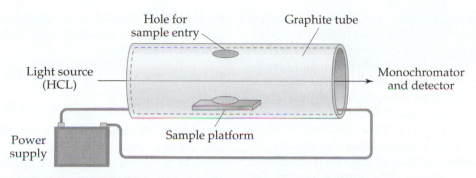

FIGURE 19.6 General design of a graphite furnace for atomic absorption spectroscopy. The role played by each of these components is described in the text.

of sample is enough to provide low limits of detection because each atom is measured many times as it absorbs light within the furnace. Although a graphite-furnace instrument typically has a considerably lower limit of detection than a laminar-flow device, it tends to be less precise than a laminar-flow device. Even small amounts of contamination from a particle of dust can also create a problem when using a graphite-furnace instrument.

19.3C Optimizing Atomic Absorption Spectroscopy

Several variables can impact the results that are obtained by atomic absorption spectroscopy. Flame temperature and sample composition are certainly important. The flame must be hot enough to atomize the sample efficiently. The flame need not excite atoms for this measurement, but some excitation is inevitable. If a flame is too hot, extensive ionization of sample atoms may occur and this will decrease the signal that is obtained from nonionized atoms.

We saw in Section 19.2C that the type of oxidant versus fuel used to produce the flame will affect the flame's temperature (see Table 19.1). An excess of oxidant is generally used to ensure that the fuel is totally consumed and doesn't leave unburned carbon particles (or "soot") in the flame. Soot particles are undesirable because they will scatter light and give an error in absorption measurements. However, some metals react with oxygen at a high temperature to give metal oxides that do not readily form atoms. Such oxides are said to be "refractory." Aluminum and molybdenum are two metals that form refractory oxides in oxygen-rich flames. This situation creates a problem because samples have less than a millisecond to atomize in a flame before they are out of the region of analysis. One way of overcoming this problem for elements that form refractory oxides is to instead use nitrous oxide, N_2O, as the oxidant. Using N_2O as the oxidant prevents free and reactive oxygen atoms from being in the flame, and thus prevents the formation of metal oxides as these oxide atoms combine with metals. Another benefit of using nitrous oxide as an oxidant is it combines with a fuel like ethylene (C_2H_2) in a highly exothermic reaction, as shown next.

$$C_2H_2 + 5\,N_2O \rightarrow 2\,CO_2 + H_2O + 5\,N_2 \qquad (19.4)$$

This fuel–oxidant reaction results in a flame with a temperature of around 3100 K (see Table 19.2) and leaves behind no free oxygen atoms or molecules that might interfere with absorption measurements.

The location and length of the light path through the flame are also important when using atomic absorption spectroscopy. The long axis of the flame, which is typically 10 cm in length, must be aligned carefully with the light beam to control and maximize this path length (see Figure 19.5 for the location of this path length in a laminar-flow furnace and Figure 19.6 for its location in a graphite furnace). The height of the burner with respect to the light path determines what part of the flame is analyzed during the absorption measurement. Recall that the flame temperature (and, therefore, the extent of atomization) varies with position in the flame, which means that the composition of atoms also varies considerably throughout the flame. Fortunately, this parameter can easily be optimized by using standard samples of the desired analyte before unknown samples are examined under a given set of conditions.

Formation of Nonvolatile Compounds. In addition to the formation of refractory oxides when excess oxygen is present in the flame, various sample components can lead to the formation of refractory solids. Calcium is an excellent example. For instance, if phosphate or sulfate ions are present in a solution they interfere with the analysis of Ca^{2+} in the same solution. Calcium sulfate and calcium phosphate are both sparingly soluble materials, but low concentrations of calcium and phosphate or sulfate anions can coexist in solution. When water is evaporated away from such a solution as the droplets of this solution enter a flame, the concentrations of these ions quickly become high enough to form solid calcium phosphate or calcium sulfate. Both of these are refractory solids that at ordinary flame temperatures do not atomize effectively.

Problems related to the formation of refractory substances in a flame can be overcome by using various chemical methods. For instance, in the case of calcium ions one can add to the samples and standards a chemical that will act as either a "releasing agent" or a "protective agent." A

releasing agent is a reagent that causes the precipitating anion (sulfate or phosphate in this case) to form a precipitate with something else. As an example, lanthanum chloride is often added to a solution in which calcium is to be measured. Lanthanum will form compounds with sulfate and phosphate that are even less soluble than the corresponding calcium compounds. Thus, the calcium is released from the phosphate or sulfate and remains in a form that can be readily measured (see Figure 19.7). A **protective agent** is a reagent that prevents the formation of the refractory compound by a different mechanism. For instance, ethylenediaminetetraacetic acid (EDTA) is often added to calcium samples to tie up calcium ions as a soluble $CaEDTA^{2-}$ complexes, which again prevents calcium ions from combining with phosphate or sulfate ions. When $CaEDTA^{2-}$ enters a flame, the EDTA readily burns away in the heat of the flame, allowing calcium to be converted into atoms that can be measured. EDTA can similarly be used as a protective agent in the analysis of many other elements that form complexes with this agent (see Chapter 9).

Hydride Generation. Another method to increase the volatility of some chemical species is to alter them before they enter a flame. Arsenic, selenium, antimony, bismuth, germanium, tin, tellurium, and lead can all be converted into volatile metal hydrides for this purpose. An example of this reaction is shown in Equation 19.5 for arsenic (present in an acidic aqueous solution in the

form H_3AsO_3), which is reacted with sodium borohydride to form the metal hydride AsH_3 before this solution enters a flame.

$$4\,H_3AsO_3 + 3\,H^+ + 3\,NaBH_4 \rightarrow 3\,H_3BO_3 + 4\,AsH_3 +$$
$$3\,H_2O + 3\,Na^+ \quad (19.5)$$

Metal hydrides like AsH_3 are volatile compounds that will enter a flame as molecules but will quickly burn away to make water and leave behind the central metal atom for analysis. Because this method removes the analyte from its matrix, detection limits can be improved by 10- to 100-fold through this approach.[7,8]

Ionization Interferences. Still another difficulty in doing quantitative analysis by flame methods is that even though conditions are usually chosen to minimize the ionization of atoms, some ionization is going to occur. The ion of an element will have a different emission or absorption spectrum than the parent atom, so this process reduces the signal that might be obtained from the atom and forms a more complex spectrum for the sample. One way of minimizing sample ionization is to use a lower flame temperature, but this approach also affects sample atomization (and excitation, in the case of atomic emission spectroscopy), so this method may not always be the best approach.

Another way of dealing with ionization effects is to add in an **ionization buffer** to the sample.[3,5] This idea can be illustrated by using sodium as an example.

$$Na \rightleftarrows Na^+ + e^- \qquad K_{i,Na} = \frac{[Na^+][e^-]}{[Na]} \quad (19.6)$$

The preceding ionization reaction can be described by an equilibrium constant that is called an *ionization constant* and whose value depends strongly on temperature. In many ways, this equilibrium constant can be treated in the same way as the reaction of a weak acid in water. The percent ionization will vary with concentration, with the ratio $[Na^+]/[Na]$ decreasing at high sodium ion concentrations. This will result in a nonconstant, increasing slope of a plot of emission intensity versus $[Na^+]$ for the aqueous solution entering the flame. We can overcome this difficulty by adding an electron source (or ionization buffer) to maintain a constant concentration of free electrons in the flame, which will help to maintain a constant ratio for $[Na^+]/[Na]$. Typically, lithium chloride is added to the sample for sodium or potassium measurements to supply such a constant concentration of electrons.

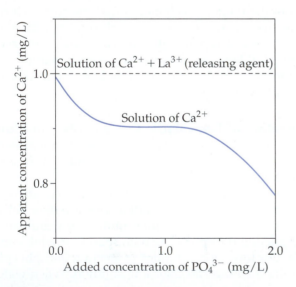

FIGURE 19.7 Effect of adding La^{3+} as a protecting agent on the apparent concentration of Ca^{2+} that is measured by atomic absorption spectroscopy for Ca^{2+} samples that contain various concentrations of phosphate, PO_4^{3-}. In the absence of some La^{3+}, the phosphate will combine with Ca^{2+} and form a sparingly soluble material and lower the number of calcium atoms that form in the flame for analysis. The presence of lanthanum prevents this reaction from occurring as the excess La^{3+} (added in this example at a concentration of 1000 mg/L) combines instead with the phosphate ions. (This graph is based on data obtained from Perkin-Elmer.)

$$Li \rightleftarrows Li^+ + e^- \qquad K_{i,Li} = \frac{[Li^+][e^-]}{[Li]} \quad (19.7)$$

If lithium is present in a sufficiently high concentration, it will maintain a constant free-electron concentration in exactly the same fashion that an aqueous buffer maintains a constant pH during the use of an acid or base.

Instrument Design.   Many features are the same for flame emission and for atomic absorption spectroscopy. This means most flame atomic absorption instruments can also be used for atomic emission spectroscopy. The features necessary for both of these methods are a flame, a monochromator, and a detector. For atomic absorption measurements, there is an additional need for a light source, such as a hollow cathode lamp. The general design of instruments for carrying out atomic emission, atomic absorption, and atomic fluorescence measurements are shown in Figure 19.8, which summarizes the similarities and differences in such devices.

Figure 19.8 indicates that all of these instruments have the monochromator placed *after* the flame or heating source. This design is important for atomic emission measurements in which only the wavelength characteristic of the analyte is to be examined. Such a design is also important for atomic absorption and atomic fluorescence measurements, because we want only a single wavelength of light to reach the detector, but a heating source such as a flame or plasma can act as a source of many wavelengths of light.

There are two sources of light that might interfere during atomic absorption measurements. The first is light that is emitted by atoms other than the analyte in the flame. The second is other wavelengths that are emitted by the hollow cathode lamp that is used as a light source. In addition, light that is emitted by the analyte in the flame will interfere when atomic absorption rather than emission is to be measured.

Light of other wavelengths can be removed by using a good monochromator, but light emitted by the analyte will have the same wavelengths as light that is emitted by the hollow cathode lamp for use in the

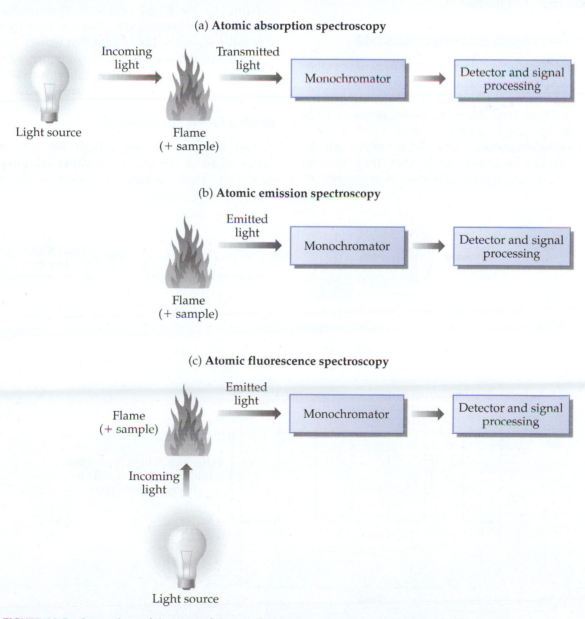

FIGURE 19.8   Comparison of the general design of instruments for performing (a) atomic absorption spectroscopy, (b) atomic emission spectroscopy, and (c) atomic fluorescence spectroscopy.

absorption measurement. Several schemes are available for dealing with this issue. The simplest approach is to use a *signal chopper*. A signal chopper is a propeller that rotates in the light path between the hollow cathode lamp and the flame. This propeller alternately opens and blocks the path of incident light that is being given off by the hollow cathode lamp, but has no effect on the light that is emitted by atoms in the flame. Therefore, the detector alternatively sees light at the characteristic wavelength that is the sum of the transmitted and emitted light when the chopper is open, but sees only emitted light when the chopper has blocked the incident light. The result is a signal at the detector that is a square wave. In this square wave, the high value is the sum of the intensity of the emitted light plus transmitted light and the low value is the intensity of only the emitted light (see Figure 19.9). The difference in these two values gives the intensity of transmitted light, which is used to calculate the true absorbance of the sample.

19.4 ATOMIC EMISSION SPECTROSCOPY

There are two approaches that are used in modern laboratories for carrying out experiments based on atomic emission spectroscopy: flame instruments and plasma instruments. These two classes of instruments differ in the methods they use to atomize samples and to excite atoms for emission measurements. One major advantage of emission instruments is that they can be designed to measure several elements simultaneously.

This is in contrast to atomic absorption instruments, which are used to measure just one element at a time.

19.4A Flame Instruments

The general design of an instrument for atomic emission spectroscopy is similar to that used for atomic absorption spectroscopy, as shown earlier in Figure 19.8. One difference is that atomic emission instruments do not require a hollow cathode lamp as a light source and do not need a chopper to distinguish between the light from this source and light that is being given off by the flame. Group IA and IIA elements are commonly determined by flame emission spectroscopy because they have fairly small values for their excitation energies. This feature means that these elements will have strong emission, allowing them to be detected at even the relatively low temperatures produced by common fuel–oxidant mixtures in flames (roughly 2200–3300 K, as indicated in Table 19.1). Specific instruments for flame emission spectroscopy can be developed to measure individual elements by examining only those wavelengths of light that would be emitted by the atoms of these elements. A common example is the use of instruments in clinical laboratories to measure sodium and potassium in blood, urine, or sweat samples.

19.4B Plasma Instruments

Plasma instruments use a high temperature plasma rather than a flame for sample atomization and excitation. These devices are used in the method of

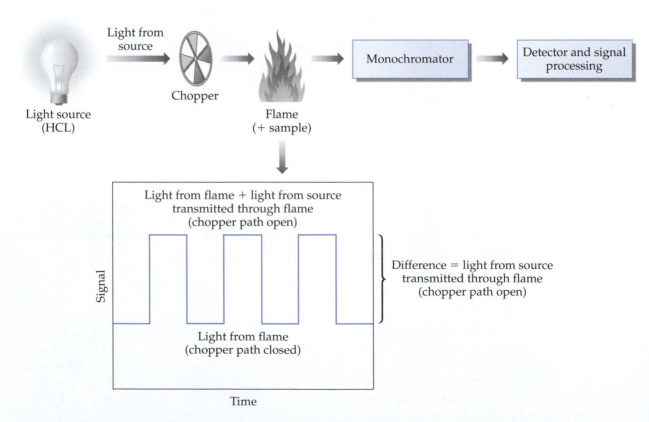

FIGURE 19.9 Use of a chopper in a laminar flow instrument to correct for light emission from the flame.

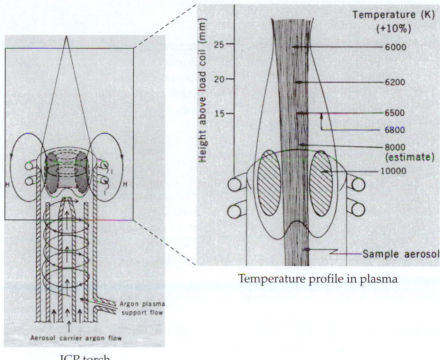

ICP torch

Temperature profile in plasma

FIGURE 19.10   The general design of a device to create an argon plasma in inductively coupled plasma (ICP) atomic emission spectroscopy (left), and the temperature profile of the plasma (right). The coils around the outside of the device have an alternating radio-frequency current I passing through them that creates a magnetic field H. This magnetic field, in turn, leads to movement and collisions of electrons and argon ions in the plasma and produces heat. (These figures are produced with permission from the AAAS from V.A. Fassel, "Quantitative Elemental Analysis by Plasma Emission Spectroscopy," *Science*, 202 (1978) 183–191.)

inductively coupled plasma atomic emission spectrometry (ICP–AES).[3,8,9] An example of this type of plasma source is given in Figure 19.10. The two major advantages of plasma instruments are the high temperatures that they can attain for emission measurements and their capability of examining many elements simultaneously. Most plasma instruments use argon as the plasma source and are powered by a radio-frequency alternating current (AC) that flows through a copper coil that is wound as an inductor around the argon flow. A spark injects electrons into the load coil region of the plasma. These electrons interact with the oscillating magnetic field produced through induction by the oscillating radio-frequency current in the load coil. Collisions between the electrons and argon atoms subsequently ionize the argon to create additional electrons which lead to further ionization events. The result is a plasma in a confined region of space that has a very high temperature. An aerosol of the sample is carried into the center of the plasma by a second stream of argon gas. A third stream of argon is passed through the narrow outer tube as a cooling sheath to keep the parts of the instrument that surround the plasma from melting. The flow of argon in these three streams is critical to generate a stable plasma.

Argon is considerably more expensive than the hydrocarbon gases like methane or acetylene and oxidants like air or oxygen that are used in atomic absorption spectroscopy. This makes ICP–AES more expensive to perform. However, argon is an inert gas that will not lead to the formation of refractory oxides. In addition, ICP-AES has limits of detection that can be decreased by 10- to 100-fold compared to those obtained with a flame as the heat source.[8,9] In addition, ICP–AES can easily be used to carry out a multielement analysis. Instruments that are used in ICP–AES have an extremely wide linear dynamic range that can cover more than six orders of magnitude in concentration.

There are three major designs for an ICP–AES instrument. These designs differ in the way in which multiple elements are handled. In one design, the instrument is set at the factory to have several slits and photomultipliers lined up at positions to monitor emission of light by specific elements. It is extremely difficult to modify these instruments to measure elements that were not considered in the original design. These devices do, however, allow the simultaneous analysis of as many as 40–50 elements. A second common design is one that allows the sequential measurement of several elements. This instrument is programmed to move its optical components to optimum positions for whatever elements are desired. These measurements take more time to carry out than with the first type of instrument, but are more flexible in terms of the range of elements that can be measured. A third design is based on the use of an array detector that can measure many elements simultaneously or that can be set for whatever elements the operator wishes to examine.

Key Words

Atomic absorption spectroscopy 459	Atomic spectroscopy 459	Inductively coupled plasma atomic emission	Protective agent 468
Atomic emission spectroscopy 459	Boltzmann distribution 462	spectrometry 471	Releasing agent 468
	Flame emission spectroscopy 460	Ionization buffer 468	
Atomic fluorescence spectroscopy 463	Graphite furnace 466	Laminar flow burner 465	
	Hollow cathode lamp 465		

Other Terms

Atomization 461	Heisenberg uncertainty principle 466	Secondary combustion zone 462	Sputtering 465
Desolvation 461			Volatization 461
Doppler effect 466	Interzonal region 461	Primary combustion zone 461	
Dissociation 461	Ionization 461		
Excitation 461	Ionization constant 468	Signal chopper 470	

Questions

WHAT IS ATOMIC SPECTROSCOPY?

1. What is meant by the term "atomic spectroscopy"? What basic requirement is needed before a sample can be examined by this method?
2. What is the difference between atomic emission spectroscopy and atomic absorption spectroscopy? Explain why both of these methods are types of atomic spectroscopy.
3. Explain why the spectrum for an analyte in atomic spectroscopy is much simpler than a typical spectrum that is acquired in molecular spectroscopy.
4. Why is the emission spectrum for hydrogen atoms simpler than it is for hydrogen molecules (H_2)?
5. Early in the twentieth century the age and temperature of a large number of stars was estimated based on emission spectroscopy. Explain how this was accomplished.

HOW IS ATOMIC SPECTROSCOPY USED IN CHEMICAL ANALYSIS?

6. What are two methods based on atomic spectroscopy that can be used to measure a specific element in a sample?
7. What is "flame emission spectroscopy"? Describe how this method was first developed.
8. Describe how atomic spectroscopy can be used to measure several elements at the same time. Name one specific type of atomic spectroscopy that can be used for this purpose.
9. Ernest Rutherford determined that the alpha particles noted in some types of radioactive decay were actually helium nuclei. Explain how he could have determined that helium nuclei were alpha particles by using atomic spectroscopy.

SAMPLE ATOMIZATION

10. What is meant by "atomization" when you are working with atomic spectroscopy? Why is atomization important to this method?
11. Explain what happens during each of the following events in sample atomization.
    (a) Desolvation
    (b) Volatization
    (c) Dissociation
    (d) Excitation
    (e) Ionization
12. Which of the events listed in Problem 11 are required for any type of atomic spectroscopy that is performed with a liquid sample? Which of these events are required in atomic emission spectroscopy but not atomic absorption spectroscopy? Which of these events are undesirable in both of these methods?
13. What is the typical amount of time that is allowed for sample atomization when you are using a flame in atomic spectroscopy? Use this information to explain why the rate and efficiency of sample atomization is important in obtaining an optimum signal in atomic spectroscopy.
14. A typical droplet that passes into the flame of an instrument for atomic absorption spectroscopy is found to have a diameter of 1.5 μm. If this droplet is formed from an aqueous solution that contains 2.5×10^{-5} M $CaCl_2$, what mass of $CaCl_2$ will remain as a solid particle after this droplet has undergone desolvation?

SAMPLE EXCITATION

15. Why do you think the emission spectrum from hydrogen atoms is simpler than the emission spectrum for atoms of any other element?
16. Explain what happens during the absorption or emission of light from sodium as its electrons move between the 3s and 3p orbitals. Which of these events would be associated with the absorption of light? Which of these events would be associated with the emission of light?
17. Calculate the change in energy that occurs as an electron moves between the 3s ground-level electronic state for sodium and its first excited state at the 3p level if the wavelength of absorbed light that leads to this transition is 589.0 nm. What frequency of light will be given off by sodium if it emits light as it returns from the 3p to the 3s level?
18. In addition to the well-known emission from sodium at 589 nm, there are fainter emissions that can be observed at 330 nm (4p → 3s transition) and 285 nm (5p → 3s), as well as 818 nm (3d → 3p) and 1140 nm (4s → 3p).
    (a) Explain why these other emission wavelengths are so much weaker than that which occurs at 589 nm for sodium.

(b) What is the energy of a photon that is emitted at each of these wavelengths? What is the frequency for each of these emission wavelengths?

(c) At what energy and wavelength would you predict the emission of light from a 5p → 3d transition to occur?

19. What are two general factors that contribute to the intensity of light emission by a sample during atomic spectroscopy?

PROPERTIES OF FLAMES

20. Explain why more than one type of flame is available for use in atomic spectroscopy. Why is the temperature of such a flame important to this method?

21. What approximate temperature is associated with each of the following fuel and oxidant mixtures when they are used to create a flame for atomic spectroscopy?
(a) Propane–air
(b) Acetylene–oxygen
(c) Hydrogen–oxygen

22. Explain why a flame is considered to be a heterogeneous system. Why is this fact important to remember when you are performing atomic spectroscopy?

23. Draw a simple diagram of the temperature profile in a flame. Label each of the following regions in your diagram: primary combustion zone, interzonal region, secondary combustion zone.

24. Discuss how the Boltzmann equation can be used to describe the relative amount of excitation that is present in a population of atoms as the temperature is varied. Explain how the relative size of these populations is related to the signal that is seen in atomic emission spectroscopy.

25. How much enhancement of the emission signal for the 3p → 3s transition of sodium do you expect if the temperature of the flame that is used to atomize and excite this sample is raised from 2000 K to 3000 K?

26. How much enhancement in the emission signal for the 3p → 3s transition of sodium do you expect if the method that is used to atomize and excite this sample is changed from a propane–air flame to an inductively coupled plasma?

ANALYTE MEASUREMENT

27. What are three general approaches that can be used to measure analytes through the use of atomic spectroscopy? What type of signal is measured in each of these approaches? Describe how this signal is related to analyte concentration.

28. In a copper measurement that is made by atomic emission spectroscopy, a blank solution gives a measured intensity at 216.6 nm of 0.2 unit. A 1.00 ppm Cu^{2+} solution that is measured under the same conditions gives an emission intensity of 18.4 units and an unknown sample produces an emission intensity of 12.7 units. What is the concentration of copper in the sample?

29. Lithium is often used as an internal standard in the measurement of sodium. A 5.0 ppm solution of Li^+ gives an emission signal of 46.7 units when measured at 671 nm, and a 5.0 ppm solution of Na^+ gives an emission signal of 35.5 units at 589 nm. A 50.0 mL aliquot of unknown solution containing Na^+ is mixed with a 5.0 mL solution containing 30.0 ppm Li^+ and is found to give an emission signal of 54.5 units at 589 nm and 21.3 units at 671 nm. What was the concentration of Na^+ in the original unknown sample?

30. A copper measurement is made by atomic absorption spectroscopy of the same sample and standard as were examined in Problem 28. A blank solution gives an absorbance reading of 0.004 units at 216.6 nm, and the 1.00 ppm Cu^{2+} standard has an absorbance of 0.305. What do you predict will be the absorbance of the unknown solution from Problem 28?

31. Atomic absorption spectroscopy is used to determine the concentration of zinc in a series of standards and an unknown sample. The following values of transmittance are measured when a zinc hollow cathode lamp is used as the light source along with a laminar flame that has a light path length of 10 cm. What is the concentration of zinc in the unknown sample?

Sample Zn Concentration (ppm)	% Transmittance
0.00	89.5
0.50	67.4
1.00	50.8
2.00	28.8
5.00	5.27
Unknown	35.6

32. Atomic absorption spectroscopy is used to measure the zinc content of an unknown sample solution by using the method of standard addition. Two 5.00 mL portions of an unknown zinc sample solution are put into 10.00 mL volumetric flasks. The first is diluted to the mark with distilled water and the second is diluted to the mark with an aqueous solution containing 1.00 ppm zinc. The first solution gives an absorbance reading of 0.386 and the second gives an absorbance of 0.497. What is the concentration of zinc in the original unknown sample solution?

33. What is "atomic fluorescence spectroscopy"? How is this method similar to atomic emission spectroscopy? How are these two methods different?

34. Explain how light that is emitted by heated atoms can be distinguished from light that is emitted from the same type of atoms by fluorescence.

35. When using atomic fluorescence spectroscopy to measure the concentration of manganese in a solution, the signal when the chopper is open is found to be 17.4 units, and the signal is 3.5 units when the chopper is closed. A solution known to contain 12.0 ppm manganese gives a signal of 28.5 units when the chopper is open and 5.7 units when it is closed. What is the manganese concentration in the solution?

36. Explain why atomic fluorescence spectroscopy can give a significantly lower limit for the measurement of an element compared to atomic absorption spectroscopy.

LAMINAR FLOW AND GRAPHITE FURNACE INSTRUMENTS

37. Describe the general design of a laminar flow instrument. Explain how this type of instrument is used for sample atomization in atomic absorption spectroscopy.

38. If the detection limit for zinc in atomic absorption spectrometry is 2 ng/mL and the flow rate of sample into a laminar flow burner is 1 mL/min, how many zinc atoms are entering the burner every second when a sample is being analyzed that has this concentration of zinc? If this signal from the

laminar flow burner is averaged over a period of 10 s, how many zinc atoms will be analyzed during this period of time?

39. What safety issues are associated with "flashback" in a laminar flow burner? How can this problem be minimized or avoided?

40. Describe the general design of a hollow cathode lamp. Explain how you would construct a hollow cathode lamp that can produce light that is selective for a particular element, such as aluminum.

41. There are several hollow cathode lamps available that can be used for multiple elements.
    (a) Explain how you would construct this type of lamp to look at zinc, aluminum, iron, nickel, and copper.
    (b) What features of atomic spectroscopy make it possible to use such a multielement hollow cathode lamp for absorbance measurements?

42. Beer's law assumes that monochromatic light is being used during an absorption measurement. Discuss why a hollow-cathode lamp can be used to meet this requirement in atomic absorption spectroscopy.

43. Explain how each of the following factors contributes to the broadening of an absorption band for an atom: the Heisenberg uncertainty principle and the Doppler effect.

44. Describe the general design of a graphite furnace. Explain how this type of instrument is used for sample atomization in atomic absorption spectroscopy.

45. What are the advantages and disadvantages of using a graphite furnace compared to a laminar flow instrument for atomic absorption spectroscopy?

OPTIMIZING ATOMIC ABSORPTION SPECTROSCOPY

46. Why is an excess of oxidant often used in the fuel mixture for atomic spectroscopy? What is a possible disadvantage of using an excess of oxidant?

47. Discuss why nitrous oxide is sometimes used in place of oxygen or air as an oxidant for a flame in atomic spectroscopy.

48. Explain why the position of the light path through the flame is important in atomic absorption spectroscopy.

49. Why does this position of this light path need to be changed and optimized each time the temperature and conditions within the flame are changed? Describe a simple procedure that you can use to optimize this position for a particular analyte.

50. What is a "refractory oxide"? Give two examples of refractory oxides. Describe how these materials can create problems when you are conducting a measurement using atomic absorption spectroscopy.

51. Suppose calcium nitrate is used to make up standard solutions for the atomic absorption measurement of calcium in unknown samples that also contain low concentrations of phosphate. What error might result during this analysis due to the sample matrix? How could this measurement be improved to minimize this error?

52. What is a "releasing agent" and how is this used in atomic spectroscopy? Give one specific example of a releasing agent.

53. What is a "protective agent" and how is this used in atomic spectroscopy? Give one specific example of a protective agent.

54. A food chemist wants to measure the calcium concentration in milk by using atomic absorption spectroscopy. A 10.00 mL aliquot of the milk is first evaporated to dryness and the organic material is destroyed by using wet ashing in nitric acid. This acidic solution is transferred to a 100.00 mL volumetric flask and diluted to volume with distilled water. A 5.00 mL aliquot of this diluted sample is transferred to a 50.00 mL volumetric flask, to which a pH 9 solution of EDTA is added to the mark on the volumetric flask. The absorbance that is measured for this final solution is found to be 0.543 when it is measured at 422.6 nm. A reagent blank gives an absorbance of 0.000 under the same conditions, and a standard having a calcium concentration of 10.0 ppm that is prepared in the same fashion gives an absorbance of 0.768.
    (a) What was the concentration of calcium in the milk?
    (b) What was the purpose of adding the EDTA to the diluted sample? How do you think the results would have changed if EDTA had not been added?

55. Explain how hydride generation can be used to increase the volatility of some analytes for their analysis by atomic spectroscopy.

56. Why is the ionization of atoms a problem in atomic spectroscopy? How does a change in temperature affect this ionization process?

57. Describe what is meant by an "ionization buffer" in atomic spectroscopy? What is the function of the ionization buffer?

58. The emission of light that is measured for potassium in atomic emission spectroscopy is greater in the presence of lithium than in its absence. Explain why this difference occurs.

59. What are two sources of light that might interfere during a measurement that is made by atomic absorption spectroscopy?

60. What is a "signal chopper"? Explain how a signal chopper makes it possible to tell the difference between transmitted and emitted light in atomic absorption spectroscopy.

61. Draw a plot that shows the time dependence of the intensity of light that reaches the detector when a moderate concentration of analyte is being measured when using a chopper during atomic absorption spectroscopy. Make a similar plot for samples that have a high or low concentration of this analyte, as well as a plot for a blank. Use these plots to explain why there is a difficulty in measuring both high concentrations and low concentrations of analyte in this type of analysis.

62. The signal that is produced by using a signal chopper during a particular atomic absorption measurement gives a light intensity of 24.5 units when the chopper is open and 16.6 units when it is closed. A blank solution gives a reading of 86.5 units when the chopper is open and 0.4 units when it is closed. What is the absorbance of the sample?

ATOMIC EMISSION SPECTROSCOPY

63. Compare and contrast the general design of a flame instrument that would be used in atomic emission spectroscopy vs. atomic absorption spectroscopy. What parts of these instruments are the same? What parts are different?

64. Describe the method of inductively coupled plasma atomic emission spectroscopy. How is this method similar to flame emission spectroscopy? How is it different?

65. Explain how the plasma is created in ICP-AES.

66. The detection limit of calcium is 1 ng/mL by a laminar flow instrument in AAS, 0.02 by a graphite furnace in AAS, 0.1 by a flame source in AES, and 0.02 by ICP-AES. Explain why these detection limits differ to such a great extent.

67. Describe two ways in which an instrument for ICP-AES could be used for multielement analysis. What are the advantages and disadvantages of each approach?

CHALLENGE PROBLEMS

68. About what wavelength do you expect to be an important emission line for neon?

69. Look up the wavelengths for important emissions for helium and sodium. Use this information to explain why the first observation of helium (from the solar spectrum) was thought to be sodium. Note that helium had not yet been discovered on Earth.

70. Doppler broadening is an important constituent of the width of an atomic emission line. The broadening caused by the Doppler effect ($\Delta\lambda_D$) is given by the following relationship,

$$\Delta\lambda_D = 7.16 \times 10^{-7}(\lambda_m)(T/M)^{1/2} \qquad (19.8)$$

where λ_m is the wavelength at the midpoint of the emission band (e.g., given in units of angstroms or nm), T is the absolute temperature (in kelvin), and M is the molar mass of the atom (in units of g/mol).

(a) How much broadening due to the Doppler effect would be expected for emission at the 589 nm line of sodium at 3000 K?

(b) How much broadening due to the Doppler effect would be expected for emission at the 330 nm line of sodium at 3000 K?

(c) How much broadening due to the Doppler effect would be expected for emission at the 589 and 330 nm lines of sodium at 6000 K?

TOPICS FOR DISCUSSION AND REPORTS

71. Locate a scientific article in which spectroscopy has been used as a tool in astronomy. Determine what type of spectroscopy was used in the study and describe how this method was used in the study.

72. Another method that can be used to tell the difference between transmitted and emitted light from a flame is an approach based on the Zeeman effect.[5,7] Locate some information on the Zeeman effect and how this method can be used to correct for problems with interferences related to emission during atomic absorption spectroscopy. Write a report on your findings.

73. Some other techniques that are used to tell the difference between transmitted and emitted light in flames are listed below.[5,7] Obtain information on one of these techniques and discuss the basis behind this approach.
(a) Continuum–source correction method
(b) Two–line correction method
(c) Smith–Hieftje correction method

74. Use a source such as References 10 and 11 or the Internet to locate an example where a laser has been used as part of a system for chemical analysis. Identify how the laser was used in the study and why this type of light source was useful for the given measurement.

75. Atomic clocks are considered to be the most accurate measures of time in the world. The operation of these clocks is based on the use of atomic spectroscopy. Locate more information on atomic clocks from sources like Reference 12. Discuss how these clocks work and how they make use of the transitions in the energy of atoms.

76. One area in which atomic spectroscopy is often employed is geology. Methods based on inductively coupled plasma as a heating source are particularly useful for such work. Locate some information from the scientific literature or the Internet on applications of ICP-AES or ICP–mass spectrometry in geology. Write a report on your findings.

77. An alternative approach for elemental analysis is to use X-ray fluorescence.[7,8] Obtain information on this method and learn how it works. Share your results with your classmates. Discuss what advantages or disadvantages this method might have compared to atomic absorption spectroscopy or atomic emission spectroscopy.

References

1. NASA/IPAC Extragalactic Database.
2. G. Tautvaisievne et al., "First Stellar Abundances in the Dwarf Irregular Galaxy IC 1613," *The Astronomical Journal*, 134 (2007) 2318–2327.
3. J. Inczedy, T. Lengyel and A. M. Ure, *International Union of Pure and Applied Chemistry-Compendium of Analytical Nomenclature: Definitive Rules 1997*, Blackwell Science, Malden, MA.
4. G. Maludzinska, (Ed.), *Dictionary of Analytical Chemistry*, Elsevier, Amsterdam, Netherlands, 1990.
5. J. D. Ingle and S. R. Crouch, *Spectrochemical Analysis*, Prentice Hall, Upper Saddle River, NJ, 1988.
6. F. Szabadvary, *History of Analytical Chemistry*, Pergamon, Press, New York, 1966.
7. G. W. Ewing, (Ed.), *Analytical Instrumentation Handbook*, 2nd ed., Marcel Dekker, New York, 1997.
8. D. A. Skoog, F. J. Holler, and T. A. Nieman, *Principles of Instrumental Analysis*, 5th ed., Saunders, Philadelphia, 1998.
9. V. A. Fassel and R. N. Kniseley, "Inductively Coupled Plasmas," *Analytical Chemistry*, 46 (1974) 1155A–1164A.
10. D. L. Andrews, *Lasers in Chemistry*, Springer-Verlag, New York, 1996.
11. J. Wilson and J. F. B. Hawkes, *Lasers: Principles and Applications*, Prentice-Hall, Englewood Cliffs, NJ, 1987.
12. D. W. Ball, "Atomic Clocks: An Application of Spectroscopy," *Spectroscopy* 21 (2007) 14–20.

Chapter 20

An Introduction to Chemical Separations

Chapter Outline

20.1 INTRODUCTION: THE GREEN REVOLUTION

The phrase "Green Revolution" describes a major change in food production that started in the 1940s. Scientists at that time were working to improve crop production in third-world countries by using new, hardier crops and better agricultural techniques. As a result, Mexico went from importing half of its wheat to exporting large amounts of this grain by the 1960s. The same techniques were then adopted by other nations as a means for increasing their food production.[1,2]

One change that appeared during the Green Revolution was an increased use of chemicals to control weeds and insects. The first chemical utilized on a large-scale for this purpose was 2,4-dichlorophenyoxyacetic acid (2,4-D), which is a herbicide used to kill broadleaf weeds (see Figure 20.1). The use of 2,4-D and other herbicides has created a dramatic increase in food production throughout the world.[3,4] This use has also caused many of these chemicals to now appear in our water, soil, and food. As a result, agencies like the U.S. Environmental Protection Agency and U.S. Food and Drug Administration now monitor and regulate the amounts of 2,4-D and other herbicides that appear in food and drinking water.[5]

Many methods are employed for the measurement of 2,4-D and other chemicals in environmental samples. This type of analysis requires techniques that can deal with a variety of complex samples, ranging from water and soil to plant and animal tissues. To help in such an analysis, it is often necessary to first isolate the analyte of interest from other sample components.[6,7] In this chapter, we will learn about extractions and chromatography, two tools that are often employed in such separations.

20.1A What Is a Chemical Separation?

The isolation of 2,4-D from an environmental sample is an example of a chemical separation. A **chemical separation** is a method that involves the complete or partial isolation of one chemical from another in a mixture of two or more substances. There are many ways chemical separations can be performed. For instance, we saw in Chapter 11 how precipitation can be used to remove metal ions from a solution for analysis. Other common separation methods you may already have used include filtration, centrifugation, and distillation.

There are several strategies that can be used to obtain a chemical separation.[8] The first strategy uses two or more chemical or physical environments (referred to as "phases") to separate chemicals. The precipitation of metal ions is an example of this type of process because it utilizes the conversion of soluble ions into a solid precipitate. Another example is distillation, in which the components of a liquid mixture are separated based on their abilities to move from the liquid into a vapor phase. These separations require that the chemicals in a sample have different equilibrium constants for their transfer

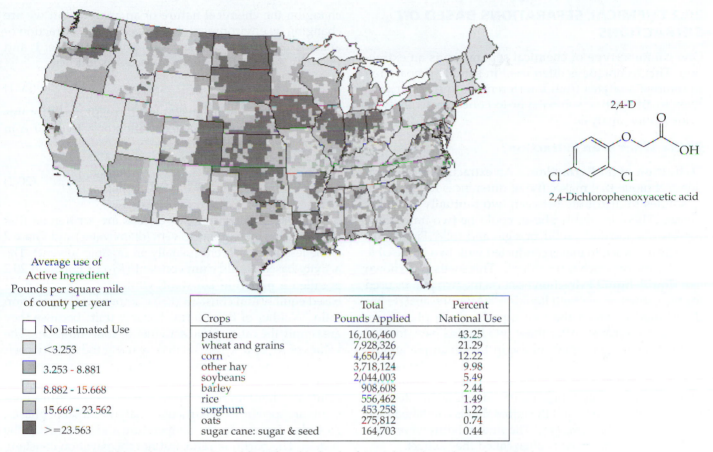

Crops	Total Pounds Applied	Percent National Use
pasture	16,106,460	43.25
wheat and grains	7,928,326	21.29
corn	4,650,447	12.22
other hay	3,718,124	9.98
soybeans	2,044,003	5.49
barley	908,608	2.44
rice	556,462	1.49
sorghum	453,258	1.22
oats	275,812	0.74
sugar cane: sugar & seed	164,703	0.44

Average use of Active Ingredient Pounds per square mile of county per year

☐ No Estimated Use
☐ <3.253
☐ 3.253 - 8.881
☐ 8.882 - 15.668
☐ 15.669 - 23.562
☐ >=23.563

FIGURE 20.1   2,4-Dichlorophenyoxyacetic acid (2,4-D) and its levels of annual use throughout the United States. (These data are from 1992 and are courtesy of the U. S. Geological Survey.)

from one phase to the other. The greater these differences are, the easier it will be to separate the chemicals.

A second strategy found in chemical separations is to use an approach where chemicals have different rates of travel through a system.[8] Centrifugation is one example of this approach, where a gravitational field is used to separate large particles (such as a precipitate) from a smaller material (e.g., a dissolved molecule or ion). A separation based on rates of travel requires the use of an external force like gravity or electrical field to move chemicals through the system. Sometimes several strategies are combined when isolating a chemical, as occurs when we precipitate a chemical (a phase-based separation) and use centrifugation to remove this precipitate from a liquid (a rate-based separation).

20.1B How Are Chemical Separations Used in Analytical Chemistry?

Many measurement techniques work well with samples that contain only a few chemicals but have problems when the sample has many substances present. This issue becomes a problem when we are examining complex samples, as would occur if we were looking for a trace chemical like 2,4-D in groundwater or food. Chemical separations are extremely valuable in this situation,

where they can be employed to remove an analyte from other agents that might interfere with its measurement. A chemical separation can also be utilized to place an analyte in a matrix that is more compatible with the analysis method. An example would be if we wanted to measure 2,4-D in a water sample, but had to first transfer the 2,4-D into an organic solvent for analysis. We can also use a chemical separation to adjust the sample volume (and, thus, the analyte concentration) to a level that is more compatible with the measurement method.

Another reason for conducting a chemical separation is to examine several analytes in a sample, as would occur if we wished to look for both 2,4-D and related herbicides in water. This type of analysis requires a separation that can isolate these chemicals from one another and pass them through a device that gives a signal related to the amount of each substance. Such a method might involve taking a sample and separating it into various fractions (e.g., water-soluble compounds and organic-soluble compounds) that can be examined separately. Alternatively, such an analysis might involve combining the separation method with a detection device as part of a single analysis system. We will see examples of both these approaches later when we discuss the methods of extraction and chromatography.

20.2 CHEMICAL SEPARATIONS BASED ON EXTRACTIONS

One common type of chemical separation is an extraction. This technique is often used in sample preparation to remove analytes from interferences, to transfer analytes to alternative solvents, or to concentrate analytes prior to their analysis.

20.2A What Is an Extraction?

Definition of an Extraction. An **extraction** is a separation technique that makes use of differences in the ability of solutes to distribute between two mutually insoluble phases. These insoluble phases could be two immiscible liquids, a liquid and a solid, or a gas and solid. Figure 20.2 shows an extraction that is conducted with two immiscible liquids, one of which is the sample. This method is known as a **liquid–liquid extraction** because the two phases used for the extraction are both liquids.[9,10] After the analyte has distributed between the two liquids, these phases are allowed to separate. After these two liquids have formed distinct layers, the upper phase in this example and its contents are removed for later use or analysis. The bottom liquid is then further extracted or discarded. The sample or phase that originally contained the chemicals of interest is known as the *raffinate* and the liquid that is combined with this sample is the *extractant*. The final mixture of analyte and the extracting phase is often called the "extract."[9]

The distribution of an analyte between the two phases in an extraction is an example of a solubility equilibrium (see Chapter 6). In this type of process we are not changing the chemical nature of an analyte, but we are changing its environment. In the case of the extraction of analyte A between two liquid phases (Phase 1 and Phase 2), we can describe this process as follows.

$$A_{Phase\ 1} \rightleftharpoons A_{Phase\ 2} \tag{20.1}$$

We can also write an equilibrium expression for this extraction based on the concentration or activity of A in each of the two liquids.

$$K_D^\circ = \frac{a_{A,Phase\ 2}}{a_{A,Phase\ 1}} \quad \text{or} \quad K_D = \frac{[A]_{Phase\ 2}}{[A]_{Phase\ 1}} \tag{20.2}$$

These equilibrium expressions are written so that Phase 1 is the more polar solvent (often water) and Phase 2 is the less polar solvent (usually an organic solvent). The activity-based equilibrium constant (K_D°) in Equation 20.2 is called a **partition constant**, while the concentration-based equilibrium constant (K_D) is known as the **partition ratio**.[9] Neither of these terms has any units because they are simply the ratio of two activities or concentrations. The values of K_D° and K_D are related by using the activity coefficients of A in each of the two phases, using an approach similar to that used for other types of equilibrium constants in Chapters 6–10. Both of these equilibrium constants are specific for a particular analyte and set of phases; they also depend on the temperature and pressure of the system. The partition ratio, but not the partition constant, can vary with the total amount of analyte in the system because this amount can affect the analyte's activity coefficients in the two phases.

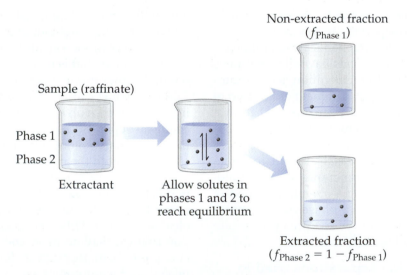

FIGURE 20.2 Example of a liquid–liquid extraction, showing the distribution of a solute between two mutually insoluble phases. The terms $f_{Phase\ 1}$ and $f_{Phase\ 2}$ refer to the fraction of a solute that remains in Phase 1 after the extraction or that is extracted into Phase 2, respectively. In this method, the lower-density liquid (often an organic solvent) will appear on top of the more dense liquid (usually water). Sometimes an additional agent is also placed in this system to bind with analytes and improve their ability to go into the extractant. An example would be the use of a complexing agent to bind metal ions and convert them into a neutral metal–ligand complex that can go into an organic phase. If such an agent is employed, it is referred to as an *extracting agent*.

Many analytes can exist in multiple forms in a solution. Examples are weak acids and chemicals that form complexes with other substances. We have already learned how we can use an analytical concentration (C_A) to represent the total amount of such an analyte, regardless of its various forms. Although each of these forms will have its own unique partition constant or partition ratio during an extraction, we can utilize analytical concentrations to describe the overall distribution of an analyte in two phases. This task is accomplished by using the **distribution ratio** (D_c), which is defined as follows.[9]

$$D_c = \frac{C_{A,\text{Phase 2}}}{C_{A,\text{Phase 1}}} \quad (20.3)$$

Like K_D, the value of D_c has no units and will depend on such things as the type of solute and phases that are being employed, as well as the temperature and pressure. However, D_c may be affected by the pH or concentration of any substances that may react with the analyte in side reactions. If an analyte exists in only one form in both Phases 1 and 2, its distribution ratio will be the same as its partition ratio, or $D_c = K_D$. These two values will not be equal, however, if the analyte has multiple forms in either of these phases. It is often possible to use our knowledge of chemical equilibrium to determine how D_c might change with a given set of conditions, as is demonstrated in the following exercise.

EXERCISE 20.1 | **Relationship Between the Distribution Ratio and Partition Ratio**

2,4-D is a monoprotic weak acid with a K_a of 1.20×10^{-3} at an ionic strength of zero. In an extraction based on water and ethyl acetate, the weak acid form (HA) of 2,4-D and its conjugate base (A^-) have separate K_D values. Only the neutral, weak acid form has any appreciable degree of extraction ($K_{D,HA} = 550$ and $K_{D,A^-} \approx 0$). Show how D_c for 2,4-D is related to the concentrations of HA and A^-, as well as to the individual partition ratios for HA and A^-.

SOLUTION

Because there are two species of 2,4-D that can exist in water (Phase 1), the analytical concentration of 2,4-D in this phase will be $C_{A,\text{Phase 1}} = [HA]_{\text{Phase 1}} + [A^-]_{\text{Phase 1}}$. We also know that essentially none of A^- enters ethyl acetate, because its partition ratio for this phase is essentially zero. Thus, the analytical concentration of 2,4-D in Phase 2 will be $C_{A,\text{Phase 2}} = [HA]_{\text{Phase 2}}$. This gives the following expression for D_c in this system.

$$D_c = \frac{C_{A,\text{Phase 2}}}{C_{A,\text{Phase 1}}} \approx \frac{[HA]_{\text{Phase 2}}}{[HA]_{\text{Phase 1}} + [A^-]_{\text{Phase 1}}}$$

We can use this expression to relate D_c to the partition ratio and acid dissociation constant for HA by first dividing the numerator and denominator by $[HA]_{\text{Phase 1}}$. We can then substitute in $K_{D,HA} = [HA]_{\text{Phase 2}}/[HA]_{\text{Phase 1}}$ and $K_a/[H^+] = [A^-]_{\text{Phase 1}}/[HA]_{\text{Phase 1}}$. The result is Equation 20.4.

$$D_c \approx \frac{[HA]_{\text{Phase 2}}/[HA]_{\text{Phase 1}}}{1 + [A^-]_{\text{Phase 1}}/[HA]_{\text{Phase 1}}} = \frac{K_{D,HA}}{1 + K_a/[H+]} \quad (20.4)$$

We will see later how this relationship can be used to determine the effect a change in pH will have on the extraction of a weak acid like 2,4-D. However, we can already predict that more 2,4-D will be extracted at a low pH, because the value of D_c for 2,4-D will increase as $[H^+]$ increases.

The value of D_c (or K_D° and K_D) reflects the ability of a solute to enter one phase versus another. If a solute has a value for D_c that is greater than 1.0, this means the solute will tend to occupy Phase 2 versus Phase 1. If D_c is less than 1.0, then the original sample (Phase 1) will contain a higher total concentration of the solute than Phase 2. One practical consequence of this relationship is that we must have a difference in the D_c values for two solutes if we are to separate these chemicals using a given set of phases. The larger this difference is in D_c, the easier it will be to separate the solutes by an extraction.[8]

General Types of Extractions. The only type of extraction we have considered up to this point has been a liquid–liquid extraction. Some other types of extractions are listed in Table 20.1. One way these methods can be grouped is according to the phases that they employ. For example, a liquid–liquid extraction uses two immiscible liquids, while a liquid and solid are the two phases present during a liquid–solid extraction. It is also possible to use more exotic types of phases such as a supercritical fluid, as described in Box 20.1.

A second way extractions can be grouped is according to the mechanism by which they separate chemicals. One possible mechanism is "partitioning," which involves a chemical entering both of the two phases being employed for the extraction. Liquid–liquid extraction is an example of a method that uses partitioning. Another mechanism for carrying out an extraction is to use "adsorption," which occurs when solutes interact with the surface of a solid. Adsorption is the separation mechanism used in a solid–phase extraction, where a bare support is the extracting phase. A solid support can also be used for partitioning if it contains a liquid coating or a chemically bonded layer on its surface.[8,10]

BOX 20.1
Supercritical Fluid Extractions

Although most extractions use a liquid (or sometimes a gas) as the extracting solvent, a *supercritical fluid* can also be used for this purpose. A supercritical fluid is a state of matter that exists above a certain critical temperature and pressure for a given chemical.[1] Under these conditions, the chemical is neither a gas nor a liquid, but has intermediate properties of both. For instance, above a pressure of 73.8 bar (72.9 atm) and a temperature of 31.1°C (a little over room temperature), carbon dioxide becomes a supercritical fluid (see Figure 20.3). In this state, carbon dioxide has a higher density and greater solvating ability than its gaseous form, but a lower density and lower viscosity than its liquid form. Furthermore, the density of the supercritical fluid can be adjusted by altering the temperature and pressure, making it possible to control the properties of carbon dioxide when it is in this state.

All of these properties have made supercritical fluids valuable as extracting solvents, especially when dealing with solid samples. The resulting extraction method is known as a *supercritical fluid extraction (SFE)*.[1,11–13] An example is the extraction of caffeine from coffee beans by supercritical carbon dioxide during the production of decaffeinated coffee. The same approach is used on a smaller scale to extract caffeine and other chemicals from solid samples for analysis. Carbon dioxide is often used in these applications because it is inexpensive and easy to convert into a supercritical fluid. Supercritical carbon dioxide is also environmentally friendly because it is nontoxic, cuts down on the use of organic solvents, and reduces the generation of chemical waste as a result

of chemical analysis. These features have created great interest in the use of supercritical fluids for applications ranging from chemical synthesis and analysis to the processing of foods and natural products.

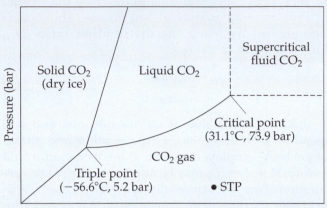

FIGURE 20.3 Phase diagram for carbon dioxide, showing the conditions in which this chemical exists as a supercritical fluid. The critical point is the pressure (73.9 bar) and temperature (31.1°C) above which carbon dioxide becomes a supercritical fluid. The triple point of carbon dioxide (or the point at which the solid, liquid, and gas forms of this chemical all exist) is included in this plot, along with the approximate location of standard temperature and pressure conditions (STP; see Chapter 6).

TABLE 20.1 General Types of Extractions

Type of Extraction	Description
Accelerated solvent extraction	Use of elevated temperature and pressure to increase the rate and extent of an extraction
Gas–solid adsorption	Use of a solid material to adsorb and separate gases in a gas mixture
Liquid–liquid extraction	Extraction of a liquid sample with a liquid phase
Microwave-assisted extraction	Use of microwave radiation to increase the rate and extent of an extraction
Solid-phase extraction	Extraction of a liquid or gas sample with a solid support that contains an adsorbing surface or chemical coating that can interact with analytes
Solid-phase microextraction	Use of uncoated or coated fiber delivered by a syringe for the extraction of analytes
Soxhlet extraction	A combined use of distillation and extraction for obtaining analytes from solid samples
Supercritical fluid extraction	Use of a supercritical fluid as the extracting phase

20.2B Using and Describing Extractions

Single-Step Extractions. The simplest type of extraction uses one step to place a sample in contact with the extracting phase. This approach is called a **single-step extraction**. If we call the sample "Phase 1" (the phase that originally contains the analyte) and the extractant "Phase 2" (the phase used to remove some analyte from the sample), it can be shown through mass balance that the fraction of A that remains in Phase 1 after one extraction ($f_{Phase\ 1,1}$) is given by Equation 20.5.

Single-Step Extraction: $f_{Phase1,1} = \dfrac{1}{1 + D_c(V_2/V_1)}$ (20.5)

In this equation, V_1 and V_2 are the volumes of Phases 1 and 2, and D_c is the concentration distribution ratio for A in these phases. The term V_2/V_1 gives the relative amount of the two phases versus one another and is known as the *phase ratio*.[9] If we know the fraction of A that remains in the sample, we can also determine the fraction that has entered Phase 2 after one extraction ($f_{Phase2,1}$), as shown below.

$$f_{Phase\ 2,1} = 1 - f_{Phase\ 1,1}$$
$$\text{or} \quad 1 = f_{Phase\ 1,1} + f_{Phase\ 2,1} \quad (20.6)$$

The following exercise illustrates how these equations can be used to predict the extent of extraction that will occur for an analyte between two liquids.

EXERCISE 20.2 **Predicting the Degree of Extraction for 2,4-Dichlorophenyoxyacetic Acid**

A 25.0 mL sample of water containing 2,4-D is adjusted to pH 4.00 and combined with 10.0 mL of ethyl acetate for a liquid–liquid extraction. What is the percent extraction of 2,4-D under these conditions for a dilute sample? (*Note*: Assume that the amount of 2,4-D is sufficiently small to make both D_c and K_D independent of its concentration.)

SOLUTION

The fraction of 2,4-D that remains in the original water sample or is transferred into ethyl acetate can be estimated by using Equations 20.5 and 20.6. This calculation requires that we know the phase ratio, as given by $V_2/V_1 = (10.0 \text{ mL}/25.0 \text{ mL})$. We can find D_c for 2,4-D by using the relationship derived in Exercise 20.1 along with the known values for K_a (1.20×10^{-3}) and K_D (550). This gives $D_c \approx (550)/(1 + 1.20 \times 10^{-3}/10^{-4.00})) = 42.3$. Placing these values into Equations 20.5 and 20.6 gives the following results.

$$f_{\text{Phase 1,1}} = \frac{1}{1 + 42.3\ (10.0 \text{ mL}/25.0 \text{ mL})}$$

$$= \mathbf{0.0558}\ (\text{or } 5.58\% \text{ 2,4-D still in the sample})$$

$$f_{\text{Phase 2,1}} = 1 - 0.05580 = \mathbf{0.9442}\ (\text{or } 94.42\% \text{ extracted})$$

This result makes sense because the value for D_c is much greater than one, which indicates that 2,4-D favors ethyl acetate over water as a solvent at pH 4.00.

Besides allowing us to determine how much solute will be extracted, Equations 20.5 and 20.6 provide clues as to how we might alter the extent of this extraction. One way we can increase the amount of extracted solute is by increasing the phase ratio (V_2/V_1). This ratio can be increased by raising the volume or the amount of Phase 2 (the extractant) versus Phase 1 (the sample). An increase in the phase ratio will give a larger denominator in Equation 20.5 and create a decrease in the fraction of solute that remains in Phase 1 ($f_{\text{Phase 1,1}}$). Another way we can alter the extraction is by changing D_c. The value of D_c can be varied by changing the phases that are used in the extraction. Changing the temperature or pressure creates a similar effect (see Figure 20.4). Another possibility is to change the distribution of the solute among its various forms, which alters D_c but not the K_D values for these species. This last approach can be accomplished by using side-reactions during extraction, as will be discussed later in this section.

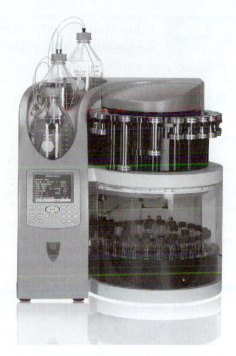

FIGURE 20.4 A commercial system used for *accelerated solvent extraction* (*ASE*). This method is also known as *pressurized solvent extraction* (*PSE*) and uses heating and elevated pressures to increase the speed and efficiency of an extraction.

(Reproduced with permission form Dionex.)

Multistep Extractions. Another way we can increase the extent of an extraction is to use several portions of Phase 2. This method is known as a **multistep extraction**. If we extract a sample several times with equal but separate volumes of a second phase, the fraction of remaining solute that is removed with each extraction step will be the same. However, the overall amount of extracted solute will increase. We can describe the overall extent of this extraction by first multiplying each of the nonextracted fractions. The result is shown in Equation 20.7, where n is the number of times the sample has been extracted with fresh portions of Phase 2 and $f_{\text{Phase 1,}n}$ is the total fraction of the solute in Phase 1 after n extraction steps.

For a Multi-step Extraction:

$$f_{\text{Phase1,}n} = \left[\frac{1}{1 + D_c(V_2/V_1)}\right]^n \tag{20.7}$$

The total fraction of solute that is transferred to Phase 2 after n steps ($f_{\text{Phase 2,}n}$) is then as follows.

$$f_{\text{Phase 2,}n} = 1 - f_{\text{Phase 1,}n} \tag{20.8}$$

Equations 20.7 and 20.8 are identical to Equations 20.5 and 20.6 when we are using only one extraction ($n = 1$). However, Equation 20.7 now indicates that as we increase the number of extractions, the size of $f_{\text{Phase 1,}n}$ will decrease (approaching a lowest possible value of zero) and the value of $f_{\text{Phase 2,}n}$ will increase (approaching a maximum value of 1.0).

EXERCISE 20.3	Using Multiple Extractions for 2,4-Dichlorophenyoxyacetic Acid

Using the same pH and sample as in Exercise 20.2, what fraction of 2,4-D will be removed from 25.0 mL water after two, three, and four extractions with fresh portions of 10.0 mL ethyl acetate?

SOLUTION

The phase ratio and value of D_c are the same as in Exercise 20.2, so we can simply place these values into Equation 20.7 along with n = 2, 3, or 4 (for two to four extraction steps) to find the fraction of 2,4-D that remains in the sample. The result for two extractions is given below.

$$f_{\text{Phase }1,2} = \left(\frac{1}{1 + 42.3\,(10.0\ \text{mL}/25.0\ \text{mL})} \right)^2$$

$$= \mathbf{0.0031} \text{ (or 0.31\% remaining in the sample)}$$

$$f_{\text{Phase }2,2} = 1 - 0.00311 = \mathbf{0.9969} \text{ (or 99.69\% extracted)}$$

It is important to note that the fraction of extracted 2,4-D is now larger than it was in Exercise 20.2 for a single-step extraction (94.42%). This trend continues for three extractions (99.98% extracted) and four extractions (greater than 99.99% extracted).

Extractions with Side Reactions. Another way we can control the degree of extraction is to use side reactions that convert an analyte into a form that is either easier or more difficult to extract. Table 20.2 shows how acid–base reactions can be used for this purpose. For example, we saw earlier that the partition ratio for a weak acid will depend on its value for K_a and the pH of its surrounding solution. This feature means we can adjust the pH to convert a weak acid from a neutral form (HA, which can be extracted by an organic solvent) to a charged form (A$^-$, which has little or no extraction by the same solvent). In this situation, the highest degree of extraction will be obtained when we use a pH that is below the pK$_a$ of a weak monoprotic acid. The opposite effect will occur for a weak monoprotic base, which will be best extracted in its neutral basic form (B) as opposed to its conjugate acid form (BH$^+$).

One way we can use these side reactions is to move a solute from its extracting phase back into a fresh portion of its original solvent. This approach is known as a *back extraction* and is often used to place an analyte into a more appropriate phase for measurement or to remove an analyte from other extracted substances. As an example, if the pH were changed from 4.00 to 8.00, Equation 20.4 predicts that D_c for 2,4-D will change from 42.3 to 0.00458 when using water and ethyl acetate as the two phases. At a phase ratio of 10.0 mL/25.0 mL, adjusting the pH of 8.0 will result in 99.82% of 2,4-D going from ethyl acetate back into water.

When dealing with analytes that are metal ions, complexation reactions can also be utilized to control the extent of an extraction. In this case, a ligand is added that will form a neutral complex with the metal ion, making it possible to extract this ion into an organic solvent. Oxine is one ligand that can be employed for this purpose (see Figure 20.5). When using such a ligand, the concentration of the ligand and the pH can both be adjusted to control the degree of metal-ion extraction, because oxine and related ligands are often weak acids or bases.[8,10]

20.2C A Closer Look at Extractions

Degree of Extraction versus Solute Purity. We now know that an increase in the number of extraction steps will increase the amount of analyte that is extracted. So why do we not always use a large number of extraction steps? The reason is that multiple extraction steps also increase the extraction of all other substances in our sample. This problem is illustrated in Figure 20.6 for two chemicals (A and B) that have D_c values of 2 and 0.1, respectively. If the phase ratio is 1.0 in a one-step extraction, 66.7% of solute A will be extracted, but only 9.1% of solute B will be removed from the sample. After two extractions, the total amount of extracted A rises to 88.9%, while B increases to 17.4%. After five extractions, over 99% of A has been extracted and 38%

TABLE 20.2 Distribution Ratio Expressions for Analytes with Acid–Base Properties

Type of Extraction	Distribution Ratio
Extraction of a Weak Monoprotic Acid (HA) $$K_{D,HA} \updownarrow \begin{array}{l} \text{HA} \xrightleftharpoons[]{K_a} \text{H}^+ + \text{A}^- \quad \text{Phase 1} \\ \hline \text{HA} \qquad\qquad\qquad\qquad \text{Phase 2} \end{array}$$	$$D_c \approx \frac{[\text{HA}]_{\text{phase 2}}}{[\text{HA}]_{\text{phase 1}} + [\text{A}^-]_{\text{phase1}}} = \frac{K_{D,HA}}{1 + K_a/[\text{H}^+]}$$
Extraction of a Weak Monoprotic Base (B) $$K_{D,B} \updownarrow \begin{array}{l} \text{B} + \text{H}_2\text{O} \xrightleftharpoons[]{K_b} \text{BH}^+ + \text{OH}^- \quad \text{Phase 1} \\ \hline \text{B} \qquad\qquad\qquad\qquad\qquad \text{Phase 2} \end{array}$$	$$D_c \approx \frac{[\text{B}]_{\text{phase 2}}}{[\text{B}]_{\text{phase 1}} + [\text{BH}^+]_{\text{phase1}}} = \frac{K_{D,B}}{1 + K_b/[\text{OH}^-]}$$ $$= \frac{K_{D,B}}{1 + K_b[\text{H}^+]/K_w}$$

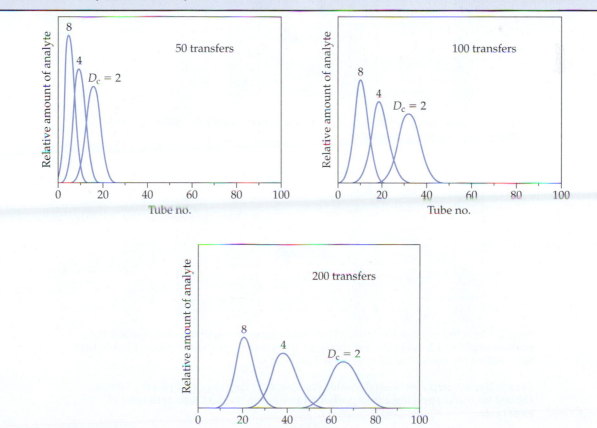

OH

8-Hydroxylquinoline (oxine)

Neutral complex of
aluminum and oxine

FIGURE 20.5   Reaction of 8-hydroxyquinoline (or oxine) with
Al³⁺, converting the aluminum ions into a neutral metal–ligand
complex that can be extracted by an organic solvent. This same
reaction can be used to precipitate Al³⁺ from an aqueous solution
for gravimetric analysis.

of B. If we continue this process, we will eventually reach a
point where essentially all of A and B have been extracted,
giving the same relative amounts we had in the original
sample! This effect is known as *coextraction* and is a problem anytime we use multiple extractions.[10]

The degree of coextraction for two chemicals will
depend on their distribution ratios. If there are equal
amounts of two chemicals in a sample, an extraction will
work best in separating these chemicals if they have a difference in D_c of at least 100-fold and D_c is large for one chemical but small for the other. A solute with a D_c =10 will be
90.9% extracted in a single step with a phase ratio of 1.0,
while only 9.1% of a solute with $D_c = 0.1$ will be coextracted.

Even better separations are obtained with larger differences in distribution ratios. This effect becomes less pronounced with multiple extractions, which provide a
greater opportunity for weakly extracted solutes to enter
the second phase. Thus, extractions involving multiple
solutes always have a compromise between the amount
of recovered analyte and the degree to which the analyte
coextracts with other solutes.

Countercurrent Extractions.   One way of improving
the purity of an analyte after an extraction while still
obtaining good recovery is to employ a **countercurrent
extraction** (see Figure 20.7). This method differs from a traditional extraction in that it uses multiple portions of *both*
Phases 1 and 2. In this system, one phase is kept in a series
of tubes or containers, each of which contains a fixed
amount of the second phase. After equilibration, each portion of the top phase is moved down one tube. The sample
is applied to the first tube of this system. As the components of this sample distribute between the two phases
and are moved with the upper second phase, they travel
from one tube to the next based on their D_c values.[8]

Table 20.3 shows some results that were obtained
with a countercurrent extraction. Solutes with small distribution ratios pass from the first tube to the last in the
fewest steps, while compounds with large D_c values take
longer to reach the last tube. The result is an extraction in

TABLE 20.3   Relative Distributions of Three Analytes (D_c = 2, 4, and 8) in a Craig Apparatus with 100 Tubes after 50,
100, and 200 Transfers (Phase Ratio = 1)

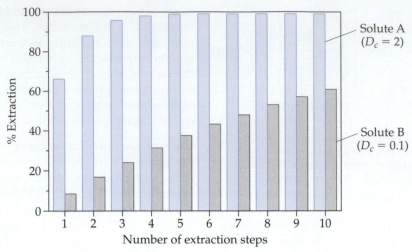

FIGURE 20.6 Illustration of the extent of solute extraction when using 1 to 10 extractions (phase ratio = 1.0) for a mixture of two solutes (A and B) with concentration distribution ratios of 2 and 0.1.

which there is an improvement in analyte recovery and fewer problems due to coextraction. One problem with this technique is that it can be time-consuming and tedious to carry out. A device was invented in the early 1940s by American chemist L.C. Craig to automate this method and overcome these problems for the isolation and study of antimalarial drugs.[14–16] This device (now known as a "Craig apparatus") uses a series of glass tubes that can hold a fixed portion of one phase while allowing the simultaneous transfer of each top phase to the next tube. The result of this process is that solutes partition between the phases in each tube, but eventually travel to the end of the apparatus where they are collected. In the next section we see how a similar but more convenient approach is used in the method of **chromatography**.

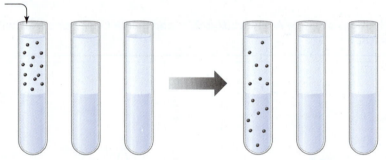

Step 1: Place the sample in the top layer (phase 1) of the first tube and equilibrate with the bottom layer (phase 2.)

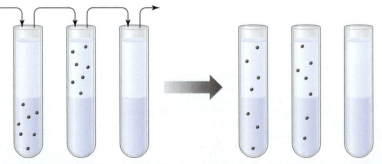

Step 2: Move the top portion (phase 1) of each tube to the right one unit. Collect the portion of phase 1 leaving the last tube and apply a fresh portion of phase 1 to the first tube. Allow the content of all tubes to reequilibrate.

Step 3: Repeat Step 2 as needed, continuing to collect the portion of phase 1 leaving the last tube and applying a frest portion of phase 1 to the first tube at the end of each cycle.

FIGURE 20.7 Use of a countercurrent extraction for the separation of chemicals. The result of this process is sometimes known as a "Craig countercurrent distribution."

20.3 CHEMICAL SEPARATIONS BASED ON CHROMATOGRAPHY

Although extractions are often applied for the isolation of a chemical from a sample or to remove interferences, there are limitations to this approach when we are separating closely related chemicals. Chromatography is another separation method that is commonly used for separating such substances and for the analysis of complex chemical mixtures.

20.3A What Is Chromatography?

Definition of Chromatography. The term "chromatography" was coined in the early 1900s by Mikhail S. Tswett (see Figure 20.8).[17–20] Chromatography is a separation technique in which the components of a sample are separated based on how they distribute between two chemical or physical phases, one of which is stationary and other of which is allowed to travel through the separation system.[9] This process can be illustrated by using the separation of 2,4-D from related herbicides in water (see Figure 20.9). To begin this separation, the sample is applied to the top of a packed tube, known as the **column**.[9] The column in this example consists of a tube that contains a coated solid support, where the coating

on the support interacts with herbicides as they pass through the system. A separate phase (a liquid, in this case) is used to apply the sample to the column and pass the sample components through to the other end. Those substances that have the weakest interactions with the fixed phase in the column will travel through more quickly than those with strong interactions, resulting in a separation.

Figure 20.9 shows that there are three main components that make up any chromatographic system. The first component is the phase that is flowing through the column and causes sample components to move toward the column's end. This component is called the **mobile phase** because it is traveling through the column. The second component is the fixed phase that is coated or bonded within the column; this is called the **stationary phase** because it always remains in the system. It is the stationary phase that is responsible for delaying the movement of compounds as they travel through the column. The third key component of a chromatographic system is the **support** onto which the stationary phase is coated or attached. This combination of stationary phase, mobile phase, and support, plus any associated equipment, makes up a device known as a *chromatograph.*[9,21]

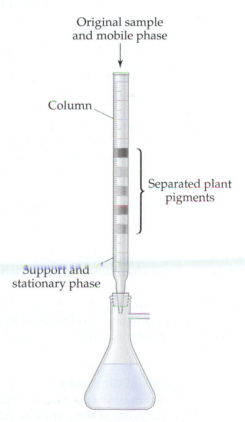

FIGURE 20.8 Mikhail Semenovich Tswett (1872–1919) and an example of a separation that he obtained using chromatography.[17–19] Tswett was a Russian botanist interested in isolating and studying plant pigments such as chlorophyll. To help him in his work, he developed a method in which a tube, or "column," was packed with a solid material and used to separate plant pigments as they were passed through the column in the presence of flowing liquid. He first described this new method to others in 1903, calling the technique "chromatography," which literally means "color writing" after the Greek words *khroma* ("color") and *graphein* ("to write"). The diagrams in (b) are adapted from papers by Tswett that described his technique.[20]

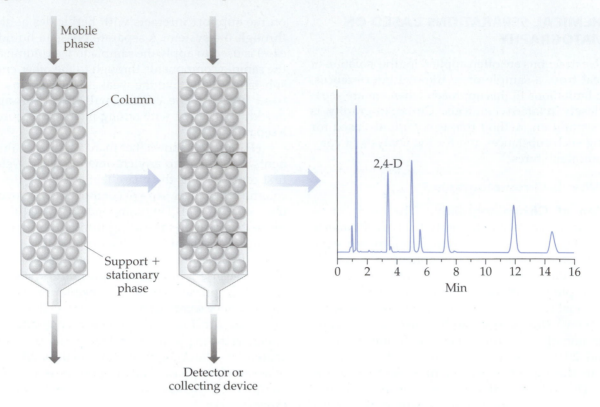

FIGURE 20.9 Separation of 2,4-D from other sample components by chromatography. The main components of the chromatographic system (the mobile phase, stationary phase, and support, which together make up the "column") are shown on the left. The plot on the right (called a chromatogram) shows the measured amount of each chemical that leaves the column after a given amount of time (or volume of applied mobile phase). (The chromatogram was provided courtesy of Sigma-Aldrich.)

Like extractions, chromatography uses two phases (a mobile phase and a stationary phase) for isolating one chemical from another. However, chromatography is also based on the different rates of travel that substances have through the system. This second feature makes chromatography similar to a countercurrent extraction (see Section 20.2C), but there is an important difference between these two methods. In particular, a countercurrent extraction has a well-defined number of contact steps between the two phases. One of these contact steps occurs each time Phase 1 and Phase 2 are combined and allowed to equilibrate in a tube. In contrast to this, solutes move back and forth between the mobile phase and stationary phase in a continuous manner as they travel through the column. This difference makes chromatography much easier to use than countercurrent extractions for the analysis and isolation of chemicals.

General Types of Chromatography. There are many different forms of chromatography. All of these forms can be classified according to their mobile phase, stationary phase, and support (see Table 20.4). The main way of categorizing chromatographic techniques is according to their mobile phase.[9] If a gas is the mobile phase, the method is called "gas chromatography" or "GC" (a technique we will discuss in Chapter 21). If the mobile phase is a liquid, the technique is known as

"liquid chromatography" or "LC" (see Chapter 22). It is even possible to use a supercritical fluid as the mobile phase, giving a method known as *supercritical fluid chromatography* or *SFC*. All of these methods can be divided into further subcategories based on their separation mechanism and type of stationary phase. For instance, the use of underivatized solid particles in GC or LC as the stationary phase gives rise to the methods of "gas–solid chromatography" and "liquid–solid chromatography."[9]

Another way chromatographic methods can be grouped is according to the type of support they utilize.[9,21] A method in which a column contains the support and stationary phase is known as *column chromatography*. If this column is packed with support particles that contain the stationary phase, the method is called *packed-bed chromatography*. If the stationary phase is instead placed directly onto the interior wall of the column, the method is known as *open-tubular chromatography*. It is also possible to have the support and stationary phase present on a flat plane (like a piece of paper, glass, or plastic); this format gives a method known as *planar chromatography*. As is indicated in Table 20.4, there are many possible combinations for the mobile phase, stationary phase, and support. This variety is what makes chromatography so valuable in separating and analyzing a large range of chemicals and samples.

TABLE 20.4   General Categories of Chromatographic Methods

Categories Based on Mobile Phase and Stationary Phase

Gas Chromatography	Type of Stationary Phase
Gas–solid chromatography	Solid, underivatized support
Gas–liquid chromatography	Liquid-coated support
Liquid Chromatography	Type of Stationary Phase
Adsorption chromatography	Solid, underivatized support
Partition chromatography	Liquid-coated or derivatized support
Ion-exchange Chromatography	Support containing fixed charges
Exclusion chromatography	Porous, inert support
Affinity chromatography	Support with immobilized biological ligand

Categories Based on Support

Name of Method	Type of Support
Packed-bed chromatography	Porous matrix consisting of a solid material (either a single polymer or small particles) packed with a column
Open-tubular chromatography	The interior wall of a column or a support coated onto this wall
Planar chromatography	A support consisting of a flat plane, such as a piece of paper, glass, or plastic

20.3B Using and Describing Chromatography

One of the great strengths of chromatography is its ability to take the individual components of a sample and isolate these from each other so they can be more easily identified or measured. This ability was illustrated in Figure 20.9 for the separation of herbicides in a water sample. Chromatography is often required in analytical methods that deal with complex samples. Such samples are a challenge for many detection techniques, which may work well for simple solutions but not for a mixture of many chemicals. At the same time, chromatography requires other techniques to help monitor the passage of chemicals through the column and to identify or quantitate these chemicals. The result is a combination of separation plus detection that makes chromatography valuable as a method for analyzing either simple or complex samples.

Describing Chromatographic Separations. The most common way of performing chromatography in analytical chemistry is to inject a small volume of sample onto a column and observe the time or volume of mobile phase it takes each sample component to pass through the system. The movement of solutes through the column in this case is referred to as "elution," and the mobile phase used to pass solutes through the column is known as the "eluent."[9,21]

The result of a chromatographic separation is shown by making a graph of the response measured by a detector at the end of the column as a function of the time or volume of mobile phase that is needed for elution. An example of such a plot, known as a **chromatogram**,[9,21] was given in Figure 20.9 for 2,4-D. Figure 20.10 shows a more general example and includes several terms that are used to describe these plots. In this plot, there is always a minimum amount of time required for any substance (even the mobile phase) to pass through the system. The term **void time** (t_M) is used to represent this time and it is measured by determining the time that is required for a totally nonbinding (or "nonretained") substance to travel through the column.[21] If a substance is bound (or "retained") by the stationary phase, it will travel slower through the column and exit at some later time. The average time required for this process is known as that substance's **retention time** (t_R).[9,21] The length of this retention time is determined by the eluting substance's structure, as well as the type of stationary phase and mobile phase being used in the chromatographic system. This relationship between the retention time and chemical structure makes t_R useful as a way of identifying peaks in chromatograms.

Elution volume is another way of describing the movement of substances in chromatography. The average volume of mobile phase it takes to move a compound through the column is called that compound's **retention volume** (V_R). Similarly, the volume of mobile phase it takes to elute a totally nonretained substance is known as the column **void volume** (V_M). The values of t_R and V_R, as well t_M and V_M, are related to each other through the flow rate (F) of the mobile phase (in units of volume per time), as shown below.[9,21]

$$V_R = t_R F \tag{20.9}$$

$$V_M = t_M F \tag{20.10}$$

For instance, suppose that 2,4-D exits a column with a retention time of 12.3 min at a flow rate of 1.25 mL/min. The void time under these same conditions is 0.84 min.

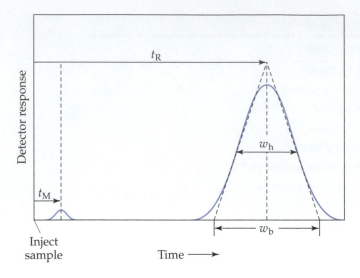

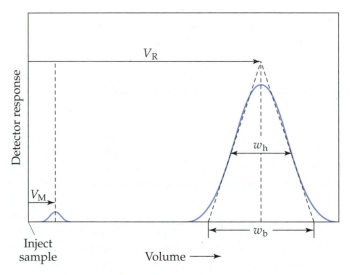

FIGURE 20.10 Examples of chromatograms plotted as a function of time the image on the top or volume of applied mobile phase the image on the bottom. The terms t_M and V_M in these plots represent the void time and void volume, while t_R and V_R represent the retention time or retention volume for the injected compounds. The terms w_b and w_h indicate the baseline width and width of half-height for the peaks shown in these chromatograms.

Using Equations 20.9 and 20.10, we get a retention volume of (12.3 min)(1.25 mL/min) = 15.4 mL for 2,4-D and a void volume of (0.84 min)(1.25 mL/min) = 1.05 mL.

As chemicals travel through a chromatographic system, the width of the region that contains each compound (also known as the compound's "peak" or "band") gradually becomes broader. This process is known as **band-broadening** and occurs even for substances that have little or no binding to the stationary phase.[21,22] Band-broadening is produced because no two individual molecules, atoms or ions (even of the same type) are likely to take exactly the same amount of time to travel through the system. Even a pure substance will give a peak in a chromatogram that has a distribution of elution times, as can be seen for 2,4-D in Figure 20.9. We discuss the processes that lead to this band-broadening later in this chapter.

Applications of Chromatography. The first piece of information chromatography can provide about a multi-component sample is the overall appearance of the chromatogram. Through the number, position, and size of the peaks in this chromatogram we obtain a fingerprint pattern or chemical profile that is characteristic of the sample being analyzed. This pattern can be useful in examining and comparing the general content of samples. We can learn even more by examining the position of each peak. This second item of information can be obtained by using the retention times or retention volumes of the eluting substance. These values, in turn, are related to the structure of the substance and are helpful in chemical identification.[17]

A third piece of information that can be obtained through chromatography is the amount of a compound that is present in the sample. Obtaining such information requires that the compound be clearly separated from other sample components. One way the amount of a compound can be measured by chromatography is by injecting standard solutions for this compound and measuring their peak heights or areas to produce a calibration curve (see Figure 20.11). This analysis is performed by using an internal standard (see Chapter 5) to correct for any variations that are produced during sample pretreatment and injection.[17]

20.4 A CLOSER LOOK AT CHROMATOGRAPHY

The separation of compounds by chromatography requires two things. First, there must be some difference in the degree of retention of the injected compounds. Second, the band-broadening must be small enough to allow the peaks produced by each compound to be easily distinguished from each other.

20.4A Analyte Retention in Chromatography

Measures of Retention. We now know that the retention time and retention volume for a chemical are related to the structure of the chemical and can be used to help identify this analyte in a chromatogram. Two closely related measures of solute retention in chromatography are the *adjusted retention time* (t_R') and *adjusted retention volume* (V_R'), which are defined as follows.[9]

$$t_R' = t_R - t_M \tag{20.11}$$

$$V_R' = V_R - V_M \tag{20.12}$$

As an example, if 2,4-D has a retention time of 12.5 min on a column with a void time of 1.2 min at 0.80 mL/min, the adjusted retention time for 2,4-D would be (12.5 min – 1.2 min) = 11.3 min. Also, the adjusted retention time and adjusted retention volume for an analyte can be calculated from each other if we know the flow rate of mobile phase through the column.

$$V_R' = t_R' F \tag{20.13}$$

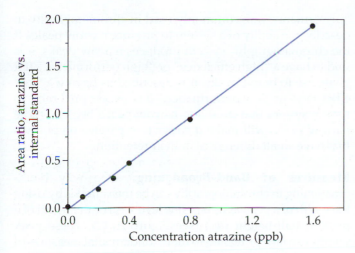

FIGURE 20.11 Calibration curve for the quantitation of atrazine in water samples by gas chromatography. Terbuthylazine was used as an internal standard. (Based on data from T.R. Shepherd, Ph.D. Dissertation, University of Nebraska-Lincoln, Lincoln, 1991.)

In our example, using 2,4-D, an adjusted retention time of 11.1 min at 0.80 mL/min would correspond to an adjusted retention volume of (11.3 min · 0.80 mL/min) = 9.0 mL.

Like t_R and V_R, the adjusted retention time and adjusted retention volume are useful as indexes of analyte retention. The value of t_R' and V_R' can also be viewed as direct measures of the strength of the interaction between the analyte and chromatographic system, because they correct for the contribution of the column void time or void volume to t_R and V_R. The same correction occurs when we use a parameter known as the **retention factor** (previously called the "capacity factor"). The retention factor is represented by the symbol k (or k'), and is calculated using either of the following relationships.[9,21]

$$k = t_R'/t_M \quad \text{or} \quad k = V_R'/V_M \qquad (20.14)$$

It is important in these calculations to use the same units on t_R' and t_M (or V_R' and V_M), which gives a value of k that has no units. Either expression in Equation 20.14 should give the same result for k, due to the fact that t_R' is directly proportional to V_R' and t_M is proportional to V_M (see Equations 20.10 and 20.13).

EXERCISE 20.4 | **Calculation and Use of Retention Factors**

A liquid chromatographic method for analyzing 2,4-D gives a retention time of 8.43 min for this compound at 1.25 mL/min. The void time of the column under these conditions is 1.38 min. What are t_R' and k for 2,4-D under these conditions? How would these values change if the flow rate were increased to 1.50 mL/min?

SOLUTION

The adjusted retention time can be found using Equation 20.11, where $t_R' = (8.43 \text{ min} - 1.38 \text{ min}) = 7.05 \text{ min}$. We can then place the values of t_R' and t_M

into Equation 20.14, giving $k = (7.05 \text{ min})/(1.38 \text{ min}) = $ **5.11**. An identical answer would be found if we were to instead use V_R' and V_M to find the retention factor.

If we increase the flow rate from 1.25 mL/min to 1.50 mL/min without any other change in the system, all compounds passing through the column will have shorter elution times. In fact, t_R, t_R', and t_M will all decrease by the ratio (1.25 mL/min)/(1.50 mL/min) = 0.80, or have 80% of their observed times at 1.25 mL/min. The retention factors for these compounds will not be affected by this change. This occurs because k is the ratio of t_R' and t_M, and both of these terms are being altered by the same relative amount as we vary the flow rate.

The value of k is unaffected by flow rate, making it a more reliable measure of retention than t_R or t_R'. Another advantage of using k is it is a relative value with no units, in which the adjusted retention time or adjusted retention volume for an analyte is compared to the void time or void volume of the column. One way of looking at this concept is to say that k represents the relative amount of time an analyte spends in the stationary phase versus the mobile phase (an idea we will revisit in the next section). In this view, the retention for a solute with a retention factor of k will be equal to $k + 1$ void volumes or void times for the column.

Factors that Affect Retention. The retention of an analyte in chromatography is usually determined by the intermolecular forces that occur between the analyte, the stationary phase, and mobile phase. We learned in Chapter 7 that intermolecular forces can occur *between* two chemical species as a result of dispersion forces, dipole–dipole interactions, hydrogen bonding, and ionic interactions. Individually these forces are much weaker than those involved in the formation of chemical bonds, but they are often important in determining a compound's boiling point, solubility, and ability to partition between two chemical or physical phases.

The distribution of an analyte between the stationary phase and mobile phase in a chromatographic column can be described by the general process,

$$A_M \rightleftharpoons A_S \qquad (20.15)$$

where A_M and A_S represent the analyte when it is present in the mobile phase or stationary phase, respectively. Equation 20.15 is similar to the reaction used in Equation 20.1 to describe an extraction. This similarity occurs because chromatography and extractions both use two phases to separate compounds. When an analyte reaches equilibrium between the stationary phase and mobile phase, we can write the following expression for its partition ratio (K_D).

$$K_D = \frac{[A]_S}{[A]_M} \qquad (20.16)$$

This particular expression is for a partition-based separation, in which $[A]_S$ and $[A]_M$ represent the effective concentrations of analyte in the stationary phase and mobile phase *at equilibrium*. You may be wondering at this point how we can have equilibrium in chromatography if the mobile phase and sample components are always moving through the system. The answer is that the applied analytes are viewed as approaching a *local equilibrium*, which occurs at the center of their chromatographic peaks. As the peak for an analyte moves through the column, the center of this peak and the location of this local equilibrium also move through the system.

Equation 20.16 shows that a substance that has strong interactions with the stationary phase (a higher value for $[A]_S$ than $[A]_M$) will have a large partition ratio. We would expect this, in turn, to result in a large retention time and retention volume. Similarly, a chemical that has only weak interactions with the stationary phase should have a small value for its partition ratio and give a small retention time or retention volume. This relationship between retention and K_D can be shown in a more quantitative fashion by using the retention factor. It is important to remember that the retention factor is a measure of the relative amount of time a compound spends in the stationary phase versus mobile phase, or $k = t'_R/t_M$. This fact means that k will be proportional to the relative moles of that compound in the stationary phase versus mobile phase at equilibrium, or $k = $ (mol A)$_S$/(mol A)$_M$. If we convert from moles to concentration by multiplying the top and bottom of this expression by the volumes of stationary phase (V_S) and mobile phase (V_M), we get Equation 20.17.

$$k = \frac{[A]_S V_S}{[A]_M V_M} = K_D(V_S/V_M) \qquad (20.17)$$

This relationship indicates k will be directly related to both the degree of that compound's interactions with the stationary phase (K_D) and the relative amount of stationary phase in the column (or the phase ratio, V_S/V_M).[9]

The separation of two chemicals by chromatography always requires that these chemicals have some difference in their retention. The most common way this situation is produced is when these chemicals have different equilibrium constants for their distribution between the mobile phase and stationary phase. A large difference in these equilibrium constants makes it easier to resolve these chemicals. It is also possible in some cases to have differences in the phase ratio for two analytes, especially if these analytes have a large difference in size. The result in either case is the analytes will have different rates of travel through the column, leading to their separation.

20.4B Chromatographic Band-Broadening

A second factor to consider in chromatography is the degree of band-broadening that occurs for analytes as they travel through a column. This factor determines the extent to which two closely eluting compounds can be discriminated from one another. The terms "efficiency" and "performance" are commonly used in chromatography to describe the ability of a system to produce narrow peaks. If the chromatographic system produces narrow peaks, it is said to have a "high efficiency" or "high performance." If it gives rise to broad peaks, it is described as having a "low efficiency" or "low performance." It is always preferable to have a system that produces narrow peaks because these narrow peaks will make it easier to separate compounds that have small differences in their retention.

Measures of Band-Broadening. One way band-broadening in chromatography can be measured is by using the width of a peak at its baseline level (w_b) or the width of a peak at half-height (w_h) (see Figure 20.12). These peak widths can also be related to more fundamental measures of the distribution in experimental results, such as the standard deviation (σ) or variance (σ^2) of a peak (see Chapter 4). Many software packages for data acquisition can automatically determine the variance or standard deviation of a chromatographic peak. It is also possible to calculate these manually from the general shape of the peak. If the peak fits a Gaussian-shaped curve, w_b and w_h can be used to find σ through the following formulas.[9,21]

$$w_b = 4\sigma \qquad (20.18)$$

$$w_h = 2.355\sigma \qquad (20.19)$$

These equations represent known relationships between the standard deviation of a Gaussian peak and the relative width of this type of peak at different

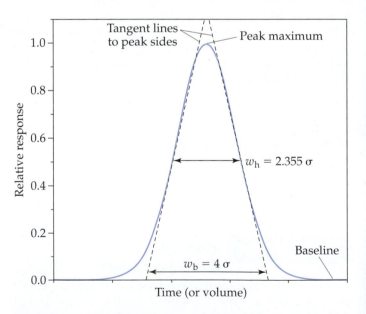

FIGURE 20.12 Method for the manual measurement of the baseline width (w_b) and width at half-height (w_h) for a Gaussian-shaped peak. The "half-height" of the peak is the height measured at half the distance between the baseline and the maximum value for the Gaussian peak. The "baseline width" is determined by drawing lines that are tangent to both sides of the peak and looking at where these lines intersect the baseline of the chromatogram. The mathematical relationships shown between w_h or w_b and the standard deviation of the peak are those predicted for a peak that follows a true Gaussian distribution.

heights. Similar relationships can be employed for other types of peaks.

Although peak widths are useful as a first approximation for describing band-broadening and column efficiency, they are highly dependent on the retention time (or retention volume) of the compound being used for their measurement. To compare the band-broadening for substances with different retention times, the **number of theoretical plates** (*N*) is used. Equation 20.20 shows the most general formula used for determining *N* in chromatography.[9,21]

General definition of *N*: $N = (t_R/\sigma)^2$ (20.20)

When using this expression, the same units must be employed for both t_R and σ, which results in a value for *N* that has no units, or that is often given simply as a "plate number."

Table 20.5 provides some typical plate numbers for common chromatographic columns. These values are sometimes given in terms of the number of plates obtained per unit length of column (such as plates per meter) to compare columns with different sizes. As we can see from Table 20.4, the value of *N* will depend on both the type of column and chromatographic method being used in a separation. It is always true that as we move to a larger value for *N* for a column, the column will be better at separating compounds. One way of illustrating this concept is to view *N* as being roughly equal to the number of times a substance will "equilibrate" between the mobile phase and stationary phase as it travels through the column. In fact, the term "plate number" was originally employed for this purpose in distillations, where a series of glass plates were used to increase the number of condensation and vaporization steps for separating two volatile chemicals. In the same manner, if a compound is able to undergo more "equilibration" steps in chromatography (as represented by a large *N*), it will become easier for the column to differentiate this compound from other substances in the sample.

If we are working with a chromatographic system that can automatically calculate the standard deviation or variance for a peak, these values can be used directly with Equation 20.20 to find *N*. The value of *N* can also be obtained based on measurements of the baseline width or the width at half-height. If we know a peak has a Gaussian shape, Equations 20.18 or 20.19 can be substituted into Equation 20.20 to get the following expressions for *N*.[9]

For a Gaussian peak: $N = 16(t_R/w_b)^2$

or $N = 5.545(t_R/w_h)^2$ (20.21)

It is important to keep in mind that these expressions are valid *only* for a Gaussian peak. If a peak has some other shape, then Equation 20.20 or an alternative relationship should be used.

EXERCISE 20.5   **Determining the Number of Theoretical Plates for a Column**

The efficiency of a new 30 m long GC column is evaluated by injecting a standard sample of pentadecane. The retention time measured for pentadecane is 10.05 min and its width at half-height is 0.064 min. Find the number of theoretical plates for this column, assuming pentadecane has a Gaussian-shaped peak. Express the answer in terms of both the total number of theoretical plates and the number of theoretical plates per meter of column length.

SOLUTION

The retention time for pentadecane is 10.05 min and its width at half-height is 0.064 min. If we assume this peak has a Gaussian shape, *N* can be calculated by using Equation 20.18.

$N = 5.545 \, (t_R/w_b)^2 = 5.545 \, (10.05 \text{ min}/0.064 \text{ min})^2$

$= \textbf{137,000 plates}$ for a 30 m long column

Another way of describing the efficiency of this system would be to say that we have (137,000 plates/30 m) = **4570 plates/m**.

TABLE 20.5   Typical Lengths, Plate Numbers, and Plate Heights for Chromatographic Columns*

Type of Column	Length, *L*	Number of Theoretical Plates, *N*	Plate Height, *H*
Gas Chromatography			
Classic packed column	2 m	3,600–4,000	0.50–0.55 mm
SCOT capillary column	15 m	15,800–27,300	0.55–0.95 mm
WCOT capillary column	30 m	43,900–480,000	0.06–0.68 mm
Liquid Chromatography [a]			
Packed column (4.6 mm inner diameter)	10–25 cm	6,000–25,500	0.01–0.04 mm
Microbore column (1.0 mm inner diameter)	25–100 cm	18,000–100,000	0.01–0.04 mm

*This table is based on information provided in C.F. Poole and S. K. Poole, *Chromatography Today*, Elsevier, Amsterdam, the Netherlands, 1991. The topics of SCOT (support-coated open tubular) columns and WCOT (wall-coated open tubular) columns are discussed in more detail in Chapter 21.

[a]The data shown for liquid chromatography are for columns used in high-performance liquid chromatography (HPLC). This information includes columns with particles that fall in the range of 5 to 20 μm.

It is often assumed that the peaks eluting from a chromatographic system are Gaussian in shape, but this assumption is not always true. Non-Gaussian peaks might occur when we are working with (1) an overloaded column, (2) a column that contains large empty spaces, (3) an analyte that has slow binding or release from the stationary phase, or (4) a column that has more than one type of site that interacts with analytes. Any of these situations can result in distortion of the observed peak shape, creating either "peak tailing" (a peak with a sharper front edge than back) or "peak fronting" (a peak with a sharper back edge than front).[9,21]

A test can be conducted to see whether a peak is really symmetrical by using a parameter known as the A/B ratio (or *asymmetry factor*).[22,23] Figure 20.13 shows how the A/B ratio is measured. First, the baseline level for the peak is determined along with the time (or volume) at which the peak's maximum response appears. Horizontal lines are drawn parallel to the baseline and at either one-half or one-tenth the distance between the baseline and the peak maximum. Next, a vertical line is drawn down from the peak top, and the width of the line segments at one-half or one-tenth height are measured on either side of this vertical line. The width obtained between the vertical line and peak's trailing edge (the "A" distance) is then divided by the width between this vertical line and the peak's front edge (the "B" distance).[23] For a Gaussian-shaped curve, the values for A and B will be the same, giving an A/B ratio of one. If peak tailing is present, an A/B ratio greater than one will be obtained. An A/B ratio less than one will be observed for a system with peak fronting.

Although N is useful for describing column efficiency, it does have the disadvantage of being dependent on column length. One way we can use N to compare columns of different sizes is to find the number of plates generated per unit column length (e.g., by expressing efficiency in terms of plates per meter). We can also use a related measure of efficiency known as the **height equivalent of a theoretical plate** or "plate height" (**HETP** or **H**).[9,21] The value of H is calculated from N by using Equation 20.22,

$$H = L/N \qquad (20.22)$$

where L is the length of the column. The value we obtain for H is in units of distance and gives the length of the column that corresponds to one theoretical plate, or one "equilibration" step of the analyte with the stationary phase. We know that a *large* number of theoretical plates helps give good column efficiencies and narrow peak widths, so a *small* plate height is also desirable. Typical plate height values for chromatographic systems are given in Table 20.5.

EXERCISE 20.6 | **Determining the Plate Height of a Chromatographic System**

Determine the plate height for the chromatographic system in Exercise 20.5. How would the plate height and plate number for this system change if the length of the column were doubled?

SOLUTION

We found earlier that N was 137,000 for a 30 m long column. We can use this information with Equation 20.22 to get the plate height for this column.

$$H = (30 \text{ m})/(137{,}000 \text{ plates})$$

$$= \mathbf{2.2 \times 10^{-4} \, m} \text{ (or } \mathbf{0.22 \, mm}\text{)}$$

The value for H is essentially independent of column length and should be unchanged as we double the column length. A twofold increase in L will lead to a twofold increase in N, assuming peak broadening due to system components other than the column is negligible.

Sources of Band-Broadening. There are many processes that can cause an initially narrow sample peak to become wider as it travels through a chromatographic system. These processes have one thing in common—they each cause individual solutes (even of the same type) to have slightly different rates of travel through the system. Many of these processes depend on how fast individual solutes move about in solution. This movement is described by *diffusion* and

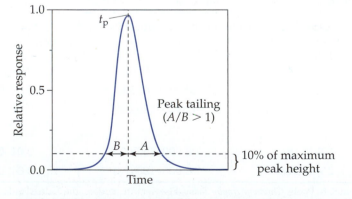

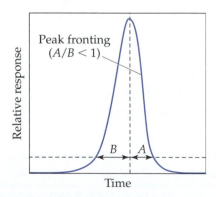

FIGURE 20.13 Determination of the asymmetry of chromatographic peaks based on A/B ratios. The distances A and B are measured from the sides of a peak to the location of the peak maximum (t_p) and along a line that (in this case) is drawn at one-tenth of the total height of the peak. Examples of both peak tailing and peak fronting are shown in this figure, along with their respective values for the A/B ratio.

mass transfer. Diffusion refers to the movement of a solute from a region of high concentration to one of lower concentration; mass transfer also describes a solute's movement from one region to the next, but in this case, the movement is not necessarily from a region of high to low concentration.[8,9,21]

The rate of a solute's diffusion depends on a term known as the **diffusion coefficient (D)**.[21] The value of D is a constant characteristic of the size and shape of a solute, as well as the temperature and type of phase in which the solute is present. Some typical diffusion coefficients are listed in Table 20.6. Large values for D, as occur for small solutes like water and benzene, represent fast motion of a substance through its surroundings. Small diffusion coefficients, such as those for proteins like hemoglobin or myosin, represent much slower movement. In chromatography a large diffusion coefficient will result in faster movement of a solute between the mobile phase and stationary phase or other regions of the system.

Figure 20.14 shows five major band-broadening processes that occur within a typical chromatographic column.[8,21,22] One of these processes is *longitudinal diffusion*, which refers to the broadening of a compound's peak due to the diffusion of solutes along the length of the column. In this process, diffusion causes molecules to move away from the peak's center (i.e., the region of highest concentration) to the regions of lower concentration at the front and back edges of the peak. This process has the greatest effect when the analyte is allowed to spend a long period of time in the mobile phase, as occurs at low flow rates.

Eddy diffusion is a band-broadening process that occurs whenever there are support particles within a column. Despite its name, this process actually has nothing to do with true diffusion. Instead, it is produced by the presence of the large number of flow paths around support particles, with each path having a slightly different length. As a sample moves through the column, some solutes will arrive at the end before others because of the different paths they take. Eddy diffusion is not affected by the mobile phase flow rate, but it does depend on the size and shape of the support particles and the effectiveness with which they are packed into the column (i.e., factors that determine the number and types of flow paths in the system).

Mobile phase mass transfer is another band-broadening process that occurs because of the presence of support particles. Band-broadening is created in this process from the different rates of travel a solute has *across* any given slice of the column. Fortunately, individual solutes also diffuse between flow channels as they move down the column, which tends to even out any band-broadening that results from differences in flow rates across the column's radius. Because mobile phase mass transfer and eddy diffusion both involve the flowstreams present within the column, they are sometimes put together when describing band-broadening in chromatography. This combination of processes is referred to as "convective mass transfer."

A fourth band-broadening process that can occur is *stagnant mobile phase mass transfer*. This process is related to the rate of diffusion or mass transfer of solutes as they go from the mobile phase outside the pores of the support to the mobile phase within the pores of the support or directly in contact with the support's surface. These two regions are called the "flowing mobile phase" and "stagnant mobile phase," respectively. Because a molecule does not travel down the column when it is in the stagnant mobile phase, it spends a longer time in the column than molecules that remain in the flowing mobile phase.

The actual interaction of a chemical with the stationary phase will also lead to band-broadening, giving a process known as *stationary phase mass transfer*. This

TABLE 20.6   Typical Diffusion Coefficients for Compounds in Gases and Liquids*

Substance (Molar Mass)	Diffusion Coefficient (cm^2/s)	Surrounding Medium
Diffusion in gases		
Water (18 g/mol)	1.020	H$_2$ (1 atm, 27°C)
Water (18 g/mol)	0.277	Air (1 atm, 30°C)
Benzene (78 g/mol)	0.096	Air (1 atm, 25°C)
Diffusion in liquids		
Ammonia (17 g/mol)	1.76×10^{-5}	Water (20°C)
Sucrose (342 g/mol)	4.59×10^{-6}	Water (20°C)
Ribonuclease (13,683 g/mol)	1.19×10^{-6}	Water (20°C)
Hemoglobin (68,000 g/mol)	6.9×10^{-7}	Water (20°C)
Myosin (493,000 g/mol)	1.1×10^{-7}	Water (20°C)

*Values taken from E.N. Fuller, P.D. Schettler, and J.C. Giddings, "A New Method for Prediction of Binary Gas-Phase Diffusion Coefficients," *Industrial and Engineering Chemistry*, 58 (1966) 19–27; and C.R. Cantor and P.R. Schimmel, *Biophysical Chemistry, Part II: Techniques for the Study of Biological Structure and Function*, Freeman, San Francisco, 1980, Chapter 10.

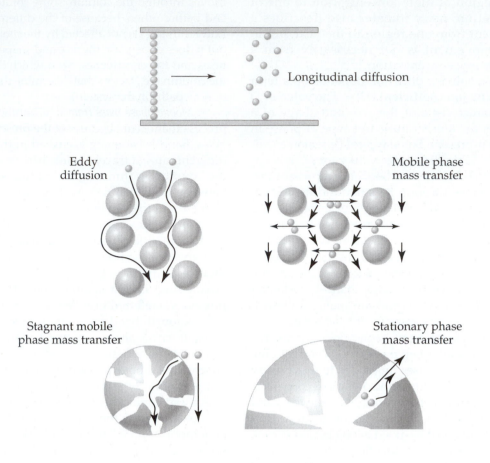

FIGURE 20.14 Band-broadening processes that can occur in chromatography.

process is related to the movement of chemicals between the stagnant mobile phase and stationary phase. Because individual solutes of the same type may spend different lengths of time in the stationary phase, they will also spend different amounts of time in the column, which creates band-broadening. Stationary phase mass transfer and stagnant mobile phase mass transfer are usually the two most important types of band-broadening in chromatography, but they are minimized when fast diffusion is present and slow flow rates are used.

In addition to the band-broadening that takes place in a column, it is possible to have broadening of a peak before or after it enters the column. This is known as *extra-column band-broadening*. This type of band-broadening might be produced as the sample leaves an injection device or as the sample passes through tubing before or after the column. Similarly, broadening can be created by a detector through which analytes pass after leaving the column. Although extra-column band-broadening is usually small compared to band-broadening within a column, it can be important when working with a small or very efficient chromatographic system.

The van Deemter Equation.   The plate height of a column is not only used to describe band-broadening and efficiency. Plate heights are also used in chromatography to see how a particular experimental factor (such as flow rate or support diameter) might affect band-broadening. Many equations based on plate heights have been developed for this purpose, the most famous of which is the **van Deemter equation.**[21,22,24]

van Deemter equation: $H = A + B/u + Cu$ (20.23)

The term u in this equation represents the *linear velocity* (a measure of the rate of travel in units of distance per time),[9,21] which is directly related to the flow rate of the mobile phase through the equation $u = (F \cdot L)/V_M$. The terms A, B, and C in Equation 20.23 are constants that represent the contributions to band-broadening due to eddy diffusion (A), longitudinal diffusion (B), and stagnant mobile phase mass transfer plus stationary phase mass transfer (C) (see Box 20.2).

Figure 20.15 gives a typical plot of the van Deemter equation for an analyte injected onto a gas-chromatographic column. This plot shows how H changes with the linear velocity (or flow rate). The relative size of the three different

BOX 20.2

A Closer Look at the van Deemter Equation

Plate heights and plate numbers have been used since 1941 to describe column efficiency in chromatography, but it was not until the mid-1950s that it was demonstrated how these values were related to the processes responsible for band-broadening in chromatographic columns. This task was first accomplished by J.J. van Deemter, F.J. Zuiderweg, and A. Klinkenberg,[25] who were working at the Shell Oil Company on the new technique of gas–liquid chromatography. They were interested in using this method to separate volatile compounds, like those found in petroleum products. As part of their work, van Deemter and his colleagues wanted to determine how various factors affected the efficiency obtained in gas–liquid chromatography, making it possible to develop better ways of separating compounds by this method.[10]

In their work, this group of scientists assumed there were several independent band-broadening processes that occurred simultaneously as a compound passed through the column. They then developed equations to describe the distribution of elution times that would be produced by each of these processes. The widths of these distributions were related to physical parameters of the chromatographic system, such as the diameter of support particles in the column and the flow rate of the mobile phase. After these equations had been developed for the separate band-broadening effects, they were combined to show the net effect of a given parameter on the overall efficiency and plate height for the column. The result obtained for a packed column is shown below.

$$H = 2\lambda d_p + \frac{2\gamma D_m}{u} + \frac{8k\,d_s^2\,u}{\pi^2(1+k)^2 D_s} \qquad (20.24)$$

In this equation, H is the measured plate height for the column and u is the linear velocity of the mobile phase,

while D_m and D_s are the diffusion coefficients for the injected compound in the mobile phase and stationary phase, respectively. The term k is the retention factor, d_p is the diameter of the support particles in the column, and d_s is the thickness of the stationary phase coating. The other terms in this equation are constants used to describe the packing structure of support particles in the column (λ) and the change in diffusion of the analyte in the presence of these particles (γ).

One way van Deemter and his coworkers used their equation was to predict how the efficiency of a column would change with the flow rate or linear velocity (u). The other parameters in Equation 20.24 were assumed to be constant and lumped together to produce what is now known as the "van Deemter equation."

$$\text{van Deemter Equation: } H = A + B/u + Cu \qquad (20.23)$$

where

$$A = 2\lambda d_p \qquad\qquad B = 2\gamma D_m$$

$$C = [8kd_s^2]/[\pi^2(1+k)^2 D_s]$$

According to the model used by van Deemter and his colleagues, the A, B, and C terms in this equation represent the band-broadening that arises from eddy diffusion (A), longitudinal diffusion (B), and a combination of stationary phase mass transfer plus stagnant mobile phase mass transfer (C). Mobile phase mass transfer was later added to modified versions of this equation by other scientists who developed improved models to describe band-broadening in chromatography. Most of these other relationships have the same general form as Equation 20.23, but include slightly different constants in the A, B, and C terms.[5]

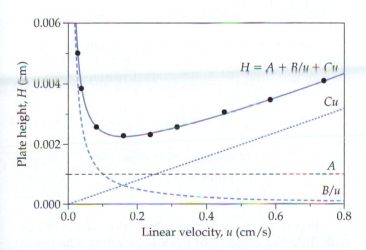

FIGURE 20.15   A typical van Deemter plot for gas chromatography. The solid line shows the overall van Deemter plot, while the dashed lines show the contributions of the various terms in the van Deemter equation to this plot.

parts of the van Deemter equation (the A-, B-, and C-terms) are also shown to indicate the amount each term contributes to the total measured plate height. This graph indicates that at low flow rates and low linear velocities the B-term (which represents longitudinal diffusion) is the most important source of band-broadening and gives the largest contribution to H. At high flow rates, the C-term (which represents stagnant mobile phase mass transfer and stationary phase mass transfer) becomes the largest contributor to the overall plate height. Eddy diffusion (which does not have any flow-rate dependence) is represented by the A-term, which gives a constant contribution to H at all linear velocities.[22]

Because these band-broadening processes are affected differently by a variation in flow rate, the overall result in Figure 20.15 is a "U"-shaped curve in which there is an "optimum linear velocity" (u_{opt}), where H has its lowest value (known as the "optimum plate height," H_{opt}). One reason we might use this plot is to determine what flow rate will give the best efficiency for a column. Another way a van Deemter plot can be employed is to compare the plate heights of columns that contain different supports. This information is valuable to scientists as they try to develop more efficient columns or compare the performance of several existing columns.

EXERCISE 20.7 | **Using the van Deemter Equation**

What are the optimum linear velocity and optimum plate height for the column in Figure 20.15? How will the plate height and number of theoretical plates for this column change as the linear velocity is altered from u_{opt} to 0.5 cm/s? How does the efficiency of this column compare to another one of the same size that produces a plate number of 350,000 at 0.5 cm/s?

SOLUTION

The minimum point in Figure 20.15 occurs at linear velocity of approximately 0.16 cm/s and a plate height of 0.0023 cm, so these are the values of u_{opt} and H_{opt}. Changing from a linear velocity of 0.16 cm/s to 0.5 cm/s will cause an increase in the plate height to about 0.0032 cm. This increase in plate height will result in a less efficient system. The new value of N can be determined by using Equation 20.22 and the given column length of 15.0 m.

$$H = L/N$$

$$3.2 \times 10^{-5} \text{ m} = (15.0 \text{ m})/N$$

$$\therefore N = \textbf{470,000}$$

This value of N is greater than the plate number of 350,000 given for the second column, so the first column is more efficient at 0.5 cm/s.

20.4C Controlling Chromatographic Separations

Measuring the Separation of Peaks.   The final success of any chromatographic separation will depend on how well the peaks of interest are separated. There are several ways to evaluate the extent of this separation. One way is to use the **separation factor (α)** between the peak of interest and its nearest neighbor. The separation factor is a measure of the relative difference in retention of two solutes as they pass through a column. The value of α is calculated as shown below, where k_1 is the retention factor for the solute that exits first from the column and k_2 is the retention factor for the second solute.[9,21]

$$\alpha = k_2/k_1 \qquad (20.25)$$

Like the retention factor, the separation factor is a unitless parameter. The separation factor becomes larger as the relative difference in retention increases between two peaks. This feature makes α useful in describing the effectiveness of a chromatographic separation. The separation

factor can also indicate whether it is feasible to resolve two compounds by a given column, where a value greater than one is needed for a separation to occur.

Another approach for describing the separation of two peaks is to use the **peak resolution (R_s)**. The value of R_s for two adjacent peaks can be calculated through the following formula.[9,21]

$$R_s = \frac{t_{R_2} - t_{R_1}}{(w_{b_2} + w_{b_1})/2} \qquad (20.26)$$

In this relationship, t_{R_1} and w_{b_1} are the retention time and baseline width (both in the *same* units of time) for the first eluting peak, while t_{R_2} and w_{b_2} are the retention time and baseline width of the second peak. This produces a unitless value for R_s that represents the number of baseline widths that separate the centers of the two peaks. An important advantage of using the peak resolution instead of the separation factor is that R_s considers both the difference in retention between two compounds (as represented by $t_{R_2} - t_{R_1}$) and the degree of band-broadening (as represented by w_{b_1} and w_{b_2}). Some practice in calculating the separation factor and peak resolution is provided in the next exercise.

EXERCISE 20.8 | **Describing the Separation of Chromatographic Peaks**

A soil scientist uses liquid chromatography to measure the amount of atrazine and its degradation product hydroxyatrazine in soil extracts. When making injections of standards, the peak due to atrazine has a retention time of 6.09 min and a baseline width of 0.21 min, while the peak due to hydroxyatrazine has a retention time of 5.71 min and a baseline width of 0.20 min. The column void time under these conditions is 0.75 min. What separation factor and peak resolution would be expected for these peaks if the atrazine and hydroxyatrazine were injected as a mixture?

SOLUTION

The values of α and R_s are found by using Equations 20.25 and 20.26, where k_2 (atrazine) = (6.09 min − 0.75 min)/(0.75 min) = 7.1$\underline{2}$ and k_1 (hydroxyatrazine) = (5.71 min − 0.75 min)/(0.75 min) = 6.6$\underline{1}$.

$$\alpha = (7.1\underline{2})/(6.6\underline{1}) = \textbf{1.08}$$

$$R_s = \frac{(6.09 \text{ min} - 5.71 \text{ min})}{(0.21 \text{ min} + 0.21 \text{ min})/2} = \textbf{1.85}$$

Figure 20.16 shows how the value of R_s changes with different degrees of peak separation. The lowest possible value for R_s is zero, which occurs when two peaks have exactly the same degree of retention and are not separated by the chromatographic system. A value for R_s greater than zero represents some degree

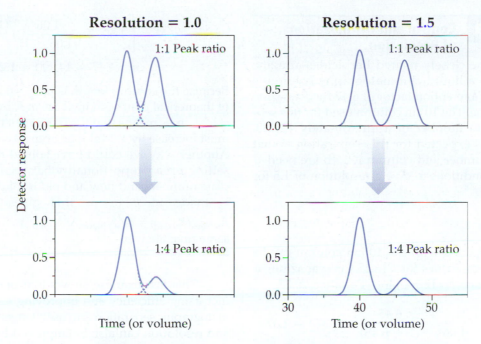

FIGURE 20.16 Degree of separation obtained between two peaks with size ratios of 1:1 or 1:4 when a resolution (R_s) of 1.0 or 1.5 is present between the peaks of similar size.

of separation between the peaks, with the extent of this separation becoming larger as R_s increases. Ideally, it is desirable to have no significant overlap between these peaks. This situation typically occurs when R_s is greater than 1.5 and is said to represent "baseline resolution." For many separations, peak resolution values between 1.0 and 1.5 are also adequate. This is especially true if the peaks are about the same size and can be measured using peak heights, which are less affected by overlap than peak areas.

Factors Affecting the Separation of Peaks. Because the resolution between two peaks is a measure of both the difference in compound retention and band-broadening, any factors that affect retention or peak widths will also affect R_s. The effects these factors will have on the resolution is given by Equation 20.27.[25]

$$R_s = \frac{\sqrt{N}}{4} \cdot \frac{(\alpha - 1)}{\alpha} \cdot \frac{k}{(1 + k)} \qquad (20.27)$$

In this equation, k is the retention factor for the second peak, α is the separation factor between the first and second peaks, and N is the number of theoretical plates for the column being used in this separation. This relationship is called the **resolution equation** of chromatography[21] and is simply a modified version of Equation 20.26 where k, α, and N have been substituted in place of t_R and w_b (see Appendix A for the derivation). This equation is useful because it shows in a quantitative fashion that the degree of a separation in chromatography will be affected by three factors: (1) the extent of band-broadening in the column (N), (2) the overall degree of peak retention (k), and (3) the

selectivity of the column's stationary phase in binding to one compound versus another (α). The change in resolution that is obtained when varying each of these parameters is shown in Figure 20.17. It is also possible to estimate how changing these parameters will affect the resolution of a real chromatographic separation, as is illustrated in the following exercise.

Resolution equation:

$$R_s = \frac{\sqrt{N}}{4} \cdot \frac{(\alpha - 1)}{\alpha} \cdot \frac{k}{(1 + k)}$$

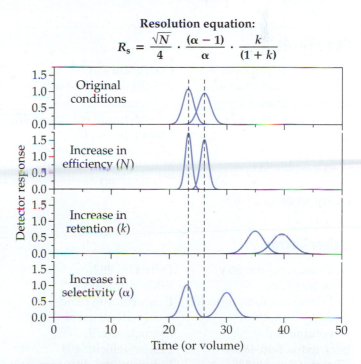

FIGURE 20.17 Effects of changes in the number of theoretical plates (N), retention factor (k), or separation factor (α) on the degree of separation that is obtained between neighboring peaks in a chromatogram.

| EXERCISE 20.9 | Control and Optimization of Resolution |

A separation of the closely related herbicides atrazine and cyanazine by a liquid chromatography column is known to produce a retention factor of 6.45 for cyanazine and 6.08 for atrazine. The 10 cm column used for this separation has a plate number of approximately 12,500. What resolution is expected for this separation? What minimum plate number and column length are needed under the same conditions to give a resolution of 1.5 for this separation?

SOLUTION

The resolution obtained with the 10 cm column can be found by placing the values for k (6.45 for cyanazine), α (6.45/6.08 = 1.06), and N (12,500) into Equation 20.27.

$$R_s = \frac{\sqrt{12,500}}{4} \cdot \frac{1.06 - 1}{1.06} \cdot \frac{6.45}{(1 + 6.45)} = 1.370 = \mathbf{1.37}$$

The same relationship can be utilized to determine the minimum value of N that is needed to increase R_s to 1.5. To do this, we can assume k and α will be constant on the old and new columns if all other conditions besides N and the column length are unchanged. We can then place these numbers along with the desired value for R_s into Equation 20.27 and solve for N.

$$R_s = 1.50 = \frac{\sqrt{N}}{4} \cdot \frac{(1.06 - 1)}{1.06} \cdot \frac{6.45}{(1 + 6.45)}$$

$$\sqrt{N} = \left[1.50 \cdot 4 \cdot \frac{1.06}{(1.06 - 1)} \cdot \frac{(1 + 6.45)}{6.45} \right]$$

$$\text{or} \quad N = 14,990 = \mathbf{15,000}$$

Because the column length is proportional to the number of theoretical plates, an increase in N by (15,000)/(12,500) = 1.20 times means that the required length of the column must increase by 1.2-fold, or from 10 cm to at least 12 cm. Another way we could have solved this problem is by setting up a proportionality between the new and old plate numbers and new and old resolutions. If no change is made in k or α, this gives the relationship $R_{s,New}/R_{s,Old} = \sqrt{N_{New}}/\sqrt{N_{Old}}$, which results in the same final answer.

This last exercise shows that one practical way of increasing efficiency and improving resolution in a chromatographic separation is to use a longer column. Efficiency and resolution can also be improved by operating under flow-rate conditions that produce smaller plate heights (like at the optimum linear velocity of the van Deemter plot), or by changing the dimensions of the support used in the column (with smaller diameter particles leading to higher efficiencies). Obtaining better resolution by increasing retention is a third possibility, which generally involves changing the mobile phase or stationary phase. Changing the selectivity of the chromatographic system is often the most difficult option for improving resolution because it requires detailed knowledge of the interactions taking place between the injected compounds and the column.

Key Words

Band-broadening 488
Chemical separation 476
Chromatogram 487
Chromatography 484
Column 485
Countercurrent extraction 483
Diffusion coefficient 493
Distribution ratio 479

Extraction 478
Height equivalent of a theoretical plate 492
Liquid–liquid extraction 478
Mobile phase 485
Multistep extraction 481
Number of theoretical plates 491

Partition constant 478
Partition ratio 478
Peak resolution 496
Resolution equation 497
Retention factor 489
Retention time 487
Retention volume 487
Separation factor 496

Single-step extraction 480
Stationary phase 485
Support 485
van Deemter equation 494
Void time 487
Void volume 487

Other Terms

A/B ratio (asymmetry factor) 492
Adjusted retention time 488
Adjusted retention volume 488
Back extraction 482
Chromatograph 485
Coextraction 483
Column chromatography 486

Diffusion 492
Eddy diffusion 493
Extra-column band-broadening 494
Extractant 478
Linear velocity 494
Longitudinal diffusion 493
Mass transfer 493
Mobile phase mass transfer 493

Open-tubular chromatography 486
Packed bed chromatography 486
Phase ratio 480
Planar chromatography 486
Raffinate 478
Stagnant mobile phase mass transfer 493

Stationary phase mass transfer 493
Supercritical fluid 480
Supercritical fluid chromatography 486
Supercritical fluid extraction 480

Questions

WHAT IS A CHEMICAL SEPARATION AND HOW IS IT USED IN ANALYTICAL CHEMISTRY?

1. What is a "chemical separation"? What are some examples of chemical separation methods?
2. What general strategy for separating chemicals is used in each of the following methods?
   (a) Distillation
   (b) Centrifugation
   (c) Extraction
   (d) Chromatography
3. What are four general reasons why a chemical separation might be used as part of an analytical procedure?
4. What is the role of the chemical separation step in each of the following procedures?
   (a) Filtering sand away from a river water sample for analysis of pollutants in the water
   (b) Precipitation of barium sulfate from a sample for a gravimetric analysis of barium
   (c) Evaporation of the water away from a protein solution, leaving behind a dried residue of the protein for analysis

WHAT IS AN EXTRACTION?

5. What is an "extraction"? How is an extraction used in chemical separations?
6. What is a "liquid–liquid extraction"? Describe how this method is carried out.
7. Define each of the following terms and state how they are used to describe the extraction of a solute.
   (a) Partition constant
   (b) Partition ratio
   (c) Distribution ratio
8. Write expressions similar to those in Equation 20.2 that describe the partition constant and partition ratio for the extraction of benzene from water into chloroform. Show how the values of this partition constant and partition ratio are related to one another.
9. For a weak monoprotic acid (HA), what happens to the value of D_c in Equation 20.4 as the pH is decreased? What value does D_c approach under these conditions? What happens to the value of D_c as the pH is increased? Explain the reasons for changes based on the acid–base properties of HA.
10. Derive an equation that relates the distribution ratio to the partition ratio and acid dissociation constants for the extraction of a weak diprotic acid (H_2A), where the only extracted form is the diprotonated species. Write your final expression in terms of K_D, the acid dissociation constants for this acid, and $[H^+]$. How does your result compare the expression shown in Equation 20.4 for a weak monoprotic acid?
11. Amitriptyline is a tricyclic antidepressant that is also a weak monoprotic base. A clinical chemist wishes to extract this drug with chloroform from a blood sample for analysis. Derive an expression that shows how the distribution ratio for this compound is related to its partition ratio between water and chloroform and its base association constant. (Note: Assume that only the neutral form of this base has any significant degree of extraction.) Write your final expression in terms of K_D, the base ionization constant for the amitriptyline, and $[OH^-]$.
12. Explain why a large difference in the distribution ratios for two chemicals makes it easier to separate these chemicals by an extraction.

13. List three different types of extraction. What phases are used in these methods?
14. What is the difference between an extraction that is based on partitioning and one that is based on adsorption?

USING AND DESCRIBING EXTRACTIONS

15. Describe the general approach that is used to conduct a single-step extraction.
16. What is the "phase ratio"? Why is this term important to consider when performing an extraction?
17. What are two ways in which the amount of extracted analyte can be varied in a single-step extraction?
18. The trace amount of chloroform in a water sample is to be extracted using pentane. The partition ratio for chloroform in these solvents is approximately 110. If a 200.0 mL water sample is extracted with 15.0 mL of pentane, what percent of chloroform will be removed after a single extraction?
19. What will happen to the percent extraction of chloroform in Problem 18 if the volume of pentane is changed to 25.0 mL? What minimum volume of pentane is needed to extract 99% of the chloroform from a 200.0 mL water sample?
20. Pharmaceutical chemists often use the distribution ratio for a drug in water versus octanol as a means for characterizing a drug's polarity and ability to cross cell membranes. This special distribution ratio is referred to as the "octanol–water partition coefficient" (K_{ow}, a term we discussed in Chapter 7). A new drug is found at pH 7.4 to be 87.5% extracted from a 100.0 mL aqueous sample after one extraction with 15.0 mL of octanol. What is K_{ow} value for this drug?
21. The caffeine in a sample of tea is extracted by combining 80.0 mL of the tea with 15.0 mL of dichloromethane. If 89% of the caffeine is extracted after a single step, what is the distribution ratio of caffeine under these conditions?
22. How does the degree of extraction change for a solute in going from a single-step extraction to a multistep extraction?
23. If 74.3% of an herbicide is extracted from a 100.0 mL food extract when using 15.0 mL of ethyl ether in a single-step extraction, what total percent of this herbicide will be extracted from a fresh 100.0 mL portion of the same food sample when using three extractions, each of which involves using 5.0 mL of ethyl ether?
24. An organic chemist wishes to use an extraction to isolate a natural product from an aqueous sample that is prepared from a plant. This chemical has a D_c value of 19.0 in the presence of water and ether.
   (a) What fraction of this chemical will be extracted from 50.0 mL of water after one step using 10.0 mL of ether?
   (b) What total fraction of this chemical will be extracted from the same water sample after two steps, each using fresh portions of 10.0 mL of ether?
   (c) How many extraction steps with 10.0 mL of ether are needed to recover at least 99% of this chemical from the original sample.
25. Describe how a side reaction can be used to alter the extent of extraction for a solute. Give two examples of how acid–base reactions can be used to alter the extent of an extraction.
26. What is a "back extraction"? Describe how a back extraction can be used to help purify a chemical from other contaminating substances in a sample.

27. A biochemist is trying to isolate a natural product that is a weak monoprotic acid. The pK_a value of this compound is estimated to be 6.21 at 25°C. Preliminary experiments indicate that this compound has a distribution ratio of 75 between water and ethyl acetate at pH 3.00, where the principal species is the protonated and neutral form. The conjugate base of this compound does not show any appreciable extraction at any pH.
    (a) What fraction of this compound will be extracted from a 30.0 mL water sample into 20.0 mL of ethyl acetate at pH 4.00?
    (b) What fraction of the compound that enters the ethyl acetate at pH 4.00 will go back into a fresh 30.0 mL portion of water at pH 8.00?
    (c) What overall fraction of the compound in the original sample will be present in the fresh portion of water after the biochemist has performed the back extraction in Part (b)?

28. A pharmaceutical company plans to use a liquid–liquid extraction to purify a drug that is a weak monoprotic base. This chemical has a pK_b of 9.52 at 25°C. The K_D for the neutral form of this chemical (B) is 210 when using water and chloroform at 25°C.
    (a) What fraction of this drug will be extracted from a 15.0 mL aqueous sample into 10.0 mL of chloroform at pH 9.00 and 25°C?
    (b) What fraction of this drug will be back extracted at 25°C from the 10.0 mL of chloroform and into a fresh 50.0 mL portion of water that is buffered at pH 2.00?
    (c) What overall fraction of the drug will be isolated from the original sample and placed into the final 50.0 mL portion of water under these conditions?

29. A clinical chemist wishes to create an extraction method for urine samples that can be used for a variety of drugs. These drugs can be divided into three general categories: (1) those that are weak acids, (2) those that are weak bases, and (3) drugs that are neither acids nor bases. Describe how using side reactions, a general scheme based on extractions can be developed to obtain separate extracted samples for each of these three classes of compounds.

30. Explain how complexation formation might be used to alter the extent of an extraction. Give an example of a binding agent that might be used to help extract some metal ions.

A CLOSER LOOK AT EXTRACTIONS

31. Describe why the purity of an extracted solute is affected as an increased number of extractions are carried out on a sample. What happens to the recovery of the solute under the same conditions?

32. Two chemicals are present in the same sample, where compound A is the analyte of interest and B is a possible interfering agent. It is known that A has a distribution ratio of 26.7 at pH 7.0 when extracted from water with toluene. Compound B has a distribution ratio of 0.24 under the same conditions.
    (a) What percent of each compound will be extracted from a 100.0 mL sample of water when using 20.0 mL of toluene in a single-step extraction?
    (b) What percent of A and B will be extracted after two, three, and four extractions, if each extraction step uses a fresh 20.0 mL portion of toluene?
    (c) Describe how the recovery and purity of compound A will change as the number of extractions increases for this sample.

33. Repeat the calculations in Problem 32 for two chemicals that have distribution ratios of 26.7 and 2.4 at pH 7.0. How do the results compare with those in Problem 32? Explain any differences that you observe. What changes could be made to improve the separation of these two chemicals?

34. What is a countercurrent extraction? How is this method performed? What are the advantages of using this method versus a simpler multistep extraction?

35. What is a Craig apparatus? How is this device used to conduct a countercurrent extraction?

WHAT IS CHROMATOGRAPHY?

36. Explain what is meant by the term "chromatography."

37. What are the three main parts of a chromatographic system? What role is played by each of these components?

38. Define the terms "column" and "chromatograph" as they relate to chromatography.

39. How are chromatography and extractions similar in the way they separate chemicals? How are these methods different?

40. Define the terms "liquid chromatography," "gas chromatography," and "supercritical fluid chromatography." What is the key difference in these methods?

41. Explain how the separation mechanism and type of stationary phase can be used to classify chromatographic methods. List two specific classes of chromatographic methods based on such a scheme.

42. What is the difference between "packed bed chromatography," "open-tubular chromatography," and "planar chromatography"?

USING AND DESCRIBING CHROMATOGRAPHY

43. What are the advantages of using chromatography with other analytical methods?

44. What is a "chromatogram"? What is the general form of a chromatogram?

45. Define each of the following terms as related to chromatography.
    (a) Void time
    (b) Void volume
    (c) Retention time
    (d) Retention volume

46. What is "band-broadening"? How does band-broadening affect the peaks that are obtained in a chromatogram? What types of information can be learned about an analyte or a sample by using chromatography? How does a chromatogram provide this information?

47. A separation of alcohols by gas chromatography gives a peak for air (representing a nonretained substance) at 0.45 min at a flow rate of 10.0 mL/min. The retention time for 2-propanol is 3.01 min. What is the void volume of this chromatographic system? What was the total retention volume for 2-propanol on this system?

48. A biochemist uses liquid chromatography to purify a protein from a cell culture sample. This protein elutes from the column with a retention volume of approximately 20.5 mL. The void volume of the column is 4.5 mL. What will be the retention time and void time of this protein and column if this separation is conducted on an inexpensive benchtop system at a flow rate of 0.050 mL/min (i.e., approximately one drop per minute)? What will the retention time and void time be if a slightly more expensive system is used to create a flow rate of 0.5 mL/min?

49. An environmental chemist obtains the following data for the injection of a series of atrazine standards onto a chromatographic system.

Atrazine Concentration (μg/L):	1.00	2.00	4.00	6.00	10.00	15.00
Relative Peak Height:	2.05	4.13	7.99	12.96	21.14	29.84

An injection of a drinking water sample under the same conditions gives a relative peak height of 5.13 for an atrazine peak. The allowable limit for atrazine in drinking water is 3μg/L. Does the concentration of atrazine in the water sample exceed this limit?

50. The following data were obtained for a series of alcohol standards injected onto a chromatographic system, where 1-propanol represents the internal standard.

	Content	
Standard Number	Methanol (mg/L)	1-Propanol (mg/L)
1	250	250
2	500	250
3	1,000	250
4	2,000	250

	Relative Peak Areas	
Standard Number	Methanol (mg/L)	1-Propanol (mg/L)
1	1,325	13,120
2	2,419	11,996
3	5,208	12,875
4	10,063	12,530

An unknown sample is analyzed under these same conditions and is found to give peaks for methanol and 1-propanol with areas of 2,258 and 12,486. Like the above standards, the final concentration of 1-propanol placed into the sample is known to be 250 mg/L, and it is known that no 1-propanol was present in the original sample. Determine the concentration of methanol that was present in the unknown blood sample.

ANALYTE RETENTION IN CHROMATOGRAPHY

51. Define each of the following terms. What advantages are there to using these parameters instead of the total retention time or total retention volume to describe the binding of an analyte to a column?
   (a) Retention factor
   (b) Adjusted retention time
   (c) Adjusted retention volume
52. Figure 20.18 shows a chromatogram that was obtained for a mixture of cholesterol-lowering drugs. The void time in this separation occurs at approximately 2.0 min.
   (a) Measure or calculate the retention times and retention volumes for peaks 1–5.

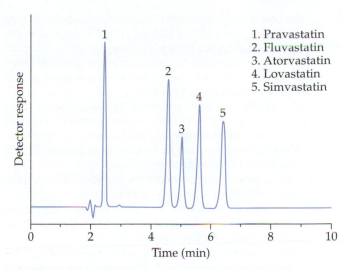

FIGURE 20.18 Separation of the cholesterol-lowering drugs (statins) by normal-phase liquid chromatography on a 4.6 mm inner diameter and 15 cm long column at 1.0 mL/min. The drugs in the injected mixture were (1) pravastatin, (2) fluvastatin, (3) atorvastatin, (4) lovastatin, and (5) simvastatin. (Reproduced with permission from Sigma-Aldrich.)

   (b) Determine the adjusted retention times and adjusted retention volumes for peaks 1–5.
   (c) Find the retention factors for peaks 1–5.
53. For the same chromatographic system as in Figure 20.18, determine how the values of k, t_R, and V_R will change for peak 5 as (1) the average mobile phase flow-rate is decreased from 1.0 to 0.75 mL/min, and (2) as the column length is increased from 15 to 25 cm.
54. A company supervising the cleanup of an explosives manufacturing plant wishes to determine the amounts of several explosives (e.g., TNT) that are present in soil taken from one of the plant's former waste sites. This analysis is to be done by extracting these substances from soil samples, followed by their separation and measurement using liquid chromatography. This gives the following data at a flow rate of 1.50 mL/min.[26]

Compound	Retention Time (min)
Nitrate (added to act as a nonretained compound)	2.00
TNT (2,4,6-trinitrotoluene)	5.00
RDX (hexahydro-1,3,5-trinitro-1,3,5-triazine)	6.15
Tetryl (methyl-2,4,6-trinitrophenylnitramine)	7.36
HMX (octahydro-1,3,4,5-tetranitro-1,3,5,7-tetrazocine)	8.35

   (a) Determine the adjusted retention time and adjusted retention volume for each compound.
   (b) What were the retention factors for TNT, RDX, Tetryl and HMX?
   (c) How would the total and adjusted retention times for TNT, RDX, Tetryl, and HMX be affected if the flow rate used in

the liquid chromatography (LC) method was accidently decreased from 1.50 to 1.00 mL/min? How would the retention factors for these same compounds be affected?

55. Why are intermolecular forces usually important in chromatography? How are these interactions related to the retention of an analyte in a column?

56. A series of phenols were separated by liquid–liquid chromatography using 1,2,3-tris(2-cyanoethoxy)propane as the stationary phase and 2,3,4-trimethylpentane as the mobile phase.
    (a) Write a general reaction that represents a phenol (P) as it distributes between these two phases.
    (b) Write a general expression for the partition ratio (K_D) for phenol P in this chromatographic system.
    (c) Write a general equation that gives the retention factor (k) for phenol P in this system. Show how the value of k is related to K_D for the phenol.

57. Why is a difference in retention needed in chromatography for a successful separation? What are some general ways in which this difference in retention can be produced?

58. Three substituted aromatic compounds are found to have retention factors of 2.32, 4.58, and 7.89 on a column that is operated under constant mobile phase and flow-rate conditions. If all of these compounds have the same phase ratio on the column, what is the relative size of their partition ratios in this chromatographic system?

59. An environmental chemist wants to use gas chromatography to measure the herbicide atrazine in water samples. This analysis is originally conducted on a 10 m long and 0.53 mm inner-diameter open-tubular column that contains a 1.20 μm thick coating of a stationary phase, which gives a retention factor of 10.5 for atrazine. The chemist later switches to another column of the same size and type that has a 2.65 μm thick coating of stationary phase (an increase in V_S of 2.2-fold), but with approximately the same mobile phase void volume. If all other conditions are kept the same, how much will the retention factor for atrazine change between the old and new column?

CHROMATOGRAPHIC BAND-BROADENING

60. What is meant by the term "band-broadening" in chromatography? What is meant by "high-efficiency" or "high-performance" when these terms are used to describe a chromatographic system?

61. Define each of the following measures of band-broadening. What are the advantages and disadvantages to using each of these terms in the description of band-broadening?
    (a) w_b
    (b) w_h
    (c) σ
    (d) N
    (e) H

62. Explain why the expressions in Equation 20.21 are true for only a Gaussian peak. What alternative expression could you use if you had a non-Gaussian peak?

63. Figure 20.19 shows a chromatogram that was obtained for a standard mixture of several naphthalene-related compounds. Using the information obtained for peaks 3–6, estimate the number of theoretical plates and plate height for this system.

64. Explain how the appearance of the chromatogram in Figure 20.19 would change if (1) the number of theoretical plates is increased, or (2) the plate height is decreased. Assume that all other conditions are kept constant to get the same retention for each injected chemical.

65. What do the terms "peak tailing" and "peak fronting" refer to in chromatography? What factors can produce peak tailing or peak fronting?

66. Define what is meant by the "asymmetry factor" for a chromatographic peak. Describe how this is determined.

67. What general peak shape (e.g., "peak fronting," "peak tailing," or "Gaussian") is represented by each of the following A/B ratios?
    (a) $A/B = 1.49$
    (b) $A/B = 0.78$
    (c) $A/B = 1.00$

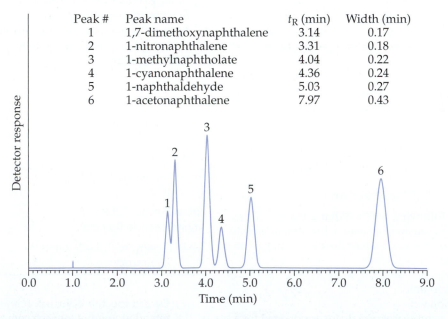

Peak #	Peak name	t_R (min)	Width (min)
1	1,7-dimethoxynaphthalene	3.14	0.17
2	1-nitronaphthalene	3.31	0.18
3	1-methylnaphtholate	4.04	0.22
4	1-cyanonaphthalene	4.36	0.24
5	1-naphthaldehyde	5.03	0.27
6	1-acetonaphthalene	7.97	0.43

FIGURE 20.19 Separation of various naphthalene-related compounds by reversed-phase liquid chromatography on a 4.1 mm inner diameter and 25 cm long column at 1.0 mL/min. The void time of this separation is at 1.02 min. The widths listed are the baseline widths of these peaks (w_b). (This chromatogram was generated using DryLab software, courtesy of LCResources.)

68. Figure 20.20 shows a set of chromatograms obtained for the separate injection of two herbicides (diquat and paraquat) onto a reversed-phase liquid chromatographic column. Use the information in this chromatogram to determine the *A/B* ratio for diquat and paraquat under these conditions. Do these peaks represent peak tailing or peak fronting? How do you think the asymmetry of these peaks would affect their separation if both diquat and paraquat were present in the same sample?

69. Define the following terms and state why they are important in chromatography.
    (a) Diffusion
    (b) Mass transfer
    (c) Diffusion coefficient

70. Describe each of the following processes and state how each contributes to chromatographic band-broadening.
    (a) Eddy diffusion
    (b) Longitudinal diffusion
    (c) Mobile phase mass transfer
    (d) Stationary phase mass transfer
    (e) Stagnant mobile phase mass transfer
    (f) Extra-column band-broadening

71. What is the van Deemter equation? What band-broadening processes are represented by each part of this equation? Explain why a plot of the van Deemer equation follows a "U"-shaped curve.

72. How will the value of the plate height in Figure 20.15 change if the linear velocity is increased from 0.16 cm/s to 0.60 cm/s? What advantages would there be in using higher linear velocities? What would be the disadvantages?

73. A food chemist using gas chromatography notices that the plate height for a particular analyte is 0.41 mm at a linear velocity of 12 cm/s, 0.25 mm at 20 cm/s, 0.32 at 50 cm/s, and 0.50 mm at 70 cm/s. What is the reason for this observed change in the plate height?

CONTROLLING CHROMATOGRAPHIC SEPARATIONS

74. What is the definition of the "separation factor" in chromatography? How is this used to describe separations? What are the advantages and disadvantages to using this parameter?

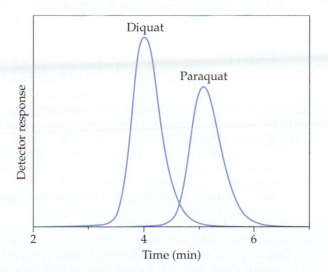

FIGURE 20.20 Separation of diquat from paraquat on a chromatographic column.

75. What is meant by "peak resolution" in chromatography? How is this determined? What are the advantages and disadvantages to using this parameter?

76. A scientist working for the U.S. Food and Drug Administration wishes to determine the degree of separation that occurs for the standard sample in Figure 20.19. Calculate the separation factor obtained for each set of neighboring peaks in this chromatogram. From this information, which particular peaks would you focus on if you were given the task of further improving this separation?

77. Calculate the resolution for each set of neighboring peaks in Figure 20.19. Based on these results, which peaks would you focus on if you were to improve this separation? How do these results compare to those obtained in Problem 76 when using separation factors to evaluate the separation of these peaks?

78. What is the "resolution equation" in chromatography? Based on this equation, what are three general approaches that can be used to improve a separation in chromatography?

79. A chemist wishes to adjust the separation in Figure 20.19 so that a better resolution is obtained between 1,7-dimethoxy-naphthaldehyde and 1-nitronaphthalene (which have retention factors of 2.1 and 2.2, and a resolution of 0.96). Determine how the resolution between these two peaks will change if each of the following changes is made separately to the chromatographic system.
    (a) An increase in the length of the column from 25.0 cm to 50.0 cm, with all other conditions remaining constant
    (b) A change in the mobile phase so the retention times for the 1-naphthaldehyde and 1-acetophenone are each changed by roughly 2.5-fold (giving *k* values of 5.2 and 5.5, respectively)
    (c) A switch to a similar, but slightly different type of LC column that has about the same plate number as the original column, but that now gives retention factors of 2.0 and 2.2 for 1-naphthaldehyde and 1-acetophenone

CHALLENGE PROBLEMS

80. It is possible to estimate the partition ratio (K_D) for a non-ionic solute between two liquids by using the solubility (S) of an analyte in each of the liquids.
    (a) Using the fact that $K_{sp,Phase\ 1} = S_{Phase\ 1}$ and $K_{sp,Phase\ 2} = S_{Phase\ 2}$ for a nonionic analyte with no significant side reactions, derive an equation that shows how K_D for this analyte will be related to $S_{Phase\ 1}$ and $S_{Phase\ 2}$.
    (b) Iodine is known to have solubilities in water and in carbon tetrachloride of 1.32×10^{-3} M and 0.115 M, respectively. What is the predicted partition ratio for iodine when it is extracted using these two solvents? How does your result compare to the measured partition ratio of 82.6?

81. Derive the relationship given in Table 20.2 between D_c and the partition ratio for a weak base. (*Hint*: See the process used in Exercise 20.1.)

82. Oxine (a chelating agent shown earlier in Figure 20.5) has two sites that can act as weak acids, as shown by the following general reactions.

$$H_2Q^+ \rightleftharpoons HQ + H^+ \qquad pK_{a_1} = 5.0$$

$$HQ \rightleftharpoons Q^- + H^+ \qquad pK_{a_2} = 9.9$$

Only the neutral form of oxine (HQ) will enter an organic solvent such as chloroform, where $K_D = 400$ in the presence of water and chloroform.[8]

(a) Derive an equation that shows how the value of D_c for oxine will change with pH.

(b) Prepare a plot of $\log(D_c)$ for oxine versus pH. Over what pH range would oxine be expected to have its highest degree of extraction by chloroform in the presence of water?

83. Prepare a spreadsheet in which the percent of extraction of two solutes (A and B) is calculated for a particular set of distribution ratios and a given phase ratio.

(a) Test your spreadsheet by using it to calculate the percent extraction of A and B under the same conditions as used in Figure 20.6. Based on your spreadsheet, after how many extractions greater than 90%, 95%, or 99% of A will be extracted? What percent of B will be extracted under these same conditions?

(b) How are the results of your calculation altered if the phase ratio is changed from 1.0 to 2.0 or 0.5? What happens when it is changed from 1.0 to 20? What does this tell you about the role that the volume of extracting solvent can play in determining the efficiency of such a liquid–liquid extraction?

(c) How are the results of your calculation altered if the phase ratio is kept at 1.0 but D_c for the two solutes is changed from values of 2.0 and 0.1 to values of 10.0 and 0.1? What happens when these values are changed to 2.0 and 0.02? What does this tell you about the importance of D_c in determining the extent of an extraction and the purity of an extracted solute?

84. The *response factor* (f) of analyte in a given detection system can be calculated by comparing the analyte's peak area ($Area_A$) to the area obtained for an equivalent amount of a reference compound ($Area_R$),[9]

$$f = [(Area_A)/(Area_R)] \, f_R \qquad (20.28)$$

where f_R (usually assigned a value of one) is a multiplication factor used to correct for any known differences in the analyte and reference compound.

(a) A drug standard containing a mixture of barbiturates was analyzed by the method of gas chromatography. Each barbiturate in this mixture was present at a concentration of 5 mg/mL. The average peak areas (in arbitrary units) obtained for multiple injections of this standard were as follows: butabarbital, 9.05; barbital, 9.01; pentobarbital, 8.24; secobarbital, 9.02; and hexobarbital, 6.32. Determine the response factor for each of these drugs, using butabarbital as the reference compound.

(b) The following data were obtained for the injection of a series of butabarbital standards onto the same system as in Part (a).

Butabarbital Concentration (mg/mL):	2.50	5.00	7.50	10.00
Average Peak Area:	4.35	9.05	12.98	18.43

An injection of a patient sample under the same conditions gave a peak area for secobarbital of 23.53. Determine the approximate concentration of secobarbital in the patient sample.

85. A simple model of chromatography can be prepared based on the Craig countercurrent distribution.[8]

$$P_{r,n} = \frac{r'}{n'(r-n)'} \, p^n \, q^{r-n} \qquad (20.29)$$

In this equation, $P_{r,n}$ is the relative amount of an analyte that will be found in tube n of the apparatus (with the first tube being tube 0, the second tube 1, and so on) after r number of transfers. The values p and q are the relative fraction of analyte that will be found in the mobile phase and stationary phase of each tube, respectively, at equilibrium. These values in turn, can be related directly to the retention factor for the analyte through Equations 20.30 and 20.31.

$$p = 1/(1+k) \qquad (20.30)$$

$$q = k/(1+k) \qquad (20.31)$$

(a) Use these equations to prepare a spreadsheet that models a Craig system with 100 tubes. Calculate and plot the fraction of three analytes ($k = 2$, 4, and 8) in each tube on this system at $r = 50$ and 100. Check your plots by comparing the results to those in Table 20.3.

(b) Use your spreadsheet to model Craig systems with 50, 100, or 200 tubes and for analytes with k values of 2, 4, or 8 after 50 and 100 transfers. How do these results compare?

(c) What observations can you make from the plots in Parts (a) and (b) about the effects of analyte retention (k) and system length (n) on these separations? How can you relate these observations to the factors that were stated in this chapter to be important in the separation of analytes by chromatography?

86. The number of theoretical plates for an asymmetric peak can be estimated as follows,

$$N = \frac{4.17 \, (t_p/w_{0.1})}{A/B_{0.1} + 1.25} \qquad (20.32)$$

where t_P is the elution time for the top of the peak, $w_{0.1}$ is the total width of the peak (in time units) at the one-tenth height level, and $A/B_{0.1}$ is the asymmetry value obtained at the one-tenth height level (e.g., as used in Figure 20.13).[27,28] Estimate the $A/B_{0.1}$ ratios and values of N for the diquat and paraquat peaks in Figure 20.20. How do these results compare to those obtained if you instead used the relationships in Equation 20.21, which apply only to peaks following a Gaussian distribution? What conclusions can you make from this comparison?

87. The average amount of time that it takes a molecule or atom of a substance to travel a particular distance through a liquid or gas (or any other medium) can be calculated using the *Stokes—Einstein equation*.[29]

$$d = (2\,Dt)^2 \qquad \text{or} \qquad t = d^2/(2D) \qquad (20.33)$$

In these equations, D is the diffusion coefficient for the substance that is moving, t is the amount of time that the substance is allowed to diffuse away from its starting position, and d is the average (or root-mean-squared) distance that is traveled by the substance in the allowed period of time t.

(a) If a typical diffusion coefficient for a small chemical in a gas at room temperature is $1 \, cm^2/s$, determine the average amount of time that it will take this chemical to travel 0.1 mm in a gas (a common radius for a gas-chromatographic (GC) open-tubular column).

(b) If a typical diffusion coefficient for a small chemical in a liquid at room temperature is $10^{-5} \, cm^2/s$, approximately how long will it take this compound to travel 0.1 mm in a liquid? From these results and those in Part (a), what can you conclude regarding the speed of diffusion and mass transfer in gas chromatography compared to liquid chromatography?

88. One way the separating power of a chromatographic system can be described is by determining how many peaks can be resolved at baseline resolution within a given section of a chromatogram. This can be determined using a quantity known as the *separation number (SN)*, which is found by the following formula,[9]

$$SN = \frac{t_{R,(z+1)} - t_{R,z}}{w_{h,z} + w_{h,(z+1)}} \qquad (20.34)$$

where $t_{R,z}$ and $t_{R,(z+1)}$ are the total retention times observed on a column for two saturated hydrocarbons that have carbon chains that are z or $(z+1)$ carbon units long, and where $w_{h,z}$ and $w_{h,(z+1)}$ are the measured widths at half-height for these peaks.

(a) A separation of n-alkanes by gas chromatography gives retention times of 15.3 min for n-heptane and 21.8 min for n-octane. The half-height widths for these peaks are 0.18 min and 0.26 min, respectively. Calculate the resolution of these two peaks and the number of theoretical plates for the column. What is the separation number in the region of the chromatogram between these two peaks?

(b) A similar but longer column is used under the same conditions, giving retention times of 30.5 min and 43.5 min for n-heptane and n-octane and peak widths at half-height of 0.25 min and 0.36 min. What is the new resolution between these peaks and plate number for the system? What is the separation number for this longer column? How does this result compare to that obtained in Part (a)? What do these results tell you about the effect of column efficiency on resolution and the ability to separate multiple chemicals in a sample?

89. Construct a spreadsheet in which Equation 20.27 is used to examine the effects of α, k, and N on R_s.

(a) Use your spreadsheet to prepare a graph in which R_s is plotted versus k over the range of 0 to 10 for a separation in which $N = 10,000$ and $\alpha = 1.1$. Over what range of retention factors is the biggest improvement in resolution obtained? What is the minimum retention factor needed to obtain baseline resolution? What do you think are the advantages and challenges of using even larger retention factors for the separation?

(b) Prepare a second graph in which R_s is plotted versus N over the range of 0 to 50,000 for a separation in which $k = 2$ and $\alpha = 1.1$. Over what range of plate numbers does the biggest improvement in resolution occur? What is the minimum value of N needed in this example to obtain baseline

resolution? What do you think are the advantages and possible challenges of using even larger plate numbers for such a separation?

(c) Construct a third graph in which R_s is plotted versus α over the range of 1.0 to 1.5 for a separation in which $k = 2$ and $N = 10,000$. Over what range of separation factors is the biggest improvement in resolution obtained? What is the smallest separation factor that is needed in this case to obtain baseline resolution? What do you think are the possible advantages and challenges of using even larger separation factors for this separation?

TOPICS FOR REPORTS AND DISCUSSION

90. Find a recent research article that describes the analysis of 2,4-D or another herbicide.
Describe how chemical separations were used in this article.

91. Chemical separation methods like extractions and chromatography are used in a wide range of industries and research laboratories. Visit a local laboratory and ask them how such methods are used at their facility. Discuss your findings with your class.

92. Using the Internet or other resources, obtain more information on one of the following extraction techniques. Write a report on this method, including a description of how it works and the types of chemicals or samples for which it is used.
(a) Solid-phase extraction
(b) Solid-phase microextraction
(c) Microwave-assisted extraction
(d) Soxhlet extraction

93. Obtain information from the Internet or other sources describing the proper use of a separatory funnel for liquid–liquid extractions. Use this information to prepare a standard operating procedure for this method.

94. There are many agents besides oxine that can be used to complex with metal ions and aid in their extraction. Other reagents that can be employed for this purpose are listed below.[8,10] For any one of these reagents, find more information its structure, properties, and use as metal binding agent. Write a report about this reagent.
(a) Cupferron
(b) 8-Mercaptoquinoline
(c) Dithizone
(d) Pyridylazonaphthol (PAN)

95. One of the oldest methods for chemical separations is a distillation. Write a report on the history of distillations[30,31] or on modern applications of this technique.[8] Compare the method you choose to extractions and chromatography in terms of the mechanism by which it separates chemicals.

References

1. R. E. Evenson and D. Gollin, "Assessing the Impact of the Green Revolution, 1960 to 2000," *Science*, 300 (2003) 151–167.

2. G. Conway, *The Doubly Green Revolution*, Cornell University Press, Ithaca, New York, 1998.

3. M. Windholz, Ed., *The Merck Index*, 10th Edition, Merck & Co., Rahway, NJ, 1983.

4. International Programme on Chemical Safety, *2,4-dichloropheoxyacetic acid (2,4-D)*, World Health Organization, Geneva, Switzerland, 1984.

5. U.S. Environmental Protection Agency Report, "2,4-D RED Facts," *EPA-738-F-05-002*, U.S. EPA, Washington, DC, 2005.

6. T. Cairns and J. Sherma, *Emerging Strategies for Pesticide Analysis*, CRC Press, Boca Raton, FL, 1992.

7. R. Grover, *Environmental Chemistry of Herbicides*, CRC Press, Boca Raton, FL, 1988.

8. B. L. Karger, L. R. Snyder, and C. Horvath, *An Introduction to Separation Science*, Wiley, New York, 1973.

9. J. Inczedy, T. Lengyel, and A. M. Ure, *International Union of Pure and Applied Chemistry—Compendium of Analytical Nomenclature: Definitive Rules 1997*, Blackwell Science, Malden, MA, Chapter 9.

10. J. Rydberg, C. Musikas, and G. R. Choppin, Eds., *Principles and Practices of Solvent Extraction*, Marcel Dekker, New York, 1992.

11. T. L. Chester, J. D. Pinkston, and D. E. Raynie, "Supercritical Fluid Chromatography and Extraction," *Analytical Chemistry*, 64 (1992) 153R–170R.

12. M. A. McHugh and V. J. Krukonis, *Supercritical Fluid Extraction: Principles and Practice*, 2nd ed., Elsevier, Amsterdam, the Netherlands, 1994.

13. L.T. Taylor, *Supercritical Fluid Extraction*, Wiley, New York, 1996.

14. N. Kresge, R. D. Simoni, and R. L. Hill, "Lyman Creighton Craig: Developer of the Counter-Current Distribution Method," *Journal of Biological Chemistry*, 280 (2005) e4–e6.

15. L. C. Craig, "Identification of Small Amounts of Organic Compounds by Distribution Studies. Application to Atabrine," *Journal of Biological Chemistry*, 150 (1943) 33–45.

16. L. C. Craig, "Identification of Small Amounts of Organic Compounds by Distribution Studies. II. Separation by Counter-current Distribution," *Journal of Biological Chemistry*, 155 (1944) 519–534.

17. M. Tswett, "Physikalisch-chemische Studien über das Chlorophyll. Die Adsorptionen," *Berichten der Deutschen Botanischen Gesellschaft*, 24 (1906) 316–323.

18. M. Tswett, "Adsorptionanalyse und chromatographische Methode. Anwendung auf die Chemie des Chlorophylls," *Berichten der Deutschen Botanischen Gesellschaft*, 24 (1906) 384–393.

19. M. Tswett, *Chromophylls in the Plant and Animal Kingdom*, Karbasnikov, Warsaw, 1910.

20. L. S. Ettre, "M.S. Tswett and the Invention of Chromatography," *LC-GC*, 21 (2003) 458–467.

21. R. E. Majors and P. W. Carr, "Glossary of Liquid-Phase Separation Terms," *LC-GC*, 19 (2001) 124–162.

22. C. F. Poole and S. K. Poole, *Chromatography Today*, Elsevier, New York, 1991.

23. B. A. Bidlingmeyer and F. V. Warren Jr., "Column Efficiency Measurement," *Analytical Chemistry*, 56 (1984) 1583A–1596A.

24. J. J. van Deemter, F. J. Zuiderweg, and A. Klinkenberg, "Longitudinal Diffusion and Resistance to Mass Transfer as Causes of Non Ideality in Chromatography," *Chemical and Engineering Science*, 5 (1956) 271–289.

25. A. S. Said, "Comparison Between Different Resolution Equations," *Journal of High Resolution Chromatography*, 2 (2005) 193–194.

26. *EPA Method 8330, Determination of Concentration of Nitroaromatics and Nitramines by High-Performance Liquid Chromatography*, U.S. Environmental Protection Agency, Washington, DC.

27. J. P. Foley and J. G. Dorsey, "Equations for the Calculation of Chromatographic Figures of Merit for Ideal and Skewed Peaks," *Analytical Chemistry*, 55 (1983) 730–737.

28. J. P. Foley and J. G. Dorsey, "A Review of the Exponentially Modified Gaussian (EMG) Function: Evaluation and Subsequent Calculation of Universal Data," *Journal of Chromatographic Science*, 22 (1984) 40–46.

29. C. R. Cantor and P. R. Schimmel, *Biophysical Chemistry, Part II: Techniques for the Study of Biological Structure and Function*, Freeman, San Francisco, 1980.

30. F. Szabadvary, *History of Analytical Chemistry*, Pergamon Press, New York, 1966.

31. A. J. Liebmann, "History of Distillation," *Journal of Chemical Education*, 33 (1956) 166–173.

Chapter 21

Gas Chromatography

Chapter Outline

21.1 INTRODUCTION: THERE'S SOMETHING IN THE AIR

News Release—Mexico City, March 31, 2006—"Mexico City a Living Laboratory for Smog Study":

Whether this city has the most polluted air in the world is a matter of debate: indignant Mexican officials lobbied to have it striken from the Guinness Book of World Records this year after it held the title two years running. What's not in question is its attraction to the hundreds of atmospheric scientists who are wrapping up a monthlong study of the reach and impact of Mexico City's pollution.... Scientists and graduate students have been working 14-hour days to measure the giant plume of gases, dust and particles that rise out of Mexico City each day and generally drift to the northeast, sometimes as far as the Gulf of Mexico. Over the course of hours, the emissions mix and are altered by sunlight to create so-called secondary pollutants—some only irritating, others carcinogenic. Using instrument readings from ground equipment, weather balloons, airplanes and NASA satellites, scientists hope to figure out how they form and how far they travel.[1]

Problems with smog and air pollution have become common in modern cities and developing nations.[2,3] Smog is formed through the reaction of nitrogen oxides with volatile organic compounds (VOCs), both of which are emitted when fuel is burned by automobiles, industrial plants, and homes (see Figure 21.1). VOCs are made up of a large group of small organic compounds with boiling points below 200°C. These low boiling points allow such compounds to easily enter the atmosphere. Once VOCs are in the air, they can react with nitrogen oxides in the presence of light to form ground-level ozone, the main component of smog.[4–7]

There has been an ongoing effort in the United States and Europe to monitor and reduce the causes of smog and air pollution. Similar programs are appearing in Mexico and other countries. These programs have helped to improve air quality in the United States,[2] but there is still a need for worldwide progress in this area.[1,3] A crucial part of this effort has been the use of analytical methods to monitor the sources of smog and to determine the levels of ozone, nitrogen oxides, and VOCs in air. In this chapter, we will examine the technique of *gas chromatography* (*GC*), which is an important tool for examining VOCs and other volatile chemicals.

21.1A What Is Gas Chromatography?

Gas chromatography (**GC**) is a type of chromatography in which the mobile phase is a gas.[8] The presence of a gas mobile phase makes GC valuable for separating substances like VOCs that occur naturally as gases or that can easily be placed into a gaseous phase. This same feature

Sources of NO$_x$ and volatile organic compounds (VOCs)

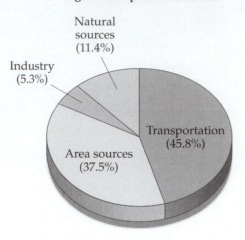

FIGURE 21.1 A haze due to smog is present over Mexico City during many days of the year. The volatile organic compounds (VOCs) and nitrogen oxides (NO$_x$) that create this smog come from such sources as automobiles and public transportation, industry, fires, and natural sources. (This graph is based on data provided in S. Guzman, "Suspiro de Vida," *EJ Magazine*, Fall 2003, http://www.ejmagazine.com/2003bsuspiro.html.)

makes GC useful for examining the many volatile chemicals that are of interest in fields like environmental testing, forensic analysis, and work in the petroleum industry.

21.1B How Is Gas Chromatography Performed?

Figure 21.2 shows a typical system for performing GC. A system that is used to perform GC is called a *gas chromatograph*.[9] The first major component of a gas chromatograph is the gas source that supplies the mobile phase. This source is typically a gas cylinder equipped with pressure regulators to deliver the mobile phase at a controlled rate. The second part of the gas chromatograph is its injection system, which often consists of a heated loop or port into which the sample is placed and converted into a gaseous form. The third part of the system is the column. The column contains the stationary phase and support material for the separation of components in a sample. This column is held in an enclosed area known as the *column oven* that maintains the temperature at a well-defined value. The fourth part of the GC system is a detector, and associated recording device, that monitors sample components as they leave the column.

A plot of the detector response versus the time that has elapsed since sample injection onto a GC system is known as a *gas chromatogram*. An example of such a plot is given in Figure 21.3. In Chapter 20, we saw how a plot like this can be used to help identify and measure the components in an injected sample. Clues to the identity of a peak can be obtained by comparing the peak's retention time to that observed for the injection of a known sample of the suspected chemical. This process can be further aided by the use of a detector that can selectively monitor or confirm the structure of the eluting substance, as we will see in Section 21.5A. Once a peak has been identified, the amount of analyte in this peak can be determined by comparing the peak's area or height to that which is obtained for the injections of standards that contain the same, or a similar, analyte. An internal standard is also often used as part of this process (see Chapter 5) to correct for variations in analyte content that might have occurred during the sample pretreatment or injection onto the GC system.

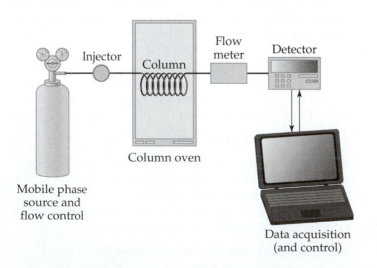

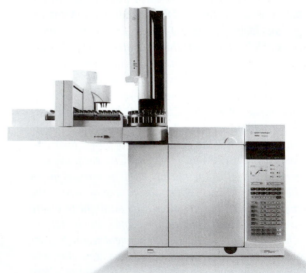

FIGURE 21.2 On the left, a diagram of a typical gas chromatograph and on the right an example of a commercial gas chromatograph. The image on the right does not show the mobile phase source but does include all other components of a typical GC system. (The image of the commercial instrument is reproduced with permission from Agilent.)

Peak label	Name	Mass	Boiling point (°C)	Peak label	Name	Mass	Boiling point (°C)
1	Ethane	30.0694	−88.6	13	Isopentane	72.1498	30
2	Ethylene	28.0536	−103.7	14	*n*-Pentane	72.1498	36.1
3	Propane	44.0962	−42.06	15	*trans*-2-Pentene	70.134	37
4	Propylene	42.0804	−47.4	16	3-Methyl-1-Butene	70.134	20
5	Isobutane	58.123	−11.7	17	1-Pentene	70.134	30
6	*n*-Butane	58.123	−0.45	18	*cis*-2-Pentene	70.134	37
7	Acetylene	26.0378	−28.1 (sublimes)	19	2,2-Dimethylbutane	86.1766	49.7
8	*trans*-2-Butene	56.1072	0.88	20	2-Methylpentane	86.1766	62
9	1-Butene	56.1072	−6.1	21	2.3-Dimethylbutane	86.1766	58
10	Isobutene	56.1072	−6.9	22	Isoprene	68.1182	34
11	*cis*-2-Butene	56.1072	3.7	23	4-Methyl-1-Pentene	84.1608	54
12	Cyclopentane	70.134	49	24	2-Methyl-1-Pentene	84.1608	62

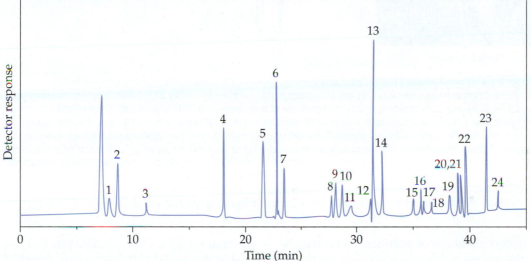

FIGURE 21.3   A gas chromatogram for the analysis of volatile organic compounds in urban air. (This chromatogram is based on data obtained from Perkin Elmer.)

EXERCISE 21.1   Analyzing Chemicals by GC

An environmental chemist wishes to use the method in Figure 21.3 to examine acetylene in air samples. Describe how we could use this method to both identify and measure acetylene.

SOLUTION

An initial identification of acetylene in an unknown sample can be made by seeing if there is a peak with a retention time around 23.4 min, the elution time for acetylene in Figure 21.3. The presence of acetylene can be confirmed by examining the suspected peak with a selective detector such as a mass spectrometer, an approach we discuss in Section 21.5A.

The amount of acetylene in an unknown sample can be measured by comparing the size of its peak to that obtained for standards that contain known amounts of acetylene. Ideally, the sample and standards should each contain a fixed level of an internal standard (such as an isotopically labeled form of acetylene when detection is performed by mass spectrometry). A calibration curve is then prepared with data obtained from the standards by plotting on the *y*-axis the peak area or height ratio for acetylene versus the internal standard (e.g., Area$_{Acetylene}$/Area$_{Internal\ Standard}$), with the amount of acetylene in the same standards being plotted on the *x*-axis. The amount of acetylene in an unknown sample can then be determined by using this curve and the unknown's peak area or height ratio versus the internal standard.

Because GC can be used for the identification, measurement, and separation of volatile chemicals, this method has been a popular chemical analysis tool for many decades. The origins of this technique go to back to shortly after World War I, when researchers began to study the selective adsorption of gases to solids for use in gas masks and the recovery of gasoline from natural gas. These same solids were then used with temperature control for analytical applications like the determination of hydrocarbons in natural gas and separation of volatile organic acids.[10] The first modern system for gas chromatography was developed in 1945–1947 by a German chemist named Erika Cremer (see Figure 21.4).[10–12] Similar devices were later

Erika Cremer

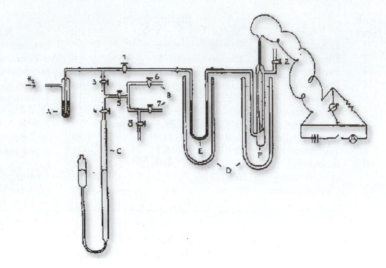

FIGURE 21.4 Erika Cremer (1900–1996) and the first modern GC system that was developed in her laboratory. Cremer began her work in gas chromatography at the University of Innsbruck in Austria during World War II. Along with creating the first prototype of a modern gas chromatograph, Cremer also developed many terms and concepts that are still used to describe GC separations. The gas chromatograph shown in this figure used a thermal conductivity detector (shown on the far right) to monitor analytes as they eluted from the column. A chromatogram was recorded on this device by using a team of four students: The first student operated the instrument, the second kept track of the time, the third watched and read off the results on the detector, and the fourth recorded these results.[11,12] (The photo is reproduced with permission and courtesy of the Archive of the University of Innsbruck; the diagram is from a 1947 dissertation by Fritz Prior, the student who worked with Erika Cremer in building the first gas chromatographic system.)

created by others that made use of partition-based separations, in which liquids were coated onto solids for use as stationary phases. The combined improvement in systems and columns for GC led to rapid growth in this field over the 1950s and 1960s.[10] Many of the techniques that were developed at that time for the separation and analysis of gases and volatile chemicals are still in use today.

21.2 FACTORS THAT AFFECT GAS CHROMATOGRAPHY

21.2A Requirements for the Analyte

Volatility and Thermal Stability. To examine a chemical by gas chromatography, it is necessary to place this chemical into a gaseous mobile phase so that the analyte can then enter and pass through the GC column. This requirement means that the injected analyte must be *volatile*, or able to go easily into the gas phase. The volatility of a chemical is related to its vapor pressure and boiling point. For instance, volatile chemicals like those that are classified as VOCs will have high vapor pressures and low boiling points. This property is what allows VOCs to enter the atmosphere and react with nitrogen oxides to form ozone. The same property also makes it easy for gas chromatography to separate and analyze such compounds.

There are several pieces of information we can use to make an educated guess as to whether a particular analyte will be volatile enough for its examination by gas chromatography. One valuable piece of information is the size of the analyte. In general, a chemical with a molar mass above 600 g/mol will have a volatility that is too low for work with GC.[13] Many smaller chemicals can be analyzed by gas chromatography, but the presence of polar functional groups can make it difficult to examine some low-mass substances. Thus, the boiling point of a chemical should always be considered when determining if the chemical can be studied by GC. A chemical with a boiling point below 500°C at 1 atm should be volatile enough for GC analysis.[9] This can be confirmed by injecting a sample of the analyte onto a GC system and seeing if it elutes from the column in a reasonable amount of time.[14]

EXERCISE 21.2 **Chemical Volatility in GC**

The boiling points and masses of several VOCs are listed in Figure 21.3. How do these values compare with the general properties that are needed to make an analyte suitable for GC?

SOLUTION

All of the VOCs in Figure 21.1 have molar masses below 600 g/mol and boiling points well below 500°C–550°C (range, –104°C for ethylene to 61°C for 2-methyl-1-pentene). These values agree with the general guidelines that were given for the analyte requirements of GC. This explains why GC works so well for such chemicals. A closer look at Figure 21.1 indicates that analytes with the lowest masses and boiling points (ethane and ethylene) elute first from the column, while analytes that have higher masses and boiling points (see peaks 19–21 and 22–24) tend to elute much later. This trend indicates that volatility is also an important factor in determining the retention of analytes in a GC system.

Besides being volatile, an analyte must have good thermal stability if it is to be examined by GC. An analyte that is not thermally stable may degrade at the high temperatures that are often used during the injection and separation of samples by GC. This type of degradation can make it difficult to detect and measure such an analyte. It is often hard to know in advance if a particular chemical will be stable enough for analysis by GC. This stability can be tested by injecting a new analyte onto a GC system and seeing if this chemical produces a chromatogram that has a single well-defined peak with good retention and detection properties. If thermal stability is a problem, such degradation can sometimes be minimized or eliminated by selecting a proper sample injection technique (see Section 21.5B) or by using chemical derivatization, as discussed in the next section.

Chemical Derivatization. Although some analytes can be injected directly onto a GC system, many chemicals are not sufficiently volatile or stable for this approach. A common solution to this problem is to change the structure of the analyte to give this chemical a more volatile or more thermally stable form. The process of altering the chemical structure of an analyte is known as **derivatization**.[15,16] Derivatization in GC typically involves replacing one or more polar groups on an analyte, such as an alcohol or amine group, with fewer polar groups. This change reduces the intermolecular interactions of the altered chemical, making it more volatile and easier to place into the gas phase. The same type of change also tends to make a compound more thermally stable.

Derivatization is not needed in GC for substances like VOCs, but it is needed for larger less volatile chemicals like cholesterol (an analyte we studied in Chapter 18). Cholesterol has a relatively low volatility due to both its mass (387 g/mol) and the presence of an alcohol group in its structure, which can lead to the formation of hydrogen bonds with neighboring molecules. These features cause cholesterol to give a broad peak with a long retention time when this analyte is injected onto a GC column (see Figure 21.5). The alcohol group of cholesterol can also lead to the degradation of this compound at high temperatures, as occurs when this alcohol group combines with a hydrogen on a neighboring carbon of cholesterol to release water.[17]

It is important to notice that each of these problems in the GC analysis of cholesterol is caused by the alcohol group of this compound. This observation suggests that we could overcome these problems by altering the alcohol group of cholesterol through derivatization. As an example, we could replace the hydrogen on this alcohol group with —Si(CH$_3$)$_3$, or trimethylsilyl (TMS).

$$CH_3$$
$$H_3C-Si-Cl + HO-R \longrightarrow H_3C-Si-O-R + HCl$$
$$CH_3 \qquad\qquad\qquad CH_3$$

Trimethylchlorosilane (TCMS, a TMS reagent)   TMS derivative

(21.1)

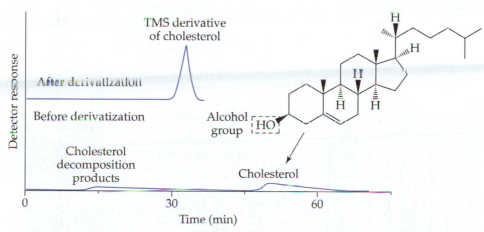

FIGURE 21.5 Analysis of cholesterol (bottom) before derivatization and (top) after derivatization on a GC column containing a nondeactivated support. Both chromatograms were obtained under the same conditions. (Reproduced with permission from W.J.A. Vanden Heuvel, "Some Aspects of the Chemistry of Gas-Liquid Chromatography," in *Gas Chromatography of Steroids in Biological Fluids*, M.B. Lipsett, (Ed.), Plenum Press, New York, 1965, pp. 277–295.)

The product of this reaction is known as a *TMS-derivative*.[15,16] The replacement of a hydrogen on the alcohol group with a bulkier and less polar TMS group means that the derivatized form of cholesterol will be more thermally stable. Although we have increased the mass of our analyte by adding the TMS group (a change in mass of 72 g/mol), we also now have a derivative that is less likely to form hydrogen bonds than underivatized cholesterol. This makes the derivative more volatile. The result of this change is an improved chromatogram that takes less time to perform and contains only a single well-defined peak for the cholesterol derivative (see top of Figure 21.5).

There are many ways in which chemicals can be derivatized for GC analysis.[15,16] Besides increasing the volatility and thermal stability of a compound, derivatization can be utilized to change the response of an analyte on certain GC detectors. As an example, derivatization can be used to place halogen atoms (I, Cl, Br, or F) onto an analyte to improve its response on an electron-capture detector, a device discussed in Section 21.5A. Derivatization can also be used to alter a GC separation by changing the positions of overlapping peaks or by preventing broad peaks caused by interactions between the column support and polar groups on an analyte.

21.2B Factors that Determine Retention in Gas Chromatography

In Chapter 20 we learned that there are two general factors that determine the ability of chromatography to separate chemicals: retention and column efficiency. As is true for any chromatographic method, the retention of a compound in GC will be determined by how much time this substance spends in the mobile phase versus the stationary phase. In GC this retention will be affected by (1) the volatility of an injected compound, (2) the temperature of the column, and (3) the degree to which the compound interacts with the stationary phase.

The low density of gases causes analytes passing through a GC column to have little or no interaction with the mobile phase. As a result, the volatility of an analyte is the main factor that causes this chemical to stay in the mobile phase during a GC separation. This fact also means that the most volatile analytes in a sample will tend to spend the most time in the mobile phase and elute the most quickly from a GC column. This idea is illustrated in Figure 21.6, for a sample that consists of a group of saturated, straight-chain hydrocarbons known as "*n*-alkanes." Compounds like these that have the same general structure, but which differ in the length of a single carbon chain, are known as *homologs* (or a "homologous series"). Volatility decreases and retention increases as we increase the size of the carbon chain in a group of homologs, as is shown in Figure 21.6 in going from peak 1 (obtained for an *n*-alkane containing a chain with 6 carbon atoms) to peak 16 (for an *n*-alkane with a chain of 40 carbons).

Temperature also plays an important role in GC separations. The effects of varying column temperature are shown in Figure 21.7. Decreasing the column temperature leads to longer retention, because this causes the injected analytes to be less volatile and spend less time in the mobile phase. Increasing the temperature produces the opposite effect, with analytes becoming more volatile and passing through the column more quickly as they spend more time in the mobile phase. This effect is the reason why GC systems contain a column oven for temperature control.

Many of the measures of retention discussed in Chapter 20 (such as t_R, V_M, and k) are affected by temperature. An alternative measure of retention that shows a smaller change with temperature is the **Kováts retention index (I)**.[18,19] The Kováts retention index for a chemical is calculated by using Equation 21.2. In this equation, the retention of an analyte on a particular column is compared to the retention seen at the same temperature and on the same column for a series

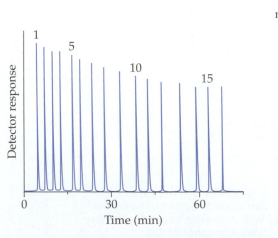

Peak number	Chemical name	Chemical structure	Boiling point (°C)
1	*n*-Hexane	$CH_3(CH_2)_4CH_3$	69
2	*n*-Heptane	$CH_3(CH_2)_5CH_3$	98
3	*n*-Octane	$CH_3(CH_2)_6CH_3$	126
4	*n*-Nonane	$CH_3(CH_2)_7CH_3$	151
5	*n*-Decane	$CH_3(CH_2)_8CH_3$	174
6	*n*-Undecane	$CH_3(CH_2)_9CH_3$	196
7	*n*-Dodecane	$CH_3(CH_2)_{10}CH_3$	216
8	*n*-Tetradecane	$CH_3(CH_2)_{12}CH_3$	254
9	*n*-Hexadecane	$CH_3(CH_2)_{14}CH_3$	287
10	*n*-Octadecane	$CH_3(CH_2)_{16}CH_3$	316
11	*n*-Eicosane	$CH_3(CH_2)_{18}CH_3$	343
12	*n*-Tetracosane	$CH_3(CH_2)_{22}CH_3$	391
13	*n*-Octacosane	$CH_3(CH_2)_{26}CH_3$	432
14	*n*-Dotriacontane	$CH_3(CH_2)_{30}CH_3$	468
15	*n*-Hexatriacontane	$CH_3(CH_2)_{34}CH_3$	498
16	*n*-Tetracontane	$CH_3(CH_2)_{38}CH_3$	525

FIGURE 21.6 GC separation of a series of *n*-alkanes. A temperature program was used for analyte elution during this separation. (Based on data from Alltech.)

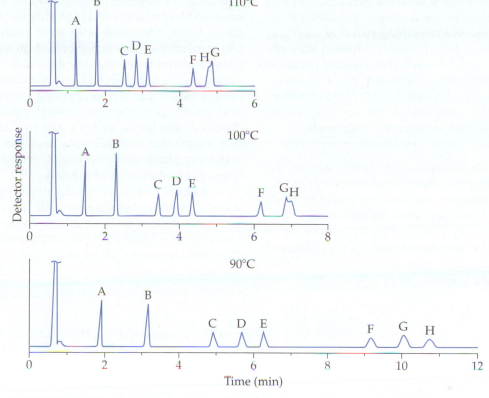

FIGURE 21.7   GC separation of an eight-component column test mixture at various temperatures. The compounds in the sample were (A) n-nonane, (B) n-decane, (C) 1-octanol, (D) n-undecane, (E) 2,6-dimethylphenol, (F) 2,4-dimethylaniline, (G) naphthalene, and (H) n-dodecane.

(Adapted with permission from J.V. Hinshaw, "Optimizing Column Temperature," *LC-GC*, 9 (1991) 94–98.)

of n-alkanes that are used in this calculation as reference compounds.[8]

$$I = 100z + 100 \cdot \frac{\log t'_{Rx} - \log t'_{Rz}}{\log t'_{R(z+1)} - \log t'_{Rz}} \quad (21.2)$$

The value of t'_{Rx} in Equation 21.2 is the adjusted retention time for the analyte (*Note*: We learned in Chapter 20 that $t'_R = t_R - t_M$ for an analyte, where t_R is the retention time of the analyte and t_M is the column void time).[20] This value is compared to t'_{Rz} (the adjusted retention time for an n-alkane that is found to elute just before the analtye) and $t'_{R(z+1)}$ (the adjusted retention time for an n-alkane that elutes just after the analyte). The values of z and $(z+1)$ in Equations 21.2 refer to the number of carbon atoms in the n-alkanes. Because these n-alkanes are homologs of each other, the n-alkane with z carbon atoms should elute more quickly than the n-alkane with $z + 1$ carbon atoms.

EXERCISE 21.3   **Calculating a Kováts Retention Index**

A sample of acenaphthene (a common chemical used in plastics) was injected onto a 25 m long × 0.3 mm inner diameter GC column at 140°C. A sample of the n-alkanes

$C_{14}H_{30}$, $C_{15}H_{32}$, and $C_{16}H_{34}$ was injected on the same column under identical conditions.[21] These chemicals had the following retention times: acenaphthene, 10.40 min; $C_{14}H_{30}$, 8.04 min; $C_{15}H_{32}$, 12.42 min; $C_{16}H_{34}$, 19.64 min.[22] The column void time was found to be 1.33 min. What was the Kováts retention index for acenaphthene on this column?

SOLUTION

The n-alkanes eluting just before and after acenaphthene were $C_{14}H_{30}$ and $C_{15}H_{32}$. These are the n-alkanes that should be used in Kováts retention index of acenaphthene, making $z = 14$ and $z + 1 = 15$. The adjusted retention time for $C_{14}H_{30}$ is $t'_{Rz} = (8.04 \text{ min} - 1.33 \text{ min})$ = 6.71 min (t'_{Rz}) and for $C_{15}H_{32}$ it is $t'_{R(z+1)} =$ (12.42 min − 1.33 min) = 11.09 min. The adjusted retention time for acenaphthene is $t'_{Rx} = (10.40 \text{ min} - 1.33 \text{ min})$ = 9.07 min. We get the following Kováts retention index when placing these values into Equation 21.2.

Kováts retention index for acenaphthene:

$$I = 100(14) + 100 \cdot \frac{\log(9.07 \text{ min}) - \log(6.71 \text{ min})}{\log(11.09 \text{ min}) - \log(6.71 \text{ min})}$$

$$= 1460$$

The digits in the hundreds and thousands places of this index tell us that acenapththene is eluting after the *n*-alkane with 14 carbon atoms. The remaining two digits indicate that, on a log scale, the adjusted retention time of acenaphthene is 60% of the way between the adjusted retention times for *n*-alkanes that contain 14 (z) and 15 ($z + 1$) carbon atoms. Thus, this index provides a quick means for comparing the retention of this analyte to that expected for these *n*-alkanes when used as reference compounds.

Temperature and compound volatility are not the only things that affect the retention of chemicals in GC. Another factor that affects retention is the extent to which an injected chemical interacts with the stationary phase. This idea is illustrated in Table 21.1 by using the Kovat's retention index to compare the retention of several model compounds on GC columns that contain two very different stationary phases: squalene and Carbowax M.

Squalane is a large nonpolar stationary phase based on a saturated hydrocarbon that has only weak intermolecular forces when it interacts with most analytes. As a result, GC separations that are performed on squalane columns give retention times that are determined mainly by the volatility of each analyte. Carbowax 20M is a polar stationary phase that can have strong intermolecular forces with polar analytes, such as through hydrogen bonding or dipole–dipole forces, as discussed in Chapter 7. This feature means that both volatility and interactions with the stationary phase can determine the retention of many analytes on a Carbowax 20M column.

A closer look at the results for the test compounds in Table 21.1 indicates that they all have similar Kováts retention indices on a squalane column, with elution times between those for *n*-pentane ($I = 500$) and *n*-heptane ($I = 700$). However, all of these test compounds show much stronger retention on a Carbowax 20M column. The

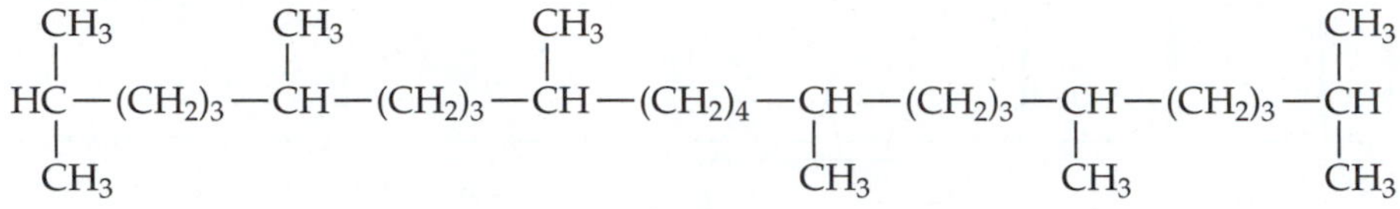

Squalane

Carbowax 20M (n ≈ 450)

TABLE 21.1   Comparison of the Retention for Some Model Compounds on Two Gas Chromatography Columns*

Name and Structure of Injected Compound	Kováts Retention Index Squalane ($I_{Squalane}$)	Kováts Retention Index Carbowax 20M ($I_{Carbowax\ 20M}$)	Difference in Retention Δ ($I = I_{Carbowax\ 20M} - I_{Squalane}$)
Benzene 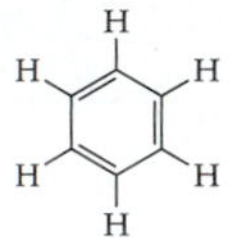	653	975	322 ($\Delta I_{Benzene} = \boldsymbol{X'}$)
1-Butanol HOCH$_2$CH$_2$CH$_2$CH$_3$	590	1126	536 ($\Delta I_{1\text{-Butanol}} = \boldsymbol{Y'}$)
2-Pentanone CH$_3$CCH$_2$CH$_2$CH$_3$	627	995	368 ($\Delta I_{2\text{-Pentanone}} = \boldsymbol{Z'}$)
1-Nitropropane CH$_2$CH$_2$CH$_3$NO$_2$	652	1224	572 ($\Delta I_{1\text{-Nitropropane}} = \boldsymbol{U'}$)
Pyridine 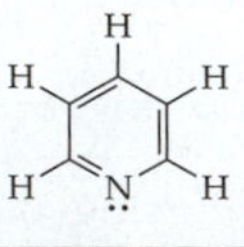	699	1209	510 ($\Delta I_{Pryidine} = \boldsymbol{S'}$)

*The numbers in this table were obtained from W.O. McReynolds, "Characterization of Some Liquid Phases," *Journal of Chromatographic Science*, 8 (1970) 685–691. The symbols shown in parentheses on the right represent the McReynolds constants for each of these particular compounds.

reason for this effect is that these compounds now have stronger interactions with the stationary phase in the Carbowax 20M column. In addition, this increase in retention is largest for analytes that can undergo strong hydrogen bonding or that have large dipole moments, such as 1-butanol, pyridine, and 1-nitropropane. This type of comparison based on the injection of a particular set of test compounds is valuable as a means for comparing the retention properties of various GC stationary phases. This topic is discussed further in Box 21.1.

21.2C Column Efficiency in Gas Chromatography

One benefit of using a gas as the mobile phase is this gives GC very high efficiency and narrow peaks (e.g., see Figures 21.3 and 21.6). These sharp and narrow peaks make it easy to measure small quantities of analytes and allow GC to separate a large number of compounds in a single run. There are several reasons for the high efficiency of GC systems. Many of these reasons are related to the fact that a gas is being used as the mobile phase. The low density of gases means that analytes in these gases can move about quickly by diffusion. This feature is important because most processes that cause peak broadening in chromatography will be reduced by the presence of fast diffusion, with longitudinal diffusion being the main exception (see Chapter 20). The presence of less peak broadening, in turn, makes it easier for the column to discriminate between the analyte and other sample components.

Low viscosity is another feature of gases that promotes high efficiency in GC. As the viscosity goes down for a mobile phase, it is possible to use a longer column. A longer column will have a larger number of theoretical plates, which will increase the resolution of separations that are performed on that column (see Chapter 20). A point will eventually be reached in which an increase in column length creates too much pressure for the mobile phase to pass through the column. Fortunately, the much lower viscosity of gases versus liquids means that much longer and more efficient columns can typically be used in GC than in liquid chromatography, a method we will discuss in Chapter 22.

21.3 GAS CHROMATOGRAPHY, MOBILE PHASES, AND ELUTION METHODS

21.3A Common Mobile Phases in Gas Chromatography

Like any type of chromatography, the mobile phase in GC is used to apply and transport compounds through the column. One important difference between GC and other chromatographic techniques is that a gas mobile phase will play little or no role in determining a compound's retention. Instead, retention is determined by the compound's volatility, the column temperature, and interactions of injected chemicals with the stationary phase. Because the main purpose of the mobile phase in GC is to simply move solutes along the column, the mobile phase in this technique is often referred to as the **carrier gas**.[8]

Examples of common carrier gases used in GC are hydrogen (H_2), helium (He), nitrogen (N_2), and argon (Ar). All of these gases are relatively inexpensive, easy to obtain, and (with the exception of hydrogen) inert and safe to use. These gases are usually provided by a standard gas cylinder, but sometimes they are supplied by a gas generator connected to the GC system. For instance, a generator might be used to isolate nitrogen from air or to

---

BOX 21.1

Comparing Gas Chromatography Stationary Phases

The compounds shown in Table 21.1 are often used to compare and evaluate the retention properties of GC stationary phases. This approach makes use of values known as *McReynolds constants*.[23] The McReynolds constants for a GC stationary phase are determined by measuring the Kováts retention indices for model compounds on both the stationary phase of interest and on a nonpolar reference stationary phase (squalane) at the same temperature. The difference in *I* for each compound on the two stationary phases (Δ*I*) is then determined by using Equation 21.3.

$$\Delta I = I_{\text{Test Stationary Phase}} - I_{\text{Squalane}} \qquad (21.3)$$

The compounds in Table 21.1 have been selected for evaluating GC columns because these compounds represent chemicals with several possible types of intermolecular forces. Benzene is a general model for compounds that can mainly interact through dispersion forces with a stationary phase. 1-Nitropropane represents compounds that have dipole moments and strong dipole-related interactions. 1-Butanol can take part in hydrogen bonding and can act as either a proton acceptor or proton donor. 2-Pentanone contains a carbonyl group, which also allows it to take part in hydrogen bonding as a proton acceptor. The aromatic nitrogen in pyridine makes it act as a base that can also form hydrogen bonds and act as a proton acceptor.

It is common for manufacturers of GC stationary phases to give the values of the McReynolds constants when they describe their products. McReynolds constants also provide a way for scientists to compare different GC stationary phases and choose those that will work best in most applications. We will see an example of such a list later when we discuss stationary phases that are recommended for gas–liquid chromatography (Section 21.4B). By convention, the value of Δ*I* that is reported in these lists is referred to as *X*′ if it was measured for benzene, *Y*′ for 1-butanol, *Z*′ for 2-pentanone, *U*′ for 1-nitropropane, and *S*′ for pyridine. The average of these Δ*I* values is also sometimes reported.[16]

produce H_2 by passing an electrical current through water, breaking the water down into oxygen and hydrogen gas.

The carrier gas should always have high purity to avoid contamination or damage to the column and GC system. Impurities like water, oxygen, organic substances, and particulate matter can be removed by passing the carrier gas through a series of traps and filters before it enters the column. The carrier-gas source should also be equipped with regulators for pressure and flow-rate control. In some cases, it is necessary to use special devices to maintain a constant flow rate as the temperature or pressure of the system is varied. This is especially important when the column conditions are changed over time, as will be discussed in the next section.

21.3B Elution Methods in Gas Chromatography

The General Elution Problem.   We saw earlier that temperature is an important factor in determining how strongly a compound will be retained in a GC column. If the same temperature is used throughout this separation, this is called an **isothermal method** (where "isothermal" means "constant temperature").[8] An isothermal method works well if the sample is relatively simple or has only a few known compounds. This is generally used for samples containing only relatively volatile analytes (e.g., low-mass compounds with boiling points below 100°C).[16] The main strength of this approach is its simplicity. The fact that no cooling down or reequilibration period is needed between samples also helps minimize the time that elapses between sample injections.

Using the same conditions throughout a GC separation is convenient, but it creates a problem when one is working with complex samples. A complex sample will probably contain chemicals with a wide range of volatilities or interactions with the stationary phase. An example is shown in Figure 21.8 for the analysis of n-alkanes that have 10 to 18 carbons in their structure ($C_{10}-C_{18}$). Some of these chemicals ($C_{10}-C_{13}$) pass through the column quickly, creating peaks with low retention and making these peaks difficult to resolve. Other chemicals in the sample (e.g., C_{18}) are well resolved, but go through the column too slowly and give broad peaks with long elution times. We would ideally prefer for all of injected chemicals to come out between these extremes, like the peaks shown for $C_{14}-C_{17}$ which are well resolved, but also pass through the column in a reasonable amount of time.

It is often difficult in chromatography or in any other separation method to find a single set of conditions that can separate all the components of a complex sample with adequate resolution and in a reasonable amount of time. This difficulty is known as the **general elution problem**. One way of dealing with this problem is to vary the separation conditions during the analysis of a sample, giving an approach called **gradient elution** (or *gradient programming*).[8] A typical method using gradient elution

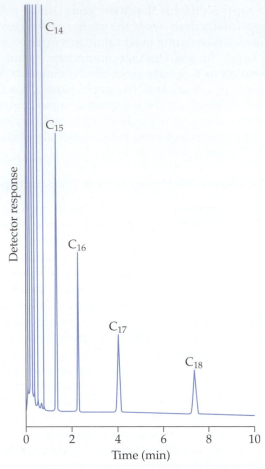

FIGURE 21.8   An example of the general elution problem as shown by the separation of the n-alkanes decane ($C_{10} = C_{10}H_{22}$) through octadecane ($C_{18} = C_{18}H_{38}$) at constant temperature.

(Reproduced with permission from S. Nygren, "Faster GC Analyses Performed by Flow Programming in Short Capillary Columns," *Journal of High Resolution Chromatography*, 2 (1979) 319–323.)

will begin with conditions that allow early eluting compounds to stay on the column longer, helping them to become better separated. The conditions are then changed over time to help other compounds also elute with good resolution and within a satisfactory amount of time.

Temperature Programming.   The most common way of performing gradient elution in GC is to vary the temperature of the column over time. This technique is known as **temperature programming**.[8] An example of temperature programming is given in Figure 21.9 for the same set of n-alkanes that were injected in Figure 21.8 under isothermal conditions. This method makes use of the known relationship between analyte retention and column temperature in GC, where an increase in temperature leads to a decrease in retention. This relationship can be used to improve the separation in Figure 21.8 by starting at a lower temperature so that the most volatile substances in the sample ($C_{10}-C_{13}$) are more strongly retained. The temperature can then be gradually increased to allow other analytes in the sample to elute

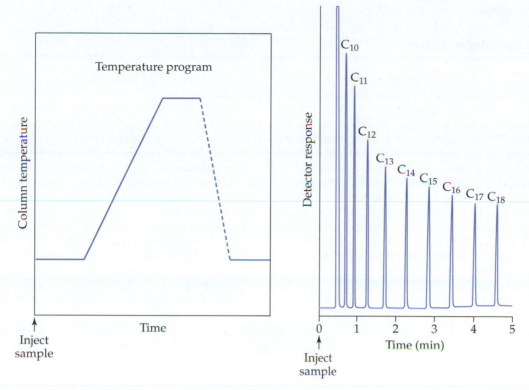

FIGURE 21.9 Separation of the *n*-alkanes decane ($C_{10} = C_{10}H_{22}$) through octadecane ($C_{18} = C_{18}H_{38}$) by temperature programming. The plot on the left shows the temperature program that was used and the plot on the right shows the resulting chromatogram.

(Reproduced with permission from S. Nygren, 'Faster GC Analyses Performed by Flow Programming in Short Capillary Columns," *Journal of High Resolution Chromatography*, 2 (1979) 319–323.)

with reasonable retention times while still being resolved from each other.

A temperature program usually begins with an initial isothermal step at a relatively low column temperature. It is during this step that the sample is injected, allowing the most volatile compounds to interact with the column and be separated. The next step in the temperature program is known as the temperature ramp; it is during this step that analytes with intermediate or high boiling points are eluted from the column. A linear change in temperature over time is generally used because of its simplicity and ability to elute compounds with a large range of volatilities. The rate at which the temperature is increased during the ramp will vary from one GC method to the next, but is often in the range of 1–30°C/min.[9] In some cases, a nonlinear change in temperature or a series of linear ramps can also be used.

After the temperature ramp has been completed, the third part of a temperature program is an isothermal step in which the temperature is held for some period of time at the upper limit of the ramp. This step is optional but useful in making sure that all analytes have time to elute and in ensuring that there are no low volatility substances that remain on the column from one sample

injection to the next. The fourth step in the temperature program is a cooling-down period in which the column is returned back to its initial temperature. It is important to allow sufficient time for this cooling process. If this is not done, the temperature of the GC column may be too high when the next sample is applied, meaning the most volatile compounds in the sample will travel too quickly through the column and will not be adequately separated. Box 21.2 shows how both isothermal elution and temperature programming have been used in important analytical applications of GC.

21.4 GAS CHROMATOGRAPHY SUPPORTS AND STATIONARY PHASES

21.4A Gas Chromatography Support Materials

Packed Columns. GC columns can be placed into one of two major categories based on the type of support that they employ: packed columns and open-tubular columns. A **packed column** is filled with small support particles that act as an adsorbent or that are coated with the desired stationary phase.[8,9] In GC a packed column is made up of a glass or metal tube that is usually 1–2 m long and a few millimeters in diameter.

BOX 21.2
Analytical Chemistry in Space

One application of GC has been its use in the study of space and other planets. For example, GC systems have recently been sent on a probe to study the atmosphere and surface of Titan and are planned for future robot probes to explore the surface of Mars.[24,25] GC systems were also present on the two U.S. Viking probes that landed on Mars in the late 1970s. The location of this device on the Viking lander is shown in Figure 21.10, which included a gas chromatograph and mass spectrometric detector. A robotic arm served as the surface sampler to acquire samples of Martian soil for analysis by the GC system, which then separated any volatile compounds in the sample prior to their detection and identification. The goal of this analysis was to determine if any organic compounds were present in Martian soil as a test for the possible existence of life on that planet.[26] Due to the large number of possible compounds that could have been present, temperature programming was employed as part of this analysis.

GC systems have also been included on probes sent by the United States and former Soviet Union to Venus.[27,28] For instance, in 1978 a Pioneer–Venus probe used a GC system to analyze the atmosphere of Venus as the probe descended toward the planet surface. A profile of the atmosphere was generated by analyzing the major gases at regular intervals during the probe's descent. Because this analysis involved a much smaller number of possible analytes than the Mars lander (major gases versus trace organic compounds), an isothermal elution method was employed. One benefit of using isothermal elution instead of temperature programming in this case was it allowed more rapid processing of samples during descent of the probe, because no reequilibration of column temperature was required at the end of each analysis.

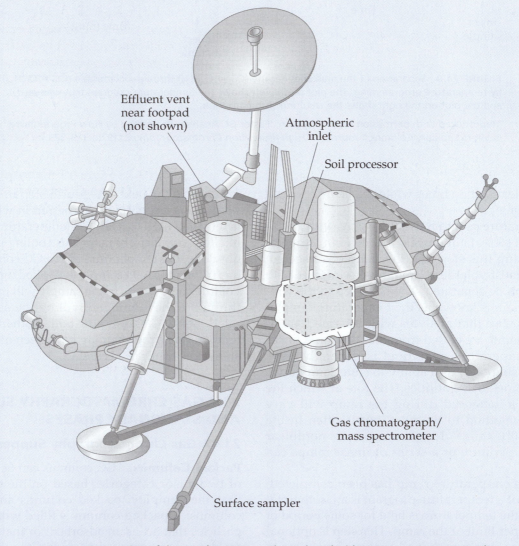

Effluent vent
near footpad
(not shown)

Atmospheric
inlet

Soil processor

Gas chromatograph/
mass spectrometer

Surface sampler

FIGURE 21.10 Location of the gas chromatograph, equipped with a mass spectrometer as the detector, on the Mars Viking lander.

(Adapted with permission from D. R. Rushneck *et al.*, "Viking Gas Chromatograph-Mass Spectrometer," *Reviews of Scientific Instrumentation*, 49 (1978) 817–834.)

Diatomaceous earth is a common support placed in packed GC columns.[16] This material is formed from fossilized diatoms (see Figure 21.11) and mainly consists of silicon dioxide or *silica* (empirical formula, SiO_2). Many other materials can also be used in packed GC columns. Diatomaceous earth and related supports are mainly used in *gas–liquid chromatography* (discussed in Section 21.4B), where they provide a surface onto which a stationary phase can be coated or attached.[8] Supports like molecular sieves or porous polymers have surfaces that can adsorb certain chemicals, allowing these materials to act as both the support and stationary phase within the GC system (a technique called *gas–solid chromatography*, which is discussed in Section 21.4B).[8]

Packed GC columns are useful when a large amount of sample must be separated. This is because the packed support particles have a big surface area that can be used with large amounts of a stationary phase. This feature makes it possible to inject relatively large samples or quantities of chemicals onto such columns. A disadvantage of packed columns is that they often have lower efficiencies than columns based on open-tubular supports (see discussion in the next section). As a result, packed columns are used in analytical methods only when a limited number of compounds must be separated.[9]

Open-Tubular Columns. An **open-tubular column** (or "capillary column") is a tube that has a stationary phase coated on or attached to its interior surface (see Figure 21.12).[8,9] This type of column in GC will generally have a length between 10 and 100 m and an inner

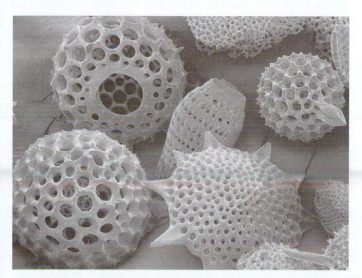

FIGURE 21.11 An image of microscopic organisms known as diatoms, whose fossilized remains make up the main component of diatomaceous earth. Diatomaceous earth is obtained from mines in the United States, Europe, Algeria, and the former Soviet Union. After being mined, this material is fused by heating, broken apart, and sorted according to particle size. This process gives a support for GC that is inexpensive and that contains small, rigid, and uniform particles. This support also has a high surface area, which allows it to be coated with a relatively large amount of stationary phase. (This image is magnified about 2500 times and is reproduced with the permission and courtesy of Dr. Nan Yao, Princeton University.)

diameter of 0.1 to 0.75 mm. A coating of polyimide is present on the outside of this column to give it better strength and flexibility for handling and storage. Open-tubular columns tend to have better efficiency and resolution, lower detection limits, and faster separations than packed columns. These properties make an open-tubular column the support of choice for most analytical applications of GC.

There are three types of open-tubular columns in GC based on how the stationary phase is placed in the column.[8] The first type is a *wall-coated open-tubular (WCOT) column*, in which a thin film of a liquid stationary phase is placed directly on the wall of the column. These columns are very efficient, but have a small sample capacity as a result of their low surface area. The second type of open-tubular column is a *support-coated open-tubular (SCOT) column*. A SCOT column has an interior wall that is coated with a thin layer of a particulate support, plus a thin film of a liquid stationary phase that is coated onto this support layer. This coating gives a SCOT column a thicker layer of stationary phase than a WCOT column, resulting in a less efficient column but one with a larger sample capacity. A *porous-layer open-tubular (PLOT) column* is the third type of open tubular column in GC. A PLOT column also contains a porous material that is deposited on the column's interior wall, but the surface of this material is now used directly as the stationary phase without any additional coating.[9] This makes PLOT columns useful in the method of gas–solid chromatography (see Section 21.4B).

Table 21.2 summarizes some important properties of open-tubular columns in GC. The material used to make the outer tube of most modern open-tubular columns is fused silica. This material gives a group of supports that are also known as *fused-silica open-tubular columns (FSOT)*.[9,16] One way these columns differ is in their inner diameters, which range from 0.10 to 0.75 mm. Open-tubular columns with small diameters have greater efficiencies and better resolving power than wider-bore columns. This is a result of the shorter distances analytes must diffuse in narrow capillaries as the analytes travel between the mobile phase and stationary phase. An advantage of wider bore open-tubular columns is they have a larger surface onto which the stationary phase can be coated and can contain a thicker coating of this stationary phase than small diameter columns. As a result, wide-bore columns have fewer problems with overloading when working with large samples.

21.4B Gas Chromatography Stationary Phases

Another major part of a GC column is the stationary phase. Three types of stationary phases are used in GC: (1) solid adsorbents, (2) liquids coated on solids, and (3) bonded phases.

Gas–Solid Chromatography. If a solid adsorbent is used as the stationary phase in GC, the resulting method is referred to as **gas–solid chromatography (GSC)**. This technique uses the same material as both

General dimensions

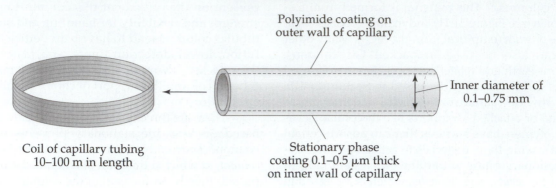

Coil of capillary tubing
10–100 m in length

Type of Stationary Phase

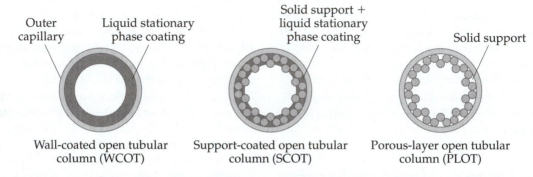

FIGURE 21.12   The top image shows typical dimensions of an open-tubular column in GC and the bottom image shows various ways in which the support and stationary phase can be placed within an open-tubular column.

the support and stationary phase, with retention occurring through the adsorption of analytes to the support's surface.[8] An example of a support for GSC is a *molecular sieve* (see Figure 21.13).[9,16] A molecular sieve is a porous material that is composed of a mixture of silica (SiO_2), alumina (Al_2O_3), water, and an oxide of an alkali or alkaline earth metal, such as sodium or calcium. When these are combined in a particular ratio, they produce a support with a series of pores with well-defined sizes and binding regions. The ability of an analyte to adsorb to this support will depend on the size of the analyte and the strength with which it interacts with the support's surface. Molecular sieves are useful in retaining such things as small hydrocarbons and gases like hydrogen, oxygen, carbon monoxide, and nitrogen.[9,16]

TABLE 21.2   Properties of Common Types of Gas Chromatography Columns*

Type of Column	Column Efficiency (plates/meter)	Maximum Sample Injection Volume (μL)
Packed columns		
2 mm inner diameter	2000	10
Open-tubular columns		
0.75 mm inner diameter (megabore)	1100	6
0.53 mm inner diameter (wide-bore)	1000	5
0.32 mm inner diameter (medium-bore)	3200	2
0.25 mm inner diameter (narrow-bore)	4000	1
0.10 mm inner diameter (ultranarrow-bore)	10000	0.5

*These results are for the following amounts of stationary phase: packed column, 5% w/w coating; 0.10 mm inner diameter open-tubular column, 0.2 μm coating; 0.25 mm inner diameter column, 1.0 μm coating; 0.32 mm inner diameter column, 1.0 μm coating; 0.53 mm inner diameter column, 5.0 μm thick coating; 0.75 mm inner diameter column, 5.0 μm thick coating.[13,18] The values in this table are approximate and will vary with column length, the amount of stationary phase, and type of stationary phase or analyte.

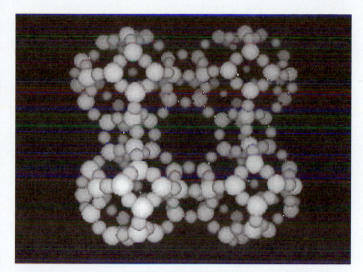

FIGURE 21.13 Structure of a molecular sieve (also known as a *zeolite*). The small circles show the positions of oxygen atoms, while the larger circles represent silicon or aluminum atoms. The arrangement of these atoms gives a material with a well-defined structure or pore size, which can be used to adsorb some gases and other chemical species. (Reproduced with permission from S. Burgmay, Bryn Mawr University.)

Other supports that can be used in GSC include organic polymers such as porous polystyrene and inorganic substances such as silica or alumina. These supports can be placed in either packed columns or in porous-layer open-tubular columns.[29] The extent to which an analyte will bind to these materials will be determined by the support's total surface area, the size of the support's pores, and the functional groups located on the support's surface. Increasing the surface area of a GSC support will increase the phase ratio and result in higher retention for analytes. Pore size is important because only compounds smaller than these pores will be able to contact the surface area within this space. The polarity of the support and its functional groups will also affect how analytes will bind to them. Nonpolar supports like porous polystyrene will have only weak intermolecular forces and be fairly nonselective in their binding. Polar supports like molecular sieves, silica, and alumina tend to have strong binding, particularly for compounds that are polar and that can form hydrogen bonds.[9,16]

Gas–Liquid Chromatography.   Another type of GC method is one in which a chemical coating or layer is placed onto the support and used as the stationary phase. This method is known as **gas–liquid chromatography (GLC)**, and it is the most common type of gas chromatography.[8] Table 21.3 lists several types of liquids that are used as stationary phases in GLC. All of these liquids have high boiling points and low volatilities, which allows them to stay within the column at the relatively high temperatures that are often used in GC for sample injection and elution. These liquids are also "wettable," which means they are easy to place onto a support in a thin, uniform layer.

Many of the stationary phases in Table 21.3 are based on a *polysiloxane*. A polysiloxane has the following general structure.[16]

$$\left[\begin{array}{c} R_1 \\ | \\ -Si-O- \\ | \\ R_2 \end{array}\right]_n \left[\begin{array}{c} R_3 \\ | \\ -Si-O- \\ | \\ R_4 \end{array}\right]_m$$

This structure consists of a backbone of silicon and oxygen atoms attached in long strings of Si—O—Si bonds. In GLC, the total size of these chains varies from molecular masses of a few thousand up to over a million,[16] which gives these polymers a low volatility. The remaining two bonds on each silicon atom are attached to side groups (R_1-R_4) that can have a variety of structures. These

TABLE 21.3   Recommended Stationary Phases for Gas–Liquid Chromatography

Stationary Phase[a]	Relative Polarity[b]
100% dimethylpolysiloxane	16 (**Nonpolar**)
5% phenyl–95% methylpolysiloxane	33
14% cyanopropylphenyl–86% methylpolysiloxane	67
50% phenyl–50% methylpolysiloxane	119
50% trifluoropropyl–50% methylpolysiloxane	146
50% cyanopropylmethyl–50% phenylmethylpolysiloxane	228
Polyethylene glycol	322 (**Polar**)

[a]These data are from the following references: R.L. Grob, *Modern Practice of Gas Chromatography*, 3rd ed., Wiley, New York, 1995; S.O. Falwell, "Modern Gas Chromatographic Instrumentation," In *Analytical Instrumentation Handbook*, G.W. Ewing, Ed., Marcel Dekker, New York, 1997, Chapter 23; H.M. McNair, "Method Development in Gas Chromatography," *LC-GC*, 11 (1993) 794–800.

[b]These rankings are based on the McReynolds constant for benzene (X'), which is used here as an overall measure of stationary phase polarity.

groups range from methyl groups (the least polar of these side chains) to cyanopropyl groups (the most polar). In some polysiloxanes only one type of side group is used throughout the entire chain, while in others, a mixture of two or more groups is employed. It is possible by altering the amount and type of these groups to produce stationary phases with a variety of polarities and specificities. This flexibility, plus their good temperature stability, has made polysiloxanes popular as stationary phases for GLC.

Bonded Phases. One difficulty with using a liquid stationary phase in GC is that even the most nonvolatile liquid will slowly vaporize or break apart and leave the column over time. This process is known as *column bleed*.[16,20] This loss of stationary phase will change the retention characteristics of the column. Column bleed can also cause some GC detectors to have a high background and a noisy signal as the stationary phase leaves the column and enters the detector. Various techniques are used to minimize column bleed. One approach is to use a stationary phase that is covalently attached to the support, resulting in a **bonded phase**.[16,20] A bonded phase can be produced by reacting groups on a polysiloxane stationary phase with *silanol groups* (general formula, —Si—OH) that are located on the surface of a silica support. A second approach for minimizing column bleed is to cross-link the stationary phase. This approach forms a *cross-linked stationary phase* that now has a larger and more thermally stable structure.[16,20] Either a bonded phase or cross-linked stationary phase will provide a more stable column that can be used at

higher temperatures than columns that contain a coated stationary phase. It is for this reason that bonded and cross-linked stationary phases are preferred in analytical applications of GC.

21.5 GAS CHROMATOGRAPHY DETECTORS AND SAMPLE HANDLING

21.5A Types of Gas Chromatography Detectors

The detector of a GC system is used to determine when something is eluting from the column, to measure the amount of this substance and, in some cases, to help identify it. The most common detectors that are employed in GC are listed in Table 21.4. Some of these detectors are general detectors that respond to a wide range of substances, while others are selective detectors that respond to only a particular set of compounds.

General Detectors. The **thermal conductivity detector (TCD)** can be used for both organic and inorganic compounds. A TCD measures the ability of the eluting carrier gas and analyte mixture to conduct heat away from a hot-wire filament, a property known as "thermal conductivity." This ability will vary as different analytes elute from the GC column, providing a means for both detecting and measuring these analytes.[9,16,20]

One common type of TCD is based on an electronic circuit known as a "Wheatstone bridge" (see Figure 21.14). A circuit consists of four resistors that are arranged in a parallel circuit, with two resistors being present in each side. When these resistors are properly balanced electronically, the voltage difference across the center of

TABLE 21.4 Properties of Common Gas Chromatography Detectors*

Detector Name	Compounds Detected	Detection Limits[a]
General detectors		
Thermal conductivity detector (TCD)	Universal—all compounds	10^{-9} g
Flame ionization detector (FID)	All organic compounds	10^{-12} g carbon
Selective detectors		
Nitrogen-phosphorus detector (NPD)	Nitrogen and phosphorus-containing compounds	10^{-14}–10^{-13} g Nitrogen 10^{-14}–10^{-13} g Phosphorus
Electron capture detector (ECD)	Compounds with electronegative groups	10^{-15}–10^{-13} g
Structure-specific detectors		
Mass spectrometry	Universal—full-scan mode	10^{-10}–10^{-9} g (Full-scan mode)
	Selective—SIM mode	10^{-12}–10^{-11} g (SIM mode)

*These data were obtained from the manufacturers of these detectors and the following references: B. Erickson, "Measuring Nitrogen and Phosphorus in the Presence of Hydrocarbons," *Analytical Chemistry*, 70 (1998) 599A–602A; D. Noble, "Electron Capture Detection for GC," *Analytical Chemistry*, 67 (1995) 439A–442A.

[a]The FID, NPD, and ECD are flow-sensitive detectors with detection limits best expressed in terms of mass of detected analyte per time (e.g., g/s); the detection limits for these devices given here are given for a one-second pulse of analyte so they may be given in units of g/s. The TCD is a concentration-sensitive detector with a detection limit that should be expressed as grams of detected analyte per unit volume; the preceding detection limit is for a 1 s interval at a carrier-gas flow rate of 6 mL/min or 0.1 mL/s.

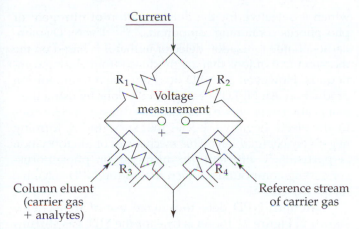

Current

R_1 R_2

Voltage
measurement

R_3 R_4

Column eluent
(carrier gas
+ analytes)

Reference stream
of carrier gas

FIGURE 21.14 A simple Wheatstone bridge circuit used in thermal conductivity detectors for GC. The symbols R_1, R_2, R_3, and R_4 represent the four resistors that are present in this circuit. The material in the exposed resisters is made up of a material such as tungsten or a tungsten-rhenium alloy that will change in its ability to carry a current as its temperature varies. This particular drawing shows a two-cell detector. A four-cell detector design can also be used in which the carrier gas/analyte mixture is passed over two of the resistors (e.g., R_2 and R_3) and the carrier-gas reference stream is allowed to pass over the other two resistors (R_1 and R_4).

TABLE 21.5	Thermal Conductivities of Representative Compounds and Carrier Gases	
	Thermal Conductivity (mW/m · K)[a]	Molar Mass (g/mol)
Inorganic compounds		
Hydrogen (H_2)	240	2.02
Helium (He)	193	4.00
Nitrogen (N_2)	34	28.01
Argon (Ar)	23	39.95
Oxygen (O_2)	37	32.00
Carbon Dioxide (CO_2)	29	44.01
Organic compounds		
Methane (CH_4)	54	16.04
Ethane (CH_3CH_3)	39	30.07
Propane ($CH_3CH_2CH_3$)	34	44.10
n-Butane ($CH_3CH_2CH_2CH_3$)	32	58.12
n-Pentane ($CH_3CH_2CH_2CH_2CH_3$)	29	72.15

[a]The listed values are for pure gases at a pressure of one atmosphere and at a temperature of 150°C. These conditions are roughly the same temperature and pressure conditions as are present within a typical thermal conductivity detector.

the circuit will be equal to zero. But if any of these resistors has changes in its electrical properties, a nonzero voltage will be produced. When this circuit is used in a TCD, at least two of the resistors consist of wire filaments that are exposed to the pure carrier gas or analyte/carrier-gas mixture that is eluting from the GC column. When a current is passed through the Wheatstone bridge, the resistors will begin to heat up. Some of this heat will be removed by the gas that surrounds the two exposed resistors. If the gases surrounding these two resistors are not the same (as will occur when analytes are eluting from the column), the amount of heat that is removed from the two resistors will differ. This causes the circuit to become unbalanced and gives an electrical signal that is related to the amount of analyte that is eluting from the column.

The carrier gas that is used with a TCD should have a thermal conductivity that is as different as possible from the thermal conductivities of any compounds that are to be detected. Table 21.5 shows the thermal conductivities for common GC carrier gases and a few representative analytes. Hydrogen and helium are often used with a TCD because they are the two carrier gases with the greatest difference in thermal conductivities versus most analytes. Helium is most commonly used with a TCD in the United States because of its safety; however, hydrogen is often used in other countries or in cases where helium is the analyte to be measured. The main advantage of a TCD is its ability to respond to any analyte that is different from the carrier gas and is present in a sufficient quantity to be measured. This makes a TCD valuable in detecting a large range of organic and inorganic compounds. A TCD is also nondestructive, which makes it possible to pass analytes onto a second type of detector for further analysis or to collect these analytes

after they exit the TCD. One disadvantage of a TCD is it will respond to impurities in the carrier gas, to a stationary phase that is bleeding from the column, or to air that is leaking into the GC system. A TCD can also be sensitive to changes in separation conditions, such as might occur during temperature programming. Another disadvantage of a TCD is it has a relatively poor lower limit of detection compared to other common GC detectors. As a result, the TCD is mostly used with relatively high concentration samples and for analytes that do not give a good response on other detectors, such as H_2, N_2, CO, H_2O, SO_2, NO_2, and CO_2.

The **flame ionization detector (FID)** is another type of general GC detector. A FID detects organic compounds by measuring their ability to produce ions when they are burned in a flame.[9,16,20] The design of a typical FID is shown in Figure 21.15. The flame in the FID is usually formed by burning the eluting compounds in a mixture of hydrogen and air. An additional amount of a "makeup gas" like nitrogen is sometimes combined with the carrier gas and hydrogen before entry into the flame; this is done to help provide a steady, optimum flow rate for detection. Positively charged ions produced by the flame are collected by a negative electrode that surrounds the flame. As these ions are produced they create a current at the electrode, thereby allowing the presence of the eluting compound to be detected.

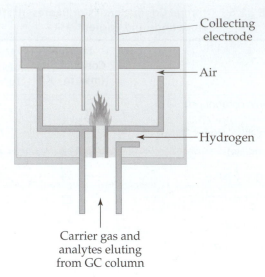

FIGURE 21.15 A flame ionization detector.

As the carrier gas and analytes are passed into a FID, they are combined with a stream of hydrogen and passed into the center of the flame while air is allowed to enter through the flame's exterior. This process creates a flame that has a hydrogen-rich area in the middle and a large amount of oxygen around its exterior. When organic analytes enter this flame, they initially encounter the hydrogen-rich region, where bonds are broken between carbon and most of the other atoms in organic compounds. This leads to the formation of CHO^+ ions that can undergo an acid–base reaction with water in the flame to produce H_3O^+. When these gas-phase H_3O^+ ions reach the collector electrode, they give a response that is related to the number of carbon atoms in the original analyte.

One advantage of using a FID for the analysis of organic compounds is that this detector gives little or no signal for many small inorganic compounds like He, Ar, and N_2 (typical carrier gases) or O_2, CO_2, and H_2O (common contaminants of carrier gases). This provides the FID with a low background signal and creates limits of detection for organic compounds that are 100- to 1000-fold lower than those that can be obtained with a TCD. The FID can also be used with temperature programming and has a large linear range, making it a good choice when sensitive detection is needed for the routine analysis of organic compounds. A disadvantage of the FID is it is a destructive detector that breaks down analytes during the process of their measurement. This prevents a FID from being connected directly to other types of detectors or techniques for compound analysis. It is possible, though, to split the carrier gas stream after the GC column so that part goes to a FID and the other part goes to a different detector.

Selective Detectors. A second group of GC detectors are those that are specific for a particular type of chemical. One example is the **nitrogen-phosphorus detector (NPD)**,

which is selective for the determination of nitrogen- or phosphorus-containing compounds.[9,16,20] The NPD is similar to a flame ionization detector in that it is based on the measurement of ions that are produced from eluting compounds. However, an NPD does not use a flame for ion production. An NPD instead generates ions by using thermal heating at or above a surface that can supply electrons to any electronegative species that surround it, forming negatively charged ions. This mechanism of ion formation is particularly efficient for nitrogen or phosphorus-containing compounds, which makes the NPD selective for such chemicals.

Modern NPD detectors make use of the scheme shown in Figure 21.16. As is true for the FID, gas streams of both hydrogen and air are mixed together with the analytes and carrier gas that are eluting from the GC column. But in an NPD the flow rate of hydrogen is kept at a level that is too slow to produce a self-sustaining flame. What happens instead is the hydrogen and analytes are passed over an electrically heated bead that contains rubidium. The heated surface of this bead causes some analytes and hydrogen to dissociate into carbon radicals and free hydrogen atoms. Nitrogen- or phosphorus-containing molecules that enter this heated region tend to break apart to form $\cdot CN$ or $\cdot PO_2$ radicals. These radicals contain unpaired electrons and are highly electronegative, so they tend to acquire additional electrons from the heated bead. The exact way in which this process happens is not totally clear, but it is known that this process results in negative ions like CN^- or PO_2^-. These ions are then collected at an electrode and create a current that is used to detect and measure an eluting analyte.

The greatest strength of the NPD is its good selectivity and low limits of detection for nitrogen- and phosphorus-containing compounds. In fact, the NPD provides the lowest available detection limits for the GC analysis of such substances. Like a FID, the NPD does not detect many common carrier gases or impurities. When

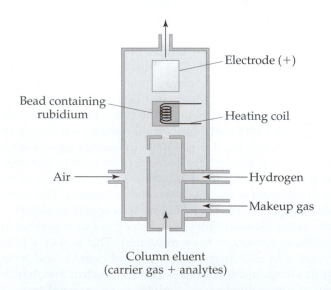

FIGURE 21.16 A nitrogen–phosphorus detector.

using an NPD, it is necessary to periodically change the heated material, because this material will slowly degrade over time. It is also important to carefully control heating of the bead and the flow rate of hydrogen through the NPD to obtain a stable signal for this detector. Several carrier gases can be used with an NPD, but the heating conditions and hydrogen flow rate need to be optimized for each of these mobile phases.

The **electron capture detector (ECD)** is another selective detector used in GC. This device detects compounds that have electronegative atoms or groups in their structure, such as halogen atoms (I, Br, Cl, and F) and nitro groups ($-NO_2$).[9,16,20] An ECD can also be used to detect polynuclear aromatic compounds, anhydrides, and conjugated carbonyl compounds, among others. The design of a typical ECD is shown in Figure 21.17. This detects compounds based on the capture of electrons by electronegative atoms or groups in the molecule. These electrons are produced by a radioactive source, such as ^3H or ^{63}Ni. Both of these sources emit beta particles (high-energy electrons) as part of their decay process. As these particles are released, some of them will collide with the carrier gas. When these collisions occur, the carrier gas takes on some of the particle's energy and creates the release of a large number of secondary electrons at lower energies.

When no analytes are eluting from the GC column, only the carrier gas is entering the detector and a steady stream of secondary electrons is produced. These electrons are passed onto a positive collector electrode and produce a measurable current. When an analyte that contains electronegative atoms elutes from the column, the atoms in this compound will capture some of the secondary electrons and will reduce the number of these electrons that reach the collector electrode. This decrease in current allows the compound to be detected.

Compounds that contain multiple halogen atoms are particularly good at capturing these secondary electrons, which gives them low limits of detection on an ECD. Substances with only one halogen atom or other types of electronegative groups will also give a response. One disadvantage of the ECD is it has a fairly narrow linear range. Another disadvantage is that a radioactive source is required. Using ^3H produces a larger number of β-particles than ^{63}Ni, but ^{63}Ni is less troublesome to work with as a radioactive hazard. Argon or nitrogen are the carrier gases of choice for this detector due to their large size versus hydrogen and helium, which makes it easier for argon and nitrogen to collide with the emitted beta particles. A small amount of methane is also sometimes included in the carrier gas to maintain the production of secondary electrons and to create a stable detector response.

Gas Chromatography/Mass Spectrometry. Another common detector used in GC is a mass spectrometer, a device we first saw in Chapter 3. The resulting combination is known as **gas chromatography/mass spectrometry (GC/MS)** and is a powerful tool for both measuring and identifying analytes as they elute from a GC column.[16] This method first converts a portion of the eluting analytes into gas-phase ions that can be separated and detected. This process may involve simply putting a charge on the original analyte, giving a "molecular ion," or it may involve breaking this molecular ion into smaller pieces known as "fragment ions" (see Chapter 3). The amount (or intensity) of ions that are produced by a chemical in mass spectrometry can be used in the measurement of a chemical. The chemical can also be identified by looking at the mass of the molecular ion (which is related to the molecular weight of the original analyte) or the types of fragment ions that are produced (which are determined by the structure of the analyte). The overall pattern of ions can also be compared to the patterns of reference compounds to identify an unknown substance.

Figure 21.18 shows a typical mass spectrometer that is used in GC/MS. Analytes that enter this device first enter an ionization chamber in which some of their molecules are converted into ions. The creation of ions is accomplished in this example by *electron-impact ionization* (*EI*), in which the eluting analytes pass through a beam of high-energy electrons (typically 70 eV). These electrons bombard some of the analyte molecules (M), causing an electron to be removed and forming a molecular ion ($M^{+\bullet}$). The molecular ion can then undergo a rearrangement or decomposition to give fragment ions.[30] Another type of ionization method that can be used in mass spectrometry is *chemical ionization* (*CI*), which makes use of a gas-phase acid to protonate the analyte and form a molecular ion with general structure MH^+. CI is a gentler ionization process than EI, often giving a larger signal for the molecular ion and fewer fragment ions.[30]

After some analyte molecules have been converted into gas-phase ions, these ions are separated from each

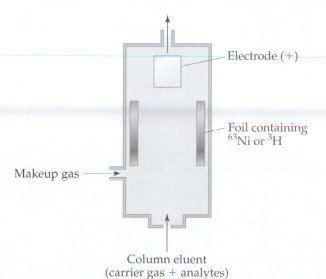

FIGURE 21.17 An electron-capture detector. Argon or nitrogen is the preferred carrier gas for this detector. This preference is due to the large size of these gases vs. hydrogen and helium, which makes it easier for argon and nitrogen to collide with the emitted beta particles.

Electrode (+)

Foil containing ^{63}Ni or ^3H

Makeup gas

Column eluent (carrier gas + analytes)

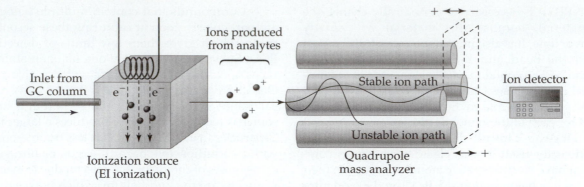

FIGURE 21.18 A transmission quadrupole mass spectrometer with an electron-impact ionization source for use in gas chromatography/mass spectrometry (GC/MS). Other mass spectrometers that can be used in GC/MS include ion traps, time-of-flight instruments, and magnetic-sector instruments.

other based on their mass-to-charge ratios. In Figure 21.18, this separation occurs in a *quadrupole mass analyzer*. This device uses a series of four parallel rods that are held at well-defined AC and DC potentials. Each opposing pair of rods is held at the same potential and any two neighboring rods are held at exactly opposite potentials. Over time, the potentials on these rods are continuously varied. This creates a varying electric field through which only ions with a certain mass-to-charge ratio will be able to pass. If the potentials on the rods are altered over time, ions with different mass-to-charge ratios can be collected and measured.[30]

There are several ways of viewing the information obtained in GC/MS. The first way is to use a *mass spectrum*.

As we saw in Chapter 3, a mass spectrum is a plot of the intensity of each ion that is detected over a range of mass-to-charge ratios. In the case of GC/MS, this plot is made for the data that are collected at a particular retention time (generally 50–600 atomic mass units). This type of plot can be used to identify a chemical in a chromatographic peak. The second type of graph that is employed in GC/MS is a *mass chromatogram*, which is a plot of the number of ions that are measured at each elution time. A mass chromatogram can be constructed by using the intensities measured for all measured ions (see Figure 21.19) or it can be prepared using only ions with particular mass-to-charge ratios. It is this last property that allows GC/MS to be utilized as a general detector or selective detector.

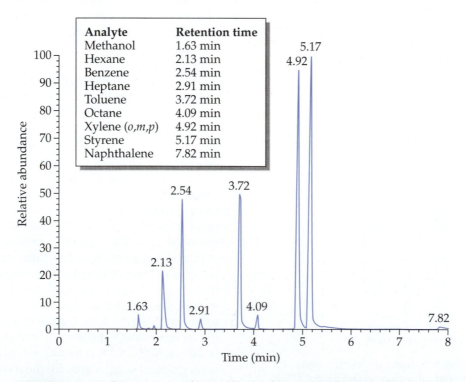

FIGURE 21.19 A chromatogram obtained by gas chromatography/mass spectrommetry (GC/MS) for a VOC standard. This sample contained 10 ppb of 11 different volatile organic compounds. (Reproduced with permission from S. Basiaga, University of Nebraska, Lincoln.)

The use of GC/MS as a general detector by collecting information on a wide range of ions is known as the *full-scan mode* of GC/MS. The full-scan mode is useful for looking at a broad range of compounds in a single analysis or for determining the identity of an unknown compound from its mass spectrum. The use of GC/MS to collect information on only a few ions is called *selected ion monitoring (SIM)*. In this approach, only a few ions characteristic of the compounds of interest are examined. SIM is employed when low detection limits are desired and when it is known in advance what compounds are to be analyzed and what types of ions they produce in the mass spectrometer. SIM provides detection limits that are roughly 100- to 1000-times lower than the full-scan mode because it monitors ions at only a few mass-to-charge ratios instead of ions covering a wide mass range. This difference allows SIM to spend more time collecting a signal for ions that are related to a particular analyte, which helps improve the limit of detection for the analyte while also avoiding interferences from other compounds in the sample by providing a means for more specific detection.

are injected separately onto this GC/MS system (see Figure 21.20). From these results, explain how we could modify this GC/MS method to perform a selective analysis for styrene.

SOLUTION

This problem requires that SIM be used. The mass spectrum for styrene contains the largest number of ions at a mass-to-charge ratio of 104 (the molecular ion), with a reasonable number of ions also being produced at a mass-to-charge ratio of 78 (a fragment ion). The mass spectrum for *o*-xylene has only a small number of ions at these mass-to-charge ratios. Thus, all that is required to modify this method is to set up the mass analyzer so that it now only monitors ions with mass-to-charge ratios of 104 (the first choice) or 78 (the second choice). Although information on other compounds in the sample will now be lost, this change should produce a technique that is more selective and has a lower detection limit for styrene.

EXERCISE 21.4 | **Selected Ion Monitoring in Gas Chromatography/Mass Spectrometry**

An analyst wishes to modify the GC/MS method in Figure 21.19 to create a selective method for for styrene that can be used in the presence of *o*-xylene. Mass spectra are collected for each of these two compounds when they

21.5B Sample Injection and Pretreatment

Gas Samples. Because GC requires that analytes enter the gas phase for analysis, gaseous analytes and samples are natural candidates for this technique. If an analyte is present as a gas at moderate-to-high concentrations, it should be possible to directly sample and inject this gas

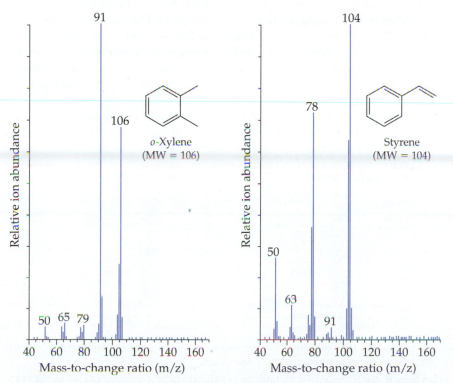

FIGURE 21.20 Mass spectra for separate samples of *o*-xylene and styrene, as obtained by gas chromatography/mass spectrometry (GC/MS) using electron impact ionization.

Sample loading position

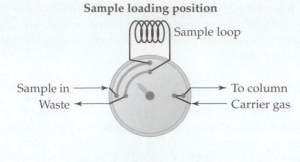

Sample loop

Sample in →

Waste ←

→ To column

← Carrier gas

Sample injection position

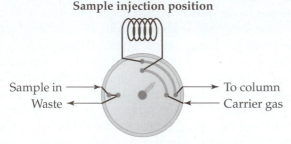

Sample in →

Waste ←

→ To column

← Carrier gas

FIGURE 21.21 A sampling valve for the injection of gas samples onto a gas chromatography (GC) system. The paths that the sample and carrier gas take through the valve are shown for both the sample loading position and the sample injection position of the valve.

onto a GC system. This approach can be performed by passing the sample through a gas-tight valve like the one shown in Figure 21.21. Instead of using this type of valve, a gas-tight syringe could also be used to inject a known volume of gas into a GC system.

For trace components of gases like the volatile organic compounds (VOCs) that are found in air, it is often necessary to collect and concentrate these analytes before they are examined by GC. One way this collection and concentration can be performed is by passing a large volume of the gas sample through a solid-phase extraction cartridge or through a liquid into which the analytes will dissolve. Another way of collecting analytes in a gas is to use a *cold trap* (or "cryogenic trap"). This second approach involves passing the gas through a hollow tube or coil that is kept at a low temperature. As the analytes enter this tube or coil, they will condense and be collected. After these

substances have been collected, they can be released and passed into the GC system by raising the temperature.

Liquid Samples. Liquid samples are frequently used in GC. For example, GC has been used to analyze VOCs in the blood of people who are routinely exposed to air pollution. Figure 21.22 shows various ways in which a liquid sample might be placed into a GC system. When using columns that can handle relatively large sample volumes (6–10 μL for packed columns or wide-bore open-tubular columns), *direct injection* can be used for a liquid sample.[20] Direct injection uses a calibrated microsyringe to apply the desired volume of sample to the system. The microsyringe is passed through a gas-tight septum and into a heated chamber where the liquid sample and its contents are vaporized and swept by the carrier gas into the column. This approach allows essentially 100% transfer of the analytes into the column. The main limitation of direct injections is that they cannot be used with many open-tubular columns, which often require smaller sample volumes than those that can be accurately delivered with a microsyringe.

Split injection is a common way of overcoming this last problem. In split injection a microsyringe is again used to inject a liquid sample into a GC system. As the sample is converted into a gas in the injection area, its vapors are divided so that only a small portion (0.01%–10%) goes into the column (the remainder going to a side vent).[20] This procedure allows the volume of sample that reaches the column to be greatly reduced. The result is a lower chance of overloading the column with the sample, especially for narrow-bore or moderate-bore open-tubular columns. Split injection works well for relatively concentrated samples, but can be difficult to use for trace analysis, because much of the sample is discarded during the injection process. Another problem with split injection is that analytes with different volatilities may not have equal fractions that enter the column. This effect can create a variable recovery for analytes and can affect their measurement.

Splitless injection is a type of direct injection that is specifically performed on a system that can also be used for split injection. This approach is again carried out with

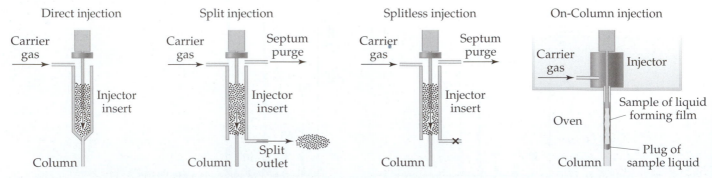

FIGURE 21.22 Common gas chromatography (GC) injection techniques for liquid samples. The dotted regions in each diagram represent the injected sample components after they have entered the vapor phase.

(Reproduced with permission from K. Grob, "Injection Techniques in Capillary GC," *Analytical Chemistry*, 66 (1994) 1009A–1019A.)

microsyringes and open-tubular columns, but the side vent of the injection system is now kept closed so that most of the injected and vaporized sample will go into the column.[2] After the analytes have entered the column, the side vent is opened to flush out any undesired vapors from the injection chamber before the next sample is applied. Because little sample is lost during its injection, this approach is better to use than split injection when performing a trace analysis. However, splitless injection does have greater difficulty with overloading the column by a sample.

Another way of applying a liquid sample to a GC system is *cold on-column injection*. In this technique a microsyringe containing the sample is passed through the "injector" and directly into the column or an uncoated precolumn. The region around the syringe is initially kept cool so that the sample can be deposited in the column as a liquid film. This liquid is later heated, causing the analytes to enter the carrier gas to begin their passage through the column.[20] Because the sample is applied directly to the column or precolumn in this method, essentially 100% of each analyte is available for measurement. This approach can have problems if there are any low-volatility substances in the sample, which will accumulate at the top of the column or precolumn and alter the properties of the GC system over time. In this situation, splitless injection is preferred because it allows such components to be removed between injections.

It is quite common for some pretreatment to be required before a liquid sample can be analyzed by GC. In Section 21.2A we saw how derivatization is often used to convert analytes into a more volatile or thermally stable form. This is one type of sample pretreatment. Another common type of pretreatment is when analytes are moved from their original sample into a solvent that is more suitable for injection onto a GC system. In particular, water and aqueous samples (such as blood) should *not* be injected directly onto most GC columns. Water binds strongly to many GC columns and can create problems with the long-term behavior and reproducibility of these columns. In addition, the water may contain dissolved solids, salts, or other nonvolatile compounds that are not suitable for injection onto a GC system.

To avoid these problems with water, aqueous samples are often pretreated before GC analysis by using liquid–liquid extraction, solid-phase extraction, or solid-phase microextraction (methods discussed in Chapter 20). Both liquid–liquid extraction and solid-phase extraction are used to transfer analytes from water to a more volatile solvent and to help remove the analytes from nonvolatile substances in the sample. These methods can also be used with solvent evaporation to reduce the final volume of the extract and increase the concentration of analyte in this extract for easier detection.[14] In solid-phase microextraction, analytes can be extracted from an aqueous sample by using a coated fiber that is later placed directly into

the GC system. This approach again removes analytes from water or nonvolatile substances in the sample. Solid-phase microextraction also reduces column overloading effects because little or no liquid is injected along with the analytes.

Headspace analysis is another technique that makes it possible to avoid the introduction of water and nonvolatile compounds into a GC system. Headspace analysis is based on the fact that volatile analytes in a liquid or solid sample will also be present in the vapor phase (or "headspace") that is located above the sample. If a portion of this headspace is collected, it can be used to measure volatile analytes without interference from other, less volatile compounds that were in the original sample.[20] This approach is often used to examine VOCs in water.

Headspace analysis can be carried out in two ways: the static method and the dynamic method (see Figure 21.23). In the static method, the sample is placed in an enclosed container and its contents are allowed to distribute between the sample and its vapor phase. After equilibrium has been reached, a portion of the vapor phase is collected and injected onto a GC system. In the dynamic method (also called the "purge-and-trap" technique), an inert gas is passed through the sample to carry away volatile compounds. This gas is then passed through a cold trap or solid adsorbent to collect and concentrate these volatile solutes for analysis. Although the dynamic technique requires more time and effort than the static method, it is more reproducible and allows better detection of compounds with relatively low volatilities.

Solid Samples. GC can also be adapted for work with chemicals that are adsorbed or held within solid samples. An example of this would be the use of GC to measure VOCs that are adsorbed by soil. A common way of handling this type of analysis is to first extract the compounds of interest from the solid material. This can be accomplished by using liquid–liquid extraction or supercritical fluid extraction (see Chapter 20). The extracted analytes are then placed into an organic solvent and treated as liquid samples, as described in the last section.

It is also possible to analyze some solids without performing an extraction. For instance, headspace analysis can be used to examine volatile compounds that are present in a solid. For less volatile substances, the method of *thermal desorption* can be employed. This method involves placing a known quantity of the solid into a chamber where the solid can be heated. As this solid is heated, its volatile components will enter the gas phase, allowing them to be trapped or placed into a GC system for testing.

The general makeup of a solid can also be examined by using a method known as *pyrolysis gas chromatography* (or *pyrolysis GC*).[16,20] This technique is useful for solid substances like plastics and polymers that are not volatile and cannot be easily derivatized into a volatile form. Pyrolysis GC involves heating a solid sample in a

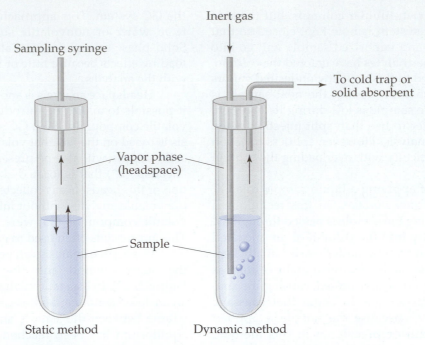

Sampling syringe

Inert gas

To cold trap or
solid absorbent

Vapor phase
(headspace)

Sample

Static method

Dynamic method

FIGURE 21.23   The static and dynamic methods of headspace analysis.
(Reproduced with permission from J. V. Hinshaw, "Headspace Sampling," *LC-GC*,
8 (1990) 362–368.)

controlled fashion to break the solid apart into smaller, more volatile chemical fragments. When these more volatile chemicals are then separated by the GC column, the result is a special type of gas chromatogram known as a *pyrogram*. The pattern of eluting chemicals in the pyrogram forms a chemical "fingerprint" for the substance that is being tested. This feature makes it possible to use a pyrogram to characterize the structure of the original solid or to identify this solid by comparing it with known standard samples.

Key Words

Bonded phase *522*
Carrier gas *515*
Derivatization *511*
Electron capture
    detector *525*
Flame ionization
    detector *523*

Gas chromatography *507*
Gas chromatography/
    mass spectrometry *525*
Gas–liquid
    chromatography *521*
Gas–solid
    chromatography *519*

General elution
    problem *516*
Gradient elution *516*
Headspace analysis *529*
Isothermal method *516*
Kováts retention index *512*
Nitrogen-phosphorus
    detector *524*

Open-tubular column *519*
Packed column *517*
Temperature
    programming *516*
Thermal conductivity
    detector *522*

Other Terms

Chemical ionization *525*
Cold on-column
    injection *529*
Cold trap *528*
Column bleed *522*
Column oven *508*
Cross-linked stationary
    phase *522*
Diatomaceous earth *519*
Direct injection *528*

Electron-impact
    ionization *525*
Full-scan mode
    (of GC/MS) *527*
Fused-silica open-tubular
    column *519*
Gas chromatograph *508*
Gas chromatogram *508*
Gradient programming *516*
Homologs *512*

Mass chromatogram *526*
McReynolds constants *515*
Molecular sieve *520*
Porous-layer open-tubular
    column *519*
Polysiloxane *521*
Pyrogram *530*
Pyrolysis gas
    chromatography *529*
Quadrupole mass
    analyzer *526*

Selected ion monitoring *527*
Silanol group *522*
Silica *519*
Split injection *528*
Splitless injection *528*
Support-coated open-
    tubular column *519*
Thermal desorption *529*
TMS-Derivative *512*
Wall-coated open-tubular
    column *519*

Questions

WHAT IS GAS CHROMATOGRAPHY AND HOW IS IT PERFORMED?

1. What is "gas chromatography" and how is it used in chemical analysis?
2. Name each of the main components of a basic GC chromatograph. Explain the function of each component.
3. Define each of the following terms: (a) gas chromatogram, (b) gas chromatograph, and (c) column oven. How is each of these used in gas chromatography?
4. Discuss how GC can be used for chemical separation and measurement.
5. The amount of morphine in a blood sample is determined by GC. First, a 2 mL sample of blood is adjusted to pH 9.9 and 20 μg of nalorphine is added as an internal standard. The morphine and nalorphine are then extracted into a toluene:hexane:isoamyl alcohol mixture, evaporated to dryness, and derivatized with N-methyl-bis-trifluoroacetamide (MBTFA). Out of a total volume of approximately 40 μL, 1 μL is then injected onto a GC system and analyzed. The following results are obtained for the sample and a series of standards. What was the concentration of morphine in the unknown sample?

Concentration (ng/mL)	Relative Peak Area	
	MBTFA-Morphine	MBTFA-Nalorphine
0	0.1	56.0
20	15.2	49.2
40	32.5	52.4
60	50.1	54.1
80	65.3	52.3
Unknown	45.8	53.9

6. The following data were obtained on a GC/MS system using toluene as a VOC standard and a fixed amount of deuterated toluene as an internal standard.

Toluene Concentration (ppb)	Relative Peak Height	
	Toluene	Deuterated Toluene
0	12	233
100	76	228
200	167	247
300	250	249
400	291	218

(a) An unknown sample that is injected under the same conditions as the standards gives a relative peak height of 178 for toluene and 229 for deuterated toluene. What is the concentration of toluene in this sample?
(b) The detector that is used in this GC system has a response factor of 0.71 of isopentane vs. toluene. If a sample gives a relative peak height of 88 for isopentane and 238 for deuterated toluene, what is the approximate concentration of isopentane in this sample?

REQUIREMENTS FOR THE ANALYTE

7. What general characteristics should a compound have if it is to be analyzed by gas chromatography?
8. Which of the following chemicals should be suitable for analysis by GC? Explain your answers. (*Note:* bp is the boiling point at 1 atm.)
   (a) Toluene (C_7H_8, bp 110.6°C)
   (b) Naphthalene ($C_{10}H_8$, bp 218°C)
   (c) Decanoic acid ($C_9H_{19}CO_2H$, bp 270°C)
   (d) Melene ($C_{30}H_{60}$, bp 380°C)
9. If the chemicals in Problem 8 were injected onto a nonpolar GC column, what would be their expected elution order based on volatility?
10. What is meant by the term "thermal stability"? Why is thermal stability important in GC?
11. Discuss how it can be determined experimentally whether a chemical has sufficient thermal stability for analysis by GC. What are some general ways in which problems with thermal stability can be overcome in GC?
12. What is meant by chemical "derivatization"? What are some reasons for the use of chemical derivatization in GC?
13. What is a "TMS derivative"? How can this make a compound more suitable for GC?
14. The structure of dicamba, a common herbicide, is shown below. Draw a reaction that shows how this compound would react with trimethylchlorosilane when this reagent is used to derivatize dicamba for analysis by GC. Explain how this reaction changes the structure of dicamba so that it is now easier to examine by gas chromatography.

Dicamba

15. The following reaction is for the derivatization of a carboxylic acid with methanol in the presence of BF_3. The result is a derivative known as a "methyl ester."

Explain why this type of derivatization might be useful in GC in terms of the changes it produces in volatility or thermal stability.

FACTORS THAT DETERMINE RETENTION IN GAS CHROMATOGRAPHY

16. Explain why chemical volatility is important in GC. In general, how does an increase in volatility affect the retention of a chemical in GC?
17. What is meant by the term "homologs"? What kind of behavior is expected when a mixture of homologs is injected onto a GC system?

18. What is the expected elution order in GC for each of the following homologous compounds?
    (a) An alcohol sample that contains 1-butanol ($CH_3(CH_2)_3OH$), ethanol (CH_3CH_2OH), methanol (CH_3OH), and 1-propanol ($CH_3(CH_2)_2OH$)
    (b) Diesel fuel that contains benzene (C_6H_6), ethylbenzene ($C_6H_5CH_2CH_3$), and toluene ($C_6H_5CH_3$)
    (c) A sample of butter that contains butyric acid ($CH_3(CH_2)_2CO_2H$) and caproic acid ($CH_3(CH_2)_4CO_2H$, also known as hexanoic acid)

19. How does temperature affect the retention of compounds on GC columns? What happens to the retention of a chemical in GC as the temperature is increased?

20. What is the Kováts retention index? How is it calculated? How is it used in GC?

21. Injection of ethane, propane, butane, pentane, hexane, heptane, and octane on a squalene column under isothermal conditions gives retention times of 3.51, 4.98, 7.30, 11.00, 16.85, 26.12, and 40.8 min, respectively. Air, which is nonretained, elutes at 1.00 min. An injection of benzene, butanol, nitropropane, pyridine, and 2-pentanone under the same conditions gives retention times of 20.00, 6.50, 4.10, 10.05, and 10.35 min. What are the Kováts retention indexes for benzene, butanol, nitropropane, pyridine, and 2-pentanone under these conditions?

22. A new GC column that is used under the same conditions as in Problem 21 gives retention times for the n-alkanes ethane (C_2H_6) through octane (C_8H_{18}) that are equal to 3.00, 3.82, 4.98, 6.62, 8.94, 12.22, and 16.80 min. Air, which is nonretained by this column, elutes at 1.00 min under these conditions. The injection of benzene, butanol, nitropropane, pyridine, and 2-pentanone gives retention times of 13.05, 5.30, 4.10, 6.60, and 14.10 min, respectively.
    (a) What are the Kováts retention indexes for benzene, butanol, nitropropane, pyridine, and 2-pentanone on this new column?
    (b) How do the Kováts retention indexes calculated for this new GC column compare to those obtained in Problem 21 for a GC column that contains squalane as the stationary phase? What information does this tell you about the retention properties of this new column?

23. A useful feature of the Kováts retention index it that a plot of I or $\log(t'_R)$ versus the number of carbon atoms in a chain for a series of homologs gives a linear relationship under isothermal conditions. Based on such a trend, what are the expected adjusted retention times and retention times for nonane (C_9H_{20}) and decane ($C_{10}H_{22}$) when using the same column and conditions that were employed in Problem 22?

COLUMN EFFICIENCY IN GAS CHROMATOGRAPHY

24. Explain how the low density and low viscosity of a gas both help to provide high efficiency in gas chromatography.

25. Explain why columns in GC can often be made much longer or used at higher flow rates than columns used in liquid chromatography.

26. Using the information provided in Table 21.2, estimate the total number of theoretical plates that would be expected for each of the following GC columns. What would be the approximate plate heights for these same columns?
    (a) A 2 m long and 2 mm inner-diameter packed column
    (b) A 20 m long and 0.32 mm inner diameter open-tubular column
    (c) A 50 m long and 0.10 mm inner diameter open-tubular column

27. A chromatogram for the analysis of volatile organic compounds in unleaded gasoline using a 100 m long column gave a separation that had 400,000 theoretical plates. Based on this information, estimate the baseline widths that would be expected for peaks due to n-pentane ($t_R = 10$ min), toluene ($t_R = 33$ min) and 1,2,4-trimethylbenzene ($t_R = 58$ min) in this separation. You may assume that each peak has a Gaussian shape.

COMMON MOBILE PHASES IN GAS CHROMATOGRAPHY

28. What is the role of the mobile phase in GC?

29. What is meant by the term "carrier gas"? What carrier gases are often used in GC?

30. Why is it important to use high-purity gases in GC? What types of impurities can be found in the carrier gas?

31. Why are pressure and flow-rate control important for the carrier gas in GC?

ELUTION METHODS IN GAS CHROMATOGRAPHY

32. In GC, what is meant by the term "isothermal method"? What are the advantages of an isothermal method?

33. What is the "general elution problem"? Explain how this problem can affect the separation of a complex sample mixture.

34. What is "gradient elution"? How does gradient elution help solve the general elution problem?

35. What is "temperature programming"? Explain why temperature programming can be used in GC to help solve the general elution problem.

36. Describe the typical steps in a temperature program for GC. Explain the purpose for each of these steps.

GAS CHROMATOGRAPHY SUPPORT MATERIALS

37. What is a "packed column"? What are typical dimensions of a packed column in GC?

38. What is diatomaceous earth? How is this used in GC?

39. Under what circumstances is it helpful to use a packed column in GC? What are the disadvantages of packed columns in GC?

40. What is an "open-tubular column"? What are the typical dimensions of an open-tubular column in GC?

41. Compare and contrast each of the following items.
    (a) Wall-coated open-tubular column
    (b) Support-coated open-tubular column
    (c) Porous-layer open-tubular column

42. What is a "fused-silica open-tubular column"? How is this used in GC?

GAS CHROMATOGRAPHY STATIONARY PHASES

43. What is "gas-solid chromatography"? What general types of stationary phases are used in gas–solid chromatography?

44. What is a "molecular sieve"? How are molecular sieves used in gas–solid chromatography?

45. What is "gas–liquid chromatography"? What general types of stationary phases are used in this method?

46. What is the general structure of a polysiloxane? Why are polysiloxanes useful as stationary phases in gas chromatography?

47. What is "column bleed"? What causes column bleed in GC?

48. Define each of the following terms. How are these items used to overcome column bleed in GC?
    (a) Bonded phase
    (b) Silanol group
    (c) Cross-linked stationary phase

TYPES OF GAS CHROMATOGRAPHY DETECTORS

49. Describe the way in which each of the following general GC detectors produces a signal. What properties of an analyte will determine its ability to produce a signal on each of these detectors?
    (a) Thermal conductivity detector
    (b) Flame ionization detector
50. Describe the way in which each of the following selective GC detectors produces a signal. What properties of an analyte will determine its ability to produce a signal on each of these detectors?
    (a) Nitrogen-phosphorus detector
    (b) Electron-capture detector
51. Which of the following compounds would you expect to provide a good response on a nitrogen-phosphorus detector?
    (a) Parathio
    (b) Imipramine
    (c) Ethylene glycol

Parathion

Imipramine

Ethylene glycol

52. Which of the following compounds would you expect to provide a good response on an electron capture detector?
    (a) DDT
    (b) Tetrahydrofuran
    (c) Freon-12

DDT

Tetrahydrofuran

Freon-12

53. What is "gas chromatography/mass spectrometry"? Explain why this approach can be used for either the general or selective detection of analytes.
54. What is a molecular ion? What is a fragment ion? What information is provided by each of these different types of ions in mass spectrometry?
55. What is "electron-impact ionization"? What is "chemical ionization"? How are each of these techniques used in mass spectrometry?
56. Explain how a quadrupole mass spectrometer can be used to separate ions with different mass-to-charge ratios.
57. Define or describe each of the following terms as related to GC/MS.
    (a) Mass spectrum
    (b) Mass chromatogram
    (c) Full-scan mode
    (d) Selected ion monitoring
58. Which common GC detector would you recommend for each of the following situations? Explain the reasons for each choice.
    (a) The analysis of urine samples at the Olympics to determine if any athletes are taking performance-enhancing drugs
    (b) The trace analysis of polychlorinated biphenyls in water samples
    (c) Determination of oxygen, carbon dioxide, and water as semi-trace components in a sample of helium

SAMPLE INJECTION AND PRETREATMENT

59. Describe how gaseous samples are typically injected or placed into a GC system.
60. Describe each of the following approaches for the injection of liquids into a GC system. What are the advantages and disadvantages for each of these approaches?
    (a) Direct injection
    (b) Split injection
    (c) Splitless injection
    (d) On-column injection
61. Why does the direct injection of water-based samples create problems with many GC columns? Discuss how this problem can be avoided through the use of an extraction.
62. What is "headspace analysis"? How is this used in GC?
63. Describe the dynamic and static methods of headspace analysis.
64. Explain how the contents of a solid sample can be analyzed by GC by using an extraction.
65. Define each of the following items and describe how they are used in GC.
    (a) Thermal desorption
    (b) Pyrolysis gas chromatography
    (c) Pyrogram
66. Discuss some of the reasons behind each of the following selections of sample injection or pretreatment methods.
    (a) In EPA method 505, GC is used for the analysis of chlorinated pesticides in drinking water following extraction of the original sample with hexane.
    (b) A company uses pyrolysis-GC to characterize the degree of branching that occurs within a preparation of the high-density polyethylene (HDPE).
    (c) Headspace analysis is used by a police laboratory for the determination of blood-alcohol levels.
    (d) Endrin (used in the past as an insecticide) tends to rearrange to form endrin aldehyde or endrin ketone when

it is applied to a GC system by splitless injection, but does not do so when PTV or cold on-column injection is used.

CHALLENGE PROBLEMS

67. Look up the boiling points (if available) and molar masses for each of the following compounds, using resources such as the *CRC Handbook of Chemistry and Physics* or the *Merck Index*.[31,32] Use this information to determine whether each of these chemicals can be examined by GC without derivatization.
    (a) Glucose
    (b) Quinoline
    (c) Erythromycin
    (d) Insulin
    (e) Triheptylamine
    (f) Coumarin

68. Some data on the thermal stability of chemicals can also be obtained from resources like the *CRC Handbook of Chemistry and Physics* and the *Merck Index*.[31,32] For example, both References 31 and 32 indicate whether a compound is thermally unstable by providing an abbreviation like "d," "dec," or "decomp" next to the information given on the compound's melting points and/or boiling points. Based on this information, determine which of the following chemicals have low thermal stabilities and give the approximate temperatures at which they begin to decompose.
    (a) Benzene
    (b) Glycine
    (c) Folic acid
    (d) *tert*-Butylamine
    (e) Tetradecane
    (f) Pyruvic acid

69. The following retention times were measured for 1-octanol, 2,4-dimethylaniline, and *n*-dodecane by GC on a column where the void time was 0.55 min.[33]

Temperature (°C)	1-Octanol	2,4-Dimethylaniline	*n*-Dodecane
80	8.50	13.74	17.16
90	5.57	8.87	10.56
100	3.86	5.92	6.86
110	2.79	4.27	4.72
120	2.10	3.12	3.31
130	1.60	2.37	2.46

Retention Time (min)

(a) Make a plot of the retention time versus column temperature T (in °C) for each of these compounds. What observations can you make from these graphs?
(b) Make a second series of graphs in which the logarithm of each compound's adjusted retention time, or $\log (t'_R)$, is plotted versus $1/T$ (now expressed in kelvin). What observations can you make from this new group of graphs?
(c) Based on the results in Part (b), discuss the relative advantages of using $\log(t'_R)$ instead of t_R as measure of retention when calculating a Kováts retention index.

70. The same compounds listed in Table 21.1 were used to examine a commercial column that contained a liquid stationary phase with the tradename "OV-17." This column gave Kováts retention indices of 772 for benzene, 748 for 1-butanol,

789 for 2-pentanone, 895 for 1-nitropropane, and 901 for pyridine. What are the McReynolds constants for OV-17?

71. A catalog lists the following data for some commercial GC columns. Based on this information, rank these columns in terms of (a) their overall polarity, and (b) their ability to retain compounds like benzene, butanol, pyridine, nitropropane, and 2-pentanone. Which columns are the most similar in your rankings? Which have the greatest differences? Explain how you arrived at your answers.

	McReynolds Constants				
Type of Column	*X'*	*Y'*	*Z'*	*U'*	*S'*
Diglycerol	371	826	560	676	854
Ethofat 60/40	191	382	244	380	333
Igepal CO-630	192	381	353	382	344
OV-1	16	55	44	65	42

72. The proper selection of a stationary phase for GC will depend on factors such as the physical nature and chemical properties of the analytes, the different chemical interactions of these analytes, and the temperatures to be used during the separation. For instance, gas–solid chromatography is generally the method of choice when working with very volatile analytes that occur as gases at room temperature, while GLC tends to work better for less volatile analytes. Explain each of the following choices of stationary phases for GC.
    (a) A petroleum chemist uses a PLOT column containing a polystyrene/divinyl benzene porous polymer for the analysis of $C_1 - C_{10}$ saturated hydrocarbons (methane through *n*-decane).
    (b) An OV-1 column (which contains 100% dimethylpolysiloxane as the stationary phase) is employed for the analysis of volatile organic compounds in air.
    (c) A chemist switches from a DB-5 column (which contains 5% phenyl-95% methylpolysiloxane as the stationary phase) to a Carbowax 20M column for analyzing a mixture of alcohols like methanol, ethanol, *n*-propanol, and 2-propanol.

73. Although the carrier gas does not affect solute retention, the choice of which gas to use as the mobile phase is important in other ways to a GC method. Factors that need to be considered in choosing a carrier gas include the type of detector that is being used and possible risks or hazards involved in using the gas.
    (a) Discuss the relative risks that would be associated with the use of hydrogen, helium, nitrogen, or argon as carrier gases. What types of chemical or physical hazards are associated with each of these gases?
    (b) What detectors discussed in this chapter can be used with each of the following gases: hydrogen, helium, nitrogen, or argon? Explain how this is related to the principles behind the operation of these GC detectors.
    (c) One advantage of using a low-mass carrier gas like hydrogen or helium is that this produces much faster rates of compound diffusion than heavier carrier gases, such as nitrogen or argon. How do you think the use of a low-mass carrier gas will affect the efficiency of a GC separation?

TOPICS FOR REPORTS AND DISCUSSION

74. The speed and efficiency of GC have made this technique popular for chemical separation and analysis in a large number of fields.[9,16] Obtain information on how GC is employed in one or more of the following areas.
   (a) Clinical chemistry
   (b) Environmental science
   (c) Forensic science
   (d) Petroleum chemistry
   (e) Polymer science
   (f) Food science

75. Contact an analytical laboratory in your area that makes use of gas chromatography. Discuss how GC is used in this laboratory and describe the types of analytes and samples it is used to examine.

76. For several years there has been interest in developing GC columns with temperature limits above those of most current GC columns. This has led to an area known as *high-temperature gas chromatography* (HTGC).[16] Look up information on the types of columns that are used in this technique. What advantages can be gained by having a column with a higher upper temperature limit? What are some applications of HTGC?

77. In *cryogenic gas chromatography*, the GC system is modified so that it can be used to perform separations that are far below room temperature.[16] What specific modifications are made in the GC system in order for this approach to be used? What types of analytes do you think would be best analyzed under these low-temperature conditions? How might the use of a lower temperature affect the types of stationary phases that can be used in the GC method? What types of stationary phases are most often used in such separations?

78. Beside temperature programming, another type of gradient elution that can be used in GC is *flow programming* (also called *pressure programming*).[8,9,16] Obtain information on this technique and discuss how it is performed. What are the advantages of this method compared to temperature programming? What are its disadvantages? Give some examples of its applications.

79. Besides the detectors that have been mentioned in this chapter, there are many other types of monitoring devices that can be used with GC columns. A few examples are listed below. Find information on one or more of these detectors. Report on the principles behind the detector's operation and compare it to other GC detectors discussed in this text. State what types of compounds can be monitored by the detector and discuss for some its typical applications in chemical analysis.[9,13,16]
   (a) Electrolytic-conductivity detector
   (b) Flame photometric detector
   (c) Photoionization detector
   (d) Thermionic emission detector
   (e) Gas density balance
   (f) Sulfur chemiluminescence detector

80. The earliest commercial systems for gas chromatography first appeared approximately 50 years ago.[34] Obtain more information on the history of GC instrumentation and write a report that summarizes developments that have occurred in this field.

81. The area of *high-speed gas chromatography* (or *fast GC*) is an area of ongoing research.[35] Describe this technique and how it is performed. What are the advantages of this approach? What are its disadvantages? How does its instrumentation differ from that used in normal GC? What are some applications of this method?

References

1. S. Enriquez, "Mexico City a Living Laboratory for Smog Study," *Los Angeles Times*, March 31, 2006, A.20.
2. J. M. Lentz and W. J. Kelley, "Clearing the Air in Los Angeles," *Scientific American*, 268 (1993) 32–39.
3. D. Cyranoski and I. Fuyuno, "Climatologists Seek Clear View of Asia's Smog," *Nature*, 434 (2005) 128.
4. S. Sillman, "Photochemical Smog: Ozone and Its Precursors," *Handbook of Weather, Climate and Water*, 1 (2003) 227–242.
5. P. B. Kelter, J. D. Carr and A. Scott, *Chemistry: A World of Choices*, WCB/McGraw-Hill, St. Louis, MO, 1999, Chapter 12.
6. *Photochemical Smog: Contribution of Volatile Organic Compounds*, Organization for Economic Co-operation and Development (OECD), Paris, 1982.
7. *Photochemical Oxidants*, World Health Organization, Switzerland, Geneva, 1979.
8. J. Inczedy, T. Lengyel, and A. M. Ure, *Compendium of Analytical Nomenclature*, 3rd ed., Blackwell Science, Malden, MA, 1997.
9. S. O. Falwell, "Modern Gas Chromatographic Instrumentation," In *Analytical Instrumentation Handbook*, G. W. Ewing, Ed., Marcel Dekker, New York, 1997, Chapter 23.
10. H. A. Laitinen and G. W. Ewing, *A History of Analytical Chemistry*, Maple Press, New York, 1977, Chapter 5.
11. L. S. Grinstein, R. K. Rose, and M. H. Rafailovich, *Women in Chemistry and Physics: A Biobibliographic Sourcebook*, Greenwood Press, Westport, CT, (1993), 129–135.
12. J. V. Hinshaw, "Handling Fast Peaks," *LC-GC*, 19 (2001) 1136–1140.
13. R. L. Grob, *Modern Practice of Gas Chromatography*, 3rd ed., Wiley, New York, 1995.
14. H. M. McNair, "Method Development in Gas Chromatography," *LC-GC*, 11 (1993) 794–800.
15. J. Drozd, *Chemical Derivatization in Gas Chromatography*, Elsevier, Amsterdam, the Netherlands, 1981.
16. C. F. Poole and S. K. Poole, *Chromatography Today*, Elsevier, Amsterdam, the Netherlands, 1991, Chapter 8.
17. W. J. A. VandenHeuvel, "Some Aspects of the Chemistry of Gas-Liquid Chromatography," In *Gas Chromatography of Steroids in Biological Fluids*, M. B. Lipsett Ed., Plenum Press, New York, (1965), 277–295.
18. E. sz. Kováts, "Gas Chromatographic Characterization of Organic Substances in the Retention Index System," *Advances in Chromatography*, Vol. 1, J. C. Giddings and R. A. Keller, Eds., Marcel Dekker, New York, 1965, Chapter 7.

19. L. S. Ettre, "Retention Index Systems. Its Utilization for Substance Identification and Liquid-Phase Characterization," *Chromatographia*, 6 (1973) 489–495.

20. J. V. Hinshaw, "A Compendium of GC Terms and Techniques," *LC-GC*, 10 (1992) 516–522.

21. W. J. A. VandenHeuvel, "Some Aspects of the Chemistry of Gas-Liquid Chromatography," In *Gas Chromatography of Steroids in Biological Fluids*, M. B. Lipsett, Ed., Plenum Press, New York, (1965), 277–295.

22. J. F. Sprouse and A. Varano, "Development of a GC Retention Index Library," *American Laboratory*, 16 (1984) 54–68.

23. W. O. McReynolds, "Characterization of Some Liquid Phases," *Journal of Chromatographic Science*, 8 (1970) 685–691.

24. S. O. Akapo, J. M. Dimandja, D. R. Kojiro, J. R. Valentin, and G. C. Carle, "Gas Chromatography in Space," *Journal of Chromatography A*, 843 (1999) 147–162.

25. M. C. Pietrogrande, M. G. Zampolli, F. Dondi, C. Szopa, R. Sternberg, A. Buch, and F. Raulin, "In Situ Analysis of the Martial Soil by Gas Chromatography: Decoding of Complex Chromatograms of Organic Molecules of Exobiological Interest," *Journal of Chromatography A*, 1071 (2005) 255–261.

26. D. R. Rushneck et al., "Viking Gas Chromatograph-Mass Spectrometer," *Reviews of Scientific Instrumentation*, 49 (1978) 817–834.

27. V. I. Oyama, G. C. Carle, F. Woeller, J. B. Pollack, R. T. Reynolds, and R. A. Craig, "Pioneer Venus Gas Chromatography of the Lower Atmosphere of Venus," *Journal of Geophysical Research*, 85 (1980) 7891–7901.

28. B. G. Gel'man, Y. V. Drozdov, V. V. Mel'nikov, V. A. Rotin, V. N. Khokhlov, V. B. Bondarev, G. G. Dol'nikov, A. V. D'yachkov, D. F. Nenarokov, L. M. Mukhin, N. V. Porshnev, and A. A. Fursov, "Chemical Analysis of Aerosol in the Venusian Cloud Layer by Reaction Gas Chromatography on Board the Vega Landers," *Document NASA-TM-88421*, National Aeronautics and Space Administration, Washington, DC, (1986), 1–6.

29. Z. Ji and R. E. Majors, "Porous-Layer Open-Tubular Capillary GC Columns and Their Applications," *LC-GC*, 16 (1998) 620–632.

30. D. A. Skoog, E. J. Holler, and T. A. Nieman, *Principles of Instrumental Analysis*, 5th ed., Harcourt Brace, Philadelphia, PA, 1998.

31. D. R. Lide, Ed., *CRC Handbook of Chemistry and Physics*, 83rd ed., CRC Press, Boca Raton, FL, 2002.

32. M. Windholz, Ed., *The Merck Index*, 10th ed., Merck & Co., Rahway, NJ, 1983.

33. J. V. Hinshaw, "Optimizing Column Temperature," *LC-GC*, 9 (1991) 94–98.

34. L. S. Ettre, "Fifty Years of GC Instrumentation," *LC-GC Europe*, 18 (2005) 416–421.

35. R. Sacks, H. Smith and M. Nowak, "High-Speed Gas Chromatography," *Analytical Chemistry*, 70 (1998) 29A–37A.

Chapter 22

Liquid Chromatography

Chapter Outline

22.1 INTRODUCTION: BATTLING A MODERN EPIDEMIC

In 2006, the U.S. health community looked back on 25 years of a battle it had fought with a modern epidemic. This battle began in 1981 when the U.S. Center for Disease Control noted an unusual number of fatal infections that were occurring in young men from the Los Angeles area. This disease was given the name "acquired immunodeficiency syndrome" or AIDS. Scientists soon realized that AIDS was linked to exposure to blood that was infected with the human immunodeficiency virus (HIV). By 1986 there was still no known cure for this disease. A large-scale effort was then undertaken to find a treatment.[1,2]

The first drug available for fighting AIDS was 3'-azido-3'-deoxythymidine, or AZT (see Figure 22.1). AZT is an analog of thymidine, one of the four nucleotides that make up DNA. When it is phosphorylated by the body, AZT inhibits replication of the HIV virus by blocking reverse transcriptase, an enzyme needed for production of the virus. AZT was first tested for use in 1985 and was approved for use in AIDS patients only two years later by the U.S. Food and Drug Administration. Although AIDS is still not curable, AZT and related drugs have helped convert this disease from one that was fatal to one that is now treatable.[3]

The development process for drugs like AZT involves a large amount of chemical analysis. For instance, measurements must be made during animal and human studies to determine the effects and appropriate doses for a new drug. Analytical techniques are also utilized by drug manufacturers to monitor the quality of their products once a drug is ready to go to market. One method that is frequently used during drug development and production is *liquid chromatography*. In this chapter we will learn about this technique and see how it is employed for the separation and analysis of chemicals.

22.1A What Is Liquid Chromatography?

Liquid chromatography (LC) is a chromatographic technique in which the mobile phase is a liquid.[4-7] This is the type of chromatography that was originally developed by Russian botanist Mikhail Tswett in 1903 (see Chapter 20). Tswett conducted this work using an apparatus similar to that shown in Figure 22.2, in which he used gravity to pass a liquid mobile phase and a mixture of plant pigments through a packed column.[8] This general approach is still used to this day for the purification of chemicals from a variety of samples. Although liquid chromatography did not become a major analytical tool until the 1960s, it is now the main type of chromatography found in analytical laboratories.[9] The popularity of LC is largely due to the ability of this method to work directly with liquid samples, which makes it valuable in such areas as food testing, environmental testing, and biotechnology.[6,7,10,11]

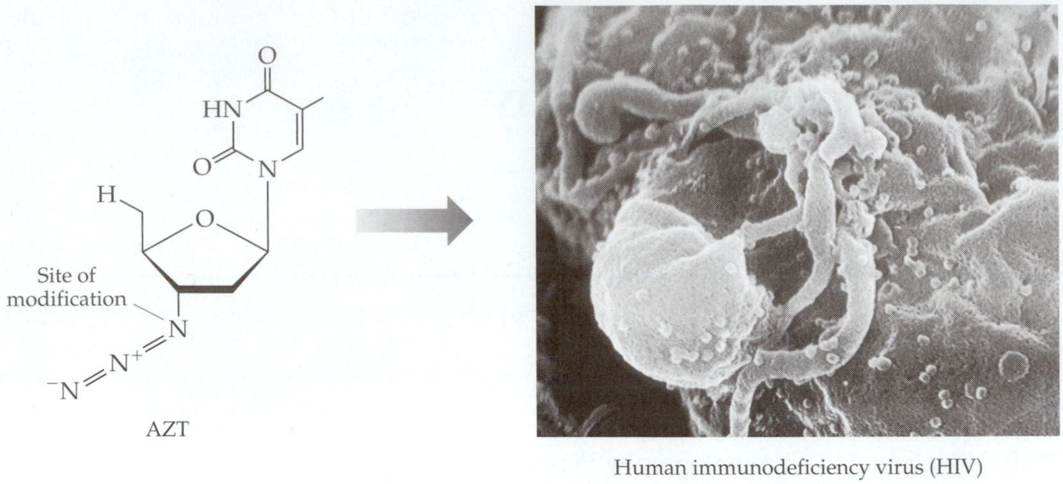

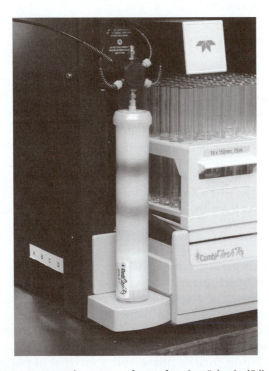

Human immunodeficiency virus (HIV)

FIGURE 22.1   Structure of AZT and a photo of the HIV virus. The location on AZT that is modified from the structure of thymidine is indicated by the label on the left. When phosphorylated, AZT acts on the enzyme reverse transcriptase, preventing the replication of HIV. The image on the right is a scanning electron microscope image of HIV-1 budding from a cultured lymphocyte (courtesy of the Centers for Disease Control).

22.1B How Is Liquid Chromatography Performed?

Liquid chromatography can be carried out in many ways, but most modern analytical laboratories use an LC system like the one shown in Figure 22.3. This system, which is used to do LC (known as a *liquid chromatograph*), typically includes a support and stationary phase enclosed in a column and a liquid mobile phase that is delivered to the column by means of a pump. For analytical applications, an injection device is used to apply samples to the column, while a detector monitors and measures analytes as they leave the column. A collection device can also be placed after the column to capture analytes as they elute.

An example of a modern LC separation is also shown in Figure 22.3. A plot of these results is made by showing the response of the detector as a function of the amount of time or volume of the mobile phase that is required for sample components to leave the column. This plot is known in LC as a *liquid chromatogram*. The retention time or volume of a peak in this plot can be employed to help identify a sample component, while the size of the peak can be related to the amount of this component in the sample.

FIGURE 22.2   A modern system for performing "classical" liquid chromatography and a typical separation obtained with such a system for a series of colored compounds that are applied to a silica column. Because of its simple system requirements and low cost, classical liquid chromatography is a popular tool for the purification of chemicals. This approach is the basis of the technique "flash chromatography" that is used by many modern synthetic chemists and that makes use of a column such as the one shown in this figure. (This figure is used with permission and courtesy of Teledyne Isco, Inc., and shows results obtained with RediSep Rf Gold™ silica column on a Combiflash® flash chromatography system.)

EXERCISE 22.1	Qualitative and Quantitative Analysis using Liquid Chromatography

A chemist working in a quality-control laboratory wishes to employ liquid chromatography to monitor the amount of caffeine in a pharmaceutical product. Describe how this chemist might use the system in Figure 22.3 to help identify the presence of caffeine and to measure the amount of caffeine in this product.

SOLUTION

The LC system and conditions that are used in Figure 22.3 give a peak for caffeine that appears at a retention time of approximately 6 min. This retention time provides a means for qualitatively determining if caffeine is in the injected sample. The amount of caffeine in the sample can be determined by using a calibration curve to compare the height or area for this particular peak to the values that are obtained for the injection of standards containing known amounts of caffeine. For best results, a fixed amount of an internal standard with similar chemical and physical properties to caffeine (but which can be detected independently of this analyte) should also be included in the standards and samples. This internal standard would be used to correct for variations that occur during the pretreatment or injection of samples and standards, as discussed in Chapter 10.

22.2 FACTORS THAT AFFECT LIQUID CHROMATOGRAPHY

Like gas chromatography (GC), liquid chromatography requires both a difference in retention and good efficiency for it to separate two given chemicals. Although LC and GC have many things in common, they also have differences in terms of their sample and analyte requirements, their formats, and the role played by the mobile phase in these methods.

22.2A Requirements for the Analyte

The first requirement that must be met before we can examine a chemical by liquid chromatography is that it must be possible to place this chemical into a liquid that can be injected onto the column. Fulfilling this requirement is usually not a problem because many chemicals are soluble in some liquid. This property has made liquid chromatography popular for the separation of biological compounds, polymers, and other chemicals that cannot be easily examined by GC or other methods. The use of a liquid as the mobile phase also allows LC to be performed at a much lower temperature than is typically used in GC, which makes LC better suited for thermally unstable compounds.

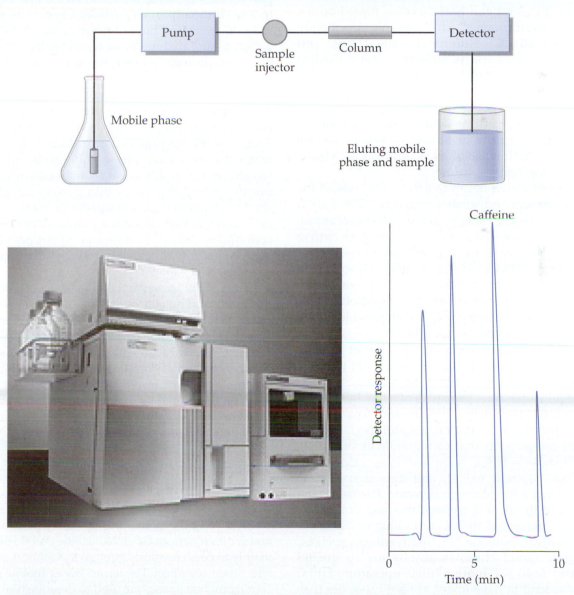

FIGURE 22.3   A modern system for performing chemical analysis by liquid chromatography and a typical separation obtained with such a system. This system is an example of high-performance liquid chromatography. (The separation shown is based on data from Alltech; the photo of the commercial high-performance liquid chromatography (HPLC) system was provided by Waters.)

A second requirement in liquid chromatography is there must be a difference in retention between the analytes to be separated. We learned in Chapter 21 that retention in GC can be altered by adjusting the temperature and type of stationary phase within the column. In LC retention can also be altered by changing the mobile phase. This difference is due to the higher density of liquids versus gases, which means that solute retention in LC will depend on the interactions of sample components with both the mobile phase and stationary phase. This feature makes LC more flexible than GC when optimizing and controlling the degree of retention that is obtained in a separation.

22.2B Column Efficiency in Liquid Chromatography

Because of the smaller plate numbers that are found in LC columns, a greater emphasis is often placed on the efficiency or "performance" for these columns. Before the mid-1960s all LC columns contained large and irregularly shaped supports similar to those used in packed GC columns. Although these supports were useful in separating some chemicals by LC, they often gave broad peaks and poor separations. These supports also had limited mechanical stability and could only be used under gravity flow or at low pressures. In this chapter we refer to the use of such supports as "classic liquid chromatography."

In the 1960s a shift began to occur in LC toward the use of more efficient and smaller supports.[9] This shift to more efficient supports, along with the inclusion of appropriate instrumentation to use such materials, gave birth to the modern technique of **high-performance liquid chromatography (HPLC)**.[4,9-11] An example of a separation that was produced by HPLC was given earlier in Figure 22.3. The presence of a more efficient support in this method produces narrower peaks, which provides better separations and lower limits of detection.

One consequence of using smaller support particles in HPLC is the need for greater pressures to pass the mobile phase through the column. Most modern HPLC columns require operating pressures of a few hundred to a few thousand pounds-per-square inch (psi, where 1 atm = 14.7 psi). These conditions require special pumps and system components that can operate at such pressures. The sample in HPLC is applied using a closed system (e.g., an injection valve), and detection is often performed using a flowthrough detector. The fast analysis times, good limits of detection, and ease of automation of HPLC make this the LC technique of choice for most analytical applications.[10,11] Disadvantages of HPLC versus classic liquid chromatography are its greater expense and need for more skilled operators. HPLC columns also tend to have lower sample capacities than those for classic liquid chromatography.

Since the original development of HPLC, there has been an ongoing effort to create even more efficient supports for this method. This trend is demonstrated in Table 22.1, where these supports have shown a consistent shift to smaller sizes.[9] These small supports help create faster mass transfer which, in turn, produces smaller plate heights, larger plate numbers, and narrower peaks. An important factor that limits the extent to which these supports can be reduced in size is the corresponding increase in pressure that is needed to pass the mobile phase through the column. Most modern HPLC systems can work at pressures up to 5000–6000 psi. Specialized systems that can operate at even higher pressures have recently been developed for work in a method known as *ultrahigh pressure liquid chromatography*.[12,13]

There have been many types of particles and support formats explored for use in HPLC (see Figure 22.4).[9] In a traditional porous particle, the mobile phase typically flows around but not through the particle. This situation means analytes must travel within the particle by means of diffusion, a relatively slow process that leads to significant band-broadening. One way of improving this process is to use *perfusion particles*, which contain larger pores that allow the mobile phase to pass both through and around the support particles.[4,9,14] This type of flow decreases the average distance solutes must diffuse to reach the stationary phase and results in less band-broadening. A similar effect is obtained with a *nonporous support* or one that has a thin porous layer or porous shell.[4,9] Another way of improving mass transfer is to use a column that contains a porous support that is one continuous bed. This latter type of support is made by using a specially prepared porous polymer and is known as a *monolithic column*.[15,16]

All of the LC separations we have considered up to this point have been based on "column chromatography," in which the support is held within an enclosed system such as a tube (see Chapter 20).[4,5] It is also possible in LC to conduct a separation in which the stationary phase is placed onto a flat surface to perform "planar chromatography" (see Box 22.1).[4,5] In this chapter, we focus on the principles and analytical applications of column chromatography, but many of these same principles also apply to planar liquid chromatography.

22.2C Role of the Mobile Phase in Liquid Chromatography

In LC the retention of solutes will depend on interactions involving *both* the mobile and stationary phases. This situation is different from GC, in which the mobile phase is used mainly to pass chemicals through the column and does not affect their retention. The terms "weak mobile phase" and "strong mobile phase" are used to describe how solutes are retained on an LC column in the presence of a given liquid.

TABLE 22.1 Change in Liquid Chromatography Supports Over the Last 50 Years*

Year(s) of Acceptance	Particle Size	Most Popular Nominal Size (μm)	Plates/15 cm (approx.)
1950s	Irregular-shaped	100	200
1967	Glass bead	50 (pellicular)[a]	1,000
1972		10	6,000
1985		5	12,000
1992		3–3.5	22,000
1998[b]		1.5 (pellicular)[a, b]	30,000
1999		5.0 (Poroshell)	8,000[c]
2000		2.5	25,000
2003		1.8	32,500

*Reproduced with permission from R. E. Majors, *American Laboratory*, October (2003) 46–54.
[a]The term "pellicular" refers to a particle that has a solid core and porous outer layer.
[b]Nonporous silica or resins.
[c]For protein MW 5700.

A **strong mobile phase** in LC is a pure solvent or a solution that quickly elutes a retained analyte from a column. This situation is created when the analyte favors staying in the mobile phase versus the stationary phase, as occurs if the analyte is more soluble in the mobile phase or has only limited interactions with the stationary phase. A **weak mobile phase** in LC is a liquid that slowly elutes a retained analyte. This second situation occurs when the analyte has better solubility in the stationary phase than the mobile phase or when the mobile phase promotes good interactions of the solute with the stationary phase. We will see several examples of strong or weak mobile phases later in this chapter.

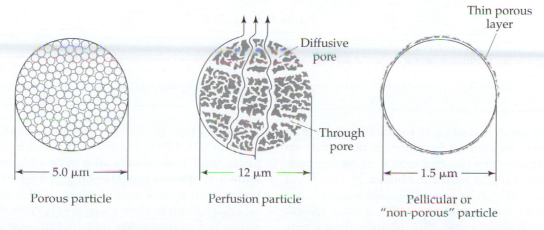

FIGURE 22.4 General structures of porous, perfusion, and pellicular particles used in high-performance liquid chromatography (HPLC). (Reproduced with permission from R.E. Majors, *American Laboratory*, October (2003) 46–54.)

BOX 22.1

Paper Chromatography and TLC

There are two main ways of using a planar support in liquid chromatography: paper chromatography and thin-layer chromatography. *Paper chromatography* is the older of these two techniques and uses paper as the support. *Thin-layer chromatography (TLC)* is more popular than paper chromatography in modern laboratories and uses a particulate support like silica that is coated on a glass or plastic sheet.[5–7]

These two methods are carried out by first applying a spot of the sample near one edge of the plane (see Figure 22.5). This edge is then placed in contact with the mobile phase, which enters and travels along the surface of the plane. Capillary action is often used to create flow of the mobile phase in these methods. As the mobile phase passes over the sample, it begins to carry the analytes with it over the support and stationary phase. The support is later removed from the mobile phase before the solvent front reaches the other edge of the plate, and the plate is examined to determine the location of analyte bands.[6,7]

In both paper chromatography and TLC, analytes that do not interact with the stationary phase will travel at the same rate as the mobile phase (or the "solvent front"). Analytes that do interact with the stationary phase will have slower rates of travel. This retention is described in planar chromatography by using the ratio of the distance traveled by a given analyte (D_s) versus the distance traveled in the same amount of time by the solvent front (D_f). This gives a value known as the **retardation factor (R_F)**.

$$R_F = D_s/D_f \qquad (22.1)$$

The value of R_F is small for analytes with strong retention and will always be between zero and one, because D_s must be less than or equal to D_f.[5,6]

TLC and paper chromatography are inexpensive and easy to carry out. Such advantages have made these techniques popular in organic chemistry for the analysis and isolation of newly synthesized compounds.[17] These methods are also used in large-scale qualitative tests during employee drug screening and in clinical chemistry to examine the amino acid composition of various samples.[18,19] Disadvantages of TLC and paper chromatography include the fact that they are usually carried out manually and have lower resolution, poorer precision, and worse limits of detection than HPLC (the method of choice for most applications involving quantitative analysis by liquid chromatography).[6,7] However, it is possible to partly overcome these disadvantages by using more efficient supports and conducting TLC as part of an automated system.[6,20]

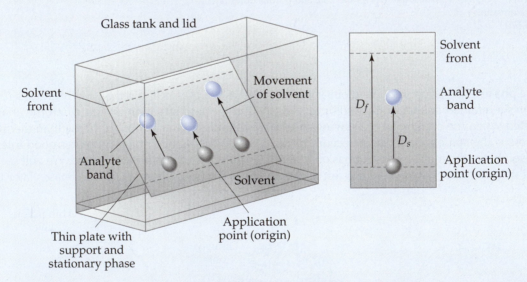

FIGURE 22.5 Typical system used to perform thin-layer chromatography (TLC).

Changing the composition of the mobile phase is an important means for altering the retention of analytes on LC columns. The use of a constant mobile phase composition for elution is called **isocratic elution**.[4,5] The liquid that is used for elution in this case is usually a mixture of a solvent that is a weak mobile phase and another solvent that is a strong mobile phase; this mixture gives a new mobile phase that produces an intermediate degree of retention for the analyte. Although isocratic elution is simple and inexpensive to perform, it does make it difficult to elute all solutes with good resolution and in a reasonable

amount of time (due to the general elution problem; see Chapter 21). Changing the composition of the mobile phase over time results in a type of gradient elution known as **solvent programming**.[5,6] This second method begins with a weak mobile phase, which allows weakly retained solutes to elute more slowly from the column. A switch is then made over time to a stronger mobile phase to allow highly retained solutes to elute more quickly from the column. By convention, the weak mobile phase in solvent programming is referred to as "solvent A," and the strong mobile phase is referred to as "solvent B."[4]

Solvent programming can be carried out in one or more steps and by using a linear change or nonlinear change in the mobile phase content over time.

22.3 TYPES OF LIQUID CHROMATOGRAPHY

A common way of grouping LC techniques is according to the mechanisms by which they separate solutes. This results in five main types of LC, as is illustrated in Figure 22.6. These categories include (1) adsorption chromatography, (2) partition chromatography, (3) ion-exchange chromatography, (4) size-exclusion chromatography, and (5) affinity chromatography.[5]

22.3A Adsorption Chromatography

General Principles. The first main type of LC is **adsorption chromatography**.[4,5] This is a chromatographic technique that separates solutes based on their adsorption to the surface of a support. In LC, this method is also known as *liquid–solid chromatography (LSC)*;[4] the equivalent method in GC is gas–solid chromatography. Adsorption chromatography uses the same material as both the stationary phase and support. In fact, many of the supports used in gas–solid chromatography are also used in liquid–solid chromatography.

The process leading to retention in adsorption chromatography is represented by Equation 22.2. This process involves the binding of an analyte (A) to the surface of a support and the competition of this analyte with the n moles of the mobile phase (M) for these binding sites.

$$A + n\,M\text{-Surface} \rightleftharpoons A\text{-Surface} + n\,M \qquad (22.2)$$

This model indicates that the retention of an analyte in adsorption chromatography will depend on the binding strength of A to the support and on the surface area of this support. This retention factor will also depend on how much mobile phase is displaced from the surface by A, the strength with which the mobile phase binds to the support, and the relative amount of mobile phase that is displaced by the analyte.[6]

The strength of a mobile phase in adsorption chromatography is characterized by a term known as the *elutropic strength* ($\varepsilon°$).[4,6,7] The elutropic strength is a measure of how strongly a particular solvent or liquid mixture will adsorb to the surface of a given support. Some examples of elutropic strengths are listed in Table 22.2 for silica and alumina supports. A liquid with a large elutropic strength will strongly adsorb to the given support, which (as indicated in Equation 22.2) will prevent the analyte from binding to the support. As

Adsorption chromatography:
solutes adsorb to
a support's surface

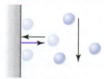

Size-exclusion chromatography:
porous support separates solutes
based on size/shape

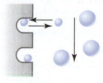

Partition chromatography:
solutes partition into a non-polar
or polar coating

Ion-exchange chromatography:
charged solutes bind to fixed charges

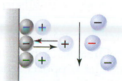

Affinity chromatography:
Solutes selectively bind to a
biologically-related ligand

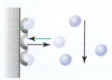

FIGURE 22.6 The five main categories of liquid chromatography, based on the mechanism of separation. The various shaded circles represent different types of solutes that are passing through the chromatographic system.

TABLE 22.2 Elutropic Strength ($\varepsilon°$) for Various Solvents on Silica and Alumina*

Solvent	$\varepsilon°$ on Silica	$\varepsilon°$ on Alumina
n-Pentane	0	0
n-Hexane	0	0.01
Carbon tetrachloride	0.11	0.17
Isopropyl ether	0.32	0.28
Chloroform	0.26	0.36
Methylene chloride	0.30	0.40
Tetrahydrofuran	0.53	0.51
Ethyl acetate	0.48	0.60
Acetonitrile	0.52	0.55
Dioxane	0.51	0.61
Isopropanol	0.60	0.82
Methanol	0.70	0.95
Water	>0.73	>0.95

*Data obtained from E. Katz *et al.*, Eds., *Handbook of HPLC*, Marcel Dekker, New York, 1998, and C.F. Poole and S.K. Poole, *Chromatography Today*, Elsevier, Amsterdam, the Netherlands, 1991.

a result, a liquid with a large elutropic strength will act as a strong mobile phase for that support, because the presence of this liquid will cause the analyte to spend more time in the mobile phase and to elute quickly from the column. For example, Table 22.2 indicates that methanol has a higher elutropic strength than n-pentane when using silica as a support material. These numbers make sense because silica is a polar support and methanol is more polar than n-pentane, giving methanol stronger binding to the surface of silica. This information also tells us that methanol is expected to be a much stronger mobile phase than n-pentane when we are using a column that contains silica in adsorption chromatography.

Stationary Phases and Mobile Phases. Silica (empirical formula, SiO_2) is the most popular support in adsorption chromatography. Because silica is polar in nature, it will most strongly retain polar compounds. A strong mobile phase for silica will also be polar. Alumina (empirical formula, Al_2O_3) is another polar material that is used in adsorption chromatography. Like silica, alumina is a general-purpose support, but can retain some polar solutes so strongly that they are irreversibly adsorbed onto its surface. Carbon-based materials are sometimes used as nonpolar supports in adsorption chromatography, giving columns that retain nonpolar solutes and that have a strong mobile phase that is nonpolar. An increase in surface area for any of these supports will generally lead to stronger analyte retention because this increases the amount of stationary phase versus mobile phase (or the phase ratio).[4,6,7]

We learned earlier that elutropic strength is used to describe the ability of a mobile phase to bind to a particular support in adsorption chromatography. For any support, liquids that have a higher elutropic strength will represent stronger mobile phases. This situation becomes more complicated when using liquid mixtures, because the elutropic strength changes in a nonlinear fashion when different solvents are combined. This effect is illustrated in Figure 22.7, where adding even a small amount of a polar solvent like isopropanol to a nonpolar solvent like hexane results in a large change in the elutropic strength on silica. This nonlinear behavior makes it difficult to use solvent programming in adsorption chromatography. It is also important to use high-purity solvents for the mobile phases in adsorption chromatography because even a trace impurity (e.g., a small amount of water contamination in a nonpolar solvent like hexane) can cause a large change in elutropic strength.

Moving to a liquid with a lower elutropic strength in adsorption chromatography will give a mobile phase that binds more weakly to the support. The result is a liquid that acts as a weak mobile phase, because the analyte will be able to easily bind to the support in the presence of this liquid, leading to high retention. If a switch is instead made to a mobile phase with a higher elutropic strength, it will be more difficult for analytes to adsorb to the support and there will be lower analyte retention. Adjustments in the mobile phase also can be used to alter the selectivity factor in adsorption chromatography by changing from one solvent mixture to another with an equivalent elutropic strength. In Figure 22.7, this could be done by moving from a 33%:67% (v/v) mixture of methyl-t-butyl ether (MTBE) in hexane to a 29%:71% mixture of tetrahydrofuran (THF) in hexane or a 8%:92% mixture of isopropanol in dichloromethane, all of which have elutropic strengths of 0.35. This will result in a chromatogram where the retention factors for many analytes are about the same, but with some shifts in retention occurring that may lead to improved resolution between closely eluting peaks.[6]

Applications. The relative low cost and widespread availability of supports like alumina and silica have made them popular as preparative tools by chemists specializing in synthetics to help purify new chemicals. Silica and alumina work well in separating chemicals that are present in an organic solvent, which will act as a weak mobile phase for these supports. Adsorption chromatography is especially useful in separating geometrical isomers and chemicals that belong to a given class of substances (as

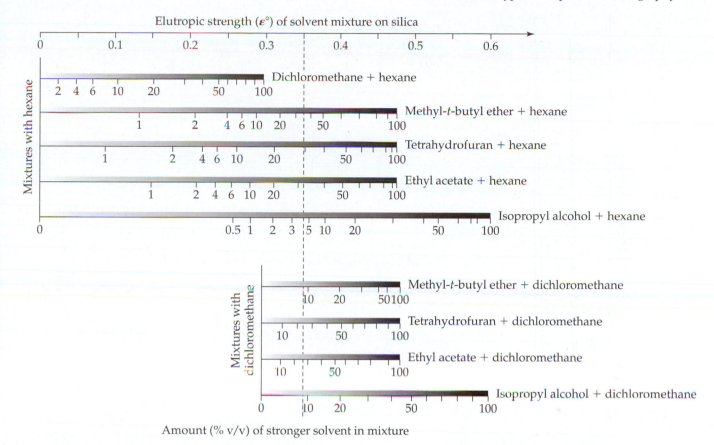

FIGURE 22.7 Solvent elutropic strengths for various mixtures of two solvents on silica. (Adapted from V.R. Meyer and M.D. Palamereva, *Journal of Chromatography*, 641 (1993) 391.)

illustrated in Figure 22.8). Such an approach might be used by a chemist specializing in synthetics to remove undesired side-products during the design of improved methods for synthesizing AZT or other drugs.[6,7]

There are some problems in the use of adsorption chromatography in analytical applications. These problems include the heterogeneous nature of the surface on silica or alumina and the ability of these surfaces to act as catalysts for some chemical reactions. These supports can also create nonreproducible retention for polar compounds and require the use of good-quality solvents for the mobile phases to give consistent elutropic strengths. Many of these difficulties are the same as those seen in gas–solid chromatography (see Chapter 21).

22.3B Partition Chromatography

General Principles. **Partition chromatography** is the second major type of liquid chromatography.[5] It is a liquid chromatographic technique in which solutes are separated based on their partitioning between a liquid mobile phase and a stationary phase coated on a solid support. The support used in partition chromatography is usually silica, but can be other materials. Originally, partition chromatography involved coating this support with a liquid stationary phase that was immiscible with the mobile phase. Most modern columns for partition chromatography employ stationary phases that are chemically bonded to the support.

There are two main categories of partition chromatography: *normal-phase chromatography* and *reversed-phase chromatography*.[4–7] The key difference in these methods is the polarity of their stationary phases. **Normal-phase chromatography** (also called "normal-phase liquid chromatography" or "NPLC") is a type of partition chromatography that uses a polar stationary phase.[4,5] Examples of such stationary phases are shown in Table 22.3 and generally contain groups that can form hydrogen bonds or undergo dipole interactions. Because NPLC has a polar stationary phase, it retains polar compounds most strongly. The weak mobile phase in NPLC is a nonpolar liquid (for example, *n*-hexane or toluene), which is used as the injection solvent. A strong mobile phase is a polar liquid, such as water or methanol.

Reversed-phase chromatography (also known as "reversed-phase liquid chromatography" or "RPLC") is the second type of partition chromatography. RPLC uses a nonpolar stationary phase, which is opposite to or "reversed" in polarity from the stationary phase that is

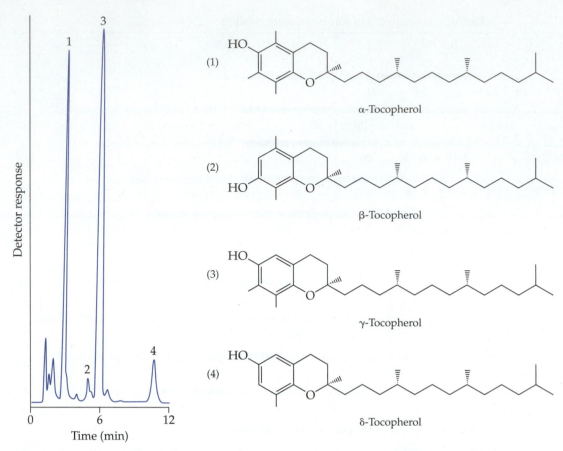

FIGURE 22.8 Separation of tocopherols in corn oil by adsorption chromatography using silica as the support and stationary phase. Notice in this case that the addition or change in position of a single methyl group causes a large change in retention for this class of analytes. This separation involved isocratic elution at 1.0 mL/min with a mobile phase that contained 0.3% isopropyl alcohol in isooctane. (Reproduced with permission from B.A. Bidlingmyer, *Practical HPLC Methodology and Applications*, Wiley, New York, 1992.)

utilized in normal phase chromatography. (*Note*: Historically, NPLC was developed before RPLC.)[4,5] Some common stationary phases used in RPLC are also shown in Table 22.3. These stationary phases typically contain saturated alkane groups like C_8 or C_{18} chains, which form a nonpolar layer on the support. RPLC is currently the most popular type of liquid chromatography.[6,10,11,21] The main reason for this popularity is that the weak mobile phase in RPLC is a polar solvent, such as water. This makes RPLC ideal for the separation of solutes in aqueous-based systems, including drugs in clinical or pharmaceutical samples.[6,7,10,11]

The retention of solutes in partition chromatography can be described by the solubility equilibrium in Equation 22.3. This process is described by a distribution constant (K_D) in the same manner used to describe the distribution of an analyte in an extraction (see Chapter 20).[7]

$$A_{\text{mobile phase}} \xrightleftharpoons{K_D} A_{\text{stationary phase}}$$

$$K_D = \frac{[A]_{\text{stationary phase}}}{[A]_{\text{mobile phase}}} \qquad (22.3)$$

The distribution constant for the analyte in the mobile phase and stationary phase of the column can be directly related to the solute's retention factor (k), as shown below,

$$k = K_D(V_S/V_M) \qquad (22.4)$$

where V_S is the volume of the stationary phase in the column, and V_M is the column void volume. This last equation indicates that the retention factor will increase in partition chromatography when we increase the tendency of this solute to enter the stationary phase (K_D) or increase the relative volume of stationary phase versus mobile phase in the column (V_S/V_M, the phase ratio).

The relative strength of a mobile phase in either NPLC or RPLC can be described by using a *solvent polarity index* (P).[22,23] According to Table 22.4, the value of P is low for a nonpolar solvent and increases as we move to more polar solvents like water. One advantage of using the polarity index to describe mobile phase strength in NPLC and RPLC is that P changes linearly as two different solvents are mixed together. This makes this index useful in

TABLE 22.3 Common Stationary Phases for Partition Chromatography

Normal-Phase Liquid Chromatography (NPLC)

Stationary Phase Name	Abbreviation	Structure
Cyanopropyl phase	CN	Support-$CH_2CH_2CH_2CN$
Aminopropyl phase	NH_2	Support-$CH_2CH_2CH_2NH_2$
Diol phase	Diol	Support-$(CH_2)_3$ OCH_2CHCH_2 with OH OH

Reversed-Phase Liquid Chromatography (RPLC)

Stationary Phase Name	Abbreviation	Structure
Octadecyl phase	C_{18}	Support-$(CH_2)_{17}CH_3$
Octyl phase	C_8	Support-$(CH_2)_7CH_3$
Cyclohexyl phase	CH	Support— (cyclohexyl)
Phenyl phase	PH	Support— (phenyl)

adjusting the mobile phase composition and analyte retention. To calculate the solvent polarity index for a mixture of two solvents, the following formula is used,

$$P_{tot} = \varphi_A P_A + \varphi_B P_B \qquad (22.5)$$

where P_{tot} is the solvent polarity index for a mixture of solvents A and B that have an individual polarity index of P_A or P_B, respectively, and φ_A or φ_B are the volume fractions of these solvents in the new mobile phase.[23] An example of this calculation is given in the following exercise.

TABLE 22.4 Solvent Polarities for Various Liquids in Partition Chromatography*

Liquid	Solvent Polarity Index, P
Carbon tetrachloride	1.56
Isopropyl ether	1.83
Chloroform	4.31
Methylene chloride	4.29
Tetrahydrofuran	4.28
Ethyl acetate	4.24
Acetonitrile	5.64
Dioxane	5.27
Isopropanol	3.92
Methanol	5.10
Water	10.2

*Data obtained form E. Katz *et al.*, Eds., *Handbook of HPLC*, Marcel Dekker, New York, 1998.

EXERCISE 22.2   Determining the Solvent Polarity Index for a Mobile Phase

A pharmaceutical chemist working with AZT begins work with an RPLC column using a 10%:90% (v/v) mixture of acetonitrile and water as the mobile phase. What is the solvent polarity index for this mixture? What would the solvent polarity index be if the mobile phase were changed to a 25%:75% (v/v) mixture of acetonitrile in water?

SOLUTION

From Table 22.4 we can get the solvent polarity indexes for acetonitrile and water, which gives us values of 5.64 for $P_{Acetonitrile}$ and 10.2 for P_{H_2O}. We are told that the volume fractions for these solvents in the mobile phase are 0.10 and 0.90, or 10% and 90% (v/v). We can then use this information with Equation 22.5 to calculate P_{tot} for this mixture.

$$P_{tot} = (0.10)(5.64) + (0.90)(10.2) = \mathbf{9.74}$$

A similar calculation for a 25%:75% mixture of acetonitrile and water gives a P_{tot} of **9.06**.

Stationary Phases and Mobile Phases. Early work with NPLC and RPLC involved the use of liquid stationary phases coated onto solid supports. It is for this reason that partition chromatography is also sometimes known as "liquid–liquid chromatography." However, the use of liquid stationary phases can lead to column bleeding, as we saw in Chapter 21 when using such stationary phases in GC. This problem can be overcome by using stationary phases that are chemically bonded to the support. These bonded phases are now widely used in partition chromatography due to their better stability and efficiency compared to liquid stationary phases. Silica is often used as the support for NPLC or RPLC columns. To place bonded phases on this support, silanol groups on the surface of silica are first treated with an organosilane that contains the desired stationary phase as a side chain. An example of such a reaction is given in Figure 22.9, in which C_{18} groups are placed on silica for use in RPLC. When we are preparing a bonded-phase support with silica, it is important to react with or cover as many silanol groups as possible. If this is not done, the result is a support that has more than one way of interacting with analytes. These "mixed-mode" interactions can create broader peaks and give lower resolution. These effects can be minimized by later reacting the silica with a small organosilane (such as trimethylchlorosilane) that can reach more silanol groups on the surface of silica. This process is known as *endcapping*.[4] Agents like triethylamine and trifluoroacetic acid can also be added to the mobile phase to prevent silanol groups from binding to analytes.[6,7,23]

A weak mobile phase for NPLC will be a solvent or solvent mixture with a low value for solvent polarity index (P_{tot}) and a strong mobile phase will be one with a high P_{tot} value. The weak mobile phase for RPLC will be one with a high P_{tot} and a strong mobile phase will be one with a low P_{tot}. If we are dealing with an analyte that is a weak acid or weak base, the pH of the mobile phase is another factor that can have a large effect on retention in partition chromatography. For instance, in RPLC the protonated neutral form of a monoprotic weak acid (HA) will elute later than its conjugate base (A$^-$). As we alter the pH, this will alter the relative amounts of these species. Because most acid–base reactions are very fast,

what will usually be observed is only one peak that has a weighted average of the retention times for the acid and conjugate base. A similar effect occurs for analytes that undergo other types of rapid reactions, like complex formation with mobile phase additives.[6]

Applications. Partition chromatography is currently the most common type of liquid chromatography used in analytical laboratories. NPLC has similar applications to those listed earlier for adsorption chromatography with silica or alumina. These applications typically involve the use of NPLC for separating analytes in organic solvents and chemicals that contain polar functional groups. Examples of chemicals for which NPLC is employed include steroids, pesticides, terpenoids, nonionic detergents, sugars, and metal complexes.[6,10,11,23]

RPLC is by far the most popular type of partition chromatography and liquid chromatography. There are several reasons for this popularity. First, RPLC separates chemicals based on their overall polarity, which makes this method useful for a broad range of substances. The fact that the weak mobile phase for RPLC is a polar solvent like water is another valuable feature that allows aqueous-based samples to be injected directly onto a reversed-phase column. This feature makes RPLC popular for examining clinical, biological, and environmental samples, as indicated by the list of possible analytes in Table 22.5.[6,10,11,23] RPLC is often employed in the pharmaceutical industry as a method for separating and analyzing drugs during their testing and development. This is illustrated in Figure 22.10, where levels of an AIDS drug and its metabolites are being measured after this agent is given to humans and animals.

22.3C Ion-Exchange Chromatography

General Principles. **Ion-exchange chromatography (IEC)** is a liquid chromatographic technique in which solutes are separated by their adsorption onto a support containing fixed charges on its surface.[4,5] Ion-exchange is a

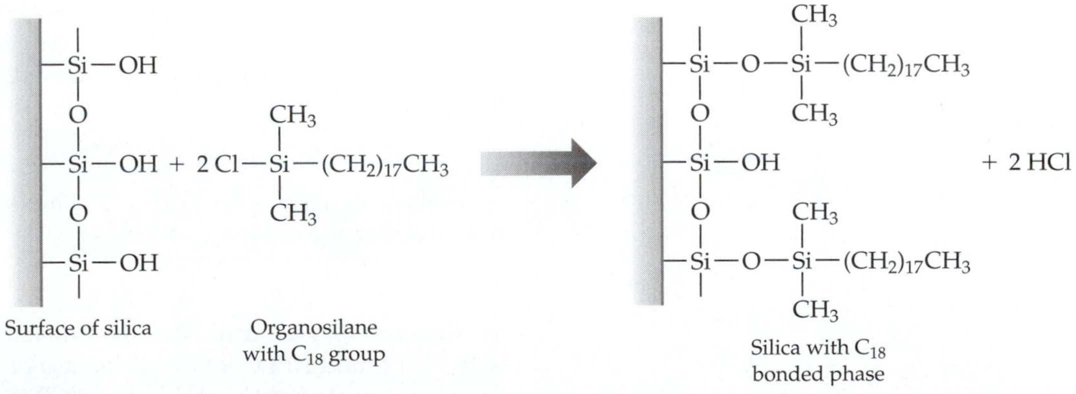

FIGURE 22.9 Reaction of silanol groups on the surface of silica with an organosilane to form a C_{18} bonded stationary phase for reversed-phase chromatography. The relatively large size of this particular organosilane will prevent it from reacting with all of the available silanol groups of the silica. Some of the remaining silanol groups can be removed by later reacting them with a smaller organosilane like trimethylchlorosilane (a reagent discussed in Chapter 21) in a process known as *endcapping*.

TABLE 22.5 Common Applications of Reversed-Phase Liquid Chromatography

Area	Analytes
Biochemistry	Amino acids, proteins, carbohydrates, lipids
Clinical chemistry	Drugs, drug metabolites, bile acids, amino acids
Environmental chemistry	Pesticides, herbicides, phenols, polychlorinated biphenyls
Food chemistry	Artificial sweeteners, antioxidants, aflatoxins, additives
Forensic chemistry	Drugs, poisons, alcohol, narcotics
Industrial chemistry	Condensed aromatics, dyes, propellants, surfactants
Pharmaceutical chemistry	Antibiotics, sedatives, steroids, analgesics

technique used routinely in industry for the removal or replacement of ions in products. A home water softener is a common example of the use of ion-exchange. Ion-exchange is also used in chromatography for the separation of charged compounds, including inorganic ions, organic ions, amino acids, proteins, and nucleic acids.[6,7,10,11,23]

A typical ion-exchange reaction is shown below, which describes the competition of a sample cation (A^+) and a competing cation (C^+) for a negatively charged ion-exchange site on a support.

$$A^+ + Support^-(C^+) \rightleftharpoons Support^-(A^+) + C^+ \quad (22.6)$$

$$K_{A,C} = \frac{[Support^-(A^+)][C^+]}{[A^+][Support^-(C^+)]} \quad (22.7)$$

A similar process can be written for the binding of a sample anion (A^-) to a support with positively charged groups in the presence of a competing anion (C^-). The equilibrium constant $K_{A,C}$ for such an ion-exchange reaction is called a *selectivity coefficient* because it describes how effectively the analyte will compete with the given competing ion for the ion-exchange sites.[4] As the size of this selective coefficient increases, we get higher retention for the analyte.[6,23]

There are several factors that will affect the retention of charged analytes on an ion-exchange column. These factors include (1) the nature and accessibility of the ion-exchange groups on the support, (2) the type and concentration of analyte ions, and (3) the nature and concentration of the competing ions in the mobile phase. The pH of the mobile phase will also be important if we have ion-exchange sites, analytes, or competing ions that are weak acids or bases, because a change in pH may affect the charges on such agents.

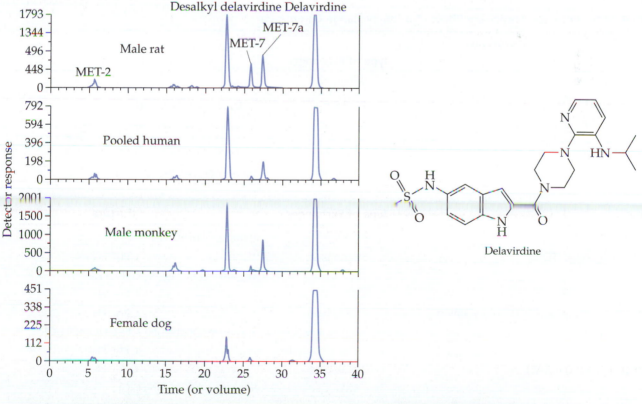

FIGURE 22.10 Use of reversed-phase chromatography for examining the concentration of the AIDS drug delavirdine and its metabolites (desalyl delavirdine, MET-2, MET-7, and MET-7a) in various biological samples. (Reproduced with permission from R. L. Voorman *et al.*, *Drug Metabolism and Disposition*, 26 (1998) 631–639.)

Stationary Phases and Mobile Phases.   There are two general types of stationary phases that are used in ion-exchange chromatography. The first type is a "cation-exchanger," which has a negatively charged group and is used to separate positive ions. The second type is an "anion-exchanger," which has a positively charged group and is used to separate negative ions. These two groups of stationary phases are used in the methods of **cation-exchange chromatography** and **anion-exchange chromatography**, respectively.[4,5] Table 22.6 shows some charged groups that are utilized as stationary phases in these methods. For cation-exchange chromatography, the stationary phase is either the conjugate base of a strong acid (e.g., a sulfonic acid) or the conjugate base of a weak acid (such as a carboxylate group). For anion-exchange chromatography, the conjugate acid of either a strong base (such as a quaternary amine) or a weak base (like a tertiary amine) is used as the stationary phase.[6,7,10,23]

Silica can be used as a support for ion-exchange chromatography if it has been modified to contain charged groups on its surface. Another support that is commonly used in ion-exchange chromatography for small inorganic and organic ions is *polystyrene*. Polystyrene is prepared by polymerizing styrene in the presence of divinylbenzene, a cross-linking agent (see Figure 22.11). The amount of divinylbenzene in this mixture determines the degree of cross-linking of the support, which affects its pore size, swelling, and rigidity. Ion-exchange sites are usually added to the polystyrene after it has been polymerized. Polystyrene and other organic polymeric supports that are used in chromatography are also referred to as "resins."[6,23]

Carbohydrate-based "gels" are another common type of support used in ion-exchange chromatography. These materials are prepared by taking a naturally occurring carbohydrate and chemically modifying it to place ionic functional groups in its structure. These supports are especially useful in the separation of biological compounds, which can have very strong, undesirable binding to organic polymer resins like polystyrene. *Agarose* is one carbohydrate gel used as a support in ion-exchange chromatography (see structure in Figure 22.12). Cross-linked dextran or cellulose gels are also sometimes used. A large number of alcohol groups are present in all of these supports, which makes these materials hydrophilic and gives them low nonspecific binding for biological molecules. These same groups are used to place ion-exchange sites on carbohydrate supports. Another valuable feature of these supports is their large pore size, which makes them attractive in work with large analytes like proteins and nucleic acids.[4,6,23]

A strong mobile phase in ion-exchange chromatography is usually a solution that contains a high concentration of competing ions. As we can see from the reaction in Equation 22.6, a high concentration of competing ions (C^+) will make it more difficult for an analyte ion (A^+) to bind to the stationary phase, leading to lower analyte retention. Changing the concentration of these competing ions is the most common way of altering the retention of analytes in

---

TABLE 22.6   Common Stationary Phases for Ion-Exchange Chromatography

Cation-Exchange Chromatography

Name	Type of Exchanger	Structure
Sulfonic Acid	Strong acid	$Support\text{-}SO_3^-H^+$
Sulfoethyl (SE)	Strong acid	$Support\text{-}O(CH_2)_2SO_3^-H^+$
Sulfopropyl (SP)	Strong acid	$Support\text{-}O(CH_2)_3SO_3^-H^+$
Carboxylic acid	Weak acid	$Support\text{-}COO^-H^+$
Carboxymethyl	Weak acid	$Support\text{-}OCH_2COO^-H^+$

Anion-Exchange Chromatography

Name	Type of Exchanger	Structure
Quaternary amine	Strong base	$Support\text{-}CH_2N(CH_3)_3^+Cl^-$
Triethylaminoethyl (TEAE)	Strong base	$Support\text{—}O(CH_2)_2N^+(CH_2CH_3)_3\,Cl^-$
Diethyl(2-hydroxypropyl) quaternary amino (QAE)	Strong base	$Support\text{—}O(CH_2)_2N^+(CH_2CH_3)_2(CH_2CHOHCH_3)\,Cl^-$
Diethylaminoethyl (DEAE)	Weak base	$Support\text{—}O(CH_2)_2N^+H(CH_2CH_3)_2\,Cl^-$
p-Aminobenzyl (PAB)	Weak base	$Support\text{—}OCH_2\text{—}C_6H_4\text{—}NH_3^+Cl^-$

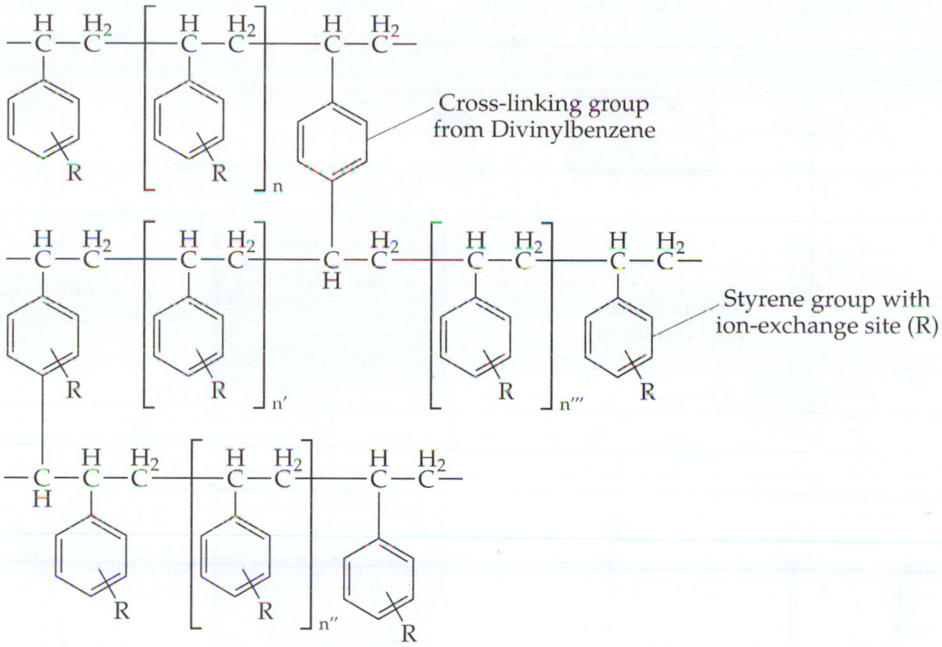

FIGURE 22.11 Structure of polystyrene. The "R" groups represent the ion-exchange sites that are placed on the polystyrene and that can appear at various positions in the styrene rings.

ion-exchange chromatography. The retention of analytes in this method will also be affected by the type of competing ion and type of ion-exchange sites that are present, which will both affect the size of the selectivity coefficient in Equation 22.7. In addition, the pH of the mobile phase can be adjusted to alter retention if we are working with analytes, competing ions, or exchange sites that are weak acids or weak bases. Adding a complexing agent to the mobile phase can also affect the charge of an analyte and alter its retention. For instance, complexation of Fe^{3+} with excess Cl^- can be used to form a $FeCl_4^-$ complex, which can then be retained by anion-exchange chromatography.[6,7,10,23]

Applications. Ion-exchange is frequently employed in removing certain types of ions from samples or solutions. As an example, ion-exchange resins are commonly utilized in water-purification systems for the production of deionized water (see Chapter 2). In this case, cation-exchange supports are used to replace the cations in water with H^+ and anion-exchange supports replace anions with OH^-, which combine to give water. Ion-exchange chromatography has also been used for many years as a

Basic structure of agarose

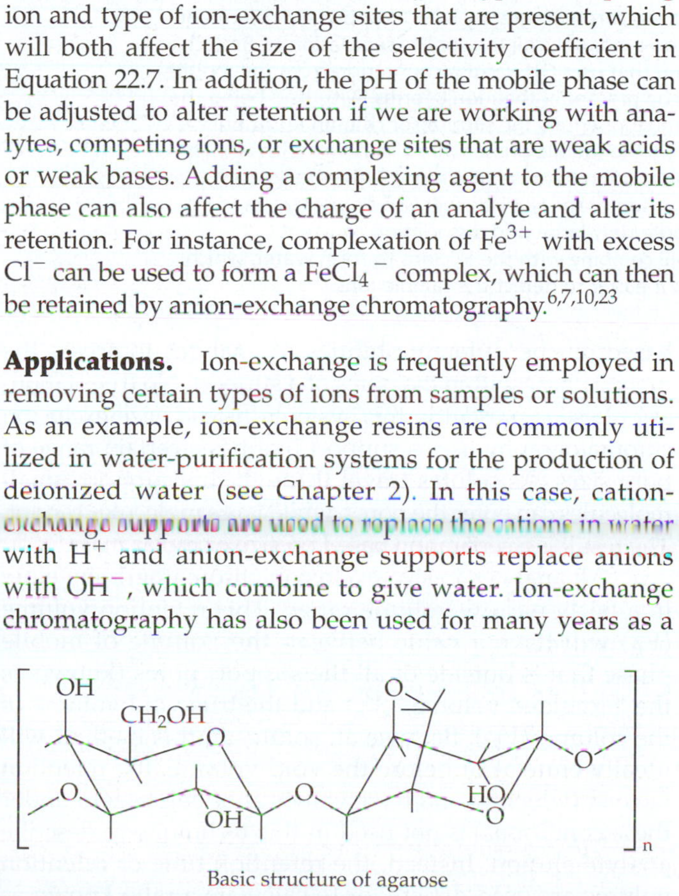

FIGURE 22.12 Structure of agarose. Agarose is a polymer made up of repeating units of D-galactose and 3,6-anhydro-L-galactose. When these polymer chains are heated and allowed to cool, they form a gel that contains a network of large pores that are useful in work with biological macromolecules like DNA and proteins.

preparative tool in biochemistry for purifying proteins, peptides, and nucleotides. In addition, ion-exchange supports are frequently employed for concentrating small inorganic and organic ions from samples like food, environmental samples, and commercial products to help analyze trace metals or ionic contaminants.[6,7,10,23]

Another application of ion-exchange chromatography is in the direct separation and analysis of samples. However, with traditional ion-exchange supports it is often necessary to use a relatively high concentration of competing ions to elute analytes from the column. This high concentration of ions can make it difficult to detect the elution of analytes if this is being monitored by a device such as a conductivity detector (see Section 22.4A). A way to overcome this problem is to use a special type of ion-exchange chromatography known as **ion chromatography (IC)**.[4] In this method, the background signal from competing ions is reduced by using a low number of charged sites for the stationary phase, which will require a lower concentration of competing ions for the elution of sample ions. This method is often used with a second column or membrane separator (of opposite charge to the first ion-exchange column), which replaces competing ions that have high conductivity with ions that have a lower conductivity. The use of a second column or membrane separator for this purpose gives a technique known as *suppressor ion chromatography*.[4]

Figure 22.13 shows a typical system for performing suppressor ion chromatography. In this example, a sample containing fluoride, chloride and bromide ions is first applied to a low-capacity anion-exchange column. These ions are eluted from this column in the presence of a mobile phase that contains a dilute solution of NaOH, with OH^- acting as a competing anion. As the analyte anions elute from this first column, they are applied to a

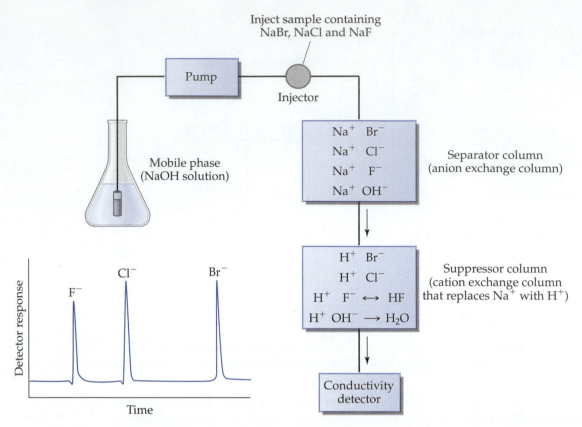

FIGURE 22.13 General design of a system for performing ion chromatography with a suppressor column or membrane. This particular example is for the use of an anion-exchange column for the separation of Br^-, Cl^-, and F^-. In the separator column (an anion-exchange column that uses OH^- as the counterion from the mobile phase), the bromide, chloride, fluoride, and hydroxide will be present in their ionic forms, with Na^+ being the main counterion for all of these species. As the separated mixture enters the suppressor column (a cation-exchange column that uses H^+ as the counterion), the Na^+ in the mobile phase will be replaced with H^+. The bromide and chloride will remain as Br^- and Cl^- ions in this column because HBr and HCl are strong acids that are almost completely dissociated in water. A significant fraction of the fluoride will also be present as F^- if the pH is not too low, because HF is a weak acid with a moderately large acid dissociation constant ($K_a = 6.8 \times 10^{-4}$). However, most of the hydroxide ions will combine with the H^+ ions to form water, which decreases the conductance of the mobile phase and makes it easier to detect the sample ions.

suppressor column or membrane that contains a cation-exchange resin with H^+ on its surface. In this suppressor column, all cations associated with the sample and mobile phase anions (mostly Na^+, in this example) are replaced by hydrogen ions. This converts the competing ion OH^- to H_2O, which is nonconductive. However, anions like F^-, Cl^- and Br^- (which are the conjugate bases of strong or moderately strong acids like HF, HCl and HBr) do not change their form and retain a high conductivity. This conversion causes the conductivity due to the competing ions to decrease without affecting the signal from sample ions, allowing better limits of detection to be obtained. A similar scheme can be performed with cations by employing a cation-exchange analytical column and an anion-exchange suppressor column or membrane.

22.3D Size-Exclusion Chromatography

General Principles. The fourth general type of liquid chromatography is **size-exclusion chromatography (SEC)**.[4,5] SEC is a LC technique that separates substances according to differences in their size. This technique is based on the different abilities of analytes to access the mobile phase within the pores of a support. No true stationary phase is present in this system. Instead, size-exclusion chromatography uses a support that has a certain range of pore sizes. As solutes travel through this support, small molecules can enter the pores, while large molecules cannot. The result is a separation based on size or molar mass.

All analytes in size-exclusion chromatography elute in a fairly narrow volume range. This retention volume (V_R) will have a value between the volume of mobile phase that is outside of all the support pores (known as the "excluded volume," V_E) and the true void volume of the column (V_M). Because all solutes in this method will ideally elute at or before the void volume, the retention factor k (which would have values at or below zero under these conditions) is not used in this technique to describe analyte elution. Instead, the retention time or retention volume are used directly or to calculate a ratio known as K_o, which is defined in Equation 22.8.[23]

$$K_o = \frac{(V_R - V_E)}{(V_M - V_E)} \tag{22.8}$$

For instance, if an analyte has a retention volume of 20.5 mL on a column with a known excluded volume of 13.1 mL and a true void volume of 24.8 mL, the value of K_o for the analyte would be K_o = (20.5 mL − 13.1 mL)/ (24.8 mL − 13.1 mL) = 0.63. We can see from Equation 22.8 that K_o is simply the fraction of the volume between V_M and V_E in which the solute elutes. For small molecules, K_o will equal or approach one. For large molecules, K_o will be equal to or approach zero. Solutes with intermediate sizes will have K_o values between these limits.

Stationary Phases and Mobile Phases. The ideal support in SEC consists of a porous material that does not interact directly with the injected solute. Many of the materials we have already discussed can be used in SEC. Carbohydrate-based supports like dextran and agarose in their underivatized form can be used in SEC for biological compounds and aqueous-based samples. A material known as polyacrylamide gel (see Chapter 23) can also be employed for such samples. Polystyrene can be used for SEC when working with samples in organic solvents, and silica containing a diol-bonded phase can be utilized when working with aqueous samples.[6,10,23]

An important feature of all these supports is they have a porous structure. The range of pore sizes in these supports determines the size of compounds that they will separate. This relationship is illustrated in Table 22.7, where the molar mass of proteins that can enter size-exclusion supports becomes smaller as the degree of cross-linking of these supports is increased and their average pore size decreases. It is also essential that the surfaces of these supports have little or no interactions with the analytes, making it possible to measure retention due to size-exclusion effects.

The mobile phase in SEC can be either a polar or nonpolar solvent. Because there is no true stationary phase, there is also no "weak" or "strong" mobile phase in this method. Instead, the selection of the mobile phase depends mostly on the solubility of the analytes and the support's stability. If an aqueous mobile phase is used in size-exclusion chromatography, the technique is called *gel filtration chromatography*. If an organic mobile phase is used, the technique is known as *gel permeation chromatography*.[4]

TABLE 22.7 Relationship Between Pore Size and Molar Mass of Proteins Separated by Size-Exclusion Chromatography*

Average Pore Size (nm)	Molar Mass of Protein (Da)
5.0	100–10,000
12.5	500–80,000
50	1,000–700,000
100	40,000–5,000,000

*This information is for globular proteins on ethylene glycol and methacrylate copolymer supports. Similar trends are seen for other size-exclusion chromatography (SEC) supports and analytes. (Data obtained from Tosoh Bioscience.)

Applications. As a preparative tool, SEC is often used with biological samples to remove small solutes from large agents like proteins. It can also be used to transfer large analytes from one solution to another or to remove salts from a sample. In analytical applications, SEC is frequently employed in the separation of biomolecules and polymers. This approach also can be utilized in estimating the molar mass of an analyte like a protein or the distribution of molar masses in a polymer.

To estimate a molar mass (or "molecular weight," MW), the size-exclusion column is first used to examine standard compounds that are similar to the analyte but have known masses. These standards should cover a range of masses that span from those totally included in the pores of the support (eluting at V_M) to those that are totally excluded from the pores (eluting at V_E). A plot of log(MW) versus K_o (or the retention time or retention volume) is then prepared for these standards, as shown in Figure 22.14. This graph is generally curved, with a linear range between V_E and V_M. This linear range makes it possible to determine MW for analytes that elute in this range by comparing their retention to that noted for the standards.[23]

EXERCISE 22.3	Estimating the Molar Mass of a Protein by Size-Exclusion Chromatography

A biochemist wishes to use the column in Figure 22.14 to estimate the molar mass of a newly isolated viral protein. This protein gives a peak that elutes at 13.5 mL under the same experimental conditions. What is K_o for this protein? What is its molar mass?

SOLUTION

The excluded volume in Figure 22.14 is 11.0 mL and the void volume is 16.0 mL, so K_o for the unknown protein is (13.5 mL − 11.0 mL)/(16.0 mL − 11.0 mL) = **0.50**. In addition, we can see from our calibration curve that a retention volume of 13.5 mL would correspond to a log(MW) value of 4.65, or a molar mass of $10^{4.65}$ = **45,000 g/mol** for the isolated protein.

22.3E Affinity Chromatography

General Principles. The fifth type of liquid chromatography is **affinity chromatography (AC)**.[4,5,24] Affinity chromatography is a liquid chromatographic method that is based on biologically related interactions. This method makes use of the selective, reversible interactions that characterize most biological systems. Examples of systems include the binding of an enzyme with its substrate, the binding of an antibody with an antigen, or of a hormone with its receptor. Biological interactions are used in affinity chromatography by immobilizing one of a pair of interacting molecules onto a solid support and placing it into a column. The immobilized molecule is known as the *affinity ligand* and represents the stationary phase in the

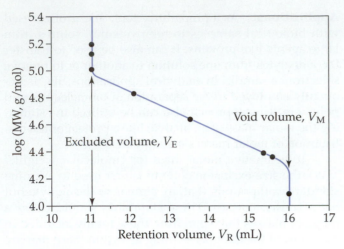

FIGURE 22.14 Typical calibration curve obtained in the separation of proteins by size-exclusion chromatography. (Based on information provided in C.F. Poole and S.K. Poole, *Chromatography Today*, Elsevier, Amsterdam, the Netherlands, 1991.)

column.[24] The column containing the immobilized ligand can then be employed as a selective adsorbent for the complementary molecule.

Affinity columns are often used in an "on/off" elution format, as shown in Figure 22.15. In this method, the sample is applied to the column in the presence of an application buffer. Because of the strong and selective nature of most biological interactions, the affinity ligand will bind to the analyte of interest during this step while allowing most other sample components to pass through as a nonretained peak. After these nonretained components have been washed from the column, a separate elution buffer is applied to release the retained analyte. This analyte is detected as it leaves the column or is collected for later use. The column and affinity ligand are then placed back into the original mobile phase, allowing them to be regenerated prior to the injection of the next sample.[23,24]

The retention of analyte (A) in an affinity column can be described by a complexation reaction in which A combines with the affinity ligand (L) to form the complex (A-L),

$$A + L \underset{}{\overset{K_A}{\rightleftharpoons}} A-L \quad K_A = \frac{[A-L]}{[A][L]} \tag{22.9}$$

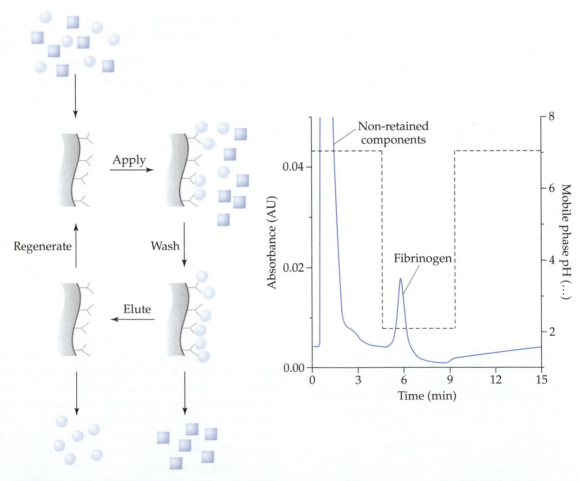

FIGURE 22.15 The "on/off" elution mode of affinity chromatography and a typical separation obtained by this method. The example given is for the analysis of fibrinogen in human plasma using antibodies as the affinity ligands. The dashed line in this example shows how the pH of the mobile phase was changed during this separation as a switch was made between a pH 7 application buffer and a pH 2 elution buffer. (Adapted with permission from J.P. McConnell and D.J. Anderson, *Journal of Chromatography*, 615 (1993) 67.)

where K_A is the association equilibrium constant for the formation of complex A–L. If a 1:1 complex is formed between A and L, the retention factor for A on the affinity column can be related to K_A and the amount of affinity ligand by Equation 22.10.[24]

$$k = K_A (m_L/V_M) \qquad (22.10)$$

The term m_L in this equation is the total moles of active ligand sites in the column and V_M is the void volume of the column. This last equation indicates that the retention factor in affinity chromatography will depend both on the strength of binding between the analyte and ligand (K_A) and the concentration of the available binding sites for an analyte on the affinity ligand (m_L/V_M).

EXERCISE 22.4 Retention in Affinity Chromatography

Antibodies are commonly used as ligands in affinity chromatography. A 10 cm long and 4.1 mm inner diameter affinity column has a void volume of 1.0 mL and contains 10 nmol of antibodies for HIV-1 reverse transcriptase. At pH 7.0 (the sample application conditions), these antibodies will bind to this enzyme with an association equilibrium constant equal to $1.0 \times 10^8\ M^{-1}$. What is the retention factor for reverse transcriptase on this column at pH 7.0?

SOLUTION

The retention factor for reverse transcriptase can be estimated by using Equation 22.10 along with the given values for m_L, V_M, and K_A. The result is shown below.

$$k = (1.0 \times 10^8\ M^{-1})(1.0 \times 10^{-8}\ mol)/(1.0 \times 10^{-3}\ L)$$
$$= 1.0 \times 10^3$$

At a flow rate of 1 mL/min, this value for k would correspond to a retention time of 1000 min or roughly 16.7 h at pH 7. This result shows us that while other solutes will pass almost immediately through this column, reverse transcriptase will bind tightly and elute only after a long time. However, we can speed up this process by applying a separate mobile phase that promotes release of this analyte, such as by altering the pH to lower the values of K_A and k.

Many of the ligands used in affinity chromatography have large equilibrium constants for their analytes. This feature results in extremely high retention factors and long retention for such columns under their sample application conditions. This is why a separate mobile phase is often used to release retained analytes from these columns. The strong binding of affinity ligands is the result of the many different types of forces that can all take part in forming the complex between the analyte and ligand (e.g., hydrogen bonding, dispersion forces, coulombic interactions, and dipole–dipole interactions,

as discussed in Chapter 6). The proper fit of the analyte with the ligand is also required for good binding to occur. This combination of many interactions gives affinity ligands strong binding as well as good selectivity.

Stationary Phases and Mobile Phases. The stationary phase in an affinity column is represented by the affinity ligand, which is the main factor determining what compounds can be separated by such a column. There are several types of affinity ligands, as shown in Table 22.8, but all can be classified into one of two categories: "high-specificity ligands" and "general ligands."[24] High-specificity ligands are compounds that bind to only one or a few very closely related molecules. Examples include antibodies for binding to foreign agents and single-stranded nucleic acids for separating and binding to complementary strands. General ligands are compounds that bind to a family or class of related molecules. High-specificity ligands tend to have large equilibrium constants in their binding to analytes and require the on/off elution format in Figure 22.15. General ligands tend to have lower equilibrium constants and can sometimes be used with isocratic elution. Although many affinity ligands are biological compounds, other binding agents, such as synthetic dyes and immobilized metal ions, can also be used in this method (see Table 22.8).

Several types of supports are utilized in affinity chromatography. Carbohydrate gels like agarose or cellulose are commonly used with affinity ligands for the purification of biological molecules. Silica can be used with affinity ligands by first converting this support into a diol-bonded phase or other form that has low nonspecific binding for

TABLE 22.8 Examples of Affinity Ligands*

Affinity Ligand	Retained Substances
Biological ligands	
Antibodies	Antigens (drugs, hormones, peptides, proteins, viruses, cell components)
Inhibitors, substrates, cofactors, coenzymes	Enzymes
Lectins	Sugars, glycoproteins, glycolipids
Nucleic acids	Complementary nucleic acids, DNA/RNA-binding proteins
Protein A/protein G	Antibodies
Non-biological ligands	
Boronates	Sugars, glycoproteins, diol-containing compounds
Triazine dyes	Nucleotide-binding proteins and enzymes
Metal chelates	Metal-binding amino acids, peptides, and proteins

*Reproduced with permission from D.S. Hage, "Affinity Chromatography." *In Handbook of HPLC*, E. Katz, R. Eksteen, P. Schoenmakers, and N. Millier, Eds., Marcel Dekker, New York, 1998, Chapter 13.

most biological agents. An important factor to consider in affinity chromatography is the method by which the affinity ligand is attached to the support, or the "immobilization method." This immobilization usually involves covalently coupling the affinity ligand to a support through amine, carboxyl, or sulfhydryl groups on the ligand. If appropriate immobilization conditions are not used, the ligand may be denatured or attached in a way that blocks its binding to the analyte. For the immobilization of small molecules, the use of a "spacer arm" between the ligand and support may also be necessary to avoid steric hindrance during binding due to close proximity of the affinity ligand to the support.[23,24]

A weak mobile phase in affinity chromatography is one that allows strong binding between the analyte and affinity ligand. This weak mobile phase is usually a solvent that mimics the pH, ionic strength and polarity of the affinity ligand in its natural environment and is known as the *application buffer*. The application buffer is typically used during the application, washing, and regeneration steps in Figure 22.15. A strong mobile phase in affinity chromatography is a solvent that can readily remove the analyte from the affinity ligand. This solvent is also called the *elution buffer*. The pH, ionic strength or polarity of the mobile phase can be altered to lower the association equilibrium constant for the analyte–ligand interaction (a method known as "nonspecific elution"), or a competing agent can be added to the mobile phase to displace the analyte from the affinity ligand (a technique called "biospecific elution").[24] Both approaches cause the analyte to spend more time in the mobile phase and leads to lower retention.

Applications.   Affinity chromatography is frequently used as a large-scale purification method for enzymes and proteins. This type of application uses columns that contain immobilized agents that can selectively bind to and retain the desired target substances in the presence of other sample components. Affinity chromatography is also used as a method for sample preparation. Examples include the use of affinity columns containing antibodies for the isolation of cellular proteins or the use of immobilized metal ions to isolate recombinant proteins containing histidine tags as part of their structure.

The selectivity of affinity chromatography has also made it appealing for use in the direct analysis of complex biological samples. One example is the use of boronate affinity columns in the measurement of glycated hemoglobin, an indicator of long-term blood sugar levels in diabetes. Affinity columns also have been used with HPLC for the measurement of hormones, proteins, drugs, herbicides, and other agents in biological and environmental samples. Another important application of affinity ligands and selective binding agents is in the area of chiral separations (see Box 22.2).

Another way affinity chromatography is sometimes used is in the study of biological interactions. For instance, as we saw in Equation 22.10, the retention of an injected analyte can be used to obtain information on the equilibrium constant between this analyte and the immobilized ligand. Other information that can be obtained from such experiments includes the number of binding sites for an analyte on the ligand and the rate of analyte–ligand binding.[24]

22.4 LIQUID CHROMATOGRAPHY DETECTORS AND SAMPLE PRETREATMENT

22.4A Types of Liquid Chromatography Detectors

There are several types of detectors available for HPLC (see Table 22.9). These include both general and specific devices that measure the refractive index, absorbance, fluorescence, conductivity, or electrochemical properties

TABLE 22.9   Properties of Common Liquid Chromatography Detectors*

Detector Name	Compounds Detected	Gradient Compatible?	Detection Limits
General Detectors			
Refractive-index detector	Universal (all compounds)	No	0.1–1 μg
UV/Vis absorbance detector	Compounds with chromophores	Yes	0.1–1 ng
Evaporative light-scattering detector	Nonvolatile compounds	Yes	10 μg
Selective Detectors			
Fluorescence detector	Fluorescent compounds	Yes	1–10 pg
Conductivity detector	Ionic compounds	No	0.5–1 ng
Electrochemical detector	Electrochemically active compounds	No	0.01–1 ng
Structure-Specific Detectors			
Mass Spectrometry	Universal (Full-scan mode)	Yes	0.1–1 ng
	Selective (SIM mode)		

*These data are for commercial instruments, as obtained from C.F. Poole and S.K. Poole, *Chromatography Today*, Elsevier, Amsterdam, the Netherlands, 1991, p. 568, and manufacturers of these detectors. The concentration limits of detection for these devices can be estimated by dividing the above values by 10–100 μL, typical injection volumes used in HPLC.

BOX 2.2
Chiral Separations

When an organic compound has four different atoms or groups attached to a carbon, this compound will have two different "chiral" forms that are mirror images of each other. Although most of the chemical and physical properties of these forms are identical, these forms can have different intermolecular interactions with other chiral compounds. This difference can be important because the proteins and peptides in our body are also chiral (being made up of L-amino acids, as opposed to D-amino acids). As a result, it is not unusual for the different chiral forms of a chemical to have different interactions with biological agents like proteins, creating large differences in the activity and toxicity of these agents in the body.[25-7]

It is possible to use chromatography to separate and examine the individual forms of a chiral compound by using a stationary phase that is also chiral (known as a *chiral stationary phase*, or *CSP*).[25-27] An example of such a separation is shown in Figure 22.16. There are many types of chemicals that can be used as CSPs. For instance, many large biological agents like carbohydrates, proteins, and peptides can be used as CSPs because they are composed of amino acids and simple sugars

that are chiral. Cyclodextrins (see Chapter 9) are one group of carbohydrates that have been used for this purpose in both liquid chromatography and gas chromatography. CSPs can also be based on synthetic organic compounds or on chiral cavities that are formed on the surface of support.[6,10,11,24-27] Another way a chiral separation might be performed is to place in the mobile phase a chiral binding agent (such as a cyclodextrin) that will have different interactions with the individual forms of a chiral analyte, leading to differences in the retention of these forms on a chromatographic column.[25,27]

Prior to the early-1990s, many drugs and food additives were used as a mixture of their individual chiral forms. However, this policy changed as the creation of new stationary phases for LC and GC made chiral separations possible for a large number of drugs and other substances. This development allowed more data to be obtained on how the different chiral forms of a drug can affect the body. Those data, in turn, led in 1992 to stricter regulations from the U.S. Food and Drug Administration concerning the production, use, and marketing of chiral drugs.[25]

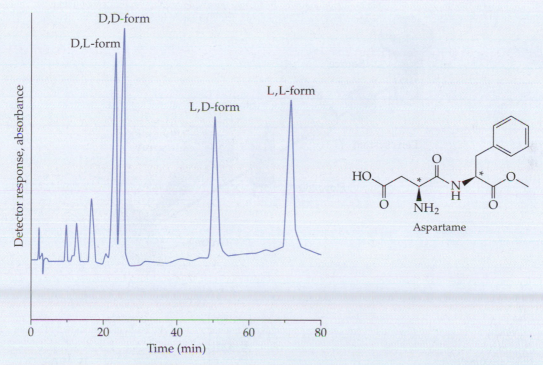

FIGURE 22.16 Chiral separation of the various forms of aspartame (the active ingredient in the artificial sweetener NutraSweet). This separation was performed on a column that used a crown ether as a chiral stationary phase. Aspartame is a dipeptide that has two carbons that can act as chiral centers (see asterisks). The presence of these two chiral centers results in four distinct chemical species: D,L-aspartame, D,D-aspartame, L,D-aspartame and L,L-aspartame. (*Note*: The L,L-form is the one used as a sweetener in commercial products.) Chemical species that have differences in only one chiral center and that produce a set of mirror-image compounds are known as *enantiomers* (e.g., D,D-aspartame and L,L-aspartame). Chemical species that have differences in their chiral centers and that are not mirror images of each other are known as *diasteriomers* (such as D,L-aspartame and D,D-aspartame). (Adapted with permission from S. Motellier and I. W. Wainer, *Journal of Chromatography*, 516 (1990) 365–373.)

of eluting analytes. Mass spectrometry can also be used for either general or selective detection in liquid chromatography.[6,23,28]

General Detectors. An **absorbance detector** is used to monitor many types of analytes in HPLC. We saw in Chapter 17 that absorbance is a measure of the ability of a substance to absorb light at a particular wavelength. Absorbance detectors for liquid chromatography make use of light in the ultraviolet or visible range and include a special sample cell, known as a "flow cell." This flow cell is designed to allow the mobile phase and analytes to pass through the detector in a continuous manner as they exit the column. Ideally, this cell should have a reasonably long path length to provide low detection limits for the analyte. The volume of this cell also needs to be kept to a minimum to avoid adding large amounts of extra-column band-broadening to the system.

The simplest type of absorbance detector for HPLC is a *fixed-wavelength absorbance detector*. This device is set to always monitor a specific wavelength. This wavelength is often 254 nm, because mercury lamps (a common light source) have intense emission at this wavelength and many organic compounds with aromatic

groups or unsaturated bonds absorb light in this range. A *variable wavelength absorbance detector* is a more flexible design that allows the monitored wavelength to be varied over a wide range (e.g., 190–900 nm). This is accomplished by adding a more advanced monochromator to the system. A third possible design is a *photodiode-array (PDA) detector* (see Figure 22.17). A PDA is an absorbance detector that uses an array of small detector cells to simultaneously measure the change in absorbance at many wavelengths. This array makes it possible to record an entire spectrum for a compound as it elutes from a column, which can be valuable in identifying overlapping peaks.[6,11,23]

Absorbance detectors can detect any compound that absorbs light at the wavelength(s) being monitored. Absorbance detectors can easily be used with gradient elution, provided that the weak and strong mobile phases do not have significant differences in their absorbances at the wavelengths being used for detection. The limits of detection for these devices are also quite good, typically being in the range of $10^{-8}M$. The main disadvantage of these devices is they require a compound to have a chromophore that can absorb at the wavelengths being monitored or that can be derivatized into a form that does absorb. These detectors also provide little information on

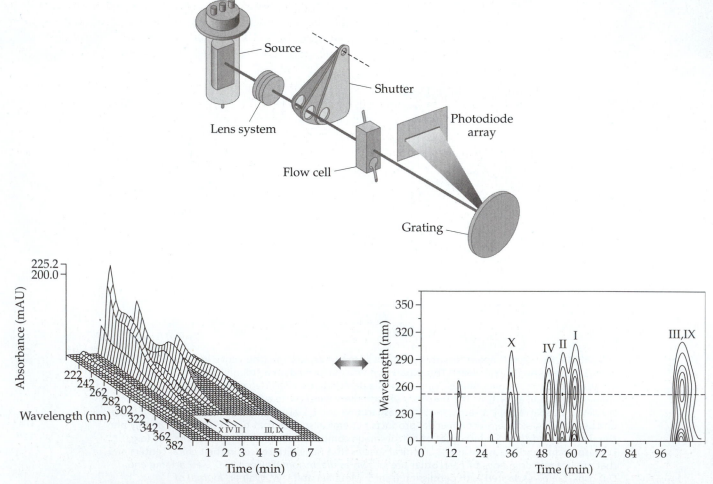

FIGURE 22.17 A photodiode array detector and examples of data that can be collected by such a detector when used with HPLC. (The schematic of the photodiode array detector is reproduced with permission from S.A. Borman, *Analytical Chemistry*, 55 (1983) 836A–842A; the bottom figures are reproduced with permission from D.G. Jones, *Analytical Chemistry*, 57 (1985) 1057A–1073A.)

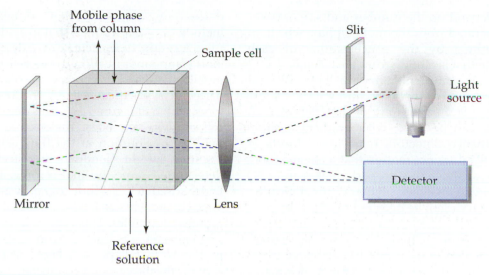

FIGURE 22.18 General design of a refractive-index detector.

the structure of an analyte. However, they are valuable in quantitating analytes once calibration curves have been obtained with standard solutions of these compounds.

A **refractive index (RI) detector** is one of the most universal detectors available for HPLC.[11,23] This detector measures the ability of the mobile phase and analytes to refract or bend light. The refractive index will change as the composition of the mobile phase changes, such as when analytes elute from a column. One type of RI detector used in HPLC is shown in Figure 22.18. In this design, light from a visible light source is passed through two flow cells, one containing mobile phase eluting from the column and the other containing a reference solution (usually pure mobile phase). These flow cells are at an angle to one another, which causes the light to be bent at their interface if there is any difference in the content and refractive index of their solutions. To increase the extent of this bending, the light is passed a second time through this interface by a mirror and then is sent on to a detector that is sensitive to the position of the light beam. As analytes elute from the column, the refractive index of the solution in the sample flow cell will be different from that in the reference flow cell. This difference in refractive index causes the light beam to be bent and produces a response at the detector.

A key advantage of an RI detector is it will respond to any compound that has a different refractive index from the mobile phase, provided that enough solute is present to give a measurable signal. This makes an RI detector useful in work where an analyte cannot be easily measured by other devices or where the nature or properties of an analyte are not yet known. For instance, RI detectors are valuable in separations of carbohydrates, which usually do not possess good chromophores for absorbance detection. One disadvantage of an RI detector is it does not have limits of detection as low as absorbance detectors or many other HPLC detectors. In addition, its signal is sensitive to changes in the mobile phase

composition and temperature, making the RI detector difficult to use with gradient elution.

A third general detector for HPLC is an **evaporative light-scattering detector (ELSD)**.[11] This detector can be used for any solute that is less volatile than the mobile phase. The way in which an ELSD works is illustrated in Figure 22.19. First, a mobile phase leaving the column is converted into a spray of small droplets. As the solvent in these droplets evaporates, small solid particles that contain the nonvolatile sample components are left behind. These particles are then passed through a beam of light, where they scatter some of the incoming light. The extent of this scattering is then measured. The degree of light

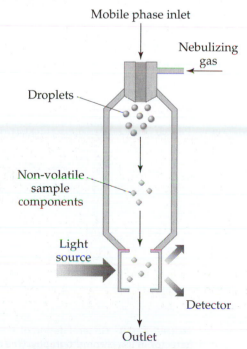

FIGURE 22.19 Operation of an evaporative light-scattering detector.

scattering will depend on the number and size of particles that are generated from the mobile phase which, in turn, is determined by the concentration of each nonvolatile solute that is eluting from the column.

An ELSD is complementary to both RI and absorbance detectors. For instance, an ELSD has a low background signal and better limit of detection than an RI detector and can be used with gradient elution. Although absorbance detectors have lower limits of detection, an ELSD does not require that a chromophore be present in the analyte. This property makes an ELSD capable of examining analytes, such as lipids and carbohydrates, that often cannot be detected by absorbance.

Selective Detectors. There are several specific detectors that can also be used in HPLC. A good example is a **fluorescence detector**.[6,11,23] This device measures the ability of chemicals to absorb and emit light at a particular set of wavelengths. Because these wavelengths are characteristic of a given chemical, this method can provide a signal that has a low background and is reasonably specific for the analyte of interest. Fluorescence can be used to selectively detect any analyte absorbing and emitting light at the given excitation and emission wavelengths. Although relatively few chemicals are fluorescent, those that do fluoresce are frequently of great importance. Examples include many drugs and their metabolites, food additives, and environmental pollutants. Fluorescence can also be used to detect analytes that are first converted to a fluorescent

derivative; compounds that can be detected this way include alcohols, amines, amino acids, and proteins.

A **conductivity detector** is a device that can monitor ionic compounds in HPLC by measuring the ability of the mobile phase and its contents to conduct a current when placed in an electrical field.[6,23] The design of such a detector is shown in Figure 22.20. This design consists of a flow cell and two electrodes. The electrodes apply an electric field to the solution in the flow cell and measure the resulting current. Conductivity detectors can be used to detect any compound that is ionic. These detectors are widely used in ion chromatography and in the analysis of ionic components of foods, industrial samples, and environmental samples. This type of device can be used with gradient elution as long as the ionic strength (and possibly pH) of the mobile phase is kept constant. It is also necessary for the background conductance of the mobile phase to be sufficiently low so that sample ions can be detected.

An **electrochemical detector** is another device used to monitor specific compounds in HPLC (see Figure 22.21).[6,11,23] This combination is known as *liquid chromatography/electrochemical detection* (or *LC/EC*). An electrochemical detector can be used to measure the ability of an analyte to undergo either oxidation or reduction. One way such a reaction can be monitored is by measuring the change in current (i.e., electron flow) that a reaction produces when present at a constant potential. Another way is to measure the change in the potential when a constant current is applied. Examples of compounds that may be detected by their reduction

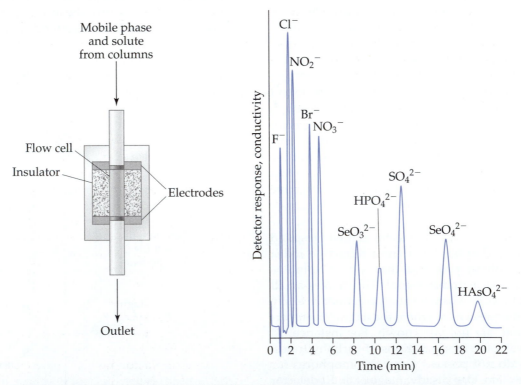

FIGURE 22.20 General design of a conductivity detector and an example of the use of this detector in ion chromatography. (The chromatogram is reproduced with permission of C.F. Poole and S.K. Poole, *Chromatography Today*, Elsevier, Amsterdam, the Netherlands, 1991.)

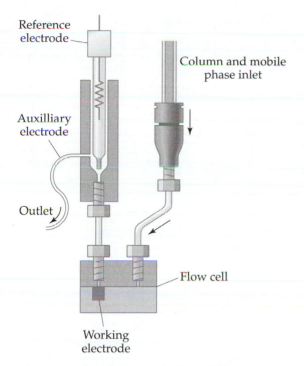

Reference electrode

Column and mobile phase inlet

Auxilliary electrode

Outlet

Flow cell

Working electrode

FIGURE 22.21 Design of an electrochemical detector for liquid chromatography. The purpose of the auxilliary electrode, working electrode, and reference electrode in such a device is described in Chapter 22 (Reproduced with permission from P.T. Kissinger, *Analytical Chemistry*, 49 (1977) 447A–456A.).

include aldehydes, ketones, oximes, conjugated acids, esters, nitriles, unsaturated compounds, aromatics, and activated halides. Compounds that may be detected by their oxidation include phenols, mercaptans, peroxides, aromatic amines, diamines, purines, and some carbohydrates. The response of an electrochemical detector depends on the extent of oxidation or reduction that occurs at the given potential of the electrode. The limit of detection can be quite low as a result of the accuracy with which electrical measurements (especially current) can be made.

Liquid Chromatography/Mass Spectrography. Another type of detector that can be employed in HPLC is a mass spectrometer. The result is a technique known as **liquid chromatography/mass spectrometry (LC/MS)**.[6,11] This is similar to gas chromatography/mass spectrography (GC/MS) (Chapter 21) in that the combined use of mass spectrometry with HPLC makes it possible to both quantitate chemicals and to identify these chemicals based on the masses of their molecular ions or fragment ions. If we look at all ions that are produced in the mass spectrometer (the "full-scan mode," as discussed in Chapter 21), LC/MS can be used as a general detection method. If we instead look at a few ions that are characteristic of a particular set of analytes ("selected ion monitoring"), LC/MS can be used for selective detection. An example is shown in Figure 22.22, where LC/MS is used during pharmaceutical studies to identify and

measure the levels of an AIDS drug and its metabolites in biological samples.

The most common way of performing LC/MS is to use **electrospray ionization (ESI)**,[28] as illustrated in Figure 22.22. In this ionization method, the sample is placed into a solvent and sprayed from a highly charged needle (3–5 kV). The solvent in the charged droplets will evaporate away quickly, giving smaller droplets with an excess of positive or negative charge. At some critical point, the coulombic forces in this droplet will exceed its cohesive forces and cause the droplet to divide (a process known as "coulombic explosion"). Eventually, the molecules in the droplet will be released as ions and enter the gas phase. These ions can be examined by a variety of mass analyzers, such as a quadrupole mass analyzer (see Chapter 21).

Electrospray ionization can be used in LC/MS to examine substances that range from small polar compounds to proteins. It can also be utilized with gradient elution methods. The use of LC/MS with electrospray ionization is particularly useful for the analysis of proteins and peptides, which tend to give ions with mass-to-charge ratios that are outside the range of many mass analyzers. In ESI this problem is overcome by the fact that many charges are often placed on one biomolecule, giving ions with reasonably low mass-to-charge ratios. As an example, a protein with a molar mass of 20,000 g/mol that carries 20 charges per molecule after ESI will give an ion with a mass-to-charge ratio of 1000, a value easy to examine with a quadrupole mass analyzer. One difficulty associated with this process is that a single protein or peptide can create many molecular ions. However, this difficulty can be addressed by computer programs designed to analyze such spectra and provide the true molar masses of proteins from such information.

22.4B Liquid Chromatography Equipment and Sample Pretreatment

An injection valve is often used in HPLC to introduce a sample into the system. Figure 22.23 shows a typical valve that is utilized for this purpose, in which a syringe applies the sample through a port and into a small sample loop. Excess sample passes through this loop and out the other side to a waste container. When the sample is injected, the position of the valve is switched so that its three internal flow channels move and connect to a different series of entry and exit ports. This switch places the sample loop in the path of the mobile phase and causes the sample to be passed onto the column. The valve is later placed back into its original position and the next sample is injected. This process can be carried out either manually or automatically.[11,23]

Another factor needed when using HPLC is a means for applying the mobile phase at a sufficiently high pressure to allow it to move through the column. This pressure is commonly generated through the use of

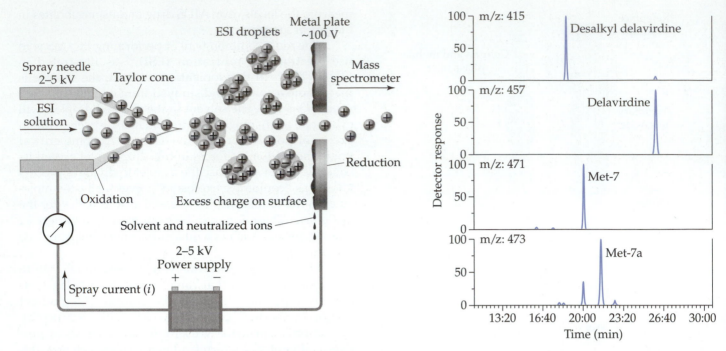

FIGURE 22.22 Scheme by which electrospray ionization (ESI) produces ions. In this method, a sample solution is pumped through a needle held at a positive voltage. This results in the production of tiny droplets with an excess of positive charge. As these droplets travel through space, the solvent evaporates, the charge begins to concentrate and the droplets start to break apart. Eventually, some of the molecules in these droplets are ejected as positively charged ions (e.g., the original molecule now associated with n H^+, giving $M(H^+)_n$) that are sent into a mass analyzer. The terms "oxidation" and "reduction" are shown to indicate that oxidation-reduction processes are occurring in this system during ion formation, making this device a type of electrochemical cell. The chromatograms on the right show the use of this method in the selected ion monitoring for the AIDS drug delavirdine and its metabolites. (The diagram of the ESI system is reproduced with permission from N.B. Cech and C.G. Enke, *Mass Spectrometry Reviews*, 20 (2001), 362; the chromatograms are reproduced with permission from R. L. Voorman, et al., *Drug Metabolism and Disposition*, 26 (1998) 631–639.)

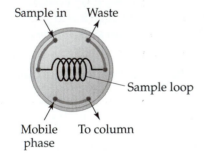

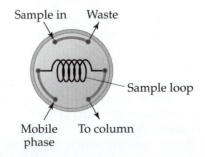

FIGURE 22.23 A six-port injection valve for HPLC.

one or more mechanical pumps. Two types of HPLC pumps are given in Figure 22.24. The first of these is a *reciprocating pump*, where a rotating cam is used to move a piston in and out of a solvent chamber; the movement of the piston causes the mobile phase to flow into the chamber and out toward the column. A reciprocating pump is often used with conventional HPLC columns and at flow rates in the mL/min range. For lower flow rates and small columns, a *syringe pump* can be employed. This pump consists of a syringe in which a plunger is pressed in by a motor, creating flow of the mobile phase out of the syringe. This design gives an even flow and works well at small flow rates, such as those in the μL/min range.[11]

Some typical dimensions for HPLC columns are shown in Table 22.10. A typical HPLC column for analytical applications is 10 or 25 cm in length and has an inner diameter of 4.1 or 4.6 mm. Longer columns or capillaries with smaller diameters can also be used to provide a larger number of theoretical plates and more efficient separations (e.g., "microbore columns" and "packed-capillary columns"). One trade-off in going to the longer, narrower columns is that we must use smaller amounts of sample to avoid overloading the system. These columns also require slower flow rates to

Syringe Pump

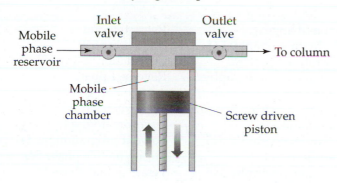

Reciprocating Pump

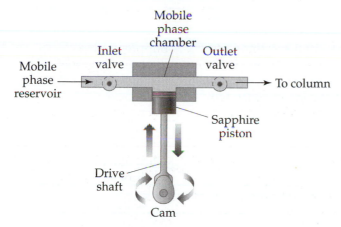

FIGURE 22.24 Design of syringe pumps and reciprocating pumps for HPLC.

computers for controlling the pumps and other system components, as well as for recording the results of the separation. It is also possible to use devices that allow most or part of the mobile phase to be recycled and used for multiple sample injections. This last feature is important because mobile phase solvents can make up a large fraction of the chemical waste that is generated by an analytical laboratory.

Because analytes examined by HPLC are injected as a solution onto the column, these analytes must first be placed into a liquid that is compatible with the mobile phase used at the beginning of the separation method (i.e., the weak mobile phase). One or more pretreatment steps may be needed to dissolve and transfer these substances into an appropriate solvent. For example, compounds that are in a nonpolar organic solvent and that are to be separated by RPLC would first have to be extracted or placed into water or another polar solvent. If solid matter is present in the sample, this matter would be removed by centrifugation or filtration prior to injection to avoid clogging within the column and chromatographic system.[11,23]

There are some situations in which analytes are derivatized in HPLC, but this is done for different reasons than in GC, because good sample volatility and thermal stability are not needed in liquid chromatography. Instead, derivatization is often employed in HPLC to improve the response of an analyte in devices such as fluorescence or electrochemical detectors. Derivatization can also be used to improve the separation of the solute from other sample components by changing the structure of the solute and altering its retention on the column. Unlike GC, derivatization in HPLC may take place either before the sample is injected ("precolumn derivatization") or after the analyte has eluted from a column ("postcolumn derivatization"). Precolumn methods can be used to either alter a solute's retention or improve its response to a detector, while postcolumn techniques are used only to improve the ability to detect an analyte.[23,29]

provide reasonable operating pressures. In some cases these slower flow rates can be an advantage, such as in LC/MS because it means less solvent will need to be removed before analyte ions can be separated and examined by the mass spectrometer.[6,23]

Other pieces of equipment can also be included as part of an HPLC system. Many HPLC systems include

TABLE 22.10 Common Column Sizes for Analytical Applications of Liquid Chromatography*

Column Type	Length	Internal Diameter (mm)	Typical Flow Rate
Conventional packed column	5–30 cm	4–5	1–3 mL/min
Microbore column	10–100 cm	1–2	0.05–0.2 mL/min
Packed capillary	20–200 cm	0.1–0.5	0.1–20 μL/min
Semipacked capillary	1–100 m	0.02–0.1	0.1–2 μL/min
Open tubular column	1–100 cm	0.01–0.075	0.05–2 μL/min

*Reproduced with permission from C. F. Poole and S. K. Poole, *Chromatography Today*, Elsevier, Amsterdam, the Netherlands, 1991.

Key Words

Absorbance detector *558*	Electrochemical detector *560*	Ion-exchange chromatography *548*	Refractive-index detector *559*
Adsorption chromatography *543*	Electrospray ionization *561*	Isocratic elution *542*	Retardation factor *542*
Affinity chromatography *553*	Evaporative light-scattering detector *559*	Liquid chromatography *537* Liquid chromatography/ mass spectrometry *561*	Reversed-phase chromatography *545*
Anion-exchange chromatography *550*	Fluorescence detector *560*	Normal-phase chromatography *545*	Size-exclusion chromatography *552*
Cation-exchange chromatography *550*	High-performance liquid chromatography *540*	Partition chromatography *545*	Solvent programming *542* Strong mobile phase *541*
Conductivity detector *560*	Ion chromatography *551*		Weak mobile phase *541*

Other Terms

Affinity ligand *553*	Gel permeation chromatography *553*	Monolithic column *540*	Suppressor ion chromatography *551*
Agarose *550*	Liquid chromatograph *538*	Nonporous support *540*	Syringe pump *562*
Application buffer *556*	Liquid chromatography/ electrochemical detection *560*	Paper chromatography *542*	Thin-layer chromatography *542*
Chiral stationary phase *557*		Perfusion particles *540*	Variable-wavelength absorbance detector *558*
Elution buffer *556*		Photodiode-array detector *548*	
Elutropic strength *543*	Liquid chromatogram *538*	Polystyrene *550*	Ultrahigh pressure chromatography *540*
Endcapping *548*	Liquid–solid chromatography *543*	Reciprocating pump *562*	
Fixed-wavelength absorbance detector *558*		Selectivity coefficient *549*	
Gel filtration chromatography *553*		Solvent polarity index *546*	

Questions

WHAT IS LIQUID CHROMATOGRAPHY AND HOW IS IT PERFORMED?

1. Define "liquid chromatography" and explain, in general, how this method is used in chemical analysis.
2. What is a "liquid chromatograph"? What are some components that are found in a liquid chromatograph?
3. What is a "liquid chromatogram"? What types of information can this provide about an analyte or sample?
4. The manufacturer of a C_{18} reversed-phase column reports the following retention factors for chemicals that it recommends as a standard test mixture: resorcinal, 0.2; acetophenone, 1.4; napthalene, 4.3; and anthracene, 9.8. A chromatogram that was actually obtained with this mixture is shown in Figure 22.25. Estimate the retention time and retention factor for each peak in this chromatogram. Use this information to identify each of these peaks.
5. A series of alkaline earth ions is examined and quantitated by cation-exchange chromatography. The injection of a standard mixture of such cations onto a column at 1.5 mL/min gives the following retention time: Mg^{2+}, 3.65 min; Ca^{2+}, 4.32 min; Sr^{2+}, 5.73 min; and Ba^{2+}, 10.14 min. The column void time under these same conditions is 0.55 min.
   (a) A series of Ca^{2+} standards under these conditions gives peak areas of 235 units at 1 ppm, 468 units at 2 ppm, 695 units at 3 ppm, and 950 units at 4 ppm. An unknown Ca^{2+} sample gives a peak area of 579 units. What is the concentration of Ca^{2+} in this sample?

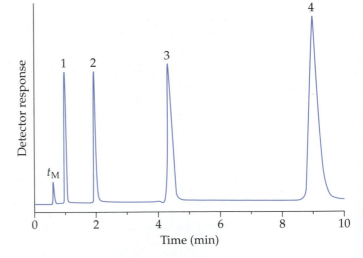

FIGURE 22.25 Separation of a typical test mixture on a C_{18} reversed-phase column.

   (b) The injection of an unknown sample onto this same column at 1.0 mL/min using the same column and mobile phase gives a peak with a retention time of 6.48 min. The column void time at this flow rate is 0.83 min. What is the most likely identity of the cation in the unknown peak?

REQUIREMENTS FOR THE ANALYTE

6. What is the first requirement that must be met before a chemical or sample can be used in liquid chromatography (LC)?

7. What role does the mobile phase play in LC with regards to analyte retention? How is this different from the role of the mobile phase in gas chromatography (GC)?

8. A pharmaceutical chemist wishes to compare two types of reversed-phase columns for the separation and analysis of the drug phenytoin in serum samples. The first column has a 3.0 mm inner diameter and is 6.0 cm long, containing a C_{18} stationary phase and 3 μm diameter support particles. The second column has an inner diameter of 4.6 mm and is 10.0 cm long, also containing a C_{18} stationary phase, but now having 5 μm diameter support particles. Both columns have similar phase ratios and provide the same retention factors for phenytoin in the presence of equivalent mobile phases.
   (a) At 1.0 mL/min, the first column has a void time of 0.34 min and gives a retention time of 30.52 min for phenytoin in solvent A and 10.27 min in solvent B. What is the retention factor for phenytoin in both of these solvents?
   (b) If the second column has a void time of 1.33 min at 1.0 mL/min, what is the expected retention time for phenytoin on this column in solvents A and B?

9. Explain why LC tends to have greater band-broadening than GC. Why are differences in the viscosities of liquids vs. gases also important to consider when comparing LC with GC?

COLUMN EFFICIENCY IN LIQUID CHROMATOGRAPHY

10. Explain what is meant by "high-performance liquid chromatography." How does this differ from "classical liquid chromatography"?

11. Describe the differences between each of the following types of supports for liquid chromatography.
    (a) Porous particle
    (b) Perfusion particle
    (c) Nonporous support
    (d) Monolithic column

12. The baseline width measured in Problem 8 for phenytoin on the first column in solvent B was 0.46 min at a flow rate of 1.0 mL/min. The baseline width measured for phenytoin on the second column under the same conditions was 2.16 min. What is the number of theoretical plates and plate height for each of these columns? Is the first or second column more efficient in this separation?

13. Use the data in Table 22.1 to estimate each of the following values.
    (a) The number of theoretical plates expected for a 25 cm long and 4.6 mm inner-diameter column that is packed with 5 μm diameter porous particles (traditional type, not Poroshell)
    (b) The value of N for a 5 cm long and 2.1 mm inner-diameter column that is packed with 2.5 μm diameter porous particles
    (c) The height equivalent of a theoretical plate for a 25 cm long and 10 mm inner-diameter column that is packed with 50 μm diameter pellicular particles
    (d) The value of H for a 3 cm long and 2.1 mm inner-diameter column that is packed with 1.5 μm diameter pellicular particles

14. Explain how LC can be used to perform either "column chromatography" or "planar chromatography."

ROLES OF THE MOBILE PHASE IN LIQUID CHROMATOGRAPHY

15. How is the role of the mobile phase in LC different from its role in GC? How is the role of the mobile phase in these two methods the same?

16. What is meant by the terms "strong mobile phase" and "weak mobile phase" in LC?

17. Define "isocratic elution" and "solvent programming." How is each of these methods performed?

18. A sample containing several proteins is to be separated on a (diethylamino) ethyl (DEAE) ion-exchange column. This sample is injected onto the column in the presence of a mobile phase that contains pH 8.2, 20 mM tris buffer. This mobile phase is then changed to pH 8.2, 20 mM tris plus 0.25 M NaCl over the course of 40 min during the separation.
    (a) Identify the weak mobile phase and strong mobile phase in this example. Which represents the "solvent A" and which is "solvent B"?
    (b) Discuss the purpose of each of the weak and strong mobile phases in this example in terms of the general elution problem.

ADSORPTION CHROMATOGRAPHY

19. What is "adsorption chromatography"? What type of stationary phase is used in this method?

20. Give a general reaction that can be used to describe the retention of an analyte by adsorption chromatography. Based on this reaction, what factors would you expect to be important in determining the retention of an analyte in this method?

21. What is an "elutropic strength"? How is this used in adsorption chromatography?

22. List two common stationary phases used in adsorption chromatography. What types of solvents are strong mobile phases for each of these stationary phases?

23. Describe how the retention of solutes can be adjusted in adsorption chromatography. Describe how the selectivity factor for a separation can be adjusted in this method.

24. An analyte is injected onto a silica column and eluted using hexane as the mobile phase. Under these conditions it elutes with a long retention time. To decrease this retention, some tetrahydrofuran is to be added to the mobile phase.
    (a) What is the approximate elutropic strength of hexane on silica?
    (b) What is the elutropic strength of tetrahydrofuran on silica?
    (c) Which of these two solvents is a stronger mobile phase on this stationary phase?
    (d) Approximately how much tetrahydrofuran, in % (v/v), must be added to hexane to change the elutropic strength of this mobile phase by 0.10 unit? By 0.20 or 0.30 unit?

25. A mixture of 20% tetrahydrofuran and 80% hexane is found to elute an analyte from a silica column within a reasonable amount of time, but the peak for the analyte has insufficient resolution from neighboring peaks. Give one other mixture of two solvents that has the same elutropic strength and that might be used in place of the tetrahydrofuran/hexane mixture as the mobile phase.

26. What are some applications for adsorption chromatography? What general types of substances are separated by this method?

PARTITION CHROMATOGRAPHY

27. What is "partition chromatography"? How does this differ from adsorption chromatography?

28. Compare and contrast reversed-phase liquid chromatography and normal-phase liquid chromatography. How are these methods the same? How are they different?

29. What are some common stationary phases in normal-phase liquid chromatography? What types of solvents are strong mobile phases in this method? What solvents act as weak mobile phases?

30. What are some common stationary phases in reversed-phase liquid chromatography? What solvents are strong and weak mobile phases in this technique?

31. Give a general reaction that can be used to describe the retention of an analyte by partition chromatography. Based on this reaction, what factors would you expect to be important in determining the retention of an analyte in this method?

32. What is a "solvent polarity index"? How is this index used in partition chromatography?

33. Find the solvent polarity index for each of the following mobile phases.
    (a) Acetonitrile
    (b) Hexane
    (c) 30% tetrahydrofuran:70% hexane
    (d) 50% Isopropanol:50% water

34. One technique that can be used in reversed-phase liquid chromatography (RPLC) and normal-phase liquid chromatography (NPLC) to help improve the selectivity of a separation is to vary the types of solvents that are combined to give a particular solvent polarity index. An environmental chemist wishes to use this approach with RPLC by using different mixtures of acetonitrile, methanol, and water as the mobile phase. One mixture that is found to give good retention for the target analyte is a mixture that contains 60% acetonitrile, 20% methanol, and 20% water (v/v).
    (a) What is the solvent polarity index of this mixture?
    (b) What mixture of only acetonitrile and water will give the same solvent polarity index as in Part (a)?
    (c) What mixture of only water and methanol will result in the same solvent polarity index as found in Part (a)?

35. Why are bonded phases common in modern RPLC and NPLC?

36. Describe how a bonded phase like a C_{18} group can be placed onto silica for use in RPLC.

37. What is "endcapping"? Why is this method important in the preparation of bonded phases for liquid chromatography?

38. The change in retention factor in going from one mobile phase to another in partition chromatography can be estimated by using the following relationships,[23]

$$\text{For RPLC:} \quad \log(k_1/k_2) = (P_{tot1} - P_{tot2})/2 \quad (22.11)$$

$$\text{For NPLC:} \quad \log(k_1/k_2) = (P_{tot2} - P_{tot1})/2 \quad (22.12)$$

where k_1 and k_2 are the retention factors for the same analyte in the presence of two mobile phases with solvent polarity indexes of P_{tot1} and P_{tot2}.
    (a) Based on your knowledge of RPLC and NPLC, explain any differences in the two preceding equations.
    (b) An analyte is found to elute from a reversed-phase column with a retention factor of 12.5 when a 20% isopropanol/80% water mixture is used as the mobile phase. What retention factor is expected for this same analyte if the mobile phase is changed to a 10% isopropanol/90% water mixture?

    (c) What mixture of isopropanol and water is needed if the analyte in Part (b) is to be eluted with a retention factor of 5.0?

39. The herbicide 2,4-dichlorophenyoxyacetic acid (2,4-D) (discussed in Chapter 21) has higher retention on a reversed phase column at an acidic pH than at a neutral pH. Explain why this difference in retention occurs. From your explanation, how do you expect the retention of 2,4-D will change on an anion-exchange column under the same conditions?

40. What are some typical applications for NPLC? What are some typical applications for RPLC? Why do you think RPLC is currently the more popular of these two methods?

ION-EXCHANGE CHROMATOGRAPHY

41. Define "ion-exchange chromatography." What is the mechanism of separation in this method?

42. What is "cation-exchange chromatography"? What is "anion-exchange chromatography"? What types of analytes are retained in each of these methods?

43. State whether you would use cation-exchange chromatography or anion-exchange chromatography for each of the following situations.
    (a) The separation and analysis of Ba^{2+}, Ca^{2+}, Mg^{2+} and Sr^{2+}
    (b) The separation and analysis of Br^-, Cl^-, NO_3^- and SO_4^{2-}
    (c) The separation of amino acids under basic conditions (pH > pI of the amino acids)

44. Write a reaction for anion-exchange chromatography that is similar to the given in Equation 20.7 for cation-exchange chromatography. Write an expression for the equilibrium constant for your reaction. How do your reaction and equation differ from those in Equation 22.7?

45. What is a "selectivity coefficient" in ion-exchange chromatography? What is the selectivity coefficient for the reaction you wrote in Problem 43?

46. What are some general factors that affect the retention of an analyte in ion-exchange chromatography?

47. List several types of supports that are used in ion-exchange chromatography. Describe the useful properties and applications for each support in such separations.

48. What type of solvent is a weak mobile phase in ion-exchange chromatography? What is a strong mobile phase? What factors can be varied to adjust the retention of analytes in this method?

49. What are some general applications for ion-exchange chromatography?

50. What is "suppressor ion chromatography"? Discuss how this type of chromatography works, using the system in Figure 22.13 as an example.

SIZE-EXCLUSION CHROMATOGRAPHY

51. Define "size-exclusion chromatography." What is the mechanism of separation in this method?

52. Explain why all analytes in size-exclusion chromatography elute within a relatively narrow range of retention volumes. What are the upper and lower limits to this range of volumes? How are these limits related to the column and its support material?

53. What properties are desired for a support in size-exclusion chromatography? What general types of supports are used in this method? Why is the pore size of the support important?

54. Explain why there is no weak or strong mobile phase in size-exclusion chromatography.
55. What is "gel filtration chromatography"? What is "gel permeation chromatography"?
56. What are some typical applications of size-exclusion chromatography?
57. A polymer sample has a retention volume of 25.2 mL on a size-exclusion column that has a reported void volume of 29.3 mL and an excluded volume of 14.1 mL. What is the average value of K_o for this polymer sample?
58. The following data were obtained for a series of protein standards injected onto a size-exclusion column. An unknown protein injected under the same conditions has an elution volume of 16.75 mL.

Protein/Analyte	Molar Mass (g/mol)	Retention Volume (mL)
Blue dextran	>10^6	11.00
Ferritin monomer	450,000	11.22
Gamma globulin	167,000	13.36
Hexokinase	104,000	14.85
Bovine serum albumin	68,000	15.92
Ovalbumin	45,000	16.97
Trypsin inhibitor	24,000	18.68
Cytochrome C	12,500	19.01

(a) What are the approximate values of V_M and V_e for this column?
(b) What is K_o for the unknown protein?
(c) What is the approximate molar mass of the unknown protein?

AFFINITY CHROMATOGRAPHY

59. What is "affinity chromatography"? What types of interactions form the basis for affinity chromatography?
60. Define or describe the following terms as related to affinity chromatography.
(a) Affinity ligand
(b) Immobilization method
(c) High-specificity ligand
(d) General ligand
(e) Application buffer
(f) Elution buffer
61. Describe the typical "on/off" elution method for affinity chromatography. What is the purpose of each step in this process?
62. What general factors determine the retention factor for an analyte in affinity chromatography? Write an equation that shows how these factors are related to k.
63. An affinity column with a void volume of 1.5 mL contains 150 nmol of protein A, an affinity ligand that binds strongly to many antibodies. The association equilibrium constant for the binding of protein A with one type of antibody (rabbit IgG) is approximately $4.0 \times 10^8 \ M^{-1}$ at pH 7.4.
(a) What is the approximate retention factor for rabbit IgG on this column at pH 7.4?
(b) At pH 7.4, how long would it take rabbit IgG to elute from the column at 1.0 mL/min?

64. When a pH 2.5 elution buffer is applied to the protein A column in Problem 63, rabbit IgG elutes within 3 min at 1.0 mL/min. What is the maximum association equilibrium constant that can be present between protein A and rabbit IgG under these conditions?
65. What is a "chiral separation"? What types of stationary phases are used in chiral separations? Why are these separations important in pharmaceutical analysis?
66. What types of supports are used in affinity chromatography?
67. Describe what type of solvent is a strong mobile phase or weak mobile phase in affinity chromatography?
68. What is meant by the terms "biospecific elution" and "nonspecific elution" in affinity chromatography? How does each of these methods work?

TYPES OF LIQUID CHROMATOGRAPHY DETECTORS

69. How does a refractive-index detector work? What are the advantages and disadvantages of this detector when used in LC?
70. Describe three types of absorbance detectors that are used in LC. What are the advantages and disadvantages of these detectors?
71. Describe how each of the following selective detectors is used in LC. What types of analytes can each of these detectors be used to monitor?
(a) Fluorescence detector
(b) Evaporative light-scattering detector
(c) Conductivity detector
(d) Electrochemical detector
72. What is electrospray ionization and how does it work? How is this method used in liquid chromatography/mass spectrometry (LC/MS)?
73. The chromatograms that are shown in Figures 22.10 and 20.22 for delaviridine and its metabolites were obtained using LC/MS operated in a general detection mode or in a selective detection mode, respectively.
(a) How are these chromatograms similar in terms of their measured response? How are they different?
(b) Explain how a detector based on electrospray-ionization mass spectrometry might be used to obtain either of these two sets of chromatograms.

LIQUID CHROMATOGRAPHY EQUIPMENT AND SAMPLE PRETREATMENT

74. Describe how samples are typically injected or placed into an LC system.
75. What are some common types of pumps used in HPLC?
76. Give some examples of sample pretreatment methods that are employed in LC.
77. Define "precolumn derivatization" and "postcolumn derivatization." How can each of these be used in liquid chromatography?

CHALLENGE PROBLEMS

78. Define the following terms.
(a) Paper chromatography
(b) Thin-layer chromatography
(c) Solvent front
(d) Retardation factor

79. A laboratory that performs clinical drug tests obtains a spot on a thin-layer chromatography (TLC) plate that travels a distance of 3.47 cm when the solvent front has traveled 4.50 cm.
    (a) What is the R_F value for the compound that makes up this spot?
    (b) The manufacturer of this TLC kit lists the following R_F values for common drugs. What is the most likely identity of the compound in the unknown sample?

Compound	R_F value
Morphine	0.14
Codeine	0.22
Quinine	0.38
Meperidine	0.53
Amitriptyline	0.63
Methadone	0.67
Pentazocine	0.73
Phenobarbital	0.77
Phencyclidine	0.79
Propoxyphene	0.81
Xylocaine	0.82
Methaqualone	0.87

80. It can be shown that R_F in planar chromatography can be related to the retention factor (k) for a solute in a column when both systems use the same stationary phase and mobile phase.

$$k = (1 - R_F)/R_F \quad \text{or} \quad R_F = 1/(1 + k) \quad (22.13)$$

An organic chemist wishes to use TLC to estimate the retention that would be expected during the separation of a series of newly synthesized compounds on a silica column. Screening of the reaction mixture by a silica TLC plate gives three major bands with migration distances of 1.23, 1.86, and 2.59 cm for a solvent front that travels 5.28 cm. What are the expected retention factors for the compounds in these bands on a silica column under the same mobile phase conditions?

81. Describe one type of LC (adsorption, partition, size-exclusion, etc.) that could be used for each of the following cases. Explain the reason for each of your answers.
    (a) The removal of salts from a protein sample prior to analysis by mass spectrometry
    (b) The analysis of morphine and its metabolites in serum samples
    (c) Isolation of a specific bacterial protein from a cell culture
    (d) Analysis of the content of nitrate and nitrite ions in drinking water

82. The retention factor k for a solute in adsorption chromatography can be described by the following equation,[6]

$$\log k = \log(V_a \cdot w/V_M) + \alpha'(S_0 - A_s\varepsilon_1) \quad (22.14)$$

where V_a is the volume of adsorbed solvent per unit mass of support, w is the weight of the support in the column,

V_M is the column void volume, α' is the support's activity parameter (related to the surface energy of the support), S_0 is free energy of adsorption for the analyte on the support, A_s is the cross-sectional area of the analyte, and ε_1 is the elutropic strength of the mobile phase. Using the previous equation, determine whether there will be an increase or decrease in k when an increase is made in the term $(V_a \cdot w/V_M)$. What happens to k when there is an increase in A_s or ε_1? How do you explain these trends?

83. Based on your knowledge of intermolecular interactions and chemical polarity (see Chapter 7), predict the order in which the following compounds will elute from a C_{18} reversed phase column at pH 8.0: octanoic acid ($C_7H_{15}COOH$), 1-octanol ($C_7H_{15}CH_2OH$), octadecanoic acid ($C_{17}H_{35}COOH$).

84. The retention of a small analyte in RPLC can be described by Equation 22.15,

$$\log k = \log k_w + a\varphi + b\varphi^2 \quad (22.15)$$

where k_w is the retention factor measured in the presence of water, φ is the volume fraction of an organic solvent in the mobile phase, and a or b are constants for a given combination of solute and solvent.[6] Analysis of the herbicide atrazine by RPLC gives the following best-fit parameters for Equation 22.15: $\log k_w = 2.42$, $a = -6.2$, $b = 4.1$.[30]
    (a) Use a spreadsheet to prepare plots of $\log k$ and k vs. φ for atrazine between $\varphi = 0.0$ and 0.6. Based on these results, will the retention factor increase or decrease for atrazine as more organic solvent is added to the mobile phase? How does this answer fit with your knowledge of reversed-phase chromatography?
    (b) At what values of φ will a retention factor of greater than 2.0 be obtained for atrazine? Under what conditions will the retention factor be less than 2.0? What is the estimated retention factor for atrazine if no organic modifier is present?

85. A scientist wishes to develop an ion chromatography system for the analysis of small cations such as Na^+, K^+, and Ca^{2+}.
    (a) The first part of this system has a strong cation exchanger that initially contains H^+ as the counterion. Give one specific stationary phase that could be used for this purpose. If a dilute solution of HCl is used as the mobile phase, will the resulting combinations of the analyte cations and competing cations with their counterions be expected to have low or high conductivity?
    (b) The second part of this system is a suppressor column that contains an anion exchanger with OH^- as the counterion. Give one specific stationary phase that could be used in such a column. After passing through this column, will the resulting combinations of the analyte cations and competing cations with their counterions be expected to have any change in their conductivity? How will this affect your ability to detect the analytes?

86. A scientists notices that the injection of standards containing the ions phosphate (PO_4^{3-}) or phosphite (PO_3^{3-}) give separate peaks when examined by ion chromatography, while injections of phosphate or monohydrogen phosphate (HPO_4^{2-}) appear to give a single peak. Explain why this is the case. (*Hint*: Consider the acid–base reactions of phosphate, as discussed in Chapter 8.)

87. The retention factor k for an analyte during biospecific elution in affinity chromatography can be described by Equation 22.16,

$$k = \frac{K_{A,A}\,(m_L/V_M)}{1 + K_{A,I}\,[I]} \tag{22.16}$$

where $K_{A,A}$ is the association equilibrium constant for the binding of the analyte to the immobilized ligand, $K_{A,I}$ is the association equilibrium constant for the binding of a competing agent in the mobile phase to the same ligand, $[I]$ is the mobile phase concentration of the competing agent, m_L is the moles of binding sites for the analyte in the column, and V_M is the column void volume.[24]

(a) If an analyte has a retention factor of approximately 250 in the absence of any competing agent (where $k = K_{A,A}(m_L/V_M)$) and the competing agent has a $K_{A,I}$ value of $1.0 \times 10^3\,M^{-1}$, what concentration of competing agent will be needed to give a retention factor of 20 for the analyte? A retention factor of 5?

(b) How do the results in Part (a) change if $K_{A,I}$ has a value of $1.0 \times 10^4\,M^{-1}$?

(c) Discuss how you might use Equation 22.16 to determine the value of $K_{A,I}$ for this system.

TOPICS FOR DISCUSSION AND REPORTS

88. Select an environmental chemical that might be found in water and obtain information on how this substance can be examined using liquid chromatography. State the type of liquid chromatography that was used for the separation or analysis of this chemical, including the types of stationary phase, mobile phase, and detection method that were employed.

89. A number of other specialized chromatographic methods are listed below.[6,7,10,11,23,24] Find more information on one or more of these methods and prepare a report. Discuss in your report how the method works, including a description of the stationary phases and mobile phases that are commonly used with this technique. Also give some examples of common applications for the method.
(a) Immobilized metal-ion affinity chromatography
(b) Ion-pair chromatography
(c) Hydrophobic interaction chromatography
(d) Fast-protein liquid chromatography
(e) Microbore liquid chromatography
(f) Process-scale liquid chromatography

90. Obtain more information on the method of ultrahigh pressure liquid chromatography.[12,13] What are the advantages of this method compared to traditional HPLC? What are some additional challenges and problems that must be faced when working with ultrahigh pressure liquid chromatography?

91. There are many detectors for liquid chromatography besides those already listed in this chapter.[6,23,25] Obtain information on one of the following detectors. Describe how the detector works and state what types of analytes it can be used to monitor.
(a) Flame photometric detector
(b) ICP atomic-emission detector
(c) Chemiluminescence detector
(d) LC-NMR
(e) LC-FTIR
(f) Optical activity detector

92. Obtain more information concerning the use of liquid chromatography in one of the following areas. For each application, state what types of stationary phases, mobile phases and detection methods are currently being used.
(a) Chiral separations
(b) Proteomics
(c) Therapeutic drug monitoring
(d) Polymer size analysis

93. Look up the material safety data sheet (MSDS) for one or more of the following solvents used in liquid chromatography. Describe any chemical or physical hazards that are associated with the solvent and give methods for its proper handling and disposal. Also, give at least one type of liquid chromatography in which the solvent might be utilized.
(a) Methyl-t-butyl ketone
(b) Hexane
(c) Acetonitrile
(d) Isopropanol

94. Obtain information on one of the following chiral stationary phases.[6,10,11,24–27] Discuss how this stationary phase is used in chiral separations and give some examples of its applications.
(a) Cyclodextrins
(b) Pirkle columns
(c) Cellulose derivatives
(d) α_1-Acid glycoprotein

95. A *microdialysis probe* is a tool explored in recent years as a means for placing real-time biological samples into an HPLC system.[31] Obtain more information on this method. Write a report on how this method works and describe some of its applications.

96. A special type of stationary phase recently developed for affinity chromatography and chiral separations is a *molecularly imprinted polymer*, or *MIP*.[24,32,33] This polymer is prepared by forming a porous polymer around a target analyte and then removing this analyte, leaving a cavity that is complementary to this target in its shape and interactions. Obtain more information on molecularly imprinted polymers. Describe to your class how they are prepared and used in chromatographic applications.

97. The topic of supercritical fluids was discussed in Chapter 20 as an alternative phase for performing extractions. Supercritical fluids can also be used as mobile phases in chromatography, giving a technique known as *supercritical fluid chromatography*.[4,23,34–36] Obtain more information on this method. Discuss what advantages it has compared with GC and LC and describe some of its applications.

References

1. A. Verghese, "AIDS at 25: An Epidemic of Caring," *New York Times*, June 4, 2006.
2. World Health Organization, *AIDS, Profile of an Epidemic*, WHO, Washington, DC, 1989.
3. Amanda Yarnell, "AZT," *Chemical & Engineering News*, 83 (2005) 48.
4. R. E. Majors and P. W. Carr, "Glossary of Liquid-Phase Separation Terms," *LC-GC*, 19 (2001) 124–162.

5. J. Inczedy, T. Lengyel, and A. M. Ure, *International Union of Pure and Applied Chemistry—Compendium of Analytical Nomenclature: Definitive Rules 1997*, Blackwell Science, Malden, MA, Chapter 9.

6. C. F. Poole and S. K. Poole, *Chromatography Today*, Elsevier, New York, 1991.

7. B. L. Karger, L. R. Snyder, and C. Horvath, *An Introduction to Separation Science*, Wiley, New York, 1973.

8. L. S. Ettre, "M.S. Tswett and the Invention of Chromatography," *LC-GC*, 21 (2003) 458–467.

9. R. E. Majors, "A Review of HPLC Column Packing Technology," *American Laboratory*, October (2003) 46–54.

10. E. Katz, R. Eksteen, P. Schoenmakers, and N. Miler, Eds., *Handbook of HPLC*, Marcel Dekker, New York, 1998, Chapter 10.

11. W. J. Lough and I. W. Wainer, *High Performance Liquid Chromatography: Fundamentals Principles and Practice*, Blackie Academic, New York, 1995.

12. J. W. Thompson, J. S. Mellors, J. W. Eschelbach, and J. W. Jorgenson, "Recent Advances in Ultrahigh-Pressure Liquid Chromatography," *LC-GC*, 24 (2006) 16–20.

13. J. E. McNair, K. C. Lewis, and J. W. Jorgenson, "Ultrahigh-Pressure Reversed-Phase Liquid Chromatography in Packed Capillary Columns," *Analytical Chemistry*, 69 (1997) 983–989.

14. N. B. Afeyan, S. P. Fulton, and F. E. Regnier, "Perfusion Chromatography: Recent Developments and Applications," *Applied Enzyme Biotechnology*, 9th (1991) 221–231.

15. M. Jacoby, "Monolithic Chromatography," *Chemical & Engineering News*, 84 (2006) 14–19.

16. F. Svec and C. G. Huber, "Monolithic Materials: Promises, Challenges, Achievements," *Analytical Chemistry*, 78 (2006) 2100–2108.

17. A. Braithwaite and F. J. Smith, *Chromatographic Methods*, Chapman and Hall, New York, 1985, Chapter 3.

18. A. I. Vogel, A. R. Tatchell, B. S. Furnis, A. J. Hannaford, and P. W. G. Smith, *Vogel's Textbook of Practical Organic Chemistry*, 5th ed., Longman, London, 1989.

19. N. W. Tietz, Ed., *Textbook of Clinical Chemistry*, Saunders, Philadelphia, PA, 1986.

20. C. F. Poole, "Progress in Planar Chromatography," *Trends in Analytical Chemistry*, 4 (1985) 209–213.

21. R. E. Majors, "Current Trends in HPLC Column Usage," *LC-GC*, 15 (1997) 1008–1015.

22. L. Rohrschneider, "Solvent Characterization by Gas-Liquid Partition Coefficients of Selected Solutes," *Analytical Chemistry*, 45 (1973) 1241–1247.

23. B. Ravindranath, *Principles and Practice of Chromatography*, Wiley, New York, 1989.

24. D.S. Hage, Ed., *Handbook of Affinity Chromatography*, CRC Press, Boca Raton, FL, 2005.

25. D. W. Armstrong, "Direct Enantiomeric Separations in Liquid Chromatography and Gas Chromatography." In *A Century of Separation Science*, H. J. Issaq, Ed., Marcel Dekker, New York, 2002, Chapter 33.

26. G. Guebitz and M. G. Schmid, "Chiral Separation Principles: An Introduction," *Methods in Molecular Biology*, 243 (2004) 1–28.

27. S. Allenmark, *Chromatographic Enantioseparations: Methods and Applications*, 2nd ed., Ellis Horwood, New York, 1991.

28. G. W. Ewing, Ed., *Analytical Instrumentation Handbook*, 2nd ed., Marcel Dekker, New York, 1997.

29. G. Lunn and L. C. Hellwig, *Handbook of Derivatization Reactions for HPLC*, Wiley-Interscience, New York, 1998.

30. J. G. Rollag, M. Beck-Westermeyer, and D. S. Hage, "Analysis of Pesticide Degradation Products by Tandem High-Performance Immunoaffinity Chromatography and Reversed-Phase Liquid Chromatography," *Analytical Chemistry*, 68 (1996) 3631–3637.

31. C. E. Lunte, D. O. Scott, and P. T. Kissinger, "Sampling Living Systems using Microdialysis Probes," *Analytical Chemistry*, 63 (1991) 773A–780A.

32. K. Haupt, "Molecularly Imprinted Polymers: Artificial Receptors for Affinity Separations." In *Handbook of Affinity Chromatography*, 2nd ed., D. S. Hage, Ed., CRC Press, Boca Raton, FL, 2005, Chapter 30.

33. M. Komiyama, T. Takeuchi, T. Mukawa, and H. Asanuma, *Molecular Imprinting: From Fundamentals to Applications*, Wiley-VCH, Weinheim, Germany, 2002.

34. R. M. Smith, S. B. Hawthorne, Eds., *Supercritical Fluids in Chromatography and Extraction*, Elsevier, Amsterdam, the Netherlands, 1997.

35. M. D. Palmieri, "An Introduction to Supercritical Fluid Chromatography. Part 1: Principles and Instrumentation," *Journal of Chemical Education*, 65 (1988) A254–A259.

36. M. D. Palmieri, "An Introduction to Supercritical Fluid Chromatography. Part 2: Applications and Future Trends," *Journal of Chemical Education*, 66 (1989) A141–A147.

Chapter 23

Electrophoresis

Chapter Outline

23.1 INTRODUCTION: THE HUMAN GENOME PROJECT

February 2001 saw one of the greatest achievements of modern science. It was at this time that two scientific papers appeared, one in the journal *Science* and the other in *Nature*, reporting the sequence of human DNA (or the "human genome").[1,2] These papers were the result of a major research effort known as the Human Genome Project, which was formally begun in 1990 under the sponsorship of the U.S. Department of Energy and the National Institutes of Health.[3]

Although it was anticipated to take 15 years to finish, this project was "completed" in about a decade. This early completion was made possible by several advances that occurred in techniques for sequencing DNA. One common approach for sequencing DNA is the Sanger method (see Figure 23.1). In the Sanger method, the section of DNA to be examined (known as the "template") is mixed with a segment of DNA that binds to part of this sequence (the "primer"). This mixture is placed into four containers that have the nucleotides and enzymes needed to build on the template. These containers also have special labeled nucleotides that will stop the elongation of DNA after the addition of a C, G, A, or T to its sequence. The DNA strands formed in each container are later separated according to their size. By comparing the length of these strands and by knowing which labeled nucleotides are at the end of each strand, the sequence of the DNA can be determined.[4]

The Sanger method was originally developed as a manual technique that took long periods of time to perform. Thus, one thing that had to be addressed early in the Human Genome Project was the creation of faster, automated systems for sequencing DNA.[5,6] Both traditional and newer systems for accomplishing this sequencing utilize a separation method known as *electrophoresis*. In this chapter we learn about electrophoresis, look at its applications, and see how improvements in this technique made the Human Genome Project possible.

23.1A What Is Electrophoresis?

Electrophoresis is a technique in which solutes are separated by their different rates of migration in an electric field (see Figure 23.2).[7–10] To carry out this method, a sample is first placed in a container or support that also contains a background electrolyte (or "running buffer"). When an electric field is later applied to this system, the ions in the running buffer will flow from one electrode to the other and provide the current needed to maintain the applied voltage. At the same time, positively charged ions in the sample will move toward the negative electrode (the cathode), while negatively charged ions will move toward the positive electrode (the anode). The result is a separation of these ions based on their charge and size. Because many biological compounds have charges or ionizable groups (e.g., DNA and proteins), electrophoresis is frequently utilized in biochemical and

DNA replication

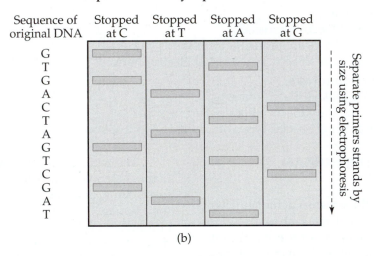

(a)

Separate and analyze primer strands

(b)

FIGURE 23.1   Sequencing of DNA by the Sanger method. This method is named after F. Sanger, one of the scientists who originally reported this technique.[4] The final DNA sequence is determined in this method by looking at the sequence of the primer strands and using the complementary nucleotides (C for G, A for T, G for C, and T for A) to describe the sequence of the original DNA.

medical research. This approach can also be adapted for work with small ions (like Cl^- or $NO_3{}^-$) or for large charged particles (such as cells and viruses).

Even though it has been known for one hundred years that substances like proteins and enzymes have a characteristic rate of travel in an electric field,[11–13] electrophoresis did not become a routine separation method until around the 1930s. One notable advance occurred in 1937 when a scientist named Arne Tiselius (Figure 23.3) used electrophoresis for the separation of serum proteins.[3,14] Tiselius conducted this separation by employing a U-shaped tube in which he placed his sample and running buffer. When he applied an electric field, proteins in the sample began to separate as they migrated toward the electrodes of opposite charge. However, the use of a large sample volume gave a series of broad and only partially resolved regions that contained different mixtures of the original proteins.[15]

The method employed by Tiselius is now known as *moving boundary electrophoresis*, because it produced a series

of moving boundaries between regions that contained different mixtures of proteins, as shown in Figure 23.3.[10,16] Today it is more common to use small samples to allow analytes to be separated into narrow bands or zones, giving a method known as *zone electrophoresis*.[8–10,16] An example of zone electrophoresis is shown in Figure 23.1, where DNA is sequenced by separating its strands of various lengths into narrow bands on a gel.

There are many ways in which electrophoresis is used for chemical analysis. These include the sequencing of DNA, as well as the purification of proteins, peptides, and other biomolecules. In clinical chemistry, electrophoresis is an important tool for examining the patterns of amino acids, serum proteins, enzymes, and lipoproteins in the body. Electrophoresis is also used in the analysis of organic and inorganic ions in foods, commercial products, and environmental samples. In addition, electrophoresis is an essential component of medical and pharmaceutical research for the characterization of

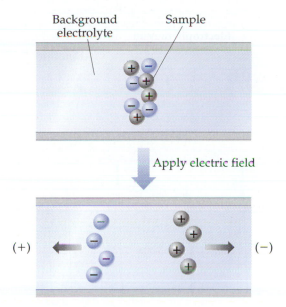

Background electrolyte   Sample

Apply electric field

(+) ←   → (−)

FIGURE 23.2 Separation of positively and negatively charged analytes in a sample by electrophoresis.

proteins in normal and diseased cells and for looking for new substances.[10]

23.1B How Is Electrophoresis Performed?

Electrophoresis can be performed in a variety of formats (see Figure 23.4). One format is to apply small amounts of a sample to a support (usually a gel) and allow the analytes in this sample to travel in a running buffer through the support when an electric field is applied.

This approach is known as *gel electrophoresis* (a method we will discuss in Section 23.3).[17–19] It is also possible to separate the components of a sample by using a narrow capillary that is filled with a running buffer and placed into an electric field. This second format is called *capillary electrophoresis* (discussed in Section 23.4).[17,19–22]

Depending on the type of electrophoresis being used, the resulting separation can be viewed in one of two ways. In the case of gel electrophoresis, the separation is stopped before analytes have traveled off the support. The result is a series of bands where the **migration distance (d_m)** characterizes the extent to which each analyte has interacted with the electric field. This approach is similar to that used to characterize the retention of analytes in thin-layer chromatography and paper chromatography (see Chapter 22). Because the migration distance of an analyte through a gel for electrophoresis will depend on the exact voltage and time used for the separation, it is common to include standard samples on the same support as the sample to help in analyte identification. The intensity of the analyte band is then used to measure the amount of this substance in the sample.

In capillary electrophoresis all analytes travel the same distance, from the point of injection to the opposite end where a detector is located. The analytes will differ, however, in the time it takes them to travel this distance, in a manner similar to what occurs in the chromatographic methods of gas chromatography (GC) and high-performance liquid chromatography (HPLC). In this situation the **migration time (t_m)** for each analyte is measured and recorded.[7] The resulting plot of detector response versus migration time is called an

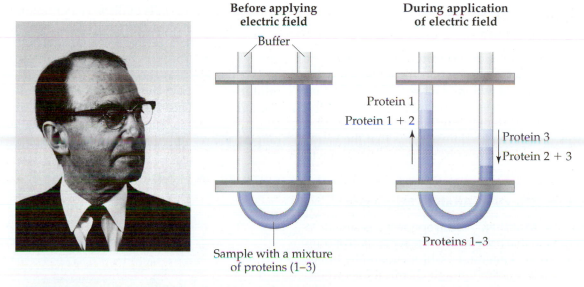

Before applying electric field

During application of electric field

Buffer

Protein 1
Protein 1 + 2

Protein 3
Protein 2 + 3

Sample with a mixture of proteins (1–3)

Proteins 1–3

FIGURE 23.3 Arne W. K. Tiselius (1902–1971), and an example of a protein separation performed by moving boundary electrophoresis. Tiselius was a Swedish scientist who won the 1948 Nobel Prize in chemistry for his early work in the field of electrophoresis. Tiselius began this research while working as a graduate student at the University of Uppsala in Sweden. He received his doctorate degree in 1930 and later returned in 1937 to the University of Uppsala as a professor of biochemistry. It is here that he explored the use of moving boundary electrophoresis to separate chemically similar proteins in blood.[3,15] Electrophoresis is still used today by clinical chemists when they examine the pattern of major and minor proteins in blood, urine, and other samples from the body.

Electrophoresis gel/support

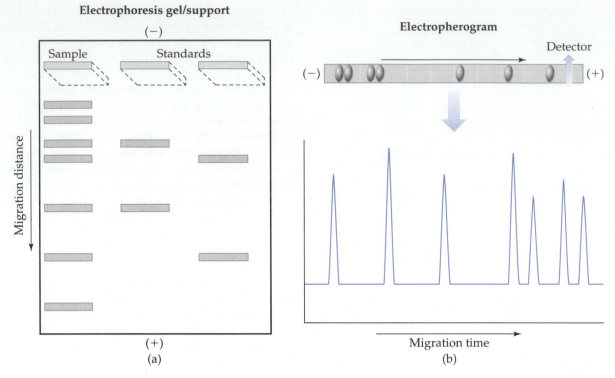

(a)

Electropherogram

(b)

FIGURE 23.4 Examples of the results produced by (a) gel electrophoresis, and (b) capillary electrophoresis.

electropherogram. The migration times in this plot can be used to help in analyte identification, while the peak heights or areas are used to determine the amount of each analyte. An internal standard is usually injected along with the sample to correct for variations during injection or small fluctuations in the experimental conditions during the separation.

23.2 GENERAL PRINCIPLES OF ELECTROPHORESIS

The separation of analytes by electrophoresis has two key requirements. The first requirement is there must be a difference in how analytes will interact with the separation system. In electrophoresis this requirement means the analytes must have different migration times or migration distances. The second requirement is that the bands or peaks for the analytes must be sufficiently narrow to allow them to be resolved.

23.2A Factors Affecting Analyte Migration

Electrophoretic Mobility. Electrophoresis is similar to chromatography in that both involve the separation of compounds by differential migration. Chromatography brings about differential migration through chemical interactions between analytes with the stationary phase and mobile phase. In electrophoresis, differential migration is produced by the movement of analytes in an electric field, where their rate of migration will depend on their size and charge.

The overall rate of travel of a charged solute in electrophoresis will depend on two opposing forces (see Figure 23.5). The first of these forces (F_+) is the attraction

of a charged solute toward the electrode of opposite charge. This force depends on the strength of the applied electric field (E, units of volts per distance) and the charge on the solute (z). The second force acting on the solute is resistance to its movement, as created by the surrounding medium. The force of this resistance (F_-) depends on the "size" of the solute (as described by its solvated radius r), the viscosity of the medium (η), and the solute's velocity of migration (v, in units of distance per time).

When an electric field is applied, a solute will accelerate toward the electrode of opposite charge until the forces F_+ and F_- become equal in size (although opposite in direction).[10,21] At this point a steady-state situation is produced in which the solute begins to move at a constant velocity. This velocity can be found by setting the expressions for F_+ and F_- equal to each other and rearranging the resulting equation in terms of v.

$$6\pi r\eta v = E z \qquad \text{or} \qquad v = \frac{E z}{6\pi r\eta} \qquad (23.1)$$

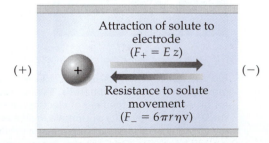

FIGURE 23.5 Forces that determine electrophoretic mobility.

To see how this velocity will be affected by only the strength of the electric field, we can combine the other terms in Equation 23.1 to give a single constant (μ),

$$v = \mu E \qquad (23.2)$$

where $\mu = z/(6 \pi r \eta)$. This new combination of terms is known as the **electrophoretic mobility**, which is represented by the symbol μ.[7,9] The value of μ is often expressed in units of $m^2/V \cdot s$ or $cm^2/kV \cdot min$ and is constant for a given analyte under a particular set of temperature and solvent conditions. The value of μ also depends on the apparent size and charge of the solute, as represented by the ratio z/r in Equation 23.1. This last feature means that any two solutes with different charge-to-size ratios can, in theory, be separated by electrophoresis.

EXERCISE 23.1	Determining the Electrophoretic Mobility for an Analyte

The apparent electrophoretic mobility for an analyte in capillary electrophoresis can be found by rewriting Equation 23.2 in the form shown.

$$\mu = \frac{v}{E} = \frac{(L_d/t_m)}{(V/L)} \qquad (23.3)$$

In this equation, V is the voltage applied to the electrophoretic system over a length L, and L_d is the distance traveled from the point of application to the detector by the analyte in migration time t_m.

A sample of several proteins is applied to a neutral-coated capillary with a total length of 25.0 cm and a distance to the detector of 22.0 cm. Two of the proteins in the sample give migration times of 15.3 min and 16.2 min when using an applied voltage of 20.0 kV. What are the migration velocities and electrophoretic mobilities of these proteins under these conditions? What will their electrophoretic mobilities and migration times be at an applied voltage of 10.0 kV?

SOLUTION

The electrophoretic mobility of the first protein can be found by substituting the known values for L_d (22.0 cm), t_m (15.3 min), V (20.0 kV), and L (25.0 cm) into Equation 23.3.

$$\text{Protein 1:} \quad \mu = \frac{(22.0 \text{ cm}/15.3 \text{ min})}{(20.0 \text{ kV}/25.0 \text{ cm})} = 1.80 \text{ cm}^2/\text{kV} \cdot \text{min}$$

A similar calculation for the second protein gives an electrophoretic mobility of **1.70 cm²/kV · min**. The lower electrophoretic mobility of the second protein makes sense because it takes longer for this protein to migrate through the system. The migration velocities for these proteins can be found by simply dividing their distance of travel by their migration times ($v = L_d/t_m$), which gives $(22.0 \text{ cm}/15.3 \text{ min}) = $ **1.44 cm/min** and $(22 \text{ cm}/16.2 \text{ min}) = $ **1.36 cm/min** for proteins 1 and 2.

If we lower the applied voltage from 20 kV to 10 kV (a twofold change), the migration times will increase and the migration velocities for these proteins will decrease (also by twofold), but their electrophoretic mobilities will remain exactly the same. This situation occurs because the electrophoretic mobility is independent of voltage and electric field strength, while migration times and velocities are not. Thus, if there is a decrease in V and E, Equation 23.3 indicates there must be a proportional decrease in v and t_m to keep μ constant.

Secondary Interactions. To obtain good separations in electrophoresis it is often necessary to adjust the conditions of this method to change the electrophoretic mobility of a solute. We can accomplish this goal by using secondary reactions that alter the charge or apparent size of the solute. If an analyte is a weak acid or weak base, for example, its net charge can be varied by changing the pH. In the case of a weak monoprotic acid, the main species at a pH well below the pK_a will be the neutral form of the acid (HA), while the dominant species at a pH much greater than the pK_a will be the negatively charged conjugate base (A^-). At an intermediate pH, we will have a mixture of these two forms and the average charge for all of these species will be somewhere between "0" and "−1." As a result, the overall observed electrophoretic mobility for such a compound (as well as for other weak acids and weak bases) can be adjusted by varying the pH.

It is also possible to use side reactions to change the effective size or charge of the analyte. This effect occurs in a method known as sodium dodecyl sulfate polyacrylamide gel electrophoresis (SDS-PAGE), which is a technique for separating proteins according to their size (see Section 23.4C). This analysis begins by first denaturing the proteins and coating them with sodium dodecyl sulfate, a negatively charged surfactant. The coating process can be thought of as a type of complexation reaction. The negative coating not only alters the overall charge but helps convert a protein into a rod-shaped structure, which alters its size and shape.[18,19]

Another approach for altering the apparent electrophoretic mobility of an analyte is to use a solubility equilibrium. As an example, we could include a second phase within the running buffer into which the analyte can partition as it moves through the system (such as through the use of micelles, a method we will examine in Section 23.4C). Because the analyte in such a system will usually have different mobilities when it is present in the running buffer or in the second phase, the partitioning of an analyte between these regions leads to a change in the analyte's rate of travel through the electrophoretic system. Physical interactions can also affect analyte migration. For instance, DNA sequencing by gel electrophoresis uses a porous support to separate DNA strands of different lengths. The same strategy is used in SDS-PAGE for protein separations.

Electroosmosis. Up until now we have examined only the direct movement of an analyte in an electric field. It is also possible for the running buffer to move in such a field. This phenomenon can occur if there are any fixed charges present in the system, such as on the interior surface of an electrophoretic system or on a support within this system (see Figure 23.6). The presence of these fixed charges attracts ions of opposite charge from the running buffer and creates an electrical double layer at the surface of the support. In the presence of an electric field, this double layer acts like a piston that causes a net movement of the buffer toward the electrode of opposite charge versus the fixed ionic groups. This process is known as **electroosmosis** and results in a net flow of the buffer and its contents through the system.[7]

The extent to which electroosmosis affects the buffer and analytes in electrophoresis is described by using a term known as the *electroosmotic mobility* (or μ_{eo}).[7] This term has the same units as the electrophoretic mobility μ. The value of μ_{eo} depends on such factors as the size of the electric field, the type of running buffer that is being employed, and the type of charge that is present on the support. This relationship is described by Equation 23.4,

$$\mu_{eo} = (\varepsilon \zeta E)/\eta \tag{23.4}$$

where E is the electric field, ε and η are the dielectric constant and viscosity of the running buffer, and ζ is the zeta potential (which represents the charge on the support).

Depending on the direction of buffer flow, electroosmosis can work either with or against the inherent migration of an analyte through the electrophoretic system. The overall observed electrophoretic mobility (μ_{Net}) for an analyte will be equal to the sum of its own electrophoretic mobility (μ) and the mobility of the running buffer due to electrosmotic flow (μ_{eo}).

$$\mu_{Net} = \mu + \mu_{eo} \tag{23.5}$$

In gel electrophoresis, electroosmotic flow is often small compared to the inherent rate of analyte migration. This is not usually true in capillary electrophoresis, where the support has a relatively large charge and high surface area compared to the volume of running buffer (see Section 23.3).

23.2B Factors Affecting Band-Broadening

The same terms used to describe efficiency in chromatography (e.g., the number of theoretical plates N and the height equivalent of a theoretical plate H) can be used to describe band-broadening in electrophoresis. Two particularly important band-broadening processes in electrophoresis are (1) longitudinal diffusion and (2) Joule heating.

Longitudinal Diffusion. You may recall from Chapter 20 that longitudinal diffusion occurs when a solute diffuses away from the center of its band along the direction of travel, causing this band to broaden over time and to become less concentrated. One factor that affects the extent of this band-broadening is the "size" of the diffusing solute, or its solvated radius. Because larger analytes have slower diffusion, they will be less affected by longitudinal diffusion than smaller substances. The rate of this diffusion will also decrease as we increase the viscosity of the running buffer or lower the temperature of the system.

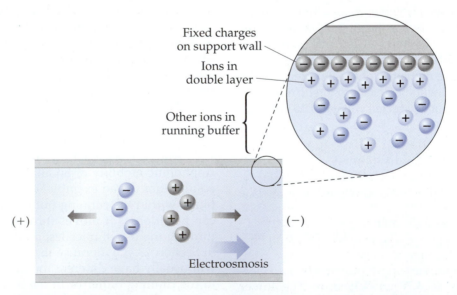

Fixed charges on support wall

Ions in double layer

Other ions in running buffer

(+)   (−)

Electroosmosis

FIGURE 23.6 The production and effects of electroosmosis. This particular example shows a support that has a negatively charged interior. Such a situation is often encountered when working with a support that is an uncoated silica capillary. The interior wall of this capillary has silanol groups at its surface, which can act as weak acids and form a conjugate base with a negative charge. The extent of electroosmosis in this case will depend on the pH of the running buffer, because this will affect the relative amount of the silanol groups that are present in their neutral acid form or charged conjugate base form.

The extent of longitudinal diffusion will depend on the amount of time that is allowed for this process to occur.[10] This time, in turn, will be affected in electrophoresis by the size of the electric field, because lower electric fields result in smaller migration velocities and longer migration times.[22] Electroosmosis will also affect the time needed for an electrophoretic separation and diffusion. If electroosmosis moves in a direction opposite to that desired for the separation of analytes, the effective rate of travel for these analytes is decreased and the time allowed for longitudinal diffusion is increased. If electroosmosis instead occurs in the same direction as analyte migration, longitudinal diffusion is decreased.

One way we can minimize the effects of longitudinal diffusion in electrophoresis is to have an analyte move through a porous support. If the pores of this support are sufficiently small, they will inhibit the movement of analytes due to diffusion and help provide narrower bands. If the pore size becomes too small, a size-based separation will also be created. Although this last feature is not always desirable, in some cases it can be an advantage, such as in the sequencing of DNA by gel electrophoresis.

Joule Heating. The most important band-broadening process in electrophoresis is often **Joule heating**.[21-23] This process is caused by heating that occurs whenever an electric field is applied to the system. According to *Ohm's law* (see Chapter 14), placing a voltage V across a medium with a resistance of R requires that a current of I be present to maintain this voltage across the medium.[10]

$$\text{Ohm's law:} \qquad V = I \cdot R \qquad (23.6)$$

As current flows through the system, heat is generated. This heat production depends on the voltage, current, and time t the current passes through the system, as shown below.

$$\text{Heat} = V \cdot I \cdot t \qquad (23.7)$$

As heat is produced, the temperature of the electrophoretic system will begin to rise. This rise in temperature will increase longitudinal diffusion and lead to increased band-broadening. In addition, if the heat is not distributed uniformly throughout the electrophoretic system, the temperature will not be the same throughout the system. An uneven temperature will lead to regions with different densities (causing mixing) and different rates of diffusion, which results in even more band-broadening. Other problems created by an increase in temperature include possible degradation of the analytes or components of the system and the evaporation of solvent from the running buffer, the latter of which can alter the pH and composition of the buffer. All of these factors lead to a loss of reproducibility and efficiency in the system.

One way Joule heating can be decreased is by using a lower voltage for the separation. A lower voltage, however, will lower the migration velocities of analytes and give longer separation times. An alternative approach is to use more efficient cooling for the system, which would allow higher voltages to be used and provide shorter separation times. Another possibility is to add a support to the electrophoretic system that minimizes the effects of Joule heating due to uneven heat distribution and density gradients in the running buffer. Examples of these approaches will be given later when we examine the methods of gel electrophoresis and capillary electrophoresis.

Another factor that affects Joule heating is the ionic strength of the running buffer. A lower ionic strength for this buffer will lower heat production, because at low ionic strengths there are fewer ions in this buffer. This lower ionic strength creates a greater resistance R to current flow at any given voltage because fewer ions are available to carry the current. We can see from Ohm's law in Equation 23.6 that as R increases a smaller current is needed at voltage V. This smaller current, in turn, will create lower heat production, as shown by Equation 23.7.

Other Factors. Eddy diffusion (a process we discussed in Chapter 20 for chromatography) is another factor that can sometimes lead to band-broadening in electrophoresis. This type of band-broadening can occur if a support is used to minimize the effects of Joule heating, a situation that creates multiple flow paths for analytes through the support. If the support interacts with analytes, band-broadening due to these secondary interactions will be introduced as well; this extra band-broadening also occurs when secondary interactions are used to adjust analyte mobility, such as complexation reactions or partitioning into a micelle. These latter effects are similar to those described in Chapter 20 for stationary phase mass transfer in chromatography. Broadening of the peaks before or after separation can be another issue when dealing with highly efficient systems, such as those used in capillary electrophoresis.

Wick flow is another source of band-broadening that occurs in gel electrophoresis.[19] In such a system, the gel is kept in contact with the electrodes and buffer reservoirs through the use of wicks. Because this support is often open to air, the presence of any Joule heating will lead to some evaporation of solvent in the running buffer from the support. As this solvent is lost, it is replenished by the flow of more solvent through the wicks and from the buffer reservoirs. This flow leads to a net movement of buffer from each reservoir towards the center of the support. The rate of this flow depends on the rate of solvent evaporation, so it will increase with the use of a high voltage or high current. The extent of this flow varies across the support, with the fastest rates occurring furthest from the center of the support.

23.3 GEL ELECTROPHORESIS

23.3A What Is Gel Electrophoresis?

One of the most common types of electrophoresis is the method of **gel electrophoresis**. This technique is an electrophoretic method that is performed by applying a sample to a gel support that is then placed into an electric field.[17–20] Typical separations obtained by gel electrophoresis were shown previously in Figures 23.1 and 23.4. In this type of system, several samples are usually applied to the gel and allowed to migrate along the length of the support in the presence of an applied electric field. The separation is stopped before analytes have left the end of the gel, with the location and intensities then being determined.

It is important to remember in gel electrophoresis that the velocity of an analyte's movement will be related to the distance it has traveled in the given separation time (as represented by the migration distance). The farther this distance is from the point of sample application, the higher the migration velocity is for the analyte and the larger its electrophoretic mobility. This migration distance will, in turn, be related to the size and charge of the analyte and can be used in identifying such a substance.

23.3B How Is Gel Electrophoresis Performed?

Equipment and Supports. Some typical systems for carrying out gel electrophoresis are shown in Figure 23.7. These systems may have a support that is held in either a vertical or horizontal position. This support contains a running buffer with ions that carry a current through the support when an electric field is applied. To replenish this buffer and its components as they move through the support or evaporate, the ends of the support are placed in contact with two reservoirs that contain the same buffer solution and the electrodes. Once samples have been placed on the support, the electrodes are connected to a power supply and used to apply a voltage across the support. This electric field is passed through the system for a given amount of time, causing the sample components to migrate. After the electric field has been turned off, the gel is removed and examined to locate the analyte bands.

The type of support we use in such a system will depend on our analytes and samples.[17,19] Cellulose acetate, filter paper, and starch are useful supports for work with relatively small molecules, like amino acids and nucleotides. Electrophoresis involving large molecules can

be carried out on agarose, a support that we discussed in Chapter 22. The resulting approach is known as "agarose electrophoresis." In addition to its low nonspecific binding for many biological compounds, agarose has a low inherent charge. Agarose also has relatively wide pores that allow it to be employed in work with large molecules, such as during the sequencing of DNA.

The most common support used in gel electrophoresis is *polyacrylamide*. This combination is often referred to as *polyacrylamide gel electrophoresis*, or *PAGE*.[17–19] Polyacrylamide is a synthetic, transparent polymer that is prepared as shown in Figure 23.8. It can be made with a variety of pore sizes that are smaller than those in agarose and of a size more suitable for the separation of proteins and peptide mixtures. Like agarose, polyacrylamide has low nonspecific binding for many biological compounds and does not have any inherent charged groups in its structure.

Sample Application. The samples in gel electrophoresis are applied to small "wells" that are made in the gel during its preparation (see Figures 23.4 and 23.7). A sample volume of 10–100 µL is then placed into one of these wells by using a micropipette. These sample volumes help provide a sufficient amount of analyte for later detection and collection, but they also create a danger of introducing band-broadening by creating a large sample band at the beginning of the separation.

A common approach to create narrow sample bands is to employ two types of gels in the system: a "stacking gel" and a "running gel."[19] The running gel is the support used for the electrophoretic separation of substances in the sample. In a vertical gel electrophoresis system, this gel is formed first and is located throughout the middle and lower section of the system (see right-hand portion of Figure 23.7). The stacking gel has a lower degree of cross-linking (giving it larger pores) and is located on top of the running gel. The stacking gel is also the section of the support in which the sample wells are located. After a sample has been placed in the wells and an electric field has been applied, analytes will travel quickly through the stacking gel until they reach its boundary with the running gel.

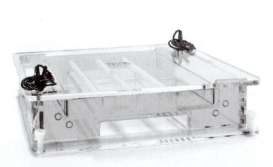

Horizontal gel electrophoresis system   Vertical gel electrophoresis system

FIGURE 23.7 Horizontal (image on the left) and vertical (image on the right) gel electrophoresis systems. (Reproduced with permission from Thermo Fisher Scientific)

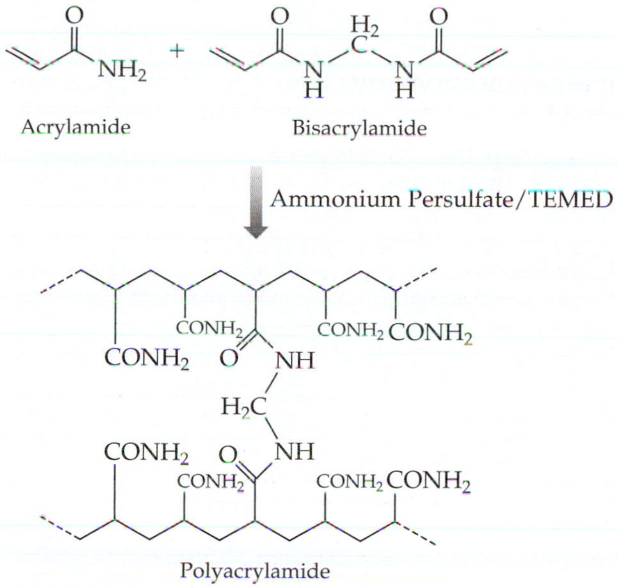

FIGURE 23.8 Preparation of a polyacrylamide gel. In this reaction, acrylamide is used as the monomer and bisacrylamide is used as a cross-linking agent. The reaction of these two agents is begun by adding ammonium persulfate, where persulfate ($S_2O_8{}^{2-}$) forms sulfate radicals ($SO_4{}^-$) that cause the acrylamide and bisacrylamide to combine. *N,N,N',N'*-Tetramethylethylenediamine (TEMED) is added to this mixture as a reagent that stabilizes the sulfate radicals. The size of the pores that are formed in the polyacrylamide gel will be related to how much bisacrylamide is used vs. acrylamide. As the amount of bisacrylamide is increased, more cross-linking occurs and smaller pores are formed in the gel. As less bisacrylamide is used, larger pores are formed, but the gel also becomes less rigid.

These substances will then travel much more slowly, allowing other parts of the sample to catch up and to form a narrower, more concentrated band at the top of the running gel. The result is a system that can use larger sample volumes without introducing significant band-broadening into the final electrophoretic separation.

Detection Methods. There are several ways analytes can be detected in gel electrophoresis. Analyte bands can be examined directly on the gel or they can be transferred to a different support for detection. Direct detection can sometimes be performed visually (when dealing with intensely colored proteins like hemoglobin) or by using absorbance measurements and a scanning device known as a *densitometer*.[9,20]

The most common approach for detection in gel electrophoresis is to treat the support with a stain or reagent that makes it easier to see the analyte bands. Examples of stains that are used for proteins are Amido black, Coomassie Brilliant Blue, and Ponceau S. These stains are all highly conjugated dyes with large molar absorptivities (see Chapter 18). Silver nitrate is used in a method known as *silver staining* to detect low concentration proteins. DNA bands can be detected by using ethidium bromide (see Chapter 2). When separating enzymes, the natural catalytic ability of these substances can be employed for their

detection, as occurs when using the fluorescent compound NAD(P)H to detect enzymes that generate this substance in their reactions.[19,20]

Another possible approach for detection in gel electrophoresis is to transfer a portion of the analyte bands to a second support (such as nitrocelluose), where they are reacted with a labeled agent. This approach is known as "blotting."[19] There are several blotting methods. One such method is a **Southern blot** (named after its discoverer Edwin Southern, a British biologist).[24] A Southern blot is used to detect specific sequences of DNA by having these sequences bind to an added, known sequence of DNA that is labeled with a radioactive tag (^{32}P) or with a label that can undergo chemiluminescence. A **Northern blot** (which was developed after the Southern blot) is similar, but is instead used to detect specific sequences of RNA by using a labeled DNA probe.[25]

Another type of blotting method is a **Western blot**.[26,27] A Western blot is used to detect specific proteins on an electrophoresis support. In this technique, proteins are first separated on a support by electrophoresis and then blotted onto a second support like nitrocellulose or nylon. The second support is then treated with labeled antibodies that can specifically bind the proteins of interest. After the antibodies and proteins have been allowed to form complexes, any extra antibodies are washed away and the remaining bound antibodies are detected through their labels, indicating whether there is any of the protein of interest present. This method is used to screen blood for the HIV virus by looking for the presence of proteins from this virus in samples.

There also has been growing interest in the use of instrumental methods for analyzing bands on electrophoresis supports. For instance, mass spectrometry is becoming a popular method for determining the molecular mass of a protein in a particular band. Such an analysis is accomplished by removing a portion of the band from the gel (or sometimes by looking at the gel directly) and examining this band by *matrix-assisted laser desorption/ionization time-of-flight mass spectrometry (MALDI-TOF MS)* (see Box 23.1). This approach makes it possible to identify a particular analyte (such as a protein) by its molecular mass even when there are many similar analytes in a sample.

23.3C What Are Some Special Types of Gel Electrophoresis?

Sodium Dodecyl Sulfate Polyacrylamide Gel Electrophoresis. Whenever a porous support is present in an electrophoretic system, it is possible that large analytes may be separated based on their size as well as their electrophoretic mobilities. This size separation occurs in a manner similar to that which occurs in size-exclusion chromatography and can be used to determine the molecular weight of biomolecules. This type of analysis is accomplished for proteins in a technique known as **sodium dodecyl sulfate polyacrylamide gel electrophoresis**, or **SDS-PAGE** (see Figure 23.10).[18,19]

BOX 23.1

Matrix-Assisted Laser Desorption/Ionization Time-of-Flight Mass Spectrometry

Matrix-assisted laser desorption/ionization time-of-flight mass spectrometry (MALDI-TOF MS) is a type of mass spectrometry in which a special matrix capable of absorbing light from a laser is used for chemical ionization. The term "MALDI" was first used in 1985 to describe the use of a laser to cause ionization of the amino acid alanine in the presence of tryptophan (the "matrix" in this case).[28] In 1988 it was shown almost simultaneously by two research groups, one in Germany and one in Japan, that MALDI-TOF MS could also be employed in work with large biomolecules, such as proteins.[29,30] The value of this method was recognized in 2002 when members of both these groups shared the Nobel Prize in chemistry for the development of this technique.

Figure 23.9 shows the typical way in which a sample is analyzed by MALDI-TOF MS. First, the sample is mixed with a matrix that can readily absorb UV light. This mixture is then placed on a holder in the MALDI-TOF instrument, where pulses of a UV laser are aimed at the sample and matrix. As the matrix absorbs some of this light, it transfers its energy to molecules in the sample, causing these to form ions. These ions are then passed through an electric field into a time-of-flight mass analyzer, where ions of different mass-to-charge ratios will travel at different velocities. The number of ions arriving at the other end is measured at various times, allowing a mass spectrum to be obtained for analytes in the sample.[31]

MALDI-TOF MS is a soft ionization approach that results in a large amount of molecular ions and few, if any, fragment ions for most analytes. This method also has a low background signal, a high mass accuracy, and can be used over a wide range of masses. These properties make MALDI-TOF MS valuable in the study and identification of proteins after they have been separated by techniques like SDS-PAGE or 2-dimensional (2-D) electrophoresis (see Section 23.3). MALDI-TOF MS can also be used to look at peptides, polysaccharides, nucleic acids, and some synthetic polymers.[31,32]

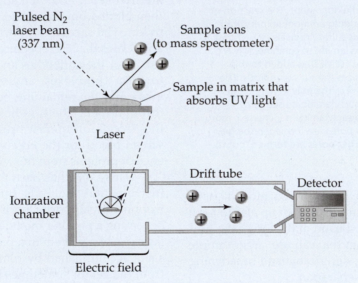

FIGURE 23.9 The analysis of a sample by MALDI-TOF MS. The individual steps in this analysis are described in the text.

In SDS-PAGE, the proteins in a sample are first denatured and their disulfide bonds broken through the use of a reducing agent. This pretreatment converts the proteins into a set of single-stranded polypeptides. These polypeptides are then treated with *sodium dodecyl sulfate (SDS)*, a surfactant with a nonpolar tail and a negatively charged sulfate group. The nonpolar end of this surfactant coats each protein, forming roughly linear rods that have an exterior layer of negative charge. The result for a mixture of proteins is a series of rods with different lengths but similar charge-to-mass ratios. Next, these protein rods are passed through a porous polyacrylamide gel in the presence of an electric field. The negative charges on these rods (from the SDS coating) cause them to all move toward the positive electrode, while the pores of the gel allow small rods to travel more quickly to this electrode than large rods.

At the end of an SDS-PAGE run, the positions of protein bands from a sample are compared to those obtained for known protein standards applied to the same gel. This comparison is made either qualitatively or by preparing a calibration curve. The calibration curve is typically prepared by plotting the log of the molecular weight (MW) for the protein standards versus their migration distance (d_m) or *retardation factor* (R_f). The retardation factor for an analyte band in SDS-PAGE is calculated by using the ratio of a protein's migration distance over the migration distance for a small marker compound (d_s), where $R_f = d_m/d_s$. The resulting plot of log(MW) versus d_m or R_f gives a curved response with an intermediate linear region for proteins with sizes that are neither totally excluded from the pores nor able to access all pores in the support.

Sample pretreatment

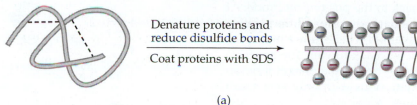

Denature proteins and
reduce disulfide bonds

Coat proteins with SDS

(a)

Protein separation

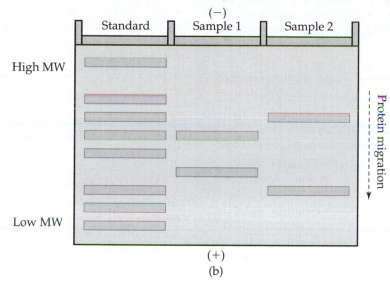

(b)

FIGURE 23.10   Preparation of proteins and their separation by sodium dodecyl sulfate polyacrylamide gel electrophoresis (SDS-PAGE).

| **EXERCISE 23.2** | **Using SDS-PAGE for Estimating the Molecular Mass of a Protein** |

The proteins in the standard in Figure 23.10 have molecular weights (from top-to-bottom) of 200, 116, 97, 66, 45, 31, 23, and 14 kDa. What are the molecular weights of the proteins in sample 1?

SOLUTION

The first band in sample 1 is at approximately the same location as the **66 kDa** band in the standard sample. The second band in sample 1 appears between the 45 kDa and 31 kDa bands in the standard, giving this second protein a mass of roughly **38 kDa**. A similar analysis for the second sample gives proteins with estimated masses of 31 and 97 kDa.

Isoelectric Focusing.   Another type of electrophoresis that often employs supports is **isoelectric focusing** (IEF).[10] IEF is a method used to separate zwitterions (substances with both acidic and basic groups, as discussed in Chapter 8). Zwitterions are separated in IEF based on their isoelectric points by having these compounds migrate in an electric field across a pH gradient. In this pH gradient, each zwitterion will migrate until it reaches a region where the pH is equal to its isoelectric point. At this point, the zwitterion will no longer have any net charge and its electrophoretic

mobility will become zero, causing the analyte to stop migrating.[1] The result is a series of tight bands, where each band appears at the point where pH = pI for a given zwitterion.

The reason isoelectric focusing produces tight bands for these analytes is that even if a zwitterion momentarily diffuses out of the region where the pH is equal to its pI, the system will tend to "focus" the zwitterion back into this region (see Figure 23.11). This focusing occurs because of the way the pH gradient is aligned with the electric field. High pH's occur toward the negative electrode, so as solutes diffuse out of their band and

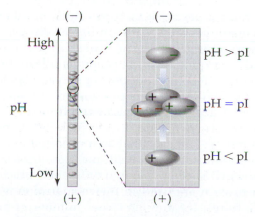

FIGURE 23.11   Isoelectric focusing.

toward this region they will take on a more negative charge and be attracted back to the positive electrode. At the same time, zwitterions that move toward the positive electrode and region of lower pH will acquire a more positive charge and be attracted back toward the negative electrode. It is this focusing property that makes it possible for IEF to separate zwitterions with only very small differences in their pI values.

To obtain a separation in IEF, it is necessary to have a stable pH gradient. This pH gradient is produced by placing in the electric field a mixture of small reagent zwitterions known as *ampholytes*. These are usually polyprotic amino carboxylic acids with a range of pK_a values.[6] When these ampholytes are placed in an electric field, they will travel through the system and align in the order of their pK_a values. The result is a pH gradient that can be used directly or by cross-linking the ampholytes to a support to keep them stationary in the system.

IEF is a valuable tool for separating proteins or other compounds that contain both positive and negative charges. These include some drugs, as well as bacteria, viruses, and cells. Applications of this method range from biotechnology and biochemistry to forensic analysis and paternity testing. IEF is particularly useful in providing high-resolution separations between different forms of enzymes or cell products. For instance, it is possible with this method to separate proteins with differences in pI values as small as 0.02 pH units.

2-Dimensional Electrophoresis. Another way gel electrophoresis can be utilized is in **two-dimensional** (or **2-D**) **electrophoresis**, which is a high-resolution technique used to look at complex protein mixtures.[19,33] In this method, two different types of electrophoresis are conducted on a single sample. The first of these separations is usually based on a isoelectric point, as accomplished by using isoelectric focusing. The second separation method (SDS-PAGE) is according to size.

A typical 2-D electrophoresis method is illustrated in Figure 23.12. First, a small band of sample is applied to the top of a support for use in isoelectric focusing. The support used in this case is typically agarose or a polyacrylamide gel with large pores. After this first separation has been finished, some proteins will have been separated based on their pI values, but there may still be many proteins with similar isoelectric points and overlapping bands. A further separation is obtained by turning this first gel on its side and placing it at the top of a second support (a polyacrylamide gel) for use in SDS-PAGE. This process gives a separation according to size, in which each band from the first separation has its own lane on the SDS-PAGE gel. The result is a series of peaks that are now separated in two dimensions (one based on pI and the other on size) across the gel. The fact that two different characteristics of each protein are used in their separation makes it possible to resolve a much larger number of proteins than is possible by either IEF or SDS-PAGE alone.

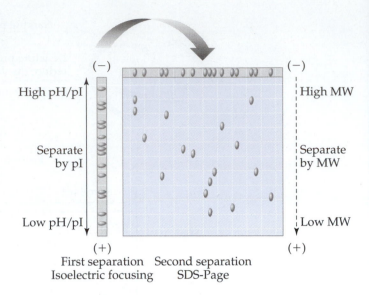

FIGURE 23.12 Two-dimensional gel electrophoresis, using a combination of isoelectric focusing and SDS-PAGE as an example.

After a 2-D separation has been finished, the protein bands can be detected using the methods discussed in Section 23.3B. Staining with Coomassie blue or silver nitrate is often used in the location and measurement of these bands. Analysis by mass spectrometry is another option. Other issues to consider are the interpretation and analysis of the many protein bands that can occur in a single sample. This analysis requires the use of computers to help image and catalog the location of each band and to correlate this information with that obtained by other methods, such as mass spectrometry.

23.4 CAPILLARY ELECTROPHORESIS

23.4A What Is Capillary Electrophoresis?

Another type of electrophoresis is the method of **capillary electrophoresis** (**CE**). CE is a technique that separates analytes by electrophoresis and that is carried out in a capillary. This method was first reported in the late 1970s and early 1980s and is sometimes known as "capillary zone electrophoresis."[23,34] CE in its current form is typically conducted in capillaries with inner diameters of 20–100 μm and lengths of 20–100 cm.[7] The use of these narrow-bore tubes provides efficient removal of Joule heating by allowing this heat to be quickly dissipated to the surrounding environment.[8,17,23] This removal of heat helps to decrease band-broadening and provides much more efficient and faster separations than gel electrophoresis (see Figure 23.13).

One reason capillary electrophoresis is more efficient than gel electrophoresis is that Joule heating is greatly reduced as a source of band-broadening. Also, capillary electrophoresis is often used with no gel or support present, which eliminates eddy diffusion and secondary interactions with the support (other than the capillary wall). The result is that longitudinal diffusion now becomes the main source of band-broadening. Under these conditions,

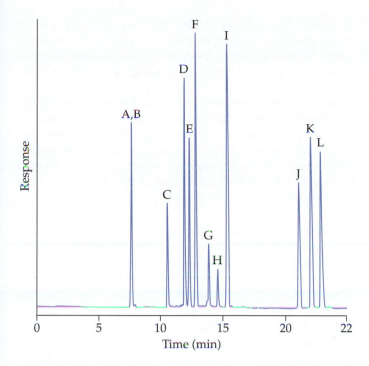

FIGURE 23.13 An early example of capillary electrophoresis, used here use for the separation of dansylated amino acids (represented by peaks A–L). (Reproduced with permission from J.W. Jorgenson and K.D. Lukacs, "Capillary Zone Electrophoresis," *Science*, 222 (1983) 266–272.)

the number of theoretical plates (N) expected for this system is given by the following equations,

$$N = \frac{\mu E L_d}{2D} \quad \text{or} \quad N = \frac{\mu V L_d}{2DL} \qquad (23.8)$$

where D is the diffusion coefficient of the analyte, μ is the electrophoretic mobility of the analyte, E is the electric field strength, L is the total length of the capillary, L_d is the distance from the point of injection to the detector, and V is the applied voltage (where $E = V/L$).[8]

Equation 23.8 shows that the value of N (representing the efficiency of the CE system) will increase as we use higher electric fields and voltages. This result makes sense because higher electric fields will cause the analyte to migrate faster and spend less time in the capillary. These shorter migration times will decrease band-broadening because less time is allowed for longitudinal diffusion. The result is a fast separation with a high efficiency and narrow peaks.

EXERCISE 23.3 The Effect of Electric Field Strength on Efficiency in Capillary Electrophoresis

The protein 1 in Exercise 23.1 has a diffusion coefficient of approximately 2.0×10^{-7} cm^2/s in its running buffer. If longitudinal diffusion is the only significant band-broadening process present during the separation of this protein by capillary electrophoresis, what is the maximum number of theoretical plates that would be expected for this protein's peak at an applied voltage of

20.0 kV and at 30.0 kV? What factors may cause lower values for N to be obtained?

SOLUTION

We can use Equation 23.8 along with the conditions given in Exercise 23.1 and the electrophoretic mobility calculated earlier for protein 1 to get the expected value for N at 20.0 kV.

$$N = \frac{\left(1.80 \text{ cm}^2/\text{kV} \cdot \text{min}\right) \cdot 20.0 \text{ kV} \cdot 22.0 \text{ cm}}{2 \cdot \left(2.0 \times 10^{-7} \text{cm}^2/\text{s}\right) \cdot (60 \text{ s/min}) \cdot 25.0 \text{ cm}}$$

$$= 1.3 \times 10^6 \text{ theoretical plates}$$

If we increase the applied voltage from 20.0 to 30.0 kV (or by 1.5-fold), Equation 23.8 indicates we will see a proportional increase of 1.5-fold in N from 1.3×10^6 to 1.9×10^6 **plates**. Factors that might give lower plate numbers include the presence of adsorption between the protein and capillary wall, extra-column band-broadening, or an increase in Joule heating as the voltage is increased.

Besides providing efficient separations, we have seen that the use of high electric fields in capillary electrophoresis also reduces the time needed for a separation. This relationship can be shown by rewriting Equation 23.3 to give the expected migration time for an analyte in terms of the electric field, the electrophoretic mobility of the analyte, and the length of the capillary.

$$t_m = \frac{L_d L}{\mu V} = \frac{L_d}{\mu E} \qquad (23.9)$$

For instance, Equation 23.9 indicates that the migration time for the protein in Exercise 23.3 will decrease by 1.5-fold (from 15.3 to 10.2 min) if we increase the applied voltage from 20.0 to 30.0 kV. The result is a situation in which we can improve both the efficiency and speed of a separation by increasing the voltage. This feature has made capillary electrophoresis popular for the analysis of complex samples, such as those used in DNA sequencing. Unfortunately, there is a limit to how high the voltage can be increased before Joule heating again becomes important. Most CE systems are capable of using voltages of up to 25–30 kV, but significant Joule heating can appear at lower voltages.

23.4B How Is Capillary Electrophoresis Performed?

Equipment and Supports. Besides being faster and more efficient than gel electrophoresis, capillary electrophoresis is easier to perform as part of an instrumental system. An example of a CE system is shown in Figure 23.14.[8,21] Along with the capillary, this system includes a power supply and electrodes for applying the electric field, two containers that create a contact between these electrodes and the solution within the capillary, an on-line detector, and a means for injecting samples onto the capillary. Because these instruments can use voltages of up to 25–30 kV, they include safety features that protect the user from the high-voltage region and that can turn off this voltage when the system is opened for maintenance or the insertion of samples and reagents.

The capillary in a CE system is typically made of fused silica. This capillary can be used directly or it can be modified to place various coatings on its interior surface. An uncoated silica capillary can lead to a significant amount of flow due to electroosmosis when working at a neutral or basic pH, due to deprotonation of the silica's surface silanol groups. One useful feature of this electroosmosis is it tends to cause all analytes, regardless of their charge, to travel in the same direction through the CE capillary. This effect means that a sample containing many types of ions can be injected at one end of the capillary (at the positive electrode), with electroosmosis then carrying these through to the other end (to the negative electrode) and past an on-line detector. This format is called the "normal polarity mode" of CE.[8] It is important to remember in this situation that a separation of ions will still occur, but that the observed mobility will now be equal to the sum of an analyte's inherent electrophoretic mobility plus the mobility created by electroosmosis (see Equation 23.5). This effect on observed mobility will, in turn, affect the observed migration time and the efficiency and resolution obtained for the separation.

Although many analytes will travel in the same direction as electroosmotic flow through a CE system, it is possible for some to have migration rates faster than electroosmosis, which will carry them in the opposite direction. The analysis of these ions in a silica capillary is performed by injecting them at the end by the negative electrode and allowing them to migrate toward the positive electrode and against electroosmotic flow. This method is known as the "reversed polarity mode" of CE.[8] In addition, electroosmotic flow can be altered by changing the pH (which changes the degree of deprotonation and charge on silica), or by placing a coating on the surface of the support. In this second case, a neutral coating helps to reduce electroosmosis while a positively charged coating will reverse the direction of this flow toward the positive rather than negative electrode.

Injection Techniques. There are two features of capillary electrophoresis that place special demands on how samples can be injected. First, the small volume of a CE capillary must be considered. A typical 50 μm I.D. × 25 cm long capillary for CE will contain only 0.5 μL of running buffer. Another factor to consider is the high efficiency of capillary electrophoresis. Both of these factors restrict the sample volumes that can be injected without introducing significant band-broadening (< 10 nL for a 0.5 μL volume capillary).[8]

There are two techniques that make it possible to inject these small sample volumes onto a CE system. The first technique is *hydrodynamic injection*, which uses a difference in pressure to deliver a sample to the capillary. This method can be carried out by placing one end of the capillary into the sample in an enclosed chamber and applying a pressure to this chamber for a fixed period of time, where the amount of injected sample will depend on the size of the pressure difference and the amount of time that this pressure is applied. Once the sample has entered the capillary, the separation is begun after the capillary

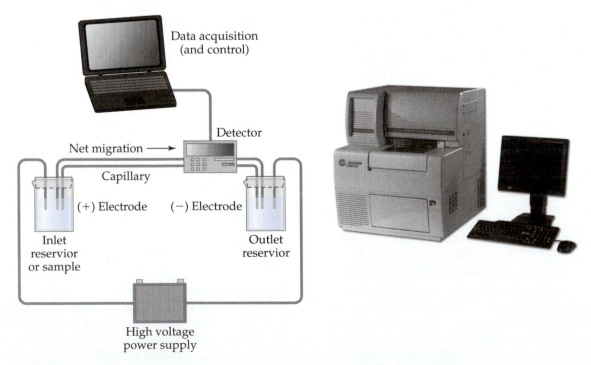

FIGURE 23.14 General design of a capillary electrophoresis system, and a commercial instrument for capillary electrophoresis. (The picture on the right is courtesy of Beckman Instruments.)

has been put back into contact with the running buffer and electrodes. A second technique that allows the injection of small sample volumes is *electrokinetic injection*. This method again begins by placing the capillary into the sample, but an electrode is also now in contact with the sample. When an electric field is applied across the capillary, electroosmostic flow and the electrophoretic mobility of the analytes cause them to enter the capillary. The amount of each analyte that is injected in this method will depend on the analyte's electrophoretic mobility, the electric field, and the time over which this field is applied.[8]

There are various methods for concentrating samples and providing narrow analyte bands in CE. One such method is *sample stacking* (see Figure 23.15).[21] Sample stacking occurs when the ionic strength (and therefore the conductivity) of the sample is less than that of the running buffer. When an electric field is applied to such a system, analytes will migrate quickly through the sample matrix until they come to the boundary between the sample and running buffer. Because the running buffer has a higher ionic strength than the sample, the rate of analyte migration decreases at this boundary. This decrease in migration rate causes the analytes to concentrate into a narrower band as they enter the running buffer. The overall effect is similar to what occurs when using stacking gels in traditional electrophoresis.

Detection Methods. Examples of detection methods that are used for capillary electrophoresis are shown in Table 23.1. Many of these methods are also used in liquid chromatography (see Chapter 22).[8,21] An important difference between detection in LC and CE is the need in CE for methods that can work with very small sample sizes. This need is a result of the small injection volumes that are required in capillary electrophoresis to avoid excessive band-broadening. Selective monitoring methods that work well for this purpose are electrochemical and fluorescence detection. Ultraviolet-visible (UV-vis) absorbance, conductance, and mass spectrometry detection are also often employed in CE.

Initial conditions

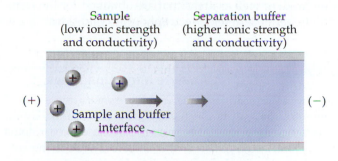

Conditions after sample stacking

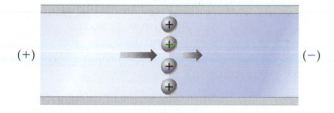

FIGURE 23.15 Principle of sample self-stacking.

Another difference between detection in LC and CE concerns how their signals vary with analyte retention or migration. In LC, all analytes pass at the same flow rate (that of the mobile phase) through the detector and spend the same amount of time in this device. This effect makes it possible to directly compare the peak areas of two analytes with different retention times. However, in capillary electrophoresis analytes with different migration times also spend different amounts of time in the detector. A correction must be made for this difference if we wish to compare the areas of two analytes in the same CE run. We can make this adjustment by using a corrected peak area (A_c), which is equal to the ratio of the measured peak area (A) for an analyte divided by its migration time.

$$A_c = A/t_m \qquad (23.10)$$

TABLE 23.1 Properties of Common Capillary Electrophoresis Detectors*

Detector Name	Compounds Detected	Detection Limits
General detectors		
Ultraviolet-visible (UV/vis) absorbance detector	Compounds with chromophores	10^{-13}–10^{-16} mol
Selective detectors		
Fluorescence detector	Fluorescent compounds	10^{-15}–10^{-17} mol
Laser-induced fluorescence detector	Fluorescent compounds	10^{-18}–10^{-20} mol
Conductivity detector	Ionic compounds	10^{-15}–10^{-16} mol
Electrochemical detector	Electrochemically active compounds	10^{-18}–10^{-19} mol
Structure-specific detectors		
Mass spectrometry	Compounds forming gas-phase ions	10^{-16}–10^{-17} mol

*These data are for commercial instruments.

This correction allows areas for different analytes to be compared, as well as areas that are obtained for the same analyte under different electrophoretic conditions.

EXERCISE 23.4 | **Correcting Peak Areas for Analyte Migration Times**

Proteins 1 and 2 in Exercise 23.1 have measured areas of 1290 and 1360 units at 20.0 kV when examined by an absorbance detector. If it is known that these two proteins have a similar response to the detector, what is the corrected area and relative amount of each protein in the given sample?

SOLUTION

The migration times for these proteins (given in Exercise 23.1) are 15.3 min and 16.2 min. Placing these data into Equation 23.10 along with the measured peak areas gives the following results.

Protein 1: $\quad A_{c,1} = (1290)/(15.3 \text{ min}) = 84.3$

Protein 2: $\quad A_{c,2} = (1360)/(16.2 \text{ min}) = 84.0$

If these proteins have a similar response to the detector, then we can say from these corrected areas that there is approximately the same amount of both proteins in the sample. If we had used the uncorrected areas for this calculation, we would have incorrectly concluded that protein 2 was present at a greater level.

Along with the various detection methods we discussed for liquid chromatography in Chapter 22, another detection approach that is used in capillary electrophoresis is **laser-induced fluorescence (LIF)**.[6,8,21] This method employs a laser to excite a fluorescent compound, allowing the detection of this agent through its subsequent emission of light. There are several advantages to using a laser as the excitation source. First, the laser is monochromatic and has a high intensity, which allows for the selective and strong excitation of a compound with an excitation spectrum that overlaps with the emission wavelength of the laser. Also, the laser beam can be focused as a very narrow beam. This feature is extremely valuable in work with the small-bore capillaries found in capillary electrophoresis. One limitation of LIF detection is it does require an analyte that is naturally fluorescent or that can be converted into a fluorescent derivative. This second option makes use of a fluorescent tag like fluorescein or rhodamine (see Chapter 18).

LIF detection with CE was used in the automated DNA sequencing systems that made early completion of the Human Genome Project possible. This detection involved the use of several fluorescent dyes, one for each of the four terminating nucleotides present during the Sanger reaction. These labeled DNA strands were then separated based on their lengths by capillary electrophoresis (see Figure 23.16). It was possible to further increase the speed of this analysis by using a single laser beam to simultaneously examine a whole array of capillaries, each sequencing a different segment of DNA. The utilization of multiple

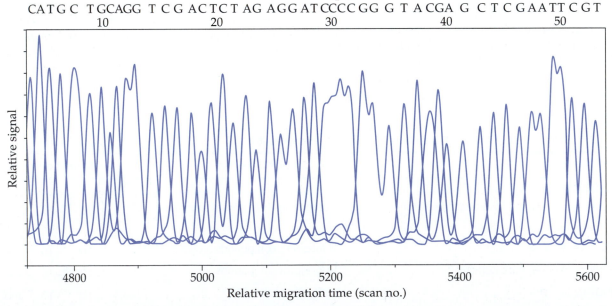

FIGURE 23.16 DNA sequencing by a capillary electrophoresis system. The top panel shows the original electrophoretic data, and the bottom panel shows the same data after they have been processed to determine the nucleotide sequence of the DNA segment being examined (as given by the symbols A, C, T, and G). (Based on data from J. Bashkin, in *Capillary Electrophoresis of Nucleic Acids*, K.R. Mitchelson and J. Cheng, Eds., Humana Press, Totowa, NJ, 2001, Chapter 7.)

capillaries in a single CE system is known as *capillary array electrophoresis* (*CAE*).[6,7,35] Such a system can examine many DNA sequences at the same time, which increases sample throughput and lowers the cost per analysis.

23.4C What Are Some Special Types of Capillary Electrophoresis?

The main capillary electrophoresis method that has been discussed up to this point is zone electrophoresis, in which differences in the charge/size ratio of analytes is the only means employed for their separation. It is possible to include other chemical and physical interactions in a CE

system to give additional types of separations. Examples include CE methods that separate analytes based on their size, isoelectric points, or interactions with additives in the running buffer. The use of CE in microanalytical systems has also been a topic of great interest (see Box 23.2).

Capillary Sieving Electrophoresis.   One useful feature of gel electrophoresis is the ability of some supports to separate analytes based on size, as occurs for proteins in SDS-PAGE. The same effect can be obtained in capillary electrophoresis by including an agent in the CE system that "sieves" the analytes, or separates them based on size. This approach is known

---

BOX 23.2

Analytical Chemistry on a Chip

The development of silicon microchips created a revolution in the computer and electronic industries. The result over the past few decades has been a continuous decrease in the size of electronic devices and an increase in their capabilities. A similar change is occurring in analytical chemistry. This change began in 1979, when methods developed for the creation of microchips where used to make a gas chromatographic system on a silicon wafer.[36] In 1990 it was proposed that all of the components of a chemical analysis could even be placed onto a miniaturized system. The resulting device is now known as a "lab-on-a-chip" or a *micro total-analysis system (μTAS)*.[37]

Capillary electrophoresis was one of the first analytical methods that was adapted for use on a microchip.[38] An example of such a device is given in Figure 23.17. There are now many reports that have used microchips for CE.[39–41] One feature that makes CE and microchips a good match is the need in CE for narrow channels to avoid the effects of Joule heating. In

addition, the elimination of Joule heating allows CE to work with short separation channels and high electric fields, as is used in Figure 23.17. The creation of an electric field to separate analytes and to generate electroosmotic flow for CE is relatively easy to obtain with a microchip by including electrodes as part of this device. The availability of detection schemes like LIF that are capable of working with small detection volumes is also valuable when placing a CE system on a microchip.[38–41]

CE is not the only analytical method that has been carried out on microchips. Other methods have included liquid chromatography, gel electrophoresis, biosensing, water analysis, flow injection analysis, and solid-phase extraction. There are several potential advantages of using microchips with these techniques. One is the small sample requirements of these devices. The ability to make these systems fully portable or disposable are additional advantages. The possibility of making microchips on a large scale and at a low cost are other attractive features.[39–41]

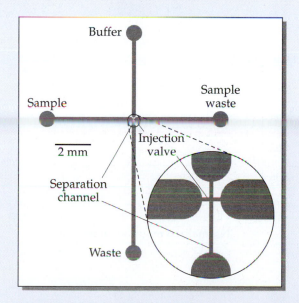

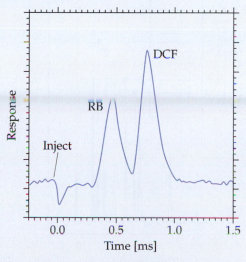

FIGURE 23.17   Design of a microchip-based system for performing electrophoresis and the use of this device in the fast separation of the dyes rhodamine B (RB) and dichlorofluorescein (DCF). (Reproduced with permission from S.C. Jacobson, C.T. Culbertson, J.E. Daler, and J.M. Ramsey, *Analytical Chemistry*, 70 (1998) 3476–3480.)

as **capillary sieving electrophoresis (CSE)**.[7] A comparison of the results of CSE and a size separation by gel electrophoresis (e.g., by SDS-PAGE) is given in Figure 23.18.

There are several ways we can perform capillary sieving electrophoresis. The first way is to place a porous gel in the capillary, like the polyacrylamide gels employed in SDS-PAGE. This method is called *capillary gel electrophoresis (CGE)*.[7] One problem with these gels is they are not always stable in the high electric fields used in capillary electrophoresis and must frequently be replaced. A second approach, and the one now used in DNA sequencing by CE, is to add to the running buffer a large polymer that can entangle with analytes and alter their rate of migration. This approach provides a system with better reproducibility and stability than those using gels, because the polymer is continuously renewed as the running buffer passes through the capillary.

Electrokinetic Chromatography.   Ordinary capillary electrophoresis works well for separating cations and anions, but it cannot be used to separate neutral substances from each other. Instead, these substances migrate as a single peak that travels with the electroosmotic flow. It is possible to employ CE with such compounds if we place in the running buffer a charged agent that can interact with these substances. This approach is called *electrokinetic chromatography*. One common way of carrying out this method is to employ micelles as additives, giving a subset of electrokinetic chromatography known as **micellar electrokinetic chromatography (MEKC)**.[7,21,42,43]

A *micelle* is a particle formed by the aggregation of a large number of surfactant molecules, such as sodium dodecyl sulfate (SDS). We saw earlier that SDS has a long nonpolar tail attached to a negatively charged sulfate group. When the concentration of a surfactant like SDS reaches a certain threshold level (known as the "critical micelle concentration"), some of the surfactant molecules come together to form micelles. If these micelles form in a polar solvent like water, the nonpolar tails of the surfactant will be on the inside of the aggregate (giving a nonpolar interior), while the charged groups at the other end will be on the outside by the solvent (see Figure 23.19).

When micelles based on SDS are placed into the running buffer of a CE system, they will be attracted toward the positive electrode. If a sample with several neutral compounds is now injected into this system, some of these neutral substances may enter the micelles and interact with their nonpolar interior. This interaction involves a partitioning process similar to that found in liquid–liquid extractions and some types of liquid chromatography (see Chapters 20 and 22), in which the micelles act as the "stationary phase." Although these neutral compounds normally travel with the electroosmotic flow through the capillary, while they are in the micelles they migrate with the micelles in the opposite direction. The result is a separation of neutral compounds based on the degree to which they enter the micelles. Micelles can also alter the migration times for charged substances through partitioning and charge interactions between the analytes and micelles.

Other Methods.   There are many other types of CE that have been explored for use in chemical analysis and separations. For example, isoelectric focusing can be carried out in a capillary, creating the method of *capillary isoelectric focusing (CIEF)*.[7,21] This technique involves the production of a pH gradient across the capillary for the separation of

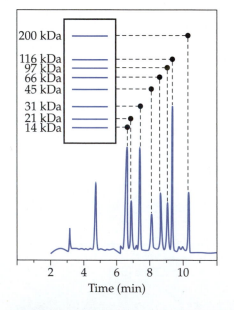

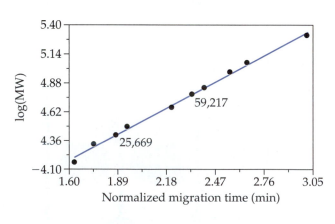

FIGURE 23.18   Comparison of capillary electrophoresis and traditional SDS-PAGE in the separation and analysis of proteins. The figure on the left shows an electropherogram obtained for a series of proteins with molecular masses ranging from 14 to 200 kDa. The inset shows an SDS-PAGE gel for the same proteins. The calibration curve on the right shows how the migration times for these proteins in CE are related to their molecular masses. (Adapted with permission from Bio-Rad Laboratories.)[8]

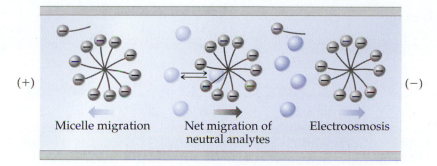

FIGURE 23.19 Micellar electrokinetic chromatography. The circles represent analytes from the injected sample. A negatively charged surfactant such as sodium dodecyl sulfate is used in this example, as represented by the circles that contain negative charges and nonpolar tails.

zwitterions. One way CIEF can be conducted is shown in Figure 23.20. In this example, the electrodes are in contact with two different electrolyte solutions: (1) the "catholyte," which is a basic solution located by the cathode, and (2) the "anolyte," which is an acidic solution located by the anode. The capillary contains a mixture of ampholytes that will create a pH gradient when an electric field is applied between these electrodes. A coated capillary is also used in this case to minimize or eliminate electroosmotic flow. When a sample (generally a mixture of proteins) is injected onto this system, its zwitterions will migrate until they reach a region where the pH is equal to their pI. Once these bands have formed, they are pushed through the capillary and past the detector by applying pressure to the system.

Another type of capillary electrophoresis occurs when biologically related agents are placed as additives in the running buffer. As analytes travel through this buffer, their overall mobility will be affected by their binding to these agents (see Figure 23.21). The result is a method known as *affinity capillary electrophoresis* (*ACE*).[7,21,44,45] One common use of ACE is in the separation of chiral analytes through the use of binding agents like cyclodextrins or proteins (see Chapters 8 and 22). This method can also be used in clinical and pharmaceutical assays and for the study of biological interactions.

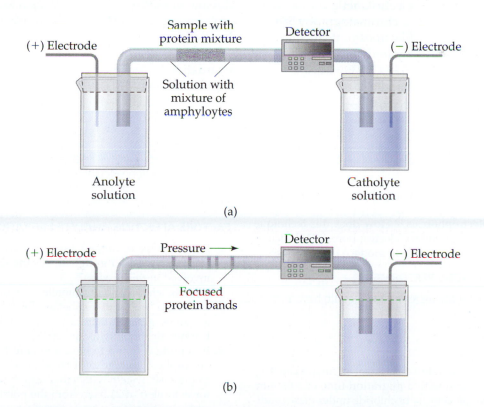

FIGURE 23.20 Capillary isoelectric focusing. The diagram in (a) shows the initial configuration of this system before an electric field is applied. The diagram in (b) shows the separation of proteins from the original sample after the electric field has been applied. These protein bands are later passed through the capillary and past the detector by applying pressure to the system.

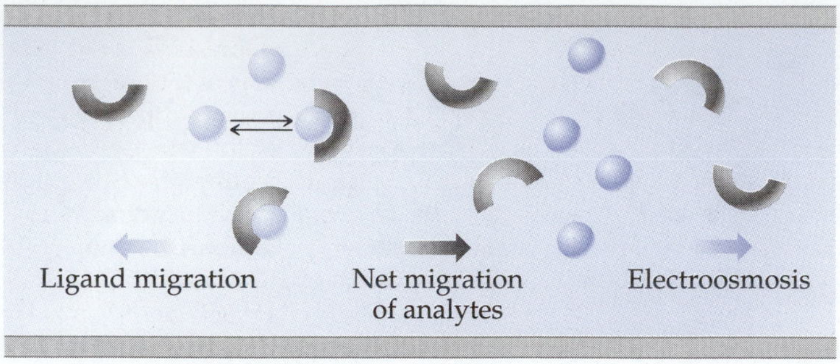

FIGURE 23.21 Affinity capillary electrophoresis. The circles represent analytes from the injected sample. The half-circles represent a binding agent for one or more of these analytes that has been added to the running buffer.

Key Words

Capillary electrophoresis 582	Electrophoretic mobility 575	Micellar electrokinetic chromatography 588	Southern blot 579
Capillary sieving electrophoresis 588	Gel electrophoresis 577	Migration distance 573	Two-dimensional electrophoresis 582
Electroosmosis 576	Isoelectric focusing 581	Migration time 573	Western blot 579
Electropherogram 574	Joule heating 577	Northern blot 579	
Electrophoresis 571	Laser-induced fluorescence 586	SDS-PAGE 579	

Other Terms

Affinity capillary electrophoresis 589	Densitometer 579	Micro total-analysis system 587	Silver staining 579
Ampholytes 582	Electrokinetic injection 585	Moving boundary electrophoresis 572	Sodium dodecyl sulfate 580
Capillary array electrophoresis 587	Electrokinetic chromatography 588	PAGE 578	Wick flow 577
Capillary gel electrophoresis 588	Electroosmotic mobility 576	Polyacrylamide 578	Zone electrophoresis 572
Capillary isoelectric focusing 588	Hydrodynamic injection 584	Retardation factor 580	
	MALDI-TOF MS 579	Sample stacking 585	
	Micelle 588		

Questions

WHAT IS ELECTROPHORESIS?

1. Define "electrophoresis" and explain how this method is used to separate chemicals.
2. What is "zone electrophoresis"? How does this technique differ from "moving boundary electrophoresis"? Which of these methods is more common in modern laboratories?

HOW IS ELECTROPHORESIS PERFORMED?

3. Define each of the following terms and explain how they are used in electrophoresis.
   (a) Migration distance
   (b) Migration time
   (c) Electropherogram
4. Chloride is found to migrate a distance of 35 cm in a capillary electrophoresis system with a migration time of 5.63 min. What is the migration velocity of chloride under these conditions? At the same velocity, what distance would chloride have traveled in 2.5 min?
5. A protein is found to migrate a distance of 3.2 cm in 30 min when 100 V is applied to a 10 cm long polyacrymide gel. What

is the migration velocity of this protein? If an applied voltage of 200 V is used instead, how long will it take the same protein to migrate a distance the entire length of the 10 cm gel?

FACTORS AFFECTING ANALYTE MIGRATION

6. Explain why a charged substance will tend to move at a constant velocity through an electric field. What forces are involved in this process?
7. What is "electrophoretic mobility"? How is this term related to the movement of a substance in an electric field? What are some general factors that affect the size of the electrophoretic mobility for an analyte?
8. A peptide is found to have a migration time of 8.31 min at an applied voltage of 10.0 kV and on a capillary electrophoresis system with a total length of 25.0 cm. The detector is located at 21.5 cm from the point of sample injection and conditions are used so that electroosmotic flow is negligible. What is the migration velocity of the peptide? What is its electrophoretic mobility?
9. How would the migration velocity and migration time for the peptide in the previous problem change if the voltage

was changed to 15.0 kV? What change, if any, would occur in the electrophoretic mobility?

10. What are some examples of secondary interactions that can affect analyte migration in electrophoresis?

11. Dicamba is a herbicide commonly used on broadleaf weeds. Its major metabolite is dichlorosalicylic acid (DCSA). Both compounds are weak acids. Dicamba has a single weak-acid group, which is a carboxylic acid group with a pK_a of 1.94. DCSA has two weak-acid groups: a carboxylic acid group with a pK_a of 2.08 and a phenol group with a pK_a of 8.60. A separation of these compounds at pH 7.4 by capillary electrophoresis (CE) gives a migration time of 2.05 min for dicamba and 2.35 min for DSCA. The same compounds had migration times of 2.06 min and 4.1 min at pH 10.0.[46] Explain why there is a large shift in the migration time for DCSA over this pH range but no significant shift in the migration time for dicamba.

12. Two chiral forms of a drug have identical electrophoretic mobilities in a CE system. However, it is possible to separate these forms when β-cyclodextrin (a complexing agent we discussed in Chapter 7) is placed as an additive into the running buffer. Based on your knowledge of chemical reactions, explain why the presence of β-cyclodextrin might lead to such a separation.

13. What is "electroosmosis"? What causes electroosmosis to occur? How does electroosmosis affect the movement of analytes in an electrophoresis system?

14. What is the "electroosmotic mobility"? What factors affect this mobility?

15. A neutral compound injected onto a capillary electrophoresis system has a migration time of 1.52 min through a 50 cm long capillary at 30.0 kV, with the detector being located 30.0 cm from the point of injection. If it is assumed that the observed electrophoretic mobility for this compound is equal to the electroosmotic mobility, what is the value of μ_{eo} under these conditions?

16. The electroosmotic mobility for a particular separation is found to be $8.3 \times 10^{-10} m^2/V \cdot s$. A 25 cm long capillary is used for this separation at an applied voltage of 20.0 kV, with the detector being located 20 cm from the point of injection. What is the expected migration time for a neutral analyte in this system (i.e., an analyte that travels through the system only due to electroosmotic flow)?

FACTORS AFFECTING BAND-BROADENING

17. Use the same equations as given in Chapter 20 for chromatography to calculate the following values. (*Note:* You can assume Gaussian peaks are present.)
    (a) The plate number of a peak in capillary electrophoresis that has a migration time of 7.30 min and baseline width of 0.12 min
    (b) The plate height of the system in Part (a) for a capillary with a total length of 35.0 cm
    (c) The plate number of a band in gel electrophoresis with a migration distance of 4.2 cm and baseline width of 2.1 mm

18. Use the equations given in Chapter 20 to calculate the following values.
    (a) The resolution of two peaks in capillary electrophoresis with migration times of 10.1 min and 10.4 min with baseline widths of 0.15 min and 0.16 min, respectively
    (b) The resolution of two bands in gel electrophoresis with migration distances of 2.3 cm and 2.6 cm with an average baseline width of 1.5 mm

19. How does longitudinal diffusion affect band-broadening in electrophoresis? How is this type of band-broadening related to the time of the separation and the size of the analytes? Why can a porous support help minimize the effects of this process?

20. What is "Joule heating"? What causes this heating? Why does Joule heating result in band-broadening in electrophoresis?

21. What are some approaches that can be used to minimize the effects of Joule heating in electrophoresis?

22. One useful tool in optimizing a separation for capillary electrophoresis is an "Ohm's law plot." This graph is prepared by plotting the measured current of the electrophoretic system at various applied voltages.[21]
    (a) Based on Equation 23.6, what information will be provided by the slope of this plot?
    (b) Deviations from linearity are often seen at high voltages in an Ohm's law plot. What do you think is usually the source of these deviations?

23. Under what conditions will eddy diffusion be present during electrophoresis?

24. What is "wick flow"? How does wick flow create band-broadening? In what types of electrophoresis can wick flow be important?

WHAT IS GEL ELECTROPHORESIS?

25. What is "gel electrophoresis"? How is this technique used for analyte identification and measurement?

26. A biochemist looking for a particular protein in a cell sample obtains the following results when using gel electrophoresis and a Western blot to compare this sample with standards containing the same protein. What is the approximate amount of this protein in the unknown cell sample?

Amount of Protein (ng)	Relative Band Area
0.0	50
5.0	560
10.0	1120
20.0	2040
Unknown sample	980

27. Figure 23.22 shows a typical result that is obtained for the analysis of a human serum sample by gel electrophoresis. Explain how this information might be used by a physician to detect both qualitative and quantitative changes in serum proteins for their patients.

HOW IS GEL ELECTROPHORESIS PERFORMED?

28. Draw a diagram of a typical gel electrophoresis system and label its main components. Explain the difference between horizontal and vertical gel electrophoresis systems.

29. List some supports that can be used in gel electrophoresis. What type of support is often used with DNA? What type is often used with proteins?

30. Describe how samples are usually applied to a support in gel electrophoresis. Explain the purpose of a "stacking gel" versus a "running gel."

31. Describe how each of the following items can be used for detection in gel electrophoresis.
    (a) Densitometer
    (b) Coomassie Brilliant Blue

(c) Silver staining

(d) Blotting

32. What is the difference between a Southern blot and a Northern blot? What is the difference between a Southern blot and a Western blot? How is each of these methods performed?

33. What is MALDI-TOF MS (matrix-assisted time-of-flight mass spectrometry)? Explain how this method can be used for identifying the contents of bands in gel electrophoresis.

WHAT ARE SOME SPECIAL TYPES OF GEL ELECTROPHORESIS?

34. Explain how SDS-PAGE (sodium dodecyl sulfate polyacrylamide gel electrophoresis), is performed. Describe why SDS-PAGE can provide information on the molecular mass of a protein.

35. The molecular mass of a protein is to be estimated by SDS-PAGE. The following migration distances are obtained for proteins of known mass on the gel: 200 kDa, 0.33 cm; 116.3 kDa, 0.57 cm; 66.3 kDa, 0.91 cm; 36.5 kDa, 1.63 cm; 21.5 kDa, 1.96 cm; 14.4 kDa, 2.24 cm. The unknown protein has a migration distance of 1.25 cm on the same gel. What is the approximate molecular weight of this protein?

36. A biochemist uses the same conditions as in the last problem to look for proteins with approximate masses of 18.5 kDa, 40.2 kDa, and 91.8 kDa. What are the expected migration distances for these proteins in this gel?

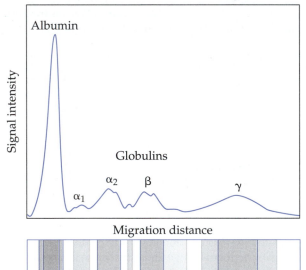

Densitometer scan of stained gel

Albumin

Signal intensity

Globulins

α_2

β

α_1

γ

Migration distance

Staining pattern of gel

FIGURE 23.22 Pattern obtained by staining of serum proteins that have been separated by gel electrophoresis. The bottom plot shows the protein bands on this gel after staining and the top tracing shows the intensity of these bands, as determined by using a densitometer. The labels on the bands refer to the types of proteins that are present in a given region or zone. The general location of these bands can be used to identify proteins in an unknown sample, while their intensity can be used to indicate the relative amount of each protein that is present. (Based on data from J.M. Anderson and G.A. Tetrault, "Electrophoresis." In *Laboratory Instrumentation*, 4th ed., M.C. Haven, G.A. Tetrault, and J.R. Schenken, Eds., Van Nostrand Reinhold, New York, 1995, Chapter 12.)

37. What is "isoelectric focusing"? Describe how this method separates analytes.

38. What is an "ampholyte"? How is an ampholyte used in isoelectric focusing?

39. What is "2-D electrophoresis"? What types of electrophoresis are often used in this method? What advantages are there in the use of 2-D electrophoresis for complex samples?

WHAT IS CAPILLARY ELECTROPHORESIS?

40. What is meant by the term "capillary electrophoresis"? How does this method differ from gel electrophoresis?

41. The amount of a nitrate in an unknown sample is to be quantitated by capillary electrophoresis. Another anion is added to all samples and standards as an internal standard (IS) prior to injection. The following results are obtained. What is the amount of the nitrate in the unknown sample?

Concentration of Nitrate (mg/L)	Peak Height	Concentration of Internal Standard (mg/L)	Peak Height
0.0	0.2	2.5	9.8
5.0	18.8	2.5	10.2
10.0	43.1	2.5	11.5
15.0	55.2	2.5	10.1
Unknown sample	15.1	2.5	9.7

42. Capillary electrophoresis and laser-induced fluorescence detection were used to determine the amount of a fluorescein-labeled peptide in a biological sample. A fixed amount of non-conjugated fluorescein was added to each sample as an internal standard. The following results were obtained in this method for a series of standards.

Concentration of Peptide (nM)	Peak Area — Peptide	Peak Area — Fluorescein
0.0	109	546
15.0	2185	598
25.0	3174	532
50.0	7046	601

The unknown sample gave measured peak areas of 4098 and 556 for the labeled peptide and fluorescein. What was the concentration of this peptide in the unknown sample?

43. What helps give capillary electrophoresis high efficiency? What processes are normally the most important in CE in determining the band-broadening of this method?

44. A small anion with a diffusion coefficient of 3.0×10^{-5} cm^2/s and an electrophoretic mobility of 1.58 cm^2/kV · min is to be analyzed at 20.0 kV by a capillary electrophoresis system. The system has a total length of 40.0 cm and the detector is located 33.0 cm from the point of injection. In the absence of electroosmotic flow, what is the maximum number of theoretical plates that can be obtained for this anion (i.e., assuming longitudinal diffusion is the only band-broadening process)? What will the migration time of the analyte be under these conditions?

45. A chemist wishes to use the plate number measured for a CE to estimate the diffusion coefficient for a new drug. This drug is injected onto a 42.5 cm long neutral coated capillary, which has no binding for the drug and produces negligible electroosmotic flow. The detector is located 38.0 cm from the point of injection. The drug has a measured migration time of 14.8 min and a baseline width of 26 s, when a voltage of 15.0 kV is applied across the capillary.
    (a) What is the number of theoretical plates for the peak due to the drug, if it is known that this peak has a Gaussian shape?
    (b) What is the diffusion coefficient for the drug (in units of cm^2/s)? What assumptions did you make in reaching your answer?

HOW IS CAPILLARY ELECTROPHORESIS PERFORMED?

46. What are the main components of a capillary electrophoresis system? How does this system differ from the equipment used in gel electrophoresis?
47. Why does the use of an uncoated silica capillary lead to electroosmotic flow in capillary electrophoresis? What is the direction of this flow? How does electroosmosis affect the apparent migration of analytes through the CE system?
48. What is the "normal polarity" mode of CE? What is the "reversed polarity" mode? In what general situations are these two modes utilized?
49. Explain why it is necessary in capillary electrophoresis to use small injection volumes. What are some difficulties in working with these small volumes? What are some advantages?
50. What is "hydrodynamic injection"? What is "electrokinetic injection"? How does each of these methods work?
51. What is "sample stacking"? Describe one way sample stacking can be accomplished in capillary electrophoresis.
52. List some general and some selective detectors that are used in capillary electrophoresis. How does this list compare to that given in Chapter 22 for liquid chromatography?
53. Two isoforms of a protein are found to elute with migration times of 20.3 min and 24.5 min from a capillary electrophoresis system. These protein peaks have measured areas of 3430 and 1235 units, respectively. What is the relative amount of each protein in the sample?
54. A drug is found to have a peak area of 11,250 units and it migrates through a 50.0 cm long capillary in the presence of an applied voltage of 15.0 kV. The detector is located at a distance of 45.0 cm from the point of injection. What will the expected area be for this sample if it is injected onto the same system, but now using an applied voltage of 20.0 kV?
55. What is "laser-induced fluorescence"? Explain why this technique is useful in capillary electrophoresis.

WHAT ARE SOME SPECIAL TYPES OF CAPILLARY ELECTROPHORESIS?

56. What is "capillary sieving electrophoresis"? What are two ways in which this method can be performed?
57. A protein is injected onto the same system used in Figure 23.18. What is the molecular weight of this protein if it has a normalized migration time of 2.65?
58. Define each of the following terms.
    (a) Electrokinetic chromatography
    (b) MEKC
    (c) Micelle
    (d) Critical micelle concentration

59. What is "capillary isoelectric focusing"? Describe one way this method can be performed.
60. What is "affinity capillary electrophoresis"? What are some applications of this method?

CHALLENGE PROBLEMS

61. Compare the capillary electrophoresis system in Figure 23.14 with the electrochemical cells that are discussed in Chapter 10.
    (a) What similarities can you find in these two types of systems? Based on this comparison, which part of the electrochemical cell would be equivalent to the capillary in a CE system? Which part of an electrochemical cell do you think would be equivalent to the gel or support in a gel electrophoresis system?
    (b) Describe how current is carried from the power supply and throughout an electrophoresis system. What is the role of the running buffer in this regard? What are the roles of the electrodes?
    (c) One tool we used in Chapter 6 for solving chemical problems was to use the method of charge balance, which says that the number of positive and negative charges in a system must be equal. And yet, in electrophoresis we use an electric field to separate analytes with positive and negative charges. Why do you think this separation is possible? (Hint: Consider your answers to Parts (a) and (b).)
62. The effect of electroosmotic flow on the overall observed electrophoretic mobility (μ_{Net}) and migration velocity (t_m) for an analyte in electrophoresis is given by the following equations,

$$v = \mu_{Net} E = \frac{(\mu + \mu_{osm})V}{L} \qquad (23.11)$$

$$t_m = \frac{L_d L}{(\mu + \mu_{osm})V} = \frac{L_d}{\mu_{Net} E} \qquad (23.12)$$

where all terms are the same as described earlier in this chapter.[21,23]
    (a) A cation has an electrophoretic mobility of 2.50 $cm^2/kV \cdot min$ on a CE system containing a 30.0 cm long coated, neutral capillary with a detector located 25.0 cm from the point of injection. What migration time and migration velocity would be expected for this cation when using an applied voltage of 15.0 kV?
    (b) What migration time and velocity would be obtained for the same cation as in Part (a) if a switch was made from the neutral capillary to a negatively charged capillary of an identical size, but that gives an electroosmotic mobility of 4.10 $cm^2/kV \cdot min$?
    (c) Repeat the calculations in Parts (a) and (b) now using an anion that has an electrophoretic mobility of -2.50 $cm^2/kV \cdot min$. Compare your results with those obtained for the cation. What does this comparison tell you about the role electroosmotic flow plays in the analysis of cations and anions in CE?
63. The effect of electroosmotic flow on the efficiency and resolution of a separation in electrophoresis is given by the equations shown,[23]

$$N = \frac{(\mu + \mu_{osm})V}{2D} \qquad (23.13)$$

$$R_s = 0.177 \, (\mu_1 - \mu_2)\sqrt{\frac{V}{D(\mu_{avg} + \mu_{osm})}} \qquad (23.14)$$

where μ_1 and μ_2 are the electrophoretic mobilities of the first and second eluting solutes, μ_{avg} is the average electrophoretic mobility of solutes 1 and 2, and D is their diffusion coefficient.

(a) The same protein as in Exercise 23.1 is examined on a CE system with a 25.0 cm long negatively charged capillary (22.0 cm to the detector) at 20.0 kV. This new capillary has an electroosmotic mobility of 3.0 cm^2/kV · min. If the protein still has an inherent electrophoretic mobility of 1.70 cm^2/kV · min and a diffusion coefficient of 2.0×10^{-7} cm^2/s, how many theoretical plates are possible for this system? How does this result compare with that in Exercise 23.1?

(b) Make a plot showing what resolution would be expected at various values for μ_{osm} in the case where the applied voltage is 10.0 kV and two analyte peaks have electrophoretic mobilities of 1.70 and 1.72 cm^2/kV · min, with an average diffusion coefficient of 2.0×10^{-7} cm^2/s . At what values of μ_{osm} versus μ_{avg} will the largest resolutions be obtained? What values of μ_{osm} will give the smallest resolutions?

64. The amount of sample applied to a capillary by hydrodynamic injection can be determined by the following form of the Hagen–Poiseuille equation,

$$\text{Sample volume} = \frac{\Delta P \, d^4 \pi t}{128 \eta \, L} \qquad (23.15)$$

where ΔP is the pressure applied across the capillary during injection, d is the inner diameter of the capillary, t is the time over which the pressure is applied, η is the viscosity of the applied solution, and L is the total length of the capillary.[8]

(a) What volume of sample would be applied to a 50 μm ID $\times$ 20 cm long capillary if a pressure of 0.5 psi is applied for 1 s to a solution with a viscosity of 0.01 poise?

(b) How much sample would be applied under the same conditions but using a 1 s pulse on a 25 μm ID $\times$ 20 cm long capillary?

65. The amount of an analyte that is applied to a capillary by electrokinetic injection is described by Equation 23.16,

$$Q = \frac{(\mu + \mu_{osm})VACt}{L} \qquad (23.16)$$

where Q is the quantity of analyte injected, μ is the electrophoretic mobility of the analyte, μ_{osm} is the mobility due to electroosmosis, V is the applied voltage, A is the cross-sectional area of the capillary, C is the analyte concentration in the original sample, t is the time over which the electric field is applied for injection, and L is the distance over which the voltage is applied.[8]

(a) Based on Equation 23.16, which types of analytes will have the largest injected quantities in electrokinetic injection: those that move with electroosmotic flow or against it?

(b) How does the value of Q change with the size of electroosmotic flow? Are small values or large values for μ_{osm} desirable in this method?

66. Compare and contrast the following analytical methods in terms of the way in which they separate analytes.

(a) Electrokinetic chromatography and reversed-phase chromatography

(b) Capillary electrophoresis and ion-exchange chromatography

(c) SDS-PAGE and capillary gel electrophoresis

(d) Affinity capillary electrophoresis and affinity chromatography

TOPICS FOR REPORTS AND DISCUSSION

67. Obtain more information on the Human Genome Project. Discuss the challenges this project presented to analytical chemists. What changes in DNA sequencing methods were made to make this project possible?

68. Now that human DNA has been sequenced, scientists have begun to examine the vast number of proteins that are encoded by this DNA. This research has lead to an area known as "proteomics." Obtain more information on proteomics and the challenges that are presented by this field to chemical analysis. Describe some analytical methods that are being used in this field.

69. Contact or visit a local hospital or biochemical laboratory. Report on how electrophoresis is used in these laboratories.

70. Compare and contrast the advantages and disadvantages for each of the following pairs of methods

(a) Gel electrophoresis versus capillary electrophoresis

(b) Gel electrophoresis versus HPLC

(c) Capillary electrophoresis versus HPLC

71. Use the Internet to obtain material safety data sheet (MSDS) information for the various chemicals that are shown in Figure 23.8 for the preparation of a polyacrylamide gel. Identify any chemical or physical hazards that are associated with these reagents.

72. Work with Northern and Southern blots in gel electrophoresis often involves the use of phosphorus-32 (^{32}P) as a radiolabel. Obtain further information on any special requirements, training or facilities that are needed for dealing with this agent. In addition, find out why phosphorus-32 is used as a label for these applications. Write a report discussing your findings.

73. There are several additional types of electrophoresis besides those that were discussed in this chapter. A few examples are listed below.[10,17] Write a report on one of these methods. Include of description of how the method separates analytes, its applications, and its advantages and disadvantages.

(a) Isotachophoresis

(b) Pulsed-field electrophoresis

(c) Dielectrophoresis

(d) Immunoelectrophoresis

74. The need for small sample sizes originally created several challenges in the design of CE instruments. This same feature has made capillary electrophoresis attractive for the analysis of samples for which only small volumes are available. For instance, it has been shown that CE can be used to analyze the content of single cells. Obtain a recent research article or review on this topic. Write a report that discusses how CE was used to analyze single cells in the subject paper.

75. Some early examples of capillary electrophoresis can be found in Referemces 23 and 34. Look up these articles and examine how electrophoresis was performed in them. How do the methods described in these papers compare to those that are now commonly used in CE, as described in this chapter?

76. *Capillary electrochromatography* is a method in which the movement of an analyte through a stationary phase is achieved through the use of electroosmotic flow rather than by only a difference in pressure.[7,47,48] Obtain more information on this method. Report on how this method works and on some of its recent applications. Discuss how this method is related to both traditional liquid chromatography and capillary electrophoresis, even though it is usually classified as a chromatographic method.

77. Four chemicals that can be used as matrices in MALDI-TOF MS are nicotinic acid, sinapinic acid, α-cyano-4-hydroxycinnamic acid, and 2,5-dihydroxybenzoic acid. Obtain more information on one or more of these chemicals and learn about how they are used in MALDI-TOF MS. What chemical or physical properties make these chemicals used in this method? What are some analytes that can be examined with these matrices?

78. Find a recent research article that describes or uses a "lab-on-a-chip" or μTAS device. Describe this device and the application for which it was employed. What advantages or disadvantages were reported for this device versus more traditional methods for chemical analysis?

References

1. J. C. Venter *et al.*, "The Sequence of the Human Genome," *Science*, 291 (2001) 1304–1351.

2. E. S Lander *et al.*, "Initial Sequencing and Analysis of the Human Genome," *Nature*, 409 (2001) 890–921.

3. *The New Encyclopaedia Britannica*, 15th ed., Encyclopaedia Britannica, Inc., Chicago, IL, 2002.

4. F. Sanger, "The Early Days of DNA Sequences," *Nature Medicine*, 7 (2001) 267–268.

5. "A History of the Human Genome Project," *Science*, 291 (2001) 1195.

6. N. J. Dovichi and J. Zhang, "How Capillary Electrophoresis Sequenced the Human Genome," *Angewandte Chemie International Edition*, 39 (2000) 4463–4468.

7. M.-L. Riekkola, J. A. Jonsson, and R. M. Smith, "Terminology for Analytical Capillary Electromigration Techniques (IUPAC Recommendations 2003)," *Pure and Applied Chemistry*, 76 (2004) 443–451.

8. T. Blanc, D. E. Schaufelberger, and N. A. Guzman, "Capillary Electrophoresis." In *Analytical Instrumentation Handbook*, 2nd ed., G. W. Ewing, Ed., Marcel Dekker, New York, 1997, Chapter 25.

9. J. Inczedy, T. Lengyel, and A. M. Ure, *Compendium of Analytical Nomenclature*, 3rd ed., Blackwell Science, Malden, MA, 1997.

10. B. L. Karger, L. R. Snyder, and C. Hovath, *An Introduction to Separation Science*, Wiley, New York, 1973, Chapter 17.

11. W.B. Hardy, "On the Coagulation of Proteid by Electricity," *Journal of Physiology*, 26 (1899) 288–304.

12. W. B. Hardy, "Colloidal Solution. The Globulins," *Journal of Physiology*, 33 (1905) 251–337.

13. L. Michaelis, "Elektrische Uberfuhrung von Fermenten," *Biochemische Zeitschrift*, 16 (1909) 81–86.

14. A. W. K. Tiselius, "A New Apparatus for Electrophoretic Analysis of Colloidal Mixtures," *Transactions of the Faraday Society*, 33 (1937) 524–531.

15. Nobel Lectures, Chemistry 1942–1962, Elsevier, Amsterdam, the Netherlands, 1964.

16. J. W. Jorgenson, "Electrophoresis," *Analytical Chemistry*, 58 (1986) 743A–760A.

17. J. C. Giddings, *Unified Separation Science*, Wiley, New York, 1991, Chapter 8.

18. L. Stryer, *Biochemistry*, Freeman, New York, 1988, Chapter 3.

19. D. S. Hage, "Chromatography and Electrophoresis," In *Contemporary Practice in Clinical Chemistry*, W. Clarke and D. R. Dufour, Eds., AACC Press, Washington, DC, 2006, Chapter 7.

20. J. M. Anderson and G. A. Tetrault, "Electrophoresis," In *Laboratory Instrumentation*, 4th ed., M. C. Haven, G. A. Tetrault, and J. R. Schenken, Eds., Van Nostrand Reinhold, New York, 1995, Chapter 12.

21. J. P. Landers, *Handbook of Capillary Electrophoresis*, CRC Press, Boca Raton, FL, 1992.

22. P. Camilleri, Ed., *Capillary Electrophoresis*, CRC Press, Boca Raton, FL, 1997.

23. J. W. Jorgenson and K. D. Lukacs, "Zone Electrophoresis in Open-Tubular Glass Capillaries," *Analytical Chemistry*, 53 (1981) 1298–1302.

24. E. M. Southern, "Detection of Specific Sequences among DNA Fragments Separated by Gel Electrophoresis," *Journal of Molecular Biology*, 98 (1975) 503–517.

25. J. C. Alwine, D. J. Kemp, and G. R. Stark, "Method for Detection of Specific RNAs in Agarose Gels by Transfer to Diazobenzyloxymethyl-Paper and Hybridization with DNA Probes," *Proceedings of the National Academy of Science USA*, 74 (1977) 5350–5354.

26. J. Renart, J. Reiser, and G. R. Stark, "Transfer of Proteins from Gels to Diazobenzyloxymethyl-Paper and Detection with Antisera: A Method for Studying Antibody Specificity and Antigen Structure," *Proceedings of the National Academy of Science USA*, 76 (1979) 3116–3120.

27. W. N. Burnette, "Western Blotting: Electrophoretic Transfer of Proteins from Sodium Dodecyl Sulfate-Polyacrylamide Gels to Unmodified Nitrocellulose and Radiographic Detection with Antibody and Radioiodinated Protein A," *Analytical Biochemistry*, 112 (1981) 195–203.

28. M. Karas, D. Bachmann, and F. Hillenkamp, "Influence of the Wavelength in High-Irradiance Ultraviolet Laser Desorption Mass Spectrometry of Organic Molecules," *Analytical Chemistry*, 57 (1985) 2935–2939.

29. M. Karas and F. Hillenkamp, "Laser Desorption Ionization of Proteins with Molecular Masses Exceeding 10,000 Daltons," *Analytical Chemistry*, 60 (1988) 2299–2301.

30. K. Tanaka, H. Waki, Y. Ido, S. Akita, Y. Yoshida, and T. Yoshida, "Protein and Polymer Analyses up to m/z 100 000 by Laser Ionization Time-of-Flight Mass Spectrometry," *Rapid Communications in Mass Spectrometry*, 2 (1988) 151–153.

31. F. Hillenkamp and J. Peter-Katalinic, Eds., *MALDI MS: A Practical Guide to Instrumentation, Methods and Applications*, Wiley, New York, 2007, 31.

32. M. J. Stump, R. C. Fleming, W.-H. Gong, A. J. Jaber, J. J. Jones, C. W. Surber, and C. L. Wilkins, "Matrix-Assisted Laser Desorption Mass Spectrometry," *Applied Spectroscopy Reviews*, 37 (2002) 275–303.

33. D. E. Garfin, "Two-Dimensional Gel Electrophoresis: An Overview," *Trends in Analytical Chemistry*, 22 (2003) 263–272.

34. F. E. P. Mikkers, F. M. Everaerts, and Th. P. E. M. Verheggen, "High-Performance Zone Electrophoresis," *Journal of Chromatography*, 169 (1979) 11–20.

35. K. R. Mitchelson and J. Cheng, Eds., *Capillary Electrophoresis of Nucleic Acids*, Humana Press, Totowa, NJ, 2001.

36. S. C. Terry, G. H. Jerman, and J. B. Angell, "A Gas Chromatographic Air Analyzer Fabricated on a Silicon Wafer," *IEEE Transactions on Electron Devices*, 12 (1979) 1880–1886.

37. A. Manz, N. Graber, and H. M. Widmer, "Miniaturized Total Chemical Analyses Systems. A Novel Concept for Chemical Sensing," *Sensors and Actuators*, B1 (1990) 244.

38. D. J. Harrison, K. Fluri, K. Seiler, Z. Fan, C. S. Effenhauser, and A. Manz, "Micromachining a Miniaturized Capillary Electrophoresis-Based Chemical Analysis System on a Chip," *Science*, 261 (1993) 895–897.

39. M. Freemantle, "Downsizing Chemistry," *Chemical & Engineering News*, 77 (1999) 27–36.

40. O. Geschke, H. Klank, and P. Telleman, *Microsystem Engineering of Lab-on-a-Chip Devices*, Wiley-VCH, Weinheim, Germany, 2004.

41. Charles S. Henry, Ed., *Microchip Capillary Electrophoresis*, Springer-Verlag, New York, 2006.

42. S. Terabe, K. Otsuka, K. Ichikawa, A. Tsuchiya, and T. Ando, "Electrokinetic Separations with Micellar Solutions and Open-Tubular Capillaries," *Analytical Chemistry*, 56 (1984) 111–113.

43. S. Terable, K. Otsuka, and T. Ando, "Electrokinetic Chromatography with Micellar Solution and Open-Tubular Capillary," *Analytical Chemistry*, 57 (1985) 834–841.

44. N. H. H. Heegaard and C. Schou, "Affinity Ligands in Capillary Electrophoresis," In *Handbook of Affinity Chromatography*, D. S. Hage, Ed., CRC Press, Boca Raton, FL, 2005, Chapter 26.

45. R. Neubert and H.-H. Ruttinger, Eds., *Affinity Capillary Electrophoresis in Pharmaceutics and Biopharmaceutics*, CRC Press, Boca Raton, FL, 2003.

46. J. Yang, X.-Z. Wang, D. S. Hage, P. L. Herman, and D. P. Weeks, "Analysis of Dicamba Degradation by Pseudomonas Maltophilia Using High-Performance Capillary Electrophoresis," *Analytical Biochemistry*, 219 (1994) 37–42.

47. S. Eeltink and W. Th. Kok, "Recent Applications in Capillary Electrochromatography," *Electrophoresis*, 27 (2006) 84–96.

48. Z. Deyl and F. Svec, Eds., *Capillary Electrochromatography*, Elsevier, Amsterdam, the Netherlands, 2001.

APPENDIX A

APPENDIX A.1 EXAMPLE OF A MATERIAL SAFETY DATA SHEET (MSDS)*

SIGMA-ALDRICH

Material Safety Data Sheet

Version 3.2
Revision date 03/25/2009
Print date 05/19/2009

1. PRODUCT AND COMPANY IDENTIFICATION

Product name : Benzene

Product number : 401765
Brand : Sigma-Aldrich

Company : Sigma-Aldrich
3050 Spruce Street
SAINT LOUIS MO 63103
USA

Telephone : +1 800-325-5832
Fax : +1 800-325-5052
Emergency phone # : (314)-776-6555

2. COMPOSITION/INFORMATION ON INGREDIENTS

Formula : C_6H_6
Molecular Weight : 78.11 g/mol

CAS-No	EC-No	Index-No	Concentration
Benzene			
71-43-2	200-753-7	601-020-00-8	-

3. HAZARDS IDENTIFICATION

Emergency overview

OSHA hazards
Flammable liquid, target organ effect, Irritant, carcinogen, mutagen

Target organs
Blood, eyes, female reproductive system, bone marrow

HMIS classification
Health hazard: 2
Chronic health hazard: *
Flammability: 3
Physical hazards: 0

NFPA rating
Health hazard: 2
Fire: 3
Reactivity hazard: 0

Potential health effects

Inhalation   May be harmful if inhaled. Causes respiratory tract irritation.
Skin   May be harmful if absorbed through skin. Causes skin irritation.

Sigma-Aldrich Corporation
www.sigma-aldrich.com

Eyes   Causes eye irritation.
Ingestion   Aspiration hazard if swallowed - can enter lungs and cause damage. May be harmful if swallowed.

4. FIRST AID MEASURES

General advice
Consult a physician. Show this safety data sheet to the doctor in attendance. Move out of dangerous area.

If inhaled
If breathed in, move person into fresh air. If not breathing give artificial respiration. Consult a physician.

In case of skin contact
Wash off with soap and plenty of water. Take victim immediately to hospital. Consult a physician.

In case of eye contact
Rinse thoroughly with plenty of water for at least 15 minutes and consult a physician.

If swallowed
Do NOT induce vomiting. Never give anything by mouth to an unconscious person. Rinse mouth with water. Consult a physician.

5. FIRE-FIGHTING MEASURES

Flammable properties
flash point   −11.0 °C (12.2 °F) - closed cup
ignition temperature   562 °C (1,044 °F)

Suitable extinguishing media
For small (incipient) fires, use media such as "alcohol" foam, dry chemical, or carbon dioxide. For large fires, apply water from as far as possible. Use very large quantities (flooding) of water applied as a mist or spray; solid streams of water maybe ineffective. Cool all affected containers with flooding quantities of water.

Specific hazards
Flash back possible over considerable distance. Container explosion may occur under fire conditions.

Special protective equipment for fire-fighters
wear self contained breathing apparatus for fire fighting if necessary.

Further information
Use water spray to cool unopened containers.

6. ACCIDENTAL RELEASE MEASURES

Personal precautions
Use personal protective equipment. Avoid breathing vapors, mist or gas. Ensure adequate ventilation. Remove all sources of ignition. Evacuate personnel to safe areas. Beware of vapours accumulating to form explosive concentration. Vapours can accumulate in low areas.

Environmental precautions
Prevent further leakage or spilage if safe to do so. Do not let product enter drains.

Methods of cleaning up
Contain spilage, and then collect with non-combustible absorbent material, (e.g. sand, earth, diatomaceous earth, vermiculite) and place in container for disposal according to local/national regulations (see section 13).

7. HANDLING AND STORAGE

Handling
Avoid inhalation of vapour or mist. Keep away from sources of ignition - No smoking. Take measures to prevent the build up of electrostatic charge.

Sigma-Aldrich Corporation
www.sigma-aldrich.com

*This MSDS is reproduced with permission and courtesy of Sigma-Aldrich.

Storage
Keep container tightly closed in a dry and well-ventilated place. Containers which are opened must be carefully resealed and kept upright to prevent leakage. Store in cool place.

8. EXPOSURE CONTROLS/PERSONAL PROTECTION

Components with workplace control parameters

Components	CAS-No	Value	Control Parameters	Update	Basis
Benzene	71-43-2	TWA	0.5 ppm	2007-01-01	USA. ACGIH threshold limit values (TLV)
Remarks	Leukemia Substances for which there is a Biological Exposure Index or Indices (see BEI section) confirmed human carcinogen: The agent is carcinogenic to humans based on the weight of evidence from epidemiologic studies. Danger of cutaneous absorption				
		STEL	2.5 ppm	2007-01-01	USA. ACGIH threshold limit values (TLV)
					Leukemia Substances for which there is a Biological Exposure Index or Indices (see BEI section) confirmed human carcinogen: The agent is carcinogenic to humans based on the weight of evidence from epidemiologic studies. Danger of cutaneous absorption
		TWA	1 ppm	1989-03-01	USA. OSHA - TABLE Z-1 Limits for Air Contaminants -1910.1000
			Sec. 1910.1028 Benzene. The final benzene standard in 1910.1028 applies to benzene except some subsegments of industry where exposures are consistently under the action level (i.e. distribution and sale of fuels, sealed containers and pipelines, coke production, oil and gas drilling and production, natural gas processing, and the percentage exclusion for liquid mixtures): for the excepted subsegments, the benzene limits in Table Z-2 apply. See Table Z-2 for the limits applicable in the operations or sectors excluded in 1910.1028.			
		STEL	5 ppm	1989-03-01	USA. OSHA - TABLE Z-1 Limits for Air Contaminants -1910.1000
			See Table Z-2 for the limits applicable in the operations or sectors excluded in 1910.1028. The final benzene standard in 1910.1028 applies to all occupational exposures to benzene except some subsegments of industry where exposures are consistently under the action level (i.e. distribution and sale of fuels, sealed containers and pipelines, coke production, oil and gas drilling and production, natural gas processing, and the percentage exclusion for liquid mixtures): for the excepted subsegments, the benzene limits in Table Z-2 apply. Sec. 1910.1028 Benzene			
		TWA	1 ppm	1993-08-30	USA. Occupational Exposure Limits (OSHA) - TABLE Z-1 Limits for Air Contaminants

---

	Value		Update	Basis
	STEL	5 ppm	1993-06-30	USA. Occupational Exposure Limits (OSHA) - TABLE Z-1 Limits for Air Contaminants
				Z37.40-1989
	TWA	10 ppm	2007-01-01	USA. Occupational Exposure Limits (OSHA) - TABLE Z2
	CEIL	25 ppm	2007-01-01	USA. Occupational Exposure Limits (OSHA) - TABLE Z2
				Z37.40-1989
	Peak	50 ppm	2007-01-01	USA. Occupational Exposure Limits (OSHA) - TABLE Z2
				Z37.40-1989
				Sec. 1910.1028. See Table Z-2 for the limits applicable in the operations or sectors excluded in 1910.1028. The final benzene standard in 1910.1028 applies to all occupational exposures to benzene except some subsegments of industry where exposures are consistently under the action level (i.e. distribution and sale of fuels, sealed containers and pipelines, coke production, oil and gas drilling and production, natural gas processing, and the percentage exclusion for liquid mixtures): for the excepted subsegments, the benzene limits in Table Z-2 apply.

Personal protective equipment

Respiratory protection
Where risk assessment shows air-purifying respirators are appropriate use a full-face respirator with-multi-purpose combination (US) or type ABEK (EN14387) respirator cartridges as a backup to engineering controls. If the respirator is the sole means of protection, use a full-face supplied air respirator. Use respirators and components tested and approved under appropriate government standards such as NIOSH (US) or CEN (EU).

Hand protection
Handle with gloves.

Eye protection
Safety glasses

Skin and body protection
Choose body protection according to the amount and concentration of the dangerous substance at the work place.

Hygiene measures
Avoid contact with skin, eyes and clothing. Wash hands before breaks and immediately after handling the product.

9. PHYSICAL AND CHEMICAL PROPERTIES

Appearance	
Form	Liquid
Colour	Colourless

Safety data

pH	no data available
Melting point	5.5 °C (41.9 °F)
Boiling point	80 °C (178 °F)
Flash point	−11 °C (12.2 °F) - closed cup
Ignition temperature	562 °C (1,044 °F)
Lower explosion limit	1.3 % (V)
Upper explosion limit	8 % (V)
Vapour pressure	221.3 hPa (166.0 mmHg) at 37.7 °C (99.9 °F) 99.5 hPa (74.6 mm Hg) at 20.0 °C (68.0 °F)
Density	0.874 g/mL at 25 °C (77 °F)
Water solubility	no data available

10. STABILITY AND REACTIVITY

Storage stability
Stable under recommended storage conditions

Conditions to avoid
Heat, flames and sparks

Materials to avoid
acids, bases, halogens, strong oxidizing agents, metallic salts

Hazardous decomposition products
Hazardous decomposition products formed under fire conditions - carbon oxides

Hazardous reactions
Vapours may form explosive mixture with air.

11. TOXICOLOGICAL INFORMATION

Acute toxicity

LD50 Oral - rat - 2,990 mg/kg

LD50 Inhalation - rat - female - 4 h - 44,700 mg/m^3

LD50 Dermal - rabbit - 8,263 mg/kg

Irritation and corrosion

Skin - rabbit - skin irritation

Eyes - rabbit - eye irritation

Sensitisation
no data available

Chronic exposure
Carcinogenicity - human - male - inhalation
Tumorigenic:Carcinogenic by RTECS criteria. Leukemia Blood:thrombocytopenia.
Carcinogenicity - rat - oral
Tumorigenic:Carcinogenic by RTECS criteria. Endocrine:tumors. Leukemia.

This is or contains a component that has been reported to be carcinogenic based on its IARC, OSHA, ACGIH, NTP, or EPA classification.

IARC: 1 - Group 1: Carcinogenic to humans (Benzene)

NTP: Known to be human carcinogen (Benzene)

Genotoxicity in vitro - human - lymphocyte
Sister chromatid exchange

Genotoxicity in vitro - mouse - lymphocyte
Mutation in mammalian somatic cells.

Genotoxicity in vitro - mouse - inhalation
Sister chromatid exchange

Laboratory experiments have shown mutagenic effects.

Developmental toxicity - rat - inhalation
Effects on embryo or fetus: extra embryonic structures (e.g., placenta, umbilical cord). Effects on embryo or fetus: fetotoxicity (except death, e.g., stunted fetus).

Developmental toxicity - mouse - inhalation
Effects on embryo or fetus: cytological changes (including somatic cell genetic material). specific developmental abnormalities: blood and lymphatic system (including spleen and marrow).

Reproductivity toxicity - mouse - intraperitoneal
Effects on fertility: pre-implantation mortality (e.g., reduction in number of implants per female; total number of implants per corpora lutea). Effects on embryo or fetus: fetal death.

Signs and symptoms of exposure

Nausea, dizziness, headache, narcosis, inhalation of high concentrations of benzene may have in initial stimulatory effect on the central nervous system characterized by exhilaration, nervous excitation and/or giddiness, depression, drowsiness, or fatigue. The victim may experience tightness in the chest, breathlessness, and loss of consciousness, tremors, convulsions, and death due to respiratory paralysis or circulatory collapse can occur in a few minutes to several hour following severe exposures.
Aspiration of small amounts of liquid immediately causes pulmonary edema and hemorrhage of pulmonary tissue. Direct skin contact may cause erythema. Repeated or prolonged skin contact may result in drying, scaling dermatitis, or development of secondary skin infections. The chief target organ is the hematopietice system. Bleeding from the nose, gums, or mucous membranes and the development of purpuric spots, pancytopenia, leukopenia, thrombocytopenia, aplastic or hyperplastic, and may not correlate with peripheral blood-forming tissues. The onset of effects of prolonged benzene exposure may be delayed for many months or years after the actual exposure has ceased., blood disorders

Potential health effects

Inhalation	May be harmful if inhaled. Causes respiratory tract irritation.
Skin	May be harmful if absorbed through skin. Causes skin irritation.
Eyes	Causes eye irritation.
Ingestion	Aspiration hazard if swallowed - can enter lungs and cause damage. May be harmful if swallowed.
Target organs	Blood. eyes. female reproductive system. bone marrow.

Additional information
RTECS: CY1400000

12. ECOLOGICAL INFORMATION

Elimination information (persistence and degradability)

Biodegradability   Result: - Readily biodegradable.

SARA 302 Components
SARA 302: No chemicals in this material are subject to the reporting requirements of SARA Title III. section302.

SARA 313 Components		
Benzene	CAS-No. 71-43-2	Revision date 2007-07-01

SARA 311/312 Hazards
Fire hazard, acute health hazard, chronic health hazard

Massachusetts right to know components		
Benzene	CAS-No. 71-43-2	Revision date 2007-07-01

Pennsylvania right to know components		
Benzene	CAS-No. 71-43-2	Revision date 2007-07-01

New Jersey right to know components		
Benzene	CAS-No. 71-43-2	Revision date 2007-07-01

California Prop. 65 components
WARNING! This product contains a chemical known in the state of California to cause cancer.

Benzene	CAS-No. 71-43-2	Revision date 2004-05-12

California Prop. 65 components
WARNING! This product contains a chemical known in the state of California to cause birth defects or other reproductive harm.

Benzene	CAS-No. 71-43-2	Revision date 2004-05-12

16. OTHER INFORMATION

Further information
Copyright 2009 Sigma-Aldrich Co. License granted to make unlimited paper copies for internal use only.
The above information is believed to be correct but does not purport to be all inclusive and shall be used only as a guide. The information in this document is based on the present state of our knowledge and is applicable to the product with regard to appropriate safety precautions. It does not represent any guarantee of the properties of the product. Sigma-Aldrich Co., shall not be held liable for any damage resulting from handling or from contact with the above product. See reverse side of invoice or packing slip for additional terms and conditions of sale.

Bioaccumulation	Leuciscus idus (golden orfe) - 3 d
	Bioconcentration factor (BCF) - 10

Ecotoxicity effects

Toxicity to fish	LC50 - Oncorhynchus mykiss (rainbow trout) - 5.90 mg/l - 98 h
	LC50 - Pimephales promelas (fathead minnow) - 15.00–32.00 mg/l - 98 h
	LC50 - Lepomis macrochirus (bluegill) - 230.00 mg/l - 96 h
	NOEC - Pimephales promelas (fathead minnow) - 10.2 mg/l - 7 d
	LOEC - Pimephales promelas (fathead minnow) - 17.20 mg/l - 7 d
Toxicity to daphnia and other aquatic invertebrates.	EC50 - Oncorhynchus mykiss (water flea) - 22.00 mg/l - 48 h
	EC50 - Daphnia magna (water flea) - 9.20 mg/l - 48 h
Toxicity to algae	EC50 - Pseudokirchneriella subcapitata (green algae) - 29.00 mg/l - 72 h

Further information on ecology
no data available

13. DISPOSAL CONSIDERATIONS

Product
Burn in a chemical incinerator equipped with an afterburner and scrubber but exert extra care in igniting as this materials is highly flammable. Observe all federal, state, and local environmental regulations. Contact a licensed professional waste disposal service to dispose of this material.

Contaminated packaging
Dispose of as unused product.

14. TRANSPORT INFORMATION

DOT (US)
UN-number: 1114   Class:3   Packing group: II
Proper shipping name: Benzene
Marine pollutant: no
Poison inhalation hazard: no

IMDG
UN-number: 1114   Class:3   Packing group: II   EMS-No: F-E, S-D
Proper shipping name: Benzene
Marine pollutant: no

IATA
UN-number: 1114   Class:3   Packing group: II
Proper shipping name: Benzene

15. REGULATORY INFORMATION

OSHA hazards
Flammable liquid, target organ effect, irritant, carcinogen, mutagen

DSL status
All components of this product are on the Canadian DSL list.

APPENDIX A.2 ESTIMATING THE NUMBER OF SIGNIFICANT FIGURES IN A RESULT

Rules for Determining the Number of Significant Figures in a Reported Result

The following rules should be followed when estimating the number of significant figures (or sig. figs.) in a reported value. These guidelines should be followed only if information is not provided on the measured or calculated precision of the value, such as determined by the propagation of errors (see Chapter 4).

Rule 1. The final value of an experimental result should be written so that it has only one figure with any appreciable uncertainty.

> *Example:* **57.3** ($\pm$0.2) (3 sig. figs.)
> **8.9381** ($\pm$0.0001) (5 sig. figs.)

(*Note*: If no uncertainty is reported for a value, it is implied that there is a variation of 0.5 units in the last number given on the right.)

Rule 2. All nonzero digits in a reported number (other than those in an exponent) are considered significant.

> *Examples:* **13.32** (4 sig. figs.)
> **3.32** $\times 10^{34}$ (3 sig. figs.)

Rule 3. All zeros that appear in a reported number between nonzero digits (other than in an exponent) are considered significant.

> *Examples:* **180,088** (6 sig. figs.)
> **6.022** $\times 10^{23}$ (4 sig. figs.)

Rule 4. Zeros that appear to the left of all nonzero digits only indicate the magnitude of a number and are not considered to be significant.

> *Examples:* 0.00**289** (3 sig. figs.)
> 0.**528** $\times 10^{9}$ (3 sig. figs.)

Rule 5. Zeros that appear to the right of all nonzero digits are significant only if there is a decimal point present.

> *Examples:* **257**,000 (3 sig.figs.)
> **2.570** $\times 10^{5}$ (4 sig. figs.)

(*Note*: This rule is important to avoid confusion when a zero at the end of a number does represent a significant figure.)

Rule 6. A number that is an integer (1, 2, 3, etc.) has an infinite number of significant figures.

> *Examples:* **1 = 1.000...**
> **2 = 2.000...**

Rule 7. When eliminating nonsignificant figures, round off the final result to the first digit that has any uncertainty in its value.

> *Examples:* 789.33 ($\pm$1) = **789**
> 2.375 ($\pm$0.2) = **2.4**

Rule 8. If the nonsignificant figures are half way between the nearest odd and even values of the first significant digit, use the even value.

> *Examples:* 88.950 ($\pm$0.5) = **89.0**
> 2385 ($\pm$20) = **2380**

Procedures for Estimating the Number of Significant Figures in a Calculated Result

The following procedures should be followed when estimating the number of significant figures in a value that results from a calculation. It should be noted that these rules are only rough guidelines and that a better approach for estimating the precision of a calculated value is to use the propagation of errors, as discussed in Chapter 4.

Addition or Subtraction: Align the numbers at the decimal point. Round the final result to the same position as the number that has its right-hand digit in the highest position versus the decimal.

> *Example:*
> | | |
> |---|---|
> | 567.29 | (significant to hundredths position) |
> | 4.018 | (significant to thousandths position) |
> | +2,881.4 | (significant to *tenths* position) |
>
> 3452.708 = **3452.7** (five significant figures to *tenths* position)

Multiplication or Division: The final result is given the same number of significant figures as the value in the calculation that has the fewest number of significant figures.

> *Example:*
> | | |
> |---|---|
> | 567.29 | (five significant figures) |
> | $\times$ 4.018 | (*four* significant figures) |
>
> 2279.3712 = **2279** (*four* significant figures)

Logarithims and Antilogarithms: For taking the logarithm (or "log") of a number, the mantissa of the log (i.e., the digits to the right of the decimal) should have the same number of significant figures as the original number. For an antilogarithm (or "antilog"), the calculated result should have the same number of significant figures as the mantissa of the original logarithm.

> *Example:* $\log(1.59 \times 10^{-9}) = $ **–8.799** (*three* significant figures in mantissa)
> antilog(-8.799) = $10^{-8.799}$
> = **1.59** $\times 10^{-9}$ (*three* significant figures)

Mixed Calculations: For calculations that involve a combination of steps, round off to the correct number of significant figures only after the final answer has been obtained. To do this, keep track of the number of significant figures that result from each step. Perform those operations that are in parentheses or brackets first. If no brackets or parentheses are present, perform the log/antilog calculations first, followed second by multiplication/division, and third by addition/subtraction.

Example: $x = 1.78 + (3.58 + 5.9789)/29.3$
$= 1.78 + (9.5589)/29.3$
$= 1.78 + 0.32624$
$= 2.10624 = $ **2.11** (*three* significant figures)

(*Note*: The underlined values in the preceding calculation are known as *guard digits*, which are used to avoid errors produced by the premature rounding of numbers.)

Other Guidelines: Because a true integer (1, 2, 3, etc.) has an infinite number of significant figures, this can safely be ignored when estimating the number of significant figures for a calculation. When you are using a chemical or physical constant (e.g., π or Avogadro's number), use a value for this constant that has at least as many, and preferably more, significant figures than the other numbers in your calculation. This avoids unnecessary rounding errors.

Example: $C = 2\pi r$
$= 2 \, (3.1416) \, (2.890 \text{ cm})$
$= 18.158 = $ **18.16** (*four* significant figures)

(*Note*: In this example, the integer "2" was not considered in determining the number of significant figures, because it has many more than any of the other numbers in the given equation. Also, if $\pi = 3.14$ had been used instead of 3.1416 in the preceding calculation, this would have improperly limited the final result to three significant figures and given an answer of 18.1.)

APPENDIX A.3 NORMALITY

In most of this textbook we have described the concentrations of solutes in terms of molarity and amount of material in units of moles. However, another well-established way we could have expressed the concentration of polyprotic acids or bases and materials that are under oxidation–reduction reactions is using units of normality. This approach requires the use of "equivalents" as a measure of amount. An "equivalent" is the amount of material that reacts with one mole of H^+ or OH^- in an acid–base reaction or with one mole of electrons in an oxidation–reduction reaction.

The normality of an acid is related to the number of moles of hydroxide ion that are required to neutralize a liter of the acid's solution. In the same manner, the normality of a base is the number of moles of hydrogen ions that are needed to neutralize a liter of the base in its solution.

For instance, sulfuric acid is a diprotic acid, so the normality (N) of a 0.0100 M solution will be double the value of its molarity, or (2 equivalents/mol) · 0.0100 M = 0.0200 N H_2SO_4. This means that the equivalent weight of sulfuric acid, in units of grams per equivalent (g/eq), is half the molecular weight, because (98.0 g/mol)/(2 eq/mol) = 49.0 g/eq. In a similar manner, the normality of a 0.100 M $KMnO_4$ solution used to titrate iron in acid solution is 0.500 N because $KMnO_4$ is a five-electron oxidizing agent (see half-reaction in Equation A3.1).

$$MnO_4^- + 8\,H^+ + 5\,e^- \leftrightarrows Mn^{2+} + 4\,H_2O \qquad (A3.1)$$

The equivalent weight of $KMnO_4$ in this case would be (158.0 g/mol)/(5 eq/mol) = 31.6 g/eq.

A danger in the use of normality as a concentration unit is that its value depends on the reaction that is being described. For instance, if a 0.100 M $KMnO_4$ solution is being used in a neutral solution, the product is MnO_2 and the oxidation state of Mn in this chemical changes by only three, as shown by the following reaction.

$$MnO_4^- + 4\,H^+ + 3\,e^- \leftrightarrows MnO_2 + 2\,H_2O \qquad (A3.2)$$

In this case, the concentration of the 0.100 M $KMnO_4$ solution in terms of normality would be 0.300 N. The equivalent weight of $KMnO_4$ in this situation is (158.0 g/mol)/(3 eq/mol) = 52.7 g/eq.

APPENDIX A.4 DERIVATION OF ERROR-PROPAGATION FORMULAS

Error Propagation in Common Mathematical Operations

The process by which random errors are carried from one number to the next through a calculation is known as the **propagation of random errors**, or **error propagation**. In Chapter 4 there were several equations provided that can be used to determine how random errors affect the results of mathematical operations like addition/subtraction, multiplication/division, exponentiation, and logarithms or antilogarithms. In this appendix, we will see how these equations were obtained and will learn what types of assumptions were made during the development of these relationships.

A General Description of Random Error Propagation

To begin our derivation of the equations in Table 4.2 we need a general way of relating the random error produced in a calculated result "y" to the random error that is present in any of the parameters that are used to determine the value of y. We start this process by using the following equation to indicate that y is a function of all the numbers ($a,b,c \ldots$) that are used in calculating the value of y.

$$y = f(a,b,c \ldots) \tag{A4.1}$$

Our goal here is to relate the precision in y, as described by its standard deviation (s_y), to the uncertainty and precision for each parameter on the right-hand side of Equation A4.1, where the precisions of a, b, $c \ldots$ are represented by the standard deviations s_a, s_b, $s_c \ldots$.

In order to meet our goal, we need to obtain some way of describing how the value of y will change when small variations (caused by random errors) are present in a, b, $c \ldots$. This is done by writing a differential equation in which the expected change in y (as given by its **deviation**, dy) is related to the deviations that may be present in each factor on the right-hand side of Equation A4.1 (da, db, $dc \ldots$). The differential equation that we will use in this case is given by Equation A4.2. Along with the terms da, db, dc, and so on, this relationship includes the partial derivatives of our original function with respect to each of our starting parameters. These partial derivatives show how the equation for calculating y is affected by changes in our initial parameters. In Equation A4.2, the partial derivatives of our function for y with respect to a, b, c, and so on, are represented by the terms $(\partial y/\partial a)$, $(\partial y/\partial b)$, and $(\partial y/\partial c) \ldots$.

$$dy = (\partial y/\partial a)\, da + (\partial y/\partial b)\, db + (\partial y/\partial c)\, dc + \ldots \tag{A4.2}$$

The next step in our derivation involves describing the precision produced in our final *mean* result (y) in terms of the random error that might be present for any *individual* result (expressed here as y_i). This random error

can be expressed in terms of the deviation in y_i (written as dy_i), which is defined as follows.

$$dy_i = (y_i - y) \tag{A4.3}$$

The same type of definition can be used to describe the random error present in any particular value of $a_i, b_i, c_i \ldots$ by using a corresponding deviation for that term (da_i, db_i, dc_i, etc).

As we go from each member in a set of results ($y_1 \ldots y_n$) to the mean result for this set (y), the net deviation in our final result (dy) should be equal to the sum of the deviations for the individual results (or $dy_1 \ldots dy_n$). This is represented mathematically by Equation A4.4.

$$dy = \sum(dy_i) \tag{A4.4}$$

Again, the same argument can be made for the parameters a, b, $c \ldots$ in setting da equal to $\sum(da_i)$, db equal to $\sum(db_i)$, dc equal to $\sum(dc_i)$, and so on.

If we want to describe precision by using standard deviations, we now need to consider how deviation terms like dy, da, db, $dc \ldots$, are related to s_y, s_a, s_b, $s_c \ldots$. This can be done by going back to the formula given by Equation 4.7 in Chapter 4 for determining the standard deviation for a group of results. A slightly modified version of this relationship is shown below in a form that is based on our calculated result (y).

$$s_y = \sqrt{\sum(y_i - y)^2/(n - 1)} \tag{A4.5}$$

In looking back at Equations A4.3 and A4.4, we can see that ($y_i - y$) was defined as being equal to dy_i and that dy was said to be equal to (dy_i). By making these substitutions into Equation A4.5, we get the following expressions.

$$s_y = \sqrt{\sum(dy_i)^2/(n - 1)} \tag{A4.6}$$

$$= \sqrt{(dy)^2/(n - 1)} \tag{A4.7}$$

or

$$(s_y)^2 = (dy)^2/(n - 1) \tag{A4.8}$$

Equation A4.8 will become useful to us later, because it provides a means of directly relating s_y to dy. By using the same general approach and types of relationships, we can also relate s_a to da, s_b to db, s_c to dc, and so on.

One small problem that we still have to deal with before we can use s_y, s_a, s_b, $s_c \ldots$ in place of dy, da, db, $dc \ldots$ is the fact that the relationship given between these factors in Equation A4.8 is based on the *squares* of these terms. This is required in going from dy to s_y, because a deviation term like dy can be either positive or negative in value, but a standard deviation like s_y must always be positive. Thus, before we can replace dy with s_y, da with s_a, and so on, in Equation A4.2 (our initial goal for this derivation), we must first square both sides of Equation A4.2. This gives us the following result.

$$(dy)^2 = [(\partial y/\partial a)(da) + (\partial y/\partial b)(db) +$$

$$(\partial y/\partial c)(dc) + \ldots]^2 \qquad (A4.9)$$

As we expand and multiply the right-hand side of the preceding expression, we will get two different types of parameter combinations. One type of combination will be the *true square terms*, like $(\partial y/\partial a)^2(da)^2$ and $(\partial y/\partial b)^2(db)^2$, which are produced whenever we multiply the same factor or group of factors by themselves. An important property of these square terms is that they will always give a product that is positive. This means that the products of the square terms will never cancel each other out when they are added together. Thus, these terms will always be important whenever we examine the propagation of random errors in calculations.

The other type of combination that we get when we multiply through the right-hand side of Equation A4.9 are the *cross terms*. These are terms that are produced when two different groups of parameters are multiplied together, with $[(\partial y/\partial a)(da)][(\partial y/\partial b)(db)]$ being one such example. Cross terms differ from the square terms in that their products can be either positive or negative in value. This is useful to know because if it is assumed that the deviations present in the individual parameters (da_i, db_i, dc_i. . .) are truly random and independent, then the products of the cross terms will tend to cancel out and their sum will approach zero; this will be especially true when the number of individual values (n) is large.

If it is true that the random errors in our starting parameters are really random and independent of one another, and if the number of our individual parameter values is sufficiently large, then we can further assume that the overall contribution of the cross terms in Equation A4.9 is essentially zero. This leaves us with just the true square terms, which allows Equation A4.9 to be simplified into the following form.

$$(dy)^2 = (\partial y/\partial a)^2(da)^2 + (\partial y/\partial b)^2(db)^2 +$$

$$(\partial y/\partial c)^2(dc)^2 + \ldots \qquad (A4.10)$$

By dividing both sides of Equation A4.10 by $(n - 1)$, we get the result shown in Equation A4.11.

$$(dy)^2/(n-1) = (\partial y/\partial a)^2(da)^2/(n-1) + (\partial y/\partial b)^2$$

$$(db)^2/(n-1) + (\partial y/\partial c)^2(dc)^2/(n-1) + \ldots \qquad (A4.11)$$

If we compare this new relationship with Equation A4.8, we see that it is now relatively easy to replace $(dy)^2/(n-1)$ with $(s_y)^2$, $(da)^2/(n-1)$ with $(s_a)^2$, and so on. This

provides us with the final relationship for random error propagation that is given in Equation A4.12.

$$(s_y)^2 = (\partial y/\partial a)^2(s_a)^2 + (\partial y/\partial b)^2(s_b)^2 + (\partial y/\partial c)^2(s_c)^2 + \ldots$$

$$(A4.12)$$

This final equation is a general relationship that shows how random errors would be expected to affect a result obtained by any type of mathematical operation. In the next few sections, we'll look at how this equation can be further simplified when working with specific types of calculations.

Random Error Propagation During Addition or Subtraction

Addition and subtraction are the first group of specific mathematical operations that we will examine with regard to the propagation of random errors. As a general example, let's consider how the final uncertainty or precision that we obtain for our result y will be affected by the uncertainty in the values of a, b, and c during the following calculation.

$$y = a - b + c \qquad (A4.13)$$

To relate the standard deviation of our result (s_y) to the standard deviations for a, b, and c (as given by s_a, s_b and s_c), we start with the following general relationship obtained from Equation A4.12.

$$(s_y)^2 = (\partial y/\partial a)^2(s_a)^2 + (\partial y/\partial b)^2(s_b)^2 + (\partial y/\partial c)^2(s_c)^2$$

$$(A4.14)$$

Next, we take derivatives of the equation to be used in our calculation (Equation A4.13) in order to determine the values of the partial derivatives $(\partial y/\partial a)$, $(\partial y/\partial b)$, and $(\partial y/\partial c)$. The results we obtain for these partial derivatives are provided in Equations A4.15 through A4.17.

$$(\partial y/\partial a) = 1 \qquad (A4.15)$$

$$(\partial y/\partial b) = -1 \qquad (A4.16)$$

$$(\partial y/\partial c) = 1 \qquad (A4.17)$$

The preceding partial derivatives are then substituted into Equation A4.14. By further simplifying this new equation, we obtain the result that was given earlier in Equation 4.6 of Chapter 4.

$$(s_y)^2 = (1)^2(s_a)^2 + (-1)^2(s_b)^2 + (1)^2(s_c)^2 \qquad (A4.18)$$

$$= (s_a)^2 + (s_b)^2 + (s_c)^2 \qquad (A4.19)$$

or

$$s_y = \sqrt{(s_a)^2 + (s_b)^2 + (s_c)^2} \qquad (4.6)$$

Random Error Propagation During Multiplication or Division

The general error-propagation expression we derived in Equation A4.12 can also be used to produce a simplified relationship for examining how the uncertainty in a result y is determined by the multiplication or division of numbers, such as is performed in the following example.

$$y = (ab)/c \qquad (A4.20)$$

The approach to relating the precision and standard deviation of y to the precision of a, b, and c is the same as we employed in the previous example for addition or subtraction. We begin by using Equation A4.12 to write a general random error propagation formula for our specific mathematical operation, as shown in Equation A4.21.

$$(s_y)^2 = (\partial y/\partial a)^2(s_a)^2 + (\partial y/\partial b)^2(s_b)^2 + (\partial y/\partial c)^2(s_c)^2 \qquad (A4.21)$$

We then use Equation A4.20 to determine the partial derivatives of y with respect to a, b, and c (see Equations A4.22–A4.24).

$$(\partial y/\partial a) = (b/c) \qquad (A4.22)$$

$$(\partial y/\partial b) = (a/c) \qquad (A4.23)$$

$$(\partial y/\partial c) = (-a\,b/c^2) \qquad (A4.24)$$

These partial derivatives are then substituted into Equation A4.21 to give Equation A4.25.

$$(s_y)^2 = (b/c)^2(s_a)^2 + (a/c)^2(s_b)^2 + (-a\,b/c^2)^2(s_c)^2 \quad (A4.25)$$

The last step involves dividing both sides of Equation A4.25 by the square of our original formula in Equation A4.20 (i.e., we divide the left-hand side by y^2 and the right-hand side by $(a^2\,b^2/c^2)$.

$$(s_y/y)^2 = (s_a/a)^2 + (s_b/b)^2 + (s_c/c)^2 \qquad (A4.26)$$

By taking the square root of both sides of this relationship, we get the final error-propagation formula for multiplication and division that was originally given in Equation 4.7 of Chapter 4.

$$(s_y/y) = \sqrt{(s_a/a)^2 + (s_b/b)^2 + (s_c/c)^2} \qquad (4.7)$$

Random Error Propagation During Exponentiation

The third type of calculation that we will consider for random error-propagation is that of exponentiation. This mathematical operation can be represented by the general formula,

$$y = a^x \qquad (A4.27)$$

where a is a variable and x is a constant. To derive an error-propagation formula for this case, we begin once again by using Equation A4.12, which has the following form for the preceding operation.

$$(s_y)^2 = (\partial y/\partial a)^2(s_a)^2 \qquad (A4.28)$$

There is really only one overall term on the right-hand side of this equation, so we can simplify this particular relationship by taking the square root of both sides, giving Equation A4.29.

$$s_y = (\partial y/\partial a)s_a \qquad (A4.29)$$

We next obtain an equation for the partial derivative $(\partial y/\partial a)$ by using our initial expression in Equation A4.27.

$$(\partial y/\partial a) = x\,a^{(x-1)} \qquad (A4.30)$$

By substituting Equation A4.30 into Equation A4.29, we get the following expression.

$$s_y = [x\,a^{(x-1)}]s_a \qquad (A4.31)$$

Finally, we can simplify Equation A4.31 by dividing both sides of this relationship by our original function ($y = a^x$). This provides us with the same result that was given in Chapter 4.

$$s_y/y = (xa^{(x-1)}/a^x)s_a \qquad (A4.32)$$

or

$$(s_y/y) = x(s_a/a) \qquad (4.12)$$

Random Error Propagation When Using Logarithms

The next type of calculation that we examine is one that involves taking the logarithm of a number. Let's suppose we're taking the natural logarithm (ln) of a, which gives y as our result.

$$y = \ln(a) \qquad (A4.33)$$

Starting again with Equation A4.12, the general expression for random error propagation in this situation can be represented by the relationship given in Equation A4.34.

$$(s_y)^2 = (\partial y/\partial a)^2(s_a)^2 \qquad (A4.34)$$

Like the case of exponentiation that we considered earlier, there is really only one overall term on the right-hand side of Equation A4.34, so we can simplify this by taking the square root of both sides. This produces the result shown in Equation A4.35.

$$s_y = (\partial y/\partial a)s_a \qquad (A4.35)$$

The second step in this process is to determine the value of the partial derivative $(\partial y/\partial a)$ for Equation A4.33. The result is given below.

$$(\partial y/\partial a) = 1/a \qquad (A4.36)$$

We then substitute this partial derivative into Equation A4.35 to produce the following relationship between s_y and s_a.

$$s_y = (s_a/a) \qquad (4.9)$$

The same general process can be used to describe the propagation of random errors when taking the base-ten logarithm (or log) of a number, as represented by Equation A4.37.

$$y = \log(a) \qquad (A4.37)$$

To determine the propagation of errors in this situation, we first change the base-ten logarithm into a natural logarithm (or ln) by using the conversion factor $\log(e)$, which approximately equals 0.434.

$$y = 0.434 \ln(a) \qquad (A4.38)$$

We then use the same approach as already described for natural logarithms to derive an error propagation formula for Equation A4.37. The resulting expression for a base-ten logarithm (see Equation 4.8) only differs from the expression derived for natural logarithms (Equation 4.9) in that it includes the log/ln conversion factor, 0.434.

$$s_y = 0.434 \, (s_a/a) \qquad (4.8)$$

Random Error Propagation When Using Antilogarithms

The final type of calculation that we examine is the conversion of a number into its antilogarithm. An example of this mathematical operation is shown in Equation A4.39.

$$y = e^a \qquad (A4.39)$$

At first glance this appears to be similar to exponentiation, which we examined previously. However, these two operations do differ in that exponentiation involves raising a variable to some constant power, while determining an antilogarithm involves taking a constant (in this case, the number "e") and raising it to a variable power, a. One consequence of this difference is that random errors propagate in different ways during these two types of calculations.

The initial error-propagation formula that we use for the case of antilogarithms is obtained by writing Equation A4.12 in the following form.

$$(s_y)^2 = (\partial y/\partial a)^2 (s_a)^2 \qquad (A4.40)$$

There is again only one combined term on the right-hand side of this relationship, so we can simplify this equation by taking the square root of both sides.

$$s_y = (\partial y/\partial a)s_a \qquad (A4.41)$$

The value of the partial derivative $(\partial y/\partial a)$ for Equation A4.39 can be described by the relationship in Equation A4.42.

$$(\partial y/\partial a) = e^a \qquad (A4.42)$$

Substituting this partial derivative into Equation A4.41 then gives us the following expression.

$$s_y = e^a s_a \qquad (A4.43)$$

Finally, if we divide Equation A4.43 by our original function $(y = e^a)$, we get the random error propagation formula that was provided earlier in Chapter 4.

$$s_y/y = (e^a/e^a)s_a \qquad (A4.44)$$

or

$$(s_y/y) = s_a \qquad (4.11)$$

A similar type of random error propagation occurs when taking a base-ten antilogarithm (or antilog) of a number, such as we see in Equation A4.45.

$$y = 10^a \qquad (A4.45)$$

To find the error-propagation equation for this case, we first modify Equation A4.45 into a form that is now written in terms of e. This is done by using the conversion factor of $\ln(10)$, or approximately 2.303.

$$y = e^{2.303a} \qquad (A4.46)$$

The procedure for determining the propagation of errors in Equation A4.43 is then the same as we used for Equation A4.39. The only difference is that the final relationship derived for the base-ten antilogarithm (Equation 4.10) contains the base-ten/base-e conversion factor of 2.303.

$$(s_y/y) = 2.303s_a \qquad (4.10)$$

608 Appendix A •

APPENDIX A.5 LEAST-SQUARES ANALYSIS

The equations given in Chapter 4 for determining the slope and intercept of a best-fit line were obtained by using a technique known as the **method of least squares**. In this method, the best-fit of any equation to a set of data is said to be obtained when the parameters for this fit give the smallest difference in the actual and predicted results. To determine this difference, we use a value that is known as the *sum of the squares of the residuals*, or the **least-squares value**. This is represented by the term $\sum(r_i)^2$, which is determined by using the following relationship,

$$\sum(r_i)^2 = \sum(y_{i,calc} - y_i)^2 \tag{A5.1}$$

where y_i represents each of the individual experimental values $(y_1 \ldots y_n)$ and $y_{i,calc}$ is the value calculated under the same set of conditions by using the best-fit parameters and corresponding equation of interest. For example, if a straight line is being fit to a set of (x_i,y_i) values, then $y_{i,calc}$ would be estimated for each point by using Equation A5.2,

$$y_{i,calc} = mx_i + b \tag{A5.2}$$

where m is the best-fit slope and b is the best-fit intercept. In this situation, Equation A5.2 can be used in place of $y_{i,calc}$ in Equation A5.1 to give the following equivalent relationship.

$$\sum(r_i)^2 = \sum(\{mx_i + b\} - y_i)^2 \tag{A5.3}$$

As already stated, one way that we can use Equation A5.3 is to determine how well a linear equation fits a set of (x,y) values. But this relationship can also be employed to help determine what values for the slope (m) and intercept (b) give the best possible fit of an equation to our data. To use least squares for this purpose, we first need to make a few simplifying assumptions. For instance, it is usually assumed that the independent variable (x) is known exactly (i.e., $s_X = 0$) and that only the dependent variable $(y_1, y_2, \ldots, y_n)$ has any variability in it. Also, it is often assumed that the error in the y values is random and that the random error in all y values follows the same type of distribution. And, finally, it is generally assumed that the size of the random error in y is roughly the same throughout the entire range of x values that is being tested.

To find the best-fit values of the slope and intercept, we begin by taking the partial derivative of $\sum(r_i)^2$ versus both m and b, as shown in Equations A5.4 and A5.5.

$$\partial[\sum(r_i)^2]/\partial m = \sum 2\,(\{m\,x_i + b\} - y_i)(x_i)$$

$$= 2\,[m(\sum x_i^2) + b(\sum x_i) - (\sum x_iy_i)] \tag{A5.4}$$

$$\partial[\sum(r_i)^2]/\partial b = \sum 2\,(m\,x_i + b - y_i)$$

$$= 2\,[m\,(\sum x_i) + n\,b - (\sum y_i)] \tag{A5.5}$$

We then find the minimum value of $\sum(r_i)^2$ (i.e., the point of best fit) by setting each of its partial derivatives equal to zero and finding the values of m and b that meet this condition.

For $\partial[\sum(r_i)^2]/\partial m$:

$$2\,[m(\sum x_i^2) + b\,(\sum x_i) - (\sum x_iy_i)] = 0 \tag{A5.6}$$

$$b = [(\sum x_iy_i) - m\,(\sum x_i^2)]/(\sum x_i) \tag{A5.7}$$

For $\partial[\sum(r_i)^2]/\partial b$:

$$2[m\,(\sum x_i) + n\,b - (\sum y_i)] = 0 \tag{A5.8}$$

$$b = [(\sum y_i) - m\,(\sum x_i)]/n \tag{A5.9}$$

Because the expressions on the right-hand sides of Equations A5.7 and A5.9 are both equal to b, we can set these two equations equal to each other and use them to solve for m. This gives rise to Equation 4.27, which was listed in Table 4.9 of Chapter 4 for calculating the slope of a best-fit line.

$$[(\sum x_iy_i) - m\,(\sum x_i^2)]/(\sum x_i) =$$

$$[(\sum y_i) - m\,(\sum x_i)]/n \tag{A5.10}$$

$$\Rightarrow n\,(\sum x_iy_i) - n\,m(\sum x_i^2) =$$

$$(\sum x_i)(\sum y_i) - m\,(\sum x_i)^2 \tag{A5.11}$$

$$\Rightarrow m[n\,(\sum x_i^2) - (\sum x_i)^2] =$$

$$n\,(\sum x_iy_i) - (\sum x_i)\,(\sum y_i) \tag{A5.12}$$

$$\therefore\ m = [n\,(\sum x_iy_i) - (\sum x_i)\,(\sum y_i)]/[n\,(\sum x_i^2) -$$

$$(\sum x_i)^2] \tag{4.28}$$

Substituting Equation 4.27 for m into Equation A5.9 then gives the following solution for b, which was given as Equation 4.28 in Chapter 4.

$$b = [(\sum y_i)(\sum x_i^2) - (\sum x_iy_i)(\sum x_i)]/[n\,(\sum x_i^2) -$$

$$(\sum x_i)^2] \tag{4.29}$$

The standard deviations in m and b can also be calculated by using the propagation of errors. For instance, the standard deviation of the slope (s_m) is given by the following equation.

$$(s_m)^2 = (\partial m/\partial y_1)^2(s_{Y1})^2 + (\partial m/\partial y_2)^2(s_{Y2})^2 + \cdots$$

$$= \sum[(\partial m/\partial y_i)^2(s_{Yi})^2] \tag{A5.13}$$

Note in this equation that we assume all x values are known exactly, or that $s_X = 0$. This is indicated by the fact that no terms relating s_m to s_X appear in Equation A5.13.

To get a more exact equation for s_m, we next need to evaluate the partial derivative $(\partial m / \partial y_i)$. We can accomplish this by going back to the equation we derived for the best-fit value for m and use this to take the desired partial derivative of m versus y_i.

$$m = [n\left(\sum x_i y_i\right) - \left(\sum x_i\right)\left(\sum y_i\right)]/[n\left(\sum x_i^2\right) - \left(\sum x_i\right)^2]$$
(A5.14)

$$(\partial m / \partial y_i) = [n\left(\sum x_i\right) - \left(\sum x_i\right)]/[n\left(\sum x_i^2\right) - \left(\sum x_i\right)^2]$$
(A5.15)

Substituting the preceding partial derivative into the propagation of error formula for s_m and assuming that $s_Y = s_{Yi}$ (that there is uniform variance throughout the graph) gives the result shown below.

$$(s_m)^2 = \sum[(\partial m / \partial y_i)^2 (s_{Yi})^2]$$

$$= \sum[([n\left(\sum x_i\right) - \left(\sum x_i\right)]/[n\left(\sum x_i^2\right) -$$

$$\left(\sum x_i\right)^2])^2 (s_{Yi})^2]$$
(A5.16)

$$(s_m)^2 = (n/[n\left(\sum x_i^2\right) - \left(\sum x_i\right)^2]) (s_Y)^2$$
(A5.17)

or

$$\therefore s_m = \sqrt{n/[n\left(\sum x_i^2\right) - \left(\sum x_i\right)^2]}(s_Y)$$
(4.31)

Using a similar treatment, the standard deviation of the intercept (s_b) can be shown to be given by the following formula.

$$(s_b)^2 = ((\sum x_i^2)/[n(\sum x_i^2) - \left(\sum x_i\right)^2]) (s_Y)^2$$
(A5.18)

or

$$s_b = \sqrt{(\sum x_i^2)/[n(\sum x_i^2) - \left(\sum x_i\right)^2]}(s_Y)$$
(4.32)

Notice that to determine either s_m or s_b, it is necessary to determine the standard deviation of all the y values (s_Y). This is given by

$$s_Y = \sqrt{\sum(y_i - y_{i,calc})^2/(n - 2)}$$
(A5.19)

where $(n - 2)$, the degrees of freedom for a linear fit. Using that fact that $y_{i,calc} = (m x_i + b)$ then gives the following relationship for s_Y, as stated earlier in Chapter 4.

$$s_Y = \sqrt{\sum(y_i - m x_i - b)^2/(n - 2)}$$
(4.30)

APPENDIX A.6 SUMMARY OF TERMS AND TESTS FOR DESCRIBING AND COMPARING DATA

Item to be Addressed	Tool or Test	Equations to Use	
Describing Experimental Results			
"How close is my result to the actual answer?"	Absolute error (e)	$e = x - \mu$	(4.1)
	Relative error (e_r)	$e_r = (x - \mu)/\mu$	(4.2)
"What is the most representative value for my results?"	Average or mean ($\bar{x}$)	$\bar{x} = \sum (x_i)/n$	(4.3)
"How reproducible is my experimental result?"	Range (R_x)	$R_x = x_{high} - x_{low}$	(4.4)
	Standard deviation (s)	$s = \sqrt{\sum (x_i - \bar{X})^2/(n - 1)}$	(4.5)
The Propagation of Errors			
"How do random errors affect my calculations?"	Error propagation	See Equations in Table 4.1	
Sample Distributions and Confidence Intervals			
"How do I describe a large set of numbers?"	Normal distribution, true mean	$y = \dfrac{1}{\sigma\sqrt{2\pi}} \cdot e^{1/2\,[(x-\mu)^2/(\sigma^2)]}$	(4.13)
	(μ) and true standard deviation (σ)		
"How precise is my estimated mean?"	Standard deviation of the mean	$s_{\bar{X}} = s/\sqrt{n}$	(4.14)
"How reliable is a mean obtained for a small data set?"	Confidence interval	$C.I. = \bar{x} + t \cdot s_{\bar{X}} \text{ (or } s)$	(4.15-4.16)
Comparing Experimental Results and Detecting Outliers			
"Is an experimental result the same as a reference value?"	Student's t-test	$t = \dfrac{|\bar{x} - \mu|}{s_{\bar{X}}}$	(4.17)
"Do two experimental means have the same value?"	Student's t-test	$s_{pool} = \sqrt{\{(n_1 - 1)\cdot(s_1)^2 + (n_2 - 1)\cdot(s_2^2)\}/(n_1 + n_2 - 2)}$	(4.18)
		$s_{\bar{X}pool} = \dfrac{s_{pool}}{\sqrt{(n_1 \cdot n_2)/(n_1 + n_2)}}$	(4.19)
		$t = |\bar{x}_1 - \bar{x}_2|/s_{\bar{X}pool}$	(4.20)
"Do two groups of experimental means have the same values?"	Paired Student's t-test	$s_d = \sqrt{\sum (d_i - \bar{d})^2/(n - 1)}$	(4.21)
		$s_{\bar{d}} = s_d/\sqrt{n}$	(4.22)
		$t = |\bar{d}|/s_{\bar{d}}$	(4.23)
"Do two experimental results have the same precision?"	F-test	$F = (s_2)^2/(s_1)^2 \text{ (where } s_2 \geq s_1)$	(4.24)
"Is a high or low value in a data set an outlier?"	Q-test	$Q = |x_o - x_n|/(x_{high} - x_{low})$	(4.25)
	T_n-test	$T_n = |x_o - \bar{X}|/s$	(4.26)
Fitting Experimental Results			
"What is the best-fit line for a set of data?"	Linear regression	See Equations in Table 4.8	
"How well does the best-fit line describe the data?"	Correlation coefficient (r)	See Equations in Table 4.8	
	Residual plots		

APPENDIX A.7 BALANCING OXIDATION–REDUCTION REACTIONS

A balanced oxidation–reduction reaction should have an equal number of atoms for each element on both sides and the same total charge for the reactant and product sides. During this process, we will often know many of the reactants and products in the oxidation–reduction reaction, which serves as a useful starting point. However, we also need to balance the atoms and charges on each side. We will now see how we can do this for reactions that occur in acidic, basic, or neutral pH solutions.

Reactions in an Acidic Solution. One approach that can be used to balance oxidation–reduction reactions involves first writing down the two half-reactions for this process, one for reduction and one for oxidation. We then proceed by balancing these two half-reactions separately and then combining them to get the overall balanced oxidation–reduction equation. This method is called the "half-reaction method." This technique is based on the conservation of mass, which means there cannot be a net gain or loss of atoms or electrons in the final balanced reaction. As a result, the number of electrons lost by the reducing agent must be the same as the number of electrons gained by the oxidizing agent.

To show how this approach is carried out for an acidic aqueous solution, we will use the oxidation of copper metal in nitric acid as an example.

$$Cu(s) + HNO_3(aq) \leftrightarrows Cu^{2+}(aq) + NO(g) \quad (A7.1)$$

Although this equation does show most of the reactants and products for this reaction, it is not yet balanced. There are no hydrogen atoms shown on the right, there are different numbers of oxygen atoms on the two sides of this equation, and the overall charge is not the same on the two sides.

The first step in balancing this equation is to identify the half-reactions by assigning oxidation numbers. This gives the following result.

Oxidation Number:

$$\begin{array}{cccc} Cu(s) & + & HNO_3(aq) & \leftrightarrows & Cu^{2+}(aq) & + & NO(g) \\ 0 & & +5 & & +2 & & +2 \end{array} \quad (A7.2)$$

We can see from this information that copper is being oxidized because its oxidation number changes from 0 to +2. Nitrogen is being reduced as it goes from an oxidation number of +5 to +2. Oxygen has an oxidation number of −2 on both sides of this reaction. This process gives use the following two initial half-reactions, which are not yet balanced.

Initial Oxidation Half-Reaction:

$$Cu(s) \leftrightarrows Cu^{2+}(aq) \quad (A7.3)$$

Initial Reduction Half-Reaction:

$$HNO_3(aq) \leftrightarrows NO(g) \quad (A7.4)$$

The second step in the half-reaction method is to balance each of the half-reactions. This, in turn, involves several smaller steps. The first of these steps is to balance the electrons in each half-reaction by adding electrons to the side for which the oxidation numbers are more positive. This requires that we add two electrons to the right of Equation A7.3 and three electrons to the left of Equation A7.4, giving the following modified half-reactions.

$$Cu(s) \leftrightarrows Cu^{2+}(aq) + 2\,e^- \quad (A7.5)$$

$$HNO_3(aq) + 3\,e^- \leftrightarrows NO(g) \quad (A7.6)$$

Next, we need to balance all atoms that are neither oxygen nor hydrogen. In this case, both the copper atoms and nitrogen atoms are already balanced, so there is nothing further to do. We then balance the oxygen atoms in the nitric acid half-reaction by attributing them to two water molecules as reaction products. The hydrogens are then balanced by placing 3 H^+ on the left of Equation A7.6, which gives the result shown below.

$$HNO_3(aq) + 3\,H^+ + 3\,e^- \leftrightarrows NO(g) + 2\,H_2O \quad (A7.7)$$

At this point, we have balanced both the charge and mass of our two half-reactions, allowing us to write them with an equal sign between their right and left sides. Students often wonder why they are entitled to put water, hydrogen ion, or hydroxide ion on either side of a balanced redox reaction. This approach is valid if the reaction is carried out in water because there is plenty of water available and it is always in equilibrium with both hydrogen ion and hydroxide ion. Often it is known whether the reaction is carried out in acidic or basic solution. This dictates whether H^+ or OH^- is to be used in balancing the half-reactions. (*Note*: It is important during this process to remember that redox half-reactions must *not* be balanced by adding H_2 or O_2 to either side, because the addition of these chemicals would result in a change in oxidation state of the H or O atoms in the reaction.) The two balanced half-reactions that we now have in our example follow.

Balanced Oxidation Half-Reaction:

$$Cu(s) \leftrightarrows Cu^{2+}(aq) + 2\,e^- \quad (A7.8)$$

Balanced Reduction Half-Reaction:

$$HNO_3(aq) + 3\,H^+ + 3\,e^- \leftrightarrows NO(g) + 2\,H_2O \quad (A7.9)$$

The third main step in the half-reaction method is to equalize the number of electrons transferred in the two half-reactions. This is required because the same number of electrons must be gained and lost in the net reaction. Therefore, we must multiply each reaction by numbers that will allow the reactions to exchange an identical number of electrons. For the two half-reactions in Equations A7.8 and A7.9, two electrons are involved in one of the half-reactions and three are involved in the other. To obtain an equal number of electrons we must multiple the half-reaction with two electrons by three and the one with two by three. This gives six electrons in each of our modified half-reactions, as shown below.

$$3\,Cu \leftrightarrows 3\,Cu^{2+} + 6\,e^- \qquad (A7.10)$$

$$2\,HNO_3 + 6\,H^+ + 6\,e^- \leftrightarrows 2\,NO + 4\,H_2O \qquad (A7.11)$$

In the fourth step to balancing an oxidation–reduction equation, we must add the half-reactions and cancel common terms to get the final oxidation–reduction reaction.

Oxidation Half-Reaction: $\quad 3\,Cu \leftrightarrows 3\,Cu^{2+} + 6\,e^-$

Reduction Half-Reaction:

$$2\,HNO_3 + 6\,H^+ + 6\,e^- \leftrightarrows 2\,NO + 4\,H_2O$$

Overall Reaction:

$$3\,Cu + 2\,HNO_3 + 6\,H^+ + 6\,e^- \leftrightarrows 3\,Cu^{2+} + 2\,NO +$$

$$4\,H_2O + 6\,e^- \qquad (A7.12)$$

The only common terms on the right and left of the overall equation in this example are the six electrons that appear on both sides. When we remove these electrons from both sides, we get the following final reaction.

$$3\,Cu + 2\,HNO_3 + 6\,H^+ \leftrightarrows 3\,Cu^{2+} + 2\,NO +$$

$$4\,H_2O \qquad (A7.13)$$

The fifth step in the half-reaction method is to check the final reaction and make sure it is balanced with respect to both mass and charge. When we do this, we find that the same number of atoms for each element are present on both sides of this equation. The total charge is also balanced, with a value of "+6" on both sides. Thus, we have confirmed that we have reached our final answer.[1]

EXERCISE A7.1 | **Balancing an Oxidation–Reduction Equation Under Acidic Conditions**

Balance the following oxidation–reduction reaction that is carried out in an acidic aqueous solution.

$$Cr_2O_7^{2-} + NO(g) \leftrightarrows Cr^{3+} + NO_3^-$$

SOLUTION

We first need to determine the oxidation numbers of our reactants and products, as shown below.

$$Cr_2O_7^{2-} + NO \leftrightarrows Cr^{3+} + NO_3^-$$
$$+6 \qquad\quad +2 \qquad\quad +3 \qquad\quad +5$$

This result tells us that nitrogen is being oxidized as we go from NO to NO_3^- and chromium is being reduced as we go from $Cr_2O_7^{2-}$ to Cr^{3+}. Next, we must balance the atoms and charge in each of these half-reactions.

Oxidation Half-Reaction:

$$NO(g) \leftrightarrows NO_3^-$$

$$NO(g) \leftrightarrows NO_3^- + 3\,e^-$$

$$NO(g) \leftrightarrows NO_3^- + 4\,H^+ + 3\,e^-$$

$$NO(g) + 2\,H_2O \leftrightarrows NO_3^- + 4\,H^+ + 3\,e^-$$

Reduction Half-Reaction:

$$Cr_2O_7^{2-} \leftrightarrows Cr^{3+}$$

$$Cr_2O_7^{2-} + 6\,e^- \leftrightarrows 2\,Cr^{3+}$$

$$Cr_2O_7^{2-} + 14\,H^+ + 6\,e^- \leftrightarrows 2\,Cr^{3+}$$

$$Cr_2O_7^{2-} + 14\,H^+ + 6\,e^- \leftrightarrows 2\,Cr^{3+} + 7\,H_2O$$

At this point, the oxidation half-reaction involves three electrons, but the reduction half-reaction involves six. To make these numbers equal, we must multiple the oxidation half-reaction by two. We can then combine these two half-reactions and eliminate common terms.

Oxidation Half-Reaction:

$$2\,NO + 4\,H_2O \leftrightarrows 2\,NO_3^- + 8\,H^+ + 6\,e^-$$

Reduction Half-Reaction:

$$Cr_2O_7^{2-} + 14\,H^+ + 6\,e^- \leftrightarrows 2\,Cr^{3+} + 7\,H_2O$$

Overall Reaction:

$$Cr_2O_7^{2-} + 2\,NO(g) + 6\,H^+ \leftrightarrows 2\,Cr^{3+} +$$

$$2\,NO_3^- + 3\,H_2O$$

In checking our answer, we see that there are two chromium atoms, two nitrogen atoms, nine oxygen atoms, and six hydrogen atoms on each side. There is also a total charge of "+4" on each side, so this oxidation–reduction equation is now balanced.

Reactions in Basic or Neutral Solutions. The previous section showed how we can balance an oxidation–reduction equation for a reaction that is taking place in water at an acidic pH. The half-reaction method can also be employed for solutions at a basic pH, using an approach similar to that described for acidic media. The key difference in these two situations is that acidic solutions have a relatively a high concentration of H^+ ions, while basic solutions in water have a higher concentration of OH^- ions. Thus, when we are balancing the hydrogen atoms for an oxidation–reduction equation at a basic pH we use OH^- instead of H^+ to balance the hydrogen atoms on either side of the equation. This idea is illustrated in the following exercise.

| **EXERCISE A7.2** | **Balancing an Oxidation–Reduction Equation Under Basic Conditions** |

Balance the oxidation–reduction reaction in Exercise A7.1 for the case where this reaction is carried out in an aqueous solution at a basic pH.

SOLUTION

We can begin this process by using the same oxidation numbers and initial half-reactions that we identified in Exercise A7.1. When we balance these half-reactions, the results are shown below.

Oxidation Half-Reaction:

$$NO(g) \leftrightarrows NO_3^-$$

$$NO(g) \leftrightarrows NO_3^- + 3\,e^-$$

$$NO(g) + 4\,OH^- \leftrightarrows NO_3^- + 3\,e^-$$

$$NO(g) + 4\,OH^- \leftrightarrows NO_3^- + 2\,H_2O + 3\,e^-$$

Reduction Half-Reaction:

$$Cr_2O_7^{2-} \leftrightarrows Cr^{3+}$$

$$Cr_2O_7^{2-} + 6\,e^- \leftrightarrows 2\,Cr^{3+}$$

$$Cr_2O_7^{2-} + 6\,e^- \leftrightarrows 2\,Cr^{3+} + 14\,OH^-$$

$$Cr_2O_7^{2-} + 7\,H_2O + 6\,e^- \leftrightarrows 2\,Cr^{3+} + 14\,OH^-$$

We again have the same problem as in Exercise A7.1, where the oxidation half-reaction involves three electrons, but the reduction half-reaction involves six. We solve this the same way as before, by multiplying the oxidation half-reaction by two. We can then combine these two half-reactions and eliminate common terms.

Oxidation Half-Reaction:

$$2\,NO + 8\,OH^- \leftrightarrows 2\,NO_3^- + 4\,H_2O + 6\,e^-$$

Reduction Half Reaction:

$$Cr_2O_7^{2-} + 7\,H_2O + 6\,e^- \leftrightarrows 2\,Cr^{3+} + 14\,OH^-$$

Overall Reaction:

$$Cr_2O_7^{2-} + 2\,NO(g) + 3\,H_2O \leftrightarrows 2\,Cr^{3+} +$$
$$2\,NO_3^- + 6\,OH^-$$

In checking this answer, we have 2 chromium atoms, 2 nitrogen atoms, 12 oxygen atoms, and 6 hydrogen atoms on each side. There is also a total charge of "–2" on each side, so this equation is balanced.

Now that you have balanced oxidation–reduction equations occurring in acidic or basic solutions, you may be wondering how to do this if the solution has a neutral pH. The answer is that at a neutral pH solution you can balance the reaction as if it were present in *either* an acidic or basic solution. Furthermore, you'll find by a close examination of the last two exercise that the balanced equations aren't really that different, but are merely different ways of expressing the same balanced oxidation–reduction equation. One of these answers (Exercise A7.1) is simply more convenient to use for solutions at a low pH, and the other (Exercise A7.2) is more functional for a solution at a high pH. There are, however, other things that can happen at a high pH or a low pH that create different behaviors to be seen in an oxidation–reduction reaction. For instance, in a basic solution the actual Cr(III) product is not Cr^{3+}, but rather a complex of Cr^{3+} with OH^-, giving $Cr(OH)_3$ or $Cr(OH)_4^-$. Also, dichromate doesn't exist at high pH, but instead becomes two moles of chromate. As a result, the overall balanced equation in the last exercise would be more appropriately written as follows.

Overall Reaction:

$$2\,CrO_4^{2-} + 2\,NO(g) + 4\,H_2O \leftrightarrows 2\,Cr(OH)_3 +$$
$$2\,NO_3^- + 2\,OH^- \quad (A7.14)$$

In addition to the half-reaction method that is described here, there are other methods for balancing redox reactions, but all these approaches result in a description of the products and reactants following the law of conservation of mass.

References

1. J. D. Carr, "Stoichiometry for Copper Dissolution in Nitric Acid: A Comment," *Journal of Chemical Education*, 67 (1990) 183.

APPENDIX A.8 DERIVATION OF BEER'S LAW

The derivation of Beer's law is based on the model shown here. The power of light in this model is given by the variable P and the distance that the light travels is represented by the variable x, where $x = 0$ at the beginning of the sample and $x = b$ (the total path length) at the end of the sample. The initial power of light entering the sample in this model is equal to P_0 and the final power of the emerging light is $P_{x=b}$.

It is assumed in this model that the analyte is uniformly distributed in the sample. It is also assumed that the rays of light that pass through the sample all have the same distance of travel in the sample. This last assumption is represented in this model through the use of a parallel beam of rays that strike a square sample cell at an angle that is perpendicular to the cell's surface.

To begin the derivation of Beer's law, let's look at what happens as this beam of light passes through a given region of the sample with a thickness of dx. We will call the power of light that enters this region P. The power of light that remains after the light has passed through the given region of the sample is given by the term $(P - dP)$, where dP is the change in the power of light due to the absorption of light by analyte in this section of the sample.

We can relate the decrease in the power of light $(-dP)$ to the power of the entering light (P) and the thickness of the sample region through the following differential equation,

$$- dP = c\, P\, dx \qquad (A8.1)$$

where c is a proportionality constant. This equation simply states that the amount of absorbed light will be directly proportional to the thickness of the layer through which the light passes. We can next rearrange this equation to combine all terms containing P (or dP) on one side and all terms related to x (including the term dx) on the other side.

$$\frac{-dP}{P} = c\, dx$$

To use this relationship to find the final intensity of light after it has passed through the *entire* sample, we can integrate both sides of this equation by going from a distance of $x = 0$ to $x = b$ on the right-hand side and from the initial power of P_0 to the final power of $P_{x=b}$ (or simply "P") on the left-hand side.

$$\int_{P_0}^{P_{x=b}} \frac{-dP}{P} = c \int_{x=0}^{x=b} dx \qquad (A8.3)$$

The integrated expression that results from this process follows.

$$- \ln(P) - \{-\ln(P_0)\} = c(b) - c(0)$$

or

$$- \ln\left(\frac{P}{P_0}\right) = c\, b \qquad (A8.4)$$

In Equation 17.11 of Chapter 17, it was stated that the ratio P/P_0 was also equal to the transmittance (T), where $T = P/P_0$. The absorbance (A) was also defined in Chapter 17 as being the base-10 logarithm of the transmittance.

$$A = - \log(T) = \log(P_0/P) \qquad (17.12)$$

Combining Equation A8.4 with Equation 17.12 gives the following equation.

$$A = c\, b \qquad (A8.5)$$

The contributions to the term c are divided into the concentration of the analyte (C, such as given n units of molarity), and the inherent molar absorptivity of the analyte (ε, with units of L/mol · cm if b has units of cm and C is in molarity). Substituting the fact that $c = \varepsilon\, C$ into Equation A8.5 gives the following equation,

Beer's law: $\qquad A = \varepsilon\, b\, C \qquad (17.13)$

which is the final form of the Beer–Lambert law (or "Beer's law") that as provided in Equation 17.13 of Chapter 17.

Effect of Polychromatic Light on Beer's Law

An assumption made in Beer's law is that the absorbance of the sample is being measured using only "monochromatic light"(i.e., light that contains only one wavelength). There is really no such thing as perfectly monochromatic light, but the monochromators of modern instruments for conducting analytical measurements based on spectroscopy come quite close in providing such light for absorbance measurements. We describe the range of wavelengths that strike the sample as $\Delta\lambda$.

When considering whether this light is "monochromatic" or "polychromatic" (i.e., containing more than one wavelength), what is important when using Beer's law is whether the molar absorptivity for the analyte changes over this wavelength range. This change in molar absorptivity ($\Delta\varepsilon$) must be small for Beer's law to give a good description of the light absorption. This usually means that the relative change in the molar absorptivity ($\Delta\varepsilon/\varepsilon$) is less than 0.01

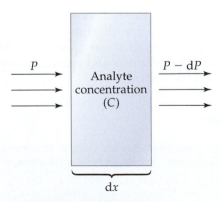

APPENDIX A.8 Model for Beer's Law Derivation.

compared to the mean value of ε over the range of wavelengths that are being used in the analysis. It is for this reason that absorbance measurements are generally made at wavelengths that correspond to a peak maximum in the absorption spectrum, because it is in the region where $(\Delta\varepsilon/\varepsilon)$ will be quite small.

This approach can be justified mathematically in the following way. First, at two wavelengths that pass through the sample (λ_1 and λ_2) the analyte of interest has molar absorptivities of ε_1 and ε_2. Also, let's say that f_1 and f_2 represent the fraction of incident radiant power at these wavelengths versus the total power of light entering the sample. We will use the terms $P_{0,1}$ and $P_{0,2}$ to represent the original incident power of light at these two wavelengths, and P_1 and P_2 will be used to represent the power of the transmitted light at these same two wavelengths. We can first write a simple expression in which the total power of transmitted light (P) is equal to the value of P_1 and P_2 if we assume that only the two given wavelengths of light are passing through the sample.

$$P = P_1 + P_2 \tag{A8.6}$$

We can next use Beer's law at each of the two wavelengths to derive the following relationships.

$$P_1/P_{0,1} = 10^{-\varepsilon_1 b C} \text{ or } P_1 = P_{0,1} \cdot 10^{-\varepsilon_1 b C} \tag{A8.7}$$

$$P_2/P_{0,2} = 10^{-\varepsilon_2 b C} \text{ or } P_2 = P_{0,2} \cdot 10^{-\varepsilon_2 b C} \tag{A8.8}$$

Substituting Equations A8.7 and A8.8 into Equation A8.6 results in the following expression.

$$P = P_{0,1} \cdot 10^{-\varepsilon_1 b C} + P_{0,2} \cdot 10^{-\varepsilon_2 b C} \tag{A8.9}$$

We can also relate the values of $P_{0,1}$ and $P_{0,2}$ to the total combined and power of the incident light (P_0) by using the following equations.

$$P_{0,1} = f_1 P_0 \tag{A8.10}$$

$$P_{0,2} = f_2 P_0 \tag{A8.11}$$

Substituting Equations A8.9–A8.11 into the relationship between the absorbance A and the ratio P/P_0 gives Equation A8.12.

$$A = -\log(P/P_0) = -\log(f_1 10^{-\varepsilon_1 b C} + f_2 10^{-\varepsilon_2 b C}) \tag{A8.12}$$

We can now see how these two wavelengths affect the slope of the Beer's law plot by taking the first derivative of the absorbance A in Equation A8.12 with respect to the concentration C.

$$dA/dC = \frac{\varepsilon_1 f_1 b 10^{-\varepsilon_1 b C} + \varepsilon_2 f_2 b 10^{-\varepsilon_2 b C}}{f_1 10^{-\varepsilon_1 b C} + f_2 10^{-\varepsilon_2 b C}} \tag{A8.13}$$

In the case where the light is "monochromatic" (or $\varepsilon_1 = \varepsilon_2$), Equation 18.13 reduces to $dA/dC = \varepsilon b$, which is the constant slope ideally seen for Beer's law in a plot of A versus C. However, if the light is polychromatic and ε_1 does not equal ε_2, the right-hand side of Equation 18.13 does not reduce to a simple combination of constants. Instead, what is produced is a change in slope with concentration that will always be concave toward the concentration axis if $\Delta\varepsilon > 0$. In addition, the extent of this curvature will become worse as the difference $|\varepsilon_1 - \varepsilon_2|$ increases.

APPENDIX A.9 DERIVATION OF THE RESOLUTION EQUATION

The resolution equation, as shown in Equation 20.27 and in Chapter 20, is a very important relationship in helping to optimize separations in chromatography or in predicting how these separations might change as experimental conditions are varied.

$$R_s = \frac{\sqrt{N}}{4} \cdot \frac{(\alpha - 1)}{\alpha} \cdot \frac{k}{(1 + k)} \quad (20.27)$$

This equation shows that the peak resolution (R_s) that is obtained between two neighboring peaks in chromatography will be affected by three factors. The first factor is the extent of band-broadening in the column, as represented by the number of theoretical plates (N). The second factor is the overall degree of peak retention, as represented by the retention factor for the second of the two peaks (k). The third factor is the selectivity of the column in retaining the compounds in the two peaks, as represented by the separation factor (α).

The form of the resolution equation that is shown in Equation 20.27 can be derived by starting with the following formula that is used to calculate R_s from experimental data, as was also provided earlier in Chapter 20.

$$R_s = \frac{t_{R_2} - t_{R_1}}{(w_{b_2} + w_{b_1})/2} \quad (20.26)$$

In this relationship, t_{R_1} and w_{b_1} are the retention time and baseline width for the first eluting peak, while t_{R_2} and w_{b_2} are the retention time and baseline width of the second peak. The first assumption we need to make to get from Equation 20.26 to Equation 20.27 is that the widths of the two neighboring peaks are approximately equal, or that $w_{b_2} \approx w_{b_1}$. This assumption means the average baseline width term $(w_{b_2} + w_{b_1})/2$ in the denominator of Equation 20.26 can be replaced by either w_{b_2} or w_{b_1}. If w_{b_2} is chosen for this substitution, we get Equation A9.1.

$$R_s = \frac{t_{R_2} - t_{R_1}}{w_{b_2}} \quad (A9.1)$$

The next step in the derivation involves using the number of theoretical plates (N) as a measure of peak width rather than w_b. We can do this by making a second assumption that both of the peaks being examined have Gaussian shapes. This assumption allows us to use the relationship given in Equation 20.21 of Chapter 20 for calculating N from w_b and t_R.

For a Gaussian Peak: $N = 16 (t_R/w_b)^2 \quad (20.21)$

A simple rearrangement of this equation makes it possible to obtain the following expressions that relate w_b to t_R and N.

$$w_b = t_R \sqrt{16/N} \text{ or } w_b = (t_R 4)/\sqrt{N} \quad (A9.2)$$

By using Equation A9.2, we can replace the value of w_{b_2} in Equation A9.1 with the equivalent term $(t_{R_2} 4)/\sqrt{N}$.

This substitution and a rearrangement of the resulting expression produces Equation A9.3.

$$R_s = \frac{t_{R_2} - t_{R_1}}{(t_{R_2} 4)/\sqrt{N}} \text{ or }$$

$$R_s = \frac{\sqrt{N}}{4} \cdot \frac{t_{R_2} - t_{R_1}}{t_{R_2}} \quad (A9.3)$$

The third step in obtaining the resolution equation is to replace each retention time in Equation A9.3 with a corresponding retention factor (k). We can accomplish this by combining the relationship $k = t'_R/t_M$, as given in Chapter 20 in Equation 20.14, with the relationship $t'_R = t_R - t_M$, as given in Equation 20.11, in which t'_R is the adjusted retention time and t_M is the void time.

$$k = t'_R/t_M \text{ or } k = (t_R - t_M)/t_M \quad (A9.4)$$

Equation A9.4 can then be rearranged in terms of t_R, which gives Equation A9.5.

$$t_R = t_M (k + 1) \quad (A9.5)$$

Using Equation A9.5, we can modify Equation A9.3 by replacing t_{R2} with $t_M (k_2 + 1)$ and t_{R1} with $t_M (k_1 + 1)$. The result is Equation A9.6, where all of the t_M terms cancel out because they appear in both the numerator and denominator.

$$R_s = \frac{\sqrt{N}}{4} \cdot \frac{t_M(k_2 + 1) - t_M(k_1 + 1)}{t_M(k_2 + 1)}$$

or

$$R_s = \frac{\sqrt{N}}{4} \cdot \frac{k_2 - k_1}{k_2 + 1} \quad (A9.6)$$

The final step in deriving the resolution equation involves replacing the retention factor for the first eluting peak (k_1) with an equivalent term based on the separation factor, α. The term α was defined by Equation 20.25 in Chapter 20 as $\alpha = k_2/k_1$, which can also be written in the following form.

$$k_1 = k_2/\alpha \quad (A9.7)$$

Rearranging this equation slightly allows one of the two retention factors (e.g., k_1) to be calculated from the other if the value of α is known. If we use Equation A9.7 to replace k_1 with k_2/α in Equation A9.6, we get the modified expression for R_s that is shown in Equation A9.8.

$$R_s = \frac{\sqrt{N}}{4} \cdot \frac{k_2 - (k_2/\alpha)}{k_2 + 1}$$

or

$$R_s = \frac{\sqrt{N}}{4} \cdot \frac{k_2(1 - 1/\alpha)}{k_2 + 1} \quad (A9.8)$$

If we now replace k_2 with "k" and use the fact that $(1 - 1/\alpha)$ can also be written as $(\alpha - 1)/\alpha$, we arrive at the final resolution equation in Equation 20.27.

APPENDIX B

APPENDIX B.1 ESTIMATED INDIVIDUAL ACTIVITY COEFFICIENTS FOR ORGANIC IONS IN WATER AT 25°C*

Type of Ion	Ion Size Parameter, a (pm)	Activity Coefficient at Ionic Strength I (M)							
		$I = 0.0005$	0.001	0.002	0.005	0.01	0.02	0.05	0.10
Charge = +1 or −1									
$(C_3H_7)_4N^+$, $(C_6H_5)_2CHCOO^-$	$a = 800$	0.976	0.966	0.954	0.932	0.911	0.886	0.848	0.816
$CH_3OC_6H_4COO^-$, $(C_3H_7)_3NH^+$, $(NO_2)_3C_6H_2O^-$	700	0.975	0.966	0.954	0.931	0.909	0.882	0.841	0.806
$C_6H_5COO^-$, $HOC_6H_4COO^-$, $ClC_6H_4COO^-$, $C_6H_5CH_2COO^-$, $(CH_3CH_2)_4N^+$, $(C_3H_7)_2NH_2^+$, $CH_2=CHCH_2COO^-$, $(CH_3)_2CHCH_2COO^-$	600	0.975	0.966	0.953	0.930	0.907	0.878	0.833	0.795
$(CH_3CH_2)_3NH^+$, $(C_3H_7)NH_3^+$, Cl_2CHCOO^-, Cl_3CCOO^-	500	0.975	0.965	0.952	0.928	0.904	0.874	0.825	0.783
CH_3COO^-, $ClCH_2COO^-$, $(CH_3)_4N^+$, $(CH_3CH_2)_2NH_2^+$, $H_2NCH_2COO^-$	450	0.975	0.965	0.952	0.928	0.903	0.872	0.821	0.776
$^+H_3NCH_2COOH$, $(CH_3)_3NH^+$, $CH_3CH_5NH_3^+$	400	0.975	0.965	0.952	0.927	0.901	0.869	0.816	0.769
$(CH_3)_2NH_2^+$, $HCOO^-$, $H_2Citrate^-$, $CH_3NH_3^+$	350	0.975	0.965	0.951	0.926	0.900	0.867	0.811	0.762
Charge = +2 or −2									
$CH_2(CH_2CH_2(COO)_2^{2-}$, $CH_2CH_2CH_2(COO)_2^{2-}$, Congo Red^{2-}	$a = 700$	0.905	0.871	0.826	0.751	0.682	0.606	0.500	0.423
$C_6H_4(COO)_2^{2-}$, $H_2C(CH_2COO)_2^{2-}$, $(CH_2CH_2COO)_2^{2-}$	600	0.904	0.870	0.824	0.747	0.675	0.595	0.482	0.400
$H_2C(COO)_2^{2-}$, $(CH_2COO)_2^{2-}$, $(CHOHCOO)_2^{2-}$	500	0.904	0.868	0.822	0.743	0.668	0.583	0.464	0.376
$C_2O_4^{2-}$, $HCitrate^{2-}$	450	0.903	0.868	0.821	0.740	0.664	0.577	0.454	0.363
Charge = +3 or −3									
$Citrate^{3-}$	$a = 500$	0.796	0.728	0.644	0.512	0.403	0.297	0.177	0.111

*This table is based on data provided in J. Kielland, "Individual Activity Coefficients of Ions in Aqueous Solutions," *Journal of the American Chemical Society*, (1937), 59, 1675–1678. The last number to the right and underlined in each activity coefficient is a guard digit (see discussion of guard digits in Chapter 2).

APPENDIX B.2 SOLUBILITY PRODUCTS FOR VARIOUS SALTS IN WATER*

Cation	Anion	Dissociation Reaction	Solubility Product, K_{sp}	$pK_{sp} = -\log(K_{sp})$
Aluminum (Al^{3+})	Hydroxide (OH^-)	$Al(OH)_3(s) \rightleftharpoons Al^{3+} + 3\,OH^-$	$[Al^{3+}][OH^-]^3 = 1.3 \times 10^{-33}$	32.89
	Phosphate (PO_4^{3-})	$AlPO_4(s) \rightleftharpoons Al^{3+} + PO_4^{3-}$	$[Al^{3+}][PO_4^{3-}] = 9.84 \times 10^{-21}$	20.01
	8-Quinolinolate[a]	$AlL_3(s) \rightleftharpoons Al^{3+} + 3\,L-$	$[Al^{3+}][L^-]^3 = 1.00 \times 10^{-29}$	29.00
	Sulfide (S^{2-})	$Al_2S_3(s) \rightleftharpoons 2\,Al^{3+} + 3\,S^{2-}$	$[Al^{3+}]^2[S^{2-}]^3 = 2 \times 10^{-7}$	6.7
Barium (Ba^{2+})	Carbonate (CO_3^{2-})	$BaCO_3(s) \rightleftharpoons Ba^{2+} + CO_3^{2-}$	$[Ba^{2+}][CO_3^{2-}] = 2.58 \times 10^{-9}$	8.59
	Chromate (CrO_4^{2-})	$BaCrO_4(s) \rightleftharpoons Ba^{2+} + CrO_4^{2-}$	$[Ba^{2+}][CrO_4^{2-}] = 1.17 \times 10^{-10}$	9.93
	Fluoride (F^-)	$BaF_2(s) \rightleftharpoons Ba^{2+} + 2\,F^-$	$[Ba^{2+}][F^-]^2 = 1.84 \times 10^{-7}$	6.74
	Phosphate (PO_4^{3-})	$Ba_3(PO_4)_2(s) \rightleftharpoons 3\,Ba^{2+} + 2\,PO_4^{3-}$	$[Ba^{2+}]^3[PO_4^{3-}]^2 = 3.4 \times 10^{-23}$	22.47
	Sulfate (SO_4^{2-})	$BaSO_4(s) \rightleftharpoons Ba^+ + SO_4^{2-}$	$[Ba^+][SO_4^{2-}] = 1.08 \times 10^{-10}$	9.97
Bismuth (Bi^+)	Oxychloride (OCl^-)	$BiOCl(s) \rightleftharpoons Bi^+ + OCl^-$	$[Bi^+][OCl^-] = 1.8 \times 10^{-31}$	30.75
Cadmium (Cd^{2+})	Iodate (IO_3^-)	$Cd(IO_3)_2(s) \rightleftharpoons Cd^{2+} + 2\,IO_3^-$	$[Cd^{2+}][IO_3^-]^2 = 2.5 \times 10^{-8}$	7.60
	Sulfide (S^{2-})	$CdS(s) \rightleftharpoons Cd^{2+} + S^{2-}$	$[Cd^{2+}][S^{2-}] = 8.0 \times 10^{-27}$	26.10
Calcium (Ca^{2+})	Carbonate (CO_3^{2-})	$CaCO_3(s) \rightleftharpoons Ca^{2+} + CO_3^{2-}$	$[Ca^{2+}][CO_3^{2-}] = 2.8 \times 10^{-9}$	8.54
	Fluoride (F^-)	$CaF_2(s) \rightleftharpoons Ca^{2+} + 2\,F^-$	$[Ca^{2+}][F^-]^2 = 5.3 \times 10^{-9}$	8.28
	Hydroxide (OH^-)	$Ca(OH)_2(s) \rightleftharpoons Ca^{2+} + S^{2-}$	$[Ca^{2+}][OH^-]^2 = 5.5 \times 10^{-6}$	5.26
	Oxalate ($C_2O_4^{2-}$)	$CaC_2O_4(s) \rightleftharpoons Ca^{2+} + C_2O_4^{2-}$	$[Ca^{2+}][C_2O_4^{2-}] = 2.32 \times 10^{-9}$	8.63 (actually a hydrate)
	Phosphate (PO_4^{3-})	$Ca_3(PO_4)_2(s) \rightleftharpoons 3\,Ca^{2+} + 2\,PO_4^{3-}$	$[Ca^{2+}]^3[PO_4^{3-}]^2 = 2.07 \times 10^{-29}$	28.68
	8-Quinolinolate[a]	$CaL_2(s) \rightleftharpoons Ca^{2+} + 2\,L^-$	$[Ca^{2+}][L^-]^2 = 7.6 \times 10^{-12}$	11.12
	Sulfate (SO_4^{2-})	$CaSO_4(s) \rightleftharpoons Ca^{2+} + SO_4^{2-}$	$[Ca^{2+}][SO_4^{2-}] = 4.93 \times 10^{-5}$	4.31
Cerium(III) (Ce^{3+})	Hydroxide (OH^-)	$Ce(OH)_3(s) \rightleftharpoons Ce^{3+} + 3\,OH^-$	$[Ce^{3+}][OH^-]^3 = 1.6 \times 10^{-20}$	19.80
Cerium(IV) (Ce^{4+})	Hydroxide (OH^-)	$Ce(OH)_4(s) \rightleftharpoons Ce^{4+} + 4\,OH^-$	$[Ce^{4+}][OH^-]^4 = 2 \times 10^{-48}$	47.7
Chromium(II) (Cr^{2+})	Hydroxide (OH^-)	$Cr(OH)_2(s) \rightleftharpoons Cr^{2+} + 2\,OH^-$	$[Cr^{2+}][OH^-]^2 = 2 \times 10^{-16}$	15.7
Chromium(III) (Cr^{3+})	Hydroxide (OH^-)	$Cr(OH)_3(s) \rightleftharpoons Cr^{3+} + 3\,OH^-$	$[Cr^{3+}][OH^-]^3 = 6.3 \times 10^{-31}$	30.20
Copper(I) (Cu^+)	Bromide (Br^-)	$CuBr(s) \rightleftharpoons Cu^+ + Br^-$	$[Cu^+][Br^-] = 6.27 \times 10^{-9}$	8.20

Cation	Anion	Dissociation Reaction	Solubility Product, K_{sp}	$pK_{sp} = -\log(K_{sp})$
	Chloride (Cl⁻)	$CuCl(s) \rightleftharpoons Cu^+ + Cl^-$	$[Cu^+][Cl^-] = 1.72 \times 10^{-7}$	6.76
	Cyanide (CN⁻)	$CuCN(s) \rightleftharpoons Cu^+ + CN^-$	$[Cu^+][CN^-] = 3.47 \times 10^{-20}$	19.46
	Hydroxide (OH⁻)	$CuOH(s) \rightleftharpoons Cu^+ + OH^-$	$[Cu^+][OH^-] = 1 \times 10^{-14}$	14.0
	Iodide (I⁻)	$CuI(s) \rightleftharpoons Cu^+ + I^-$	$[Cu^+][I^-] = 1.27 \times 10^{-12}$	11.90
	Sulfide (S²⁻)	$Cu_2S(s) \rightleftharpoons 2\,Cu^+ + S^{2-}$	$[Cu^+]^2[S^{2-}] = 2.5 \times 10^{-48}$	47.60
	Thiocyanate (SCN⁻)	$CuSCN(s) \rightleftharpoons Cu^+ + SCN^-$	$[Cu^+][SCN^-] = 1.77 \times 10^{-13}$	12.75
Copper(II) (Cu^{2+})	Carbonate (CO₃²⁻)	$CuCO_3(s) \rightleftharpoons Cu^{2+} + CO_3^{2-}$	$[Cu^{2+}][CO_3^{2-}] = 1.4 \times 10^{-10}$	9.86
	Chromate (CrO₄²⁻)	$CuCrO_4(s) \rightleftharpoons Cu^{2+} + CrO_4^{2-}$	$[Cu^{2+}][CrO_4^{2-}] = 3.6 \times 10^{-6}$	5.44
	Hydroxide (OH⁻)	$Cu(OH)_2(s) \rightleftharpoons Cu^{2+} + 2\,OH^-$	$[Cu^{2+}][OH^-]^2 = 2.2 \times 10^{-20}$	19.66
	Phosphate (PO₄³⁻)	$Cu_3(PO_4)_2(s) \rightleftharpoons 3\,Cu^{2+} + 2\,PO_4^{3-}$	$[Cu^{2+}]^3[PO_4^{3-}]^2 = 1.40 \times 10^{-37}$	36.85
	Sulfide (S²⁻)	$CuS(s) \rightleftharpoons Cu^{2+} + S^{2-}$	$[Cu^{2+}][S^{2-}] = 6.3 \times 10^{-36}$	35.20
Iron(II) (Fe^{2+})	Hydroxide (OH⁻)	$Fe(OH)_2(s) \rightleftharpoons Fe^{2+} + 2\,OH^-$	$[Fe^{2+}][OH^-]^2 = 4.87 \times 10^{-17}$	16.31
	Carbonate (CO₃²⁻)	$FeCO_3(s) \rightleftharpoons Fe^{2+} + CO_3^{2-}$	$[Fe^{2+}][CO_3^{2-}] = 3.13 \times 10^{-11}$	10.50
	Sulfide (S²⁻)	$FeS(s) \rightleftharpoons Fe^{2+} + S^{2-}$	$[Fe^{2+}][S^{2-}] = 6.3 \times 10^{-18}$	17.20
Iron(III) (Fe^{3+})	Hydroxide (OH⁻)	$Fe(OH)_3(s) \rightleftharpoons Fe^{3+} + 3\,OH^-$	$[Fe^{3+}][OH^-]^3 = 2.79 \times 10^{-39}$	38.55
	Phosphate dihydrate	$FePO_4 \bullet 2H_2O(s) \rightleftharpoons Fe^{3+} + PO_4^{3-} + 2\,H_2O$	$[Fe^{3+}][PO_4^{3+}] = 9.91 \times 10^{-16}$	15.00
Lanthanum (La^{3+})	Fluoride (F⁻)	$LaF_3(s) \rightleftharpoons La^{3+} + 3\,F^-$	$[La^{3+}][F^-]^3 = 7 \times 10^{-17}$	16.2
	Hydroxide (OH⁻)	$La(OH)_3(s) \rightleftharpoons La^{3+} + 3\,OH^-$	$[La^{3+}][OH^-]^3 = 2.0 \times 10^{-19}$	18.70
	Phosphate (PO₄³⁻)	$LaPO_4(s) \rightleftharpoons La^{3+} + PO_4^{3-}$	$[La^{3+}][PO_4^{3-}] = 3.7 \times 10^{-23}$	22.43
Lead (Pb^{2+})	Carbonate (CO₃²⁻)	$PbCO_3(s) \rightleftharpoons Pb^{2+} + CO_3^{2-}$	$[Pb^{2+}][CO_3^{2-}] = 7.4 \times 10^{-14}$	13.13
	Chloride (Cl⁻)	$PbCl_2(s) \rightleftharpoons Pb^{2+} + 2\,Cl^-$	$[Pb^{2+}][Cl^-]^2 = 1.70 \times 10^{-5}$	4.77
	Fluoride (F⁻)	$PbF_2(s) \rightleftharpoons Pb^{2+} + 2\,F^-$	$[Pb^{2+}][F^-]^2 = 3.3 \times 10^{-8}$	7.48
	Hydroxide (OH⁻)	$Pb(OH)_2(s) \rightleftharpoons Pb^{2+} + 2\,OH^-$	$[Pb^{2+}][OH^-]^2 = 1.43 \times 10^{-15}$	14.84
	Sulfate (SO₄²⁻)	$PbSO_4(s) \rightleftharpoons Pb^{2+} + SO_4^{2-}$	$[Pb^{2+}][SO_4^{2-}] = 2.53 \times 10^{-8}$	7.60
	Sulfide (S²⁻)	$PbS(s) \rightleftharpoons Pb^{2+} + S^{2-}$	$[Pb^{2+}][S^{2-}] = 8.0 \times 10^{-28}$	27.10
Magnesium (Mg^{2+})	Ammonium phosphate	$Mg(NH_4)PO_4(s) \rightleftharpoons Mg^{2+} + NH_4^+ + PO_4^{3-}$	$[Mg^{2+}][NH_4^+][PO_4^{3-}] = 2.5 \times 10^{-13}$	12.60
	Carbonate (CO₃²⁻)	$MgCO_3(s) \rightleftharpoons Mg^{2+} + CO_3^{2-}$	$[Mg^{2+}][CO_3^{2-}] = 6.82 \times 10^{-6}$	5.17

Cation	Anion	Dissociation Reaction	Solubility Product, K_{sp}	$pK_{sp} = -\log(K_{sp})$
	Fluoride (F⁻)	$MgF_2(s) \rightleftharpoons Mg^{2+} + 2\ F^-$	$[Mg^{2+}][F^-]^2 = 5.16 \times 10^{-11}$	10.29
	Hydroxide (OH⁻)	$Mg(OH)_2(s) \rightleftharpoons Mg^{2+} + 2\ OH^-$	$[Mg^{2+}][OH^-]^2 = 5.61 \times 10^{-12}$	11.25
	8-Quinolinolate[a]	$MgL_2(s) \rightleftharpoons Mg^{2+} + 2\ L^-$	$[Mg^{2+}][L^-]^2 = 4.0 \times 10^{-16}$	15.40
Potassium (K⁺)	Perchlorate (ClO₄⁻)	$KClO_4(s) \rightleftharpoons K^+ + ClO_4^-$	$[K^+][ClO_4^-] = 1.05 \times 10^{-2}$	1.98
Silver (Ag⁺)	Bromide (Br⁻)	$AgBr(s) \rightleftharpoons Ag^+ + Br^-$	$[Ag^+][Br^-] = 5.35 \times 10^{-13}$	12.27
	Carbonate (CO₃²⁻)	$Ag_2CO_3(s) \rightleftharpoons 2\ Ag^+ + CO_3^{2-}$	$[Ag^+]^2[CO_3^{2-}] = 8.46 \times 10^{-12}$	11.07
	Chloride (Cl⁻)	$AgCl(s) \rightleftharpoons Ag^+ + Cl^-$	$[Ag^+][Cl^-] = 1.77 \times 10^{-10}$	9.75
	Chromate (CrO₄⁻)	$Ag_2CrO_4(s) \rightleftharpoons 2\ Ag^+ + CrO_4^{2-}$	$[Ag^+]^2[CrO_4^{2-}] = 1.12 \times 10^{-12}$	11.95
	Cyanide (CN⁻)	$AgCN(s) \rightleftharpoons Ag^+ + CN^-$	$[Ag^+][CN^-] = 5.97 \times 10^{-17}$	16.22
	Iodide (I⁻)	$AgI(s) \rightleftharpoons Ag^+ + I^-$	$[Ag^+][I^-] = 8.52 \times 10^{-17}$	16.07
	Sulfide (S²⁻)	$Ag_2S(s) \rightleftharpoons 2\ Ag^+ + S^{2-}$	$[Ag^+]^2[S^{2-}] = 6.3 \times 10^{-50}$	49.20
	Thiocyanate (SCN⁻)	$AgSCN(s) \rightleftharpoons Ag^+ + SCN^-$	$[Ag^+][SCN^-] = 1.03 \times 10^{-12}$	11.99
Strontium (Sr²⁺)	Fluoride (F⁻)	$SrF_2(s) \rightleftharpoons Sr^{2+} + 2\ F^-$	$[Sr^{2+}][F^-]^2 = 4.33 \times 10^{-9}$	8.36
Zinc (Zn²⁺)	Fluoride (F⁻)	$ZnF_2(s) \rightleftharpoons Zn^{2+} + 2\ F^-$	$[Zn^{2+}][F^-]^2 = 3.04 \times 10^{-2}$	1.52
	Carbonate (CO₃²⁻)	$ZnCO_3(s) \rightleftharpoons Zn^{2+} + CO_3^{2-}$	$[Zn^{2+}][CO_3^{2-}] = 1.46 \times 10^{-10}$	9.94
	Hydroxide (OH⁻)	$Zn(OH)_2(s) \rightleftharpoons Zn^{2+} + 2\ OH^-$	$[Zn^{2+}][OH^-]^2 = 3 \times 10^{-17}$	16.5
	Sulfide (S²⁻)	$ZnS(s) \rightleftharpoons Zn^{2+} + S^{2-}$	$[Zn^{2+}][S^{2-}] = 1.6 \times 10^{-24}$	23.80[b]

*These K_{sp} values were determined from *Lange's Handbook of Chemistry*, 15th ed., J. A. Dean, (Ed.), McGraw-Hill, New York, 1999. The listed values were acquired at temperatures between 18° and 25°C.

[a]This reagent is also known as 8-hydroxyquinoline.

[b]This value is for the α form of ZnS. The β form has a K_{sp} equal to 2.5×10^{-22}.

APPENDIX B.3 ACID DISSOCIATION CONSTANTS IN WATER

Acid	Acid Dissociation Reaction	Acid Dissociation Constant, K_a (25°C)[a]	$pK_a = -\log(K_a)$
Acetic acid	$CH_3COOH \rightleftarrows CH_3COO^- + H^+$	1.75×10^{-5}	4.757
Acetylsalicylic acid	$C_6H_4(C_2H_3O_2)COOH \rightleftarrows C_6H_4(C_2H_3O_2)COO^- + H^+$	3.3×10^{-4}	3.48
Adipic acid	$HOOC(CH_2)_4COOH \rightleftarrows HOOC(CH_2)_4COO^- + H^+$	3.71×10^{-5}	4.43
	$HOOC(CH_2)_4COO^- \rightleftarrows {}^-OOC(CH_2)_4COO^- + H^+$	3.89×10^{-6}	5.41
Alanine[b]	${}^+H_3NCH(R)COOH \rightleftarrows {}^+H_3NCH(R)COO^- + H^+$	4.57×10^{-3}	2.34
	${}^+H_3NCH(R)COO^- \rightleftarrows H_2NCH(R)COO^- + H^+$	1.35×10^{-10}	9.87
4-Aminopyridine (conj. acid)[c]	$C_5H_5NNH_3^+ \rightleftarrows C_5H_5NNH_2 + H^+$	4.26×10^{-10}	9.37
Ammonium	$NH_4^+ \rightleftarrows NH_3 + H^+$	5.62×10^{-10}	9.25
Arginine[b]	${}^+H_3NCH(RH^+)COOH \rightleftarrows {}^+H_3NCH(RH^+)COO^- + H^+$	1.51×10^{-2}	1.82
	${}^+H_3NCH(RH^+)COO^- \rightleftarrows H_2NCH(RH^+)COO^- + H^+$	1.02×10^{-9}	8.99
	$H_2NCH(RH^+)COO^- \rightleftarrows H_2NCH(R)COO^- + H^+$	7.9×10^{-13}	12.1
Ascorbic acid	$C_6H_8O_6 \rightleftarrows C_6H_7O_6^- + H^+$	7.94×10^{-5}	4.10
	$C_6H_7O_6^- \rightleftarrows C_6H_6O_6^{2-} + H^+$	1.62×10^{-12}	11.79
Asparagine[b]	${}^+H_3NCH(R)COOH \rightleftarrows {}^+H_3NCH(R)COO^- + H^+$	6.92×10^{-3}	2.16
	${}^+H_3NCH(R)COO^- \rightleftarrows H_2NCH(R)COO^- + H^+$	1.86×10^{-9}	8.73
Aspartic acid[b]	${}^+H_3NCH(RH)COOH \rightleftarrows {}^+H_3NCH(RH)COO^- + H^+$	1.02×10^{-2}	1.99
	${}^+H_3NCH(RH)COO^- \rightleftarrows H_3NCH(R^-)COO^- + H^+$	1.26×10^{-4}	3.90
	${}^+H_3NCH(R^-)COO^- \rightleftarrows H_2NCH(R^-)COO^- + H^+$	1.0×10^{-10}	10.00
Benzoic acid	$C_6H_5COOH \rightleftarrows C_6H_5COO^- + H^+$	6.28×10^{-5}	4.202
Boric acid	$B(OH)_2OH \rightleftarrows B(OH)_2O^- + H^+$	5.79×10^{-10}	9.237
Bromoacetic acid	$BrCH_2COOH \rightleftarrows BrCH_2COO^- + H^+$	2.05×10^{-3}	2.69
Butanoic acid	$C_3H_7COOH \rightleftarrows C_3H_7COO^- + H^+$	1.52×10^{-5}	4.82
Carbonic acid	$H_2CO_3 \rightleftarrows HCO_3^- + H^+$	4.46×10^{-7}	6.351
	$HCO_3^- \rightleftarrows CO_3^{2-} + H^+$	4.69×10^{-11}	10.329
Chloric acid[d]	$HClO_3 \rightleftarrows ClO_3^- + H^+$	Strong acid in water	—
Chloroacetic acid	$ClCH_2COOH \rightleftarrows ClCH_2COO^- + H^+$	1.40×10^{-3}	2.85
Chromic acid	$H_2CrO_4 \rightleftarrows HCrO_4^- + H^+$	1.8×10^{-1}	0.74
	$HCrO_4^- \rightleftarrows CrO_4^{2-} + H^+$	3.20×10^{-7}	6.49
Citric acid	$HOOC(CH_2COOH)COOH \rightleftarrows HOOC(CH_2COOH)COO^- + H^+$	7.45×10^{-4}	3.128
	$HOOC(CH_2COOH)COO^- \rightleftarrows {}^-OOC(CH_2COOH)COO^- + H^+$	1.73×10^{-5}	4.761
	${}^-OOC(CH_2COOH)COO^- \rightleftarrows {}^-OOC(CH_2COO^-)COO^- + H^+$	4.02×10^{-7}	6.396

Acid	Acid Dissociation Reaction	Acid Dissociation Constant, K_a (25°C)[a]	$pK_a = -\log(K_a)$
Cysteine[b]	$^+H_3NCH(RH)COOH \rightleftharpoons {}^+H_3NC(RH)COO^- + H^+$	$2.\underline{0} \times 10^{-2}$	1.7
	$^+H_3NCH(RH)COO^- \rightleftharpoons {}^+H_3NCH(R^-)COO^- + H^+$	$4.3\underline{7} \times 10^{-9}$	8.36
	$^+H_3NCH(R^-)COO^- \rightleftharpoons H_2NCH(R^-)COO^- + H^+$	$1.8\underline{2} \times 10^{-11}$	10.74
1,6-Diaminohexane (conj. acid)[c]	$^+H_3N(CH_2)_6NH_3^+ \rightleftharpoons H_2N(CH_2)_6NH_3^+ + H^+$	1×10^{-10}	10.0
	$H_2N(CH_2)_6NH_3^+ \rightleftharpoons H_2N(CH_2)_6NH_2 + H^+$	8×10^{-12}	11.1
Dimethylamine[c]	$(CH_3)_2NH_2^+ \rightleftharpoons (CH_3)_2NH + H^+$	1.7×10^{-11}	10.77
Ethylamine[c]	$CH_3CH_2NH_3^+ \rightleftharpoons CH_3CH_2NH_2 + H^+$	2.3×10^{-11}	10.63
Ethylenediaminetetraacetic acid	$H_6EDTA \rightleftharpoons H_5EDTA^+ + H^+$	$1.\underline{0}$	0.0
	$H_5EDTA^+ \rightleftharpoons H_4EDTA + H^+$	$3.\underline{2} \times 10^{-2}$	1.5
	$H_4EDTA \rightleftharpoons H_3EDTA^- + H^+$	$1.0\underline{2} \times 10^{-2}$	1.99
	$H_3EDTA^- \rightleftharpoons H_2EDTA^{2-} + H^+$	$2.1\underline{4} \times 10^{-3}$	2.67
	$H_2EDTA^{2-} \rightleftharpoons HEDTA^{3-} + H^+$	$6.9\underline{2} \times 10^{-7}$	6.16
	$HEDTA^{3-} \rightleftharpoons EDTA^{4-} + H^+$	$6.4\underline{6} \times 10^{-11}$	10.19
Formic acid	$HCOOH \rightleftharpoons HCOO^- + H^+$	1.77×10^{-4}	3.75
Glutamic acid[b]	$^+H_3NCH(RH)COOH \rightleftharpoons {}^+H_3NCH(RH)COO^- + H^+$	$6.9\underline{2} \times 10^{-3}$	2.16
	$^+H_3NCH(RH)COO^- \rightleftharpoons {}^+H_3NCH(R^-)COO^- + H^+$	$5.0\underline{1} \times 10^{-5}$	4.30
	$^+H_3NCH(R^-)COO^- \rightleftharpoons H_2NCH(R^-)COO^- + H^+$	$1.1\underline{0} \times 10^{-10}$	9.96
Glutamine[b]	$^+H_3NCH(R)COOH \rightleftharpoons {}^+H_3NCH(R)COO^- + H^+$	$6.4\underline{6} \times 10^{-3}$	2.19
	$^+H_3NCH(R)COO^- \rightleftharpoons H_2NCH(R)COO^- + H^+$	$1.0\underline{0} \times 10^{-9}$	9.00
Glycine[b]	$^+H_3NCCH_2COOH \rightleftharpoons {}^+H_3NCH_2COO^- + H^+$	$4.4\underline{7} \times 10^{-3}$	2.35
	$^+H_3NCH_2COO^- \rightleftharpoons H_2NCH_2COO^- + H^+$	$1.6\underline{6} \times 10^{-10}$	9.78
Histidine[b]	$^+H_3NCH(RH^+)COOH \rightleftharpoons {}^+H_3NC(RH^+)COO^- + H^+$	$2.\underline{5} \times 10^{-2}$	1.6
	$^+H_3NCH(RH^+)COO^- \rightleftharpoons {}^+H_3NCH(R)COO^- + H^+$	$1.0\underline{7} \times 10^{-6}$	5.97
	$^+H_3NCH(R)COO^- \rightleftharpoons H_2NCH(R)COO^- + H^+$	$5.2\underline{5} \times 10^{-10}$	9.28
Hydrobromic acid[d]	$HBr \rightleftharpoons Br^- + H^+$	Strong acid in water	—
Hydrochloric acid[d]	$HCl \rightleftharpoons Cl^- + H^+$	Strong acid in water	—
Hypochlorous acid	$HOCl \rightleftharpoons OCl^- + H^+$	2.95×10^{-8}	7.53
Hydrofluoric acid	$HF \rightleftharpoons F^- + H^+$	6.8×10^{-4}	3.17
Hydrogen cyanate	$HOCN \rightleftharpoons OCN^- + H^+$	3.3×10^{-4}	3.48
Hydrogen cyanide	$HCN \rightleftharpoons CN^- + H^+$	4.93×10^{-10}	9.31

Acid	Acid Dissociation Reaction	Acid Dissociation Constant, K_a (25°C)[a]	$pK_a = -\log(K_a)$
Hydrogen sulfide	$H_2S \rightleftharpoons HS^- + H^+$	9.5×10^{-8}	7.02
	$HS^- \rightleftharpoons S^{2-} + H^+$	1×10^{-14}	14
Hydrogen thiocyanate	$HSCN \rightleftharpoons SCN^- + H^+$	8×10^{-2}	1.1
Hydroiodic acid[d]	$HI \rightleftharpoons I^- + H^+$	Strong acid in water	—
Hypobromous acid	$HOBr \rightleftharpoons OBr^- + H^+$	2.06×10^{-9}	8.69
Hypoiodous acid	$HOI \rightleftharpoons OI^- + H^+$	2.3×10^{-11}	10.64
Hypophosphorus acid	$H_3PO_2 \rightleftharpoons H_2PO_2^- + H^+$	6.3×10^{-2}	1.2
Iodoacetic acid	$ICH_2COOH \rightleftharpoons ICH_2COO^- + H^+$	7.5×10^{-4}	3.12
Isoleucine[b]	$^+H_3NCH(R)COOH \rightleftharpoons {}^+H_3NCH(R)COO^- + H^+$	4.79×10^{-3}	2.32
	$^+H_3NCH(R)COO^- \rightleftharpoons H_2NCH(R)COO^- + H^+$	1.74×10^{-10}	9.76
Lactic acid	$CH_3CH(OH)COOH \rightleftharpoons CH_3CH(OH)COO^- + H^+$	1.35×10^{-4}	3.87
Leucine[b]	$^+H_3NCH(R)COOH \rightleftharpoons {}^+H_3NCH(R)COO^- + H^+$	4.68×10^{-3}	2.33
	$^+H_3NCH(R)COO^- \rightleftharpoons H_2NCH(R)COO^- + H^+$	1.82×10^{-10}	9.74
Lysine[b]	$^+H_3NCH(RH^+)COOH \rightleftharpoons {}^+H_3NCH(RH^+)COO^- + H^+$	1.70×10^{-2}	1.77
	$^+H_3NCH(RH^+)COO^- \rightleftharpoons H_2NCH(RH^+)COO^- + H^+$	8.51×10^{-10}	9.07
	$H_2NCH(RH^+)COO^- \rightleftharpoons H_2NCH(R)COO^- + H^+$	1.52×10^{-11}	10.82
Malic acid	$HOOCCH_2CH(OH)COOH \rightleftharpoons HOOCCH_2CH(OH)COO^- + H^+$	3.9×10^{-4}	3.40
	$HOOCCH_2CH(OH)COO^- \rightleftharpoons {}^-OOCCH_2CH(OH)COO^- + H^+$	7.8×10^{-6}	5.11
Malonic acid	$HOOCCH_2COOH \rightleftharpoons HOOCCH_2COO^- + H^+$	1.49×10^{-3}	2.83
	$HOOCCH_2COO^- \rightleftharpoons {}^-OOCCH_2COO^- + H^+$	2.03×10^{-6}	5.69
Methionine[b]	$^+H_3NCH(R)COOH \rightleftharpoons {}^+H_3NCH(R)COO^- + H^+$	6.61×10^{-3}	2.18
	$^+H_3NCH(R)COO^- \rightleftharpoons H_2NCH(R)COO^- + H^+$	8.32×10^{-10}	9.08
Methylamine	$CH_3NH_3^+ \rightleftharpoons CH_3NH_2 + H^+$	2.70×10^{-11}	10.657
Nitric acid[d]	$HNO_3 \rightleftharpoons NO_3^- + H^+$	Strong acid in water	—
2-Nitrobenzoic acid	$C_6H_4(NO_2)COOH \rightleftharpoons C_6H_4(NO_2)COO^- + H^+$	6.6×10^{-3}	2.18
3-Nitrobenzoic acid	$C_6H_4(NO_2)COOH \rightleftharpoons C_6H_4(NO_2)COO^- + H^+$	3.5×10^{-4}	3.46
4-Nitrobenzoic acid	$C_6H_4(NO_2)COOH \rightleftharpoons C_6H_4(NO_2)COO^- + H^+$	3.6×10^{-4}	3.44
2-Nitrophenol (o-nitrophenol)	$C_6H_4(NO_2)OH \rightleftharpoons C_6H_4(NO_2)O^- + H^+$	6.8×10^{-8}	7.17
3-Nitrophenol (m-nitrophenol)	$C_6H_4(NO_2)OH \rightleftharpoons C_6H_4(NO_2)O^- + H^+$	5.3×10^{-9}	8.28
4-Nitrophenol (p-nitrophenol)	$C_6H_4(NO_2)OH \rightleftharpoons C_6H_4(NO_2)O^- + H^+$	7.0×10^{-8}	7.15

Acid	Acid Dissociation Reaction	Acid Dissociation Constant, K_a (25°C)[a]	pK_a = –log(K_a)
Nitrous acid	$HNO_2 \rightleftharpoons NO_2^- + H^+$	4.6×10^{-4}	3.37
Oxalic acid	$HOOCCOOH \rightleftharpoons HOOCCOO^- + H^+$	5.90×10^{-2}	1.23
	$HCOOCOO^- \rightleftharpoons {}^-OOCCOO^- + H^+$	6.40×10^{-5}	4.19
Perchloric acid[d]	$HClO_4 \rightleftharpoons ClO_4^- + H^+$	Strong acid in water	—
Phenol	$C_6H_5OH \rightleftharpoons C_6H_5O^- + H^+$	1.28×10^{-10}	9.89
Phenylalanine[b]	${}^+H_3NCH(R)COOH \rightleftharpoons {}^+H_3NCH(R)COO^- + H^+$	6.31×10^{-3}	2.20
	${}^+H_3NCH(R)COO^- \rightleftharpoons H_2NCH(R)COO^- + H^+$	4.90×10^{-10}	9.31
Phosphoric acid	$H_3PO_4 \rightleftharpoons H_2PO_4^- + H^+$	7.11×10^{-3}	2.148
	$H_2PO_4^- \rightleftharpoons HPO_4^{2-} + H^+$	6.34×10^{-8}	7.198
	$HPO_4^{2-} \rightleftharpoons PO_4^{3-} + H^+$	4.22×10^{-13}	12.375
Phosphorus acid	$H_3PO_3 \rightleftharpoons H_2PO_3^- + H^+$	1.0×10^{-2} (18°C)	2.00
	$H_2PO_3^- \rightleftharpoons HPO_3^{2-} + H^+$	2.6×10^{-7} (18°C)	6.59
Phthalic acid	$HOOCC_6H_4COOH \rightleftharpoons HOOCC_6H_4COO^- + H^+$	1.12×10^{-3}	2.950
	$HOOCC_6H_4COO^- \rightleftharpoons {}^-OOCC_6H_4COO^- + H^+$	3.90×10^{-6}	5.408
Piperazine	${}^+H_2NC_4H_8NH_2^+ \rightleftharpoons HNC_4H_8NH_2^+ + H^+$	2.76×10^{-6}	5.56
	$HNC_4H_8NH_2^+ \rightleftharpoons HNC_4H_8NH + H^+$	1.48×10^{-10}	9.83
Piperidine	$C_5H_{10}NH_2^+ \rightleftharpoons C_5H_{10}NH + H^+$	7.53×10^{-12}	11.123
Proline[b]	${}^+H_3NC(R)COOH \rightleftharpoons {}^+H_3NC(R)COO^- + H^+$	1.12×10^{-2}	1.95
	${}^+H_3NC(R)COO^- \rightleftharpoons H_2NC(R)COO^- + H^+$	2.29×10^{-11}	10.64
Propanoic acid	$C_3H_7COOH \rightleftharpoons C_3H_7COO^- + H^+$	1.34×10^{-5}	4.87
Pyridine	$C_5H_5NH^+ \rightleftharpoons C_5H_5N + H^+$	5.62×10^{-6}	5.25
Pyrophosphoric acid	$H_4P_2O_7 \rightleftharpoons H_3P_2O_7^- + H^+$	1.4×10^{-1}	0.85
	$H_3P_2O_7^- \rightleftharpoons H_2P_2O_7^{2-} + H^+$	3.2×10^{-2}	1.49
	$H_2P_2O_7^{2-} \rightleftharpoons HP_2O_7^{3-} + H^+$	1.7×10^{-6}	5.77
	$HP_2O_7^{3-} \rightleftharpoons P_2O_7^{4-} + H^+$	6×10^{-9}	8.22
Salicylic acid	$C_6H_4(OH)COOH \rightleftharpoons C_6H_4(OH)COO^- + H^+$	1.07×10^{-3}	2.97
	$C_6H_4(OH)COO^- \rightleftharpoons C_6H_4(O^-)COO^- + H^+$	3.98×10^{-14}	13.40
Serine[b]	${}^+H_3NCH(R)COOH \rightleftharpoons {}^+H_3NCH(R)COO^- + H^+$	6.46×10^{-3}	2.19
	${}^+H_3NCH(R)COO^- \rightleftharpoons H_2NCH(R)COO^- + H^+$	6.17×10^{-10}	9.21
Succinic acid	$HOOC(CH_2)_2COOH \rightleftharpoons HOOC(CH_2)_2COO^- + H^+$	6.21×10^{-5}	4.207
	$HOOC(CH_2)_2COO^- \rightleftharpoons {}^-OOC(CH_2)_2COO^- + H^+$	2.32×10^{-6}	5.635

Acid	Acid Dissociation Reaction	Acid Dissociation Constant, K_a (25°C)[a]	$pK_a = -\log(K_a)$
Sulfuric acid[d]	$H_2SO_4 \rightleftharpoons HSO_4^- + H^+$	Strong acid in water	—
	$HSO_4^- \rightleftharpoons SO_4^{2-} + H^+$	1.03×10^{-2}	1.987
Sulfurous acid	$H_2SO_3 \rightleftharpoons HSO_3^- + H^+$	1.54×10^{-2}	1.81
	$HSO_3^- \rightleftharpoons SO_3^{2-} + H^+$	1.02×10^{-7}	6.91
Tartaric acid	$HOOC(CHOH)_2COOH \rightleftharpoons HOOC(CHOH)_2COO^- + H^+$	1.05×10^{-3}	2.98
	$HOOC(CHOH)_2COO^- \rightleftharpoons {}^-OOC(CHOH)_2COO^- + H^+$	4.57×10^{-5}	4.34
Threonine[b]	${}^+H_3NCH(R)COOH \rightleftharpoons {}^+H_3NCH(R)COO^- + H^+$	8.13×10^{-3}	2.09
	${}^+H_3NCH(R)COO^- \rightleftharpoons H_2NCH(R)COO^- + H^+$	7.94×10^{-10}	9.10
Trichloroacetic acid	$CCl_3COOH \rightleftharpoons CCl_3COO^- + H^+$	1.99×10^{-1}	0.70
Trimethylamine[c]	$(CH_3)_3NH^+ \rightleftharpoons (CH_3)_3N + H^+$	1.6×10^{-10}	9.80
Tryptophan[b]	${}^+H_3NCH(R)COOH \rightleftharpoons {}^+H_3NCH(R)COO^- + H^+$	4.27×10^{-3}	2.37
	${}^+H_3NCH(R)COO^- \rightleftharpoons H_2NCH(R)COO^- + H^+$	4.68×10^{-10}	9.33
Tyrosine[b]	${}^+H_3NCH(RH)COOH \rightleftharpoons {}^+H_3NCH(RH)COO^- + H^+$	5.75×10^{-3}	2.24
	${}^+H_3NCH(RH)COO^- \rightleftharpoons H_2NCH(RH)COO^- + H^+$	6.46×10^{-10}	9.19
	$H_2NCH(RH)COO^- \rightleftharpoons H_2NCH(R^-)COO^- + H^+$	3.39×10^{-11}	10.47
Valine[b]	${}^+H_3NCH(R)COOH \rightleftharpoons {}^+H_3NCH(R)COO^- + H^+$	5.13×10^{-3}	2.29
	${}^+H_3NCH(R)COO^- \rightleftharpoons H_2NCH(R)COO^- + H^+$	1.91×10^{-10}	9.72
Water	$H_2O \rightleftharpoons OH^- + H^+$	1.01×10^{-14}	13.997

[a]Most of the K_a values in this appendix were obtained from NIST Database 46. Some values were obtained from J. A. Dean, Ed., *Lange's Handbook of Chemistry, 15th ed.*, McGraw-Hill, New York, 1999, and the *CRC Handbook of Chemistry and Physics, 81st ed.*, CRC Press, Boca Raton, FL, 2000. Values that are underlined are guard digits.

[b]The symbol "R" represents the side chain of the given amino acid; the full structures of the amino acids are provided in Table 8.8 of Chapter 8.

[c]The K_a and pK_a values shown for these bases are for the corresponding conjugate acids.

[d]The K_a values for these acids are greater than or near to $10^{1.74} = 55.5$, the approximate K_a for H_3O^+.

APPENDIX B4.1 FORMATION CONSTANTS FOR METAL IONS WITH VARIOUS CHELATING AGENTS

Metal Ion	Chelating Agent and Log K_f for Metal Ion[a]			
	EDTA	Trien	CDTA	EGTA
Al^{3+}	16.4		19.5	13.90
Ba^{2+}	7.88		8.58	8.30
Ca^{2+}	10.65	1.4	13.1	10.86
Cd^{2+}	16.5	10.7	19.7	16.5
Co^{2+}	16.45	10.9	19.7	12.3
Cu^{2+}	18.78	20.06	22.0	17.7
Fe^{2+}	14.30	7.72	18.9	11.8
Fe^{3+}	25.1		30.0	20.5
Hg^{2+}	21.5	24.7	24.8	22.9
Li^+	2.95			
Mg^{2+}	8.79	1.4	11.0	5.28
Mn^{2+}	13.89	4.87	17.5	12.2
Na^+	1.86			
Ni^{2+}	18.4	14.0	20.2	13.5
Ra^{2+}	7.07			
Sr^{2+}	8.72		10.5	8.42
Ti^{3+}	21.3			
V^{2+}	12.7	10.7		
Zn^{2+}	16.5	11.95	19.3	12.6

Abbreviations: EDTA, ethylenediaminetetraacetic acid; trien, triethylenetetramine; CDTA, *trans*-1,2-diaminocyclohexanetetraacetic acid; EGTA, ethyleneglycoldiaminetetraacetic acid.

[a]These values are from NIST Standard Reference Database 46—*NIST Critically Selected Stability Constants for Metal Complexes Database, vol. 8.0*, NIST, Gaithersburg, MD, 2004.

[b]The formation constants for additional metal ions with EDTA can be found in Table 9.6 of Chapter 9.

APPENDIX B4.2 OVERALL FORMATION CONSTANTS (β) FOR METAL IONS WITH VARIOUS LIGANDS

Ligand	Metal Ion	Value of $\log(\beta_n)$ for Ligand with Metal Ion[a]					
		$\beta_1 = K_{f1}$	β_2	β_3	β_4	β_5	β_6
Ammonia (NH_3)	Ag^+	3.40	7.40				
	Cu^{2+}	4.13	7.61	10.48	12.59	12.54	10.54
	Cd^{2+}	2.60	4.5	6.04	6.92	6.6	4.9
	Co^{2+}	2.05	3.62	4.61	5.31	5.43	4.75
	Hg^{2+}	8.80	17.50	18.5	19.4		
	Ni^{2+}	2.72	4.88	6.54	7.67	8.33	8.31
	Zn^{2+}	2.27	4.61	7.01	9.06		
Chloride (Cl^-)	Ag^+	2.9	4.7	5.0	5.9		
	Hg^{2+}	6.7	13.2	14.1	15.1		
Cyanide (CN^-)	Ag^+		21.1	21.8	20.7		
	Au^+		38.3				
	Cd^{2+}	6.0	11.1	15.6	17.9		
	Hg^{2+}	18.0	34.7	38.5	41.5		
	Ni^{2+}				31.3		
	Pb^{2+}				10		
	Zn^{2+}				16.7		
Fluoride (F^-)	Al^{3+}	6.1	11.15	15.0	17.7	19.4	19.7
	Be^{2+}	5.1	8.8	11.8			
	Cr^{3+}	4.4	7.7	10.2			
	Fe^{3+}	5.2	9.2	11.9			
Hydroxide (OH^-)	Al^{3+}				33.3		
	Ca^{2+}	1.3					
	Cd^{2+}	4.3	7.7	10.3	12.0		
	Ce^{4+}	13.3	27.1				
	Fe^{3+}	11.0	21.7				
	Zn^{2+}	4.4		14.4	15.5		
	Zr^{4+}	13.8	27.2	40.2	53		
Thiocyanate (SCN^-)	Ag^+	7.6	9.1	10.1			
	Au^+		25				
	Au^{3+}		42				
	Cu^{2+}	1.7	2.5	2.7	3.0		
	Fe^{3+}	2.3	4.2	5.6	6.4	6.4	
	Hg^{2+}		16.1	19.0	20.9		
	Ni^{2+}	1.2	1.6	1.8			

Ligand	Metal Ion	Value of $\log(\beta_n)$ for Ligand with Metal Ion[a]					
		$\beta_1 = K_{f1}$	β_2	β_3	β_4	β_5	β_6
Thiosulfate ($S_2O_3^{2-}$)	Zn^{2+}	0.7	1.04	1.2	1.6		
	Ag^+	8.82	13.5				
	Cu^+	10.3	12.2	13.8			
	Hg^{2+}	29.86	32.26				

[a]These values are from A. Ringbom, *Complexation in Analytical Chemistry*, Krieger Publishing, Huntington, NY, 1979, and NIST Standard Reference Database 46—*NIST Critically Selected Stability Constants for Metal Complexes Database*, vol. 8.0, NIST, Gaithersburg, MD, 2004.

APPENDIX B.5 STANDARD POTENTIALS FOR REDUCTION HALF-REACTIONS

Half-Reaction	Standard Potential $E°$ (V) vs. SHE
Aluminum	
$Al^{3+} + 3\,e^- \rightleftharpoons Al$	−1.71
Arsenic	
$H_3AsO_4 + 2H^+ + 2\,e^- \rightleftharpoons HAsO_2 + 2\,H_2O$	+0.58
(or $H_3AsO_4 + 2H^+ + 2\,e^- \rightleftharpoons H_3AsO_3 + H_2O$)	
Barium	
$Ba^{2+} + 2\,e^- \rightleftharpoons Ba$	−2.90
Bismuth	
$Bi^{3+} + 3\,e^- \rightleftharpoons Bi$	+0.31
$BiOCl + 2\,H^+ + 3\,e^- \rightleftharpoons Bi + Cl^- + H_2O$	+0.16
Bromine	
$2\,BrO_3^- + 12\,H^+ + 10\,e^- \rightleftharpoons Br_2 + 6\,H_2O$	+1.52
$Br_2 + 2\,e^- \rightleftharpoons 2\,Br^-$	+1.09
Calcium	
$Ca^{2+} + 2\,e^- \rightleftharpoons Ca$	−2.76
Cadmium	
$Cd^{2+} + 2\,e^- \rightleftharpoons Cd$	−0.40
Cerium	
$Ce^{4+} + e^- \rightleftharpoons Ce^{3+}$	+1.72
$Ce^{4+} + e^- \rightleftharpoons Ce^{3+}\ (1.0\ M\ HClO_4)$	+1.70
$Ce^{4+} + e^- \rightleftharpoons Ce^{3+}\ (1.0\ M\ HNO_3)$	+1.61
$Ce^{4+} + e^- \rightleftharpoons Ce^{3+}\ (0.5\ M\ H_2SO_4)$	+1.44
$Ce^{4+} + e^- \rightleftharpoons Ce^{3+}\ (1.0\ M\ HCl)$	+1.47
Chlorine	
$HClO + H^+ + e^- \rightleftharpoons \tfrac{1}{2}Cl_2 + H_2O$	+1.63
$Cl_2 + 2\,e^- \rightleftharpoons 2\,Cl^-$	+1.40
Chromium	
$Cr_2O_7^{2-} + 14\,H^+ + 6\,e^- \rightleftharpoons 2\,Cr^{3+} + 7\,H_2O$	+1.36
$CrO_4^{2-} + 4\,H_2O + 3\,e^- \rightleftharpoons Cr(OH)_3 + 5\,OH^-$	−0.12
$Cr^{3+} + 2\,e^- \rightleftharpoons Cr^{2+}$	−0.41
$Cr^{2+} + 2\,e^- \rightleftharpoons Cr$	−0.56

Half-Reaction	Standard Potential $E°$ (V) vs. SHE
Cobalt	
$Co^{2+} + 2\ e^- \rightleftharpoons Co$	−0.28
Copper	
$Cu^{2+} + 2\ e^- \rightleftharpoons Cu$	+0.34
$Cu^{2+} + e^- \rightleftharpoons Cu^+$	+0.16
Fluorine	
$F_2 + 2\ e^- \rightleftharpoons 2\ F^-$	+2.87
Gold	
$Au^+ + e^- \rightleftharpoons Au$	+1.68
$Au^{3+} + 3\ e^- \rightleftharpoons Au$	+1.42
Hydrogen	
$2\ H^+ + 2\ e^- \rightleftharpoons H_2(g)$	0.000... (Reference)
$2\ H_2O + 2\ e^- \rightleftharpoons H_2(g) + 2\ OH^-$	−0.83
Indium	
$In^{3+} + 3\ e^- \rightleftharpoons In$	−0.34
Iodine	
$IO_3^- + 6\ H^+ + 5\ e^- \rightleftharpoons \frac{1}{2} I_2 + 3\ H_2O$	+1.20
$I_2 + 2\ e^- \rightleftharpoons 2\ I^-$	+0.53
Iron	
$FeO_4^{2-} + 8\ H^+ + 3\ e^- \rightleftharpoons Fe^{3+} + 4\ H_2O$	+ 2.20
$Fe^{3+} + e^- \rightleftharpoons Fe^{2+}$	+0.77
$Fe^{3+} + e^- \rightleftharpoons Fe^{2+}$ (1 M HClO$_4$)	+0.77
$Fe^{3+} + e^- \rightleftharpoons Fe^{2+}$ (1 M HNO$_3$)	+0.75
$Fe^{3+} + e^- \rightleftharpoons Fe^{2+}$ (0.5 M H$_2$SO$_4$)	+0.68
$Fe^{3+} + e^- \rightleftharpoons Fe^{2+}$ (1 M H$_3$PO$_4$)	+0.44
$Fe^{3+} + e^- \rightleftharpoons Fe^{2+}$ (1 M HCl)	+0.73
$Fe^{2+} + 2\ e^- \rightleftharpoons Fe$	−0.41
$Fe^{3+} + 3\ e^- \rightleftharpoons Fe$	−0.04
Lead	
$PbO_2 + SO_4^{2-} + 4\ H^+ + 2\ e^- \rightleftharpoons PbSO_4 + 2\ H_2O$	+1.68
$PbO_2 + 4\ H^+ + 2\ e^- \rightleftharpoons Pb^{2+} + 2\ H_2O$	+1.46
$Pb^{2+} + 2\ e^- \rightleftharpoons Pb$	−0.13

Half-Reaction	Standard Potential $E°$ (V) vs. SHE
Lithium	
$Li^+ + e^- \rightleftharpoons Li$	−3.04
Magnesium	
$Mg^{2+} + 2\,e^- \rightleftharpoons Mg$	−2.36
Manganese	
$MnO_4^- + 4\,H^+ + 3\,e^- \rightleftharpoons MnO_2 + 2\,H_2O$	+1.68
$MnO_4^- + 8\,H^+ + 5\,e^- \rightleftharpoons Mn^{2+} + 4\,H_2O$	+1.51
$Mn^{3+} + e^- \rightleftharpoons Mn^{2+}$	+1.51
$MnO_2 + 4\,H^+ + 2\,e^- \rightleftharpoons Mn^{2+} + 2\,H_2O$	+1.21
$MnO_4^- + 2\,H_2O + 3\,e^- \rightleftharpoons MnO_2 + 4\,OH^-$	+0.59
Mercury	
$Hg^{2+} + 2\,e^- \rightleftharpoons Hg$	+0.85
$Hg_2^{2+} + 2\,e^- \rightleftharpoons 2\,Hg$	+0.80
$Hg_2Cl_2 + 2\,e^- \rightleftharpoons 2\,Hg + 2\,Cl^-$	+0.268
$Hg_2Cl_2 + 2\,e^- \rightleftharpoons 2\,Hg + 2\,Cl^-$ (0.1 M KCl)	+0.334
$Hg_2Cl_2 + 2\,e^- \rightleftharpoons 2\,Hg + 2\,Cl^-$ (saturated KCl)	+0.242[b]
$Hg_2Cl_2 + 2\,e^- \rightleftharpoons 2\,Hg + 2\,Cl^-$ (saturated NaCl)	+0.236
Nickel	
$Ni^{2+} + 2\,e^- \rightleftharpoons Ni$	−0.23
Nitrogen	
$NO_3^- + 3\,H^+ + 2\,e^- \rightleftharpoons HNO_2 + H_2O$	+0.94
Oxygen	
$O_3 + 2\,H^+ + 2\,e^- \rightleftharpoons O_2 + H_2O$	+2.07
$H_2O_2 + 2\,H^+ + 2\,e^- \rightleftharpoons 2\,H_2O$	+1.78
$O_2 + 4\,H^+ + 4\,e^- \rightleftharpoons 2\,H_2O$	+1.23
$O_2 + 2\,H^+ + 2\,e^- \rightleftharpoons H_2O_2$	+0.68
$O_2 + 2\,H_2O + 4\,e^- \rightleftharpoons 4\,OH^-$	+0.40
Phosphorus	
$H_3PO_4 + 2\,H^+ + 2\,e^- \rightleftharpoons H_3PO_3 + H_2O$	−0.28
Plutonium	
$Pu^{4+} + e^- \rightleftharpoons Pu^{3+}$	+0.97
$Pu^{3+} + 3\,e^- \rightleftharpoons Pu$	−2.00

Half-Reaction	Standard Potential $E°$ (V) vs. SHE
Ruthenium	
$Ru^{3+} + e^- \rightleftharpoons Ru^{2+}$	+0.25
Silver	
$Ag^+ + e^- \rightleftharpoons Ag$	+0.80
$AgCl + e^- \rightleftharpoons Ag + Cl^-$	+0.222
Sodium	
$Na^+ + e^- \rightleftharpoons Na$	−2.71
Sulfur	
$S + 2\,H^+ + 2\,e^- \rightleftharpoons H_2S(g)$	+0.17
$S + 2\,H^+ + 2\,e^- \rightleftharpoons H_2S(aq)$	+0.14
Thallium	
$Tl^{3+} + 2\,e^- \rightleftharpoons Tl^+$	+1.25
$Tl^+ + e^- \rightleftharpoons Tl$	−0.34
Tin	
$HSnO_2^- + H_2O + 2\,e^- \rightleftharpoons Sn + 3\,OH^-$	−0.91
$Sn^{4+} + 2\,e^- \rightleftharpoons Sn^{2+}$	+0.14
$Sn^{2+} + 2\,e^- \rightleftharpoons Sn$	−0.14
Uranium	
$U^{3+} + 3\,e^- \rightleftharpoons U$	−1.8
Vanadium	
$VO_2^+ + 2\,H^+ + e^- \rightleftharpoons VO^{2+} + H_2O$	+1.00
$VO^{2+} + 2\,H_2O + e^- \rightleftharpoons V^{3+} + H_2O$	+0.34
$V^{3+} + e^- \rightleftharpoons V^{2+}$	−0.26
$V^{2+} + 2\,e^- \rightleftharpoons V$	−1.2
Zinc	
$Zn^{2+} + 2\,e^- \rightleftharpoons Zn$	−0.76

[a]The standard potentials are all shown for the reduction half-reaction at 25°C versus the standard hydrogen electrode. The chemicals listed in bold include the element that is being oxidized or reduced in the given half-reaction. The abbreviation "SHE" stands for the standard hydrogen electrode. The listed values were obtained from J. A. Dean (Editor), *Lange's Handbook of Chemistry, 15th Ed.*, McGraw-Hill, New York, 1999, or the *CRC Handbook of Chemistry and Physics, 81st Edition*, CRC Press, Boca Raton, FL, 2000. In most cases, the listed values have been rounded to the nearest 0.01 V; standard potentials listed with more significant digits are those that correspond to common reference electrodes.

[b]This potential represents the saturated calomel electrode.

APPENDIX C

USING A SPREADSHEET FOR DATA ANALYSIS

A spreadsheet was defined in Chapter 2 as a program that can be used to record, analyze, and manipulate data. A spreadsheet is valuable in carrying out repetitive calculations or in constructing graphs (e.g., see Figure 2.8 in Chapter 2 and Figure 12.17 in Chapter 12). A spreadsheet is constructed by placing a series of entries based on text, numbers, or formulas into a table. These entries can then be used in calculations or to make graphs. In this Appendix we discuss the basic features of most spreadsheet programs and see how a spreadsheet can be used to carry out calculations or to construct a graph. We use the spreadsheet program Microsoft Excel as the basis of this discussion, but the same general features we discuss for this program can also be found in other types of spreadsheet. Further help on the use of spreadsheets is given by accessing the "Help" menu of the program. Additional information on Microsoft Excel can also be found by using resources such as References *1–4*.

BASIC DESIGN OF A SPREADSHEET

The main body of a spreadsheet is laid out in a grid that contains a series of boxes, known as "cells," in which you make individual entries. The columns in this grid are represented at the top of the grid by a series of letters, starting with A and proceeding on up as needed. The rows in the grid are represented by a series of numbers that are shown along the side of the grid. The location of a specific cell is identified through the column and row in which that cell is located. For example, a cell that is located in column B and in row 2 would be referred to as cell "B2." The current location of the cursor on this spreadsheet can be seen in the "name bar," which is located to the left and just above the table. The contents of the cell are shown just to the right of the name bar in a "formula bar," which is identified by the symbol "f_x" in Microsoft Excel 2007.

Scroll bars are located at the bottom and side of the table to allow you to move from one area of the table in a spreadsheet to another. This is a useful feature because some spreadsheets you may create for data analysis can involve cells that are located over a large number of columns and rows. A menu bar is at the top of the spreadsheet and contains several menus that can be accessed to carry out data analysis or to construct graphs. For instance, in Microsoft Excel 2007 these menus include "Home," "Insert," "Page Layout," "Formulas," "Data," "Review," and "View." Below the menu bar is a tool bar that shows the various options that are available under the selected menu. There is also a menu for opening, closing, printing, and saving files in the upper left-hand corner by the circular logo in Microsoft Excel 2007. In the upper right corner there is an option to access a help menu by the circle that has an enclosed question mark.

ENTERING DATA, TEXT, AND FORMULAS INTO CELLS

There are four basic types of entries that can be made into a cell in a spreadsheet: numbers, text, functions, and formulas (i.e., combinations of numbers, functions, and sometimes text). You can select the cell that you want to use for data entry by pointing and clicking at that cell with the mouse or by moving to that cell with the cursor. The active cell (i.e., the one into which any entered data will be placed) has a boldface box around it to allow you to easily identify it in the spreadsheet. The entry can either be made directly into the cell or by placing the entry into the formula box after the desired active cell has been selected.

Text and numbers can be entered in the normal fashion in a cell. A function must be entered in a specific format that will depend on the type of spreadsheet you are using. For Microsoft Excel, functions begin with the symbol "@" and are followed by a name or abbreviation that represents a particular type of function. (*Note*: in recent versions of Excel, the symbol "=" can also be used in place of "@" at the start of the name for a function.) Some examples of common functions that are used in data analysis are listed below (see your particular spreadsheet program for many other examples).

@AVERAGE(values)	Gives the mean or average of the selected values
@STDEV(values)	Gives the standard deviation of the selected values
@MEDIAN(values)	Gives the median for the selected values
@LOG10(value)	Gives the base-10 logarithm of the selected value
@LOG(value)	Gives the natural logarithm of the selected value

For each function, a value or set of values must be selected to place in parentheses after the function name. These values are the numbers that are evaluated in finding the result for the particular function. The values that are used here can be selected by pointing the cursor or mouse to a cell that contains them as a function is entered, or by entering these values directly into the function.

A formula can be constructed by using a combination of numbers, functions, and sometimes text. This type of entry should begin with a "=" or "+" sign to tell the spreadsheet that a formula rather than a text entry is being made. The entry is then continued by inserting in a series of numbers, functions, and signs for mathematical operations (e.g., "+" or "−" for addition or subtraction; "*" or "/" for multiplication or division; "^" for exponentiation or to raise a value to some power). As an example, the following formula would be written to represent the addition of a constant value (4.52) to the base-10 logarithm for the result of 2.0 divided by 2.5.

$$+4.52+@LOG10(2.0/2.5)$$

One extremely useful feature of formulas and functions in spreadsheets is that the "value" used in these calculations can be made to refer to an entry in a particular cell, rather than entering the numerical value directly into the formula or function. For instance, a function similar to the preceding one can be written as follows,

$$+C8+@LOG10(B9/D9)$$

in which the values that appear in cells C8, B9, and D9 are now used in finding the final result of this formula. A formula or function can also be set up in this way to use the value from a cell that has a particular relative location versus the cell with formula (e.g., two cells to the right, as entered through the use of the cursor or mouse). A value from a specific cell that always has the same absolute location in the spreadsheet can also be used in the formula by using "$" signs in the cell name (e.g., by using A2 to refer to the absolute location of cell A2). This feature of being able to use data and values from other cells in formulas and calculations is what makes spreadsheets particularly useful in analyzing tables of large quantities of numbers or data.

CREATING A GRAPH USING A SPREADSHEET

There are many types of graphs that can be created using a spreadsheet program. A few examples include a bar graph, a line graph, a scatter plot, and a pie graph. Most of the graphs that we use in this textbook can be created in Microsoft Excel by using a Scatter plot. To prepare this type of plot, you first need to enter a series of x and y values for the plot. One way this can be done is by creating two adjacent columns of data, one for the x-values and the other for the corresponding y-values. These values should then be selected with the mouse and cursor. The "Insert Menu" should then be selected, and "Scatter" can be selected for the type of plot that will be prepared from these data under the option labeled "Charts." If you have highlighted the x- and y-values in advance, a chart that contains these values in a plot should appear. You also have option of first selecting the type of chart and then later selecting the x- and y-values that will be used by pointing to the chart box and using the right mouse button to go to "Select Data." This second approach is useful when the x- and y-values are not in adjacent columns or if you want to make several plots on the same graph using data from different parts of the spreadsheet. In this situation, each set of data is referred to by the spreadsheet program as a different "Series" of numbers in the plot. The same general method can be used to make other types of plots using the spreadsheet.

Once your basic graph has been created, you can edit it and change its appearance by using the options that appear when you click on the graph using the right mouse button. Some of these options allow you to change the appearance of the overall chart area, the type of chart that is being used, and the data that are used in the plot. Options are also available for selecting the format of the numbers that appear on the x- and y-axes, the labels for these axes, the number and location of tick marks, and the presence or absence of grid marks. The formatting for the axes can be changed by clicking on the axis to which a change is to be made and selecting the "Format Axis" option. The type of markers and whether a line is shown (or not) for a particular series of data can be selected by clicking on the desired series of data in the plot and choosing the "Format Data Series" option.

In Chapter 4 we discuss how linear regression can be carried out on a set of data to obtain a best-fit line. The equations that can be used for this purpose are provided in that chapter and an example of how a spreadsheet can be constructed to use these equations is given in the problems at the end of that chapter. However, linear regression is also available as a built-in function for spreadsheet programs. In the case of Microsoft Excel 2007, linear regression on the data in a selected plot can be carried out by going to the "Layout" menu at the top of the page (*Note*: The graph must be selected in order for this menu option to appear), followed by selection of the option "Trendline" and then "Linear Trendline." This places a best-fit line for the data in the plot. To obtain more information on this fit, you can select "More Trendline Options" at the bottom of the "Trendline" option and choose "Display Equation on chart" and "Display R-squared value on chart." These selections will cause the equation for the best-fit line to appear in the plot as well as the coefficient of determination (r^2). The correlation coefficient (r) can be found by simply taking the square root of the r^2 value.

References

1. S. Copestake, *Excel 2002 in Easy Steps*, Barnes & Noble, Warkwickshire, UK, 2003.
2. E. J. Billo, *Microsoft Excel for Chemists*, 2nd ed., Wiley, New York, 2001.
3. R. De Levie, *How to Use Excel in Analytical Chemistry and in General Scientific Data Analysis*, Cambridge University Press, Cambridge, UK, 2001.
4. J. Cronan, *Microsoft Office Excel 2007 QuickSteps*, McGraw Hill, New York, 2007.

SELECTED ANSWERS

Chapter 1

1.6

a. Sample = coal, analyte = sulfur, matrix = group of all substances in the coal

b. Sample = drug tablet, analyte = drug, matrix = group of all substances in the tablet

c. Sample = fumes emitted by industrial plant, analyte = carbon monoxide, matrix = group of all substances in the fumes

1.8

a. Major components

b. Major component

c. Minor component

1.14

a. Qualitative analysis would be used initially to determine if these drugs are present in the samples above an accepted cut off limit. If any samples are found to give positive results for the drugs, they would probably be reanalyzed using quantitative analysis to determine the amount of drug that is present.

b. A structural analysis could be used to provide information on the structure of the compound, which could then be used to identify this chemical. Property characterization might also be used for this purpose.

c. Quantitative analysis would be utilized in determining the amount of drug that is present in the product.

d. This is an example of spatial analysis because it seeks to obtain information on how a particular analyte is distributed in space along a river bed. This work would also involve quantitative analysis as the amount of pollutant is measured at different locations in the river.

Chapter 2

2.12

a. Oxidizing agent, corrosive

b. Corrosive

c. No appreciable hazardous properties

d. Carcinogen, toxin, dangerous for the environment

e. Flammable, harmful or irritant

f. Flammable, can be explosive

2.14 Acetonitrile: flammability – flammable liquid (flash point below 100° F); reactivity – stable; health – may be harmful if inhaled or absorbed

Sodium borohydride: flammability – combustible if heated; reactivity – unstable or may react violently if mixed with water; health – may be harmful if inhaled or absorbed; special – water reactive

2.31

a. 6,855 m   b. 5.76×10^5 Pa   c. 59 kg

d. 193 km/hr   e. 9200 J   f. 94.6 L

2.32

a. 37.0°C or 310.2 K

b. –273.15°C or –459.67°F

c. 293 K or 68°F

d. –4°F or 253 K

2.33

a. 25.8 pg or 0.0258 ng

b. 125 μL or 0.125 mL

c. 150 kg/mol or 150 kDa

d. 589 nm or 0.589 μm

e. 600 MHz or 0.600 GHz

f. 25 kV or 0.025 MV

2.35 Estimated % transmittance = 53.7%; estimated absorbance = 0.270

2.38

a. 9   b. 7   c. 3   d. 5 (for miles)   e. 3 (for cm)   f. 3

2.39

a. 1.52 M b. 7.463   c. 6.63×10^{-34} J · s

d. 5.52 ns   e. 8.37   f. 0.165

2.44

a. 143.321   b. 2.453   c. 0.984

d. –5.697   e. 1.29×10^{-3} f. 1.5×10^{-7}

2.47

a. 1.789 g Cu

b. 0.0148 mol/L NaOH (or 0.0148 M)

c. 342.3025 g/mol $C_{12}H_{22}O_{11}$

d. 11 g/cm^3

Chapter 3

3.5 The change in the gravitational pull from that which occurs at 340 ft above sea level to that which occurs at 5,280 ft above sea level would account for the error.

3.12

a. Macroanalytical balance or semimicrobalance

b. Precision balance

c. Microbalance or ultramicrobalance

3.13 The precision balance has a resolution of 200,000; this balance would be used to weigh the calcium carbonate. The resolution of the analytical balance is 800,000 and this balance would be used to weigh the EDTA.

3.17 5.3083 g

3.28

a. Volumetric pipet

b. Micropipet

c. Several devices might be used here, such as a 2.00 mL volumetric pipet, a micropipette, or a Mohr pipet

d. Several devices might be used here, such as a micropipette, a Mohr pipet or a serological pipet

e. Syringe

f. Buret

3.32

a. 50.008 mL

b. 50.003 mL

3.35 This solution contains 161 mol ethylene glycol and 278 mol water, so water is the solvent and ethylene glycol is the solute.

3.37

a. 12.5% (v/v) acetonitrile, 25.0% (v/v) methanol

b. 70.3% (w/w) iron, 17.6% (w/w) chromium, 8.03% (w/w) nickel, 2.01% (w/w) manganese, 1.01% (w/w) silicon, 0.803% (w/w) carbon, 0.257% (w/w) other elements

c. 5.0 g/L dissolved solids

3.39

a. 5.0×10^{-6} g/L or 5.0 ppm Cu^{2+}

b. 8.3×10^{-6} g/L or 8.3 ppm Be^{+2}

c. 1.7×10^{-4} g/L or 0.17 parts-per-thousand $NaIO_3$

3.44

a. 1.014 M b. 0.02285 M

c. 0.1242 M d. 0.1081 M

3.45 1.87 M or 1.94 m formaldehyde

3.48 Individual concentrations, 0.0108 M acetic acid and 0.00642 M acetate; analytical concentration of acetic acid plus acetate, 0.0172 M

3.58 120 mL

3.59 0.425 mM creatinine

3.62 50.1 μM at 4°C, 49.7 μM at 45°C

Chapter 4

4.2

a. Random error   b. Systematic error   c. Systematic error

4.6 Result, absolute error and relative error (%): 117 μmol/mL, 5 μmol/mL, 4.$\underline{1}$%; 119 μmol/mL, 3 μmol/mL, 2.$\underline{5}$%; 111 μmol/mL, 11 μmol/mL, 9.$\underline{0}$%; 115 μmol/mL, 7 μmol/mL, 5.$\underline{7}$%; 120 μmol/mL, 2 μmol/mL, 1.$\underline{6}$%

4.10

a. 89.$\underline{3}$ mg

b. 279.0 nm

c. 134.9 s

4.15

a. Range, 34.1 min − 32.8 min = 1.3 min; standard deviation, 0.55 min

b. Range, 0.24% − 0.19% = 0.05%; standard deviation, 0.022%

c. Range, 0.01018 M − 0.00998 M = 0.0002 M; standard deviation, 0.00010 M

4.17 Mean, 99.8 mg/dL; range, 112 mg/dL − 88 mg/dL = 24 mg/dL; standard deviation, 5.6 mg/dL; relative standard deviation, 5.7%

4.23

a. 3.05 (± 0.04)   b. 10.18 (± 0.07)

c. 77,600 (± 1,200)   d. 65 (± 6)

e. 43.6 (± 0.4)   f. 6.9 (± 0.3) $\times 10^6$

4.25

a. 0.30458 (± 0.00017)   b. 6.0 (± 0.3) $\times 10^{-4}$

c. 242 (± 6)   d. 1.696 (± 0.002)

e. 0.186 (± 0.004)   f. 2.53 (± 0.02)

4.28

a. 0.147 (± 0.003)   b. 186 (± 4)

c. 30.069 (± 0.002)   d. 6.0 (± 0.3) $\times 10^{-4}$

e. −0.34 (± 0.03)   f. 1.4 (± 0.9) $\times 10^{-12}$

4.30 0.0636 (± 0.0003) M

4.33 −6.5 (± 0.2) $\times 10^4$ J/mol

4.38 $\bar{x}$ = 10.4 mg/dL, $s_{\bar{x}}$ = 0.3 mg/dL

4.42 90% C.I. = 3.0 ± 0.3 μg/L (n = 5)

4.49 t (0.11) < t_c (2.36), results are the same at the 95% confidence level

4.51 t (1.8) < t_c (1.94), results are the same at the 90% confidence level

4.53 t (6.8) > t_c (2.18), results are not the same at the 95% confidence level

4.55 t (1.7) < t_c (2.78), results are the same at the 95% confidence

4.58

a. t (1.37) < t_c (2.31), mean results are the same at the 95% confidence

b. The new employee obtained a result with a larger degree of variation. The level of precision was not the same at the 95% confidence level.

4.60 F (30.3) > F_c (6.39), there is a significant improvement in the precision at the 95% confidence

4.64 No points can be rejected

4.69 Slope = 0.0062, intercept = 0.0022

Chapter 5

5.10 Absolute error = 0.019% (w/w); relative error = 2.2%

5.12 Percent recovery = 98.7%

5.15 t (3.$\underline{6}$) > t_c (2.36), results are not the same at the 95% confidence

5.20 Standard deviation = 0.6 mg/mL, relative standard deviation = 4%

5.24 The RSD from 15 to 35 mg/L ranges from 2.5 to 8%; the concentration range that gives an RSD at or below 5% is approximately 18 to 47 mg/L.

5.26

a. Interoperator precision

b. Day-to-day precision

c. Within-run precision

5.30 LOD = 0.02 ng/mL; LOQ = 0.08 ng/mL

5.33 The dynamic range extends from the lower limit of detection (insufficient data to estimate in this case) up to at least 20.00 mg/L. The linear range for data that fits within 10% of a best-fit line extends from the lower limit of detection up to approximately 12.50 mg/L.

5.38 The calibration sensitivity (or slope) at 7.50 mg/L is approximately 0.022 L/mg. This value is consistent throughout the linear range, but begins to decrease at concentrations that are above the linear range.

5.45 Total analysis time = 35 min

5.51 Using an allowable range of $\bar{x}$ ± 2 s would correspond to values of 8.61 to 9.58 in the control chart. The values which are outside of this range are those obtained on Days 7 and 11. The control samples on these days would need to be reanalyzed and possible corrective action taken before the assay could be used to examine samples and standards.

Chapter 6

6.6

a. Pure oxygen gas at a pressure of one bar (formerly 1 atm)

b. Pure sodium chloride crystals

c. Pure liquid methanol

d. A 1.00 M solution of NaCl in water

e. A 1.00 M solution of methanol in water

f. Pure helium gas at a pressure of one bar (formerly 1 atm)

6.10 Activity coefficients = 0.734

6.13 $I = 0.00163\ M$

6.14 I for NaCl solution = 0.100 M; I for Na_2SO_4 solution = 0.300 M

6.17 The individual activity coefficients for Na^+ and Cl^- are both approximately 0.85 in this case.

6.21

a. Activity coefficient for H^+ = 0.93; activity coefficient for NO_3^- = 0.92

b. Activity coefficient for K^+ = 0.86; activity coefficient for OH^- = 0.87

c. Activity coefficient for Ba^{2+} = 0.53; activity coefficient for Cl^- = 0.84

6.22

a. Mean activity coefficient = 0.92$\underline{5}$

b. Mean activity coefficient = 0.86$\underline{5}$

c. Mean activity coefficient = 0.61$\underline{8}$

6.26 $\gamma = 1.16$

6.34

a. $K° = (a_{H_3O^+})(a_{HSO_4^-})/(a_{H_2SO_4})(a_{H_2O})$

b. $K° = (a_{Zn(NH_3)2+})/(a_{Zn^{2+}})(a_{NH_3})$

c. $K° = (a_{Pb^{2+}})(a_{Cl^-})^2/(a_{PbCl_2})$

6.35

a. $K = [H_3O^+][HSO_4^-]/[H_2SO_4]$

b. $K = [Zn(NH_3)^{2+}]/[Zn^{2+}][NH_3]$

c. $K = [Pb^{2+}][Cl^-]^2$

6.36 $K = (3.6 \times 10^{-4})(1.0 \times 10^{-4})/(2.0 \times 10^{-4}) = 1.8 \times 10^{-4}$

6.39 $\Delta G° = -9{,}140\ J/mol$

6.42 $Q = 1.1 \times 10^{-12}$; this value is smaller than the equilibrium constant, so some of the solid will dissolve to form more aqueous ions.

6.46 $C_{Phosphoric\ acid} = [H_3PO_4] + [H_2PO_4^-] + [HPO_4^{2-}] + [PO_4^{3-}]$

6.49 $2\,[Ca^{2+}] + [Na^+] + 2\,[Mg^{2+}] = [Cl^-]$

6.52

a. Roots for x, $-0.33\underline{8}$ and $-0.037\underline{0}$

b. Roots for x, $0.42\underline{4}$ and $-0.42\underline{6}$

c. Roots for x, $3.3\underline{4}$ and $0.15\underline{5}$

d. First divide all terms by x, and then obtain the roots for x, $32.\underline{5}$ and $-0.0049\underline{4}$

6.55

a. $x = -1.8$ b. $x = -2.5 \times 10^{-3}$

c. $x = 0.77$ d. $x = 0.0779$

Chapter 7

7.9 All three of these ions are precipitated with hydroxide as $Al(OH)_3$, $Fe(OH)_3$, and $Cr(OH)_3$, as occurs during the addition of a dilute ammonia solution. If concentrated NaOH is added, however, aluminum and chromium dissolve by forming $Al(OH)_4^-$ and $Cr(OH)_4^-$ while iron hydroxide remains insoluble. The addition of H_2O_2 to the solution oxidizes the chromium to dichromate ($Cr_2O_7^{2-}$), which is precipitated as the yellow solid $BaCrO_4$ after the addition of $BaCl_2$.

7.15

a. From most polar to least: dimethylsulfoxide > acetone > chloroform > benzene; dimethylsulfoxide and acetone will mix best with water, while benzene will mix best with octane

b. Acetone can undergo hydrogen bonding, dipole-dipole interactions and interactions involving dispersion forces; benzene can undergo interactions involving dispersion forces; chloroform can undergo dipole-dipole interactions and interactions involving dispersion forces; dimethylsulfoxide can undergo dipole-dipole interactions, hydrogen bonding and interactions involving dispersion forces

7.16

a. From most polar to least: acetic acid > propanol > propane > cyclohexane

b. Acetic acid and propanol are the most soluble in water; cyclohexane is the most soluble in octanol

c. Acetic acid can undergo hydrogen bonding, dipole-dipole interactions, and interactions involving dispersion forces; cyclohexane and propane can undergo interactions involving dispersion forces; n-propanol can undergo hydrogen bonding and interactions involving dispersion forces

7.23 $S = 2.79 \times 10^{-2}\ M$

7.26

a. $K°_{sp} = (a_{Ag^+})(a_{Br^-})$ _or_ $K_{sp} = [Ag^+][Br^-]$

b. $K°_{sp} = (a_{Sr^{2+}})(a_{F^-})^2$ _or_ $K_{sp} = [Sr^{2+}][F^-]^2$

c. $K°_{sp} = (a_{Ca^{2+}})^3(a_{PO_4^{3-}})^2$ _or_ $K_{sp} = [Ca^{2+}]^3[PO_4^{3-}]^2$

d. $K°_{sp} = (a_{Mg^{2+}})(a_{NH_4^+})(a_{PO_4^{3-}})$ _or_ $K_{sp} = [Mg^{2+}][NH_4^+][PO_4^{3-}]$

7.28

a. $S = 1.3 \times 10^{-12}\ M$ for the α-form of ZnS or $1.6 \times 10^{-11}\ M$ for the β-form

b. $S = 8.9 \times 10^{-14}\ M$

7.35

a. $[C_6H_6] = 0.0229\ M$ b. Volume = 2.04 mL

7.37 $S = 0.477\ M$; volume = 1.05 L

7.41 $C_{solute} = 2.09 \times 10^{-4}\ M$

7.57 $Ni(dmg)_2$ precipitates are known to form small crystals that are extremely difficult and slow to filter. The second, more patient, student had a precipitate that was made up of much larger and more easily filtered crystals.

7.60

a. $Q = 5.6 \times 10^{-5} > K_{sp} = 2.53 \times 10^{-8}$ (precipitation will occur); sulfate is the limiting reagent so the mass of $PbSO_4$ formed is $(0.010\ mol/L)(100/150)(0.150\ L)(303.3\ g/mol) = 0.15\ g$

b. $Q = [Pb^{2+}][Cl^-]^2 = \{(0.080)(50/250)\}\{(0.0050)(200/250)\}^2 = 2.56 \times 10^{-7}$; this is less than the K_{sp} of 1.7×10^{-5} so precipitation will not occur.

7.63 Mass precipitated = 0.25 g

7.68 $[Ag^+] = 7.3 \times 10^{-7}\,M$ in pure water; $[Ag^+] = 5.4 \times 10^{-11}\,M$ in 0.010 M KBr

7.70

a. $[CuI] = 1.17 \times 10^{-6}\,M$ b. $[CuI] = 2.54 \times 10^{-11}\,M$

7.73

a. pH = 2.81   b. $[Fe^{3+}] = 8.8 \times 10^{-20}\,M$

Chapter 8

8.7

a. $K_a^{\circ} = (a_{H_3O^+})(a_{OCN^-})/(a_{HOCN})(a_{H_2O})$

b. $K_a^{\circ} = (a_{H_3O^+})(a_{C_2H_5COO^-})/(a_{C_2H_5COOH})(a_{H_2O})$

c. $K_a^{\circ} = (a_{S^{2-}})(a_{H_3O^+})/(a_{HS^-})(a_{H_2O})$

d. $K_a^{\circ} = (a_{CH_3COOH_2^+})(a_{CH_3COO^-})/(a_{CH_3COOH})^2$

8.9

a. $H_3PO_4 > H_2PO_4^- > HPO_4^{2-}$

b. $HClO > HBrO > HIO$

c. $H_3PO_2 > H_3PO_3 > H_3PO_4$

8.13

a. $K_b = (a_{RNH_3^+})(a_{OH^-})/(a_{RNH_2})(a_{H_2O})$

b. $K_b = (a_{HClO})(a_{OH^-})/(a_{ClO^-})(a_{H_2O})$

c. $K_b = (a_{RCOOH})(a_{OH^-})/(a_{RCOO^-})(a_{H_2O})$

d. $K_b = (a_{CH_3OH_2^+})(a_{CH_3O})/(a_{CH_3OH})^2$

8.15

a. Methylamine is slightly stronger than ethylamine, which is considerably stronger than ammonia

b. $PO_4^{3-} > HPO_4^{2-} > H_2PO_4^-$

c. Piperidine >> pyridine > 2,2′-bipyridine

8.17 Autoprotolysis is the act of one molecule donating a hydrogen ion to another molecule of the same compound.

For water: $H_2O + H_2O \rightleftarrows H_3O^+ + OH^-$

For methanol: $CH_3OH + CH_3OH \rightleftarrows CH_3OH_2^+ + CH_3O^-$

8.19

a. $3.7 \times 10^{-11}\,M$ b. $1.96 \times 10^{-10}\,M$ c. $5.40 \times 10^{-8}\,M$

8.24

a. $HCOO^-$, $K_b = 5.71 \times 10^{-11}$

b. $C_6H_5O^-$, $K_b = 7.89 \times 10^{-5}$

c. $HCrO_4^-$, $K_b = 5.6 \times 10^{-14}$

d. CrO_4^{2-}, $K_b = 3.16 \times 10^{-8}$

8.28 At pH 6.8, $[H^+] = 1.\underline{6} \times 10^{-7}\,M$; at pH 5.2, $[H^+] = 6.\underline{3} \times 10^{-6}\,M$

8.30

a. pH = 5.20   b. $pK_b = 7.084$

c. pOH = 11.332   d. pCl = 3.82

8.38

a. pH = 1.52   b. pH = 11.78

c. pH = 2.10   d. pH = 11.7

Each of these calculations assumes complete dissociation is present and that the activity coefficients for the resulting species are all 1.00.

8.40 $[HCl] = 0.480\,M$; pH = 0.319

8.43 These solutions are probably too dilute to ignore the contribution due to the autoprotolysis of water. If this autoprotolysis is considered, the results are as follows:

a. pH = 7.064   b. pH = 6.20

c. pH = 6.97   d. pH = 7.011

8.47

a. pH = 2.76   b. pH = 3.96

c. pH = 1.60   d. pH = 3.02

8.49 pH = 1.63

8.52

a. pH = 3.75   b. pH = 7.43

8.56

a. pH = 4.16   b. pH = 4.24   c. pH = 8.91

8.57

a. pH = 4.61

b. After 25 mL NaOH, moles conjugate base = 2.5×10^{-4} mol

After 50 mL NaOH, moles conjugate base = 5.0×10^{-4} mol

After 100 mL NaOH, moles conjugate base = 9.0×10^{-4} mol

c. After 25 mL NaOH, pH = 6.76

After 50 mL NaOH, pH = 7.27

After 100 mL NaOH, pH = 10.70

8.61 $[H_3PO_4] = 6.16 \times 10^{-3}\,M$, $[H_2PO_4^-] = 4.38 \times 10^{-2}\,M$

8.62 pH = 10.81

8.65

a. At pH 7.0, $[Na_2HPO_4]/[NaH_2PO_4] = 0.634$

At pH 6.5, $[Na_2HPO_4]/[NaH_2PO_4] = 0.200$

At pH = 7.5, $[Na_2HPO_4]/[NaH_2PO_4] = 2.00$

b. Total concentration of the phosphate species will be 0.20 M.

c. Equal volumes (500 mL each) of the two solutions are needed. For $Na_2HPO_4 \cdot H_2O$, the mass would be 17.8 g, and for $NaH_2PO_4 \cdot 2H_2O$ the mass is 15.6 g.

8.71 The curve reaches a maximum at the pK_a of the acid making up the buffer. The curve would show a maximum at pH = 6.99 for imidazole because that is the pH which equals the pK_{a1} of imidazole.

8.75 Fraction of $H_2A^{2+} = 0.27$, fraction of $HA^+ = 0.73$, faction of A = 0.00011

8.77

a. Fraction of $NH_3 = 0.98$

b. $[NH_3] = 0.024\underline{6}\,M$

c. Fraction of $NH_3 = 0.0056$; there is much more free ammonia at pH 11 than at pH 7.

8.81

a. pH = 2.37   b. pH = 11.50

c. pH = 4.67   d. pH = 9.78

8.88

a. pI = 5.44   b. pI = 6.29

c. pI = 5.63   d. pI = 5.32

Chapter 9

9.8

a. Mg^{2+} is the Lewis acid and OH^- is the Lewis base

b. Ag^+ is the Lewis acid and Cl^- is the Lewis base

c. Fe^{3+} is the Lewis acid and $EDTA^{4-}$ is the Lewis base

d. CH_3COOH is the Bronsted acid and NH_3 is the Bronsted base. These are also the Lewis acid and base because the H^+ from acetic acid accepts an electron pair from ammonia to form ammonium.

9.12 Reaction 1 shows what really happens as water molecules are substituted by ammonia. Reaction 2 is a simplified representation and emphasizes what most people regard as the important changes that are occurring in this reaction.

9.15

a. $K_f = [BaOH^+]/[Ba^{2+}][OH^-]$

b. $K_f = [Cu(NH_3)_2^{2+}]/[Cu^{2+}][NH_3]^2$

c. $K_f = [Ni(CN)_4^{2-}]/[Ni^{2+}][CN^-]^4$

d. $K_f = [FeF_6^{3-}]/[Fe^{3+}][F^-]^6$

9.20 $\beta_4 = 8.6 \times 10^{17}$

9.21 $K_{f1} = 59$, $K_{f2} = 6.7$, $K_{f3} = 2.5$, $K_{f4} = 0.2$

9.22 $[Ni^{2+}] = 1.6 \times 10^{-6}\,M$

9.24

a. $C_{Cu(II)} = [Cu^{2+}] + [Cu(NH_3)^{2+}] + [Cu(NH_3)_2^{2+}] + [Cu(NH_3)_3^{2+}] + [Cu(NH_3)_4^{2+}] + [Cu(NH_3)_5^{2+}] + [Cu(NH_3)_6^{2+}]$

b. $\alpha_{Cu2+} = 1/(1 + \beta_1[NH_3] + \beta_2[NH_3]^2 + \beta_3[NH_3]^3 + \beta_4[NH_3]^4 + \beta_5[NH_3]^5 + \beta_6[NH_3]^6)$

$\alpha_{Cu(NH_3)_4^{2+}} = \beta_4[NH_3]^4/(1 + \beta_1[NH_3] + \beta_2[NH_3]^2 + \beta_3[NH_3]^3 + \beta_4[NH_3]^4 + \beta_5[NH_3]^5 + \beta_6[NH_3]^6)$

c. $\alpha_{Cu2+} = 2.2 \times 10^{-9}\ \alpha_{Cu(NH_3)_4^{2+}} = 0.857$

9.36 $Pb^{2+} + EDTA^{4-} \rightleftarrows Pb(EDTA)^{2-}\quad K_f = 1.0 \times 10^{18}$

Both nitrogen atoms of EDTA and its four carboxylate oxygens can bond simultaneously to a large metal ion such as Pb^{2+} to form a highly stable complex.

9.38 Mass = 3.7223 g, concentration of EDTA = 0.0269 M

9.40

a. $\alpha_{EDTA4-} = 1.9 \times 10^{-3}$

b. The principle form at pH 7.5 is $HEDTA^{3-}$

9.46

a. $K'_f = 2.6 \times 10^{10}$ b. $K'_f = 1.5 \times 10^{17}$

c. $K'_f = 2.0 \times 10^{14}$ d. $K'_f = 1.8 \times 10^{-2}$

9.49 $K'_f = 0.029$ at pH 5, $K'_f = 444$ at pH 10; the value at pH 10 is closest to the true formation constant

9.56 $K_A = 1.6 \times 10^9\,M$

Chapter 10

10.7

a. Zn/Zn^{2+} and Ag^+/Ag

b. Pb/Pb^{2+} and PbO_2/Pb^{2+}

c. I_2/I^- and $S_2O_6^{2-}/S_4O_6^{2-}$

10.11

a. Cl, 0   b. Au, 0

c. Ca, +2; O, –2   d. K, +1; S, –6; O, –2

e. O, –2; Fe, 8/3   f. H, +1; O, –1

g. F, –1; Xe, +4   h. N, –3; H, +1; Cr, +6; O, –2

i. H, +1; O, –2; C, –2

10.14

a. Fe is reduced from +3 to +2; Al is oxidized from 0 to +3; O is always at an oxidaton number of –2

b. Co is reduced from +3 to +2; Sn is oxidized from 0 to +2; O is at an oxidation number of –2 and H is +1

c. One O is reduced from –1 to –2, and the other oxygen is oxidized from –1 to 0; H is at +1

d. F is reduced from 0 to –1; Br is oxidized from –1 to 0

10.15

a. No

b. Yes, Cl is reduced from 0 to –1 and I is oxidized from +1 to 0

c. No

d. Yes, Zn is oxidized from 0 to +2 and Cu is reduced from +2 to 0

10.18 Fe^{2+} is oxidized according to the half-reaction:

$Fe^{2+} \rightleftarrows Fe^{3+} + 1\,e^-$

Cr(VI) is reduced according to the half-reaction:

$Cr_2O_7^{2-} + 14\,H^+ + 6\,e^- \rightleftarrows 2\,Cr^{3+} + 7\,H_2O$

10.20

a. Sn^{2+} is oxidized and Hg(II) is reduced. Two electrons are transferred.

$Sn^{2+} \rightleftarrows Sn^{4+} + 2\,e^-$ and $2\,HgCl_2 + 2\,e^- \rightleftarrows Hg_2Cl_2 + 2\,Cl^-$

b. I_2 is reduced and S(IV) is oxidized. Two electrons are transferred.

$I_2 + 2\,e^- \rightleftarrows 2\,I^-$ and $2\,S_2O_3^{2-} \rightleftarrows S_4O_6^{2-} + 2\,e^-$

c. Fe^{2+} is oxidized and O(–1) is reduced. Two electrons are transferred.

$Fe^{2+} \rightleftarrows Fe^{3+} + 1\,e^-$ and $H_2O_2 + 2\,H^+ + 2\,e^- \rightleftarrows 2\,H_2O$

d. Br^- is oxidized and Br(V) is reduced. As written, five electrons are transferred.

$2\,Br^- \rightleftarrows Br_2 + 2\,e^-$ and $BrO_3^- + 6\,H^+ + 5\,e^- \rightleftarrows \frac{1}{2}\,Br_2 + 3\,H_2O$

10.21

a. $K^o = (a_{Sn4+})(a_{Hg_2Cl_2})(a_{Cl^-})^2/\{(a_{Sn2+})(a_{HgCl_2})^2\}$

b. $K^o = (a_{I^-})^2(a_{S_4O_6^{2-}})/\{(a_{I_2})(a_{S_2O_3^{2-}})^2\}$

c. $K^o = (a_{H_2O})^2(a_{Fe3+})^2/\{(a_{H_2O_2})(a_{Fe2+})^2(a_{H^+})^2\}$

d. $K^o = (a_{Br_2})^3(a_{H_2O})^3/\{(a_{Br^-})^5(a_{BrO_3^-})(a_{H^+})^6\}$

10.24 $K = 1.95 \times 10^5$

10.28 The ability to undergo reduction is also the ranking of the strength of these chemicals as oxidizing agents:

$O_3 > Cl_2 > Cr^{3+} > Zn^{2+} > Na^+$

10.31

a. $E^o_{Net} = +0.17\,V$ b. $E^o_{Net} = -1.10\,V$ c. $E^o_{Net} = +0.27\,V$

The order of decreasing tendency for these reactions to occur is c > a > b

10.35 $E^o_{Net} = 1.53\,V$, $\Delta G^o = -2.95 \times 10^5\,J$ or 295 kJ

10.38 $K = 1.5 \times 10^{37}$

10.40

a. $Zn + Cu^{2+} \rightleftarrows Cu + Zn^{2+}$ $K = 1.7 \times 10^{37}$

b. $[Zn^{2+}] = 0.050\,M$, $[Cu^{2+}] = 3.2 \times 10^{-39}\,M \approx 0\,M$

Mass of copper metal = 0.159 g, mass of zinc metal = 0.837 g

10.48

a. $E^o_{Cell} = +2.48\,V$ b. $E^o_{Cell} = +0.74\,V$

c. $E^o_{Cell} = +0.36\,V$ d. $E^o_{Cell} = +0.36\,V$

10.54 $E^o_{Cell} = +0.242\,V$ (based on a calomel electrode containing saturated KCl)

10.55

a. Cathode half-reaction: $Cu^{2+} + 2\,e^- \rightleftarrows Cu$

Anode half-reaction: $H_2 \rightleftarrows 2\,H^+ + 2\,e^-$

b. $E^o_{Cell} = +0.34\,V$

10.59

a. $E = E^o - \{RT/(1\,F)\}\ln\{a_{Cu^+}/a_{Cu2+}\}$

b. $E = E^o - \{RT/(2\,F)\}\ln\{1/a_{Hg2+}\}$

c. $E = E^\circ - \{RT/(1\ F)\}\ \ln\{1/a_{Tl^+}\}$

d. $E = E^\circ - \{RT/(2\ F)\}\ \ln\{a_{Sn^{2+}}/a_{Sn^{4+}}\}$

10.63

a. $E = 0.222 - 0.05916\ \log\{a_{Cl^-}\}$

b. $E = -0.83 - 0.02958\ \log\{(a_{H_2})(a_{OH^-})^2\}$

c. $E = +0.94 - 0.02958\ \log\{(a_{NO_2^-})(a_{OH^-})^2/(a_{NO_3^-})\}$

d. $E = +1.63 - 0.02958\ \log\{(a_{Cl_2})/[(a_{HOCl})^2(a_{H^+})^2]\}$

10.66 $E_{Cell} = +1.09$ V

10.68 $3\ Ag^+ + Bi_{(s)} \rightleftarrows Bi^{3+} + 3\ Ag_{(s)}$ $E_{Net} = +0.41$ V

10.69

a. $Ni^{2+} + 2\ e^- \rightleftarrows Ni$ and $Co^{2+} + 2\ e^- \rightleftarrows Co$

b. E° for the nickel half-reaction is –0.23 V (vs SHE) and E° for the cobalt half-reaction is –0.28 V (vs SHE); because the nickel E° value is more positive, we might expect based on only this information that nickel would be the cathode and cobalt would be the anode.

c. $E_{Cell} = -0.01$ V

d. The negative sign on the answer to Part (c) tells us that we have incorrectly chosen the nickel electrode as the cathode. Actually, the cell potential is +0.01 V, with cobalt being the cathode and nickel being the anode.

10.75

a. $E_{Cell} = +0.62$ V

b. Error = 0.03 V

10.76 The reactions that are directly affected by a change in pH are (a), (c) and (d).

10.77 If the pH increases by one unit, the activity of OH^- will increase by ten-fold. This change will result in the log term of the equation changing by 0.011 V per pH unit.

10.79 At pH 2.00, $E_{Cell} = 1.15$ V; at pH 4.00, $E_{Cell} = 0.97$ V

10.84

a. Fe^{2+}

b. $Fe(OH)_2$

c. Fe^{3+}

d. FeO_4^{2-}

10.86

a. $E_{Cell} = +0.90$ V; the cathode is the platinum in the cerium solution and the anode is the platinum in the iron solution.

b. $E_{Cell} = 0.69$ V

c. $E_{Cell} = 0.71$ V

Chapter 11

11.10

a. Dry ashing would melt the chocolate and burn it to CO_2 to leave behind the iron and other metal oxides.

b. Sodium carbonate fusion will dissolve the silicate to form water-soluble sodium silicate.

c. Wet ashing will keep the dissolved mercury in solution rather than the mercury being volatilized away in the other high temperature methods.

11.14

a. Medium or course porosity

b. Fine or medium porosity

c. Fine or medium porosity

11.20 $2\ Mg(NH_4)PO_4 \cdot 6\ H_2O \rightarrow Mg_2(P_2O_7) + 13\ H_2O + 2\ NH_3$

11.30 The final solid being weighed must have a mass greater than 0.0500 g, and the original sample must have a mass of at least 1.40 g.

11.32

a. Use the hydrolysis of urea ($H_2N–CO–NH_2$) to gradually increase the pH of a solution that contains some hydrogen phosphate (or the more acidic species, phosphoric acid, etc.). As the pH increases, more phosphate species will be present as PO_4^{3-} for use in precipitation.

$H_2N–CO–NH_2 + H_2O \rightarrow CO_2 + 2\ NH_3$

$NH_3 + H_2O \rightleftarrows NH_4^+ + OH^-$

$HPO_4^{2-} \rightleftarrows PO_4^{3-} + H^-$

b. Hydrolysis of sulfamic acid:

$NH_2SO_3H + H_2O \rightarrow NH_4^+ + H^+ + SO_4^{2-}$

c. Hydrolysis of dimethyloxalate:

$(COOCH_3)_2 + H_2O \rightarrow 2\ CH_3OH + (COOH)_2$

11.34 0.0040 mol biacetyl and 0.0080 mol hydroxylamine are needed

11.40

a. Gravimetric factor = 0.5884

b. Gravimetric factor = 0.5224

c. Gravimetric factor = 0.7877

d. Gravimetric factor = 0.8534

11.43 Purity of original silver metal = 91.18%

11.45 Mass of NaCl = 0.3368 g; mass of NaBr = 0.0286 g

11.48 $Fe^{3+} + 3\ OH^- \rightarrow Fe(OH)_3$

$2\ Fe(OH)_3 \rightarrow Fe_2O_3 + 3\ H_2O$

11.49 mass Fe_2O_3 = 20.5 g

11.53 $[Ni^{2+}]$ = 0.0228 M

11.54

a. Gravimetry factor = 0.05873

b. mass of Al = 0.009708 g

11.58

a. mass of C = 0.03717 g (or 0.00309 mol); mass of H = 0.00554 g (or 0.00550 mol); mass of O = 0.0110 g (or 0.00069 mol)

b. Empirical formula, C_5H_8O

11.65 0.0574 g ammonium carbonate; 0.0298 g sodium carbonate; 0.0093 g NaCl

Chapter 12

12.5 Titration error = –0.73 mL

12.11 Mass of aspirin = 1.0431 g

12.12 Ammonia content = 1.64% (w/w)

12.14 Molar mass = 176.0 g/mol

12.17 Percent absorbic acid = 70.06% (w/w)

12.24 Nitrogen content = 1.44% (w/w); Protein content = 8.25% (w/w)

12.30 For the smallest relative error, one should use about 40 mL of a 50 mL buret for the analysis. Concentrated HCl is about 12 M, so the diluted HCl is about 12(10/250) = 0.48 M. The mass of TRIS needed for the analysis is 2.3 g.

12.40 Titration error = –0.57 mL

12.41 Thymol blue changes from yellow to blue over a pH range of 8.0–9.6. This is the same range of pH for phenolphthalein so this student's error should be no worse than for a student using phenolophthalein.

12.46
  a. Equivalence point = 32.13 mL
  b. pK_a = 3.44

12.49 [HA] = 0.1326 M; K_a = 8.0 × 10^{-6} or pK_a = 5.10

12.50
  a. $K = [NH_4^+][F^-]/[NH_3][HF] = (K_{b,NH_3} K_{a,HF})/K_w$
  b. $K = 1.2 × 10^6$, which is much smaller than K for HF titrated with NaOH ($K_{a,HF}/K_w = 6.8 × 10^{10}$)

12.52
  a. $[HNO_3]$ = 15.82 M
  b. At beginning of titration, pH = 0.10; at equivalence point, pH = 7.00
  c. At 10.00 mL, pH = 0.38; at 25.00 mL, pH = 0.89; at 40.00 mL, pH = 12.34

12.55
  a. At 0% titration, pH = 12.90; at 100% titration, pH = 7.00
  b. At 50% titration, pH = 12.35

12.58 Original sample, pH = 2.73; at 50% titration, pH = 3.17; at 100% titration, pH = 7.32; at 30.00 mL titrant, pH = 11.33

12.60 K_a = 5.47 × 10^{-7}

12.62 At 0 mL titrant, pH = 12.45; at 10 mL titrant, pH = 10.42; at 20 mL titrant, pH = 2.15; at equivalence point, pH = 6.10

12.66
  a. At 0% titration, pH = 2.71; at 50% titration, pH = 4.43; at 100% titration, pH = 4.92; at 150% titration, pH = 5.41; at 200% titration, pH = 9.21
  b. At 35% titration, pH = 4.70; at 70% titration, pH = 4.80; at 135% titration (35% between first and second equivalence points), pH = 5.14; at 170% titration (70% between first and second equivalence points), pH = 5.78

12.68
  a. Concentration of base = 0.0862 M
  b. K_b = 2.1 × 10^{-4}, pK_b = 3.67

12.72 $[H_2SO_4]$ = 0.03096 M, [Oxalic acid] = 0.02179 M

12.74 $[H_2SO_3]$ = 0.02272 M, $[H_2SO_4]$ = 0.06442 M

12.78
  a. At midpoint of phenolphthalein color change (pH = 8.8), titration error = –0.0026 mL (–0.013%); at start of range (pH = 8.0), titration error = –0.00025 mL (–0.001%)
  b. At midpoint of methyl red color change (pH 5.3), titration error = –0.00125 mL (–0.005%); at midpoint of thymolphthalein color change (pH 10.0), titration error = 0.025 mL (0.10%). Phenolpthalein and methyl red both will give endpoints with only small titration errors, but thymolphthalein will give a noticeably late end point.

Chapter 13

13.7
  a. $Ca^{2+} + EDTA^{4-} → CaEDTA^{2-}$
  b. $[Ca^{2+}]$ = 4.10 × $10^{-3}M$
  c. Water hardness = 411 ppm $CaCO_3$

13.9
  a. $Ag^+ + Cl^- → AgCl_{(s)}$
  b. $[Ag^+]$ = 0.00912 M

13.14
  a. pH 3–5.25
  b. pH 2.0–4.1, or pH 10.4–10.6 in presence of auxiliary agent
  c. pH 8.2–10.2 (pH 8.2–9.1 in presence of an auxiliary agent)

13.16 $[Ca^{2+}]$ = 0.01797 M, $[Mg^{2+}]$ = 0.00614 M

13.17 Mass $Na_2H_2EDTA \cdot 2H_2O$ = 14.8896 g

13.21 $[Ni^{2+}]$ = 0.01521 M, $[Mg^{2+}]$ = 0.01883 M

13.24 $[CN^-]$ = 0.1491 M

13.28
  a. pH 8.0–10.5   b. pH 9.5–10.6   c. pH 7.5–10.5

13.30 $\alpha_{Fe^{3+}}$ = 1.0 × 10^{-8}

13.32 $[Zn^{2+}]$ = 0.02294 M, $[Ni^{2+}]$ = 0.04656 M

13.36 The calcium complex of murexide at pH 8 is CaH_3I, which is red-orange. The nickel complex is yellow, so the color change at the endpoint is yellow to red-orange. The reaction occurring at the endpoint is as follows:
$HEDTA^{3-} + NiH_3I → NiEDTA^{2-} + CaH_3I + H^+$

13.40 $[Al^{3+}]$ = 3.38 × 10^{-2} M

13.43 $[NH_2OH]$ = 4.53 × 10^{-3} M

13.45 $[SO_4^{2-}]$ = 0.1894 M

13.47
  a. $Ca^{2+} + EDTA^{4-} → CaEDTA^{2-}$
  Volume titrant needed = 29.76 mL
  b. At 0 mL titrant, pCa = 1.90; at equivalence point, pCa = 6.17; at 10.00 mL titrant, pCa = 2.23; at 30.00 mL titrant, pCa = 8.48

13.55 $[Br^-]$ = 4.00 × 10^{-3} M

13.60 Silver content = 98.38% (w/w)

13.64 $[Cl^-]$ = 0.373 M

13.69
  a. $Ag^+ + Cl^- → AgCl_{(s)}$
  Volume titrant needed = 100.00 mL
  b. At 0% titration, pAg = 1.30; at 50% titration, pAg = 1.90; at 100% titration, pAg = 4.88; at 10.00 mL excess titrant, pAg= 6.95

Chapter 14

14.6
  a. Charge = 0.0840 C, which corresponds to 5.24 × 10^{17} electrons
  b. Mass of copper = 2.77 × 10^{-5} g

14.11 I = 3.5 × 10^{-14} A

14.29
  a. Class one electrode
  b. Metallic indicator electrode
  c. Class two electrode

14.30
  a. $[Fe^{2+}]/[Fe^{3+}]$ = 2.45
  b. $[Fe^{2+}]$ = 0.0542 M, $[Fe^{3+}]$ = 0.0221 M

14.47 $[Na^+]$ = 0.0625 M

14.48 $[Na^+]$ must be greater than 5 × 10^{-7} M

14.53 [Unknown 1] = $1.16 \times 10^{-3}\ M$, [Unknown 2] = $1.26 \times 10^{-5}\ M$, [Unknown 3] = $1.73 \times 10^{-1}\ M$

Chapter 15

15.10 COD = 4,810 mg O_2/L

15.11 $[Fe^{3+}]$ = 0.02310 M, $[Fe^{2+}]$ = 0.08816 M

15.15

  a. mass $K_2Cr_2O_7$ = 11.7676 g

  b. Volume H_2SO_4 = 106 mL

15.21 $[Fe^{2+}]$ = 0.01735 M, $[Fe^{3+}]$ = 0.00761 M

15.29 The $E^{o'}$ for this indicator is 0.85 V and the color change should occur in a range that is (0.05916 V)/2 below and above this value, so it should change over the range 0.82 to 0.88 V.

15.38

  a. $Fe^{2+} \rightleftarrows Fe^{3+} + e^-$
  $Ce^{4+} + e^- \rightleftarrows Ce^{3+}$

Titration reaction: $Fe^{2+} + Ce^{4+} \rightleftarrows Fe^{3+} + Ce^{4+}$

Volume of titrant = 40.00 mL

  b. At 0% titration, E_{Cell} = +0.27 V; after 10.00 ml of cerium is added, E_{Cell} = +0.43 V

  c. At equivalence point, E_{Cell} = +0.84 V; after 20.00 mL of excess titrant, E_{Cell} = +1.18 V

15.43 The titration error in this case is only +0.062 μL, or 3.1 ppm for the titration of Fe^{2+} with 0.0050 N Ce^{4+}

15.49 Volume = 0.793 mL

15.51 Content of uranium =26.05% (w/w)

15.57

  a. $Fe^{2+} \rightleftarrows Fe^{3+} + e^-$
  $MnO_4^- + 8\ H^+ + 5\ e^- \rightleftarrows Mn^{2+} + 4\ H_2O$

Titration reaction:
  $MnO_4^- + 5\ Fe^{2+} + 8\ H^+ \rightleftarrows Mn^{2+} + 5\ Fe^{3+} + 4\ H_2O$

Volume of titrant = 15.00 mL

  b. At 50% titration, E_{Cell} = +0.46 V

  c. At equivalence point, E_{Cell} = +1.17 V

  d. After a 10.00 mL excess of titrant, E_{Cell} = +1.29 V

15.66 Moles of dichromate reduced by organics = 2.13×10^{-4} mol

15.72 Volume NaS_2O_3 = 0.02224 L or 22.24 mL

15.75 Concentration of titrant = 0.00494 g/mL, water content = 0.775% (w/w)

15.77 Iodine number = 16.1

Chapter 16

16.5 $[Pb^{2+}]$ = $4.88 \times 10^{-3}\ M$, $[Cu^{2+}]$ = $3.66 \times 10^{-2}\ M$

16.7 Copper content = 22.08% (w/w)

16.14 No, Ag^+ requires one electron per ion to be reduced to silver metal, while Ni^{2+} requires two electrons per ion to be reduced to nickel metal. Therefore, the original concentration of Ag^+ must have been double the original concentration of Ni^{2+} for this situation to occur.

16.17 Vitamin C content = 4.20% (w/w)

16.27 Only the relative levels of the three soluble oxidation states can be determined in this case (because no data for standards is provided), but these results do indicate that CrO_4^{2-}, Cr^{3+} and Cr^{2+} were all initially present at identical concentrations.

16.30 $[Ag^+]/[Cu^{2+}]$ = 1.14

16.30 LOD for $[Cd^{2+}]$ = $8.6 \times 10^{-5}\ M$

16.36 The top 15 feet of the lake are in equilibrium with atmospheric oxygen. This no longer occurs below the "thermocline" mark at 15 feet. Water below that level is more highly reduced and has less oxygen. This effect becomes more pronounced as the depth increases from 15 to 75 feet below the surface.

16.38

  a. Mass of cadmium = 2.3×10^{-9} g

  b. Charge = 8.0×10^{-6} C

  c. Fraction of original cadmium that was reduced and later reoxidized = 0.0046

Chapter 17

17.4 510 nm

17.5 The strongest absorption occurs at 525 nm and the weakest absorption occurs at 700 nm.

17.14

  a. v = 2.249×10^8 m/s   b. v = 1.989×10^8 m/s

  c. n = 1.0005   d. n = 1.3588

17.15 t = 8.3×10^{-10} s

17.17

  a. 8.56×10^{13} s^{-1}, infrared light

  b. 1.57×10^6 m^{-1}, visible light

  c. 0.14 nm, X-ray

  d. 1.83×10^1 cm^{-1}, infrared light (far IR)

17.19 A wavelength of 2.5 μm is in the infrared range and 400 nm is in the visible range. The number of photons at 2.5 μm that would have the equivalent energy of one photon at 400 nm would be 6.25, but you can't have just a fraction of a photon, so at least 7 such photons would be need to supply at least as much energy as one 400 nm photon.

17.20 Difference in frequency = 1.197×10^{12} s^{-1}; difference in energy = 7.93×10^{-22} J

17.26

  a. 475 nm is in the green-blue region of visible light

  b. 250 nm is in the ultraviolet region

  c. 675 nm is in the red-orange region of visible light

  d. 1000 nm is in the infrared region

17.32 For the first solution, 450–500 nm light is blue to green-blue, so if this light is absorbed the object will appear as the complementary color, which is red-orange. If the solution absorbs 250–300 nm light, this light is in the ultraviolet and the object will appear colorless.

17.39

  a. F = 2.25×10^{-8}; this fraction is far too small to be an important issue when using sunlight for remote sensing.

  b. Reflection of light by particulate matter in the atmosphere would be an important factor.

17.42 At 30°, θ_2 = 19.25°; at 45°, θ_2 = 27.79°; at 60°, θ_2 = 34.82°

17.53 Concentration of Unknown #1 = $3.94 \times 10^{-4}\ M$; concentration of Unknown #2 = $1.10 \times 10^{-3}\ M$

17.57

Transmittance	%Transmittance	Absorbance
0.156	15.6	0.807
0.358	35.8	0.446
0.561	56.1	0.251
0.689	68.9	0.162
0.780	78.0	0.108
0.056	5.6	1.250

17.59 $\epsilon = 124$ L/mol·cm

17.60 $C = 5.34 \times 10^{-5} M$

17.64 Unknown concentration $= 2.94 \times 10^{-4} M$

17.68

a. Approximately 0.3 to 0.4 for 2.5% and 0.1 to 1.3 for 3%

b. Approximately 0.3 to more than 2.0 for 2.5% and 0.20 to more than 2.0 for 3%

17.72

a. The absorbance at 465 nm is low and on the side of a peak, while 645 nm is a peak maximum; therefore, a plot made using data obtained at 645 nm will have a larger slope and better linearity than a plot made using data obtained at 465 nm.

b. The absorbance at 645 nm is low but 660 nm is a peak maximum; therefore, a plot made using data obtained at 660 nm will have a larger slope.

17.75 The measured absorbance will be 1.23 instead of 1.30, which is a relative error of 5.4%.

Chapter 18

18.8 In order from easiest to most difficult: d > c > b > a (not possible)

18.10 $C = 3.6 \times 10^{-9} M$

18.12 $C = 1.58 \times 10^{-2} M$

18.23

a. $[Fe^{2+}] = 1.73 \times 10^{-4} M$

b. If some other species absorbs at the given wavelength, the measured absorbance will be too high and will be interpreted inaccurately as a higher concentration of Fe^{2+}.

18.24 Volume = 541 L; it is assumed the dye is distributed uniformly throughout the pond and that no other solute in the pool absorbs significantly at 450 nm.

18.25 $C = 14.1 \times 10^{-3} M$

18.28 Concentration of caffeine in brewed coffee $= 3.37 \times 10^{-3} M$

18.34

a. $[Cu^{2+}] = 3.53 \times 10^{-2} M$

b. The analyte by itself has a very small absorbance but the copper-trien complex absorbs quite strongly. Past the endpoint when only more trien is added, there is no increase in absorbance; this result means trien has no measurable absorbance at the given wavelength.

18.38 $[Q] = 7.65 \times 10^{-4} M$, $[P] = 4.69 \times 10^{-4} M$

18.56 Double bonds absorb at about 1650 cm^{-1} (6.0 μm); alkanes do not absorb here.

18.57

a. Ketones, such as acetone, show strong absorption near 1700 cm^{-1}. There is no such absorption present in this case, so the solvent must be the hydrocarbon mixture.

b. The absorption at 2950 cm^{-1} is due to a C-H stretch and the absorption at 1450 cm^{-1} is from a C-C stretch, which are present in both vegetable and mineral oil. The absence of a peak near 1720 cm^{-1}, which is characteristic of an ester group, shows that the sample can not be vegetable oil. Thus, it must be mineral oil.

18.74 $C = 6.45 \times 10^{-5} M$

Chapter 19

19.14 mass of $CaCl_2 = 4.9 \times 10^{-18}$ g

19.17 $\Delta E = 3.37 \times 10^{-19}$ J, $v = 5.09 \times 10^{14}$ s^{-1} (or Hz)

19.21

a. 2267 K   b. 3342 K   c. 3080 K

19.25 Based on the Boltzmann equation, the fraction of sodium atoms in the 3p versus 3s orbitals at these temperatures will be 1.5×10^{-5} and 8.8×10^{-4}, respectively. As a result, the enhancement in signal intensity due to just this effect will be $(8.8 \times 10^{-4})/(1.5 \times 10^{-5})$, which is a 59-fold increase in intensity.

19.28 Concentration of $Cu^{2+} = 0.687$ ppm

19.29 Concentration of Na^+ in original solution = 10.8 ppm

19.30 $A = 0.209$

19.31 Zn^{2+} concentration = 1.62 ppm

19.35 Manganese concentration = 7.3 ppm

19.38 Number of zinc atoms $= 3.0 \times 10^{11}$ atoms/s or 3.0×10^{12} atoms in 10 s

19.51 Add La^{3+} to form lanthanum phosphate or EDTA to form CaEDTA, which both prevent the formation of calcium phosphate (*Note*: the EDTA will burn away in the flame)

19.54

a. Original calcium concentration in milk = 707 ppm

b. The EDTA was added as a protective agent to assure that calcium phosphate would not form in the flame. Milk is rich in phosphate, which is a serious concern in this measurement. If EDTA had not been added there is a good chance that a low result would have been obtained.

19.62 $A = 1.038$

Chapter 20

20.10 $D_c = K_{D,H2A}[H^+]^2/([H^+]^2 + K_{a1}[H^+] + K_{a1}K_{a2})$

20.18 89.2% extraction

20.21 $D_c = 43.2$

20.23 86.8% extraction

20.27

a. 98.0% extraction

b. 55.6% back extraction

c. 54.5% overall back extraction into water

20.32

a. Compound A, 84.2% extraction; compound B, 4.6% extraction

b. Compound A, 97.5% extraction ($n = 2$); 99.6% extraction ($n = 3$); 99.9% extraction ($n = 4$); compound B, 8.95% extraction ($n = 2$); 13.1% extraction ($n = 3$); 17.1% extraction ($n = 4$)

c. The recovery of compound A increases but its purity decreases as more extractions are conducted on the sample.

20.47 Void volume = 4.5 mL, total retention volume for 2-propanol = 30.1 mL

20.49 Atrazine concentration = 2.41 μg/L, which is below the allowable limit

20.54

a. TNT, t'_R = 3.00 min, V'_R = 4.50 mL; RDX, t'_R = 4.15 min, V'_R = 6.22 mL; Tetryl, t'_R = 5.36 min, V'_R = 8.04 mL; HMX, t'_R = 6.35 min, V'_R = 9.52 mL

b. TNT, k = 1.50; RDX, k = 2.08; Tetryl, k = 2.68; HMX, k = 3.18

c. The total and adjusted retention times would all increase by 1.5-fold but the retention factors would not be affected.

20.59 The retention factor will increase by 2.2-fold.

20.67

a. Peak tailing

b. Peak fronting

c. Gaussian peak

20.73 A linear velocity of 12 cm/s is below the optimum linear velocity in the van Deemter curve, 20 cm/s is near the optimum, and 50 cm/s or 70 cm/s are above this optimum value.

20.77 Peaks 1 and 2, R_s = 0.97; peaks 2 and 3, R_s = 3.65; peaks 3 and 4, R_s = 1.39; peaks 4 and 5, R_s = 2.63; peaks 5 and 6, R_s = 8.4

Chapter 21

21.5 Morphine concentration = 54.5 ng/mL

21.8 All of these compounds have molar masses below 600 g/mol and boiling points below 500°C at 1 atm, so it is possible (in theory) to analyze each by GC. However, some derivatization may be required for decanoic acid, and melene would be expected to strong retention on a GC column due to its large size and high boiling point.

21.9 The approximate order of elution, from first to last, would be as follows: toluene, naphthalene, decanoic acid, and melene. This assumes that decanoic acid has sufficient thermal stability for the GC analysis.

21.18 The approximate order of elution, from first to last, would be as follows:

a. methanol, ethanol, 1-propanol and 1-butanol

b. benzene, toluene and ethylbenzene

c. butyric acid and caproic acid

21.21 Kováts retention indexes: benzene, I = 639; butanol, I = 370; nitropropane, I = 246; pyridine, I = 478; 2-pentanone, I = 485

21.27 n-Pentane, w_b = 0.063 min; toluene, w_b = 0.21 min; 1,2,4-trimethylbenzene, w_b = 0.37 min

21.51 a. Parathion and b. imipramine

21.52 a. DDT and c. Freon-12

Chapter 22

22.5

a. Concentration of Ca^{2+} = 2.47 ppm

b. This peak is due to Ca^{2+}

22.8

a. In solvent A, k = 88.8; in solvent B, k = 29.2

b. In solvent A, t_R = 119 min; in solvent B, t_R = 40.2 min

22.12 Column 1, N = 7980, H = 7.5 μm; column 2, N = 5540, H = 18 μm. Column 1 is more efficient than column 2.

22.18 a) Weak mobile phase = pH 8.2, 20 mM Tris buffer (solvent A); strong mobile phase = pH 8.2, 20 mM Tris plus 0.25 M NaCl (solvent B)

22.25 A mixture of 2% isopropyl alcohol: 98% hexane or a mixture of 10% ethyl acetate: 90% dichloromethane are two possibilities.

22.33

a. P = 5.64   b. P = 0.1

c. P_{tot} = 1.35   d. P_{tot} = 7.06

22.38

a. The difference in the signs on the right reflects the difference in the polarities of the stationary phase in RPLC and NPLC.

b. k = 25.8

c. Approximately 33% isopropanol: 67% water

22.39 2,4-D is a weak acid that will have its neutral, acidic form as the principle species at a low pH. This species is what leads to the higher retention on a reversed phase column at an acidic pH versus a neutral pH. This same pH dependence will also lead to a decrease in retention for 2,4-D on an anion-exchange column as the pH is decreased.

22.43

a. Cation-exchange chromatography

b. Anion-exchange chromatography

c. Under these conditions, the amino acids will have a net negative charge; use anion-exchange chromatography

22.57 K_o = 0.730

22.58

a. V_M = 11.00 mL, V_e = 19.01 mL

b. K_o = 0.718

c. Molar mass = 69,000 g/mol

22.63

a. k = 4.0 × 10^4 b. t_R = 60,000 min or 1000 h

Chapter 23

23.4 v = 6.22 cm/min, distance in 2.5 min = 15.6 cm

23.8 v = 2.59 cm/min, μ = 6.47 cm^2/kV · min

23.11 The main species for dicamba throughout the given pH range is the conjugate base, A$^-$. The pK_a of dicamba is sufficiently low that the relative amount of this conjugate base versus dicamba is not significantly altered by the change in pH. However, the second pK_a for DCSA does occur in the range over which the pH is being altered. This means the relative amounts of the acid–base forms of DCSA are changing, which alters the apparent mobility that is observed for this compound.

23.15 μ_{eo} = 32.9 cm^2/kV · min

23.17

a. N = 59,200   b. H = 5.9   c. N = 6,400

23.18

a. R_s = 1.94   b. R_s = 2.00

23.26 Protein amount = 9.1 ng

23.35 Molecular weight = 52 kDa

23.41 Amount of nitrate = 4.3 mg/L

23.44 N = 434,500; migration time = 41.8 min

23.53 The relative amount of the first versus second isoform is 3.35.

GLOSSARY

A/B Ratio. A parameter used to see whether a chromatographic peak is symmetrical; also known as the *asymmetry factor*.

Absolute Error (e). The difference between an experimental result (x) and the result's true value (τ), where $e = x - \tau$; used as a measure of the accuracy of an experimental result.

Absorbance (A). A value used to describe the absorption of light by matter and that is equal to the negative value of the base-10 logarithm of the transmittance (T), where $A = -\log(T)$; also called the *optical density*.

Absorbance Detector. A device that is used to measure the ability of a substance to absorb light at a particular wavelength.

Absorption. The transfer of energy from an electromagnetic field (as possessed by light) to a chemical entity (e.g., an atom or molecule).

Absorption Spectrum. A plot of the intensity of light that is absorbed (or transmitted) by a sample at various wavelengths, frequencies, or energies.

Absorptivity (a). A term used for the proportionality constant in Beer's law when this constant is expressed in units other than L/mol · cm; also called the *extinction coefficient*.

Accelerated Solvent Extraction (ASE). A method that uses heating and elevated pressure to increase the speed and efficiency of an extraction; also known as *pressurized solvent extraction*.

Accepted SI Unit. An important or common measurement unit that is related to but not directly derived from fundamental SI units.

Accuracy. The degree of agreement between an experimental result and its true value.

Acid (Arrhenius model). A chemical that results in an increase in hydrogen ions in aqueous solution.

Acid (Brønsted–Lowry model). A chemical that donates a proton (or hydrogen ion) to another chemical (a base).

Acid Dissociation. The process by which an acid undergoes dissociation to form hydrogen ions.

Acid Dissociation Constant. An equilibrium constant used to describe the strength of an acid; this equilibrium constant is related to the ability of the acid to dissociate and transfer a hydrogen ion (or proton) to a given base, which is often water; also known as the *acidity constant* or K_a.

Acid–Base Indicator. A chemical or mixture of chemicals that changes color over a known range of pH.

Acid–Base Reaction. In the Brønsted–Lowry model, a process that involves the transfer of the hydrogen ion from an acid to a base.

Acid–Base Titration. A titration in which the reaction of an acid with a base is used for measuring an analyte.

Acidic Solution (in water). An aqueous solution with a pH less than 7.0.

Acidity Constant. See *Acid dissociation constant*

Activity (a). See *Chemical activity*

Activity Coefficient, Concentration-Based (γ_c). A term used to relate the activity (a) of a chemical to its concentration (c) in a solution; this is a unitless term that is used in the relationship $a = \gamma_c (c/c°)$ where $c°$ is the concentration of the substance of interest in a standard state. (Note: $c°$ is generally equal to one

molar when molarity is the concentration unit used for the chemical of interest, so this is not always written as part of the relationship between a, γ_c, and c.)

Acute Exposure. Exposure to a single concentrated dose of a chemical or other substance.

Acute Toxin. A chemical that causes a harmful effect after a single exposure.

Adjusted Retention Time (t'_R). A measure of solute retention in chromatography, as calculated from t_R through the relationship $t'_R = t_R - t_M$, where t_M is the column void time.

Adjusted Retention Volume (V'_R). A measure of solute retention in chromatography, as calculated from V_R through the relationship $V'_R = V_R - V_M$, where V_M is the column void volume.

Adsorption. A process by which a chemical interacts with a surface; a process by which impurities can stick to the surface of a crystal.

Adsorption Chromatography. A chromatographic technique that separates solutes based on their adsorption to the surface of a support; also known as *liquid–solid chromatography* or *LSC*.

Adsorption Indicator. A charged dye used as an indicator for a precipitation titration; the adsorption of this indicator to precipitate particles, which possess a slight charge opposite to that of the dye, is used to signal the end point.

Affinity. See *Association constant*

Affinity Capillary Electrophoresis (ACE). A type of capillary electrophoresis in which biologically related agents are placed as additives in the running buffer.

Affinity Chromatography (AC). A liquid chromatographic method that is based on biologically related interactions.

Affinity Ligand. The immobilized molecule that is used in affinity chromatography as the stationary phase.

Agarose. A carbohydrate gel that consists of a polymer made up of repeating units of D-galactose and 3,6-anhydro-L-galactose.

Agarose Electrophoresis. A type of electrophoresis in which agarose is used as a support material.

Aliquot. A portion of a stock solution or sample that is taken for the preparation of a second, less concentrated solution.

Allergen. A substance that may produce an allergic reaction.

Alpha error. See *Type 1 error*

Alternate Hypothesis (H_1). In statistical tests, an initial hypothesis in which the model and experimental results are thought to be different at a given confidence level.

Alternating Current. A current in which the direction of the movement of electrons reverses at a regular rate.

Alumina. A material with the empirical formula Al_2O_3; a polar support employed in adsorption chromatography.

Ames Test. A method that uses microorganisms to test for the ability of a chemical to produce mutations in DNA.

Ampere (A). The fundamental unit of electric current in the SI system; defined as the constant current that produces a force of 2×10^{-7} newton per meter of length when maintained in two straight parallel conductors of infinite length and negligible circular cross section that are placed one meter apart in a vacuum.

Amperometry. A method of electrochemical analysis in which the current passing through an electrochemical cell is measured at a fixed potential.

Amphiprotic. A substance that can act as either an acid or a base.

Ampholytes. A mixture of small zwitterions that is used in isoelectric focusing to produce a stable pH gradient.

Amplitude. The intensity of a wave of light (or electromagnetic radiation), as measured by the height of the crests of this wave.

Analog Display. A display of a signal that is shown as having a continuous range of values.

Analysis. The act of performing a chemical measurement or the actual method used to examine a sample or analyte within that sample; also referred to as an *assay* or *determination* of the analyte in a sample.

Analyte. The particular chemical or substance that is being measured or studied within a sample.

Analytical Balance. A balance with an enclosed weighing area, typically having a readability of 0.1 milligram or less.

Analytical Chemistry. The science of chemical measurements; the field of chemistry that deals with the use and development of tools and processes for examining and studying chemical substances.

Analytical Concentration. The total concentration of a chemical in a solution, regardless of the chemical's final form or number of species; also called *total concentration*.

Analytical Method. The specific approach that is used to perform a chemical analysis; also referred to as the *analytical technique*.

Analytical Technique. See *Analytical method*

Angle of Incidence (θ_i). The angle at which light strikes a boundary between two regions that have different values for their refractive index; this angle is measured versus a line known as the "normal" that is perpendicular to the boundary between the two regions.

Angle of Reflection (θ_r). The angle at which light is reflected off of a boundary between two regions that have different values for their refractive index; this angle is measured versus a line known as the "normal" that is perpendicular to the boundary between the two regions.

Anion-Exchange Chromatography. A type of ion-exchange chromatography that uses a positively-charged group to separate negative ions.

Anode. The electrode in an electrochemical cell at which oxidation occurs.

Anodic Stripping Voltammetry. A combination of coulometry and voltammetry in which the working electrode is first set at a potential that is suitable for reduction of the analyte, followed later by a scan in potential in a positive direction to reoxidize the reduced analyte and measure the resulting current.

Anolyte. An acidic solution located by the anode in capillary isoelectric focusing.

Antibody. A protein produced by the body's immune system that has the ability to specifically bind to a foreign agent, such as a bacterial cell, virus, or protein from another organism.

Application Buffer. The weak mobile phase in affinity chromatography, which allows strong binding between the analyte and affinity ligand; this weak mobile phase is usually a solvent that mimics the pH, ionic strength, and polarity of the affinity ligand in its natural environment.

Argentometric Titration. A titration method that uses Ag^+ as a titrant.

Arithmetic Mean (average, $\bar{x}$). The sum of a series of values divided by the total number of values in that set; used to provide a single representative number for a group of observations; also called the *mean*.

Ascarite. A material made up of sodium hydroxide adsorbed onto a clay-based material.

Ashing. Pretreatment of a sample by dry or wet methods that converts metals in the sample into metal ions in a solution.

Ashless Paper. A paper that when burned will leave little or no ash, as often used in a gravimetric analysis.

Asphyxiant. A chemical that interferes with the transport of oxygen in the body.

Aspirator. A device that is connected to a faucet and uses the flow of water from the faucet to create a difference in pressure, typically used to draw liquid through a funnel for the collection of a precipitate.

Assay. See *Analysis*

Association Constant (K_a). A term used with biological ligands in place of the term *formation constant*; also called the *affinity* of a biological ligand.

Asymmetry Factor. See *A/B ratio*

Atomic Absorption Spectroscopy (AAS). A type of atomic spectroscopy that examines light that is absorbed by atoms.

Atomic Emission Spectroscopy (AES). A type of atomic spectroscopy that examines light that is emitted by atoms.

Atomic Fluorescence Spectroscopy (AFS). A type of atomic spectroscopy that examines light that is emitted by atoms after these atoms have first been excited by absorbing a photon.

Atomic Force Microscope. An instrument that is used to perform atomic force microscopy.

Atomic Force Microscopy (AFM). A method in which a small cantilever with a sharp tip is passed over the surface of the sample, where the slight deflections of this tip are measured to create an image of the surface.

Atomic Mass (atomic weight). The number of grams that are contained in one mole of a particular type of atom. (Atomic mass = gram atoms/mol)

Atomic Spectroscopy. Measurement of the wavelength or intensity of light that is emitted or absorbed by free atoms.

Atomization. The process of converting a chemical compound into its constituent atoms.

Autoprotolysis. An acid–base reaction in which the same chemical acts as both the acid and base.

Autoprotolysis Constant. An equilibrium constant used to describe the autoprotolysis of a chemical; for water, the autoprotolysis constant is also known as K_w.

Autotitrator. A system that is designed for use in automating an acid–base titration or other type of titration; this system is capable of precisely delivering various amounts of titrant to a sample while an electrode or other device is used to measure a response (such as pH) for the sample/titrant mixture.

Auxiliary Electrode. An additional electrode that is used in some electrochemical cells to pass current and to provide a complementary half-reaction to that taking place at the working electrode, thus providing a complete electrical circuit without running the risk of changing the properties of the reference electrode over time; also known as a *counter electrode*.

Auxiliary Ligand. An additional complexing agent that is added to reduce side reactions during a complexometric titration.

Average. See *Arithmetic mean*

Back Extraction. A type of extraction in which a solute is allowed to distribute from its extracting phase back into a fresh portion of its original solvent.

Back Titration. A technique in which an excess of a known amount of a reagent is added to react with an analyte in a sample; the amount of reagent that remains after this process is then determined by using a titration, with the difference between the original amount of reagent and the amount that is titrated then being used to determine how much analyte was present in the sample.

Balance. A precision weighing instrument that is used to measure small masses.

Band-Broadening. A process that occurs as chemicals travel through a chromatographic system or other separation device, in which the width of the region that contains each chemical gradually becomes broader.

Base (Arrhenius model). A chemical that results in an increase in hydroxide ions in an aqueous solution.

Base (Brønsted–Lowry model). A chemical that can accept a proton (or hydrogen ion) from another chemical (an acid).

Base Ionization Constant. An equilibrium constant used to describe the strength of a base; this equilibrium constant is related to the ability of the base to accept a hydrogen ion (or proton) from a given acid, which is often water; also known as the *basicity constant*, *protonation constant*, or K_b.

Baseline Resolution. A situation in which there is no significant overlap between two peaks in a chromatographic or electrophoretic separation.

Basic Solution (in water). An aqueous solution with a pH greater than 7.0.

Basicity Constant. See *Base ionization constant*

Battery. A collection of electrochemical cells.

Beer–Lambert Law. See *Beer's Law*

Beer's Law. An equation used to relate the absorbance (A) of a homogeneous sample to the concentration (C) of a dilute absorbing analyte in this sample as given by $A = \epsilon\, b\, C$, where C is in units of M (or mol/L), b is the *path length* (in cm), and the term ϵ is the molar absorptivity (in units of L/mol $\cdot$ cm); also known as the *Beer–Lambert law*.

Beer's Law Plot. A plot of the measured absorbance versus the concentration of the absorbing analyte.

Bell-Shaped Curve. See *Normal distribution*

Best-Fit Line. A line that produces the best possible description and smallest overall deviations from the individual data points in a plot.

Beta Error. See *Type 2 error*

Biamperometry. A detection method used in the Karl Fisher titration that uses two electrodes, each of which has a controlled potential such that current will only flow when both iodine and iodide are present in solution.

Bidentate Ligand. A ligand that has two binding sites for a metal ion.

Biohazard. A biological substance that presents a health hazard.

Bioluminescence. Chemiluminescence produced by a chemical reaction of a biological agent.

Biospecific Elution. An elution technique used in affinity chromatography in which a competing agent is added to the mobile phase to displace the analyte from the affinity ligand.

Biotechnology Grade. Chemicals and solvents that have been purified and prepared for use in biotechnology; often used in molecular biology, electrophoresis assays, and in DNA/RNA or peptide sequencing and synthesis.

Blank Sample. A sample that contains no analyte.

Blotting. An approach for detection in gel electrophoresis in which a portion of an analyte band is transferred to a support such as nitrocelluose, where the analytes are reacted with a labeled agent.

Boltzmann Distribution. The relative distribution of chemical species in different energy states as a function of temperature as given by the following equation,

$$\frac{N_i}{N_0} = \frac{P_i}{P_0} e^{-\Delta E/(k\cdot T)}$$

where N_i and N_0 are the number of excited and ground-state chemical species, P_i and P_0 are integers that describe the number of possible ways these two energy levels can come about, ΔE is the energy difference between these two states per atom (or per molecule), k is the Boltzmann constant (with a value of 1.38×10^{-23} J/K), and T is the absolute temperature (in kelvin).

Bonded Phase. A stationary phase that is covalently attached to a support material.

Borosilicate Glass. A type of glass that contains a higher percentage of boron oxide and a lower percentage of sodium oxide than ordinary soda–lime glass; examples are Pyrex and Kimex.

BP Grade. A reagent chemical that meets or exceeds specifications set by the British Pharmacopeia (BP).

Bragg Equation. An equation used to predict the angles at which constructive interference will be observed for X-ray diffraction by the atoms in a crystal as given by $n\lambda = 2d\,\sin(\theta)$, where λ is the wavelength of the X rays passed through the crystal, d is the interplanar distance (or "lattice spacing") between atoms in the crystal, and n is the order of diffraction for the observed constructive interference band.

Brønsted–Lowry Model. A model for describing acids and bases that defines an acid as a chemical that donates a proton (or hydrogen ion) to another chemical, and a base as a chemical that accepts a proton from another chemical (an acid).

Buffer Capacity. The moles of strong acid or strong base that must be added per liter of buffer to produce a pH change of 1.0 units.

Buffer Index (β). The moles of a strong acid or strong base that are required to produce a given change in pH per unit volume.

Buffer Solution. A mixture of an acid and its conjugate base, which gives a solution that will tend to keep the same pH, even when small amounts of acid, base or water are added to it.

Buoyancy. A force that works in opposition to gravity when you are weighing an object; the size of this force will depend on both the density of the material that is being weighed and the density of the medium that surrounds it.

Buoyancy Correction. The process of adjusting for the effect of buoyancy of the measured mass of an object; this requires some knowledge of the density of the sample, the density of weights used to calibrate the balance, and the density of surrounding medium.

Buret. A volumetric device used to accurately measure and deliver variable amounts of a liquid; this device consists of a graduated glass tube with an opening at the top for the addition of a liquid and a stopcock at the bottom for the precise delivery of this liquid into another container.

Calibration. The use of standards and their response to determine the amount of an analyte in a sample; this task is often accomplished by making a plot of the signals given by a method for standards that contain known amounts of the analyte.

Calibration Curve. A plot of the signals given by a method for standards that contain known amounts of the analyte.

Calibration Sensitivity. The slope at a particular point in a calibration curve.

Calomel Electrode. A common reference electrode for potentiometry based on a mercury/mercury chloride electrode.

Candela (cd). The fundamental unit of luminous intensity in the SI system; defined as the luminous intensity measured in a given direction from a source that emits monochromatic radiation with a frequency of 540×10^{12} hertz and that has a radiant intensity of 1/638 watt per steradian in the observed direction.

Capacity (of a balance). The largest mass that can be reliably measured by a particular balance; also known as the *maximum load* of a balance.

Capacity Factor. See *Retention factor*

Capillary Array Electrophoresis (CAE). The utilization of multiple capillaries in a single capillary electrophoresis system.

Capillary Column. See *Open-tubular column*

Capillary Electrophoresis (CE). A type of electrophoresis that is performed using a narrow capillary that is filled with a running buffer, also known as *CZE*.

Capillary Gel Electrophoresis (CGE). The use of a porous gel in capillary electrophoresis.

Capillary Isoelectric Focusing (CIEF). Isoelectric focusing carried out in a capillary electrophoresis system.

Capillary Sieving Electrophoresis (CSE). A type of capillary electrophoresis in which an agent is included in the separation that can separate analytes based on their size.

Capillary Zone Electrophoresis (CZE). See *Capillary electrophoresis*

Carcinogen. A substance that causes cancer.

Carrier Gas. The mobile phase in gas chromatography.

Cathode. The electrode in an electrochemical cell at which reduction occurs.

Catholyte. The basic solution located by the cathode in capillary isoelectric focusing.

Cation-Exchange Chromatography. A type of ion-exchange chromatography that uses a negatively charged group to separate positive ions.

Cerate. A strong oxidizing agent with the formula Ce^{4+} that is often used in redox titrations.

Certified ACS Grade. A chemical that meets or exceeds specifications set by the American Chemical Society (ACS).

Certified Reference Material (CRM). A material that has documented values for its chemical content or physical properties.

Charge. The integral of electrical current over time.

Charge Balance Equation. An approach for solving chemical equations that makes use of the fact that the sum of all positive and negative charges in a closed system should be zero.

Charging Current. A current that is produced when the applied potential of an electrode is changed, as created by a charging of the electric double layer of the electrode solution interface.

Chelate. The type of complex that forms between a metal ion and a chelating agent; from the Greek word *chele* for the "claw" of lobsters and crabs.

Chelate Effect. The tendency of chelating agents to give more stable complexes with metal ions than monodentate ligands.

Chelating Agent. A ligand that has multiple interactions with a metal ion.

Chemical Activity. A measure of the difference in chemical potential between a substance in some given state (μ) versus a standard reference state for that same substance ($\mu°$), where $a = e^{(\mu - \mu°)/(RT)}$, R being the ideal gas law constant and T the absolute temperature; this is a unitless term also known as *activity* or *relative activity*.

Chemical Equilibrium. A situation that occurs when the forward and reverse rates of a reaction are equal.

Chemical Hazard. Any chemical that is a physical or a health hazard; also called *hazardous chemical*.

Chemical Hygiene Plan (CHP). A collection of standard operating procedures used to promote safety in a laboratory.

Chemical Identification. The use of chemical testing to identify an unknown substance in a sample.

Chemical Ionization (CI). A method of ionization in mass spectrometry that makes use of a gas-phase acid to protonate the analyte and form a molecular ion with general structure MH^+.

Chemical Kinetics. The field of chemistry that is concerned with the rates of chemical processes.

Chemical Oxygen Demand (COD). The use of a strong oxidizing agent such as dichromate to determine the equivalent amount of oxygen that would have been consumed by organic compounds in a water sample.

Chemical Potential (μ). A measure of the amount of energy per mole that is available through the reaction of a chemical when it is present in a particular state; generally expressed in units of J/mol, this is also known as the *partial molar Gibbs energy*.

Chemical Separation. A method that involves the complete or partial isolation of one chemical from another in a mixture of two or more substances.

Chemical Thermodynamics. The field of chemistry that is concerned with the changes in energy that take place during chemical reactions or phase transitions and the overall extent to which such processes can occur.

Chemiluminescence. The emission of light as a result of a chemical reaction.

Chiral Stationary Phase. A stationary phase for chromatography that is chiral and that is used in the separation of chiral chemicals.

CHN Analyzer. An instrument that is designed to determine the carbon, hydrogen, and nitrogen content of a sample.

Chromatogram. A graph of the response in chromatography that is measured by a detector at the end of the column as a function of the time or volume of the mobile phase that is needed for elution.

Chromatograph. An instrument that is used to perform chromatography.

Chromatography. A separation technique in which the components of a sample are separated based on how they distribute between two chemical or physical phases, one of which is stationary and other of which is allowed to travel through the separation system.

Chromophore. The portion of a molecule that has properties that allow the molecule to absorb light.

Chronic Exposure. Exposure to a repeated, long-term dose of a chemical or other substance.

Chronic Toxin. A chemical that causes a harmful effect after long-term exposure.

Class A Glassware. Glassware that has met specifications for "Class A" volume measurements, as given by the American Society for Testing and Materials (ASTM), where Class A glassware has an accuracy that is generally twice as good as it is for Class B glassware.

Class B Glassware. Glassware that has met specifications for "Class B" volume measurements, as given by the American Society for Testing and Materials (ASTM), where Class B glassware has an accuracy that is generally half of that for Class A glassware.

Class One Electrode. A metal electrode that is in contact with a solution that contains metal ions of that element; also known as an *electrode of the first kind*.

Class Three Electrode. A metal electrode that is in contact with a salt of its metal ion (or a complex of this metal ion) and a second, coupled reaction involving a similar salt (or complex) with a different metal ion; also known as an *electrode of the third kind*.

Class Two Electrode. A metal electrode that is in contact with a slightly soluble salt of that metal and that is in a solution containing the anion of this salt; also known as an *electrode of the second kind*.

Classical Liquid Chromatography. A type of liquid chromatography that makes use of a support that consists of relatively large and irregularly shaped particles.

Classical Method. An analytical technique that produces a result by using only experimentally determined quantities such as a measured mass or volume, along with known atomic or molecular weights and well-defined chemical reactions (e.g., a titration or gravimetric method).

Coagulation. A process by which particles stick together due to the electrostatic attractions of oppositely charged ions on neighboring particles.

Coefficient of Determination (r^2). A calculated factor that is used to determine the degree of agreement between a best-fit line and experimental data; this number is equal to the square of the correlation coefficient and gives a value that is always between zero and one, with one indicating a perfect fit and zero representing a random relationship between the best-fit line and data.

Coefficient of Variation (CV). See *Relative standard deviation*

Coextraction. The extraction of more than one solute under a given set of conditions.

Cold On-Column Injection. An injection technique for gas chromatography in which a microsyringe containing the sample is passed through the "injector" and directly into the column or an uncoated precolumn. The region around the syringe is initially kept cool so the sample can be deposited as a liquid film. This liquid is later heated, placing analytes into the carrier gas for separation.

Cold Trap. A way of collecting analytes from a gas that involves passing the gas through a hollow tube or coil that is kept at a low temperature. As the analytes enter the tube or coil, they condense and are collected. After these substances have been collected, they are then released by raising the temperature; also known as a *cryogenic trap*.

Colloid. See *Colloidal dispersion*

Colloidal Dispersion. A system containing a dispersion of smaller particles with sizes between 1 nm and 1 μm; also called *colloid*.

Colorimetry. An analytical method in which the analyte is combined with a reagent that will form a colored product; the color of this product is then compared to the color of standards, making it possible to determine the amount of analyte that is present in the sample or to simply see if the analyte is present above a certain level as part of a screening assay.

Column. A tube or enclosed container that holds the stationary phase for a chromatographic system.

Column Bleed. The loss of a stationary phase from a chromatographic column over time.

Column Chromatography. A type of chromatography in which a column contains the support and stationary phase.

Column Oven. An enclosed area in a gas chromatographic system that is used to hold the column and maintain this column at a well-defined temperature.

Combination Electrode. A pH electrode that contains both a reference electrode and an indicator electrode.

Combustible. A term describing a substance that is relatively easy to burn.

Combustion Analysis. A method in which the combustion of a sample is used to measure the relative amount of carbon, hydrogen, and other elements in a sample.

Common Ion Effect. An effect that occurs when an additional source is present for one or more ions in a dissolved salt; this effect often results in a lower solubility for the dissolved salt.

Comparative Weighing. A weighing procedure in which both the sample of interest and a reference weight are measured on the same balance, with their difference in weight and the known mass of the reference being used to determine the mass of the sample.

Competitive Binding Immunoassay. An immunoassay that involves the incubation of analyte in the sample with a fixed amount of a labeled analyte analog (containing an easily measured tag) and a limited amount of antibodies that binds to both the native analyte and labeled analog.

Complex. The product of a complex formation reaction.

Complex Formation. A reaction in which there is reversible binding between two or more distinct chemical species, such as EDTA and as Fe^{3+}.

Complexometric Titration. A titration that involves complex formation.

Compound Electrode. A device in which the modification of a pH electrode or other type of ion-selective electrode allows for the measurement of other analytes.

Compressed Gas. A gas that is held in an enclosed container at an elevated pressure.

Concentration. The amount of a substance within a given volume or mass of solution.

Concentration-Dependent Equilibrium Constant (K). An equilibrium constant that is written in terms of chemical concentrations.

Concentration Distribution Ratio. See *Distribution ratio*

Conditional Formation Constant (K'_f). An equilibrium constant that describes complex formation under a given set of reaction conditions; also called an *effective stability constant*.

Conditional Potential. See *Formal potential*

Conductance. The reciprocal of resistance.

Conductivity Detector. A device that can monitor ionic compounds by measuring the ability of a solution and its contents to conduct a current when placed in an electrical field.

Confidence Interval (C.I.). A range that follows an experimental number to express the degree of certainty that can be placed in that result; this is expressed using the following form,

$$\text{C.I.} = \bar{x} \pm ts_{\bar{x}} \text{ (or } s)$$

where $\bar{x}$ is the mean result that is being reported, t is the Student's t-value for the given number of measurements and desired confidence level, s is the standard deviation for the overall group of results, and $s_{\bar{x}}$ is the standard deviation for the mean.

Confidence Level. The degree of probability, or certainty, that is desired in stating that two values are either the same or that a measured result falls within a specified range of values.

Confidence Limit. The range of values that follows a mean in a confidence interval.

Conjugate Acid. An acid that is produced through the reaction of its parent base with another acid.

Conjugate Base. A base that is produced through the reaction of its parent acid with another base.

Constant Current Coulometry. A type of coulometry in which the current is maintained at a constant level during the analysis.

Constant Potential Coulometry. A type of coulometry in which the potential is maintained at a fixed value during the analysis.

Constructive Interference. A type of interference in which intersecting waves have crests and troughs that combine to give an overall observed amplitude that is increased.

Control Chart. A graph that uses the results obtained with a control material to follow the performance of an analytical method over time.

Control Material. A substance that is analyzed periodically by an analytical method to determine whether the procedure is working in a consistent manner.

Controlled Potential Electrolysis. A type of electrolysis in which the potential is controlled during the analysis.

Convective Mass Transfer. A term used to describe the combined effects of mobile-phase mass transfer and eddy diffusion in chromatography.

Coordinate Bond. The type of bond that forms when one chemical shares a pair of electrons with a metal ion; also known as a *coordinate covalent bond* or *dative bond*.

Coordination Covalent Bond. See *Coordinate bond*

Coordination Sphere. A metal ion and its surrounding ligands in a complex.

Coprecipitation. The presence of ions and other impurities in the structure of a growing crystal or precipitate.

Correlation Chart. A graph in which the results of a new method are plotted versus those obtained by a reference technique.

Correlation Coefficient (r). A calculated parameter that is used to judge the goodness of fit between a best-fit line and experimental data; the numerical value of r is always between -1 and 1, with a value of 1 or -1 (or $|r| = 1$) representing perfect agreement between the data and best-fit line, and a value of zero representing no correlation between the best-fit line and data.

Correlation Study. A technique for evaluating the accuracy of an analytical method by taking a group of several samples and measuring the amount of analyte in each by using both the method of interest and a second established technique.

Corrosive. A chemical that causes the destruction of living tissue at the site of contact.

Coulometric Titration. A type of titration in which the titrant is generated by means of coulometry and in the presence of the analyte.

Coulometry. A technique that uses a measure of charge for chemical analysis.

Counter Electrode. See *Auxiliary electrode*

Countercurrent Extraction. A special type of extraction that uses multiple portions of both the original sample solvent and extracting phase.

Craig Apparatus. A device used to perform a countercurrent extraction that makes use of a series of glass tubes that can hold a fixed portion of one phase while allowing the simultaneous transfer of each top phase to the next tube.

Craig Countercurrent Distribution. The result of a countercurrent extraction, generally as performed on a Craig apparatus.

Crest. The region of maximum intensity in a wave.

Critical Micelle Concentration. The threshold concentration of a surfactant above which the surfactant molecules come together to form micelles.

Critical Value. In a statistical test, the maximum or minimum cutoff value to which a test statistic is compared to see if the model and experimental value of interest can be said to be different at the selected confidence level and given degrees of freedom.

Cross-Linked Stationary Phase. A stationary phase that has been cross-linked on a chromatographic support.

Cyrogenic Trap. See *Cold trap*

Crystal. A solid that is relatively pure and has a high degree of order in its structure.

Crystal Growth. The process by which molecules or ions are added to existing nuclei of a precipitating chemical.

Crystallization. The use of precipitation to form crystals.

Cumulative Formation Constant. See *Overall formation constant*

Cupellation. See *Fire assay*

Current. The amount of electrical charge that flows through a conducting medium in a given amount of time.

100% Current Efficiency. A condition in which all the electrons that are passed through an electrochemical cell are used to oxidize or reduce the analyte.

Cyclic Voltammetry. A type of voltammetry in which the potential is scanned back and forth in a linear fashion over time.

Cyclic Voltammogram. A plot of the measured current versus applied potential as obtained by cyclic voltammetry.

Cyclodextrin. A cyclic polymer of glucose that is formed by certain types of bacteria; this results in a cone-shaped structure that has a nonpolar interior and polar upper and lower edges that are ringed by alcohol groups.

Daniell Cell. An electrochemical cell that makes use of the ability of Cu^{2+} ions in an aqueous solution to be reduced by zinc metal and form copper metal plus dissolved Zn^{2+} ions, resulting in a current.

Dative Bond. See *Coordinate Bond*

Day-to-Day Precision. The variation in results that is obtained over several days when using a particular analytical method.

DC Voltammetry. A type of voltammetry in which the potential is gradually increased from zero to a more negative value; also called *direct current voltammetry.*

Debye–Hückel Limiting Law (DHLL). A simplified version of the extended Debye–Hückel law for use at low ionic strengths; this equation is given by $\log(\gamma) = -0.51 z^2 I^{1/2}$, where z is the change on an ion, γ is the activity coefficient of the ion, and I is the ionic strength.

Degrees of Freedom (f). A statistical quantity that describes the number of values in a data set that are required to define the overall population of results; in calculations involving multiple variables, the degrees of freedom is equal to the total number of values minus the number of fitted parameters.

Deionization. See *Deionized water*

Deionized Water (DI water). Water prepared by deionization, where cations or anions in the water are exchanged for hydrogens ions (H^+) and hydroxide ions (OH^-), most of which combine to form water.

Demasking Agent. An agent that is used to release a metal ion from a masking agent so that the metal ion can be measured.

Densitometer. A scanning device used to detect analytes in gel electrophoresis and other analytical methods that use a planar support.

Density (ρ). The mass (m) per unit volume (V) of a substance, where $\rho = m/V$.

Dependent Variable (y). A measured or calculated quantity whose value is examined as a function of the independent variable in an experiment; an example is the set of the values that are plotted on the y-axis of a linear graph.

Derivatization. The process of altering the chemical structure of an analyte.

Derived SI Unit. A unit of measure that can be obtained by combining the fundamental units of the SI system.

Descartes' Law. See *Snell's law*

Desiccant. A material that absorbs water vapor from the air.

Desiccator. A container used to store samples in a dry environment.

Desolvation. The removal of solvent from a sample, such as by evaporating or burning away this solvent in a flame.

Destructive Interference. A type of interference in which intersecting waves have crests and troughs that cancel each other to give an overall observed amplitude that is decreased.

Determination. See *Analysis*

Deuterium Lamp. A light source often used in UV-vis spectroscopy that consists of two inert electrodes across which a high voltage is imposed in a quartz bulb that is filled at a low pressure with D_2.

Diatomaceous Earth. A material made up of fossilized diatoms and that mainly consists of silicon dioxide; used as a support material in packed GC columns.

Diatomite. See *Diatomaceous earth*

Dichromate. A strong oxidizing agent with the formula $Cr_2O_7^{2-}$ that is often used in redox titrations.

Dielectric Constant (ϵ) A measure of the degree to which a solvent or material will allow an electrostatic force from one charged body (such as an ion) to affect another; sometimes used as a rough indicator of the polarity for a chemical.

Diffraction. A process by which a wave, such as light, spreads around an object in its path.

Diffraction Pattern. A pattern caused by wave diffraction that results in regions with constructive or destructive interference; an image that is produced when a crystal of a chemical is exposed to X rays; in which this image depends on the spacing and distances of the atoms in the crystal.

Diffuse Reflection. A type of reflection that occurs when the boundary between two regions with different refractive indices is irregular instead of smooth, causing light to be reflected in many directions and to not retain its original image.

Diffusion. A process by which a solute moves away from a region of high concentration to one of lower concentration.

Diffusion Coefficient (D). A term used to describe the rate of a solute's diffusion; the value of D is a constant characteristic of the size and shape of a solute, as well as the temperature and type of phase in which the solute is present.

Digital Display. A display of a signal that is shown as having a discrete and fixed number of possible values.

Dilution. A solution prepared by adding more solvent to a reagent or sample; the process of preparing a more dilute solution.

Dimensional Analysis. A process in which the units on a set of numbers used in a calculation are recorded and compared to ensure the final result is expressed in the desired fashion.

Dimethylglyoxime (dmg). An organic binding agent that is often used for the precipitation and analysis of nickel ions.

Diode Array Detector. A spectrophotometer in which the monochromator has an entrance slit but no exit slit, allowing light of many wavelengths to enter the sample and to be detected simultaneously by an array of small diode detectors.

Dipole Moment. A measure of the polarity for a chemical, as given in units of debye (D).

Dipole–Dipole Interaction. An intermolecular force that takes place between two chemicals that have permanent dipole moments.

Direct Coulometry. A type of coulometry in which the analyte is what is being oxidized or reduced.

Direct Current. A current in which the direction of electron movement always proceeds in the same direction.

Direct Current Voltammetry. See *DC voltammetry*

Direct Injection. An injection method in gas chromatography that uses a calibrated microsyringe to apply the desired volume of liquid to the system.

Direct Titration. A titration in which the amount of analyte is determined by combining it directly with the titrant while the appearance of the end point for this titration is monitored.

Direct Weighing. A weighing procedure in which an object is placed on a balance pan and the mass is recorded directly from the balance display.

Dispersion Force. An intermolecular force that occurs when the movement of electrons in one molecule creates a temporary dipole moment, which induces another temporary but complementary dipole moment in a neighboring molecule; also called *London forces* or *van der Waals forces*.

Displacement Titration. A titration in which one metal ion (the analyte) displaces another metal ion from EDTA; this method is used in the case where the second metal ion has a suitable indicator available for its detection.

Dissociation. A process by which an ionic substance dissolves by forming ions; the breaking of chemical bonds in a solute.

Dissociation Constant (K_D). A term that is equal to the reciprocal of the association constant for the binding of a biological ligand with a target compound.

Dissolved Inorganic Gases. Contaminants in water that include dissolved gases like carbon dioxide and oxygen.

Dissolved Inorganic Solids. Contaminants in water that consist of various ions, minerals, and metals.

Dissolved Organics. Contaminants in water that consist of a variety of organic compounds from either natural sources or human-related activities.

Distillation. See *Distilled Water*

Distilled Water. Water that is prepared by distillation, in which the water is heated to boiling, with the steam that is given off then being recondensed and used as purified water.

Distribution Ratio (D_c). The ratio of the analytical concentrations for a solute in two phases as this solute distributes between these phases at a given pressure and temperature; also known as the *concentration distribution ratio*.

Dixon's Test. See *Q-test*

Doppler Effect. An effect that creates a distribution in the observed energies and wavelengths of light in the absorption or emission spectrum for atoms, due to some atoms moving toward the detector while others move away from the detector.

Double-Beam Instrument. A device in spectroscopy in which the original beam of light is split so that half the light goes through a reference solution while the other half passes through the sample, minimizing errors that are caused by drift in the intensity of the lamp or in the response of the detector.

Double-Layer Current. See *Charging current*

Dry Ashing. A method of sample preparation in which a weighed portion of a sample is heated to red heat in a porcelain dish that is open to the air.

Dynamic Range. The largest possible range that can be used by an analytical technique, extending from the lower limit of detection to the upper limit of detection.

Eddy Diffusion. A band-broadening process that occurs whenever there are support particles within a column, as produced by the presence of the large number of flow paths around support particles.

Effective Stability Constant. See *Conditional formation constant*

Effluent. The mixture of mobile phase and analytes that elutes from the end of a column or chromatographic system.

Electric Potential. The work required a per-unit charge to move a charged particle from one point to another, or from one chemical to another; also known as the *potential*.

Electrochemical Analysis. The use of electrochemistry for the analysis of chemicals.

Electrochemical Cell. A system or device in which the oxidation and reduction processes of an oxidation–reduction reaction occur in different locations, with electrons flowing from one location to the other through an external circuit.

Electrochemical Detector. A device that can be used to measure the ability of an analyte to undergo either oxidation or reduction; this device might measure the change in current that a reaction produces when present at a constant potential, or it might measure the change in the potential that this reaction creates when a constant current is applied to the system.

Electrochemical Reaction. See *Oxidation–reduction reaction*

Electrochemistry. The study of electrochemical reactions and their applications.

Electrode. A conducting material at which one of the half-reactions in an electrochemical cell is taking place.

Electrode of the First Kind. See *Class one electrode*

Electrode of the Second Kind. See *Class two electrode*

Electrode of the Third Kind. See *Class three electrode*

Electrodeposition. See *Electrogravimetry*

Electrogravimetry. A type of gravimetric analysis where a dissolved analyte is converted into a solid by either oxidation or reduction in such a way that the product is tightly attached to an inert electrode for later use in a mass measurement, providing a direct measure of the amount of analyte that was in the sample; also known as *electrodeposition*.

Electrokinetic Chromatography. A type of capillary electrophoresis in which a charged agent is placed into the running buffer to interact with analytes.

Electrokinetic Injection. An injection technique employed in capillary electrophoresis in which an electric field is applied across the capillary, allowing electroosmostic flow, and the electrophoretic mobility of the analytes to cause them to enter the capillary.

Electrolysis. The flow of current and the reaction that it creates in an electrolytic cell.

Electrolyte. The solution of ions that surrounds an electrode in an electrochemical cell.

Electrolytic Cell. An electrochemical cell in which an external power source is used to apply an electric current and cause a particular oxidation–reduction reaction to occur.

Electromagnetic Radiation. A wave of energy that propagates through space with both electrical- and magnetic-field components; also referred to as *light*.

Electromotive Force (emf). A term used to describe the potential of an electrochemical cell in which no appreciable current is flowing.

Electron-Capture Detector (ECD). A selective detector used in GC that is based on the capture of electrons by electronegative atoms or groups in an analyte.

Electron-Impact Ionization (EI). A method for the creation of ions in mass spectrometry by which analytes are passed through a beam of high-energy electrons; these electrons bombard some of the analyte molecules (M), causing electrons to be removed from the analyte to form both molecular ions ($M^{\cdot+}$) and fragment ions.

Electronic Balance. A balance that uses an electrical mechanism to determine the mass of an object.

Electronic Laboratory Notebook (ELN). A digital record of a laboratory experiment in which text can be combined directly with graphs, structures, images, and other computer-based sources of information.

Electroosmosis. Movement of the running buffer in electrophoresis, as caused by the presence of fixed charged in the system and the creation of an electrical double layer at the support's surface.

Electroosmotic Mobility (μ_{eo}). The observed mobility due to electroosmosis in electrophoresis.

Electropherogram. A plot of detector response versus migration time in electrophoresis.

Electrophoresis. A technique in which solutes are separated by their different rates of migration in an electric field.

Electrophoretic Mobility (μ). A constant used in electrophoresis to relate the velocity of migration (v) for a charged solute to the strength of the applied electric field (E), where $v = \mu E$.

Electrospray Ionization (ESI). An ionization method used in mass spectrometry, in which the sample is placed into a solvent and sprayed from a highly charged needle; the solvent in the charged droplets evaporates away quickly, giving smaller droplets with an excess of positive or negative charge that will eventually cause each droplet to divide and molecules in the droplets to be desorbed as ions that enter the gas phase.

Eluent. The mobile phase that is used to pass solutes through a column in chromatography.

Elution. The movement of solutes through a column in chromatography.

Elution Buffer. The strong mobile phase in affinity chromatography, which acts to readily remove the analyte from the affinity ligand.

Elutropic Strength (ϵ^{o}). A measure of the strength with which a mobile phase is adsorbed to a solid support.

Emission. The release of light by matter.

Emission Spectrum. A plot of the intensity of light that is emitted by matter at various wavelengths, frequencies, or energies.

Emulsion. A suspension of small droplets of one liquid in another immiscible liquid.

Endcapping. The treatment of silica with a small organosilane to react with and cover silanol groups.

End Point. The experimental estimate of the equivalence point in a titration.

Enzyme Electrode. A compound electrode that uses enzymes to convert analytes into products that can be measured by potentiometry; also known as an *enzyme substrate electrode*.

Enzyme Substrate Electrode. See *Enzyme electrode*

EP Grade. A reagent chemical that meets or exceeds specifications set by the European Pharmacopeia.

Equilibrium Constant. A ratio used to describe the relationship of the activities or concentrations of the products versus reactants for a reaction at equilibrium; usually represented by the symbol K.

Equivalence Point. The point in a titration curve at which exactly enough titrant has been added to react with all of the analyte.

Equivalent. A measure of chemical content that describes the amount of a chemical that is available for a specific type of reaction.

Eriochrome Black T. A metallochromic indicator used to signal the end point during the titration of Ca^{2+} by EDTA.

Error Assessment. The process of identifying all sources of errors that can occur in a method and in determining how to correct these errors.

Error Propagation. The way in which errors are carried through a calculation or a series of experimental steps.

Ethylenediamine. A common chelating agent, which has the formula $H_2NCH_2CH_2NH_2$.

Ethylenediaminetetraacetic Acid (EDTA). A common chelating agent for metal ions.

Etiological Agent. A microorganism or related toxin that can cause human disease.

Evaporative Light-Scattering Detector (ELSD). A detector used in liquid chromatography to monitor solutes that are less volatile than the mobile phase; this detector converts the mobile phase leaving the column into a spray of small droplets that are allowed to evaporate and leave behind small solid particles that contain nonvolatile sample components; these particles are then passed through a beam of light, where they scatter some of the incoming light and are measured.

Excitation. An increase in the energy of a chemical species such as an atom or molecule, placing this species into an excited state.

Experimental Mean ($\bar{x}$). The arithmetic mean, or average, that is calculated for a group of experimental values; this approaches the true mean (μ) as the number of values in the data set becomes larger.

Explosive. A substance that can cause a sudden, violent chemical reaction with the release of gas and heat.

Extended Debye–Hückel Equation. An equation that relates the activity coefficient for an ion to the ionic strength of its solution, the charge on the ion, and three adjustable parameters (a, A, and B).

Extinction Coefficient. See *Absorptivity*

Extra-Column Band-Broadening. Broadening of a peak before or after it enters a chromatographic column.

Extractant. The phase that is combined with a sample and used to extract its solutes.

Extracting Agent. A substance that is placed in an extraction system to bind with analytes and improve their ability to go into the extracting phase.

Extraction. A separation technique that makes use of differences in the distributions of solutes between two mutually insoluble phases.

Extreme value. See *Outlier*

Fajans Method. A precipitation titration for the determination of chloride by its reaction with silver ions and that makes use of a negatively-charged dye such as fluorescein and related derivatives as an adsorption indicator.

Faradaic Current. A current that is created by the oxidation or reduction of the analyte or some other electroactive species.

Faraday Constant (F). A constant that gives the charge that is present in one mole of electrons (or elementary charge), equal to approximately 9.6485×10^4 C/mol.

FCC Grade. A reagent chemical that meets or exceeds specifications set by the Food Chemicals Codex (FCC).

Ferroin. A common redox indicator consisting of an iron ion that is complexed with three molecules of 1,10-phenanthroline; this indicator has a change in color as it undergoes a one-electron reduction process.

Figures of Merit. The properties used to characterize an analytical method.

Filter. A porous structure that forms a barrier to solids, but that allows liquid to pass through.

Filtrate. The liquid that passes through a filter during the process of filtration.

Filtration. A process by which a filter is used to physically separate a solid material from a liquid.

Fire Assay. A technique used since ancient times for determining the purity of gold and silver; also known as *cupellation*.

Fixed-Wavelength Absorbance Detector. An absorbance detector that always monitors a specific wavelength.

Flame Emission Spectroscopy (FES). A type of atomic emission spectroscopy that uses a flame as the atomization and excitation source.

Flame Ionization Detector (FID). A general gas chromatography detector that detects organic compounds by measuring their ability to produce ions when these compounds are burned in a flame.

Flammable. A term describing a material that is easy to ignite.

Flow-Injection Analysis. A method in which samples are injected sequentially into a flowing stream of a reagent, which then reacts with the contents of these samples to give a change in signal, such as through the formation of a colored product.

Fluorescence. The emission of light by a sample after this sample has become electronically excited by the absorption of a photon, with the light emission being due to a spin-allowed transition such as a singlet–single transition.

Fluorescence Detector. A device that measures the ability of chemicals to absorb and emit light at a particular set of wavelengths.

Fluorescence Quantum Yield (ϕ_F). The efficiency of fluorescence by a chemical, as given by the ratio of the number of fluoresced photons divided by the number of absorbed photons.

Fluorescence Spectroscopy. A method that uses fluorescence to characterize or measure chemicals.

Fluoride Ion-Selective Electrode. A solid-state ion-selective electrode used for the measurement of fluoride.

Fluorometer. An instrument that is used to perform fluorescence spectroscopy and makes use of simple filters for wavelength selection.

Flux. A fusing agent that is used in the method of fusion to pretreat and dissolve samples of rocks or metals for the analysis of metal ions.

Formal Potential ($E°'$). The expected potential for a given redox couple when the activities of the species undergoing oxidation or reduction are exactly 1.0, and when using a specific type of solution or electrolyte; also called a *conditional potential*.

Formal Solution. See *Formality*

Formality (F). A concentration unit equal to the moles of an ionic compound that are present per liter of solution; for instance, a solution that contains 1.0 mole of such a solute in a final solution volume of 1.0 L is referred to as a 1.0 F, or 1.0 *formal*, solution.

Formation Constant (K_f). The equilibrium constant for a given step during a complex formation process; also known as a *stability constant*.

Formula Weight (FW). The number of grams that are contained in one mole of an ionic substance (FW = grams ionic substance/mol); see *Molar mass*.

Fourier Transform Infrared Spectroscopy. A type of infrared spectroscopy in which an interferometer is used to cause positive and negative interference to occur at sequential wavelengths as a moving mirror changes the path length of an incoming light beam, and Fourier transform is used to convert this information into a spectrum.

Fraction of Species Equation. An equation that shows how the fraction of a chemical species will change as a given parameter is varied for a chemical system.

Fraction of Species Plot. A graph that shows how the fraction of a chemical species will change as a given parameter is varied for a chemical system.

Fraction of Titration. The ratio of the moles of titrant that have been added at any given point in the titration versus the moles of analyte that were originally present in the sample.

Frequency (ν). The number of waves that pass by a given point in space in a specified amount of time; the number of cycles of an event that occur per unit time.

Fresnel Equation. A relationship that gives the fraction of light that will be reflected as it enters the boundary at a right angle (an expanded form of this equation is used for work at other angles).

Fritted Glass Crucible. See *Sintered glass crucible*

F-Test. A statistical test for comparing the standard deviations or variances for experimental results; this is determined by using a test statistic known as the F-value, which is calculated by using $F = (s_2)^2/(s_1)^2$ where s_2 and s_1 are the two standard deviations being compared, with the larger of these two terms always appearing in the top of this ratio.

Full-Scan Mode. The use of a mass spectrometer, such as in gas chromatography/mass spectography, as a general detector by collecting information on a wide range of ions.

Fundamental (base) SI Unit. A basic unit of measure in the SI system with which other units of measurement can be described; the current SI base units include the meter, kilogram, second, mole, kelvin, ampere, and candela.

Fused-Silica Open-Tubular (FSOT) Columns. An open-tubular column in which the outer tube is made of fused silica.

Fusion. A sample pretreatment method that involves melting a flux, which is used to dissolve samples of rocks or metals for the analysis of metal ions.

Galvanic Cell. An electrochemical cell in which an oxidation–reduction reaction occurs spontaneously, resulting in a flow of electrons; also called a *voltaic cell*.

Gas Chromatogram. A plot of the detector response versus the time that has elapsed since sample injection onto a gas chromatographic system.

Gas Chromatograph. A system that is used to perform gas chromatography.

Gas Chromatography (GC). A type of chromatography in which the mobile phase is a gas.

Gas Chromatography/Mass Spectrometry (GC/MS). The combined use of gas chromatography with a mass spectrometer.

Gas–Liquid Chromatography. A gas chromatography method in which a chemical coating or layer is placed onto the support and used as the stationary phase.

Gas-Sensing Electrode. A compound electrode that has been modified for the analysis of a gas.

Gas–Solid Adsorption. Use of a solid material to adsorb and separate gases in a gas mixture.

Gas–Solid Chromatography. This technique involves the use of the same material as both the support and stationary phase, with retention being based on the adsorption of analytes to the support's surface.

Gaussian Curve. See *Normal distribution*

Gel Electrophoresis. An electrophoretic method that is performed by applying a sample to a gel support that is then placed into an electric field.

Gel Filtration Chromatography. A type of size-exclusion chromatography that uses an aqueous mobile phase.

Gel Permeation Chromatography. A type of size-exclusion chromatography that uses an organic mobile phase.

General Conference on Weights and Measures (CGPM). An international conference that is held every few years to update the SI system; also known in French as the *Conférence Générale des Poids et Mesures*.

General Elution Problem. A problem that arises in chromatography and other separation methods when working with a complex sample, where it is often difficult to find a single set of conditions that can separate all sample components with adequate resolution and in a reasonable amount of time.

General Method. A procedure that can detect a wide range of compounds; also called a *universal method*.

Glass-Membrane Electrode. An indicator electrode that uses a thin glass membrane for selectively detecting the desired ion.

Globar. A light source used in infrared spectroscopy that consists of a heated, inert rod that is made of silicon carbide.

Gooch Crucible. A crucible that contains a porous bottom, but is used with a glass-fabric disk as a filter.

Good Buffers. A group of zwitterionic buffers that are named after Norman E. Good, who first proposed the use of such agents in biological research.

Good Laboratory Practices (GLPs). A set of guidelines that promote proper work and conduct within the laboratory.

Gradient Elution. An elution method in chromatography in which the separation conditions are changed during the analysis of a sample; also known as *gradient programming*.

Gradient Programming. See *Gradient elution*

Gran Plot. A type of graph that can be used to locate and provide an accurate estimate of the equivalence point of a titration by plotting a special function of pH versus the volume of titrant to give a linear response for the titration curve.

Graphite Furnace. A device for sample atomization in atomic absorption spectroscopy that uses a hollow, cylindrical piece of graphite that is heated electrically by passing a current through the cylinder.

Gravimetric Analysis. An analytical method that uses only measurements of mass and reaction stoichiometry to determine the amount of analyte in a sample; also known as *gravimetry*.

Gravimetric Factor. A conversion factor often employed in calculations for a gravimetric analysis, in which the conversion factor is used to multiply the measured mass of a precipitate to obtain the mass of desired analyte.

Gravimetric Titration. A titration where the mass of titrant is measured; also known as a *weight titration*.

Gravimetry. See *Gravimetric analysis*

Gravitational Acceleration Constant (g). A measure of the pull of gravity on an object at a particular location; on the earth the value for g varies with both altitude and latitude.

Guard Digit. An extra nonsignificant figure that is carried with a number through a calculation until the final result is obtained, used as a means to protect against rounding errors.

Half-Reaction. A chemical reaction that is written to show electrons among either the products or reactants.

Half-Wave Potential ($E_{1/2}$). The potential half-way up the wave in a plot of current versus applied potential, as obtained by direct current voltammetry or a related technique.

Hazardous Chemical. See *Chemical hazard*

Headspace Analysis. A technique based on the fact that volatile analytes in a liquid or solid sample will also be present in the vapor phase that is located above the sample; if a portion of this headspace is collected, it can be used to measure volatile analytes without interference from other, less volatile compounds that were in the original sample.

Height Equivalent of a Theoretical Plate (H). A measure of efficiency in chromatography and other separation methods, as determined by taking the ratio of the number of theoretical plates (N) and the length of the separation system (L), where $H = L/N$; also known as the *plate height*.

Heisenberg Uncertainty Principle. A principle that states it is not possible to know with perfect knowledge both the lifetime of an atom in an excited state (Δt) and the uncertainty in the energy that is associated with this excited state (ΔE).

Hematopoietic Toxin. A chemical that damages the formation or development of blood (the *hematopoietic system*).

Henderson–Hasselbalch Equation. An equation named after Lawrence J. Henderson and Karl A. Hasselbalch that describes how a change in the ratio of the amounts of a conjugate base and acid will affect the pH for a solution that contains these chemicals and that is at equilibrium.

Henry's Law. A relationship used to describe the solubility of a gas in a liquid; this relationship can be expressed as $C_{solute} = K_H P_{solute}$, where C_{solute} is the saturated concentration of the gas in a particular liquid, P_{solute} is the partial pressure of the gas in equilibrium with the solution, and K_H is a proportionality factor known as *Henry's law constant*.

Henry's Law Constant. See *Henry's law*

Hepatotoxin. A chemical that causes damage to the liver (the *hepatic system*).

Heterogeneous Material. A material with a composition that varies from one point to the next within its structure.

High-Performance Liquid Chromatography (HPLC). The modern instrumental form of liquid chromatography that makes use of small and efficient supports, which often require the use of high pressures.

Hold-Up Time. See *Void time*

Hold-Up Volume. See *Void volume*

Hollow Cathode Lamp. The light source commonly used in atomic absorption spectroscopy, which contains a small, hollow cylindrical piece of the metal/element that the lamp will be used to help analyze in a flame; this cylinder acts as a cathode and is bombarded with positive ions such as Ar^+ or Ne^+ that cause some atoms to be dislodged from the cathode, to be excited, and to later emit light.

Homologs. Compounds that have the same general structure, but which differ in the length of a single carbon chain.

HPLC Grade. Chemicals that have been purified and prepared for use in high-performance liquid chromatography (HPLC).

Hydrated Radius. The size of an ion plus the shell of water that surrounds it when the ion is in an aqueous solution.

Hydration Layer. The shell of water that surrounds an ion when the ion is in an aqueous solution.

Hydrodynamic Injection. An injection technique employed in capillary electrophoresis that uses a difference in pressure to deliver a sample to the capillary.

Hydrogen Bonding. An intermolecular interaction in which a hydrogen atom is shared in a noncovalent bond between molecules that contain atoms such as nitrogen or oxygen.

Hydrogen Lamp. A light source often used in ultraviolet/visible spectroscopy that consists of two inert electrodes across which a high voltage is imposed in a quartz bulb that is filled at a low pressure with H_2.

Hygroscopic. A term used to describe a substance that absorbs or attracts moisture from the air.

Hypothesis. A statement that describes the initial guess for the outcome of an experiment.

Ignition. The use of a flame in a gravimetric analysis to remove filter paper after a precipitate has been collected on this filter paper for weighing.

Immiscible. A term used to describe two liquids that do not dissolve to any appreciable extent in each other.

Immobilization Method. A term used in affinity chromatography for the method by which the affinity ligand is attached to the support.

Immunoassay. An analytical method that uses an antibody as a reagent.

Immunoglobulin. An antibody.

Incident Radiant Power (P_0). The original intensity of light striking a boundary between two media or entering a sample.

Inclusion. A process by which a precipitate may include some ions that are different from the desired cation or anion in the precipitate, but that have similar sizes and charges.

Independent Variable (x). A measured or calculated quantity whose values are chosen (either arbitrarily or by design) when performing an experiment; an example is the set of values that are plotted on the x-axis of a linear graph.

Index of Refraction. See *Refractive index*

Indicator Electrode. The electrode in potentiometry that gives a potential related to the activity and concentration of the analyte.

Indirect Titration. A method that indirectly measures an analyte through the effect that this analyte has on the concentration of another chemical (such as a metal ion) that can be titrated in a solution.

Individual Chart. A control chart that is used in situations where just one measurement per analyte is performed on a sample; also known as a *Levey–Jennings chart*.

Inductively Coupled Plasma Atomic Emission Spectrometry (ICP–AES). A type of atomic emission spectroscopy that uses a high temperature plasma for sample atomization and excitation.

Inflammable. A term used to describe a substance that is relatively easy to burn.

Inflection Point. A point in a graph in which the slope goes through a maximum or minimum and changes from an increasing to a decreasing value.

Infrared Spectroscopy. A spectroscopic method that uses infrared light to study or measure chemicals; also called *IR spectroscopy*.

Instrumental Method. An analytical technique that uses an instrument-generated signal for detecting the presence of an analyte or determining the amount of an analyte in a sample.

Intercept. The point on an axis that is intersected by a best-fit line or set of data, usually referring to the intersection at the y-axis; for a linear relationship, the intercept is represented by the term b in the equation $y = mx + b$, where m is the slope, x is the independent variable, and y is the dependent variable.

Interference. The combination of waves to give either an overall increase in the observed amplitude (constructive interference) or a decrease in the observed amplitude (destructive interference).

Interference Plot. A graph in that the apparent amount of analyte that is measured by a method is plotted versus the amount of a second substance that has been added to the sample.

Interferometer. A device that causes positive and negative interference to occur at sequential wavelengths as a moving mirror changes the path length of the light beam.

Interlaboratory Precision. The variation obtained with a single analytical method and sample, but by different laboratories.

Intermolecular Forces. Noncovalent, electrostatic interactions that cause separate but neighboring molecules or chemical species to attract or repel one another.

Internal Standard. An added substance that is not present in the original sample, but with similar properties to the analyte and the ability to be detected separately from this compound.

International Bureau of Weights and Measures (BIPM). The organization responsible for maintaining the SI system; also known in French as the *Bureau International des Poids et Mesures.*

Interoperator Precision. The variation obtained with a single analytical method and sample, but by different analysts.

Interzonal Region. The intermediate region in a flame where the flame temperature reaches its maximum and a local thermal equilibrium is reached.

Iodimetric Titration. A group of redox titrations that involve the use of iodine as a titrant, reagent, or analyte; also known as *iodimetry.*

Iodimetry. See *Iodometric titration*

Iodine Number. A measure of the degree of unsaturation in organic compounds such as fats, described as the grams of I_2 that react per 100 g of fat.

Ion Chromatography (IC). A special type of ion-exchange chromatography in which the background signal due to competing ions is reduced by using a low number of charged sites for the stationary phase to lower the concentration of competing ions that will be needed to elute sample ions; this method is also often used with a second column or membrane separator (of opposite charge to the first ion-exchange column) to replace competing ions that have high conductivity with ions that have a lower conductivity.

Ion-Exchange Chromatography (IEC). A liquid chromatographic technique in which solutes are separated by their adsorption onto a support containing fixed charges on its surface.

Ionic Interaction. A type of intermolecular force in which electrostatic attraction occurs between two ions with opposite charges, or repulsion occurs between two ions with the same type of charge.

Ion Product. The product of the concentrations or activities of individual ions that are formed as an ionic solid dissolves in a solvent.

Ion-Selective Electrode. An indicator electrode that can respond to individual types of anions or cations.

Ionic Strength, Concentration-Based (I_c). A term used to represent the amount of charge due to all of the ions present in a solution; this is calculated using the formula $I_c = \frac{1}{2} \Sigma (c_i z_i^2)$, where c_i is the concentration of ion i, and z_i is the charge on this same ion; I_c has the same units as the concentration term (c_i) that is used in its calculation.

Ionization Buffer. An easily ionizable species that is added to a flame and sample to prevent the ionization of other species in the flame.

Ionization Constant. An equilibrium constant that describes the ionization of a chemical species, such as in a flame.

IR Spectroscopy. See *Infrared spectroscopy*

Irritant. A noncorrosive chemical that causes reversible inflammation (swelling and redness) on contact with living tissue.

Isobestic Point. The point of intersection in the spectra for two absorbing species, representing a place where these species have an identical molar absorptivity.

Isocratic Elution. The use of a constant mobile-phase composition for elution in chromatography.

Isoelectric Focusing (IEF). An electrophoretic method used to separate zwitterions based on their isoelectric points by having these compounds migrate in an electric field across a pH gradient.

Isoelectric Point (pI). The pH at which a zwitterionic compound has a net charge of zero.

Isoionic Point. The pH that arises when only the neutral form of a zwitterionic compound is placed into a solution.

Isothermal Method. A method that is performed at a constant temperature.

Jones Reductor. A reductor that consists of a column that contains amalgamated zinc as a reducing agent; also known as a *zinc reductor.*

Joule Heating. Heating that occurs when an electric field is applied to an electrophoretic system.

Junction Potential. A potential that is present whenever two solutions or regions exist in an electrochemical cell that have different chemical compositions.

Kelvin (K). The fundamental unit of temperature in the SI system; the kelvin temperature scale is set so that the lowest possible temperature (absolute zero) is given a value of 0 K and the triple point of water (where the gas, solid, and liquid forms of water all exist in equilibrium) is at 273.16 K.

Kilogram (kg). The fundamental unit of mass in the SI system; defined as the mass of a platinum–iridium cylinder that is kept as the international standard for the kilogram at the International Bureau of Weights and Measures.

Kimex. See *Borosilicate glass*

Kjeldahl Method. An analytical technique that uses a digestion, distillation, and a back titration to measure the nitrogen content in organic samples.

Kováts Retention Index (I). A measure of retention in gas chromatography, as determined by comparing a compound's retention on a given column to the retention seen under identical conditions for *n*-alkanes.

K_w. The autoprotolysis constant of water.

Lab-on-a-Chip. See *Micro total-analysis system*

Laboratory Grade. See *Technical grade*

Laboratory Information Management System (LIMS). A computer software package for collecting data from instruments in a laboratory and processing this information into a suitable form for a report.

Laboratory Notebook. A record of the procedures that are used by a scientist, the experimental results that are obtained, and the conclusions that are reached from these experiments.

Laboratory Waste Management. The procedures for disposing of or handling used chemicals.

Laminar Flow Burner. A type of burner used in atomic absorption spectroscopy in a spray of sample droplets is mixed with a fuel and oxidant, and burned in a long narrow flame.

Laser-Induced Fluorescence (LIF). A method that employs a laser to excite a fluorescent compound, allowing the detection of this agent through its subsequent emission of light.

Le Châtelier's Principle. A principle that states that when a change in stress is placed on a system at equilibrium (such as a change in reactant or product concentrations), the system will respond to partially relieve this stress (e.g., by creating more products or reactants).

Least-Squares Analysis. A method for determining the best-fit parameters between an equation and a data set, as obtained by minimizing the sum of the square of the residuals between the equation and the data.

Leveling Effect. The reaction of water with a strong acid to form H_3O^+, or the reaction of water with a strong base to form OH^-; the result is that water tends to equalize the strength of both strong acids and strong bases when they are placed into this solvent.

Levey–Jennings Chart. See *Individual chart*

Lewis Acid. A chemical that can accept a pair of electrons from another substance.

Lewis Base. A chemical that can donate a pair of electrons to another substance.

Ligand. A chemical that shares an electron pair with a metal ion; from the Latin word *ligare,* which means "to bind" or "to tie."

Light. Another term for "electromagnetic radiation," or a wave of energy that propagates through space with both electrical- and magnetic-field components.

Limit of Detection (LOD). The lowest or highest amount of analyte that can be detected by a method.

Limit of Quantitation (LOQ). The smallest or largest amount of analyte that can be measured within a given range of accuracy and/or precision.

Limiting Diffusion Current. The current measured as the plateau in a plot of current versus applied potential, as obtained by direct-current voltammetry or a related technique.

Linear Range. The portion of a method's range that gives a linear dependence between its response and the concentration or measured property for an analyte.

Linear Regression. A procedure in which a given set of (x,y) values is fit to a linear equation, which has the general form $y_{i,calc} = mx_i + b$, where m is the slope of the best-fit line, b is the intercept, x_i is a given x-value in the data set, and $y_{i,calc}$ is predicted value that is given by the best-fit line for x_i.

Linear Titration Curve. A titration in which the measured response makes use of a value such as absorbance that is directly related to the concentration or activity of an analyte.

Linear Velocity (u). A measure of the rate of travel in units of distance per time.

Liquid Chromatogram. A plot of the detector response versus the time that has elapsed since sample injection onto a liquid chromatographic system.

Liquid Chromatograph. A system that is used to perform liquid chromatography.

Liquid Chromatography (LC). A chromatographic technique in which the mobile phase is a liquid.

Liquid Chromatography/Electrochemical Detection (or LC/EC). A method that combines liquid chromatography with an electrochemical detector.

Liquid Chromatography/Mass Spectrometry (LC/MS). The use of liquid chromatography with a mass spectrometer.

Liquid Junction Potential. A junction potential formed at the boundary between two solutions with different compositions.

Liquid–Liquid Chromatography. A type of partition chromatography that involves the use of a liquid stationary phase that is coated onto a solid support.

Liquid–Liquid Extraction. An extraction in which the two phases (the extractant and raffinate) are both liquids.

Liquid–Solid Chromatography (LSC). See *Adsorption chromatography*

Logarithmic Titration Curve. A titration in which the measured response makes use of a logarithmic expression of concentration or activity (such as pH).

London Forces. See *Dispersion force*

Longitudinal Diffusion. Broadening of a chemical's peak due to the diffusion of this chemical along the length of a chromatographic column or separation system.

Lower Limit of Detection. The smallest amount of analyte that can be examined by a method.

Luminescence. The emission of light from an excited-state chemical.

Luminometer. An instrument used to measure chemiluminescence.

Major Component. A substance that makes up more than 1% of the total composition for a sample.

Major-Component Analysis. Measurement of one or more major components in a sample; also called *major constituent*.

Major Constituent. See *Major component*

MALDI-TOF MS. See *Matrix-assisted laser desorption/ionization time-of-flight mass spectrometry*

Masking Agent. A ligand that is added to prevent a titrant from reacting with a particular substance.

Mass. The quantity of matter in an object; this property is constant regardless of the object's location.

Mass Balance. An application of the law of conservation of mass, which states that matter is neither created nor destroyed as a result of an ordinary chemical reaction.

Mass Balance Equation. An equation that shows how the total concentration of a chemical is related to the concentrations of its various species.

Mass Chromatogram. A plot of the number of ions that are measured by mass spectrometry at a given elution time from a chromatographic column.

Mass Spectrometer. An instrument in which ions are separated (or analyzed) based on their mass-to-charge ratios; this is used to measure the masses of individual atoms or molecules by first converting these atoms or molecules into ions.

Mass Transfer. The movement of a solute from one region to another.

Material Safety Data Sheet (MSDS). A set of one or more sheets that, by law, must be sent with each chemical substance that is produced by a manufacturer or that is imported for distribution; items in an MSDS include (1) a list of the chemicals found in the material and their common names; (2) information on the chemical and physical properties of the material; (3) health hazards that are associated with the material; (4) the maximum allowable limit of exposure to the substance, chemical; (5) an indication as to whether the chemical is a known carcinogen or has been found to be a potential carcinogen; and (6) any precautions to follow for the safe handling and use of the material.

Matrix. The entire group of chemicals and substances that makes up a sample; also referred to as the *sample matrix*.

Matrix-Assisted Laser Desorption/Ionization Time-of-Flight Mass Spectrometry (MALDI-TOF MS). A method used for ionization and volatilization in mass spectrometry, in which a sample is mixed with a matrix that can readily absorb ultraviolet light, followed by exposure of this mixture to pulses of an ultraviolet laser that transfers some of its energy to molecules in the sample and forms ions.

Maximum Load (of a balance). See *Capacity*

McReynolds Constants. Constants used to compare the retention properties of stationary phase in gas chromatography, as determined by measuring the Kováts retention indices for model compounds on both the stationary phase of interest and on a nonpolar reference stationary phase (squalane) at the same temperature.

Mean. See *Arithmetic mean*

Mean Activity Coefficient ($\gamma_\pm$). An activity coefficient that is a weighted average of the activity coefficients for both the negatively and positively charged ions in solution.

Measuring Pipet. See *Mohr pipet*

Mechanical Balance. A balance that uses a mechanical approach for determining the mass of an object.

Median (x_m). The value that occurs in the exact center of a group of numbers when these are arranged in increasing order; if there is an odd number of values, this value will be the one with an equal number of results that are higher and lower in the data set; if there is an even number of results, then the median is the average of the two central results.

Meker Burner. A special type of burner that provides a uniform hot flame over a relatively wide area.

Meniscus. The curved upper surface of a liquid.

Metal Coordination Complex. See *Metal-ligand complex*

Metallic Indicator Electrode. An electrode made of an inert metal that is used in an electrochemical cell to oxidize or reduce another substance; this type of indicator electrode is made from a material such as platinum, palladium, or gold.

Metal-Ligand Complex. A complex formed between a metal ion and ligand that involves the creation of a coordinate bond; also known as a *metal coordination complex*.

Metallochromic Indicator. A chemical that has a change in its color or its fluorescence properties when it is free in solution or complexed to a metal ion.

Meter (m). The fundamental unit of length in the SI system; defined as the distance traveled by light in a vacuum in $1/299,792,458$ of a second. (*Note*: This definition also fixes the speed of light at a value of $299,792,458$ m/s).

Method Validation. The process of characterizing an analytical technique and proving it will fulfill its intended purpose.

Micellar Electrokinetic Chromatography (MEKC). A subset of electrokinetic chromtogrpahy that employs micelles as running buffer additives.

Micelle. A particle formed by the aggregation of a large number of surfactant molecules, such as sodium dodecyl sulfate (SDS).

Microorganisms. Organisms such as bacteria and algae; often found as contaminants in water.

Micropipet. A pipet that is used to deliver extremely small volumes, typically on the order of 0.1 to 5000 μL; these pipets have disposable tips that can easily be changed between samples and can often be adjusted to deliver solutions over a wide range of volumes; also known as a *pipetter*.

Micro Total-Analysis System (μTAS). A miniaturized system that contains all of the components needed for a chemical analysis; also called a *lab-on-a-chip*.

Microwave-Assisted Extraction. Use of microwave radiation to increase the rate and extent of an extraction.

Migration Distance (d_m). The distance that an analyte travels on a support in a given amount of time, such as in separations based on gel electrophoresis.

Migration Time (t_m). The time required for an analyte to travel a given distance, such as used in separations based on capillary electrophoresis.

Minor Component. A substance that makes up 0.01%–1% of the total composition for a sample.

Minor Component Analysis. Measurement of one or more minor components in a sample.

Miscible. A term used to describe two liquids that can form a stable solution when they are mixed in any proportion.

Mobile Phase. The phase in chromatography that flows through the column and that causes sample components to move towards the end of a column.

Mobile-Phase Mass Transfer. A band-broadening process that results from the different rates of travel a solute has across any given slice of a chromatographic column.

Model. The component of a statistical technique that represents the result, method, or predicted behavior to which an experimental value is being compared.

Mohr Method. A method for detecting the end point in an argentometric titration by using the reaction of Ag^+ with chromate to form a red precipitate of silver chromate.

Mohr Pipet. A pipet that contains many calibrated markings to allow it to measure and deliver a variety of volumes within this calibrated range; like a volumetric pipet, a Mohr pipet is designed "to deliver" the desired volume of solvent through only the process of natural draining, and not with blowing or any forced delivery; also known as a *measuring pipet*.

Molal Solution. See *Molality*

Molality (m). A concentration unit equal to the moles of a solute that are present per kilogram of solvent; for instance, a solution that contains 1.0 mole of a solute in 1.0 kg of solvent is referred to as a 1.0-*m*, or 1.0-molal, solution.

Molar Absorptivity. A term used for the proportionality constant in Beer's law when this constant is expressed in units of L/mol · cm.

Molar Mass (molecular weight, MW). The number of grams that are contained in one mole of a substance (MW = grams chemical/mol); although this is often used to describe both molecular and ionic compounds, some use it only to refer to molecular substances; also called *formula weight*.

Molar Solution. See *Molarity*

Molarity (*M*). A concentration unit equal to the moles of a solute that are present per liter of solution; for instance, a solution that contains 1.0 mole of a solute in a final solution volume of 1.0 L is referred to as a 1.0 *M*, or 1.0 molar, solution.

Mole (mol). The fundamental unit for describing the amount of a substance in the SI system; defined as the number of individual entities of a substance that is equal to the number of carbon atoms in 0.012 kg of carbon-12.

Molecular Luminescence Spectroscopy. Use of the emission of light from an excited-state chemical to study molecules.

Molecular Occlusion. See *Occlusion*

Molecular Sieve. A porous material that is composed of a mixture of silica (SiO_2), alumina (Al_2O_3), water, and an oxide of an alkali or alkaline earth metal, such as sodium or calcium; when these materials are combined in a particular ratio, they produce a support with a series of pores with well-defined sizes and binding regions; also known as a *zeolite*.

Molecular Spectroscopy. The examination of the interactions of light with molecules.

Molecular Weight (MW). See *Molar mass*

Monochromatic Light. Light that contains only one wavelength.

Monodentate Ligand. A chemical like ammonia, water, or Cl^- that can donate only one pair of electrons to a metal ion; also known as a *simple ligand*.

Monolithic Column. A column that contains a support that is one continuous bed rather than a packed bed made up of many small particles; this type of support is made by using a specially prepared porous polymer as the support.

Monoprotic Acid. An acid that can donate only one hydrogen ion to a base.

Monoprotic Base. A base that can accept only one hydrogen ion from an acid.

Moving Boundary Electrophoresis. An electrophoretic method that produces a series of moving boundaries between regions that contain different mixtures of analytes.

Multistep Extraction. A type of extraction in which several steps are used to place the sample in contact with an extracting phase.

Mutagen. A substance that causes a change in DNA.

Nephelometry. A technique in which the intensity of light that is scattered by a solution is compared to the original intensity of this light, with the scattered light being measured at a right angle to the incoming light.

Nephrotoxin. A substance that causes damage to the kidneys (i.e., the nephritic system).

Nernst Equation. An expression used for a reversible half-reaction to relate the reduction potential under nonstandard conditions (E) to the activities of the reactants and products and the half-reaction's standard reduction potential ($E°$).

Nernst Glower. A light source used in infrared spectroscopy that consists of a heated, inert rod that is made of a mixture of rare-earth oxides.

Neurotoxin. A chemical that creates an adverse effect on the central nervous system.

Neutral Solution (in water). An aqueous solution with a pH equal to 7.0.

NF Grade. A reagent chemical that meets or exceeds specifications set by the National Formulary (NF).

NFPA (National Fire Prevention Association) Label. A means for identifying chemical hazards that is usually drawn as a diamond with four colored areas (blue = overall health risk, red = flammability, yellow = reactivity with other substances, white = reactivity with water or the ability to oxidize other compounds).

Nitrogen-Phosphorus Detector (NPD). A detector used in gas chromatography for the determination of nitrogen- or phosphorus-containing compounds, based on the measurement of ions that are produced from such analytes.

Noise. A term used in method validation and statistical testing that refers to the random variation in a signal.

Nonporous Support. A support for liquid chromatography that has a thin porous layer or porous shell.

Nonspecific Elution. An elution technique used in affinity chromatography in which the pH, ionic strength, or polarity of the mobile phase is altered to lower the association equilibrium constant for the analyte–ligand interaction.

Normal Distribution. A common mathematical model used for describing random errors and various types of experimental measurements; also known as a *bell-shaped curve* or *Gaussian curve*.

Normal-Phase Chromatography. A type of partition chromatography that uses a polar stationary phase; also known as *normal-phase liquid chromatography* or *NPLC*.

Normal-Phase Liquid Chromatography (NPLC). See *Normal-phase chromatography*

Normal Polarity Mode. A method used in capillary electrophoresis during which a sample is injected at one end of the capillary (the positive electrode when using an uncoated silica capillary) and is carried by electroosmosis to a detector near the other end of the capillary (the negative electrode, in this case).

Normality (N). A concentration unit equal to the number of gram-equivalent weights (or equivalents) of a solute per liter of solution; for instance, a solution that contains 1.0 equivalent of a solute in 1.0 L of a solution is referred to as a 1.0 N, or 1.0 normal, solution.

Northern Blot. A blotting method that is used to detect specific sequences of RNA through their binding with a labeled DNA probe.

Nucleation. A process by which small particles (or "nuclei") are formed of a precipitating chemical.

Null Hypothesis (H_0). In statistical tests, an initial hypothesis in which the model and experimental results are thought to be indistinguishable at a given confidence level.

Number of Observations (n). The total number of measured or observed values within a group of results; also sometimes known as the *sample size*.

Number of Theoretical Plates (N). A measure of band-broadening in chromatography and other separation methods; the most general formula used for determining N in chromatography is $N = (t_R/\sigma)^2$, where t_R is the retention time of a solute, and σ is the standard deviation for the solute's peak; also known as the *plate number*.

Occlusion. The trapping of solvent or molecular impurities inside of a precipitate; also called *molecular occlusion*.

Octanol–Water Partition Ratio (K_{ow}). A term used to rank chemicals based on their relative polarity by comparing their ability to dissolve in water versus 1-octanol.

Ohm's Law. A relationship between the potential (E), current (I), and resistance (R) in an electrical or electrochemical system, as given by following formula $E = I \cdot R$.

Open-Tubular Chromatography. A type of column chromatography in which the stationary phase is placed directly onto the interior wall of the column.

Open-Tubular Column. A chromatographic column consisting of a tube that has a stationary phase coated on or attached to its interior surface; also known as a *capillary column*.

Optical Density. See *Absorbance*

Optimum Linear Velocity (u_{opt}). The linear velocity in chromatography or a separation method that produces the smallest plate height and best efficiency for that system.

Optimum Plate Height (H_{opt}). The smallest plate height for chromatography or other type of separation system.

Ostwald-Folin Pipet. A pipet that is used to measure and deliver a single, specific volume of a liquid to another container, such as a volumetric flask; this is similar to a volumetric pipet, but is now marked "To Deliver/ Blow Out," meaning that it delivers the indicated volume only when the last bit of its contents is blown out with a pipet bulb.

Ostwald Ripening. A technique for obtaining precipitates with both larger and purer particles in which a precipitate is heated in its original solution to a temperature that is near the boiling point of the solution.

Outer-Sphere Association Constant (K_{os}). An equilibrium constant that describes the association of a metal ion complex with a ligand that is present in the layer of molecules that is immediately next to this complex.

Outlier. A data point that does not fit the general trend observed for a group of results that are obtained under supposedly identical conditions; also known as an *outlying value* or *extreme value*.

Outlying Value. See *Outlier*

Overall Analysis Time. The time needed for all steps in the preparation and analysis of a sample.

Overall Formation Constant (β_n). The product of the formation constants for *n*-stepwise reactions between a metal (M) and ligand (L) leading to the complex $M(L)_n$, where $\beta_n = K_{f1} K_{f2}...K_{fn}$; also known as a *cumulative formation constant*.

Oxidant. See *Oxidizing agent*

Oxidation. A process in which a chemical loses one or more electrons.

Oxidation Half-Reaction. A half-reaction in which electrons are one of the products.

Oxidation Number. The charge that an element in a chemical would have if this element existed as a solitary ion but still possessed the same number of electrons that it has in the given chemical; also called the *oxidation state*.

Oxidation–Reduction Reaction. A reaction that involves the oxidation and reduction of chemicals; also called an *electrochemical reaction*.

Oxidation State. See *Oxidation number*

Oxidation–Reduction Titration. See *Redox titration*

Oxidizer. A substance that readily yields oxygen to support the combustion or oxidation of other chemicals.

Oxidizing agent. A chemical that can readily undergo reduction, causing other chemicals to be oxidized.

Packed-bed Chromatography. A type of column chromatography in which the column is packed with support particles that contain the stationary phase.

Packed Column. A chromatographic column that is filled with small support particles that act as an adsorbent or that are coated with the desired stationary phase.

PAGE. See *Polyacrylamide gel electrophoresis*

Paired Student's *t*-Test. A statistical test that is used to compare two sets of identical samples that are analyzed by different methods with similar standard deviations for their results.

Paper Chromatography. A type of planar chromatography that uses paper as the support.

Parallax Error. An error produced when reading a calibrated mark or scale when it is viewed at any nonperpendicular angle, producing readings that may be either too high or low in value.

Partially Miscible. A term used to describe two liquids that give a difference in the volume for their mixture versus the total volumes of the two liquids before they were combined.

Partial Molar Gibbs Energy. See *Chemical potential*

Particulates. Large suspended particles that are found as contaminants in water.

Partition Chromatography. A liquid chromatographic technique in which solutes are separated based on their partitioning between a liquid mobile phase and a stationary phase coated on a solid support.

Partition Constant ($K_D°$). An equilibrium constant that gives the ratio of the activities for a solute in two phases as this solute distributes between these phases at a given pressure and temperature.

Partition Ratio (K_D). An equilibrium constant that gives the ratio of the concentrations for a solute in two phases as this solute distributes between these phases at a given pressure and temperature.

Partitioning. A process by which a chemical enters both of two separate phases, such as occurs in a liquid–liquid extraction.

Parts-per-Billion (ppb). A means of expressing chemical content in which there is one part of the analyte or chemical of interest for every 10^9 parts of a mixture; often used to describe weight-per-weight, volume-per-volume, and weight-per-volume ratios.

Parts-per-Million (ppm). A means of expressing chemical content in which there is one part of the analyte or chemical of interest for every 10^6 parts of a mixture; often used to describe weight-per-weight, volume-per-volume, and weight-per-volume ratios.

Parts-per-Thousand (‰). A means of expressing chemical content in which there is one part of the analyte or chemical of interest for every one thousand parts of a mixture; often used to describe weight-per-weight, volume-per-volume, and weight-per-volume ratios.

Parts-per-Trillion (ppt). A means of expressing chemical content in which there is one part of the analyte or chemical of interest for every 10^{12} parts of a mixture; often used to describe weight-per-weight, volume-per-volume, and weight-per-volume ratios.

Path Length (*b*). The distance light must travel through a sample.

Peak Fronting. A condition in which a chromatographic peak has a sharper back edge than front.

Peak Resolution (R_s). A parameter used to describe the separation of two peaks in chromatography; the value of R_s for two adjacent peaks can be calculated through the following formula,

$$R_s = \frac{t_{R_2} - t_{R_1}}{(w_{b_2} + w_{b_1})/2}$$

where t_{R_1} and w_{b_1} are the retention time and baseline width (both in the same units of time) for the first eluting peak, and t_{R_2} and w_{b_2} are the retention time and baseline width of the second peak.

Peak Tailing. A condition in which a chromatographic peak has a sharper front edge than back.

Peptization. The conversion of a solid precipitate into a colloid suspension.

Percent Recovery. A measure of an analytical method's accuracy, as determined by calculating the percent change in the measured amount of an analyte in a sample after a known amount of the same analyte has been spiked into the sample.

Percent Relative Error (e_r, %). A measure of accuracy that is obtained by dividing the absolute error of an experimental result (x) by the true value (τ) and multiplying the final answer by 100, where e_r (%) = 100 ($x - \tau$)/τ.

Percent Transmittance (% T). The percent of light that is transmitted by a sample, where % T = 100 (T) and T is the measured transmittance.

Perfusion Particles. A type of support used in liquid chromatography that contains large pores that allow the mobile phase to pass both through and around the support particles.

Permanganate. A strong oxidizing agent with the formula MnO_4^- that is often used in redox titrations and chemical analysis.

pH (*notational definition*). A working definition of pH, given in this text as pH = $-\log (a_{H^+}) \approx -\log([H^+])$, where a_{H^+} is the activity of hydrogen ions (or hydronium ions in an aqueous solution), and [H^+] is the corresponding concentration of these ions.

pH Electrode. An indicator electrode that is selective for the detection of hydrogen ions.

pH Scale. A scale used to describe the pH of water and aqueous samples.

Phase Ratio. A ratio that gives the relative amount of one phase versus another in an extraction or chromatographic method.

Phenolphthalein. An acid–base indicator that is widely used in acid–base titrations.

Phosphorescence. The emission of light by a sample after this sample has become electronically excited by absorbance of a photon, with the light emission occurring after the excited electron undergoes an intersystem crossing into a triplet state; this situation requires a spin-forbidden transition from the triplet state to a singlet state for light emission.

Phosphorescence Spectroscopy. A spectroscopy technique that utilizes phosphorescence to characterize or measure chemicals.

Photodiode-Array Detector (PDA). An absorbance detector that uses an array of small detector cells to simultaneously measure the change in absorbance at many wavelengths.

Photoelectrical Effect. The ejection of electrons from certain materials when these materials are hit with photons of light.

Photon. A term used to describe an individual particle of light.

Pipetter. See *Micropipet*

pK_a. A value equal to $-\log(K_a)$, where K_a is an acid dissociation constant.

pK_w. A value equal to $-\log(K_w)$, where K_w is the autoprotolysis constant of water.

Planar Chromatography. A type of chromatography in which the support and stationary phase are present on a flat plane, such as on a piece of paper, glass, or plastic.

Planck's Constant (h). A proportionality constant used in Planck's equation to relate the energy and frequency for a photon of light; this constant has a value of approximately 6.626×10^{-34} J·s, regardless of the type of light or photons that are being examined; also known as the *Planck constant*.

Planck's Equation. An equation that relates the energy (E_{Photon}) and frequency (ν) for a photon of light through a proportionality constant (h), where $E_{Photon} = h\nu$.

Plate Height. See *Height equivalent of a theoretical plate*

Plate Number. See *Number of theoretical plates*

pM. A value equal to $-\log[M^{n+}]$ for a solution of ion M^{n+}.

Poison. A substance that can kill, injure, or impair a living organism.

Polyacrylamide. A synthetic polymer that is often used as a support in gel electrophoresis.

Polyacrylamide Gel Electrophoresis (PAGE). An electrophoretic method in which a polyacrylamide gel is used as the support.

Polychromatic Light. Light containing a mixture of two or more wavelengths.

Polydentate Ligand. A ligand that has multiple binding sites for metal ions.

Polymerase Chain Reaction (PCR). A method for increasing the amount of given sequences of DNA, as conducted by using a series of cycles in which the DNA strands are separated, combined with primers and appropriate reagents, and converted back into double-stranded DNA.

Polyprotic Acid. An acid that can donate more than one hydrogen ion to a base.

Polyprotic Base. A base that can accept more than one hydrogen ion from an acid.

Polysiloxane. A type of chemical that consists of a backbone of silicon and oxygen atoms attached in long strings of Si—O—Si bonds, with the remaining two bonds on each silicon atom being attached to side groups that can have a variety of structures; a common type of stationary phase in gas chromatography.

Polystyrene. A material that is made by polymerizing styrene in the presence of divinylbenzene.

Porous-Layer Open-Tubular Columns (PLOT). A type of open-tubular column in gas chromatography that contains a porous material that is deposited on the interior wall of a column, with the surface of this material being used as the stationary phase.

Postprecipitation. A process by which impurities collect on a precipitate after it has been formed but is still standing in its original solution.

Potassium Hydrogen Phthalate (KHP). A common primary standard for acid–base titrations.

Potential. See *Electric potential*

Potentiometer. A device used to measure the potential difference between two electrodes in an electrochemical cell; also known as a *voltmeter*.

Potentiometric Titration. The use of a potential measurement to follow the course of a titration.

Potentiometry. A technique for electrochemical analysis that is based on the measurement of a cell potential with essentially zero current passing through the system.

Potentiostat. A device that controls the difference in potential that is applied to an electrochemical cell.

Pourbaix Diagram. A plot of potential versus pH that is used to show the principal species for an element that will be present under a given set of reaction conditions.

Powder Diffraction. The use of X-ray crystallography to examine a powder that is prepared from small crystals of a chemical.

Precipitant. A precipitating agent that is added during a gravimetric analysis to form a weighable solid whose mass can be used to calculate the mass of any analyte that is present.

Precipitate. A solid that is formed as a dissolved chemical leaves a solution that contains more than the solubility limit of the chemical.

Precipitation. A process that occurs when a portion of a dissolved chemical leaves a solution to form a solid.

Precipitation from Homogeneous Solution. A technique in which a precipitating agent is formed slowly in the solution after it has been stirred and made homogeneous.

Precipitation Titration. A titration method in which the reaction of a titrant with a sample leads to an insoluble precipitate.

Precision. The variation in individual results obtained under identical conditions.

Precision Balance. A balance with an open weighing area, typically having a readability of 1 mg or greater; also referred to as a *top-loading balance.*

Precision Plot. A diagram that gives a visual representation of how the precision of a method changes with the concentration or measured property of an analyte.

Preoxidation. A sample pretreatment process used to oxidize and convert an analyte to a higher oxidation state.

Prereduction. A sample pretreatment process used to reduce and convert an analyte to a lower oxidation state.

Pressurized Solvent Extraction (PSE). See *Accelerated solvent extraction*

Primary Combustion Zone. The region in the center and the bottom of a flame where the combustion of a sample begins.

Primary Standard. A pure substance that is stable during storage, can be weighed accurately, and undergoes a known reaction with the solution it is used to characterize.

Primary Standard Solution. A solution that is prepared using a chemical that is a primary standard.

Procedure. The entire group of operations (including sample preparation, measurement, and data handling) that is used for a chemical analysis method; also known as a method's *protocol.*

Property Characterization. Measurement of some specific chemical or physical property of an analyte (e.g., its color, crystal shape, and mechanical strength, or its ability to interact with light, electrons, or other chemicals).

Protective Agent. A reagent that prevents the formation of a refractory compound in a flame by creating an alterative chemical species that can burn and easily dissociate into atoms.

Protocol. See *Procedure*

Protonation Constant. See *Base ionization constant*

Pyrex. See *Borosilicate glass*

Pyrogens. Pieces of bacterial cell walls and lipopolysaccharides from bacteria that can be present as contaminants in water; these may hinder the growth of cells or tissues in biological assays.

Pyrogram. A type of gas chromatogram produced in pyrolysis gas chromatography, which provides a picture or fingerprint pattern of the volatile compounds that are given off as the test substance is heated.

Pyrolysis Gas Chromatography. A method used in gas chromatography that involves heating a solid sample in a controlled fashion to break the solid apart into smaller, more volatile chemical fragments that are separated by a column.

Pyrophoric. A term used to describe a chemical that can ignite in the presence of air or moisture.

Q-Test. A statistical test for detecting outliers; this is performed by taking the absolute difference between the suspected outlier's value (x_0) and its nearest neighboring value (x_n), and comparing this difference to the total range of values that are present in the data set ($x_{high} - x_{low}$) by using the ratio $Q_{exp} = |x_0 - x_n| / (x_{high} - x_{low})$; also known as *Dixon's test.*

Quadratic Equation. An equation where the highest order term for "x" is x^2; this type of equation is often written in the general form $0 = A x^2 + B x + C$, where A, B, and C are constants.

Quadratic Formula. A formula used to solve for the roots of a quadratic equation.

Quadrupole Mass Analyzer. A type of mass spectrometer that uses a series of four horizontal rods that are held at well-defined AC and DC potentials. Each opposing pair of rods is held at the same potential and any two neighboring rods are held at exactly opposite potentials. Over time, the potentials on these rods are continuously varied. This creates a varying electric field through which only ions with a certain mass-to-charge ratio will be able to pass; also known as a *transmission quadrupole mass spectrometer.*

Qualitative Analysis. A measurement in which the goal is to determine whether or not a particular analyte is present in a sample above a certain minimum level.

Quality Control. The process of monitoring the routine performance of a method; the use of chemical or physical measurements to determine whether the composition or properties of a particular product or raw material are of satisfactory quality for sale or further use.

Quantify. The act of measuring the amount of an analyte within a sample.

Quantitate. The act of measuring the amount of an analyte within a sample.

Quantitative Analysis. A measurement in which the goal is to provide a numerical value for the amount of an analyte within a sample.

Quartz Crystal Microbalance (QCM). A device that uses a thin oscillating quartz crystal as a sensor for chemicals in samples; if chemicals adsorb to the surface of this crystal, it will change its frequency of vibration, which provides a signal that is related to the mass of the deposited material.

Radiant Power (P). The intensity of light, expressed in units of watts.

Radioactive. A term used to describe a material that emits ionizing radiation.

Radiocontrast Agent. A substance used to absorb X rays.

Raffinate. The sample or phase in an extraction that originally contains the chemicals of interest.

Raman Scattering. An effect in which the scattering of light by molecules involves a small change in the wavelength of the light.

Raman Spectroscopy. A method that uses Raman scattering to study or measure chemicals.

Random Error. An error that results from random variations in experimental data.

Random Sample. A sample that is acquired by arbitrarily taking part of a material and testing this portion to obtain information on the entire contents of the material.

Range (R_x). The difference between the largest and smallest value in a set of data; used to describe the variation within a group of numbers.

Rayleigh Scattering. The scattering of photons of light by particles such as atoms or molecules that are much smaller than the wavelength of light (particle diameter $< 0.05 \lambda$).

Reaction Quotient (Q). A ratio of the activities or concentrations of the products versus reactants in a chemical reaction, as given under nonequilibrium conditions.

Readability (of a balance). The size of the smallest division in mass that is displayed on a balance's display.

Reciprocating Pump. A pump in which a rotating cam is used to move a piston in and out of a solvent chamber, creating flow of mobile phase or solvent into and from this chamber.

Redox Couple. A pair of two different oxidized and reduced forms of the same element in an oxidation–reduction reaction.

Redox Indicator. An indicator that has a change in color as it undergoes an oxidation–reduction reaction.

Redox Titration. A titration that makes use of an oxidation–reduction reaction; also known as an *oxidation–reduction titration*.

Reducing Agent. A chemical that is easily oxidized, causing other chemicals to be reduced; also called a *reductant*.

Reductant. See *Reducing agent*

Reduction Half-Reaction. A half-reaction in which electrons are one of the reactants.

Reduction. A process in which a chemical gains one or more electrons.

Reductor. A device used for sample prereduction that consists of a column that contains an insoluble form of a reducing agent.

Reflection. A process that occurs whenever light encounters a boundary between two regions that have different refractive indices, where at least part of the light changes its direction of travel and returns to the medium in which it was originally traveling.

Reflection Grating. A grating that contains a polished and reflective surface that has a series of parallel and closely spaced steps cut into its surface.

Refraction. A process in which the direction of travel for light is changed as it passes through a boundary between two regions that have different refractive indices.

Refractive Index (n). The ratio of the velocity of light in a true vacuum (c) versus the velocity of light in a particular medium (v), where $n = c/v$; also known as the *index of refraction*.

Refractive Index Detector. A detector used in liquid chromatography that measures the ability of the mobile phase and analytes to refract or bend light.

Regular Reflection. See *Specular reflection*

Relative Activity. See *Chemical activity*

Relative Error (e_r). A measure of accuracy that is obtained by dividing the absolute error of an experimental result (x) by the true value (τ), where $e_r = (x - \tau)/\tau$.

Relative Standard Deviation (RSD). A measure of precision that is equal to the standard deviation for a number divided by that number's value; also known as the *coefficient of variation* or *CV*.

Releasing Agent. A reagent that causes a precipitating anion to form a precipitate with some chemical agent other than the analyte.

Remote Sensing. The use of an analytical instrument to examine a distant sample.

Reprecipitation. A technique used to improve the purity of a precipitate; in this method, the original precipitate is removed from the rest of its original solution, redissolved in a pure solvent, and later precipitated again from this new solution.

Representative Sample. A sample that reflects the overall composition and properties of the original material.

Reproductive Toxin. An agent that creates damage to the reproductive system.

Residual Plot. A graph for visually examining data to detect any deviations from a best-fit line that is prepared by plotting the difference, or residual, in the experimental and calculated y values ($y_i - y_{calc}$) versus the corresponding x values.

Resistance. The resistance to the flow of current in the presence of an applied electrical potential.

Resolution (of a balance). The maximum load of a balance divided by the readability of the balance; a measure of the number of distinct masses that can be determined by a balance.

Resolution Equation. An equation used to show the relationship between peak resolution, peak retention, efficiency, and selectivity in chromatography.

Response. The signal for an analytical method when measuring a given amount of analyte or property.

Retardation Factor (R_f). A measure of retention in gel electrophoresis, where the migration distance of an analyte (d_m) is divided by the migration distance for a small marker compound (d_s), where $R_f = d_m/d_s$.

Retardation Factor (R_F). In planar chromatography, the ratio of the distance traveled by a given solute (D_s) to the distance traveled over the time by the mobile phase in the solvent front (D_f), as given the relationship $R_F = D_s/D_f$.

Retention Factor (k). A measure of solute retention in chromatography that is calculated using $k = t'_R/t_M$ or $k = V'_R/V_M$, where t'_R is the adjusted retention time of the solute, t_M is the column void time, V'_R is the adjusted retention volume for the solute, and V_M is the column void volume; also called the *capacity factor*.

Retention Time (t_R). The average time required for a given retained substance to pass through a chromatographic system; also known as the *total retention time*.

Retention Volume (V_R). The average volume of mobile phase required to pass a given retained substance through a chromatographic system; also known as the *total retention volume*.

Reversed-Phase Chromatography. A type of partition chromatography that uses a nonpolar stationary phase; also known as *reversed-phase liquid chromatography* or *RPLC*.

Reversed-Phase Liquid Chromatography (RPLC). See *Reversed-phase chromatography*

Reversed Polarity Mode. A method used in capillary electrophoresis during which a sample is injected at one end of the capillary (the negative electrode when using an uncoated silica capillary) and migrates against electroosmosis to a detector near the other end of the capillary (the positive electrode, in this case).

Robustness. The ability of an analytical technique to provide a consistent response when small variations are made in its experimental conditions.

Rounding Error. An error that is produced when a number is rounded off too early in a calculation.

Running Buffer. The background electrolyte in electrophoresis.

Running Gel. A support used for an electrophoretic separation of substances in a sample.

Salt Bridge. A device that allows current to flow between two electrodes while preventing the mixing of their electrolytes.

Salting Coefficient. A constant used to describe the effect of ionic strength on the activity coefficient for a neutral agent.

Salting-Out Effect. A decrease in the solubility of a neutral compound as the ionic strength is increased.

Sample. The portion of a material that is taken for chemical analysis.

Sample Matrix. See *Matrix*

Sample Stacking. A method for concentrating samples and providing narrow analyte bands in capillary electrophoresis, which makes use of a sample that has a lower ionic strength (and lower conductivity) than that of the running buffer.

Sample Throughput. The number of samples that can be processed by an analytical method in a given period of time.

Sample Size. See *Number of observations.*

Sampling Error. An error created when only a portion of a nonuniform substance is used for an analysis.

Sampling Plan. The specific approach used when acquiring a sample.

Sandwich Immunoassay. An immunoassay that uses two different types of antibodies that each bind to the analyte of interest; the first of these two antibodies is attached to a solid-phase support and is used for extraction of the analyte from samples, while the second antibody contains an easily measured tag and serves to place a label onto the analyte for measurement.

Saturated Calomel Electrode. A calomel electrode that contains a saturated solution of KCl.

Saturated Solution. A solution in which the dissolved concentration of a solute is equal to its maximum solubility at equilibrium.

Scanning Probe Microscopy (SPM). A type of microscopy that uses a physical probe to scan a surface and create an image of a sample.

Scattering. The change in travel of one particle (such as a photon) due to its collision with another particle (e.g., an atom or molecule).

Screening Assay. An analytical technique that is designed to determine whether or not an analyte is present in a sample above a certain minimum level; a method that is used to decide whether further tests and more rigorous assays need to be performed on a sample.

SDS-PAGE. See *Sodium dodecyl sulfate polyacrylamide gel electrophoresis*

Second (s). The fundamental unit of time in the SI system; defined as the amount of time equal to 9,192,631,770 periods of the radiation corresponding to the transition between the two hyperfine levels of the ground state of cesium-133.

Secondary Combustion Zone. The outer cone of a flame, where oxygen from the surrounding air can lead to additional combustion.

Secondary Standard. A reagent or reagent solution that is characterized by using a primary standard.

Secondary Standard Solution. A solution that is prepared using a chemical that is a secondary standard.

Selected Ion Monitoring (SIM). The use of gas chromatography/mass spectrography to collect information on only a few ions.

Selectivity. The ability of an analytical method to detect and discriminate between an analyte and other chemicals in a sample; often referred to as *specificity.*

Selectivity Coefficient ($K_{A,C}$). A type of equilibrium constant used to describe an ion-exchange reaction; a measure of an analytical method's specificity, as determined by comparing the relative signal that is produced by one substance versus another in that method.

Sensitivity. A measure of how the response of an analytical method changes as the amount of analyte or a sample property is varied.

Separation Factor (α). A measure of the relative difference in retention of two solutes as they pass through a column, as determined by the ratio $\alpha = k_2/k_1$, where k_1 is the retention factor for the solute that exits first from the column, and k_2 is the retention factor for the second solute.

Separation Method. An approach used to remove one type of chemical from another.

Serological Pipet. A pipet that contains many calibrated markings to allow it to measure and deliver a variety of volumes within this calibrated range; this pipet is marked "To Deliver/Blow Out," meaning that it delivers the indicated volume only when the last bit of its contents is blown out with a pipet bulb.

Shewhart Chart. A type of control chart that is often used in industrial laboratories and in situations where enough time and material are available to perform several measurements on every sample.

SI System (International System of Units). A system that provides a set of uniform standards for describing such things as mass, length, time, and other measurable quantities; also known in French as the *Système Internationale d'Unités.*

Signal. The net change in response between a blank and a sample known to contain the analyte.

Signal Chopper. A propeller that rotates in the light path between the lamp and flame in an instrument for atomic spectroscopy; the chopper alternately opens and blocks the path of incident light that is being given off by the lamp but does not affect light from the flame.

Signal-to-Noise Ratio (S/N). The ratio of the "noise" (or random variation in the blank signal) versus the "signal" for a sample (or the net change in response measured between the blank and sample).

Significant Figures (sig figs). The digits that can be used to reliably record or report a number.

Silanol Group. A chemical group with the general formula $-Si-OH$ that is located on the surface of silica.

Silica. A material composed of silicon dioxide (empirical formula, SiO_2), which is often used as a solid support in chromatography.

Silver Reductor. See *Walden reductor*

Silver/Silver Chloride Electrode. A common reference electrode that consists of a silver wire that is coated with silver chloride.

Silver Staining. A staining method that uses silver nitrate for the detection of low-concentration proteins.

Simple Ligand. See *Monodentate ligand*

Single-Beam Instrument. A device used in spectroscopy that has a single path for the light to take through the instrument.

Single-Crystal Diffraction. The use of X-ray crystallography to examine a single crystal of a chemical.

Single-Pan Mechanical Balance. See *Substitution balance*

Single-Step Extraction. An extraction in which one step is used to place a sample in contact with an extracting phase.

Sintered Glass Crucible. A crucible that has a porous glass filter in the bottom that can be used to directly collect a precipitate; also called *fritted glass crucible*.

Size-Exclusion Chromatography (SEC). A liquid chromatographic technique that separates substances according to differences in their size.

Slope. The change in the dependent variable versus the change in the independent variable for a graph; for a linear relationship the slope has a constant value and is represented by the term m in the equation $y = mx + b$, where b is the intercept of the line on the y-axis, x is the independent variable, and y is the dependent variable.

Snell's Law. A relationship that predicts the angle of refraction for light based on the angle of incidence of the light and the refractive indices for the original medium in which the light was traveling and the new medium it is entering; also known as *Descartes' law*.

Sodium Carbonate Fusion. A way to pretreat and dissolve solid samples for analysis by weighing the powdered sample and mixing it with solid sodium carbonate in a platinum crucible prior to heating of this mixture to a red-hot temperature.

Sodium Dodecyl Sulfate. A surfactant with a nonpolar tail and a negatively charged sulfate group, used in methods such as SDS-PAGE.

Sodium Dodecyl Sulfate Polyacrylamide Gel Electrophoresis (SDS-PAGE). An electrophoretic method for the separation of proteins based on their size, in which the proteins are first denatured and their disulfide bonds broken through the use of a reducing agent. The resulting single-stranded polypeptides are then treated with sodium dodecyl sulfate and passed through a porous polyacrylamide gel in the presence of an electric field for their separation.

Sodium Ion-Selective Electrode. An indicator electrode that contains a glass membrane and is relatively selective for the measurement of sodium ions.

Solid-Phase Extraction. Extraction of a liquid or gas sample with a solid support that contains an adsorbing surface or chemical coating that can interact with analytes.

Solid-Phase Microextraction. Use of uncoated or coated fiber delivered by a syringe for the extraction of analytes.

Solid-State Ion-Selective Electrode. An ion-selective electrode in which the sensing element is a crystalline material or a homogeneous pressed pellet.

Solubility. The maximum concentration or amount of a chemical that can be placed into a solvent to form a stable solution.

Solubility Constant. An equilibrium constant that describes the placement of a solid chemical into a solution.

Solubility Equilibrium. A process in which a chemical enters a solution and reaches equilibrium between its dissolved and solid states.

Solubility Product. An equilibrium constant that describes the conditions needed to saturate a solution with a dissolved ionic solid; represented by the term K_{sp}.

Solute. A substance that is dissolved within another substance (the solvent) to produce a solution.

Solution. A uniform mixture of one substance (the solute) within another (the solvent).

Solvent. A substance that dissolves another substance (the solute) to produce a solution.

Solvent Polarity Index (P). A measure of mobile phase strength in partition chromatography (for example, in normal-phase liquid chromatography (NPLC) or reverse-phase liquid chromatography (RPLC).

Solvent Programming. A type of gradient elution in which the composition of the mobile phase is changed over time.

Southern Blot. A blotting method used to detect specific sequences of DNA, based on the binding of these sequences to an added, known sequence of DNA that is labeled with a radioactive tag or with a label that can undergo chemiluminescence.

Soxhlet Extraction. A combined use of distillation and extraction for obtaining analytes from solid samples.

Spatial Analysis. A study of the way in which a particular analyte is distributed throughout a matrix.

Speciation. The ability to discriminate between different forms of an analyte.

Specificity. See *Selectivity*

Spectrochemical Analysis. The use of spectroscopy to identify a sample or measure chemicals in a sample.

Spectrofluorometer. An instrument that is used to perform fluorescence spectroscopy and that involves the use of a sophisticated monochromator.

Spectrometer. An instrument that is used to collect a spectrum.

Spectrometry. The use of spectroscopy to measure a spectrum.

Spectrophotometric Titration. The use of absorption spectroscopy to follow the course of a titration when the analyte, titrant, or product of the titration reaction has significant absorption of visible or ultraviolet light.

Spectroscopy. A technique that uses light to obtain quantitative or qualitative information on the chemical or physical properties of a sample.

Spectrum. The pattern that is observed when light is separated into its various colors, or spectral bands; also used in mass spectrometry to describe a plot of ion intensity versus the mass-to-charge ratio for an ion; plural form, *spectra*.

Specular Reflection. A type of reflection that occurs when light strikes a boundary between two regions that is a flat plan, causing the light to be reflected in a well-defined manner and to retain its original image; also known as *regular reflection*.

Spiked Recovery Study. A study conducted by taking a typical sample and spiking a portion of this sample with a known amount of the analyte; the amounts of analyte in the original and spiked samples are then measured, with the difference in these values then being compared to the amount added to the spiked sample.

Split Injection. A sample injection method in gas chromatography (GC) in which a microsyringe is used to place a sample into the heated injection port of the GC system, but with only part of the resulting vapors being allowed to enter the column.

Splitless Injection. A type of direct injection in gas chromatography that is specifically performed on a system that can also be used for split injection. This approach is performed with microsyringes and open-tubular columns, but the side vent of the injection system is now kept closed so that most of the injected and vaporized sample will go into the column. After the analytes have entered the column, the side vent can be reopened to flush out any undesired vapors from the injection chamber before the next sample is applied.

Spontaneously Combustible. A term used to describe a material that can ignite spontaneously.

Spreadsheet. A computer program used to record, analyze, and manipulate data.

Sputtering. The collision of positive ions such as Ar^+ with a hollow cathode, causing some metal atoms at this cathode to be dislodged and enter into the surrounding gas phase.

Stability Constant. See *Formation constant*

Stacking Gel. A gel with a relatively low degree of cross-linking and large pores that is located on top of the running gel in a gel electrophoresis system.

Stagnant Mobile-Phase Mass Transfer. A band-broadening process that is related to the rate of diffusion or mass transfer of solutes as they go from the mobile phase outside the pores of the support to the mobile phase within the pores of the support or directly in contact with the support's surface.

Standard. A material known to contain the analyte of interest, with the analyte being present in a known amount or producing a known property in the material.

Standard Addition Method. A technique used to determine the concentration of an analyte in a sample by measuring the signals for the unknown sample and for portions of this sample that have been spiked with known amounts of the analyte.

Standard Cell Potential (E^o_{Cell}). The potential that develops between an anode and a cathode when all of the components in an electrochemical cell are in their standard states.

Standard Chemical Potential (μ^o). A measure of the amount of energy per mole that is available through the reaction of a chemical when it is present in its standard state, generally expressed in units of J/mol.

Standard Deviation (s or σ). A measure of the variation within a set of values, which is calculated by using the following relationship,

$$s = \sqrt{\Sigma(x_i - \overline{x})^2/(n-1)}$$

where n is the number of values within the set, $\overline{x}$ is the arithmetic mean for this set of values, and x_i represents any individual value; the symbol s is used to represent the estimated standard deviation for a small group of results, while the symbol σ is used to describe the overall standard deviation of a large population of values.

Standard Deviation of the Mean ($s_{\overline{x}}$). A factor that is used to describe the precision of an experimentally determined mean, as given by the following formula, $s_{\overline{x}} = s/\sqrt{n}$ where (s) is the standard deviation of the entire data set and n is the number of

values in this set; also known as the *standard error* or *standard error of the mean*, *SEM*.

Standard Electrode Potential (E^o). The potential expected under standard conditions for a given half-reaction compared to a standard hydrogen electrode.

Standard Error. See *Standard deviation of the mean*

Standard Error of the Mean (SEM). See *Standard deviation of the mean*

Standard Hydrogen Electrode. An inert platinum electrode at which the half-reaction involves the reduction of hydrogen ions to form hydrogen gas, which is assigned a reference value of exactly 0.000 V for determining the standard reduction potentials of all other half-reactions.

Standard Operating Procedure (SOP). A specific set of instructions that describes how a particular method or task should be performed.

Standard Potential. See *Standard electrode potential*

Standard Solution. A solution with an accurately known concentration, making it suitable for use as a standard in a chemical analysis.

Standard State. The reference form for a chemical that is said to have an amount of inherent energy per mole that is equal to the standard chemical potential; the standard state for a chemical has an activity of exactly one and is based on either the pure form of a chemical or a particular concentration of this chemical.

Standardization. The process of determining the concentration of a titrant or some other type of standard solution.

Starch Indicator. A visual indicator that is employed in redox titrations that have iodine as the titrant or the analyte, in which iodine (probably in the form of the linear I_3^- ion) forms a complex with starch and gives an intense blue color.

Stationary Phase. A fixed phase in a chromatographic column or system that is responsible for delaying the movement of chemicals as they travel through the column or system.

Stationary-Phase Mass Transfer. A band-broadening process that is related to the movement of chemicals between the stagnant mobile phase and the stationary phase.

Stepwise Formation Constant. The formation constant for an individual reaction in a stepwise reaction process.

Stepwise Reaction. A reaction with multiple steps, often used to describe a series of complex formation reactions.

Stock Solution. A reagent solution that is used to make other, less concentrated solutions for use in an assay.

Stoichiometry. The number of moles of reactants versus products that appear in a balanced chemical reaction.

Stray Light. Light that reaches a detector without going through a sample.

Strong Acid. An acid that undergoes essentially complete dissociation to form hydrogen ions (or protons) in a given solvent.

Strong Base. A base that undergoes an essentially complete reaction as it accepts hydrogen ions (or protons) from an acid.

Strong Mobile Phase. A mobile phase in chromatography that quickly elutes a retained solute from a column.

Structural Analysis. A chemical measurement in which the goal is to determine features such as the molar mass, elemental composition, functional groups, and/or structure of the analyte.

Student's *t*-test. A statistical method that uses the Student's *t*-value to compare an experimental value with a known value or to compare two experimental values; for the comparison of an experimental mean with a known value, this is done by calculating the ratio $t = |\bar{x} - \mu| / s_{\bar{x}}$, where $\bar{x}$ is the experimental mean being tested, μ is the true value for the same data set, and $s_{\bar{x}}$ is the standard deviation of the mean.

Student's *t*-value (*t*). A mathematical factor used to correct for the greater uncertainty that is present when working with small data sets rather than with larger sets; this factor is commonly used in calculating both confidence intervals and in the comparison of the experimental values; see *Student's t-test* and *Confidence intervals.*

Substitution Balance. A balance that has a single pan and a set of removable counterweights on one side of a balance beam, and a counterweight of known mass on the other side; the mass of an object on the pan is determined by adding or removing counterweights until this side is in balance with the known mass on the other side.

Successive Approximations. A method for solving equations, in which estimates are placed into the equation and used to obtain new estimates for the actual answer.

Supercritical Fluid. A state of matter that exists above a certain critical temperature and pressure for a given chemical; under these conditions, the chemical is neither a gas nor a liquid, but has intermediate properties of both.

Supercritical Fluid Chromatography. A type of chromatography that uses a supercritical fluid as the mobile phase.

Supercritical Fluid Extraction. A type of extraction that uses a supercritical fluid as one of the phases.

Supernatant Liquid. See *Supernate*

Supernate. The liquid that is in contact with a precipitate; also called a *supernatant liquid.*

Supersaturated Solution. A solution in which the concentration of a dissolved chemical is temporarily greater than its maximum solubility at equilibrium.

Support. The material onto which the stationary phase is coated or attached in a chromatographic system.

Support-Coated Open-Tubular Columns (SCOT). A type of open-tubular column in gas chromatography that has an interior wall coated with a thin layer of a particulate support; a thin film of a liquid stationary phase is then coated onto this layer of support material.

Suppressor Ion Chromatography. A type of ion chromatography in which the use of a second column or membrane separator (of opposite charge to the first ion-exchange column) replaces competing ions that have high conductivity with ions that have a lower conductivity.

Surface Analysis. The analysis of a material's outermost layers.

Syringe. A device made of a graduated glass or plastic barrel with a flange and an open needle that takes a gas or liquid into the barrel and uses a metal, glass, or plastic plunger to push out and dispense this gas or liquid; typically used to measure and dispense volumes of 0.5–500 μL.

Syringe Pump. A pump that consists of a syringe in which a plunger is pressed in by a motor to create flow of mobile phase or solvent out of the syringe.

Systematic Error. An error that results in a constant bias of the results from the true answer.

Taring (to "tare"). A procedure that involves first placing the weighing container onto the balance and having the balance electronically reset its display (by pressing a "tare" button) so that it reads zero when the container is present.

Technical Grade. Chemicals of reasonable purity for cases where no official standards exist for quality or impurity levels; often used in manufacturing or general laboratory applications; also known as *Laboratory Grade.*

Temperature Programming. A method of performing gradient elution in gas chromatography in which the temperature of the column is varied over time.

Temporal Analysis. See *Time-dependent analysis*

Teratogen. A substance that leads to the production of non-hereditary birth defects.

Test Statistic. A factor that is calculated in a statistical test to see whether two values agree or differ in their values at the desired confidence level; one example is the Student's *t*-value, which can be used as a test statistic for comparing an experimental result to a true value or to another experimental result.

Tetradentate Ligand. A ligand that has four binding sites for a metal ion.

Thermal Conductivity. A property that describes how fast heat can be transferred through a material, such as a gas or gas mixture.

Thermal Conductivity Detector (TCD). A general detector for gas chromatography that measures the ability of the eluting carrier gas and analyte mixture to conduct heat away from a hot wire filament.

Thermal Desorption. A method used in gas chromatography (GC) for examining volatile compounds in a solid. This method is conducted by placing a known quantity of the solid into a chamber where it is heated. As the analytes go into the gas phase, they are trapped or placed into a GC system for testing.

Thermobalance. An instrument for performing a thermogravimetric analysis, which includes a high-quality analytical balance along with a furnace for heating the sample in a controlled fashion.

Thermocouple. A heat-sensing detector used in infrared spectroscopy in which the junction of two different wire conductors generates an electrical voltage that depends on the temperature difference between the ends of two wires.

Thermodynamic Equilibrium Constant ($K°$). An equilibrium constant that is written in terms of only chemical activities.

Thermogravimetric Analysis (TGA). A technique in which the mass of a sample is measured as the temperature of the sample is varied; also called *thermogravimetry.*

Thermogravimetric Curve. A graph made by thermogravimetric analysis, in which the measured mass of the sample is plotted versus temperature.

Thermogravimetry. See *Thermogravimetric analysis*

Thin-Layer Chromatography (TLC). A type of planar chromatography in which a particulate support like silica is coated on a glass or plastic sheet.

Thiosulfate. A reagent with the formula $S_2O_3^{2-}$ that is often used in iodometric titrations and chemical analysis methods involving iodine.

Time-Dependent Analysis. A study of how one or more analytes varies in a sample over time; also called *temporal analysis*.

Titrant. The reagent that is combined with the analyte during a titration.

Titration Curve. A plot of the measured response versus the amount of added titrant.

Titration Error. The difference in the amount of titrant that is needed to reach the true equivalence versus the end point.

Titration. A procedure in which the quantity of an analyte in a sample is determined by adding a known quantity of a reagent that reacts completely with the analyte in a well-defined manner; also known as *titrimetric analysis*.

Titrimetric analysis. See *Titration*

TMS-Derivative. The product of a derivatization reaction that takes place between a chemical and trimethylsilyl (TMS).

T_n-test. A statistical test for the detecting of outliers; this is performed by taking the absolute difference between the suspected outlier's value (x_o) and the data set's mean value ($\bar{x}$), and comparing this difference to the standard deviation of the entire data set (s) by using the ratio $T_n = |x_o - \bar{x}|/s$.

"To Contain" (TC). A marking found on some types of volumetric glassware indicating that they are designed to contain the amount of liquid for which they have been calibrated.

"To Deliver" (TD). A marking that indicates a volumetric device will deliver the correct measured volume when the contents of the device are released by allowing the device to drain, without any blowing or forced delivery, into the desired receptacle.

"To Deliver/Blow Out." A marking that indicates a volumetric device will deliver the measured volume only when the last bit of its contents is blown out with a pipet bulb or by some other means of forced delivery.

Top-Loading Balance. See *Precision balance*

Total Concentration. See *Analytical concentration*

Total Retention Time. See *Retention time*

Total Retention Volume. See *Retention volume*

Toxicology. The study of how chemicals affect living organisms.

Trace Analysis. Measurement of one or more trace components in a sample.

Trace Component. A substance that makes up less than 0.01% (100 parts-per-million) of the total composition for a sample.

Trace Metal Grade. Chemicals prepared to have low levels of trace metals; used for the preparation of reagents and samples for trace-metal analysis.

Transfer Pipet. See *Volumetric pipet*

Transmission. The passage of electromagnetic radiation through matter with no change in energy taking place.

Transmission Grating. A grating in which diffraction is produced by having light pass through a grating composed of a series of small slits that create constructive and destructive interference of the light.

Transmission Quadrupole Mass Spectrometer. See *Quadrupole mass analyzer*

Transmittance (T). The relative amount of light that is transmitted through a sample.

Tridentate Ligand. A ligand that has three binding sites for a metal ion.

Trough. The region of minimum intensity in a wave.

True Mean (μ). The actual arithmetic mean for a group of values, as obtained for a large number of values or in the absence of any random error.

Tungsten/Halogen Lamp. A tungsten lamp in which a small amount of iodine is also present in the lamp.

Tungsten Lamp. A light source that uses a heated tungsten wire.

Turbidimetry. A technique for examining the degree of light that is scattered by a sample by measuring the decrease in power of light that makes it through the sample without being scattered.

Turbidity. The "cloudiness" of a solution due to the presence of a light-scattering substance such as a precipitate.

Two-Dimensional Electrophoresis (2-D electrophoresis). A method in which two different types of electrophoresis are performed on a single sample; the first of these separations is usually based on isoelectric focusing and the second on sodium dodecyl sulfate polyacrylamide gel electrophoresis.

Type 1 Error. An error that occurs in a statistical test when it is concluded that the model and experimental value are not the same when they really are equivalent; also known as *alpha error*.

Type 2 Error. An error that occurs in a statistical test when it is concluded that the model and experimental value are the same when the result is part of an entirely different set of data; also known as *beta error*.

Ultra-high-Pressure Liquid Chromatography. A type of liquid chromatography that is performed at pressures greater than the range of 5000–6000 psi that is normally used in high-performance liquid chromatography.

Ultraviolet-visible Spectroscopy. See *UV-vis spectroscopy*

Universal Method. See *General method*

Unsaturated Solution. A solution in which the final concentration of an added solute is given by the total amount that was added to the solution.

Upper Limit of Detection. The largest amount of analyte that can reliably be measured by a method.

USP Grade. A reagent chemical that meets or exceeds specifications set by the United States Pharmacopeia (USP).

UV-vis Absorbance Spectrometer. An instrument that is used to examine the absorption of light in UV-vis spectroscopy.

UV-vis Spectroscopy. A type of spectroscopy that is used to examine the ability of an analyte to interact with ultraviolet or visible light through absorption.

Van Deemter Equation. An equation used in chromatography to show the relationship between the plate height (H) for a system and the linear velocity (u); this equation has the general form $H = A + B/u + Cu$, where A, B, and C are constants of the system.

Van der Waals Forces. See *Dispersion force*

Variable-Wavelength Absorbance Detector. A type of absorbance detector that allows the monitored wavelength to be varied over a wide range.

Variance (V or σ^2). The square of the standard deviation for a population of results; used to describe the variation in a group of results.

Void Time (t_M). The time required for a totally nonbinding (or nonretained) substance to travel through the column during chromatography, also known as the *hold-up time*.

Void Volume (V_M). The volume of mobile phase it takes to elute a totally nonretained substance from a chromatographic system; the volume of mobile phase within a chromatographic system, also known as the *hold-up volume*.

Volatization. The conversion of a chemical from a solid or liquid phase into the gas phase, such as in the presence of heat.

Volhard Method. A titration method that involves the titration of Ag^+ with thiocyanate to give solid AgSCN; the indicator is Fe^{3+}, which reacts with the first bit of excess thiocyanate after the equivalence point to form $FeSCN^{2+}$, a red soluble complex.

Voltaic Cell. See *Galvanic cell*

Voltammetry. A method in which a current is measured as the potential is changed as a function of time.

Voltammogram. A plot of current versus applied potential that is obtained in voltammetry.

Voltmeter. See *Potentiometer*

Volume. The amount of space that is occupied by a three-dimensional object; the official base unit of volume in the SI system is the cubic meter (m^3), but chemists generally use the related unit of the liter, which is equal to $1 \times 10^{-3} m^3$ or a thousand cubic centimeters ($1000 cm^3$).

Volume-per-Volume (v/v). A measure of the chemical content of a gas or liquid mixture, determined by taking the volume of the specific gas or liquid of interest and dividing this by the total volume of the mixture; often expressed in terms of a percent or some related ration, such as ppm, ppb, or ppt.

Volumetric Analysis. A procedure in which volume measurements are used for characterizing a sample.

Volumetric Flask. A piece of glassware that is used to prepare and dilute solutions to a specific volume (generally in the range of 1–2000 mL); this consists of a round, flat-bottomed lower region for mixing and holding solutions, a long upper neck with a calibrated mark, and an opening at the top where an inert stopper can be securely placed for mixing the solution within the flask.

Volumetric Pipet. A pipet (or pipette) that is used to measure and deliver a single, specific volume of a liquid to another container, such as a volumetric flask; also known as a *transfer pipet*.

Volumetric Titration. A titration in which volume measurements are used for characterizing a sample.

Walden Reductor. A reductor that consists of a column that contains silver granules as a reducing agent; also known as a *silver reductor*.

Wall-Coated Open-Tubular Column (WCOT). A type of open-tubular column in gas chromatography in which a thin film of a liquid stationary phase is coated or placed directly on the wall of the column.

Water Hardness. The total concentration of calcium and magnesium ions in the water, as expressed in units of milligrams of $CaCO_3$ per liter of water.

Water-Reactive. A term used to describe a chemical that will react with water to become flammable or give off large quantities of flammable or toxic substances.

Wavelength (λ). The distance between any two neighboring crests in a wave.

Wavenumber ($\overline{\nu}$). A value equal to the reciprocal of the wavelength (λ), where $\overline{\nu} = 1/\lambda$.

Weak Acid. An acid that only partially dissociates to form hydrogen ions (or protons) in a given solvent.

Weak Base. A base that undergoes only a partial reaction as it accepts hydrogen ion (or protons) from an acid.

Weak Mobile Phase. A mobile phase in chromatography that slowly elutes a retained solute from a column.

Weighing. The process of determining the mass or weight of a substance.

Weight. A measure of the pull of a force, such as gravity, on an object.

Weight by Difference. A weighing procedure in which the mass of a sample is determined by taking the difference between the mass of its container and the mass of the container plus the sample.

Weight-per-Volume (w/v). A measure of the chemical content of a mixture, determined by taking the mass of the analyte or substance of interest and dividing this by the total volume of the mixture; this can be expressed as a percent or as a related ratio (ppm, ppb, or ppt) if the density of the solution is approximately equal to 1.0 g/mL, as is true for dilute aqueous solutions near room temperature.

Weight-per-Weight (w/w). A measure of the chemical content of a mixture, determined by taking the mass of the analyte or substance of interest and dividing this by the total mass of the mixture; often expressed in terms of a percent or some related ration, such as ppm, ppb, or ppt.

Weight Titration. See *Gravimetric titration*

Western Blot. A blotting method used to detect specific proteins, based on the transfer of these proteins onto a support like nitrocellulose or nylon followed by the treatment of this support with labeled antibodies that can specifically bind the proteins of interest.

Wet Ashing. A technique in which a weighed amount of sample is placed in a concentrated acid and heated to the boiling point of the acid, usually in a porcelain dish; this is done in such a way as to oxidize any organic material so that it is lost as CO_2 but the mineral components remain behind dissolved in the acid.

Wheatstone Bridge. An electronic circuit that consists of four resistors that are arranged in a parallel circuit, with two resistors being present in each side; when these resistors are properly balanced electronically, the voltage difference across the center of the circuit will be equal to zero, but if any of these resistors changes in its electrical properties, a nonzero voltage will be produced.

Wick Flow. A source of band-broadening in gel electrophoresis due to evaporation of solvent from the wicks, such as in the presence of Joule heating.

Within-Day Precision. The variation obtained for an analytical method during a single day, usually over the course of several runs.

Within-Run Precision. A measure of the variation of an analysis for a single sample during one analysis session, or "run."

Working Electrode. The electrode in an electrochemical cell at which the reduction (or oxidation) of the analyte is carried out.

X-Ray Crystallography. A method for examining the structure of a chemical based on the diffraction pattern that is produced when a crystal of this chemical is exposed to X rays.

X-Ray Imaging. An imaging method based on the exposure of a sample to X rays.

Zeolite. See *Molecular sieve*

Zinc Reductor. See *Jones reductor*

Zone Electrophoresis. An electrophoretic method that uses small amounts of sample to allow analytes to be separated into narrow bands or zones.

Zwitterion. A chemical that possesses a net charge of zero, but contains groups with an equal number of negative and positive charges.

INDEX

Note: Page numbers in italics denote graphic material. A page number with an appended italicized 't' denotes material contained in a table. A page number with an appended italicized 'b' denotes material contained in a box.

A

Common Physical Constants[a]

Symbol	Name[b]	Value	Uncertainty ($\pm$ 1 S.D.)
c	Speed of light in a vacuum	$2.997\ 924\ 58 \times 10^8$ m/s	Exact value
F	Faraday constant	$9.648\ 533\ 99 \times 10^4$ C/mol	$\pm 0.000\ 000\ 24 \times 10^4$ C/mol
h	Planck constant	$6.626\ 068\ 96 \times 10^{-34}$ J $\cdot$ s	$\pm 0.000\ 000\ 33 \times 10^{-34}$ J $\cdot$ s
k	Boltzmann constant	$1.380\ 6504 \times 10^{-23}$ J/K	$\pm 0.000\ 0024 \times 10^{-23}$ J/K
N_A	Avogadro's number	$6.022\ 141\ 79 \times 10^{23}$ mol^{-1}	$\pm 0.000\ 000\ 30 \times 10^{23}$ mol^{-1}
R	Gas constant	$8.314\ 472$ J/(mol $\cdot$ K)	$\pm 0.000\ 015$ J/(mol $\cdot$ K)
		or $8.205\ 746 \times 10^{-2}$ L $\cdot$ atm/(mol $\cdot$ K)	$\pm 0.000\ 014 \times 10^{-2}$ L $\cdot$ atm/(mol $\cdot$ K)

[a]These values are the 2006 CODATA recommended values, as provided by the U.S. National Institute of Standards and Technology (NIST). The uncertainties listed are given in terms of $\pm$ one standard deviation ($\pm$ 1 S.D.).

[b]Avogadro's number, N_A, is officially listed by the NIST as the "Avogadro constant" and R is the "molar gas constant"; F is also called "Faraday's constant," h is commonly known as "Plank's constant" and k is called "Boltzmann's constant."